Characterization and Failure Analysis of PLASTICS

ASM International
The Materials Information Society

www.asminternational.org

First printing, December 2003
Second printing, June 2004
Digital printing, February 2010

Comments, criticisms, and suggestions are invited and should be forwarded to ASM International.

ASM International staff who worked on this project include Steve Lampman, Editor; Bonnie Sanders, Manager of Production; Nancy Hrivnak, Jill Kinson, and Carol Polakowski, Production Editors; and Scott Henry, Assistant Director of Reference Publications.

Library of Congress Cataloging-in-Publication Data

Characterization and failure analysis of plastics.
p. cm.
Collection of articles from ASM International handbooks.
Includes bibliographical references and index.
1. Plastics—Fracture. I. ASM International.

TA455.P5C463 2003
620.1'9236—dc22

2003057732

ISBN-13: 978-0-87170-789-5
ISBN-10: 0-87170-789-6
SAN: 204-7586

ASM International®
Materials Park, OH 44073-0002
www.asminternational.org

Printed in the United States of America

Preface

This book is collection of *ASM Handbook* articles on how engineering plastics are characterized and understood in terms of properties and performance. It approaches the subject of characterization from a general standpoint of engineering design, materials selection, and failure analysis. These basic activities of the engineering process all require clear understanding of plastics performance and properties by various methods of physical, chemical, and mechanical characterization.

The first section introduces the fundamental elements of engineering plastics and how composition, processing, and structure influence their properties and performance. The second section contains articles on material selection and design, where the requirements of a plastic part are synthesized and analyzed in terms of function, shape, process, and materials. The next sections then cover the important physical, chemical, and mechanical properties of plastics.

The last section covers failure analysis, which is the ultimate stage of characterization in the life of a part, but really only the penultimate stage in the overall engineering process. Failure analysis, in a broad sense, is another iteration of the design process, as it can provide important information on product and process improvements. Thus, it closely ties together with the characterization of properties and performance plastics during design and materials selection.

This book would not have been possible without the original contributions from the authors of the Handbook articles. Thanks are extended to them.

Steve Lampman
May 2003

Contents

Introduction

Engineering Plastics: An Introduction

AN ENGINEERING PLASTIC may be defined as a synthetic polymer with mechanical properties that enable its use in the form of a load-bearing shape. Polymers, which constitute the major portion of an engineering plastic, are made up of extremely large molecules formed from polymerization of different monomers. Engineering plastics all have, as their principal constituent, one or more synthetic polymer resins and almost universally contain additives. Additives, which have much smaller molecules than polymers, provide color, flexibility, rigidity, flame resistance, weathering resistance, and/or processibility. They can be grouped into two main categories: (a) those that modify the characteristics of the base polymer by physical means, including plasticizers, lubricants, impact modifiers, fillers, and pigments and (b) those that achieve their effect by chemical reactions, including flame retardants, stabilizers, ultraviolet absorbers, and antioxidants.

The basic structure of polymers influences the properties of both polymers and the plastics made from them. An understanding of this basic structure permits the user to understand which polymers may be acceptable for a certain application and which may not. The chemical structure of a polymer is very important because it dictates so many polymer properties. Much of the processing used to create engineering plastics is directed toward optimizing the properties that might be attainable using the basic structure of the polymer. For example, special processing techniques are used to produce polymer fibers. Such fibers have substantially greater stiffness and strength along their length than do the unoriented polymers from which they are manufactured. This is because special processing has been used to orient the covalent bonds of an appropriate long-chain polymer in the lengthwise (axial) direction of the fiber. The design of such processing would not be possible without an understanding of chemical structure.

This introductory article describes the various aspects of chemical structure that are important to an understanding of polymer properties and, thus, their eventual effect on the end-use performance of engineering plastics. This article also includes some general information on the classification and naming of polymers and plastics. The other articles provide more specific details on how plastics are characterized and evaluated during the various stages of engineering from design to failure analysis. Materials evaluation or characterization is a basic engineering activity that is done during design, manufacture, service, and failure analysis. For example, it begins during the design phase, when designers must select an appropriate material and process to achieve a given function and shape. This process of design involves a complex series of steps in evaluating the alternatives and interrelationships of materials, processes, shape, and function (Fig. 1) (Ref 1). The first stage of design is conceptual, where materials and processes are considered in broad terms. Initial selection may be either a "materials-first" approach or a "process-first" approach. In the material-first approach, the designer begins by selecting a material class and narrowing it down. Then, manufacturing processes consistent with the selected material are considered and evaluated. Chief among the factors to consider are production volume and information about the size, shape, and complexity of the part. With the process-first approach, the designer begins by selecting the manufacturing process, guided by the same factors. Then, materials consistent with the selected process are considered and evaluated, guided by the performance requirements of the part.

Some level of materials evaluation is done during any stage of engineering. It begins at the conceptual stage of design, where the typical ranges of key properties are compared for general categories of materials (such as metal, plastics, ceramic, or composites). For example, general comparisons of some common properties are given in Table 1 (Ref 2) for metals, ceramics, and polymers. The precision of property data needed during the conceptual stage of design is more comparative in terms of key physical principles for the shape or function. Refinements and more detailed specification of materials, process, shape, and function are achieved during the stages of detailed design. The overall design process is also iterative (Fig. 2), because it may be necessary to reexamine alternatives during the earlier stages of design. In this sense, failure analysis can also be viewed as an extension of an iterative design process, because failure analysis is, or should be, another feedback loop that can influence the conceptual or detailed evaluation of materials, processes, shape, and function.

With these general concepts in mind, this book provides a collection of articles on the performance and characterization of plastics. The first section contains articles on the evaluation and characterization of engineering plastics during the stage of design and materials selection. The other sections contain articles on the physical, chemical, thermal, and mechanical characteristics and analysis of plastics. The last section contains articles on the failure analysis of plastics. This approach is meant to cover the overall characterization of plastics from the beginning stages of design to the last stage when a plastic component reaches the end of its useful life either by unintended interruption of service (i.e., failure) or by intentional removal from service to prevent failure. In a way, design and failure analysis are complementary activities in reverse (Fig. 3) (Ref 3). Design is the process of synthesizing and analyzing conditions into the reality of an actual or hypothetical component. In contrast, failure analysis is the dissection of an actual component in order to synthesize and understand the significance of a hypothetical design in a given failure. In this sense, analysis and synthesis of engineering factors are prominent in different areas of each process, although the individual steps within the processes contain both.

Polymer Structure

A polymer structure contains many (*poly-*) repeats of some simpler chemical unit, called a *mer.* Another term often used in place of mer unit is monomer unit, but this term is also used to indicate the basic chemical compound from which the polymer is polymerized. For example, the polymer polyethylene is produced from the monomer ethylene, although the mer unit of polyethylene is distinct from the source monomer (Fig. 4). For this reason, the term mer unit is preferred when referring to the basic repeat unit of a polymer.

Polymer properties are primarily dictated by the polymer structure, which in turn is influenced by basic chemical composition, morphology, and processing. The polymer structure can be divided into that which is within the mer unit, within the molecule, and between molecules. The repeating mer units of polymers are held together by covalent bonds. Covalent bonds are stronger than the metallic bonds that hold metals together, but weaker than ionic bonds (Table 2). In comparison to metals, intermetallics, and ceramics and glasses, polymers also have a very low coordination number (CN), which is defined as the number of cation/anion (i.e., positive-ion/negative-ion) near neighbors. The very low CNs of polymers, in addition to the prevalence of light atoms such as carbon and hydrogen as the backbone of most polymers, tends to result in lower density relative to metals and ceramics. The localized nature of electrons in polymers also renders them good electrical insulators and poor thermal conductors.

Unlike either metallic or ionic bonds, covalent bonds are very directional in character. This means that the atoms in the molecule are oriented with fixed bond angles between atoms in a polymer molecule (dictated by the chemical and electronic structures of the atoms involved). Depending on the nature of the covalent bonds, the mer units may form one-, two- (rarely), or three-dimensional molecules. For example, polyethylene is a long-chain molecule that forms when the double bond between carbon atoms in the ethylene molecule (C_2H_4) is replaced by a single bond between adjacent carbon atoms (Fig. 4). Three-dimensional covalent bonding is typified by cross linking that occurs when thermoset plastics are cured. Differences in chemical bonding are the reasons for the differences between thermoplastics and thermosets. Thermoplastics are invariably composed of long, individual molecules that are bonded to each other by secondary chemical bonds, which are much weaker than the primary covalent bonds that hold the molecules together. On the other hand, thermosets are invariably composed of some type of three-dimensional covalently bonded network structure.

Mer Structure

The structure within a mer involves the elements, their bonding, the flexibility of the mer, the bulkiness of the mer, the side groups, and possible geometric isomerism (i.e., different structural arrangement of elements in a polymer compound). The elements and bonding within the mer represent the most basic and unchangeable aspect of the structure of a particular polymer. The various elements of polymers are discussed in more detail in the section "Chemical Composition and Structure" in this article, while this section briefly introduces the general structure and strength of the bonds within a mer unit. Table 3 (Ref 6) gives a list of bond energies for bonds that occur commonly in polymers. Bond strength has a dramatic influence on important properties, such as thermal decompo-

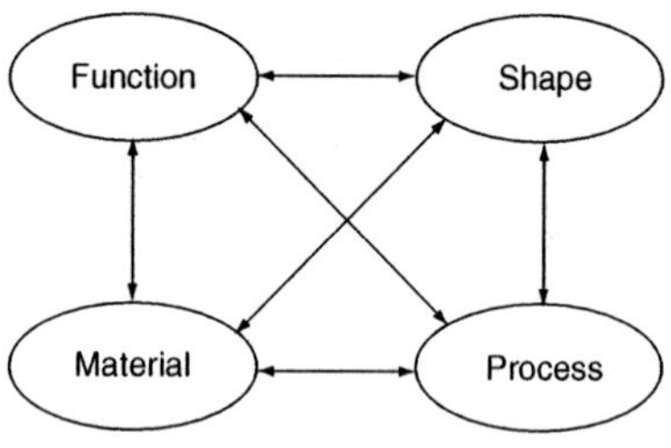

Fig. 1 Interrelated factors involved in the design process. Source: Ref 1

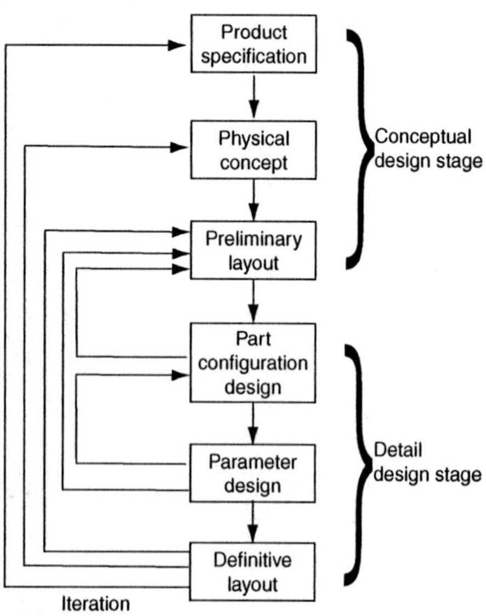

Fig. 2 Stages and steps in the iterative process of design. Source: Ref 1

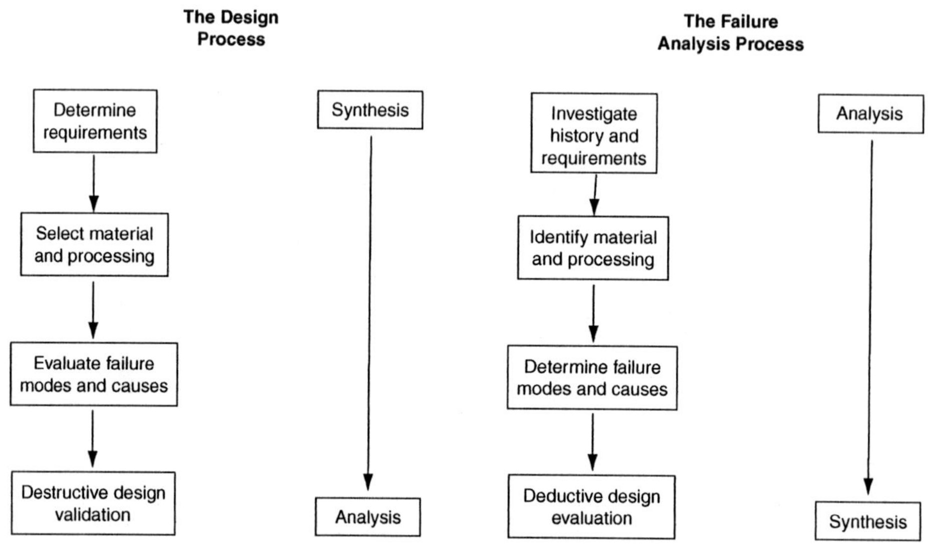

Fig. 3 General steps and the roles of synthesis and analysis in the processes of design and failure analysis. Source: Ref 3

Table 1 General comparison of properties of metals, ceramics, and polymers

Property (approximate values)	Metals	Ceramics	Polymers
Density, g/cm^3	2–22 (average ~8)	2–19 (average ~4)	1–2
Melting points	Low (Ga = 29.78 °C, or 85.6 °F) to high (W = 3410 °C, or 6170 °F)	High (up to 4000 °C, or 7230 °F)	Low
Hardness	Medium	High	Low
Machinability	Good	Poor	Good
Tensile strength, MPa (ksi)	Up to 2500 (360)	Up to 400 (58)	Up to 140 (20)
Compressive strength, MPa (ksi)	Up to 2500 (360)	Up to 5000 (725)	Up to 350 (50)
Young's modulus, GPa (10^6 psi)	15–400 (2–58)	150–450 (22–65)	0.001–10 (0.00015–1.45)
High-temperature creep resistance	Poor to medium	Excellent	...
Thermal expansion	Medium to high	Low to medium	Very high
Thermal conductivity	Medium to high	Medium, but often decreases rapidly with temperature	Very low
Thermal shock resistance	Good	Generally poor	...
Electrical characteristics	Conductors	Insulators	Insulators
Chemical resistance	Low to medium	Excellent	Good
Oxidation resistance	Generally poor	Oxides excellent; SiC and Si_3N_4 good	...

Source: Ref 2

sition. As can be seen from Table 3, it is dependent on both the elements and type of bonds involved.

Flexibility and Bulkiness of the Mer. The flexibility and bulkiness of a mer unit influence interactions between molecules, such as crystallization. This directly affects some important properties of the polymer chain or the network that is built from it. The flexibility of the mer unit is largely determined by the type of bonds involved in the backbone of the unit. Replacing *single (saturated) bonds* with *double (unsaturated) bonds* reduces the flexibility of the unit. Aromatic rings (discussed in the section "Chemical Composition and Structure" in this article) or cyclic groups in the backbone also reduce flexibility and add bulkiness. Furthermore, aromatic rings and many cyclic groups are more chemically stable than are double bonds.

Bulkiness can also be increased by adding large, inflexible side groups to the mer. However, even though bulky, inflexible side groups can increase mer bulkiness, flexible side groups may have almost the opposite effect. One significant possible effect of side groups is their role in producing secondary transitions in polymers, as discussed further in the section "Properties of Polymers" in this article.

Geometric Isomers. An isomer is a compound, radical, ion, or nuclide that contains the same number of atoms of the same elements, but differs in structural arrangement and properties. This structural feature occurs only in those thermoplastic polymers with a double bond or cyclic structure in their backbone chains. In such polymers two totally different types of mer can occur: they are cis forms and trans forms, which cannot be converted into one another by bond rotation. Figure 5 shows these two configurations for the mer of polyisoprene. The two forms are known as geometric isomers of each other, and they produce noticeably different properties. In a given polymer, property differences depend on factors such as mer bulkiness and in resulting interactions between molecules. In this example, the molecule or atom, *R*, is placed on an unsaturated carbon chain in either a cis or trans position. In the cis position, the unsaturated bonds lie on the same side of the chain. In the trans position, they are on opposite sides. The difference between these two possibilities is important in butadiene rubbers. The cis structure makes the molecule tend to coil rather than remain linear. This coiling is believed to be responsible for the elasticity observed in elastomers (e.g., rubber).

Table 2 Bond energies for various materials

Bond type	Material	Bond energy kJ/mol	Bond energy kcal/mol
Ionic	NaCl	640	153
	MgO	1000	239
Covalent	Si	450	108
	C (diamond)	713	170
Metallic	Hg	68	16
	Al	324	77
	Fe	406	97
	W	849	203
van der Waals	Ar	7.8	1.8
	Cl_2	31	7.4
Hydrogen	NH_3	35	8.4
	H_2O	51	12.2

Source: Ref 5

Table 3 Bond energies for common bonds in polymers

Bond	Bond energy kJ/mol	Bond energy kcal/g · mol
C–C	350	83
C–H	410	99
C–F	440	105
C–Cl	330	79
C–O	350	84
C–S	260	62
C–N	290	70
N–N	160	38
N–H	390	93
O–H	460	111
C=C	200	147
C=C	810	194
C=O	715	171
C=N	615	147

Source: Ref 6

Polymer Structure

The mer unit defines the chemical composition of a polymer, but complete information about the chemical structure of a polymer has several variations of how the mers combine to form a polymer. These variations in structure within the molecule may involve stereoisomerism, branching, molecular weight and distribution, end groups and impurities, and copolymerization. Polymer size is quantified primarily by molecular weight (MW), molecular-weight distribution (MWD), and branching. These factors are briefly described in this section with more details given in the next article, "Effects of Composition, Processing, and Structure on Properties of Engineering Plastics" in this book.

Stereoisomers. In addition to possible geometric isomers of the mer unit, there is another type of isomerism possible when mer units are bonded. If a polymer model is constructed in three dimensions, taking into account tetrahedral bonding, it is found that in some cases two or more different chain configurations can be produced that are consistent with the structural model but cannot be converted into each other without breaking and reforming covalent bonds. This is shown in Fig. 6 for a simple vinyl. A common example of stereoisomerism is with polypropylene (PP), where the $-CH_3$ side group may arrange itself in isotactic form (all side groups on the same side of the chain), syndiotactic form (side groups in regular alternating sides of the chain), or atactic (random) form. Isotactic and syndiotactic polymers are stereoregular and have properties that are different from one another and from the atactic form of the same polymer. They differ significantly in properties largely because of the changes produced in such structural factors as intermolecular bonding and crystallinity. For example, an atactic polymer tends to be a rubbery amorphous material, while an isotactic polymer is more crystalline with more stiffness and melting temperatures.

Branching. Many thermoplastic polymers are composed almost completely of linear chains. Others have chains with branches. These branches can have short or long lengths and can occur rarely or frequently along a chain. For a polymer of a given MW, a more highly branched polymer has a lower density and a lower degree of entanglement. The increased number of chains from branching increases the amount of free volume in the polymer. Chain ends reduce packing efficiency, and the additional free volume available offers sites into which the polymer can be displaced under stress.

Branching interferes with intermolecular bonding and has a significant effect on rheology and crystallinity. Branching lowers dimensional stability and reduces the glass-transition temperature (T_g) with other major factors (i.e., MW and MWD) being constant. At a particular molecular weight, branching may also lead to a decrease in the melting temperature (T_m) of thermoplastics. Increased branching in polymers also decreases their ability to conduct heat. The increase in free volume from branching lowers the efficiency of thermal conduction due to a more tortuous path

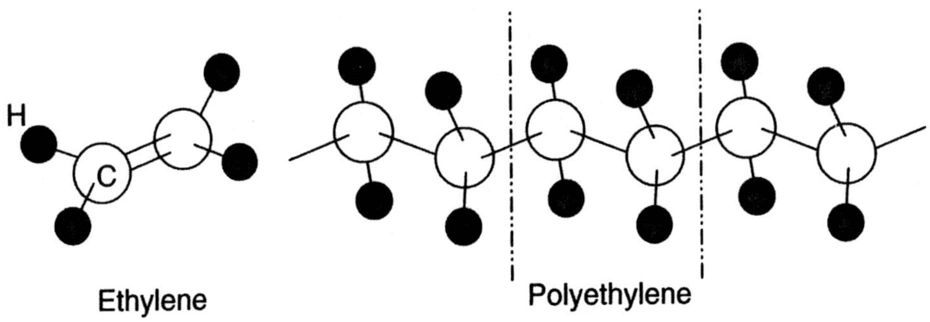

Fig. 4 Ethylene and polyethylene. Source: Ref 4

for heat conduction along primary valence bonds.

Polyethylene (PE) is a good example of how branching influences properties of thermoplastics. Polyethylene is produced in four principal grades: high density (HDPE), low density (LDPE), linear low density (LLDPE), and ultrahigh molecular weight (UHMWPE). Structurally, these grades differ in the degree and type of branching on the main chain and in overall molecular weight. At a particular molecular weight, branching leads to a decrease in T_m. Therefore, the orientation of high-molecular, linear chains can lead to an exceptionally high T_m. For example, UHMWPE, with almost perfect chains, displays the highest T_m of the different PE grades with a T_m of about 150 °C (300 °F) and a crystallinity exceeding 70%. On the other extreme, LDPE has randomly displaced branches and a T_m of about 100 °C (212 °F) and crystallinity of less than 50%.

Molecular Weight and Distribution. Because thermoplastic polymers are composed of long molecules, they can vary in molecular weight. The molecular weight can be just barely enough to qualify the materials as a polymer (rather than an oligomer), or it may represent hundreds of thousands of mer units. Except for a few cases, the molecular weight represents an average; almost all polymers have a molecular-weight distribution. Polymer samples with the same average molecular weight can have very different molecular-weight distributions. A formulation having a broader molecular-weight distribution has more chains at both the high and low end of the molecular-weight spectrum. In plastics with a broad distribution of molecular weights, the average molecular weight can be calculated in several different ways (see the next article "Effects of Composition, Processing, and Structure on Properties of Engineering Plastics" in this book).

Molecular weight and molecular-weight distribution are useful in characterizing the properties of plastic. For any given polymer, the lower its molecular weight, the more flexible it will be as there are a greater number of chain ends per unit volume for short chain species. Another important consequence of high molecular weight is its effect on crystallinity. In contrast to the typical crystalline structures of low-molecular-weight materials (such as metals), most polymers are amorphous (noncrystalline). Amorphous polymers do not have sharp melting points; instead, they pass from hard glassy structure below the T_g to a viscous liquid state or a rubbery structure above the T_g (Fig. 7a) (Ref 8). Structurally, the molecular chains in an amorphous polymer are randomly arranged in three dimensions (Fig. 8) (Ref 7). Examples of amorphous polymers include polyvinyl chloride (PVC), polymethyl methacrylate (PMMA), and polycarbonate (PC).

Nonetheless, some polymers can exhibit limited crystallinity, when the polymer chains arrange themselves into an orderly structure. In general, simple polymers with little or no side branching or strong hydrogen bonds (as in nylon) crystallize more easily, whereas crystallization is inhibited in heavily cross-linked polymers and in polymers containing bulky side groups. As noted, amorphous polymers exhibit a T_g, when the amorphous regions become mobile. In contrast, semicrystalline polymers exhibit both a T_g and a melting temperature T_m. At the latter temperature, the ordered crystalline regions melt and become disordered random coils. While the magnitude of the T_g of a polymer depends only on the inherent flexibility of the polymer chain, the magnitude of T_m is also a function of the attractive forces between chains. For both amorphous and crystalline polymers, the T_g goes up with the number-average molecular weight (Fig. 7).

Many other properties can be characterized by molecular weight. For example, elongation at break for acrylic samples with different molecular weights can be reduced to a single curve when weight-average molecular weight is used. Plastics with narrow molecular weights are preferred for low warpage in thin-wall injection molding, film extrusion, and rotational molding. Plastics with moderately low molecular weights are suitable for high-speed processing, such as high draw-down rate extrusion, high-speed calendering, and injection molding. Most processing conditions require materials with high molecular weights. This is especially true for extrusion and blow molding, which require sufficient melt strength for the extrudate to support itself as it exits from the die.

End Groups and Impurities. Regardless of the simplicity or complexity of the mer unit, the end of a polymer chain must be different from any section within the chain. For example, the end of the PE chain, which is made up of $-CH_2$ units, will be a $-CH_3$ unit. Depending on the

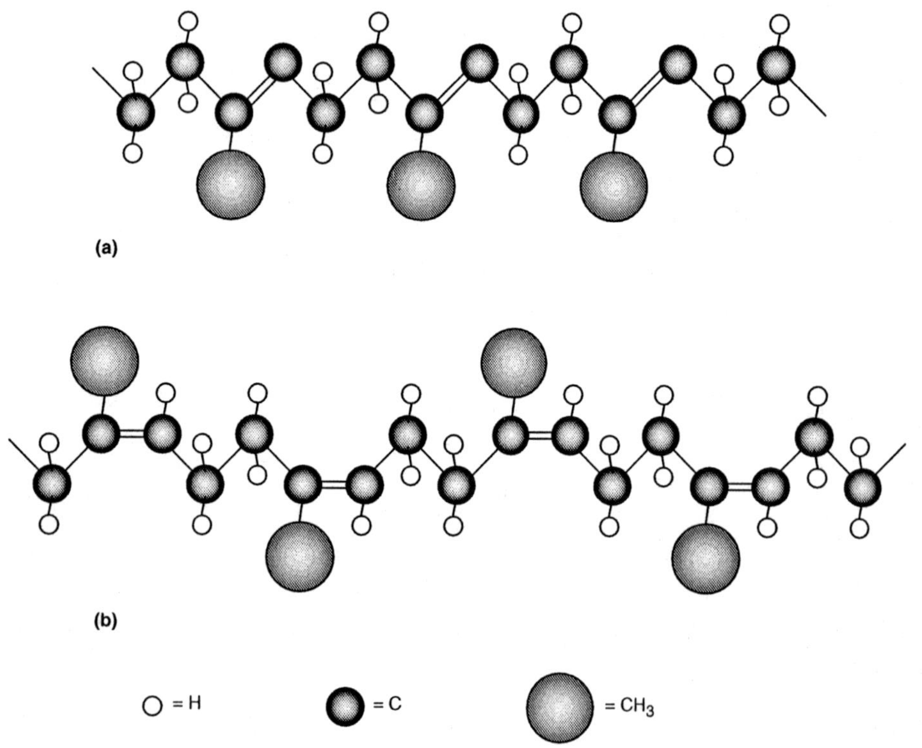

Fig. 5 Geometric isomers of polyisoprene. (a) Cis-polyisoprene (natural rubber). (b) Trans-polyisoprene (gutta percha). Source: Ref 4

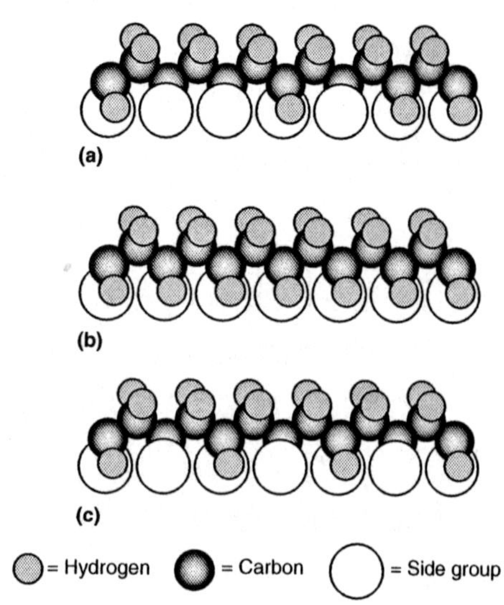

Fig. 6 Stereoisomers in a simple vinyl polymer. (a) Atactic (random arrangement of side groups). (b) Isotactic (all side groups on same side). (c) Syndiotactic (regularly alternating side groups). Source: Ref 7

polymerization process, the end group may be fairly similar to, or very different from, the chain. End groups will thus have either somewhat or very different chemical properties from the rest of the chain.

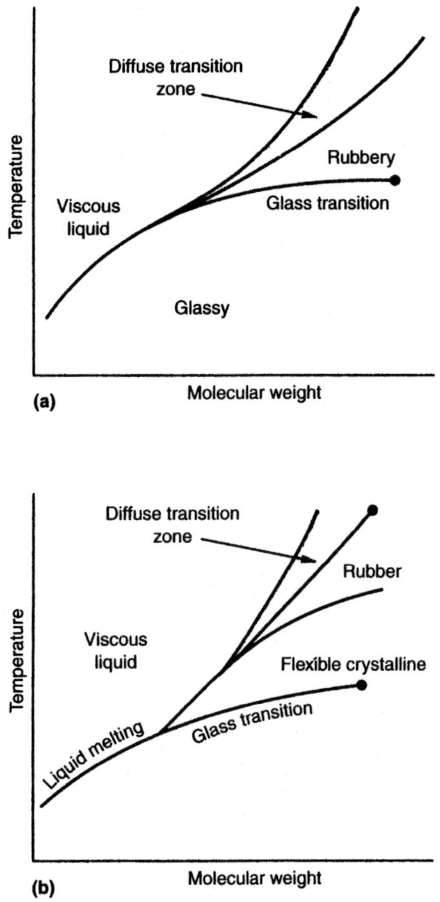

Fig. 7 Influence of molecular weight and temperature on the physical state of polymers. (a) Amorphous polymer. (b) Crystalline polymer. Source: Ref 8

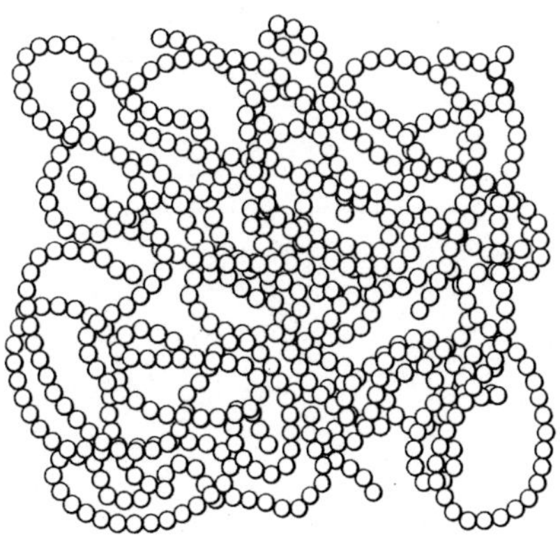

Fig. 8 Polymer structure. The spheres represent the repeating units of the polymer chain, not individual atoms. Source: Ref 7

It is also possible to have an impurity polymerized into the polymer chain. Such impurities will, of course, also have different chemical properties from the rest of the chain and thus may act as sites for decomposition, cross linking, or other chemical reactions.

Copolymerization. In many cases, two (or more) mers are combined to make a copolymer. This can be done in four different ways (Fig. 9). The mers can be polymerized together alternately to form an alternating copolymer, in a random manner to produce a random copolymer, or in blocks to produce a block copolymer. Another possibility is that one polymer can be grafted onto the other to form a graft copolymer. Because many copolymer properties are between those of the two polymers, this is a way to improve, for example, the impact resistance of a brittle polymer (which is the purpose of adding butadiene to PS).

Structure between Polymer Molecules

Important structural aspects from interactions between polymer molecules include secondary bonding, crystallinity, and cross linking. These basic structural features of polymer materials can be influenced by the internal structure of the individual polymers chains in a material and by the interactions (or bonding) between polymer chains. For example, thermoplastics are invariably composed of long, individual molecules that are bonded to each other by secondary chemical bonds, which are much weaker than the primary covalent bonds that hold the molecules together. Their overall structure is generally amorphous, but some thermoplastics can become partly crystalline. The extent of crystallization of thermoplastics depends on the internal features of the individual polymer chains. These internal features include stereoisomerism, branching, molecular weight, and molecular-weight distribution, as previously noted.

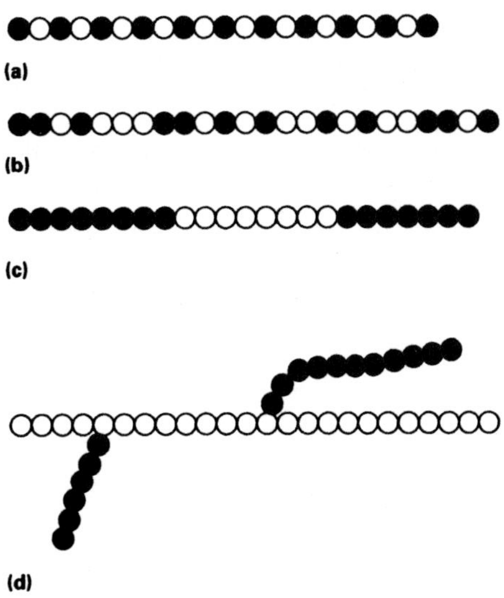

Fig. 9 Types of copolymers. (a) Alternating. (b) Random. (c) Block. (d) Graft

In general, the bonds between polymer molecules can be either weaker secondary bonds (i.e., van der Waals bond, hydrogen bonds) or stronger primary bonds (covalent bonding). These differences in chemical bonding are the reasons for the differences between thermoplastics and thermosets. Weak secondary bonds account for the behavior of thermoplastics, which are typified by low melting temperatures, low stiffness, and low strength exhibited by many polymers. The weaker secondary bonds are relatively easy to disrupt (with moderate heat, for example) without rupturing the bonds within an individual polymer molecule. Thermoplastic materials often melt upon heating, but return to their original solid condition when cooled.

In contrast, thermosets are invariably composed of some type of three-dimensional cross linking of polymers chains. The cross linking of thermoset plastics often involves primary (covalent) bonding, but sometimes cross linking may occur from hydrogen bonds, which are a stronger form of secondary bond (Table 2). This results in three-dimensional networks of cross-linked molecular chains. Most engineering thermosets involve cross linking by covalent bonding. When these types of bonds are present, an increase in temperature does not lead to plastic deformation. Thermoset plastics change chemically during processing and do not melt upon reheating. Rather, they will remain strong until they break down chemically (depolymerize) via charring or burning.

Some polymers appear to be midway between thermoplastics and thermosets. These materials can be reformed somewhat, but not completely, with the application of heat. Their properties are midway between the two extremes because their bonding is midway between. These polymers have long, individual molecules that are lightly cross linked to each other by covalent bonds or perhaps hydrogen bonds. This cross linking may have been done intentionally—to improve the stiffness or temperature resistance of the polymer, for example—or it may have happened unintentionally because of some degradation process, such oxidation or weathering.

For example, cross linking can be used to produce high-performance composite matrices that can be molded as thermoplastics and subsequently cross linked to produce varying degrees of thermosetting properties. Similarly, elastomers or rubbers can use different degrees of cross linking to vary the properties from those of an art gum eraser to those of a hard industrial rubber. Elastomers differ from thermoplastics and thermosetting polymers in that they are capable of rubbery behavior and are capable of very large amounts of recoverable deformation (often in excess of 200%). Structurally, these materials consist of networks of heavily coiled and heavily cross-linked polymer chains. For example, polyisoprene (Fig. 5) is a synthetic rubber with the same basic structure as natural rubber, but lacking the impurities found in natural rubber. The addition of sulfur to this compound and the application of pressure and a temperature of approximately 160 °C (320 °F) cause sulfur cross links to form. As the degree of cross linking increases, the rubber becomes harder. This particular process is known as *vulcanization.* A schematic of these cross-linking arrangements is shown in Fig. 10.

Secondary bonds occur from coulombic attraction between adjacent molecules or atoms, and the secondary bonds may hold adjacent macromolecules together along the length of the polymer chain (Fig. 11). In thermoplastics, such bonds can have a tremendous effect on its properties, because these are the only bonds that occur between the molecules of a thermoplastic. Even in the case of a thermoset, secondary bonds have an influence on solvent resistance and electrical properties, for example.

The weakest form of secondary bond is the dispersion bond, which arises from the internal fluctuations of electron clouds in an atom. Dispersion bonds can occur even between nonpolar atoms such as helium, which condenses at low temperatures. Dispersion bonding also occurs in hydrocarbon polymers such as PE. Secondary bonds from molecular dipoles (Fig. 12) are stronger than dispersion bonds. These types of interactions occur between induced dipoles, between induced dipoles and polar molecules, and between polar molecules.

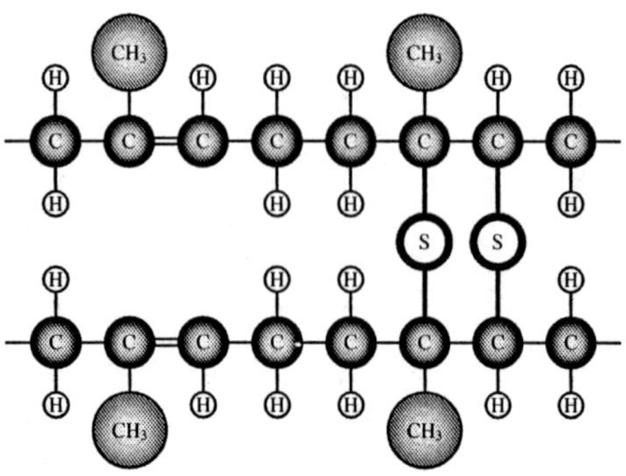

Fig. 10 Cross linking in polyisoprene. Source: Ref 4

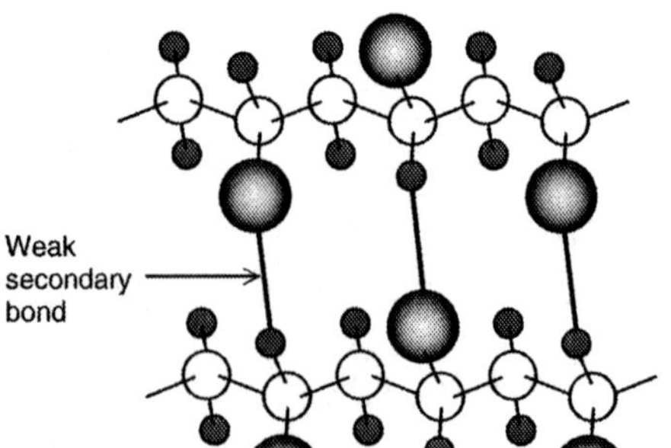

Fig. 11 Secondary bonding between two polymer chains. Source: Ref 4

Dipoles occur because atoms such as oxygen, chlorine, and fluorine are much more electronegative than the atoms to which they are bonded, such as carbon and hydrogen. Polar groups include C–O, C–Cl, C–F, O–H, and N–H. In each case, either carbon or hydrogen is at the more electropositive end of the bond. On the other hand, C–C and C–H are approximately nonpolar groups. Strong bonds result from the interaction of such preexisting electrical dipoles within a polymer with an atom or another dipole on another polymer molecule.

Hydrogen Bonds. The strongest of all such secondary bonds in polymers is the hydrogen bond. In this case, a dipole is formed from hydrogen bonded to a more electronegative element such as oxygen or nitrogen. This bond then interacts with an electronegative element such as oxygen, nitrogen, chlorine, or fluorine bonded elsewhere, forming a secondary bond that can have up to 10% of the strength of a primary covalent bond. Hydrogen bonds can occur in thermoplastics (such as nylon), or they can be the cross link bonds in some thermosets.

Crystallinity is not only possible in polymers, some thermoplastic polymers have substantial crystallinity. Such polymers are termed semicrystalline because the degree of crystallinity never reaches 100%; they include such important thermoplastics as PE and nylons (or polyamides, PA). Most polymers, however, have either very little or no true crystallinity and are generally referred to as noncrystalline or amorphous. As noted, thermoset polymers are seldom crystalline, because cross linking inhibits the mobility of individual chains.

The degree of crystallinity in a thermoplastic polymer can have a tremendous influence on its properties Crystallinity is an important feature of the structural strength of many polymers and is used in some thermoplastics to produce higher temperature resistance than would otherwise be obtainable. The polymer having the most flexible chains generally has the highest degree of crystallinity; this is the reason PE has the highest degree of crystallinity of any polymer. Anything that improves the ability of the chains to pack into a regular crystalline array improves crystallinity. Thus, the polymer having the greatest chain regularity also tends to have the higher crystallinity. Also, polymers without bulky side groups have substantially higher crystallinity than those with such groups. For these reasons, isotactic polystyrene (PS) has some crystallinity, while atactic PS is completely amorphous. However, even isotactic PS shows much less crystallinity than polypropylene (PP), which has a much smaller side group. Linear polymers have higher crystallinity than branched polymers. Strong dipoles in a mer also generally improve crystallinity.

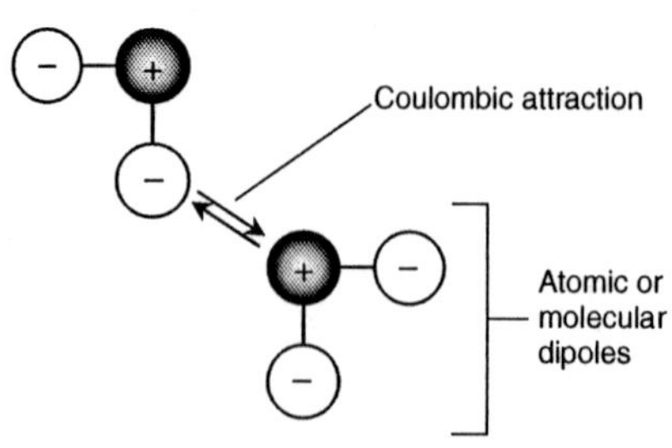

Fig. 12 Secondary bonding between two molecular dipoles. Source: Ref 4

Special processing techniques are often used to produce, increase, or direct crystallinity in polymers. The processing of fibers, for example, is aimed at producing highly oriented, crystalline regions to yield stiffness and strength in the fiber direction. Processing techniques are also used to enhance crystallization. The amount of crystalline fraction and the size of crystalline regions can be affected by the addition of nucleating agents, or seed particles, which can be small, inorganic particles. Plastics with seeds contain a higher crystalline fraction with small domains. Crystallinity is also affected by the temperature gradient in processing. A high mold temperature reduces temperature gradients and the amount of crystallization, whereas a low mold temperature increases the crystallization rate. A high melt pressure in molding can also reduce dwell time in the barrel, reducing the temperature loss, which tends to decrease the amount of crystallization.

The cooling temperature rate also affects the amount of crystallinity. Generally, the maximum crystallization rate is observed at about 0.9 of the T_m, measured in absolute temperature. For a material cooled at approximately the T_m, sufficient crystallinity will develop. If the material has a high T_g and the cooling process takes place below it, amorphism can increase. Material with a tendency to crystallize will exhibit gradual crystallization and postshrinkage when stored at temperatures above the T_g. During crystallization, the crystalline polymer packs all of the low-molecular-weight components and impure species into the interstices between the crystalline regions, leaving these as contaminated boundaries of lower strength and modulus. Shrinkage during crystallization may leave stresses and voids in these interstices, weakening them even more. The surface between crystalline regions and amorphous interstices is the weak interface at which cracking is most likely to begin.

For crystalline material, control of crystallinity is generally more important than control of molecular weight in changing mechanical properties. For these materials, the property can be correlated with density, which in turn is related to crystallinity. One primary example is PE, which in the commercial market is classified according to density. Hydraulic stress during injection-molding flow and calendering aligns the polymer molecules parallel to each other and favors crystallization. In these cases, tensile strength in the machine direction is generally higher. During tension measurement, elongation can reach several times the original length if necking occurs. In the necking region, the unoriented polymer chains are transformed into thin, oriented chains, resulting in a single, sharp-moving neck. Polyethylene and polyethylene terephthalate (PET) are known to exhibit necking.

The recently developed liquid crystal polymers are one extreme of such aligned polymers. Because of rigid molecules, these materials tend to align themselves in melts or solutions. By properly aligning them with stress during the solidifying stage, high tensile strength in one direction can be obtained. In some cases, the strength can be higher than that of steel.

Cross Linking. For thermosets, a major structural influence on properties is the number and type of cross links. In network polymers such as epoxies, the network is produced by the joining of many short chains. Nonetheless, the actual length of these short chains can vary considerably, with the network polymer becoming stiffer as the chains become shorter. The number of cross links formed also influences the final properties, with stiffness increasing as concentrations of cross links increase. In a thermoplastic, any cross links that are produced have a dramatic effect on properties, because such cross links change the thermoplastic nature of the material and may also destroy crystallinity.

Chemical Composition and Structure

Polymer structures can contain many different elements, but very few have more than four chemical elements. Nonetheless, the mer unit of many polymers and the way these mer units are bonded together to form a macroscopic polymer can be extremely complex. This is because several different types of bonds can occur and combine, and because the elements involved can be arranged in many different ways.

Most common polymers are made from compounds of carbon, although polymers can be made from inorganic chemicals, such as silicates and silicones. Carbon is common in the backbone of many polymer structures because of its unique ability to form extensive, stable covalent bonds with itself. Polymers and other compounds based on the chain-forming properties of carbon are called organic compounds. Although most polymers are organic, some inorganic polymers do exist. For example, many ceramic glasses could be considered inorganic polymers. However, because such inorganic glasses have very different properties from organic polymers, largely because they have ionic as well as covalent bonding, they are not usually treated as polymers.

In polymers and other molecules, the most common type of bond between carbon atoms is one in which each atom is bonded in a perfectly symmetrical three-dimensional arrangement to four neighboring atoms. Such a bond is known as a tetrahedral bond. When pure carbon is bonded together solely with tetrahedral bonds, the result is the form of carbon known as diamond. Most of the carbon atoms found in the backbone of polymer molecules are bonded together with tetrahedral bonds.

Another important type of bond that occurs between carbon atoms in polymers is the double bond. In such bonds, the carbon atom makes four bonds, as occurs with the tetrahedral bond, but two of these are between the same two carbon atoms. Although this yields a strong bond, it is more subject to chemical attack than are two separate single bonds.

Carbon also forms these same two types of bonds with elements other than itself. Thus, carbon atoms in polymers will be bonded in some combination of single and double bonds that adds up to four bonds per carbon atom.

Another element very prevalent in polymers is hydrogen. Unlike carbon, hydrogen can make only one bond with another element. Thus, hydrogen is never part of the backbone of a polymer, because a continuous backbone requires that each atom therein be bonded to at least two other atoms. However, hydrogen is the most common side or pendant attachment to the atoms of a polymer backbone, and most polymers contain many hydrogen atoms in their structures.

Hydrocarbon Polymers. Carbon and hydrogen form the structure of many polymers, known as hydrocarbon polymers, that are important commodity thermoplastics. In many of these, the mer unit is very simple, with the simplest of all being that of the synthetic polyethylene (Fig. 4). This simple mer unit is covalently bonded into long linear or branched chains. Polyethylene is an important commodity thermoplastic. Note that the PE structure is shown as a combination of two identical $-CH_2-$ units. Why is the mer unit not shown as a single $-CH_2-$ unit? Strictly speaking, this would be the correct mer unit for PE. However, because PE is actually polymerized from the compound ethylene and almost all other polymers have at least two atoms in their backbone chain, the mer unit of PE is normally shown as comprising two carbon and four hydrogen atoms (Fig. 4).

Other common hydrocarbon polymers (Fig. 13) have more complex mer structures than PE. For example, a slightly more complex mer unit is found in PP, which is used as a commodity thermoplastic in items such as medicine bottles, syringes, textile fibers, and packaging films and is also used as an engineering plastic. As previously noted, the polymer structures may also have several variations. Polyethylene, for example, can be low density (LDPE) or high density (HDPE) depending on the extent of chain branching and orientation. Polymers may also have atactic (random), isotatic (one-sided), or syndiotactic (regular alternation) arrangement of side groups.

Carbon-Chain Polymers. For reasons that are explained later in this article, most engineering plastics are not based on hydrocarbon polymers. Only a few elements other than carbon and hydrogen occur frequently in polymers. Of these, the ones that commonly occur in pendant groups on the side of the polymer backbone are chlorine, fluorine, oxygen, and nitrogen. Polytetrafluoroethylene (PTFE) is one of the simplest nonhydrocarbon carbon-chain thermoplastics. Its mer (Fig. 14) resembles that of PE with fluorine substituted for hydrogen. Polymethyl methacrylate (PMMA) is a carbon-chain polymer with a more complex mer unit (Fig. 14).

Heterochain Polymers. Two elements other than carbon that occur fairly often in the backbone of polymers are oxygen, which forms two bonds with other elements, and nitrogen, which forms three. Sulfur also occurs in the backbone of some polymers and, like oxygen, can form two bonds with other elements. Silicon occurs in the backbone of a specialized group of polymers known as silicones. Like carbon, silicon can form four tetrahedral bonds, but it does not form long chains and three-dimensional structures as easily as does carbon.

Polymers that have two or more elements in their backbones are known as heterochain polymers. For reasons that are described later in this article, heterochain polymers are often stronger and have higher temperature resistance than carbon-chain polymers. An important heterochain polymer that is used extensively as an engineering plastic is nylon 6/6 (Fig. 15).

Silicon and oxygen make up the backbone of the silicones, but even these inorganic-chain polymers invariably have carbon in their pendant groups. The most common silicone is polydimethyl siloxane (PDMS) (Fig. 15). Silicones

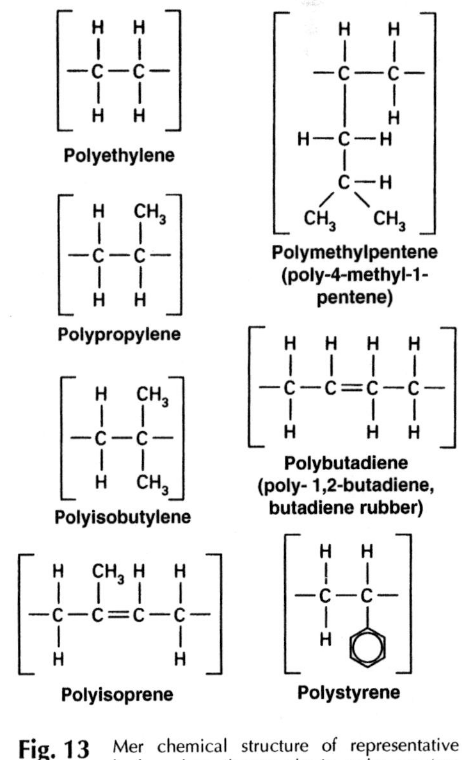

Fig. 13 Mer chemical structure of representative hydrocarbon thermo-plastic polymers (see Table 6 for glass-transition temperatures)

are generally not used as engineering plastics, but rather as adhesives, sealants, lubricants, and elastomers. The structures of other heterochain polymers are given in Fig. 15.

Polymers Containing Aromatic Rings. In addition to the various elements that may be found in polymers, a specialized chemical feature occurs in many important polymers. This is the aromatic ring, originally so called because it occurs in many compounds that have a distinctive aroma. It is also known as the benzene ring or phenyl group. It is a ring of six carbon atoms with alternating double and single bonds between them (Fig. 16). It represents a very special structure in organic chemistry because the positions of the double and single bonds actually resonate back and forth, with the result that each bond in the ring has characteristics midway between that of a double and single bond.

Such aromatic rings can occur either bonded into the backbone of polymers or attached as a side group. They can be very important to the properties of the polymer. For reasons that are described later in this article, high-temperature thermoplastic polymers almost invariably have such rings in their backbone. Several important high-temperature thermoplastics are shown in Fig. 17. Because the aromatic ring is composed of only carbon and hydrogen, it can also occur in fairly simple hydrocarbon polymers, such as the important commodity thermoplastic PS.

Polymer Names

Even for the experienced, it is not always easy to decipher the meaning of the names given to polymers. This is because a given polymer may have as many as four different types of names assigned to it: a systematic name, a chemical name, a customary name, and a commercial name. It is also quite common to abbreviate the names of polymers. Table 4 is a list of polymer abbreviations compiled from ASTM D 4000 (Ref 9, 10). The abbreviations in bold type are standard abbreviations listed in ASTM D 4000.

The systematic name is that assigned according to nomenclature rules adopted by the International Union of Pure and Applied Chemistry. Such a name is unique to the specific polymer and completely specifies the chemical structure of the simplest mer unit that can be described for the polymer. The systematic name for PE is poly(methylene), that for PS is poly(1-phenylethylene), and that for PVC is poly(1-chloroethylene). Although naming polymers by such a system seems to be a good approach, systematic names are not widely used. This is because the nomenclature rules are quite complicated, many of the resulting names are quite lengthy, and other names have simply become accepted.

The chemical name is used by polymer chemists in most of their descriptions. In some cases, this name is the same as the systematic name, and sometimes it is a shortened version of the systematic name, which lumps together several slightly different polymers under one term. The chemical name is invariably a name that resembles a systematic name in that it is composed of the "poly-" prefix followed by a chemical group. Polyethylene, polystyrene, and polyvinyl chloride are all examples of such names.

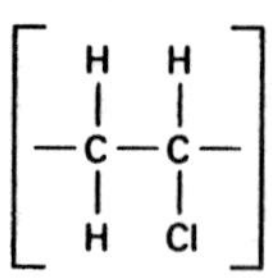

Polyvinyl chloride (vinyl)

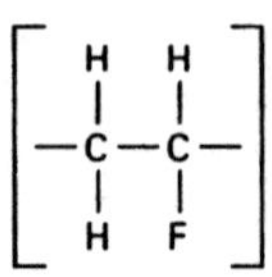

Polyvinyl fluoride

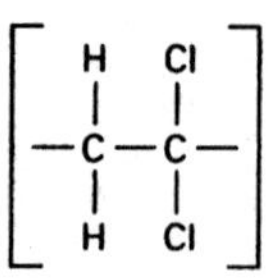

Polyvinylidene chloride

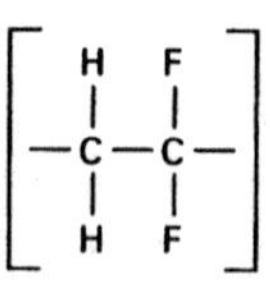

Polyvinylidene fluoride

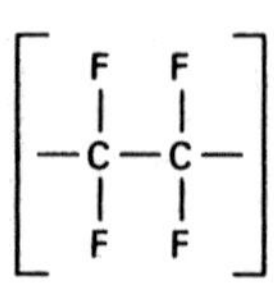

Polytetrafluoroethylene

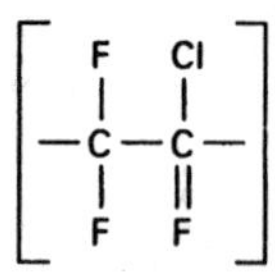

Polychlorotrifluoroethylene

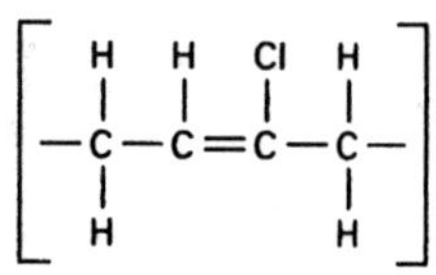

Polychloroprene (chloroprene rubber, or neoprene)

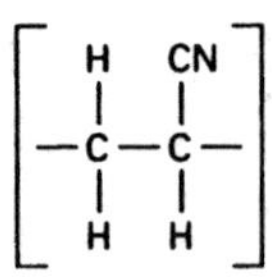

Polyacrylonitrile

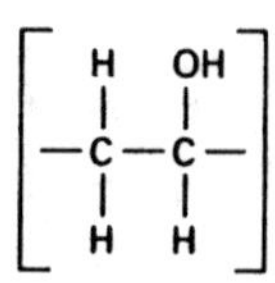

Polyvinyl alcohol

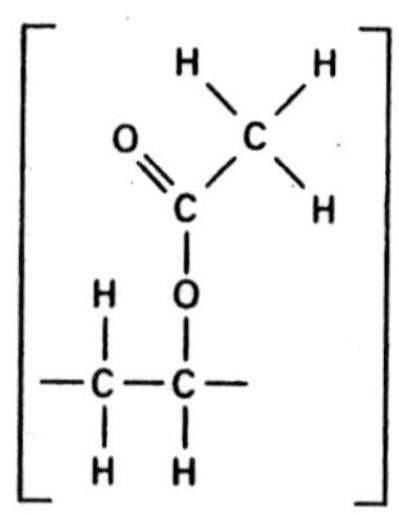

Polvinyl acetate

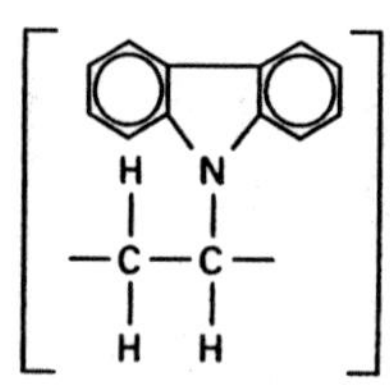

Polyvinyl carbazole

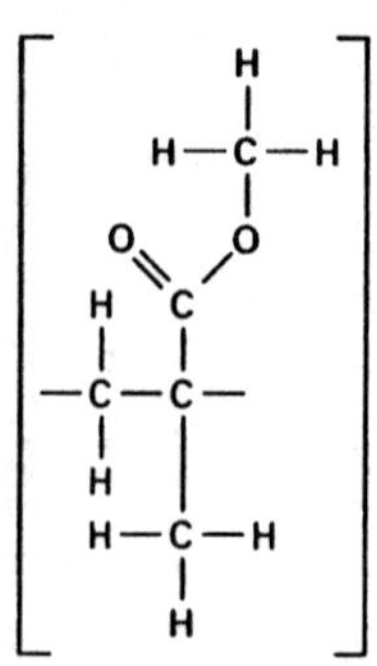

Polymethyl methacrylate

Fig. 14 Mer chemical structure of representative nonhydrocarbon carbon-chain thermoplastic polymers

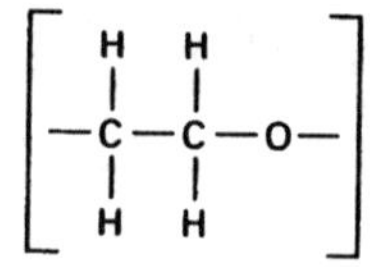

Polyethylene oxide

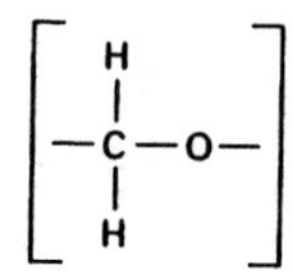

Polyoxymethylene

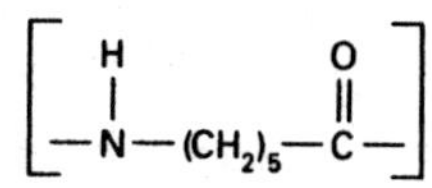

Polyamide (Nylon 6)

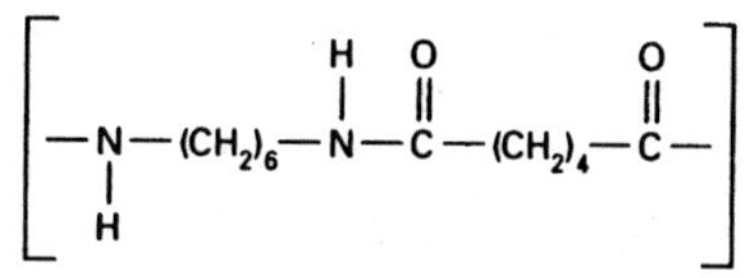

Polyamide (Nylon 6/6)

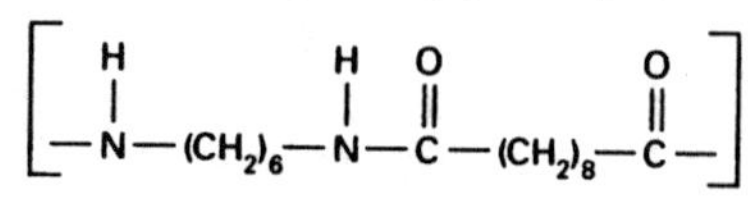

Polyamide (Nylon 6/10)

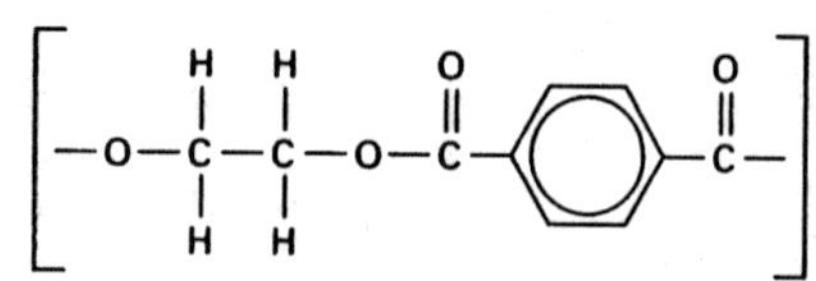

Polyethylene terephthalate

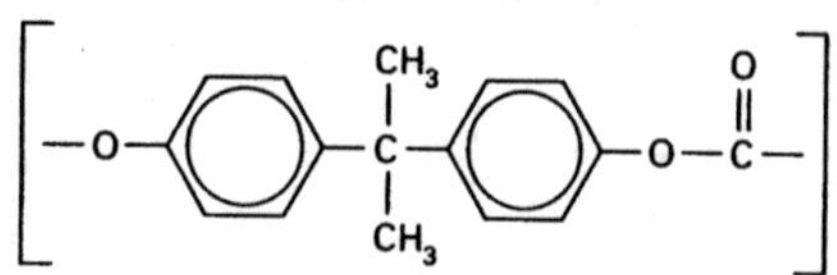

Polycarbonate

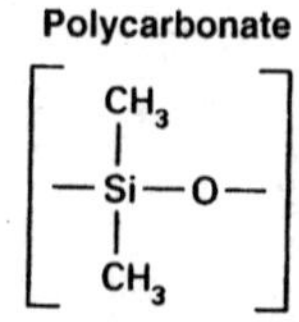

Polydimethyl siloxane (silicone rubber)

Fig. 15 Mer chemical structure of representative heterochain thermoplastic polymers

The chemical name is commonly used by polymer scientists. These names are based on the names of the mer unit of the polymer or, for complex polymers, on the name of one or more prominent chemical groups that make up the polymer. Figure 18(a) and (b) lists chemical groups that may be involved in the naming of polymers. This book generally refers to polymers by their chemical names or, for groups of polymers, by chemical family names.

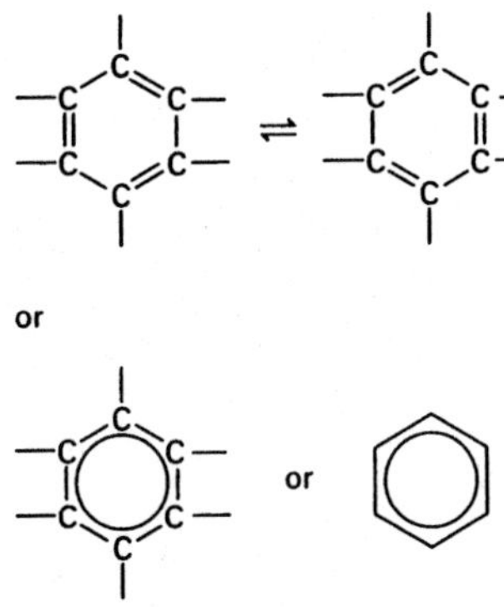

Fig. 16 Carbon ring structure of the phenyl group (also known as benzene ring or aromatic ring)

The customary name (or common name) often lumps together even more polymers than does the chemical name. Such names are unpredictable, being derived from early marketing terms for the material, modified chemical names, or other sources. They are often used in a generic sense to describe a group of polymers without using proprietary commercial names. Such names include vinyl, acrylic, and nylon.

The commercial name is assigned by the company marketing the polymer and is usually proprietary. A given polymer may have several different commercial names, because several different companies may market the same polymer, and the same commercial name may refer to several different polymers. However, some of these names, such as nylon, have been allowed to become generic and are now used as customary names.

Properties of Polymers

This introductory article cannot cover all polymer properties, nor can it discuss all of the structural influences on any given property. Instead, the most important properties of polymers and the most significant influences of structure on those properties are covered. The next article in this book, "Effects of Composition, Processing, and Structure on Properties of Engineering Plastics," discusses properties in more detail. Other articles cover specific properties or characteristics more thoroughly with particular emphasis on the performance of plastic products.

Thermal Properties

Thermal properties include dimensional stability, thermal decomposition, thermal expansion, and thermal conductivity. The thermal characteristics that are important in the application of engineering plastics are listed in Table 5.

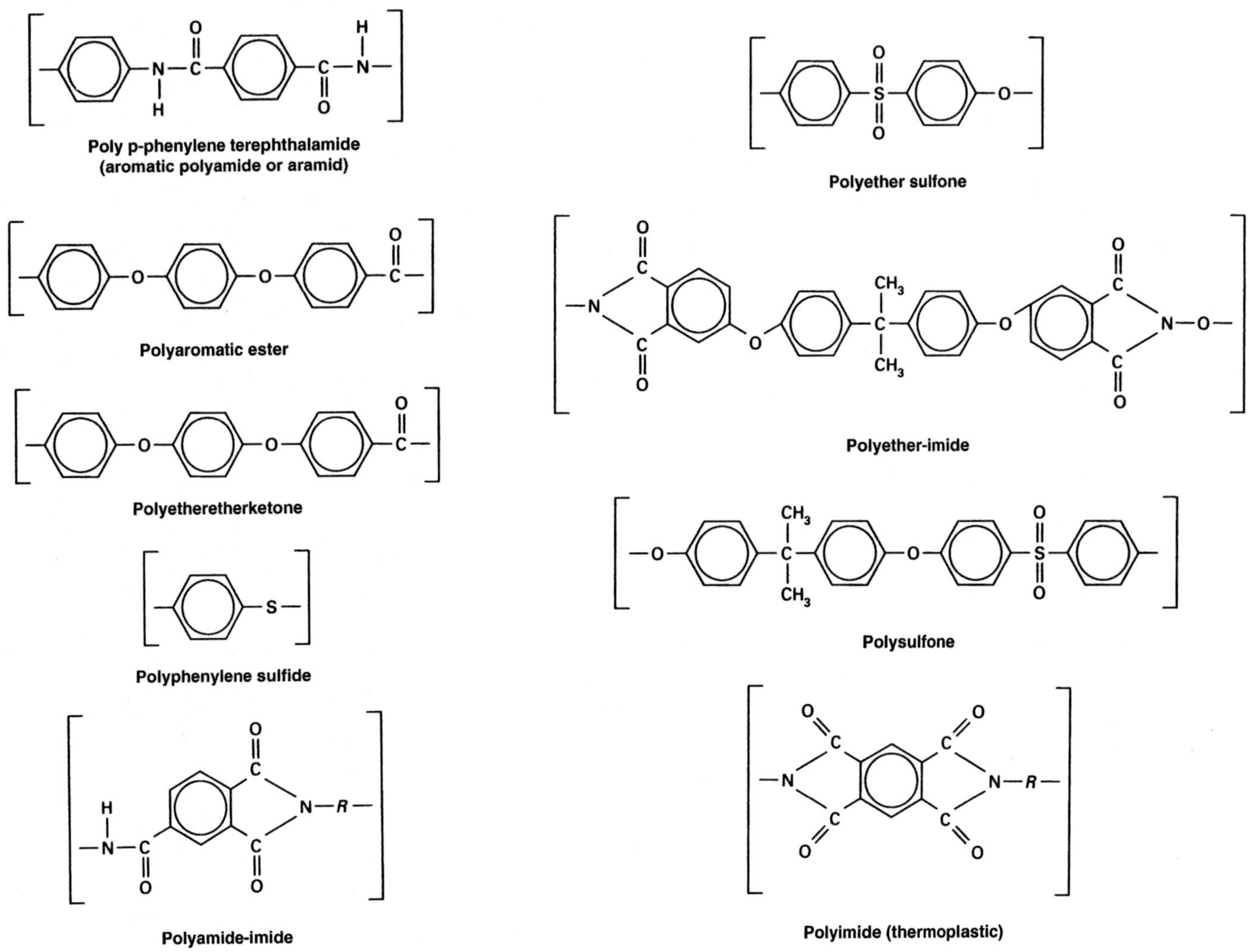

Fig. 17 Mer chemical structure of representative thermoplastic polymers for high-temperature service

Dimensional stability is the most important thermal property for the majority of polymers because a polymer cannot be used at a temperature above which it loses dimensional stability. For most thermoplastic polymers, the main determinant of dimensional stability is the T_g of the polymer (Table 6). Because of the partially or completely noncrystalline nature of polymers, they undergo a transition as a function of temperature that is not seen in fully crystalline materials. This T_g is a measure of the temperature at which the noncrystalline portions of the polymer change from a glassy state (at low temperature) to a rubbery state (at higher temperatures). This is the most important temperature that can be specified for most polymers because in all but highly crystalline polymers it represents the temperature above which the polymer loses most of its stiffness and thus its dimensional stability.

Glass-transition temperatures are influenced by moisture absorption and the intentional addition of plasticizers. Absorbed moisture invariably lowers the T_g, and the more moisture is absorbed, the lower the transition temperature. This is consistent with the role of water as a plasticizer, which is why absorbed moisture can reduce the strength of plastics. Plasticizers are low-molecular-weight additives that lower strength and T_g. The lowering of transition temperatures by plasticizers can be quantitatively described by various mixing formulas (Ref 11, 12), which can be quite useful for predicting the loss of properties due to absorbed moistures.

There is much argument about the character of the glass transition, which occurs in the noncrystalline regions of the polymer. It may be a second-order phase transformation that is severely influenced by kinetics, or it may be a purely kinetic process. The actual temperature at which loss of dimensional stability is noted depends on the rate of testing. For example, if a polymer is heated at a moderate rate, a loss of stiffness and dimensional stability will be observed at a temperature near the listed T_g for the polymer. If, however, it is heated very rapidly, such a loss in dimensional stability will not be noted until a higher temperature is reached. (Of course, in a given application it is possible that the gradual change in properties as a function of temperature may make the polymer unusable even at a temperature below the T_g).

The change in properties at the glass transition occurs not at a distinct temperature, but over a range of temperatures. Thus, the T_g specified for a polymer actually represents roughly the center of a transition region. In a thermoplastic polymer such as PS, the change that occurs gradually over the T_g region eventually leads to a complete loss of dimensional stability. In a network polymer such as epoxy, the change is less severe, but nonetheless produces significant softening and loss of mechanical properties.

One way to understand the reason for the substantial change in properties at the T_g is to focus on the expansion that occurs in the polymer as temperature is increased. The free volume, which may be thought of as room inside the polymer, gradually increases until cooperative rotational motion of five to ten mer units is possible. At this point the polymer can deform in response to an applied stress, for example, much more easily than it could at a lower temperature. Clearly, the flexibility and bulkiness of the mer unit and the cohesive energy between molecules strongly influence the temperature at which this can occur. The more flexible and less bulky the mer unit, the easier it is for the cooperative rota-

Table 4 Abbreviations and names of plastics

Abbreviation(a)	Plastic family name(b)
ABA	Acrylonitrile-butadiene-acrylate
ABS	Acrylonitrile-butadiene-styrene
ACS	Acrylonitrile-styrene and chlorinated polyethylene
AES	Acrylonitrile-styrene and ethylene propylene rubber
AMMA	Acrylonitrile-methyl methacrylate
ARP	Aromatic polyester
ASA	Acrylonitrile-styrene-acrylate
CA	Cellulose acetate (acetate)
CAB	Cellulose acetate (butyrate)
CAP	Cellulose acetate propionate
CE	Cellulose plastics, general
CF	Cresol formaldehyde
CMC	Carboxymethyl cellulose
CN	Cellulose nitrate (celluloid)
CP	Cellulose propionate (propionate)
CPVC	Chlorinated polyvinyl chloride
CPE	Chlorinated polyethylene
CS	Casein
CTA	Cellulose triacetate (triacetate)
CTFE	Polymonochlorotrifluoroethylene
DAP	Poly(diallyl phthalate)
DMC	Dough molding compound (usually polyester)
EC	Ethyl cellulose
EAA	Ethylene-acrylic acid
EEA	Ethylene-ethyl acrylate
EMA	Ethylene-methacrylic acid
EP	Epoxy, epoxide
EPD	Ethylene-propylene-diene
EPM	Ethylene-propylene polymer
ETFE	Ethylene-tetrafluoroethylene copolymer
EVA (EUAC)	Ethylene-vinyl acetate
EVOH, EVAL, EVOL	Ethylene-vinyl alcohol
FEP	Fluorinated ethylene propylene copolymer
FEP	Tetrafluoroethylene-hexafluoropropylene copolymer
FF	Furan formaldehyde
HDPE	High-density polyethylene
HIPS	High-impact polystyrene
LDPE	Low-density polyethylene
IPS	Impact styrene
LLDPE	Linear low-density polyethylene
MBS	Methacrylate-butadiene styrene
MDPE	Medium-density polyethylene
MF	Melamine-formaldehyde (melamine)
PA	Polyamide (some nylons)
PAI	Polyamide-imide
PARA	Polyaryl amide
PB	Polybutene-1
PBT (PBTP, TMT)	Polybutylene terephthalate, (polyester)
PC	Polycarbonate
PCT	Poly-(1,4-cyclohexyl-enediaminemethylene terephthalate)
PCTFE	Polychlorotrifluoroethylene
PE	Polyethylene
PEBA	Polyether block amide
PEEK	Polyetheretherketone
PEEKK	Polyetheretherketoneketone
PEG	Polyethylene glycol
PEI	Polyether-imide
PEK	Polyetherketone
PEO	Polyethylene oxide
PESV (PES)	Polyether sulfone
PET (PETP)	Polyethylene terephthalate, (polyester)
PETG	Glycol modified polyethylene terephthalate comonomer
PF	Phenol-formaldehyde (phenolic)
PFA	Perfluoro alkoxy alkane
PI	Polyimide
PIB	Polyisobutylene
PMMA (PMM)	Polymethyl methacrylate, (acrylic)
PMMI	Polymethylmethacrylimide
PMP	Poly(4-methyl pentene-1)
POM	Polyoxymethylene (acetal), Polyacetal, polyformaldehyde
POP	Polyphenylene oxide
PP	Polypropylene plastics
PPE	Polyphenylene ether
PPG	Polypropylene glycol
PPO	Polyphenylene oxide
PPO	Polypropylene oxide
PPS	Polypropylene sulfide
PPOX	Polypropylene oxide
PPS	Polyphenylene sulfide
PPSU	Polyphenylene sulfone
PS	Polystyrene (styrene)
PSU (PS)	Polysulfone
PTFE	Polytetrafluoroethylene
PUR	Polyurethane (urethane)
PVA	Polyvinyl acetal
PVAC	Polyvinyl acetate
PVAL (PVA)	Polyvinyl alcohol
PVB	Polyvinyl butyral
PVC	Polyvinyl chloride
PVDC	Polyvinylidene chloride
PVDF	Polyvinylidene fluoride
PVF	Polyvinyl fluoride
PVFM	Polyvinyl formal
PVK	Polyvinylcarbazole
PVP	Polyvinyl pyrrolidone
P4MP1	Poly-4-methyl pentene-1
RF	Resorcinol-formaldehyde
SAN	Styrene-acrylonitrile
SB	Styrene-butadiene
SI	Silicone plastics
SMA	Styrene-maleic anhydride
SMS	Styrene/α-methylstyrene
TEEE	Thermoplastic elastomer, ether-ester
TEO	Thermoplastic elastomer-olefinic
TES	Thermoplastic elastomer-styrenic
TPEL	Thermoplastic elastomer
TPES	Thermoplastic polyester (general)
TPS	Toughened polystyrene
TPUR	Thermoplastic polyurethane
UF	Urea-formaldehyde (Urea)
UP	Unsaturated polyester
UPVC	Unplasticized PVC
VLDPE	Very-low-density polyethylene
XPS	Expanded polystyrene

(a) Abbreviations in bold are standard symbols in ASTM D 4000. (b) Common names or common short version of full name are in parenthesis.
Sources: Ref 9, 10

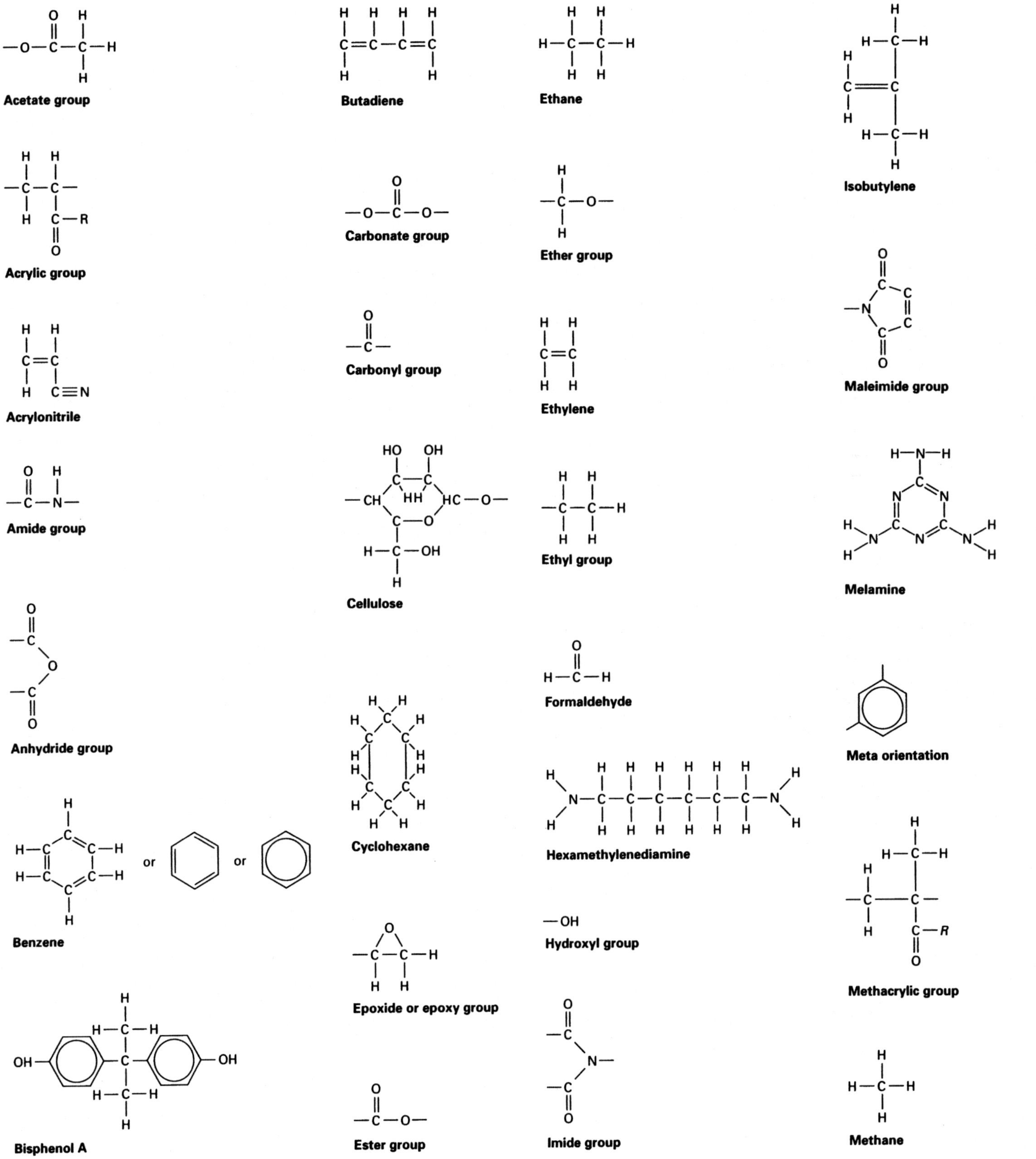

Fig. 18(a) Chemical groups in the naming of polymers. Acetate group to methane tion to occur and thus the lower the T_g. However, if the polymer molecules are bonded to one another by strong secondary bonds, the bonding will interfere with such motion, even if the chain is very flexible and not very bulky. This, of course, is what gives thermosets higher average T_gs than thermoplastics. Those thermoplastics with the highest T_gs have stiff, bulky chains and strong intermolecular hydrogen bonding between chains.

The crystalline portion of a semicrystalline polymer has a thermodynamic T_m similar to those found in other crystalline materials. In some semicrystalline polymers this may be the most important transition temperature. If high crystallinity (roughly 50% or higher) can be obtained, it may permit a polymer to be used above its T_g. High crystallinity can be attained (with difficulty) only in thermoplastics. However, if substantial crystallinity can be obtained, loss of dimensional stability will not occur at T_g because the crystalline regions will not undergo a glass transition and thus will restrict the deformation of the noncrystalline regions. Thus, in such polymers it is possible to extend the region of acceptable dimensional stability above T_g. If crystallinity is quite high (say 80% or more), this may extend the short-term use temperature almost to the T_m.

Substantially crystalline polymers in the temperature range between T_g and T_m are referred to as leathery, because they are made up of a combination of the rubbery noncrystalline regions and the stiff, crystalline regions. Thus, PE, PP, and other polymers are still useful at room temperature, and PA is useful to moderately elevated temperatures, even though these temperatures are above their respective T_gs.

As with T_g, T_m is increased by a decrease in chain flexibility, an increase in bulkiness, or an increase in the strength of intermolecular bonding. However, for a crystalline polymer, decreases in chain flexibility and increases in bulkiness may need to be limited because these factors adversely influence crystallinity. In a crystalline polymer, dimensional stability increases with added crystallinity because this decreases the portion of the polymer that is influenced by T_g.

Numerous examples of the influence of structure on T_g and T_m can be noted in Table 6. Polyethylene, for example, has a T_g of either about −100 or −20 °C (−150 or −5 °F). It has a T_m of 115 °C (240 °F) for the less-crystalline, low-density version, and 137 °C (280 °F) for the more highly crystalline, high-density version. The difference is due to increased intermolecular bonding in the more highly crystalline, high-density polyethylenes (HDPEs). Polyethylene is flexible, not bulky, and has only weak dispersion bonds between chains. Polystyrene with its bulky, aromatic side groups (Fig. 13), has a T_g of about 100 °C (212 °F) and a T_m (for the little crystallinity that occurs) of 240 °C (465 °F). It also is held together by dispersion bonds only. Polycarbonate (PC) has two aromatic rings in its backbone (Fig. 15). This produces a very stiff, bulky mer. Furthermore, its heterochain structure permits hydrogen bonding between molecules. The T_g of PC is 150 °C (300 °F), and its T_m is 265 °C (510 °F).

The heterochain thermoplastics (Fig. 17) have the highest values for T_g and T_m. These high-temperature polymers have inflexible and bulky rings and cyclic structures and are all heterochain polymers having many sites for intermolecular

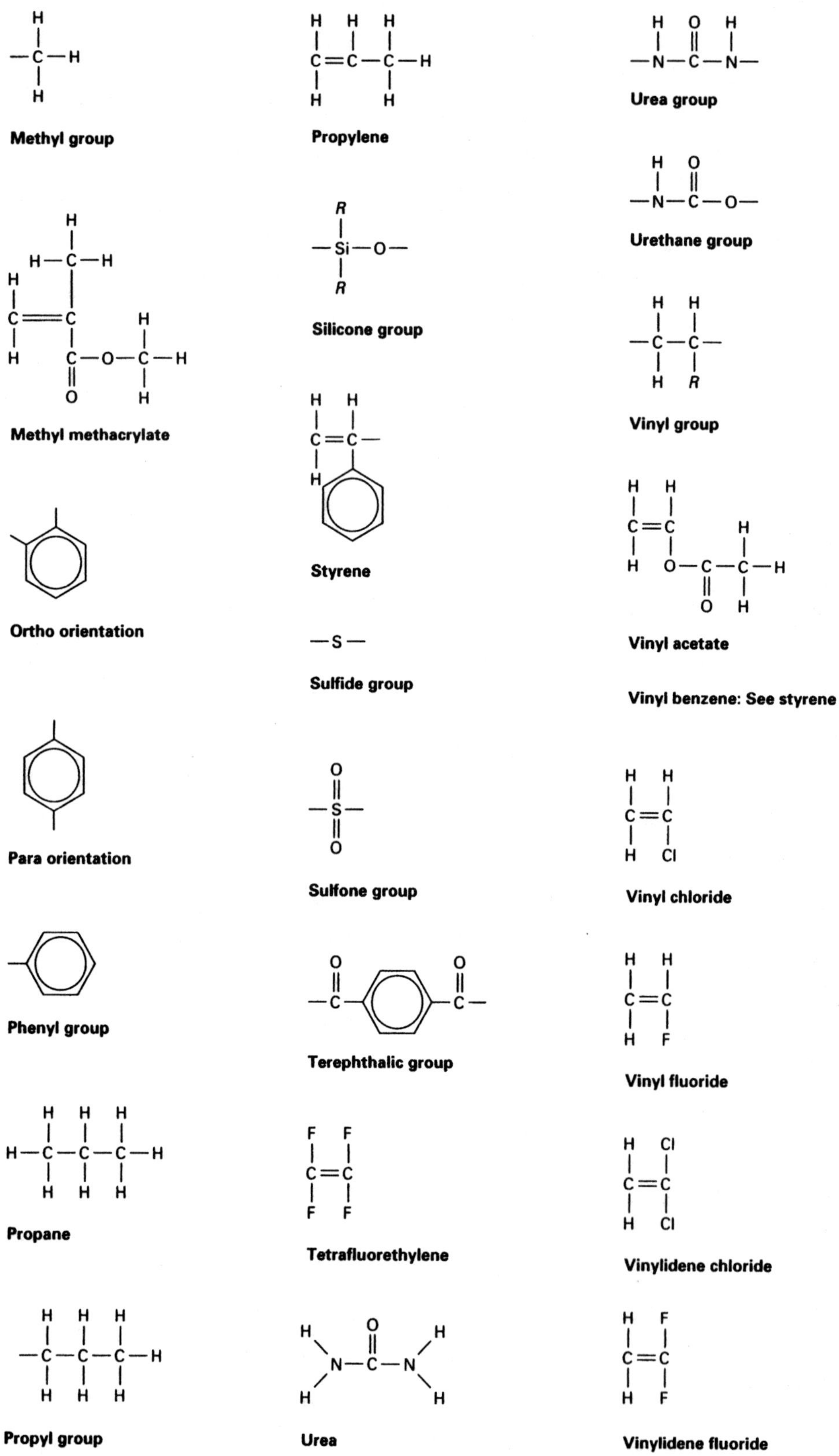

Fig. 18(b) Chemical groups in the naming of polymers. Methyl group to vinylidene fluoride

hydrogen bonding. It should be noted that the flexible ether and sulfide linkages included in most of these polymers do lower the T_g, but are added intentionally to give the chain enough flexibility so that the polymer can be processed and, in some cases, so that high crystallinity can be attained. Crystallinity is used to extreme effect in the aramid fiber poly (*p*-phenylene terephthalamide), shown at the top of Fig. 17, to produce a highly oriented, crystalline structure whose extremely strong hydrogen bonding gives it not only a high T_g but also a T_m that is actually above its decomposition temperature.

Flexibility and bulkiness are also used to modify the T_g of thermosets. For example, flexibilizers that usually contain fairly long segments of $-CH_2-$ units are added to epoxies to make them less brittle; they also lower the T_g of the cured resin. On the other hand, the epoxies with the highest T_gs are cross linked from both resins and curing agents that are relatively inflexible and bulky. Because thermosets are covalently cross linked, secondary bonding has only a small influence upon the T_g. However, the cross-link density of the thermoset has a dramatic effect on the T_g, and in many cases much effort is spent in the formulation and cure of thermoset resins to ensure that they achieve a high cross-link density.

In addition to the T_g and T_m, polymers can undergo other transition temperatures. These include phase changes in the crystalline phase as well as various transitions in the noncrystalline regions. The latter are usually due to side-group motion, but may also result from motion of some subunit of the chain itself. These transitions can have an influence on properties, but the influence is usually on properties other than dimensional stability.

Structural factors originating within the molecule also have an influence on dimensional stability. Different stereoisomers have different T_gs and T_ms and may have very different percentages of crystallinity. Branching interferes with intermolecular bonding and crystallinity and thus lowers dimensional stability. Increases in molecular weight increase T_g and T_m somewhat, but the ease of crystallization also decreases. Thus, increased molecular weight may have an adverse effect on the dimensional stability of crystalline polymers. Copolymerization usually produces a T_g somewhere between the two mers, or a double T_g. However, the influence of copolymerization on the T_m is much more dramatic. In many cases, copolymerization causes the T_m to drop so low that crystallinity is totally destroyed.

Thermal Decomposition. For applications having moderate thermal requirements for the polymer, thermal decomposition may not be an important consideration. However, if the polymer is one offering dimensional stability to high temperatures, it is possible that its processing and/or service temperatures may approach its decomposition temperature. The thermal decomposition temperature of the polymer is largely determined by the elements and bonding within the mer unit. Thermal decomposition occurs when the primary covalent bonds of the polymer are ruptured. The decomposition temperature, as well as the general chemical resistance of the polymer, is thus increased by stronger bonds as well as by the inclusion of the mer of elements and bonds that are not easily attacked by chemicals or other agents.

Table 3 lists the strengths of common bonds in polymers. To a first approximation, the higher the energies of the bonds within the mer, the higher the thermal decomposition temperature. However, there are several complications to this approximation. For example, because a double bond is less stable than two single bonds, rupture of the bond to produce two single bonds is relatively easy. Inclusion of a double bond into a ring or cyclic structure, however, greatly strengthens both double and single bonds.

Thermal Expansion. In a thermoset, the ease or difficulty of thermal expansion is dictated for the most part by the degree of cross linking, as well as the overall stiffness of the units between cross links. Less-flexible units are also more resistant to thermal expansion. Influences such as secondary bonding have much less effect on the thermal expansion of thermosets.

In a thermoplastic, thermal expansion is controlled less by the stiffness of the chains than by the strength of the secondary bonds between molecules. For example, thermoplastics held together by strong hydrogen bonds generally expand less than those held together by dispersion bonds. However, thermal expansion is also greatly reduced by crystallinity, and the absence or presence of substantial crystallinity may greatly alter the thermal expansion of a polymer. Thus, any factors that interfere with crystallinity, such as branching or copolymerization, may increase the thermal expansion coefficient as well.

Table 5 Thermal properties of selected plastics

Material	Heat deflection temperature at 1.82 MPa (0.264 ksi)		UL Index		Thermal conductivity		Coefficient of thermal expansion, 10^{-5}/K
	°C	°F	°C	°F	W/m · K	Btu · in./h · ft² · °F	
Acrylonitrile-butadiene-styrene (ABS)	99	210	60	140	0.27	1.9	5.3
ABS-polycarbonate (ABS-PC) alloy	115	240	60	140	0.25	1.7	3.5
Diallyl phthalate (DAP)	285	545	130	265	0.36	2.5	2.7
Polyoxymethylene (POM)	136	275	85	185	0.37	2.6	3.7
Polymethyl methacrylate (PMMA)	92	200	90	195	0.19	1.3	3.4
Polyarylate (PAR)	155	310	...	...	0.22	1.5	3.1
Liquid crystal polymer (LCP)	311	590	220	430	...	...	0.5
Melamine-formaldehyde (MF)	183	360	130	265	0.42	2.9	2.2
Nylon 6	65	150	75	165	0.23	1.6	2.5
Nylon 6/6	90	195	75	165	0.25	...	4.0
Amorphous nylon 12	140	285	65	150	0.25	...	7.0
Polyarylether (PAE)	160	320	160	320	...	...	3.0
Polybutylene terephthalate (PBT)	...	...	120	250	...	...	4.5
PC	129	265	115	240	0.20	...	3.8
PBT-PC	129	265	105	220	...	...	2.8
PEEK	...	...	250	480	0.25	1.7	2.6
Polyether-imide (PEI)	210	410	170	340	0.22	1.5	3.1
Polyether sulfone (PESV)	203	395	170	340	...	...	5.5
PET	224	435	140	285	0.17	...	1.5
Phenol-formaldehyde (PF)	163	325	150	300	0.25	1.7	1.6
Unsaturated polyester (UP)	279	535	130	265	0.12	0.8	1.6
Modified polyphenylene oxide alloy (PPO)	100	212	80	175	...	...	3.8
Polyphenylene sulfide (PPS)	260	500	200	390	0.17	...	3.0
Polysulfone (PSU)	174	345	140	285	0.26	1.8	3.1
Styrene-maleic anhydride terpolymer (SMA)	103	215	80	175	...	...	...

UL, Underwriters' Laboratory

Any cross linking has a substantial effect on the thermal expansion of a thermoplastic. In a noncrystalline thermoplastic, the thermal expansion coefficient is reduced. In a crystalline thermoplastic, however, the decreased expansion due to cross linking may be partially offset by loss of crystallinity.

Thermal conductivity is also dependent on primary and/or secondary bonding, in that heat is conducted more easily through a polymer that is strongly bonded. Thus, thermosets usually have higher thermal conductivities than do thermoplastics.

In general, however, the thermal conductivity of polymers is low, and polymer structure does not alter the value very much. Thermal conductivity can be increased by adding metallic fillers or electrically insulating fillers such as alumina, if electrical conductivity is undesirable. Likewise, thermal conductivity is decreased by foaming with air or some other gas, as is done to make styrofoam coffee cups.

Table 6 Glass-transition temperatures (T_g), and melting temperatures (T_m) of representative thermoplastic polymers

Chemical name	T_g °C	T_g °F	T_m °C	T_m °F
Hydrocarbon thermoplastics (Fig. 13)				
Polyethylene				
HDPE	−90 or −20	−130 or −5	137	280
LDPE	−110 or −20	−165 or −5	115	240
Polypropylene				
Atactic	−18	0	176	350
Isotactic	−10	15	176	350
Polyisobutylene	−70, −60	−95, −75	128	260
Polyisoprene				
Cis: natural rubber	−73	−100	28	80
Trans: gutta percha	...	...	...	...
Polymethyl pentene (poly-4-methyl-1-pentene)	29	85	250	480
Polybutadiene (poly-1,2-butadiene, butadiene rubber)				
Syndiotactic	−90	−130	154	310
Isotactic	−90	−130	120	250
Polystyrene				
Atactic	100, 105	212, 220	(a)	(a)
Isotactic	100, 105	212, 220	240	465
Nonhydrocarbon carbon-chain thermoplastics (Fig. 14)				
Polyvinyl chloride (vinyl)	87	190	212	415
Polyvinyl fluoride	−20	−5	200	390
Polyvinylidene chloride	−17	1	198	390
Polyvinylidene fluoride	−35	−30	...	...
Polytetrafluoroethylene	−97, 126	−140, 260	327	620
Polychlorotrifluoroethylene	45	115	220	430
Polychloroprene (chloroprene rubber, or neoprene)	−50	−60	80	175
Polyacrylonitrile	104, 130	220, 265	317	600
Polyvinyl alcohol	85	185	258	495
Polyvinyl acetate	29	85	...	...
Polyvinyl carbazole	150, 208	300, 405	...	...
Polymethyl methacrylate				
Syndiotactic	3	35	105, 120	220, 250
Isotactic	3	35	45	115
Heterochain thermoplastics (Fig. 15)				
Polyethylene oxide	−67 to −27	−90 to −15	62–72	145–160
Polyoxymethylene	−85	−120	175	345
Polyamide				
Nylon 6	50	120	215	420
Nylon 6/10	40	105	227	440
Polyethylene terephthalate	69	155	265	510
Polycarbonate	150	300	265	510
Polydimethyl siloxane (silicone rubber)	−123	−190	−54	−65
Thermoplastic polymers for high-temperature service (Fig. 17)				
Poly *p*-phenylene terephthalamide (aromatic polyamide or aramid)	375	705	~640(c)	~1185(c)
Polyaromatic ester	...	...	421	790
Polyether ether ketone	143	290	334	635
Polyphenylene sulfide	85	185	285	545
Polyamide-imide	277–289	530–550	(d)	(d)
Polyether sulfone	225	435	(d)	(d)
Polyether-imide	215	420	(d)	(d)
Polysulfone	193	380	(d)	(d)
Polyimide (thermoplastic)	280–330	535–625	(d)	(d)

(a) Polymer is generally 95% or more noncrystalline. Any T_m given is for remaining crystalline portion or for crystalline version. (c) T_d = 500 °C (930 °F). *R* contains at least one aromatic ring. (d) Polymer is generally 95% or more noncrystalline. Any T_m given is for remaining crystalline portion or for crystalline version.

Mechanical Properties

The general mechanical behavior a polymer may be that of a fiber, plastic, or elastomer (Fig. 19). The use depends on the relative strength of its intermolecular bonds and structural geometry. Noncrystalline polymers with weak intermolecular forces are usually elastomers or rubbers at temperatures above their T_g. In contrast, polymers with strong hydrogen bonds and the possibility of high crystallinity can be made into fibers. Polymers with moderate intermolecular forces are plastic at temperatures below T_g. Some polymers, such as nylon, can function both as a fiber and as a plastic. Other polymers, such as isotactic polypropylene, lack hydrogen bonds, but because of their good structural geometry, they can serve both as a plastic and as a fiber.

Because of the partially or completely noncrystalline structure of polymers, they undergo a change in mechanical behavior that is not seen in fully crystalline materials. At temperatures well below T_g, plastics exhibit a high modulus and are only weakly viscoelastic. At temperatures above T_g, there is drastic reduction of modulus. Therefore, the T_g is the most important temperature that can be specified for most polymers because in all but highly crystalline polymers, it represents the temperature above which the polymer loses most of its stiffness, as previously noted in the section "Dimensional Stability" in this article. Mechanical properties are also affected by molecular weight. Most material manufacturers provide grades with different molecular weights. High-molecular-weight materials have high-melt viscosities and low-melt indexes. For a commercial product, a melt index is generally an inverse indicator of molecular weight.

When molecular weight is low, the applied mechanical stress tends to slide molecules over each other and separate them. The solid, with

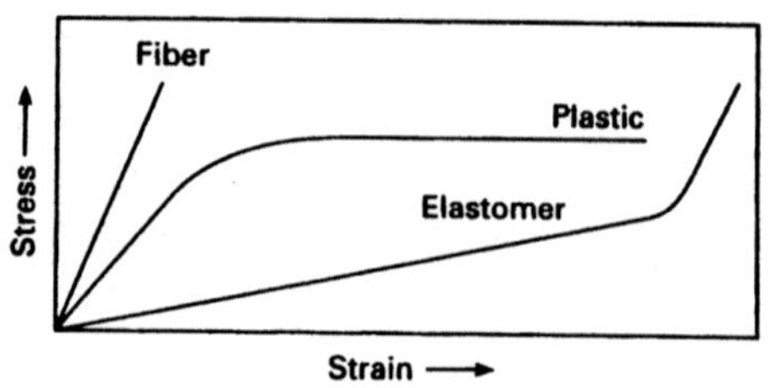

Fig. 19 Typical stress-strain curve for a fiber, plastic, and elastomer

very little mechanical strength, has negligible structural value. With a continuing increase in molecular weight, the molecules become entangled, the attractive force between them becomes greater, and mechanical strength begins to improve. It is generally desirable for materials manufacturers to make plastics with sufficiently high-molecular weights to obtain good mechanical properties. For PS this molecular weight is 100,000, and for PE this value is 20,000. It is not desirable to increase molecular weight further because melt viscosity will increase rapidly, although there are occasional exceptions to this rule. The yield strength of PP decreases when molecular weight increases. High molecular weight and branching reduce crystallinity. Polymers with high intermolecular interaction, such as hydrogen bonding, do not require high molecular weight to achieve good mechanical properties.

Several different types of mechanical properties are used to characterize polymers, but three important properties of load-bearing polymers (plastics) are usually stiffness, strength, and toughness. These three properties are briefly described in the following paragraphs with more details in other articles.

Stiffness. The same factors that influence thermal expansion dictate the stiffness of a polymer. Thus, in a thermoset, the degree of cross linking and the overall flexibility of the units is most important. In a thermoplastic, crystallinity and secondary bond strength control stiffness. A typical modulus-temperature curve is shown in Fig. 20. At temperatures below T_g, most plastic materials have a tensile modulus of about 2 GPa (0.3×10^6 psi). If the material is semicrystalline (at least 50% crystalline), a small drop in modulus is generally observed at T_g, while a large drop is seen at T_m. The T_g is primarily associated with amorphous, rather than crystalline resins or cross-linked thermosets. Resins that are partially crystalline have at least a 50% amorphous region, which is the region that has a T_g. If the material is amorphous, a single decrease is usually seen at temperatures near T_g. At even higher temperatures, there is another similar drop in modulus, and the plastic flows easily as a high-viscosity liquid. At this condition, the plastic can be processed by extrusion or molding.

Strength. The concept of strength is much more complex than that of stiffness. Many different types of strength exist, including short- and long-term strengths, static or dynamic strengths, and impact strength. Some strength aspects are intertwined with those of toughness, as well. Because of this complexity, this section provides a simplified overview of strength in order to point out the most important influences on it. It is also important to point out the importance of specific strength. Engineering plastics are not as strong as metals, but due to the lower density of plastics, the specific strengths of structural plastics are higher than those of metallic materials. This is shown in Table 7, which compares the range of mechanical properties of plastics with those of other engineering materials. These data show that glass-filled plastics have strength-to-weight ratios that are twice those of steel and cast aluminum. In addition to glass fillers, other types of additives (such as plasticizers, flame retardants, stabilizers, and impact modifiers) can also modify the mechanical properties of plastics.

The short-term yield strength of a polymer is largely controlled by the bonding that holds the polymer together. In a thermoplastic, both the intrachain covalent bonding and the interchain secondary bonding contribute to strength. Crystallinity is also very important: if it is substantial, the molecules will extend between the regions and into noncrystalline regions. Thus, the crystalline regions work cooperatively and increase the yield strength of the material, while also restricting the deformation in the noncrystalline regions. Unless crystallinity is impeded, increased molecular weight generally increases yield strength. Cross linking increases short-term yield strength substantially, but has an adverse effect on toughness. Of course, in thermosets, increased cross-link density increases short-term yield strength.

Long-term rupture strengths in thermoplastics are increased much more by increased secondary bond strength and crystallinity than by increased intrachain covalent bond strength. Fatigue strength is similarly influenced, and all factors that influence thermal dimensional stability also influence fatigue strength. This is because substantial heating is often encountered in fatigue. Short-term failure strengths, in most cases, and impact strengths, invariably, are determined by the factors that control toughness.

Toughness. Like strength, this is a complex topic and is simplified for this discussion. Even the definition of toughness is complex; definitions of a tough material range from one having a high elongation to failure to one in which a lot of energy must be expended to produce failure. For this discussion, the latter definition is used.

For high toughness, a polymer needs both the ability to withstand load and the ability to elongate substantially without failure. It may appear that factors contributing to high stiffness will thus be required, but this is incorrect because of the inverse relationship between flaw sensitivity and toughness. The higher the stiffness and yield strength of a material, the more flaw sensitive it becomes. However, because some load-bearing capacity is required to provide toughness, high toughness is achieved by a trade-off of factors.

Because crystallinity increases both stiffness and yield strength, an increase in crystallinity usually decreases toughness. This is true below the T_g in a mostly noncrystalline polymer and below or above the T_g in a substantially crystalline polymer. However, above the T_g in a polymer having only moderate crystallinity, increased crystallinity improves toughness. An increase in molecular weight from low values increases toughness, but with continued increases, toughness begins to drop.

Cross linking produces some dimensional stability and improves toughness in a noncrystalline polymer above the T_g, but high levels of cross linking lead to embrittlement and a loss of toughness. This is one of the problems encountered in thermosets for which an increase in the T_g is desired. Increased cross linking or stiffening of the chain segments increases the T_g, but also decreases toughness, sometimes to an unacceptable degree.

One of the classic ways to increase toughness is to blend, fill, or copolymerize a brittle polymer with a tough one. While some loss in stiffness is usually encountered, the result can be a very satisfactory combination of properties. Copolymerization to produce toughened regions

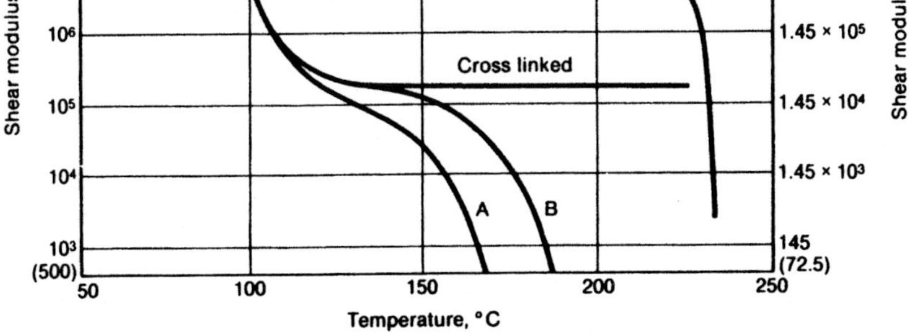

Fig. 20 Shear modulus versus temperature for crystalline isotactic polystyrene (PS), two linear atactic PS materials (A and B) with different molecular weights, and lightly cross-linked atactic PS

is the principle used to produce impact-resistant PS and acrylonitrile-butadiene-styrene.

Toughness may also be influenced dramatically by secondary transitions. For example, PC has a T_g of 150 °C (300 °F) yet is quite tough at room temperature. This results from a low-temperature secondary transition that occurs in PC and gives the polymer some degree of rubbery character, even below its T_g. Some secondary transitions produce deleterious effects, however. If they occur at approximately the required use temperature, they may cause an unanticipated change in properties during use. Toughness may decrease in the vicinity of a transition temperature.

Chemical Properties

Chemical properties are numerous, and this section briefly introduces solubility, permeability, and chemical resistance. The latter category includes a wide range of properties, such as environmental resistance, radiation resistance, and so forth. Generally, plastics exhibit excellent resistance to many forms of chemical attack and are better than many metals, especially in weak acids or alkalis. They are, however, attacked by strong oxidizing acids. Thermoplastics can also be dissolved by various organic solvents. As molecular weight increases, solubility in a particular solvent decreases. Cross linking, even in slight amounts, may make the plastic insoluble. More crystalline polymers exhibit higher chemical resistance, because the denser packing of the chain molecules makes it difficult for a solvent or other chemical substance to penetrate. Fuels, fats, oils, and even water may cause some plastics to swell and soften. This is of particular importance for materials used in gaskets and seals.

The solubility of the polymer in various solvents and the tendency for a solvent to diffuse into and/or swell a given polymer are important considerations for many applications. The mutual solubility of a polymer and a given solvent are strongly influenced by the elements and bonding within the mer and, to a lesser extent, by the bonding between polymer molecules. This is because "like dissolves like," which means that a polymer will not dissolve in a solvent unless the chemical structure of its mer unit is fairly similar to that of the solvent. Other factors, such as interactions between molecules (which themselves depend largely on the chemical structure of the mer) also influence solubility, although not as dramatically. It should be noted that solubility will, in turn, affect other properties, such as permeability.

Plasticization of polymers is a very important aspect of solubility. A plasticizer is a chemical added to a polymer to improve its processing characteristics or to alter its physical and/or mechanical properties. A plasticizer generally lowers the temperature resistance of a polymer as well as its hardness, stiffness, and tensile strength. It may, however, also increase the toughness of the polymer. In some polymers, plasticizers are required to bring the processing temperature below the decomposition temperature.

Plasticizers must form a homogeneous mixture with the polymer at processing temperatures without chemically degrading it and without separating out as the mixture cools. Thus, they must be compatible with the polymer and have a fairly high molecular weight and low volatility.

More plasticizers are used in PVC than in any other polymer. PVC plastic pipe illustrates the properties of PVC without plasticizers, while vinyl in raingear and upholstery illustrate the properties produced by heavy plasticization.

Permeability. Secondary bonding is one of the most important influences on polymer permeability to gases or other small molecules. If a molecule interacts strongly with a polymer, it will not be readily able to diffuse through it. Although this depends on a complex interaction between the polymer and the diffusant, an increase in the polarity of the polymer usually increases the interactions with the diffusant, thus reducing permeability. Usually, strong polar or hydrogen bonding in a polymer interferes with the permeability of polar molecules, while dispersive bonding has little influence. This is the reason PE and highly crystalline hydrocarbon polymers have limited solubility in most solvents and yet are completely permeable to most gases. An example of this is the escape of onion odors from a plastic bag.

However, the higher the crystallinity and/or density of the polymer, the lower the permeability, because the free volume through which the molecule must diffuse is reduced. Cross linking usually reduces permeability, unless crystallinity is destroyed in the cross-linking process.

Table 7 Range of mechanical properties for common engineering materials

Material	Elastic modulus		Tensile strength		Maximum strength/density		Elongation at break, %
	GPa	10^6 psi	MPa	ksi	$(km/s)^2$	$(kft/s)^2$	
Ductile steel	200	30	350–800	50–120	0.1	1	0.2–0.5
Cast aluminum alloys	65–72	9–10	130–300	19–45	0.1	1	0.01–0.14
Polymers	0.1–21	0.02–30	5–190	0.7–28	0.05	0.5	0–0.8
Glasses	40–140	6–20	10–140	1.5–21	0.05	0.5	0
Copper alloys	100–117	15–18	300–1400	45–200	0.17	1.8	0.02–0.65
Moldable glass-filled polymers	11–17	1.6–2.5	55–440	8–64	0.2	2	0.003–0.015
Graphite-epoxy	200	30	1000	150	0.65	1.3	0–0.02

The solubility of the diffusant in the polymer also influences permeability. Solubility usually reduces permeability because a molecule that is interacting with the polymer does not simply diffuse through it. Of course, if the solubility is high enough, eventually the solvent will pass through to the other side.

Chemical Resistance. Although resistance to attack by chemicals, environments, and radiation depends on the chemical nature and bonding in the mer, it often depends even more on weak links in the polymer chain. Such weak links include chemical defects in the chain, branch points, and polymer end groups. Such weak links often have a much greater chemical effect than their concentration would indicate. For example, polytetrafluoroethylene (PTFE), which is otherwise very stable, decomposes by depolymerization that is initiated by an unzipping from its end (Ref 13). Such weak links influence all types of environmental resistance, including resistance to temperature, ultraviolet radiation, ozone, and others.

The specialized chemical degradation problem known as environmental stress cracking and crazing is produced by a combination of factors, including solubility and polymer toughness. The active agent must dissolve in the polymer and wet the surface of a flaw to reduce its surface energy. With reduced surface energy, the flaw, when stressed, may then more easily propagate to failure. In the case of environmental stress crazing, the solvent dissolves some of the lower-molecular-weight-material in the polymer, producing crazes that act as flaw sites for stress cracking.

Aging and weathering of plastics depend on the nature of the environment and the incident radiation. Most plastics oxidize and degrade if kept for long periods at elevated temperatures in the presence of air. Sunlight is also damaging, because ultraviolet radiation can cause polymer degradation unless stabilizers are added.

Electrical and Optical Properties

Important electrical properties include dielectric constant, dielectric strength, dispersion, and conductivity. Optical properties are briefly discussed in this section.

Dielectric Properties. Because polymers are good insulators, they may be able to store electrical charge effectively, thus serving as good dielectrics. The dielectric constant of a polymer is improved significantly by the existence of permanent dipoles within the polymer. However, if the permanent dipoles are bulky, the polymer may only be useful as a dielectric at low frequencies. This is because at higher frequencies the dipoles cannot keep up with changes in field and become unable to store charge. The polymer is said to undergo dispersion.

Dielectric strength is greatly influenced by internal and external impurities. Polytetrafluoroethylene has excellent, small permanent dipoles combined with a nonstick surface that does not gather surface impurities. It is viewed as an excellent dielectric material at low frequencies even though its small dipoles do not store as much charge as bulkier dipoles. Because dielectric breakdown can also occur by mechanical or thermal collapse, dielectric strength is improved by increasing the basic mechanical strength of the polymer (such as by adding fiber reinforcements to PTFE) and/or by increasing its thermal dimensional stability.

Conductivity. In most cases, polymers make poor electrical conductors. This is because the primary chemical bonding in most polymers is covalent, and thus there are no free electrons or ions to conduct charge. Specialized polymers that have sufficient charge carriers to be semiconductors or conductors have been created, but are often brittle, inflexible, insoluble materials with no commercial possibilities. While some advances are being made in creating conductive polymers, most conductivity is produced by adding a conductive second phase to the polymer.

Optical properties such as color, clarity, transparency, and so forth, may not seem to be very important properties, but if the polymer is to be used as a window in a jet aircraft, for example, such properties become very important. When transparency is required, inclusions, voids, and all crystallinity must be avoided. This is because the change of refractive index at the boundary of such a region would interfere with the passage of light.

Both the refractive index of the polymer and its color are dictated by the details of chemical bonding. Most polymers are colorless and thus can often be colored as desired. Other optical properties are often influenced more by macroscopic morphologies and flaws than by the basic structure of the polymer.

Other Properties

There are many other types of properties that may be important to a polymer application but are not covered in this article. For example, the molten properties of a polymer are very important to processing. Melt properties of a true thermoplastic are influenced by mer flexibility and bulkiness, by isomerism, by branching, by molecular weight, and by molecular-weight distribution. For example, the blow-molding resin that is used to produce PE bottles is a linear resin having a high average molecular weight but a broad molecular-weight distribution. A linear resin with high average molecular weight ensures that the resin is strong and tough enough in finished form, while the low-molecular-weight chains act as a lubricant in the melt and allow the resin to flow easily. Without this broad molecular-weight distribution, even a resin with much lower average molecular weight could not be blow molded successfully.

Engineering Thermoplastics

Any list identifying engineering thermoplastics is partly subjective, because certain thermoplastics are only marginally load bearing and others are upgraded to structural capability by reinforcing the neat (unmodified) resin with fibers. The following thermoplastic resins are briefly described:

- Acetals (AC)
- Polyamides (PA), specifically nylons
- Polyketones
- Polycarbonates (PC)
- Polyether-imides (PEI)
- Polyether sulfones (PES or PESV, with the latter the preferred ASTM abbreviation)
- Polysulfones (PSU)
- Polyphenylene ether blends (PPE) and polyphenylene oxide (PPO)
- Polyphenylene sulfides (PPS)
- Polyethylene terephthalates (PET)
- Polybutylene terephthalates (PBT)
- Acrylonitrile-butadiene-styrenes (ABS)

These materials by no means constitute the totality of the engineering thermoplastic family, but they do represent a broad cross section of properties and applications. Table 8 lists properties of these materials.

Acetals (AC) are highly crystalline plastics that are strong, rigid, and have good moisture, heat, and solvent resistance. Acetals are based on formaldehyde polymerization technology to produce either homopolymers (from polymerization of a single monomer) or copolymers.

Melting points of the homopolymer acetals are higher than those of the copolymers (175 °C, or 350 °F, versus 165 °C, or 330 °F), and the homopolymers are harder, have higher resistance to fatigue, are more rigid, and have higher tensile and flexural strength with generally lower elongation (Table 8). Some high-molecular-weight homopolymer grades are extremely tough and have higher elongation than copolymers. Homopolymer grades are available that are modified for improved hydrolysis resistance to 80 °C (180 °F), similar to that of copolymer materials.

The copolymers remain stable in long-term, high-temperature service and offer exceptional resistance to the effects of immersion in water at high temperatures. Neither type resists strong acids, and the copolymers are virtually unaffected by strong bases. Both types are available in a wide range of melt-flow grades.

Both the homopolymers and copolymers are available in several unmodified and glass-fiber-reinforced injection-molding grades. Both are available in grades filled with polytetrafluoroethylene (PTFE) or silicone, and the homopolymer is available in chemically lubricated low-friction formulations. The acetals are also available in extruded rod and slab form for machined parts.

The properties of acetals make them suitable for a diverse range of applications, including:

- Materials-handling conveyors
- Automotive components (e.g., fuel-handling components and instrument panel components)
- Appliances (e.g., housings, gears, and bearings)
- Plumbing components (e.g., shower heads, ball cocks, and faucet underbodies)
- Consumer products (e.g., toys, sporting goods, and soap dispensers)

Acrylic plastics comprise a broad array of polymers and copolymers in which the major monomeric constituents belong to two families of ester-acrylates and methacrylates. These are used singly or in combination. Hard, clear acrylic sheet is made from methyl methacrylate, whereas molding and extrusion pellets are made from methyl methacrylate copolymerized with small percentages of other acrylates or methacrylates. The use of additives and modifiers during the polymerization process allows the production of different types of acrylic plastic sheets and molding compounds, each of which is formulated to enhance a specific set of properties. Most types are available in colorless form and also in a variety of transparent, translucent, and opaque colors. Grades per ASTM D 788 differ in molecular weight and in their principal properties, particularly flow rate, heat resistance, and toughness.

Straight (unmodified) grades of acrylic plastic are noted for their outstanding optical properties and weatherability. Colorless acrylic plastic is as transparent as the finest plate glass and is capable of giving almost complete transmittance of visible light. Acrylic plastics have outstanding resistance to the effects of sunlight and exposure to the elements over long periods of time. They do not yellow significantly, nor do they undergo any significant changes in physical properties. Most of the transparent, translucent, and opaque colors of acrylic have the same outstanding resistance to weathering.

Impact-modified acrylic grades, depending on the modifier used, have toughnesses up to 20 times that of unmodified acrylics. The butadiene-modified grades have the greatest toughness, but are not as transparent as the acrylic-modified grades. In addition to toughness, the acrylic-modified grades resist changes due to weathering better than do most thermoplastics. Sheet extruded from acrylic-base impact-modified grades has excellent thermoforming characteristics and can be rigidified by applying glass-reinforced polyester to the second surface with a spray gun or by using a closed-mold process.

Acrylonitrile-butadiene-styrene (ABS) consists of a rubberlike toughener (polybutadiene particles) suspended in a continuous phase of styrene-acrylonitrile. This versatile amorphous resin family is divided into three classifications:

- *Standard grades* are grouped by impact strength: medium, high, or very high.
- *Specialty grades* are heat resistant, platable, flame resistant, or transparent.
- *Alloyed grades* include alloys of ABS with polyvinyl chloride (ABS-PVC), polycarbonate (ABS-PC), nylon (ABS-PA), and styrene-maleic anhydride (ABS-SMA)

All grades are fabricated primarily by injection molding or extrusion.

One of the major advantages of ABS is its excellent toughness, as indicated by the relatively high Izod impact strength of many grades. Although ABS is notch sensitive, it is much less so than many other plastics, including PC and PA (nylon). In addition to good impact strength at room temperature, ABS retains significant impact strength at temperatures as low as –40 °C (–40 °F). This has led to its use in applications such as drain, waste, and vent pipes and pipe fittings, camper tops, and truck-bed liners.

ABS products are very resistant to chemical attack, and most also have good environmental stress-cracking resistance. ABS is resistant to acids (except concentrated oxidizing acids), alkalis, salts, essential oils, and a wide range of food and pharmaceutical products. It is attacked by many solvents, however, including ketones and esters.

In addition to the applications mentioned previously, medium-impact ABS has long been used for refrigerators (door liners, shelves, crisper drawers) because of its excellent environmental stress-cracking resistance and appearance. ABS is also used extensively in automotive applications. High-impact and heat-resistant grades and ABS alloys are used in instrument panels, armrests, interior trim panels, seat-belt retainers, glove-compartment doors, and liftgates, and plating grades are used in wheel covers, grilles, headlight bezels, mirror housings, and decorative trim. Other applications include appliances, business and consumer electronics, luggage, packaging, and telecommunications.

Polyamides (PAs), or nylons, were the first of the thermoplastic resins, originally developed as high-strength textile fibers. These semicrystalline plastics are available in compositions for molding and extruding, for solution and fluidized-bed coatings, and for casting.

Nylon 6/6 is the most widely used nylon plastic because of its overall balance of properties. The second most widely used is nylon 6. Other commercial nylon grades include 4/6, 6/10, 6/12, 11, and 12. Both the nylon 6 and nylon 6/6

Table 8 Properties of selected thermoplastic and thermosetting engineering plastics

Material	Tensile strength		Tensile modulus			Flexural strength		Flexural modulus		Notched impact strength		Rockwell hardness	Specific gravity
	MPa	ksi	GPa	10^6 psi	Elongation, %	MPa	ksi	GPa	10^6 psi	J/m	ft · lbf/in.		
Engineering thermoplastics													
Acetal													
Copolymer	60.7	8.8	2.8	0.41	60	89.6	13	2.6	0.375	69	1.3	R80	1.41
Homopolymer	68.9	10	3.6	0.52	40	97.2	14.1	2.8	0.410	75	1.4	R94	1.42
Polyamides													
Nylon 6	80.7	11.7	2.6	0.38	30–100	108	15.7	2.7	0.39	32–53	0.6–1.0	R119	1.12–1.14
Nylon 6/6	94.5	13.7	1.59–3.79	0.23–0.55	15–60	114–117	16.5–17.0	2.8–3.1	0.41–0.45	29–53	0.55–1.0	R120	1.13–1.15
PEEK	91.7	13.3	1.1	0.16(a)	50	170	24.7	3.6	0.53	85	1.6	...	1.32
Polycarbonate	62–72.4	9–10.5	2.3	0.34	110–125	76–103	11–15	2–2.3	0.30–0.34	640–850	12–16	M62–70	1.2
PEI	105	15.2	3.0	0.43	...	152	22.0	3.3	0.48	53	1.0	M109	...
PES	84.1	12.2	2.6	0.38	...	129	18.7	2.6	0.375	75	1.4	M88	...
PSU	70.3	10.2	2.48	0.36	75	106	15.4	2.69	0.390	64	1.2	M69	1.24
PPE	53.8	7.8	2.5	0.36	50–60	88.3	12.8	2.5	0.36	267	5.0	R115	1.06–1.10
PPS (neat)(b)	65.5	9.5	...	...	...	96	13.9	3.8	0.55	16	0.30	R120	1.35
PPS (40 wt% glass)	138	20	...	...	...	160	23.2	12	1.7	58	1.09	R123	1.6
PET (neat)(b)	62.1	9.0	...	...	...	...	...	2.8	0.40	26.7	0.5	...	...
PET (30% glass fiber)	150	22	...	...	2.2	235	34.0	8.96	1.30	95	1.8	R120	1.68
PBT	52	7.5	...	...	...	82.7	12	2.3	0.34	53	1.0	R117	...
ABS													
Medium impact	45	6.5	2.5	0.36	...	76	11	2.8	0.4	160	3.0	R108–118	1.03–1.07
High impact	39	5.6	2.2	0.32	...	66	9.5	2.2	0.32	270	5.0	R102–113	1.01–1.05
Very high impact	32	4.7	1.8	0.26	...	54	7.8	1.8	0.26	400	7.5	R90–100	1.01–1.04
Engineering thermosets													
Aminos													
UF (cellulose filled)	38–48	5.5–7.0	...	...	0.5–1.0	75–110	11–16	9.6–10.3	1.4–1.5	14–18	0.27–0.34	M110–120	1.5
MF (cellulose filled)	48–55	7–8	...	...	0.6–0.9	75–110	11–16	7.6	1.1	16–19	0.30–0.35	M120	1.5
PUR (unfilled)(c)	24	3.50	...	...	110	...	...	0.518	0.075	213.5	4.0	...	1.01
PUR (20% glass flakes)(c)	32	4.70	...	...	20	...	...	1.38	0.200	107	2.0	...	1.15
Unreinforced polyesters													
Orthophthalic	55	8.0	3.45	0.50	2.1	85	12	3.45	0.50	...	...	...	...
Isophthalic	75	11	3.38	0.49	3.3	131	19	3.59	0.52	...	...	Barcol 40	...
BPA fumerate	40	6	2.83	0.41	1.40	110	16	3.38	0.49	...	...	Barcol 34	...
Reinforced polyesters(d)													
Orthophthalic	152	22	5.5	0.8	...	220	32	6.9	1.0	...	...	...	...
Isophthalic	193	28	11.7	1.7	...	240	35	7.6	1.1	571	10.7	Barcol 45	...
BPA fumerate	124	18	11.0	1.6	...	160	23	9.0	1.3	640	12	Barcol 40	...
Unreinforced epoxy(e)	42.7–82.7	6.2–12	2.7–3.4	0.395–0.50	1.35–5.7	103–131	15–19	2.48–2.93	0.360–0.425	...	...	...	...
Phenolics													
Cellulose filled	...	...	6.9–9.7	1.0–1.4	...	70–90	10–13	...	...	16–58.7	0.30–1.10	M100–110	1.35–1.45
Mineral filled	...	...	10.3–17.2	1.5–2.5	...	70–90	10–13	...	...	16.24	0.30–0.45	M105–115	1.50–1.70
Glass fiber filled	...	...	17.2–20.7	2.5–3.0	...	80–140	12–20	...	...	21–800	0.40–15.0	M110–120	1.75–2.10
Unreinforced polyimide	38.6	5.6	3.9	0.57	...	176	25.5	4.0	0.58	53.37	1.0	...	1.32

(a) Tensile modulus at 150 °C (300 °F). (b) Values for neat PPS and PET would not appear on supplier data sheet because both are reinforced for engineering/structural applications. (c) Values listed are for reaction injection molded polyurethane, both unfilled and filled (20% glass flakes parallel to the flow direction of the mold-filling process). Data supplied by Mobay Corp. (d) Reinforced with 40 wt% glass fibers. (e) Typical property value ranges for DGEBA epoxy (refer to text) cured/hardened with aliphatic amine, Lewis acid (boron trifluoride monoethylamine), anhydride, or aromatic amine. Source: Ref 14 grades are supplied neat or reinforced (30 to 35 vol% glass fiber).

Key characteristics of nylons are their resistance to oils and greases. Other characteristics include outstanding resistance to solvents, bases, fatigue, repeated impact, and abrasion; a low coefficient of friction; high tensile strength and toughness; barrier properties; creep resistance; and retention of properties over a wide temperature range, from –60 to 110 °C (–75 to 230 °F). Mechanical properties of nylons are listed in Table 8. Limitations of nylons are high moisture pickup, with resulting changes in dimensional and mechanical properties; high mold shrinkage; and notch sensitivity, unless suitably blended for toughness. (Toughened unreinforced nylon 6/6 has a notched Izod impact strength of 907 J/m, or 17 ft · lbf/in.)

Nylons display a low coefficient of friction when they contact many other materials, so they are frequently used in journal bearings, bushings, gears, and cams. Sliding parts often require no lubrication. Additives such as molybdenum disulfide, graphite, or PTFE resin are sometimes employed to enhance the natural lubricity of nylon.

Polybutylene terephthalate (PBT), like PET, is a semicrystalline thermoplastic polyester. This polyester is characterized by low moisture absorption, excellent electrical properties, broad chemical resistance, lubricity, and durability. Although PBT is most commonly processed by injection molding, other processing options include structural foam molding, fiber and nonfilament spinning, nonwoven-fabric formation, thermoforming, blow molding, and profile, film, and sheet extrusion.

PBT has good tensile strength, ranging from 50 MPa (7.5 ksi) for neat grades to 170 MPa (25 ksi) for glass-reinforced grades. Corresponding flexural modulus values range from 2.30 to 10.3 GPa (0.340 to 1.5 × 10^6 psi). Notched Izod impact strength ranges from 55 to 910 J/m (1 to 17 ft · lbf/in.). The addition of flame retardants, reinforcements, impact modifiers, minerals, and other polymers can enhance flammability resistance and other properties. Properties of neat PBT are listed in Table 8.

For applications in which friction is a consideration, such as gears and bearings, the intrinsic lubricity, smooth surface, and low coefficient of friction of PBT against itself helps it resist abrasion and eliminates the need for lubrication.

The chemical resistance, thermal stability, and hydrolytic stability of PBT make it suitable for automobile grilles, body panels, fenders, wheel covers, and components for door handles, mirrors, and windows. It is also used in under-the-hood distributor caps, rotors, ignition components, parts for headlamp systems, windshield-wiper assemblies, water pumps, and brake systems. Nonautomotive applications include materials-handling components, electrical/electronic components, lawn and garden products, and housewares.

Polycarbonates (PCs) are amorphous thermoplastics that are characterized by a combination of toughness, transparency, heat and flame resistance, and dimensional stability. As shown in Table 8, PCs are noted for high notched Izod impact strength, 640 to 850 J/m (12 to 16 ft · lbf/in.), and good retention of impact strength at temperatures as low as –50 °C (–60 °F). The insulating and other electrical characteristics of PCs are excellent and are almost unchanged by temperature and humidity conditions. One exception is arc resistance, which is lower than that of many other plastics.

Polycarbonates are generally unaffected by greases, oils, and acids. Water at room temperature has no effect, but continuous exposure in hot (65 °C, or 150 °F) water causes gradual embrittlement. The resins are soluble in chlorinated hydrocarbons and are attacked by most aromatic solvents, esters, and ketones, which cause crazing and cracking in stressed parts. Polycarbonates are supplied in neat and glass-fiber-reinforced grades and can be processed by all thermoplastic processing methods. Injection molding, sheet and profile extrusion, blow molding, and foam molding are the most frequently used. Other processing methods include rotational molding and coextrusion with other polymers. Applications for PCs include:

- Components for business machines and telecommunication equipment
- Appliance parts
- Automotive components
- Sporting equipment
- Food and beverage containers and microwave cookware
- Medical components and devices

Polyether-imides (PEIs) are amorphous thermoplastics that have high heat resistance, high strength and modulus, excellent electrical properties that remain stable over a wide range of temperatures and frequencies, and excellent processibility. Unmodified PEI resins are transparent and characterized by inherent flame resistance and low smoke generation.

PEI resins are available in an unreinforced grade for general-purpose injection molding, blow molding, foam molding, and extrusion, in four glass-fiber-reinforced grades (10, 20, 30, and 40% glass), in carbon-fiber-reinforced grades, in bearing grades, and in several high-temperature grades.

The high T_g of PEIs (217 °C, or 420 °F) and the high-performance strength and modulus characteristics at elevated temperatures are provided by the very rigid imide groups in the chemical structure. The ether linkages confer flexibility to the molecular chain, providing good melt flow during processing and good practical toughness in the end products. The high T_g allows PEIs to be used intermittently at 200 °C (390 °F) and permits short-term excursions to even higher temperatures.

Unreinforced PEI is one of the strongest engineering thermoplastics. At 180 °C (355 °F), tensile strength and flexural modulus remain in excess of 41 MPa (6.0 ksi) and 2.06 GPa (0.300 × 10^6 psi), respectively. Higher strength and stiffness at elevated temperatures up to the T_g are achieved with glass or carbon-fiber reinforcement. The exceptionally good long-term resistance to creep at high temperatures and stress levels has allowed reinforced PEI resins to replace metal and other materials in an increasing number of structural applications.

PEIs are primarily used in the automotive, electrical/electronic, packaging, aircraft, industrial, and medical markets. Other uses are as appliances and hardware, and in the field of fluid engineering.

Polyether sulfone (PES) is an amorphous thermoplastic that belongs to the sulfone family, which also includes polyarylsulfone and polysulfone. Although the chemical properties and some physical properties of this family are similar, PES is more desirable for applications that make use of its superior thermal stability and mechanical properties. Properties of PES can be retained at temperatures up to 200 °C (390 °F) for thousands of hours. It is inherently flame resistant and emits very low levels of toxic fumes when burned.

Processing by conventional injection molding, extrusion, or blow molding techniques is possible with PES. Sheet and film can be vacuum formed. Highly filled PES compounds (30 wt% chopped glass fibers plus other additives) can be compression molded.

Mechanical strength, resistance to oils and gasoline at elevated temperatures, low flammability, and low emissions of toxic gases and smoke make PES highly attractive in demanding automotive applications. Its features are important for applications in which safety standards are stringent and are becoming more so.

PES can be molded to close tolerances, and it provides significant savings compared to traditional metals. Applications include fuse housings, water pumps, turbochargers, supercharger parts, and car heater fans and bearing cages where prolonged high-temperature resistance is vital. Other applications include electrical and electronic components, aircraft radomes and other aerospace components, medical equipment, fluid-handling parts and equipment, and consumer items. Properties of PES are listed in Table 8.

Polyethylenes (PEs), which represent one of largest-volume plastics in use, are very stable and extremely durable polymers characterized by good chemical resistance and excellent mechanical properties. The resin design parameters that determine end-use properties are density, molecular weight, and molecular-weight distribution. The influence of each is illustrated in Table 9. By carefully changing the balance of these parameters, different products can be manufactured to provide high-performance properties specific to certain application requirements.

For example, HDPE resin is produced with nominal density in the 0.944 to 0.954 g/cm^3 range, which optimizes tensile strength, load-bearing strength, and barrier properties while slightly sacrificing some toughness and mechanical properties. This loss in toughness and mechanical properties is more compensated by the use of high-molecular-weight (HMW) polymers. High-density polyethylene resins with less than 200,000 molecular-weight (MW) units are considered general-purpose commodity grades. These grades typically favor easier flow properties, balanced with moderate end-use physical properties.

The HDPE polymer grades with MW in the range of 200,000 to 500,000 are considered high-performance, high-molecular-weight HDPEs (HMW HDPEs). The combination of high molecular weight and high density provides high stiffness, abrasion resistance, chemical resistance, and extended product service life in critical environmental applications. High-molecular-weight resins also provide excellent environmental stress-cracking resistance, high tensile strength, and practical toughness with excellent impact resistance at temperatures as low as −50 °C (−60 °F). Their excellent melt strength allows the high draw ratios necessary for reducing wall thickness of finished products.

Polyethylene terephthalate (PET) is part of a family of thermoplastic polyesters that also includes polybutylene terephthalate. Essentially all PET products offered commercially are reinforced with short glass fibers, minerals, or glass/mineral combinations. Proprietary modifier packages are added in order to achieve acceptable PET crystallization rates at conventional mold temperatures. Injection molding is the principal fabrication technique for this family of thermoplastics.

Reinforced PET grades are available at glass-fiber loadings of 15 to 55%, corresponding to a flexural modulus range of 5.79 to 16.9 GPa (0.840 to 2.45 × 10^6 psi). Glass-fiber/mineral blends at levels of 35 to 40% are also offered to satisfy applications that require a high degree of dimensional stability. Flame-retardant grades are available at glass-fiber loadings of 30 and 43%, or with glass-fiber/mineral blend levels of 45%. Impact-modified products have also been developed in which the notched Izod is increased from 95 to 230 J/m (1.87 to 4.4 ft · lbf/in.) for a 30% glass-fiber level. Properties of neat and reinforced PET are listed in Table 8.

Table 9 Basic polymer parameters and their influence on resin properties

Properties	Density	Molecular weight	Molecular-weight distribution
Environmental stress cracking resistance	Decreases	Increases	Broadens
Impact strength	Decreases	Increases	Narrows
Stiffness	Increases	...	...
Hardness	Increases	...	...
Tensile strength	Increases	...	...
Permeation	Decreases	...	...
Warpage	Decreases	...	Broadens
Abrasion resistance	...	Increases	...
Flow processibility	...	Decreases	Broadens
Melt viscosity	...	Increases	Narrows
Copolymer content	Decreases	...	...

Current applications for PET include the industrial, automotive, and electrical industries. In the industrial market, the combination of high stiffness and low moisture absorption permits the use of reinforced PET in structural applications such as furniture chair arms and frames, pump housings, and hand tools. Automotive applications include structural components (e.g., luggage racks, mirror backs, door-latch mechanisms, grille supports) and electrical parts (e.g., head-lamp reflectors, lamp sockets, alternator housings, and ignition rotors). Electrical components made from PET are primarily composed of flame-retardant grades.

Polyketones are partially crystalline thermoplastics that can be used at high temperatures. They also have excellent chemical resistance, high strength, and excellent resistance to burning. Although they require high melt temperatures, polyketones can be extruded and injection molded with standard processing equipment. Commercially available polyketones include:

- Polyaryl ether ketones (PAEK or PEK), repeating ether and ketone groups combined by phenyl rings
- Polyether ether ketones (PEEK), repeating monomers of two ether groups and a ketone group
- Polyether ketone ketones (PEKK), repeating monomers of one ether group and two ketone groups

Polyketones are available in neat as well as glass- and carbon-fiber-reinforced forms. The reinforced grades are noted for their strength retention at elevated temperatures. For example, a 30 vol% glass-fiber-reinforced polyketone has a heat-deflection temperature of 325 °C (619 °F) at 1.82 MPa (0.264 ksi) and a tensile strength exceeding 30 MPa (4.4 ksi) at 250 °C (480 °F). Polyketone resins are useful in a broad spectrum of applications that require their unique combination of properties. For example, low flammability, low smoke generation, chemical resistance, high-temperature resistance, and strength make PAEK materials suitable for aircraft/aerospace applications such as engine components, cabin interior material, air ducts, and nonstructural exterior parts. Other applications include wire and cable in the electrical/electronics market, pump components in the chemical-processing market, backup seals in the downhole oil equipment market, and bearing surfaces in the industrial equipment market. Properties of the PEEK resin system are listed in Table 8.

Polyphenylene ether (PPE) materials are actually alloys, or blends, that contain high-impact polystyrene and additives in various proportions. Glass-reinforced grades, heat-resistant grades (containing nylon), and platable grades are also available.

The PPE blends are characterized by their outstanding moisture resistance, high strength and heat resistance, and excellent dielectric properties over a wide range of frequencies and temperatures. The addition of rubber-modified high-impact polystyrene increases the impact strength considerably (values as high as 530 J/m, or 10 ft · lbf/in., can be achieved). Properties of modified PPE materials are listed in Table 8.

Modified PPE resins are suitable for extrusion, blow molding, and thermoforming processes. However, injection molding is the most commonly used processing method. Structural foam materials can be molded on standard injection-molding equipment, which heightens their cost effectiveness.

Most of the markets for PPE resins are similar to those for other specialty thermoplastics: business machines, automobiles, televisions, and appliances. Hydrolytic stability is an important factor in the selection of PPE resins for pumps, impellers, shower heads, chemical-process equipment, and filter bodies. Metal-plated modified PPE also performs well in enclosures shielded from electromagnetic interference and radiofrequency interference, which are used in the automotive and appliance industries.

Polyphenylene sulfide (PPS) is a crystalline, high-performance thermoplastic that is characterized by outstanding high-temperature stability, inherent flame resistance, and resistance to diverse chemical environments. A wide range of injection-molding grades of PPS are available, including:

- A series of compounds that contain various glass-fiber levels, recommended for mechanical applications requiring high strength and impact resistance and for electronic applications requiring good insulating characteristics
- A series of compounds that contain various mineral fillers plus glass-fiber reinforcement, suitable for electrical applications requiring high arc resistance and low arc tracking, for current-carrying parts in electrical assemblies, and for microwave ovenware and appliance components

Unreinforced PPS resins are available as powders for slurry coating and electrostatic spraying. The resin coatings are suitable for food-service applications as well as for chemical-processing equipment. Properties of neat and reinforced PPS are listed in Table 8.

Injection-moldable PPS compounds require processing temperatures of 300 to 360 °C (575 to 675 °F). Mold temperatures can range from 40 to 150 °C (100 to 300 °F) to control the crystallinity. Cold-molded parts deliver optimal mechanical strength, and hot-molded highly crystalline parts provide optimal dimensional stability at high temperatures.

Polypropylene (PP). Reinforced polypropylenes, although based on a commodity thermoplastic, have become contenders in the engineering resin field in recent years. Advances in filler and reinforcement technologies and an attractive cost-performance balance are two major reasons. Polypropylene is readily combined with mineral fillers such as talc, mica, and calcium carbonate, as well as with glass and carbon fibers. Although 50 wt% is the maximum concentration usually used, concentrates are available with higher percentages of filler/reinforcement. Table 10 compares typical mechanical properties of PP and other thermoplastics containing different percentages of glass filler.

Polypropylene is commonly produced either as a homopolymer or copolymer (in which the comonomer is ethylene). Polypropylene in both forms is very resistant to moisture; has good chemical resistance to acids, alkalies, and solvents; and can be processed by extrusion, injection molding, and blow molding. Copolymerization with ethylene improves the toughness of PP, as well as the flexibility, but slightly reduces heat resistance. Basic physical properties of homopolymer and copolymer PP are contained in Table 11. The automotive, appliance, consumer products, and medical markets represent application areas for PPs.

Polystyrene (PS) is one of the oldest commercially produced thermoplastic polymers, having been introduced in the 1930s. The homopolymer, known as crystal PS, is a brilliant, clear, noncrystalline plastic with excellent

Table 10 Room-temperature mechanical properties of selected thermoplastics with glass filler

Thermoplastic	Glass fiber content, wt%	Tensile strength(a)		Tensile elongation at break(a), %	Tensile modulus(a)		Flexural strength(b)		Flexural modulus(b)		Izod impact strength notched(c)		Compressive strength(d)	
		MPa	ksi		kPa	psi	MPa	ksi	GPa	10^6 psi	J/m	ft · lbf/in.	MPa	ksi
Styrene	...	46	6.7	2.2	320	46	97	14.0	3	0.45	11	0.2	97	14.0
	20	76	11	1.0	760	110	107	15.5	7	0.96	53	1.0	111	16.1
	30	93	13.5	1.0	900	130	117	17.0	8	1.22	53	1.0	120	17.4
	40	103	15	1.0	1100	160	121	17.5	10	1.47	59	1.1	122	17.7
Styrene-acrylonitrile (SAN)	...	72	10.5	3.0	390	56	103	15.0	4	0.55	27	0.5	103	15.0
	20	90	13	2.0	860	125	129	18.7	8	1.10	53	1.0	134	19.5
	30	107	15.5	1.5	1000	145	155	22.5	10	1.52	53	1.0	141	20.5
	40	119	17.2	1.5	1240	180	161	23.4	12	1.80	53	1.0	148	21.5
Acrylonitrile-butadiene-styrene (ABS)	...	48	7	8.0	210	30	72	10.5	3	0.38	240	4.5	69	10.0
	20	90	13	3.0	620	90	117	17.0	6	0.80	80	1.5	86	12.5
	30	105	15.2	3.0	690	100	128	18.5	7	1.00	75	1.4	107	15.5
	40	110	16	2.0	1030	150	145	21.0	9	1.30	69	1.3	118	17.1
Flame-retardant ABS	...	40	5.8	5.1	240	35	83	12.0	2	0.33	213	4.0	52	7.5
	20	76	11	2.0	510	74	107	15.5	5	0.71	64	1.2	97	14.0
Polypropylene (PP)	...	32	4.7	15.0	130	19	41	6.0	2	0.30	27	0.5	34	5.0
	20	59	8.5	3.0	380	55	55	8.0	4	0.60	43	0.8	41	6.0
	30	62	9	3.0	450	65	59	8.5	6	0.80	59	1.1	45	6.5
	40	69	10	2.0	520	75	62	9.0	7	1.00	69	1.3	48	7.0
Glass-coupled PP	...	32	4.7	15.0	130	19	41	6.0	2	0.30	27	0.5	41	6.0
	20	76	11	3.0	410	60	83	12.0	4	0.60	75	1.4	69	10.0
	40	97	14	2.0	550	80	131	19.0	7	1.00	85	1.6	90	13.0
Polyethylene (PE)	...	30	4.3	9.0	100	15	38	5.5	2	0.22	69	1.3	28	4.0
	20	48	7	3.0	410	60	62	9.0	4	0.55	75	1.4	34	5.0
	30	69	10	2.0	590	85	76	11.0	6	0.80	91	1.7	48	7.0
	40	76	11	2.0	760	110	86	12.5	7	1.00	91	1.7	55	8.0
Acetal (AC)	...	61	8.8	60.0	280	41	90	13.0	3	0.37	69	1.3	36	5.2
	20	83	12	2.0	830	120	110	16.0	7	1.00	53	1.0	83	12.0
	30	90	13	1.8	930	135	114	16.5	8	1.20	43	0.8	83	12.0
Polyester	...	55	8	200.0	280	40	88	12.8	2	0.34	11	0.2	90	13.0
	20	117	17	5.0	690	100	152	22.0	6	0.85	80	1.5	110	16.0
	30	131	19	4.0	1030	150	179	26.0	8	1.20	96	1.8	124	18.0
	40	152	22	3.0	1380	200	207	30.0	10	1.50	107	2.0	138	20.0
Flame-retardant polyester	...	61	8.9	60.0	280	40	101	14.7	3	0.38	48	0.9	100	14.5
	30	131	19	3.0	1100	160	176	25.5	9	1.30	69	1.3	124	18.0
Nylon 6	...	81	11.8	200.0	280	40	103	15.0	3	0.40	53	1.0	90	13.0
	20	128	18.5	3.0	690	100	159	23.0	6	0.80	80	1.5	148	21.5
	30	155	22.5	3.0	900	130	186	27.0	8	1.10	117	2.2	159	23.0
	40	185	26.8	2.0	970	140	207	30.0	9	1.30	160	3.0	159	23.0
Flame-retardant nylon 6	...	85	12.3	60.0	290	42	110	16.0	3	0.40	53	1.0	90	13.0
	30	152	22	3.0	900	130	228	33.0	9	1.35	91	1.7	16	2.3
Nylon 6/6	...	79	11.4	300.0	130	19	103	15.0	1	0.19	53	1.0	34	4.9
	20	138	20	3.0	830	120	193	28.0	6	0.85	64	1.2	159	23.0
	30	179	26	2.0	1030	150	259	37.5	9	1.30	107	2.0	165	24.0
	40	214	31	2.0	900	130	293	42.5	11	1.60	139	2.6	172	25.0
Flame-retardant nylon 6/6	...	67	9.7	35.0	130	19	90	13.0	1	0.18	53	1.0	34	4.9
	30	148	21.5	2.0	830	120	172	25.0	7	1.00	85	1.6	159	23.0
Nylon 6/12	...	61	8.8	150.0	200	29	86	12.5	2	0.29	53	1.0	76	11.0
	20	124	18	4.0	690	100	193	28.0	6	0.90	59	1.1	131	19.0
	30	152	22	4.0	900	130	221	32.0	8	1.10	128	2.4	152	22.0
Polycarbonate (PC)	...	62	9	110.0	240	34.5	93	13.5	2	0.34	160	3.0	86	12.5
	10	90	13	5.0	480	70	110	16.0	4	0.60	107	2.0	124	18.0
	20	110	16	5.0	620	90	138	20.0	6	0.80	117	2.2	138	20.0
	30	131	19	4.0	900	130	165	24.0	8	1.20	128	2.4	145	21.0
	40	152	22	3.5	1170	170	193	28.0	10	1.40	144	2.7	148	21.5
Polysulfone (PSU)	...	70	10.2	75.0	250	36	106	15.4	3	0.39	32	0.6	97	14.0
	20	131	19	3.0	620	90	138	20.0	5	0.75	64	1.2	138	20.0
	30	148	21.5	3.0	830	120	155	22.5	7	1.00	75	1.4	155	22.5
	40	165	24	2.0	1240	180	172	25.0	9	1.25	107	2.0	172	25.0
Polyphenylene sulfide (PPS)	40	138	20	1.5	1410	205	234	34.0	12	1.80	80	1.5	172	25.0

(a) ASTM D 638 test method. (b) ASTM D 790 test method. (c) ASTM D 256 test method with 6.35 mm 1/4 in.) bar. (d) ASTM D 695 test method stiffness and processibility. However, the low impact strength of crystal PS limits its use. High-impact polystyrene (HIPS) was developed in the early 1950s to meet the demand for a tougher resin. Most commercial PSs have a weight-average molecular weight in the range of 150,000 to 350,000.

HIPS resins are known for their ease of processing, excellent dimensional stability, good impact strength, and high rigidity. Relative disadvantages of HIPS are poor high-temperature properties, poor oxygen barrier properties, low light (UV) stability, and lower chemical resistance than most crystalline polymers. The largest single use for HIPS is in packaging and disposables, specifically for food packaging or food service. For example, typical extrusion and thermoforming applications include dairy containers, vending and portion cups, lids, plates, and bowls. In other areas, injection-molded products such as flatware, closures, safety razors, and pens account for large volumes of various HIPS grades.

Polysulfone (PSU) is a clear, rigid, amorphous thermoplastic with properties and processing characteristics similar to those of PES. The primary difference is its continuous service temperature of 160 °C (320 °F). Properties of PSU are given in Table 8.

Polysulfone can be used in a variety of applications, particularly in molded and extruded items that require excellent hydrolytic stability, resistance to high temperatures, and resistance to prolonged exposure to steam or hot water. Its high heat-deflection temperature, combined with excellent hydrolytic stability and an ability to retain mechanical properties in hot and wet environments, makes it suitable for medical and food-service applications that require repeated hot-water cleaning or sterilization. Microwave cookware also represents a significant market for PSU. Electrical and electronic applications are a growing market, especially injection-molded printed circuit boards and connectors. Polysulfone has also been used as a membrane support for reverse osmosis, ultrafiltration, and gas separation.

Polyvinyl chloride (PVC) has been used commercially for more than 50 years, since flexible (plasticized) PVC was introduced in the mid 1930s. A rigid engineering grade of PVC became useful in the early 1950s for piping on naval vessels. Applications for rigid PVC have steadily grown to include its use as an engineering thermoplastic; it is now the second most commonly used plastic material, in terms of volume, after PE.

PVC offers a number of unique features:

- *Low combustibility:* PVC has low combustibility because of its halogen content (57%). While other materials often use halogen-containing additives to achieve low combustibility, PVC offers naturally low combustibility without additives that can sometimes cause problems due to migration. In fact, PVC itself has been used as an additive in other polymer systems to reduce combustibility.
- *Toughness:* PVC compounds are usually ductile and tough. They can be designed to be virtually unbreakable, with a notched Izod impact strength of greater than 0.5 J/mm (>10 ft · lbf/in.) at −40 °C (−40 °F).
- *Weatherability:* Properly designed and processed PVC compounds have outstanding weatherability, including good color and impact retention, good tensile and flexural strength retention, and no loss in modulus (stiffness). For example, rigid vinyl exterior window profiles and house siding installations have accumulated more than 30 years of weathering history with good color and physical property retention.
- *Outstanding dimensional control:* PVC compounds can be designed to have either high or low melt viscosity to meet processing and property requirements. High-melt-viscosity compounds are typically used for good dimensional control in extruded profiles.
- *Low melt viscosity:* In injection molding and sheet extrusion, PVC must have an excellent melt flow to fill large, complex molds or wide extrusion dies. PVC can be designed to lower its molecular weight to promote flow while retaining excellent physical properties. However, PVC is somewhat difficult to injection mold because of its limited processing window. It recrystallizes when cooled, with the crystallites forming physical cross links. These physical cross links effectively make PVC a very-high-molecular-weight polymer at room temperature, giving an outstanding balance of melt flow and physical properties.

PVC is readily modified to attain enhanced properties using compounding additives that are available from several industries that supply the vinyl industry. Rubbery materials, blended with PVC to enhance toughness, are based on rubbers such as butadiene (ABS, acrylonitrile-butadiene-styrene, or MBS, methacrylate-butadiene-styrene, and nitrile rubber), butylacrylate (acrylic and modified acrylic modifiers), and ethylene (chlorinated polyethylene, or CPE, and ethylene/vinyl acetate, or EVA). Other blending ingredients, such as methyl methacrylate (MMA) copolymer and styrene-acrylonitrile (SAN) copolymers, are used as processing aids to reduce melt fracture during PVC processing. Other polymeric ingredients, such as α-methyl styrene-acrylonitrile, styrene-maleic anhydride (S/MA), and glutarimide acrylic copolymer, are used in blends and alloys to increase the softening temperature of PVC. Chlorinated polyvinyl chloride (CPVC) compounds have PVC-like properties, except for an increased softening temperature.

Polyvinyl chloride itself is sometimes used as an additive resin to ABS or impact PS alloys to enhance flame-retardant properties. Blends and alloys are available as balanced compounds containing all other necessary ingredients and need only be processed by extrusion, injection molding, or other processes to achieve the desired properties.

Table 11 Physical properties of polypropylene

Property	Homopolymer	Copolymer	ASTM test method
Tensile strength, MPa (ksi)	31–41 (4.5–6.0)	21.4 (3.1)	D 638
Elongation, %	100–600	300	D 638
Flexural modulus, GPa (10^6 psi)	1.2–1.7 (0.170–0.250)	0.9 (0.130)	D 790
Notched Izod impact strength, J/m (ft · lbf/in.)	20–53 (0.4–1.0)	763 (14.0)	D 256
Deflection temperature under load, °C (°F)			
At 1.82 MPa (0.264 ksi)	50–60 (120–140)	43 (110)	D 648
At 0.45 MPa (0.066 ksi)	110–120 (225–250)	85 (185)	D 648

Source: product data sheets, Quantum Chemical Corporation, USI Division

Engineering Thermosets

As noted, engineering thermosets are polymers with three-dimensional networks of cross-linking bonds between chains. They are known as network polymers, or cross-linked thermosets. Thermoset resins may be either wet (solution, dispersion) or dry (powder), and they may be compounded with catalysts, accelerators, lubricants, fillers, and other processing additives. Catalysts cause cross linking, whereas accelerators promote and modify the curing reaction. Lubricants aid in processing and facilitate mold release. Basic thermoset resins are generally filled and/or reinforced. A filler may be fibrous (e.g., wood flour, glass fiber, carbon fiber) or in flake or granular form (e.g., mica, talc, calcium carbonate). Depending on the end use, combinations of fillers are often used. Fillers provide reinforcement and extend the resin. Other additives, such as pigments and colorants, can also be used.

Curing of thermosets involves the application of elevated temperature and pressure for a given time period to form the cross-linking chemical bonds. Once the cross-linked molecular network forms during this curing process, reapplication of temperature and pressure, even in excess of cure requirements, will not melt-flow the resin system out of shape. Network polymers do not have real glass-transition temperatures, and they degrade (depolymerize) at elevated temperatures.

Common examples include Bakelite and polyester resins used in fiberglass and epoxy adhesives. In Bakelite, cross links form by means of phenol rings, which are integral parts of each chain. The structure of Bakelite is shown in Fig. 21. Starting materials and representative chemical structures for several important families of thermosets are shown in Fig. 22. Six common thermosets are briefly described in this section. They are not the totality of engineering thermosets, but they do represent the range of properties and applications. The six thermosets described in this section are categorized according to their service-temperature capabilities:

- *Low-temperature thermosets:* The aminos, polyurethanes, and unsaturated polyesters used at temperatures under approximately 120 °C (250 °F)
- *Medium-temperature thermosets:* The epoxies and phenolics, used at approximately 120 to 260 °C (250 to 500 °F)
- *High-temperature thermosets:* The polyimides, used at temperatures above approximately 260 °C (500 °F)

There is overlap between these categories; the thermal performance of a resin may be equivalent to that of some resins in an adjacent group.

Amino resins are formed by the controlled reaction of formaldehyde with compounds containing the NH_2 amino group. The most widely used of the amino resins are those made with urea (urea-formaldehyde) and melamine (melamine-formaldehyde). They are supplied as liquid or dry resins and filled molding compounds. Applying heat in the presence of a catalyst converts the material into a hard, rigid, abrasion-resistant solid that has high resistance to deformation under load.

Both urea and melamine molding compounds can be compression, transfer, or injection molded. Molding temperatures for ureas are approximately 140 to 170 °C (280 to 340 °F); for melamine, they are 155 to 170 °C (310 to 340 °F). Compression molding pressures for both materials can vary from approximately 14 to 40 MPa (2 to 6 ksi).

Melamines are superior to urea in resistance to heat, boiling water, and normal acids and alkalis. They also exhibit better performance when cycled between wet and dry conditions. Moldings of both melamines and ureas swell and shrink slightly in varying moisture conditions. Baking molded parts accelerates postmold shrinkage and improves dimensional stability.

Cellulose-filled urea resins are used in circuit breakers, receptacles, and other electrical wiring devices; bases for toasters and other appliances; and consumer products such as buttons, knobs, handles, piano keys, and camera parts. Cellulose-filled melamine resins are principally used for dinnerware, utensil handles, food-service trays, and housings for electric shavers and mixers. Industrial melamine compounds are used for such items as meter blocks, connector plugs, automotive and aircraft ignition parts, and switch housings. Some of these products contain glass fiber or mineral reinforcement. In liquid form, both urea and melamine resins are also used as baked-enamel coatings, particle-board binders, and paper and textile treatment materials. Typical property values are shown in Table 8.

Polyurethane resins (PURs) are usually formed by the reaction of a diisocyanate with a polyol. The material is supplied as flexible and rigid foams, as elastomers, and as a liquid for coatings.

The flexible foams use toluene diisocyanate (TDI) or polymethylene diphenylene isocyanate (PMDI). The largest-volume use is in furniture and bedding. In addition, auto seats, carpet underlay, fabric thermal interlining, and packaging use flexible foam extensively.

The rigid foams are formulated mostly with PMDI and are used as insulation foam for building construction, for the transport of cold fuels and food products, and in furniture.

The elastomers can be used for applications requiring superior toughness, superior resistance to tear and abrasion, and cold-temperature impact and flexibility. Their major shortcoming is low resistance to steam, fuels, strong acids, and bases.

The coating form of PUR is based on the TDI formulation and is used in applications requiring abrasion resistance, skin flexibility, fast curing, good adhesion, and chemical resistance.

Reaction injection molding has recently gained importance in the automotive industry for producing fascia, door panels, and fenders from solid PUR reinforced with up to 20 wt% glass fibers or glass flakes. Highly reactive liquid systems are metered and impingement mixed under high pressure, injected into a mold, and then cured in the mold. A new internal mold release technology has increased productivity and the surface quality of the finished parts. Properties of reaction-injection-molded PUR are listed in Table 8.

Thermoset polyester resins are widely used in transportation, construction, electrical, and consumer products. They are generally produced from the reaction of an organic alcohol (a glycol) with a saturated (isophthalic) and an unsaturated (maleic or fumaric) organic acid. The polyester is then dissolved in a liquid reactive monomer such as styrene, and the solutions are sold as polyester resins. Some polyesters are

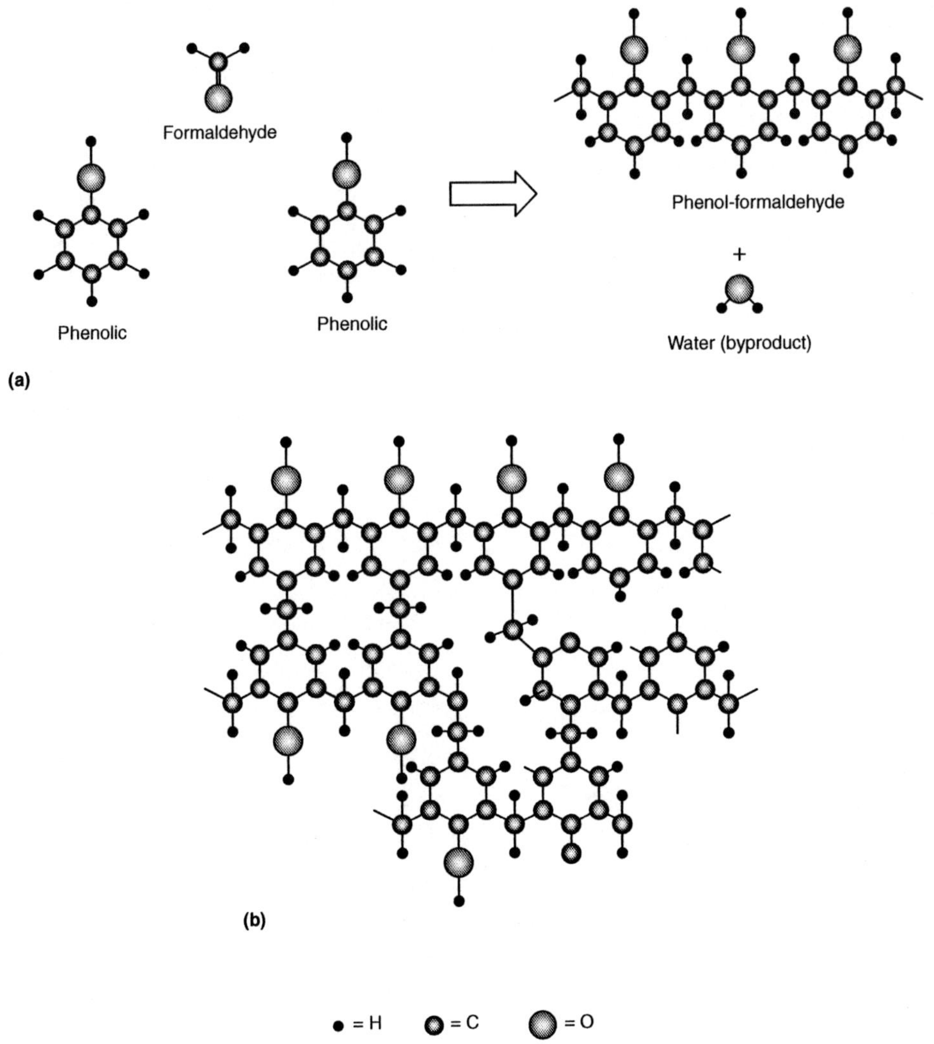

Fig. 21 Structure of a phenol formaldehyde. (a) Two phenol rings join with a formaldehyde molecule to form a linear chain polymer and molecular by-product. (b) Excess formaldehyde results in the formation of a network, thermosetting polymer due to cross linking. Source: Ref 4

supplied as pellets or granular solids. Polyesters are often premixed with glass fiber to form bulk molding compounds (BMCs) or sheet molding compounds (SMCs). Polyester resins with glass-fiber reinforcements can be formulated to provide different mechanical, thermal, electrical, and flammability properties. Table 8 compares the mechanical properties of unreinforced and reinforced thermoset polyesters.

Because of their low cost, ease of processing, and good performance characteristics, unsaturated polyesters are the most extensively used type of thermoset resin. Unsaturated polyesters are generally combined with chopped, continuous, or woven glass fibers, as well as filler and additives, to alter the properties for specific applications.

The versatility of thermoset polyesters allows them to be used in a broad variety of processes. By selection of the appropriate cross-linking initiator, they can be cured at any point from room temperature to 175 °C (350 °F). Resin and glass fibers are combined at the mold in hand lay-up, spray-up, filament winding, pultrusion, and resin transfer molding. Both BMCs and SMCs, as well as other molding compounds, are used as input materials for compression, injection, and transfer molding processes. However, because the fibers are not preplaced in these three molding operations, fiber orientation caused by molding compound flow can produce variable anisotropy in the finished parts. Properties of glass-reinforced polyesters depend on the type of polyester (see Table 8), the glass content (generally from 30 to 70 wt%), and the type and form of glass used.

Epoxy resins are unique among thermosetting resins because of their low shrinkage during curing and their combination of excellent properties (notably adhesion, chemical resistance, and electrical and thermal properties). Epoxies are used in coatings, adhesives, composites, electronics, building materials, and civil engineering applications. Reinforced epoxy structures provide high strength-to-weight ratios, and some can be used at temperatures as high as 260 °C (500 °F).

The diglycidyl ether of bisphenol A (DGEBA), which is based on the condensation of bisphenol A and epichlorohydrin, continues to be the most common type of epoxy resin. Low-molecular-weight epoxies are liquid and are usually cured by amines, carboxylic acid anhydrides, and Lewis acid and base catalysts. Higher-molecular-weight epoxies are cured through their hydroxyl groups. Aliphatic epoxies can be produced by the epoxidation of glycols, polyols, vegetable oils, polyesters, and polyethers. Cycloaliphatic epoxies are produced by the peracetic oxidation of olefins. Epoxy novolacs are epoxidized phenolic novolacs. Properties of unreinforced DGEBA epoxy are given in Table 8.

Epoxy resins are amenable to a variety of formulating techniques. In a typical formulation, the resin component contains epoxy resin(s) and epoxide-containing reactive diluents, while the curative component consists of hardener(s), catalysts, and accelerators. Nonreactive diluents, resin modifiers, fillers, reinforcers, colorants (pigments, dyes), flow additives (thixotropic agents, viscosity suppressants), processing aids (antifoam agents, mold release agents), and other property-regulating additives (adhesion promotors, surfactants, fire retardants) are commonly added to either or both components.

Epoxide-containing reactive diluents are basically low-viscosity epoxy resins or monoepoxides, such as butyl glycidyl ether. Some commercial DGEBA resins are prediluted with reactive diluents. Resin modifiers, such as polyesters,

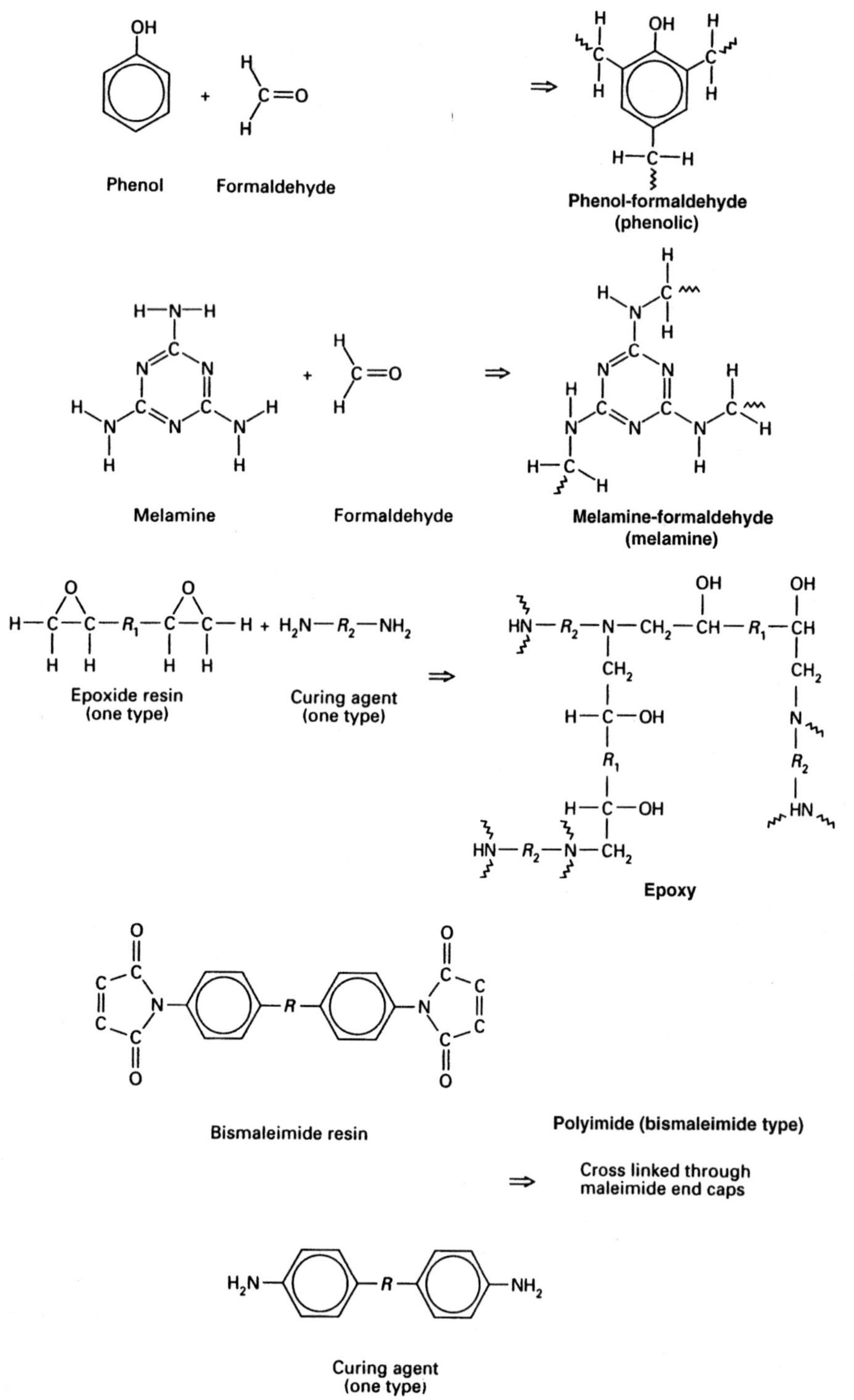

Fig. 22 Chemical structure of representative thermoset plastics

polyurethanes, silicones, acrylics, and butadiene-acrylonitrile polymers, may be included in the formulation to impart special properties, such as flexibility, impact strength, and adhesion. Nonreactive diluents reduce viscosity and cost and increase the pot life. Reinforcing fibers, such as glass, carbon, and aramid, considerably improve mechanical properties and make epoxies suitable in many structural applications. Typical epoxy fillers are powdered metals (for electrical/thermal conductivity), alumina (thermal conductivity), mica (electrical resistance), graphite powders (lubrication), and silica, talc, and calcium carbonate (cost reduction).

The properties of epoxy resins vary over a wide range, depending on the composition and processing of the formulation and the final shape and service environment of the part. Liquid resins and hardeners form low-viscosity systems that can be cured at temperatures from –40 to 200 °C (–40 to 390 °F), depending on the curing agent. Epoxies wet and adhere well to many substrates. They tend to lose mechanical properties when exposed to high temperatures and high humidity (120 °C, or 250 °F, and 95% relative humidity) for extended periods. However, this limitation is highly formulation dependent.

Epoxies are used in coatings, electronics/electrical insulation, composites, construction, and adhesives. Epoxy coatings are noted for their toughness, excellent adhesion, corrosion resistance, and chemical resistance. Marine and maintenance coatings are generally cured at room temperature by polyamido amines. Beverage container coatings are generally DGEBA resins that are modified to produce waterborne systems and are cured by melamines. Solvent-based and waterborne container coatings are both used. In automotive applications, electrodeposited epoxides provide corrosion protection, and epoxy powder coatings are used in under-the-hood applications. Solid epoxies are used in pipe, industrial, and appliance coatings, notably as powder coatings requiring high-temperature curing.

Epoxies cure without giving off volatiles, and their low shrinkage during cure makes them ideal as lightweight, high-strength replacements for metals in many structural applications, especially in the aircraft, aerospace, and automotive industries. The DGEBA, along with brominated epoxies, is used in laminated circuit boards. Carbon-fiber-reinforced epoxy composites are used in the aircraft and aerospace industries. Epoxies are also useful as encapsulating materials for electrical and electronic devices, and they provide outstanding properties in adhesives, grouts, and construction materials.

Phenolic resins are formulated from the reaction between phenol and formaldehyde. The main resin types are:

- *Single-stage resole resins,* which do not liberate ammonia during or after molding, are preferred for applications in which metal corrosion or odor may be a concern. In addition, they show good resistance to stress cracking in parts that are wet on one side and dry on the other.
- *Two-stage novolac resins* are the most widely used and offer wider molding latitude, better dimensional stability, and longer shelf life than resole materials.

Phenolic resins are available in flake, powder, and liquid (solution emulsion) form to meet a variety of mechanical and electrical requirements. They exhibit dimensional and thermal stability and have outstanding load-bearing capabilities at elevated temperatures. Phenolics can be molded by compression, transfer, and injection molding to close tolerances at low cost.

Phenolic resin thermosets include unfilled resin and filled resin systems. For the latter, fillers include glass, carbon, and nylon fiber; wood and cotton flock; aluminum powder; rubber; cellulose fabric; and minerals. Properties of various filled phenolics are listed in Table 8.

Phenolics find application in foundry molds and cores, plywood and particle board bonding, brake linings, insulation, abrasives, coatings, varnishes, and laminates. Filled and reinforced resole and novolac resins are used as engineering plastics in electrical (wiring devices, heavy electrical switch gear, circuit breakers, connectors), appliance (knobs, handles, toasters, steam irons), and automotive (brake system, transmission thrust washer, water pump housing, solenoids, starter commutators) applications. The growth in applications for phenolic resins is due to the weight and cost savings inherent in metal replacement and parts consolidation. Phenolics have replaced thermoplastics where creep resistance and thermal stability are required in downsized parts or hostile service environments.

Hybrids of the novolacs are used as impregnating resins with glass, carbon, and graphite cloth for tape wrapping or hand lay-up of aerospace components, rocket nozzle ablatives, and insulation liners. Chopped-fiber molding compounds are used mostly in the automotive, appliance, and electrical component markets. General characteristics of these materials that make them suited for the aforementioned applications are high service temperatures, good electrical properties, excellent moldability, superior dimensional stability, and relatively good moisture resistance.

Thermoset polyimides are characterized by the imide structure, which has exceptional thermal and oxidative properties. Thermoset polyimides have high elongation and toughness, which are particularly advantageous in thin-film products.

Molded polyimide parts and laminates are inherently resistant to combustion. Glass-fiber-reinforced polyimide moldings have very high flexural strength and modulus. Deformation under load is extremely low, and creep is almost nonexistent, even at high temperatures. Graphite-reinforced polyimides used for high-temperature aerospace applications retain their properties up to 315 °C (600 °F), which is the highest service temperature of any polymeric material.

Polyimide parts are fabricated by conventional injection, transfer, extrusion, and compression molding methods. Applications include aerospace and electronics, and polyimides are making inroads into the industrial market. Polyimide moldings and laminates are used in jet-engine parts, computers, photocopy machines, and integrated circuit chips. Polyimide films are employed in electric motors and in insulation for aircraft and missile wire cables. Polyimide coatings find uses in electronic and electrical devices, while polyimide foams find applications in space vehicles. Properties of unreinforced polyimide are given in Table 8.

ACKNOWLEDGMENT

Significant portions of this article were adapted from the article "Polymer Science for Engineers" by Linda Clements, in *Engineering Plastics, Engineered Materials Handbook,* Volume 2, ASM International, 1988, p 48–62.

REFERENCES

1. H.W. Stoll, *Product Design Methods and Practices,* Marcel Dekker, 1999, p 40, 148
2. V. John, *Introduction to Engineering Materials,* 3rd ed., Industrial Press, 1992
3. B.A. Miller, Materials Selection for Failure Prevention, *Failure Analysis and Prevention,* Vol 11, *ASM Handbook,* 2002, p 24
4. M.L. Weaver and M.E. Stevenson, Introduction to the Mechanical Behavior of Nonmetallic Materials, *Mechanical Testing and Evaluation,* Vol 8, *ASM Handbook,* ASM International, 2000, p 13–25
5. W.D. Callister, *Materials Science and Engineering—An Introduction,* 4th ed., John Wiley & Sons, 1997, p 21
6. F. Rodriguez, *Principles of Polymer Systems,* McGraw Hill and Hemisphere, 1982
7. W.G. Moffatt, G.W. Pearsall, and J. Wulff, *Structure,* Vol 1, *The Structure and Properties of Materials,* John Wiley & Sons, 1964, p 104
8. A. Kumar and R. Gupta, *Fundamentals of Polymers,* McGraw-Hill, 1998, p 30, 337, 383
9. "Standard Classification System for Specifying Plastic Materials," D 4000-95a, *Plastics,* Vol 08.03, *Annual Book of ASTM Standards,* ASTM
10. J. Brydson, *Plastic Materials,* 7th ed., Butterworth-Heinemann, 1999, p xxiv
11. L.E. Nielson, *Mechanical Properties of Polymers,* Van Nostrand Reinhold, 1962
12. F.N. Kelly and F. Bueche, *J. Polym. Sci.,* Vol 50, 1961, p 549
13. C. Hall, *Polymer Materials: An Introduction for Technologists and Scientists,* Macmillan, 1981
14. J.R. Davis, Guide to Materials Selection, *Engineered Materials Handbook, Desk Edition,* ASM International, 1995, p 135

Effects of Composition, Processing, and Structure on Properties of Engineering Plastics*

PLASTICS are so prevalent in our lives that it is easy to overlook the vast differences in their properties and how specialized many polymers have become. Consider the differences between aramid bulletproof vests and the polyurethane foam used in pillows. Why can plates made of crystallized polyethylene terephthalate be microwaved successfully while plastic film wrap (polyvinylidene chloride) has poor elevated-temperature properties? Consider how different polycarbonate is from plastic foam (expanded polystyrene); why is one plastic suitable for motorcycle helmets and the other for disposable coffee cups?

The answers to these questions lie in the chemical nature and morphology (structure) of the polymer chains and additions such as fillers, colorants, reinforcing agents, thermal stabilizers, plasticizers, and other modifying agents or additives. The preceding article in this book introduces the basic concepts of polymer structure and properties. This article describes in more detail the importance of chemical composition and morphology to mechanical properties and reviews basic plastic processing techniques. Table 1 and Fig. 1 show the structures and transition temperature of selected polymers.

The difference between plastics and metals or ceramics is that plastics can be melted at relatively low temperatures and formed into a variety of shapes. Advantage can be taken of their non-Newtonian flow behavior in selecting a suitable molding or finishing process. Atoms can be specifically selected to design a polymer with the desired properties through a fundamental understanding of how submolecular, molecular, intermolecular, and supermolecular forces behave. A polymer scientist can custom polymerize a plastic to meet specific application requirements.

This article describes in more detail the fundamental building-block level, atomic, then expands to a discussion of molecular considerations, intermolecular structures, and finally supermolecular issues. An explanation of important physical properties, many of which are unique to polymers, follows, and the final section discusses processing techniques.

Composition

Submolecular Structure

As noted in the preceding article, most engineering plastics are based on organic (carbon-base) polymers, where the carbon atom plays a critical role in developing final properties. Hydrogen, oxygen, nitrogen, fluorine, and chlorine are among the many atoms that are built into polymer structures in order to tailor specific properties. Table 2 lists common atoms found in plastics and gives both the electronegativity (relative tendency to attract electrons) of the atom and the number of unpaired electrons present in the outer shell. The number of unpaired electrons governs the number of covalent bonds the atom will form. The electronegativities of the constituent atoms that make up the polymer control its polarity. This, in turn, regulates the ability of the polymer to form the secondary bonds (e.g., hydrogen bonds) that have marked effects on the final thermomechanical properties.

Carbon is of fundamental importance as the most basic building block of most polymers in use. Carbon contains six valence electrons, two of which are located in the inner, most protected orbital, and all or four of which are in the outer orbital. It is the presence of four outer orbital electrons (exactly halfway between zero and eight) that causes carbon to be a neutral atom. Consequently, the electronegativity of carbon is 2.5. Metal atoms tend to be large, with a propensity to lose electrons when forming bonds; thus, their electronegativities are lower than 2.5. Elements that tend to gain electrons have electronegativities greater than 2.5. Carbon atoms share electrons when forming bonds with other carbons and, while the resultant materials can vary dramatically from diamond to graphite to hydrocarbon polymers such as polyethylene, the neutral carbon-carbon covalent bonds are stable to heat and ultraviolet (UV) light exposure.

Because carbon can form four bonds, it may bond more than once with other carbon atoms. As shown in Fig. 2, carbon-carbon single bonds are relatively stable. While carbon-carbon double bonds are shorter (as evidenced by their greater bond dissociation energy), they are more subject to attack by atmospheric oxygen. Consequently, polymers, such as polyisoprene and polybutadiene, are usually compounded with antioxidants. Carbon-carbon triple bonds are even more sensitive to oxygen attack. Although these are rarely found individually in commercial polymers, alternating triple and single bonds (called conjugated triple bonds) impart electrical conductivity to polymers, such as polyacetylene. Conjugated double bonds are more rigid. Rings of carbon-carbon single bonds, such as found in cyclohexane, assume nonplanar configurations. In contrast, rings of conjugated carbon-carbon double bonds, which occur in benzene, phenyl groups, and phenylene groups, are rigid and planar. As is discussed later in this article, these groups impart rigidity to polymers such as polystyrene (PS) and polycarbonate (PC).

Attaching other elements to a carbon atom introduces polarity, which changes the balance of the electron cloud. This can be regarded as either reducing the stability of an all-carbon material, or increasing its reactivity. Introducing polarity to the molecule through electronegativity differences between atoms has significant effects on thermal properties such as melting temperature (T_m) and mechanical properties such as Young's modulus (E). The presence of polar bonds produces higher thermal and mechanical properties in engineering plastics compared to those in nonpolar materials. Figure 2 presents chemical groups commonly found in plastics and the bond dissociation energies (E_d) for selected groups.

Hydrogen. Because the electronegativity of hydrogen, 2.1, is only slightly more electropositive than carbon, carbon-hydrogen bonds are almost as stable as carbon-carbon bonds. In the

*Adapted from A.-M.M. Baker and C.M.F. Barry, "Effects of Composition, Processing, and Structure on Properties of Engineering Plastics," *Materials Selection and Design*, Volume 20, *ASM Handbook*, ASM International, 1997, pages 434 to 456

absence of atmospheric oxygen, carbon-hydrogen bonds have good thermal and UV stability. Materials containing aliphatic (i.e., noncyclic) carbon-hydrogen bonds, such as polyethylene (PE) and polypropylene (PP), are marked by low surface energies, low adhesion, and low coefficients of friction. This makes PP automobile bumpers difficult to paint and is why printing inks adhere poorly to untreated PE bags. Aromatic carbon-hydrogen bonds (for example, in a benzene ring) are stabilized by resonance and are more stable than aliphatic carbon-hydrogen bonds.

Hydrogen can also bond to elements other than carbon, such as oxygen, in the case of the common hydroxyl group , –OH. Due to the electronegativity of oxygen, the hydroxyl group is more polar and less balanced, making this bond more highly reactive than the previous bonds considered.

Oxygen. With an electronegativity of 3.5, oxygen introduces significant polarity to polymers. It is a unique atom in that it has two pairs of readily available unbonded electrons that can form fairly strong hydrogen bonds with neighboring molecules. These unbonded electron pairs also impart high surface energy to oxygen-containing polymers. Thus, such polymers have higher mechanical properties and provide better adhesion than nonpolar hydrocarbon polymers.

Table 1 Properties of selected commodity and engineering plastics

Common name	Tensile strength, MPa	Glass transition temperature (T_g), °C	Melting temperature (T_m), °C
Low-density polyethylene (LDPE)	10–12	–120	110
High-density polyethylene (HDPE)	26–33	–120	135
Linear low-density polyethylene (LLDPE)	15–32	–120	125
Isotactic polypropylene (PP or i-PP)	31–37	–10	165
Cis-1,4-polyisoprene, natural rubber	. . .	–67	15–50
Trans-1,4-polyisoprene, gutta percha or balata	. . .	–71	56–65
Polybutadiene:			
1,4-cis	21(a)	–102	. . .
1,4-trans	14(a)	–107	. . .
1,2-isotactic	10(a)	–15	. . .
1,2-syndiotactic	11(a)	–15	90
Poly-(4-methyl-1-pentene) (TPX)	28	55	245
Atactic-polystyrene (PS or a-PS)	50	100	. . .
Syndiotactic-polystyrene (s-PS)	41	100	270
Polymethylacrylate	. . .	0	. . .
Polymethyl methacrylate			
PMMA	70	100, 105	. . .
i-PMMA	. . .	45	160
Polyvinyl chloride (PVC)	55	80, 87	212
Polyvinylidene chloride (PVDC)	. . .	–17	198
Polyvinyl fluoride (PVF)	66–131	–20	200
Polyvinylidene fluoride (PVDF or PVF_2)	48	–35	171
Polychlorotrifluoro-ethylene (PCTFE)	30–39	45, 52	220
Polytetrafluoroethylene (PTFE)	17–21	126	327
Polyvinyl acetate (PVAC)	Soft	29	. . .
Polyvinyl alcohol (PVOH)	83–152	85	$T_d < T_m$
Polyacrylonitrile (PAN)	. . .	104	$T_d < T_m$
Polyoxymethylene (POM or polyacetal)	70	–50	175
Polyethylene oxide (PEO)	13–22	–55	66
Polypropylene oxide	. . .	–62	65
Polyamide 11 (nylon 11)	38	. . .	185
Polyamide 12 (nylon 12)	45	. . .	175
Polyamide 4/6 (nylon 4/6)	100	. . .	295
Polyamide 6/6 (nylon 6/6)	80	60	264
Polyamide 6/10 (nylon 6/10)	55	40	215
Polycarbonate (PC)	62	150	. . .
Polyethylene terephthalate (PET)	72	69	265
Polybutylene terephthalate (PBT)	52	60	232
Polyether imide (PEI)	105	215	. . .
Polyamide-imide (PAI)	152	275	. . .
Polyimide (PI)	72–118	310–365	. . .
Polysulfone (PSU or PSF)	70	195	. . .
Polyarylether sulfone (PAS)	70	220	. . .
Polyether sulfone (PES)	90	230	. . .
Polyphenylene sulfide (PPS)	70	85	288
Polyether ketone (PEK)	110	155	365
Polyether ether ketone (PEEK)	92	143	334
Polyether ketone ketone (PEKK)	102	156	338
Polyether ether ketone ketone (PEEKK)	100	167	360
Polyether ketone ether ketone ketone (PEKEKK)	118	170	381
Polyphenylene oxide (PPO)	72	220	. . .
Modified polyphenylene oxide (PPO/PS)	55	140	. . .
Polydimethyl siloxane (PDMS)	. . .	–123	–85 to –65

(a) When vulcanized. Source: Ref 1–6

Hydrogen bonds are further discussed in the section on intermolecular arrangements.

Carbon and oxygen are the components of several major functional groups shown in Fig. 2. The stability of the –C–O–C– ether bond is dependent on attached groups. Because aromatic ethers have a resonating system that includes the two electron pairs from the oxygen, the larger extended structure is stabilized through resonance. This contributes to the high thermal stability and high heat-distortion temperatures of engineering plastics such as polysulfones (PSUs) and polyether ketones (PEKs). In contrast, the bond of a hydrogen to an atom adjacent to the oxygen in an aliphatic ether (referred to as the α-hydrogen) is destabilized in the presence of the oxygen. Thus, polymers—such as polyvinyl acetals and cellulosics—exhibit instability because their –O–CH_2–O– linkages are particularly sensitive to acid hydrolysis.

The carbonyl group of ketones, esters, and carbonates (shown in Fig. 2) strongly absorbs UV light in the 2800 to 3200 Å range, thus leading to polymer instability and poor outdoor aging characteristics. The ester group may hydrolyze and degrade upon exposure to water; manufacturers capitalize on this reactivity to produce polyvinyl alcohol (PVOH) from polyvinyl acetate (PVAC). Polyvinyl alcohol is a water-soluble, film-forming polymer that finds extensive use in applications ranging from photographic film to packaging. Polyvinyl acetate is not water soluble and is used in adhesives, textile applications, and latex paint.

Nitrogen, with an electronegativity of 3.0, generally forms strong bonds with carbon and, as in the case of oxygen, the unbonded electron pair generates a highly polar molecule available to form secondary bonds. The presence of both oxygen and nitrogen in the amide, urea, and urethane groups leads to strong hydrogen bonding and high sensitivities to water in the corresponding polymers. An alternative bond that nitrogen can form with carbon is an extremely rigid triple bond. This nitrile group is instrumental in generating high-modulus, heat-resistant engineering plastics such as styrene-acrylonitrile (SAN) copolymers and acrylonitrile-butadiene-styrene (ABS).

Fluorine is the most electronegative of all elements, with an electronegativity of 4.0. Its small atomic radius means that the carbon-fluorine bond length is very short. The strong bonds it forms with carbon impart low surface energy to fluoropolymers and allow them to be used for nonwetting applications such as nonstick cookware. The carbon-fluorine bond is also low in friction, which is suitable for high-lubricity applications such as mold lubricants and self-lubricating gears and bearings. This bond is extremely stable to heat, UV light, and chemical exposure, making it appropriate for high-temperature plastics and elastomers. Table 3 highlights the effects of different degrees of fluorination on maximum-use temperature from PE to polytetrafluoroethylene (PTFE). It is evident that the reduction in fluorine content generates thermal instability, but does result in a more eas-

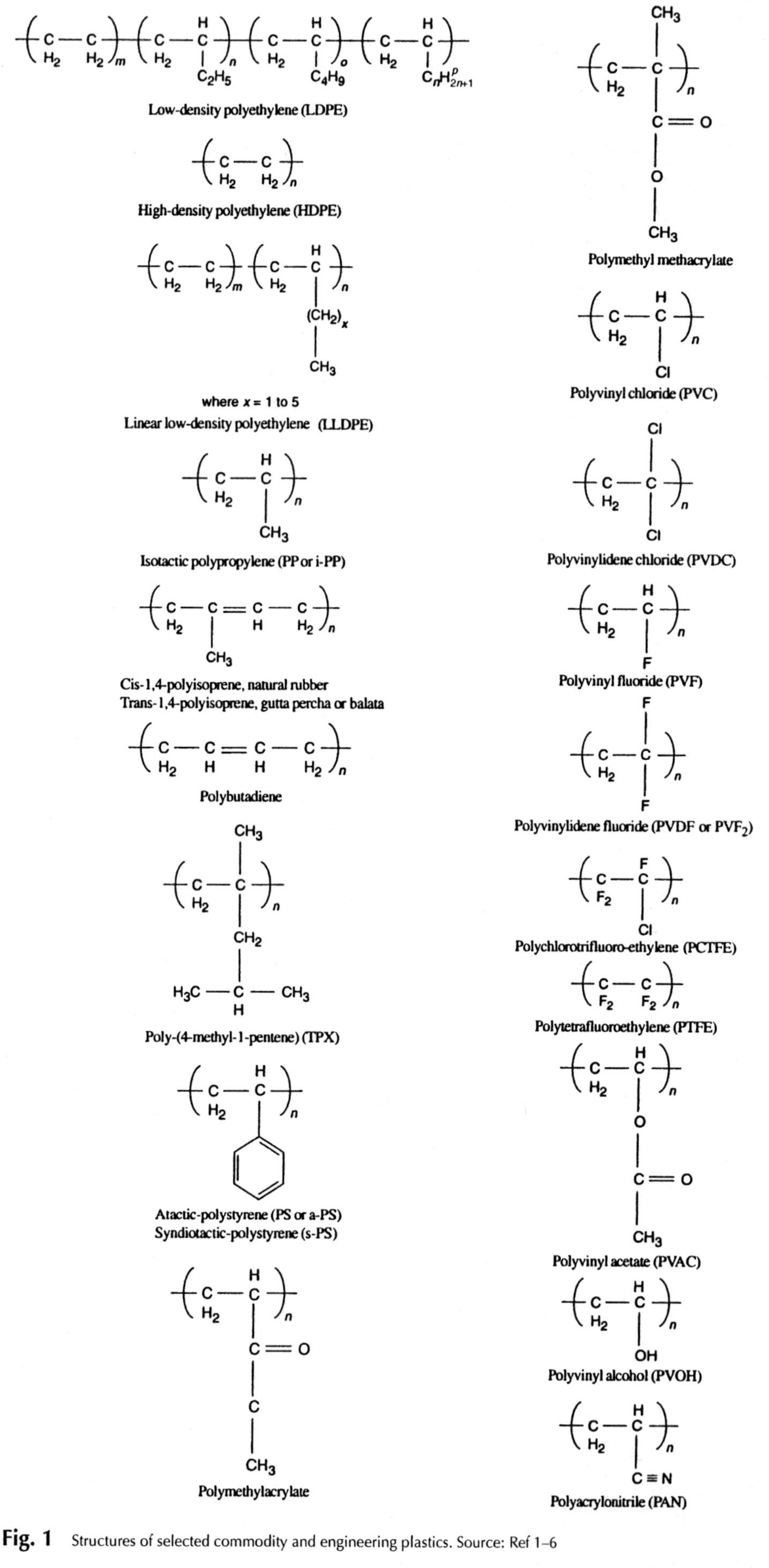

Fig. 1 Structures of selected commodity and engineering plastics. Source: Ref 1–6

ily processed polymer. Highly fluorinated plastics such as PTFE are not melt processible by traditional methods.

Chlorine. While chlorine has seven valence electrons like fluorine, its larger atomic radius reduces its electronegativity to 3.0. Thus, chlorine bonds less strongly to carbon than does fluorine. The presence of such a large and electronegative atom generates polarity that has a marked effect on mechanical properties such as stiffness. A nonpolar molecule, PE, has a tensile modulus of 175 to 280 MPa (25 to 40 ksi) and a T_m of 105 to 110 °C (220 to 230 °F). Polyvinyl chloride (PVC), which substitutes a single chlorine atom onto the PE structure, has a tensile modulus of 2400 to 6500 MPa (350 to 945 ksi) and a glass-transition temperature (T_g) (amorphous) of 75 to 105 °C (165 to 220 °F).

Molecular Structure

Polymer molecules contain multiple repeat units called mers. The number of repeat units

Table 2 Number of covalent bonds formed and electronegativities of atoms commonly found in plastics

Atom	Total number of electrons	Number of unpaired electrons	Number of covalent bonds formed	Electro-negativity(a)
H	1	1	1	2.1
C	6	4	4	2.5
N	7	3	3	3.0
O	8	2	2	3.5
F	9	1	1	4.0
Si	14	4	4	1.8
P	15	3	3 or 5	2.1
S	16	2	2 or 6	2.5
Cl	17	1	1	3.0
Br	35	1	1	2.8

(a) Electronegativity data from Ref 7

Table 3 Continuous service temperature as a function of degrees of fluorine substitution on polyethylene

Name	Continuous service temperature, °C	Repeat structure
PE	60–75	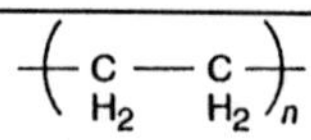
PVF	100–120	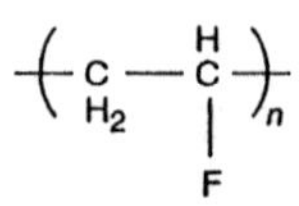
PVDF	150	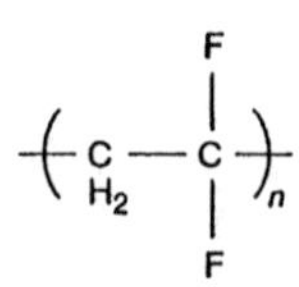
PTFE	250	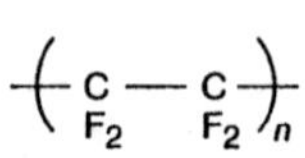

Source: Ref 10

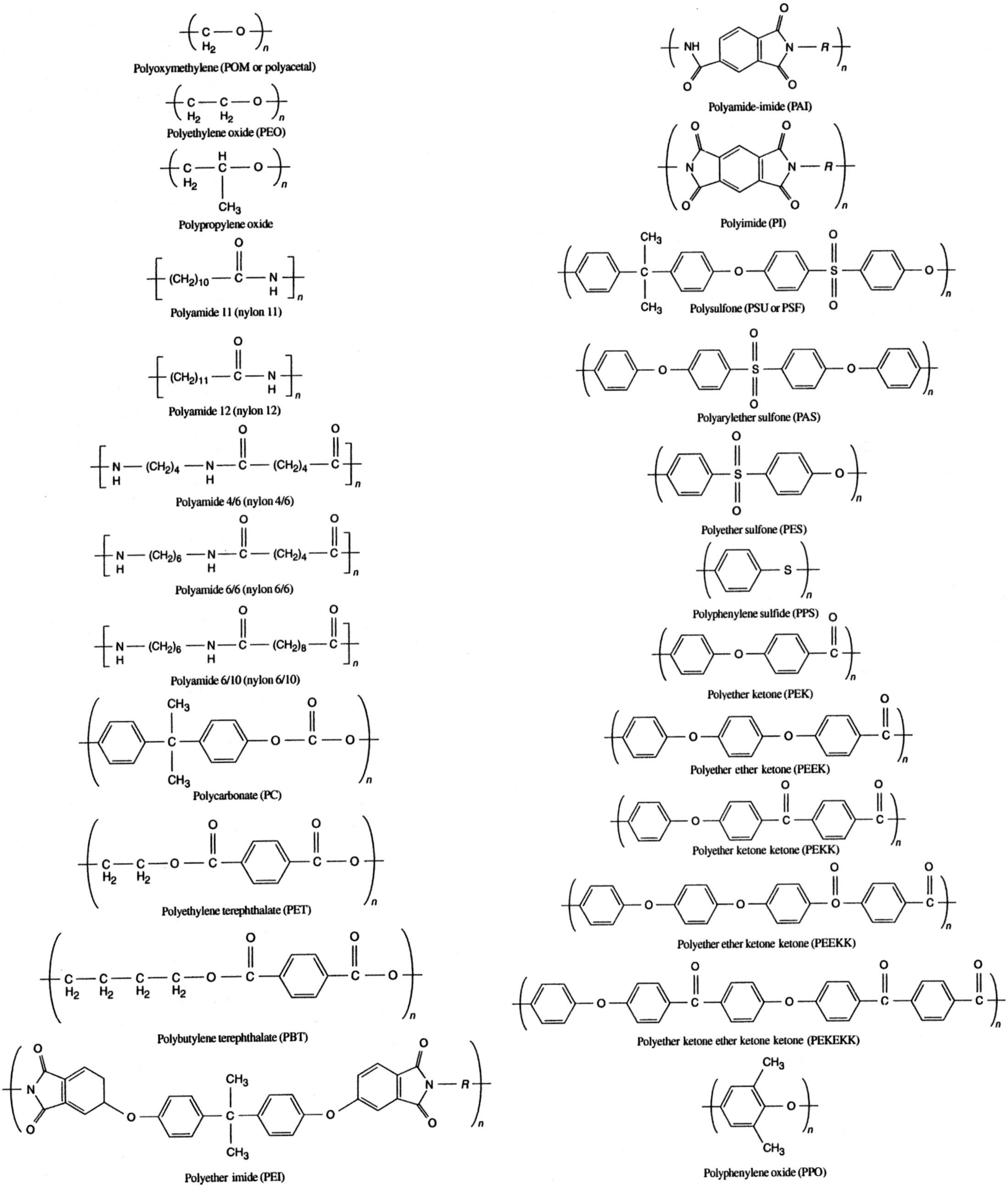

Fig. 1 (continued)

can be varied, and this strongly affects the thermal, mechanical, and rheological properties of plastics, as shown in Table 4. Polymer size is quantified primarily by molecular weight (MW), molecular-weight distribution (MWD), and branching.

Molecular weight is generally defined as either number average ($\overline{M}_n$) or weight average ($\overline{M}_w$) depending on whether the length of each molecule is averaged according to numbers of molecules present at that length (as in the case of $\overline{M}_n$) or whether large molecules are more heavily considered (as in the case of $\overline{M}_w$). Equations 1 and 2 define $\overline{M}_n$ and $\overline{M}_w$, respectively, as:

$$\overline{M}_n \equiv \frac{\sum_{j=1}^{\infty} w_i}{\sum_{i=1}^{\infty} N_i} = \frac{\sum_{i=1}^{\infty} M_i N_i}{\sum_{i=1}^{\infty} N_i} \quad \text{(Eq 1)}$$

$$\overline{M}_w \equiv \frac{\sum_{i=1}^{\infty} M_i w_i}{\sum_{i=1}^{\infty} w_i} = \frac{\sum_{i=1}^{\infty} M_i^2 N_i}{\sum_{i=1}^{\infty} M_i N_i} \quad \text{(Eq 2)}$$

where w_i is the weight of polymer species i, N_i is the number of moles of species i, and M_i is the molecular weight of that species. If a polymer system has 7 moles of 20,000 MW species and 5 moles of 60,000 MW species, then the $\overline{M}_n$ and $\overline{M}_w$ can be calculated as follows, according to Eq 3 and 4, respectively:

$$\overline{M}_n = \frac{(7 \cdot 20{,}000) + (5 \cdot 60{,}000)}{(7 + 5)} \cong 37{,}000 \quad \text{(Eq 3)}$$

$$\overline{M}_w = \frac{(7 \cdot 20{,}000^2) + (5 \cdot 60{,}000^2)}{(7 \cdot 20{,}000) + (5 \cdot 60{,}000)} \cong 47{,}000 \quad \text{(Eq 4)}$$

Because in the case of $\overline{M}_w$ the higher MW fractions of a polymer contribute more heavily, $\overline{M}_w$ is always greater than or equal to $\overline{M}_n$. $\overline{M}_n$ can be measured by methods that depend on end-group analysis or colligative properties such as osmotic pressure, boiling-point elevation, or freezing-point depression. $\overline{M}_w$ can be measured by light-scattering techniques or ultracentrifugation, both of which depend on the mass of species present (Ref 12).

As shown in Fig. 3, many physical and mechanical properties vary significantly as a function of MW, up to a threshold value, whereupon they level off asymptotically at higher MWs. Molecular entanglement can be dramatically demonstrated by the relationship of melt viscosity, η, to $\overline{M}_w$; melt viscosity being a measure of the tendency of the material to resist

Table 4 Effect of molecular weight on polyethylene

Number of $-CH_2-CH_2-$ units	Molecular weight (MW)	Softening temperature, °C	Character of polymer at 25 °C
1	30	−169(a)	Gas
6	170	−12(a)	Liquid
35	1,000	37	Grease
140	4,000	93	Wax
250	7,000	98	Hard wax
430	12,000	104	Plastic
750	21,000	110	Plastic
1350	38,000	112	Plastic

(a) Melting point. Source: Ref 11

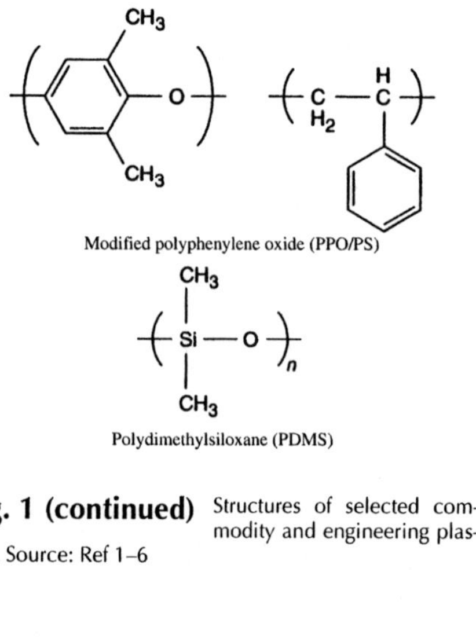

Fig. 1 (continued) Structures of selected commodity and engineering plastics. Source: Ref 1–6

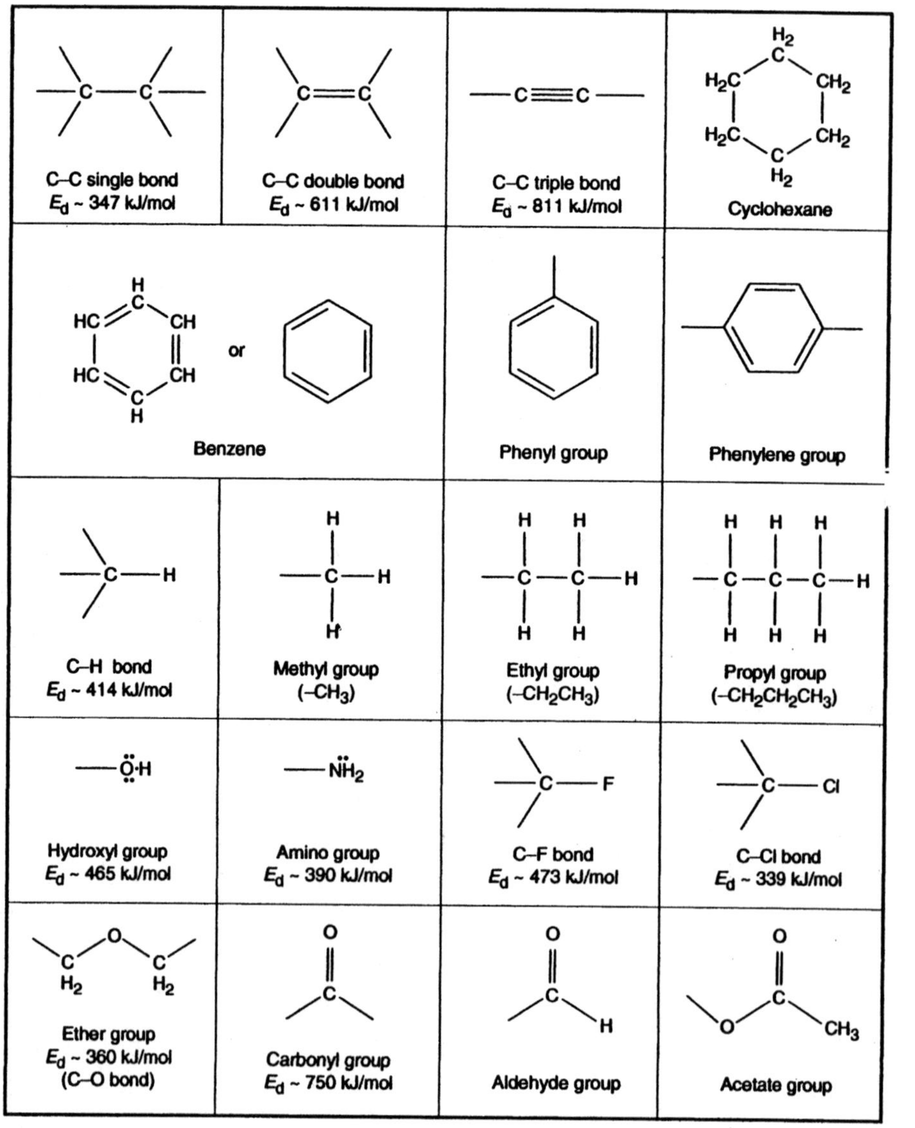

Fig. 2 Chemical groups and some bond dissociation energies (E_d) used in plastics. Adapted from Ref 8; dissociation energies from Ref 9

flow. Below a critical $\overline{M}_w$, denoted as $\overline{M}_c$, there is little chain entanglement, and the melt viscosity increases linearly with $\overline{M}_w$ until it reaches the M_c threshold. At this point, the melt viscosity increases as an exponential function of $\overline{M}_w$, with the exponent approximating 3.4 for many polymers, as shown in Fig. 4. In the elevated-slope region, molecular entanglements inhibit molecular slippage. The increased occurrence of physical chain entanglements associated with higher MWs accounts for the elevation of melt viscosity.

Below M_c the chains are short enough to align in the direction of flow and to slip past each other with relative ease. Once the critical length has been achieved, entangled polymers offer more resistance to the stresses inducing flow. This property, associated with the high MWs of engineering plastics, dramatically distinguishes them from Newtonian rheological behavior as is further explored in the section "Thermal and Mechanical Properties" in this article.

This concept of M_c can be related to mechanical properties intuitively. The degree of intermolecular attractive forces is limited by the chain length. For example, at low MWs (below M_c), chain disentanglement can occur. Above a certain size (greater than M_c), the system is highly entangled and has maximized its intermolecular bonding such that it is now limited by the strength of the chain backbone. Most industrial engineering plastics have MWs well above M_c so that moderate changes in MW will not appreciably affect properties such as yield stress or modulus.

$\overline{M}_n$ finds relevance in relating properties that depend on small molecules (such as environmental stress cracking resistance), while $\overline{M}_w$ is well suited for relating properties that depend on intermolecular attractions, because as chain length increases, the number of intermolecular bonds per molecule also increases. This is important when the property of interest measures the ability of a material to disentangle chains.

The breadth of MW range in a sample can be represented by a polydispersity index, which is equal to the ratio of $\overline{M}_w$ to $\overline{M}_n$. A material with a broad range of MWs (i.e., a high polydispersity index or broad MWD) will melt at lower temperatures than the equivalent material with a narrow range of MWs because the components with lowest MW will melt first.

Recent use of metallocene catalysts during polymerization has resulted in greater control over MWD. The narrow MWD linear low-density polyethylenes (LLDPEs) have better strength and heat-sealing properties because the lower MW components are no longer present. However, the lack of shorter polymer chains increases melt viscosity to such a degree that processing problems are often encountered. The use of blends of high- and low-MW LLDPE generates a bimodal MWD that produces a balance of good strength and ease of processing.

Chain branching also has a significant effect on flow properties. For a polymer of a given

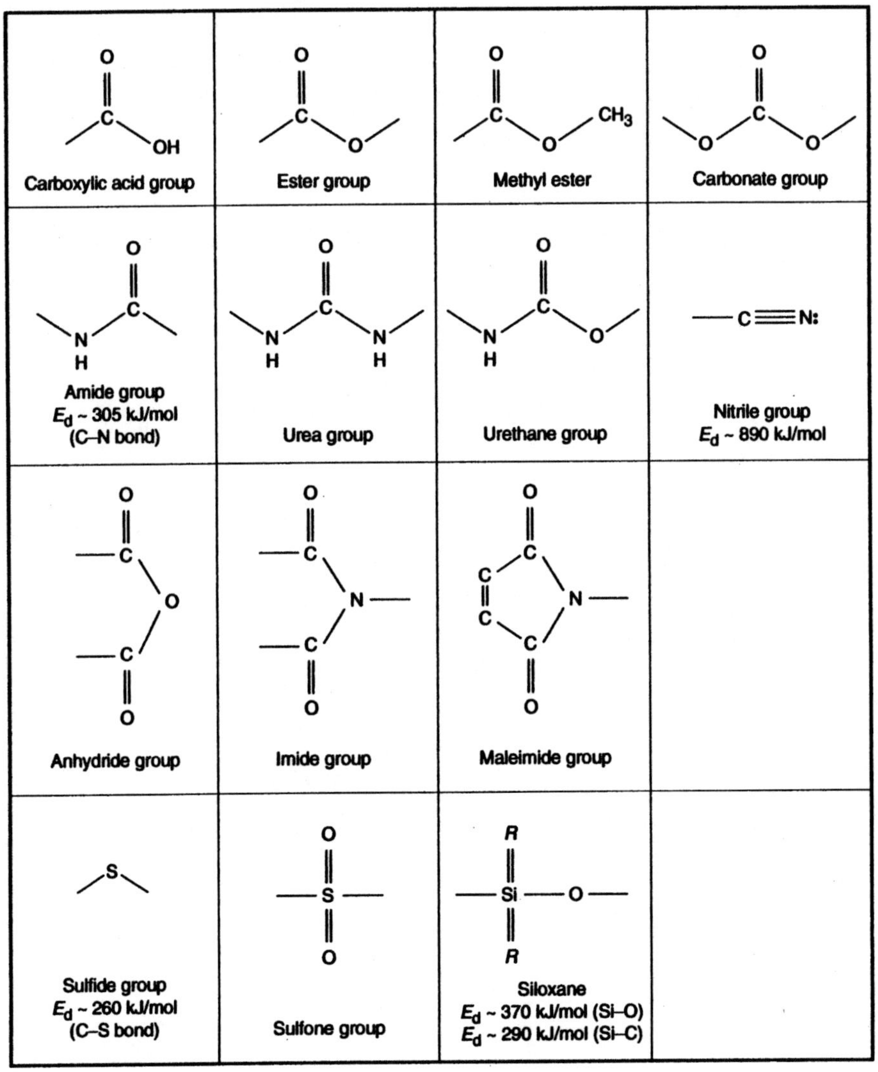

Fig. 2 (continued)

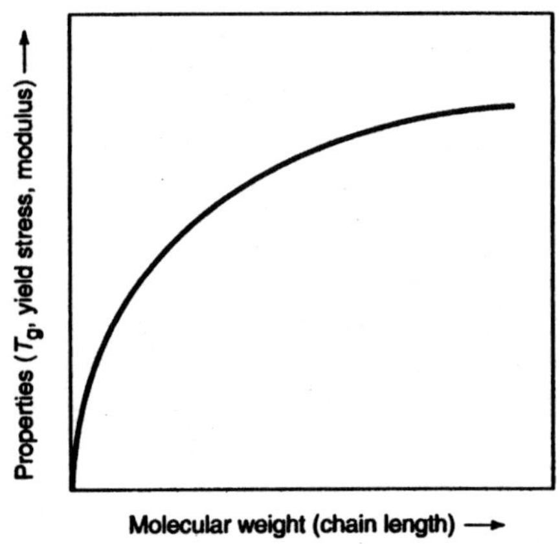

Fig. 3 General influence of molecular weight on polymer properties. Source: Ref 13

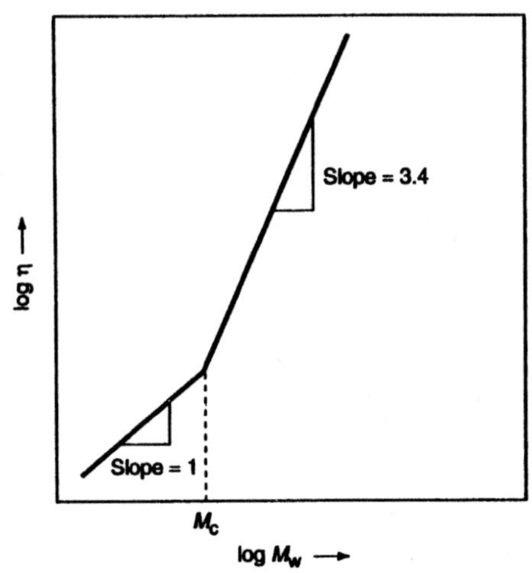

Fig. 4 Viscosity dependence on molecular weight exhibiting M_c. Source: Ref 14

MW, the more highly branched the structure is, the lower its density will be and the lower the degree of entanglement. Moreover, for any given polymer, the lower its MW, the more flexible it will be as there are a greater number of chain ends per unit volume for short chain species. Chain ends reduce packing efficiency, and the additional free volume available offers sites into which the polymer can be displaced under stress. Once the MW is greater than the M_c the end-group concentration change is insignificant for further MW increases, and the mechanical properties plateau when the total intermolecular attractions are greater than the strength of the polymer backbone.

In addition to MW and chain branching, repeat units can be added in either a random or ordered fashion. In atactic polymers, such as PS and polymethyl methacrylate (PMMA), the mers are added randomly. In contrast, the repeat units of isotactic and syndiotactic polymers are ordered. Because the side chains of atatic polymers are randomly oriented as shown in Fig. 5(a), they inhibit crystallization (as is discussed later in this article). In isotactic polymers (Fig. 5b) the side chains all extend from the same side of the backbone, while in syndiotactic polymers they alternate sides (Fig. 5c). This regularity facilitates crystallization.

Inherent Flexibility. Before expanding the scope of consideration to include interactions between neighboring molecules, it is important to appreciate the inherent flexibility of the backbone of any given molecule. In this discussion it is first assumed that every carbon-carbon bond segment is completely free to assume any position as long as the equilibrium requirement that the carbon-carbon bond angle be maintained at 109° is met. A random conformation that might occur is shown in Fig. 6.

Inclusion of the hydrogen atoms (which fill the valence electron requirements of carbon) in the spatial consideration introduces limitations to the flexibility. The hydrogen atoms impose restrictions on the number of energetically viable positions that the chain can assume. Figure 7 plots an example of different energetically favorable conformations for the case of ethane (C_2H_6). It considers what happens as one carbon is rotated around the carbon-carbon bond and demonstrates the effects of trying to force the hydrogen atoms of one carbon atom to be spatially close to the hydrogen atoms of an adjacent carbon atom.

Figure 8 dramatically demonstrates the effects of the replacement of two hydrogens by carbon-carbon triple bonds. For example, there are fewer atoms surrounding the central carbon of methylacetylene, which allows greater freedom of rotation for the carbon-carbon single bond. Of course, the greater electron density of the carbon-carbon triple bond does restrict the motion of that bond. Consideration of neopentane shows the resulting reduction of degrees of freedom when substituent hydrogens are replaced by the considerably more bulky methyl ($-CH_3$) groups. Introduction of the electronegative oxygen-containing side groups further increases stiffness of the backbone by reducing flexibility not only due to the size of this side group but because of electrical repulsion as well.

This concept accounts for the flexibility of rubbers, such as cis-1,4-polybutadiene (Table 1 and Fig. 1), that have double bonds on their main chain. The double bond eliminates two hydrogen atoms, and the additional free volume results in additional flexibility. One of the most flexible polymers, polydimethylsiloxane (PDMS), has a flexible ether linkage on the main chain and nonpolar side groups, which accounts for its lack of rigidity. The ether oxygen only forms bonds with two carbon atoms, and the lack of hydrogen atoms means that ether linkages are surrounded by ample free volume. This promotes ease of rotation. Flexibility is also introduced by ether linkages due to the smaller atomic radius of oxygen atoms compared to

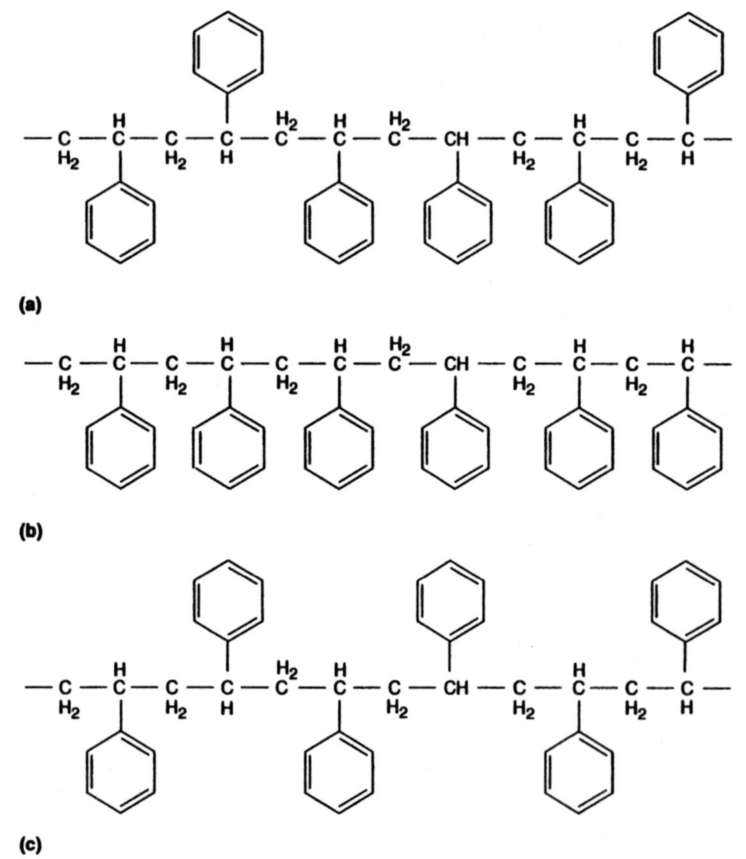

Fig. 5 Tacticity in polymers as shown by (a) atactic, (b) isotactic, and (c) syndiotactic polystyrene

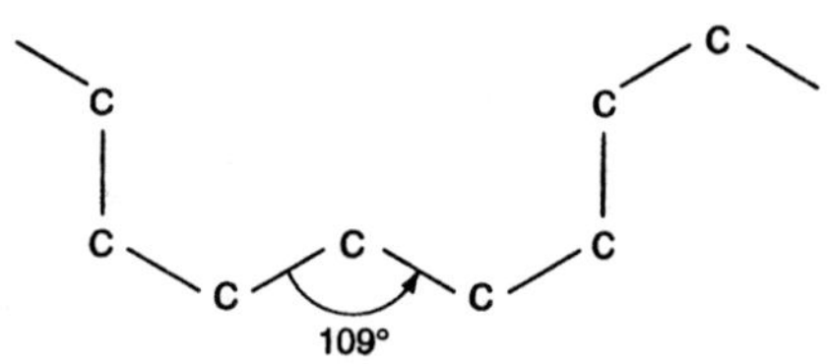

Fig. 6 Random formation of carbon-carbon bond segments. Bond angle is 109°

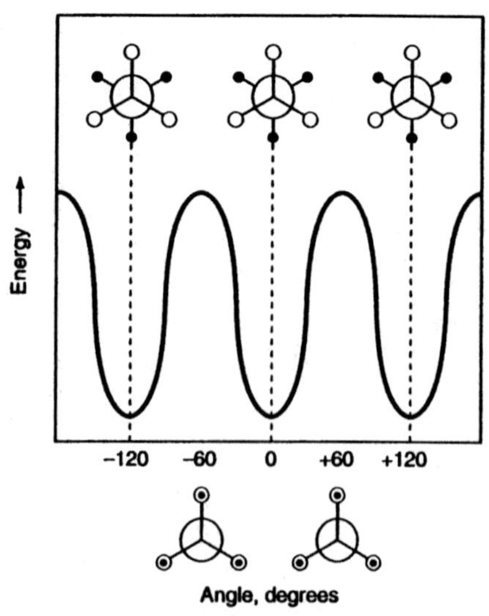

Fig. 7 Steric hindrance of ethane. Source: Ref 15

those of carbon atoms. The structure of PDMS, one of the few commercially significant polymers without carbon on its backbone, is shown in Table 1 and Fig. 1. Cis-1,4-polybutadiene is used for flexible hose, gasketing, and rubber footwear application, while PDMS is used for embedding electrical components, high-temperature gaskets, and rubber-covered rollers for laminators.

Ring structures on the backbone reduce flexibility. The presence of a phenylene group combined with the resonance among the adjacent oxygen structures of polyethylene terephthalate (PET) explains its rigidity. This is one reason why PET is well suited for the manufacture of thin-walled soda bottles. Materials such as polyacrylonitrile (PAN) and PVC are all rigid due to electrical repulsion (the nitrile group is highly polar) as well as steric hindrance (chlorine is a large atom). The repeating unit structures for PET, PAN, and PVC are shown in Table 1 and Fig. 1.

These molecular factors account in part for the elevated T_g and elastic moduli values of engineering polymers; these are discussed in the subsection "Solid Engineering Plastics" in the section "Thermal and Mechanical Properties" in this article. Polyacrylonitrile is used principally for synthetic (acrylic) fibers because its rodlike molecules form highly crystalline bundles and the high degree of hydrogen bonding provides high mechanical, thermal, and chemical resistance. With its low cost and relatively high modulus, PVC is used for water and gas pipes, window frames, siding, gutters, and identification and credit cards.

Main chain restrictions to rotation are important when considering inherent flexibility, but side chains and their morphology also play an important role. Side-chain contributions to molecular flexibility are affected by three characteristics:

- Presence of branching in the side chain
- Length of the side chain
- Polarity of the side chain

Branched side chains are even bulkier than their linear counterparts and offer greater steric hindrance. Steric hindrance is the restriction of free rotation due to spatial limitations imposed by the presence of atoms. This reduced flexibility is manifested as higher gas transition and wetting temperatures. Cyclic side groups stiffen the molecule, although this stiffening effect is diminished as the cyclic group occurs further from the main backbone. This can be seen when considering the phenyl group, which introduces significant steric hindrance, is present on increasingly long side chains. Table 5 indicates the effect of length of side chain on thermal transitions. When the phenyl group is pendant to the main chain, as in the case of PS, the T_m is 240 °C (465 °F). Locating further away by one carbon atom reduces T_m to 208 °C (405 °F), and when it is two carbon units away the T_m is only 160 °C (320 °F).

There is an interesting limitation to the lowering of thermal transition temperatures by increasing the length of purely aliphatic (linear chains, without rings) side chains. It occurs when the reduction of stiffening through increased intermolecular distance is offset by side-chain crystallization. After the aliphatic side chain reaches eight to ten carbon units in length, side-chain crystallization can occur, which again increases T_g and T_m. This is demonstrated in Table 6 and Fig. 9 for a series of polyolefins (saturated polymers containing only carbon and hydrogen). While PE has a T_m of 137 °C (280 °F), introduction of a one-carbon side chain in PP increases the T_m to 176 °C due to limited chain mobility. However, longer side chains increase the free volume enough to reduce T_m until the side-chain length exceeds eight to ten carbons. These long side chains then have sufficient mobility to crystallize and again increase T_m.

Electrical repulsion between polar side chains disrupts random coil formation of the backbone and imposes what is known as "rigid-rod" conformation. This occurs in engineering thermoplastics such as PTFE and PAN.

Intermolecular Considerations

Intermolecular arrangements are governed by both spatial considerations (such as order and distance to neighboring molecules) and by the presence of attractive forces between molecules. Intermolecular order is defined as either amorphous, crystalline, or oriented, as shown in Fig. 10.

Amorphous Versus Semicrystalline. While amorphous materials assume random, three-dimensional structures, semicrystalline polymers have very ordered, tightly packed three-dimensional arrangements connected by

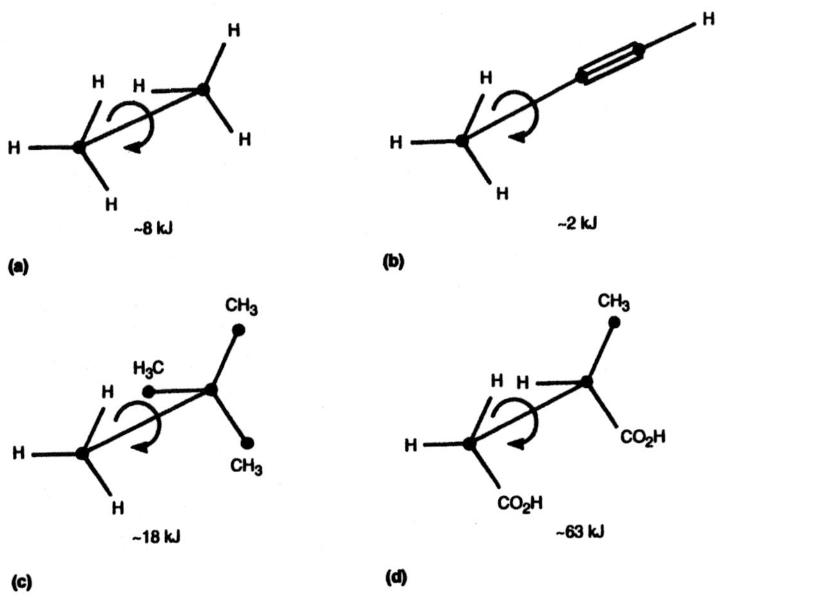

Fig. 8 Rotational energy barriers as a function of substitution. (a) Ethane. (b) Methylacetylene. (c) Neopentane. (d) Methylsuccinic acid. Source: Ref 16

Table 5 Effect of side-chain length on glass transition and melting temperatures

Side chain structure	Glass transition temperature (T_g), °C	Melting temperature (T_m), °C
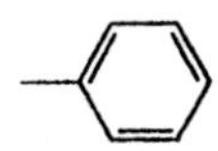	83	240
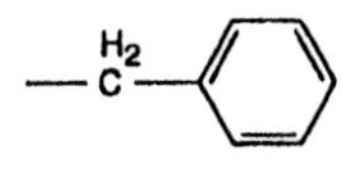	60	208
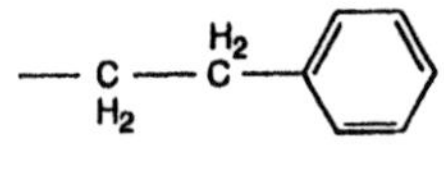	10	160

Source: Ref 17

Table 6 Effect of length of aliphatic side chain on glass transition and melting temperatures of polyolefins

Olefin	Number of carbons in side chain	Glass transition temperature (T_g), °C	Melting temperature (T_m), °C
PE	0	−122	137
PP	1	−19	176
Poly-(1-butene)	2	−24	120
Poly-(1-pentene)	3	−47	70
Poly-(1-hexene)	4	−50	−55
Poly-(1-heptene)	5	...	−40
Poly-(1-octene)	6	−60	−38
Poly-(1-dodecene)	10	...	45
Poly-(1-octadecene)	16	...	70

Source: Ref 18

amorphous regions. In the melt or solution, the chains of all polymers, except liquid crystalline polymers (LCPs), exhibit random or amorphous configurations. Liquid crystalline polymers form randomly arranged rodlike bundles. Upon cooling of the melt or evaporation of the solvent, some polymers remain amorphous whereas others crystallize. The state is determined by the regularity and flexibility of the polymer structure and the rate at which the melt is cooled or the solvent evaporated.

Polymers, such as atactic PS, atactic PMMA, atactic PP, and PVC, have large side chains or pendant groups added at irregular intervals. Because these groups prevent such polymers from forming crystalline regions, polymers with irregular structures are usually amorphous. When the pendant group or side chain is small enough, such as in PVOH and PAN, the side group can be tucked into ordered structures resulting in polymers that are semicrystalline. Moreover, regular addition of even large side groups permits the formation of tightly packed regions. Consequently, isotactic PP and syndiotactic PS are semicrystalline polymers, whereas the atactic forms are amorphous.

Because chain mobility is required to form ordered structures, polymers with regular, but rigid, structures cannot crystallize under normal processing conditions. Polycarbonate can crystallize if annealed at sufficiently high temperatures for long periods of time; however, under typical processing conditions PC is amorphous. In contrast, the structure of PE is so flexible that crystallization occurs even when the polymer melt is quenched (cooled rapidly).

Amorphous polymers exhibit a T_g that is the temperature at which the amorphous regions become mobile. In contrast, semicrystalline polymers exhibit both a T_g and a T_m. At this latter temperature, the ordered crystalline regions melt and become disordered random coils. While the magnitude of the T_g of a polymer depends only on the inherent flexibility of the polymer chain, the magnitude of T_m is also a function of the attractive forces between chains.

Although the degree of crystallinity in a given polymer varies with the processing conditions, the maximum degree of crystallinity depends on the polymer structure. Polymers such as PE, PP, polyoxymethylene (POM), and nylon 6/6 have regular, flexible structures that permit high levels of crystallization. As indicated in Table 7, increased branching that reduces the regularity of the polymer structure and its density also decreases the degree of crystallinity. The molecular architecture of these grades, shown in Fig. 11, explains why high-density polyethylene (HDPE) can achieve the highest level of crystallinity. Because the linear molecule is unimpeded by the random branches found in low-density polyethylene (LDPE), it can assume a tightly packed crystalline form.

The influence of crystallinity is best illustrated through the properties presented in Table 7 for PEs of various degrees of crystallinity. As shown in Table 7, the T_m, modulus, and hardness increase with crystallinity.

Orientation. Oriented polymers are often confused with semicrystalline polymers. In the case of oriented polymers, localized regularity is induced by mechanical deformation and is limited to small areas. Straining of polymers can result in stretched areas of parallel, linear, partially ordered structures as shown in Fig. 10(c). This uniaxial orientation results from forming processes such as fiber spinning, pipe and profile extrusion, and flat-film extrusion. The polymer chains can also be aligned parallel and perpendicular (transverse) to the primary direction of flow as shown in Fig. 10(d). Blown-film extrusion and blow molding inherently produce this biaxial orientation. In contrast, in the production of PET sheet, uniaxial orientation occurs during extrusion while the biaxial orientation is induced during a secondary stretching operation. Biaxial orientation is also the underlying concept of shrink-wrap films that revert to their amorphous conformations when enough heat is applied to reverse the induced orientation. Rotomolding and other low-shear processes produce little orientation.

Intermolecular Attractions. Secondary intermolecular attractive forces that promote crystallinity include London dispersion forces, dipole forces (either induced or permanent),

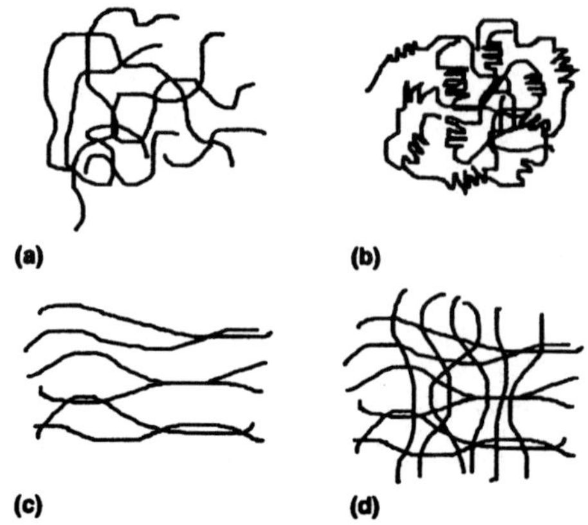

Fig. 10 Intermolecular order in polymers. (a) Amorphous. (b) Semicrystalline. (c) Uniaxial orientation. (d) Biaxial orientation

Table 7 Properties of polyethylenes of varying degrees of crystallinity

Property	Low density	Medium density	High density
Density range, g/cm^3	0.910–0.925	0.926–0.940	0.941–0.965
Crystallinity, approximate %	42–53	54–63	64–80
Melting temperature (T_m), °C	110–120	120–130	130–136
Hardness, Shore D	41–46	50–60	60–70
Tensile modulus, MPa	97–260	170–380	410–1240

Source: Ref 19

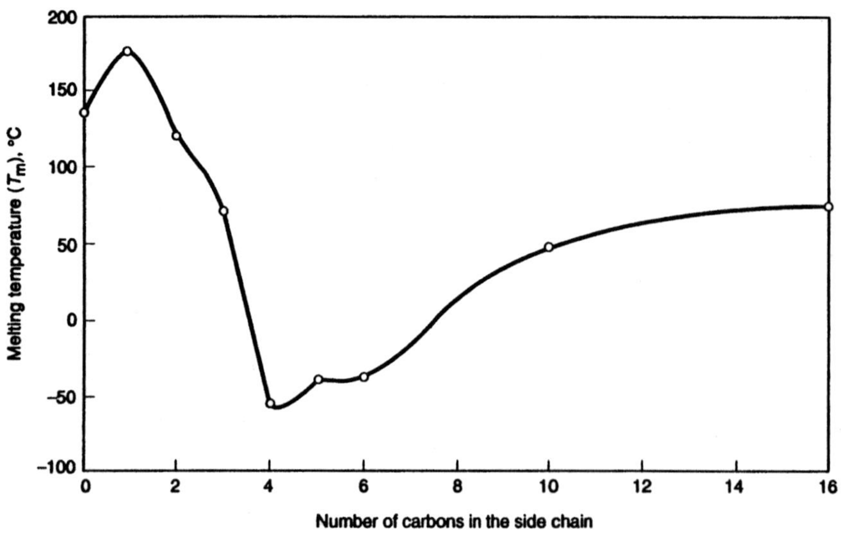

Fig. 9 The effect of aliphatic side chain on the melting temperature of polyolefins

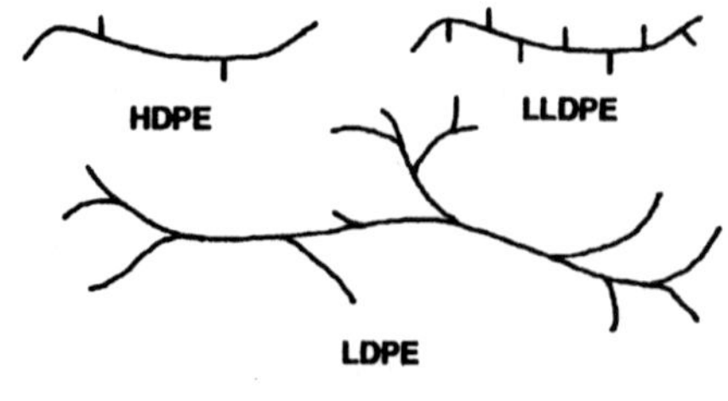

Fig. 11 Molecular architecture of high-density (HDPE), low-density (LDPE), and linear low-density (LLDPE) polyethylenes. Source: Ref 20

hydrogen bonding, and ionic bonding. These secondary bonds do not actually connect two atoms through equally shared electrons the way that a primary covalent bond does; therefore, the energy required to break secondary bonds is less than the 300 to 420 kJ/mol (Ref 21) strength of covalent bonds. The interatomic distance of covalent bonds is quite short, generally between 1 and 2 Å (Ref 21). When primary, or covalent, bonds join adjacent polymer chains, the polymer is cross linked.

London dispersion forces are the weakest of the secondary bonds with energies of 4 to 8 kJ/mol and an intermolecular distance of 3 to 5 Å (Ref 21). They are the only secondary interactions in linear, nonpolar hydrocarbons and fluoropolymers. The mobility of the valence electron clouds in these polymers results in transient states of electrical imbalance, and this momentary polarity draws two molecules together. London dispersion forces also provide significant intermolecular attractions in polar polymers, such as PVC, nylons, and PET, which form other secondary bonds.

Dipole Forces. In the presence of a polar molecule, an induced dipole can be set up in a neighboring molecule. Dipoles are the result of a covalent bond between atoms of differing electronegativities, and the resulting polarity accounts in large part for high thermal and mechanical properties of polar polymers such as PVC. These forces result in an intermolecular attraction of 4 to 21 kJ/mol (Ref 21) and often control solubility.

Hydrogen bonding occurs when the electron pair of an electronegative atom is shared by a hydrogen. The typical length of these bonds is 3 Å, with strengths of 6 to 25 kJ/mol (Ref 21). The degree to which hydrogen bonding occurs is related to the number of hydrogen bonding sites available, which in turn is related to the MW of the molecule. The facility with which they are formed is aided by regular, crystalline structures. Hydrogen bonding accounts for the high strengths of aliphatic polyamides such as nylon 6/6 and is so strong in aromatic polyamides, such as aramid fibers, that the polymers degrade before they melt.

Ionic bonding is less common than hydrogen bonding. Ionic bonding is the binding force that results from the electrostatic attraction of positively and negatively charged ions. Ionomers, with their equal numbers of positively and negatively charged ions, have high T_m and moduli. Overall, they are electrically neutral. Bond strengths are of the order of 42 to 84 kJ/mol (Ref 21). These bonds are easily eliminated by polar liquids (of high dipole moments) such as water because surface ions are readily extricated when in contact with these liquids.

Cross linking is the creation of a three-dimensional network by forming covalent bonds between polymer chains as shown in Fig. 12. While the degree of cross linking can vary, highly thermoset systems are typically rigid. Upon exposure to elevated temperatures, cross-linked polymers cannot melt and flow. The covalent bonds that form the three-dimensional network prevent melting and also do not permit dissolution in solvents. Thermoset systems, such as unsaturated polyester, epoxy, thermoset polyurethanes, polyureas, phenol formaldehyde, and melamine formaldehyde, are shaped and cross linked during processing. Representative structures are shown in Fig. 13.

Normally thermoplastic resins, such as PE, can also be cross linked after the shaping operation. Cross linking of PE does not introduce many cross links because PE is quite unreactive. The few cross links that form actually reduce regularity and therefore crystallinity. Thus, the modulus of semicrystalline thermoplastics is not increased upon cross linking, although hot creep is reduced. Hot creep is the deformation of plastics exposed to stress and elevated temperatures for prolonged periods. Polyethylenes used in wire coating are frequently cross linked for this reason.

Supermolecular Considerations

Supermolecular considerations include copolymerization, polymer blends, plasticization, incorporation of additives, and foaming.

Copolymerization. Copolymers are polymer molecules that contain several different repeat units. Usually two monomers are polymerized into one of four different configurations: random, alternating, block, or graft (Fig. 14). In random copolymers the units are distributed randomly along the polymer chains, whereas with alternating copolymers every second repeat unit is the same. Block copolymers also contain alternating segments of each monomer, but the segments are usually several repeat units long. Graft copolymers consist of a main chain composed of only one repeat unit with side chains of the second monomer.

The properties and processing characteristics of copolymers are often very different from those of the component polymers. Processing and properties can also vary with the ratio of the components and their arrangement within the copolymer. Block and graft copolymers can form two-phase systems similar to those observed with immiscible polymer blends.

Examples of random copolymers are ethylene propylene rubber (EPR), polystyrene-co-acrylonitrile (SAN), and fluorinated ethylene propylene (FEP). Ethylene propylene rubber is an amorphous elastomer, whereas PE and PP are semicrystalline plastics. Because acrylonitrile (which as PAN is difficult to process) is the minor component of SAN, it increases the melt temperature and stiffness of the PS without affecting its processibility. Fluorinated ethylene propylene is a melt-processible copolymer, while its major component, PTFE, is not.

Typical block copolymers are polyether-amides, hard-segment/soft-segment polyurethanes, and "styrenic" elastomers (for example, styrene-butadiene-styrene, or SBS, and styrene-ethylene-butylene-styrene, or SEBS). Graft copolymers are present in impact-modified polystyrene (HIPS) and ABS terpolymers. Alternating copolymers have, until recently, been laboratory curiosities.

Polymer Blends. While copolymers are mixtures of monomers that were joined together during polymerization, polymer blends are mixtures of polymer chains. The component polymers may be miscible, immiscible, or partially miscible. In the case of miscible blends, the polymers mix on a molecular level to produce a single phase. The most prominent example of this is modified polyphenylene oxide, which is a blend of polyphenylene oxide (PPO) with either PS or HIPS. Such systems exhibit a single T_g, and the mechanical properties are not affected by processing any differently than homopolymers.

With immiscible blends, the polymers cannot mix on a molecular level and therefore separate into two phases that exhibit the transition temperatures of the component polymers. In partially miscible blends, intermolecular attractions between the component polymers produce two phases that are not as sharply separated as those of immiscible blends. These blends exhibit transition temperatures that are shifted from those of the component polymers. The properties of both immiscible and partially miscible blends are sensitive to composition and processing conditions. For immiscible latex systems, such as HIPS and ABS, the size of the rubbery phases and the degree of grafting between the rigid and rubbery phases is determined during the polymerization process. However, for mechanically blended systems, the morphology is determined during the blending process and can be altered during injection molding.

Immiscible and partially miscible blends can be made compatible to provide better adhesion between the two phases. Typically, a third component, such as a block copolymer or reactive copolymer, is added to the blend to form a link between the phases.

Plasticizers are small molecules that are added to plastics to reduce viscosity during processing and to increase the flexibility of the finished product. Plasticizers such as phthalates are typically incorporated into vinyl compositions to produce flexible PVC automotive upholstery, raincoats, and luggage. Water and solvents are used as temporary plasticizers during the processing of polymers such as cellulosics and PAN.

Additives can produce significant changes in the properties and processibility of polymers. Some additives such as colorants, antioxidants,

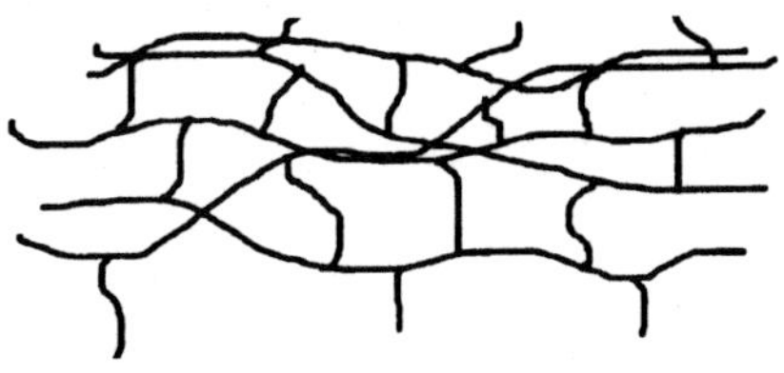

Fig. 12 Cross-linked polymer

and thermal stabilizers, do not affect the mechanical properties, but may influence viscosity during processing. In contrast, mineral fillers and glass or carbon fibers affect both mechanical properties and processibility. Fillers such as talc, calcium carbonate, and silica often reduce cost and increase the modulus, melt viscosity, and the deflection temperature under load. While fibers can significantly improve mechanical properties, their performance depends on orientation and fiber length, both of which can be affected by processing.

Foams. In foamed plastics a dispersed gaseous phase is incorporated into the plastic from the physical introduction of air or nitrogen, the degradation of chemical blowing agents, or the addition of microballoons (hollow glass or plastic microspheres) to the polymer. This gas phase reduces the weight and thermal conductivity of the plastic. While the resulting foams are classified many ways, they can generally be divided into open-cell and closed-cell foams. The individual cells (gas phases) of closed-cell foams are separated, whereas in open-cell foams these cells interconnect. Consequently, closed-cell foams are typically buoyant and are frequently used for life jackets, buoys, and other flotation devices. Foamed plastics can be made from either thermoplastic or thermoset polymers, and the modulus of the base polymer determines the flexibility of the foam. Because the walls of flexible foams collapse when pressure is applied, these materials easily dissipate mechanical and acoustic energy. This makes flexible foams particularly suitable for packaging, cushions, padding, and related applications. In contrast, high-modulus polymers produce rigid foams with a high ratio of load-bearing strength to weight. These foams typically find applications in airplane wings and automotive parts.

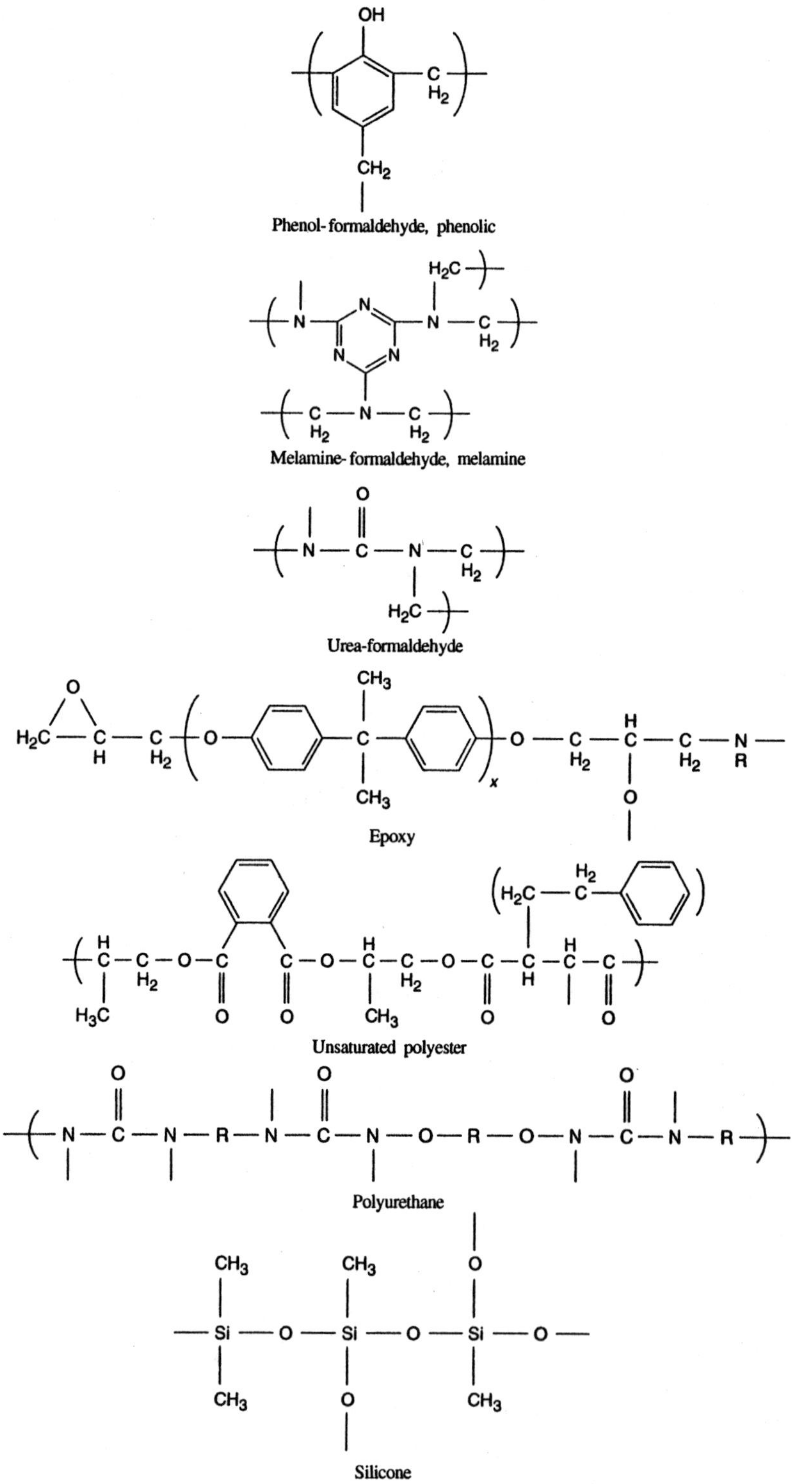

Fig. 13 Representative structures of thermoset plastics. Ref 8

Thermal and Mechanical Properties

Solid Engineering Plastics

A typical plot of stress versus strain behavior for an engineering thermoplastic is shown in Fig. 15. This classic relationship is character-

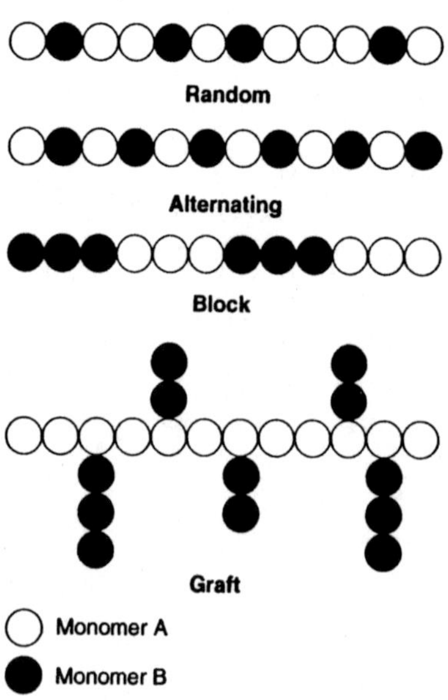

Fig. 14 Copolymer configurations. Source: Ref 22

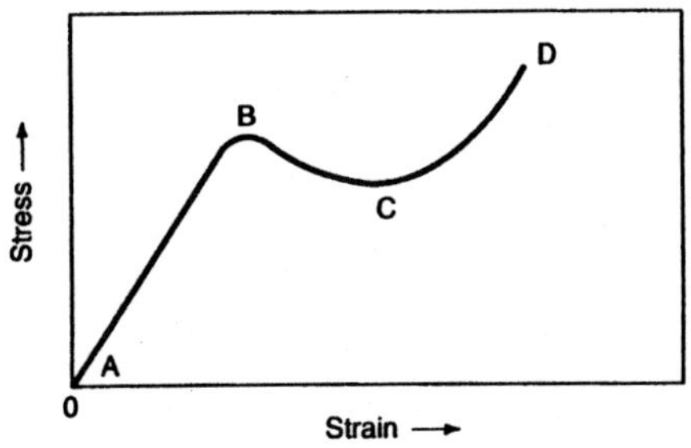

Fig. 15 Typical stress-strain curve for a polymer

ized by a linear region (shown as segment AB), which is called the linear viscoelastic region. In this region the polymer chains stretch and disentangle in response to the stress being imposed. The ratio of stress to strain (the slope) is known as either Young's modulus or the elastic modulus. Behavior in this region is like that of a purely elastic, ideal solid, governed by Hooke's law:

$$\sigma = E\varepsilon \quad \text{(Eq 5)}$$

where σ is stress, ε is strain, and the proportionality constant E is known as the spring constant or as stated earlier, Young's modulus or the elastic modulus.

Beyond this point, known as the yield point (shown as point B in Fig. 15), increased strain can be achieved with reduced stress. Secondary bonds are broken, and the strain is now irreversible. Permanent deformations such as necking begin to occur. Prior to point B, removal of stress allows the material to recover its original dimensions.

Eventually, at point C, the slope increases due to mechanically induced orientation of the polymer chains. This orientation in the direction of the imposed stress effectively increases the strength of the material. Finally, the breaking point (D) is achieved where the ultimate, or breaking, stress and strain are defined. For tensile properties, the stress is often referred to as tensile strength whereas the strain is elongation.

The stress-strain behavior presented in Fig. 15 varies strongly as a function of both strain rate and temperature. At very high strain rates, the molecules do not have adequate time to disentangle from each other and physically respond to the imposed stress. High-speed testing, known as impact testing, yields a high modulus response and low ultimate strains. In cases where the stress is imposed very slowly, the polymer chains have adequate time to disentangle and deform.

Temperature also plays an important role. At very low temperatures, polymer molecules do not have much thermal energy or mobility. Therefore, they exhibit higher moduli and lower ultimate strains than at higher temperatures. At elevated temperatures, the molecules are more

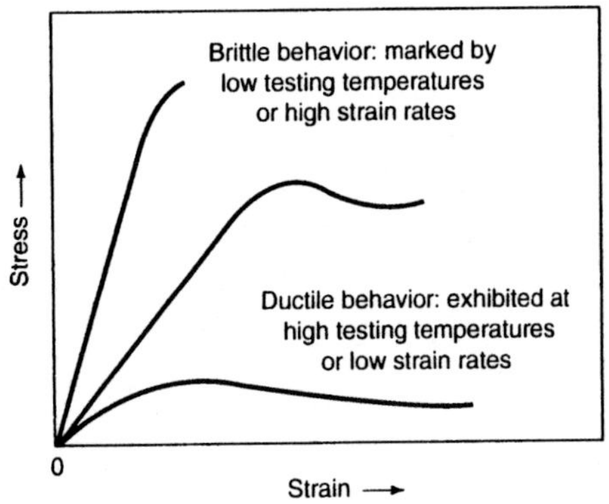

Fig. 16 Mechanical behavior of a plastic tested under different temperatures and strain rates

flexible and can distort and orient in response to the stress imposed by testing. Figure 16 shows the response of the same engineering plastic to different strain rates and different temperatures.

Figure 17 highlights the mechanical behavior of different plastics. "Strong" and "weak" are distinguished by differences in ultimate stress values, while "hard" and "soft" are differentiated by Young's moduli differences (the slope of the linear region). "Brittle" refers to a low ultimate strain, and "tough" is generally related to a large area under the stress-versus-strain curve. This definition of tough can be misleading because reinforced plastics have low ultimate strains, but are almost unbreakable.

The classic relationship of elastic modulus to temperature for polymers is presented in Fig. 18. The glassy state is characterized by limited motion of small segments of the molecule, one to four atoms in length. Behavior in this region is like that of a purely elastic, ideal solid. In the leathery region, the modulus decreases by up to three orders of magnitude for amorphous polymers. The temperature at which the polymer behavior changes from glassy to leathery is known as the T_g. This corresponds to approximately 2.5% free volume, which is the unoccupied space between molecules. The rubbery plateau has a relatively stable modulus. As temperature is further increased, rubbery flow begins, but motion does not yet involve entire molecules. In this region, deformations begin to become nonrecoverable as permanent set takes place. There is little elastic recovery in the liquid flow region, and these viscous materials, if ideal, would obey Newton's law:

$$\sigma = \eta\dot{\varepsilon} \quad \text{(Eq 6)}$$

where σ is stress, $\dot{\varepsilon}$ is strain rate, and the proportionality constant η is referred to as viscosity. The transition from the rubbery plateau to liquid flow occurs at the T_m. At this temperature, entire molecules are in motion.

Effects of Structure on Thermal and Mechanical Properties. Because free volume is generally associated with end-group concentration, T_g is a function of MW, particularly $\overline{M}_n$. Higher MWs mean longer chains, typically reduced relative concentration of end groups, and a reduction in the associated free volume. This leads to greater opportunity for molecular entanglements, which behave as physical (albeit temporary) cross links and thus drives the onset of T_g to higher temperatures. Addition of plasticizer is a means of reducing the overall "effective" MW through the incorporation of typically low MW entities into the plastic. While unimolecular plasticizers provide significant increases of free volume, which allows for enhanced rota-

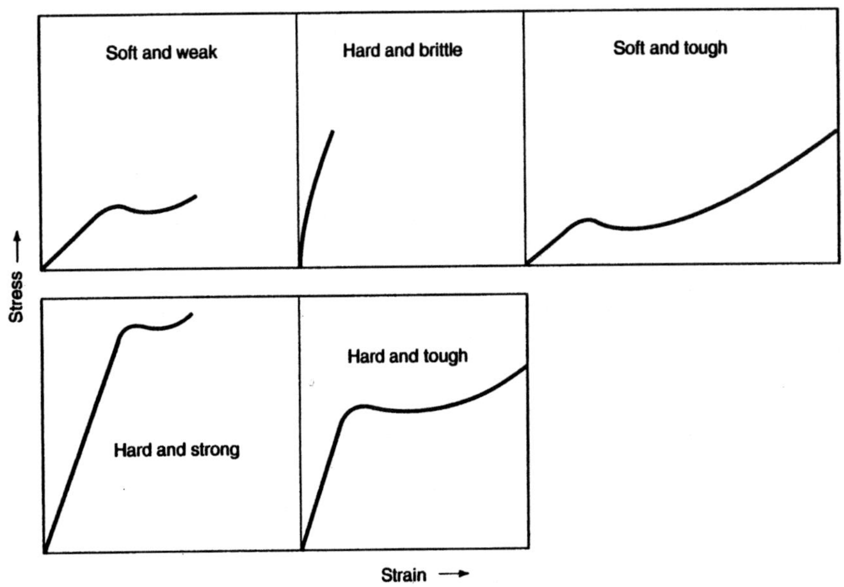

Fig. 17 Tensile stress-strain curves for several types of polymeric materials. Source: Ref 23

tional degrees of freedom for the plasticized polymer, more permanent polymeric plasticizers with their greater MW and internal plasticizers (flexible segments incorporated into the polymer) permit far less mobility. Consequently, the latter two must be added in larger amounts to achieve the same effects as produced with unimolecular plasticizers.

Increasing polarity in the polymer produces stronger attractive forces between molecules. As shown in Fig. 19, this so stiffens the polymer that the onset of T_g can be delayed. Because more thermal energy is required to overcome the stronger polar attractive forces of the molecules, T_m is increased. Thus, the nonpolar HDPE, the moderately polar POM, and the highly polar nylon 6/6 exhibit T_gs of −120, −50, and 60 °C (−185, −60, and 140 °F), respectively, while their T_ms increase as 135, 175, and 264 °C (275, 345, and 505 °F), respectively.

Figure 20 presents the effect of crystallinity on the modulus-temperature relationship. At T_m the crystal structure is overwhelmed by thermal motion of the chains, and flow occurs. Increasing the degree of crystallinity does not affect the T_g, which involves much smaller structural components than the crystal lattice. However, polymers with higher degrees of crystallinity do require higher temperatures in order to melt. Higher degrees of crystallinity lead to higher T_m and rubbery plateaus, which occur at higher moduli.

High MWs extend the rubbery region as increased entanglements serve to postpone flow or deformation. In the extreme case of numerous covalent bonds linking molecules together, cross-linked polymers never exhibit the transition from the rubbery plateau into the flow regime. The covalent bond cross links preclude flow, and the rubbery plateau simply extends until the decomposition temperature, at which point the covalent bonds are broken down. As the degree of cross linking is increased, the onset of the rubbery plateau occurs at increasingly higher moduli, as shown in Fig. 21.

Molten Engineering Plastics

Newton's law, given by Eq 6, only applies to ideal, viscous materials. A plot of log viscosity versus log shear rate for polymer melts (Fig. 22) exhibits three different regions of behavior. Regions A and C are Newtonian in that the viscosity is invariant with shear rate. Region A is often referred to as the "lower Newtonian plateau" and represents the viscosity at rates of shear that are low enough to allow the molecules to remain randomly tangled. As the rate of shear increases, region B is entered where molecules are now starting to align in the direction of flow. Because these aligned molecules offer less resistance to flow, viscosity is reduced. Finally, in region C, often referred to as the "upper Newtonian plateau," the molecules are aligned as much as possible and further increases in shear rate are no longer able to further reduce resistance to flow. This is the minimum viscosity that the molecules can achieve at a given temperature.

Processes such as extrusion and injection molding generate shear rates that are within region B, where the viscosity versus shear rate relationship is often approximated by the power law, given in Eq 7:

$$\eta = k\dot{\gamma}^{n-1} \qquad \text{(Eq 7)}$$

where η is the viscosity, $\dot{\gamma}$ is shear rate, k is a material constant called the consistency index, and n is a constant called the power-law index. Power-law indexes approximate the shear sensitivity of a polymer; values for common polymers are given in Table 8. Polymers that have very stiff backbones, such as PC and PS, tend to

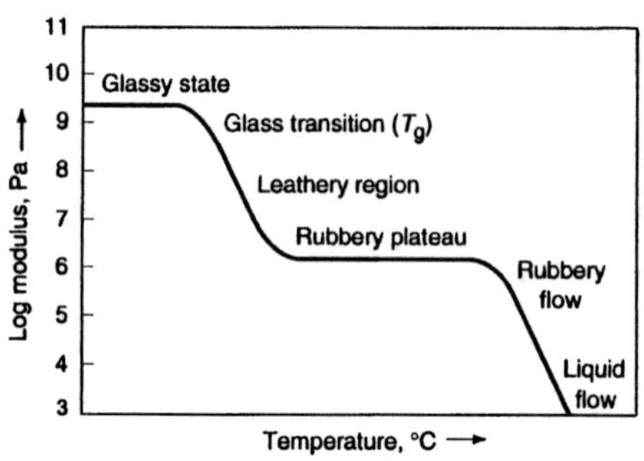

Fig. 18 Thermal dependence of elastic modulus for a typical polymer. Source: Ref 24

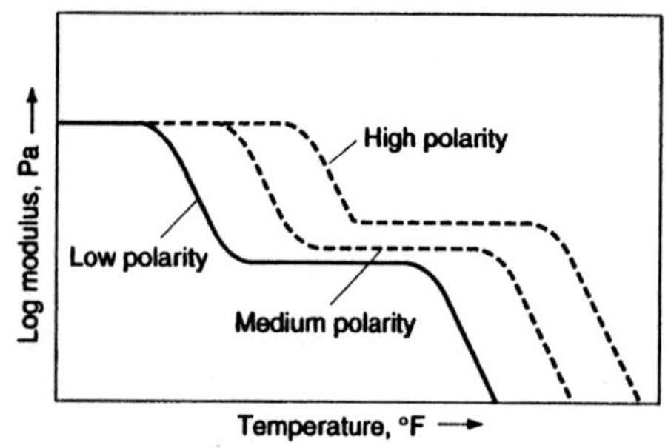

Fig. 19 Effect of temperature on modulus for polymers with different polarities. Source: Ref 25

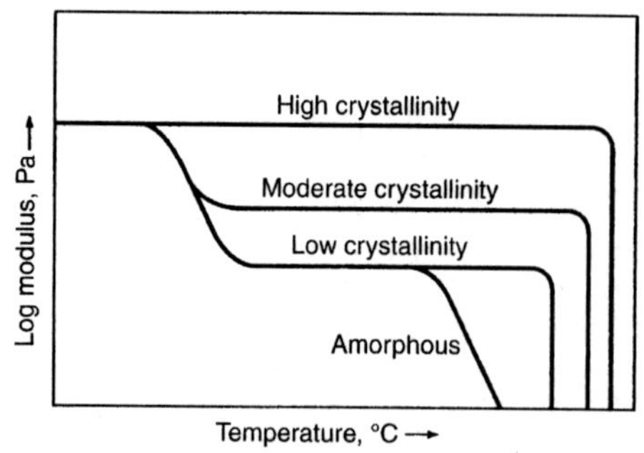

Fig. 20 Effect of temperature on modulus at various degrees of crystallinity. Source: Ref 26

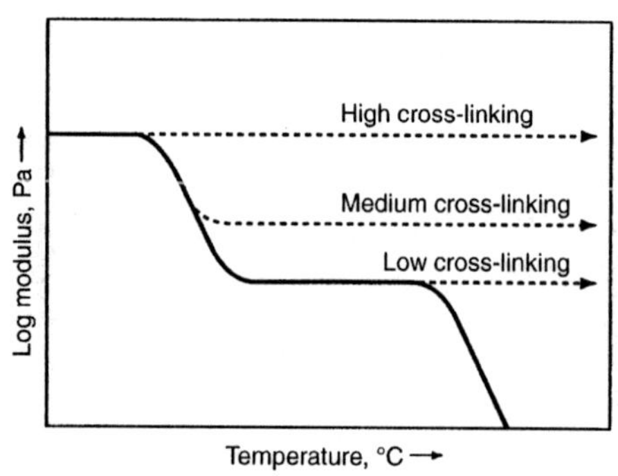

Fig. 21 Effect of temperature on modulus for different degrees of cross linking. Source: Ref 25

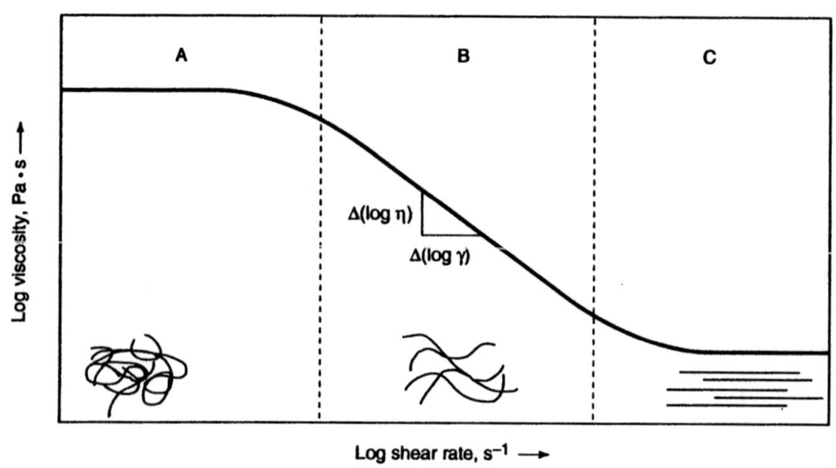

Fig. 22 General pseudoplastic behavior. Source: Ref 27

exhibit lower Newtonian plateaus that extend to shear rates of 1000 s^{-1} or more. Consequently, as discussed in the section "Processing" in this article, shear thinning does not often reduce the viscosity of these polymers during extrusion.

Viscoelasticity

Mechanical analogs to purely elastic Hookean solid behavior and purely viscous Newtonian melt behavior help describe why polymers have intermediate (viscoelastic) properties, which are time dependent. Most commonly, a spring is used to model Hookean behavior, and a dashpot (representing a piston in a viscous material similar to hydraulic fluid) represents viscous behavior. These models and their concomitant stress and strain behaviors are shown in Fig. 23(a) and (b).

Application of a deforming force (i.e., pulling) on the spring results in an immediate stretching and thus an immediate strain. Once the force is released, the spring immediately recovers its initial length. Pulling with twice the force results linearly in twice the strain. The case of the dashpot, however, is significantly different. When the "piston" has a force applied to it, it slowly starts to move (no instant displacement as in the case of the spring), and when the force is released, the dashpot stays in its new conformation. Once a force causes an ideal viscous polymer melt to flow, it remains in its new position.

Two models, combining the spring and the dashpot either in series or parallel, have been developed that attempt to better describe real polymer flow behavior. These models, Maxwell and Voigt, are named after their creators and are shown in Fig. 23(c) and (d). Figure 24, very similar to Fig. 18, shows which mechanical analogs model different regions of the log modulus versus temperature curve. The behavior shown in the Voigt model helps to explain the action known as creep. Creep occurs when, under a static load for extended periods of time, increased strain levels slowly develop, as in the case of a refrigerator that after many years distorts a linoleum floor. The Maxwell model describes stress relaxation, which occurs when polymers are subjected to a constant strain environment. Over time, the molecules relax and orient themselves to the strained position, thereby relieving stress. This occurs in applications such as threaded metal inserts into plastic parts and threaded plastic bottle caps.

Table 8 Sample power-law indexes (n) for common plastics

Polymer	n
LDPE	0.35
LLDPE	0.60
HDPE	0.50
PP	0.35
PS	0.30
ABS	0.25
PMMA	0.25
PVC	0.30
PC	0.70
PET	0.60
PBT	0.60
Nylon 6	0.70
Nylon 6/6	0.75

Source: Ref 28

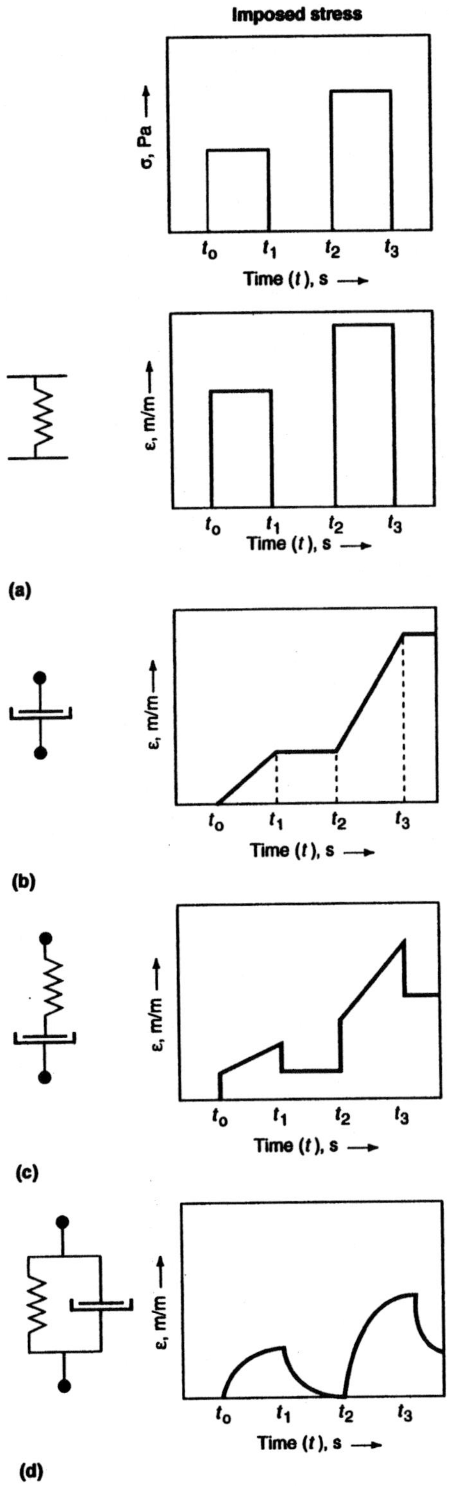

Fig. 23 Mechanical models and typical behavior. (a) Ideal Hookean solid ($\sigma = E\varepsilon$; spring model; elastic response). (b) Ideal viscous Newtonian liquid ($\sigma = \eta\dot{\varepsilon}$; dashpot model). (c) Maxwell's mechanical model for a viscoelastic material. (d) Voigt's mechanical model for a viscoelastic material. Source: Ref 29

Properties of Engineering Plastics and Commodity Plastics

Engineering plastics generally offer higher moduli and elevated-service temperatures compared to the lower-cost, high-volume, commodity plastics such as PE, PP, and PVC. These improved properties are due to chemical substituents, inherently rigid backbones, and the presence of secondary attractive forces as discussed earlier in this article. Engineering thermoplastics (e.g., POM, PC, PET, and polyetherimide, or PEI) are polymerized from more expensive raw materials, and their processing requires higher energy input compared to that of commodity plastics, which is why the engineering thermoplastics are more expensive.

Structures of Commodity Plastics. It is interesting to note the T_m elevation of HDPE from LDPE. The effect of the branched structure on density and morphology enables the high-density version to form more tightly packed crystalline regions that require more thermal energy to overcome the cohesive forces keeping the plastic from melting. Substituting a methyl group in place of a hydrogen, in the case of PP, increases T_m and tensile strength further above that of HDPE. In this case, steric hindrance due to the additional size of the methyl group stiffens the chain and restricts rotation. The substitution of a large and highly electronegative chlorine atom in PVC prevents crystallization and also increases the onset of T_g, both due to steric hindrance effects and to the attractive polar forces generated. Polar attractive forces are so extensive that the tensile strength can be seen to increase to 55 MPa (8 ksi). Polystyrene is amorphous and transparent due to the atactic positioning of the pendant phenyl group, whose randomness destroys crystallinity. The tensile strength of PS is less than that of PVC due to the lack of the highly polar pendant group.

Structures of Engineering Plastics. Phenylene and other ring structures (Table 1 and Fig. 1) attached directly into the backbone often stiffen the polymer significantly, imparting elevated thermal properties and higher mechanical properties such as increased strength. Polyoxymethylene is essentially PE with an ether substitution, but it has a much higher T_m (200 versus 135 °C, or 390 versus 275 °F, for HDPE) because of its polarity. Both of these features promote a highly crystalline morphology. The high dimensional stability, good friction and abrasion characteristics, and ease of processing of this polymer make it a popular engineering plastic for precision parts.

Polycarbonate has an extended resonating structure because of the carbonate linkage. It has such a stiff backbone that crystallization is impeded, and the resultant amorphous structure is transparent, much like PET. Physical properties of PET, however, depend strongly both on its degree of crystallinity, which is governed by degree of orientation imparted during processing, and on its annealing history. The high strength, ease of processing, and clarity of PET make it ideal for soda bottles and polyester fibers. Polycarbonate has high strength, stiffness, hardness, and toughness over a range of –150 to 135 °C (–240 to 275 °F) and can be reinforced with glass fibers to extend elevated-temperature mechanical properties. The high impact strength of high-MW PC makes it suitable for applications such as motorcycle helmets. The carbonate linkage of PC causes a susceptibility to stress cracking.

Polyether-imide has both imide groups and flexible ether groups, resulting in high mechanical properties but with enough flexibility to allow processing. Its highly aromatic (presence of benzene rings) structure allows it to be used for specialty applications.

Polyether ether ketone (PEEK), PPO, and polyphenylene sulfide (PPS) also rely on backbone benzene rings to yield high mechanical properties at elevated temperatures. Both sulfur and oxygen are electronegative atoms, creating dipole moments that promote intermolecular attractions and thus favorably affect elevated-temperature properties.

While the composition of thermoset plastics vary widely, the three-dimensional structure produced by cross linking prevents melting and hinders creep. Overall properties such as stiffness and strength are determined by the flexibility of the polymer structure and the number of cross links (cross-link density). Because epoxies, phenolics, and melamine formaldehyde contain aromatic rings, they are typically rigid and hard. Epoxies are used for adhesives, assorted electronics applications, sporting goods such as skis and hockey sticks, and prototype tooling for injection molding and thermoforming. Melamine formaldehyde is easily colored and so is often found in household and kitchen equipment, electronic housings, and switches. In contrast, phenolics are naturally dark colored and are limited to electronic and related applications where aesthetics are less important. Silicones with their flexible ether linkages are softer and often used as caulking and gasket materials. Thermoset polyurethanes vary widely from flexible to relatively rigid, depending on the chemical structure between urethane groups. Unsaturated polyesters are used for potting and encapsulating compounds for electronics and in glass-fiber-reinforced molding compounds.

This discussion of the major commodity and engineering plastics is by no means complete. It is meant rather to include concepts touched on earlier in evaluating structures in relation to their resultant properties.

Electrical Properties

Volume and/or surface resistivity, the dielectric constant, dissipation factor, dielectric strength, and arc or tracking resistance are considered important electrical properties for design. These properties relate to structural considerations such as polarity, molecular flexibility, and the presence of ionic impurities, which may result from the polymerization process, contaminants, or plasticizing additives. Table 9 shows some typical electrical property values for selected plastic materials.

Volume resistivity is a measure of the resistance of an insulator to conduction of current. Most neat polymers have a very high resistance to flow of direct current, usually 10^{15} to 10^{20} $\Omega \cdot$ cm compared to 10^{-6} $\Omega \cdot$ cm for copper. Electrical conductivity in normally insulating polymers results from the migration of ionic impurities and is affected by the mobility of these ionic species. Generally, plasticizers with their increased mobility and high relative concentration of end groups reduce resistivity and therefore increase electrical conductivity. Because absorption of water increases the mobility of ionic species, this also reduces volume resistivity. Thus, the volume resistivity of nylon 6/6 is reduced by four decades when the polymer absorbs water at ambient conditions. Addition of antistatic agents decrease surface resistivity because the polar additives migrate to the surface of the polymer and absorb humidity. In contrast, conductive fillers, such as carbon black powders and aluminum flake, can form three-dimensional pathways for conduction through insulating polymer matrices. Finally, highly conjugated polymers such as polyacetylene and polyaniline provide sufficient electron movement to reach semiconductor conductivity. For full conductivity, they rely on dopants.

Dielectric Constant and Dissipation Factor. In the presence of an electric field, polymer molecules will attempt to align in that field. The dielectric constant (or permittivity), ε or ε', is a measure of this polarization. While the dielectric constant varies from 1 for a vacuum (where nothing can align) to 80 for water, the values for polymers (shown in Table 9) are generally so low that most polymers are insulators. The dielectric constant also varies with temperature, rate or frequency of measurement, polymer structure and morphology, and the presence of other materials in the plastic. The dielectric constant of polymers typically peaks at the major thermal transition temperature (T_g and/or T_m) and then decreases because of random thermal motions in the melt. As shown in Fig. 25(a), the dielectric constant decreases abruptly as frequency increases. This occurs between 1 Hz and 1 MHz and is a result of the inability of the dipoles to align with the high-frequency electric fields. The dielectric loss, ε'', is a measure of the energy lost to internal motions of the material, and as shown in Fig. 25(b), peaks where the dielectric constant changes abruptly. The dissipation factor, tan δ, which is given by:

$$\tan \delta = \frac{\varepsilon''}{\varepsilon'} \quad \text{(Eq 8)}$$

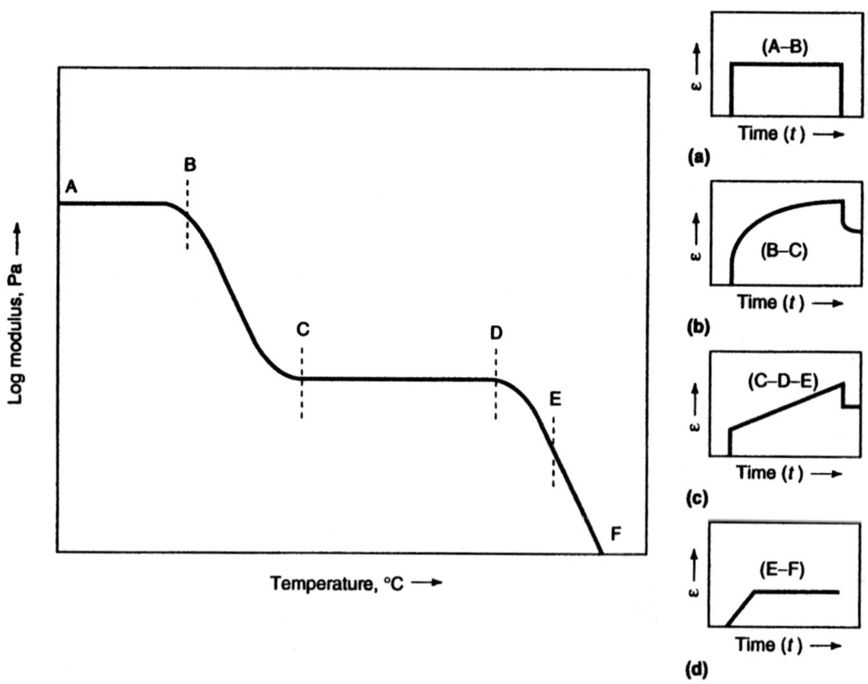

Fig. 24 Thermal dependence of elastic modulus for polystyrene. (a) Glassy region corresponding to Hookean solid behavior. (b) Leathery region corresponding to Voigt model behavior. (c) Rubbery plateau region corresponding to Maxwell model behavior. (d) Liquid flow region corresponding to Newtonian liquid behavior. Source: Ref 30

is a measure of the internal heating of plastics. Thus, little heating should occur in insulators ($\tan \delta < 10^{-3}$), whereas high-frequency welding necessitates that $\tan \delta$ be much greater (Ref 32).

Because polymer molecules are typically too long and entangled to align in electric fields, the dielectric constant usually arises from shifting of the electron shell of the polymer and/or alignment of its dipoles in the field. For nonpolar polymers, such as PTFE and PE, only electron polarization occurs and the dielectric constant can be approximated by:

$$\varepsilon = n^2 \quad \text{(Eq 9)}$$

where n is the optical refractive index of the polymer. These values vary little with frequency, and changes occurring with increased temperatures are caused by changes in free volume of the polymer. In contrast, the dielectric constants of polar polymers, such as PVC and PMMA, are greater than n^2 and change substantially with temperature and frequency. Backbone flexibility or ease of rotation of polar side groups allows some polymers to orient quickly and easily. If the electric field alternates slowly enough, the molecule may be able to align or orient in the field depending on its flexibility and mobility. Consequently, relatively flexible polymers, such as PVC and PMMA, exhibit greater decreases in dielectric constant with increased frequency than polymers, such as PEI and PSU, that have rigid backbones. The additional free volume and mobility of the plasticized PVC allows the molecules to align with minimal delay; as shown in Table 9, this doubles the dielectric constant at low frequencies.

Dielectric Strength. As the electric field applied to a plastic is increased, the polymer will eventually break down due to the formation of a conductive carbon track through the plastic. The voltage at which this occurs is the breakdown voltage, and the dielectric strength is this voltage divided by the thickness of the plastic. The dielectric strength decreases with the thickness of the insulator because this prevents loss of internal heat to the environment. Dielectric strength is increased by the absence of flaws.

Arc Resistance. In contrast to the dielectric strength, arc resistance is the ability of a polymer to resist forming a carbon tracking on the surface of the polymer sample. Because these tracks usually emanate from impurities surrounding electrical connections, arc resistance is measured by the track times. Polymers, such as PC, PS, PVC, and epoxies (which have aromatic rings, easily oxidized pendant groups, or high surface energies), are prone to tracking (Ref 33) and exhibit typical track times of 10 to 150 s (Ref 34). However, polyesters may have better tracking resistance than phenolics because of the heteroatomic backbone that disrupts the carbon track. Nonpolar aliphatic compounds or those with strongly bound pendant groups usually have better arc resistance; thus, the tracking times for PTFE, PP, PMMA, and PE are greater than 1000 s (Ref 33).

Optical Properties

Transparency, opacity, haze, and color are all important characteristics of plastics. Optical clarity is achieved when light is able to pass relatively unimpeded through a polymer sample. This is usually defined by the refractive index, n, which is shown in Fig. 26 and given by:

$$n = \frac{\sin \alpha}{\sin \beta} \quad \text{(Eq 10)}$$

where α is the angle of incident light and β is the angle of refracted light. While n for most polymers is 1.40 to 1.70, it increases with the density of the polymer and varies with temperature. In order for a material to be clear, light has to be transmitted with minimal refraction. Unstressed, homogeneous, amorphous polymers, such as PS, PMMA, and PC, exhibit a single refractive index and thus are optically clear. However, when these polymers are severely oriented, and therefore stressed, the areas with different refractive indexes produce birefringence in the molded products. Because amorphous, but heterogeneous, systems, such as the immiscible polymer blends ABS and HIPS, typically exhibit a refractive index for each polymer phase, they are usually opaque or translucent. Semicrystalline polymers, such as HDPE and nylon 6/6, effectively have two phases, the amorphous and crystalline regions. Consequently, semicrystalline polymers are usually not transparent. Finally, introduction of any nonpolymeric phases, such as fillers or fibers, into the plastic material induces opacity because these phases have their own refractive indexes.

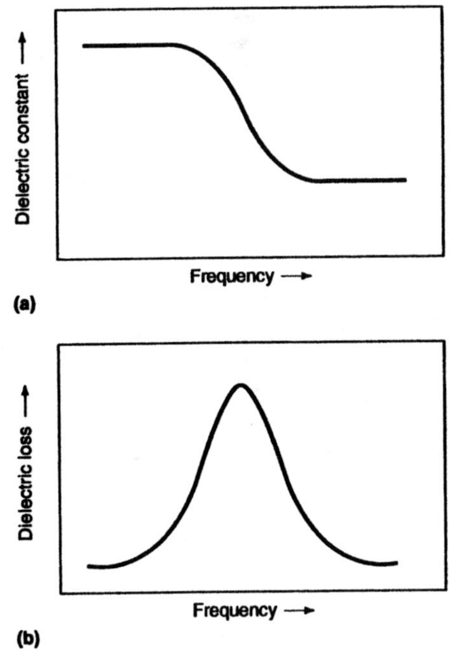

Fig. 25 Frequency dependence of the (a) dielectric constant and (b) dielectric loss. Source: Ref 31

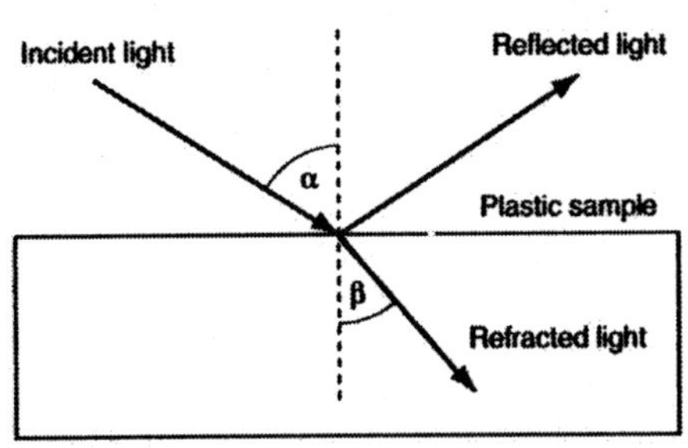

Fig. 26 Light refracted by a plastic sample

Table 9 Electrical properties of selected plastics

Plastic	Surface resistivity, Ω	Volume resistivity, Ω · cm	Dielectric strength, kV/mm	Dielectric constant		Dissipation factor	
				At 50 Hz	At 10^6 Hz	At 50 Hz	At 10^6 Hz
LDPE	10^{13}	$>10^{16}$	>70	2.3	2.3	2×10^{-4}	2×10^{-4}
PTFE	10^{17}	$>10^{18}$	60–80	2.1	2.1	2×10^{-4}	2×10^{-4}
PS	10^{14}	...	...	2.6	...	0.5×10^{-4}	2.5×10^{-4}
PMMA	5×10^{13}	$>10^{15}$	30	3.7	2.6	0.060	0.015
PVC	...	$>10^{15}$	20–40	3.5	2.7	0.003	0.002
Plasticized PVC	...	10^{15}	28	6.9	3.6	...	...
POM	10^{13}	10^{15}	70	...	3.7	0.0015	0.0055
Nylon 6/6	...	10^{15} (dry) 10^{11} (wet)	40 (dry)	4.0 (dry) 6.0 (wet)	3.4	0.02 (dry) 0.20 (wet)	...
PET	6×10^{14}	2×10^{14}	60	3.4	3.2	0.002	0.021
PBT	5×10^{13}	5×10^{13}	>45	3.0	2.8	0.001	0.017
PC	$>10^{15}$	$>10^{16}$	>80	3.0	2.9	0.900	11
Modified PPO	10^{14}	$>10^{15}$	22	2.7	2.6	4×10^{-4}	9×10^{-4}
PAI	5×10^{18}	2×10^{15}	23	...	3.9	...	0.030
PEI	...	7×10^{15}	24	3.15	3.05	0.0015	0.0064
PSU	3×10^{16}	5×10^{16}	20	3.15	3.10	0.001	0.005
PEEK	...	5×10^{16}	19	3.20	...	0.003	...

Source: Ref 4

Optical clarity can also be controlled by polymerization techniques. When the refractive indexes of multiphase systems are matched, these plastics can be optically clear, but usually only over narrow temperature ranges. Neat poly-(4-methyl-1-pentene) (TPX) is clear because the bulky side chains produce similar densities (0.83 g/cm^3), and thus similar refractive indexes, in the amorphous and crystalline regions of the polymer. Matching of refractive indexes of PVC and its impact modifier is often used in transparent films for food packaging. Domains (second phases) that are smaller than the 400 to 700 nm wavelengths of visible light will not scatter visible light and thus do not reduce clarity. In impact-modified polymers, the minor rubbery phase is usually dispersed as particles with diameters greater than 400 nm, so most of them are opaque. However, when the domains have diameters less than 400 nm or when the two phases form concentric rings whose width is too narrow to scatter visible light, the blends are clear.

When crystals are smaller than the wavelength of visible light, they will also not scatter light and the plastic will be optically clear or translucent. These crystal sizes can be controlled by quenching, use of nucleating agents, stretching, and copolymerization. In quenching, the plastic melt is rapidly cooled below the transition temperature of the polymer. The resultant reduction in thermal mobility of the polymer molecules limits crystal growth because the molecules are not able to form ordered structures. While quenching is more easily accomplished with thin parts and films, nucleating agents can reduce crystal size in a wider range of parts. The agents are small particles at which the crystallization process can begin. Consequently, many such sites competing for polymer chains will reduce the average crystal size. Stretching also promotes clarity because the mechanical stretching can break up large crystals, and the resultant thinner films are more liable to transmit light without refraction. Finally, copolymerization can reduce the regularity of the polymer structure enough to inhibit formation of large crystals. As noted, the structural regularity that is required of a polymer is to pack into tightly ordered crystallites, and randomization of the structure results in smaller areas capable of being packed together.

The surface character of processed parts also controls optical properties. Smooth surfaces reflect and transmit light at limited angles, whereas rough surfaces scatter the light. Consequently, smooth surfaces produce clear and glossy products while rough surfaces appear dull and hazy. Surface character is usually controlled by processing.

Unmodified polymers are usually clear to yellowish in color. Other colors are produced by dispersing pigments or dyes uniformly within the plastic. Poor dispersion can produce the marbled or speckled appearances favored for cosmetic cases. However, degradation of polymers will produce yellowing or browning of the plastic. Polymers such as PVC, which are particularly subject to degradation, are also discussed in the section "Processing" in this article.

Chemical Properties

Solubility is the ease with which polymer chains go into solution and is a measure of the attraction of the polymer to solvent molecules. The old adage of "like dissolves like" can be explained by considering the balance of forces that occur during dissolution of the polymer. Solubility is determined by the relative attraction of polymer chains for other polymer chains and polymer chains for solvent molecules. If the polymer-solvent interactions are strong enough to overcome polymer-polymer interactions, dissolution occurs; otherwise, the polymer remains insoluble. Swelling can be considered as partial solubility because the solvent molecules penetrate the polymer, but they cannot completely separate the chains.

When solvents and polymers have similar polarities, the polymer will dissolve in or be swollen by the solvent. Because longer chains are more entangled, higher MW hinders dissolution. Semicrystalline polymers are much harder to dissolve than similar amorphous materials. The tightly packed crystalline regions are not easily penetrated because the solvent molecules must overcome the intermolecular attractions. Elevated temperatures, which increase the mobility of solvent molecules and polymer chains, facilitate dissolution. The presence of cross links completely prevents dissolution, and such polymers merely swell in solvents.

Plasticizers must be soluble in the polymer to prevent migration to the surface (blooming) and extraction by solvents. Consequently, the relatively expensive primary plasticizers for PVC closely match the solubility of the polymer, while less expensive secondary plasticizers are less compatible with the PVC.

Permeability is a measure of the ease with which molecules diffuse through a polymer sample. The low densities of polymers compared with metals and ceramics allow enhanced permeation of species such as water, oxygen, and carbon dioxide. If there are strong interactions between the polymer and the migrating species, adsorption will be high, but permeation may be low as the migrating species is delayed from diffusing. For example, the electronegative chlorine atoms substitution in polyvinylidene chloride (PVDC) enhances adsorption of oxygen, nitrogen, carbon dioxide, and water while its tightly packed chain arrangement restricts diffusion of these species. Thus, PVDC films (commonly used as plastic wrap) are extremely valuable in food-packaging operations. As shown in Fig. 27, permeability can also be inhibited by the addition of platelike fillers, which increase the distance that water must travel in order to pass completely through the plastic.

Environmental stress cracking occurs when a stressed plastic part is exposed to a weak solvent, often moisture. The stress imparts strain to the polymer, which allows the solvent to penetrate and either extract small molecules of low $\overline{M}_n$ or to plasticize and weaken the polymer. The stress then causes fracture at these weak areas. Polymers that are exposed to UV light are particularly susceptible to environmental stress cracking. Resistance is enhanced when the permeability of the polymer to water is low.

Processing

Most thermoplastic processing operations involve heating, forming, and then cooling the polymer into the desired shape. This section briefly outlines the most common plastics manufacturing processes. The factors that must be considered when processing engineering thermoplastics are also discussed. These include melt viscosity and melt strength; crystallization; orientation, die swell, shrinkage, and molded-in stress; polymer degradation; and polymer blends.

Overview of the Major Thermoplastics Processing Operations. Although there are a number of variants, the major thermoplastics processing operations are extrusion, injection molding, blow molding, calendering, thermoforming, and rotational molding. Characteristics of each of these processes are described briefly in the paragraphs that follow. Additional information is provided in the article "Design and Selection of Plastics Processing Methods" in this book.

Extrusion is a continuous process used to manufacture plastics film, fiber, pipe, and pro-

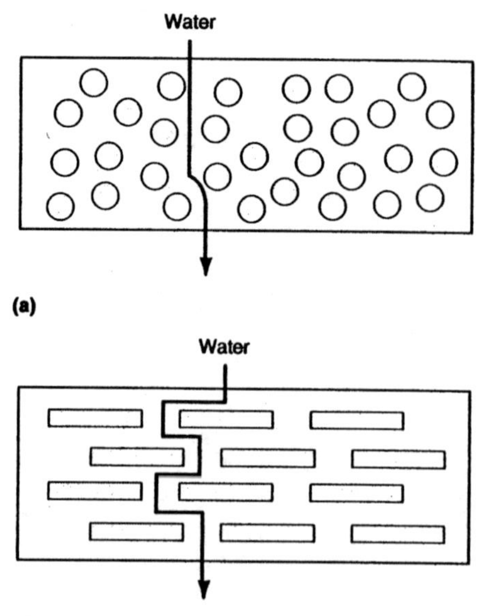

Fig. 27 Barrier pigment effect. Water passes relatively unobstructed through a polymer with spherical additives (a), but must travel around platelike fillers (b). Source: Ref 35

files. The single-screw extruder is most commonly used. In this extruder, a hopper funnels plastic pellets into the channel formed between the helical screw and the inner wall of the barrel that contains the screw. The extruder screw typically consists of three regions: a feed zone, a transition or compression zone, and a metering or conveying zone (see Fig. 10 in the article "Design and Selection of Plastics Processing Methods" in this book). The feed zone compacts the solid plastic pellets so that they move forward as the solid mass. As the screw channel depth is reduced in the transition zone, a combination of shear heating and conduction from the heated barrel begins to melt the pellets. The fraction of unmelted pellets is reduced until finally in the metering zone a homogeneous melt has been created. The continuous rotation of the screw pumps the plastic melt through a die to form the desired shape.

The die and ancillary equipment produce different extrusion processes. With blown-film extrusion, air introduced through the center of an annular die produces a bubble of polymer film; this bubble is later collapsed and wound on a roll. In contrast, flat film is produced by forcing the polymer melt through a wide rectangular die and onto a series of smooth, cooled rollers. Pipes and profiles are extruded through dies of the proper shape and held in that form until the plastic is cooled. Fibers are formed when polymer melt is forced through the many fine, cylindrical openings of spinneret dies and then drawn (stretched) by ancillary equipment. In extrusion coating, low-viscosity polymer melt from a flat-film die flows onto a plastic, paper, or metallic substrate. However, in wire coating, wire is fed through the die and enters the center of the melt stream before or just after exiting the die. Finally, coextrusion involves two or more single-screw extruders that separately feed polymer streams into a single die assembly to form laminates of the polymers. Typical extrusion pressures range from 1.5 to 35 MPa (0.2 to 5 ksi).

While single-screw extruders provide high shear and poor mixing capabilities, they produce the high pressures needed for processes such as blown and flat-film extrusion. Screw designs are changed to improve mixing, to shear gel (unmelted polymer) particles, and to provide more efficient melting. The latter designs are particularly critical to the extrusion of PE films where partially melted polymer particles are not desirable.

In addition to single-screw extruders, twin-screw extruders are available. While twin-screw extruders use two screws to convey the polymer to a die, the configuration of the screws produce different conveyance mechanisms. Intermeshing twin-screw extruders transfer the polymer from channel to channel, whereas nonintermeshing twin-screw extruders—like single-screw extruders—push the polymer down the barrel walls. In addition, intermeshing corotating twin-screw extruders tend to move the polymer in a figure-eight pattern around the two screws. Because this produces more shear and better mixing, corotating twin-screw extruders are well suited to mixing and compounding applications. Intermeshing counterrotating twin-screw extruders channel the polymer between the two screws. Twin-screw extruders also permit tighter control of shear because twin screws are usually not a single piece of metal, but two rods on which component elements are placed. Consequently, screw profiles can be "programmed" to impart specific levels of shear.

In contrast to the single- and twin-screw extruders, ram extruders have no screw, but merely use a high-pressure ram to force the polymer through a die. This provides for minimal shear and much higher pressures than available in single-screw extruder. However, ram extrusion is a batch operation, not a continuous operation.

Injection molding is a batch operation used to rapidly produce complicated parts. Plastic pellets are fed through a hopper into the feed zone of a screw and melted in much the same way as occurs in a single-screw or ram extruder. However, rather than being forced through a die, in an injection-molding machine the melt is accumulated and subsequently forced under pressure into a mold by axial motion of the screw. This pressure is typically quite high and for rapid injection and/or thin-walled parts can exceed 100 MPa (14.5 ksi). Once the part has cooled sufficiently, the mold is opened, the part ejected, and the cycle recommences. The use of multiple-cavity molds allows for simultaneous production of a large number of parts, and often little finishing of the final part is required. Polymer from multiple plasticating units (extruders) can also be injected sequentially into the same mold to form "coinjected" parts. In gas-assisted injection molding, gas is injected into the melt stream and accumulates in thicker sections of the part, whereas in foam processes the introduced gas forms small pockets (cells) throughout the melt.

Blow molding operations generate hollow products, such as soda bottles and automobile fuel tanks. The three basic processes are continuous extrusion, intermittent extrusion, and injection blow molding. In continuous-extrusion blow molding, a tube of polymer is continuously extruded. Pieces of this tube (called parisons) are cut off, inserted into the mold, and stretched into the cavity of the blow mold by air pressure. Although intermittent extrusion blow molding is similar, the tube of plastic is injected from the extruder rather than continuously extruded. In the injection-blow-molding process, a plastic preform, which for bottles resembles a test tube with threads, is injection molded. Then this preform is brought to the forming temperature (either as part of the cooling from injection molding or after being reheated) and expanded into the blow mold. Stretch blow molding is a variant of the blow-molding process, in which the preform is stretched axially by mechanical action and then expanded in the transverse direction to contact the walls of the mold.

Calendering uses highly polished precision chromium rolls to transform molten plastic continuously into sheet (>0.25 mm, or 0.01 in.) or film (≤0.25 mm, or 0.01 in.) for floor coverings. This process can also be used to coat a substrate, for example, cords coated with rubber for automotive tire use (Ref 36). Usually an extruder provides a reservoir of plastic melt, which is then passed between two to four calender rolls whose gap thickness and pressure profiles determine the final gage of the sheet being formed. Chill rolls are used to reduce the sheet temperature, and a windup station is generally required to collect the sheet product.

Thermoforming operations are used to produce refrigerator liners, computer housings, food containers, blister packaging, and other items that benefit from its low tooling costs and high output rates. In this process, infrared or convection ovens heat an extruded or calendered sheet to its rubbery state. Mechanical action, vacuum pressure, and/or air pressure force the heated sheet into complete contact with cavity of the thermoforming mold.

Rotational molding, or rotomolding, involves charging a polymeric powder or liquid into a hollow mold. The mold is heated and then cooled while being rotated on two axes. This causes the polymer to coat the inside of the mold. Because rotomolding produces hollow parts with low molded-in stresses, it is often used for chemical containers and related products where environmental stress crack resistance is required. It can also be used for hollow parts with complicated geometries that cannot be produced by blow molding.

Melt viscosity and melt strength are major factors to be considered when choosing a resin and a processing operation. While flexible polymers are generally less viscous than polymers with more rigid structures, MW, MWD, and additives are used to tailor plastics for specific processes. Resins are typically rated by their melt index, which is the flow of the melt (in grams per 10 min) through a geometry and under a load specified by ASTM D 1238 (Ref 37). Although this generates the flow at very low shear rates, it is an indication of the melt viscosity of the plastic. Extrusion-blow-molding processes require that the melt index be below 2 g per 10 min, whereas other extrusion processes require somewhat greater flow. In contrast, high-melt-index resins (6 to 60 g per 10 min) are necessary in extrusion coating, injection molding, and injection blow molding.

Low-viscosity polymers such as nylon 6/6 tend to leak (drool) from the nozzles of injection-molding machines, so they require special nozzles for injection molding. Aliphatic nylons exhibit narrow melting ranges and so need special screws in which the transition zone is relatively short, typically two or three turns (flights). Molecular weight distribution also factors into the extrusion of relatively low-viscosity polymers such as PEs. A wider MWD provides easier processing, but is detrimental to final properties such as strength and heat sealing. Narrower

MWDs, particularly with linear polymers such as HDPE and LLDPE, often necessitate changes to extruder.

High-viscosity polymers, such as PC and PSU, typically require high injection pressures and clamping tonnages. If, however, the pressure required to fill the cavity exceeds the maximum injection pressure for the press, then the cavity is underfilled. When the injection pressure is greater than clamp pressure (tonnage), then the melt can force its way through the parting line (where the mold opens to eject the finished part) and damage the mold. The former problem is common in high-speed or thin-wall injection molding of PC and other high-viscosity resins. While increasing processing temperatures does decrease the melt viscosity, increased plasticating (screw) speeds do not reduce viscosity much due to the rigid backbones of PC and PSU, which extend the lower Newtonian plateau beyond the shear rates typical of plasticating units. However, high shear is still produced during injection and can break the polymer chains, which lowers mechanical properties, such as the impact strength of PC. High-flow resins (melt index >40 g per 10 min) are available, but these generally exhibit lower MWs with the corresponding changes in properties. Other high-flow resins, which are usually immiscible blends of the primary polymer with a higher-flow plastic or additive, also affect final thermomechanical properties.

Very-high-MW or very rigid structures produce polymers that are not truly melt processible. In high-MW materials such as ultrahigh-molecular-weight polyethylene (UHMWPE) and PTFE, the intermolecular attraction and excessive chain length do not allow the materials to melt. Heat will soften these polymers, but they are usually processed as slurries in which a solvent or oil carries the unmolten polymer particles. Because this requires excessive pressure, PTFE is often processed using a ram extruder. Ultrahigh-molecular-weight polyethylene needs less pressure, but is also processed on ram or twin-screw extruders to prevent excessive shearing (as is discussed later in this article). The high MW (~10^6 Daltons, Ref 38) of the PMMA used for Plexiglas (trademark of Rohm and Haas Corp.) sheet does not permit melt processing, but rather the sheet is cast (polymerized) from the monomer (molding grade PMMA resins have MWs in the range of 60,000 Daltons, Ref 38).

The very inflexible structures of polyimides and aromatic polyamides do not permit melt processing. While polyimides are cast, more flexible variations, such as PEI and polyamide-imide (PAI) are melt processible. Similarly, copolymers and other variants of PTFE are melt processible. In both cases, the properties of the melt-processible polymers are less than those of the originals. Polyphenyl oxide is barely processible. However, blends of PPO with PS or HIPS are.

Additives such as processing aids and colorants can severely alter the viscosity of a polymer. It is not unusual for the same polymer compounded in different colors to have very different flow characteristics. Fillers and fibers typically increase melt viscosity. High loadings of fine particulate fillers, such as carbon black and titanium dioxide, can alter the low shear-rate behavior of the plastic; because these materials exhibit yield stresses, more force or pressure is required to initiate movement of the molten polymer. Regrind (processed polymer from runners and sprues) is often recombined with the virgin resin. However, because the regrind usually has a lower MW than the virgin resin, the flow characteristics of the mixture differ from those of the neat polymer.

Control of viscosity is critical in several processes. In coextrusion, the polymers must form layers and not mix with each other. Thus, the maximum viscosity difference for multi-manifold dies is 400 to 1, whereas it is 2 or 3 to 1 for feed blocks where the molten layers are in contact longer. In gas-assisted injection molding, the polymer viscosity determines where the bubble will form. Viscosity also allows polymer flow in rotary molding and extrusion coating.

Melt strength is the ability of the molten polymer to hold its shape for a period of time. Because long entangled polymer chains produce melt strength, these resins are high-MW polymers (with the related low-melt index values). However, polymers, such as PS, PET, and some nylons, which do not permit sufficient entanglement, always have low melt strength. Consequently, the processing equipment must accommodate this. Fiber extrusion lines usually place the extruder two or three floors above the windup units and draw the low-melt-strength fibers with gravity. This technique has also been used in blown-film extrusion of nylons. Polystyrene and PET are generally processed using flat-film extrusion so that the melt flows from the die to chill rollers that support the melt. As discussed previously, biaxially oriented PET films are then produced by heating the flat film to its rubbery state and stretching it on a center frame. Low-melt-strength polymers must always be injection blow molded.

Sheet materials used for thermoforming require hot strength to prevent excessive sagging of the rubbery polymeric sheet during heating. While this strength is also related to the MW and MWD, it incorporates the transition temperatures of the polymer. Because amorphous polymers exhibit broad transitions from their T_g to the molten state, they are easily thermoformed. The sharper melting transitions of polymers, such as PP, PET, and nylons, provide narrow processing temperature ranges and tend to be either too solid to form or too molten and sag. Broadening of the MWD of PP and copolymerization of PET have produced grades of these resins suitable for thermoforming. There are also special techniques that use the ductility of PP to thermoform parts.

Crystallization has two components: nucleation and crystal growth. Nucleation is the initiation of crystallization at impurities in the polymer melt and is enhanced by rapid cooling rates and nucleating agents. Crystal growth is favored by slower cooling rates (which allows the molecules enough thermally induced mobility to assume a crystalline structure). Although the maximum crystallinity occurs if the polymer is held at 0.9 T_m (K), the degree of crystallinity developed is a function of the temperatures achieved and how long the molten plastic is kept warm. Consequently, because rapid cooling produces no crystallinity or many small crystallites, it is used to produce optically clear PE-blown film and blow-molded PET bottles. Slower cooling or annealing—which produces fewer, but larger, crystals—is not always favored because mechanical properties such as impact strength are adversely affected. Moreover, while the intermolecular bonding that occurs in a crystalline polymer results in improved mechanical and thermal properties, the desire for crystalline, stress-annealed parts is balanced by economics, which usually dictate that plastics be cooled as rapidly as possible to reduce production time.

The volumetric changes (tight molecular packing) associated with crystallization produce shrinkage in plastics products. Consequently, the semicrystalline plastics shrink far more than amorphous plastics, and the degree of shrinkage varies with the cooling rate. Typical shrinkage values are presented in Table 10, but the incorporation of additives—such as fillers and glass fibers, which interrupt or enhance crystallinity—can affect shrinkage. Because flexible polymers, such as aliphatic nylons and PP, exhibit high levels of shrinkage, particularly in thick cross sections, they reduce shrinkage during extrusion by utilizing the high pressures of ram extruders to process the polymers slightly below their melting temperatures.

Crystallinity can also vary through the thickness of a part with the rapidly cooled outside surfaces and the slowly cooled core having different levels of crystallinity. This effect, which varies with polymer type and processing conditions, can alter plastic properties. With flexible polymers, such as PP, crystallization occurs throughout the thickness. However, at relatively slow injection speeds and low mold temperatures, relatively rigid polymers, such as syndio-

Table 10 Typical shrinkage values for selected polymers

Polymer	Shrinkage, mm/mm	
	Polymer	Polymer with 30% glass fiber
HDPE	0.015–0.040	0.002–0.004
PP	0.010–0.025	0.002–0.005
PS	0.004–0.007	...
ABS	0.004–0.009	0.002–0.003
POM	0.018–0.025	0.003–0.009
Nylon 6/6	0.007–0.018	0.003
PET	0.020–0.025	0.002–0.009
PBT	0.009–0.022	0.002–0.008
PC	0.005–0.007	0.001–0.002
PSU	0.007	0.001–0.003
PPS	0.006–0.014	0.002–0.005

Source: Ref 39

tactic PS, PPS, and PEK, produce layers of amorphous polymer at the surface and core of the part with a semicrystalline region between these layers (Ref 40). At high temperatures, these polymers behave more like PP.

Orientation. Different levels of orientation—and the related phenomena of die swell, shrinkage, and molded-in stress—are introduced during processing. Because gravity is the only force acting on the melt during rotational molding, very little orientation occurs in this process. Uniaxial orientation results from pipe, profile, flat-film and fiber extrusion, and calendering, whereas blow molding and blown-film extrusion induce biaxial orientation. While the actual orientation in injection molding varies with the mold design, the high flow rates generally align the polymer molecules in the direction of flow. Thermoforming also orients the polymer chains according to the design of the product.

Die swell is the expansion of the polymer melt that occurs as the extruded melt exits the die. This occurs when the aligned polymer chains escape the confines of the die and return to their random coil configuration. Die swell is dependent on processing conditions, die design, and polymer structure. It typically increases with screw speed (output rate) and decreases with higher melt temperatures and longer die land lengths. Increased MW, which produces more entanglement, also increases die swell.

Melt Fracture. At high extrusion rates, the polymer surface may also exhibit sharkskin or melt fracture. When the shear stress during extrusion exceeds the critical shear stress for the polymer, a repeating wavy pattern known as sharkskin occurs. In high-MW polyolefins this may disappear as the shear rate reaches the stick/slip region where the defect is present, but not visible. At even higher speeds, the polymer surface breaks up again in the defect known as melt fracture. This is particularly important in continuous and intermittent extrusion blow molding where these high-MW polymers are used; the output rates for continuous extrusion blow molding are typically below the critical shear rate, while those for intermittent extrusion blow molding place the process in the stick/slip region.

Shrinkage. Although shrinkage results from the volumetric contraction of the polymer during cooling, it is influenced by the relaxation of oriented polymer molecules. During processing the polymers align in the direction of flow, and their relaxation causes swelling perpendicular to this direction. Consequently, shrinkage in the direction of flow is usually much greater than transverse to flow. Addition of fillers and fibers, which also align in the flow, reduces shrinkage because they prevent the aligned molecules from relaxing. While rapid cooling can prevent the aligned polymer chains from relaxing, these chains contribute to molded-in stress.

Molded-in stress is the worst in regions where the polymer chains are highly aligned and not allowed to relax. Thus, processes with high levels of orientation produce the greatest molded-in stress. The stressed areas are points of attack for chemicals and sources of future breaks and cracks. Annealing will remove some of these stresses and is routinely required for some polymers such as PSUs. Because processes such as thermoforming and injection blow molding do not actually melt the plastic, but shape it at lower temperatures, the stretching produces high levels of molded-in stress. Usually the gate region of an injection-molded part will have the highest stresses, and consequently gate location is an important consideration in part design and failure analysis.

Polymer Degradation. Polyvinyl chloride, other chlorine-containing polymers, fluoropolymers, and POM tend to degrade under normal processing conditions. The dehydrochlorination of PVC occurs relatively easily and requires tightly controlled processing conditions. Hydrochloric acid formed during the degradation of PVC is not only corrosive to the equipment, but it catalyzes further degradation. The remaining polymer becomes increasingly rigid and discolored due to the formation of conjugated carbon-carbon double bonds. A similar reaction occurring in fluoropolymers produces the equally corrosive hydrofluoric acid. In contrast, POM depolymerizes from the ends of the polymer in an action called "unzipping"; this produces formaldehyde, which further catalyzes the depolymerization. To prevent or minimize degradation of PVC (or other chloropolymers and fluoropolymers), stabilizers are added to the plastic. With POM, copolymerization with cyclic ethers (such as ethylene oxide) or incorporation of blocking groups at the ends of the polymers (end capping) prevents unzipping.

Because many engineering polymers were produced by condensing two components to produce water, the presence of water during melt processing reverses this reaction. Thus, chains are broken, the MW is reduced, and properties decrease. In addition, water migrates to the surface of the part, resulting in the visual defect known as splay. While water uptake varies with the polarity and storage conditions of the plastic, most engineering plastics require drying before processing. Of the polymers shown in Table 11, only HDPE, PP, and rigid PVC are usually processed without some drying. While undried ABS and PMMA will not exhibit chain scission, they are typically dried to prevent splay. The remaining polymers in Table 11 are subject to chain scission and visual defects. Control of the water content in PET is of major importance for clarity of blow-molded bottles.

The combination of temperature and shear can also degrade plastics. The long entangled polymer chains of UHMWPE are easily severed in single-screw extruders. Heat-sensitive polymers such as PVC also degrade when the viscous dissipation from shear raises the melt temperature above the degradation temperature. Because counterrotating twin-screw extruders have positive material conveying characteristics, uniform residence time, and uniform temperature distributions, they are used for extruding materials such as rigid PVC. Ultra-high-molecular-weight polyethylene is often processed on twin-screw extruders or ram extruders (which have little shearing action). While shear can be a problem in extrusion processes, it is usually greatest in injection molding where polymer is forced at high velocities through small orifices. As indicated in Table 11, the processing temperatures and maximum shear conditions vary from polymer to polymer. However, as mentioned previously, when forcing highly viscous melts through thin channels, these maximum values are easily exceeded. Excess shear rates produce chain scission, whereas excess shear stress tends to produce cracking and related defects in the plastics product.

Table 11 Water absorption, processing temperatures, and maximum shear conditions for selected polymers

Polymer	Water absorption, %	Processing temperatures, °C	Maximum shear stress, MPa	Maximum shear rate, 10^3 s^{-1}
HDPE	<0.01	180–240	0.20	40
PP	0.01–0.03	200–260	0.25	100
PMMA	0.10–0.40	240–260	0.40	40
PVC, rigid	0.04–0.40	140–200	0.20	20
ABS	0.20–0.45	200–260	0.30	50
POM	0.25–0.40	190–230	0.45	40
Nylon 6/6	1.00–2.80	270–320	0.50	60
PET	0.10–0.20	280–310	0.50	...
PBT	0.08–0.09	220–260	0.40	50
PC	0.15	280–320	0.50	40
PS	0.30	310–340	0.50	50

Source: Ref 8, 39

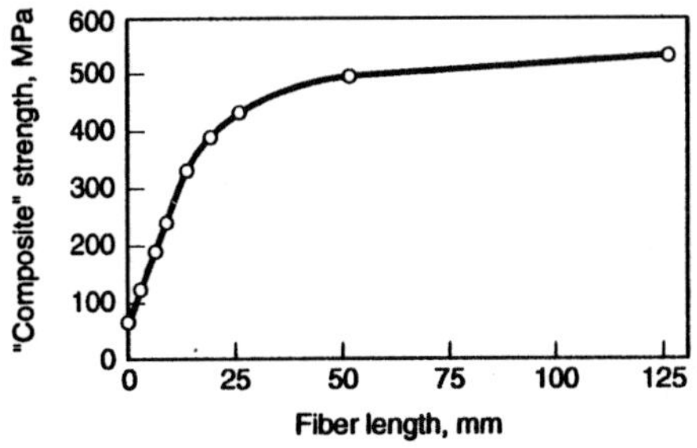

Fig. 28 Effect of fiber length on material strength. Source: Ref 41

When continuous-glass fibers or glass mats are processed using traditional thermoset processing techniques, the glass fibers usually remain unbroken. However, the discontinuous glass fibers commonly added to engineering resins are often broken during plastication and molding. As shown in Fig. 28, the fiber length is critical to the strength of the "composite." Reduction of the fiber length below a critical value results in a rapid decrease in strength. Consequently, glass fibers are often compounded into polymers using the controlled shear of twin-screw extruders. Special nonreturn valves (at the end of screws in injection-molding machines) also minimize fiber degradation.

Blends. The properties of immiscible and partially miscible blends depend on their processing conditions. Some are engineered so that one phase migrates to the air interface and governs surface properties. In immiscible polyblends, morphology is very sensitive to temperature and shear. These determine the size of the domains and whether the domains are spherical, elongated, or laminar. Phases may elongate in the flow direction.

ACKNOWLEDGMENT

Major portions of this article are based on the seminal text, *Polymer Structure, Properties and Applications,* by R. Deanin of University of Massachusetts at Lowell's Plastics Engineering Department.

REFERENCES

1. J.A. Brydson, *Plastics Materials,* 7th ed., Butterworth Heinemann, 1999
2. R.J. Cotter, *Engineering Plastics Handbook of Polyarylethers,* Gordon and Breach, 1995
3. R.D. Deanin, *Polymer Structure, Properties and Applications,* Cahners Books, 1972
4. H. Dominghaus, *Plastics for Engineers: Materials, Properties, and Applications,* Hanser Publishers, 1988
5. F. Rodriguez, *Principles of Polymer Systems,* 3rd ed., Hemisphere Publishing, 1989
6. J.H. Schut, Why Syndiotactic PS Is Hot, *Plast. Technol.,* Feb 1993, p 26–30
7. R.D. Deanin, *Polymer Structure, Properties and Applications,* Cahners Books, 1972, p 27
8. L.L. Clements, Polymer Science for Engineers, *Engineering Plastics,* Vol 2, *Engineered Materials Handbook,* ASM International, 1988, p 56–57
9. F. Rodriguez, *Principles of Polymer Systems,* 3rd ed., Hemisphere Publishing, 1989, p 23
10. H. Dominghaus, *Plastics for Engineers: Materials, Properties, and Applications,* Hanser Publishers, 1988, p 34, 347
11. R.D. Deanin, *Polymer Structure, Properties and Applications,* Cahners Books, 1972, p 54
12. S.L. Rosen, *Fundamental Principles of Polymeric Materials,* 2nd. ed., John Wiley & Sons, 1993, p 53, 54, 59
13. R.D. Deanin, *Polymer Structure, Properties and Applications,* Cahners Books, 1972, p 55
14. J.M. Dealy and K.F. Wissbrun, *Melt Rheology and its Role in Plastics Processing; Theory and Applications,* Van Nostrand Reinhold, 1990, p 369
15. R.D. Deanin, *Polymer Structure, Properties and Applications,* Cahners Books, 1972, p 130
16. J.A. Brydson, *Plastics Materials,* 5th ed., Butterworths, 1989, p 61
17. R.D. Deanin, *Polymer Structure, Properties and Applications,* Cahners Books, 1972, p 141
18. R.D. Deanin, *Polymer Structure, Properties and Applications,* Cahners Books, 1972, p 138
19. S.L. Rosen, *Fundamental Principles of Polymeric Materials,* 2nd ed., John Wiley & Sons, 1993, p 45
20. S.L. Rosen, *Fundamental Principles of Polymeric Materials,* 2nd ed., John Wiley & Sons, 1993, p 46
21. F. Rodriguez, *Principles of Polymer Systems,* 3rd ed., Hemisphere Publishing, 1989, p 23–24
22. W. Michaeli, *Plastics Processing, an Introduction,* Hanser Publishing, 1992, p 19
23. C.C. Winding and G.D. Hiatt, *Polymeric Materials,* McGraw-Hill, 1961
24. R.D. Deanin, *Polymer Structure, Properties and Applications,* Cahners Books, 1972, p 89
25. R.D. Deanin, *Polymer Structure, Properties and Applications,* Cahners Books, 1972, p 342
26. R.D. Deanin, *Polymer Structure, Properties and Applications,* Cahners Books, 1972, p 240
27. C. Rauwendaal, *Polymer Extrusion,* 2nd ed., Hanser Publishers, 1990, p 182
28. C. Rauwendall, *Polymer Extrusion,* 2nd ed., Hanser Publishers, 1990, p 218
29. M.M. McKelvey, *Polymer Processing,* John Wiley & Sons, 1962, p 26, 30
30. J.M.G. Cowie, *Polymers: Chemistry & Physics of Modern Materials,* 2nd ed., Blackie Academic and Professional, 1991, p 248
31. R.D. Deanin, *Polymer Structure, Properties and Applications,* Cahners Books, 1972, p 109
32. W. Michaeli, *Plastics Processing, an Introduction,* Hanser Publishing, 1992, p 59
33. C.C. Ku and R. Liepins, *Electrical Properties of Polymers: Chemical Principles,* Hanser Publishers, 1987, p 181–182
34. A.B. Strong, *Plastics: Materials and Processing,* Prentice-Hall, 1996, p 144
35. M.J. Austin, Inorganic Anti-Corrosive Pigments, *Paint and Coating Testing Manual,* J.V. Koleste, Ed., ASTM, 1995, p 239
36. W. Michaeli, *Plastics Processing, an Introduction,* Hanser Publishing, 1992, p 159
37. "Standard Test Method for Melt Flow Rates of Thermoplastics by Extrusion Plastometer," D 1238, *Annual Book of ASTM Standards,* Vol 08.01, ASTM
38. J.A. Brydson, *Plastics Materials,* 5th ed., Butterworths, 1989, p 399
39. *Modern Plastics Encyclopedia '92,* McGraw-Hill, 1992, p 378–428
40. Y. Ulcer, M. Cakmak, J. Miao, and C.M. Hsiung, Structural Gradients Developed in Injection Molded Syndiotactic Polystyrene (S-PS), *Annual Technical Conference of the Society of Plastics Engineers,* 1995, p 1788
41. P.K. Mallick, *Fiber-Reinforced Composites,* Marcel Dekker, 1988, p 83

Materials Selection and Design of Engineering Plastics

General Design Guidelines*

TO ENSURE the proper application of plastics, one must keep in mind three factors that determine the appropriate end-use: material selection, processing, and design. In plastics design, perhaps more so than in the design of other materials, an appropriate selection of material and processing conditions for end-use applications is essential. Plastics properties are highly influenced by the method of processing and the process conditions. If these two factors are understood, a wise choice in the actual design can be made. However, the problem is not simple because there are thousands of different resins and blends available to industry as component materials recognized by Underwriters' Laboratories (UL).

Design engineers, whether designing a new plastics product or replacing a product formerly made of another material, have four major concerns (Ref 1):

- Designing products that can be built as easily and economically as possible
- Ensuring product reliability
- Simplifying product maintenance and extending product life
- Ensuring timely delivery of materials or components

Good design essentially means that all of the above factors have been considered in detail in order to manufacture a part economically with available manufacturing methods while meeting end-use performance requirements. End-use properties are more important in plastics than in metals because most properties are a function of time and the environment, not just of the usual properties, such as strength, load, modulus, or other mechanical property considerations. The importance of applying engineering design principles to plastics has not always been fully utilized. Plastics are governed by the same physical laws and the same rules for good design as other materials, with the additional requirement that many properties, such as modulus, strength, and creep are also highly dependent on time, temperature, and environment (humidity). These principles can be applied intelligently only if the physical laws are understood and data on pertinent properties of the materials are available.

Defining End-Use Requirements

The properties to be considered depend not only on the material, but also on the application itself. Thus, it is necessary to know what performance is expected of the end product and under what circumstances. Lacking concrete data, the designer must often make reasonable estimates of end-use requirements. A logical way to accomplish this is to use a checklist that enumerates the anticipated use conditions of the article to be designed. Such a list includes the following factors (Ref 2, 3):

General information

- What is the function of the part?
- How does the assembly operate?
- Can the assembly be simplified by using a plastic?
- What tolerances are necessary?
- Can a number of functions be combined in a single molding to eliminate future assembly operations and simplify design?
- What space limitations exist?
- What service life is required?
- Is light weight desirable?
- Are there acceptance codes and specifications to be met?
- Do analogous applications exist?

Structural applications

- How is the part stressed in service?
- What is the magnitude of the stress?
- What is the stress versus time relationship?
- How much deflection can be tolerated in service?

Environment

- What are the operating temperatures, types of chemicals or solvents, humidity conditions, and service life in the expected environment?

Appearance

- What are the style, shape, color, surface finish, and decoration elements?

Economic factors

- What is the cost of the existing part, and the cost estimate of the part if made of plastic?
- Are faster assemblies and elimination of finishing operations possible?
- Will redesign of the part simplify the assembled product and thus give rise to savings in installed cost?

Manufacturing options

- Should the proposed design be machined, injection molded, or extruded, considering the number of parts to be made, the design geometry, and tolerances?
- If injection molding is chosen, how can mold design contribute to part design?
- In subsequent assembly operations, can the properties of the plastic be used further, as in snap fits or spin welding?

Part Geometry

After the preliminary study, the designer has to define the part geometry. This usually progresses through several stages, beginning with preliminary drawings and sketches that indicate the basic design and functions. More detailed sketches show appropriate wall thickness, ribs, radii, and other structures, based on end-use considerations.

Processing and Tolerances. The most versatile of the plastics forming processes is injection molding, which accounts for about 33% of all plastics consumed. It is the one process that can produce three-dimensional parts. Extrusion accounts for another 33% of plastics conversion, but only two dimensions can be controlled because it is a continuous process. A final 33% of forming uses blow molding (10% of all plastics), thermoforming, compression molding, coating, laminating, rotational molding, and all others. However, injection molding is the most important, and, except for compression or transfer molding of thermosets, is usually the most accurate in maintaining dimensional tolerances. In most cases, tolerances can be expected to be within a few thousandths of an inch.

Tolerances are determined to a large degree by the type of process used in manufacture. For melts, the melt pressure and tooling have a great influence. Blow molding, extrusion, thermoforming, and rotational molding are low-pressure processes (zero to a few hundred psi). Hence, the tolerances are usually only good to a few hundredths of an inch. Injection molding uses melt pressures up to 140 MPa (20 ksi). Thus, tooling is more rigid, and the melt that is compressible can be packed to compensate for shrinkage due to thermal contraction as the part cools. In injection molding for parts less than an

*Adapted from article by Nick R. Schott, General Design Guidelines, *Engineering Plastics,* Volume 2, *Engineered Materials Handbook,* ASM International, 1988, p 707–710

inch in length, the tolerances can be expected to be in the ten thousandths of an inch. In the case of plastic optical lenses, tolerances down to micrometers or Angstrom units can be expected.

These are exceptions, and an understanding of what is typically expected from industry can be gained by reviewing Ref 4. This publication shows that tolerances are specified as either fine or commercial. Figure 1 shows typical tolerances for a hypothetical part, with assumptions as specified. The commercial values shown represent common production tolerances at the most económical level. The fine values represent closer tolerances that can be held, but at a greater cost. Tighter tolerances can be accomplished by changing molding conditions, such as decreasing the melt temperature, increasing the injection pressure, decreasing the mold temperature, and so on. The relationship between the variables is complex, but essentially concerns the packing of extra molecules of plastics into the cavity before the melt has frozen in the gate area of the cavity. One should also keep in mind that mineral fillers decrease shrinkage because inorganic materials such as glass, mica, talc, and calcium carbonate have much lower coefficients of expansion. Also, because crystalline materials have more efficient molecular packing than amorphous materials such as polystyrene (PS) and polyvinyl chloride (PVC), the highest level of shrinkage should be expected for crystalline polymers such as nylons, polyethylene (PE), and polypropylene (PP).

Nominal Wall Thickness of Molded Parts. Most molded plastics have a nominal wall thickness of 3.2 mm (1/8 in.) or less. A lower limit is 0.50 to 0.75 mm (0.20 to 0.30 in.). A thinner wall causes the melt front to freeze in the mold cavity before the cavity is filled. The many factors that come into play include the melt temperature, injection rate, type of tool steel, mold temperature, size of runner system,

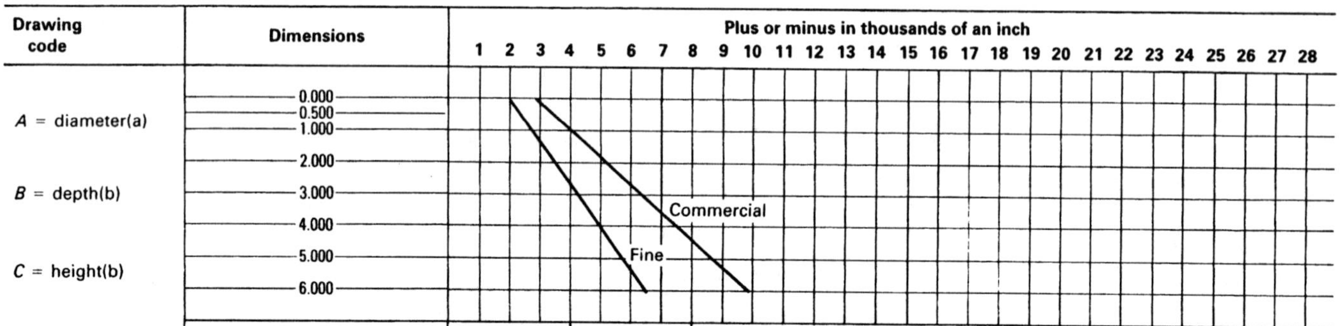

Drawing code	Dimensions	Comm. ±	Fine ±
	6.000 to 12.000 for each additional inch add (inches)	0.003	0.0015
D = bottom wall(b)	. . .	0.003	0.002
E = side wall(c)	. . .	0.003	0.002
F = hole size diameter(a)	0.000–0.125	0.002	0.001
	0.125–0.250	0.002	0.0015
	0.250–0.500	0.003	0.002
	≥ 0.500	0.003	0.002
G = hole size depth(d)	0.000–0.250	0.002	0.002
	0.250–0.500	0.003	0.002
	0.500–1.000	0.004	0.003
Draft allowance per side(d)	. . .	1°	½°
Flatness(c)	0.000–3.000	0.005	0.003
	3.000–6.000	0.007	0.004
Thread size (class)	Internal	1B	2B
	External	1A	2A
Concentricity(c)	. . .	0.005	0.003
Fillets, ribs, corners(e)	. . .	0.015	0.015
Surface finish(f)	. . .	. . .	. . .
Color stability(f)	. . .	. . .	. . .

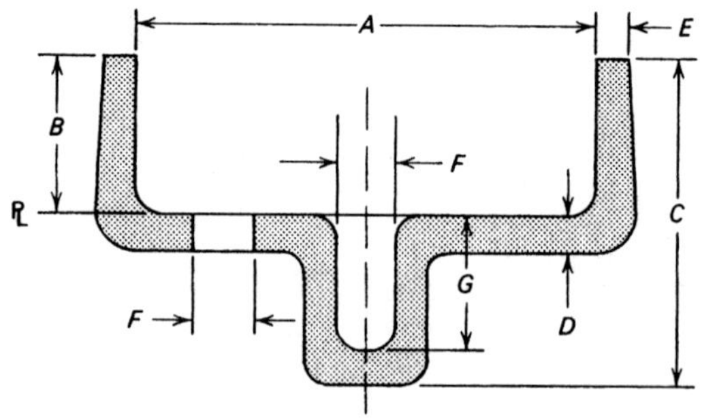

Reference notes

(a) These tolerances do not include allowance for aging characteristics of material.
(b) Parting line must be taken into consideration.
(c) Part design should maintain a wall thickness as nearly constant as possible. Complete uniformity in this dimension is impossible to achieve.
(d) Care must be taken that the ratio of the depth of a cored hole to its diameter does not reach a point that will result in excessive pin damage.
(e) These values should be increased whenever compatible with desired design and good molding technique.
(f) Customer-molder understanding is necessary prior to tooling.

Fig. 1 Typical tolerances for an injection-molded part with a 3.2 mm (1/8 in.) wall section. All dimensions in inches. Source: The Society of The Plastics Industry, Inc.

resin type, filler, and many others. Typical wall thickness values are shown in Table 1.

A wall thickness that is much over 3.2 mm (0.125 in.) poses several problems. First, the cycle time for molding becomes unduly long, and the manufacture of the part becomes uneconomical. As a rule of thumb, one can say that most articles are molded in cycle times of 2 or 3 min or less. The cooling time makes up about 75% of the molding cycle and thus has a great influence on production economics. Excessive wall thickness or nonuniform wall thickness causes other problems, such as sink marks, which are surface depressions due to excessive shrinkage, and internal voids, which are flaws that cause structural weakness. One may purposely foam a plastics melt and introduce microvoids. In foaming, a chemical or physical blowing agent is introduced; the gas causes the melt to expand and fill the cavity. Typical density reductions are 0 to 30%. Wall thicknesses are increased up to 12.7 mm (0.5 in.), or even thicker, but the surface finish can be poor in foamed parts and a painting operation is often necessary. Improvements in the smoothness of the part surface, as molded, have been achieved through nucleating the molding compound and applying external molding techniques, such as gas counterpressure, as well as through a better understanding of the injection-molding process. A better understanding of the relationship between molding parameters and surface quality has resulted in an increasing number of equipment modifications, such as hydraulic boosters, which are used to inject molten polymer at a faster rate to attain higher weight reduction in large parts.

Stress concentration in corners can lead to premature part failure. The article "Impact Loading and Testing" in this book describes both this issue and the relative notch sensitivity of neat, filled, and reinforced polymers.

Strength of Plastics

Many engineers are familiar with steel and wood and can "think and feel" in terms of these materials. Steel has a modulus of 210 GPa (30×10^6 psi), while hardwoods have a modulus of about 2.8 GPa (0.400×10^6 psi). Unfilled plastic materials have moduli that are usually less than 3.45 GPa (0.500×10^6 psi) at room temperature. The modulus is, of course, highly temperature dependent and also is affected by environmental factors, such as humidity. The modulus can be increased by the use of fillers and reinforcements. In injection molding, only short fibers are considered. They are typically 3.2 mm (0.125 in.) long and vary in concentrations from 0 to 40 or 50% maximum. A critical fiber length is necessary for reinforcement. Some fiber breakage will occur in processing, and the mechanical properties will deteriorate as the material is reground and reprocessed. The modulus of the fiber-reinforced composite can be estimated by a simple rule of mixtures:

$$E_c = f_1 E_f + f_2 E_p$$

where E_c is the modulus of the composite, E_f is the modulus of the fiber, E_p is the modulus of the plastic, and f_1, f_2 are the volume fractions of each component. Most plastics have a tensile strength of less than 35 MPa (5 ksi). Reinforcements can increase these values by up to one order of magnitude.

It is well known that stiffness is a product of modulus and moment of inertia. Stiffness can be increased by proper design of ribs, by foaming the plastic, by selecting a higher modulus resin, and by using a resin with reinforcements.

Cost Estimating Plastics Parts

It is important to have an idea of product cost before the product is off the drawing board. Dym (Ref 5) gives a simplified method of quickly estimating part costs. Two basic components determine the value of a plastic part, the cost of the resin, and the cost of processing. The cost of the material is calculated by multiplying the cost per cm^3 by the volume of the part, expressed in cm^3. Production rates in plastics molding are determined by maximum part thickness because the cycle time (pieces per hour) is determined by heat transfer (see Fig. 2). Molding costs are determined from production rate data and machine size (see Table 2 for an example). The size of the machine is determined by the size of a shot, expressed as either cm^3 ($in.^3$) of material or cm^3 (oz) of polystyrene, and by the clamping force that keeps the two halves of the mold closed. Shot sizes vary from less than 30 cm^3 (1 oz) to about 15,000 cm^3 (500 oz), while the tonnage varies from 180 to 44,500 kN (20 to 500 tonf) or larger for the biggest machines. Most molds have multicavities to make many parts in the same cycle. The optimal number of cavities can be estimated as reported by Boller (Ref 6). As a rule of thumb, Dym suggests that the number of cavities be determined by the parts that can be produced in 200 molding hours and ordered in a 45 day reorder frequency: number of cavities = 45 day requirement/200 h × pieces per hour.

The material per cavity times the number of cavities plus the weight of the sprue and runner system gives the shot size. The clamping force is estimated by calculating the projected area of each cavity, sprue, and runner system. As a rule of thumb, for each square unit of projected area, a clamping force of 14 to 28 N/mm^2 (1 to 2 $tonf/in.^2$) is used. For more viscous materials, 41 to 69 N/mm^2 (3 to 5 $tonf/in.^2$) is used. Thus, from Table 2, one can estimate the manufacturing costs. Costs can also be estimated for other processing methods. Each process is assigned a

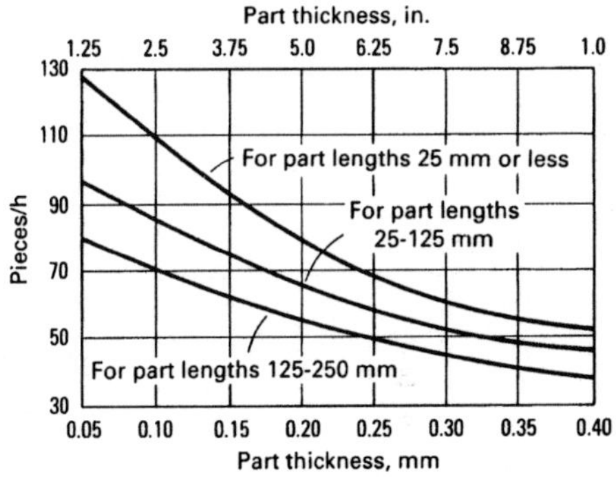

Fig. 2 Cycle time in injection molding as a function of part thickness. Source: Ref 5

Table 1 Suggested wall thickness

Thermoplastic	Minimum, any article mm	Minimum, any article in.	For small articles mm	For small articles in.	Average for most articles mm	Average for most articles in.	For large to maximum articles mm	For large to maximum articles in.
Acrylics	0.64	0.025	0.89	0.035	2.4	0.093	3.2–6.4	0.125–0.250
ABS	0.75	0.030	1.3	0.050	2.3	0.090	3.2–6.4	0.125–0.250
PA	0.38	0.015	0.64	0.025	1.5	0.060	2.4–3.2	0.092–0.125
PC	0.75	0.030	1.3	0.050	2.3	0.090	3.2–4.7	0.125–0.187
PE	0.89	0.035	1.3	0.050	1.6	0.062	2.4–3.2	0.093–0.125
PP	0.89	0.035	1.3	0.050	1.6	0.062	3.2–6.4	0.125–0.250
PS	0.75	0.030	1.3	0.050	1.6	0.062	3.2–6.4	0.125–0.250
Polyvinyls	1.6	0.062	2.4	0.093	2.4	0.093	3.2–6.4	0.125–0.250

ABS, acrylonitrile-butadiene-styrene; PA, polyamide; PC, polycarbonate

Table 2 Machine capacity in relation to cost per hour

Cost/h, 1983 $	Capacity kN	Capacity tons	Capacity cm^3	Capacity $in.^3$
18	445	50	81.1	4.95
23	670	75	162	9.9
25	890	100	213	13.0
28	1110	125	267	16.3
30	1335	150	324	19.8
32	1780	200	374	22.8
34	2225	250	533	32.5
37	2670	300	640	39.0
40	3115	350	852	52.0
43	3560	400	959	58.5
46	4005	450	1065	65.0
49	4450	500	1600	97.5
54	5340	600	1865	113.8
58	6230	700	2556	156
65	7120	800	2917	178
72	8010	900	3195	195
80	8900	1000	4392	268

Source: Ref 5

cost factor, as shown in Table 3. The cost factor times the material cost reflects the estimated purchase price of the part.

Structure, Properties, Processing, and Applications

For a good product design in plastics, the design engineer must understand the interdependence of polymer structure, material properties, processing method, and end-use application. A fundamental understanding of polymer structure allows one to understand and predict many of the properties of polymers. In turn, the thermal and rheological properties dictate the processing method. The act of processing will itself influence the properties of the plastic part. Shrinkage, warpage, density, strength, toughness, and many other properties are affected by processing. Also, the properties are highly anisotropic and depend on the temperature and shear history of the material. Because end-use properties are affected by so many variables, it is extremely difficult to get reliable data that can be used in design. Many resin companies have databases that they will share with their customers, but always with a disclaimer because many factors are beyond their control. Computerized databases can be effective at making material selection easier. Similarly, computer use in design allows optimization of part design, mold design, and process conditions. The cost benefits are substantial because computer simulations preclude expensive trial-and-error methods for tooling.

Table 3 Cost factors for various plastics processes

Process	Cost factor	
	Overall	Average
Blow molding	1⅜–5	1⅛–3
Calendering	1½–5	2½–3½
Casting	1½–3	2–3
Centrifugal casting	1½–4	2–4
Coating	1½–5	2–4
Cold-pressure molding	1½–5	2–4
Compression molding	1⅜–10	1½–4
Encapsulation	2–8	3–4
Extrusion forming	1 1/16–5	1⅛–2
Filament winding	5–10	6–8
Injection molding	1⅛–3	1 3/16–2
Laminating	2–5	3–4
Matched-die molding	2–5	3–4
Pultrusion	2–4	2–3½
Rotational molding	1¼–5	1½–3
Slush molding	1½–4	2–3
Thermoforming	2–10	3–5
Transfer molding	1½–5	1¾–3
Wet lay-up	1½–6	2–4

Source: Ref 7

Thus, plastic product design follows the same engineering principles and guidelines that are used with other materials. Use of a checklist makes the task systematic and rational. The designer must be aware that plastic properties change with time, temperature, and environment. Short-term properties can be used to screen potential candidate materials, while long-term data are required to predict performance for in-service use. Manufacturability depends on the interaction between product design and mold design. Economic benefits will accrue with part or function consolidation as a plastics part replaces metal or other engineering materials. Also, many secondary operations, such as painting, cutting, drilling, and punching, are eliminated and thus contribute to cost and labor savings.

Furthermore, since part cost is determined by material and manufacturing costs, one finds that plastics replace many other materials because they are more energy efficient. Function in design is determined by the volume of a part. Because plastics have densities of around 1.0 g/cm^3, they show a high specific strength and modulus and a low cost relative to other materials when compared on a volume rather than a weight or density basis.

General guidelines in plastics product design consist of a series of rules. These rules are based on the behavior of the plastic melt during processing and the behavior of plastics materials in service. Adherence to the guidelines leads to economical manufacturing and good part performance in the end-use application.

REFERENCES

1. Compounding Lines, *Imagineering News,* Vol 2 (No. 1), RTP Company, June 1987
2. "Design Handbook for DuPont Zytel Nylon Resin," E.I. Du Pont de Nemours & Co.
3. G.R. Moore and D.E. Kline, *Properties and Processing of Polymers for Engineers,* Prentice-Hall, 1984, p 175–180
4. *Standards and Practices of Plastics Molders and Plastics Molded Parts Buyers Guide,* The Society of the Plastics Industry, 1965
5. J.B. Dym, Cost Estimating of Plastic Parts, *Plast. Des. Forum,* Vol 8, Nov–Dec 1983, p 51
6. W.A. Boller, Specifying the Optimum Number of Cavities, *Mod. Plast.,* Vol 51, Nov 1974, p 74
7. D.V. Rosato, *Injection Molding Handbook,* Van Nostrand Reinhold, 1986, p 805, 866

Design with Plastics*

THE KEY to any successful part development is the proper choice of material, process, and design matched to the part performance requirements. The ability to design plastic parts requires knowledge of material properties—performance indicators that are not design or geometry dependent—rather than material comparators that apply only to a specific geometry and loading. Understanding the true effects of time, temperature, and rate of loading on material performance can make the difference between a successful application and catastrophic failure. Examples of reliable material performance indicators and common practices to avoid are presented in this article. Simple tools and techniques for predicting part performance (stiffness, strength/impact, creep/stress relaxation, and fatigue) integrated with manufacturing concerns (flow length and cycle time) are demonstrated for design and material selection.

Engineering plastics are now used in applications where their mechanical performance must meet increasingly demanding requirements. Because the marketplace is more competitive, companies cannot afford overdesigned parts or lengthy, iterative product-development cycles. Therefore, engineers must have design technologies that allow them to create productively the most cost-effective design with the optimal material and process selection.

The design-engineering process involves meeting end-use requirements with the lowest cost, design, material, and process combination (Fig. 1). Design activities include creating geometries and performing engineering analysis to predict part performance. Material characterization provides engineering design data, and process selection includes process/design interaction knowledge. In general, the challenge in designing with structural plastics is to develop an understanding not only of design techniques, but also of manufacturing and material behavior.

Engineering thermoplastics exhibit complex behavior when subjected to mechanical loads. Standard data sheets provide overly simplified, single-point data that are either ignored or, if used, are probably misleading. Some databases provide engineering data (Ref 1) over a range of application conditions, and knowledge-based material-selection programs have been written (Ref 2). A methodology for optimal selection of materials and manufacturing conditions to meet part performance needs is described in this article. Simple tools and techniques for the initial prediction of part performance, leading to the optimal selection of materials and process conditions, are discussed. Related coverage is provided in the articles "Effects of Composition, Processing, and Structure on Properties of Engineering Plastics" and "Design and Selection of Plastics Processing Methods" in this book.

Mechanical Part Performance

There are a wide variety of part performance requirements. Some, such as flammability, transparency, ultraviolet stability, electrical, moisture, and chemical compatibility, as well as agency approvals, are specified as absolute values or simplified choices. However, mechanical requirements such as stiffness, strength, impact, and temperature resistance cannot be specified as absolute values. For example, a part may be required to have a certain stiffness—maximum deflection for a given loading condition. The part geometry (design) and the material stiffness combine to produce the part stiffness. Thus, it is impossible to select a material without some knowledge of the part design. Similarly, the part may be required to survive a certain drop test and/or a certain temperature/time/loading condition. Again, it is impossible to select a material or design a part by using traditional, inadequate, single-point data such as notched Izod or heat-distortion temperature (HDT). In addition, it is important to consider the effects of the design and material selection of a part on its fabrication. Considerations such as flow and cycle time should be quantitatively included in the design and material-selection process. Simple yet extremely useful tools and techniques for the initial prediction of part performance are presented in this article.

The design process for thermoplastic part performance can be divided into two categories based on time-independent and time-dependent material behavior (Fig. 2). For time-independent material behavior, elastic material response is used to predict the displacement of a part under load. The maximum load occurs when the strength of the material is reached as fully plastic yielding for ductile materials or brittle failure for glass-filled materials. Time-dependent material behavior becomes important for three types of loading: monotonic loading at a given strain rate until failure occurs, constant load for a period of time, or cyclic load. In the first case, strain-rate-dependent material behavior becomes important; for constant load or displacement, time-dependent deformation or stress relaxation becomes an important design consideration; for cyclic loading, fatigue failure is an important consideration. In the next five sections, stiffness, strength, impact, creep/stress relaxation, and fatigue behavior are related to part performance. More details of these important design issues can be found in Ref 3.

Part Stiffness. Many thermoplastic parts are platelike structures that can be treated as a simply supported plate, possibly reinforced with ribs. A procedure intended to provide quick, approximate solutions for the stiffness of laterally loaded rib-stiffened plates has been developed (Ref 4). The computer program employs the Rayleigh-Ritz energy method and is capable of including the geometric nonlinearities associated with the large displacement response typical of low-modulus materials such as thermoplastics. The program allows the user to input the important parameters of specific plate structures (length, width, thickness, number of ribs, rib geometry), the boundary conditions (simply supported), clamped, point supported), and the loading (central point, uniform pressure, torsion loading). With the capability of multiple rib pattern definitions, the user can quickly determine the load-deflection response for different designs to select the one that is most effective for the specific application. This tool has been validated with finite-element results. An example demonstrating the prediction of the nonlinear load-displacement response is shown in Fig. 3.

Strength and Stiffness of Glass-Filled Plastic Parts. An accurate characterization of the strength and stiffness of glass-filled thermoplastics is necessary to predict the strength and stiffness of components that are injection molded with these materials. The mechanical properties of glass-reinforced thermoplastics are generally measured in tension using end-gated, injection-molded ASTM type I (dog-bone) specimens (Ref 3). However, the gating and the

*Adapted from G.G. Trantina, "Design with Plastics," *Materials Selection and Design,* Volume 20, *ASM Handbook,* ASM International, 1997, pages 639 to 647

direction of loading of these molded specimens yields nonconservative stiffness and strength results caused by the highly axial orientation of glass that occurs in the direction of flow (and loading) during molding.

Previous studies (Ref 5) have shown that injection-molded, glass-reinforced thermoplastics are anisotropic; that is, stiffness and strength values in the cross-flow direction are substantially lower than in the flow direction. The tensile stiffness and strength were measured by using dog-bone specimens that were cut in both the flow and cross-flow directions from edge-gated plaques of various thicknesses. The ratio of the cross-flow/flow tensile modulus and strength of 30% glass-filled polybutylene terephthalate (PBT), 30% glass-filled modified polyphenylene oxide (M-PPO), and 50% glass-filled (long glass fibers) nylon are plotted versus specimen thickness in Fig. 4 and 5. It is important to note the strong dependence of the cross-flow/flow ratio on specimen thickness and the small values of this ratio for small specimen thicknesses. These data clearly indicate that material selection and design for glass-filled materials that are based on injection-molded bars of a given thickness could be totally misleading—cross-flow properties could be only 50% of flow properties (small specimen thicknesses), and unless the thickness of the specimen is the same as the thickness of the part, the data could not be used for predicting part performance. However, for most parts (thickness less than 4 mm, or 0.16 in.) with glass loadings of 30% or greater, a simple mold-filling analysis coupled with an anisotropic stress analysis with the cross-flow stiffness of 60% of the flow stiff-

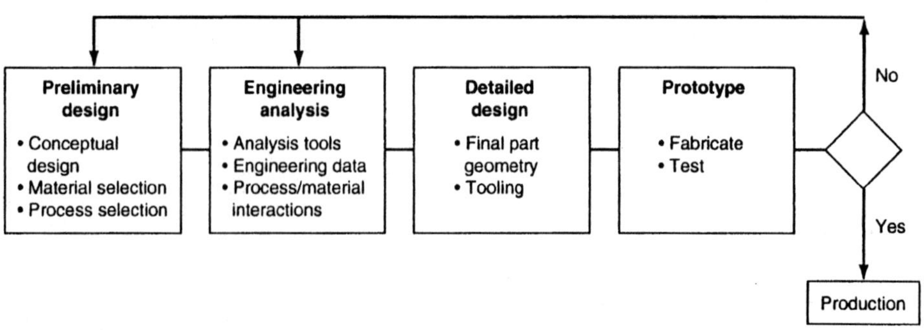

Fig. 1 Design-engineering process. The goal is to meet the end-use requirements the first time with low cost.

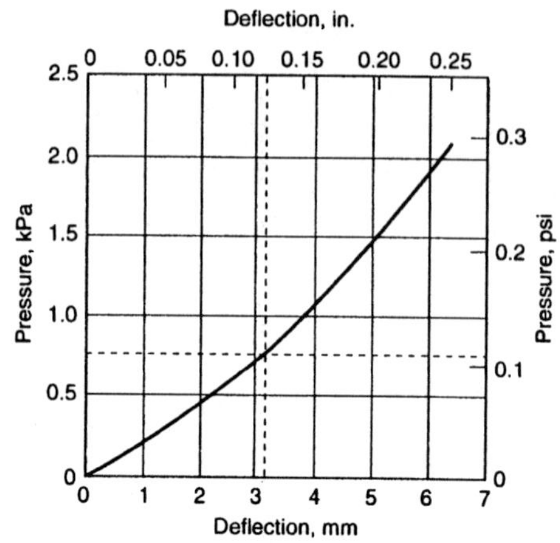

Fig. 3 Nonlinear pressure-deflection response for a 254 by 254 mm (10 by 10 in.) plate with a thickness of 2.5 mm (0.1 in.) and a material with a modulus of 2350 MPa (340 ksi)

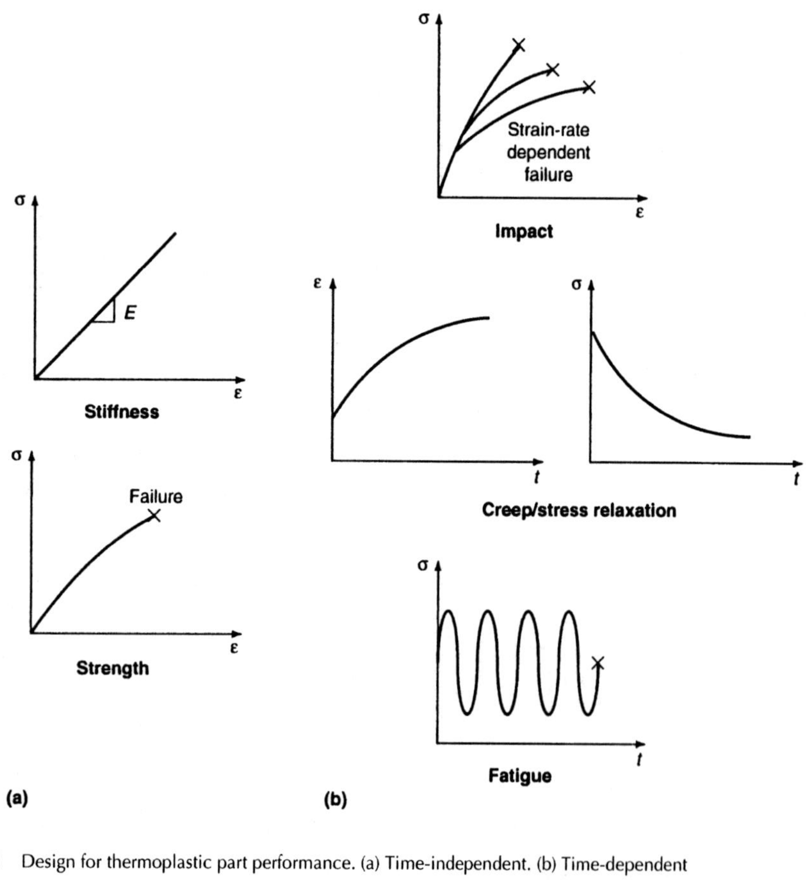

Fig. 2 Design for thermoplastic part performance. (a) Time-independent. (b) Time-dependent

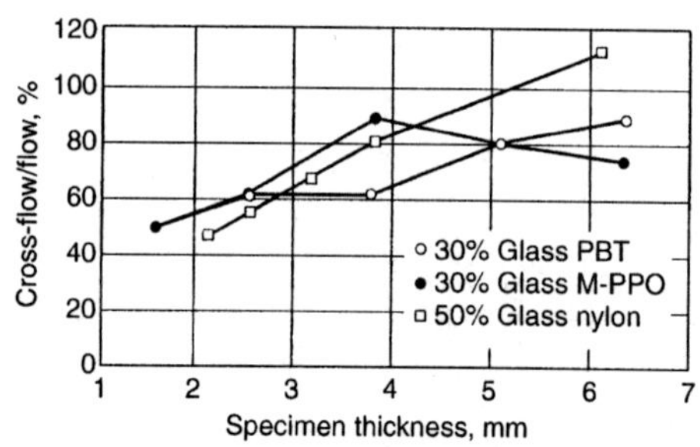

Fig. 4 Ratio of cross-flow/flow tensile modulus as a function of specimen thickness. PBT, polybutylene terephthalate; M-PPO, modified polyphenylene oxide

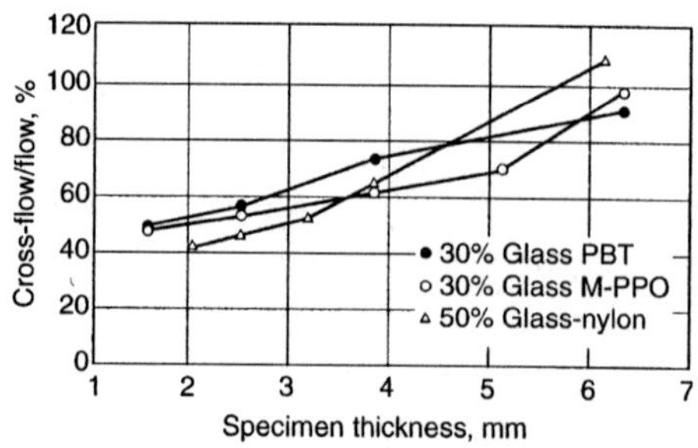

Fig. 5 Ratio of cross-flow/flow ultimate stress as a function of specimen thickness. PBT, polybutylene terephthalate; M-PPO, modified polyphenylene oxide

ness provides a reasonable prediction of part performance (Ref 3).

Part Strength and Impact Resistance. A number of test methods such as Izod (notched beam) and Gardner/Dynatup (disk) are available for measuring impact resistance (Ref 3). Such tests should not only measure the amount of energy absorbed, but also determine the effects of temperature on energy absorption. Additionally, they should be able to identify strain-rate-dependent transitions from ductile to brittle behavior. They should be applicable to a wide variety of geometric configurations. Unfortunately, these techniques provide only geometry-specific, single-point data for a specific temperature and strain rate. Also, each test provides a different ductile/brittle transition. Energy absorption, however measured, is made up of many complex processes involving elastic and plastic deformation, notch sensitivity, and fracture processes of crack initiation and propagation.

The prediction of strength and impact resistance of plastic parts is probably the most difficult challenge for the design engineer. Tensile stress-strain measurements as a function of temperature and strain rate provide one piece of useful information. Most unfilled engineering thermoplastics exhibit ductile behavior in these tensile tests, with increasing strength (maximum stress) as displacement rate increases and/or temperature decreases. However, stress-state effects must be added to the tensile behavior because the three-dimensional stress state created by notches, radii, holes, thick sections, and so forth increase the potential of brittle failure.

Ductile-to-brittle transitions in the fracture behavior of unfilled thermoplastics occur with increasing strain rates, decreasing temperatures, and increasingly constrained stress states. Figure 6 shows three common mechanical test techniques: uniaxial tension, biaxially stressed disks (usually clamped on the perimeter and loaded perpendicularly with a hemispherical tup), and notched beams loaded in bending. These three tests provide uniaxial, biaxial, and triaxial states of stress. Typical part geometries and loadings exhibit combinations of these states of stress. Thus, no one test is sufficient for part design and material selection. Furthermore, there are two competing failure modes: ductile and brittle (Fig. 6). With increasingly constrained stress states (uniaxial → biaxial → triaxial), the tendency for brittle failure tends to increase. Brittle failure occurs when the brittle failure mechanism occurs prior to ductile deformation (Fig. 6).

The calculation and measurement of the ductility ratio (Ref 6) is a method to characterize the ductility of a material for a relatively severe state of stress, for example, a beam with a notch radius of 0.25 mm (0.010 in.). The ductility ratio is defined as the ratio of the failure load in the notched-beam geometry (P_{failure}) to the maximum ductile, load-carrying capability in an unnotched-beam geometry where the height of the unnotched beam is equal to the net section height of the notched-beam geometry:

$$\text{Ductility ratio} = \frac{P_{\text{failure}}}{P_{\text{ductile}}} \quad \text{(Eq 1)}$$

where:

$$P_{\text{ductile}} = \frac{\sigma_f b h^2}{l} \quad \text{(Eq 2)}$$

and σ_f is the strength at appropriate rate and temperature, b is the beam thickness, h is the beam height, and l is the beam span.

This ductile load limit can be determined experimentally or with this plastic-hinge calculation assuming fully developed plasticity over the entire cross section and perfectly plastic material behavior. A ductility ratio of 1.0 corresponds to a ductile failure, while ductility numbers less than 1.0 correspond to varying levels of brittle behavior. Ductility ratios can be plotted as a function of strain rate at different temperatures to create fracture maps such as the one shown for polycarbonate (PC) in Fig. 7. This information is useful for material-selection and initial part design considerations.

Creep/Stress Relaxation—Time/Temperature Part Performance. Polymers exhibit time-dependent deformation (creep and stress relaxation) when subjected to loads. This deformation is significant in many polymers, even at room temperature, and is rapidly accelerated by small increases in temperature. Hence, the phenomenon is the source of many design problems. Development and application of methods are needed for predicting whether a component will sustain the required service life when subjected to loading, as the useful life of the part could be terminated by excessive deformation or even rupture. For most practical applications of polymers, predictive methods must account for part geometry, loading, and material behavior.

A common measure of heat resistance is the heat-distortion temperature (HDT). For this test, bending specimen 127 by 12.7 mm (5 by 0.5 in.) with a thickness ranging from 3.2 to 12.7 mm (0.125 to 0.5 in.) is placed on supports 102 mm (4 in.) apart, and a load producing an outer fiber stress of 0.46 or 1.82 MPa (66 or 264 psi) is applied. The temperature in the chamber is increased at a rate of 2 °C/min (3.6 °F/min). The temperature at which the bar deflects an additional 0.25 mm (0.010 in.) is called the HDT or sometimes the deflection temperature under load (DTUL). Such a test, which involves vari-

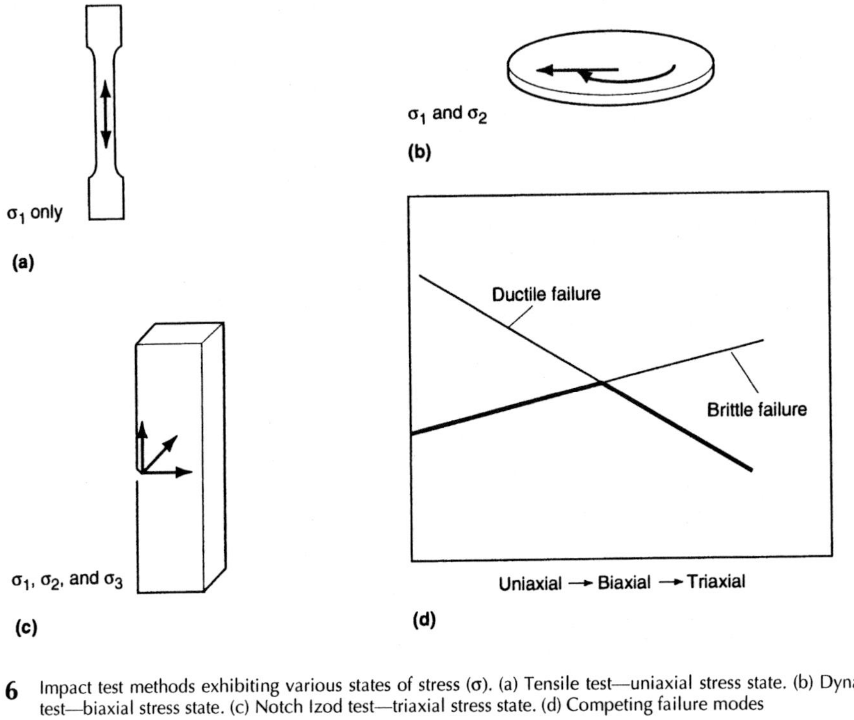

Fig. 6 Impact test methods exhibiting various states of stress (σ). (a) Tensile test—uniaxial stress state. (b) Dynatup test—biaxial stress state. (c) Notch Izod test—triaxial stress state. (d) Competing failure modes

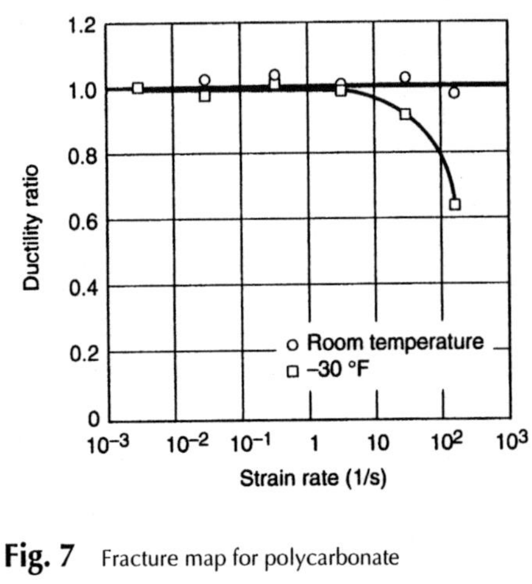

Fig. 7 Fracture map for polycarbonate

able temperature and arbitrary stress and deflection is of no use in predicting the structural performance of a thermoplastic at any temperature, stress, or time. In addition, it can be misleading when comparing materials. A material with a higher HDT than another material could exhibit more creep at a lower temperature. Also, some semicrystalline materials exhibit very different values of HDT at 0.46 and 1.82 MPa (66 and 264 psi). For example, with PBT, the HDT at 0.46 MPa (66 psi) is 154 °C (310 °F), and the HDT for 1.82 MPa (264 psi) is 54 °C (130 °F). The question of which HDT to use for comparison with another material that has the same HDT for both stress levels naturally arises. Another approach that is often used to account for the change in material modulus with temperature is the use of dynamic mechanical analysis (DMA) data (Ref 7). Although this approach may be a more useful indication of instantaneous modulus variation with temperature than HDT, it is unable to account for the time-dependent nature of most applications. For purposes of predicting part performance and for material selection, tensile creep data are the desired measurements.

To be useful for preliminary part design and material selection, creep data must be converted to simple information such as "deformation maps." A simple method is summarized where linear elastic part deformations are simply magnified by the use of deformation maps thus accounting for time and temperature effects.

A deformation map is produced directly from creep data (Ref 8). For a given temperature, T, the measured time-dependent strain, $\varepsilon(t)$ is divided by the applied stress, σ, to determine the creep compliance, J, as:

$$J(T,t) = \frac{\varepsilon(t)}{\sigma} \qquad \text{(Eq 3)}$$

The creep compliance is then normalized by dividing by the room temperature (T_0), instantaneous ($t \rightarrow 0$), elastic compliance, $J(T_0, 0)$:

$$\hat{J} = \frac{J(T,t)}{J(T_0, 0)} \qquad \text{(Eq 4)}$$

Because $J(T_0, 0)$ is the inverse of the room-temperature elastic modulus, E:

$$\hat{J} = J(T,t) \cdot E \qquad \text{(Eq 5)}$$

and

$$\hat{E} = \frac{1}{\hat{J}} \qquad \text{(Eq 6)}$$

Thus, a deformation map in time and temperature space can be produced from creep data with lines of constant compliance and modulus (Fig. 8). Thus, the design process involves calculating the linear elastic part deformation using E and then magnifying that deformation by $\hat{J}$ for the time of loading and ambient temperature. When a constant displacement is applied to a part, the calculated linear elastic stress using E is then reduced by $\hat{E}$ for the time of interest and ambient temperature. Thus, the deformation map provides a simple method to predict the time-dependent performance of plastic parts. As shown in Fig. 8, the deformation map provides the material response that can be combined with a linear elastic, time-independent analysis (in this case a finite-element stress analysis) to predict the time-dependent deformation. Validation of this approach is demonstrated by comparing it to experimentally measured part deformations (Fig. 8).

Fatigue-Cycle-Dependent Part Performance. An understanding of the deformation and fracture behavior of plastics subjected to cyclic loading is needed to predict the lifetime of structures fabricated from thermoplastics. This fatigue behavior is of concern because failure at fluctuating load levels can occur at much lower levels than failure under monotonic loading. A significant amount of information exists on the fatigue behavior of plastics. Unfortunately, very little has been documented about the application of this understanding to the prediction of the fatigue behavior of plastic parts.

There are two distinct approaches to treating and measuring the fatigue of polymers. The first approach is the traditional measurement of the number of cycles to failure (N) as a function of the fluctuating load or stress (S), that is, S-N. The "load" that is controlled is the minimum and maximum force or displacement in tension or bending. The fluctuations have a certain frequency and waveform. From a design viewpoint, it is difficult to predict part performance with these data because an enormous number of variables must be taken into consideration as well as various environmental conditions and a wide variety of materials.

The second approach to treating the fatigue of plastics is cyclic crack propagation. The use of fracture mechanics in cyclic fatigue involves the measurement of the amount of crack growth per cycle as a function of the stress-intensity factor. The fundamental addition here is the treatment of the crack length and thus an improved understanding of a fatigue mechanism. However, the same large number of variables that apply to the traditional fatigue (S-N) approach apply to the crack-propagation approach. In addition, the design engineer is challenged with determining the initial or inherent flaw size.

Even though cycle-dependent part performance is not well understood, a general design-engineering approach can be applied to the fatigue of plastic parts. First, for material selection an awareness of the fatigue performance of numerous plastics is necessary. Materials should be compared under identical test conditions to determine their relative fatigue performance. This preliminary selection should be based on the general assessment of the relative fatigue performance, taking into account the overall severity of the part loading. Next, the part loading conditions should be determined and related to the appropriate laboratory data. This task is probably the most important, yet the most difficult due to the large number of variables involved. Establishing whether the part will experience load-controlled or displacement-controlled cyclic loadings is possibly the most significant factor. Next, the effects of frequency, waveform, and load level and type must be assessed to determine if part temperature will increase, leading to thermal fatigue, or if mechanical failure will occur with little or no temperature increase. Other conditions that should be considered or matched from the laboratory specimen to the component include environmental effects (e.g., temperature), stress state, stress concentrations, and mean stress. Finally, appropriate laboratory tests or full-scale component tests should be conducted. These laboratory tests must be carefully planned to achieve correspondence to the actual service conditions.

Fracture mechanics can be used to provide an approach to predicting the fatigue lifetime of components. The important additional feature is an understanding of crack growth through meas-

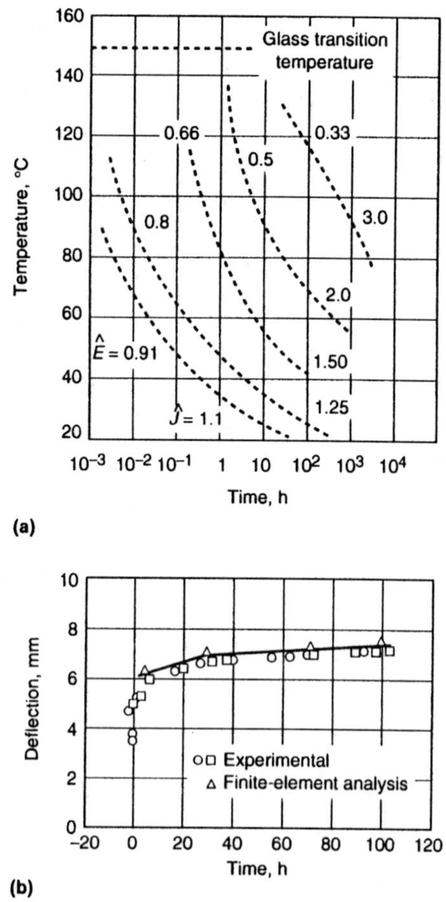

Fig. 8 Deformation map (a) used to predict PC part deformation at 82 °C (180 °F). $\hat{E} = E(T,t)/2350$ MPa. (b) Comparison of cathode-ray-tube housing creep prediction

urement of the amount of crack growth per cycle (da/dN) as a function of the cyclic range of stress-intensity factors (ΔK). Despite the fact that plastics are time-dependent materials, and that linear fracture mechanics only apply strictly to elastic materials, it appears that crack-propagation rates in many polymers can be correlated with ΔK.

During the fatigue process, the stress amplitude ($\Delta\sigma$) usually remains constant and failure occurs as the result of crack growth from an initial, subcritical size to a critical size related to the fracture toughness (K_c) of the material. The lifetime of a component is thus dependent on the

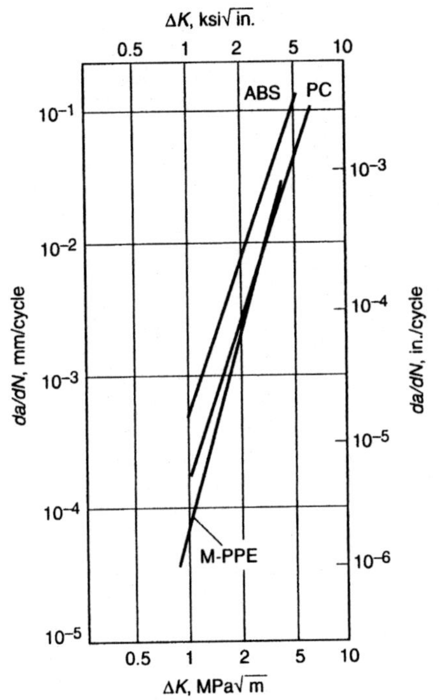

Fig. 9 Fatigue-crack-propagation behavior. ABS, acrylonitrile-butadiene-styrene; PC, polycarbonate; M-PPE, modified polyphenylene ether

initial crack size, the rate of crack growth, and the critical crack size. The relation takes the power-law form:

$$\frac{da}{dN} = A\,\Delta K^n \qquad \text{(Eq 7)}$$

where A and n are material constants varying with temperature, environment, and frequency. The stress-intensity factor range is given as:

$$\Delta K = Y(\Delta\sigma)\sqrt{a} \qquad \text{(Eq 8)}$$

where Y is a crack and structural geometry factor and a is crack length. Typical crack-propagation curves for a number of plastics (Ref 9) are shown in Fig. 9.

Fatigue lifetime of plastic parts can be calculated for design purposes by integrating the crack-growth rate expression (Eq 7) after substitution of Eq 8:

$$\frac{da}{dN} = AY^n\,\Delta\sigma^n\,a^{n/2} \qquad \text{(Eq 9)}$$

Assuming that the geometry factor Y does not change as the crack grows, this equation can be integrated to give the number of cycles to failure (N_f) that is necessary for the crack to grow from its initial size a_i to the critical size a_f. For $n \neq 2$:

$$N_f = \frac{2}{(n-2)AY^n\,\Delta\sigma^n}\left(\frac{1}{a_i^{(n-2)/2}} - \frac{1}{a_f^{(n-2)/2}}\right) \qquad \text{(Eq 10)}$$

This expression can be used to predict the fatigue lifetime of a component with an initial defect of known size.

The fatigue lifetime (number of cycles to failure) of a part is strongly dependent on the applied load. S-N curves have been generated for a number of thermoplastics (Ref 10) at room temperature with a standard tensile specimen with a net cross section of 12.7 by 3.2 mm (0.5 by 0.125 in.). The tensile load was varied from a very small load (nearly zero) to various maximum loads (stresses). A sinusoidal waveform with a frequency of 5 Hz was used. Very little or no specimen heating occurred. By choosing S-N curves for the same materials—polycarbonate (PC), modified polyphenylene ether (M-PPE), and acrylonitrile-butadiene-styrene (ABS)—whose fatigue-crack-propagation behavior is displayed in Fig. 9, the S-N data can be combined with the crack-propagation data to compute the initial crack lengths (Eq 10). The final crack length a_f is computed from the fracture toughness of these materials. Thus, over the range of stresses for the S-N curves, the initial crack lengths can be computed. Ideally, these crack lengths would be independent of applied stress level. However, while there is some variation, the average crack length was computed and used in Eq 10 to "predict" the measured S-N data from the crack-growth-rate data. These results are shown in Fig. 10 for PC, M-PPE, and ABS. These data and this approach indicate the similarity of the S-N and crack-growth-rate methods of predicting part lifetime and suggest a method of utilizing both types of data.

Manufacturing Considerations

Flow Length Estimation. The ability to manufacture plastic parts using the injection-molding process is governed by the material behavior, part geometry, and processing conditions. Estimating the flow length of the resin into a mold of a given thickness is an important manufacturing consideration for the design engineer. One example of a generic tool (Diskflow) is capable of analyzing radial flow and quantifying effects of material, geometry, or process changes (Ref 11). This tool is composed of a numerical flow analysis, automatic mesh generator, and menu-driven pre- and postprocessors. No knowledge of simulation techniques is required, though a knowledge of injec-

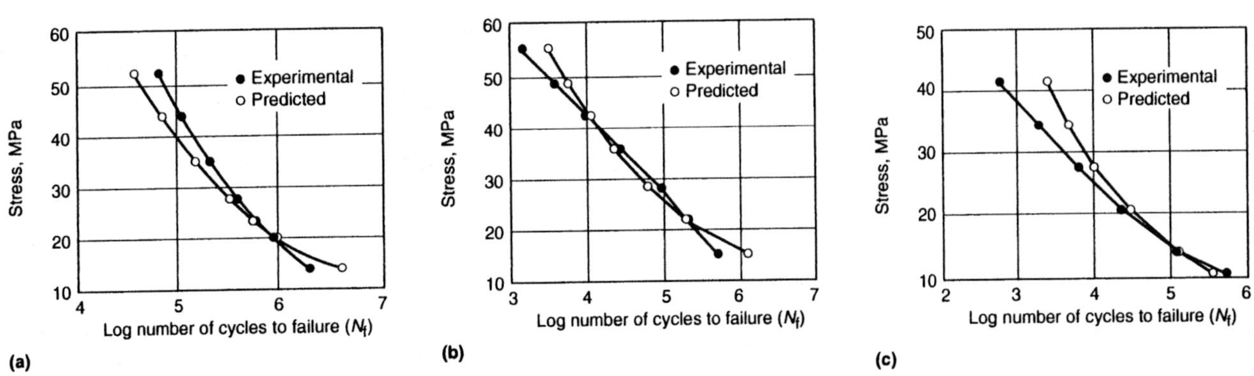

Fig. 10 S-N data compared to crack-growth prediction. (a) Polycarbonate (PC); a_i = 0.013 mm (0.5 mil). (b) Modified polyphenylene ether (M-PPE); a_i = 0.32 mm (12.5 mil). (c) Acrylonitrile-butadiene-styrene (ABS); a_i = 0.23 mm (9 mil)

tion molding is needed when interpreting the results.

For flow-length estimation, an initial flow rate is assumed constant subject to some user-specified maximum pressure limit that mimics the capability of a molding machine. As the mold fills at a constant volumetric flow rate, the injection pressure rises due to the increasing flow resistance. When the injection pressure attains the user-specified maximum, the analysis switches over to a second phase in which the injection pressure is maintained at a constant value and the flow rate is allowed to vary; the flow rate eventually decays to zero, at which point a final flow length is attained. The flow length may be defined as the farthest distance that a polymeric material travels in a mold of some nominal wall thickness given a set of processing conditions. The flow-length capability examines the feasibility of manufacturing a desired design: if the distance from the gate to the corner of the part is greater than the predicted flow length, then the part may not be manufacturable. Figure 11 shows the dependence of flow length on wall thickness for a maximum injection pressure of 103.5 MPa (15 ksi) for PC. This information is useful for assessing manufacturability in the early stages of design and material selection.

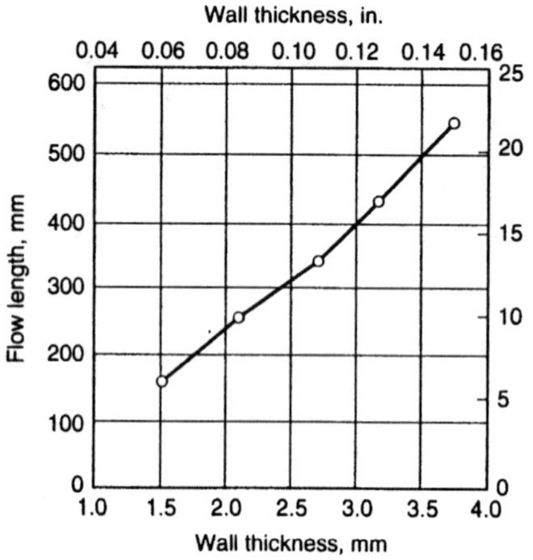

Fig. 11 Flow length versus wall thickness predicted by Diskflow mold-filling analysis. Material, unfilled PC; mold temperature, 82 °C (180 °F); melt temperature, 335 °C (635 °F); maximum injection pressure, 103.4 MPa (15 ksi)

Cycle Time Estimation. The molding of thermoplastics consists of injecting a molten polymer into the cooled mold cavity. The injected resin is held in the cavity until the part solidifies (by heat transfer). The time for the melt to cool until it solidifies to the extent that the part can be removed from the mold and retain its dimensions is generally the majority of the total cycle time. The large impact of the cooling time on the total processing cost is obvious.

During the cooling phase, heat conduction is the prime mechanism of heat transfer. The development of a simplified mold-cooling program allows designers and molders to evaluate materials and process parameters in a rapid, convenient, and cost-efficient manner. Plastic parts are usually thin, and thus a one-dimensional, transient heat-conduction analysis is adequate to approximate the cooling of the real part. The main assumption is that the mold surface is kept at a constant temperature throughout the cooling phase. Comparing calculated minimum cooling times for different material part geometries (i.e., thickness) and processing conditions help optimize the material-selection process.

Thermal material properties are strong functions of temperature. Because the thermoplastic material experiences a wide range of temperatures during the cooling phase, temperature-dependent material data such as specific heat and thermal conductivity are used for the computations. To perform the analysis the injection temperature, mold temperature, ejection temperature, material, and thickness must be chosen. The program uses a one-dimensional finite-difference scheme to calculate temperature through the thickness as a function of time. When the center of the plate reaches the specified ejection temperature, the analysis is stopped and the results are displayed graphically. By performing the analysis for a range of part thicknesses, cooling-time curves can be produced (Fig. 12). These curves can then be used to estimate cycle times in the early stages of material selection and design.

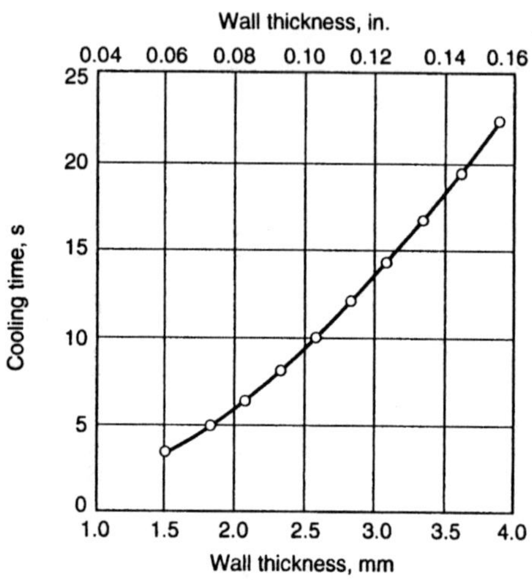

Material	Unfilled PC
Thickness, mm	1.62–3.81
Thermal conductivity, W/m · K	0.270
Specific heat, W · s/kg · K	1791
Melt temperature, °C	300
Mold temperature, °C	82
Ejection temperature, °C	112

Fig. 12 In-mold cooling time versus wall thickness predicted from one-dimensional, transient mold cooling analysis

Design-Based Material Selection

Design-based material selection (Ref 12, 13) involves meeting the part performance requirements with a minimum system cost while considering preliminary part design, material performance, and manufacturing constraints (Fig. 13). Some performance requirements such as transparency, Food and Drug Administration (FDA) approval, or flammability rating are either met by the resin or not. Mechanical performance such as a deflection limit for a given load are more complicated requirements. Time-

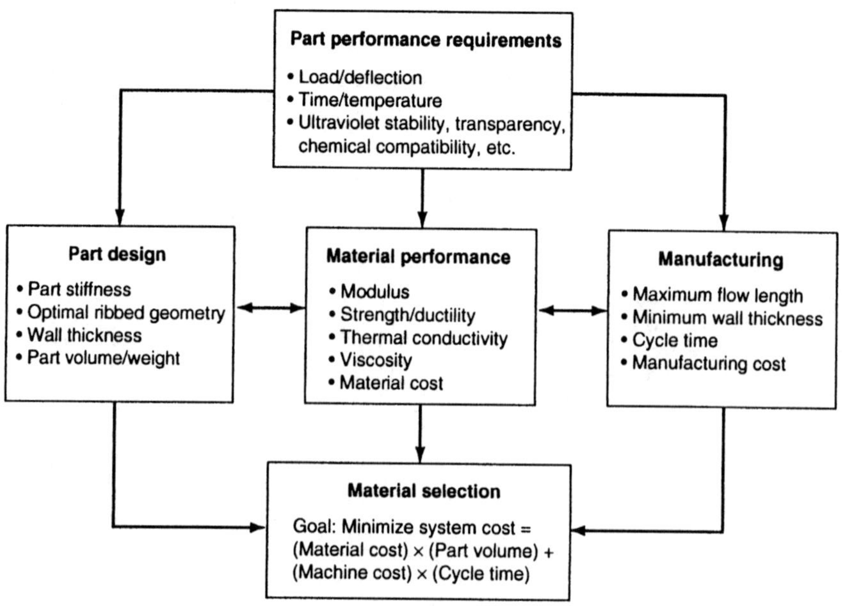

Fig. 13 Design-based material-selection process

and temperature-reduced stiffness of the material is determined from the deformation map. Part design for stiffness involves meeting the deflection limit with optimal rib geometry and part thickness combined with the material stiffness. This part geometry can be used to compute the part volume that when multiplied by the material cost provides the first part of the system cost. The second half of the system cost is the injection-molding machine cost multiplied by the cycle time. This total system cost is a rough estimate used to rank materials/designs that meet the part performance requirements. In addition, the manufacturing constraint of flow length for the part thickness must be considered. The entire process is summarized in Fig. 13.

Example 1: Materials Selection for Plate Design. A simple example is presented to illustrate the design-based material-selection process. A 254 by 254 mm (10 by 10 in.) simply supported plate is loaded at room temperature with a uniform pressure of 760 Pa (0.11 psi). The maximum allowable deflection is 3.2 mm (0.125 in.). Using a modeling program, the nonlinear load-displacement response of the plate can be computed. Through iteration, it is determined that a PC plate with a thickness of 2.5 mm (0.1 in.) satisfies the requirements (Fig. 3). From Fig. 11, the flow length is 320 mm (12.5 in.). Thus, the plate could be filled with a center gate or from the center of an edge. From Fig. 12, the in-mold cooling time is 10 s. The volume of the plate is 0.00016 m^3 (10 $in.^3$).

A second design can be produced by designing a rib-stiffened plate. Again, through iteration, a 1.5 mm (0.060 in.) thick plate with 10 ribs in each direction with a rib height of 4.5 mm (0.18 in.) and a rib thickness of 1.5 mm (0.060 in.) would meet the deflection requirement. From Fig. 11, the flow length is about 175 mm (7 in.). Thus, because a center-gated plate would have a flow length of 175 mm (7 in.), the part would probably fill if the ribs would serve as flow leaders to aid the flow. However, it is generally not recommended to push an injection-molding machine to its limits because this will exaggerate inconsistencies in the material and the process. A more thorough three-dimensional process simulation should be performed to determine the viability of this design before it is chosen. From Fig. 12, the in-mold cooling time is about 4 s, a considerable savings (6 s/part) in cycle time as compared to the plate with no ribs. In addition, the volume of the ribbed plate is 0.00013 m^3 (8 $in.^3$), a savings of 20% on material as compared to the plate with no ribs. The system cost of the ribbed plate is computed to be 73% of the plate with no ribs (Fig. 1). Because the ribs would produce a constrained, three-dimensional stress state, consideration of impact would be important for high rates of loading and low temperature (Fig. 7). The fracture map shows a tendency for brittle behavior with PC at low temperature and high loading rates for notched or constrained geometries.

If time/temperature performance were added to this example as a requirement, the optimal material may change or the initial design would need to be modified. If the same load were applied to the plate for 1000 h at a temperature of 79 °C (175 °F), the PC plate would exhibit a deformation as if its material stiffness were about 40% of the room-temperature modulus (Fig. 8). Simply increasing the thickness of the plate with no ribs to 3.5 mm (0.136 in.) would provide a design that would meet the deflection requirements. The penalty would be a 40% increase in material usage and an additional 8 s added to the cycle time. Choosing a material with more temperature resistance or initial stiffness is an option.

Example 2: Materials Selection for an Electrical Enclosure. The usefulness of this process can be demonstrated through another design example. In this case, a very simple five-sided box is chosen. The box is used as an electrical enclosure and must meet flammability requirements. This limits the number of candidate materials to examine more closely. Also, this enclosure is not painted, and therefore the resin must be unfilled to maintain acceptable aesthetics. It is unribbed to minimize sink marks on the exposed surfaces. Finally, it must support a uniform load across its surface without deflecting more than 2.5 mm (0.10 in.). The enclosure is a 300 mm wide by 450 mm long by 100 mm high (12 by 18 by 4 in.) box (Fig. 14).

A series of analyses is performed using three resins to see how they perform under different conditions. These resins are representative of what is currently used in electrical enclosures (computer housings, office equipment, etc.). They are an unfilled M-PPO resin, an unfilled ABS resin, and an unfilled PC-ABS resin blend.

To examine the relative performance of each resin, the application requirements are varied in loading, environment, and manufacturing. First, the uniform load is varied from 150 to 1200 Pa (0.02 to 0.17 psi). Next, the ambient temperature the enclosure must withstand for 1000 h under load is varied from 20 to 80 °C (68 to 175 °F). Finally, the gating scenario is changed from edge gated to center gated to multiple gates.

Using a center-gated box at 40 °C (105 °F) for 1000 h, the uniform load is varied from 150 to 1200 Pa (0.02 to 0.17 psi). For each resin the optimal wall thickness is determined to support the load at the lowest variable system cost for each loading case. Figure 15(a) compares the normalized cost of the enclosure for each resin

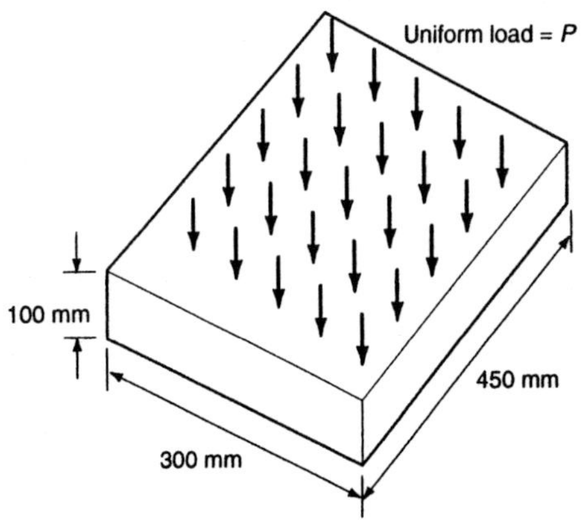

Fig. 14 Geometry of enclosure example

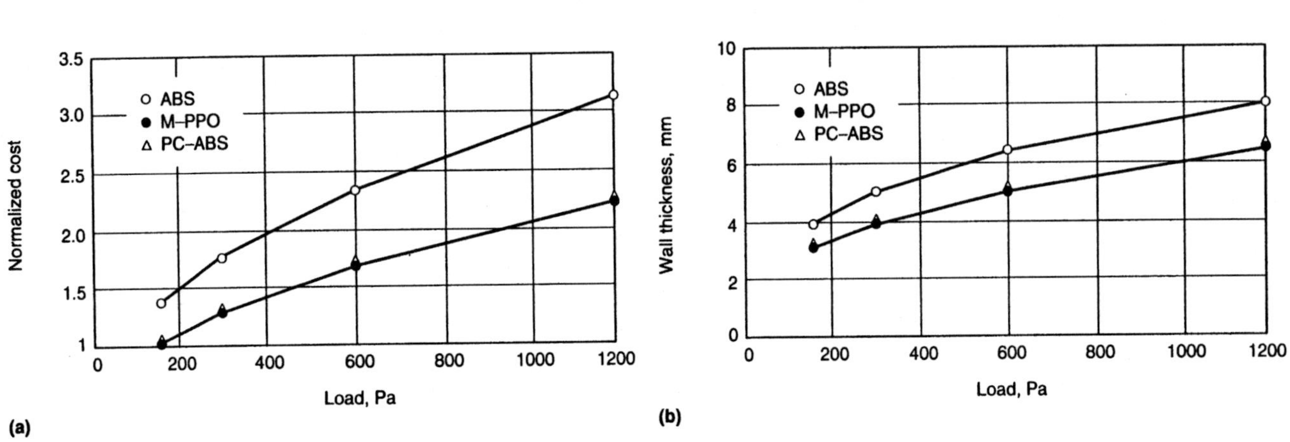

Fig. 15 Loading variation for 40 °C (105 °F) and 1000 h. ABS, acrylonitrile-butadiene-styrene; PC, polycarbonate; M-PPO, modified polyphenylene oxide

as the load is increased. As can be seen from this graph, the PC-ABS and M-PPO are virtually equivalent in cost, while the ABS is about 30% more expensive. While this may seem counterintuitive (ABS is less expensive per pound than PC-ABS or M-PPO), it is easily explained by examining Fig. 15(b), wall thickness versus loading. At this elevated temperature and long time (40 °C, or 105 °F, 1000 h), the ABS requires significantly more material to support the required load within the specified 2.5 mm deflection than either the PC-ABS or the M-PPO. This added material far outweighs the price advantage of ABS.

The cooling time is another factor that will increase the variable system cost of the ABS resin enclosure. As the wall thickness increases, the time to cool the part to ejection temperature will increase. The cooling time is also influenced by the thermal properties of each resin. Figure 16 contains a graph of the cooling time versus wall thickness for the three example materials based on one-dimensional transient heat-transfer analyses. The wall thickness for each resin to support 600 Pa (0.09 psi) at a deflection of no more than 2.5 mm (0.10 in.) is indicated on the graph. From this graph, it can easily be seen that, in this case, the cooling time for each resin will be very different.

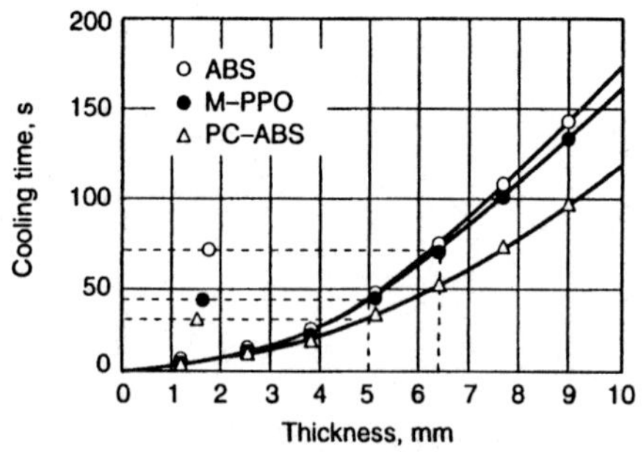

Fig. 16 Cooling time versus wall thickness. ABS, acrylonitrile-butadiene-styrene; PC, polycarbonate; M-PPO, modified polyphenylene oxide

Using a center-gated box that must support a 300 Pa (0.04 psi) load within a 2.5 mm (0.10 in.) deflection of 1000 h, the temperature was varied from 20 to 80 °C (68 to 175 °F). Figure 17(a) compares the normalized cost of these three resins as the temperature is increased. Initially, at 20 °C (68 °F) these resins have very similar variable system costs. As the temperature increases, the creep performance of each resin decreases. Figure 17(b) shows the creep modulus for each resin as the temperature changes. The creep modulus of the ABS resin decreases rapidly as temperature increases. The M-PPO maintains its stiffness longer, but eventually decreases rapidly while the PC-ABS performs better, because of the high creep resistance of the PC component of the blend.

The wall thickness to support the load must increase as temperature increases because the creep modulus decreases. This, in turn, increases the part volume and the cooling time, affecting the variable system cost. As the temperature increases, the cost rises to high levels (ABS at 80 °C, or 175 °F, 1000 h). If the application must withstand these temperature extremes, a higher-performance thermoplastic may be a better choice.

The process to manufacture this enclosure can influence how the enclosure will be designed and what material will be used. Using a box that must support a 150 Pa (0.02 psi) load within a 2.5 mm (0.10 in.) deflection in a 40 °C (105 °F) environment for 1000 h, the gating scenario is varied choosing three common configurations (Fig. 18): edge gate, center gate, and four gates. The minimum flow length necessary to fill the part is determined for each case based on the geometry of the enclosure and the gate position. The minimum wall thickness to allow each material to achieve this flow length, determined using the radial flow injection-molding simulation, is then used as a lower bound on the thickness optimization and is shown in Fig. 19(b). Figure 19(a) details the normalized cost versus minimum flow length (i.e., gating scenario). Initially, as the flow length increases (from four gates to center gate) the normalized cost does not change. The wall thickness necessary to support the load within the specified deflection is greater than the minimum wall thickness dictated by the flow-length constraint. As the flow length increases from the center-gated to the edge-gated case, the normalized cost increases because the wall thickness is now dictated by the manufacturing constraint rather than the loading condition. The gate placement now dictates the wall thickness that is necessary to fill the part.

There are other considerations that a design engineer can use to help determine the best material for an application. The strength of a resin over a range of temperatures may aid the engineer in determining if the part will fail under load. The impact performance of the resin, as indicated by the ductility ratio, can also be quite important. While it only indicates the impact performance for one specific geometry, and cannot be used in design, it does provide useful comparative information.

Conclusions

Material selection and engineering design of plastic parts can be a difficult task when there is a lack of effective and efficient design methods and the associated material data. However, methods are available to improve the design process by providing more accurate and effective predictive techniques. Fracture maps indicate the relative ductility of a material as a function of temperature and strain rate for a relatively severe stress state. A range of test data for different stress states from tensile tests, disk tests, and notched beams is used to predict part deformation and potential ductile-to-brittle behavior. For time-dependent deformation, such as creep or stress relaxation, deformation maps

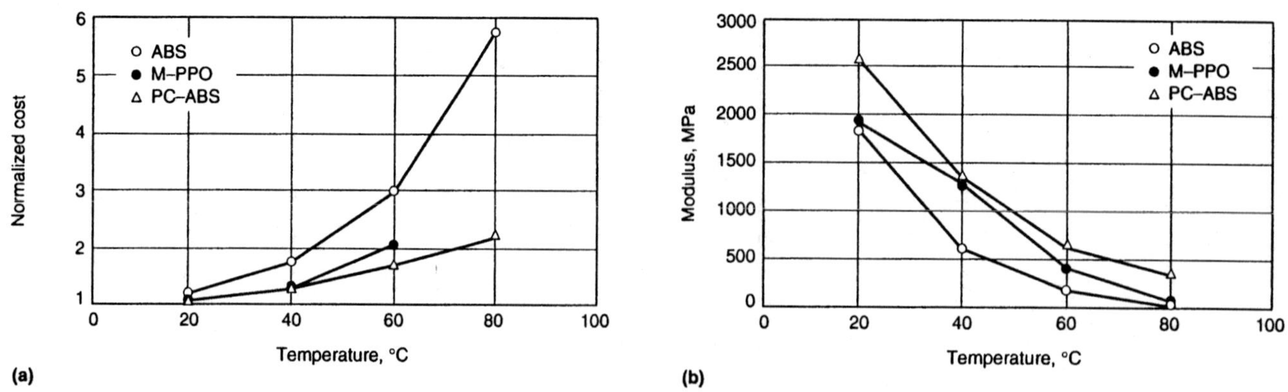

Fig. 17 Temperature variation. ABS, acrylonitrile-butadiene-styrene; PC, polycarbonate; M-PPO, modified polyphenylene oxide

can be combined with linear elastic calculations of part deformation to predict the time- and temperature-dependent deformation of the part. The cross-flow stiffness and strength of injection-molded glass-filled materials is sometimes only 50% of the stiffness and strength in the flow direction, especially for thin-walled parts. This must be accounted for in predicting part stiffness and strength. For predicting lifetime of parts subjected to cyclic loading, the combination of *S-N* data and crack-growth-rate data is useful because it provides two options: to use the *S-N* data directly or to use the initial defect size with the crack-growth-rate data. In either case, with the vast number of parameters that affect fatigue behavior, having more information is useful. The design methods and material data summarized here describe some effective and efficient techniques to select materials and design plastic parts.

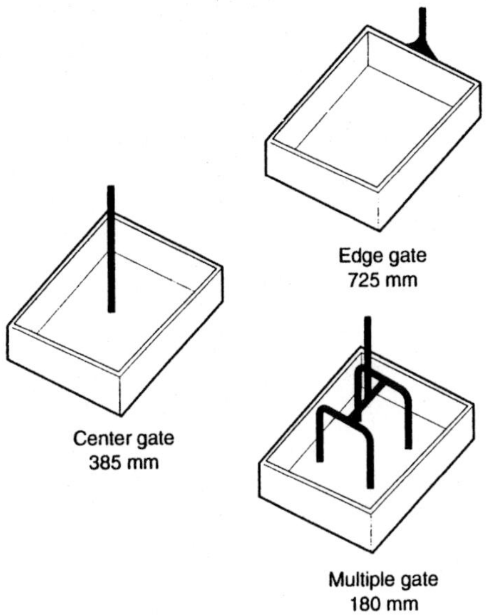

Fig. 18 Examples of gating scenarios

REFERENCES

1. G.G. Trantina and D.A. Ysseldyke, An Engineering Design System for Thermoplastics, *1989 ANTEC Conf. Proc.*, Society of Plastics Engineers, p 635–639
2. E.H. Nielsen, J.R. Dixon, and M.K. Simmons, "GERES: A Knowledge Based Material Selection Program for Injection Molded Resins," ASME Computers in Engineering Conference (Chicago), American Society of Mechanical Engineers, July 1986
3. G.G. Trantina and R.P. Nimmer, *Structural Analysis of Thermoplastic Components*, McGraw-Hill, 1994
4. K.C. Sherman, R.J. Bankert, and R.P. Nimmer, Engineering Performance Parameter Studies for Thermoplastic, Structural Panels, *1989 ANTEC Conf. Proc.*, Society of Plastics Engineers, p 640–644
5. G. Ambur and G.G. Trantina, Structural Failure Prediction with Short-Fiber Filled Injection Molded Thermoplastics, *1988 ANTEC Conf. Proc.*, Society of Plastics Engineers, p 1507
6. J.T. Woods and R.P. Nimmer, Design Aids for Preventing Brittle Failure in Polycarbonate and Polyetherimide, *1996 ANTEC Conf. Proc.*, Society of Plastics Engineers, p 3182–3186
7. M.P. Sepe, Material Selection for Elevated Temperature Applications: An Alternative to DTUL, *1991 ANTEC Conf. Proc.*, Society of Plastics Engineers, p 2257–2262
8. O.A. Hasan and G.G. Trantina, Use of Deformation Maps in Predicting the Time-Dependent Deformation of Thermoplastics, *1996 ANTEC Conf. Proc.*, Society of Plastics Engineers, p 3223–3228
9. R.W. Hertzberg and J.A. Manson, *Fatigue of Engineering Plastics*, Academic Press, 1990
10. G.G. Trantina, Material Properties for Part Design and Material Selection, *1996 ANTEC Conf. Proc.*, Society of Plastics Engineers, p 3170–3175
11. D.O. Kazmer, Development and Application of an Axisymmetric Element for Injection Molding Analysis, *1990 RETEC Conf. Proc.*
12. G.G. Trantina, P.R. Oehler, M.D. Minnichelli, Selecting Materials for Optimum Performance, *Plast. Eng.*, Aug 1993, p 23–26
13. P.R. Oehler, C.M. Graichen, and G.G. Trantina, Design-Based Material Selection, *1994 ANTEC Conf. Proc.*, Society of Plastics Engineers, p 3092–3096

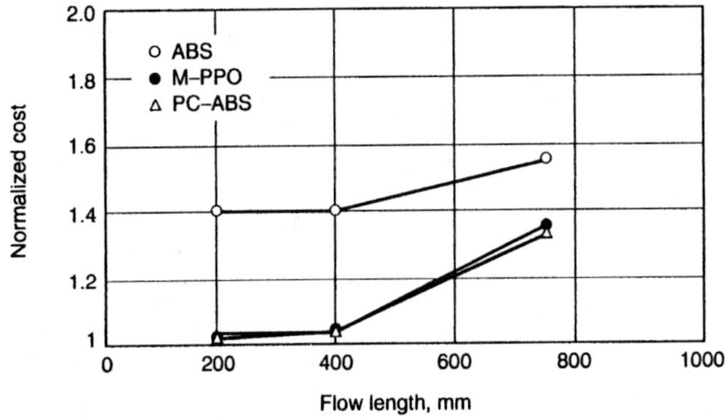

(a)

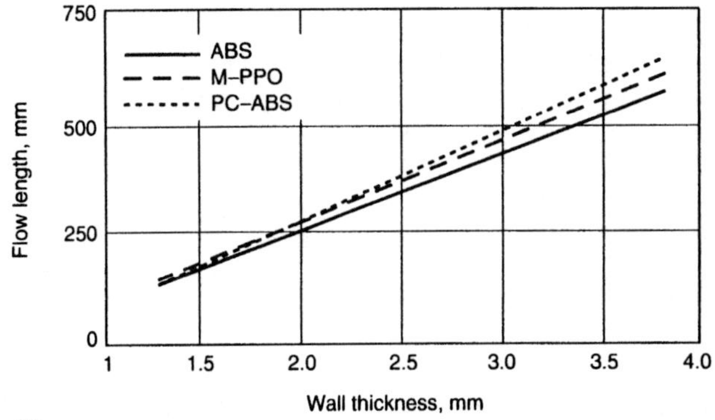

(b)

Fig. 19 Gating variations. ABS, acrylonitrile-butadiene-styrene; PC, polycarbonate; M-PPO, modified polyphenylene oxide

Design and Selection of Plastics Processing Methods

THE PRODUCTION of quality plastic parts is influenced by a number of factors, as shown in Fig. 1. These factors determine whether a plastic part meets functional requirements and is durable enough to survive years of use. In the design phase, the key factors to consider are:

- Plastic material(s) to be used
- Product shape and features
- Production process
- End-use applications

The product designer must also consider that the plastic molding or forming process influences the plastic part performance. The physical, mechanical, and chemical properties of the material can be affected by the molding/forming process. The part designer needs to understand the rudiments of plastic processing methods in order to select a plastic material, define the specific shape of the part, and define the process used to manufacture the plastic product.

This article describes key processing methods and related design, manufacturing, and application considerations for plastic parts; it includes discussion of materials and process selection methodology for plastics. Because plastics properties are highly influenced by the methods of processing and the process conditions, appropriate design for end-use applications requires proper material selection and process selection.

Plastics Processing Methods*

The primary plastics processing methods are:

- Injection molding
- Extrusion
- Thermoforming
- Blow molding
- Rotational molding
- Compression molding/transfer molding
- Composites processing
- Casting

*Adapted from Edward A. Muccio, Design for Plastics Processing, *Materials Selection and Design,* Volume 20, *ASM Handbook,* ASM International, 1997, pages 793 to 803

Other plastics processing methods exist, but most are variants of these processes. Table 1 lists characteristics and capacities of processing methods used for thermoplastic and thermoset parts.

Plastics processing is a form conversion process. The material that enters the process as plastic pellets or powder is basically the same material that exits the process as a plastic part. The plastic process converts the shape of the plastic material. However, this simple explanation of plastic processing needs to be slightly modified. Although the plastic entering the process is the same plastic exiting the process, the properties of the plastic material may be affected by the rigorous activities that occur during the process. The resulting properties of the plastic part may be different from the properties of the plastic material as defined by the plastic material manufacturer.

Each processing method can have a different effect on the final properties. Following is a brief description of the primary plastic processing methods and a summary of how each process influences part design and the properties of the plastic part.

Injection Molding

Injection molding, and all its variants, is the most popular process for producing plastic products. Designers prefer the injection molding process because, in addition to being fast and cost effective, it allows the designer the opportunity to create true three-dimensional part shapes. (Many plastic processes, such as extrusion, blow molding, thermoforming, and rotational molding, do not allow the designer to control all surfaces of the plastic part being manufactured. One surface is a function of the process, not the product design; some examples include the inside of a hollow container produced by blow or rotational molding, the length of an extruded profile, and the outer surface of a thermoformed part produced on a female mold.)

Product designers desire control over all aspects of the design of a product, and injection molding allows this to occur. Additionally, injection molding allows the designer to incorporate product design features such as holes, snaps, color, texture, and symbolization that might demand secondary operations if the design were manufactured using materials such as metal, wood, or ceramic.

The injection molding process involves several steps:

- Feed and melting of the plastic pellets
- Metering of the plastic melt
- Injection of the plastic melt into the mold
- Cooling and solidifying of the plastic in the mold
- Ejection or removal of the molded part from the mold

The following description of these steps is based on the processing required to mold a simple part such as the polystyrene poker chip shown in Fig. 2.

Feed and Melting of the Plastic Pellets. The polystyrene, in the form of pellets, is fed into the throat of the injection molding machine (Fig. 3). Initially, the plastic pellets are heated by the electric heater bands; however, the shear and friction created by turning the injection molding machine screw will provide the majority of the energy required to melt the plastic.

As the screw turns, the plastic pellets melt, and the melted material is conveyed toward the discharge end of the injection unit.

Metering of the Plastic Melt. As the plastic melt is conveyed forward through the barrel of the molding machine, it is allowed to pass through a nonreturn valve that prevents the plastic melt from traveling rearward or back through the valve. The plastic melt that moves through the valve and in front of the screw will push the screw rearward. This rearward motion of the screw, while the screw is turning, creates more shear and facilitates the melting of the plastic pellets. The amount of plastic melt that is allowed to move through the valve and reside in front of the screw is defined by a limit switch or stopping point assigned by the molding technician. The plastic melt in front of the screw will be the material that is injected into the mold to produce the plastic parts.

Injection of the Plastic Melt into the Mold. Injecting the plastic melt into the closed mold requires high pressures (between 35 and 205 MPa, or 5 and 30 ksi, on the plastic material) and often high speeds. The specific values for injecting the plastic melt are a function of the melt viscosity of the plastic material, the mold design, and the plastic product design. To allow the injection of plastic into the mold, the part designer must consider design features such as the wall thickness and gate type and location.

Wall thickness, the thickness of the major portion of the wall of the plastic part, depends on the melt characteristics (melt viscosity) of the plastic. A plastic part with thin walls (<1 mm, or 0.04 in., thick) will usually require higher molding pressures than a plastic part with a wall thickness of about 4 mm (0.16 in.). Thicker wall sections (>6 mm, or 0.25 in., thick) may result in poor part quality and molding defects such as underfill or sink marks.

Gate Type and Location. The gate (Fig. 4) is the point where the plastic melt is allowed to enter the cavity to form the part. The gate is designed to cool or freeze after the cavity has been filled and packed with plastic. This cooling prevents any plastic melt from exiting the filled cavity.

Cooling and Solidifying of the Plastic in the Mold. Plastic materials are thermal insulators; that is, they tend *not* to absorb or release thermal energy at a rapid rate. The plastic part designer must avoid thick wall sections to avoid cooling problems in the mold. Specifically, parts with thicker wall sections require a longer cooling time within the mold; additionally, the thick sections may distort, have sink marks, or contain voids (Fig. 5).

To avoid these problems, the designer must strive for a nearly constant thickness of every section of the part. This *nominal* thickness must meet the application requirements of the part, ensure nearly uniform cooling, and be fillable by the plastic material selected. As an example, a plastic material manufacturer may suggest a nominal wall thickness of 4.5 mm (0.18 in.) for a specific plastic. The plastic part designer is not bound to make all the walls this thickness, but should design the wall to average this dimension. The wall thickness may vary, but only at a reasonable rate of change (Fig. 6).

Ejection or Removal of the Molded Part from the Mold. To allow an injection-molded part to be removed from the mold requires that the part designer consider ejection surfaces and draft.

Ejection surfaces on the part provide an allowance for ejector pins to push the part out of the mold (Fig. 2). The ejector pins or other mold components such as inserts and slides will leave

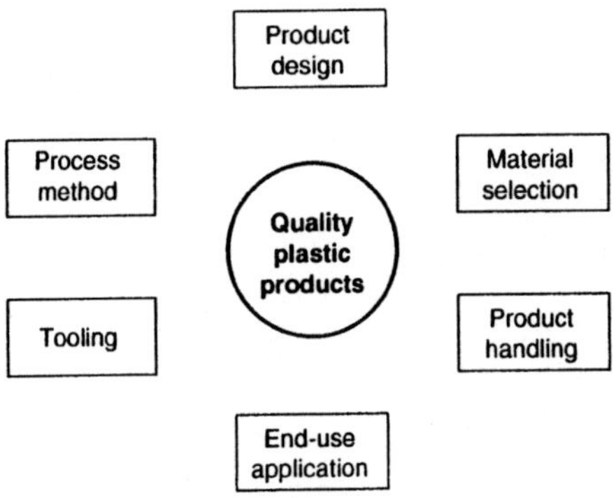

Fig. 1 Key factors in the development and production of quality plastic parts

Table 1 Thermoplastics and thermoset processing comparison

Process	Process pressure		Maximum equipment pressure		Maximum size		Pressure limited	Ribs	Bosses	Vertical walls	Spherical shape	Box sections	Slides/cores	Weldable	Good finish, both sides	Varying cross section
	MPa	ksi	MN	tonf	m^2	ft^2										
Thermoplastics																
Injection	15–45	2–7	30	3370	0.75	8.0	y	y	y	y	n	n	y	y	y	y
Injection compression	20	2.9	30	3370	1.5	16	y	y	y	n	n	n	y	y	y	y
Hollow injection	15	2.2	30	3370	2.0	20	y	y	y	y	n	y	y	y	y	y
Foam injection	5	0.7	15	1690	3.0	30	y	y	y	...	n	y	y	y	y	y
Sandwich molding	20	2.9	30	3370	1.5	16	y	y	y	y	n	n	y	y	y	y
Compression	20	2.9	30	3370	1.5	16	y	y	y	y	n	n	y	y	y	y
Stamping	20	2.9	30	3370	1.5	16	y	n	n	n	n	n	n	y	y	n
Extrusion	n/a	n/a	n/a	n/a	n/a	n/a	n/a	y	n	n/a	n	y	n	y	y	y
Blow molding	1	0.15	10	1120	2.0	20	n	n	n	y	y	y	y	y	n	n
Twin-sheet forming	1	0.15	10	1120	6.0	65	n	n	n	y	y	y	n	y	n	n
Twin-sheet stamping	1	0.15	30	3370	6.0	65	n	n	n	n	n	y	n	y	y	n
Thermoforming	0.1	0.015	n/a	n/a	...	...	n	n	n	y	n	n	y	y	n	n
Filament winding	0	0	n/a	n/a	...	...	n/a	y	n	y	y	y	n	y	n	y
Rotational casting	0.1	0.015	n/a	n/a	...	...	n	n	n	y	y	n	n	y	n	n
Thermoset plastics																
Compression																
Powder	60	8.7	30	3370	0.5	5	y	y	y	y	n	n	y	n	y	y
Sheet molding compound	6–20	0.85–3	30	3370	4–5	45–55	y	y	y	y	n	n	y	n	y	y
Cold-press molding	1	0.15	30	3370	...	...	n	n	y	y	n	n	n	n	y	y
Hot-press molding	5	0.75	30	3370	6.0	65	y	n	y	y	n	n	n	n	y	y
High-strength sheet molding compound	4–10	0.60–1.5	30	3370	3.0	30	y	y	y	y	n	n	n	n	y	y
Prepreg	0.5–5	0.07–0.75	30	3370	6.0	65	y	n	n	y	n	n	n	n	y	y
Vacuum bag	0.1	0.015	n/a	n/a	...	...	n	n	y	y	n	y	n	n	n	y
Hand lay-up	0	0	n/a	n/a	...	...	n	n	y	y	n	y	n	n	n	y
Injection																
Powder	100	14.5	10	1120	0.1	1.1	y	y	y	y	n	n	y	n	y	y
Bulk molding compound	30	4.5	30	3370	1.0	11	y	y	y	y	n	n	y	n	y	y
ZMC	30	4.5	30	3370	1.0	11	y	y	y	y	n	n	y	n	y	y
Stamping	3	0.45	30	3370	6.0	65	y	n	n	y	n	n	n	n	y	n
Reaction injection molding	1	0.15	10	1120	...	...	y	y	y	n	n	y	n	y	y	
Resin transfer molding	0.1	0.015	10	1120	...	...	n	y	n	y	n	y	n	n	y	y
High-speed resin transfer molding or fast resinject	2	0.3	30	3370	...	...	n	y	n	y	n	y	n	n	y	y
Foam polyurethane	0.5	0.07	n/a	...	...	...	n	y	y	y	y	y	n	n	y	y
Reinforced foam	1	0.15	30	3370	3.0	30	y	y	y	y	n	y	n	n	y	y
Filament winding	n/a	n/a	n/a	n/a	...	...	n/a	y	n	y	y	y	n	n	(a)	y
Pultrusion	n/a	n/a	n/a	n/a	n/a	n/a	n/a	y	n	n/a	n	y	n	n	y	y

Note: y, yes; n, no; n/a, not applicable. (a) One side of filament-wound article will exhibit a strong fiber pattern.

a witness mark on the plastic part, which the plastic part designer needs to respect. Often the part specification will include a note that states "knockout witness to be flush to or 0.125 mm (0.005 in.) below the molding surface."

Draft is the angle in the wall design that facilitates ejection from the mold (Fig. 7).

Details and design considerations for injection molding include shrinkage, postmold shrinkage, and size and location of holes and other features.

Shrinkage occurs because the plastic melt volume is greater than the solid volume, and the plastic melt is packed into the mold under high pressures. Shrinkage needs to be understood in order to produce plastic parts with a high degree of dimensional stability. Different plastics experience different amounts of shrinkage (Table 2). Additives affect the shrinkage rate. Rate and direction of flow of the melt into the mold can influence shrinkage and may cause the same material to exhibit two different types of shrinkage depending on the part geometry.

Postmold Shrinkage. It is best to have any shrinkage occur while the plastic part is constrained by the mold. Shrinkage that occurs outside the confines of the mold after the part is ejected, known as postmold shrinkage, may be uncontrolled and/or unpredictable. The result could be a major dimensional problem for an injection-molded part. Postmold shrinkage is a function of both the plastic material and the process. Several semicrystalline plastics tend to exhibit a higher potential for postmold shrinkage.

If the injection molding process is not optimized, it can contribute to postmold shrinkage. For example, consider an injection molding process that has the plastic melt in the barrel at 260 °C (500 °F) and a mold temperature of 82 °C (180 °F). The desire for productivity gains, that is, output of more parts per hour, leads to cooling the mold to 38 °C (100 °F) and speeding up the cycle. The result of this process change may not be immediately visible. While output gains may be achieved, the lower mold temperature may cause a higher degree of molded-in (residual) stress. This increased stress may be relieved after the part is removed from the mold. Over the next hours, days, or weeks, the relieving of the stress may manifest itself as postmold shrinkage.

Holes and Other Features. Injection-molded part features can be expressed as a function of the nominal wall thickness (*T*) as shown in Fig. 8 and 9.

Extrusion

The extrusion of plastic material is, surprisingly, the process that utilizes the most plastic material, even more than injection molding. One reason for this great material consumption is that extrusion is one of the few continuous plas-

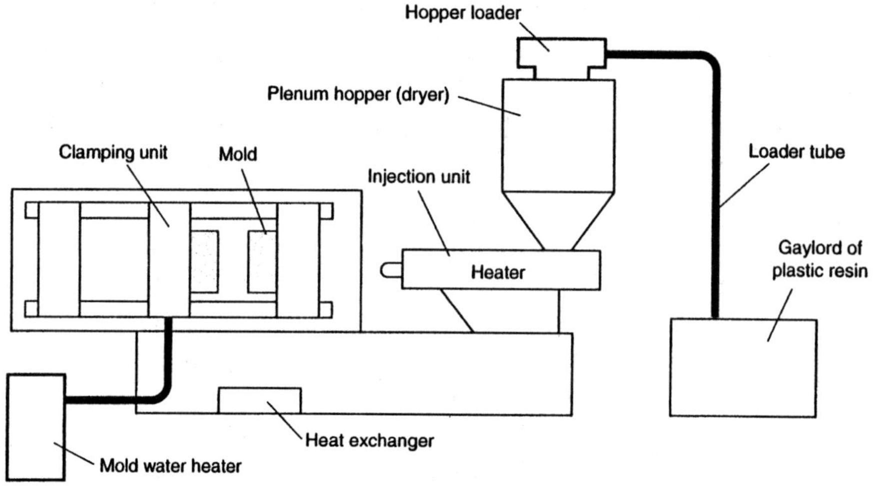

Fig. 3 Injection molding machine

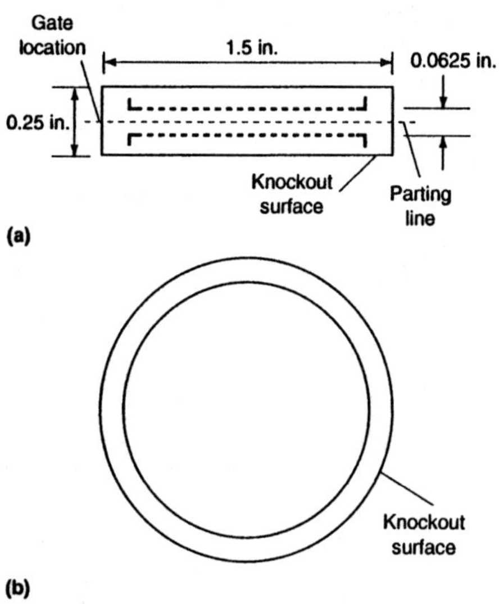

Fig. 2 Polystyrene poker chip. (a) Side view. (b) Bottom view

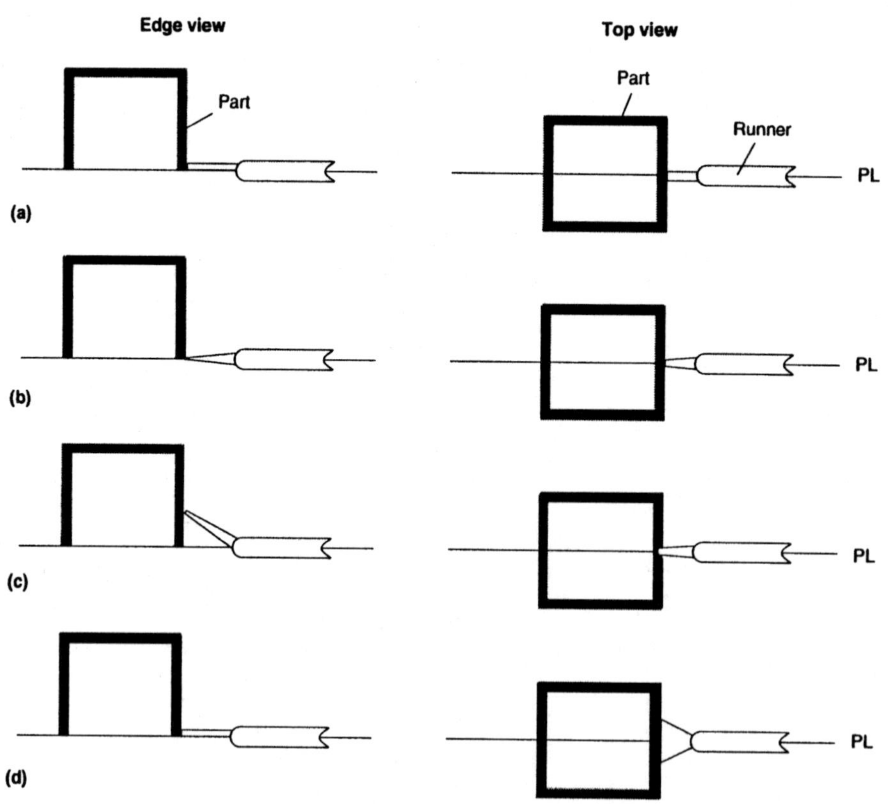

Fig. 4 Types of injection molding gates. (a) Tab gate. (b) Pinpoint gate. (c) Sub gate. (d) Fan gate. PL, parting line

tics processes. Other plastics processes are batch processes, relying upon repetition. Extrusion of plastic material is continuous, and the plastic product is cut and formed in a secondary process. Another reason is that extrusion is used to compound and produce the plastic pellets used in most other thermoplastic processing operations. For example, most plastic pellets used in the injection molding process are produced in an extruder at the plant of the material manufacturer.

The extruded product is designed as a two-dimensional cross-section shape, which is extruded in the third dimension. The third dimension is usually controlled by a cutoff operation. As an example, polyvinyl chloride (PVC) pipe is designed as two simple concentric circles. A die is fabricated, and the plastic melt is extruded through the die on a continuous basis (Fig. 10a). The length of the pipe is defined and created by cutting the continuous extrudate to the desired length.

Types of extruded parts can be categorized as:

- *Sheet* is a flat extruded profile greater than 0.0004 mm (0.010 in.) thick.
- *Film* is a flat extruded profile less than 0.0004 mm (0.010 in.) thick. Blown film (Fig. 10b) is a volume product used for trash bags, packaging, and wrappings. Cast film is a high-volume, high-tolerance product used for carrier material in the printing and audio/video recording industry.
- *Profile* is a shaped extruded profile.
- *Fiber* is a cylindrical or tubular profile less than 0.0004 mm (0.010 in.) thick.

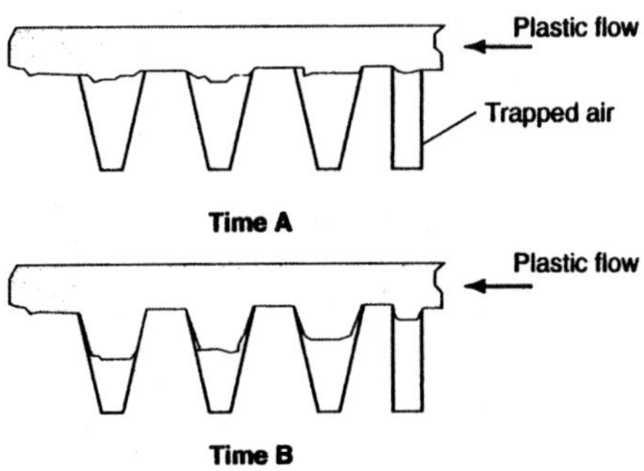

Fig. 5 Problems in cooling and solidification caused by the rib fill rate for an injection-molded part

Details and design considerations for extruded parts include die swell and orientation.

Die swell (Fig. 11) is the phenomenon where an extrudate swells to a size greater than the die from which it came. As the plastic exits the die, it tends to swell. This is associated with the reduction in pressure as well as the nature of the polymer itself. Die swell has to be considered by the product designer as well as the die designer in order to produce extrusions that meet the customer requirements.

Orientation is the phenomenon where the polymer molecules are aligned as a result of the high degree of laminar flow as well as the pulling of the extrusion takeoff apparatus. Orientation is often desirable, if controlled, because it can improve the properties of the extruded product. Biaxial orientation is orientation in two directions and improves strength in film materials.

Orientation also allows an extruded product to shrink when exposed to heat. Shrink-wrap materials for packaging and dunnage have become very important products that incorporate this phenomenon of shrinking due to controlled orientation and heating.

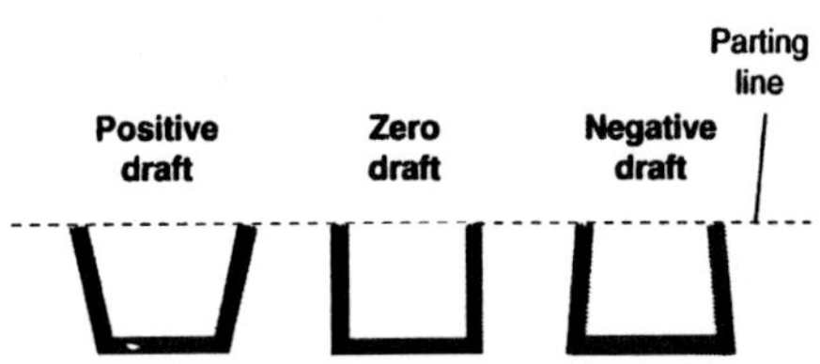

Fig. 7 Types of draft in plastic injection-molded parts

Thermoforming

Thermoforming, also referred to as vacuum forming, forms plastic sheet into shapes. The plastic sheet is placed into a clamp frame to hold it securely on all edges. The sheet material is placed into the clamp frame manually, robotically for high-volume processing, or continuously if the sheet material is produced by an in-line extruder. Thermal energy, usually in the form of convection and radiant heat from electrical heating elements, is applied for a sufficient amount of time to soften (not melt) the plastic sheet. Once the sheet is sufficiently softened, a mold is brought in contact with the sheet, and a vacuum is applied that draws the softened sheet onto the mold. After the sheet cools, it will retain the shape of the mold when the mold is removed. The thermoforming process sequence is shown in Fig. 12.

Historically, thermoforming has been considered a one-sided process; that is, the softened

Table 2 Shrinkage of selected plastic materials

Material	Shrinkage, %
Amorphous plastics	
Acrylic	0.6
Polycarbonate	0.6
Acrylonitrile-butadiene-styrene (ABS)	0.6
Polycarbonate (40% glass filled)	0.3(a) 0.5(b)
Semicrystalline plastics	
Polyethylene	2.0
Polypropylene	2.0
Nylon 6/6	1.5
Nylon (40% glass filled)	0.8(a) 0.3(b)

(a) Flow direction. (b) Transverse direction

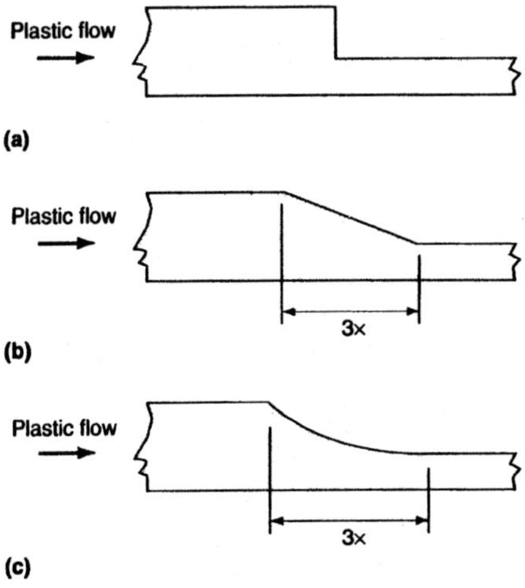

Fig. 6 Wall transitions in a plastic part. (a) Poor (sharp) transition. (b) Better (gradual) transition. (c) Best (smooth) transition

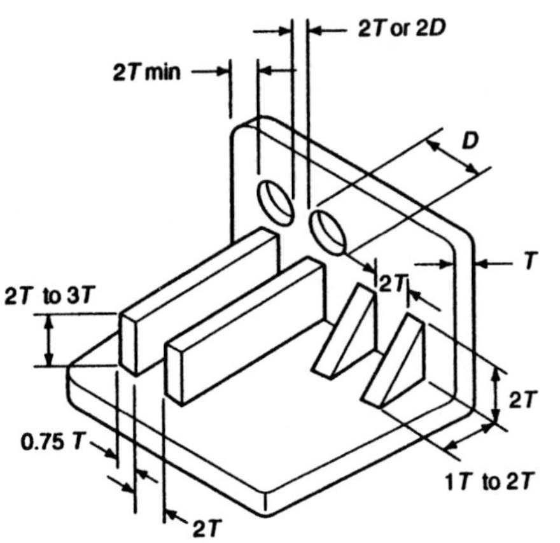

Fig. 8 Good design practice for holes and projections in injection-molded parts

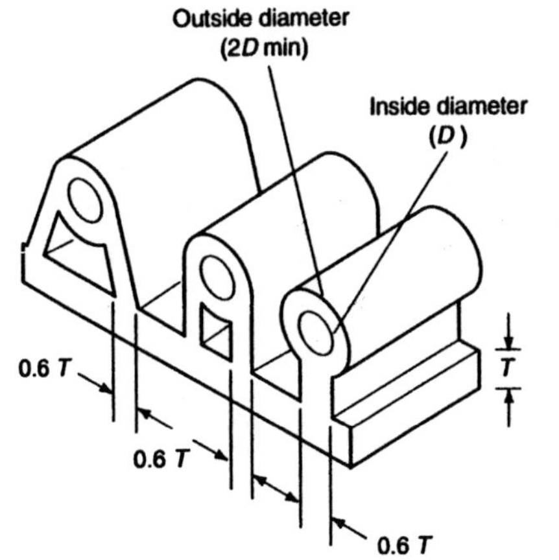

Fig. 9 Boss configurations for injection-molded plastic parts

sheet will either conform to a male mold with the inside becoming the critical surface and the outside the noncritical surface, or conform to a female mold with the outside becoming the critical surface and the inside the noncritical surface. This one-sided approach to thermoforming was satisfactory for decades when the process was used primarily for simple packaging parts. Process advancements in the mid-1990s have enabled the production of thermoformed parts that have two critical sides and sufficient dimensional accuracy to allow them to be used in key automotive, building, and construction applications. This dimensional control is accomplished by having two dies or molds, one forming either side of the sheet.

Typical Thermoformed Parts. The majority of thermoformed products are produced for the packaging market; however, broader applications include:

- Blister packages
- Foam food containers
- Refrigerator and dishwasher door liners
- Auto interior panels
- Tub/shower shells, which are later fiber reinforced
- Pickup truck bed liners
- Internally lighted acrylic and cellulose acetate butyrate (CAB) signs

The thermoforming process offers some unique tooling advantages over other conventional plastic processes, primarily because the thermoform molds are relatively simple in design and construction as well as lower in cost.

Prototypes produced using the thermoforming process can be made quickly by using simple molds made from inexpensive materials, such as wood, plaster, and epoxy. Many designers will insist on a product design review that includes the development of one or more thermoformed prototype parts.

Blow Molding

Blow molding has historically been associated with simple geometries such as bottles and containers. However, there were significant developments in the blow-molding process and its variants in the 1980s and 1990s. These developments allow the blow molding of more complex shapes such as air ducts and automobile fuel tanks.

Basic blow molding equipment (Fig. 13a) is essentially a profile extruder attached to a blowing station. The extruder produces a tube referred to as a parison. The parison can be controlled in both its size and shape. At the blowing station (Fig. 13b), the mold captures the parison and seals it by pinching either end. A blow pin is then inserted into the parison, and air is introduced at about 700 kPa (100 psi). The air causes the pinched parison to expand and take the shape of the mold.

This basic process results in a product that is dimensionally defined on the exterior surfaces. The interior surfaces are not controlled as they do not contact a mold surface. As a result, the wall thickness of a conventionally blow-molded part may vary. The nature of the conventional blow molding process also does not lend itself to incorporating design features such as holes, sharp corners, and narrow ribs.

Rotational Molding

Like blow molding, rotational molding produces a hollow product. Unlike blow molding,

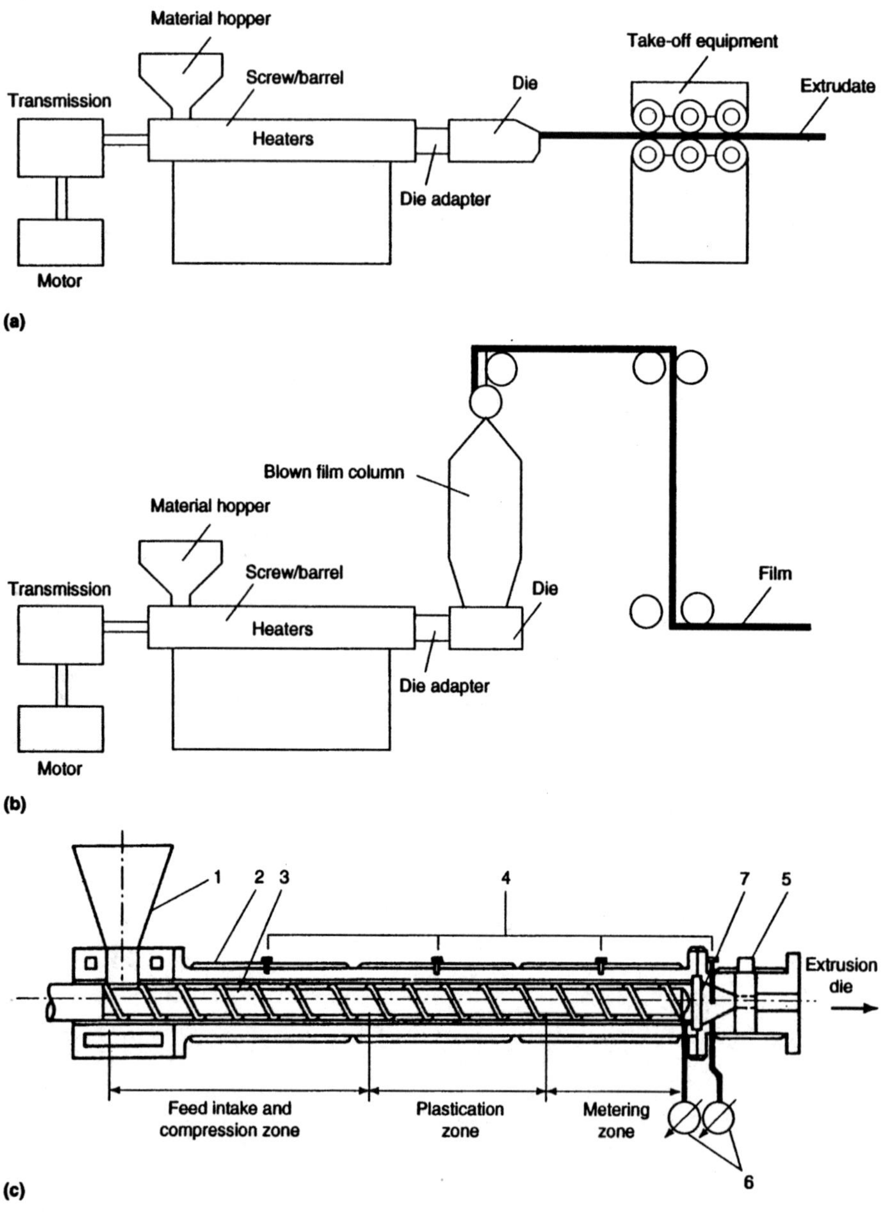

Fig. 10 Extrusion processes. (a) Profile/sheet extrusion. (b) Blown film extrusion. (c) Construction arrangement of the plastication barrel of an extruder. 1, feed hopper; 2, barrel heating; 3, screw; 4, thermocouples; 5, back-pressure regulating valve; 6, pressure-measuring instruments; 7, breaker plate and screen pack

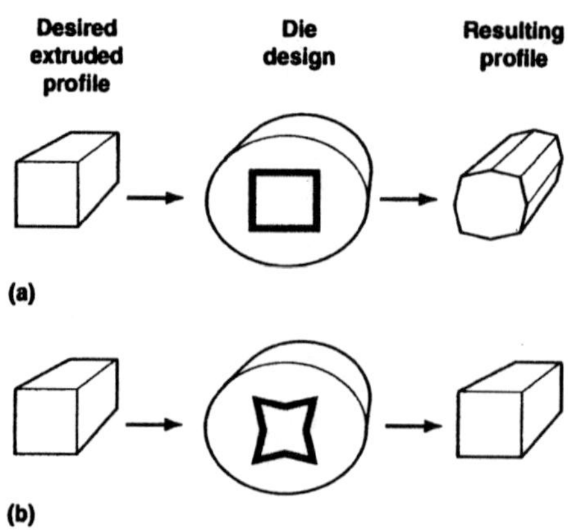

Fig. 11 Die swell in extrusion. (a) Incorrect die design for intended profile. (b) Correct die design

however, rotational molding is a relatively slow process that begins with plastic in the form of a powder, not a parison. The advantage of rotational molding is that it can produce large objects, with capacities from 1 to more than 500 gal. Additionally, the wall thickness is a function of how much plastic powder is placed into the mold, and thick (2.5 to 12 mm, or 0.10 to 0.4 in.) wall sections can be formed.

The rotational molding process uses a mold made of sheet metal or cast aluminum. Because rotational molding is a low-pressure process, tooling can be lower in strength than that used for the other molding processes.

Another advantage of rotational molding over other plastic processes is that it results in a very low-stressed product. Since the rotational process is low in pressure, and the plastic is not forced through narrow channels, it does not induce a significant amount of internal stress. The result is a high degree of dimensional stability in the final product.

Processing Sequence. The plastic powder is placed directly into the mold by the operator. The mold is attached to the rotational process equipment where it passes through three distinct process stages: loading, heating, and cooling.

Loading is the stage of the process where the plastic powder is loaded into the mold and the mold is attached to the process equipment. After loading is completed, the mold begins to rotate along three axes. Although the rotation speed is relatively slow (<20 rpm), it is sufficient to force the plastic powder to the mold walls.

Heating is the stage of the process where heat, usually generated by a natural-gas-fired heater, brings the rotating mold to a temperature high enough to fuse the plastic powder.

Cooling is the stage of the process where the mold, which is still rotating, is allowed to cool, and the fused plastic in the mold solidifies. Once the contents of the mold are sufficiently cooled, the mold is separated and the product removed.

Processing Systems. Advances in the rotational molding equipment have resulted in two distinct styles of processing systems:

- *A conventional rotational system* has the mold located on the end of one of several arms. Several molds are at various stages of the process simultaneously.
- *A clam-shell system* (Fig. 14) has the mold located on a rotating apparatus that is housed within a chestlike chamber. The clam-shell system has a fixed and compact footprint. The chamber environment changes to heat as well as cool the mold. Especially good for small rotational-molded parts, the clam-shell system provides superior control and safety.

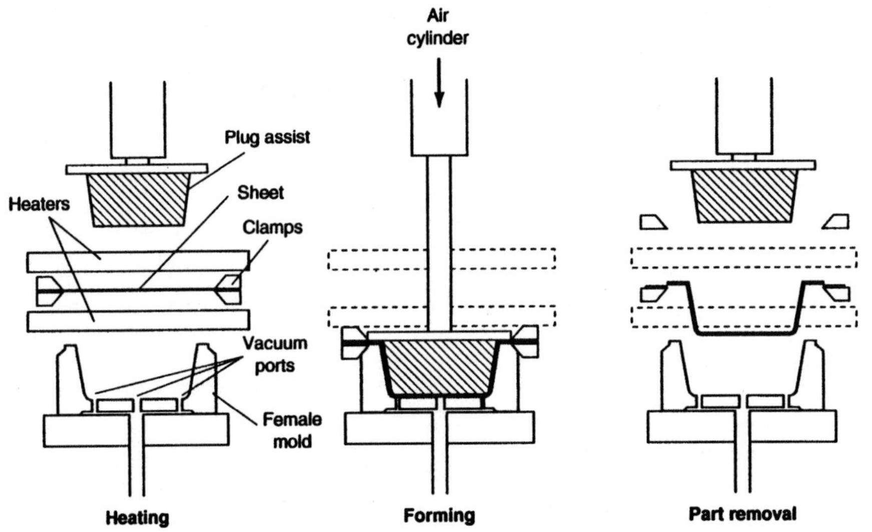

Fig. 12 Thermoforming (vacuum forming)

Compression Molding and Transfer Molding

Compression molding and transfer molding (a variant of compression molding) are two of the oldest plastic processing methods.

Compression molding is a simple process that offers the manufacturer an excellent method for producing low stress plastic parts. As shown in Fig. 15, the plastic material, usually in the form of a powder or preformed pill, is placed in the cavity or female section of the mold. Since most compression molding is associated with thermosetting plastics, the mold temperature

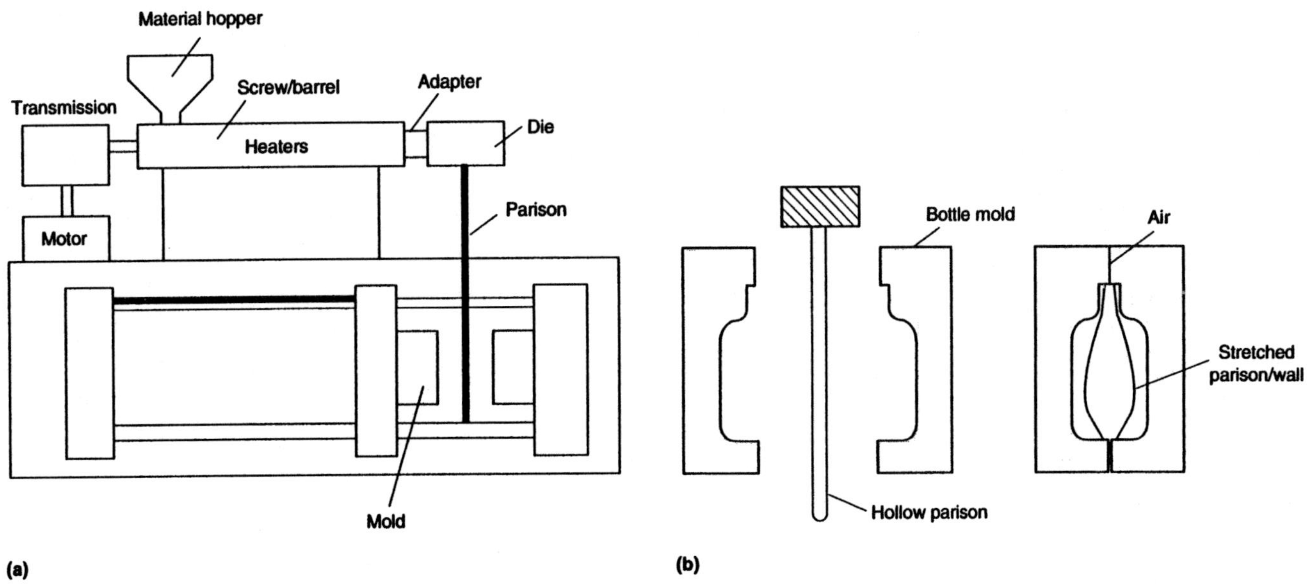

Fig. 13 Blow molding. (a) Equipment configuration. (b) Sequence at blowing station for a bottle mold

will be relatively high (149 to 177 °C, or 300 to 350 °F). The male or core portion of the mold is located on the upper portion of the mold. After the plastic material is loaded, the male portion of the mold is lowered and compresses the heated plastic in the cavity. The heat and pressure cause the material to flow and fill the cavity details. In many compression molding processes, excess plastic forms a flash that will be removed in a secondary operation. The plastic material will then be allowed to cure or set, and the plastic part will be ejected from the mold.

Because the thermosetting plastic flows only a short distance and the flow rate is relatively slow, little shear stress is developed in the process. This low stress level will result in low internal or molded-in stress in the part, which thus will have a high degree of dimensional stability.

Transfer molding is a variant of the compression molding process and is a precursor to the modern injection molding process. The main difference between compression and transfer molding is in the method by which the plastic material enters the mold. As stated previously, compression molding requires the material to be preloaded into the mold in the form of either a powder or a preform. Transfer molding utilizes a transfer ram or plunger where powder or preform is loaded into a chamber above the mold. The material in this chamber is forced through a sprue and runner/gate system to fill the mold. The transfer process allows the plastic to enter the mold in a molten or fluid state; thus, the process is adaptable to insert or overmolding. The downside of the transfer process is that there is often a need for secondary operations to separate the product from the gates/runners. There is also a certain amount of waste with the sprue and runners.

The ability to overmold or insert mold product allows many unique products to be transfer molded. These include electrical outlets, semiconductors such as integrated circuits, plastic handles for products such as knives, and plastic exteriors for metal parts.

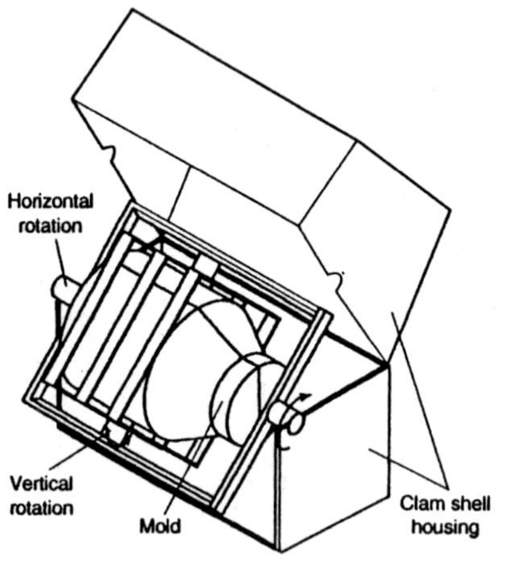

Fig. 14 Rotational molding equipment (clam-shell type)

Composites Processing

The term *composite* applies to plastic materials that are reinforced with glass, mica, metal, carbon fibers, or other materials. Composite materials provide the plastics part designer and plastics processor with the opportunity to customize a plastic compound by adjusting the type and volume of reinforcements added to a specific plastic matrix. Such composites usually exhibit synergistic behavior. Synergism, in the context of plastic materials, is when the strength of the composite is greater than the sum of the strength of the individual matrix and reinforcements. Additionally, composite plastics also distribute loads better than conventional plastics and result in superior strength-to-weight ratios. Other advantages of composite plastics are that they can be used to fabricate large parts and they offer excellent product durability and superior chemical resistance.

Although the development of composite materials and fabrication methods have made great strides, there are few high-volume processes available to manufacturers. Additionally, there are few design rules available to composite product designers. Products designed for conventional processes, such as injection molding, blow molding, extrusion, and thermoforming, can utilize a series of design guidelines based on an understanding of the materials and processes involved. Since composite product fabrication was limited, these design guidelines are still being developed. Some other disadvantages of composite processing include high product cost, high equipment cost (and limited availability), and low production rates.

Composite production processes include casting, reaction injection molding (RIM), structural reaction injection molding (SRIM), resin transfer molding (RTM), matched metal molding, filament winding, and pultrusion. Some of these processes are also used for noncomposite materials; casting is described separately in the section "Casting" in this article.

Reaction injection molding is a variant of injection molding (described previously in the section "Injection Molding" in this article). It is used for molding polyurethane, epoxy, and other liquid chemical systems. Mixing two to four components in the proper ratio is accomplished by a high-pressure impingement-type mixing head, from which the mixed material is delivered into the mold at low pressure, where it reacts (cures).

Structural reaction injection molding is in its technological infancy. As a consequence, the process is undergoing dramatic material and process improvements and rapid industrial growth. In many ways, SRIM is the natural evolution of two more established molding processes: RIM (described previously), and RTM (described later). Like RIM, SRIM uses the fast polymerization reactions of RIM-type polymers, its intensive resin mixing procedures, and its rapid resin reaction rates. Like RTM, SRIM also employs preforms preplaced in the cavity of a compression mold to obtain optimal composite mechanical properties. A schematic of the SRIM process is shown in Fig. 16.

The ability of SRIM to fabricate large, lightweight composite parts consisting of all types of precisely located inserts and selected reinforcements is an advantage that other competitive manufacturing processes find difficult to match. In addition, large SRIM parts can often be molded in 2 to 3 min with clamping pressures as low as 700 kPa (100 psi). Thus, the capital requirements of SRIM are relatively low, allowing the economical manufacture of parts with annual production volumes below 10,000 units. These advantages, when coupled with the concurrent development of a large family of commercially available SRIM resins, have led to predictions of a high SRIM annual growth rate.

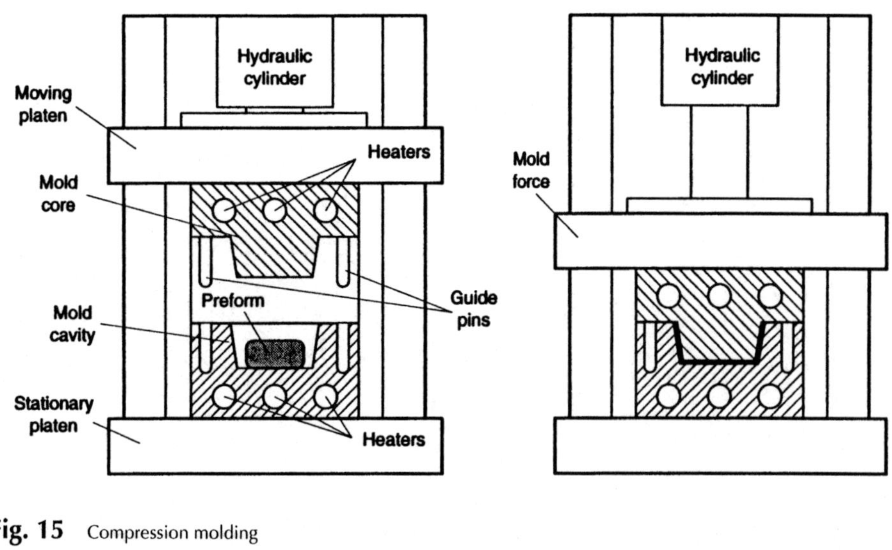

Fig. 15 Compression molding

Applications of SRIM. The first SRIM part commercially produced was the cover of the spare tire well in several automobiles produced by General Motors. Other SRIM automotive structural parts include foam door panels, sunshades, instrument panel inserts, and rear window decks. Nonautomotive SRIM applications include seat shells for the furniture market and satellite dishes for home entertainment centers. Structural reaction injection molding is an attractive process for the economical production of large, complex structural parts.

Resin transfer molding is a process by which catalyzed resin is transferred or injected into an enclosed mold in which reinforcement has been placed (Fig. 17). The fiberglass reinforcement is usually woven, nonwoven, or knitted fabric. Resin transfer molding is primarily used for prototyping, low-volume production of large, relatively complex parts, or low-to-medium-volume production of small, simple parts.

Prototyping. Resin transfer molding is an excellent process choice for making prototype components. Unlike processes such as compression molding and injection molding, which require tools and equipment approaching production level to accurately simulate the physical properties achievable in the production-level component, RTM allows representative prototypes to be molded at low cost. It should be noted that in some cases, RTM can be used to prototype components designed for other processes; the RTM component will typically have properties that exceed those of the production-level product.

When prototyping with RTM, less reactive resins are generally used, allowing long fill times and easier control of the vents. Tooling is usually low-cost epoxy, but could be made with an impervious material that would contain the resin. Prototype preforms are made by cut-and-sew methods, and any foam cores used are machined to shape. Sizes can range from small components to very large, complex three-dimensional structures. Resin transfer molding provides two finished surfaces and controlled thickness, and it requires no auxiliary vacuum or autoclave equipment. Other processes used for prototyping such as hand lay-up and wet molding give only a single finished surface, and dimensions in the thickness direction are controlled.

Matched metal molding, also known as matched-die press forming, is the most common and probably the most widely used composite-forming system. This is because forming presses are readily available from very simple, hand-operated small presses to fairly sophisticated, computer-controlled hydraulic systems. For simple forming operations, standard heated platen presses that have generally been used for flat panel molding have proven to be adequate. However, in operations where the control of deformation rate and pressure history are important, high-quality stamping presses are used.

The dies used in this forming method are generally made of metal, which can be internally heated and/or cooled. When metals are used, the dies are generally designed to fixed gap (thickness) of close tolerance. High pressures can easily be applied to the workpiece. A disadvantage of this forming method is that when there is a thickness mismatch between the formed piece and the premachined cavity, nonuniform pressure is produced on the part resulting in nonuniform consolidation.

When heating or cooling is desired, the dies usually have such a high heat content that heat-transfer times are long. Finally, matched-die fabrication costs are high because of the requirement that the two close-tolerance die halves have to match. Substituting an elastomeric material for one of the die halves will usually reduce the tooling cost and enable the application of a more uniform consolidation pressure than in an all-metal die set.

Pultrusion (Fig. 18) is a composite process that has many similarities to extrusion. Pultrusion begins with strands of reinforcement, usually glass or carbon fibers, that have been wetted in a resin tank. The resin used is most often an epoxy or polyester. The next step in the process is to pull the resin-soaked strands through a heated shaping die. The die may be in the shape of a rod, tube, I-beam, or other geometric shapes. After the resin is cured and pulled through the die, the resulting profile has a high strength-to-weight ratio and is very durable, especially in a chemical environment.

Applications for pultruded products include structural beams for electrical and chemical environments (for example, ladders for use near electrical wires), poles, and shafts. The future of the pultrusion process may be in space applications where long structural components for space stations could be manufactured in space, as required, eliminating the need to transport long beams from earth.

Filament winding is primarily used to manufacture large structural containers or tanks. The process involves several spools of reinforc-

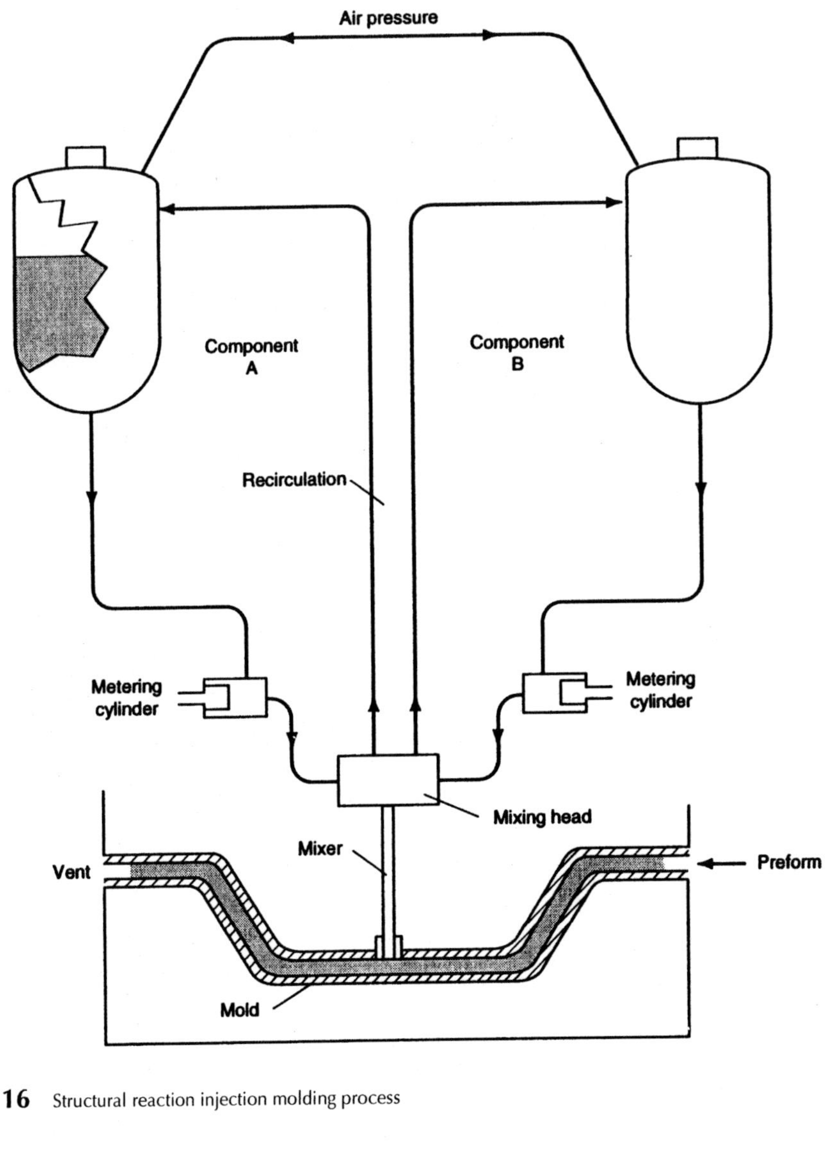

Fig. 16 Structural reaction injection molding process

ing materials such as glass or carbon strands. The strands are directed into a resin bath of either polyester or epoxy. The wetted strands are wound over a turning mandrel (Fig. 19) in different patterns to provide different strengths. After the resin has cured, the filament-wound part is removed from the mandrel and machined or assembled as required.

Applications for filament-wound composites include gasoline storage tanks, septic tanks, large-diameter drainage pipes, chemical storage systems, and sporting equipment such as golf club shafts and bike frames.

Casting

Casting is the process of pouring liquid plastic into a mold. The recent development of a wide variety of casting resins and rapid tooling fabrication has allowed the casting process to be considered as a viable process for both prototyping and low-volume production. Typical casting resins include casting acrylic, casting polycarbonate, epoxy, polyurethane, and polyester. The casting process produces plastic parts with the lowest level of internal stress and a high degree of dimensional stability.

Design Features and Process Considerations

A design feature is an aspect of the shape of a product. Principal features incorporated into the design of a plastic part are:

- *Walls* are the predominant features of the shape of a product. The nominal wall for a plastic part is the platform on which all other design features reside. Most other design features are configured and sized as a function of the size of the nominal wall (Fig. 20).
- *Projections* are design features that rise from the nominal wall. They include ribs, gussets, threads, snap-fits, and bosses (Fig. 21).
- *Depressions* are design features that enter into or reside within the nominal wall. They include through holes, blind holes, and slots (Fig. 22).

Design for assembly (DFA), or design for manufacturing and assembly (DFMA), is a design methodology that embraces the concept that well-engineered assemblies will take advantage of the high functionality of materials, such as plastics, and integrate several design features, such as snaps, alignment features, and locks, to facilitate assembly. These integral features help to eliminate conventional assembly components such as screws, washers, and nuts. Additionally, the product designers that utilize DFA techniques go beyond the design of the individual part to consider the optimization of the design of several parts in an assembly (Fig. 23). General DFA concepts are described in the articles "Introduction to Manufacturing and Design" and "Design for Manufacture and Assembly" in *Materials Selection and Design,* Volume 20 of the *ASM Handbook.*

Design for Optimal Properties and Performance. While many product designers understand that it is possible to degrade or lower material properties through processing operations, opportunities to optimize properties of the plastic and the plastic product are often overlooked. The most important step in the optimizing process is to understand how the proper-

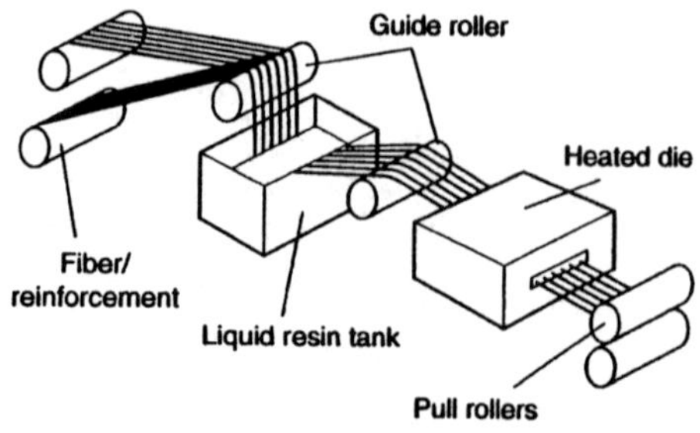

Fig. 18 Pultrusion

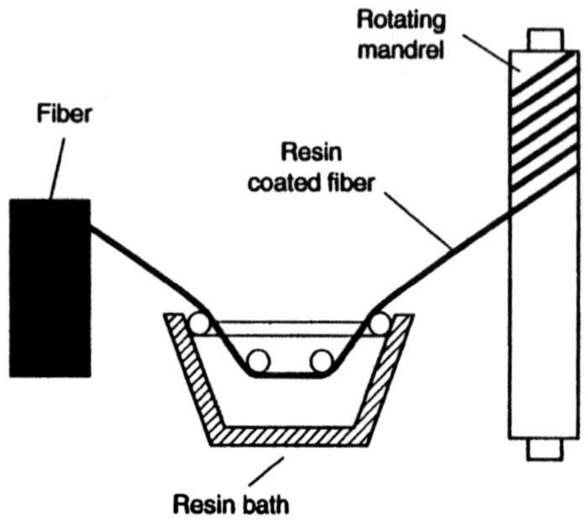

Fig. 19 Filament winding

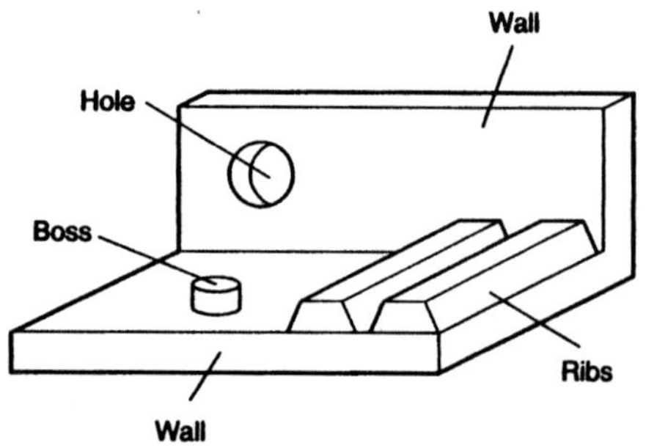

Fig. 20 Plastic part wall components

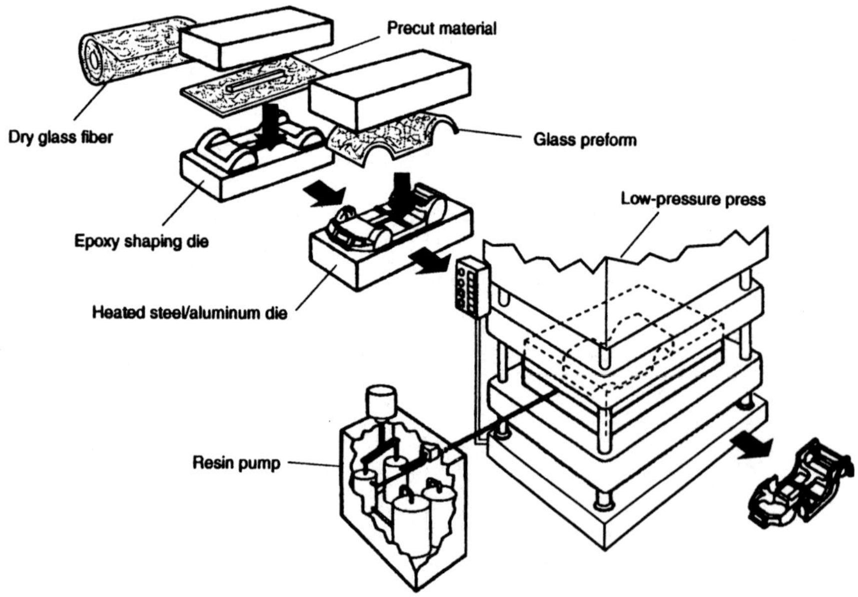

Fig. 17 High-speed resin transfer molding process

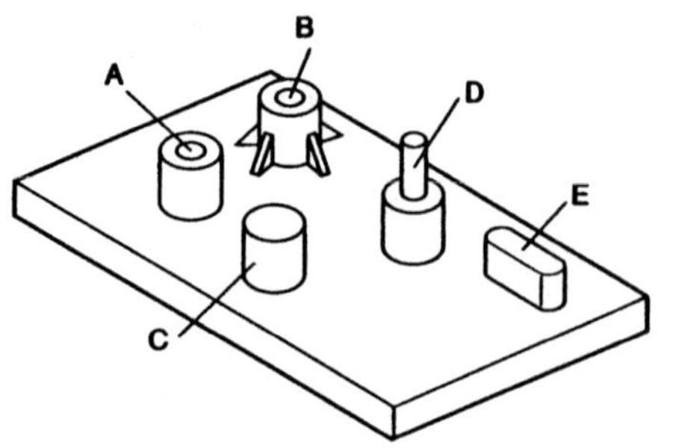

Fig. 21 Boss designs for plastic parts. A, hollow boss; B, gussetted boss; C, solid boss; D, stepped boss; E, elongated boss

ties of a plastic will vary when processed and to alter the product design to avoid or manage this variation.

The best resource for determining these variations is direct experience. Certainly, material suppliers can provide generalizations as to the effects of processing; however, there are simply too many variables (design, process equipment, additives, etc.) for detailed and reliable property variations to be published.

An example of such a property variation is a polycarbonate part designed for high-impact strength that has a wall section that is too thick. The polycarbonate shows impact degradation at a critical thickness value; thus the impact properties of the resulting plastic product may be much lower than those published by the material manufacturer.

A second example is a polyacetal part molded with a low mold temperature to improve the cycle time, resulting in an induced stress level and postmold shrinkage. The product may be dimensionally correct when measured directly after molding but too small when inspected 1 week later after delivery to the customer.

A third example is a polyphenylene sulfide part processed with a mold temperature of 93 °C (200 °F), the highest temperature the molder's water controller can safely attain. As a result, the polymer does not achieve optimal crystallinity, and the part performance suffers in its end-use application. While water-heated molds may be acceptable for many plastics, the molder did not recognize that an electric or oil-heated mold would have allowed a more appropriate mold temperature of 121 °C (250 °F) to maximize the crystallinity of the polymer.

Other Plastics Design and Processing Considerations

Stress in plastic parts is most often related to shear stress, especially for molded or extruded parts. During molding, a plastic part is subjected to a rigorous process environment. When melted plastic is forced into a closed mold, three major stress contributors are present: high pressure, high speed, and constricted flow areas.

The stress level in a part depends on the processing method used. The three major stress-producing components are present in injection molding; thus injection-molded parts usually have a high degree of molded-in stress. Compression molding usually has only the high-pressure component and thus produces parts with much lower molded-in stress.

Casting involves none of the major stress-contributing factors and, therefore, results in parts with the lowest levels of molded-in stress.

Stress levels, in a plastic part, affect the properties and performance of the product. In general, as the molded-in stress level increases, dimensional stability decreases, chemical properties lower, mechanical properties decrease, and optical properties diminish. A good example of this relationship is the compact disk (CD). A CD needs to be optically pure and mechanically strong. For efficient production, the best processing method for CDs is high-volume injection molding. However, by its nature, the injection molding process will lower the required properties, especially because the disks are extremely thin and will promote high shear stress. This problem was solved, in large part, by the development of a low-viscosity plastic (polycarbonate) that required lower molding pressures to fill the mold and thus lowered the resulting molded-in stress in the CDs.

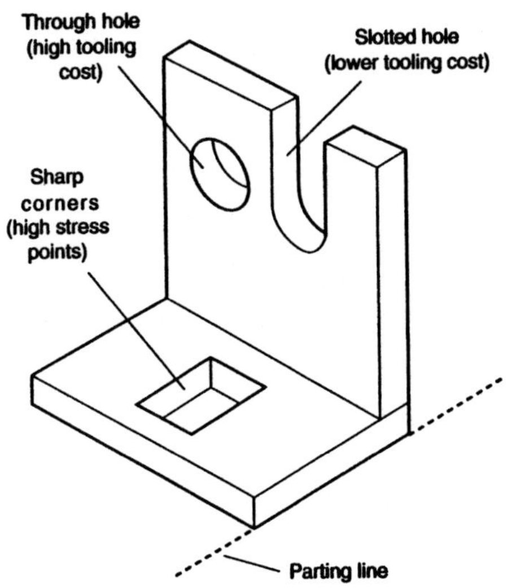

Fig. 22 Depression designs for plastic parts

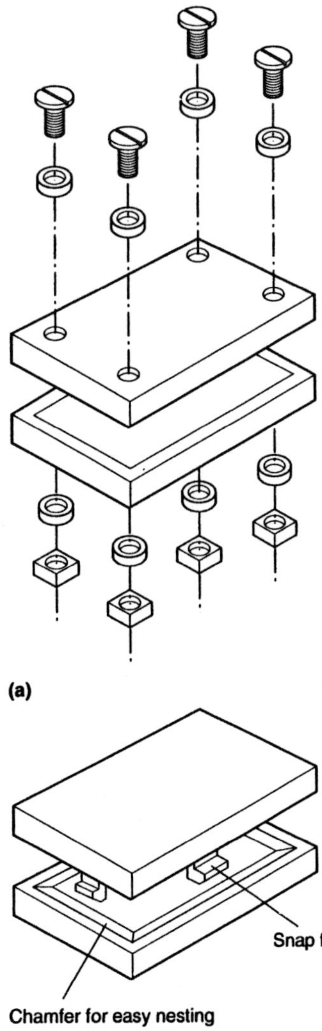

Fig. 23 Designing for product assembly. (a) Original design. (b) Improved design for ease of assembly

Published versus Actual Product Properties. It is technically naive to believe that the material manufacturer's published properties will be duplicated in the final product. Published properties are derived from parts that are molded and tested under highly controlled conditions. The environment is controlled to either "dry as molded" or 50% relative humidity conditions. The samples tested are usually ASTM/ISO test specimens that do not have the design features that often can serve to lower properties in molded products.

The property effects of additives such as colorants and regrind are often not published simply because there are too many variables involved. Also, time-related properties such as weatherability and creep often are interpretive, and published data should serve only as guidelines for the designer.

It is important that the product design team along with the manufacturing team develop and test the plastic products prior to production. Using the manufacturer's published data as a reference, the product tests should always emulate the end-use application environment as closely as is practical.

End-Use Concerns. Even well-designed and manufactured plastic products can fail because one or more aspects of the end-use application was not considered. The longer the list of end-use environmental conditions that are assessed and respected, the more successful the plastic product will be.

Many new plastic product designers tend to focus on details, but overlook some obvious end-use considerations. Examples include:

- Effects of household chemicals (for example, milk, water, cleaners, makeup, and makeup removers) on plastics used for consumer applications
- Shipping and handling factors (for example, high compartment temperatures and damage from dropped boxes)
- Time-related issues such as creep, stress relaxation, and ultraviolet weatherability

Materials-Selection Methodology

Selecting a plastic material for a specific end-use application is a challenge. It requires a thorough understanding of the end-use application, a knowledge of available plastic materials and their properties, and a methodology to sort and select all the data to make a prudent decision.

Understanding the End-Use Application. As noted previously in the section "Other Plastics Design and Processing Considerations," it is important for the designer to understand the intended application of a part including physical loads that will be applied, chemical resistance and exposure factors, and temperature. However, a designer must investigate the end-use environment even further if a good material selection is to be made.

For example, a design specification from a furniture maker noted that their products typically experience temperatures between 16 and 27 °C (60 and 80 °F). This is a narrow range, but the furniture maker understood that their customer base would always properly control the environment of the office where the furniture was to be located. However, the shipping and storage of the furniture was overlooked. Temperatures in the back of a semitrailer truck or a warehouse can be significantly different from those in the ultimate end-use environment. Temperatures below freezing and over 43 °C (110 °F) are certainly possible. The plastic materials selected for this application must be able to withstand the shipping and storage as well as the end-use application.

In another example, a manufacturer carefully studied the end-use requirements for a new tail-light lens and designed and processed the part to meet the requirements. The plastic lenses were molded in Texas and shipped (by truck) to the automaker in Detroit. The automaker was dismayed to discover upon delivery that more than 20% of the snap tabs used to hold the lens in its assembly were cracked or broken. The cause of the problem was the cyclical loading that occurred during shipping, which was different from any loading tested for the end-use application of the part.

A third and final example involves plastic material selection for a simple coat hook. The design engineer assessed the end-use application and determined that a 10 lb coat hanging for 8 to 10 h on the hook would be no problem. However, it was later discovered that the same 10 lb coat left on the hook for 30 days resulted in unacceptable deformation of the hook. The designer failed to consider the time-related phenomenon of creep.

Understanding the Properties of the Plastic Material. Thousands of plastic materials are available today, and the list is growing rapidly. It is challenging for designers to select the right plastic material, that is, one that meets all the property, processing, and cost requirements and goals. The material-selection process is further complicated by the fact that much of the information published about plastic materials focuses on their positive attributes. Very few plastics manufacturers overtly comment on the weaknesses or drawbacks of their materials.

Selecting a plastic material requires an understanding of the balance of properties; that is, how well a material meets the overall property requirements for a particular application and any effects that different properties may have on one another. For example, for a plastic material considered for a roof gutter system on a residential building, the properties that might be assessed include water resistance, impact resistance, temperature resistance (resistance to heat and cold), and weatherability (ability to withstand the rigors of outdoors).

Unfortunately, however, a simple list of attributes, although an excellent starting point, is not adequate to differentiate among the thousands of plastic materials available in order to make a proper selection. The selection process must have more depth. Consideration must go well beyond attributes and consider the specific variable or number value of each property. Once that is determined, there must be an assessment of interaction.

A more detailed property assessment might appear as follows:

Property	Requirement
Water resistance	Must not absorb >0.05% water when exposed to 100% humidity
Physical properties	Must be measured at 50% relative humidity
Impact resistance	Izod notched impact strength must be >133 J/m (>2.5 ft · lbf/in.) from −17 to 66 °C (−20 to 150 °F)
Temperature resistance	Temperature range is −40 to 71 °C (−40 to 160 °F)
Weatherability	<5% degradation in physical properties after 15 yr ultraviolet exposure (UVA and UVB)

It is clear that the process of identifying the end-use application requirements is more challenging than it may initially appear.

Other end-use properties that should not be overlooked include:

- Cost (material, labor, overhead, and yield)
- Processability (easy or difficult, inexpensive or costly)
- Maintenance (low or high, durability)
- Ease of assembly

While some of these attributes are a direct function of the product shape and design, the material-selection aspect needs to be considered. As an example, when considering ease of assembly, it may be important to have a plastic material that can be bonded using solvents. This requirement would eliminate many plastics from the list that might otherwise have been candidates.

Material-Selection Matrix. In order to simplify the material-selection process, many product designers utilize a material-selection matrix (Fig. 24). The matrix allows a direct comparison between the end-use properties desired and the actual properties available from the candidate materials. Additionally, the matrix helps prioritize and sort these properties to make the final selection.

Developing a material-selection matrix involves a number of steps:

1. Identify as many material properties and attributes as possible that are required to meet the demands of the application.
2. Assign a rating to each property and attribute (where 9 is critical, 6 is desirable, and 3 is optional).
3. List the candidate materials.
4. Rank the materials, relative to one another, as to how well they meet each property/attribute requirement. (If there are four materials being compared, the material that best meets the property/attribute requirement receives a 4. The material that least meets the property/attribute requirement receives a 1.)
5. Multiply the property rating by the material ranking.
6. Add the products of step 5.

The material with the highest sum is the top candidate. In Fig. 24, for example, PVC is the top candidate for the application represented by this matrix. This matrix analysis tool also can be readily adapted to compare processes as a means for optimizing process selection.

Desired properties	Rating	PE	PP	PVC	ABS	PS	
Impact strength	9	3	4	5	2	1	Rank
		27	36	45	18	9	Rating × rank
Low cost	6	5	4	3	2	4	Rank
		30	24	18	12	24	Rating × rank
Flexural modulus	3	1	2	3	4	5	Rank
		3	6	9	12	15	Rating × rank
Low flammability	9	2	3	5	4	1	Rank
		18	27	45	36	9	Rating × rank
Tensile strength	6	2	3	4	5	1	Rank
		12	18	24	30	6	Rating × rank
Total (sum of products)		90	111	141	108	63	

Fig. 24 Example of a material-selection matrix. PE, polyethylene; PP, polypropylene; PVC, polyvinyl chloride; ABS, acrylonitrile-butadiene-styrene; PS, polystyrene

Function and Properties Factors in Process Selection*

Manufacturing process selection is a critical step in product design. Failure to select a viable process during initial design stages can dramatically increase development costs and timing.

The first step in selecting an appropriate process for the function and properties of the specific part is to establish accurate functional requirements. Once these are established, the materials that meet these requirements can be selected. The next step is to select those processes that can handle the material to produce the required properties in the part. This may involve selecting processes to provide the maximum achievable physical properties of the material in one direction or location. Process selection must then be refined, allowing for such factors as size, shape, surface finish, and cost. Final processing detail is then established during design development through the fine tuning of the selection procedure and production trials.

This section provides an overview of various process effects and how they affect the functions and properties of the part. This is preceded by a brief discussion of functional requirements in process selection. For example, if a major functional requirement is for resistance to creep under high loads, it is probable that a long fiber-reinforced plastic is necessary; this would immediately eliminate such processes as injection and blow molding.

Establishing Functional Requirements

The functional requirements of the part must be understood before material and process selection is attempted. All requirements should be listed, and the maximum and minimum property limits should be established. At this stage, these limits must be adequate for the part rather than being set higher than necessary. It is a widely held belief that a safety factor is built into the selection process by specifying higher product performance requirements than are actually required. This is a serious mistake because overspecifying the property requirements can lead to the selection of inefficient processes and expensive materials.

The principal functional requirements, or critical requirements, are those functions that must be achieved for the part to work. A clear understanding of the critical requirements is necessary before making material and process decisions. However, for a designer using plastics for the first time, establishing critical requirements can be very difficult. For example, if a designer has been designing wood boats for an extended period of time, it would be easy to overlook the fact that not dissolving in water is a critical material requirement.

Critical requirements can be:

- Optics (windows, lamp lenses)
- Electricity (wiring, connectors)
- Temperature resistance (cookware, electrical components)
- Molecular structure (microwave cookware, coaxial cable)
- Chemical and water resistance (food contact parts, storage tanks, fuel containers, piping, boat hulls)
- Nontoxicity (food contact parts, living areas)
- Impact resistance (car bumpers, instrument panels, football helmets, tool housings)
- High flexural modulus (car structures, boat hulls, electrical component housings, pallets); can be achieved by material or design
- Low flexural modulus (pads, balls, ski boots, shoes)
- Resilience (seat pads, springs, flexible car fascias)
- High flexural and tensile strengths (piping, pressure vessels, storage silos, vehicle and aircraft structures)
- Light weight (aircraft parts, luggage)

The critical requirements can also be related to the surface quality or shape of the component, such as its appearance; feel; ability to contain fluids or air; ability to reflect or absorb light, heat, or sound energy; abrasion resistance; or adhesion capability.

Once a list of critical requirements is established, other desirable properties that would enhance the product and increase its sales potential should be listed. For example, a microwave cooking dish would have the critical properties of not being affected by microwaves, not tainting hot food, and not being softened or distorted by contact with hot food. Desirable properties would be dishwashing and freezer temperature resistance, nonstick capabilities, an easily cleaned surface, and a decorative appearance. The critical items usually define the materials and processes that can be used, while the desirable items are used to fine tune the material and secondary process selection.

Properties Considerations and Processing

Descriptions of numerous material properties that should be considered are provided below and precede the discussion of thermoplastic and thermosetting process effects on product function. Polymer structure and plastics properties have been discussed in detail in previous articles in this book. Some general property considerations are described here in terms of processing effects.

The toughness of polymers, or resistance to impact, varies with molecular structure and the type of stress application. Care must be taken in relating flexibility to toughness, but generally, a more rubbery character gives higher elongation to break and better impact resistance values, although such materials would have lower stiffness. Figure 25 shows the variation of notched impact with flexural modulus for a typical range of acrylonitrile-butadiene-styrene (ABS) resins. Some stiffness can be recovered by adding fibrous reinforcement, but the reinforcement can also affect impact resistance.

The relationship between toughness and stiffness must also be considered during material selection. Generally, a longer, higher-molecular-weight thermoplastic polymer will be tougher than a shorter, lower-molecular-weight polymer of the same chemical type. The higher-molecular-weight polymer will also provide higher melt viscosity and will be more difficult to use in processes involving flow, such as injection molding. Such high melt viscosity, on the other hand, is advantageous in such processes as blow molding or vacuum forming.

The toughness of a polymer can be increased by adding rubbery particles as a second noncontinuous phase. Such particles disrupt crack propagation. Butadiene has this effect in toughened polystyrene and ABS.

Some thermoplastics have natural toughness because of their molecular shape. All high-molecular-weight polymers are entangled. Such materials can suffer sudden losses of properties when annealed by heating or by secondary operations such as ultrasonic welding, which apparently relaxes the molecules so that they lose their impact resistance.

In thermosetting resins, which are cross linked in their final form, the molecule is infinitely large. During processing, however, its size can be as small as that of the monomer. Some of the liquid resins used to produce thermosets can have very low viscosities and can be ideal for such processes as resin transfer molding and pultrusion, in which the resin must flow through preplaced glass reinforcements. For other processes, such as extra-high-strength molding compound and prepreg compression molding, the resin must be thickened by reaction to stay on the glass reinforcements. With sheet molding compound, bulk molding compound, or the spe-

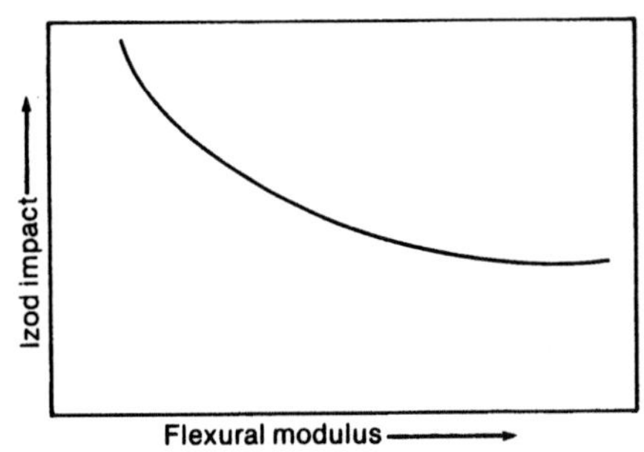

Fig. 25 Notched impact strength versus flexural modulus of ABS

*Adapted from Derek Gentle, Function and Properties Factors in Process Selection, *Engineering Plastics*, Volume 2 of the *Engineered Materials Handbook*, ASM International, 1988, pages 279 to 287

cially formulated ZMC, the resin must be chemically thickened so that it can carry fillers and glass reinforcement during process flow.

Shrinkage effects that occur during the processing of both thermoplastic and thermosetting resins are defined below.

Thermoplastic processing normally involves heating plastics granules, powders, or sheet to above the melting point of the polymer and then forcing the melt into a cooled mold until it is solid enough to handle. During this process, three shrinkage steps can take place. First, as the material phase goes from melt (liquid) to solid, a change in contraction rate occurs.

Second, some polymers crystallize below the melting point, and there is a volume change upon crystallization. Crystalline regions have a lower coefficient of expansion than amorphous regions. Therefore, crystalline polymers shrink up to five times as much as fully amorphous polymers. To be crystalline, the molecules must have the correct shape to be able to line up physically with each other and lie parallel in the crystalline areas without the stearic hindrance of side chains, for example. Polyolefins, polyamides, and polyesters tend to be crystalline, but atactic polycarbonates, polystyrenes, and acrylics are amorphous. Crystallization starts below the melting point, but can continue long after the product cools to room temperature—even for days or weeks—if room temperature is above the glass-transition temperature, T_g. This can cause unexpected warpage.

Third, normal thermal contraction during cooling will also take place. Contraction is anisotropic because of orientation. If the part is constrained during cooling, as in injection molding, stresses can be built in that will dissipate elastically as the part leaves the tool or that will appear when the part is in service, causing part distortion. This is most likely to occur if the part is subjected to high temperatures in service.

Rate of cooling and the temperature of the part upon extraction from the tool will also affect the amount of shrinkage, especially with a crystalline polymer. Thick sections will shrink at a different rate than thin sections and will be at a higher temperature leaving the tool. Shrinkage occurring outside the tool is less constrained.

Differential cooling in the tool, thickness differences, and ribs can all cause stresses between different areas of the part. These factors can also cause part distortion upon removal from the tool. Cooling fixtures can be used to constrain the part until it reaches room temperature after it is removed from the tool. The part can also be left in the tool longer, or the tool can be run at a different temperature to try to reduce warpage.

For some crystalline materials, nucleating agents can be used to increase the rate of crystallization. This will have the added benefit of reducing crystal size and giving more, better dispersed crystals.

The high pressures used in such processes as injection molding can both reduce some of the effects of shrinkage, by packing out the mold, and cause distortion of the product, by increasing internal stresses. With glass-fiber-reinforced materials, shrinkage of the resin away from the surface during molding can leave the glass fibers proud (raised above a surrounding area), resulting in an unacceptable surface finish. A similar effect is found with thermosetting materials. Some of the worst effects of shrinkage can be overcome by using blowing agents or high-pressure air to avoid sinkage over ribs and bosses or by using processes that reduce pressure gradients in the polymer melt.

Thermosetting reactions generally result in a volume loss. The degree of shrinkage varies with the type of reaction, the temperature at which the reaction takes place, and the type of bonds being formed.

After high-temperature reactions, the thermal contraction that occurs upon cooling can cause internal stresses and sometimes cracking of the more brittle resins. This can also cause loss of adhesion to reinforcing fibers.

Resin shrinkage also leaves glass fibers above the surface. By allowing the reaction to take place at the visible surface first, the mold surface is replicated; shrinkage of the resin occurs progressively toward the back surface, which will have a very poor appearance. Heating the face of the tool to a higher temperature can result in this effect. Part distortion often occurs as one surface is still reacting and shrinking after the other has solidified. Such effects can occur with most of the thermoset processes, but most easily with hand lay-up.

Shrinkage in thermosets is reduced by adding fillers, by adding a thermoplastic resin to absorb the monomer and expand during the heating cycle (as in the sheet molding compound process), by gas formation during the reaction (as in polyurethane reaction injection molding), or by adding a blowing agent.

The addition of fillers can significantly reduce the apparent shrinkage of both thermoplastic and thermosetting resins. This reduced shrinkage produces a much more stable part as molded. With higher filler loading in thermoplastics, the part is much less likely to distort upon removal from the tool or to warp in service. This increased stability is demonstrated with most processing methods, even thermoforming.

Fillers also increase the flexural modulus, but in most cases they also significantly reduce impact resistance. Filler shape is very important; platelike or fibrous fillers have greater stiffening effects and usually worsen impact resistance. They can also orient in the melt flow direction, giving anisotropic physical properties, which can lead to poor impact resistance in the cross-flow direction.

Rounded (particulate) fillers tend to have less significant effects. Some round calcite fillers have been shown to increase both stiffness and impact resistance in polyolefins. This effect may be related to the natural surface chemistry of the filler or to modification of the polymer crystalline structure.

Surface treatment of the fillers to increase (or in some cases to decrease) the adhesion of the polymer to the filler particles can improve impact resistance in comparison to untreated filler. This improvement is due to a reduction of the stress-concentration effect of the filler.

Fiber Reinforcement. The addition of glass, carbon, inorganic, or high-tensile organic fibers to a polymer will have a dramatic effect on its physical properties. These properties can vary from being similar to those of the base polymer, at low loadings, to approaching those of the reinforcement, at high loadings.

The form of the fiber is very important and has a significant effect on final physical properties. It can be very short, as with milled glass fiber, which would be less than 0.5 mm (0.020 in.) in length; short chopped to about 2 mm (0.08 in.); long chopped to 10 to 50 mm (0.4 to 2 in.); or continuous.

Glass and mineral reinforcement can also be used in flake form. Mica is one natural form of mineral flake.

Although this section uses glass as a primary reinforcement example, it is recognized that the use of inorganic fibers or whiskers or the various forms of carbon and organic fibers can provide better physical properties. Generally, adding reinforcement increases the stiffness of the plastic part. The greater the glass content, the greater the flexural modulus. This effect is true no matter what form of glass is used.

The advantage of long glass lengths is that with higher glass loadings, some physical properties, such as tensile and flexural strength, become more related to those of the reinforcement. With thermosetting resins loaded in excess of 60 wt% glass, the resin becomes simply a binder to hold the glass together. At lower glass contents and with short glass lengths, physical properties such as tensile and flexural strength still relate largely to the base polymer.

With shorter glass lengths, the impact resistance of the product varies inversely with stiffness. The use of polyurethane reaction injection molding for large automotive panels has always been a problem because adding milled or flake glass to increase stiffness becomes effective only at about 12 wt%, exactly the level at which the impact resistance is decreased dramatically, as shown in Fig. 26.

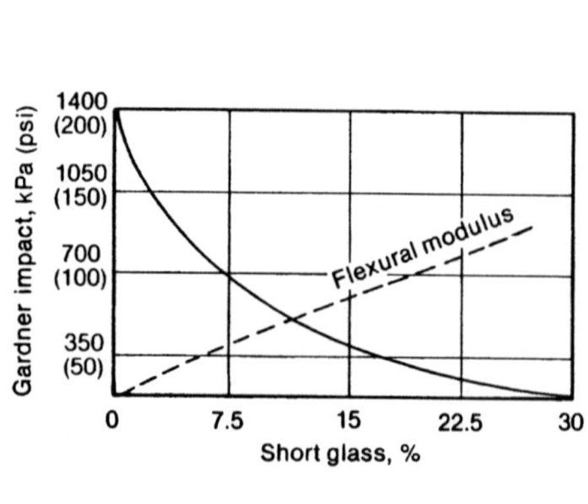

Fig. 26 Effect of short glass content in polybutylene terephthalate on Gardner impact values measured at 20 °C (70 °F)

With long glass, an apparent increase in impact resistance can be measured. This depends to some extent on the measurement system, but incorporation of long, chopped or continuous glass into either thermoplastic or thermosetting polymers can result in products with very high impact resistance. This occurs only if the elongation to break of the polymer matrix is high enough for the forces to be transferred to the reinforcement, as shown in Fig. 27.

With most amorphous thermoplastics, the flexural modulus stays relatively constant up to the region of the glass temperature or softening point, so that the material is useful up to that point. The addition of glass increases stiffness, but at the region of the polymer softening point, the properties of the material drop dramatically whether reinforced or not. Therefore, the useful temperature range of amorphous thermoplastics is unaffected by reinforcement.

On the other hand, crystalline thermoplastics tend to decrease slowly in stiffness with temperature and tend to be very susceptible to creep under load. Unreinforced crystalline plastics, therefore, have a useful range ending very much below their melting points. Reinforcement has a dramatic effect in that it increases the overall stiffness and creep resistance right up to the melting point, as shown in Fig. 28. By way of example, the heat-deflection temperature of polypropylene is 60 °C (140 °F), while it is 150 °C (300 °F) when reinforced with glass. Reinforcement is therefore much more useful with crystalline thermoplastics than with amorphous thermoplastics, in which a loss of impact properties tends to outweigh by far any gains in stiffness.

Many processes that rely on the polymer melt to carry the reinforcement as it flows into the mold will result in oriented reinforcing fibers. This can cause part distortion and weakness in the cross-fiber dimension. In structural components requiring maximum strength from the reinforcement, the fibers need to be placed in specific areas and directions in the part. This is difficult to achieve with processes that flow the polymers and reinforcement to fill the tool. The reinforcing fibers must be placed using a process such as filament winding, resin transfer molding, or hand lay-up and vacuum bagging.

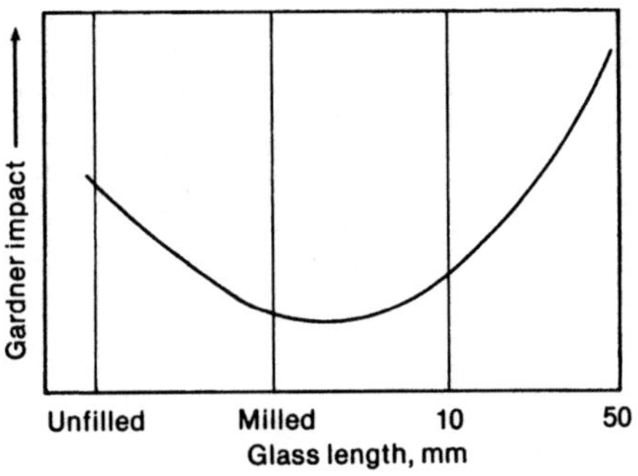

Fig. 27 Effect of glass length on Gardner impact strength

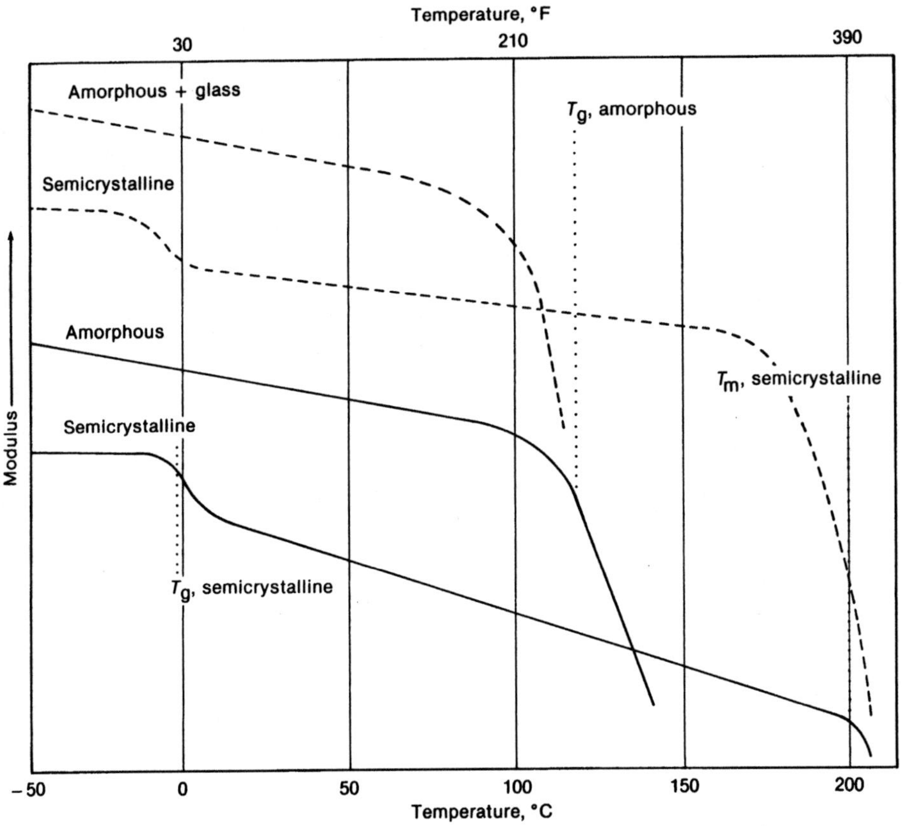

Fig. 28 Temperature versus modulus for different material types

Reinforcement Limitations. The ability of the various processes to handle reinforcements is defined in Table 3. Process and materials effects on surface finish, dimensional stability, stability of properties at elevated temperatures, and impact resistance properties are also given in Table 3. Effects are shown as directional only. With the very wide range of materials available, the overlap of performance among different processes is considerable. A compression-molded continuous-glass-reinforced thermoset could, for example, have a lower flexural modulus than an injection-molded unreinforced thermoplastic, depending on the choice of polymer.

Process Effects on Molecular Orientation

Polymer molecules in the melt form flow in a non-Newtonian manner and tend to line up in the direction of flow. Molecules in thermoplastic polymers are much longer than those in the unreacted liquid resin precursors to thermosets. The orientation effect in thermoplastics melt flow, such as occurs in injection molding, is much greater than that which occurs in such thermoset processes as reaction injection molding. Orientation of the thermoplastics molecule in the melt is frozen in as the melt solidifies upon cooling. This molecular orientation affects the properties of the product. For example, tensile strength in the direction of orientation is much greater than that in the cross-flow direction.

Molecules and fibrous reinforcement line up in the direction of flow when forced through an orifice. The material is stretched from higher pressure to lower pressure. Such effects occur in an extrusion nozzle or an injection gate. On the other hand, as the material goes through a tubular extrusion nozzle, it is also possible to orient the molecules or reinforcement perpendicular to the flow by constricting and then expanding the tube inside the nozzle while keeping the thickness constant. This stretches the extrusion radially and causes the molecules to orient around the tube.

Similar effects occur as the polymer flows out of the injection gate. If flow into the part is basically in one direction, it will orient in that direction. This takes place when a plate mold is filled from a film gate at one of the short sides.

However, if the melt is injected at the center of a disk or from the long side of a plate mold through a pin gate, the orientation of the polymer and the reinforcement can change. As the flow front expands, there is some tendency for orientation to occur across the flow. In some cases, the orientation at the surface will be different from that at the center, where the flow becomes more constrained as the mold fills. Mold temperature, part thickness, molecular and reinforcement length, and gate design all affect the orientation and therefore the shrinkage and built-in stresses. The more restricted the flow, the greater the orientation in the flow direction. Thin-wall parts are more oriented than thick-wall parts, as shown in Fig. 29.

An extreme example of orientation is that which occurs when polypropylene film is stretched in one direction. Although the film becomes very strong in that direction, it is easily fibrillated because of low mechanical properties in the direction perpendicular to the stretch. The resulting product is widely used in carpeting, ropes, and rot-proof string. Similar stretching is used in the production of high-tensile polyamides and polyester fibers for tires.

Thermoplastic Process Effects on Properties

As Table 3 indicates, the different processes are capable of producing parts with different physical properties. A brief discussion of the effects of individual thermoplastic processes on part and material properties follows.

Injection Molding. Parts can be produced with ribs, varying thickness, and superb surfaces, using almost all thermoplastics materials. Orientation of molecules and reinforcement occurs during the process. High pressures, nonuniform polymer shrinkage, and orientation can lead to warpage and sinkage over ribs and bosses. Warpage is most apparent with crystalline materials and with large, rather flat parts. Methods of controlling these effects are described below.

In injection molding, plastics granules are softened and forced under pressure into a cold mold through small orifices, or gates. Pressure is maintained on the material after injection is complete so as to reduce sinkage of the ribs and bosses as the material cools. Pressure is higher at the gates because it will not transfer effectively through the compressible and rapidly cooling melt. The additional packing pressure leads to a higher density of material near the gates and causes internal stresses. These stresses tend to be partially relieved when the part is removed from the tool, resulting in warpage.

The plastics melt must flow from the gates, through the narrow gap between cooled mold surfaces, to the edge of the tool. As the material flows, the gap becomes narrower as some of the melt solidifies at the mold surface. The pressure, flow rate, and distance between the mold faces must be great enough, and the material viscosity low enough, to fill the mold before the solidify-

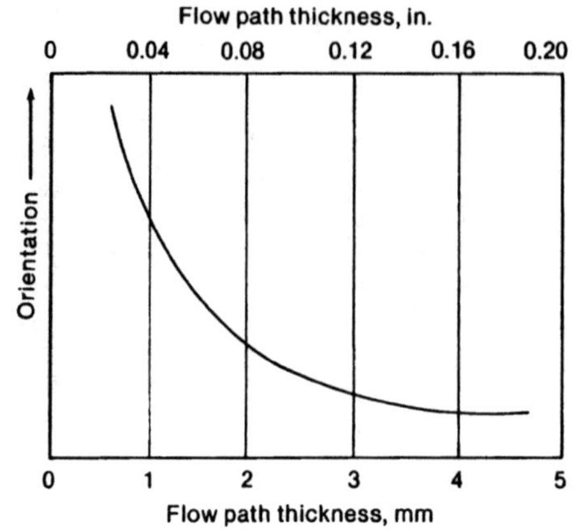

Fig. 29 Flow path thickness versus orientation

Table 3 Process reinforcement capabilities and selected properties

Process	Reinforcement: Type	Reinforcement: %	Limitations to reinforcement	Properties: Surface finish (++ to −−)	Properties: Flexural modulus, GPa	Properties: Flexural modulus, 10^6 psi	Properties: Temperature resistance (trend)	Properties: Tendency to warp (high to low)
Thermoplastics								
Injection	None	...	Longer than 2 mm (0.08 in.) is	++	2.0	0.3	Low	High
	Short glass	40	difficult; glass increases	+	4.0	0.6	to	High
	Long glass	50	stiffness but lowers impact	−−	7.0	1.0	high	Very high
Injection compression	None	...	Can handle glass up to 50 mm (2	++	2.0	0.3	Low	Slight
	Short glass	40	in.), but is better with 10 mm	+	4.0	0.6	to	Slight
	Long glass	60	(0.4 in.) or less	−	7.5	1.2	high	High
Hollow injection	None	...	Single large gate allows use of up	++	2.0	0.3	Low	Slight
	Short glass	40	to 10 mm (0.4 in.), but easier	+	4.0	0.6	to	Slight
	Long glass	50	with 2 mm (0.08 in.)	−−	7.0	1.0	medium	High
Foam injection	None	...	Glass tends to increase stiffness	−	2.5	0.4	Low	Slight
	Short glass	20	but spoils cell structure,	−−	5.0	0.7	to	High
	Long glass	40	surface, impact	−−	8.0	1.2	medium	High
Sandwich molding	Glass in core	40	<10 mm (<0.4 in.) is acceptable with development	++	3.5	0.5	Low	High
Compression	Very long glass	40	Glass orientation in ribs	−−	5.0	0.7	Medium	Slight
Stamping	None	...	Single thickness only, but up to	+	0.5	0.1	Low	High
	Very long glass	40	70% if cloth used	−	9.5	1.4	Medium	High
	Continuous	70		−	14.0	2.0	High	High
Blow molding	None	...	Very difficult	+ to −−	1.0	0.2	Low	High
Thermoforming	None	...	Very difficult	+	1.0	0.2	Low	Very high
Filament winding	Continuous	70	Shape limited	−−	40.0	6.0	High	Low
Rotational casting	None	...	Difficult	+	1.0	0.2	Low	Slight
Thermosetting								
Compression	Long glass (SMC)	30	Very long glass limits material	+	7.0	1.0	High	Very low
	+ very long (HMC)	50	flow and surface finish	−	15.0	2.1	High	Low
Injection	Long glass (BMC)	20	Difficult to handle if longer than	++	5.5	0.8	High	Very low
	Long glass (ZMC)	30	30 mm (1.2 in.); glass breakage	++	7.0	1.0	High	Very low
Stamping	Very long glass	70	No ribs or bosses	−	15.0–35.0	2.2–5.0	High	Slight
Reaction injection molding (RIM)	None	...	Reduced impact as glass loading	++	0.5	0.1	Low	Slight
	Short or flake	20	increases	+	2.0	0.3	Medium	High
Resinject (resin transfer molding)	Very long glass	55	Glass bulk limits maximum	+	12.5–27.5	1.8–4.0	High	Low
Foam polyurethane	None	...	Simple shapes and only very open	+ to −−	0.3	<0.1	Low	Slight
	Very long glass	30	glass structures	−−	7.0	1.0	Low	Slight
Filament winding	Continuous	70	Limited cross strength	−−	40.0	6.0	High	Low
Hand lay-up/vacuum bagging	Very long glass/cloth	30	...	++ to −−	7.0	1.0	High	Slight
Cold press	...	40	...	+ to −	11.0	1.5	High	Low

ing material closes off the flow path. For each material and part thickness, there is a maximum practical flow length from a gate.

The higher the pressure and the narrower the flow path, the greater the orientation. As the gap freezes off, the orientation becomes greater. Therefore, the orientation at the center of the part wall is much higher than that at the surface. For the same reason, orientation is highest near the gates. The gates should not be in areas that are likely to suffer impact or other stresses, such as chemical attack.

The maximum practical thickness of the parts is about 4 mm (0.16 in.); above this thickness, cooling time becomes excessive. The minimum normal thickness for injection molding is about 1 mm (0.04 in.); below this level, the part cools before the tool is filled, and orientation is excessive. Polystyrene drinking glasses, for example, will always split in the direction of flow when squeezed.

The largest readily available injection presses have about a 27 MN (3000 tonf) clamping force, which restricts part size to about 1 m^2 (10 ft^2) or less for more difficult and filled materials. The flow length of the plastics from any one gate is limited to about 500 mm (20 in.) with a 3 mm (0.12 in.) wall thickness. Therefore, multiple gates must be used for large parts. Gate design and position are very important for reducing part warpage and add to the complexity of orientation effects.

The strength and modulus values of parts produced by injection molding are limited by the inability of the process to handle reinforcement longer than a few millimeters without breaking the fibers or blocking the injection system. Although fillers and short fiber reinforcements can be added, this tends to produce stiffer parts having greater resistance to load at elevated temperature but much lower impact resistance.

Some specially formulated materials have been produced that contain glass approximately 10 mm (0.4 in.) in length. These materials can be used to a limited extent with injection molding, but are better suited to injection/compression processes.

The surface finish of injection-molded parts replicates the mold surface because it cools in contact with the surface, except over ribs and bosses. Part design must be aimed at keeping ribs and bosses away from the back side of visible surfaces, reducing material in the rib root, or using amorphous plastics that shrink less than crystalline types. With filled or reinforced materials, the surface tends to be dull and to show flow marks.

Cycle time is approximately 1 min. Therefore, injection molding is the most useful thermoplastics processing method, provided the size limitation, lack of ability to use long glass reinforcement, tendency toward warpage in flat parts, and sinkage over ribs can be designed around.

Injection compression molding is sometimes known as coining. The plastic melt is injected into the tool, which is held to a slightly greater opening than the ultimately desired part thickness. As the amount of injected material approaches the desired part weight, the tool is closed to compress the material and to fill out the tool.

It is important for surface quality that the tool closure start before injection stops and that the injection be completed before the tool is fully closed. This ensures that the material flow front does not stop.

Because material flows into the tool with the tool surfaces farther apart than normal, pressure requirements and orientation effects are less. Rate of injection can be higher because the flow path is more open.

As the tool is closed down to the final part thickness, the melt is squeezed to the edges of the tool. Orientation is less because the final melt is not being forced through a narrow channel by high pressure from the gate. Packing around the gate is eliminated because the injection is stopped before the tool is full. Flash is reduced because there is no sudden pressure peak, such as occurs in normal injection molding when tool fill is completed.

Longer glass, up to 50% of 50 mm (2 in.) long, can be handled if properly formulated because the lower injection pressures and larger gates allow the fibers to pass through more easily. With lower built-in stresses and less orientation, parts tend to exhibit much lower warpage when removed from the tool and less distortion and stress cracking in service.

Injection compression molding is most useful for large-area parts (up to 1.5 m^2, or 16 ft^2), and for reinforced components requiring minimum warpage. Sinkage over ribs is as bad or worse than with conventional injection molding because packing additional melt into ribs and bosses is not practical. However, the ability to add reinforcement could overcome the need to use ribs.

Although it is not widely used, injection compression molding does offer the opportunity to overcome some of the size, orientation, and reinforcement limitations of normal injection molding. Injection compression molding avoids the pressure peak obtained during normal injection. This allows larger parts to be made on the same tonnage machine as smaller parts. Internal stresses are lower because of a more even pressure distribution.

Flash is minimized, but a vertical flash or telescoping tool is necessary. This would normally have only one large, centrally located gate. Orientation is nearly eliminated. The need to use a vertical flash tool for this process limits its ability to be used for many parts, because of part shape. This is discussed in the section "Size, Shape, and Design Detail Factors in Process Selection" in this article.

Hollow injection molding was developed in the 1980s. High-pressure gas is injected into the polymer melt flow at the nozzle of the machine or at the gates of a hot manifold system. The gas flows through the areas of lowest viscosity at the hotter center of the melt. Polymer injection is stopped before the part is full, which allows the gas to fill out the molten areas. Final filling of the part is by gas pressure.

The molten areas must be designed to form a continuous path from the gate and along the ribs so that the gas pressure can be effective to the extremities of the part. Ribs must normally be widened at the root to allow for air passage.

Rib sinkage is reduced or eliminated by this process. Because the pressure peak is also eliminated, internal stresses and flashing are reduced, thus reducing warpage and finishing costs. Pressure on the mold is also reduced; therefore, much lower machine clamping tonnage is necessary, and the production of longer parts should be possible.

Surface finish is similar to that found in normal injection molding, with the added advantage of reduced rib sinkage. The process appears able to handle similar materials as normal injection molding. Limitations on reinforcement are similar to those of normal injection molding.

Hollow injection may require heavier wall sections than normal injection molding, and it may be useful to consider the process as being between normal injection molding and foam injection with an improved surface. Various alternative processes using the melt stream injection of liquid or solid blowing agents have been proposed.

Foam Injection Molding. Foamed thermoplastic parts can be produced by adding a heat-activated blowing agent to the plastics granules or by injecting gas into the polymer melt in the injection molding machine. Foaming does not occur while the melt containing the gas (produced either by the blowing agent or from gas injection) is under high pressure in the injection machine barrel. When the melt is injected into the mold, the trapped gas can expand to produce a foam.

There are many competing foam injection methods in which foaming is controlled by varying the speed of injection or by maintaining pressure on the melt as it fills the tool. Each method has its advantages, such as better foam structure, lower density, improved surface finish, or improved dimensional stability. Various machines have been developed to handle the differing processes. Because of the lower density of the foamed plastics, foaming gives a higher stiffness-to-weight ratio. On the other hand, to achieve foaming, the part thickness must be at least 4 mm (0.16 in.), and for densities below about 0.7 g/cm^3, a minimum thickness of about 6 mm (0.24 in.) is necessary to achieve a reasonable foam structure.

Cycle times are much longer than with other processes because of the greater part thickness. This is sometimes balanced by feeding more than one tool from each injection unit.

Typical foam parts have a surface made up of collapsed cells; this gives a swirl pattern somewhat similar to wood. A major advantage of the process is that the foaming action completely fills even large ribs and bosses, leaving a flat surface. This is an excellent system for articles requiring a massive internal rib and boss system for stiffness or attachment, provided a high gloss finish is not required.

The impact resistance of a foamed plastic is lower than that of the solid material; this reduces its usefulness in high-impact situations when using rigid plastics. Foamed resilient materials, however, can be used to produce massive impact-absorbing elements in such applications as automotive bumpers. Although the added thickness of foamed parts provides additional part stiffness, properties such as creep resistance, temperature resistance, and chemical resistance are similar or slightly worse than with the solid material.

Reinforcement is possible and slightly longer glass can be used with some process variations as compared to normal injection molding, but surface quality and impact resistance usually suffer significant reductions. With some of the process variations, the processing pressures are much lower than with normal injection molding, allowing very large parts of up to 2 m^2 (22 ft^2) to be produced on machines with a clamping force of about 9 MN (1000 tonf). This low pressure also tends to result in low internal stresses and very accurate dimensions. Lower-cost tooling is possible with the low pressures, but mold surface temperature control is critical to achieving dimensional accuracy.

A small amount of blowing agent is sometimes added to normal injection molding plastics to reduce rib sinkage. This is effective, but it is difficult to control the side effects.

Sandwich molding is used to produce parts with a skin of one material and a core of a different material. The skin material is normally unfilled and chosen for its good surface characteristics, while the core material is usually foamed to eliminate sinkage or reinforced to increase stiffness.

There are slight variations in the way sandwich molding is achieved, but the basic process relies on two injection units connected, through a switchable valve, to the gate system of a tool in a single clamp unit. The skin material is injected, and this is immediately followed by injection of the core material. The core material pushes the skin material to the extremities of the mold, laying down a solidifying layer of skin on the cooled mold surface as it passes.

In principle, sandwich molding can have the advantages of either a foam molding or a reinforced material injection, along with the perfect surface of unfilled plastics. With foaming, the part thickness must be similar to that used in foam injection. With a reinforced core, the total thickness must be about 1 mm (0.04 in.) thicker than that for normal injection molding. Most properties are as one would expect from the core materials, but impact is more related to the physical properties of the skin material.

Size limitations are similar to those found in normal injection molding, but multiple gating is difficult because the flow fronts always consist of skin material. Therefore, any weld areas are surrounded by pure skin material. Part corners also tend to be filled with skin material.

Mold shrinkage differences between skin and core materials can cause excessive distortion of the part near the skin-rich corners where the sandwich is not balanced. Careful selection of material to balance mold shrinkage of the core and skin materials must also consider thermal and moisture expansion effects while the part is in service.

Compression molding is one of the few thermoplastics processing methods that allows the use of very long or continuous reinforcement. There are two major variations of compression molding. One is flow molding, in which the heated plastic moves in three dimensions under the pressure exerted by the cold mold to fill the mold, carrying any reinforcement with it. Ribs and bosses are filled with plastic and reinforcement, but little true control of reinforcement orientation can be achieved even though oriented continuous fiber is used in the starting material. The other variation of compression molding is stamping, which is the deformation of a heated sheet of plastic under the pressure of a cold mold with minimal flow of material or change in reinforcement orientation. Only single-thickness parts are possible.

Flow forming is used for semistructural components containing heavy ribs and bosses in applications in which surface appearance is not important. Vehicle bumper structures and battery trays are typical components. Pressures of about 20 MPa (2.8 ksi) are necessary to flow and consolidate the part using commercially available material.

Stamping can be used for unfilled plastics where sheets are heated to above their temperature of crystallization (or above the T_g for amorphous materials) and formed in matched metal dies.

Cold forming can be carried out below T_g for amorphous polymers, or below the melt temperature, T_m, for crystalline polymers. Polypropylene can be rolled at room temperature. Some parts produced by cold forming have very high orientation of molecules in the plane of the thickness (biaxial orientation), because the parts are stretched in both length and breadth during forming. This makes them much stronger than would be expected from normal material properties. These products, such as margarine containers, tend to have very poor resistance to elevated temperature, although manipulating amorphous/crystalline changes with materials such as polyethylene terephthalate can produce crystalline parts with resistance to high temperature by stamping amorphous sheet in hot tools. Similar results can be achieved with forging, blow molding, and thermoforming. At higher forming temperatures, little orientation occurs, and parts have heat resistance almost up to the forming temperature. Cold forming is primarily used for large, single-thickness parts, such as vehicle fender liners.

Stamping is also used with heavily reinforced sheet to produce stiff, single-thickness components that replace metal stampings. Its main benefit is weight reduction and resistance to corrosion. With carbon fiber cloth reinforcement of high-temperature-resistant crystalline polymers, such as polysulfones and polyetheretherketones, this process is used for aircraft parts. With polymers such as polypropylene, it is used for car and truck belly pans. Some retention of fiber placement and orientation is possible.

With both flow forming and stamping, the process used to produce the sheet is critical to the final physical properties of the part. Sheet can be produced by laminating plastic films with interlayers of glass mat. Almost any form of glass mat can be used. The ability to use plastic films with different molecular and crystalline compositions in the center of the laminate can be an effective way of controlling crystallization in the subsequent stamping process. Incorporating both surface films (even metal) and glass veils is also possible. Equipment used for lamination can be large platen presses for batch production or rollers and double steel belt laminators for continuous production.

With continuous processes, the plastic films can be extruded straight into the laminating equipment to eliminate or reduce the troublesome heating stage of the lamination.

Plastic laminates can also be produced by similar methods that start with powder instead of film. With these various lamination processes, the glass tends to remain where it is placed, and the fiber bundles do not separate into individual glass monofilaments.

Sheet can also be produced by weaving precoated fibers into a fabric, which can be heated and stamped directly or preconsolidated into a sheet.

Glass-thermoplastic sheets for flow forming or stamping have been produced using papermaking processes. These composite sheets normally contain glass between 10 and 30 mm (0.4 and 1.2 in.) in length that has been separated into monofilaments during the wet processing stage.

Such materials can be used to produce a lightweight part by stamping without full consolidation. Such parts have specific gravities as low as 0.2, with good impact and stiffness, but comparatively low tensile strength.

When fully consolidated by stamping, the composite typically has a fairly good surface appearance because of the fineness of the glass. Physical properties tend to be very homogeneous, even after flow forming. Impact resistance, when fully consolidated, is different from that of composites formed from normal lamination processes.

Thermoforming is the forming of heated plastic sheet by the application of air pressure (pressure forming) or a vacuum between the heated sheet and the tool. The atmospheric pressure forces the sheet onto the tool, where it cools and retains the tool shape (vacuum forming).

It is impractical to form reinforced sheet, because of the low pressures involved and the tendency of the sheet to tear. Orientation and crystallization effects can be used to strengthen or modify the physical properties of the plastic material, but such techniques are of interest only for low-cost food containers and similar items.

Conventional blow molding cannot handle reinforcement. The only effect on the physical properties of the material caused by the processing is the tendency to orient the molecules in the direction of the flow through the extrusion head. This can ultimately lead to failure of the product due to splitting in the direction of flow when subjected to impact or stress corrosion.

In injection blow molding, an injection-molded preform is used instead of an extruded parison. This technique is particularly useful as a precursor for stretch blow molding, in which blowing is carried out at lower temperatures and with mechanical means or preform shapes to ensure that biaxial orientation takes place. Stiffness and strength are significantly increased by this process, as are other properties, such as resistance to the transfer of gases. Most carbonated beverage containers make use of this biaxial effect.

By using a thin outer layer of a low-viscosity polymer in a coextruded parison, it has been claimed that short glass reinforcement can be added to blow-molded materials to increase stiffness. The outer layer acts as a lubricant.

Adding an incompatible polymer to the base plastic material in blow molding can vary such properties as vapor transmission. The additive polymer tends to form platelets parallel to the part wall because of such processing forces as interfacial tension or viscosity. These platelets act as an interior barrier and could theoretically be used to modify stiffness and impact resistance. Care must be taken so that delamination does not occur, especially in the pinch-off area.

Filament Winding. By coating filaments with thermoplastic resins, filament winding can be carried out either by remelting the resin after the winding is complete or by melting the resin as the winding takes place. Fiber coating can be carried out by a variety of methods and supplied as coated, or preferably impregnated, fiber to the winding equipment. Cross-head extrusion coating is one method by which the resin could be applied on the machine itself. With filament winding, the fibers can be placed as desired if the geometric requirements of the process allow this.

Physical properties are very high, combining the strength of the fibers with the toughness of high-molecular-weight thermoplastics. Failure modes would tend to be by delamination at high temperature if the resin softens. Filament winding or thermoplastic pultrusion can be used as a precursor to stamping or compression molding as a way of producing preforms with the glass accurately oriented.

In rotational casting, a plastics powder or paste is placed in a hollow metal mold, which is then heated and rotated so that the plastic melt coats the inside of the mold. The mold is then cooled while still rotating.

Parts produced in this way are difficult to reinforce because the fibers tend to separate from the plastics. Internal stresses are very low, but there is risk of polymer degradation due to exposure to air. The process is used for decorative nonstructural parts or as an alternative to blow molding.

Thermosetting Process Effects on Properties

Thermosetting plastics can be divided into three major categories. The first group is unreinforced materials used to produce comparatively small components with good dimensional stability, stiffness, electrical properties, heat resistance, and comparatively low material costs. The manufacturing process used is usually injection molding, transfer molding, or compression molding, none of which greatly affects properties. The starting material is usually a powder of a partially reacted condensation product with a high filler loading of wood flour, cellulose powder, mineral, or thermoplastic to modify the properties and to reduce shrinkage.

Typical resins are phenolic, urea-formaldehyde, melamine, and polyimide. The products typically produced are electrical components, handles for cookware, automotive transmission components, and bathroom fittings.

The second category of thermosetting plastics consists of thermosets that result from the direct reaction of two or more liquid components in a mold to yield a foam or solid product. Foam molding, electrical potting, and reaction injection molding are typical processes in this category. Materials used are typically urethanes, silicones, and epoxies.

Processing has very little effect on the physical properties of the resultant product; chemical content and temperature of reaction are the most important factors. However, a gray area is encountered when long glass reinforcement is preplaced in the mold before the chemical reactants, which is more typical of the third category of thermosetting plastics.

The third category consists of those thermosetting plastics commonly referred to as fiberglass reinforced. Although the resin or chemical type does have a significant effect on the final properties of the part, processing plays an even more important role in determining the glass-resin relationship and therefore the physical properties of the final component.

In compression molding, the reactive resins and glass fibers are mixed outside the mold, allowed to react sufficiently to provide an intermediate product that can be handled (B-stage), placed in a heated matched metal mold (or die) and formed under high pressure until the reaction is complete. There are many versions of compression molding that give a variety of different physical properties, but typically it is difficult to control the positioning of glass because of material flow in the mold. Descriptions of the main compression molding processes and their major attributes follow.

Prepreg Molding. Glass cloth or mat is preimpregnated with a reactive resin mixture that is allowed to reach its B-stage, at which point the reaction stops, or is stopped, before cross linking starts. This prepreg can be cut to shape and laminated with other prepreg sheets to give the desired glass orientation. Pressure and heat to complete the reaction are then applied in a compression molding die or in an autoclave or oven using a vacuum bag if cure temperatures are low enough and only a few parts are required.

Glass content can be very high with this process, and glass positioning is very good. Strength depends on the choice of fabric and skill of placement. Very little movement occurs during molding. Ribs and bosses can be formed only by preplacement of the prepreg. Forming pressures must not be too high, or the glass will break where it crosses, reducing strength significantly.

The prepreg molding process is effective for producing very-high-strength composite structures, but it is very slow. If placement of the prepreg is not perfect, the result will be resin-rich areas on corners, at the part periphery, and in ribs and bosses. The process is mainly used in the aerospace industry, using epoxies or more exotic polymers. Usually, no fillers are used in order to achieve maximum strength. Resin shrinkage can be high, and surface finishing requires sanding and other labor-intensive operations.

Extra-high-strength molding compound (XMC) is a form of prepreg sheet produced by filament winding in an X-pattern onto a large drum. The looser glass weave allows some glass movement in the mold in order to fill out corners and part edges.

Mat molding is an early, lower-cost prepreg version based on chopped-glass mat. Variations of mat molding employ polyester, epoxy, or vinyl ester resins, and loose chopped glass to provide easier glass movement into ribs and bosses, as well as filler to reduce cost. Typical problems are lack of control of glass orientation and separation of glass and resin. Surface finish is poor because of resin shrinkage. Later changes involved thickening the resin so that it would carry the glass better and including additives to reduce shrinkage. Development progressed through low-profile and low-shrink resins to produce sheet molding and other molding compounds.

Sheet molding compound (SMC) is a mixture of resins, catalysts, fillers, chopped glass, thermoplastics additives to stop shrinkage, and thickening agents. Allowed to age until thickened with MgO or $Mg(OH)_2$, the sheet is stacked in the center of the hot mold, and it flows to fill the cavity under maintained high pressure. Cycle time is as low as 1 min.

With SMC, there is very little internal stress and no part shrinkage. Its comparatively low strength is due to the high filler content, low glass content, lack of control of glass placement, and maximum glass length of about 50 mm (2 in.). Some shorter glass, about 20 mm (0.8 in.), is added to obtain better flow into ribs and bosses, but orientation tends to be in the height of the rib or boss, which is the direction of flow. Lack of glass along the ribs and around the bosses leads to rib cracking under load and to bosses splitting when used for screw attachment.

If material flows around a hole in the part, the two flow fronts may not merge to give full strength. The glass tends to orient along the weld, leaving the weld line without glass. Therefore, this is an area of weakness that will soon show cracking in service. With SMC, it is often preferable to cut out the hole instead of molding it in.

Sheet molding compound has been specifically developed as a material and process to provide even distribution of glass fibers, no warpage, good surface finish in a part, and stiffness. One of its main uses is for exterior automotive body panels.

Bulk molding compound (BMC), dough molding compound, and thick molding compound are different physical forms of SMC that provide the mixture of resins, glass fillers, and so on, in the form of a log, lump, or slab rather than sheet. Physical properties are similar to those obtained with SMC.

High-strength SMC (HMC) is really derived more from prepreg and mat molding than from SMC, but SMC is commonly used to describe the compression molding of any material in sheet form. High-strength SMC is aimed at strength rather than surface finish and contains some continuous swirl, or continuous oriented glass, in addition to chopped glass.

Glass content is 50% more than in HMC, and filler content is reduced from that in SMC. Glass placement is still limited, but can at least be concentrated in one direction. It is commonly used for bumper beams, with the continuous glass oriented along the bumper for maximum flexural strength.

The use of vinyl ester resin instead of polyester is common in HMC because it produces a tougher product. When thickened to carry chopped glass, it can take up to 70 wt% glass. Even without continuous glass, a flexural modulus of more than 15 GPa (2.18×10^6 psi) is possible with formulations designed to flow into ribs and bosses and containing only 25 mm (1 in.) chopped glass.

Thermoset stamping describes a process for molding a high glass content vinyl ester resin sheet, produced on SMC-type equipment, with the glass oriented in preselected directions. The sheet is cut to fit the mold cavity and compression molded with minimal flow and glass movement. Thermoset stamping is similar to prepreg in principle, but is aimed at highly automated operation. The physical properties approach those of hand laid-up prepreg parts.

Injection molding uses materials similar to those used for compression molding. Bulk molding compound can be injection molded by using a warm barrel on a plunger-type injection molding machine and injecting into a hot mold. This product eliminates the surface porosity that causes a problem with painted SMC.

Glass fiber length starts at 25 mm (1 in.), which is further reduced during injection. Glass tends to orient more because of the longer flow lengths. Both of these effects cause the impact resistance of injection-molded parts to be lower than that of SMC parts.

ZMC is a French development, based on injection molding, that reduces glass breakage. With the use of specially formulated BMC, this method appears to overcome most of the disadvantages of injection molding without losing the advantages. Development is ongoing. Glass orientation is random at best, as in SMC.

Reaction injection molding (RIM) is a process that uses the second category of thermosetting resins (discussed previously). Liquid reactants are mixed at the entrance to the mold and react in the mold to form a solid or microcellular product. The development of RIM has taken place primarily with urethanes and ureas, which yield very tough, flexible products. The largest usage is for automotive bumper covers and spoilers.

Reinforced RIM (RRIM) is possible using glass that is in short chopped fiber form or flake form. Glass fibers cause part warpage because of fiber orientation and loss of impact resistance. Glass flake causes some loss of impact resistance, but increases stiffness and heat resistance and results in less warpage than fibers.

The recent development of polyurea formulations for RIM opens possibilities for higher stiffness and resistance to distortion at elevated temperatures. Low molding pressures promise the ability to produce components in excess of 2 m^2 (26 ft^2), but reaction speed limits flow time. The reinforcement of urethane foams with continuous glass mat (structural RIM), placed in the tool before injection, is a recent technique used to produce lightweight, stiff, impact-absorbing semistructural components.

Resin transfer molding (RTM) uses premixed liquid resins that are slowly injected into the mold, which is preloaded with glass fibers. The glass fibers can be laid in the mold by hand or as a preform. Very large components, such as sports car bodies or boats, have been made using this room-temperature process. Resins are formulated for slow room-temperature reaction and are usually polyester based. At its most primitive form, resins are gravity fed into the mold, although metering pumps are normally used. Vacuum is sometimes applied to the mold to aid filling.

Hand-laid fiber reinforcement can be placed at will, but the bulky nature of such hand lay-up does impose a practical upper glass loading of about 50 wt% because of the difficulty of closing the tool. Preforms involve a similar problem, and spray-up preforms in particular are very bulky and can hinder mold closure, especially if there are vertical sections in the part.

The ability to place foam sections in the tool to produce box sections is a significant advantage. Maximum flexural modulus is about 7 MPa (1 ksi), although some oriented areas could be much greater, but only in one direction. Resin transfer molding is a slightly improved version of hand lay-up and vacuum bagging techniques, which have similar glass content limitations.

With hand lay-up it is difficult to force more than 30 wt% by weight glass into the resin if mat is used. Cloth can increase the loading significantly. Bagging can result in contents as high as 60 wt%. The excess resin used to ensure wet-out of the glass tends to produce resin-rich areas on the back side of the molding. Both processes are difficult to control, but are widely used for low-volume production or very large parts.

Spray-up of resin and glass in one step is faster than hand lay-up, but produces parts with lower strength because of the limited glass length and content. It does not allow glass orientation.

Cold press molding is a technique that allows the use of low-cost resin tools or metal-faced resin tools, but results in a part with two good sides. Glass fiber mat, cloth, or preform is laid on one side of the tool, premixed liquid resin is poured onto the glass, and the tool is closed under pressure until the resin is cured.

Excess resin is squeezed out during the process, leaving glass loadings as high as 70%, but usually limited to about 40% by the addition of filler for improved surface appearance. Cold press molding is basically a low-cost alternative to SMC, and it yields a product superior to that produced by hand lay-up. Excess resin and glass must be sawed off after molding.

High-speed resin injection and high-speed RTM are two of the many names being used for a combination of RIM and RTM technology. High-speed high-pressure RIM injection/mixing units are used to inject fast-reacting resins into molds containing glass preforms. New, fast resins have been developed for this process, based on acrylic, polyurea, polyester, and other materials of mixed chemistry. Difficulties include pushing the reactive mixture through the glass before its viscosity becomes too high, bubbles caused by turbulence, and maintaining glass position.

Maximum glass loading depends on the resin, but is normally between 40 and 50%. An open glass structure is necessary. To make use of the high speeds, automatically formable glass systems are required. Glass preforms are mostly produced from continuous swirl mat. Accurate placement of the preform so as to avoid resin-rich edges or glass in the seal surface of the mold is still a difficulty.

Filament winding is perhaps the best automatic process for achieving high glass loadings in the desired orientation, but it does have geometric limitations. Application of the preimpregnated fiber to the surface of the tool requires that it be kept in tension. This is not possible on a concave surface. Some apparently ideal shapes, such as spheres, can be produced only if the mandrel is left inside.

It is a highly suitable process for producing symmetrical parts requiring very high strength in one direction, such as pressure tubes, or for wrapping pressure vessels. The process is also useful for producing continuous oriented fiber preforms for use in compression molding or RTM, probably in combination with other preform processes. This may well be its major outlet with the increasing requirement for areas of oriented glass in large structural components.

Process Combinations. It is becoming increasingly difficult to divide the various methods available for producing fiberglass-reinforced materials into clearly defined processes. Processes for bringing resin and glass together are being combined; therefore, there are no longer clearly defined boundaries related to physical properties. Size, shape, speed, and finish may become the important selection parameters.

Size, Shape, and Design Detail Factors in Process Selection*

Part size is limited by process pressure and available equipment, whereas the ability to achieve particular shape and design detail is dependent on the way the process operates.

As a rule of thumb, it can be said that the lower the processing pressure, the larger the part that can be produced. Other restrictions are the size of the equipment that is available, the length of flow, and the material reaction time.

Generalizations that are basically interrelated and that usually hold true are:

- The more automated and mechanized the process the greater the number of restrictions for producing a large part
- Conversely, the more hand labor involved in production, the larger the part that can be produced
- The slower the reaction, the larger the part that can be produced

In order to handle materials that react (that is, convert from liquid or melt to solid) very fast, it is necessary to use a mechanical process such as injection molding, which as a process is still limited by the time available to fill the mold before the material solidifies; thus, high pressures are used to increase the speed of mold filling.

With most labor-intensive methods, such as hand lay-up, very slow-reacting thermosetting resins are used, and there is virtually no limit on size. With some processes, size is limited only by the size of equipment that is available or can be produced.

*Adapted from Derek Gentle, Size, Shape, and Design Detail Factors in Process Selection, *Engineering Plastics*, Volume 2 of the *Engineered Materials Handbook*, ASM International, 1988, p 288–292

Each process has certain characteristics, which determine whether:

- Ribs and bosses are feasible, depending on whether one or both sides of the part reproduce the tool surface.
- The sequence of material injection and tool closure allows deep vertical sections in the surface wall.
- Material viscosity is high enough to allow the use of slides and cores in the tool without their being gummed up with material flowing into the slide mechanism.
- Hollow sections or containers are feasible.
- Hollow or foam-filled box sections can be produced to increase section stiffness.

This section illustrates the thinking that goes into the selection of processes for size and shape factors. The information herein should be considered as a guideline only. No sharp distinction has been made between processes. Furthermore, processes are always being combined and modified.

Injection molding of thermoplastics and compression molding of thermosetting plastics will continue to be the most useful, cost-effective ways of producing high-quality, high-volume parts. Other processes should be considered, however, when these basic methods fail to satisfy specific requirements in terms of physical properties, size, shape, or overall cost effectiveness related to production volume.

Part Size Factors in Process Selection

Pressure applied to the material during processing varies with material type, material viscosity, time available for material flow, and other parameters. With simple hand lay-up of polyester resin and glass fiber, pressure is limited to the use of hand rollers to ensure that the glass is adequately wet out. With cold press molding using similar materials, 1 MPa (0.15 ksi) pressure is applied to squeeze the resin through the glass. Further pressure increments are necessary for prepreg molding, high-speed RTM, and SMC, all of which can be polyester-glass combinations.

When high pressures are used in processing, the maximum part size is directly related to the force available to hold the mold together. In injection molding, this is usually known as clamp tonnage, press tonnage, or machine size. This force is limited, although machines of up to 100 MN (10,000 tonf) have been produced. The largest injection molding machines in general use are 30 MN (3000 tonf). As processing pressures are about 40 MPa (6 ksi), the maximum projected area of a part produced by this method is about 0.75 m^2 (8 ft^2). Although actual size varies with the type of material, wall thickness, and ability of the machine controls to predict mold fill, this is the order of part size possibility with the process.

Shape and Design Detail Factors in Process Selection

Both shape and design details are very process related. The ability to mold ribs, for example, may depend on the material flow during a process, or on the flowability of resin that is reinforced with glass. The ability to produce hollow shapes depends on the ability to use removable cores, including air, meltable or soluble solids, and sand. Pseudohollow shapes can also be produced using cores that remain in the part, such as foam inserts in RTM.

Another consideration when selecting a material or process for a part is its ability to be joined to another piece to complete the total part, which may require the use of welding or adhesives. Welding is usually a preferable method because it does not require the additional cost and complexity of a secondary material. Welding of thermoplastics is generally feasible, but it must be remembered that normally the weld will not contain reinforcement and may be an area of weakness. With thermosetting materials, welding is not feasible, and therefore adhesives or mechanical fasteners must be used.

Shape and Design Detail in Thermoplastics Processing

Table 1 gives an overview of the sizes and shapes that can be produced using commonly available processes. This list of processes is larger than that appearing in Table 3 because slight process modifications can have a considerable effect on achievable part size, while hardly affecting physical properties. Descriptions of each process cited in the table follow.

Injection molding is a high-pressure process that is limited by the clamp force capability of generally available equipment, to a maximum projected area (0.75 m^2 or 8 ft^2). Projected area is the area of a plan view of the part in the plane of the platen surface (perpendicular to the applied clamp force). This assumes that part thickness is about 3 mm (0.12 in.), which is an average thickness for a large injection-molded part.

At this thickness, the flow length from a gate will be about 600 mm (24 in.), making multiple gates necessary for large parts. The process details described later indicate the types of gates to use to avoid surface marks. Much greater wall thickness can be used in injection molding, allowing larger parts to be produced, but this results in excessively long molding cycles.

It must be realized that the sizes and pressures quoted in this section are for guidance only in comparing processes. With the wide range of materials available and with the ability to vary processing conditions, much larger parts can be produced by injection molding, but such parts are exceptions rather than customary production practice.

As the molten material is injected into a closed mold, the material solidifies in the basic image of both mold surfaces. Ribs and bosses can be produced, and there is no difficulty in producing vertical walls. All surfaces must have a slight draft angle to allow easy part extraction. Thus, the walls cannot be precisely vertical, but they can be within a variance of one or two degrees, depending on the surface finish. A grained surface, for example, requires a greater draft angle to stop the grain from forming an undercut.

Undercuts can be produced using slides and cores in the mold, as can ribs that are not parallel to the mold-opening direction. The sequence in which slides and cores are removed depends on their position and action, but obviously requires careful planning to prevent the mold from locking up.

Although undercuts can be produced, true hollow shapes and box sections cannot be made by injection molding, except in the mold-opening direction. For example, a bottle shape cannot be produced because it is impossible to remove to mold from the inside. However, because the material is thermoplastic, a bottle shape could be produced by injection molding two halves and welding the halves together.

Injection molding produces parts with a good finish on each side and allows production of a variety of cross sections, but one must expect and allow for the molded-in stresses that can occur when different thicknesses are present. Injection molding remains the most efficient method for high-volume production of small- to medium-size thermoplastic components.

In injection compression molding, material is injected into a partially open mold, which is then closed to force the material into the extremities of the mold. Because there is no high-pressure peak as fill is completed, the maximum processing pressure on the mold is about half that of normal injection molding. Therefore, the size of part that can be produced on the same equipment is twice that produced with injection molding. However, one must be careful to ensure that the platen size of the machine is large enough to accommodate the larger mold.

The telescoping, vertical flash type of tool used for injection compression molding makes the use of slides and cores more difficult mechanically. It also affects vertical sections of the part, in that the mold closing does not change the distance between the walls on vertical sections. This can interfere with material flow or result in wall thickness variation in the finished part. Injection compression is most often used for large, flat components with vertical sections that are restricted to the edges and that have no undercuts. In the case of molded openings, this process does not suffice; a posttrimming operation is needed.

Overall, injection compression processing results are similar to those of normal injection molding. However, there are very few machines set up to allow the simultaneous injection and mold closure that is necessary for injection compression.

Hollow Injection Molding. The lower pressures of this technique allow the production of parts of up to 2 m^2 (22 ft^2), provided that sufficiently large ribs are designed in to allow for polymer melt and gas flow to the mold cavity extremities from a central gate. The use of multiple gating can be complicated. Even with one gate, it is difficult to design the rib system to avoid trapping air between the ribs.

Apart from its lower pressures, the ability to produce large ribs and bosses without bad sinks on the surface is the major advantage of the process. Because ribs and bosses will be hollow to some extent, they may not give as much stiffness and retention as expected.

This process gives low ribs with some box section properties, but there is a definite limit to the ratio of wall thickness to hollow section. Thicker cross sections may well be finished with some gas in the section and may act as two walls with a gap in the middle. Most properties are similar to those of normal injection molding.

Foam Injection Molding. Most foam molding equipment and molds are designed for low pressures, not high pressures. There are many alternative foam processes, most of which are capable of producing parts up to about 3 m^2 (30 ft^2), but there is not much equipment available for this size. Rather, available equipment is primarily intended for producing parts up to about 2 m^2 (20 ft^2) on low-tonnage machines.

The advantage of the foam injection molding process is the capability for producing thick components with high section stiffness, multiple ribs and bosses, and a good, flat surface without sink marks. The comparatively rough surface caused by collapsing gas bubbles is normally disguised by the use of graining and painting. Varying cross sections are not a problem, provided that material flow is considered during mold design. This process is widely used for business machine housings and for massive products, such as transport pallets.

Sandwich molding uses lower pressures than normal injection molding does because part thickness is greater to accommodate the skin/core combination, especially if the core is foamed. A limitation is imposed by the use of a single gate, which is necessary on large, visible parts to avoid trapped webs of skin material.

It is important to avoid major disruptions of the melt flow, such as by holes, ribs, and bosses, because they can cause the core material to burst through the skin. Other factors are similar to those of normal injection molding.

Compression molding is mainly used with continuous or very long fiber reinforced material that is heated in sheet form until the polymer melts to give a lofted, fibrous rectangle. This hot mass is placed on one half of a cold mold, which is then closed, forcing the material to flow and fill the mold.

The high pressure of about 20 MPa (3 ksi) that is used to form well-consolidated parts limits its practical size to about 1.5 m^2 (16 ft^2). Causing the material to flow too far can result in glass-polymer separation and insufficient glass in deep ribs.

Compression molding is used for semi-structural, heavy parts with ribs, bosses, and complex shapes. Hollow shapes, such as spheres and box sections, are not possible.

Stamping is similar to compression molding when using reinforced material, except that the starting sheet is blanked such that it almost fills the tool. The sheet is heated just to the melting point, and not enough to loft. When used with unfilled material, this process is similar in that a sheet is softened to be formable without melting. With this process, there is very little flow of material. Therefore, ribs, bosses, and vertical walls are to be avoided. For the same reason, holes should be cut out after molding, and variable wall thickness is not viable.

Extrusion is used to produce either very wide sheet or complex, two-dimensional sections, including tubes and multiple hollow sections. Postforming of the simpler shapes, such as round tubes, is possible. They can be end-to-end welded to form such products as vinyl window frames. Sheet is used for further processing by thermoforming or thermoplastics stamping.

In blow molding, the round, extruded tube, or parison, is trapped in a hollow mold while it is still hot, or it is reheated. High-pressure air is then blown into the trapped parison to force it out against the cold walls of the mold. It solidifies under the air pressure on the mold walls. When removed from the mold, the excess tube, flattened by the mold sealing surface, is cut off, and the material is recycled.

Equipment size availability and the type of material used are both important factors affecting part size. Maximum size depends on the size and melt strength of the parison. High-molecular-weight, high-density polyethylene (HDPE) is one of the easiest materials to blow mold because of its very high melt viscosity. Large fuel storage tanks of about 2 m (6.5 ft) in height by 1 m (3 ft) across in width have been produced in Europe.

The hollow shapes produced by the process obviously do not allow the use of bosses, but internal ribs can be produced by forcing the material to form internal webs over rods located inside the tool.

Undercuts are possible, although difficult, to achieve, because the part must be sprung off the tool. Tube attachments can be welded to the outside of the part using spin or hot-plate welding, but because a reactive force is not possible from the inside of the tank, weld flanges should be as large as possible.

Box sections and solid areas can be produced by forcing the molten sides of the parison together with the tool. Much higher clamp forces are necessary to achieve this. Large, flat objects with some beam stiffness can be formed by flattening the parison before closing the tool, which converts the process into a combination of pressure thermoforming and twin-sheet forming. The inside surface of blow-molded parts tends to be very irregular. Therefore, it must be realized that when open-topped articles are formed by blow molding and separation, the inside will have a poor appearance.

Twin-sheet forming is very similar to blow molding, except that instead of using a hot tube as the starting point, it uses two plastic sheets that are heated, trapped in the mold, and blown onto the mold surfaces. Care must be taken to keep the hot sheets apart in the mold. This process is used for producing very large, flat components, usually on modified vacuum forming equipment. Size is limited only by the width of sheet that is available.

Twin-sheet stamping uses conventional hydraulic compression molding machines to form two heated sheets, usually of reinforced plastics, around a foam core. The core provides the reactive pressure to keep the sheets in contact with the mold surface until cold. Elastic bladders or high-pressure air can be used instead of foam. The process is used only for producing flat, panellike structures with areas of plastic-to-plastic contact and foam-filled box sections. It provides greater stiffness than a single-sheet method and uses materials that cannot be processed with direct air pressure.

Thermoforming is often referred to as vacuum forming, but it also includes pressure forming, in which high-pressure air assists the atmospheric pressure used by vacuum forming. Available equipment, comparative simplicity, and low cost make vacuum forming suitable for the production of very large, single-thickness parts. Heated plastic sheet is forced onto the mold surface as air is evacuated from the space between the sheet and the mold.

A single-thickness part can be formed to quite complex shapes. Even undercuts can be produced by using hand-removed mold inserts, but the use of ribs, bosses, hollow shapes, variation in wall thickness, and box sections is not possible.

Wall thickness variation does occur with the process, but not by choice. The material thickness varies as it is stretched over the tool. One surface, the tool side, will match the tool surface, but the other side will be rounded over styling lines because of the material thickness. In spite of this rounding, the back surface is usually used as the visible part surface, because it is free of the various tool marks that come from using low-cost tooling. Both vacuum and pressure forming are used for high-volume production of low-cost food containers, trays, and cups, using both solid and foamed starting materials.

Filament winding is ideally suited for producing an item with one axis of symmetry, such as cones, tubes, and pressure vessels. With pressure vessels, the mandrel cannot be removed from inside the part; instead, it becomes the inner surface. With thermoplastics, filament winding is often the first fiber placement stage before compression molding. Most shapes can be produced with sophisticated equipment, provided that the fibers can be kept in tension.

In rotational casting, polymer powder or paste is placed inside a rotating, heated, hollow tool, where it melts or gels on the inner surface. The still-rotating tool is cooled, usually with water spray, until the plastic solidifies. Properties are similar to those obtained with blow molding, except that ribs cannot be formed, although they can sometimes be cast in by placing preforms in the tool. This, however, is very difficult because it interferes with powder movement.

Shape and Design Detail in Thermoset Processing

Compression molding covers a wide range of material forms and processing pressures.

Powder compression molding places a B-stage, or partially reacted, polymer in powder form (mixed with particulate filler) in a heated compression tool (telescoping). Pressure is applied as the tool closes. The very high pressures and type of material limit this process to fairly small components, such as electrical fittings, low-cost dinnerware, and handles for cookware. Complex shapes (including ribs, bosses, and vertical walls) and varying part thicknesses are both possible, but hollow sections are not. Also, because welding is not possible with thermosets, adhesives or mechanical fasteners are used for joining.

Sheet molding compound (SMC) has a process pressure requirement that ranges from 6 to 20 MPa (0.85 to 3 ksi) to form components with adequate surface and physical properties. Maximum size is limited by machine availability to about 2 m^2 (20 ft^2). Developed specifically for replacement of metal surface panels in appliances and automobiles, this process is not normally used for components smaller than 0.09 m^2 (1 ft^2).

Because of very low shrinkage, ribs and bosses are easy to produce, but such is not the case with undercuts using slides and cores because of the telescoping action of the tool and the risk of getting material into the mechanism. Hollow sections and welding are not possible, but both sides of the part have a good finish, and varying cross sections are not a problem.

Cold-press molding involves placing a glass fiber preform in the tool, pouring in premixed resin (usually polyester), closing the tool, and allowing the reaction to occur. Ribs and bosses can be formed only if these sections of the tool are carefully stuffed with glass fiber when laying in the preform. Parts can be as large as the low-tonnage press that is needed to separate the low-cost tooling. Surface quality varies with the quality of the tool, but suffers from the use of unfilled resins.

Low resin viscosity and low-cost tooling limit the use of cores and slides, although hand-removed inserts can be used with this slow process if enough release agent is used.

Hot-press molding is similar to cold-press molding, except that a heated, matched-metal mold is used. The reaction is faster, and therefore more pressure is required to force the resin to fill the tool before it becomes too viscous. This process has some limits due to press size and availability.

Shape limitations are similar to those encountered in cold-press molding, except that low-profile additives can be used with the resin to improve surface finish, and hand-removed mold inserts are more difficult to handle because the process is both hotter and faster.

High-strength SMC (HMC) contains less filler and no low-profile additives, compared to conventional SMC. It usually contains oriented, continuous glass fibers, but can use random, chopped glass. The material can be molded at lower pressures, so somewhat larger components can be produced on the same equipment, but there is a limit to part complexity due to the oriented glass. Lower resin viscosity demands care with use of slides and cores. As with all high-pressure processes, hollow sections are not possible.

Prepreg use is similar to HMC, except that the glass reinforcement, in the form of cloth or chopped mat, will not go into ribs or bosses. Prepregs generally have highly oriented, continuous fibers in a matrix. Flow is minimal and orientation of prepreg plies is closely controlled. Very large parts of single thickness can be produced on fairly simple, light-construction presses, which must be strong enough to open the tool.

A vacuum bag process is used with prepreg sheet or liquid resins and fiberglass cloth or mat. The resin-glass composite is laid up on a one-sided mold and covered with a plastic film; a vacuum is then drawn between the film and the tool. Air pressure consolidates the resin-glass.

Depending on the resins used, reaction can occur at room temperature over several hours, or in an oven or autoclave. The autoclave gives higher pressures, but restricts size to that of the autoclave diameter. The air or oven cure sets almost no limit on size. Shape restrictions are similar to those in cold-press molding, but only one good side is produced.

Hand lay-up is a slow, labor-intensive process that can be used to produce components of quite large size. As with the vacuum bag process, foam sections can be incorporated to give stiffening box sections, and forms of ribs and bosses can be produced, but only on the face side of the part. Back surface quality is very poor, but for parts of 1 m^2 (10 ft^2) or larger, hand lay-up is a very useful but slow process.

The spray-up process uses chopped glass and a thermosetting resin, which are sprayed onto a gel-coated form. Only one side is smooth. However, spray-up is an important process for making boats, bathroom enclosures, truck parts, and low-volume recreational vehicles.

Injection molding is an alternative to compression molding when using powder and bulk molding compound (BMC), which is equivalent to SMC.

Powder molding uses very high injection pressures to fill the tool. It is used only for small parts when a more automated process than compression molding of powder is desired. This method does use a closed mold, which makes the use of slides and cores somewhat easier.

Bulk molding compound (BMC) or the modern version, ZMC, is used instead of SMC largely because of the nonporous surface that is obtained. Because of the high pressures and the difficulty of flowing glass through the mold, the largest feasible part at this time is less than 1 m^2 (10 ft^2).

Although similar to SMC in terms of shape factors, the process does benefit from using a closed-mold system, which simplifies the use of slides and cores for producing undercuts. Lack of readily available equipment is a serious drawback to the use of BMC and, especially, of ZMC.

Thermoset stamping is a modern equivalent of prepreg molding. It uses very high glass content resin sheet, with the glass carefully oriented in the sheet. The sheet is blanked to fit the mold cavity, and compression molded at comparatively low pressure, with minimal flow.

Large, single-thickness parts similar to steel in depth of draw are produced. Ribs, bosses, and hollow sections are not possible.

With reaction injection molding, part size is limited by the size of the presses available and the flow length of the resin reactants. Ribs, bosses, undercuts, and varying wall thickness are all possible. The greatest problem is trapped air bubbles associated with ribs and bosses. Materials cannot be welded. The use of this process is limited by the low modulus of the material.

Resin transfer molding, or resinject, is similar to cold-press molding except that after the glass preform is laid in the mold and the mold is closed, the resin is injected into the mold at a very slow rate. The very low pressures involved, with gravity often providing the filling pressure, allows the use of very low-cost, lightweight tooling. Part size is limited only by the ability to produce a double-sized mold. Fairly complex undercuts can be produced using hand-removed mold inserts. The advantage of resinject over hand lay-up is that both sides of the part have a reasonable surface. Foam inserts can also be used to provide stiffening box sections. Care must be taken to ensure progressive resin fill to avoid air entrapment, which can make the use of varying wall thickness impracticable.

High-speed resin transfer molding uses high-speed mixing pumps to inject the resins. Standard RIM equipment is sometimes used. A major difference between this process and RIM is that the resin is usually injected near the center of the part instead of at the edge. The high pressures that can be used in order to inject some of the newer, fast-reacting resins can make the use of foam inserts difficult. In addition, the higher production speeds make the stuffing of glass into ribs and bosses impractical. With larger parts, flow paths must be carefully calculated to avoid air entrapment. This is especially necessary if foam cores are used.

There is a continuous range of process capabilities between slow-speed and high-speed RTM, depending on the balance of speed and part complexity required. With all RTMs, it is preferable to use foam inserts at the edge, unless the edge can be cut to size, because placing glass exactly to the edge is not only very difficult, it can lead to thick parts if it is trapped in the tool flash or in resin-rich edges. Wrapping round foam makes it easier to place the glass.

In foam urethane molding, foam pressures are much higher than is generally thought and do require some form of press or mold closure device. Generally, the thinner the part, the higher the fill pressure. This process is used to produce massive sections, but in some self-skinning versions, the process is similar to RIM in molding requirements and pressures. Modulus varies from very low to high. Almost any shape, even a sphere, can be produced, but not hollow objects. Obviously, the process is used for different applications, based on most materials and processes discussed in this section.

Reinforced Foam. Urethane foam can be reinforced by placing glass preform in the tool, before injecting or pouring in the mixed foam reactants. As a process, this is very similar to cold-press molding or RTM. The resulting product has a high section modulus, due to thickness at low weight, and good impact resistance. Putting glass into ribs and bosses is a problem that is found in other processes using preforms. Part size is limited to press availability, but very large parts can be produced using open pouring, which avoids the problem of having foam flow too far through the glass.

Filament Winding. Like thermoplastics filament winding, this process is very specialized and would not usually be compared to other processes when considering size and shape factors for process selection, although a large variety of sizes and shapes are possible. Very complex three-dimensional parts can be produced using computer-controlled six-axis machines. This process can produce rail cars, plane fuselages, and large-diameter tanks, as well as tubing, pipes, and spherical pressure vessels.

Pultrusion, which is similar to extrusion, would only be selected when constant sections are required.

SELECTED REFERENCES

- J.F. Agassant, P. Avenas, J. Sergent, and P.J. Carreau, *Polymer Processing: Principles and Modeling,* Hanser Gardner Publications, 1991
- R.D. Beck, *Plastic Product Design,* 2nd ed., Van Nostrand Reinhold, 1980
- M.L. Berins, Ed., *Plastics Engineering Handbook of the Society of the Plastics Industry, Inc.,* 5th ed., Van Nostrand Reinhold, 1991
- J.-M. Charrier, *Polymeric Materials and Processing: Plastics, Elastomers, and Composites,* Hanser Gardner Publications, 1990
- *Engineering Plastics,* Vol 2, *Engineered Materials Handbook,* ASM International, 1988
- S. Levy and J.H. DuBois, *Plastic Product Design Engineering Handbook,* 2nd ed., Chapman and Hall, 1985
- E. Miller, *Plastic Product Design Handbook,* Part A, Marcel Dekker, 1981
- P. Mitchell, Ed., *Plastic Part Manufacturing,* Vol 8, *Tool and Manufacturing Engineers Handbook,* Society of Manufacturing Engineers, 1996
- E.A. Muccio, *Plastic Part Technology,* ASM International, 1991
- E.A. Muccio, *Plastics Processing Technology,* ASM International, 1994
- *Polymer Engineering Principles: Properties, Process, and Tests for Design,* Hanser Gardner Publications, 1993
- P.A. Tres, *Designing Plastic Parts for Assembly,* 2nd ed., Hanser Gardner Publications, 1995

Physical, Chemical, and Thermal Analysis of Plastics

Physical, Chemical, and Thermal Analysis of Thermoset Resins*

THERMOSETTING RESINS are unique among engineering materials. Using these materials to fabricate hardware depends on a chemical process, in contrast to the more conventional metallic techniques, such as machining, heat forming, and mechanical fastening. Thermoset processing is further complicated in that its chemical reactions convert the materials into an infusible, insoluble material. This dictates that chemical methods must be used to control fabrication and that physical, chemical, and thermal testing must be performed on the starting materials rather than the finished product.

The quality of thermoset hardware can only be assured by either the postfabrication testing of tagalong test specimens in conjunction with production hardware or by being certain that the proper raw materials were used and processed correctly. The first procedure only proves that the manufacturer has made a good or bad test specimen. It says little about the quality of the product. As a consequence, the latter approach of materials and processing quality control has been pursued. Instrumentation also permits a better understanding of thermoset resin formulations and their processing. This basic control concept is the basis of this article.

Because chemical reactions are involved in thermoset processing, three factors should be considered to ensure successful hardware fabrication. First, the user must ascertain that the starting raw material has been properly formulated. The term formulation is used because all thermoset materials consist of at least two reactive components, which must be present in the proper ratios. Next the user must be assured that the material will cure or react properly within predefined processing conditions, followed by an assurance that proper processing has been conducted.

Methods of characterization and quality control continue to improve since the first inception of this control concept through an Air Force contract (Ref 1). Numerous procedures and methods of physical, chemical, and thermal analysis for thermosetting polymer systems have evolved from this work and subsequent efforts. Most important among these are chromatographic and rheological determinations. Chromatography ensures that the proper formulative constituents are present in the correct amounts and has proved invaluable in the identification of the components in new formulations. Rheology ensures that the resin will process according to a predescribed pattern. Other useful tools are infrared spectroscopy for the qualitative and quantitative analysis of raw materials and, in certain cases, extent of cure; thermal analysis techniques, including differential scanning calorimetry and thermogravimetric and thermomechanical analyses for chemical reactivity and extent of chemical reaction; dielectric analysis for in-process control and analysis; and several other techniques for specialized tests.

Full, in-depth treatment of these methods and current instrumentation is beyond the scope of this publication. The main purpose is to give sufficient detail to permit the reader to understand a particular test technique and its value to the thermoset resin field. Additional information on specific techniques is available in the references cited in this article. In addition, many instrument suppliers are willing to offer assistance with specific problems.

Epoxy resins are emphasized in the examples that follow because they dominate the airframe and aerospace industries. Intense Department of Defense and National Aeronautics and Space Administration interests have resulted in many published and proprietary studies that focus on these matrices. However, other polymer systems are discussed where appropriate.

Chemical Composition Characterization

Thermoset systems consist of one or more resins and are cured using various curing agents and catalysts. The chemical structure, functionality, and composition of these components have a broad effect on ultimate polymer properties, such as dimensional stability, moisture and solvent resistance, and mechanical strength. Once the resin formulation is established, subtle changes can be made for reasons that range from raw material shortages to environmental or safety regulations. Because resin matrix consistency is essential to the reliability and reproducibility of materials, factors that contribute to the uniformity of these resin systems include the type, purity, and concentration of individual chemical components, as well as mix homogeneity. Solvent separation of individual resin constituents by some form of chromatography is the major characterization technique used for analyzing uncured thermoset systems. Chromatography, combined with various spectroscopic methods and elemental analysis for molecular structure identification, forms the backbone of quality assurance testing schemes for these materials.

Chromatography. The individual constituents of a thermoset can include resin monomers, curing agents, catalysts, stabilizers, plasticizers, fillers, cross-linked or branched polymers, and microgels. The technique or combination of techniques selected for a specific separation will be driven by the nature of the material and the end-use of the resultant data. One of the most versatile general techniques for resin and polymer separation analysis is chromatography (Ref 2–4).

Chromatographic techniques accomplish separation of a resin mixture by the interaction of soluble sample components with a flowing fluid or mobile phase and a solid, stationary phase. These methods can be divided into two major categories—gas and liquid chromatography—and subdivided further according to the type of stationary phase, as shown in Fig. 1.

High-performance liquid chromatography (HPLC) utilizes a liquid mobile phase and a solid stationary phase. The separation mechanism is based on the way in which sample molecules distribute themselves between each phase and the time spent in each. A liquid chromatographic system is shown in Fig. 2.

The liquid mobile phases used for thermoset separations are generally organic solvents or mixtures of organic solvents with water. Common ones are tetrahydrofuran, chloroform, methanol, and acetonitrile. The solvent is driven through the system by means of high-pressure constant-flow pumps. A resin sample is dissolved in solvent, injected into the chromato-

*Adapted from Deborah K. Hadad and Clayton A. May, Physical, Chemical, and Thermal Analysis of Thermoset Resins, *Engineering Plastics,* Volume 2, *Engineered Materials Handbook,* ASM International, 1988, pages 517 to 532 graph, and swept through a column packed with fine, solid particles via the mobile phase. The presence of each molecular fraction is sensed by one or more detection devices (Fig. 2, parts 5 and 6), and the total separation is displayed on a video screen or strip chart as a chromatogram. Figure 3 shows the HPLC separation of a commercial polyimide (PI) resin system. Peak locations are associated with the chemical structure of the individual formulative components. The areas under the peaks are proportional, with calibration, to the amounts of each component.

The most difficult aspect of HPLC is solvent and column selection. Because the separating power of the HPLC method depends on interactions between sample molecules, column packings, and solvents, it follows that the choice of these elements has a significant effect on the ultimate quality of component separation. By combining knowledge of the basic material characteristics obtained from other sources (manufacturer's data, other chemical analyses, previous experience) with general selection guides (Fig. 4), a choice can be made for initial test parameters. Refinements are then made based on subsequent results. A successful separation involves an interdependency among resolution, sample capacity (mass loading), and the time or speed at which the separation can be performed. By varying experimental parameters, it is possible to enhance one of these qualities at the expense of the other two. The optimal separation always involves a compromise among the three. Although HPLC has several subbranches, only two are discussed as applied to the analysis of thermoset materials: gel permeation chromatography and liquid-solid chromatography.

Gel permeation chromatography (GPC), or size exclusion separation, provides component segregation based on one physical parameter—molecular size. It is a form of liquid chromatography in which the component molecules are separated by their permeation into a porous packing gel. While the small molecules diffuse quickly into the pores and are temporarily retained, large species may be excluded from some or all of the pores and will be swept through the column first, followed by others of decreasing size (Fig. 5). The range of molecular sizes that can be separated is controlled by the distribution of varying pore sizes within the porous gel. The effective size in solution is closely related to molecular weight. Molecular weight distribution affects many characteristic physical properties of the cured material, such as tensile strength, brittleness, impact strength, toughness, chemical resistance, cure time, and melt viscosity. Subtle batch-to-batch differences in this distribution can cause significant differences in end-use material properties. Because of this, GPC has been extensively used for the quality control of incoming materials (Ref 5–15).

The best technique for studying neat resins is not necessarily the best for separating resin mixtures. It was found very early in the study of

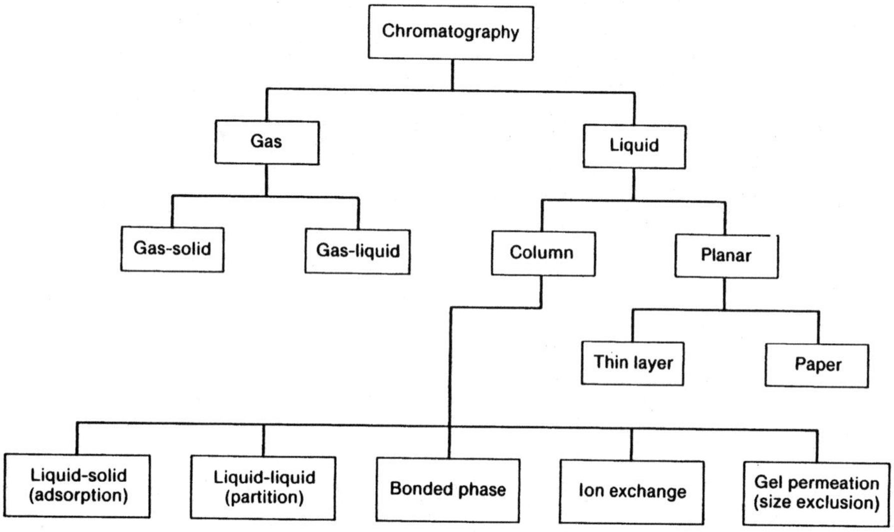

Fig. 1 Classification of chromatographic techniques

Fig. 2 Liquid chromatographic system. 1, solvent delivery pumping system; 2, autosampling/injection system; 3, interactive controller for solvent mixing and system automation; 4, ultraviolet detectors; 5, differential refractometer; 6, data-handling computer. The column is not visible in this view.

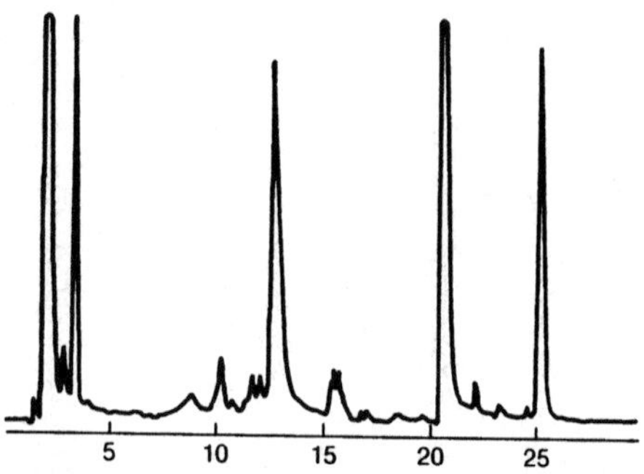

Fig. 3 HPLC chromatogram for PMR-15 polyimide

thermosets that tetrahydrofuran, which is a good solvent for many thermoset mixtures, produces poor separations in the analysis of mixtures made with Ciba-Geigy resin MY-720 and the curing agent diaminodiphenylsulfone (DDS), which is commonly used in commercial epoxy prepreg matrices (Ref 1, 16, 17). Figure 6 illustrates this behavior. As a smaller molecule, the curing agent DDS should elute after the MY-720 monomer. By using the weaker solvent chloroform, DDS is effectively separated from the MY-720 components, but this involves the resultant reduction in resolution of the higher molecular weight region and increased analysis time. This illustrates the compromises discussed previously when attempting to optimize a chromatographic separation.

Gel permeation chromatography has been used to study the composition of resin formulations (Ref 1, 17), materials in the electronics field (Ref 18–21), resin structure (Ref 22–25), aging and resin advancement (Ref 26–28), and thermoset cure kinetics (Ref 29–31). It has also been used as an on-line process control technique (Ref 32).

Liquid-Solid Chromatography (LSC). The most successful techniques for analyzing thermoset systems have been those that use LSC. This method separates individual components by their affinity for a stationary phase that is more polar than the mobile phase, as shown in Fig. 7. When this stationary phase is less polar, the technique is called reverse-phase LSC. Attachment or adsorption to this phase increases as the polarity and number of sample functional groups increase. Competition between sample and solvent molecules for sites on the stationary surface and the multiple interactions between functional groups on the same molecule with these sites account for the unique power of reverse-phase LSC. Because of its superior strength in separating complex mixtures, this method has been used almost exclusively in all quantitative HPLC applications for thermosets.

One of the major reasons LSC has been so successful is the practice of using a solvent programming technique called gradient elution. This method is usually used on complex materials where optimal resolution is desired. With this technique, the mobile phase strength (polarity) is increased during a separation by mixing two or more solvents at a programmed rate. Strongly attached molecules on the stationary phase of the column are then swept out of the column faster. The improved and consistent

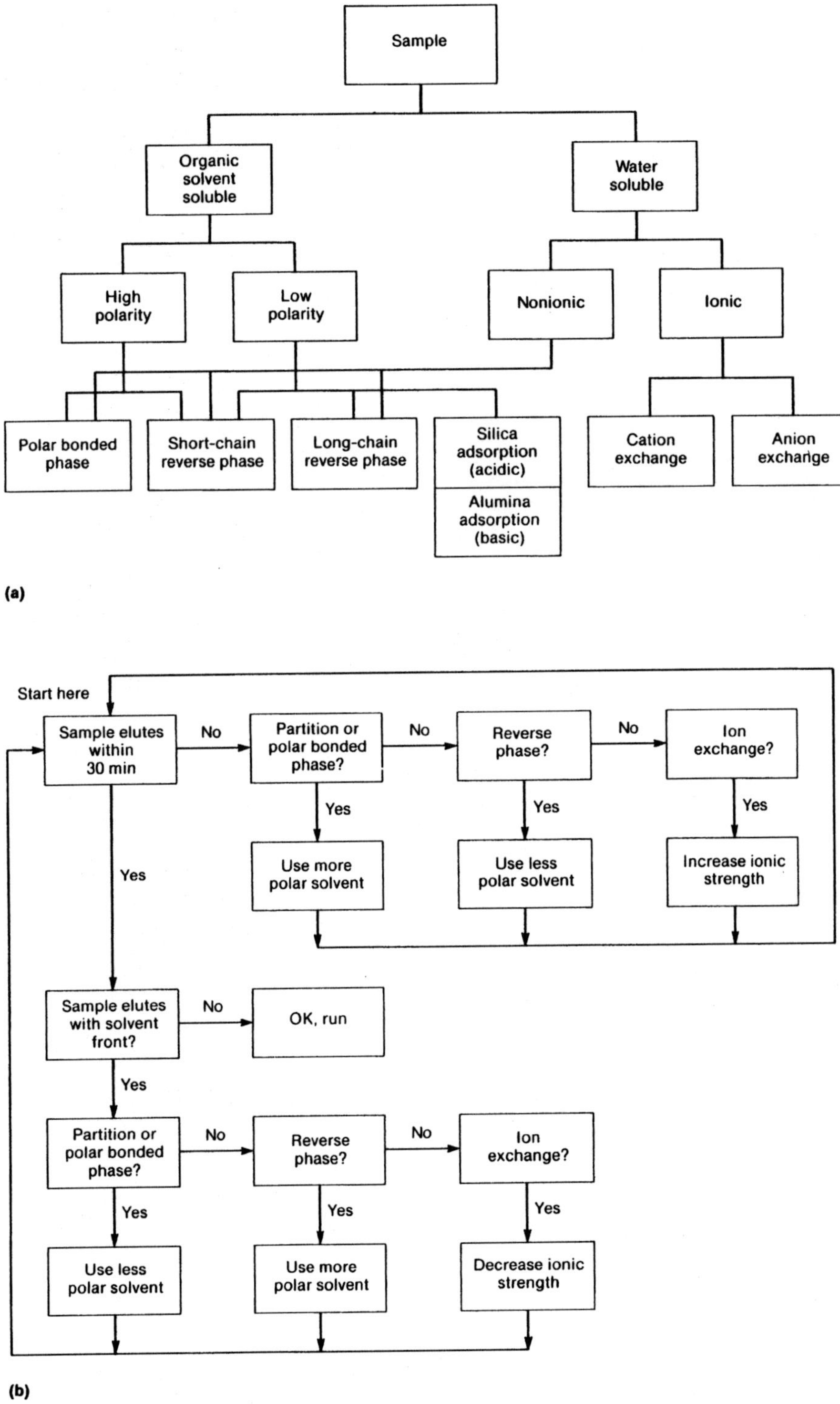

Fig. 4 HPLC method sorbent (a) and solvent (b) selection guides. Source: Alltech Associates

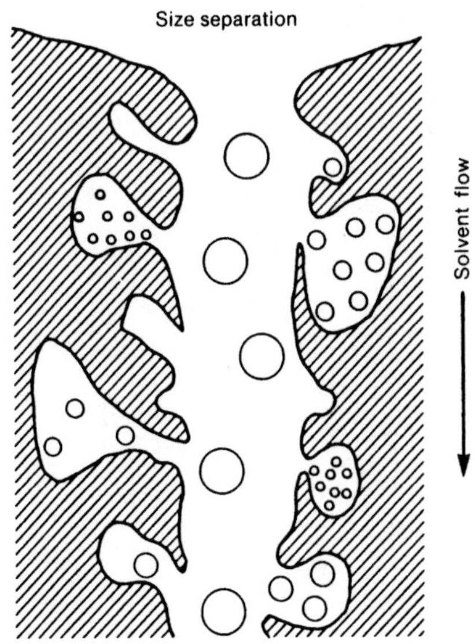

Fig. 5 Functioning of gel permeation chromatography

peak shapes in this mode afford the high precision needed for routine, quantitative analyses.

A shorter, simpler alternative to gradient elution is the isocratic method. With this technique, a mobile phase of constant solvent strength is used, which consists of a solvent mixture of constant composition. This technique offers definite advantages over other methods because solvent purity is not as critical as in gradient work. All size exclusion separations are isocratic. This technique generally requires more time and tends to produce broad peaks toward the end of the separation time.

High-performance liquid chromatography is a proven, powerful tool for separating complex adhesives (Ref 33–34), neat resins (Ref 35–42), printed circuit board materials (Ref 43), curing agents (Ref 44), and advanced composite resin systems (Ref 1, 17, 45–48) into their individual components. Therefore, this analytical method is ideally suited to quality control applications.

Quality control testing using liquid chromatography varies from a simple qualitative visual match between a sample chromatogram and that of a control to a complete quantitative analysis with the attendant statistical treatment of data and retention of a historical database. Table 1 lists the important parameters that are critical for the development of an accurate and repeatable quantitative method. The importance of these parameters was examined and verified as part of an Air Force sponsored round-robin test program (Ref 49).

Because of its ability to follow the changing composition of the sample with time, HPLC is an excellent tool for monitoring the aging of a resin system (Ref 50–52). A variety of reaction products form during aging, while the main reactants decrease, as shown in Fig. 8 and Table 2. These same techniques can also be used to study the effects of moisture (Ref 53), impurities (Ref 54), stoichiometry (Ref 55), and accelerated aging (Ref 56) on cure kinetics.

As with any analytical test technique, HPLC methods must be tailored for individual materials. An experienced analyst can choose the HPLC mode that is best suited to the separation of the particular composition. Refining the techniques for quantitative analysis demands time and care in setting parameters and statistically evaluating the resultant data.

Thin-layer chromatography (TLC) is one of the simplest, least expensive forms of chromatography. A thin, sorbent layer applied to a support material such as glass plate or quartz rod serves as the packed column, or stationary, phase. A closed container holding the development solvent represents the mobile, or liquid, phase. A drop of sample solution is applied to the sorbent, which is then placed in the developing chamber. The individual components of a mixture are separated as the mobile phase migrates upward by capillary action. These components advance at various distances up the sorbent, depending on their solubility and affinity for the sorbent material. When the solvent approaches the top of the plate or rod, the support medium is removed, the solvent is allowed to evaporate, and the locations of the various component spots are determined by any of several methods. Because of this mechanism, TLC can be used to ensure that a resin formulation has remained unaltered (Fig. 9), and it therefore serves as a quality control test in the same way as HPLC. Some of the earliest studies of thermosets using TLC involved its use for quantitative component analyses (Ref 57, 58).

Thin-layer chromatography resembles high-performance liquid chromatography in that the same solid phases are commercially available for use in effecting a separation; other phases, such as alumina, can also be used. Surface area, porosity, ambient conditioning, and type of binders are all factors that influence the capacity, efficiency, reproducibility, and selectivity of TLC sorbents.

The effectiveness of a TLC separation is largely determined by the solvents used and the way in which development is carried out. Systematic studies are available that describe the selection of these parameters (Ref 59), and increased resolution and sensitivity have been realized by complex developing modes. Multiple development increases the apparent sorbent length by utilizing several redevelopments of the same support. Continuous development uses a constant solvent flow along the sorbent, and excellent work has been done using this technique (Ref 60). More complicated developing modes are *x-y* or two-dimensional development, programmed multiple development (Ref 61), gradient elution, high-performance radial (Ref 62), and overpressure layer chromatography (Ref 63).

As with the other chromatographic techniques, detectors are needed to locate the positions of various compounds after a thin-layer separation. Traditional detection methods include ultraviolet absorption, charring, color reaction, and densitometry. One relatively new TLC detector involves the use of a flame ionization detector (FID) (Ref 64–66). This system combines an unusual detection device with an equally unusual support medium (Fig. 10). Separation is achieved on a reusable thin-layer rod

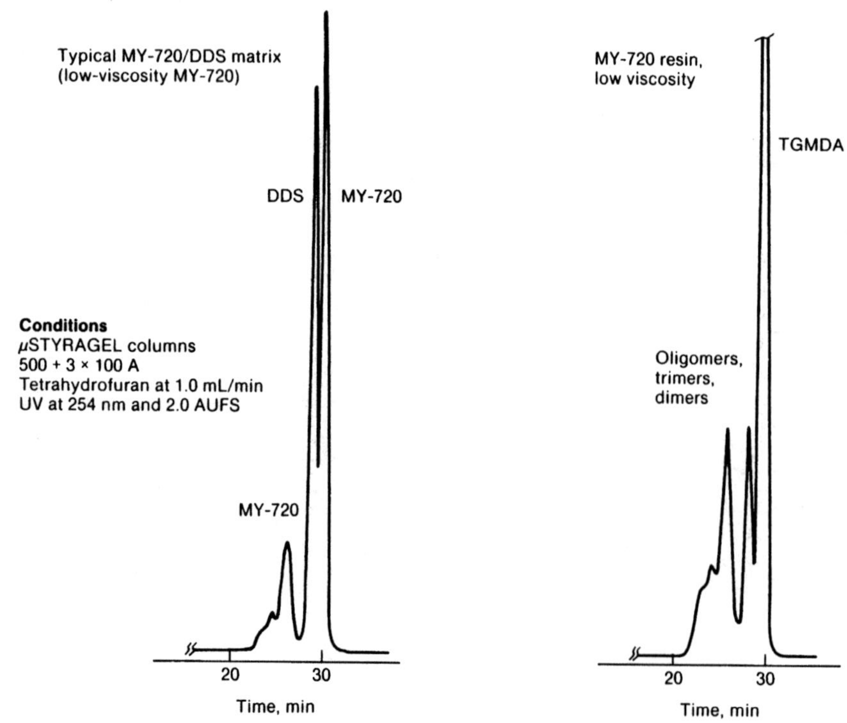

Fig. 6 Solvent effects on GPC separation of MY-720/DDS

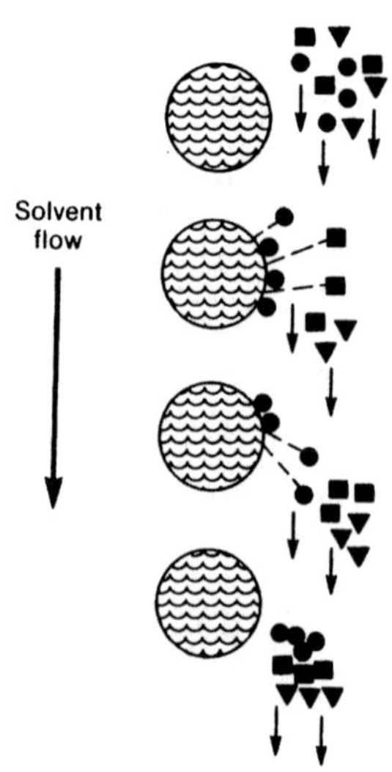

Fig. 7 Liquid-solid separation by affinity in liquid-solid chromatography

rather than a plate. The rod is conventionally developed, but the detection technique is automatic, requires no visualization reagents, and provides quantitative separation. This method has been successfully used on a simulated resin system (Ref 67). The example chromatogram shown in Fig. 11 resembles a conventional HPLC chromatogram.

Thin-layer chromatography is capable of providing quantitative information for thermosets that is at least comparable to that from the more popular separation techniques. Because of their inherent benefits (lower instrument/operating costs, simplicity, high sample throughput), TLC analyses are ideal not only for repetitive quality control applications but also for any cost-conscious analytical laboratory.

Infrared (IR) spectroscopy involves the study of molecular vibrations. A continuous beam of electromagnetic radiation is passed through or reflected off the surface of a sample, which may be a solid, a liquid, or a gas. Individual molecular bonds and bond groupings vibrate at characteristic frequencies and selectively absorb infrared radiation at matching frequencies. Thus, the amount of radiation absorbed or passed through unchanged depends on the chemical composition of the sample, and the resultant curve is known as the infrared spectrum. Figure 12 shows the infrared spectrum for a nitrile phenolic resin.

Because of the uniqueness of the infrared spectrum of a material, the identification of unknown materials, as well as chromatographic or wet chemical fractions, is one of the most powerful functions of IR spectroscopy. In addition, the progress of most thermoset reactions can be followed by monitoring appropriate functional group absorption peaks.

Infrared spectroscopy is extensively used to determine the type and amount of curing agents used in thermoset systems. As with other absorption techniques, peak areas are proportional to the concentration of the absorbing species. The sulfone (SO_2) doublet at 1148/cm (2920/in.) for the curing agent DDS is used to analyze this material quantitatively. The infrared stack plot (Fig. 13) and calibration curve (Fig. 14) are shown for a percentage DDS determination. The value of this analytical method for screening neat resins, curing agents, and mixed systems and for following the cure of resin-hardener mixes is obvious and accounts for its extensive use in thermoset resin analysis.

The application of IR spectroscopy to chemical problems has been expanded by Fourier transform infrared (FTIR) spectroscopy (Ref 68), which has become a commonplace technique because of the availability of relatively low-cost digital computers. Rapid analysis times and high resolution are two advantages of FTIR over conventional IR analysis. All radiation frequencies are incident to the sample throughout the scan, and the resulting signal, called an interferogram, is a plot of intensity

Table 1 Steps for developing quantitative HPLC procedures

Test parameter	Comments and effects
Sample preparation	Highly purified solvents Physical removal of fillers and scrim Typical concentration range: 15–60 g (0.5–2 oz) resin injected Storage of prepared resin solutions not advisable over 12–24 h (must be determined for each material)
Column equilibration	For chromatogram reproducibility and analytical precision Temperature control to eliminate solvent compressibility, flow, and solubility fluctuations
Analytical sequence	For the autosampling of a large number of samples, precise time sequencing for equilibration, test run, and reequilibration of the column; produces a consistent test environment Run blank prior to sample series to verify purity of mobile phase(s). Run samples in duplicate. For a large number of samples, run blanks during the course of the series to check for contamination buildup and to clean column.
Detector linear concentration response	Determine linear operating range of detector, specifically when using ultraviolet radiation Sample concentration versus peak area
Standard calibration solution	No specified set of instrument parameters will yield identical sample chromatograms between different instruments. Absolute quantitative results are not possible strictly by instrumental electronic response. This parameter as it relates to quantitative evaluation is determined by an operator. Every polymer mixture has its own response characteristic; thus, a generalized method for many materials is not possible. Standard solution containing the components of a mixture is used to determine response factors that remain constant regardless of the instrument used. For a large number of samples, the standard solution should be run during the course of the series.
Integration techniques	Single largest error-producing parameter in HPLC analysis Must evaluate computer parameters and specify conditions that produce consistent integration for each fraction in a mixture.

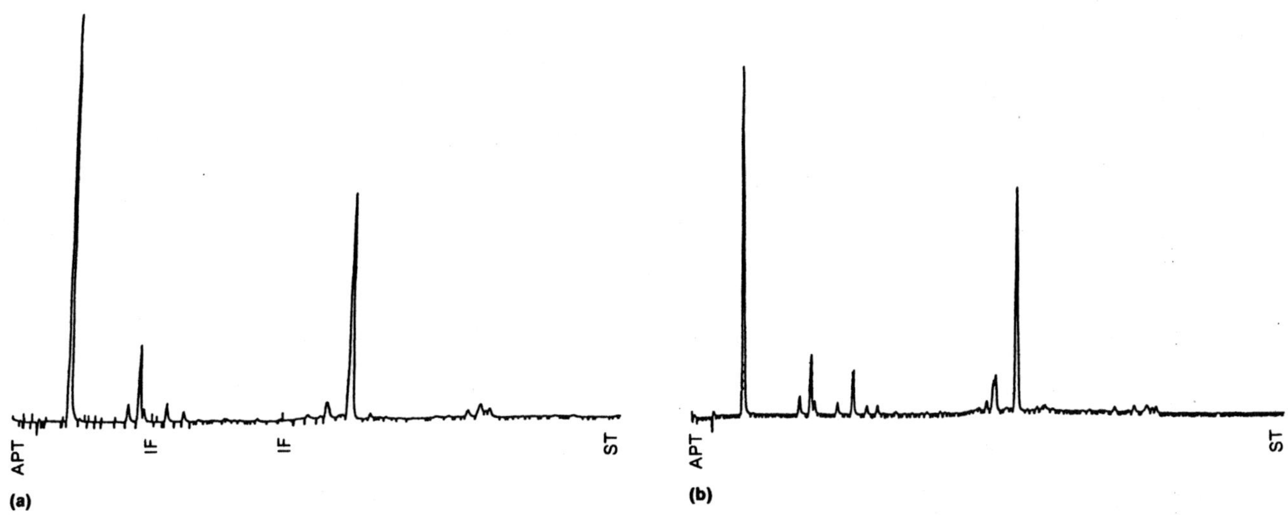

Fig. 8 HPLC chromatograms showing the ambient aging of Fiberite 934 epoxy. (a) Control. (b) Aged 65 days at ambient temperature

versus time. The final infrared spectrum is obtained by calculating the Fourier transform of the interferogram in the frequency domain.

Because of the relationship between chemical reactions and successful thermoset processing and high-quality finished hardware, kinetic studies of thermoset resin systems have increased. This result is due to the importance of understanding cure mechanisms and assessing the degree of cure/fractional conversion based on specific reactions. Elevated temperatures are required for many thermoset cures, and modification of FTIR spectrometer cells to provide the necessary environment is essential. The real-time chemical reactions of several epoxy systems have been investigated, and extent of conversion versus time curves have been constructed at several temperatures (Ref 69). Similar FTIR techniques have also been used to study thermoset curing processes (Ref 70–75).

Thermosets are usually studied by transmission of the infrared radiation through a resin sample. However, for more difficult applications, several alternative FTIR techniques are available. If the resin is not in the neat form or if it cannot be dissolved from interfering materials such as fibers or fillers, a surface analysis technique may be needed that involves some form of internal reflectance to increase the strength of the signal (Ref 76, 77). Surface analysis techniques include attenuated total reflectance, specular reflectance, diffuse reflectance, and photoacoustic spectroscopy.

Fourier transform infrared spectroscopy is a powerful technique for studying degree of cure, aging, effect of processing on resin chemistry (cure kinetics), structural characterization of polymer surfaces, degradation products, and oxidative stability. It is used in quality control (Ref 78–83); for studying composite weathering (Ref 81–83), the epoxy-fiber interphase (Ref 84, 85), and resin-water interactions (Ref 86, 87); and for real-time multicomponent analysis in the production environment (Ref 88, 89).

Other Techniques. Chromatography and IR spectroscopy are the most common techniques for chemical composition analysis, but there are many other less used methods, such as nuclear magnetic resonance spectroscopy, ultraviolet spectroscopy, gas chromatography, and combination methods. Although these methods require more complex and expensive equipment as well as highly trained operators, they can be important in special cases for the identification and quantitative analysis of unreacted resin systems. In some cases, they can be used for routine quality control testing and for monitoring resin impurities, aging, B-staging, and cure kinetics.

Processing Characterization

The way in which a resin system reacts is determined not only by the types of compounds present but also by the processing conditions (time, temperature, pressure) used to cure them. For most industrial applications, the formulation is fixed. The processing variables allow tailoring of the ultimate cured properties of the system. Processing parameters are as important as the initial chemical composition of the resin being cured. Processes such as rate, kinetics, and cessation of chemical reactions, as well as flow characteristics, are of vital significance and have been discussed in the literature (Ref 69, 90–93).

Processing characterization can be divided into two categories. The first studies the thermal properties of reactive thermoset systems. The second utilizes these thermal characteristics as the basis for monitoring and control during processing. "Smart" processing computer programs have been developed that require thermal, kinetic, rheological, and heat-transfer data for individual thermoset formulations to control fabrication cycles.

Thermal analysis measures chemical or physical changes as a function of temperature. This dependency allows access to processing and performance information relating to resins and fiber-reinforced composites and can be used

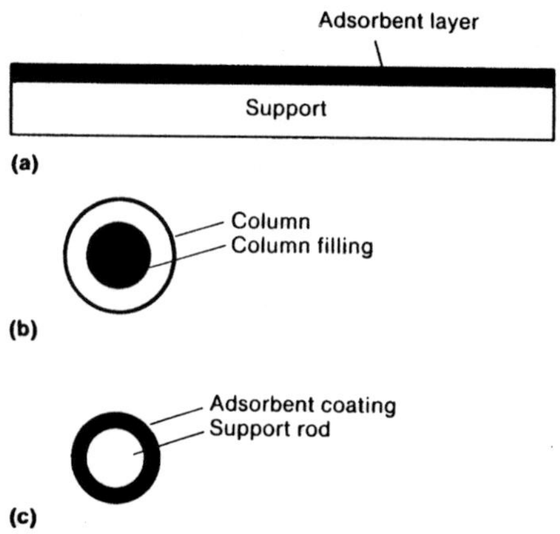

Fig. 10 Geometry of chromatographic separation methods. (a) TLC plate. (b) HPLC column. (c) TLC rod. Source: Ref 66

Table 2 Aging effects on HPLC data for Fiberite 934

	Curing agent, %		Major resin, %		Minor resin, %		Total unreacted, %	
Aging, days	Resin	Prepreg	Resin	Prepreg	Resin	Prepreg	Resin	Prepreg
0	30.30	26.80	51.00	46.20	13.00	11.40	94.3	84.4
7	29.05	23.35	46.20	40.60	12.90	10.20	88.15	75.10
14	27.45	21.05	45.50	38.70	12.80	9.80	86.90	71.70
30	25.30	13.80	43.10	30.40	10.30	5.60	78.70	49.80

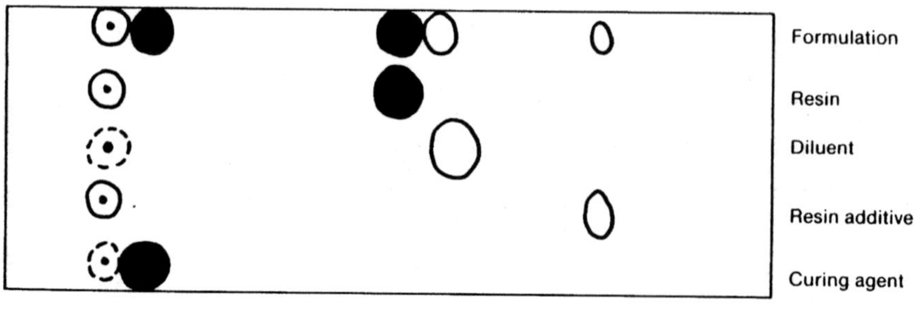

Fig. 9 TLC separation of an epoxy system and its components

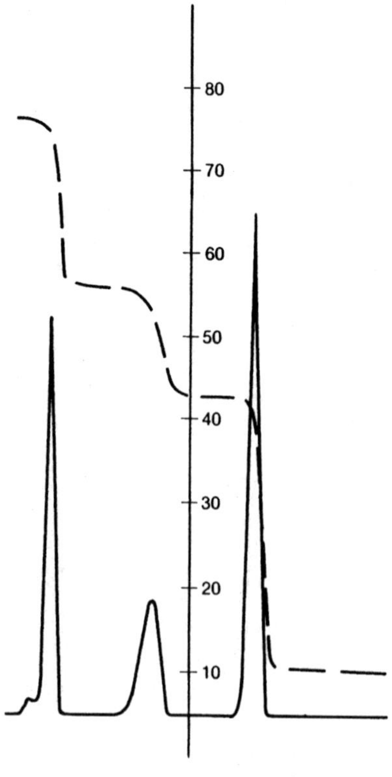

Fig. 11 TLC-FID separation of polymer mixture

for quality assurance, process control, and new material process development. Gel points, glass-transition temperatures (T_g), expansion/contraction properties, reaction rates and cure kinetics, effects of individual and combinations of components, polymer stability, and material life predictions can all be determined by thermal analyses.

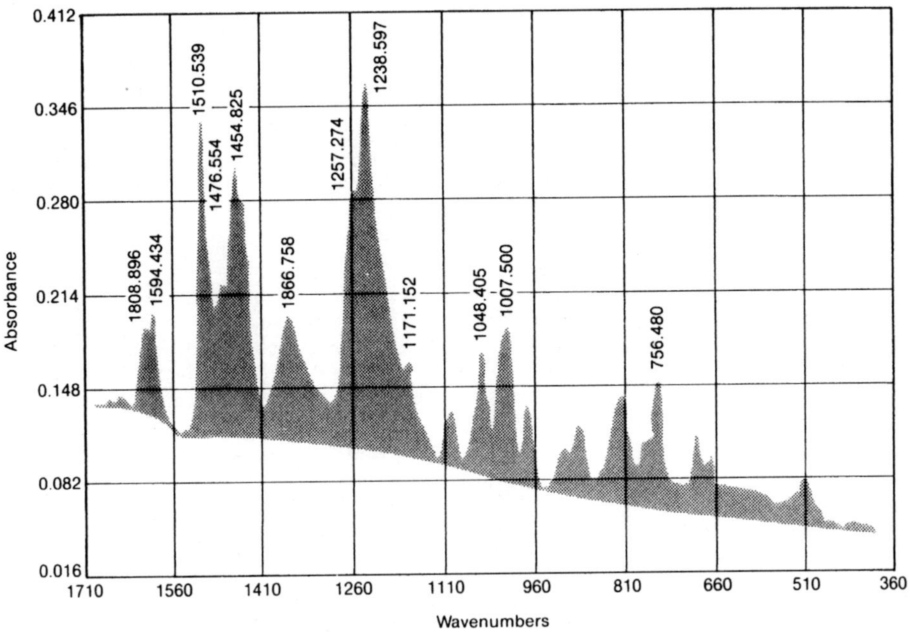

Fig. 12 Infrared spectrum of nitrile phenolic resin

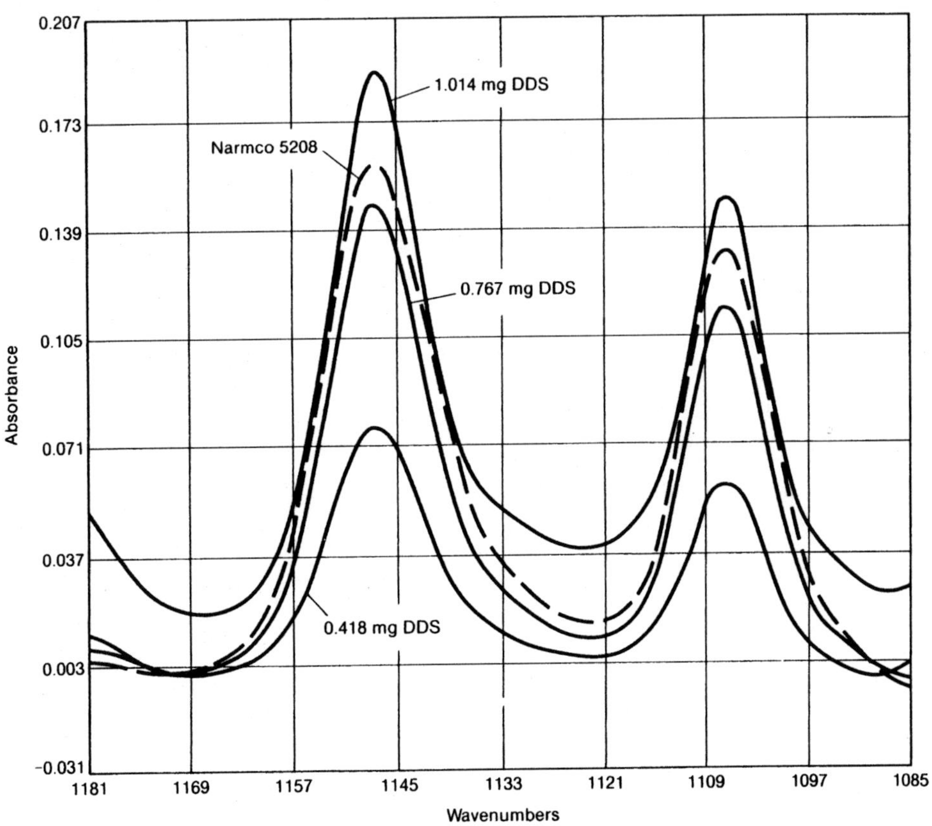

Fig. 13 Infrared stack plot for percentage DDS determination. Solvent: 10% THF, 90% $CHCl_3$

The four thermal analysis techniques used most frequently are differential scanning calorimetry, thermogravimetric analysis, thermomechanical analysis, and rheological analysis. These techniques as applied to thermosets are described in Ref 90. General information on thermal analysis and its applications is also available in the literature (Ref 94–107).

Differential Scanning Calorimetry (DSC). When a thermoset cures, the resultant chemical reaction gives off heat (exotherm) or absorbs energy (endotherm) as a function of both time and temperature. Differential scanning calorimetry measures the temperature differences between a sample and an inert reference material. These differences are recorded as a function of the sample temperature, with the area under the resultant output curve (thermogram) being directly proportional to the total energy q transferred into or out of the sample. The ordinate of the thermogram, therefore, is proportional to the rate of heat transfer, dq/dt, at any given time. A combination of dynamic and isothermal experiments can provide information on reaction rates, cure rates, specific heat, and degree of cure. A dynamic DSC curve typical of the thermoset resins used in some advanced composites and adhesives is shown in Fig. 15. Critical points on the curve are:

- T_g, the subambient glass transition temperature of the uncured resin
- T_i, the initiation temperature or onset of reaction indicating the beginning of polymerization
- T_m, a minor exotherm peak temperature associated with accelerator effects
- T_{exo}, the major exotherm peak temperature
- T_f, the final temperature, indicating the end of heat generation and completion of the cure

Several thermal characteristics affect the quality of hardware made from thermoset systems. These characteristics include temperature gradient control, heat-up rate during processing, and extent of cure. The type and number of competing chemical reactions, heat of reaction, thermal conductivity, and specific heat of a material at various stages of reaction produce temperature variations during a cure cycle that directly affect the final degree of cure. This is particularly true in thick laminates where slow heat-removal rates can drastically influence processing. Therefore, to optimize hardware fabrication, it is essential to understand the kinetic behavior of the reactive system being processed.

Control of resin advancement in raw material and the degree of cure after processing are also prerequisites for repeatable, reliable, high-quality final products. Differential scanning calorimetry has been used for quality control and degree of cure studies of molding compounds (Ref 108, 109), printed circuit board prepregs (Ref 110–112), powder paints (Ref 109), an

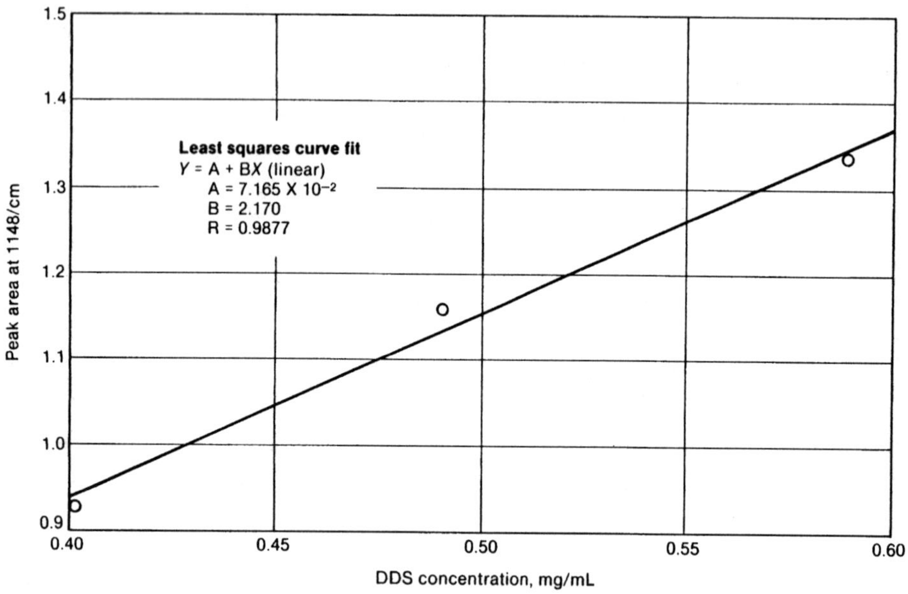

Fig. 14 Infrared standard calibration plot for percentage DDS determination

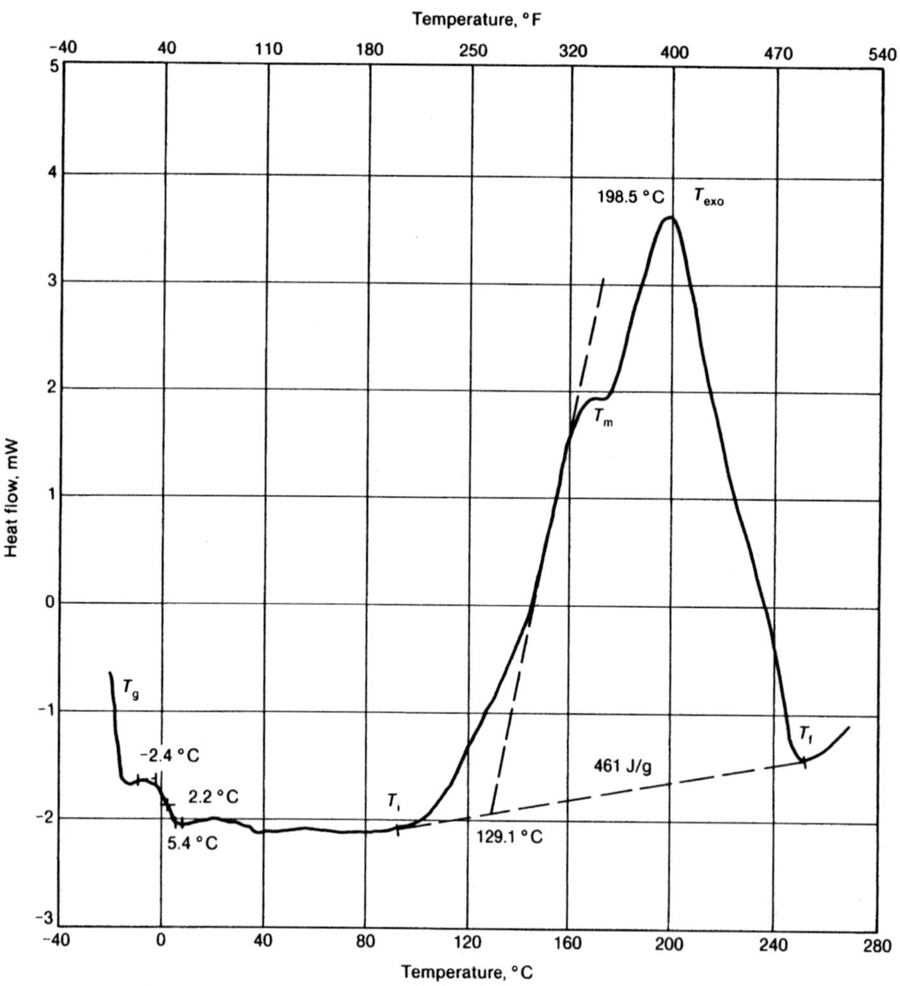

Fig. 15 DSC thermogram of Fiberite 934 epoxy, 4.89 mg (0.075 gr), 10 °C/min (18 °F/min)

ambiently cured field repair system (Ref 113), graphite-reinforced prepreg resin matrices (Ref 114, 115), and film adhesives (Ref 116).

The subject of chemical kinetics and the way in which kinetic parameters are obtained is a complex one. The effects of fillers (Ref 117), impurities (Ref 30), and catalysts (Ref 118, 119) on overall reaction have been studied, as well as the reaction kinetics of a commercial adhesive (Ref 120).

The long-term integrity of a thermoset material is influenced by a number of time-dependent factors. These include moisture and solvent diffusion, viscoelastic deformation, fatigue, chemical reactions, and the generic category of aging, which includes physical, chemical, and mechanical aging (Ref 121).

The physical aging of amorphous polymers has been described in detail in Ref 122. During some processing procedures, a material can be trapped in a nonequilibrium thermodynamic state. For example, when a polymer is rapidly cooled (quenched) to below its T_g, polymer chains are frozen before they can react. This results in excess free volume instead of a tight, dense network. Physical aging is the natural process of reaching equilibrium, and it leads to densification and embrittlement of the polymer. This type of aging in polymers is manifested by changes in relaxation times. These changes have been studied in thermosets by using DSC (Ref 122–124).

Chemical aging involves cross-linking reactions, and the results are similar to those of physical aging. If a thermoset resin is incorrectly processed, the chemical reaction is interrupted prior to cross linking. The resultant unreacted species will continue to cross link slowly over a long period of time, continuously changing

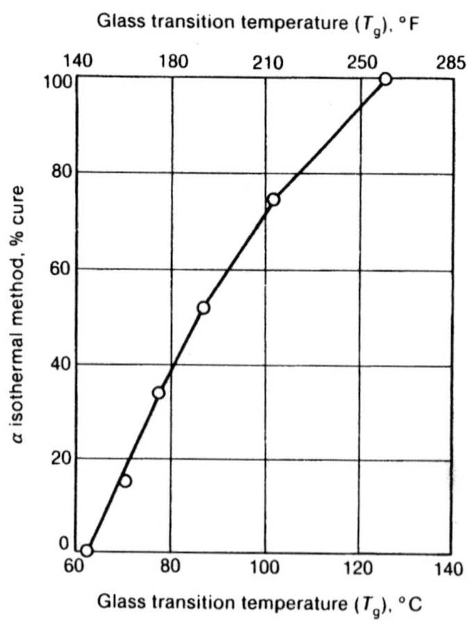

Fig. 16 Correlation of T_g with degree of cure by isothermal DSC of epoxy-glass laminate. Source: Ref 126

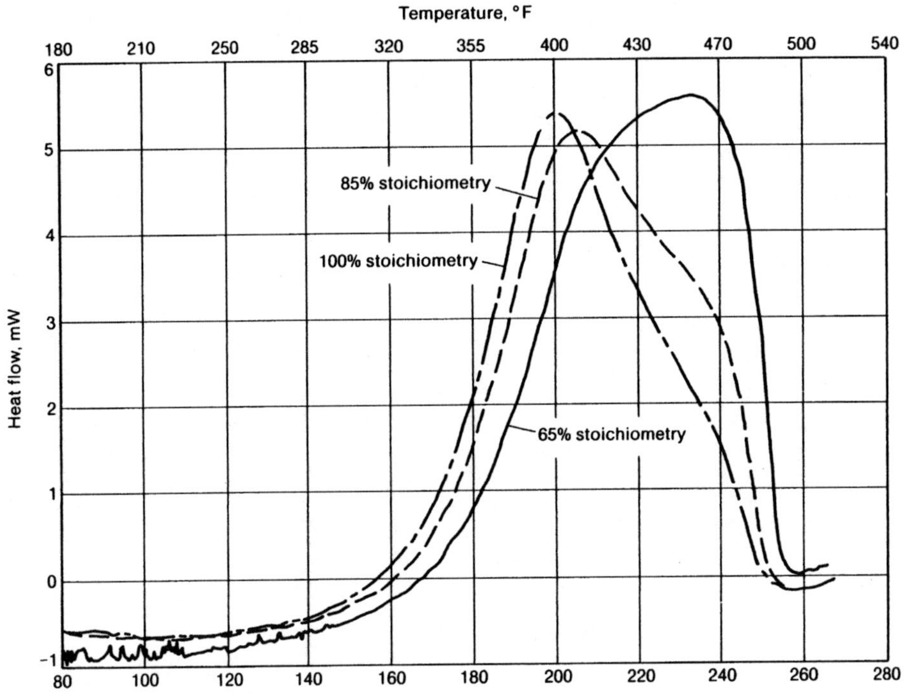

Fig. 17 Effect of resin-curing agent stoichiometry on DSC profiles at 3 °C/min (5 °F/min), 14 × 10³ cps

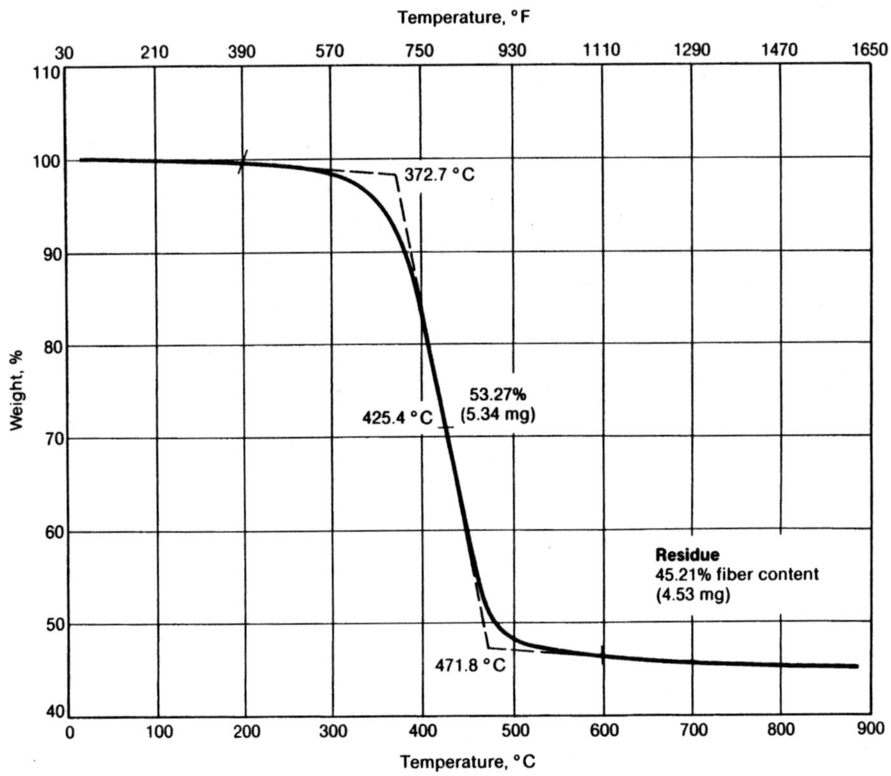

Fig. 18 Typical TGA curve for fiberglass-vinyl ester prepreg

material properties. These changes can be followed by using DSC.

The effects of aging can often be catastrophic. Undercure of the resin matrix can result in hardware failure. By using a residual DSC exotherm technique, one researcher found a dramatic relationship between degree of cure and bond performance in an epoxy adhesive (Ref 125). At an 85 to 90% cure level, the failure mode for a single lap test changed from cohesive to adhesive.

The T_g is also an indicator of degree of cure, as shown in Fig. 16, as well as a measure of degradation (Ref 127). This degradation presumably reflects a change in cross-link density.

Differential scanning calorimetry can be used to study the effects of reactant ratio or stoichiometry. Figure 17 shows variations from 65 to 100% stoichiometry for mixtures composed of MY-720/DDS. As stoichiometry increases, the exotherm peak temperatures decrease, and the thermal curves take on a more Gaussian shape. As stoichiometry decreases (corresponding to a decrease in DDS concentration), the shape of the curve exhibits a profile closer to that associated with homopolymerization of the resin alone.

Thermogravimetric analysis (TGA) involves measurement of the weight gain or loss of a material as a function of temperature and time, and it utilizes an extremely sensitive electronic microbalance. A typical weight loss curve is shown in Fig. 18. In addition to the normal decomposition profile, there is the added benefit of obtaining the amount of fabric or filler left behind as the residue. This applies for fiberglass and other fabrics and fillers that do not oxidize or form other compounds that cause a weight gain. This method has been used as an alternative to conventional muffle furnace techniques (Ref 128).

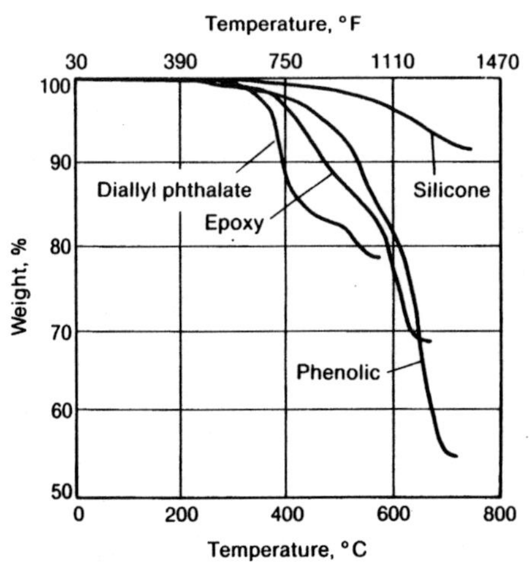

Fig. 19 TGA comparison of encapsulating materials, 20 to 30 mg (0.3 to 0.5 gr), 10 °C/min (18 °F/min), air at 40 mL/min. Source: R.E. Thomas, Motorola Semiconductor Products Division

One of the most important applications of TGA is the assessment of the thermal stability of a material. This can be done to obtain relative comparisons between different materials or as an accelerated means for lifetime predictions. Where the loss of additives such as plasticizers or antioxidants can damage a structure, decomposition profiles are excellent indicators of change. A comparison of the thermal decomposition of encapsulating materials using TGA is shown in Fig. 19. Absolute classification of thermal stability is difficult, however, because of the interaction of various aging phenomena. Because decomposition mechanisms are often diffusion controlled, sample geometry and fillers can affect the observed test results. Therefore, the data obtained on small test specimens may not be extrapolated to larger structures. This type of information should be used judiciously as a guide for further studies until TGA or other thermal techniques are developed that give better correlation.

Current kinetic models that predict material life are in the early stages of development. Predictions of material longevity require a relationship between time-to-failure and experimental variables that induce failure. Because the failure of polymer systems and composite materials is complex and involves multiple failure modes, it is important that accelerated tests model each of the relevant processes in such a way as to describe the combined effect of competing modes. The best technique to date for accurately predicting the lifetime of polymers is the factor-jump method (Ref 129–131). Experiments at very slow heating rates and low isothermal temperatures minimize the differences between actual and extrapolated service conditions.

Thermogravimetric analysis can also be used to determine moisture, volatile, and filler contents, to study the effects of additives, and to obtain separation of some components (for example, rubber from carbon black). In an attempt to determine the exact mechanisms of polymer degradation, TGA has been coupled with spectroscopic techniques to clarify degradation pathways and to identify additive components (Ref 132, 133).

Thermomechanical analysis (TMA) measures variations in the vertical displacement of a probe resting on top of a sample and is used to obtain physical property changes as a function of temperature and/or time. Some of the properties obtained using TMA are compression, expansion, and tension properties, which include expansion or shrinkage under tension, single-fiber properties, dilatometry involving volumetric expansion of a material within a confining medium, and isothermal kinetic measurements.

Figure 20 shows the typical expansion behavior of a PI resin casting. Cured thermosets typically exhibit two linear regions. The first is associated with the glassy state and is followed by a change to a second linear region of higher slope associated with the rubbery state because of T_g.

The coefficient of thermal expansion and T_g of a thermoset are closely related to the degree of cure of that resin. Fully cured materials have higher T_gs and sometimes lower expansion coefficients than under-cured or partially cured materials. Many fabrication processes induce cured-in stresses. Figure 21 shows a typical TMA profile for a material exhibiting stress relief. Thermal cycling or annealing above T_g will smooth the curve but will not elevate T_g. Ideally, T_g is observed as an abrupt change in the slope of the linear expansion versus temperature curve. However, because relaxation often occurs near T_g, the transition can be broad, depending on such factors as the material, cure state, internal stresses, and test conditions.

Because of the critical dimensional stability requirements of multilayer printed circuit boards, TMA is extensively used for determining and controlling the thermal expansion behavior (Ref 134–135) and delamination resistance (Ref 136) of these materials. Thermomechanical analysis is one of the standard test techniques for studying thermoset resins because T_g and the coefficient of thermal expansion are strongly influenced by resin composition, additives, solvents, moisture, and degree of cure.

Dynamic mechanical analysis (DMA) measures the ability of a material to store and dissipate mechanical energy upon deformation, and it follows changes in both elastic (stiffness), or storage modulus, and viscous (toughness), or loss modulus, properties. These quantities can be mathematically combined to give, in effect, a measure of the shear or flexural moduli of the material. In the case of liquids and semiliquids, the same quantities can be combined to give the apparent viscosity of the material. Instrumentation is available for measuring both liquids and semiliquids, as well as solid samples. The measurement can be made isothermally in a dynamic temperature scan and generally at different frequency and strain levels.

Rheology is the study of the flow behavior of a material and is generally applied to liquids or semiliquids. A typical rheological curve is shown in Fig. 22 for the dynamic cure of a PI prepreg. The initial drop in viscosity is associated with the softening and flowing of the resin. The peak appears when the resin hardens because of increased chain extension and stiffness as imidization takes place. The resin goes through a second melt stage as the imidized resin softens, and then viscosity rapidly increases as cure continues to completion.

The curing of a thermoset system involves a complex, multistep mechanism leading to a

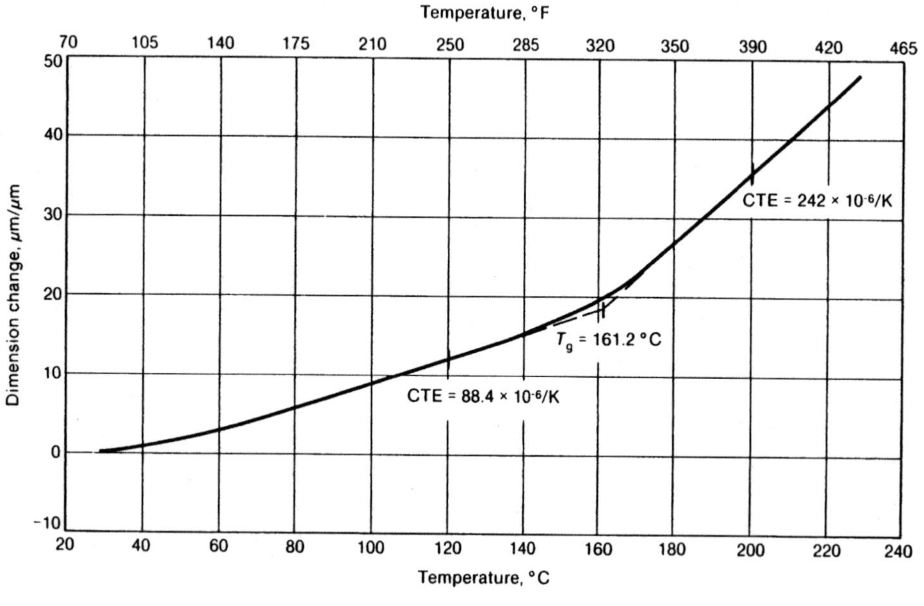

Fig. 20 Typical TMA curve for a fiberglass-polyester prepreg, 2 mm (0.08 in.), 10 °C/min (18 °F/min). CTE, coefficient of thermal expansion

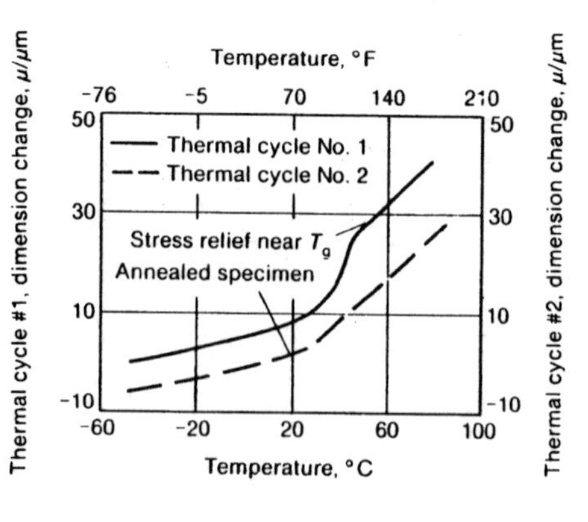

Fig. 21 TMA profile exhibiting stress relief; epoxy casting, 4.19 mm (0.16 in.), 5 °C/min (9 °F/min)

molecular network of infinite molecular weight. The gel point is the point at which a viscous liquid becomes an elastic gel; this marks the beginning of the infinite network. From a processing standpoint, this point and the flow behavior leading up to it are important characteristics. Flow behavior affects the way in which a material can be processed, and gelation marks the point at which processing flexibility ends. Other thermal techniques, such as DSC and TGA, do not detect this physical change, because chemical reactions continue unchanged following gelation. Cross-link density, T_g, and ultimate physical properties continue to increase after gelation until the reaction is complete. These characteristics are studied using DMA, and because DMA measures mechanical properties dynamically, the possibility exists for obtaining rapid information on end-product performance.

The key relationships between the process of cure and the physical properties of the cured state of thermosets have been studied (Ref 137, 138). These relationships are shown in a generic time-temperature-transformation (TTT) diagram (Fig. 23) depicting the four material states encountered during cure: liquid, elastomer (gelled rubber), ungelled glass, and gelled glass. Critical processing information can be obtained from TTT diagrams, such as the time-temperature dependence of flow, reaction kinetics, gelation, and vitrification (initiation into the ungelled glass state). This type of information is quite useful to the manufacturing engineer for developing appropriate cure cycles (Ref 139, 140). Appropriate time-temperature values for B-staging, debulking, dwells (devolatilization), pressure application points (compaction), and final conditions for cure cycles can be optimized.

The gel point of a thermoset can be empirically assigned as the point at which the shear modulus, G', is equal to the loss modulus, G'' (Ref 141). Figure 24 shows these curves for a commercial prepreg. The viscosity is increasing rapidly at this point. This modulus crossover point is more precise and operator independent than conventional gel-point determinations.

In the past, rheological tests were performed exclusively on neat resins or resins removed from the reinforcement. Some doubt was always present regarding the one-to-one correlation between the viscosity data thus obtained and the way in which a reinforced material would perform during composite fabrication. The possibility always existed of changing the resin when removing the sample. Dissolving the resin from its reinforcement poses problems in solvent removal because even a small level of residual solvents will significantly alter the viscosity profile. Heating to remove trace solvents or the resin itself can advance the matrix and alter its behavior. Simply scraping a resin sample from the reinforcement is tedious and often contaminates the sample with fiber or filler. In addition, neat resin exhibits near-Newtonian flow characteristics during the early stages of cure, while flow is non-Newtonian in the presence of fibers having large surface areas and relatively polar surfaces. As a result, the viscous-state behavior exhibited during the manufacturing process may differ sharply from that observed in the rheological test chamber. To overcome these problems, techniques have been developed to measure the apparent viscosity of the resin in the presence of fibers (Ref 139, 142, 143).

Rheological analysis has been used to study the processing of printed circuit boards (Ref 144) and the effects of moisture on structural thermoset systems (Ref 145, 146). It has also been used as a quality control tool (Ref 147, 148).

Predictive Modeling. Historically, the fabrication of advanced thermoset composite hardware involved processes derived by trial and error. However, cure cycle development can be accomplished in a more scientific and cost-effective manner if the chemical and thermal behavior of the curing resin system is thoroughly understood. This understanding is evolving through the use of mathematical models. These models predict the extent of conversion and the viscosity behavior as a function of time and temperature and offer almost unlimited potential for cure-cycle development and real-time process control of thermosets. In addition, mathematical modeling is useful for quality control applications, as a tool design aid, and as a viscosity predictor in cure-cycle control systems (Ref 149–161).

Cure Monitoring. Viscosity is frequently used to correlate physical behavior with typical processing parameters such as time and temperature. Indirect methods are required for monitoring the physical changes that occur during a production cure because direct measurement of viscosity is not possible. Many years ago, it was shown that electrical and physical measurements are analogous because they are governed by similar mathematical relationships (Ref 162). Therefore, electrical property measurements during cure should reflect the physical and therefore the chemical changes in the curing thermoset. This monitoring method is a well-established technique, and dynamic dielectric analysis (DDA) provides one method for real-time process control.

Dynamic dielectric analysis measures dielectric changes as a function of the molecular mobility of a resin. Most organic resins are polar, and their dipoles will orient in an alternating electrical field to a degree that relates to resin rheology. When the resin is a liquid, the

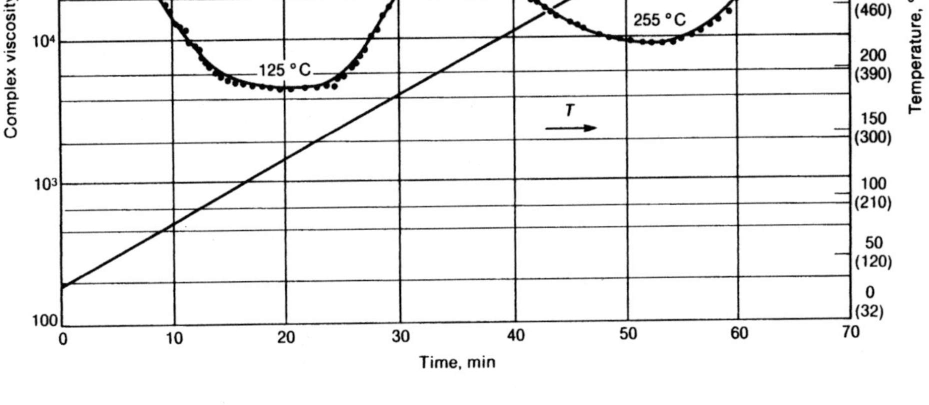

Fig. 22 Typical viscosity profile for LARC-160 PI resin

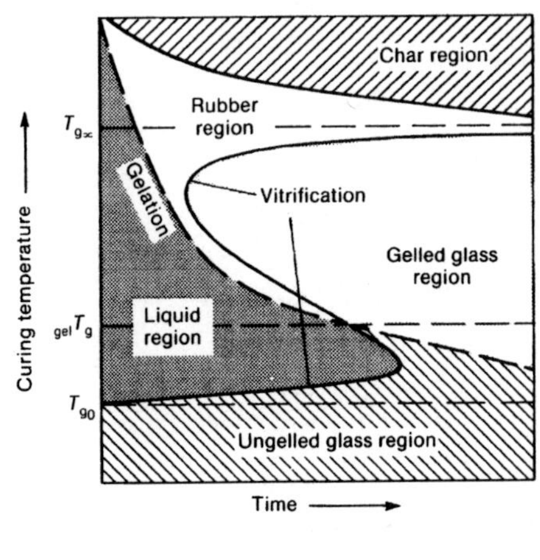

Fig. 23 Time-temperature-transformation diagram. Source: Ref 137, 138

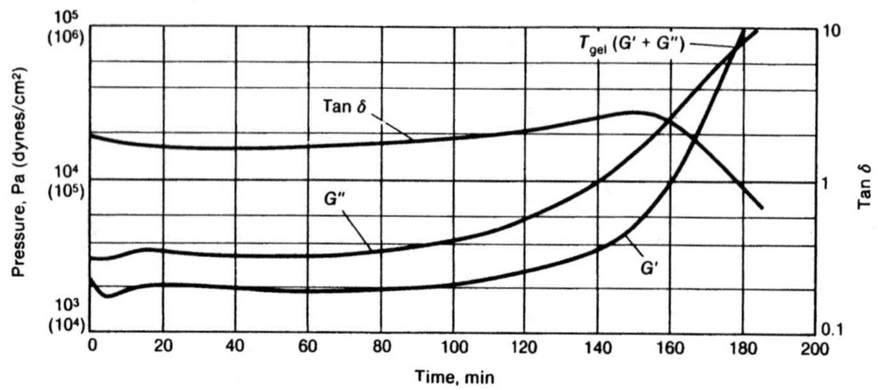

Fig. 24 Gel time from the viscosity curve of Narmco 5208 1300 epoxy prepreg; isothermal at 124 °C (255 °F). *G'*, shear modulus; *G"*, loss modulus

dipoles move quite easily. As the resin cures, it becomes increasingly difficult for these dipoles to align in the field. When final cure is reached and the polymer network is rigid, no dipole movement is possible.

For DDA to be a valid technique, the dielectric signal must correlate to the bulk viscosity. Early investigators employed a dielectric dissipation curve of a material obtained using embedded parallel plate electrodes. Because the spacing between the electrodes can change during cure because of resin shrinkage or application of pressure, planar interdigitized printed circuit probe designs with integral temperature-monitoring devices were developed (Ref 163, 164). This miniature probe allows measurement of both temperature and dielectric properties in the same localized area. The probe combines small size with built-in amplification, providing high signal-to-noise ratio and the ability to obtain property measurements at frequencies as low as 1 Hz. Because the electrode geometry is fixed and the manufacture of integrated devices is very precise, the data obtained using these probes are very reproducible.

There is no question that DDA is a valid and practical monitoring technique (Ref 164–171). By utilizing mathematical models and the appropriate instrumentation, the total automation of a resin-curing process based on intrinsic material properties should be achieved in the near future (Ref 165–172). With appropriate background knowledge of the chemical and thermal behavior of a thermoset, DDA can be used to monitor the extent of reaction (material advancement, aging), reaction rates, point of minimum viscosity, completion of cure, and effects of moisture. These properties, in turn, can be used for the quality control assessment of processibility and as a basis for totally automated, closed-loop process control.

Other Techniques. The simplest form of cure monitoring measures the processing parameters (time, temperature, pressure) that affect a material property, rather than the property itself. In the past, a cure cycle was developed empirically, and cure records of these parameters only satisfied the fabricator that parts had been cured according to a cycle originally developed on one lot of material. Computer-controlled equipment is available that is capable of handling and storing large quantities of data in real time to make it easily adaptable to process control applications. Thus, it is possible to combine the results from the physical and chemical characterization of a material with the thermal response of fabrication tools in order to develop and control "smart" cure cycles.

There are three levels of control for curing thermosets. The first level regulates the cure cycle based on the temperature of the reaction vessel (oven, press, autoclave, and so on). This is the oldest and least effective method. The second level controls by part temperature. This is the most common technique. The last level, based on primary resin properties, is called α control, where α represents the extent of chemical conversion. This method utilizes monitoring/control techniques such as dielectric analysis or ultrasonic monitoring.

By developing process specifications based on the chemical and physical cure characteristics of a resin and utilizing a control system that is able to incorporate these characteristics, consistent part fabrication can be realized with predictable engineering properties.

REFERENCES

1. D.K. Hadad, J.S. Fritzen, and C.A. May, "Exploratory Development of Chemical Quality Assurance and Composition of Epoxy Formulations," AFML-TR-76-112 and AFML-TR-77-217, Air Force Materials Laboratory, June 1976 and Jan 1978
2. E. Heftmann, *Chromatography,* 3rd ed., Van Nostrand Reinhold, 1975
3. *A User's Guide to Chromatography: Gas, Liquid, TLC,* Regis Chemical Co., 1976
4. E.L. Johnson and R. Stevenston, *Basic Liquid Chromatography,* Varian Associates, Inc., 1978
5. G. Fallick and J. Cazes, "High Performance Liquid Chromatography," AMMRC MS 77-2, Army Materials and Mechanics Research Center, Jan 1977, p 159–175
6. "Know More About Your Polymer," Waters Associates, Inc., March 1981
7. "When You Can't Change a Part, Reliability is All Important," Waters Associates, Inc., Jan 1978
8. "Polymer Testing Saves Money in Electronics," Waters Associates, Inc., April 1976
9. "A New Way to Tell Good from Bad," Waters Associates, Inc., Feb 1975
10. F.N. Larsen, Gel Permeation Chromatographic Analysis of Commercial Epoxy Resins, *Proc. Sixth International Seminar on GPC,* 1968, p 111
11. J. Ekmanis and S. Church, Simple Test of Incoming Resins Rates Batch to Batch Quality Level, *Plast. Des. Process.,* March 1977, p 30–34
12. G. Fallick and J. Cazes, Gel Permeation Chromatography for Problem Solving and Quality Control, *Mod. Plast.,* Dec 1977, p 62–66
13. T.J. McCrary, Jr., Quality Assurance Programs via Size Exclusion Chromatography. The WStatistic Concept, *Am. Lab.,* Jan 1985, p 86–90
14. T.J. McCrary, Jr., Quality Assurance Programs via Size Exclusion Chromatography: Part Two, WStatistic Regression Equations for Physical Property Correlations, *Am. Lab.,* Feb 1985, p 98–105
15. D.K. Hadad, Chemical Quality Assurance of Epoxy Resin Formulations by Gel Permeation, Liquid, and Thin Layer Chromatography, *SAMPE J.,* July/Aug 1978, p 4–10
16. D.J. Crabtree, Chromatographic Analysis of Epoxy Resins, *Liquid Chromatography of Polymer and Related Materials,* J. Cazes, Ed., Marcel Dekker, 1977, p 63–77
17. G.L. Hagnauer and I. Setton, Compositional Analysis of Epoxy Resin Formulations, *J. Liq. Chrom.,* Vol 1 (No. 1), 1978, p 55–73
18. T.D. Zucconi and J.S. Humphrey, Jr., A Comparison of Gel Permeation Chromatography and Liquid Chromatography for Epoxy Resin Analysis, *Poly. Eng. Sci.,* Vol 16 (No. 1), 1976, p 11–14
19. T.E. Baker and T.E. Judge, Control and Characterization of Adhesives in the Electronics Industry, *Adhes. Age,* April 1980, p 15–20
20. E.A. Eggers and J.S. Humphrey, Jr., Applications of Gel Permeation Chromatography in the Manufacture of Epoxy-Glass Printed Circuit Laminates, *J. Chromatogr.,* Vol 55, 1971, p 33–44
21. D.K. Hadad, Liquid Chromatographic

Characterization of Printed Circuit Board Materials, *Liquid Chromatography of Polymer and Related Materials III,* Vol 19, J. Cazes, Ed., Marcel Dekker, 1981, p 157–167

22. M.G. Rogers, The Structure of Epoxy Resins Using NMR and GPC Techniques, *J. Appl. Polym. Sci.,* Vol 16, 1972, p 1953–1958
23. H. Batzer, and S.A. Zahir, Studies in the Molecular Weight Distribution of Epoxide Resins. II. Chain Branching in Epoxide Resins, *J. Appl. Polym. Sci.,* Vol 19, 1975, p 601–607
24. H. Batzer and S.A. Zahir, Studies in the Molecular Weight Distribution of Epoxide Resins. III. Gel Permeation Chromatography of Epoxide Resins Subject to Postglycidylation, *J. Appl. Polym. Sci.,* Vol 19, 1975, p 609–617
25. H. Batzer and S.A. Zahir, Studies in the Molecular Weight Distribution of Epoxide Resins. IV. Molecular Weight Distribution of Epoxide Resins Made from Bisphenol A and Epichlorohydrin, *J. Appl. Polym. Sci.,* Vol 21, 1977, p 1843–1857
26. J.G. Hendrickson, The Analysis of Adhesive Resins in Aerospace Applications, *Proc. Nat. SAMPE Conf.,* 1969, p 541–549
27. I.L. Kalnin, H. Meisters, and H.J. Notarius, "Characterization of Epoxy Resin Advancement in Fiber Reinforced Composite Prepregs," paper presented at the 26th Annual Technology Conference on Reinforced Plastics, Section 14-A, 1-14, SPE Inc., Composites Division, 1971
28. R.D. Nuss, Effect of Different Catalysts on an Identical Thermosetting Epoxy-Anhydride Resin System Evaluated Through the B-Staged Curing Conditions by Gel Permeation Chromatography, *Liquid Chromatography of Polymers and Related Materials,* J. Cazes, Ed., Marcel Dekker, 1977, p 79–91
29. I. Antal, L. Fuzes, G. Samay, and L.C. Sillag, Kinetics of Epoxy Resin Synthesis on the Basis of GPC Measurements, *J. Appl. Polym. Sci.,* Vol 26, 1981, p 2783–2786
30. G.L. Hagnauer, P.J. Pearce, B.R. La Liberte, and M.R. Roylance, Cure Kinetics and Mechanical Properties of a Resin Matrix. Effects of Impurities and Stoichiometry, *Chemorheology of Thermosetting Polymers,* C.A. May, Ed., ACS Symposium Series 227, American Chemical Society, 1983, p 25–47
31. G.L. Hagnauer and P.J. Pearce, SEC Analysis of Epoxy Resin Cure Kinetics, *Proc. ACS Symposium on SEC,* American Chemical Society, March 1983
32. R.A. Mowery, E.N. Fuller, and R.K. Bade, On-Line Process Size-Exclusion Chromatography, *Am. Lab.,* May 1982, p 61–67
33. J.R. Woodlee, Optimization in Reverse Phase Liquid Chromatography as Applied to an Epoxy Adhesive, *Proc. Nat. SAMPE Symp.,* Vol 30, 1985, p 471–478
34. C.E.M. Morris, P.J. Pearce, and R.G. Davidson, Characterization of Two Nitrile-Epoxy Structural Adhesives, *J. Adhes.,* Vol 15, 1982, p 1–12
35. J.J. King, R.N. Castonguay, and J.P. Zizzi, HPLC Evaluation of MY-720, *Proc. Nat. SAMPE Tech. Conf.,* Vol 13, 1981, p 53–63
36. J.J. King and R.N. Castonguay, HPLC Evaluation of MY-720 II, *Proc. Nat. SAMPE Symp.,* Vol 27, 1982, p 163–177
37. C.A. Cobuzzi, J.J. King, and M.A. Chaudhari, HPLC Evaluation of MY-720 III, *Proc. Nat. SAMPE Symp.,* Vol 28, 1983, p 877–892
38. M.A. Chaudhari and J.J. King, Characterization of MY-720 IV, *Proc. Nat. SAMPE Tech. Conf.,* Vol 15, 1983, p 676–687
39. C.A. Cobuzzi, J.J. King, and M.A. Chaudhari, Characterization of MY-720 V, *Proc. Nat. SAMPE Symp.,* Vol 29, 1984, p 1261–1276
40. M.A. Chaudhari, C.A. Cobuzzi, and J.J. King, Characterization of MY-720 VI, *Proc. Nat. SAMPE Symp.,* Vol 16, 1984, p 565–576
41. G.L. Hagnauer, "HPLC and GPC Analysis of EPON 828 Epoxy Resins," AMMRC TR 79-59, Army Materials and Mechanics Research Center, Nov 1979
42. G.L. Hagnauer, Analysis of Commercial Epoxies by HPLC and GPC, *Ind. Res. Dev.,* April 1981, p 128–133
43. C.M. Tung, P.J. Dynes, and C.L. Hamermesh, "Improved Quality Control of Printed Circuit Board B-Stage Epoxy Resins," paper presented at the 23rd IPC Annual Meeting, April 1980
44. G.L. Hagnauer and D.A. Dunn, Dicyandiamide Analysis and Solubility in Epoxy Resins, *J. Appl. Polym. Sci.,* Vol 26, 1981, p 1837–1846
45. J.F. Carpenter, "Quality Control of Structural Nonmetallics," Report N00019-76-C-0138, Naval Air Systems Command, June 1977
46. D. Crozier, G. Morse, and Y. Tajima, The Development of Improved Chemical Analysis Methods for Epoxy Resins, *SAMPE J.,* Sept/Oct 1982, p 17–22
47. G.L. Hagnauer and D.A. Dunn, Quality Assurance of Epoxy Resin Prepregs, *ANTEC '84,* Society of Plastics Engineers, May 1984, p 330–333
48. G.L. Hagnauer and D.A. Dunn, Quality Assurance of An Epoxy Resin Prepreg Using HPLC, *Proc. Nat. SAMPE Tech. Conf.,* Vol 12, Oct 1980, p 648–655
49. "Air Force HPLC Round Robin Test Program," Final Report, Fourth Workshop on Physiochemical Characterization Methods, Rockwell Science Center, Aug 1979
50. G.L. Hagnauer, J.M. Murray, and B.W. Bowse, "HPLC Monitoring of Graphite-Epoxy Prepreg Aging," AMMRC TR 79-33, Army Materials and Mechanics Research Center, May 1979
51. P.J. Pearce, R.G. Davidson, and C.E.M. Morris, Aging and Performance of Structural Film Adhesives. I. A Comparison of Two High Temperature Curing, Epoxy-Based Systems, *J. Appl. Polym. Sci.,* Vol 27 (No. 11), 1982, p 4501–4516
52. P.J. Pearce, R.G. Davidson, and C.E.M. Morris, Aging and Performance of Structural Film Adhesives. II. Comparison of Two Nitrile-Epoxy Systems, *J. Appl. Polym. Sci.,* Vol 28 (No. 1), 1983, p 283–294
53. Z.N. Sanjana, W.H. Schaefer, and J.R. Ray, "Effect of Moisture on the Relative Reaction Rates of a Graphite Epoxy Prepreg," paper presented at the 35th Annual Technology Conference on Reinforced Plastics, Section 12-D, 1-7, Composites Institute, Society of the Plastics Industry, 1980
54. G.L. Hagnauer and P.J. Pearce, The Effects of Impurities on the Hydrolytic Stability and Curing Behavior of an Epoxy Resin, *Organic Coat. Appl. Polym. Sci.,* Vol 46, 1982, p 580–584
55. D.K. Hadad and M.R. Dusi, "Characterization of the Processing Behavior of Advanced Composite Materials Containing Variations in TGDDM Viscosity," paper presented at the 187th ACS National Meeting (St. Louis, MO), American Chemical Society, April 1984
56. G.L. Hagnauer, B.R. La Liberte, and D.A. Dunn, Isothermal Cure Kinetics of an Epoxy Resin Prepreg, *Organic Coat. Appl. Polym. Sci.,* Vol 46, 1982, p 646–650
57. R.G. Weatherhead, Thin-Layer Chromatography of Epoxide Resins, *Analyst,* Vol 91, 1966, p 445–448
58. H.L. Spell and R.D. Eddy, The Characterization of Epoxy Resins with the Combined Techniques of Thin Layer Chromatography and Infrared Spectroscopy, *ACS Organic Coat. Preprints,* Vol 24, 1964, p 267–272
59. J.C. Touchstone and F.D. Murrell, *Practice of Thin Layer Chromatography,* John Wiley & Sons, 1978
60. Thin-Layer System Raises Chrom Capabilities, *Ind. Res. Dev.,* Aug 1978
61. J.A. Perry and L.J. Glunz, Programmed Multiple Development: The Intrinsic Grid, *Ind. Res. Dev.,* Oct 1977, p 117–120
62. D.M. Kent and R. Vitek, Quantitation and Correlation, *Ind. Res. Dev.,* May 1978, p 99–102
63. J.M. Newman, Overpressure Layer Chromatography, *Am. Lab.,* April 1985, p 52–63
64. A.D. Woyewoda and J.C. Sipos, More with Less, *Ind. Res. Dev.,* Oct 1978
65. R.G. Ackman, Flame Ionization Detection Applied to Thin-Layer Chromatogra-

phy on Coated Quartz Rods, in *Methods in Enzymology,* Vol 12, 1981, p 205–252
66. H.O. Ranger, TLC Separations in the Third Dimension, *Am. Lab.,* Nov 1981, p 146–151
67. D.K. Hadad, "New Developments in Chromatographic Characterization and Quality Assurance of Composite Materials," paper presented at the 72nd Annual AICE Meeting (San Francisco, CA), American Institute of Chemical Engineers, Nov 1979
68. J.L. Koenig, Application of Fourier Transform Infrared Spectroscopy to Chemical Systems, *Appl. Spec.,* Vol 29 (No. 4), 1975, p 293–309
69. J.B. Enns, "The Cure of Thermosetting Epoxy/Amine Systems," Ph.D. thesis, Princeton University, 1982
70. J.F. Sprouse, Analysis of Curing Processes of Composites Using FTS-IR, *Organic Coat. Plast. Chem.,* Vol 40, 1979, p 934
71. R.E. Sacher, L.M. Chow, and J.M. Sloan, "Evaluation of the Epoxy System for the Repair of Fuel Tank M109," AMMRC-TR-83-55, Army Materials and Mechanics Research Center, Sept 1983
72. T. Provder, C.M. Neag, G. Carlson, C. Kuo, and R.M. Holsworth, Cure Reaction Kinetics Characterization of Some Model Organic Coatings Systems by FT-IR and Thermal Mechanical Analysis, *Anal. Calorim.,* Vol 5, 1984, p 377–393
73. A. Gupta, M. Cizmecioglu, D. Coulter, R.H. Liang, A. Yavrovian, F.D. Tsay, and J. Moacanin, The Mechanism of Cure of Tetraglycidyl Diaminodiphenyl Methane with Diaminodiphenyl Sulfone, *J. Appl. Polym. Sci.,* Vol 28, 1983, p 1011–1024
74. E.T. Mones and R.J. Morgan, FTIR Studies of the Chemical Structure of High Performance Composite Matrices, *Div. Poly. Chem.,* Vol 22 (No. 2), 1981, p 249–250
75. L.J. Buckley and D.K. Roylance, Kinetics of a Sterically Hindered Amine Cured Epoxy Resin System, *SAMPE Quart.,* Oct 1982, p 8–13
76. N.J. Harrick, *Internal Reflectance Spectroscopy,* John Wiley & Sons, 1967
77. P.A. Wilks, Jr., Internal Reflection Spectroscopy, *Am. Lab.,* June 1980, p 92–101
78. M.K. Antoon, K.M. Starkey, and J.L. Koenig, Applications of Fourier Transform Infrared Spectroscopy to Quality Control of the Epoxy Matrix, *Composite Materials: Testing and Design,* STP 674, American Society for Testing and Materials, 1979, p 541–552
79. A.K. Rogers, Y.A. Tajima, and R.C. Young, Material Characterization and Specification Development for 350 F Curing Epoxy-Graphite Materials, *Proc. Nat. SAMPE Symp.,* Vol 27, 1982, p 277–291
80. M.K. Antoon, B.E. Zehner, and J.L. Koenig, Spectroscopic Determination of the In-Situ Composition of Epoxy Matrices in Glass Fiber-Reinforced Composites, *Polym., Compos.,* Vol 2 (No. 2), 1981, p 81–87
81. J.F. Sprouse, "Fourier Transform Infrared Spectroscopy as a Method for Studying Weathering of Glass-Fiber Epoxy Composites," Proc. TTCP-3 Crit. Review, AMMRC-MS-77-2, Army Materials and Mechanics Research Center, 1977, p 43–54
82. B.M. Halpin, J.F. Sprouse, and G.L. Hagnauer, "Characterization of Epoxy Resins, Prepregs, and Composites Using HPLC and FTS-IR," paper presented at the 33rd Annual Technology Conference on Reinforced Plastics, Composites Institute, Society of the Plastics Industry, 1978
83. D. Roylance and M. Roylance, Weathering of Fiber-Reinforced Epoxy Composites, *Polym. Eng. Sci.,* Vol 18 (No. 4), 1978, p 249–254
84. L.T. Drzal, *SAMPE J.,* Vol 19 (No. 5), 1983, p 7
85. A. Graton, FT-IR of the Polymer-Reinforcement Interphase in Composites, *Poly. Preprints,* Vol 25 (No. 2), 1984, p 163–164
86. R.L. Levy, "Mechanism by Which Moisture Influences the Elevated Temperature Properties of Epoxy Resins," AFML-TR-199, Air Force Materials Laboratory, Dec 1976
87. M.K. Antoon and J.L. Koenig, Fourier Transform Infrared Study of the Reversible Interaction of Water and a Crosslinked Epoxy Matrix, *J. Polym. Sci.,* Vol 19, 1981, p 1567–1575
88. F.C. Hewitt, K.S. Morris, and A.J. Rein, Evolution of FTIR from the Laboratory to the Production Environments, *Am. Lab.,* Dec 1985, p 32–39
89. M.S. Roth and D. O'Donnell-Leach, FTIR Analyzer for Real-Time Multi-component Analysis of Process Streams: Part One, *Am. Lab,* Dec 1985, p 40–52
90. R.B. Prime, Thermosets, *Thermal Characterization of Polymer Materials,* E. Turi, Ed., Academic Press, 1982, p 435–569
91. J.D. Keenan, "Structure and Dynamic Mechanical Properties of TGDDM/DDS Epoxy, Graphite Fibers, and Their Composites," M.S. thesis, University of Washington, 1978
92. H.S. Chu, "Processing-Structure-Property Relations for High-Performance Amine-Cured Epoxy Polymers," M.S. thesis, University of Washington, 1982
93. C.A. May, Ed., *Chemorheology of Thermosetting Polymers,* ACS Symposium Series 227, American Chemical Society, 1983
94. W.W. Wendlandt, *Thermal Methods of Analysis,* John Wiley & Sons, 1974
95. P.F. Levy, Thermal Analysis: An Overview, *Am. Lab.,* Jan 1970
96. R.L. Blaine, "Thermal Analysis in the Electronics Industry," paper presented at the Du Pont Educational Seminar, Palo Alto, CA, E.I. Du Pont de Nemours, June 1974
97. R.L. Blaine, Using Thermal Analysis as a Process Development and Quality Control Tool in Circuit Manufacturing, *Insul. Circuits,* March 1976, p 37–42
98. D. Frisch and R. Ciccarone, Thermal Analysis for Evaluating Laminates, *Circuits Manuf.,* Vol 17 (No. 7), July 1977, p 54–58
99. R.L. Hassel, Quality Control of Thermosets, *Ind. Res. Dev.,* Vol 20 (No. 10), 1978, p 160–163
100. W.P. Brennan and R.B. Cassel, Thermal Analysis in the Electrical and Electronics Industries, *Am. Lab.,* Jan 1979, p 80–88
101. P.F. Levy, R.L. Blaine, P.S. Gill, and J.D. Lear, Thermal Analysis: Advances in Instrumentation, *Am. Lab.,* June 1979, p 79–88
102. R.H. Wehrenberg II, Thermal Analysis: The Hot New Technique for Testing Plastics, *Mech. Eng.,* Sept 1979, p 78–83
103. R. Riesen and H. Sommerrauer, Curing of Reaction Molding Resins Studied by Thermoanalytical Methods, *Am. Lab.,* Vol 15 (No. 1), Jan 1983, p 30–37
104. P.S. Gill, Thermal Analysis Developments in Instrumentation and Applications, *Am. Lab.,* Jan 1984, p 39–49
105. W.P. Brennan and M.P. DiVito, Recent Advances in Thermal Analysis Instrumentation, *Am. Lab.,* Jan 1985, p 68–79
106. G. Dugan, Thermal Analysis Supports Chemical R & D, Product Quality Control, *Res. Dev.,* June 1985, p 98–102
107. M.P. DiVito, W.P. Brennan, and R.L. Fyans, Thermal Analysis: Trends in Industrial Applications, *Am. Lab.,* Jan 1986, p 82–95
108. T.A.M.M. Maas, Optimization of Processing Conditions for Thermosetting Polymers by Determination of the Degree of Curing with a Differential Scanning Calorimeter, *Polym. Eng. Sci.,* Vol 18 (No. 1), 1978, p 29–32
109. S.J. Swarin and A.M. Wims, A Method for Determining Reaction Kinetics by Differential Scanning Calorimetry, *Anal. Calorim.,* 1976, p 155–177
110. L.T. Pappalardo, DSC Evaluation of B-Stage Epoxy-Glass Prepregs for Multilayer Boards, *Soc. Plast. Engr.,* Vol 20, 1974, p 13–16
111. L.T. Pappalardo, DSC Evaluation of Epoxy and Polyimide-Impregnated Laminates (Prepregs), *J. Appl. Polym. Sci.,* Vol 21, 1977, p 809–820
112. Z.N. Sanjana and R.N. Sampson, Measuring the Degree of Cure of Multilayer Circuit Boards, *Insul. Circuits* 1981, p 87–92
113. T.M. Donnellan, "The Curing of a Bisphenol A Type Epoxy Resin with 1,8-Diamino-p-Methane," NADC-83146-60, Naval Air Systems Command, 1982

114. J.M. Barton, "A Thermoanalytical Study of the Cure Characteristics of an Epoxy System: BSL 913," Technical Report 76138, Royal Aircraft Establishment, 1976
115. M.R. Dusi, "Chemorheological Characterization and Processing Science of an Epoxy/Amine Thermosetting Matrix," M.S. thesis, San Jose State University, 1984
116. B.G. Parker and C.H. Smith, Evaluating Cure and Shelf Life of Epoxy Prepregs and Film Adhesives, *Mod. Plast.*, Dec 1979, p 58–60
117. A. Dutta and M.E. Ryan, Effect of Fillers on Kinetics of Epoxy Cure, *J. Appl. Polym. Sci.*, Vol 24, 1979, p 635–649
118. R.J. Morgan, C.M. Walkup, and T.H. Hoheisel, Characterization of the Cure of Carbon Fiber/Epoxy Composite Prepregs by Differential Scanning Calorimetry, *J. Compos. Technol. Res.*, Vol 7, 1985, p 17–19
119. N.S. Schneider, J.F. Sprouse, G.L. Hagnauer, and J.K. Gillham, DSC and TBA Studies of the Curing Behavior of Two Dicy-Containing Epoxy Resins, *Polym. Eng. Sci.*, Vol 19 (No. 4), 1979, p 304–311
120. W.J. Sichina, Characterization of Autocatalyzed Thermosets by Differential Scanning Calorimetry, *Proc. Nat. SAMPE Symp.*, Vol 30, 1985, p 610–623
121. W. Huffered, "Application of Rate Theory to Accelerated Durability Testing of Structural Adhesives," AFML-TR-79-4199, Air Force Materials Laboratory, 1980
122. L.C.E. Struik, *Physical Aging in Amorphous Polymers and Other Materials,* Elsevier, 1978
123. Z.H. Ophir, "Structure-Property Relationships in Solid Polymers: I—Segmented Polyurethanes and II—Epoxy Thermosets," Ph.D. Thesis, Princeton University, 1979
124. E.S.W. Kong, Physical Aging and Its Effects on the Mechanical and Physical Properties of Graphite/Epoxy Composites, *Organic Coat. Appl. Sci.*, Vol 46, 1982, p 568–573
125. C.L. Brett, *J. Appl. Polym. Sci.*, Vol 20, 1976, p 1431–1440
126. A.P. Gray, *Perkin-Elmer TAAS 2,* Perkin-Elmer Corp., 1972
127. T.R. Manley, *J. Macromol. Sci.-Chem.*, Vol A8 (No. 1), 1974, p 53–64
128. J.F. Moellmer, Measuring Resin Contents of PC Laminates with Thermal Gravimetric Analysis, *Insul. Circuits,* Aug 1980, p 29
129. J.H. Flynn, *Aspects of Degradation and Stabilization of Polymers,* H.H.G. Jellinek, Ed., Elsevier, 1978, p 573–603
130. J.H. Flynn, Degradation Kinetics Applied to Lifetime Predictions of Polymers, *Polym. Eng. Sci.*, Vol 20 (No. 10), 1980, p 675–677
131. J.H. Flynn, Thermogravimetric Analysis Kinetics, *Div. Polym. Chem.*, Vol 22 (No. 1), 1981, p 310–312
132. B. Shushan, C. Williamson, and R.B. Prime, Applications of a Fully Computer Controlled Thermogravimetric-Tandem Triple Quadrupole Mass Spectrometer System (TGA/MS/MS), *ANTEC '84,* Society of Plastics Engineers, 1984, p 319–322
133. D.C. Sabatelli, G. Lavigne, J. Tanaka, and J.F. Johnson, Polymer Curing Studies Using Combined TGA-GC-FTIR-MS Techniques, *ANTEC '84,* Society of Plastics Engineers, 1984, p 311–315
134. W.M. Jensen, Controlled Thermal Expansion in Printed Wiring Boards, *SAMPE J.*, Jan/Feb 1983, p 58–59
135. W.J. Sichina, P.S. Gill, and K.F. Baker, Thermal Analysis for the PC Shop, *Circuits Manuf.*, Aug 1985, p 39–46
136. J.W. Lula, Testing for Delamination Resistance of Multilayers, *Insul. Circuits,* July 1980, p 61–63
137. J.K. Gillham, *Developments in Polymer Characterization,* Vol 3, J.V. Dawkins, Ed., Applied Science, 1982
138. J.B. Enns and J.K. Gillham, *J. Appl. Polym. Sci.*, Vol 28, 1983, p 2567
139. M.R. Dusi, M.G. Maximovich, and R.M. Galeos, Physiorheological Characterization of a Carbon/Epoxy Prepreg System, *J. Appl. Polym. Sci.*, Vol 30, 1985, p 1847–1857
140. R.J. Hinrichs, Rheological Cure Transformation Diagrams for Evaluating Polymer Cure Dynamics, *Chemorheology of Thermosetting Polymers,* C.A. May, Ed., ACS Symposium Series 227, American Chemical Society, 1983, p 187–199
141. C.M. Tung and J.P. Dynes, Relationship between Viscoelastic Properties and Gelation in Thermosetting Systems, *J. Appl. Polym. Sci.*, Vol 27, 1982, p 569–574
142. M.G. Maximovich and R.M. Galeos, Rheological Characterization of Advanced Composite Prepreg Materials, *Proc. Nat. SAMPE Symp.*, Vol 28, 1983, p 568–580
143. L.D. Lauer, Dynamic Mechanical Analysis of Epoxy Composite Prepregs, *SAMPE Quart.*, Oct 1983, p 31–35
144. W. Engelmaier and M.B. Roller, Temperature-Viscosity-Time Profiles Support Empirical Rules Governing Multilayer Printed Wiring Board Lamination, *Insul. Circuits,* April 1975, p 43–47
145. R. Hinrichs and J. Thuen, Environmental Effects on the Control of Advanced Composite Materials Processing, *SAMPE J.*, Nov/Dec 1979, p 12–21
146. J. Thuen and R. Hinrichs, Structural Adhesives Rheological Behavior Response to Process-Environmental Variations, *SAMPE J.*, Sept/Oct 1980, p 6–17
147. K.G. Kibler, Characterization of Composition Variations in a Structural Adhesive, *SAMPE Quart.*, April 1982, p 39–45
148. D.E. Jackson, D.L. Paradis, and D.L. Hawkins, Rheological Characterization of a Toughened Epoxy Adhesive System as a Quality Control Tool, *Proc. Nat. SAMPE Symp.*, Vol 27, 1982, p 310–319
149. C.A. May, M.R. Dusi, J.S. Fritzen, D.K. Hadad, M.G. Maximovich, K.G. Thrasher, and A. Wereta, Jr., Process Automation: A Rheological and Chemical Overview of Thermoset Curing, *Chemorheology of Thermosetting Polymers,* C.A. May, Ed., ACS Symposium Series 227, American Chemical Society, 1983, p 1–24
150. R.P. White, *Polym. Eng. Sci.*, Vol 14 (No. 1), 1974, p 50
151. M.B. Roller, *Polym. Eng. Sci.*, Vol 15 (No. 6), 1975, p 406
152. J.D. Keenan, SAMPE Educational Workshop, Society for the Advancement of Material and Process Engineering, June 1980
153. M.R. Dusi, D.K. Hadad, and A. Wereta, Jr., "Viscosity Predictions for Composite Reliability," paper presented at the ACS Chemical Industrial Engineer Symposium (New York), American Chemical Society, Aug 1981
154. M.R. Dusi, C.A. May, and J.C. Seferis, Predictive Models as an Aid to Thermoset Resin Processing, *Organic Coat. Appl. Sci.*, Vol 47, 1982, p 635–638
155. Y.A. Tajima and D.G. Crozier, Calculating the Viscosity of an Epoxy Resin During Cure, *Proc. Nat. SAMPE Symp.*, Vol 29, 1984, p 1277–1284
156. Y.A. Tajima and D.G. Crozier, Chemorheology of an Amine Cured Epoxy Resin, *ANTEC '84,* Society of Plastics Engineers, 1984, p 274–277
157. G.P. Schmitt and J.P. Wiley, A Statistical Model for Viscosity Changes During Epoxy Resin Cure, *ANTEC '84,* Society of Plastics Engineers, 1984, p 270–273
158. M.B. Roller, Rheology of Curing Thermosets: Critique and Review, *ANTEC '84,* Society of Plastics Engineers, 1984, p 268–269
159. R.D. Sudduth, A Simplified Analytical Approach to Calculate the Dynamic Gel Temperature, *Proc. Nat. SAMPE Symp.*, Vol 30, 1985, p 649–659
160. A.C. Loos and G.S. Springer, Curing of Epoxy Matrix Composites, *J. Compos. Mater.*, Vol 17, 1983, p 135–169
161. D.J. Boll and R.E. Hoffman, The Use of Dynamic Gel Temperatures to Develop Cure Cycles, *Proc. Nat. SAMPE Symp.*, Vol 29, 1984, p 1411–1421
162. D.K. Cheng, *Analysis of Linear Systems,* Addison-Wesley, 1961
163. J.L. Meyer, J.E. Shidler, E.S. Harrison, R.A. Edwards, and D.R. Fick, "Advanced Composites In-Process Controls/Inspection," Eleventh Quarterly Report, Contract F33615-77-C-5217, Air Force Materials Laboratory, 1980

164. S.D. Senturia, N.F. Sheppard, Jr., H.L. Lee, and S.B. Marshall, Cure Monitoring and Control with Combined Dielectric/ Temperature Probes, *SAMPE J.*, July/Aug 1983, p 22–26
165. W.E. Baumgartner and T. Ricker, Computer Assisted Dielectric Cure Monitoring in Material Quality and Cure Process Control, *SAMPE J.*, July/Aug 1983, p 6–16
166. Z.N. Sanjana and J.R. Ray, The Use of Dielectric Analysis to Characterize Composite Prepreg and Cure, *Div. Polym. Chem.*, Vol 22 (No. 2), 1981, p 225–256
167. N.F. Sheppard, M.C.W. Coln, and S.D. Senturia, A Dielectric Study of the Time-Temperature-Transformation (TTT) Diagram of DGEBA Epoxy Resins Cured with DDS, *Proc. Nat. SAMPE Symp.*, Vol 29, 1984, p 1243–1250
168. D.E. Kranbuehl, S.E. Delos, P.K. Jue, T.P. Jarvie, and S.A. Williams, Dynamic Dielectric Characterization of Thermosets and Thermoplastics Using Intrinsic Variables, *Proc. Nat. SAMPE Symp.*, Vol 29, 1984, p 1251–1260
169. D.E. Kranbuehl, S.E. Delos, E. Yi, J. Mayer, T. Hou, and W. Winfree, Correlation of Dynamic Dielectric Measurements with Viscosity in Polymeric Resin Systems, *Proc. Nat. SAMPE Symp.*, Vol 30, 1985, p 638–648
170. Z.N. Sanjana and R.L. Selby, Monitoring Cure of Epoxy Resins Using a Microdielectrometer, *Proc. Nat. SAMPE Symp.*, Vol 29, 1984, p 1233–1242
171. I.D. Maxwell and R.A. Pethrick, Dielectric Studies of Water in Epoxy Resins, *J. Appl. Polym. Sci.*, Vol 28, 1983, p 2363–2379
172. J.W. Lane et al., Dielectric Modeling of the Curing Process, *Polym. Eng. Sci.*, Vol 26 (No. 5), 1986, p 346–353

Physical, Chemical, and Thermal Analysis of Thermoplastic Resins*

DURING THE PROCESSING OF THERMOPLASTICS, polymer chains are sheared, twisted, distorted, stretched, and subjected to a vast array of flow histories. As a result of time and temperature, the macromolecular network will eventually undergo stress relaxation. The attendant distortion, warpage, and dimensional instabilities are directly related to the degree of "cruelty" suffered in processing. The consequences of processing at all stages must be addressed when discussing the thermomechanical properties of both thermoplastic and thermosetting resin systems.

Polymers are viscoelastic; that is, they respond to stress as if they were a combination of elastic solids and viscous fluids, but not always in a stable 50-50 proportion. The balance of the storage and loss components of a polymer are important in determining its melt processibility and functionality, or solids behavior. Unfortunately, the very essence of viscoelastic behavior also hinders the use of typical testing methods for predicting end-use properties.

The plastics industry has used well-established, short-term test methods for attempting to predict long-term behavior. A practical and accurate method for predicting the useful lifetime of plastics and elastomers in structural applications is thus critical with the development of new materials and demanding applications. For example, short-term data generated via instrumented impact testing provide reliable information, whereas long-term mechanical behavior, including fatigue and creep, must be qualified. This requires the use of some prediction methodologies.

This article addresses some established protocols in characterizing thermoplastics, whether they are homogeneous resins, alloyed or blended compositions, or highly modified thermoplastic composites. The information herein is applicable to all these major resin groupings; no attempt has been made to contrast one against another.

Molecular Weight Determination from Viscosity

The principal methods of molecular weight (MW) determination are based on viscosity measurements and chromatography. The latter includes gel permeation chromatography (GPC) and high-performance liquid chromatography (HPLC), which are discussed in the section "Chromatography" in this article. There are also several other methods in determining MW, such as:

- Number average MW using vapor pressure (ASTM D 3592, discontinued) and membrane osmometry (ASTM D 3750, discontinued)
- Weight number average using light scattering (ASTM D 4001)

These methods are not discussed in this article.

This section describes MW determination by viscosity measurements. The relationship between MW and viscosity (η) is:

$$\eta = K\,(MW_V)^a$$

where K is a constant, MW_V is the viscosity average MW. The exponent, a, varies from 0.5 to 1 for solution viscosity. For melt viscosity, a = 3.4, and so melt viscosity is more sensitive to MW changes. No information on MW distribution is given from viscosity measurements. It can be a good tool for assessing degradation (e.g., heat, hydrolysis) as part of failure analysis.

Current ASTM volumes include more than 20 different protocols for determining the viscosity of a polymeric solution or melt. From these viscosity measurements, mathematical relationships are employed to determine the molecular weight of the polymer. Several categories of test methods are available for making these determinations.

Solution Viscosity. The traditional approach for determining only the molecular weight of a resin, but not the molecular weight distribution, involves dissolving the polymer in a suitable solvent. However, the more structurally complicated macromolecules require the use of hostile solvents, tedious sample preparations, and costly time delays to obtain limited, single data point values. For example, the solution viscosity determination of polyvinyl chloride (PVC), according to ASTM D 1243-95 (Ref 1), requires either a 1 or 4% concentration in cyclohexanone or dinitrobenzene, while polyamides (PAs), or nylons, require formic acid (Ref 2). Other engineering polymers might require tetrahydrofuran, dimethylformamide, dimethylsulfoxide, or other equally hostile solvents (many of these solvents are also used in GPC analyses).

Various viscosity values as a function of polymer concentration are shown in Fig. 1. However, these values are only indications of molecular weight and do not reflect the elastic component of the polymer. Although the viscosity of the base resin might be useful knowledge, the value for PVC in particular is quite limited because vinyls are the most modified base resins. The ASTM D 3591-97 method (Ref 3) is recommended for determining the logarithmic viscosity of a PVC compound and for assessing the consequences of processing using the molecular weight of the base resin.

Brookfield Viscosity. Several ASTM documents are based on the inexpensive Brookfield

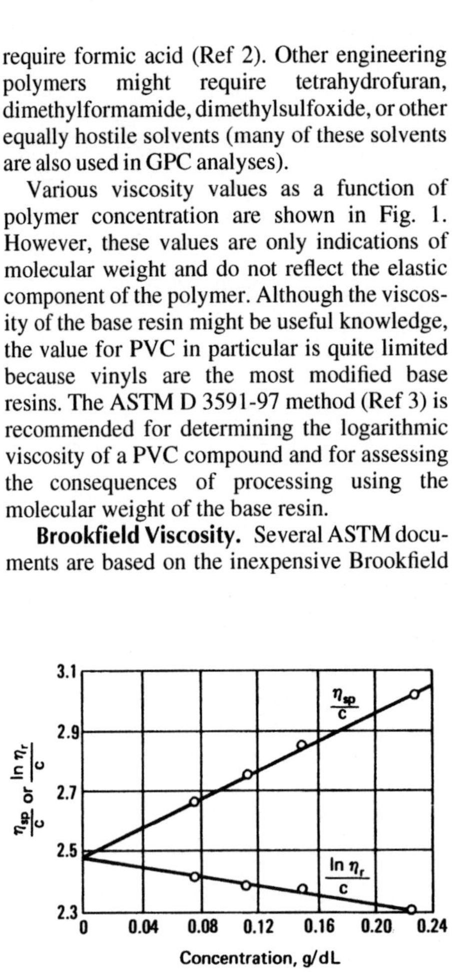

Common name	Recommended name	Symbol and defining equation
Relative viscosity	Viscosity ratio	$\eta_r = \eta/\eta_0 \simeq t/t_0$
Specific viscosity	...	$\eta_{sp} = \eta_r - 1 = (\eta - \eta_0)/\eta_0 \simeq (t - t_0)/t_0$
Reduced viscosity	Viscosity number	$\eta_{red} = \eta_{sp}/C$
Inherent viscosity	Logarithmic viscosity number	$\eta_{inh} = (\ln \eta_r)/C$
Intrinsic viscosity	Limiting viscosity number	$[\eta] = (\eta_{sp}/C)_{c=0} = [(\ln \eta_r)/C]_{c=0}$

Fig. 1 ASTM solution viscosity relationships

*Adapted from Stephen Burke Driscoll, Physical, Chemical, and Thermal Analysis of Thermoplastic Resins, *Engineering Plastics*, Volume 2, *Engineered Materials Handbook*, ASM International, 1988, pages 533 to 543

viscometer. The document that is specific to nylon solutions is ASTM D 789-98 (Ref 2), while ASTM D 1824-95 (Ref 4) gives protocol for vinyl plastisols and organosols. However, like the solution techniques, the Brookfield viscometer determines only the viscous component of the resin, which can be quite sensitive to temperature.

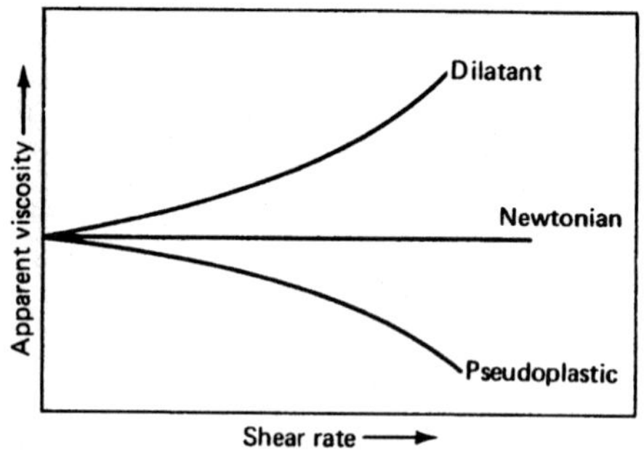

Fig. 2 SPE rheology primer

An advantage of the Brookfield viscometer is that various spindles and steady rotational speeds can be used to determine quickly and easily the fundamental rheological behavior of the solution, including the Newtonian, dilatant, pseudoplastic, or Bingham response, as shown in Fig. 2 (Ref 5). However, offsetting this advantage is the insensitivity of solution techniques to subtle changes in molecular weight. This concern is discussed in greater detail later.

Torque Rheometry. Because base PVC resin is never used alone, and because many other polymers are extensively modified with additives, including fillers and lubricants, two additional ASTM documents have been adopted for measuring important rheological characteristics using the torque rheometer.

The absorption of a plasticizer by a vinyl homopolymer, and more specifically, the identification of the drying rate of a Henschel versus ribbon blender type resin, have been cited in ASTM D 2396-94 (Ref 6). Once the formulation has been set, the torque rheometer can be used to assess the influence of the molecular weight and of fillers, lubricants, and other various additives. Figure 3 illustrates the data generation of this isothermal, steady-shear time sweep type of test, including time to reach peak and equilibrium torques.

The underlying premise that must be remembered when using the torque rheometer is that only the motor load is being measured, not the actual viscosity of the polymer solution or melt. Consequently, the determination of molecular weight is only an approximation or relative ranking. Figures 4 to 6 illustrate the effect of changing the molecular architecture of the base resin and the amount or loading level of filler or lubricant, but not the type or particle size or shape (Ref 7).

Unlike solution techniques, the torque rheometer cannot measure the actual or apparent viscosity, only the motor load imposed by the material being evaluated. Both the solution and torque systems are limited to viscosity measurements and are not capable of assessing the equally important elastic component.

Melt Flow Rate. One of the most common physical properties routinely reported on manufacturer resin data sheets or product bulletins is the melt index (MI) (for polyethylene) or melt flow rate (MFR) (for all other thermoplastic resins, alloys, and composites). ASTM D 1238-98 (Ref 8) cites the average flow (g/10 min) of a

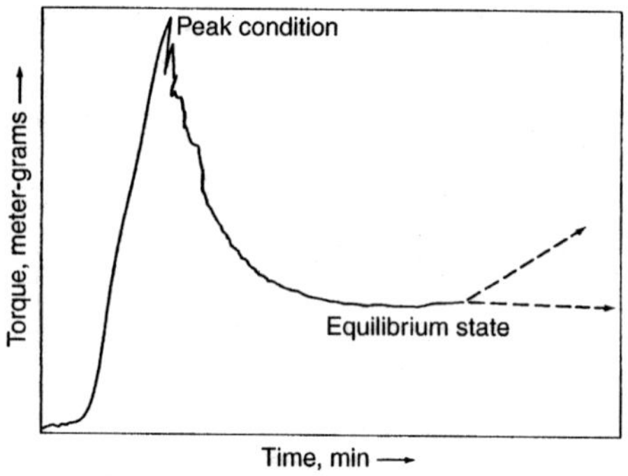

Fig. 3 Torque rheometry

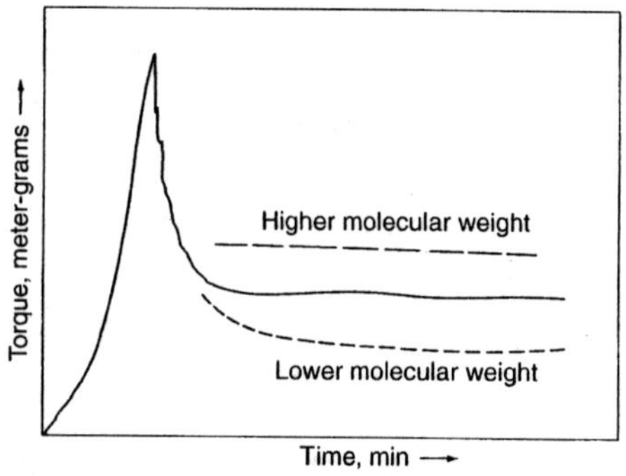

Fig. 4 Torque rheometry, function of molecular weight

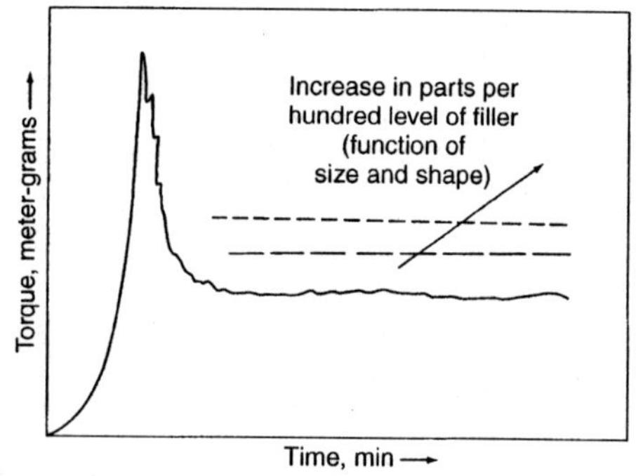

Fig. 5 Torque rheometry, function of parts per hundred of filler

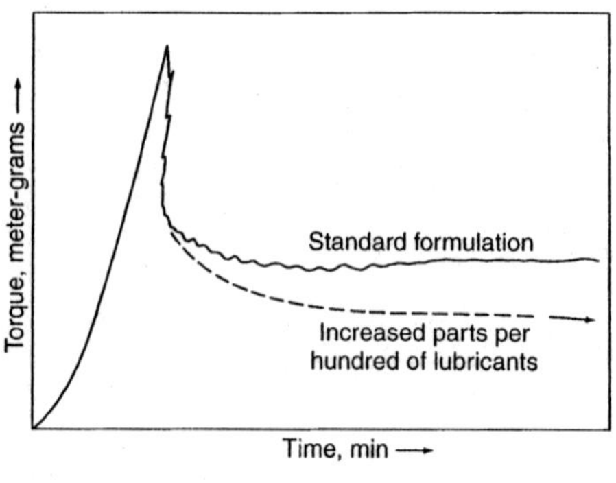

Fig. 6 Torque rheometry, function of parts per hundred of lubricant

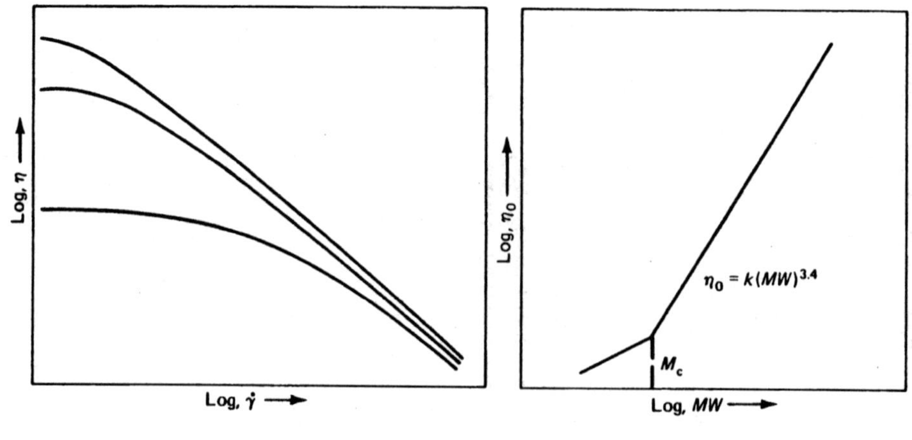

Fig. 7 Relationship of molecular weight to zero-shear viscosity

thermoplastic material through a standardized orifice under standardized conditions (temperature and dead load). The actual steady-shear rate for many materials is about 5 reciprocal seconds. ASTM D 3364-94 is similar to ASTM D 1238-98 but it uses a capillary die that is three times longer (Ref 9).

Again, the measured flow is only an indirect indication of the molecular weight, which is inversely proportional to the measured flow of the thermoplastic material. The lower the molecular weight of the polyethylene, for example, the greater the melt index. The lower the melt flow rate of the resin, the higher the molecular weight or bulk-average molecular weight of many resins blended together. The MI or MFR is only an inverse indication of the "overall" molecular weight and does not indicate anything about the equally important molecular weight distribution.

Capillary. By varying the orifice used in the extrusion plastometer, different shear rates can be obtained. Dividing the monitored shear stress by the shear rate generates single data point viscosity. Multiple evaluations of the material at different shear stresses or rates combine to give a viscosity versus shear rate profile. Attempts to measure the percent die swell have not been totally successful. ASTM D 3835-96 (Ref 10) does caution that the barrel pressure drop cannot be ignored for short capillaries of large diameters (very small length/diameter, or *L/D,* ratios). Additionally, the pressure drop at the entrance to the capillary must be measured when the *L/D* ratio is less than 40 to 1. Finally, the Rabinowitsch correction is necessary for calculating the shear rate at the capillary wall for non-Newtonian fluids.

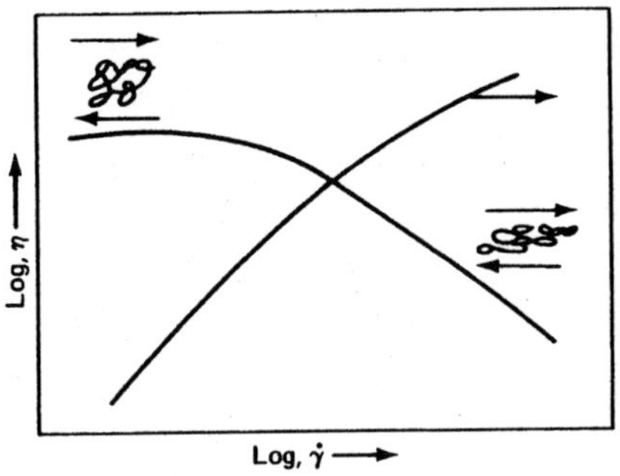

Fig. 8 Steady-shear rheometry

The Use of Cone and Plate and Parallel Plate Geometries in Melt Rheology

Determining the viscoelastic properties of a polymer melt can easily be categorized into two broad areas: steady-shear rheometry and dynamic oscillatory measurements. In both cases, the same test fixtures or tooling can be used to generate important rheological information about the material, including the effect of macromolecular structure on processibility and prediction of functional properties.

Steady-Shear Rheometry. Historically, cone and plate geometries have been used to ensure a uniform shear field on the material being tested. However, the principal limitation is that the particulates or reinforcements in filled materials might be trapped at the center of the cone. This could affect the critical gap setting, causing erroneous readings. A second limitation is that automatic temperature sweeps are quite impractical because of the constant need to adjust the gap setting. Offsetting these two disadvantages, however, is the inherent ability to shear at extremely low rates in order to extrapolate confidently to the "zero-shear" viscosity of the material. Figure 7 illustrates the reason very high molecular weight products need extremely low steady-shear rates in order to attain a flat, Newtonian plateau for extrapolating to zero shear. A low molecular weight resin might establish its Newtonian plateau at only 0.1 reciprocal seconds. A medium molecular weight product might require at least 0.01 reciprocal seconds of shear rate to stabilize, while a very high molecular weight resin might demand an unusually low shear rate of 10^{-4} to 10^{-6}/s.

Such low rates are now easily obtained by new-generation rheometers, which are equipped with precision air bearings. Concurrent with the development of a floating actuator/motor assembly is their unique sensitivity to measure subtle perturbations during rotation of the test fixtures. At very low shear rates, the first normal force developed is quite minimal and only becomes significant at higher shear rates. At low shear rates, the viscosity, η, is quite high. However, the viscosity decreases as the normal force increases with increasing shear rate due to chain entanglements that cannot be stress relieved within the time frame at that particular shear rate (Fig. 8). Stress-relaxation time is the reciprocal of the shear rate.

Extensional Rheometry. Important research has illustrated the enhanced three-fold sensitivity of extensional viscosity, compared to steady-shear methods. Extensional deformation is more representative of fiber spinning, blow molding, and some extrusion operations (Ref 11).

Dynamic Mechanical Rheometry. As reviewed elsewhere, a new generation of critically sensitive test protocols was introduced (in the 1970s) based on dynamic mechanical testing using conventional shear geometries, including cone and plate and parallel plates. These new protocols can be used to measure the rheological properties of materials, including powders, solutions, melts, and solids (Ref 12–15).

In principle, a small amount of material is dynamically oscillated over a fixed arc or amplitude. Response to this deformation is continuously monitored, and the data is reduced to the in-phase (elastic or storage) and out-of-phase (viscous or loss) components.

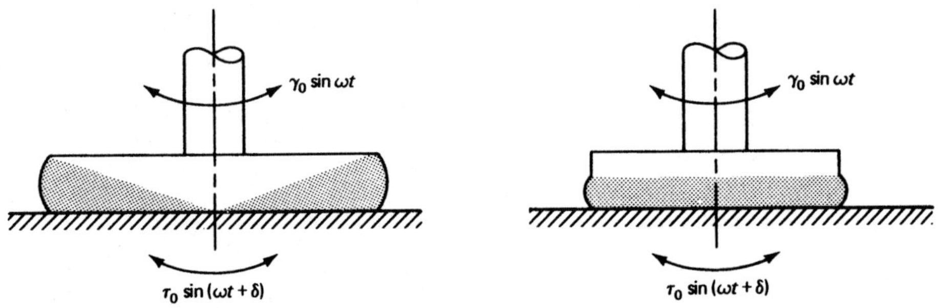

Fig. 9 Cone and plate (left) and parallel plate (right) geometries

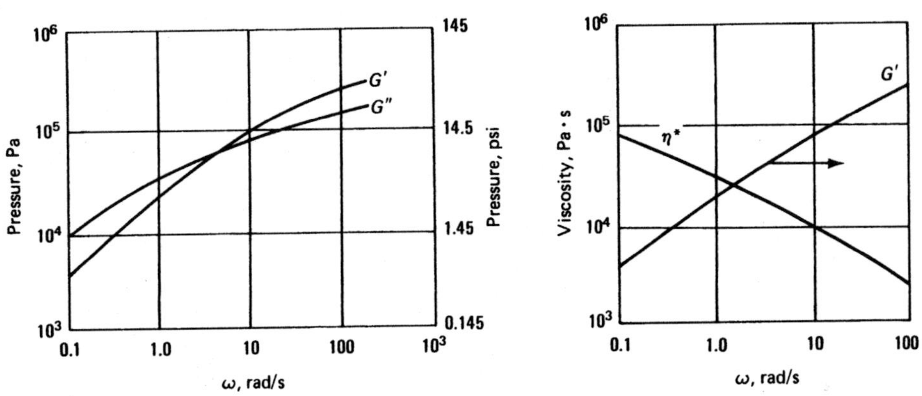

Fig. 10 Rheological profile of high-density polyethylene (HDPE)

The advantages of dynamic mechanical techniques for measuring the rheological properties of a solution or melt include: fast test capability for thermally sensitive materials, ease of sample preparation and subsequent cleanup, and the ability to use different test geometries to maximize the output signal and thereby realize the maximum sensitivity of the instrument. For example, conventional melt viscosity measurements are made using either cone and plate or parallel plate (disk) geometries (Fig. 9). Solid materials are also routinely and accurately characterized using many specific geometries, which are discussed in the following section.

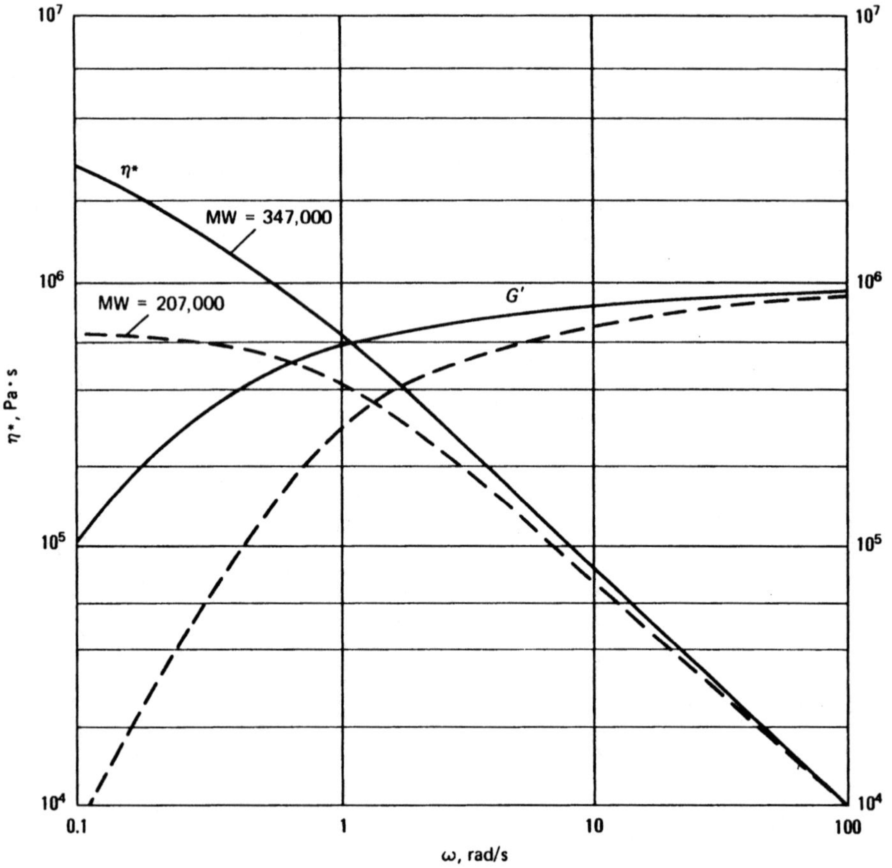

Fig. 11 Degree of polymerization of polybutadiene rubber at 25 °C (77 °F)

ASTM standards have a series of protocols for determining the rheological properties of solutions and melts. For example, ASTM D 4440-95a addresses the need to identify the viscoelastic behavior of either thermoplastic or uncured thermosetting resins, while ASTM D 4473-95a is used to monitor continuously the cure behavior of a thermoset or a vulcanizable elastomer.

Briefly, the elastic component in shear, G', relates to many manufacturing considerations, including surface appearance and die swell, which affect tool and die design, while G'', the viscous component in shear, relates to the resistance of a material to flow. The ratio of G'' to G', known as tan δ, indicates elastic memory and recovery in the melt phase, as well as impact resistance and creep behavior in the solid regime. The complex modulus, G^*, is determined by using vector analysis, and the complex viscosity, η^*, is G^* divided by the dynamic frequency, ω, in radians/s or Hz. The units for the moduli are Pa, while complex viscosity is expressed in Pa · s.

Figure 10 depicts an ASTM D 4440-95a complex melt viscosity profile of high-density polyethylene (HDPE) at 190 °C (375 °F) and 10% strain amplitude. Although the G' and G'' are the two independent variables, quite often only one is shown with the complex melt viscosity.

Specific tests can be designed to assess the contributions of base resins, fillers, and other additives. For example, a frequency sweep is quite useful in identifying the consequences of changes in molecular weight and molecular weight distribution. Figure 11 depicts the change in complex viscosity as a function of increased degree of polymerization (DP) or molecular weight (MW). Note that both the η^* and G' increase with increases in molecular weight. The true sensitivity of melt viscosity to MW is observed by the relationship $\eta^* = k(MW)^{3.4}$. This means that only a slight change in MW corresponds to a significant change in melt viscosity, unlike conventional solution

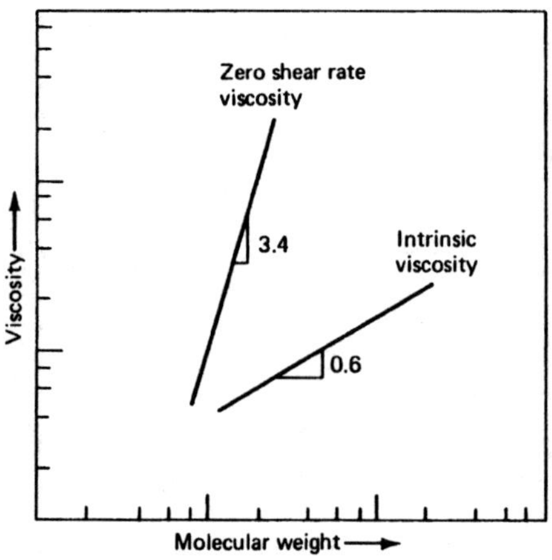

Fig. 12 Sensitivity of solution versus melt rheometry to molecular weight

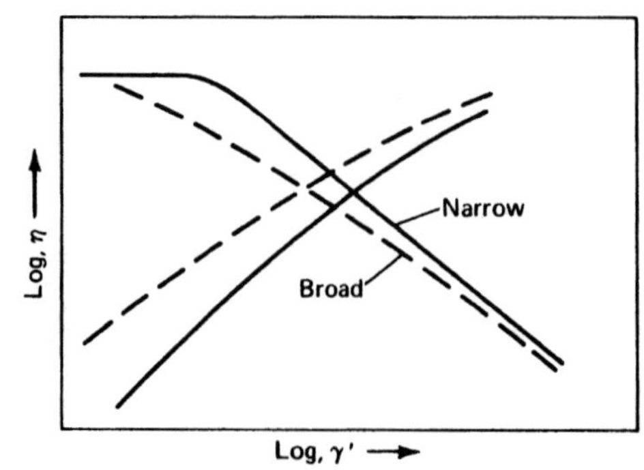

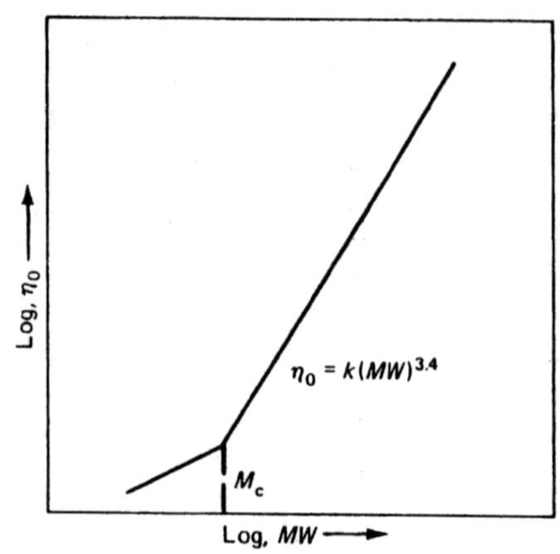

Fig. 13 Narrow versus broad molecular weight distribution

techniques, which are dramatically less sensitive ($\eta = k(MW)^{0.6}$, as shown in Fig. 12.

A second example of the unique contribution of melt viscosity measurements using dynamic mechanical techniques is the relationship between η^* and G' and molecular weight distribution. Figure 13 shows that the complex melt viscosity decreases more immediately, although gradually, with a broadening of the distribution, and that the G' increases. The same trend can be observed for short-chain branching in the polymer architecture, which is discussed in the section "Chromatography." Unlike the previously cited techniques, the observed values fingerprint the "total" rheological nature of the polymer and do not simply generate a single, average, or bulk, value.

A considerable body of work has been reported by many, including Saini and Shenoy (Ref 16), contrasting the quality of steady-shear versus dynamic oscillatory evaluations. For most unfilled homopolymers, the steady-shear viscosity, as a function of shear rate, and the complex melt viscosity, as a function of dynamic oscillation, do superimpose nicely. In other words, there is no compromise of the quality of data generated by either technique. In fact, most viscoelastic analyses in the United States since the 1970s have been in the parallel plate, dynamic oscillatory mode (ASTM D 4440-95a). This testing protocol continues because of its sensitivity and ease of operation.

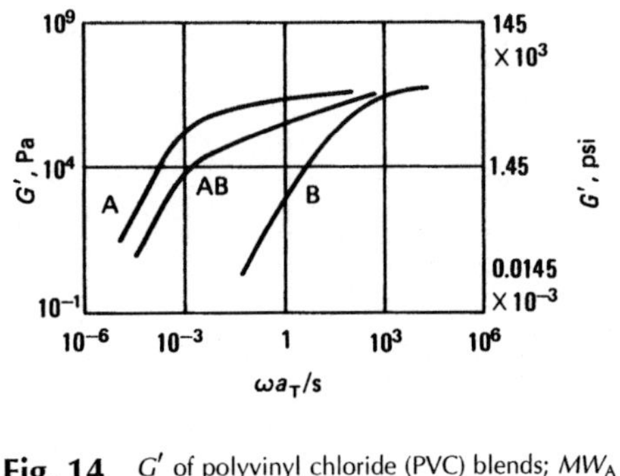

Fig. 14 G' of polyvinyl chloride (PVC) blends; $MW_A = 58 \times 10^4$; $MW_B = 5.9 \times 10^4$

Small amounts of high molecular fractions are known to have disproportionate contribution on the complex viscosity (Ref 17, 18). Park (Ref 19) used ASTM D 4440-95a to illustrate that blending two PVC-base resins with quite different molecular weights resulted in a much higher melt viscosity than predicted (Fig. 14). Park concluded that because of the large crystallinity or supermolecular structural contribution, the higher molecular weight fraction of PVC has a relatively strong influence on the melt flow properties of the blend (Ref 19).

Other important test modes include time sweeps to monitor the thermal stability of a polymer, strain sweeps to assess the contribution of filler size and shape, and temperature sweeps to generate the viscosity sensitivity of the polymer. The latter is also used to develop a time-temperature master curve, which can be invaluable in determining long-term functional properties of the polymer (Fig. 15).

These tests are not limited to laboratory "batch" studies alone. These same ASTM test protocols have been incorporated into online/real-time manufacturing schemes. The importance of compounding technology cannot be minimized. Changes in compounding equipment (single-screw versus twin-screw extruders or Banbury intensive mixers) affect the rheological behavior of the polymer. Changes in processing parameters alter both processibility and performance of the end product. With on-line rheometry, changing the processing conditions allows immediate assessment of the rheological response. Increasing the screw speed, reducing the residence time, or varying the shear history and the dispersional uniformity changes product quality. Investigative work has demonstrated the practicality of on-line rheometry.

Dynamic Mechanical Properties of Solids. When discussing the rheological behavior of a polymer, it is also important to consider the viscoelastic response of the solid. The most common experimental protocol is a temperature sweep (Ref 15). This shows important thermal and functional property transitions and generates critical information, including modulus as a function of temperature, indications of deflection temperature under load (DTUL), the maximum service temperature, and trends of impact and creep behavior.

As mentioned previously, the use of dynamic mechanical protocol allows for testing many different geometries, including torsional shear of bars and rods, dynamic tension of films and fibers, dynamic compression of foams and elastomers, and dynamic three-point bending of very high modulus or friable compositions (Fig. 16). Regardless of the geometry selected (primarily for user convenience), the mathematical relationships are similar. The shear and tension-compression approaches can be combined to determine Poisson's ratio:

	Shear	Tension-compression
Storage modulus	$G' = \tau'/\gamma_0$	$E' = \sigma'/\varepsilon_0$
Loss modulus	$G'' = \tau''/\gamma_0$	$E'' = \sigma''/\varepsilon_0$
Loss factor	$\tan\delta = G''/G'$	$\tan\delta = E''/E'$

Figure 17 illustrates the dynamic mechanical properties of a series of coated polyethylene terephthalate (PET) film. A dynamic mechanical test in tension provides a sensitive analysis of the quality of coatings on thin substrates. Uniformity and thickness of the coatings are crucial to the performance of the coated films. These factors are largely controlled by the wetting properties and the rheology of the coating during application. In Fig. 17, the effect of the coating on the mechanical loss behavior was very pronounced. The uncoated sample exhibited the relaxation behavior characteristic of PET, with the glass transition at 120 °C (250 °F), and a sec-

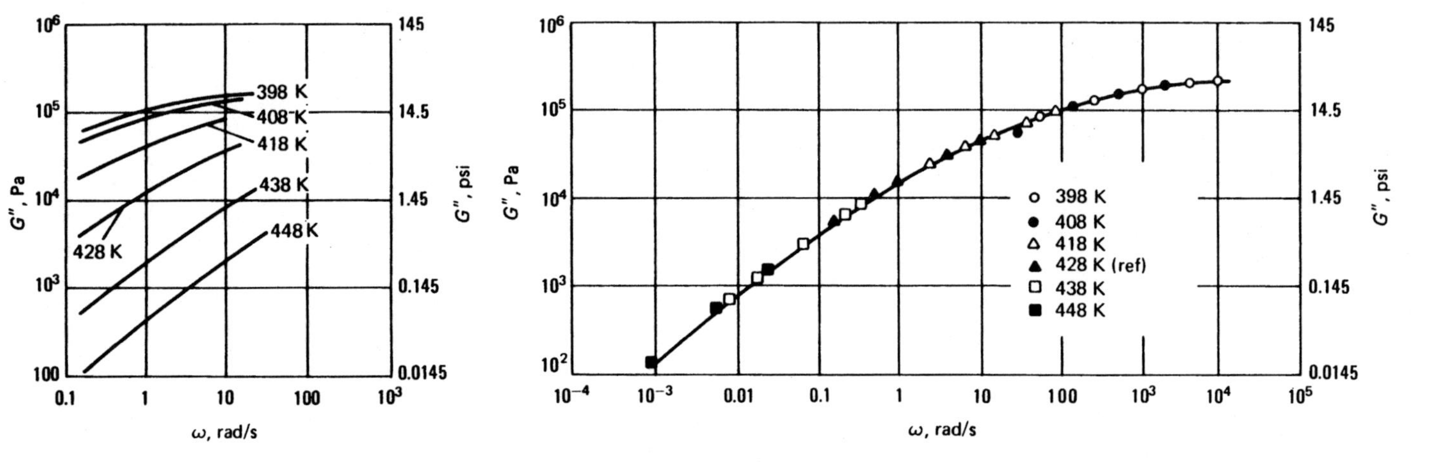

Fig. 15 Development of polyvinyl chloride (PVC) master curve

ondary transition at −60 °C (−75 °F) (at 6.28 rad/s). The sample with a good-quality coating showed a third transition at 50 °C (120 °F), while the intensity of the transition was reduced substantially with the poor-quality coating. The poor quality appeared to be related to the uneven wetting of the substrate.

The dynamic compression of a polyurethane foam is noted in Fig. 18. The dynamic mechanical properties of soft urethane foam can be determined conveniently in the compression mode. The analysis is important for both foams and elastomers; their stiffness and mechanical damping characteristics can be observed over the entire significant-temperature range. As shown in Fig. 18, the urethane sample studied had only one relaxation process, its glass transition at 25 °C (77 °F). The proximity of its glass-transition temperature to room temperature can account for the long relaxation time of the foam observed at room temperature in recovery from large deformations. Large deformations attainable with a minimum force and a slow recovery from the deformation are the key properties required for ear plugs, for example. The temperature response of a polymer construction in three-point bending is shown in Fig. 19. In all cases, it is important to note that ASTM does caution that modest thermal ramps or gradients be used, below a dynamic frequency of 10 rad/s or 1.6 Hz to ensure that important transitions are not masked by testing too quickly.

Figure 20 illustrates a typical modulus versus temperature analysis, with important regions, including the glass transition (T_g) and the secondary beta peaks. As the material is heated through the T_g region, there is a dramatic decrease in the modulus, or stiffness (rigidity), of the polymer. Concurrent with the decrease in G', there is a maximum in the tan δ response. At lower temperatures, the influence of additives, such as impact modifiers, and of compounding schemes, can be identified.

Figure 21 illustrates the contribution of additional rubber content in an acrylonitrile-butadiene-styrene (ABS) terpolymer. As the amount and type of rubber change, there is a corresponding decrease in rigidity, a modest T_g decrease at −90 °C (−130 °F) (butadiene peak), and an increase in the tan δ value (improved impact, but more creep). From a practical approach, the ASTM D 3763-02 (Ref 20) instrumented impact test provides a more detailed analysis of the impact event. Consequently, the use of instrumented impact along with the ASTM D 4065-95 (Ref 15) solid properties investigations will provide more meaningful information on the nature of the impact event, such as whether the impact was brittle or ductile (Fig. 22), or was a tear-fracture or a punched hole (Fig. 23).

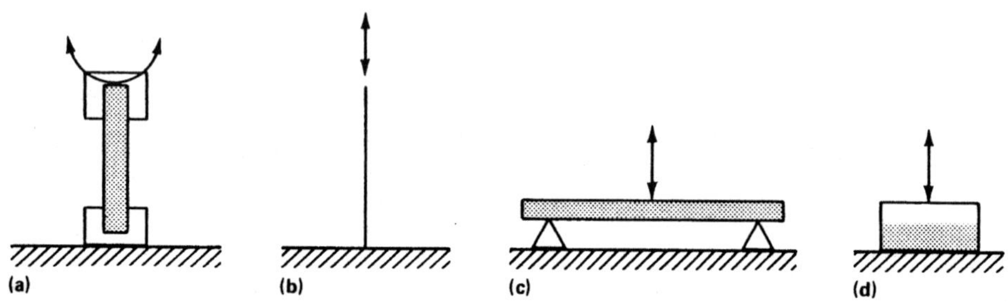

Fig. 16 Dynamic mechanical properties of solids. (a) Torsion. (b) Tension. (c) Bending. (d) Compression

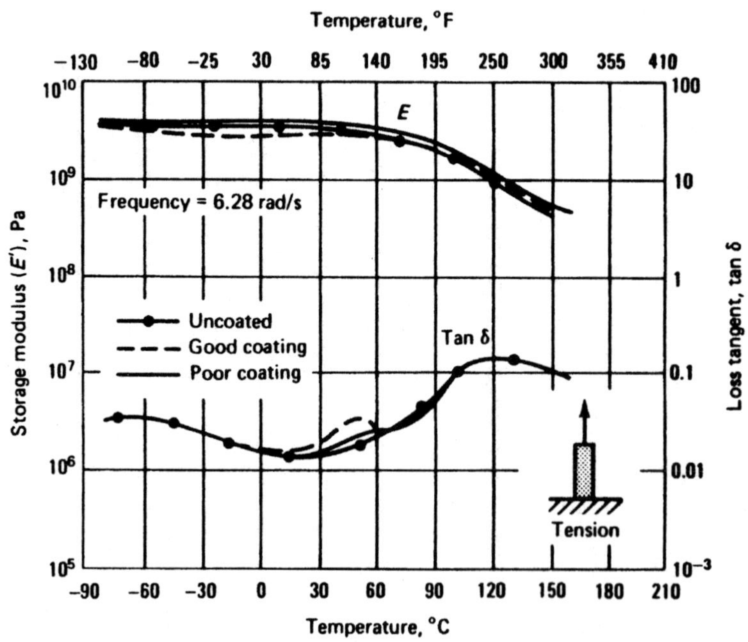

Fig. 17 Dynamic mechanical properties of polyethylene terephthalate (PET) film as a function of temperature; 0.05 mm (0.002 in.) thick specimen, 6.28 rad/s frequency

Chromatography

The need for sensitive molecular weight determinations is very important because it directs processing efforts while explaining the functional properties of the composition. Certainly there are as many disagreements or uncertainties as there are agreements on these relationships. According to Bikales (Ref 22), the more generally recognized relations include:

- Melt viscosity is related to the weight-average molecular weight, being directly proportional to $MW^{3.4}$.
- Tensile strength appears related to an average of the number average and the weight average; however, tensile impact strength appears to be the most closely related to the weight average.
- Elongation at break increases with molecular weight, but is not clearly related to the number average.
- T_g is related to the molecular weight through the equation $T_g = T_g - (K/M)$.
- Stress relaxation has been shown to be related to the molecular weight distribution of polymers, but with different types of dependence for various regions of behavior (Ref 22).

Liquid chromatography allows determination of MW and molecular weight distribution (MWD) of polymers. The methods are referred to as gel permeation chromatography (GPC) and high-performance liquid chromatography (HPLC). The analysis of polymers by conventional gas chromagraphy/mass spectrometry is normally not possible because polymers are typ-

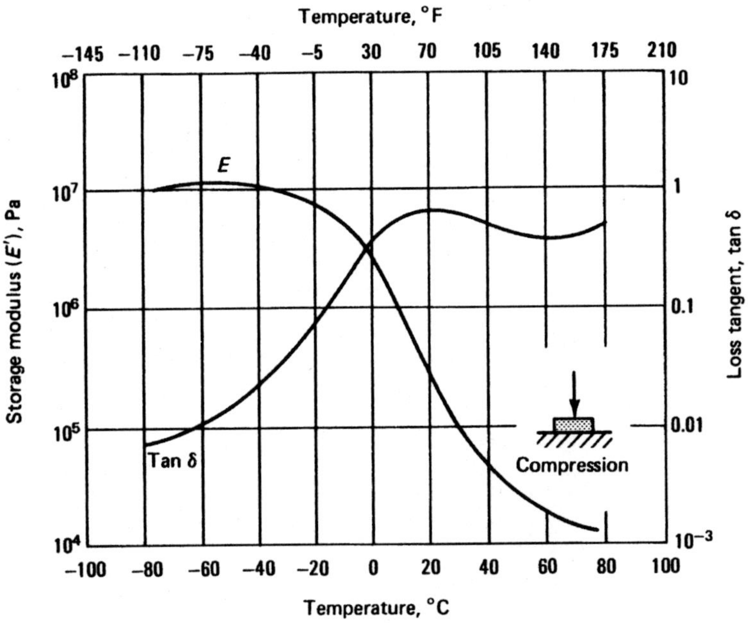

Fig. 18 Dynamic mechanical properties of a cylindrical urethane foam sample; 6.60 mm (0.26 in.) diam, 6.28 rad/s frequency

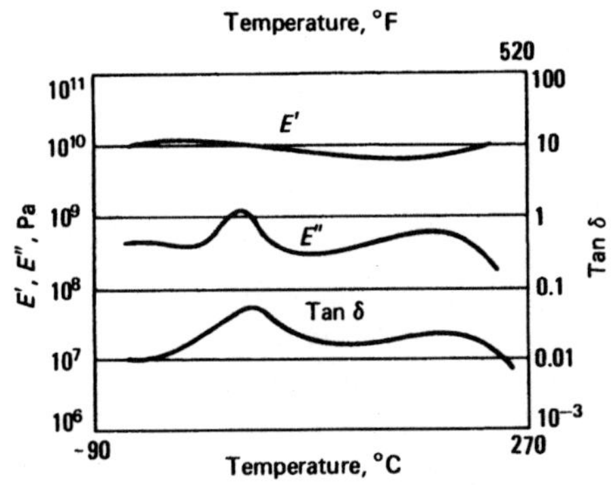

Fig. 19 Dynamic mechanical properties in three-point bending

ically high-molecular-weight materials that do not vaporize.

Gel permeation chromatography (GPC), also known as size-exclusion chromatography (SEC), separates sample molecules on the basis of their physical size. It utilizes a liquid mobile phase and a solid stationary phase, whereas gas chromatography uses an inert gas mobile phase and solid or liquid stationary phase. The stationary phase is a gel with pores of a particular average size. Molecules that are too large to permeate the pores move directly through the separation column and appear first in the chromatogram. Small molecules permeate the pores and follow a long path through the pore matrix; therefore, they have longer retention times. A size-exclusion chromatogram of a mixture containing polystyrenes is shown in Fig. 24 (Ref 23). It can be seen that the components elute in order of decreasing molecular weight.

Size-exclusion chromatography is the preferred method for separating components with high molecular weights (2000 to 2,000,000 amu), particularly those that are nonionic. It has been used in the analysis of epoxies, polyesters, polyolefins, polystyrenes, polyurethanes, polyvinyl alcohol, polyvinyl chloride, proteins, and carboxymethylcellulose (Ref 24). It can determine average molecular weight and MWD. Fast, efficient separations can often be achieved by selecting appropriate conditions for the separation problem at hand. Separation is based on hydrodynamics; the larger the molecule, the less time spent in pores (less time to elute). Past ASTM protocols (ASTM D 3536, Ref 25 and ASTM D 3593, Ref 26) have been replaced by ASTM D 5296-97 (Ref 27), which covers HPLC of polystyrene.

Modern liquid column chromatography, now called HPLC or simply liquid chromatography (LC), is made possible by technical advances in

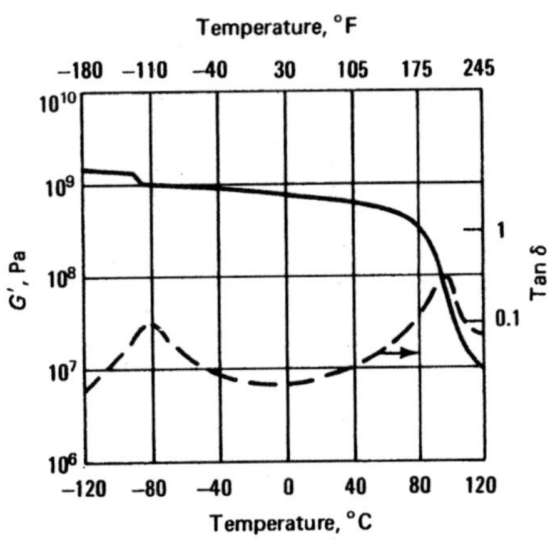

Fig. 20 Solid properties of high-impact polystyrene

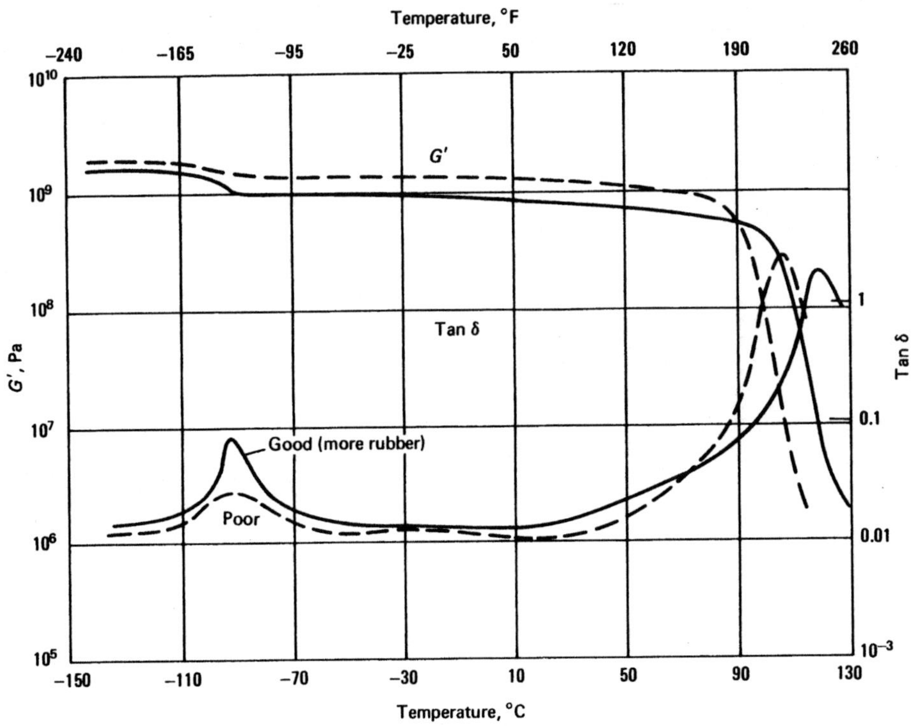

Fig. 21 Dynamic mechanical properties of two acrylonitrile-butadiene-styrene (ABS) terpolymers, 1 Hz

equipment, columns, and column-packing materials. As a result of these advances, LC is an important technique in analytical chemistry. It enables the user to perform rapid, efficient separations of complex mixtures of organic, inorganic, pharmaceutical, and biochemical compounds. At the time of publication, there are several modes of LC (bonded-phase, liquid-liquid partition, liquid-solid adsorption, ion-exchange, ion-pair, and size-exclusion), which can be employed with a single apparatus. The wide variety of available LC solvents adds to the selectivity that can be attained. Therefore, modern LC instrumentation (the liquid chromatograph) offers diversified approaches to separation problems and analysis of volatile and nonvolatile compounds.

Of continued interest is the commercialization of high-speed liquid chromatography (HSLC), which offers a threefold to fivefold reduction in analysis time, compared to conventional high-performance (HPLC) instruments, plus lower solvent consumption and enhanced analytical detectability. Complete details are noted in Ref 28.

It must be remembered that in all chromatographic studies, only the soluble portion of the polymer (the uncross-linked portion of the thermosetting resin) can be detected. Thus, there is a self-limiting availability of resin for analysis as the structure development proceeds.

Thermoanalysis

Thermoanalytical techniques include three distinctive, but complementary operations: differential scanning calorimetry (DSC), thermogravimetric analysis (TGA), and thermomechanical testing (TMT), each of which is briefly described with more details in the next article "Thermal Analysis and Thermal Properties" in this book.

Differential scanning calorimetry. measures the heat energy (calories) that a sample either absorbs or gives off at any given temperature. This technique is useful for measuring the T_g and melt temperatures, T_m, of a material, as well as the onset of thermal decomposition of blowing agents or other materials. Figure 25 is a representative DSC thermogram. Typical DSC examples are given in Fig. 26 and 27.

Thermogravimetric analysis, which measures the weight loss or gain versus a constantly increasing temperature, is especially useful in determining the concentration of an additive in a plastics formulation (including lubricant, filler, or reinforcements) or of other constituents. It is important to remember that the sample heating rate is often quite fast and that the sample might not always be at the observed temperature. Thus, important information on decomposition temperatures might be only qualitative, rather than quantitative in nature. Examples from Ref 29 illustrate the value of TGA as an analytical instrument for characterizing polymers. Figure 28 shows the composition of a nylon 6/6 lightly modified with molybdenum disulfide. Figure 29 depicts the moisture versus resin-glass content in a nylon molding compound. Figure 30 illustrates the differential decomposition of an acetal/fluorocarbon alloy. It is noteworthy that there must be a significant difference in the

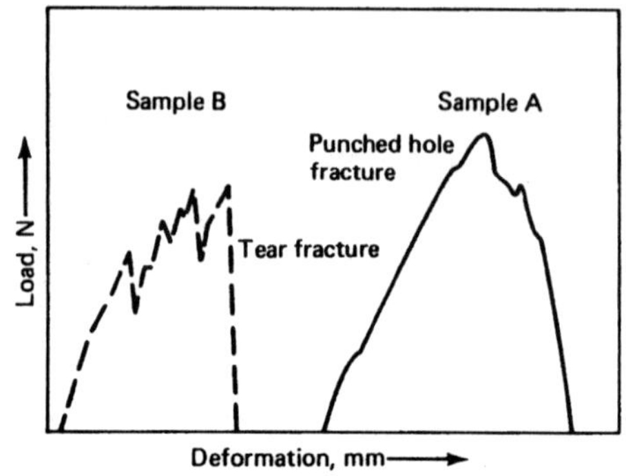

Fig. 23 Tear versus punched-hole fracture of acrylonitrile-butadiene-styrene (ABS) at 8 km/h (5 miles/h), 25 °C (77 °F). Source: Ref 21

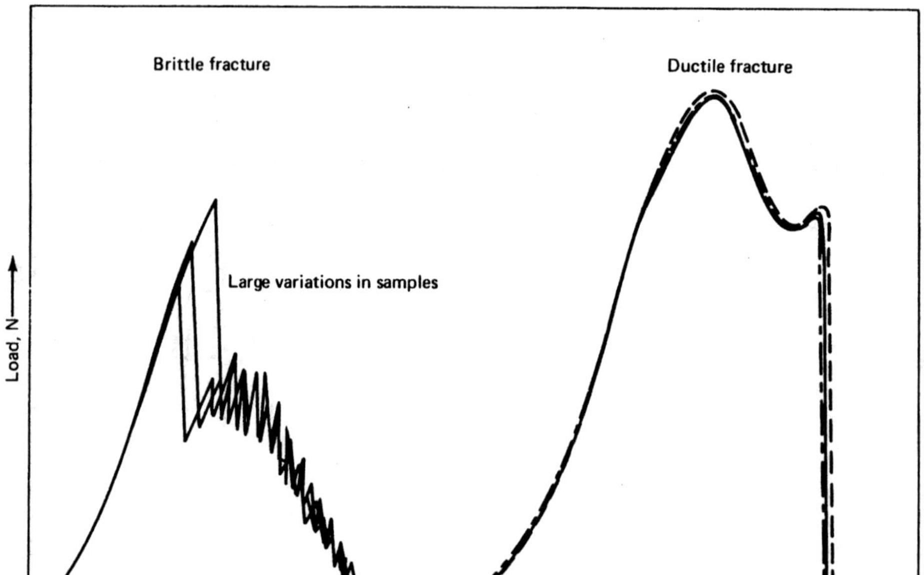

Fig. 22 Brittle versus ductile impact failure. Source: Ref 21

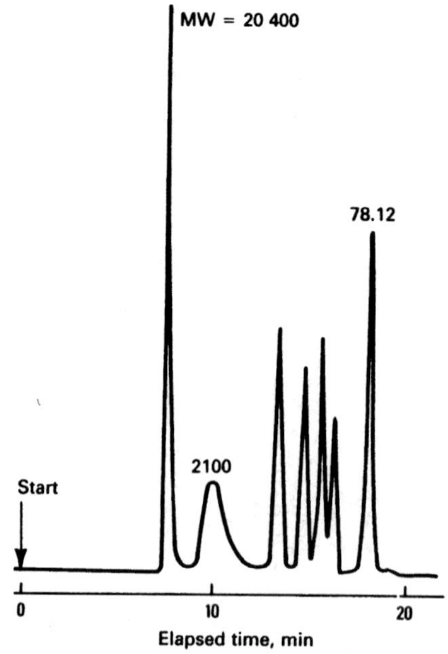

Fig. 24 Size-exclusion chromatogram. Mixture of (in order of elution) polystyrene (MW = 20,400 amu), polystyrene (MW = 2100 amu), dioctyl phthalate (390.6), dibutyl phthalate (278.3), diethyl phthalate (222.2), dimethyl phthalate (194.2), and benzene (78.12). Source: Ref 23

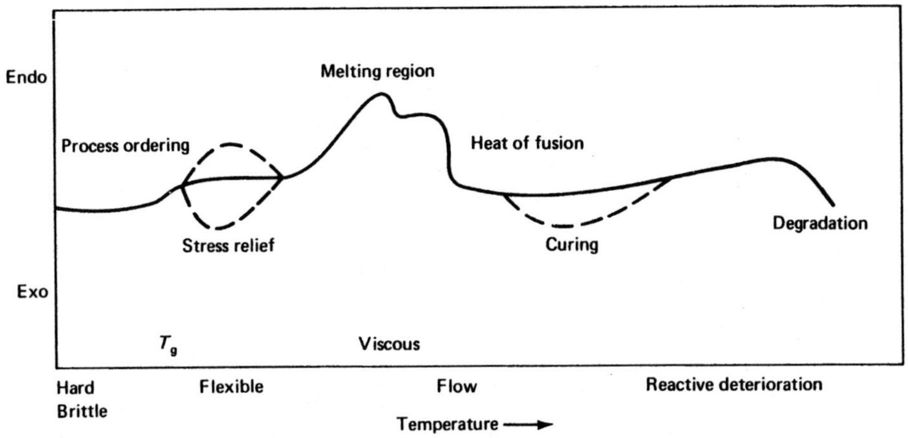

Fig. 25 Differential scanning calorimetry thermogram

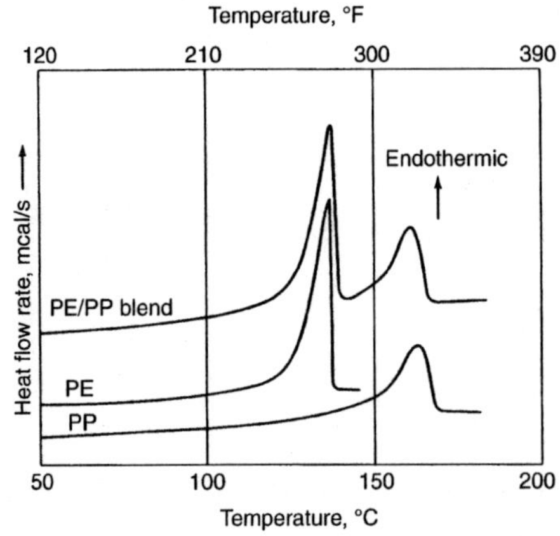

Fig. 27 Differential scanning calorimetry (DSC) of polyethylene/polypropylene blend 10 mcal/s range; 20 °C/min (36 °F/min). PE, polyethylene; PP, polypropylene. Source: Ref 29

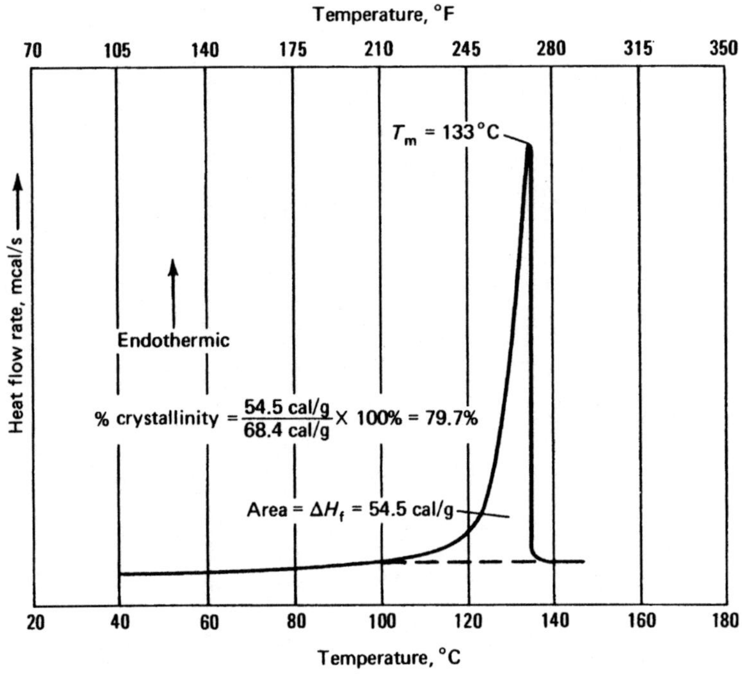

Fig. 26 Melting point and percent crystallinity of high-density polyethylene (HDPE) 10 mcal/s range; 10 °C/min (18 °F/min), 7.1 mg (1.5 gr). Source: Ref 29

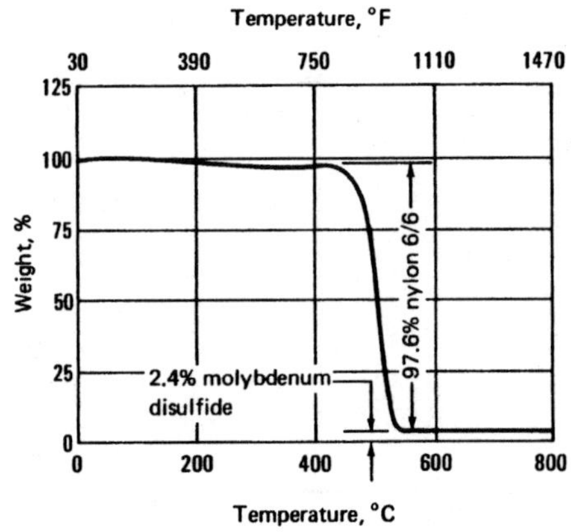

Fig. 28 Thermogravimetric analysis (TGA) of reinforced nylon 6/6; 40 °C/min (70 °F/min) in air

decomposition temperatures in order to appreciate subtleties in composition. If the two temperatures were too similar, it would not be possible to detect small variations in composition.

Thermomechanical testing measures the physical expansion/contraction of a material, as well as changes in modulus. The penetrometer actually detects the softening of the material T_g. Figure 31 shows the heat-deflection temperature of a series of materials.

When an analytical laboratory conducts a series of tests on an unknown material, in addition to running conventional infrared (IR) scans to identify composition, a GPC study characterizes the molecular weight and distribution of the polymer, DSC identifies percent crystallinity and T_g, TGA identifies weight composition, and TMT substantiates thermal regions and changes in properties over many temperatures. There are numerous ASTM documents that are germane to thermal analysis. Typical documents are cited in Ref 30 and 31.

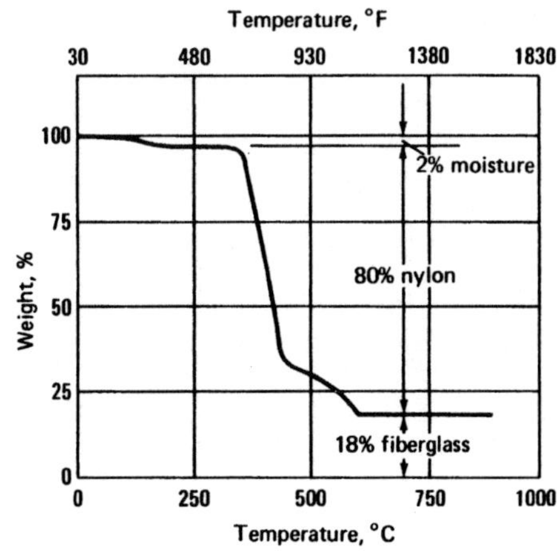

Fig. 29 Thermogravimetric analysis (TGA) of reinforced nylon; 80 °C/min (145 °F/min) in air

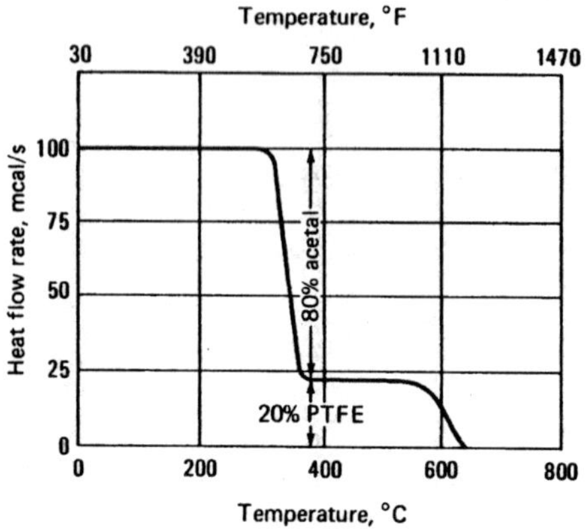

Fig. 30 Thermogravimetric analysis (TGA) of acetal/fluorocarbon blend; 40 °C/min (70 °F/min) in air

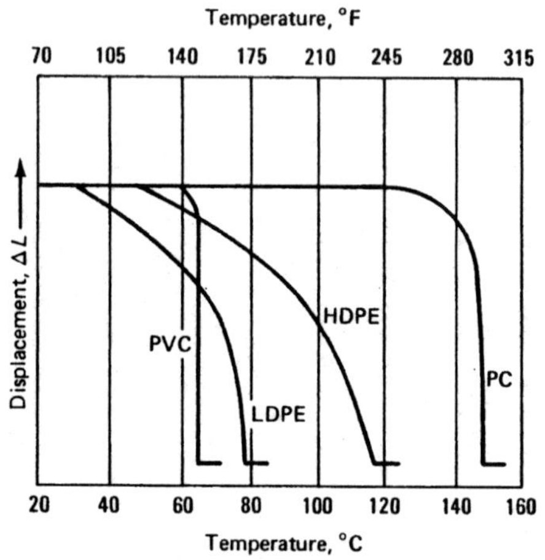

Fig. 31 Heat-deflection temperature at 1.8 MPa (0.264 ksi) of thermoplastics according to thermomechanical testing (TMT); 5 °C/min (9 °F/min) in flexure. PC, polycarbonate

REFERENCES

1. "Test Method for Dilute Solution Viscosity of Vinyl Chloride Polymers," D 1243-95, *Plastics (I),* Vol 8.01, *Annual Book of ASTM Standards,* American Society for Testing and Materials
2. "Specification for Nylon Injection Molding and Extrusion of Materials," D789-98, *Plastics (I),* Vol 8.01, *Annual Book of ASTM Standards,* American Society for Testing and Materials
3. "Recommended Practice for Determining Logarithmic Viscosity Number of Poly (Vinyl Chloride) (PVC) in Formulated Compounds," D 3591-97, *Plastics (II),* Vol 8.02, *Annual Book of ASTM Standards,* American Society for Testing and Materials
4. "Test Method for Apparent Viscosity of Plastisols and Organosols at Low Shear Rates by Brookfield Viscometer," D1824-95, *Plastics (I),* Vol 8.01, *Annual Book of ASTM Standards,* American Society for Testing and Materials
5. SPE Rheology Primer, *SPE J.,* Vol 27, Dec 1971, p 25
6. "Recommended Practice for Powder-Mix Test of Poly(Vinyl Chloride) (PVC) Resins Using a Torque Rheometer," D 2396-94, *Plastics (II),* Vol 8.02, *Annual Book of ASTM Standards,* American Society for Testing and Materials
7. S.B. Driscoll, "The Rheology of PVC," paper presented at SPE RETEC (Chicago, IL), Society of Plastics Engineers, Sept 1986
8. "Test Method for Flow Rates for Thermoplastics by Extrusion Plastometer," D 1238-98, *Plastics (I),* Vol 8.01, *Annual Book of ASTM Standards,* American Society for Testing and Materials
9. "Test Methods for Flow Rates of Poly (Vinyl Chloride) and Rheologically Unstable Thermoplastics," D 3364-94, *Plastics (II),* Vol 8.02, *Annual Book of ASTM Standards,* American Society for Testing and Materials
10. "Test Methods for Measuring the Rheological Properties of Thermoplastics with a Capillary Rheometer," D 3835-96, *Plastics (II),* Vol 8.02, *Annual Book of ASTM Standards,* American Society for Testing and Materials
11. H. Munstedt, A New Universal Extensional Rheometer for Polymer Melts, *Proc. Society of Rheology,* Oct 1978
12. "Recommended Practices for Testing Polymeric Powders and Powder Coatings," D 3451-92, Vol 6.02, *Annual Book of ASTM Standards,* American Society for Testing and Materials
13. "Standard Practice for Rheological Measurement of Polymer Melts Using Dynamic Mechanical Procedures," D 4440-95a, *Plastics (III),* Vol 8.03, *Annual Book of ASTM Standards,* American Society for Testing and Materials
14. "Standard Practice for Measuring the Cure Behavior of Thermosetting Resins Using Dynamic Mechanical Procedures," D 4473-95a, *Plastics (III),* Vol 8.03, *Annual Book of ASTM Standards,* American Society for Testing and Materials
15. "Standard Practice for Determining and Reporting Dynamic Mechanical Properties of Plastics," D 4065-95, *Plastics (II),* Vol 8.02, *Annual Book of ASTM Standards,* American Society for Testing and Materials
16. D.R. Saini and A.V. Shenoy, Dynamic and Steady-State Rheological Properties of Linear-Low Density Polyethylene Melt, *Polym. Eng. Sci.,* Vol 24 (No. 15), Oct 1984, p 1215–1218
17. S.B. Driscoll et al., "On-Line Monitoring of Quality Control During Compounding," paper presented at SPE INTEC (Newark, NJ), Society of Plastics Engineers, Nov 1985
18. S.B. Driscoll et al., "The Consequences of Compounding on Rheological Properties," paper presented at SPE RETEC (Akron, OH), Society of Plastics Engineers, Nov 1986
19. I. Park, The Effect of Molecular Weight Blending on PC Melt Rheology, *Proc. SPE ANTEC,* Vol 26, Society of Plastics Engineers, 1980
20. "Standard Test Method for High Speed Puncture Properties of Plastics Using Load and Displacement Sensors, D 3763-02, *Plastics (II),* Vol 8.02, *Annual Book of ASTM Standards,* American Society for Testing and Materials
21. S.B. Driscoll, Variable Rate Impact Testing of Polymers, *Instrumented Impact Testing of Plastics and Composite Materials,* STP 936, S.L. Kessler, Ed., American Society for Testing and Materials, 1986, p 163–186
22. N.M. Bikales, Ed., *Characterization of Polymers,* Wiley-Interscience, 1971, p 57
23. R.W. Yost, L.S. Ettre, and R.D. Conlon, "Practical Liquid Chromatography: An Introduction," Perkin-Elmer Corp., 1980
24. M.J. Kelly, Liquid Chromatography, *Materials Characterization,* Vol 10, *Metal Handbook,* American Society for Metals, 1986, p 654
25. "Test Method for Molecular Weight Averages and Molecular Weight Distribution of Polystyrene by Liquid Exclusion Chromatography (Gel Permeation Chromatography-GPC)" D 3536, *Annual Book of ASTM Standards,* American Society for Testing and Materials
26. "Test Method for Molecular Weight Averages and Molecular Weight Distribution of Certain Polymers by Liquid Size Exclusion Chromatography (Gel Permeation Chromatography-GPC) Using Universal Calibration," D 3593, *Annual Book of ASTM Standards,* American Society for Testing and Materials
27. "Test Method for Molecular Weight Averages and Molecular Weight Distribution of Polystyrene by High Performance Size-Exclusive Chromatography," D 5296-97, *Plastics (III),* Vol 8.03, *Annual Book of ASTM Standards,* American Society for Testing and Materials
28. M.W. Dong and J.L. DiCesare, High Speed Liquid Chromatography: Application in Quality Control Laboratory of Plastic Materials, *Plast. Eng.,* Vol 39 (No. 2), Feb 1983, p 25–28
29. W.P. Brennan, "Characterization and Quality Control of Engineering Thermoplastics by Thermal Analysis," Perkin Elmer Corp.
30. "Standard Test Method for Heats of Fusion and Crystallization of Polymers by Thermal Analysis," D 3417, *Annual Book of ASTM Standards,* American Society for Testing and Materials
31. "Standard Test Method for Transition Temperatures of Polymers by Thermal Analysis," D 3418, *Annual Book of ASTM Standards,* American Society for Testing and Materials

Thermal Analysis and Thermal Properties*

THERMAL ANALYSIS provides a powerful tool for researchers and engineers in determining both unknown and reproducible behavioral properties of polymer molecules. During the past three decades, the thermal analysis of polymers has progressed from a capability possessed by a few organizations to an essential characterization methodology for all organizations pursuing polymer research and product development.

This article covers the thermal analysis and thermal properties of engineering plastics with respect to chemical composition, chain configuration, and/or conformation of the base polymers; processing of the base polymers with or without additives; and the response to chemical, physical, and mechanical stresses of base polymers as unfilled, shaped articles or as components of composite structures. This article also summarizes the basic thermal properties used in the application of engineering plastics, such as thermal conductivity, temperature resistance, thermal expansion, specific heat, and the determination of glass-transition temperatures. Typical thermal properties for various plastics are summarized in Tables 1 and 2.

Glass Transition Temperature

In general terms, the glass-transition temperature (T_g) of a plastic is a threshold temperature below which the plastic is hard and glassy and above which the plastic becomes rubbery. The T_g is also a measure of the onset of long-range molecular movement in the plastic. Because the transition from glass to rubber is not a thermodynamic transition, but rather a manifestation of viscoelasticity, the exact value of the T_g depends on the method used to measure it and the rate at which the temperature is changed during the measurement. For this reason, these parameters must be specified when reporting T_g measurements and when comparing data of different plastics.

The change in properties at the glass transition occurs not at a distinct temperature but over a range of temperatures. Thus, the T_g specified for a polymer actually represents roughly the center of a transition region. In a thermoplastic polymer, for example, the change that occurs gradually over the T_g region eventually leads to a complete loss of dimensional stability. In a network polymer such as epoxy, the change is less severe but nonetheless produces significant softening and loss of mechanical properties.

The importance of the glass transition as a material property can be understood in terms of the loss of rigidity that accompanies the transition. A drop of several decades in the modulus from a common value of 10^{10} Pa (10^{11} dynes/cm^2) at T_g is usually observed above the transition temperature. Hence, the T_g, as the key controlling parameter in the time/temperature-dependent viscoelastic behavior of a polymer, is often an important factor in determining the usefulness of a given polymer.

Another way to understand the substantial change in properties at the T_g is to focus on the expansion that occurs in the polymer as temperature is increased. It is said that the free volume, which may be thought of as room inside the polymer, gradually increases until cooperative rotational motion of five to ten mer units is possible. At this point the polymer can deform in response to an applied stress, for example, much more easily than it could at a lower temperature. Clearly, the flexibility and bulkiness of the mer unit and the cohesive energy between molecules strongly influence the temperature at which this can occur. The more flexible and less bulky the mer unit, the easier it is for the cooperative rotation to occur, and thus the lower the T_g. However, if the polymer molecules are bonded to one another by strong secondary bonds, the bonding will interfere with such motion, even if the chain is very flexible and not very bulky. This, of course, is what gives thermosets a higher average T_g than thermoplastics. Thermoplastics with a high T_g have stiff, bulky chains and strong intermolecular hydrogen bonding between chains.

As a function of temperature, the modulus of any thermoplastic or thermoset may be generally described by three stages of behavior in the T_g region, as identified in Fig. 1. At temperatures below the T_g of the given polymer, the polymer exhibits a high modulus, which decreases very slightly with increasing temperature until the vicinity of the glass transition is reached. This stage is referred to as the glassy plateau. As the polymer is heated through the glass transition, the modulus of the polymer typically drops in value by three decades. This is commonly called the transition zone. Finally, at temperatures above the T_g, the modulus continues to drop until the physical integrity of the polymer is lost (a melting process for semicrystalline polymers; complete liquidlike flow above T_g for linear, amorphous polymers; or rubberlike behavior for cross-linked systems). This region of behavior above the transition is called the rubbery plateau, in reference to the high degree of molecular motion possible at these temperatures.

All polymers, ranging from simple, linear thermoplastics to filled, cross-linked resins, exhibit these different regimes of behavior with temperature. The differences stem from the specific locations of the transition temperatures and the magnitude of the respective variations with temperature. Linear, amorphous polymers, such as polystyrene (PS) or polymethyl methacrylate (PMMA), have moduli curves that closely approximate the type of thermal behavior shown in Fig. 1. The primary effect of crystallinity in linear thermoplastics is the mediation of the modulus change at T_g, and maintenance of the rubbery plateau into higher temperatures, as indicated in Fig. 1. Cross-linking a polymer produces the same qualitative effects on the temperature dependence of the modulus as does crystallinity. In both cases, long-range segmental motion necessary for complete, liquidlike flow is restricted (by chemical linkages, in the case of thermosets, and by crystallites that act as virtual cross-links below the melting point of the polymer, in the case of semicrystalline polymers).

Semicrystalline Polymers

T_g and T_m for Semicrystalline Polymers. For most polymers, the glass-transition temperature is the threshold limit for service and is not exceeded during application. In some cases,

*Adapted from Thermal Analysis and Properties, *Engineered Materials Handbook Desk Edition,* ASM International, 1995, pages 367 to 392

however, semicrystalline plastics are exempt from the concern of exceeding the T_g because their crystalline melting point (T_m), which is always above their T_g, is their temperature limit. The crystalline portion of a semicrystalline polymer has a T_m similar to those found in other crystalline materials. For semicrystalline polymers, this may be the most important transition temperature. If high crystallinity (roughly 50% or higher) can be obtained, it may permit a polymer to be used above its T_g. It is difficult to obtain high crystallinity in polymers; this is really possible only in thermoplastics. However, if substantial crystallinity can be obtained, loss of dimensional stability will not occur at T_g because the crystalline regions will not undergo T_g and will restrict the deformation of the noncrystalline regions. Thus, in such polymers it is possible to extend the region of acceptable dimensional stability above the T_g. If crystallinity is quite high (say 80% or more), this may extend at least the short-term use temperature almost to the T_m.

In fact, substantially crystalline polymers in the temperature range between T_g and T_m are referred to as leathery, because they are made up of a combination of the rubbery noncrystalline regions and the stiff, crystalline regions. Thus, polyethylene (PE), polypropylene (PP), and other polymers are still useful at room temperature, and polyamide (PA) is useful to moderately elevated temperatures.

As with the T_g, the T_m is increased by a decrease in chain flexibility, an increase in bulkiness, or an increase in the strength of intermolecular bonding. However, for a crystalline polymer, decreases in chain flexibility and increases in bulkiness may need to be limited because these factors adversely influence crystallinity. In a crystalline polymer, dimensional stability increases with added crystallinity because this decreases the portion of the polymer that is influenced by the T_g.

Crystallinity. In crystalline polymers, the mobility of polymer segments is reduced considerably. The effect is an increase in rigidity, modulus, and hardness and a decrease in solvent solubility. Impact strength sometimes increases with crystallinity at temperatures above the T_g because the crystals act as cross links. At a very high degree of crystallinity the impact strength usually decreases. The tensile strength of crystalline materials generally shows a small decrease when the temperature increases above the T_g. Tensile properties decrease only at temperatures near the T_m. Materials with this behavior have favorable applications between the T_g and T_m because they are ductile. Materials with an excessive crystalline fraction become brittle at temperatures below the T_g.

During crystallization, the crystalline polymer packs all of the low-molecular-weight components and impure species into the interstices between the spherulites, leaving these as contaminated boundaries of lower strength and modulus. Shrinkage during crystallization may further leave stresses and voids in these inter-

Table 1 Thermal properties of selected resins

Thermoplastic resins	Heat deflection temperature at 1.82 MPa (0.264 ksi)		UL index		Thermal conductivity		Coefficient of thermal expansion, 10^{-6}/°C
	°C	°F	°C	°F	W/m · K	Btu/ft · h · °F	
Acrylonitrile-butadiene-styrene (ABS)	99	210	60	140	0.27	0.16	53
ABS-polycarbonate alloy (ABS-PC)	115	240	60	140	0.25	0.14	35
Diallyl phthalate (DAP)	285	545	130	265	0.36	0.21	27
Polyoxymethylene (POM)	136	275	85	185	0.37	0.22	37
Polymethylmethacrylate (PMMA)	92	200	90	195	0.19	0.11	34
Polyarylate (PAR)	155	310	...	...	0.22	0.125	31
Liquid crystal polymer (LCP)	311	590	220	430	...	...	5
Melamine-formaldehyde (MF)	183	360	130	265	0.42	0.24	22
Nylon 6	65	150	75	165	0.23	0.13	25
Nylon 6/6	90	195	75	165	0.25		40
Amorphous nylon 12	140	285	65	150	0.25		70
Polyarylether (PAE)	160	320	160	320	...	...	30
Polybutylene terephthalate (PBT)	...	...	120	250	...	...	45
PC	129	265	115	240	0.20		38
PBT-PC	129	265	105	220	...	...	28
Polyetheretherketone (PEEK)	...	...	250	480	0.25	0.14	26
Polyether-imide (PEI)	210	410	170	340	0.22	0.125	31
Polyether sulfone (PESV)	203	395	170	340	...	...	55
Polyethylene terephthalate (PET)	224	435	140	285	0.17		15
Phenol-formaldehyde (PF)	163	325	150	300	0.25	0.14	16
Unsaturated polyester (UP)	279	535	130	265	0.12	0.07	16
Modified polyphenylene oxide alloy (PPO mod)	100	212	80	175	...	...	38
Polyphenylene sulfide (PPS)	260	500	200	390	0.17		30
Polysulfone (PSU)	174	345	140	285	0.26	0.15	31
Styrene-maleic anhydride (S/MA) terpolymer	103	215	80	175	...	...	...

Thermoset resins (neat)	Heat deflection temperature at 1.82 MPa (0.264 ksi)		Continuous service temperature		Thermal conductivity		Coefficient of thermal expansion, 10^{-6}/°C
	°C	°F	°C	°F	W/m · K	Btu/ft · h · °F	
Allyl diglycol carbonate	60–90	140–190	100	212	0.2	0.115–0.120	80–140
Bismaleimide resins	...	...	315(a) 230(b)	600(a) 450(b)	...	...	30–50
Epoxy resins	45–290	115–550	120–290	250–550	0.17–0.2	0.10–0.12	45–65
Melamine-formaldehyde	150	298	100	210	...	...	...
Phenolic resins	120–175	250–350	120–175	250–350	0.12–0.24	0.072–0.144	25–60
Polyester resins	50–205	120–400	120–150	250–300	0.17–0.22	0.10–0.13	55–100
Polyimide resins	305–360	580–680	260–315	500–600	0.10–0.34	0.058–0.20	25–80
Polyurethane (cast)	50–205	120–400	90–120	190–250	0.17–0.21	0.10–0.12	70–100
Silicone resins	...	...	260	500	0.22	0.13	80–300
Urethane elastomer	...	...	90	190	0.07–0.30	0.04–0.178	100–200
Urethane rigid foam	...	...	160	325	0.06–0.12	0.033–0.067	80

(a) Short-term continuous service temperature. (b) Long-term continuous service temperature

Table 2 Glass-transition and melting temperatures of selected thermoplastic and thermoset resins

Chemical name	Glass-transition temperature (T_g)		Melting temperature (T_m)	
	°C	°F	°C	°F
Hydrocarbon thermoplastics				
Polyethylene				
HDPE	–90 or –20	–130 or –5	137	280
LDPE	–110 or –20	–165 or –5	115	240
Polypropylene				
Atactic	–18	0	176	350
Isotactic	–10	15	176	350
Polyisobutylene	–70, –60	–95, –75	128	260
Polyisoprene				
Cis: natural rubber	–73	–100	28	80
Trans: gutta percha	...	...	...	...
Polymethylpentene (poly-4-methyl-1-pentene)	29	85	250	480
Polybutadiene (poly-1,2-butadiene, butadiene rubber)				
Syndiotactic	–90	–130	154	310
Isotactic	–90	–130	120	250
Polystyrene				
Atactic	100, 105	212, 220	(a)	(a)
Isotactic	100, 105	212, 220	240	465
Nonhydrocarbon carbon-chain thermoplastics				
Polyvinyl chloride (vinyl)	87	190	212	415
Polyvinyl fluoride	–20	–5	200	390
Polyvinylidene chloride	–17	1	198	390
Polyvinylidene fluoride	–35	–30	...	...
Polytetrafluoroethylene	–97, 126	–140, 260	327	620
Polychlorotrifluoroethylene	45	115	220	430
Polychloroprene (chloroprene rubber, or neoprene)	–50	–60	80	175
Polyacrylonitrile	104, 130	220, 265	317	600
Polyvinyl alcohol	85	185	258	495
Polyvinyl acetate	29	85	...	...
Polyvinyl carbazole	150, 208	300, 405	...	...
Polymethyl methacrylate				
Syndiotactic	3	35	105, 120	220, 250
Isotactic	3	35	45	115
Heterochain thermoplastics				
Polyethylene oxide	–67 to –27	–90 to –15	62 to 72	145 to 160
Polyoxymethylene	–85	–120	175	345
Polyamide				
Nylon 6	50	120	215	420
Nylon 6/10	40	105	227	440
Polyethylene terephthalate	69	155	265	510
Polycarbonate	150	300	265	510
Polydimethyl siloxane (silicone rubber)	–123	–190	–54	–65
High-temperature thermoplastics				
Poly *p*-phenylene terephthalamide (aromatic polyamide or aramid)	375	705	~640(b)	~1185(b)
Polyaromatic ester	...	...	421	790
Polyetheretherketone	143	290	334	635
Polyphenylene sulfide	85	185	285	545
Polyamide-imide	277–289	530–550	(a)	(a)
Polyether sulfone	225	435	(a)	(a)
Polyether-imide	215	420	(a)	(a)
Polysulfone	193	380	(a)	(a)
Polyimide (thermoplastic)	280–330	535–625	(a)	(a)
Thermoset resins				
Amino resins (melamine-formaldehyde)	None(b)	None(b)	...	...
Bismaleimide	230–345(c)	450–650(c)	...	...
Epoxy resins	60–175	140–350	...	...
Phenolic resins	300	570	...	...
Polyester resins	110	230	...	...
Polyimide resins	315–370(c)	600–700(c)	...	...
Polyurethane (cast)	135	275	...	...
Polyurethane (elastomer)	...	...	...	...
Silicone	–125	–193	...	...
Urethane rigid foam	...	...	...	...

(a) Polymer is generally 95% or more noncrystalline. Any T_m given is for remaining crystalline portion or for crystalline version. (b) T_d = 500 °C (930 °F). *R* contains at least one aromatic ring. (c) Based on private communication, American Cyanamid Co. (d) Dry. Source: Ref 1–3 and product information sheets

stices, weakening them even more. The surface between spherulites and amorphous interstices is the weak interface at which cracking is most likely to begin.

The amount of crystalline fraction and the size of crystalline spherulites can be affected by the addition of nucleating agents, or seed particles, which can be small, inorganic particles. Plastics with seeds contain a higher crystalline fraction with small domains. Crystallinity is also affected by the temperature gradient in processing. A high mold temperature reduces temperature gradients and the amount of crystallization, whereas a low mold temperature increases the crystallization rate. A high melt pressure in molding also can reduce dwell time in the barrel, reducing the temperature loss, which tends to decrease the amount of crystallization.

The cooling temperature rate also affects the amount of crystallinity. Generally, the maximum crystallization rate is observed at about 0.9 of the T_m, measured in absolute temperature. For a material cooled at approximately the T_m, sufficient crystallinity will develop. If the material has a high T_g and the cooling process takes place below it, amorphism can increase. Material with a tendency to crystallize will exhibit gradual crystallization and postshrinkage when stored at temperatures above the T_g.

For crystalline material, control of crystallinity is generally more important than control of molecular weight in changing mechanical properties. For these materials, the property can be correlated with density, which in turn is related to crystallinity. One primary example is PE, which in the commercial market is classified according to density. Hydraulic stress during injection molding flow and calendering aligns the polymer molecules parallel to each other and favors crystallization. In these cases, tensile strength in the machine direction is generally higher. During tension measurement, elongation can reach several times the original length if necking occurs. In the necking region, the unoriented polymer chains are transformed into thin, oriented chains, resulting in a single, sharp-moving neck. Polyethylene and polyethylene terephthalate (PET) are known to exhibit necking.

The recently developed liquid crystal polymers are one extreme of such aligned polymers. Because of rigid molecules, these materials tend to align themselves in melts or solutions. By properly aligning them with stress during the solidifying stage, high tensile strength in one direction can be obtained. In some cases, the strength can be higher than that of steel.

Structural and Test Effects

High-Temperature Thermoplastics. The most impressive T_g and T_m for thermoplastics are for high-temperature thermoplastics (Table 2). These high-temperature polymers have inflexible and bulky rings and cyclic structures; they are all heterochain polymers having many sites for intermolecular hydrogen bonding. It should be noted that the flexible ether and sulfide linkages included in most of these polymers do lower the T_g, but they are added intentionally to give the chain enough flexibility so that the polymer can be processed and, in some cases, so that high crystallinity can be attained. Crystallinity is used to extreme effect in the aramid fiber poly(*p*-phenylene terephthalamide) to produce a highly oriented, crystalline structure whose extremely strong hydrogen bonding gives it not only a high T_g but also a T_m that is above its decomposition temperature.

Thermosets. Flexibility and bulkiness are also used to modify the T_g of thermosets. For example, flexibilizers that usually contain fairly long segments of $-CH_2-$ units are added to epoxies to make them less brittle; they also lower the T_g of the cured resin. On the other hand, the epoxies with the highest T_g are cross linked from both resins and curing agents that are relatively inflexible and bulky. Because thermosets are covalently cross linked, secondary bonding has only a small influence on the T_g. However, the cross-link density of the thermoset has a dramatic effect on the T_g, and in many cases much effort is spent in the formulation and cure of thermoset resins to ensure that they achieve a high cross-link density.

Other Molecular Factors and Transitions. In addition to the T_g and T_m, polymers can undergo other transition temperatures. These include phase changes in the crystalline phase as well as various transitions in the noncrystalline regions. The latter are usually due to side group motion, but many also result from motion of some subunit of the chain itself. These transitions can have an influence on properties, but the influence is usually on properties other than dimensional stability.

Structural factors originating within the molecule also have an influence on dimensional stability. Different stereoisomers have different T_g and T_m and may have very different percentages of crystallinity. Branching interferes with intermolecular bonding and crystallinity and thus lowers dimensional stability. Increases in molecular weight increase T_g and T_m somewhat, but the ease of crystallization also decreases. Thus, increased molecular weight may have an adverse effect on the dimensional stability of crystalline polymers. Copolymerization usually produces a T_g somewhere between the two mers, or a double T_g. However, the influence of copolymerization on the T_m is much more dramatic. In many cases, copolymerization causes the T_m to drop so low that crystallinity is totally destroyed.

Methods of Determining T_g. As previously noted, T_g is not an inherently thermodynamic property; therefore, it depends in part on the test method. There are many different experimental techniques for detecting the T_g in polymers. The methods are usually related to the measurement of volume, heat capacity changes with temperature, or some material property related to the abrupt increase in molecular mobility associated with the glass transition. The most commonly encountered techniques include dilatometry, differential scanning calorimetry (DSC), and physical yielding. As mentioned, the glass transition is manifested by a change in the slope of the extensive thermodynamic variables, (e.g., volume or entropy) when the transition is traversed.

Monitoring the specific volume of a polymer with temperature, or dilatometry, is one of the oldest methods used for detecting the T_g in polymers. A polymer sample is placed within a dilatometer and immersed in a confining liquid. The volume change with temperature of the polymer sample is determined by following the volume of the liquid polymer assembly and subtracting the contributions of the confining liquid. The glass temperature is indicated by the temperature at which the volume of the polymer sample undergoes a change in slope.

A much easier and more widely practiced method for detecting the T_g is DSC. In DSC, the heat capacity of the polymer sample is measured relative to that of a reference material. The glass transition of a polymer is manifested by a stepwise change in the heat capacity of the sample (ASTM D 3418). The actual location of the glass transition depends on the rate of the measurement process, because relaxation rate effects become significant at about the T_g. For this reason, rate information is generally specified with DSC results.

The type of test methods used to determine the T_g that perhaps have the most direct relevance for the design engineer are based on physical property changes rather than thermodynamic ones. Engineering plastics are usually specified by softening or deflection temperatures (ASTM D 1525 and D 1637) in regard to their effective working T_g. In these methods, a standard form such as a polymer dog-bone bar is placed under a load by means of a penetration probe or some clamping configuration, and the T_g is defined as the temperature above which noticeable yielding of the polymer to the imposed load (as indicated by surface penetration or bending of the sample) becomes apparent at conventional time scales.

Generally, the T_g values obtained from DSC are lower than those obtained by dynamic mechanical methods. Many other variables relating to morphology, which in turn depend on the sample thermal history and orientation, also affect T_g measurements. For instance, the T_g of PMMA can be observed at 110 and 160 °C (230 and 320 °F) using a dilatometric and rebound elasticity technique, respectively (Ref 1). The T_g of PET can be detected at 65 to 105 °C (150 to 220 °F), depending on the degree of crystallinity and orientation (Ref 2). Hence, it is not surprising that the different factors affecting T_g became the subject of a number of reviews (Ref 1–3). Prediction of T_g values can be complicated even further in stereoisomorphic polymers (Ref 4). For PMMA, the effect of tacticity on T_g is pronounced, with syndiotactic polymers having the greatest effect, isotactic having the least, and atactic being between these two (Ref 2). In the case of PP, stereoisomorphism hardly affects the

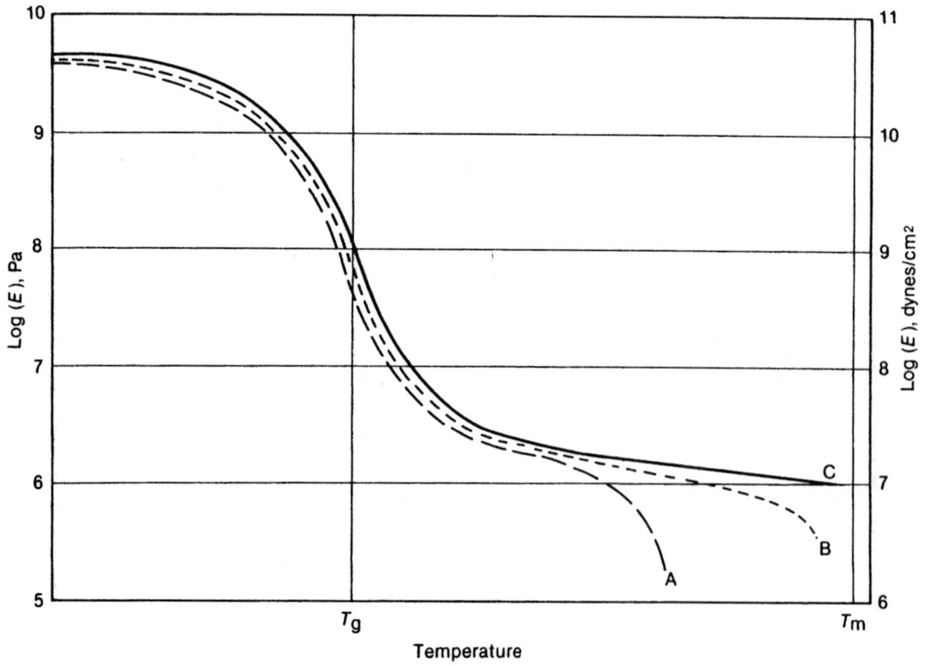

Fig. 1 Temperature dependence of the modulus, *E*, of polymers. Examples of idealized behaviors exhibited by an amorphous thermoplastic (A), a semicrystalline thermoplastic (B), and a thermoset (C)

T_g. Differential scanning calorimetry data on syndiotactic, atactic, and isotactic samples indicate T_g values of –4, –6, and –18 °C (25, 20, and 0 °F), respectively (Ref 4, 5). In isomorphic copolyesters exhibiting oxygen-methylene isomorphism, the effect of composition on T_g is minimal (Ref 6, 7).

Variation with Molecular Weight. The T_g has been shown to increase with the molecular weight of the polymer until a limiting value in T_g is achieved (Ref 8). The Fox-Flory equation:

$$T_g = T_{g\infty} - C/\overline{M}_N \quad \text{(Eq 1)}$$

relates the T_g of a polymer of a given molecular weight, M_N, to a constant, C, and the limiting glass temperature, $T_{g\infty}$. The rationale for this relationship is that chain ends contribute finite free volume to the polymer system, and the excess free volume introduced by the greater number of chain ends in lower-molecular-weight polymers has a proportional effect on the T_g of the polymer. The T_g of the monomer and that of the asymptotic value achieved by a chain of infinitely high-molecular-weight form the range of T_g as a function of molecular weight.

It is generally desirable for material manufacturers to make plastics with sufficiently high molecular weights to obtain good mechanical properties. For PS this molecular weight is 100,000 and for PE this value is 20,000. It is not desirable to increase molecular weight further because melt viscosity will increase rapidly, although there are occasional exceptions to this rule. The yield strength of PP decreases when molecular weight increases. Studies of morphology indicate that high molecular weight and branching reduce crystallinity. Polymers with high intermolecular interaction, such as hydrogen bonding, do not require high molecular weight to achieve good mechanical properties. With low molecular weight, viscosity is very low, which is commonly observed for PA.

Molecular weight and molecular-weight distribution are useful in characterizing plastic properties. In plastics with a broad distribution of molecular weights, the average molecular weight can be calculated in several different ways. The T_g goes up with the number-average molecular weight. Elongation at break for acrylic samples with different molecular weights can be reduced to a single curve when weight-average molecular weight is used. Plastics with narrow molecular weights are preferred for low warpage in thin-wall injection molding, film extrusion, and rotational molding. Plastics with moderately low molecular weights are suitable for high-speed processing, such as high draw-down rate extrusion, high-speed calendering, and injection molding. Most processing conditions require materials with high molecular weights. This is especially true for extrusion and blow molding, which require sufficient melt strength for the extrudate to support itself as it exits from the die.

Chemical structure affects the T_g of polymers in two ways: chain flexibility and substituent effects. The T_g increases with the stiffness of the backbone of the polymer chain. The effect of a higher energy barrier to the main chain rotational movement necessary for short-range and long-range segmental motion associated with T_g is demonstrated by a comparison of PE (CH_2–CH_2) and poly(*p*-phenylene). The T_g of the former is approximately –100 °C (–150 °F), while the T_g of poly(*p*-phenylene) lies above the decomposition temperature and hence exhibits no experimentally observable transition temperature. Side-group substitution in polymers affects the T_g by superimposing additional steric effects on the main chain characteristics of the polymer. The presence of polar groups (which increase intermolecular interactions such as hydrogen bonding) and an increase in the size of the substituent group both tend to raise the T_g by increasing the energy requirements necessary for chain rotation to occur. A comparison of PE (CH_2–CH_2), PP (CH_2–$CH(CH_3)$), and PMMA [CH_2–$CH(CO_2CH_3)$] with increasing side group sizes shows respective T_g of –85, –20, and 6 °C (–120, –4, and 45 °F). The introduction of long, flexible side groups, however, has the effect of decreasing T_g, due to free volume and chain flexibility effects.

Cross Linking. The influence of cross linking a polymer may be considered from two different perspectives. At lower degrees of cross linking, the presence of the cross links effectively raises the molecular weight of the polymer. In the system styrene-divinyl benzene, the T_g rose from about 77 °C (170 °F) to about 157 °C (315 °F) as the divinyl benzene content of the polymer was increased from 0 to 15 mol% (Ref 9). At high degrees of cross linking, the increase in T_g becomes nonlinear as the rotational freedom of the average chain length between cross links decreases with increased cross link density.

Copolymer Composition. In systems of random copolymers, the effect of copolymer composition on the T_g may be described by:

$$T_{G_c} = \frac{T_{g_1} + (KT_{g_2} - T_{g_1})W_2}{1 - (1 - K)W_2} \quad \text{(Eq 2)}$$

in which T_{g_1} and T_{g_2} are the glass-transition temperatures of the respective homopolymers, K is a constant, and W_2 is the weight fraction of comonomer 2. Equations of this form are equivalent to those proposed by Gordon and Taylor (Ref 10) and Wood (Ref 11).

Copolymerization is also frequently used to change the properties of plastics. For example, copolymerization with vinyl acetate increases the processibility and thermal stability of polyvinyl chloride (PVC). Copolymerization of acrylonitrile with styrene increases the T_g.

Plasticizers are low-molecular-weight compounds that are often compounded into high-molecular-weight polymers to improve processibility, impact strength, and elongation. Plasticized plastics generally have high melt indexes. Plasticizers lower melt viscosity and processing temperature. Fundamentally, they function by broadening the molecular-weight distribution and increasing the low-molecular-weight fraction of the total composition. Plasticizers are essentially nonvolatile solvents. When plasticizers are not entirely compatible, the transition from rigid to leathery properties can take place over a wider temperature range, which can be useful in certain applications. Plasticizers do not work effectively for crystalline materials because only the amorphous region is accessible to plasticizers. Sometimes they increase both the mobility of polymer chains and the crystallinity, thereby causing an antiplasticizing effect. Although a plasticizer does not decrease the T_m of a material as much as comonomers do, it depresses the T_g more.

Plasticization by a Diluent. The addition of a soluble, generally low-molecular-weight component to a polymer results in the plasticization of the polymer, as evidenced by a lowering of the T_g of the system. Plasticized PVC is a primary example of a rigid polymer that is rendered rubber-like by the addition of a diluent, commonly dioctyl phthalate. The lowering of the T_g by a diluent has shown good correlation to relationships of the types advocated by Gordon et al. and Couchmen (Ref 12, 13). The equations take the form:

$$T_{g_{12}} = \frac{W_1T_{g_1} + W_2T_{g_2}K}{W_1 + W_2K} \quad \text{(Eq 3)}$$

in which $T_{g_{12}}$ is the T_g of the plasticized polymer system, W_1 and W_2 are the mole or weight fractions of components 1 and 2, T_{g_1} and T_{g_2} are the component glass temperatures, and $K = \Delta C_{p_2}/\Delta C_{p_1}$, the ratio of the incremental heat capacities at T_g (Ref 13) where C_p is the specific heat at constant pressure. Alternately:

$$X_1\,\Delta C_{p_1} \ln\frac{T_{g_1}}{T_{g_{12}}} + X_2\,\Delta C_{p_2} \ln\frac{T_{g_{12}}}{T_{g_2}} = 0 \quad \text{(Eq 4)}$$

where $X_{1,2}$ are either the mole or mass fractions of the respective component (Ref 13). These relationships apply only to mutually soluble polymer-diluent systems in which the diluent may be a low-molecular-weight compound or another polymer. For incompatible systems, the components remain immiscible and phase separately into distinct domains for which the T_g of the respective phases remain the same. Hence, the occurrence of separate and distinct T_g in multicomponent polymer systems is useful as an indication of incompatibility or immiscibility among the components.

Moisture Effect on T_g

The effect of absorbed moisture on the T_g is invariably to lower it. This is consistent with the role of water as a plasticizer. As a rule of thumb, the more water absorbed, the lower the T_g. For this reason, a nonpolar plastic such as PS is less affected than, for example, PMMA. The lowering of T_g is sometimes quantitatively discussed in terms of several mixing formulas (Ref 14, 15). Currently, the most often used expression is:

$$T_g = \frac{X_1 C_{p_1} T_{g_1} + X_2 C_{p_2} T_{g_2}}{X_1 C_{p_1} + X_{p_2} C_{p_2}} \quad \text{(Eq 5)}$$

In this expression, T_g, and T_{g_1}, and T_{g_2} are the glass-transition temperatures of the polymer mixture, polymer 1, and diluent 2, respectively. The expressions C_{p_1} and C_{p_2} are the discontinuities in the heat capacities at the glass transitions of the components. This expression was first derived by Gordon (Ref 16) for polymer blends and was based on the Gibbs-DiMarzio entropy theory (Ref 17). Couchmen provided an alternative derivation, based on a purely thermodynamic exposition (Ref 18). The extension of the Couchman approach to plastic-diluent systems, especially epoxy-water systems, has been carried out by Karasz et al. (Ref 19–21). Couchman's derivation (but not the result) has recently been criticized by Goldstein (Ref 22). The discussion is important to an understanding of the glassy state.

These relationships can be quite useful for predicting the loss of properties due to moisture. Such a relationship is shown in Fig. 2. Because the modulus falls precipitously at the glass transition, these data give an absolute upper temperature limit. Most conservative design requires all application temperatures to be remote from the glass-transition region. The relationship between T_g and the amount of absorbed water can be affected by many factors, such as additives, thermal pretreatments, presence or absence of fillers or reinforcements, and, in thermosets, amount and type of curative, degree of cure, and so forth.

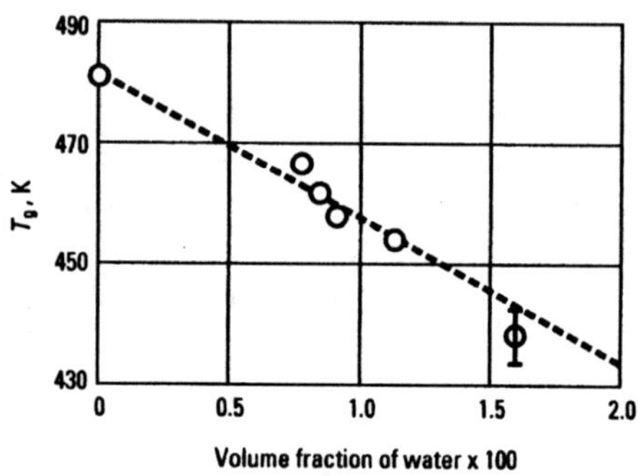

Fig. 2 Glass-transition depression data (calculated). Curve as predicted by Eq 5. Source: Ref 19

Measuring the T_g of moisture-containing resins is not accomplished without a good deal of care, because water is often lost during the measurement. This can be especially serious for high-T_g polymers, such as those used in high-technology applications. Examples of this are given in Table 3, with water loss being measured by thermogravimetric analysis (TGA). Essentially all the absorbed water may be lost unless proper precautions are taken. Measurements of T_g are often carried out by DSC (as was done by Karasz); DSC techniques are particularly adaptable to preventing loss of moisture during T_g measurements.

A method that satisfactorily measures the T_g in water-saturated thermoplastics and thermosets that do not have an excessively high cross link density is to seal the plastic and a small amount of water in a high-pressure DSC pan and then measure in a normal manner (Ref 23). The pan contains three phases (liquid and gaseous forms of water and polymer) and thus has but one degree of freedom by Gibb's phase rule, namely temperature, because this is necessarily the independent variable in a measurement of T_g. Results of using this method are given in Table 4, where they are contrasted with measurements in standard DSC analysis. The plastic, in principle, remains fully saturated with moisture during the run. A T_g measured by this technique is thus a "worst case" value of a plastic that is fully moisture saturated. A major drawback is that no values of T_g of intermediate saturations can be obtained. Other DSC techniques seem to be satisfactory as well. Simply sealing a moisture-containing plastic into a DSC high-pressure pan may be adequate, especially if the plastic sample is large in comparison to the vapor space. A drawback of the DSC method is that it generally fails to give measurable T_g for resins having a very high cross-link density, in particular some of the aerospace epoxy resins.

Perhaps the most popular method of T_g measurement is dynamic mechanical analysis (DMA). The glass transition and other relaxations are clearly distinguished, and the shear or tensile complex moduli are measured and have a clear connection to the moduli of interest for engineering design. In Fig. 3, representative dynamic mechanical data are given, with some of the commonly used measures of T_g pointed out. To date, it is not possible to run experiments in an autoclave to prevent loss of moisture. Nonetheless, there is much to recommend this technique as a routine screening method for the T_g of moisture-containing plastics and composites.

In one common mode of operation, the temperature is increased in jumps of 5 to 10 °C (9 to 18 °F) and is held for 2 min, after which the dynamic mechanical parameters are measured. This technique tends to give conservative values for the T_g of dry plastics but should not be used to determine the T_g for moisture-containing plastics, particularly if the T_g is above 100 °C (212 °F). In many instances, the plastic specimen is completely dried out by this technique. A better technique is to ramp the temperature, as in the DSC and DMA techniques, at 10 °C/min (18 °F/min), for example. A further improvement (Ref 24) is to enclose the specimen in a polytetrafluoroethylene (PTFE) bag containing oil saturated with water. Thermomechanical analysis (TMA) is also a recognized method for measuring T_g (Ref 25).

Another method of assessing T_g of composite materials by modulus measurement is used in the aerospace industry. A number of specimens of a size suitable for measuring flexural modulus are placed in a humidity chamber until they reach saturation. Then the flexural modulus is determined for individual specimens at increasing temperatures in oil baths. The modulus-versus-temperature curve is plotted, and the T_g is identified by a rapid drop of modulus on the curve. Although rather tedious and time-consuming, this method is very satisfactory for advanced composite structures, for which performance is critical. Shear modulus (G_{12}) also can be determined, using ±45° tension tests.

Epoxy Resins. It is likely that more work has been done on the effect of moisture on the T_g of epoxy resin systems than on any other plastic system. Of particular interest is the system based on tetraglycidyl methylenedianiline (TGMDA)/diamino diphenyl sulfone (DDS). These materials are the principal epoxy matrix resin systems currently used in advanced composite aircraft/aerospace applications. As shown in Table 5, TGMDA/DDS can absorb as much as 6.5 wt% water. This absorbed water results in a dramatic drop in T_g (Ref 26–28). The reduction of the T_g resulting from the absorbed moisture is also given in Table 5 and corresponds with the 13 to 15 °C/wt% (25 to 30 °F/wt%) water content, as predicted by Ellis and Karasz. Also, the amount and rate of moisture absorption of a typical TGMDA/DDS laminate were found to increase with periodic exposure to thermal spikes (Ref 29), such as those experienced on a supersonic aircraft. The absorptivity coefficient of a graphite-epoxy laminate was shown to double with such an exposure.

Not all epoxy resin systems absorb as much water as the TGMDA/DDS system, because the amount of water absorbed by an epoxy resin depends on the polarity of the epoxy resin system. The effect of moisture on T_g depends on the amount of moisture absorbed, which in turn depends on the chemical structure of the cured resin.

Table 3 Water losses during temperature scans (thermogravimetric tests)

Resin curing agent or plastic	Beginning water content, wt %	Water loss, wt %
Epon resin 826/diamino-diphenyl sulfone	2.28	–0.91 at 40 °C/min (70 °F/min) –2.02 at 10 °C/min (18 °F/min)
Polycarbonate	0.32	–0.2 at 40 °C/min (70 °F/min) –0.39 at 10 °C/min (18 °F/min)
Polysulfone	0.57	–0.25 at 40 °C/min (70 °F/min) –0.52 at 10 °C/min (18 °F/min)

Thermal Analysis

Thermal analysis describes the techniques used in characterizing materials by measuring a physical or mechanical property as a function of temperature or time at a constant temperature or as a function of temperature. This dependency allows access to processing and performance information relating to resins and fiber-reinforced composites and can be used for quality assurance, process control, and new material-process development. Gel points, T_g, expansion/contraction properties, reaction rates and cure kinetics, effects of individual and combinations of components, polymer stability, and material life predictions can all be determined by thermal analyses.

The four thermal analysis techniques used most frequently are DSC, TGA, TMA, and rheological analysis. In general thermal characterization practice, the DSC analysis technique may reveal an initial endotherm, assigned to the T_g, and either a second endotherm, indicating the T_m, or a pronounced exotherm that indicates a decomposition temperature. Because the T_g occurs at a temperature below the T_m or the decomposition point of a polymer, DSC analysis can be used to determine both the T_g and the T_m or the decomposition point. The TMA analysis method is primarily used to obtain T_g data. The TGA analysis method is normally used to obtain the onset temperature of initial polymer weight loss, as well as the extent of oxidative effects (in an air environment) or char formation (in an inert environment).

General characterizations that illustrate the results normally obtained using DSC, TMA, and TGA methods to screen the thermal properties of polymers are presented below. General information on thermal analysis and its application also are available in a variety of publications (Ref 30–43). To establish a relationship between various thermal properties, seven engineering plastics from the Society of Plastics Engineers "Resinkit" were evaluated by DSC, TMA, and TGA (see Table 6 and Fig. 4). There is good correlation between DSC and TMA transition temperatures.

Differential Scanning Calorimetry

Differential scanning calorimetry measures the energy absorbed (endotherm) or produced (exotherm) as a function of time or temperature. It is used to characterize melting, crystallization, resin curing, loss of solvents, and other processes involving an energy change. Differential scanning calorimetry may also be applied to processes involving a change in heat capacity, such as the glass transition. A schematic of a DSC thermogram is shown in Fig. 5.

An experimental analysis related to DSC is differential thermal analysis (DTA), in which temperature differentials are measured. This information on relative heat capacities, presence of solvents, changes in structure (that is, phase changes, such as melting of one component in a resin system), and chemical reactions. However, heat flow is not measured by the DTA method.

In the DSC method, the sample and reference are placed in thin metal (aluminum) pans, with the thermocouple sensors below the pans. Differential scanning calorimetry measurements can be made in two ways: by measuring the electrical energy provided to heaters below the pans necessary to maintain the two pans at the same temperature (power compensation) or by measuring the heat flow (differential temperature) as a function of sample temperature (heat flux). In the DTA method, the sensor thermocouple is placed either directly in the sample or close to the sample. The endothermic or exothermic heats of transition can be quantitatively measured by DSC but not by DTA. In short, DSC measures heat flow, while DTA measures temperature differentials.

DSC and DTA of Thermoplastics. Engineering thermoplastics have been characterized by DSC and DTA (Ref 44–52). With these methods, the following physical properties have been determined (Ref 53–55):

- T_g, and T_c (the temperature at which crystallization occurs at a maximum rate)
- Exothermic heat of polymerization or cure
- T_m
- Heat of fusion
- Exothermic heat of stress relaxation
- Specific heat as a function of temperature
- Thermal and oxidative stability
- Heat of volatilization of residual solvents

This information provides the engineer with differences between a potentially successful and a potentially inadequate sample, differences resulting from thermal or processing histories, the presence of undesirable contaminants, or changes in formulation. A typical plot of two thermoplastics and a blend is shown in Fig. 6.

Differential scanning calorimetry is used to characterize a wide variety of effects on the performance of plastics. As a few examples of the utility of DSC, it can determine the effect of a plasticizer on the melting point of nylon 11 (Fig. 7), the amount of PE in impact PC (Fig. 8), and the crystallinity of polyolefins (Fig. 9). All of these effects are important aspects of plastic performance.

DSC of Thermoset Resins. When a thermoset cures, the resultant chemical reaction gives off heat (exotherm) or absorbs energy (endotherm) as a function of both time and temperature. Differential scanning calorimetry

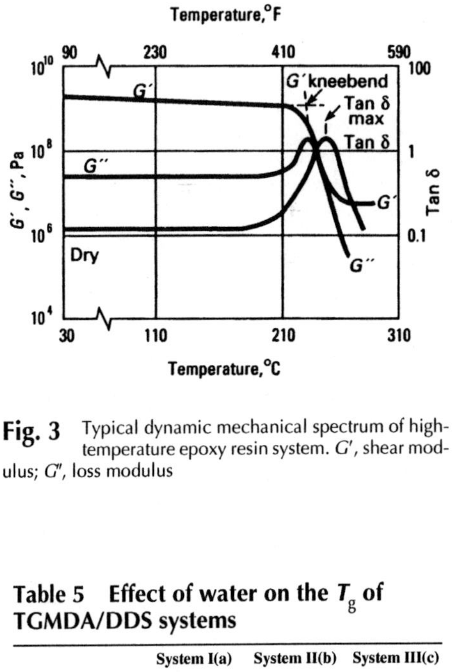

Fig. 3 Typical dynamic mechanical spectrum of high-temperature epoxy resin system. G', shear modulus; G'', loss modulus

Table 4 Differential scanning calorimetry comparison of Tg results from sealed and unsealed pans

	T_g, unsealed pan				T_g, sealed pan, resin and water	
	Dry		Wet			
Resin curing agent or plastic	°C	°F	°C	°F	°C	°F
Epon resin 826(a)/Epon curing agent Y(b)	167	33	134 (1st scan)	275	125	257
Epon resin 826/methylenedianiline	165	330	...	...	122	252
Epon resin 826/Jeffamine D-230(b)	92	200	88	190	62	145
Epon resin 826/Jeffamine D-400(b)	50	120	...	...	30	85
Polycarbonate	148	300	139	280	132	270
Polysulfone	184	365	184	365	158	315

(a) Tradename of Shell Chemical Company. (b) Tradename of Texaco Chemical Company

Table 5 Effect of water on the T_g of TGMDA/DDS systems

	System I(a)	System II(b)	System III(c)
Moisture gain, wt%	6.5(d)	5.5(d)	5.0(e)
Glass transition temperature			
Dry, °C (°F)	246 (475)	175 (350)	200 (390)
Wet, °C (°F)	144 (290)	112 (235)	140 (285)
°C/wt% water absorbed	15.7	11.5	12.0

(a) TGMDA, tetraglycidyl methylenedianiline; DDS, diaminodiphenyl sulfone. (a) NARMCO 5208 (Ref 29). (b) TGMDA/32 phr DDS/$BF_3 \cdot H_2NCH_3$ (Ref 27). (c) TGMDA/50 phr DDS (Ref 26). (d) Immersion in water at 71 °C (160 °F). (e) Immersion in water at 60 °C (140 °F) measures the temperature differences between a sample and an inert reference material. A combination of dynamic and isothermal experiments can provide information on reaction rates, cure rates, specific heat, and degree of cure. A dynamic DSC curve typical of the thermoset resins used in some advanced composites and adhesives is shown in Fig. 10. Critical points on the curve are:

- T_g, the subambient glass-transition temperature of the uncured resin
- T_i, the initiation temperature or onset of reaction, indicating the beginning of polymerization
- T_m, a minor exotherm peak temperature associated with accelerator effects
- T_{exo}, the major exotherm peak temperature
- T_f, the final temperature, indicating the end of heat generation and completion of the cure

Several thermal characteristics affect the quality of hardware made from thermoset systems. These characteristics include temperature gradient control, heatup rate during processing, and extent of cure. The type and number of competing chemical reactions, heat of reaction, thermal conductivity, and specific heat of a material at various stages of reaction produce temperature variations during a cure cycle that directly affect the final degree of cure. This is particularly true in thick laminates where slow heat-removal rates can drastically influence processing. Therefore, to optimize hardware fabrication, it is essential to understand the kinetic behavior of the reactive system being processed.

Control of resin advancement in raw material and the degree of cure after processing are also prerequisites for repeatable, reliable, high-quality final products. Differential scanning calorimetry has been used for quality control and degree of cure studies of molding compounds (Ref 57, 58), printed circuit board prepregs (Ref 59–61), powder paints (Ref 58), an ambiently cured field repair system (Ref 62), graphite-reinforced prepreg resin matrices (Ref 63, 64), and film adhesives (Ref 65). The subject of chemical kinetics and the way in which kinetic parameters are obtained is a complex one. The effects of fillers (Ref 66), impurities (Ref 67), and catalysts (Ref 68, 69) on overall reaction have been studied, as well as the reaction kinetics of a commercial adhesive (Ref 70).

The long-term integrity of a thermoset material is influenced by a number of time-dependent factors: moisture and solvent diffusion, viscoelastic deformation, fatigue, chemical reactions, and the generic category of aging, which includes physical, chemical, and mechanical aging (Ref 71).

Physical aging is the natural process of reaching equilibrium, and it leads to densification and embrittlement of the polymer. This type of aging in polymers is manifested by changes in relaxation times. These changes have been studied in thermosets by using DSC (Ref 72–74).

Chemical aging involves cross linking reactions, and the results are similar to those of physical aging. If a thermoset resin is incorrectly processed, the chemical reaction is interrupted prior to cross linking. The resultant unreacted species will continue to cross link slowly over a long period of time, continuously changing material properties. These changes can be followed by using DSC.

Thermogravimetric Analysis

Thermogravimetric analysis involves measurement of the weight gain or loss of a material as a function of temperature and time, and it utilizes an extremely sensitive electronic microbalance. Typically, a polymer sample is examined from room temperature to above its decomposition or pyrolysis temperature in nitrogen (thermal stability). Often, a platinum pan or quartz boat is used for high-temperature studies to 1000 °C (1830 °F). Sample size may vary from 1.0 to 100 mg (0.0154 to 1.54 gr). The oxidative stability of polymers in air or oxygen can also be determined by TGA.

One of the most important applications of TGA is the assessment of the thermal stability of a material. This can be done to obtain relative comparisons between different materials or as an accelerated means for lifetime predictions. Where the loss of additives such as plasticizers or antioxidants can damage a structure, decomposition profiles are excellent indicators of change.

Thermogravimetric analysis can also be used to determine moisture, volatile, and filler contents, to study the effects of additives, and to obtain separation of some components (for example, rubber from carbon black). In an attempt to determine the exact mechanisms of polymer degradation, TGA has been coupled with spectroscopic techniques to clarify degradation pathways and to identify additive components (Ref 75, 76).

Thermogravimetric Analysis of Thermosets. A typical weight-loss curve of a thermoset composite is shown in Fig. 11. In addition to the normal decomposition profile, there is the added benefit of obtaining the amount of fabric or filler left behind as the residue. This applies for fiberglass and other fabrics and fillers that do not oxidize or form other compounds that cause a weight gain. This method has been used as an alternative to conventional muffle furnace techniques (Ref 77).

A comparison of the thermal decomposition of encapsulating materials using TGA is shown in Fig. 12. Absolute classification of thermal stability is difficult, however, because of the interaction of various aging phenomena. Because decomposition mechanisms are often diffusion controlled, sample geometry and fillers can affect the observed test results. Therefore, the data obtained on small test specimens may not be extrapolated to larger structures. This type of information should be used judiciously as a guide for further studies until TGA or other thermal techniques are developed that give better correlation.

Current kinetic models that predict material life are in the early stages of development. Pre-

Table 6 Thermal characterization of Society of Plastics Engineers (SPE) reference plastics

Polymer	SPE identification number	Onset temperature: DSC, °C	Onset temperature: DSC, °F	Onset temperature: TMA, °C	Onset temperature: TMA, °F	Source of DSC of TMA transition	ASTM D 256 Izod impact, J/m	ASTM D 256 Izod impact, ft · lbf/in.	TGA: Extrapolated onset temperature, °C	TGA: Extrapolated onset temperature, °F	TGA: wt% at 600 °C (1110°F)
PVC, flexible	29	65	150	41	105	T_g	270	5.0	274	525	5.7
PVC, rigid	30	120	250	102	215	T_g	20	0.4	278	530	7.9
ABS, transparent	5	99	210	59	140	T_g, SAN	130	2.5	407	765	0
ABS, high impact	7	108	225	85	185	T_g, SAN	430	8.0	422	790	4
Nylon 6	16	183	360	155	310	T_m	160	3.0	439	820	1.2
Nylon 6/6	15	218	425	165	330	T_m	110	2.1	433	810	2.4
PET	18	151	305	130	265	T_g	40	0.7	517	960	28.3

DSC, differential scanning calorimetry; TMA, thermomechanical analysis; TGA, thermogravimetric analysis; PVC, polyvinyl chloride; ABS, acrylonitrile-butadiene-styrene; PET, polyethylene terephthalate. Experimental conditions: Heating rate = 10 °C/min (18 °F/min); N_2 flow = 50 cm^3/min (3 in.^{3}/min) in DSC, TGA, and TMA; weight = 14–21 mg (0.21–0.32 gr) in DSC and 27–36 mg (0.42–0.55 gr) in TGA; in TMA, 5.0 g (0.18 oz) applied load, and height = 1.3–1.7 mm (0.05–0.07 in.)

dictions of material longevity require a relationship between time-to-failure and experimental variables that induce failure. Because the failure of polymer systems and composite materials is complex and involves multiple failure modes, it is important that accelerated tests model each of the relevant processes in such a way as to describe the combined effect of competing modes. Experiments at very slow heating rates and low isothermal temperatures minimize the differences between actual and extrapolated service conditions.

Thermal Stability of Thermoplastics. The relative thermal stability of polymers measured by TGA is shown in Fig. 13. Based on the onset temperature of thermal degradation, the polymers are ranked in order of stability: polyimide (PI) stability is greater than that of PTFE, which is greater than that of high-density polyethylene, which is greater than that of PMMA, which is greater than that of polyvinyl chloride (PVC).

The composition of silica- and carbon-filled PTFE was determined by TGA (Fig. 14). PTFE is decomposed and volatilized in nitrogen, while the carbon filler is volatilized in air at 600 °C (1110 °F). The inorganic residue is silica.

A summary of polymer thermal properties as determined by thermal analysis and limited oxygen index (LOI) (ASTM D 2863) is given in Table 7. According to this standard, LOI is "the minimum concentration of oxygen, expressed as volume percent, in a mixture of oxygen and nitrogen that will just support flaming combustion of a polymer initially at room temperature." The logarithm of the heat of combustion varied linearly with LOI for the polymers studied. The T_g, T_m, T_p, and combustion temperature did not relate to the LOI.

Screening High-Performance Thermoplastics. Thermogravimetric analysis is an effective method for screening the stability of high-performance polymers in oxidative and inert environments. However, the TGA method has a limited degree of capability for predicting polymer thermo-oxidative stability at very high temperatures (>300 °C, or 570 °F). It is particularly useful for assessing the short-term thermo-oxidative stability ranking of a series of polymers prepared from one identical monomer and one variable monomer. The TGA method is questionable at best for predicting longer-term thermo-oxidative stability (>24 h) at high temperatures.

The use of TGA as a tool for assessing the high-temperature stability of aromatic/heterocyclic polymers predated state-of-the-art DSC and TMA techniques by approximately 10 years. Therefore, prior to the availability of other thermal characterization methods, TGA represented the major thermal analysis method for screening thermal (samples tested in an inert atmosphere) and oxidative (samples tested in air or enriched-oxygen atmospheres) behavior. The determination of thermal stability by TGA in an inert environment is frequently used to assess char yield. Char yield is defined as an area of constant weight retention, following an initial period of weight loss, normally ascribed to stable carbonaceous residue formation after polymer decomposition.

Example: Typical DSC and TGA Screening Results. Thermal characterization of two commercially available PIs and one polybenzimidazole (PBI) was conducted to obtain screening data and to determine initial thermal performance (Ref 79). In this study, DSC and TGA analyses were performed to determine, respectively, the T_g and initial oxidative weight loss of selected polymers. Specifically, as-cast and postcured film samples of Avimid N, Celazole PBI, and Eymyd L-30N polyimide were characterized. The chemical composition and background information for these polymers are listed in Table 8.

The results of the DSC and TGA thermal property screening analyses are presented in Table 9. These results illustrate the thermal analysis data normally obtained when one characterizes highly aromatic/heterocyclic polymers such as PI and PBI in a film sample state. Important comparative information shown in Table 9 is:

- The T_g value for postcured specimens determined by DSC consistently occurred ≥30 °C

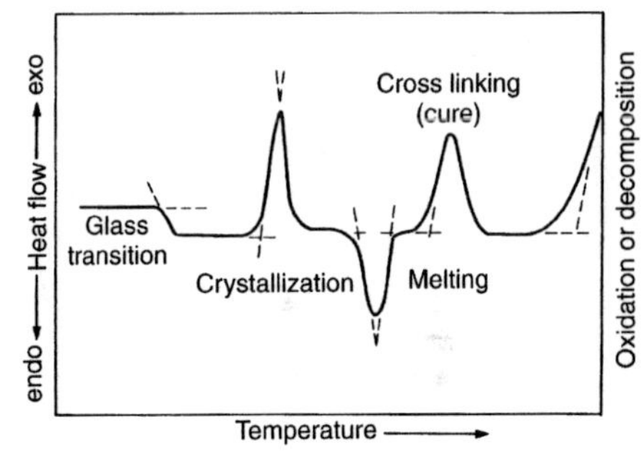

Fig. 5 Differential scanning calorimetry thermogram

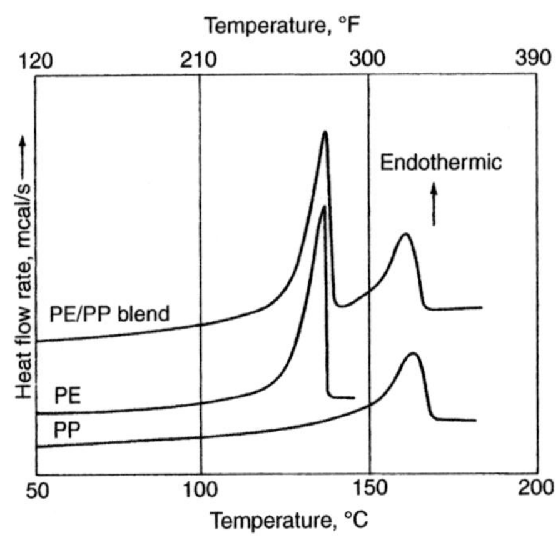

Fig. 6 Differential scanning calorimetry thermogram of polyethylene/polypropylene blend, 10 mcal/s range, 20 °C/min (36 °F/min) heating rate. PE, polyethylene; PP, polypropylene. Source: Ref 56

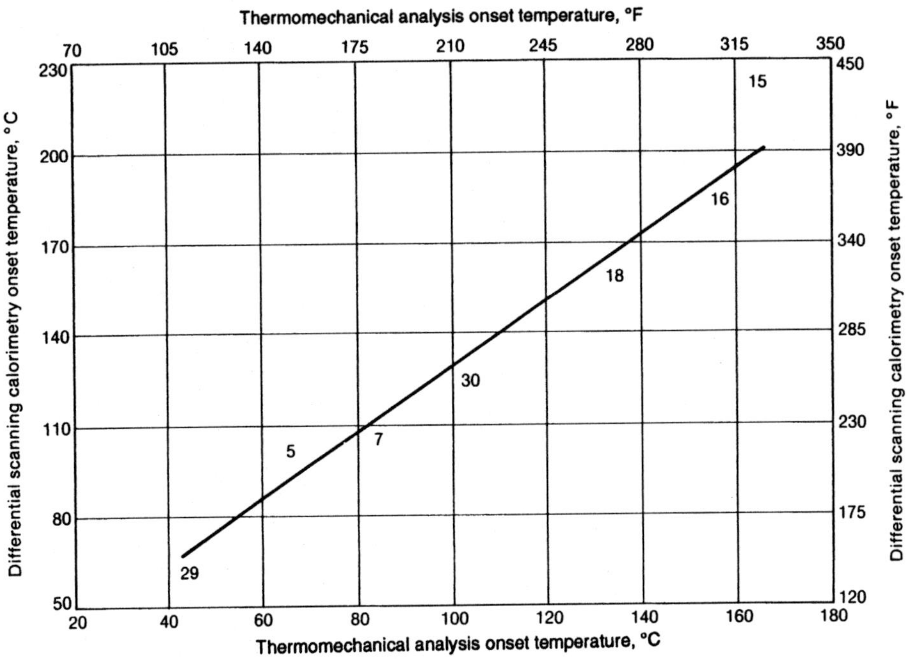

Fig. 4 Thermal analysis of Society of Plastics Engineers (SPE) reference plastics. Identification numbers are tied to SPE resin kit (see Table 6)

(≥55 °F) below the initial weight-loss temperature determined by TGA.

- The effect of postcure of PIs on the initial weight loss in air by TGA was minimal (~10 °C, or 18 °F); postcure increased the initial weight-loss temperature of PBI to a larger extent (by 30 °C, or 55 °F).

A similar thermal analysis screening study was conducted on experimental PIs in which one aromatic diamine, Ethacure 300, was reacted with three aromatic diahydrides: hexafluoropropane dianhydride (6-FDA), pyromellitic dianhydride (PMDA), and benzophenonetetracarboxylic acid dianhydride (BTDA). These PIs were postcured and analyzed by DSC and TGA to give the results presented in Table 10.

The DSC and TGA data presented in Table 10 represent expected trends in thermal behavior similar to those summarized in Table 9 for commercially available polymers. Noteworthy data trends from the thermal analyses are:

- The T_g values determined by DSC follow an expected pattern of thermal stability. The dianhydrides used normally follow the order of PMDA > 6-FDA > BTDA in descending temperature when the dianhydrides are combined with a single aromatic ring, such as Ethacure 300, to give a very inflexible (or stiff) PI repeat unit.
- The initial weight-loss temperature determined by TGA is again higher than the T_g values obtained (Table 9); the lower thermal stability is to be expected because of the methylthio substituents on the Ethacure 300 monomer versus the unsubstituted amines used in the commercial polymers given in Table 8.

For illustrative purposes, the TGA tracings of Ethacure 300/6-FDA and Ethacure 300/PMDA are presented in Fig. 15 and 16, respectively. These TGA results were determined in nitrogen instead of air, but are considered to be representative of expected trends in gross thermal stability. Commercially available and experimental thermoplastic high-performance PIs exhibit similar and classical behavior in DSC and TGA screening characterization.

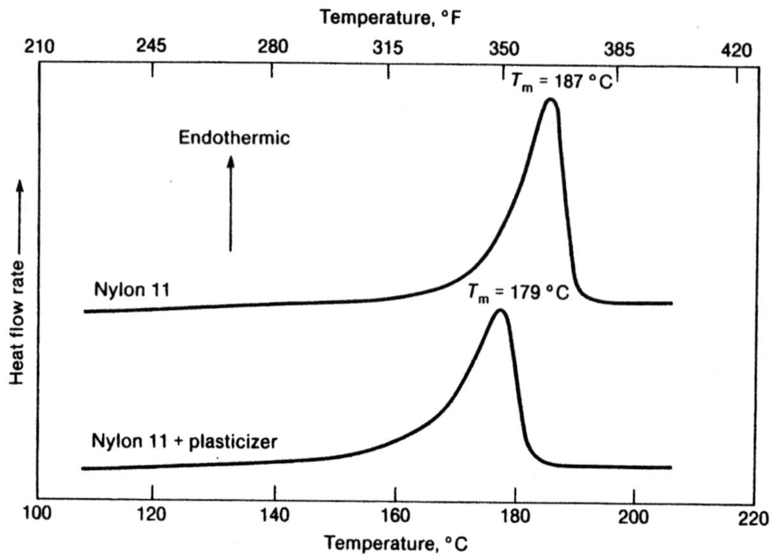

Fig. 7 Differential scanning calorimetry determination of the effect of a plasticizer on T_m of nylon 11. Range, 0.0024 W (10 mcal/s); heating rate, 20 °C/min (36 °F/min); weight, 6.8 mg (0.105 gr), both samples. Source: Ref 51

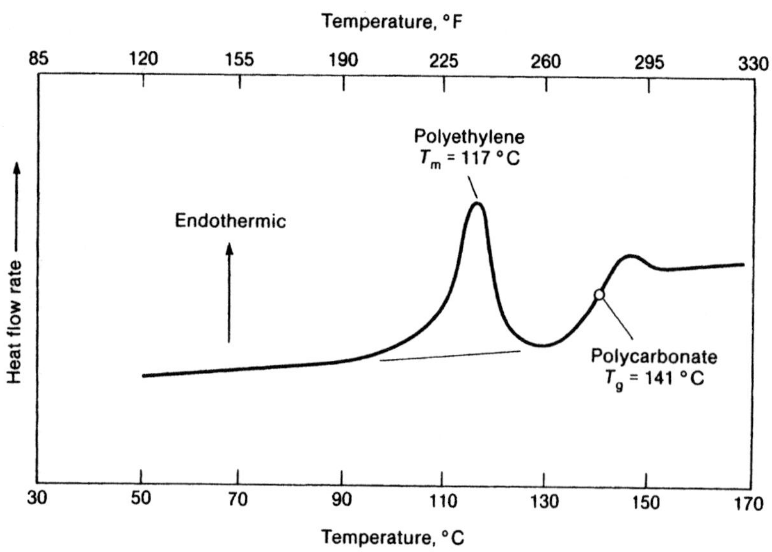

Fig. 8 Differential scanning calorimetry determination of polyethylene in impact polycarbonate. Range, 0.00048 W (2 mcal/s); heating rate, 20 °C/min (36 °F/min); weight, 23 mg (0.355 gr). Source: Ref 51

Thermomechanical Analysis

Thermomechanical analysis measures the dimensional change of a plastic as a function of time or temperature. The thermomechanical properties that have been measured are the T_g, softening point, coefficient of linear thermal expansion, heat-deflection temperatures (HDT), creep moduli, creep relaxation, degree of cure, viscoelastic behavior, and dilatometric properties.

Heat-Deflection Curves. ASTM has developed thermomechanical tests that approximate the strength and T_g of plastics, for example, the vicat softening temperature and HDT under load (DTUL) test method. Vicat softening (ASTM D 1155) and HDTs (ASTM D 648) of plastics have been determined by TMA at the high stresses of 10.3 and 1.82 MPa (1.5 and 0.264 ksi), respectively. Figure 17 shows heat-deflection curves for several thermoplastics.

Creep Modulus. Generalized tensile stress-strain curves for plastics are related to polymer properties (Fig. 18). Based on this generalization and the room-temperature TMA creep modulus, as well as the percent of creep recovery, a scheme has been developed for ranking commercial polymers (Fig. 19). The polymers are categorized by their mechanical properties: hard tough, hard brittle, soft tough, and soft weak. There is a good correlation between the TMA properties and the known tensile properties of these commercial polymers.

Thermal Expansion of Thermosets. Cured thermosets typically exhibit two linear regions. The first is associated with the glassy state and is followed by a change to a second linear region of higher slope associated with the rubbery state because of T_g.

The coefficient of thermal expansion and T_g of a thermoset are closely related to the degree of cure of that resin. Fully cured materials have higher T_g and sometimes lower expansion coefficients than undercured or partially cured materials. Many fabrication processes induce cure-in stresses. Thermal cycling or annealing above T_g will smooth the curve but will not elevate T_g. Ideally, T_g is observed as an abrupt change in the slope of the linear expansion versus temperature curve. However, because relaxation often occurs near T_g, the transition can be broad, depending on such factors as the material, cure state, internal stresses, and test conditions.

Rheology is the study of the flow behavior of a material and is generally applied to liquids or semiliquids. A typical rheological curve for the dynamic cure of a PI prepreg shows an initial drop in viscosity associated with the softening and flowing of the resin. The peak appears when the resin hardens because of increased chain extension and stiffness as imidization takes place. The resin goes through a second melt stage as the imidized resin softens, and then viscosity rapidly increases as cure continues to complction.

The curing of a thermoset system involves a complex, multistep mechanism leading to a molecular network of infinite molecular weight. The gel point is the point at which a viscous liquid becomes an elastic gel; this marks the beginning of the infinite network. From a processing standpoint, this point and the flow behavior leading up to it are important characteristics. Flow behavior affects the way in which a material can be processed, and gelation marks the point at which processing flexibility ends. Other thermal techniques, such as DSC and TGA, do not detect this physical change, because chemical reactions continue unchanged following gelation. Cross-link density, T_g, and ultimate physical properties continue to increase after gelation until the reaction is complete. These characteristics are studied using DMA, and because DMA measures mechanical properties dynamically, the possibility exists for obtaining rapid information on end-product performance.

The key relationships between the process of cure and the physical properties of the cured state of thermosets are shown in generic time-temperature-transformation (TTT) diagrams, which depict the four material states encountered during cure: liquid, elastomer (gelled rubber), ungelled glass, and gelled glass. Critical processing information can be obtained from TTT diagrams, such as the time-temperature dependence of flow, reaction kinetics, gelation, and vitrification (initiation into the ungelled glass state). This type of information is quite useful to the manufacturing engineer for developing appropriate cure cycles (Ref 85, 86). Appropriate time-temperature values for B-staging, debulking, dwells (devolatilization), pressure application points (compaction), and final conditions for cure cycles can be optimized.

The gel point of a thermoset can be empirically assigned as the point at which the shear modulus, G', is equal to the loss modulus, G''. The viscosity is increasing rapidly at this point. This modulus crossover point is more precise and operator independent than conventional gel-point determinations. The loss modulus represents the out-of-phase relation between stress-strain response of viscoelasticity materials such as plastics.

In the past, rheological tests were performed exclusively on neat (unreinforced) resins or resins removed from the reinforcement. Some doubt was always present regarding the one-to-one correlation between the viscosity data thus obtained and the way in which a reinforced material performed during composite fabrication. The possibility always existed of changing the resin when removing the sample. Dissolving the resin from its reinforcement poses problems in solvent removal, because even a small level of residual solvents will significantly alter the viscosity profile. Heating to remove trace solvents or the resin itself can advance the matrix and alter its behavior. Simply scraping a resin sample from the reinforcement is tedious and often contaminates the sample with fiber or filler. In addition, neat resin exhibits near-Newtonian flow characteristics during the early stages of cure, while flow is non-Newtonian in the presence of fibers having large surface areas and relatively polar surfaces. As a result, the viscous-state behavior exhibited during the manufacturing process may differ sharply from that observed in the rheological test chamber. To overcome these problems, techniques have been developed to measure the apparent viscosity of the resin in the presence of fibers (Ref 85, 87, 88).

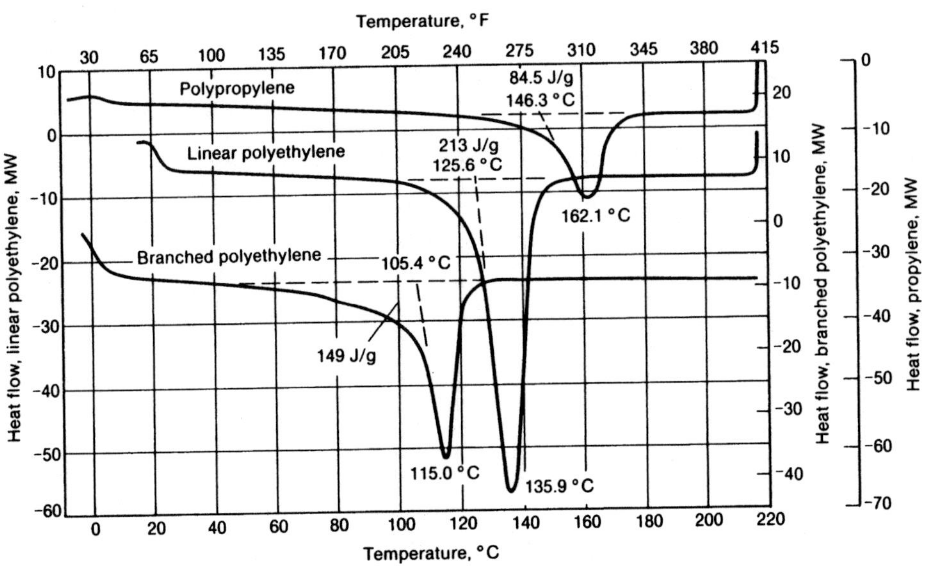

Fig. 9 Polyolefin melting profiles. Source: Ref 55

Thermal Properties

The key thermal properties often considered in the application of engineering plastics include:

- Long-term temperature resistance
- Heat-deflection temperature
- Thermal conductivity
- Thermal expansion coefficients

Typical values are summarized in Table 1, while this section describes the factors affecting these properties.

Long-term temperature resistance is the temperature at which the part must perform for the life of the device. One of the most common measures of long-term temperature resistance is the thermal index determined by the Underwriters' Laboratories. In this test, standard test specimens are exposed to different temperatures and are tested at varying intervals. Failure is said to occur when property values drop to 50% of their initial value. The property criterion for determining the long-term use temperature depends on the application.

The most common change that takes place during high-temperature exposure is an oxidation reaction, which decreases the molecular weight of the polymer. This reduction in molecular weight often is first evidenced by a reduction in physical strength and, most frequently, in impact strength, as the plastic embrittles. As the degradation reaction continues, other physical properties drop off, and eventually the electrical properties are affected. For this reason, the long-term temperature resistance is often rated differently for different applications, such as those requiring impact or other mechanical properties as opposed to those requiring only electrical insulation.

Heat-deflection temperature (also known as deflection temperature under load) is an often misused characteristic. In the standard ASTM test (D 648), the HDT is the temperature at which a 125 mm (5 in.) bar deflects 0.25 mm (0.010 in.) when a load is placed in the center. It is typically reported at both 0.44 and 1.82 MPa (0.064 and 0.264 ksi) stresses. Thus, it can give an insight into the temperature at which a part would begin to deflect under load. It should not be used as a measure of the thermal stability of the material.

Because it is a measure of the rigidity of a material, the HDT can be influenced by the addition of glass fibers. Softening or relaxation is also a function of the crystallinity of the plastic, that is, the uniform compactness of the molecular chains forming it. Glass-fiber reinforcements increase the HDT in a crystalline material such as PA to a greater extent than in an amorphous material (which has no pattern in molecular distribution) such as PC, as shown in Table 11.

Thermal Conductivity. A knowledge of the thermal conductivity and diffusivity of a polymer, be it a solid thermoplastic, a foam, or a thermoset resin, is essential to processing the material into its final configuration and to establishing appropriate applications of the material (such as polymeric foams as insulating structures). In thermoplastic processing, heat energy is either transferred or exchanged in the material to effect sufficient fluidity such that the polymer can be shaped and/or oriented appropriately. Alternatively, defining the rate of heat transfer becomes rudimentary to the control of reaction processes in the case of thermosetting resins, in which a balance of reaction time and extent of cure must be achieved. The transfer of heat into and out of a polymeric part often involves elements of convection and radiation, as well as conduction. However, because the former two processes are also dependent on fluid dynamics, such as shear heating, and part geometry, they are not considered intrinsic thermal properties.

The mechanism for thermal conduction in polymers is based on agitation or molecular movement across intramolecular or intermolecular bonds. In this context, structural changes that result in an increase in the effective frequency of contact, or that decrease interbond path lengths, increase thermal conductivity. Conversely, factors causing increased disorder or free volume in polymers usually result in a decrease in thermal conductivity. The presence of crystallinity in polymers results in improved packing of molecules and usually increases the conductivity (Ref 89).

Thermal conductivity does not vary significantly among neat plastics. The organic plastics are basically very good insulators. Consequently, to improve thermal conductivity, plastics filled with mineral or conductive materials must be used. Figure 20 shows the positive effect of adding various amounts of glass to nylon 6/6 in particular. Composites based on conductive flakes with high aspect ratios have also been explored as high-conductivity materials. For example, composites of nylon 6/6, or polybutylene terephthalate (PBT) with up to 35% aluminum flakes, were made, and their heat-transfer effectiveness reached 80 to 95% of that of the pure metal (Ref 90).

For semicrystalline polymers, the total conductivity is assumed to be the sum of contribution from the crystalline and amorphous phases (Ref 91). The crystalline phase contribution is expected to be greater than that of the amorphous contribution because of the greater degree of order and packing density achieved in the crystalline phase. The dependence of thermal conductivity on molecular weight of the polymer has been addressed by several authors (Ref 91–93). Hansen et al., for example, found that the thermal conductivity of linear PE increases proportionally to the square root of the weight-average molecular weight (Ref 91). However, increased branching in polymers decreases their ability to conduct heat. The conductivity values of PEs of differing degrees of branching are given in Table 12.

The increased number of chain ends introduced by branching increase the amount of free volume in the polymer. This and the more tortuous path for heat conduction along primary valence bonds lower the efficiency of thermal conduction. Increasing the size of the substituent group on a polymer has an analogous effect, as shown by a comparison of PS, PVC, and PE in Table 12. However, it is noteworthy that, in addition to the bulk effect of the substituent groups, the much higher conductivity values of the PEs are due in part to the increased

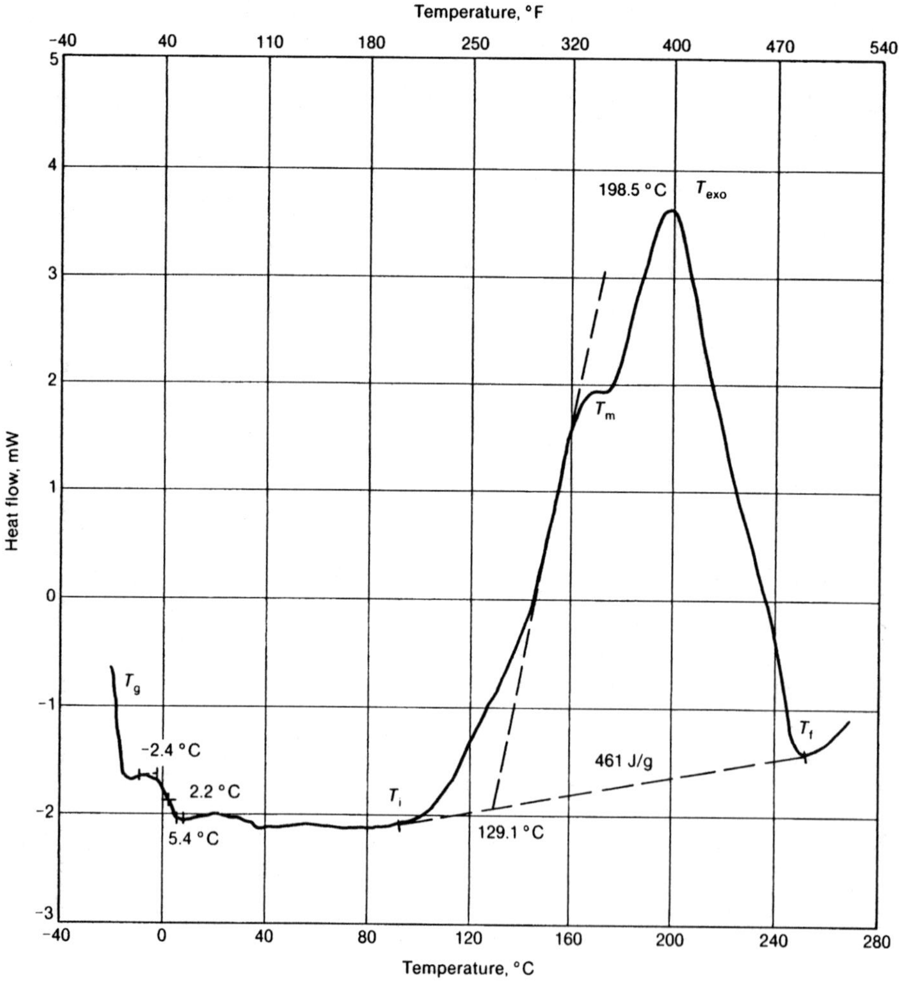

Fig. 10 Differential scanning calorimetry thermogram of Fiberite 934 epoxy, 4.89 mg (0.075 gr), 10 °C/min (18 °F/min) heating rate

degree of crystallinity in these polymers, compared to amorphous PS.

The thermal conductivity of cross-linked systems has been studied as a function of cross-link density, as well as filler content. The latter is particularly important because thermoset resins are generally used as composite structures containing either fillers or reinforcement agents. It has been shown that thermal conduction in thermoset resins is increased by the degree of cross linking achieved (Ref 94, 95). The increase in conductivity due to increased cross-link density is caused by an effective increase in the molecular weight of the resin, thus providing primary valence bonds as conductivity paths among chains through the cross-linking points.

The increase in thermal conductivity of the polymer depends on the concentration and type of fillers and reinforcements used, as shown in Table 12. Conversely, polymeric foams exhibit marked decreases in heat conduction because gaseous fillers are incorporated in the foam structure. Increasing the number of closed cells in the foam minimizes heat conduction by convection, further improving the insulating character of foamed polymeric parts.

The thermal conductivity of a polymer is also affected by its processing history. Orientation increases the thermal conductivity of polymers in the direction of stretch due to improved alignment of conduction paths (Ref 96). Finally, compression of plastics can increase the thermal conductivity by increasing the packing density of the molecules (Ref 97).

Thermal Expansion. The coefficient of thermal expansion (CTE) is an important factor in many applications involving two different materials. Because plastics have wide variations in thermal expansion, stresses are created whenever one material is connected to or encapsulated in another material. This expansion varies from material to material and is affected by the amount and type of fillers or reinforcements.

Figure 21 shows the effect of glass additions to several materials. The CTE varies, depending on polymer structure, and is generally anisotropic in nature. Parts molded with oriented molecules expand differentially, as long as the T_g is not reached. Above the T_g, the polymer tends to expand isotropically, and hysteresis is noted in the expansion curve upon cooling. Figure 22 demonstrates the change in expansion due to stress relaxation when the sample initially exceeds the T_g, in two different runs.

The CTE is also an important parameter for the selection of polymers for high-precision engineering applications. In particular, parts intended for use over a wide temperature range must have dimensional tolerances that take into account the thermal expansion characteristics of the polymer used. Certain grades of engineering thermoplastics, such as filled PBT, have CTEs that approach those of metals, while most polymers typically exhibit much higher CTEs. Representative values for common polymers measured at room temperature are shown in Table 13. Like the heat capacity, the thermal expansion of a polymer is an increasing function of the temperature with different behaviors above and below the T_g. Thermal-expansion curves of polymers, with temperature, undergo a change in the slope at the T_g while exhibiting linear dependencies above and below the transition. Coefficient of thermal expansion values for polymers are generally several times larger above the T_g than below it.

Mold Shrinkage. Another dimension-related thermal property, often considered by design engineers who use thermoplastic or thermosetting resins, is the total volume contraction associated with the solidification of the polymer from the melt. A prominent example is mold shrinkage in injection molding. Typical ranges of mold-shrinkage values have been established that characterize different types and grades of polymers, but mold shrinkage is a function not only of the volume change associated with the temperature of the polymer but also of additional intrinsic polymer properties (such as the possible intervention of crystallization) and extrinsic parameters (such as mold fill and clamping pressures).

Representative values of mold shrinkage for some common polymers are given in Table 13. Mold shrinkage tends to be greatest for flexible, crystallizable polymers, such as PE, because of the volume and hence density difference between the crystalline and amorphous phases. On the other hand, glassy polymers, such as PS, exhibit smaller dimensional changes upon cool-

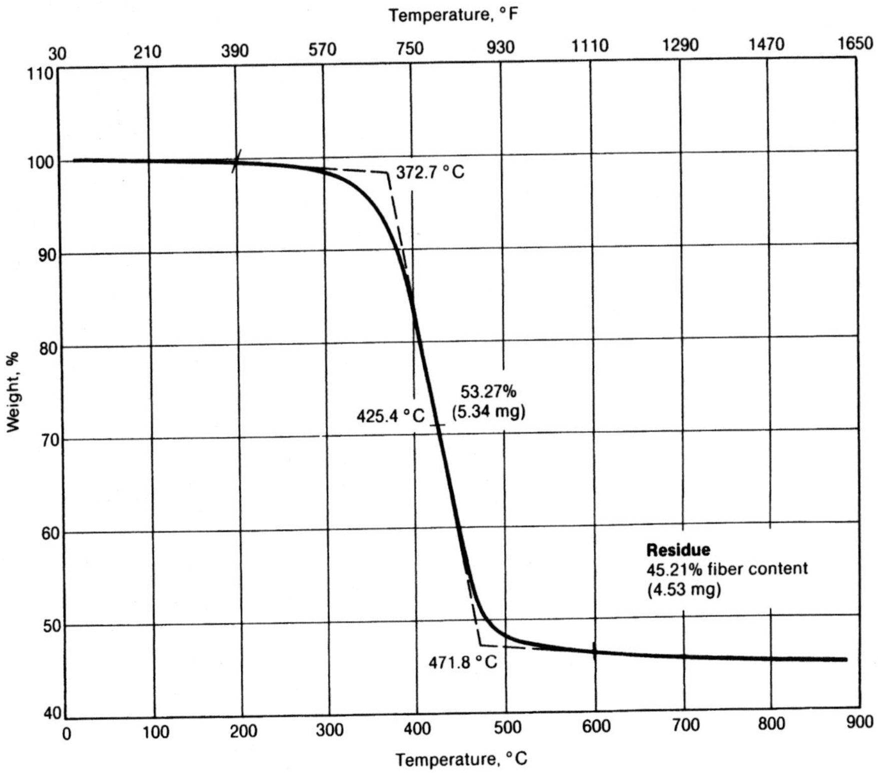

Fig. 11 Typical thermogravimetric analysis curve for fiberglass-vinyl ester prepreg

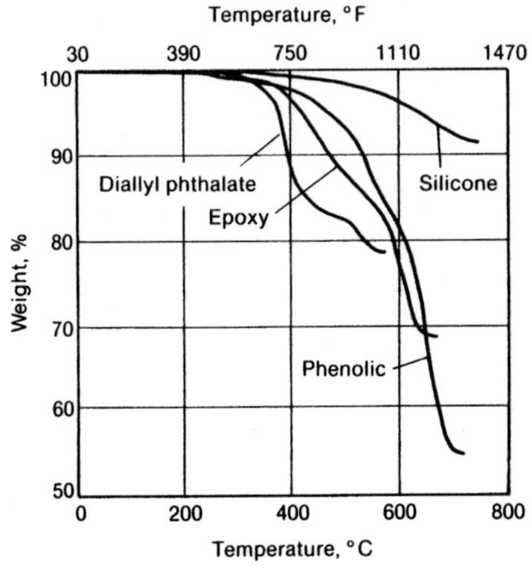

Fig. 12 Thermogravimetric analysis of encapsulating materials, 20 to 30 mg (0.3 to 0.5 gr), 10°C/min (18 °F/min), air at 40 mL/min. Courtesy of Motorola Semiconductor Products Division

ing because of the absence of crystallization. Besides the possibly large volume effects due to recrystallization, mold shrinkage becomes proportional to the thermal expansivity of a polymer. The addition of fillers to the polymer matrix, particularly those with effective wetting characteristics, generally decreases the CTE of the composite material and lowers the mold-shrinkage values. Mold-shrinkage values are useful guides to dimensional tolerance limits that must be built into molds to compensate for thermal contraction of the polymer, but additional dimensional variations can result from inappropriate processing parameters. Examples include sink marks from low mold-fill pressures and flashing from excessive pressures. Analogously, volume changes associated with specific fabrication techniques depend on the thermomechanical history of the process, as well as the thermal behavior of the polymer.

Specific heat or heat capacity, of polymers arises from the various degrees of freedom with which the chain molecules can take part as the temperature of the system is raised. Primary contributions to the experimentally observed heat capacity of a polymer include lattice vibrations, lower-frequency group vibrations, chain or segmental rotation, chain defects, and macroscopic contributions from hole and surface defects (Ref 98). Because of the many different types of contributing processes and the strong dependence of the microstructure in polymers on thermal history, exact theoretical calculations of the specific heat are extremely difficult to obtain and not very accurate.

The heat capacities of polymers increase monotonically with temperature and exhibit an incremental jump at the T_g. It has been proposed that the incremental change in the heat capacity at T_g is constant and corresponds to 2.75 cal/mole-bead-K (Ref 99). A bead is defined as the smallest structural unit that can take part in motion above the T_g. The specific heats of polymers generally range in value from 1250 to 2510 J/kg · K (0.3 to 0.6 cal/g · °C) at ambient temperatures. The specific heats of metals, in comparison, generally range one magnitude lower in value. Representative examples are shown in Table 14.

The presence of crystallinity in polymers causes a decrease in the heat capacity. The orderly packing of polymer molecules lowers the range of motions that give rise to the observed heat capacity (Ref 98). The heat capacity of a polymer has been reported to decrease with increasing molecular weight, although the changes are generally small (Ref 100). The heat capacities of most thermoset systems, such as epoxies and phenolic resins at normal degrees of cross linking, have values within the range of linear thermoplastic. The presence of fillers or reinforcement agents generally increases or decreases the heat capacity of the composite material by an amount proportional to the type and concentration of the filler phase relative to the polymeric matrix.

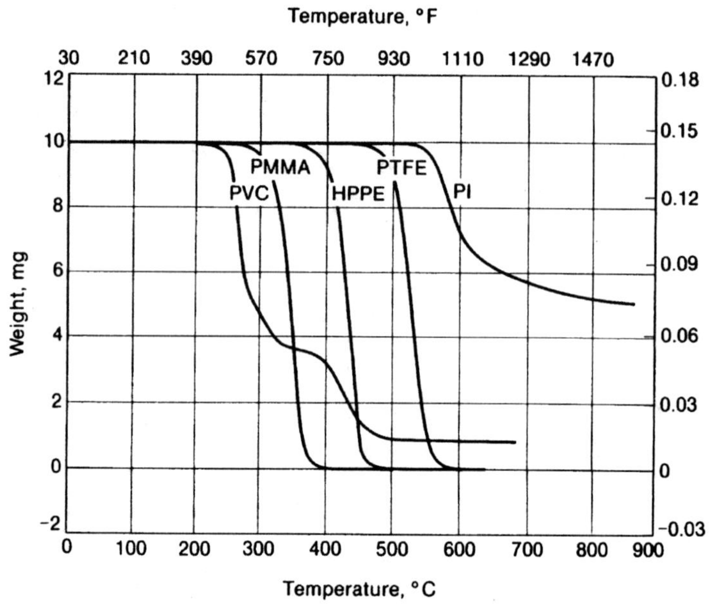

Fig. 13 Relative thermal stability of polymers by thermogravimetric analysis; 10 mg (0.15 gr) at 5 °C/min (9 °F/min) in nitrogen. PVC, polyvinyl chloride; PMMA, polymethylmethacrylate; HPPE, high-pressure polyethylene; PTFE, polytetrafluoroethylene; PI, polyimide

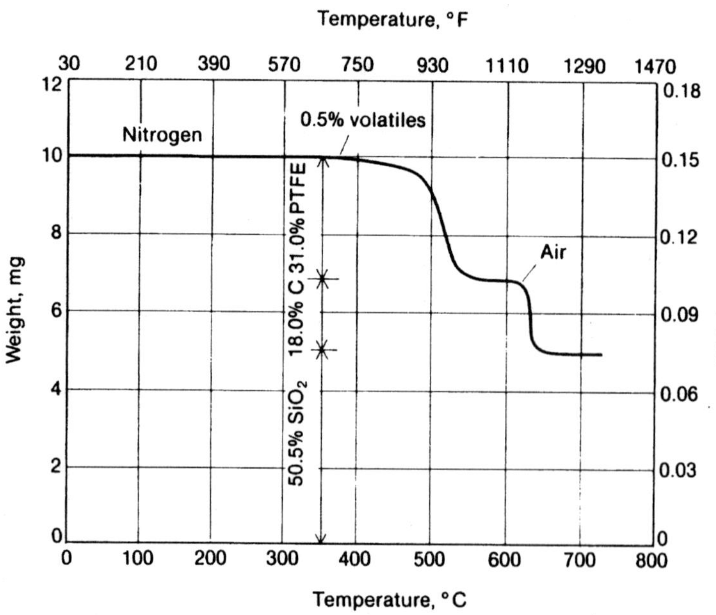

Fig. 14 Thermogravimetric analysis of silica and carbon-filled polytetrafluoroethylene (PTFE); 10 mg (0.15 gr) at 5 °C/min (9 °F/min). Source: Ref 55

Determination of Service Temperature*

Relying on the glass-transition and melting temperatures (T_g and T_m, respectively) of the

*Adapted from Shari Duzac, Thermal Degradation: Determination of Service Temperature, *Engineering Plastics*, Volume 2, *Engineered Materials Handbook*, ASM International, 1988, pages 568 to 570

plastic may not be sufficient, because plastic degradation results from the specific and combined effects of heat and chemical reactants, such as oxygen and ozone. Various aspects of this degradation may be important in determining the suitability of the plastic for a given application.

Testing to identify an important area that may limit the applicability of engineering plastics is discussed in this section. This limit is the service temperature, which is the maximum safe temperature to which the plastic can be exposed.

Service Temperature

The service temperature of a material indicates its ability to retain a certain property, whether electrical or physical, when exposed to elevated temperatures for an extended period of time. Service temperature is therefore an important property when considering the end-use applications of a plastic.

The service temperature, or relative thermal index (RTI), of a plastic is critical to its proper selection. There are generally three RTIs that are used to characterize the properties of plastic materials: electrical, mechanical with impact, and mechanical without impact. The electrical RTI is assigned based on destructive testing of the plastic material using a dielectric strength test. The mechanical with impact RTI is assigned based on the test results of monitoring the degradation of the tensile impact or Izod impact. Lastly, the mechanical without impact RTI is assigned based on the test results of the tensile strength or flexural strength tests.

Many techniques are available for estimating the thermal life expectancy of plastics. The method discussed in this article is used by Underwriters' Laboratories and is outlined in IEEE Std 101-1972.

Test Program. The RTIs are based on an aging program, from which the test performance of the material at lower temperatures is predicted, based on results at higher temperatures. A minimum of 5000 h of thermal aging is necessary before an RTI can be assigned. The final temperature ratings that result from these investigations are critically dependent on the control material selected, specimen thickness, and type of property being evaluated. Test samples should be in a stressed position to ensure maximum deterioration.

Control Material. The RTI depends on the comparison of the thermal aging characteristics of one material of proven field service history at a particular temperature level with those of another material with no field service history. Therefore, one of the most important steps in the program is to select a suitable control material that is as similar as possible to the new or candidate material. The control material should already have been assigned an RTI as a result of the same procedure. Any reformulation of a plastic should require RTI requalification. Because only a small quantity of plastic resin is used in its raw form, even small changes in the amount of flame retardants, molding process additives, and fillers can create major changes in property characteristics.

Selection of Aging Temperatures. A minimum of four aging temperatures should be selected for the thermal aging program. The temperatures may be different for each of the three properties under investigation, namely, the dielectric, impact, and mechanical strengths. It may be useful to review the aging data of the control material to estimate the appropriate oven conditioning temperatures for the candidate material.

The separation between oven temperatures should be at least 10 °C (18 °F) to minimize the effects of the temperature fluctuations of the ovens. The lowest temperature should be approximately 20 °C (35 °F) higher than the expected RTI. For example, if the expected RTI of the candidate material is 140 °C (280 °F), the four aging temperatures should be 160, 170, 180, and 190 °C (320, 340, 355, and 375 °F).

Conducting the Program. Initially, for each property and thickness being evaluated under the program, one set of at least five test specimens is subjected to the tests to establish the starting value, or 100% property retention value. For each oven aging temperature, five sets of test specimens are placed in the air-circulating ovens. At the end of the first, second, and third cycles, an additional set is added.

Generally, samples are conditioned for a specified test cycle, with the highest temperature being assigned a test cycle of 3 days. The second highest oven temperature is assigned a test cycle of 7 days, and the third, 14 days. The lowest test temperature is assigned a test cycle of 28 days.

Usually, some of the original specimens are removed from the oven and subjected to the applicable tests only at the end of the third cycle. Assuming that these specimens do not show the end-of-life value, namely, 50% of original property retention, the test is to be repeated after every third cycle until 50% retention is reached.

When this 50% retention point is achieved, the groups of specimens that were placed in the oven at delayed times are removed from the oven and tested. A performance analysis provides a more accurate determination of the time to 50% property retention. It is important to note that at least one additional data point should be obtained that shows less than 50% of the initial property value to confirm the end-of-life value.

Reviewing End-of-Life Data. The five specimens tested per cycle are used to calculate an average value of the particular property for the test cycle and oven aging temperature. The average values are plotted on a graph in which the x-axis represents time, in hours, and the y-axis represents the property value. The best-fitting curve is drawn through the data, and the 50% property retention level is determined.

The test data can best be analyzed using a computer. In this case, a second- or third-order polynomial fit is attempted through the mean data. The best-fit plot then serves as a basis for calculating the 50% property retention level for that particular material property and oven temperature.

Table 7 Thermal and oxidative properties of selected polymers

Polymer	T_g (softens)		T_m (melts)		T_p (pyrolysis)		T_c (combustion)		ΔH		Limiting oxygen index
	°C	°F	°C	°F	°C	°F	°C	°F	kJ/g	10^3 Btu/lb	
Nylon 6	50	120	215	420	431	810	450	840	39	16.8	20.8
Nylon 6/6	50	120	265	510	403	755	530	990	32	13.8	20.8
Polyester	85	185	255	490	433	810	480	900	24	10.3	20.5
Acrylic	100	212	>220	>430	290	555	>250	>480	32	13.8	18.2
Polypropylene	−20	−4	165	330	469	875	550	1020	44	18.9	18.6
Modacrylic	<80	<175	>240	>465	273	525	690	1275	...	...	29.5
Polyvinyl chloride	<80	<175	>180	>355	>180	>355	450	840	21	9.0	38
Polyvinylidene chloride	−17	1	195	385	>220	>430	532	995	11	4.7	60
Polytetrafluoroethylene	126	260	>327	>620	400	750	560	1040	4	1.7	95
Aramid honeycomb core	275	525	375	705	410	770	>500	>930	30	12.9	29.4
Aramid	340	645	560	1040	>590	>1095	>550	>1020	...	...	29
Polybenzimidazole	>400	>750	...	...	>500	>930	>500	>930	...	...	41

Source: Ref 78

Four such plots and computer analyses are required for each thickness, material, and property tested. Figure 23 shows an example data set. After completing each set of aging tests, the dielectric strength test should be repeated at maximum and minimum operating temperatures, plus 20 °C (35 °F).

Determination of Lifeline. The use of the Arrhenius equation to represent the dependence of the life of the material on temperature is assumed as the functional basis for analyzing the life test data. The Arrhenius equation for a chemical reaction rate is given by:

$$K = A \exp\left(\frac{-E}{RT}\right) \quad \text{(Eq 6)}$$

where K is the specific reaction rate, E is the activation energy of the reaction, R is the gas constant, T is the absolute temperature (in degrees Kelvin), and A is the frequency factor (assumed constant).

An adaptation of Eq 6 to represent insulation life, y, which is assumed to be inversely proportional to the chemical reaction rate, leads to:

$$\log_{10}(\text{life}) = \log_{10} y$$

$$= \text{Constant} + \left(\frac{1}{2.303}\right)\left(\frac{E}{RT}\right) \quad \text{(Eq 7)}$$

Equation 2 has the algebraic form:

$$Y = a + bX \quad \text{(Eq 8)}$$

where Y is the $\log_{10} y$, X equals $1/T$, a is a constant, and b equals $E/2.303R$, another constant. The constants a and b can be estimated by fitting the experimental data in the form of $\log_{10} y$ (=Y) versus $1/T$ to Eq 8, This can be done by graphing the data on semilog paper and visually fitting the best straight line through the points. It can

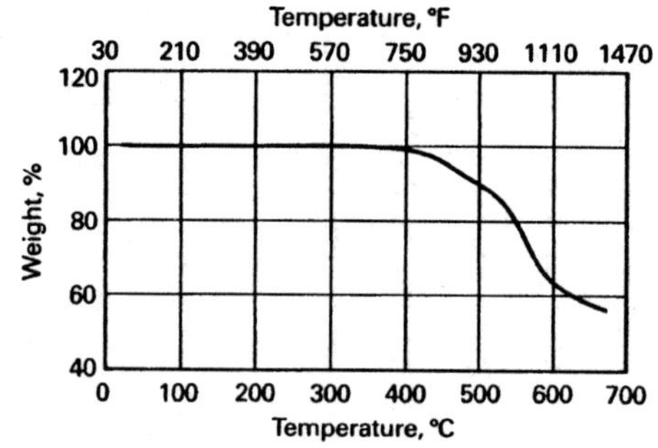

Fig. 15 Thermogravimetric analysis tracing of post-cured Ethacure 300/6-FDA (hexafluoropropane dianhydride) at 10 °C/min (18 °F/min) in nitrogen

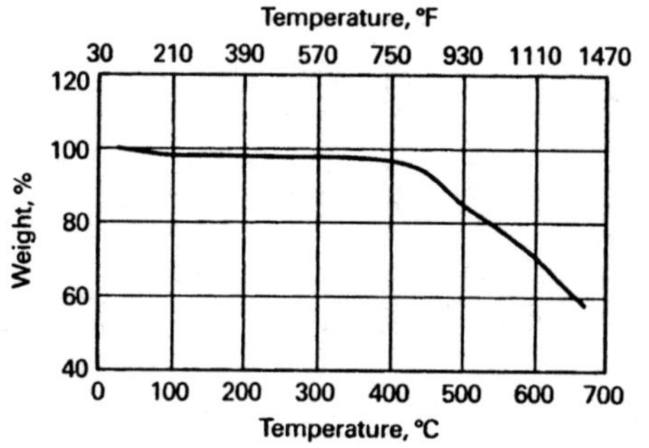

Fig. 16 Thermogravimetric analysis tracing of postcured Ethacure 300/PMDA (pyromellitic dianhydride) at 10 °C/min (18 °F/min) in nitrogen

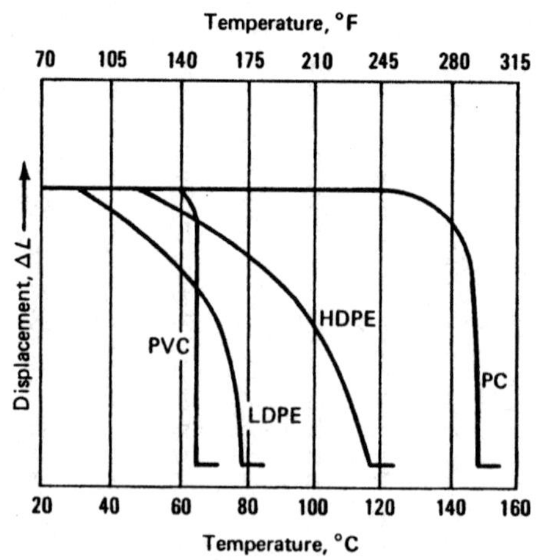

Fig. 17 Heat-deflection temperature per ASTM D 648 at 1.8 MPa (0.264 ksi) of thermoplastics according to thermomechanical analysis; 5 °C/min (9 °F/min) in flexure. PVC, polyvinyl chloride; LDPE, low-density polyethylene; HDPE, high-density polyethylene; PC, polycarbonate

Table 8 Summary of key polymers

Polymer tradename; type of material	Vendor	Chemical constituents(a)	Ref
Avimid N; polyimide	DuPont	95 MPDA:5 PPDA/6-FDA	80
Celazole PBI; polybenzimidazole	Hoescht-Celanese	Constituents can vary	81
Eymyd L-30N; polyimide	Ethyl Corporation	4-BDAF/PMDA	82
None; polyimide	None (experimental)	Ethacure 300/6-FDA	79
None; polyimide	None (experimental)	Ethacure 300/PMDA	79
None; polyimide	None (experimental)	Ethacure 300/BTDA	79

(a) MPDA, metaphenylene diamines; PPDA, paraphenylene diamine; 6-FDA, hexafluoropropane dianhydride; BTDA, benzophenonetetracarboxylic acid dianhydride; PMDA, pyromellitic dianhydride

Table 9 Thermal characterization results obtained on commercially available polyimides (PIs) and polybenzimidazoles (PBIs)

See Table 8 for specific polymer information. All measurements made on a DuPont 993 thermal analyzer equipped with appropriate differential scanning calorimetry (DSC) and thermogravimetric analysis (TGA) modules. Analyses performed in air at a 5 °C/min (9 °F/min) heatup

	First significant endotherm or T_g obtained on postcured film using DSC(a)		First significant weight-loss temperature in air using TGA			
			As-cast		Postcured(a)	
Polymer candidate	°C	°F	°C	°F	°C	°F
Avimid N	400	752	440	824	450	842
PBI	360	680	400	752	430	806
Eymyd L-30N	410	770	430	806	440	824

(a) The Avimid N, PBI, and Eymyd L-30N film samples were postcured for 2 h in an air-circulating oven at 370 °C (700 °F).

Table 10 Thermal characterization results obtained on experimental polyimide polymers

All measurements made on a DuPont 993 thermal analyzer equipped with appropriate differential scanning calorimetry (DSC) and thermogravimetric analysis (TGA) modules

	First significant endotherm or T_g in air obtained on postcured film DSC(a)		First significant weight-loss temperature in nitrogen obtained on postcured film using TGA(a)	
Polymer candidate	°C	°F	°C	°F
Ethacure 300/PMDA	335	635	390	734
Ethacure 300/6-FDA	314	597	390	734
Ethacure 300/BTDA	306	583	340	644

(a) See Table 8 for polymer information.

also be done more precisely by using the method of least squares. Figure 24 shows an example of a lifeline.

Determining the RTI. The lifelines of the control and candidate materials are plotted on the same graph. The time that corresponds to the already assigned RTI for the control material is determined for that particular property and test thickness. Usually, this time is approximately 20,000 to 100,000 h. This time then becomes the correlation time and is used to determine the corresponding temperature on the lifeline for the candidate material. Figure 25 shows an example of RTI determination. Temperature ratings are assigned in 5 °C (9 °F) increments up to 130 °C (265 °F), 10 °C (18 °F) increments up to 180 °C (355 °F) (except for 155 °C, or 310 °F), and 20 °C (35 °F) increments for greater than 180 °C (355 °F).

Thermal Properties of Thermoplastics

Representative examples of different types of engineering thermoplastics are discussed in this section primarily in terms of structure and thermal properties. The properties of thermoplastic polymers, with emphasis on their thermal properties, were reviewed by Shalaby and Bair (Ref 2, 101).

Polyethylene is produced in four principal grades: high density (HDPE), low density (LDPE), linear low density (LLDPE), and ultra-high molecular weight (UHMWPE). Structurally, these grades differ in the degree and type of branching on the main chain and in overall molecular weight. At a particular molecular weight, branching leads to a decrease in PE T_m. Therefore, UHMWPE, with almost perfect chains, displays the highest T_m, which decreases progressively from HDPE to LLDPE to LDPE.

The orientation of high-molecular, linear chains can lead to an exceptionally high T_m. Thus, gel-spun UHMWPE may exhibit a T_m of about 150 °C (300 °F) and a crystallinity exceeding 70%. On the other extreme, a low-molecular-weight LDPE with randomly displaced branches may have a T_m of about 100 °C (212 °F) and crystallinity of less than 50%. Other grades of PE melt between these two extremes, primarily depending on branching and molecular weight.

Polypropylene. Engineering thermoplastic grades of PP are primarily made of stereoregular, isotactic chains that crystallize in the helical conformation. Small amounts of atactic segments are usually present in all commercially available PP. Thermal and physical properties are affected by the weight fraction of the atactic components, which is usually about 5% or less. It has been shown by DSC that the amorphous atactic and isotactic PP display T_g values of –6 and –18 °C (21 and 0 °F), respectively (Ref 5).

Polybutylene (PB) homopolymer is made by the polymerization of 1-butene to chains that are primarily isotactic, like those of PP. However, PB differs from PP in that the solid polymer exists in four crystalline modifications, the most stable of which melts at 125 to 130 °C (260 to 265 °F). The other forms melt below 125 °C (260 °F). The thermo-oxidative stability of PB is similar to that of PP. On the other hand, filled PB displays higher low-temperature impact strength than does PP.

Poly (4-methylpentene) (PMP). Engineering thermoplastic grades of PMP are largely based on isotactic chains, which can pack into a three-dimensional structure to provide materials having 40 to 65% crystallinity. The higher degree of crystallinity is usually achieved by annealing shaped articles. The polymer is characterized by a T_g of 30 to 40 °C (85 to 105 °F) and a T_m of 245 °C (475 °F). Its high T_m and high transparency distinguish PMP from PP and PB, which share some of its other physical and chemical characteristics.

Polystyrene. Commercial grades of PS homopolymer are made by free-radical polymerization to produce an amorphous material with atactic chains. The polar, bulky phenyl groups of PS chains are responsible for stiffness, restricted mobility, and, hence, high T_g of about 100 °C (212 °F). The high T_g of PS makes it one of the most important engineering plastics.

Because of its amorphous nature, PS can be easily melt processed at temperatures well below its ceiling temperature. Extrusion and injection molding of PS can be achieved, typically at 180 to 230 °C (360 to 450 °F) and 180 to 260 °C (360 to 500 °F), respectively. The low specific heat (1170 J/kg · K, or 0.28 cal/g · °C) and coefficient of linear thermal expansion (6 to 8×10^{-5}/K) make PS one of the most useful injection-molding resins.

Typically, PS exhibits a low mold shrinkage of 2 to 6×10^{-3} mm/mm, which is much lower than those of crystalline polyolefins such as PP and PE. A key drawback of unfilled, molded PS articles is their low impact strength, a situation that may be corrected by mixing with rubber-based impact modifiers.

Styrene copolymers usually feature some correction of undesirable PS properties, while leaving its desirable ones practically intact. Most of the differences between PS and its copolymers are pertinent to basic changes in the thermal properties of the homopolymer. Thus, to increase the impact strength of PS, copolymers of styrene with variable amounts of butadiene are produced to have (among other properties) a lower T_g than the homopolymer. Copolymers of styrene with acrylonitrile are known to have better chemical and solvent resistance than PS.

The acrylonitrile-butadiene-styrene (ABS) copolymers are particularly suitable for applications requiring heat resistance, flame resistance, a high HDT (about 110 °C, or 230 °F), and a high degree of transparency. ABS copolymers are used as high-performance electroplating and structural-foam grades.

Polyvinyl Chloride and Related Polymers. This family of vinyl polymers includes PVC, polyvinylidene chloride (PVDC), copolymers of vinyl and vinylidene chlorides (VC-

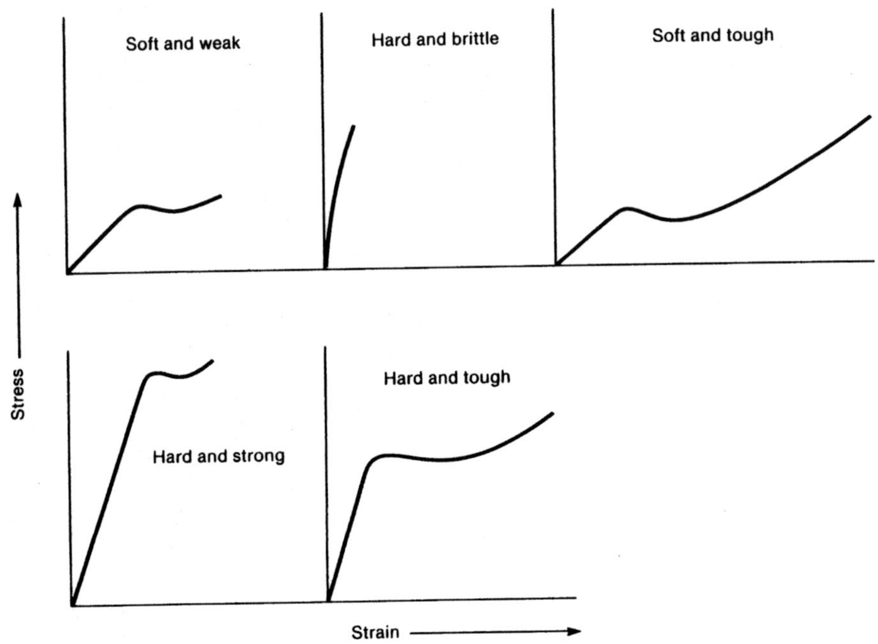

Fig. 18 Tensile stress-strain curves for several types of polymeric materials. Source: Ref 83

VDC), vinyl chloride-vinyl acetate copolymers (VC-VA), polyvinyl formal (PVFM), and polyvinyl butyral (PVB).

Polyvinyl chloride, as a commercial grade polymer, is largely an atactic, amorphous material with a T_g of 75 to 105 °C (165 to 220 °F). The high T_g of PVC is associated with the polarity and bulkiness of the chlorogroup. In the liquid state, the chain polarity becomes responsible for the high melt viscosity of the polymer. This, in turn, makes it difficult to melt process PVC without causing thermally induced dehydrohalogenation. Thus, PVC is usually compounded with stabilizers to minimize its thermal dehydrogenation, or with plasticizers, to reduce its melt viscosity and increase compliance of certain shaped articles. Copolymerization of vinyl chloride with a suitable comonomer to achieve internal plasticization of the PVC chain resulted in a family of commercially viable copolymers. Because of its propensity to generate hydrogen chloride, PVC is known for its exceptional flame resistance. Thermal properties of various vinyl polymers are compared in Table 15.

Fluoropolymers. The most important thermoplastic members of the fluoropolymer family are polytetrafluoroethylene (PTFE), poly-chlorotrifluoroethylene (PC-TFE), poly(ethylene-co-tetrafluoroethylene) (PE-TFE), poly(ethylene-co-chlorotrifluoroethylene) (PE-CTFE), and polyvinylidene fluoride (PVDF). Because of their fluorinated chains, these polymers exhibit excellent thermal stability, flame resistance, low conductance, chemical and solvent resistance, high surface and volume resistivity, and water repellency. The small size of the fluoro-groups and high polarity of the C–F bond permit tight packing of the polymer chains in the solid state. Thus, fluoropolymers generally exhibit high T_g. Crystalline members of the fluoropolymer family also melt at relatively higher temperatures, compared to other addition-type thermoplastics. The degree of crystallinity in these polymers approaches 75%, as in the case of PTFE. A comparison of thermal properties is given in Table 16.

With the exception of PTFE, the fluoropolymers described previously can easily be melt processed using conventional techniques. They display excellent melt stability, although the generation of trace amounts of the corrosive hydrofluoric acid may be encountered at elevated temperatures. Because of its high T_m and melt viscosity, PTFE is usually fabricated by sintering (cold pressing) the virgin polymer particles at 360 to 380 °C (680 to 715 °F). Thus, granular PTFE is molded into billets, sheets, and rings through preforming, sintering, and cooling. Rods and thick-wall tubes are made by ram extrusion.

Commercial grades of polymethyl methacrylate (PMMA) are amorphous materials that exhibit a T_g of about 105 °C (220 °F), using DSC. PMMA displays excellent clarity (92% transmission) and the desirable properties of a useful ET. These include a density of 1.18 to 1.19 g/cm^3, linear coefficient of thermal expansion of 6×10^{-5}/K at −30 to 30 °C (−22 to 85 °F), thermal conductivity of 0.20 W/m · K (1.36 Btu · in./h · ft^2 · °F), and specific heat of 1470 J/kg · K (0.35 cal/g · °C). However, the ceiling temperature of PMMA is relatively low, compared to other engineering thermoplastics, and care must be taken to avoid excessive processing temperatures. Despite its low ceiling temperature, PMMA offers low flame resistance.

Nitrile resins (NRs) are copolymers based on 70% acrylonitrile, 20 to 30% styrene (or methylmethacrylate, MMA), and 0 to 10% butadiene. The NRs are amorphous, and their T_g depend on their compositions. Most commercial grades form viscous liquids above 200 °C (390 °F) and can be processed by conventional melt processing methods between 200 and 205 °C (390 and 400 °F). Most NR products are characterized by a high degree of toughness and excellent barrier properties, which are attributed to the butadiene and nitrile components, respec-

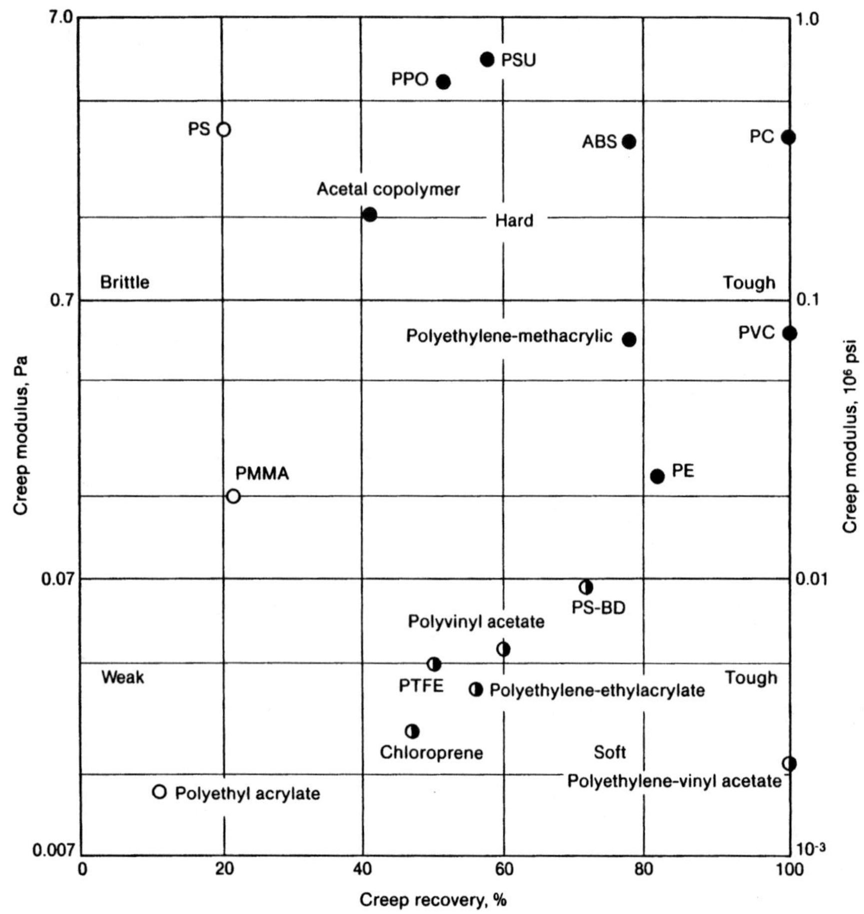

Fig. 19 Properties of commercial polymers according to thermomechanical analysis. See "Abbreviations and Symbols" in this book for definitions of abbreviations. Source: Ref 84

Table 11 Heat-deflection temperature versus glass content for selected engineering plastics

Material	Glass content							
	0%		15%		30%		40%	
	°C	°F	°C	°F	°C	°F	°C	°F
PBT, crystalline	54	130	190	375	207	405	204	400
PA, crystalline	90	195	243	470	249	480	249	480
PC, amorphous	129	265	146	295	146	295	146	295

PBT, polybutylene terephthalate; PA, polyamide; PC, polycarbonate tively. Typically, molded articles made at 455 kPa (66 psi) from a styrene or MMA terpolymer display an HDT of 75 to 77 °C (165 to 170 °F), while those made from MMA terpolymer have an HDT of 80 to 95 °C (180 to 200 °F). Film made of NR may have an oxygen or carbon dioxide permeability of 0.8 and 1.6 $cm^3/in.^2$/day at 50% relative humidity and 73 °C (165 °F). Because of their high nitrile content, NRs can undergo cycloaddition reaction at high temperatures, leading to partially aromatic segments and, hence, improved thermal stability.

Modacrylics are polymers made by the copolymerization of 25 to 85% acrylonitrile and 75 to 15% of a second comonomer. The most common type of modacrylic is based on acrylonitrile and vinyl chloride and is used primarily for fiber production by melt or solution spinning. A typical modacrylic fiber shows no distinct T_m because it is essentially amorphous and softens at 190 to 240 °C (375 to 465 °F).

Acetal Polymers (ACs). Structurally pure ACs are made of $-(CH_2O)-$ repeat units and (OH) end groups and undergo thermal depolymerization by an unzipping mechanism. Thus, commercial grades of AC are stabilized by end capping the (OH) groups or by incorporating a small fraction of ethylene oxide units in the polymer chain. These polymers are crystalline, and their molded articles are usually distinguished by their high rigidity, dimensional stability, and fatigue endurance. This is not surprising, because ACs are known to have a T_m of about 163 °C (325 °F), specific heat of 1470 J/kg · K (0.35 cal/g · °C), coefficient of thermal expansion of 8×10^{-5}/K, and HDT of 170 °C (338 °F) at 455 kPa (66 psi). Although the polymer chains are highly oxygenated, AC retains only 0.8% of water at equilibrium because of its high crystallinity. The intrinsic thermal instability of AC makes its flammability properties unsatisfactory for certain uses.

Polyamides. Nylon 6 and nylon 6/6, which are made by ring-opening and step-growth polymerizations, respectively, are by far the most important PAs used as engineering thermoplastics. However, applications based on nylon 12 and three step-growth polymers, namely nylon 6/10, nylon 6/I (from hexamethylene diamine and isophthalic acid), and nylon MXD/6 (from *m*-xylene diamine and adipic acid), are increasing steadily. The excellent properties of PAs are attributed to their high crystallinity, high melting temperatures, moderate T_g, slow melt viscosities, moderate to high thermal stability, excellent frictional properties, and resistance to solvents. The thermal properties of PAs have been discussed frequently in the literature (Ref 2, 102–104). Key thermal properties are given in Table 17.

Nylon 6 has less thermal stability than the step-growth nylons because it has the tendency to undergo thermal depolymerization by chain unzipping. Nylon 12 is more stable thermally than nylon 6 because of a more difficult generation of a 13-member ring by chain unzipping. Accordingly, nylon 6 is the least resistant of these polymers in terms of flame resistance, although nylons are generally characterized among the fair-to-poor performers.

Because of their amide-bearing chains, some of the properties of nylons are sensitivity to moisture content, or the relative humidity, of the surrounding environment. The primary effect of water on nylon properties is manifested through depression of the T_g. Lowering the T_g by plasticization with water is usually reflected in some loss of mechanical properties and increase in toughness. Nylons are particularly useful in the production of self-lubricating bearings, films, and textile fibers. Because of their low melt viscosity and polarity, they are well suited for compounding with fillers to form several types of structural composites.

Polyesters. As engineering thermoplastics, polybutylene terephthalate (PBT) and polyethylene terephthalate (PET) are the most important polyesters. Both are made by step-growth polymerization and are used extensively in the plastics and fibers industries. Polycyclohexane dimethylene terephthalate (PCHDMT), a step-growth polymer, is developed primarily for use in the fibers industry. Polycyclohexane dimethylene terephthalate was formerly of limited use as a molding resin due to its high T_m, but this problem has been addressed through copolymerization, which produces more melt-processible products. A fourth polyester that is made by ring-opening polymerization is polycaprolactone (PCL). Because of its low T_m of 60 to 64 °C (140 to 150 °F), PCL has not been used to any great extent as a primary engineering thermoplastic. However, it is used as an intermediate in the polyurethane (PUR) industry.

With the exception of PCL, these polyesters display sufficient thermal stability to make them quite useful for melt processing into several types of shaped articles. Their hydrophobic nature and high degree of crystallinity make polyesters less sensitive to hydrolytic degradation than might be anticipated on the basis of their chemistry. Key thermal and related properties are given in Table 18.

In terms of flame resistance, polyesters can be categorized as poor to fair. Additives, particularly those containing phosphorus, have been used successfully to reduce their flammability (Ref 105).

Polycarbonates are primarily based on the carbonic acid esters of bisphenol A (BPA). For special applications, small amounts of polyhy-

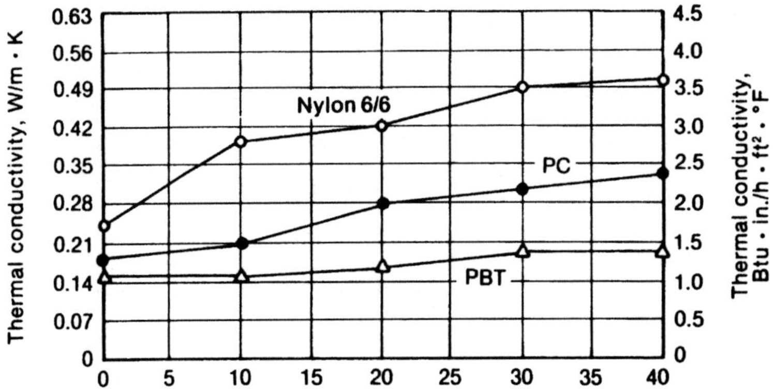

Fig. 20 Effect of glass addition on thermal conductivity. PBT, polybutylene terephthalate; PC, polycarbonate

Table 12 Thermal conductivities of polymers and other materials

Material	Thermal conductivity at 20 °C (68 °F) W/m² · K	Btu · in./s · ft² · °F
Polymers		
Polyethylene		
Low density	35	2.5
Medium density	35–42	2.5–2.9
High density	46–52	3.2–3.6
Polyvinyl chloride	13–29	0.9–2.0
Polystyrene	10–14	0.7–1.0
Epoxy resin (Shell 828, diethanolamine), filled		
20 wt% mica	18	1.3
30 wt% mica	24	1.7
40 wt% mica	33	2.3
50 wt% mica	39	2.7
Polyurethane		
20% closed cell	3.5	0.2
90% closed cell	1.7	0.1
Acetal copolymer	20	0.001
Polypropylene	20	0.001
Expanded polystyrene	3	0.0002
Other materials		
Copper	40,000	2.8
Aluminum	24,000	1.7
Steel	5000	0.35
Granite	350	0.02
Crown glass (75 wt% silica)	90	0.006

Source: Ref 4

dric phenols are mixed with BPA. Because of their highly aromatic nature, PCs are characterized by a high degree of hydrophobicity (unfilled PC typically absorbs 0.15 to 0.18% water as a 3.2 mm, or 1/8 in., thick bar for 24 h), as well as high T_g and melt strength. A typical PC, such as poly[2,2-bis-(4-phenylene)propane carbonate] has a T_g of 150 °C (300 °F) and a T_m of 220 to 230 °C (430 to 445 °F). Although PC can be obtained in a crystalline form (by annealing at 180 °C, or 355 °F, for 24 h), the relatively small difference between high T_g and T_m provides a narrow crystallization window and a lower tendency to crystallize under usual processing conditions, compared to other engineering thermoplastics. High melt strength, high T_g, and low tendency to crystallize make PC useful in blow-molding applications. High thermal transition temperatures, the intrinsic thermal stability of the polymer chain, and polymer hydrophobicity make PC useful in a broad range of applications.

Some of the key properties that distinguish PC as an exceptional engineering thermoplastic are:

- High impact strength, which may be related to ability of the polymer to efficiently absorb mechanical stresses below the T_g
- Dimensional stability over a wide range of temperatures due to high T_g and modulus, thereby permitting use at –50 to 130 °C (–60 to 265 °F) and 1.82 MPa (0.264 ksi) (with a typical heat-deflection temperature of 130 to 140 °C, or 265 to 285 °F)
- Low mold shrinkage and creep resistance, which consistently allow precision molding to a tolerance of 0.002 mm/mm
- Ease of conversion to transparent articles under conventional molding conditions because of the tendency to remain practically amorphous after melt processing

However, PC is subject to occasional solvent stress-crazing problems.

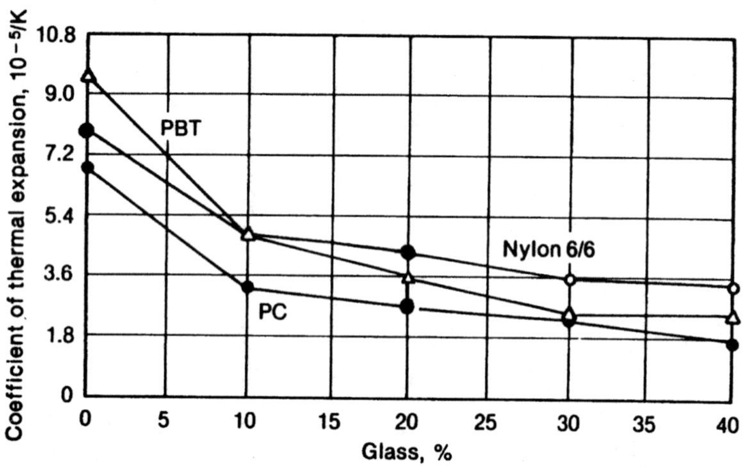

Fig. 21 Effect of glass addition on coefficient of thermal expansion. PBT, polybutylene terephthalate; PC, polycarbonate

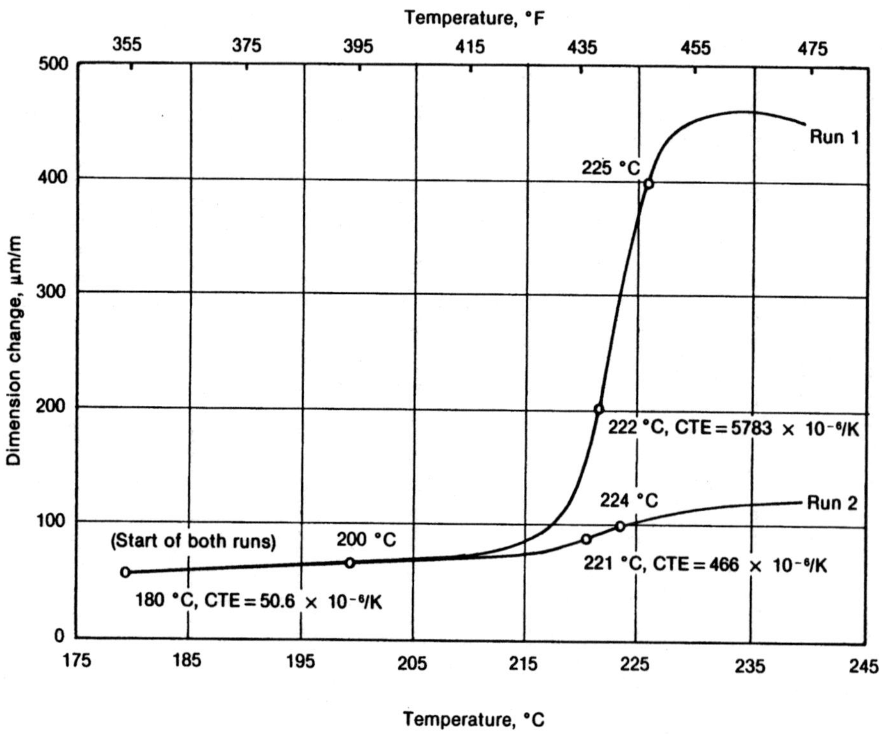

Fig. 22 Thermal analysis of oriented plastic. CTE, coefficient of thermal expansion

Table 13 Coefficients of linear thermal expansion for various polymers and other materials

Material	Coefficient of linear thermal expansion, 10^{-5}/K	Mold shrinkage, μm/m
Polyethylene		
Low density	10–20	20–40
High density	10–20	20–40
Polypropylene	2–20	10–30
Nylon 6/6	10	20
Polystyrene	6–8	2–6
Polycarbonate	7	–7
Polybutylene terephthalate		
Unfilled	6–10	9–20
Filled, glass fiber	3	2–8
Epoxy resin		
Unfilled	4–7	...
Filled, mica	2–6	2
Zinc	3.5	...
Copper	1.7	...
Silver	1.9	...

Table 14 Specific heats of various materials

Material	Specific heat at room temperature	
	J/kg · K	Cal/g · °C
Polyethylene		
Low density	2300	0.55
High density	1850	0.44
Polypropylene		
Atactic amorphous	2350	0.56
Crystalline isotactic	1800	0.43
Nylon 6/6	1670	0.40
Polystyrene	1170	0.28
Zinc	380	0.09
Copper	380	0.09
Silver	250	0.06

Substantial improvement of certain mechanical properties can be achieved by filling PC with 10 to 40% glass fiber. Because of their aromatic components and thermal stability, filled and unfilled PCs are relatively more flame resistant than most halogen-free thermoplastic resins.

Aromatic Ethers. Polyaryl ether and methyl-substituted phenylene oxide resins are the commercial forms of the aromatic ethers family of polymers. Although the latter resins are proprietary compositions, they are known to be based on mixtures of aromatic polyethers and other thermoplastic resins. The aromatic polyether chains are made by the oxidative coupling of phenolic monomers, such as dimethylphenol. Because of the aromatic and steric requirements about their rigid chains, aromatic polyethers are hydrophobic materials, are essentially amorphous, and undergo glass transition above 100 °C (212 °F), depending on the chain composition. They have excellent thermo-oxidative stability and are more flame resistant than most of the halogen-free thermoplastic resins. Because of their high T_g, shaped articles made of aromatic polyethers using conventional melt-processing techniques usually display excellent dimensional stability and high resistance to creep.

Because of their good dielectric properties, high thermo-oxidative stability, and low tendency to absorb water, this class of aromatic polymers is widely used in electrical applications. A typical commercial grade of polyaryl ethers displays a T_g of about 160 °C (320 °F), an HDT of 150 °C (300 °F) (at 1.82 MPa, or 0.264 ksi), a coefficient of thermal expansion of 3.6×10^{-3}/K, and a water absorption of 0.25% after 24 h on a 3.2 mm (⅛ in.) thick specimen. Modified phenylene oxide resins exhibit some changes in these properties as a result of compounding with more traditional thermoplastic resins. Additional changes can be observed upon filling these with glass.

Polyetheretherketone (PEEK) and related polyaromatic ketones (PAK), unlike other aromatic polymers, are crystalline. PEEK, which is commercially available; can be made from Ph–O–Ph–O–Ph–COCl by the Friedel-Crafts reaction. A sample of PEEK having a molecular weight of 2.4×10^5 dalton was reported to have a T_g of about 144 °C (290 °F) and T_m of about 342 °C (650 °F) (Ref 105). Although PEEK has a high T_m, it is easily melt processible in the vicinity of 375 °C (710 °F) and thus is used as a thermoplastic matrix for fiber-reinforced composites. It has been noted that the development of ultimate properties may be influenced by the rate of cooling from the melt through the glass

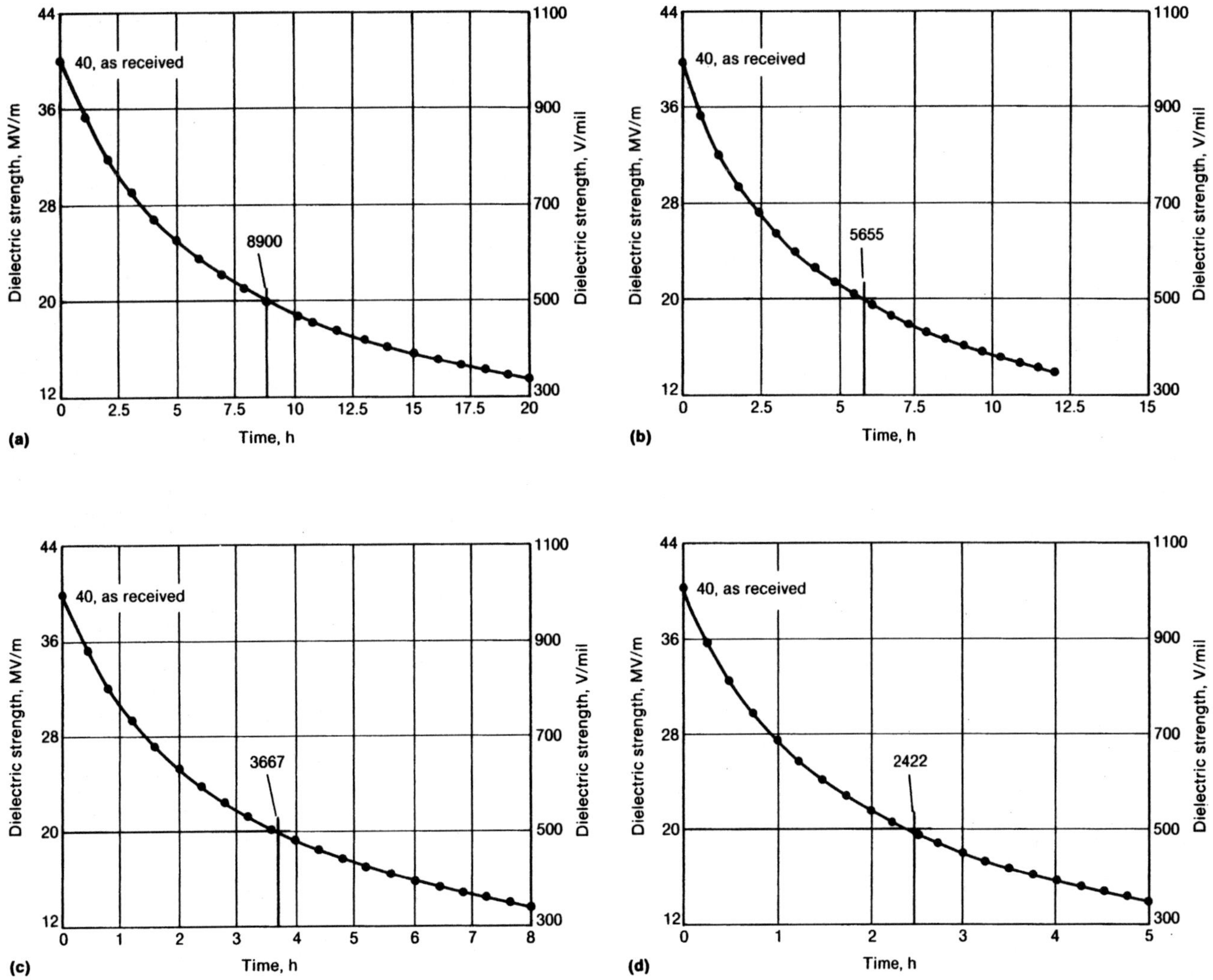

Fig. 23 50% determination of 0.80 mm ($^1/_{32}$ in.) specimen aged at four temperatures. (a) 160 °C (320 °F). (b) 170 °C (340 °F). (c) 180 °C (355 °F). (d) 190 °C (375 °F)

transition into the solid state. Thermo-oxidative decomposition studies of PEEK indicated that prolonged heating (1 to 10 h) at 375 °C (710 °F) results in about 1 to 10% weight loss (Ref 106). This weight loss is associated with the formation of benzoquinone. Polyetheretherketone has a flame resistance comparable to that of polyarylether.

In a review by Mullins and Woo (Ref 107), the synthesis and properties of different types of PAK were reported. A variety of high-molecular-weight polymers having T_g values of 151 to 216 °C (300 to 420 °F) and T_m values of 271 to 486 °C (520 to 905 °F) were discussed.

Aromatic Sulfones. The chains of these polymers consist of partially or fully aromatic building blocks interlinked with sulfonyl groups. The three major commercial forms are polysulfone (PSU) with isopropylidene biphenyl between the sulfonyl groups, polyether sulfone (PESU) having sulfonyl and ether groups interlinking *p*-phenylene groups, and polyphenylene sulfone (PPSU) consisting of biphenylene groups interconnected with ether and sulfonyl groups. Because of the steric requirements about the main chains of these polymers and the inherent stiffness of these highly aromatic structures, this class of polymers is noted for high T_g, lack of crystallinity, high thermal and thermo-oxidative stabilities, good flame resistance, excellent dimensional stability, and good creep resistance, impact strength, and hydrolytic stability. Most of the desirable properties of the aromatic sulfone polymers are associated with their high T_g. Nevertheless, these polymers can be easily processed at 340 to 395 °C (640 to 740 °F) using conventional molding equipment. Further modification can be achieved by compounding with glass fibers. Table 19 gives property values.

Cellulosics are derivatives of cellulose that are made by alkylating or acylating the natural polymer to render it thermoplastic. Most of the commercially available thermoplastic, cellulose derivatives have less desirable properties as molding or extrusion resins, compared to the majority of synthetic polymers discussed in this section.

The major thermoplastic cellulosics are ethyl cellulose (EC), cellulose acetate (CA), cellulose acetate-butyrate (CAB), cellulose acetate-propionate (CAP), and cellulose nitrate (CN). The latter polymer, CN, has limited use as a compression molding resin (processed at 85 to 120 °C, or 185 to 250 °F) because it is a potential explosive. The rest of the cellulosics are crystalline polymers that can be molded (by compression or injection) or extruded (usually as sheets) at temperatures close to their T_m to avoid excessive thermal decomposition. Because of their tendency to thermally degrade to highly flammable gases, their flame resistance can be rated as poor. Of all the cellulosics, EC, CAB, and CAP are favored as thermoplastic resins because of their moderate T_m and hence better melt processibility compared to CA and CN. Ethyl cellulose is most widely used to produce molded articles with high impact strength at low temperature. Some thermal properties are described in Table 20.

Although no accurate values for the T_g of cellulosics could be found, the heat-deflection temperature data in Table 20 may be used to predict moderate to high T_g for these polymers. This is not surprising in view of their rigid, ring-containing main chains. Because of their highly oxygenated chains, their water absorption is relatively higher than that of most synthetic thermoplastics.

Thermoplastic elastomers (TEs) and elastoplastics are copolymers that share common properties with elastomers and traditional thermoplastics. The discussion here is limited to materials whose properties approach those of engineering thermoplastics. The chains of typical TE and elastoplastic materials consist of hard and soft components. Chains of elastoplastic polymers are predominantly made of hard components. The polymers behave like compliant, or toughened, thermoplastics with limited elastomeric properties. If the chains contain a high fraction of soft components, or segments, the polymers display elastomeric properties without having covalent cross links. This is because the balance of the polymer, consisting of hard components, will either associate or aggregate intermolecularly and provide quasi-cross-links under ambient conditions. Above certain temperatures, the aggregates dissociate,

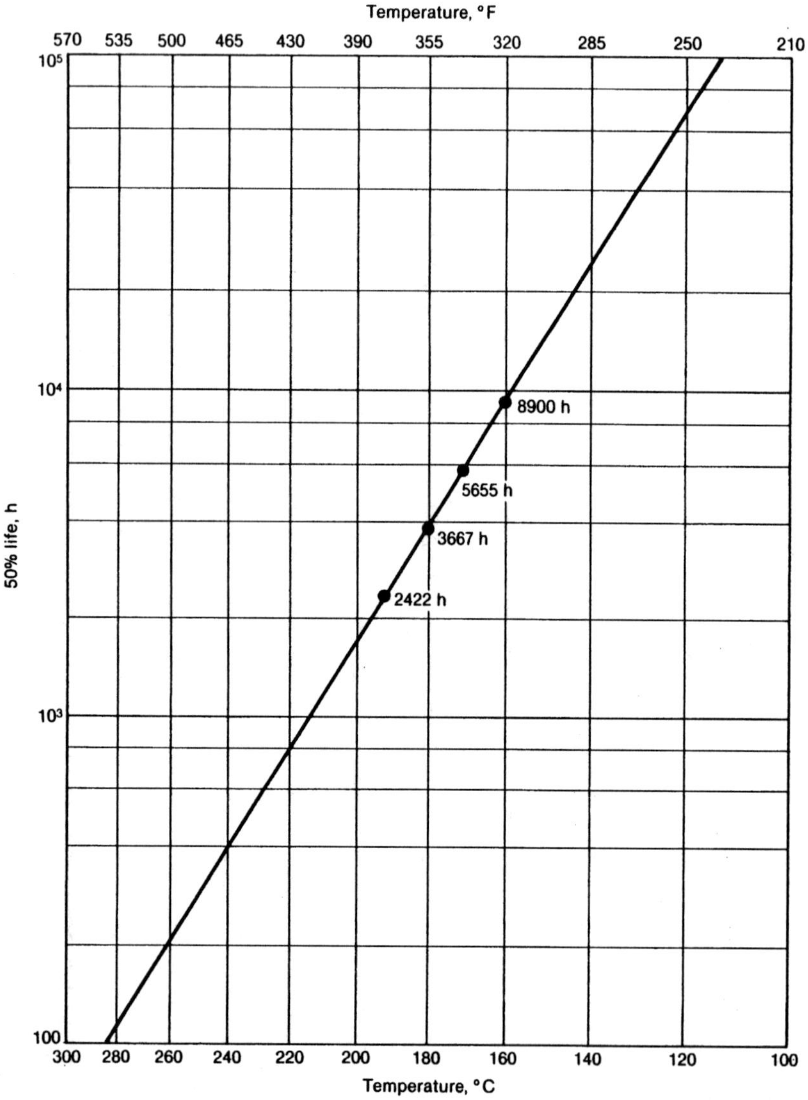

Fig. 24 Lifeline of material XYZ

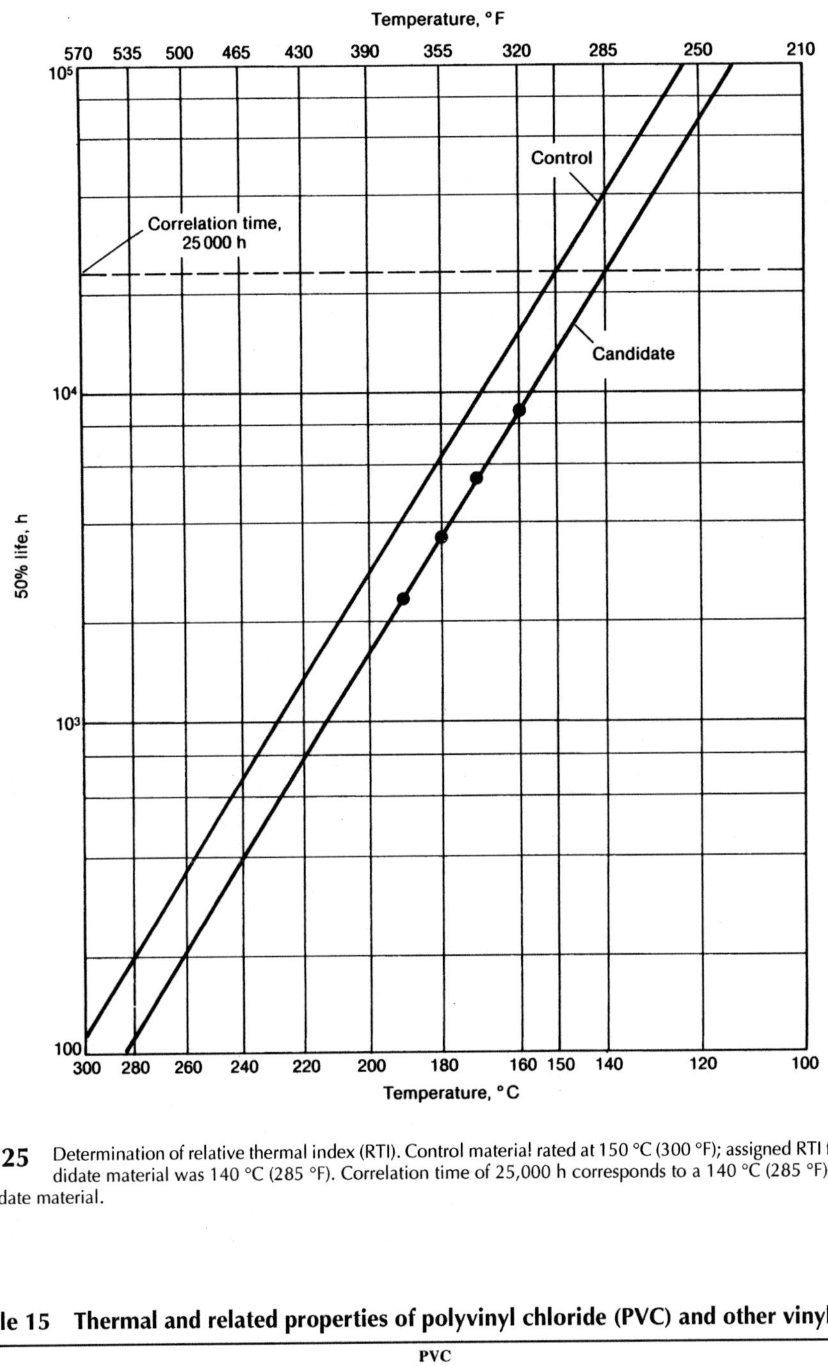

Fig. 25 Determination of relative thermal index (RTI). Control material rated at 150 °C (300 °F); assigned RTI for candidate material was 140 °C (285 °F). Correlation time of 25,000 h corresponds to a 140 °C (285 °F) RTI for candidate material.

and the polymer can undergo unidirectional viscous flow to be processed like conventional thermoplastics (Ref 108). When the hard-soft ratio (H/S) of any member of this class of polymers is low (usually <0.5), the product can be described as a TE. A polymer with high H/S (normally ≥0.5) behaves like an elastoplastic. The hard components are typically made of crystallizable, high-T_g and/or polar (capable of association) short or long segments, blocks, or grafts. Long segments, blocks, or grafts of highly flexible sequences constitute the soft component of TE and elastoplastic (Ref 108).

Major elastoplastics and the hard members of the TEs include ethylene-propylene block copolymers (EP-BL), based primarily on PP; hard styrene-butadiene block copolymers (SB-BL), with a major PS component; hard, hydrogenated styrene-butadiene block copolymers (H-SB-BL), having a major PS fraction; segmented copolymers of poly(polyoxybutylene terephthalate) and PBT (POBT-PBT); segmented copolymers of polyoxybutylene glycol and nylon 12 (POB-N) interconnected with amino-carboxylate groups (Ref 111); hard polyether urethanes made from polyoxybutylene glycol and methylene diphenylisocyanate (MDI); and hard polyester urethanes based on polycaprolactone and MDI. Some of the properties of these copolymers, including key thermal data, are given in Table 21. The data indicate the great flexibility in tailoring segmented or block copolymers to attain a broad range of thermal properties and hydrophilicity.

The key commercial engineering thermoplastic resins include triblock H-SB-BL (with PS terminal blocks), POBT-PBT with over 50% PBT hard segments, and POB-N with more than 50% nylon 12 hard segments. There have also been a number of elastoplastic copolyesters noted in the patent literature (Ref 109–112). In addition to their use as molding resins, a few grades of POBT-PBT and POB-N can be converted to strong fibers under conventional extrusion conditions. In terms of thermal and thermo-

Table 15 Thermal and related properties of polyvinyl chloride (PVC) and other vinyl polymers

Property	PVC: Rigid	PVC: Plasticized	PVC: 30% glass filled	Chlorinated PVC	PVDC	PVFM	PVB
T_g, °C(°F)	75–105 (170–220)	(a)	75–105(b) (170–220)	110 (230)	...	105 (220)	49 (120)
T_m, °C(°F)	(c)	...	...	(c)	210 (410)	(c)	(c)
Molding temperature, °C(°F)							
Compression	140–205 (285–400)	140–195 (285–385)	...	170–205 (350–400)	104–175 (220–350)	150–175 (300–350)	140–160 (280–320)
Injection	150–215 (300–415)	160–195 (320–385)	130–210 (270–405)	160–225 (325–440)	150–205 (300–400)	150–205 (300–400)	120–170 (250–340)
Heat deflection temperature, at 1.82 MPa (0.264 ksi), °C(°F)	140–170 (285–340)	...	155 (310)	202–234 (395–450)	130–150 (265–300)	150–170 (300–340)	...
Water absorption, 24 h at 3.2 mm (⅛ in.) thick, %	0.04–4.0	0.15–0.75	0.008	0.02–0.15	0.1	0.5–3.0	1.0–2.0

PVDC, polyvinylidene chloride; PVFM, polyvinyl formal; PVB, polyvinyl butyral. (a) Variable: can be lower than 75–105 °C (165–220 °F) depending on type and concentration of plasticizer. (b) Irrespective of the filler. (c) Amorphous oxidative stability, copolymers containing polyether or unsaturated moieties have poor performance. For example, their flame resistance characteristics are usually unsatisfactory. The other types of copolymers have fair stability and poor to fair flame resistance.

Table 16 Thermal properties of typical thermoplastic fluoropolymers

Property	Material				
	PVDF	PTFE	PC-TFE	1/1 PE-TFE	1/1 PE-CTFE
Density, g/cm³	1.78	2.2	2.13	1.70	1.68
T_g, °C (°F)	−45 (−50)	127 (260)	45 (115)	...	...
T_m, °C (°F)	170 (340)	327 (620)	218 (425)	270 (520)	245 (475)
Thermal conductivity, at 20–30 °C (68–95 °F) W/m² · K (Btu · in./s · ft² · °F)	...	0.18 (2.5)	0.18 (2.5)	...	...
Specific heat, at 40 °C (105 °F), J/kg · K (cal/g · °C)	...	960 (0.23)	...	...	...
Coefficient of thermal expansion, 10^{-5}/K	...	5.5	7	...	...

PVDF, polyvinylidene fluoride; PTFE, polytetrafluoroethylene; PC-TFE, polychlorotrifluoroethylene; PE-TFE, poly(ethylene-co-tetrafluoroethylene); PE-CTFE, poly(ethylene-co-chlorotrifluoroethylene); 1/1 mole ratio

Table 17 Thermal properties of representative polyamides

Property	Nylon 6	Nylon 12	Nylon 6/6	Nylon 6/10
T_g, °C (°F)	50–70 (120–160)(a)	46 (115)	57–80 (135–175)(a)	50 (120)
T_m, °C (°F)	225 (440)	180 (360)	265 (510)	219 (425)
Melt-processing temperature, °C (°F)	225–290 (440–550)	180–270 (360–525)	270–325 (520–620)	230–290 (450–550)
Specific heat, J/kg · K (cal/g · K)	1670 (0.4)	1260 (0.3)	1670 (0.4)	1670 (0.4)
Coefficient of thermal expansion, 10^{-5}/K	8.3	10.0	8.0	9.0
Heat-deflection temperature, at 455 kPa (66 psi), °C (°F)	185 (365)	145 (293)	190 (374)	165 (330)
Water absorption, 24 h, 3.2 mm (⅛ in.), %	1.3–1.9	0.25	1.5	0.4

(a) Observed range is attributed to variable sample water content: T_g increases with dryness.

Table 18 Thermal and related properties of polyester films

Property	PCL	PBT	PET	PCHDMT(a)
Density, g/cm³	...	1.31–1.38	1.38–1.41	1.23
T_g, amorphous, °C (°F)	40 (105)	60–70 (140–160)	78–80 (170–175)	85–95 (185–205)
T_m, °C (°F)	64–70 (150–160)	225–235 (440–455)	260–265 (500–510)	293 (560)(a)
T_c, °C (°F)	...	...	125–180 (260–355)	...
Heat-deflection temperature, at 345 kPa (50 psi), °C (°F)	...	...	158 (315)	165 (330)
Water absorption, at 25 °C (77 °F), 24 h immersion, %	...	...	0.55	0.33

PCL, polycaprolactone; PBT, polybutylene terephthalate; PET, polyethylene terephthalate; PCHDMT, polycyclohexane dimethylene terephthalate.
(a) With 75% cyclohexane dimethylene

Table 19 Properties of aromatic sulfone polymers

Properties	PSU	PESV	PPSU
T_g, °C (°F)	190 (375)	220–230 (430–445)	...
Heat-deflection temperature, at 1.82 MPa (0.264 ksi), °C (°F)	175 (345)	200 (397)	205 (400)
Izod impact strength, notched, J/m (ft · lbf/in.)	65 (1.2)	75 (1.4)	...

PSU, polysulfone; PESV, polyether sulfone; PPSU, polyphenylene sulfone

Thermal Properties of Thermosets

Engineering thermosets are resin systems that chemically fuse and bond with the application of elevated temperature and pressure for a given time period. The reapplication of temperature and pressure, even in excess of cure requirements, will not melt-flow the resin system out of shape. This is due to the cross-linked molecular network of the thermoset polymer, which forms in the curing process.

This section discusses the thermal and related properties of nine thermoset resin systems. The resin types are divided into three groups by low, medium, and high service temperature capabilities. The categories are based on general performance characteristics of the resin types, and they exhibit some overlap. Additional information on the thermal analysis of thermosets is contained in Ref 101.

Although neat thermoset resins are seldom used, their properties are important because resin characteristics have a strong influence on composite thermal properties. The addition of fillers and fibers can improve the properties of thermosets, but oriented fibers can cause anisotropy. These effects are not explicitly considered in this section.

Low-Temperature Resin Systems

The amino resin system is formed by an addition reaction of formaldehydes and compounds containing amino groups ($-NH_2$). The most widely used of the amino resins are those made with urea and melamine. They are supplied as liquid or dry resins and filled molding compounds. Applying heat in the presence of a catalyst converts the material into a hard, rigid, abrasion-resistant solid, with high resistance to deformation under load.

Melamines are superior to urea in resistance to normal acids and alkalies, heat, and boiling water. They also exhibit better performance when cycled between wet and dry conditions. Moldings of both melamines and ureas swell and shrink slightly in varying moisture conditions. Baking molded parts accelerates postmold shrinkage and improves dimensional stability.

In liquid form, both urea and melamine resins are also used as baked-enamel coatings, particle board binders, and paper and textile treatment materials. Typical property values are shown in Table 22.

Polyurethane resin systems are usually formed by the reaction of a diisocyanate with a polyol. The material is supplied as flexible and rigid foams, as elastomers, and as a liquid for coatings. Flexible foams use toluene diisocyanate (TDI), or polymethylene diphenylene isocyanate (PMDI).

Table 20 Properties of cellulose derivatives

Property	EC	CA	CAB	CAP	CN
T_m, °C (°F)	135 (275)	230 (445)	140 (60)	190 (90)	...
Molding temperature, °C (°F)					
Compression	120–200 (250–390)	125–215 (260–420)	125–200 (265–390)	130–205 (265–400)	85–95 (185–200)
Injection	175–260 (350–500)	170–255 (335–490)	170–250 (335–480)	170–270 (335–515)	...
Coefficient of thermal expansion, molded, 10^{-6}/K	100–200	80–180	110–170	110–170	80–120
Heat deflection temperature, at 1.82 MPa (0.264 ksi), molded, °C (°F)	45–88 (115–190)	45–90 (111–195)	45–95 (113–202)	45–110 (111–228)	60–70 (140–160)
Water absorption, 24 h at 3.2 mm (⅛ in.) thick, %	0.08–1.8	1.7–6.5	0.9–2.2	1.2–2.8	1.0–2.0

EC, ethyl cellulose; CA, cellulose acetate; CAB, cellulose acetate-butyrate; CAP, cellulose acetate-propionate; CN, cellulose nitrate

Table 21 Properties of thermoplastic elastomers and elastoplastics

Property	EP-BL(a)	SB-BL	H-SB-BL	POBT-PBT	POB-N	Polyether urethane(a)	Polyester urethane(a)
T_g, °C (°F)	...	(b)	(b)	...	65–75 (148–172)	50–70 (120–160)	50–70 (120–160)
T_m, °C (°F)	163–165	...	...	145–228	...	...	...
Injection molding temperature, °C (°F)	175–245 (350–475)	150–220 (300–425)	175–195 (350–380)	170–260 (340–500)	170–250 (340–480)	210–225 (410–440)	205–225 (400–435)
Extrusion temperature, °C (°F)	195–245 (380–475)	190–205 (370–400)	165–195 (330–380)	170–260 (340–500)	190–240 (370–460)	190–210 (370–410)	195–225 (380–440)
Mold (linear) shrinkage, mm/mm	0.015–0.021	0.001–0.022	0.003–0.022	0.003–0.014	...	0.005–0.015	0.008–0.012
Coefficient of thermal expansion, 10^{-6}/K	...	67–140	...	90–190	210–230	...	...
Specific gravity	0.90–0.98	0.9–1.2	0.9–1.20	1.15–1.25	1.0–1.02	1.15–1.28	1.14–1.21
Water absorption, 24 h at 3.2 mm (⅛ in.) thick, %	0.28	0.009–0.39	0.1–0.42	0.3–1.6	1.0–1.3	0.3	...

EP-BL, ethylene-propylene block copolymers; SB-BL, styrene-butadiene block copolymers; H-SB-BL, hydrogenated SB-BL; POBT-PBT, poly(polyoxybutylene terephthalate) and polybutylene terephthalate; POB-N, polyoxybutylene glycol and nylon 12. (a) High hardness grade. (b) Amorphous polymer, T_g varies with composition

Table 22 Thermal and related properties of amino resins

Thermal and related properties	Melamine-formaldehyde: No filler	Melamine-formaldehyde: Cellulose filler	Urea-formaldehyde, alpha cellulose filler
Cure process parameters			
Mold pressure, MPa (ksi)	13.8–34.5 (2.0–5.0)	10.3–41.4 (1.5–6.0)	13.8–55.2 (2.0–8.0)
Mold temperature, °C (°F)	150–165 (300–330)(a)	145–180 (290–360)(b)	150–260 (300–500)(c)
Mold shrinkage, mm/mm	0.011–0.012	0.006–0.008	0.007
T_g, °C (°F)	None(d)	...	...
Cured material properties			
Water absorption, 24 h, 3.2 mm (⅛ in.) thick, %	0.30–0.503	0.34–0.80	0.60
Specific gravity(e)	1.48	1.45–1.52	1.48–1.50
Continuous service temperature, °C (°F)	100 (210)	120 (250)	75 (170)
Heat deflection temperature at 1.82 MPa (0.264 ksi), °C (°F)	150 (298)	130 (266)	130–135 (266–275)
Flammability rating	...	...	94V-0
Burning rate	Self-extinguishing	Self-extinguishing	Self-extinguishing
Specific heat, kJ/kg · K (Btu/lb · °F)(e)	...	...	1.68 (0.40)
Coefficient of thermal expansion, 10^{-6}/K (in./in./°F × 10^{-6})(e)	...	45 (25.0)	27–29 (14.9–16.0)
Thermal conductivity, W/m · K (Btu/ft · h · °F)(e)	...	0.265–0.314 (0.156–0.185)	0.285–0.409 (0.168–0.241)

(a) At 21–35 MPa (3–5 psi). (b) At 10–42 MPa (1.5–6 ksi). (c) At 14–55 MPa (2–8 ksi). (d) Based on private communication, American Cyanamid Company. (e) At room temperature

The elastomers can be used for applications requiring superior toughness, superior resistance to tear and abrasion, and cold-temperature impact and flexibility. Their major shortcoming is low resistance to steam, fuels, strong acids, and bases. Property values are shown in Table 23.

The allyl resin system is a family of esters with a basic allyl radical. Allyl esters based on monobasic and dibasic acids are available as low-viscosity monomers and thermoplastic prepolymers. They are used in the preparation of reinforced thermoset molding compounds and high-performance transparent articles. Allyl resin can be processed by all modern thermoset techniques. The most common allylic resin system is diallyl phthalate (DAP). Another is allyl diglycol carbonate. Property values are shown in Table 24.

The molding compounds based on allyl prepolymers are reinforced with fibers (glass, asbestos, acrylic, orlon, polyester) and are filled with particulate materials (mineral) to improve properties. Glass fiber imparts maximum mechanical properties, acrylic fiber provides the best electrical properties, and polyester fiber improves impact resistance and strength in thin sections. Particulate fillers affect flow characteristics. Compounds can be made in a wide range of colors because the resin is essentially colorless.

Thermoset polyester resins are generally produced from the reaction of an organic alcohol (a glycol) with both a saturated (isophthalic) and an unsaturated (maleic or fumaric) organic acid. The polyester is then dissolved in a liquid reactive monomer such as styrene, and the solutions are sold as polyester resins. Some polyesters are supplied as pellets or granular solids. Polyesters are often premixed with glass fiber to form bulk molding compounds or sheet molding compounds. Polyester resins with glass-fiber reinforcements can be formulated to provide different mechanical, thermal, electrical, and flammability properties.

Because of their low cost, ease of processing, and good performance characteristics, unsaturated polyesters are the most extensively used type of thermoset resin. Unsaturated polyesters are generally combined with chopped, continuous, or woven glass fibers, as well as filler and additives, to alter the properties for specific applications. Property values are shown in Table 25.

Table 23 Thermal and related properties of polyurethane resins

Thermal and related properties	Polyurethane resin (cast)	Urethane elastomer	Urethane rigid foam
Cure process parameters			
Mold pressure, MPa (ksi)	...	5.2–13.8 (0.75–2.0)	...
Mold temperature, °C (°F)(a)	...	145–205 (293–400)	...
Mold shrinkage, mm/mm	...	0.009–0.030	...
T_g, °C (°F)	135 (275)	...	...
Cured material properties			
Water absorption, 24 h, 3.2 mm (⅛ in.) thick, %	0.20–0.60	0.70–0.90	<1%
Specific gravity(b)	1.10–1.50	1.11–1.25	0.56–0.64
Continuous service temperature, °C (°F)(b)	90–120 (190–250)	90 (190)	160 (325)
Heat deflection temperature, at 1.82 MPa (0.264 ksi), °C (°F)	50–205 (120–400)	...	...
Specific heat, kJ/kg · K (Btu/lb · °F)(b)	1.3–2.3 (0.30–0.55)	1.7–1.9 (0.40–0.45)	1.4 (0.33)
Coefficient of thermal expansion, 10^{-6}/K (in./in./°F × 10^{-6})(b)	70–100 (39–56)	100–200 (56–111)	80 (45)
Thermal conductivity, W/m · K (Btu/ft · h · °F)(b)	0.17–0.21 (0.100–0.121)	0.07–0.30 (0.041–0.178)	0.06–0.12 (0.033–0.067)

(a) At 5.2–14 MPa (0.750–2 ksi). (b) At room temperature

Table 24 Thermal and related properties of allyl resins

Thermal and related properties	Allyl diglycol carbonate neat resin	Diallyl phthalate (DAP)	
		Glass-fiber filled	Mineral filled
Cure process parameters			
Mold pressure, MPa (ksi)	...	3.4–27.6 (0.5–4.0)	3.4–27.6 (0.5–4.0)
Compression mold temperature, °C (°F)	130–160 (270–320)	145–195 (290–380)(a)	130–165 (270–330)(a)
Mold shrinkage, mm/mm	...	0.001–0.005	0.005–0.007
Cured material properties			
Water absorption, 24 h, 3.2 mm (⅛ in.) thick, %	0.20	0.12–0.35	0.20–0.50
Specific gravity(b)	1.30–1.40	1.61–1.85	1.65–1.68
Continuous service temperature, °C (°F)	100 (212)	150–205 (300–400)	150–205 (300–400)
Heat deflection temperature, at 1.82 MPa (0.264 ksi), °C (°F)	60–90 (140–190)	165–260 (325–500)	165–260 (325–500)
Burning rate, mm/min (in./min)	8.9 (0.35) to self-extinguishing	Self-extinguishing to nonburning	Self-extinguishing to nonburning
Specific heat, kJ/kg · K (Btu/lb · °F)(b)	2.3 (0.55)	1.26–1.33 (0.30–0.32)	1.26 (0.30)
Coefficient of thermal expansion, 10^{-6}/K (in./in./°F × 10^{-6})(b)	80–140 (45–79)	10–35 (5.5–20)	10–42 (5.5–2.3)
Thermal conductivity, W/m · K (Btu/ft · h · °F)(b)	0.199–0.210 (0.115–0.120)	0.20–0.60 (0.12–0.36)	0.29–1.02 (0.168–0.600)

(a) At 3.5–23 MPa (0.5–4 ksi). (b) At room temperature

Medium-Temperature Resin Systems

Epoxy resin systems include formulations such as diglycidyl ether of bisphenol A (DGEBA), multifunctional epoxies, and aliphatic epoxies. Use temperatures up to 230 to 260 °C (450 to 500 °F) can be tolerated for the latter two types of resin systems for short periods. Reinforced epoxy structures provide high strength-to-weight ratios and good thermal and electrical properties.

Filament winding and machine or hand lay-up processes with glass fiber/fabric, carbon, graphite, quartz, and aramid fibers are used to fabricate advanced aircraft fuselages, wings, and control panels; rocket motor cases; rocket nozzle structural shells; and commercial pressure vessels, tanks, and pipe. The same reinforcements are used in molding compounds, hand lay-ups, and fiber/fabric prepreg composites to match pressure, temperature, service life, weight, and cost requirements for different applications. Table 26 shows specific property values for epoxy neat resin, as well as for short-glass-fiber-reinforced molding compound.

Phenolic resin systems are formulated from the reaction between phenol and formaldehyde. The two main resin types are resoles and novolacs. Two-stage phenolic resins (novolacs) are used for general-purpose molding compounds, while hybrids of the novolacs are used as impregnating resins with glass, carbon, and graphite cloth for tape wrapping or hand lay-up of aerospace components, rocket nozzle ablatives, and insulation liners. Chopped-fiber molding compounds are used mostly in the automotive, appliance, and electrical component markets. General characteristics of these materials that make them suited for the aforementioned applications are high service temperatures, good electrical properties, excellent moldability, superior dimensional stability, and relatively

Table 25 Thermal and related properties of polyester resins

Thermal and related properties	Neat resin	Resin and 10 to 40 wt% chopped glass fiber
Cure process parameters		
Mold pressure, MPa (ksi)	...	3.4–13.8 (0.5–2.0)
Mold temperature, °C (°F)	...	140–175 (280–350)(a)
Mold shrinkage, mm/mm	...	0.001–0.012
T_g, °C (°F)	110 (230)	...
Cured material properties		
Water absorption, 24 h, 3.2 mm (⅛ in.) thick, %	...	0.06–0.28
Specific gravity(b)	1.10–1.46	1.6–2.1
Continuous service temperature, °C (°F)	120–150 (250–300)	150–175 (300–350)
Heat deflection temperature, at 1.82 MPa (0.264 ksi), °C (°F)	50–205 (120–400)	190–205 (375–400)
Flammability rating	...	94V-0
Specific heat, kJ/kg · K (Btu/lb · °F)(b)	1.3–2.3 (0.30–0.55)	1.0 (0.25)
Coefficient of thermal expansion, 10^{-6}/K (in./in./°F × 10^{-6})(b)	55–100 (31–55)	20–33 (11–18)
Thermal conductivity, W/m · K (Btu/ft · h · °F)(b)	0.17–0.22 (0.10–0.13)	0.4–0.6 (0.24–0.38)

(a) At 3.5–14 MPa (0.5–2 ksi). (b) At room temperature

Table 26 Thermal and related properties of epoxy resins

Thermal and related properties	Neat resin	Short glass fiber molding compound
Cure process parameters		
Mold pressure, MPa (ksi)	...	2.07–2.28 (0.30–0.33)
Mold temperature, °C (°F)	...	150–165 (300–330)(a)
Mold shrinkage, mm/mm	0.001–0.004	0.001–0.005
T_g, °C (°F)	60–175 (140–347)	125 (259)
Cured material properties		
Water absorption, 24 h, 3.2 mm (⅛ in.) thick, %	0.080–0.150	0.05–0.20
Specific gravity(b)	1.11–1.40	1.60–2.00
Continuous service temperature, °C (°F)	120–290 (250–550)	150–260 (300–500)
Heat deflection temperature, at 1.82 MPa (0.264 ksi), °C (°F)	45–290 (115–550)	150–275 (300–525)
Flammability rating	...	94V-0
Specific heat, kJ/kg · K (Btu/lb · °F)(b)	1.05 (0.25)	0.80 (0.19)
Coefficient of thermal expansion, 10^{-6}/K (in./in./°F × 10^{-6})(b)	45–65 (25–36)	11–35 (6–19.5)
Thermal conductivity, W/m · K (Btu/ft · h · °F)(b)	0.17–0.20 (0.10–0.12)	0.17–0.40 (0.10–0.24)

(a) At 2.1–35 MPa (0.3–5 ksi). (b) At room temperature

Table 27 Thermal and related properties of phenolic resins

Thermal and related properties	Neat resin	Chopped glass fiber molding compound
Cure process parameters		
Mold pressure, MPa (ksi)	17–26 (0.25–4.0)	1.9–27.6 (0.28–4.0)
Mold temperature, °C (°F)	130–160 (270–320)(a)	140–175 (280–350)(a)
Mold shrinkage, mm/mm	0.010–0.012	0.0001–0.0040
Tg, °C (°F)	300 (572)	...
Cured material properties		
Water absorption, 24 h, 3.2 mm (⅛ in.) thick, %	0.010–0.20	0.03–1.20
Specific gravity(b)	1.23–1.32	1.65–1.95
Continuous service temperature, °C (°F)	120–175 (250–350)	175–290 (350–550)
Heat deflection temperature, at 1.82 MPa (0.264 ksi), °C (°F)	120–175 (250–350)	150–315 (300–600)
Flammability rating	94V-1	94V-0
Specific heat, kJ/kg · K (Btu/lb · °F)(b)	1.4–1.8 (0.34–0.42)	0.85–1.25 (0.2–0.3)
Coefficient of thermal expansion, 10^{-6}/K (in./in./°F × 10^{-6})(b)	25–60 (13.8–33.3)	8–20 (4.4–11.4)
Thermal conductivity, W/m · K (Btu/ft · h · °F)(b)	0.12–0.24 (0.072–0.144)	0.32–0.60 (0.19–0.35)

(a) At 1.7–28 MPa (0.25–4 ksi). (b) At room temperature

good moisture resistance. Table 27 shows typical property values.

Phenolic resin thermosets include unfilled resin and filled resin systems. For the latter, fillers include glass, carbon, and polyamide fiber, wood and cotton flock, aluminum powder, rubber, cellulose fabric, and minerals.

Silicones. In a direct process for the production of chlorosilane intermediates, either methyl chloride or chlorobenzene is used as the starting material. Different combinations of chlorosilanes are initially subjected to hydrolysis and neutralization of the chlorosilane monomers. The crude silicone polymer produced by hydrolysis is equilibrated to stabilize it into a useful form, then devolatilized.

Usually, silicone resins are formulated to provide a three-dimensional network of siloxane (Si–O). The predominant organic groups attached to the silicon atom are hydrocarbon radicals such as methyl, phenyl, and vinyl. Heat, frequently combined with the action of metal catalysts, such as tin or platinum compounds, further condenses the polymer to form a rigid thermoset material.

Silicones are best characterized by: thermal and oxidative stability at high temperatures, up to 260 °C (500 °F); flexibility at –75 °C (–100 °F); excellent electrical properties, including resistance to corona breakdown; general inertness, exhibited as resistance to weathering, ozone, and many chemicals; general noncorrosiveness to other materials (with the exception of some construction adhesives in contact with ferrous alloys in enclosed, moist environments); inherent nonflammability and self-extinguishing properties; lubricity; unusual surface properties (such as low surface tension of the fluid resin and the capability of preventing other materials from sticking); and very low water absorption.

Silicone resin products are usually formulated for specific applications. The resin may be combined with other ingredients, such as micas and silica fillers, glass and carbon fibers, pigments, and other additives that impart special properties. Properties are given in Table 28.

High-Temperature Resin Systems

Polyimide Resins. The most common PI resin system is PMR-15, developed and licensed for production by the National Aeronautics and Space Administration (NASA) Lewis Research Center. It is available as both a liquid and a miscible powder.

The chemistry of PMR-15 begins with a solution of three monomers in a low-boiling alcohol, usually methanol. The solution is then used to impregnate fiber (yarn, fabric, or braid), or it is mixed with chopped fiber and other fillers. The alcohol solvent is easily evaporated, after which the material is heated in the 80 to 205 °C (180 to

Table 28 Thermal and related properties of silicone resins

Thermal and related properties	Neat resin	Silica-filled heat-vulcanizing molding compound	Silica-filled, carbon-fiber-reinforced, two-part room-temperature-vulcanizing molding compound
Cure process parameters			
Mold temperature, °C (°F)	...	165 (330)(a)	Room temperature(b)
Mold shrinkage, mm/mm	0–0.006	0.0030–0.0067	0.002–0.006
T_g, °C (°F)	−125 (−193)	...	...
Cured material properties			
Water absorption, 24 h, 3.2 mm (⅛ in.) thick, %	0.12(c)	0.10	...
Specific gravity(d)	0.99–1.50	1.86–1.88	1.46
Continuous service temperature, °C (°F)	260 (500)	175–260 (350–500)	205 (400)
Heat deflection temperature at 1.82 MPa (0.264 ksi), °C (°F)	...	225–345 (435–650)	...
Flammability rating	Self-extinguishing	...	...
Specific heat, kJ/kg · K (Btu/lb · °F)(d)	...	0.80–0.84 (0.19–0.20)	1.1 (0.27)
Coefficient of thermal expansion, 10^{-6}/K (in./in./°F × 10^{-6})(d)	80–300 (44–166)	55–30 (13–18)	250 (140)
Thermal conductivity, W/m · K (Btu/ft · h · °F)(d)	0.22 (0.13)	0.37–0.49 (0.22–0.29)	0.34–0.49 (0.20–0.29)

(a) In compression mold. (b) For 24 h. (c) For 7 days at 25 °C (77 °F). (d) At room temperature

Table 29 Thermal and related properties of polyimide resins

Thermal and related properties	Neat resin	50% glass fiber and resin (molding compound)
Cure process parameters		
Mold pressure, MPa (ksi)	1.4–17.2 (0.2–2.5)	1.4–6.9 (0.2–10)
Mold temperature, °C (°F)	290–315 (500–600)(a) 315 (600)(c)	175–250 (350–480)(b)
Mold shrinkage, mm/mm	0.0126	0.0005–0.0040
T_g, °C (°F)	315–370 (600–698)(d)	...
Cured material properties		
Water absorption, 24 h, 3.2 mm (⅛ in.) thick, %	0.24–0.40	0.20
Specific gravity(c)	1.19–1.43	1.60–1.95
Continuous service temperature, °C (°F)	260–315 (500–600)	250–260 (480–500)
Heat deflection temperature, at 1.82 MPa (0.264 ksi), °C (°F)	305–360 (582–680)	290–350 (550–660)
Burning rate	Nonburning	...
Specific heat, kJ/kg · K (Btu/lb · °F)(e)	1.05–1.5 (0.25–0.35)	0.63–1.13 (0.15–0.27)
Coefficient of thermal expansion, 10^{-6}/K (in./in./°F × 10^{-6})(e)	25–80 (12.7–45)	10–27 (5.5–15.1)
Thermal conductivity, W/m · K (Btu/ft · h · °F)(e)	0.10–0.34 (0.058–0.20)	0.34–0.49 (0.20–0.29)

(a) At 1.4–17 MPa (0.2–2.5 ksi). (b) At 1.4–70 MPa (0.2–10 ksi). (c) Postcure. (d) Dry. (e) At room temperature

400 °F) range to form low-molecular-weight (1500 average) PI chains (imidization).

At this stage, the material is thermoplastic, in that the molecular chains are capped at either end with a norbornenyl group, which can be obtained at temperatures above 290 °C (550 °F) to form chemical bonds (two per end cap) to neighboring chains. It is important to note that no further volatiles are generated beyond the imidization stage. The imidized prepreg or molding compound is ready to be molded, using pressure and temperature (270 °C, or 520 °F) to remelt the resin to the desired shape. In the 290 to 315 °C (550 to 600 °F) range, the resin rapidly acquires a permanent set. The final cure or set is the result of intermolecular bonds that tie the PI chains into a three-dimensional network. Because no volatiles are evolved during this final cure, dense and void-free molded parts are produced reliably.

The molded neat resin has a specific gravity of 1.3 and exhibits retention of good mechanical properties, such as Underwriters' Laboratories room-temperature flexural strength of 125 to 140 MPa (18 to 20 ksi) up to 290 °C (550 °F), with only a 35 to 50% decrease from room-temperature properties. The T_g of the fully cured material is typically 320 to 330 °C (610 to 630 °F). NASA studies have shown excellent thermo-oxidative stability with the retention of mechanical integrity for up to 1000 h of continuous exposure to 315 °C (600 °F) air.

Polyimide resin systems are used for electrical and low friction products in the aerospace industry, as well as in office equipment. They are good bearing materials, exhibiting low friction, high wear resistance, low creep, and dimensional stability. Thus, they are used for self-lubricated parts, such as bearings, bushings, thrust washers, wear rings, and seals. Properties are summarized in Table 29.

PMR-15 exhibits good chemical resistance, except when exposed to strong alkalies. It is self-extinguishing when the ignition source is removed, with low smoke generation and a high char yield (70%) that forms an insulating barrier against further flame spread.

Bismaleimide (BMI) resin systems are derived from a variety of different starting chemical compounds, with some commonality in the final resin, a 4,4′-bismaleimido-diphenylmethane (MDAB). This BMI is synthesized from 4,4′-diaminodiphenylmethane and maleic acid anhydride, with the synthesis being followed by cyclodehydration. It is the main building block of almost all commercially available BMI resins. MDAB is a yellow, high-melting-temperature, fine powder that contains no free methylene dianiline (MDA).

These resin systems exhibit a high T_g relative to postcure temperature, are dimensionally stable at elevated temperatures (>260 °C, or 500 °F), and have low flammability characteristics. In addition, they handle well in processing as a hot melt and exhibit good humidity resistance, good toughness, and excellent mechanical properties at both ambient and elevated temperatures. They are amenable to both hot-melt and solvent-resin impregnation of fibers or fabrics. They can be cast, filament wound, or molded into component shapes. For curing, a standard autoclave or compression molding process is used to apply pressure and heat, typically 700 kPa (100 psi) and 175 °C (350 °F), for 1 to 2 h for the component cure. A free-standing postcure at 250 °C (482 °F) for up to 6 h is also recommended.

Applications include aircraft primary structures (wing skins and ribs, and helicopter firewalls), high-performance structural adhesives, and printed circuit boards. Property values are shown in Table 30.

Table 30 Thermal and related properties of bismaleimide resins

Thermal and related properties	Neat resin	68.3 vol % T300 carbon fiber and resin	57.7 vol % E-glass fiber and resin
Cure process parameters			
Mold pressure, MPa (ksi)	0.59–0.69 (0.085–0.10)	...	...
Mold temperature, °C (°F)	175–190 (350–375)(a)	210 (410)(b)	210 (410)(b)
	230–245 (450–475)	210 (450)(b)(c)	210 (450)(b)(c)
T_g, °C (°F)	230–345 (450–650)(d)	265 (510)	255 (490)
Cured material properties			
Specific gravity	1.23–1.29	1.60	2.00
Continuous service temperature, °C (°F)			
Short term	315 (600)	315 (600)	315 (600)
Long term	230 (450)	230 (450)	230 (450)
Flammability rating	Low	...	...
Coefficient of thermal expansion, 10^{-6}/K (in./in./°F × 10^{-6}) at room temperature	30–50 (17–27)	...	...

(a) At 490–700 kPa (85–100 psi). (b) For 5 h. (c) Postcure. (d) Dry

ACKNOWLEDGMENT

The information in this article is largely taken from the following articles in *Engineering Plastics,* Volume 2, *Engineered Materials Handbook,* ASM International, 1988:

- R.C. Allen and R.S. Bauer, Moisture-Related Failure, p 761–769
- L.L. Clements, Polymer Science for Engineers, p 48–62
- S.B. Driscoll, Physical, Chemical, and Thermal Analysis of Thermoplastic Resins, p 533–543
- R.J. Jones, Characterization of Temperature Resistance, p 559–567
- R.E. Laramee, Thermal and Related Properties of Engineering Thermosets, p 439–444
- R.R. Maccani, Characteristics Crucial to the Application of Engineering Plastics, p 68–73
- A.T. Riga and E.A. Collins, Analysis of Structure, p 824–837
- S.W. Shalaby and P. Moy, Thermal and Related Properties of Engineering Thermoplastics, p 445–459.

REFERENCES

1. H.G. Elias, *Macromolecules,* Vol 1, Plenum, 1977, p 407
2. S.W. Shalaby, chapter 3, *Thermal Characterization of Polymeric Materials,* E.A. Turi, Ed., Academic Press, 1981
3. K.R. Beck, R. Krosmeyer, and R.I. Kunz, *J. Chem. Ed.,* Vol 61, 1984, p 668
4. D.R. Burfield, *J. Chem. Ed.,* Vol 64, 1987, p 875
5. D.R. Burfield and Y. Doi, *Macromolecules,* Vol 16, 1983, p 702
6. S.W. Shalaby and A. Kafrawy, *J. Polym. Sci., Polym. Chem. Ed.,* 1988, in press
7. S.W. Shalaby, U.S. Patent 4,190,720, 1980
8. T.G. Fox and P.J. Flory, *J. Appl. Phys.,* Vol 21, 1950, p 581
9. K. Ueberreiter and G. Kanig, *J. Chem. Phys.,* Vol 18 (No. 4), 1950, p 399
10. M. Gordon and J.S. Taylor, *J. Appl. Chem.,* Vol 2, 1952, p 493
11. L.A. Wood, *J. Polym. Sci.,* Vol 28, 1958, p 319
12. J.M. Gordon, G.B. Rouse, J.H. Gibbs, and W.M. Risen, *J. Chem. Phys.,* Vol 66 (No. 11), 1977, p 4971
13. P.R. Couchmen and F.G. Karasz, *Macromolecules,* Vol 11, 1978, p 117
14. L.E. Nielson, *Mechanical Properties of Polymers,* Van Nostrand Reinhold, 1962
15. F.N. Kelly and F. Bueche, *J. Polym. Sci.,* Vol 50, 1961, p 549
16. J.M. Gordon, G.B. Rouse, J.H. Gibbs, and W.M. Rosen, *J. Chem. Phys.,* 1977, p 4971
17. J.H. Gibbs and E.A. DiMarzio, *J. Chem. Phys.,* Vol 20, 1958, p 373
18. P.R. Couchman and F.E. Karasz, *Macromolecules,* Vol 11, 1978, p 117
19. P. Moy and F.E. Karasz, The Interactions of Water with Epoxy Resin, *Water Interactions in Polymers,* S.P. Rowland, Ed., Symposium Series 127, American Chemical Society, 1980
20. T. Ellis and F.E. Karasz, *Polymer,* Vol 25, 1984, p 664
21. G. Brinke, F.E. Karasz, and T. Ellis, *Macromolecules,* Vol 16, 1983, p 244
22. M. Goldstein, *Macromolecules,* Vol 18, 1985, p 277
23. R.C. Allen, *Proceedings of the 18th SAMPE Technical Conference,* Society for the Advancement of Material and Process Engineering, 1986, p 583
24. D.J. Boll, W.D. Bascom, and B. Motiee, *Compos. Sci. Technol.,* Vol 24, 1985, p 253
25. E.L. McKague, J.D. Reynolds, and J.E. Halkias, Thermomechanical Testing of Plastics for Environmental Resistance, *J. Test. Eval.,* Vol 1 (No. 6), Nov 1973, p 468–471
26. C.E. Browning, *Polym. Eng. Sci.,* Vol 18, 1978, p 16–24
27. A. Apicella, L. Nicolais, G. Astarita, and E. Drioli, *Polymer,* Vol 20 (No. 9), 1979, p 1143–1148
28. E.L. McKague, J.D. Reynolds, and J.E. Halkias, *J. Polym. Sci.,* Vol 22, 1978, p 1643–1654
29. E.L. McKague, J.E. Halkias, and J.D. Reynolds, *J. Compos. Mater.,* Vol 9, 1975, p 2–9
30. W.W. Wendlandt, *Thermal Methods of Analysis,* John Wiley & Sons, 1974
31. P.F. Levy, Thermal Analysis: An Overview, *Am. Lab.,* Jan 1970
32. R.L. Blaine, "Thermal Analysis in the Electronics Industry," paper presented at the Du Pont Educational Seminar, Palo Alto, CA, E.I. Du Pont de Nemours, June 1974
33. R.L. Blaine, Using Thermal Analysis as a Process Development and Quality Control Tool in Circuit Manufacturing, *Insul. Circuits,* March 1976, p 37–42
34. D. Frisch and R. Ciccarone, Thermal Analysis for Evaluating Laminates, *Circuits Manuf.,* Vol 17 (No. 7), July 1977, p 54–58
35. R.L. Hassel, Quality Control of Thermosets, *Ind. Res. Dev.,* Vol 20 (No. 10), 1978, p 160–163
36. W.P. Brennan and R.B. Cassel, Thermal Analysis in the Electrical and Electronics Industries, *Am. Lab.,* Jan 1979, p 80–88
37. P.F. Levy, R.L. Blaine, P.S. Gill, and J.D. Lear, Thermal Analysis: Advances in Instrumentation, *Am. Lab.,* June 1979, p 79–88
38. R.H. Wehrenberg II, Thermal Analysis: The Hot New Technique for Testing Plastics, *Mech. Eng.,* Sept 1979, p 78–83
39. R. Riesen and H. Sommerrauer, Curing of Reaction Molding Resins Studied by Thermoanalytical Methods, *Am. Lab.,* Vol 15 (No. 1), Jan 1983, p 30–37
40. P.S. Gill, Thermal Analysis Developments in Instrumentation and Applications, *Am. Lab.,* Jan 1984, p 39–49
41. W.P. Brennan and M.P. DiVito, Recent Advances in Thermal Analysis Instrumentation, *Am. Lab.,* Jan 1985, p 68–79

42. G. Dugan, Thermal Analysis Supports Chemical R & D, Product Quality Control, *Res. Dev.*, June 1985, p 98–102
43. M.P. DiVito, W.P. Brennan, and R.L. Fyans, Thermal Analysis: Trends in Industrial Applications, *Am. Lab.*, Jan 1986, p 82–95
44. J.E. Mark, A. Eisenberg, W. Graessley, L. Mandelkern, and J. Koenig, "Physical Properties of Polymers," paper presented to the American Chemical Society, Washington, D.C., 1984
45. T. Smith, Physical Properties of Polymers—An Introductory Discussion, *Polym. Eng. Sci.*, Vol 13 (No. 3), 1973, p 161
46. J. Haslam and H.A. Willis, *Identification and Analysis of Plastics,* Van Nostrand, 1967
47. F. Billmeyer, *Textbook of Polymer Science,* 2nd ed., Wiley-Interscience, 1981
48. R.J. Young, *Introduction to Polymers,* Chapman-Hall, 1981
49. W. Greive and A.T. Riga, *Instrumental Analysis of Plastics,* American Society for Testing and Materials, Nov 1986; also, Oct 1987
50. W.P. Brennan, "What is a T_g? A Review of the Scanning Calorimetry of the Glass Transition," Perkin Elmer, No. 7, March 1973
51. W.P. Brennan, Thermal Analysis: Useful Tool for Quality Control in a Complex Era, *Mod. Plast.*, Vol 56 (No. 1), 1979, p 98
52. P. Levy, Thermal Analysis—An Overview, *Am. Lab.*, Jan 1970
53. A.T. Riga, Inhibitor Selection for Vinyl Monomers by DSC, *Polym. Eng. Sci.*, Vol 18 (No. 12), 1976, p 836
54. A.T. Riga, Thermal Analysis as an Aid to Monomer Plant Design, *Polym. Eng. Sci.*, Vol 15 (No. 5), 1975, p 349
55. "Instrument Systems," E.I. Du Pont de Nemours & Company, Inc., 1987
56. W.P. Brennan, "Characterization and Quality Control of Engineering Thermoplastics by Thermal Analysis," Perkin Elmer Corporation
57. T.A.M.M. Maas, Optimization of Processing Conditions for Thermosetting Polymers by Determination of the Degree of Curing with a Differential Scanning Calorimeter, *Polym. Eng. Sci.*, Vol 18 (No. 1), 1978, p 29–32
58. S.J. Swarin and A.M. Wims, A Method for Determining Reaction Kinetics by Differential Scanning Calorimetry, *Anal. Calorim.*, 1976, p 155–177
59. L.T. Pappalardo, DSC Evaluation of B-Stage Epoxy-Glass Prepregs for Multilayer Boards, *Soc. Plast. Eng.*, Vol 20, 1974, p 13–16
60. L.T. Pappalardo, DSC Evaluation of Epoxy and Polyimide-Impregnated Laminates (Prepregs), *J. Appl. Polym. Sci.*, Vol 21, 1977, p 809–820
61. Z.N. Sanjana and R.N. Sampson, Measuring the Degree of Cure of Multilayer Circuit Boards, *Insul. Circuits,* 1981, p 87–92
62. T.M. Donnellan, "The Curing of a Bisphenol A Type Epoxy Resin With 1,8-Diamino-p-Methane," NADC-83146-60, Naval Air Systems Command, 1982
63. J.M. Barton, "A Thermoanalytical Study of the Cure Characteristics of an Epoxy System: BSL 913," Technical Report 76138, Royal Aircraft Establishment, 1976
64. M.R. Dusi, "Chemorheological Characterization and Processing Science of an Epoxy/Amine Thermosetting Matrix," M.S. Thesis, San Jose State University, 1984
65. B.G. Parker and C.H. Smith, Evaluating Cure and Shelf Life of Epoxy Prepregs and Film Adhesives, *Mod. Plast.*, Dec 1979, p 58–60
66. A. Dutta and M.E. Ryan, Effect of Fillers on Kinetics of Epoxy Cure, *J. Appl. Polym. Sci.*, Vol 24, 1979, p 635–649
67. G.L. Hagnauer, P.J. Pearce, B.R. La Liberte, and M.R. Roylance, Cure Kinetics and Mechanical Properties of a Resin Matrix, Effects of Impurities and Stoichiometry, *Chemorheology of Thermosetting Polymers,* C.A. May, Ed., ACS Symposium Series 227, American Chemical Society, 1983, p 25–47
68. R.J. Morgan, C.M. Walkup, and T.H. Hoheisel, Characterization of the Cure of Carbon Fiber/Epoxy Composite Prepregs by Differential Scanning Calorimetry, *J. Compos. Technol. Res.*, Vol 7, 1985, p 17–19
69. N.S. Schneider, J.F. Sprouse, G.L. Hagnauer, and J.K. Gillham, DSC and TBA Studies of the Curing Behavior of Two Dicy-Containing Epoxy Resins, *Polym. Eng. Sci.*, Vol 19 (No. 4), 1979, p 304–311
70. W.J. Sichina, Characterization of Autocatalyzed Thermosets by Differential Scanning Calorimetry, *Proc. Nat. SAMPE Symp.*, Vol 30, 1985, p 610–623
71. W. Huffered, "Application of Rate Theory to Accelerated Durability Testing of Structural Adhesives," AFML-TR-79-4199, Air Force Materials Laboratory, 1980
72. L.C.E. Struik, *Physical Aging in Amorphous Polymers and Other Materials,* Elsevier, 1978
73. Z.H. Ophir, "Structure-Property Relationships in Solid Polymers: I—Segmented Polyurethanes and II—Epoxy Thermosets," Ph.D. thesis, Princeton University, 1979
74. E.S.W. Kong, Physical Aging and Its Effects on the Mechanical and Physical Properties of Graphite/Epoxy Composites, *Organic Coat. Appl. Sci.*, Vol 46, 1982, p 568–573
75. B. Shushan, C. Williamson, and R.B. Prime, Applications of a Fully Computer Controlled Thermogravimetric-Tandem Triple Quadrupole Mass Spectrometer System (TGA/MS/MS), *ANTEC '84,* Society of Plastics Engineers, 1984, p 319–322
76. D.C. Sabatelli, G. Lavigne, J. Tanaka, and J.F. Johnson, Polymer Curing Studies Using Combined TGA-GC-FTIR-MS Techniques, *ANTEC '84,* Society of Plastics Engineers, 1984, p 311–315
77. J.F. Moellmer, Measuring Resin Contents of PC Laminates with Thermal Gravimetric Analysis, *Insul. Circuits,* Aug 1980, p 29
78. D. Price, A. Horrocks, and M. Tunc, Textile Flammability, *Chem. Brit.*, Vol 23 (No. 3), 1987, p 235
79. TRW, Inc., unpublished research
80. H.H. Gibbs, *Proc. Nat. SAMPE Symp.*, Vol 17, III-B-6, 1972
81. G.M. Moelter, R.F. Tetreault, and M.J. Hefferon, *Polym. News,* Vol 9, 1983, p 134–138
82. R.J. Jones, M.K. O'Rell, and J.M. Hom, U.S. Patent 4,111,906, 1978
83. C.C. Winding and G.D. Hiatt, *Polymeric Materials,* McGraw-Hill, 1961
84. A.T. Riga, Heat Distortion and Mechanical Properties of Polymers by Thermal-Mechanical Analysis, *Polym. Eng. Sci.*, Vol 14 (No. 11), 1974, p 764
85. M.R. Dusi, M.G. Maximovich, and R.M. Galeos, Physiorheological Characterization of a Carbon/Epoxy Prepreg System, *J. Appl. Polym. Sci.*, Vol 30, 1985, p 1847–1857
86. R.J. Hinrichs, Rheological Cure Transformation Diagrams for Evaluating Polymer Cure Dynamics, *Chemorheology of Thermosetting Polymers,* C.A. May, Ed., ACS Symposium Series 227, American Chemical Society, 1983, p 187–199
87. M.G. Maximovich and R.M. Galeos, Rheological Characterization of Advanced Composite Prepreg Materials, *Proc. Nat. SAMPE Symp.*, Vol 28, 1983, p 568–580
88. L.D. Lauer, Dynamic Mechanical Analysis of Epoxy Composite Prepregs, *SAMPE Q.*, Oct 1983, p 31–35
89. M. Hattori, *Kolloid Z.*, Vol 185, 1962, p 27
90. D.I. Cullen, M.S. Zawojski, and A.L. Holbrook, *Plast. Eng.*, Vol 44 (No. 1), 1988, p 37
91. D. Hansen and C.C. Ho, *J. Polym. Sci.*, Vol A3, 1965, p 659
92. M. Hattori, *Bull. Univ. Osaka Prefect. Ser.*, Vol A9 (No. 1), 1960, p 51
93. D. Hansen, R.C. Kantayya, and C.C. Ho, *Polym. Eng. Sci.*, Vol 6 (No. 3), 1966, p 260

94. K. Ueberreiter and S. Nans, *Kolloidn. Zh.*, Vol 123, 1951, p 92
95. K. Ueberreiter and E. Otto-Laupenmuhlen, *Kolloidn. Zh.*, Vol 133, 1953, p 26
96. K. Eigermann, *Kunstoffe*, Vol 51 (No. 9), 1961, p 512
97. J.S. Fox and M.J. Imber, *Appl. Polym. Sci.*, Vol 12, 1968, p 571
98. B. Wunderlich and H. Baur, *Adv. Polym. Sci.*, Vol 7, 1970, p 151
99. B. Wunderlich, *J. Phys. Chem.*, Vol 64, 1960, p 1052
100. V. Sochava, O.N. Trapeznikova, *Dokl. Akad. Nauk SSSR*, Vol 113, 1957, p 784
101. S.W. Shalaby and H.E. Bair, chapter 4, *Thermal Characterization of Polymeric Materials*, E.A. Turi, Ed., Academic Press, 1981
102. S.W. Shalaby and E.M. Pearce, *Int. J. Polym. Mater.*, Vol 3, 1974, p 81
103. S.W. Shalaby, E.A. Turi, and W.H. Wenner, *Thermal Methods in Polymer Analysis*, S.W. Shalaby, Ed., Franklin Institute Press, 1978, p 35
104. E.M. Pearce, S.W. Shalaby, and R.H. Barker, chapter 9, *Polymer Flammability*, Lewis, Atlas, and Pearce, Ed., Plenum Press, 1975
105. P.J. Koch, E.M. Pearce, E.M. Lapham, and S.W. Shalaby, *J. Appl. Polym. Sci.*, Vol 19, 1975, p 227
106. R.B. Prime and J.C. Seferis, *J. Polym. Sci. C, Polym. Lett.*, Vol 24, 1986, p 641
107. M.J. Mullins and E.P. Woo, *J. Macromol. Sci., Rev. Macromol. Chem. Phys.*, Vol C-27 (No. 2), 1987, p 313
108. S.W. Shalaby, *Encyclopedia of Medical Devices and Instrumentation*, J.E. Webster, Ed., John Wiley & Sons, 1988
109. G.A. Weaver, L. Price, R. Britt, and S.W. Shalaby, U.S. Patent 4,578,451, 1986
110. S.W. Shalaby, E.S. Schipper, and D.F. Koelmel, U.S. Patent 4,433,161, 1984
111. S.W. Shalaby and D.D. Jamiolkowski, U.S. Patent 4,608,428, 1986
112. S.W. Shalaby and D.F. Koelmel, U.S. Patent 4,543,952, 1985

Environmental and Chemical Effects

ENVIRONMENTAL EFFECTS on polymeric materials cover a broad range of different behaviors. For example, plasticization, solvation, and swelling, occur from the diffusion of the chemicals into the polymer. These chemical alterations of the molecular structure are not necessarily irreversible, as the effects can be reversed if the chemical is removed from the material (for example, by evaporation). Environmental stress cracking also can occur without significant absorption of an environmental reagent by the polymer. Environmental conditions can promote brittle fracture in normally ductile plastics at levels of stress or strain well below those that would usually cause failure at all. Actual degradation of a polymer, which reduces molecular weight and therefore mechanical properties, does not need to be particularly pervasive in order to be problematic. For example, degradation of a thin surface layer of material on a plastic part can facilitate premature failure or brittle failure under conditions where ductile failure would normally occur. All of this is further exacerbated by the effects of changing temperature and strain rate. All of these conditions may need to be considered in preventing or determining cause of failure in a polymeric material.

Chemical Susceptibility*

Chemical reaction kinetics of polymers are similar to those of small molecules, in spite of the enhanced resistance of crystalline, cross-linked, and stiffened polymers. A polymer is a giant molecule that differs from conventional, small molecules, such as ethane $[H(CH_2)_2H]$, predominantly by size. For example, high-density polyethylene (HDPE) $[H(CH_2)_nH]$, where n is greater than 2000, may contain numerous methylene (CH_2) groups joined together by covalent (electron-shared) bonds in a continuous chain. Polymers also usually comprise a mixture of molecules of different molecular weights and are said to be polydisperse. In the case of HDPE molecules, molecular weights can range from 14,000 to 140,000 or more. An average molecular weight of 5000 for HDPE can be cited.

The terminating ends of a polymer also have different chemical structure than that of the repeating mers. For example, the terminal groups in HDPE and other alkane polymers (saturated hydrocarbons) are methyl (CH_3), but the terminal groups of other polymeric molecules, such as polyesters, may be hydroxyl (OH) or carboxylic (COOH) groups. The chemistry of these terminal groups may differ from that of the repeating units (mers), and the effect of this difference is inversely related to the molecular weight or chain length of the polymer. Thus, the effect of such groups is of particular importance in very-low-molecular-weight polymers (oligomers).

This effect is one of several that prevent the characteristic properties of polymers from being evident unless the chain exceeds a critical or threshold length. This length varies from polymer to polymer, but it may be designated as a molecular weight of 1000. This threshold value is higher for nonpolar molecules such as those in HDPE, which has low intermolecular attractive forces (van der Waals forces) between polymer chains. This value is lower for polar molecules such as nylons, which have high intermolecular attractive forces between polymer chains. In any case, it is important to note that polymeric properties are not evident until the polymer chain is long enough to achieve strength by chain entanglement.

Many polymers, such as HDPE, are called homopolymers because they consist of sequences of identical repeating units in the polymer chain. Other polymers or macromolecules that comprise two or more differing repeating units in the chain are called copolymers. Thus, if the repeating units in a homopolymer are A, the polymer could be designated as $(A)_n$. However, if A and B repeating units are present in the chain, the macromolecule is a copolymer, which can be designated as $(AB)_n$.

The chemical susceptibility of a copolymer is based on the susceptibility of each specific component present. Thus, the chemical susceptibility of the copolymer of vinyl chloride (CH_2:CHCl) and vinyl acetate (CH_2:CH($OOCCH_3$)) is related to that of vinyl chloride and vinyl acetate. The extent of this susceptibility is related to the ratio of these components.

Many polymers and random copolymers are amorphous, and their thermal behavior is much like that of glass. However, homopolymers, which consist of repeating units with regular structure, may have a high degree of crystallinity. In contrast to amorphous polymers, which are characterized by glass-transition temperature (T_g) values, crystalline polymers are characterized by melting point (T_m) values, as well as T_g values for the amorphous areas present.

Absorption and Transport

Small polar molecules, such as water and ethanol, are readily absorbed by polar polymers, such as cellulose, proteins, and nylon. Amorphous polymers absorb these small polar molecules more readily than crystalline polymers, and the rate varies inversely with the degree of crystallinity. The absorption by polar polymers increases as the concentration of the absorbate increases. This rate is lowered by the presence of nonpolar groups in the polymer and is independent of the concentration of polar absorbates in nonpolar polymers, such as polyethylene (Ref 1).

The diffusion of liquids is related to polymer structure and temperature and is independent of chain length but is inversely related to the size of the absorbate. The rate of diffusion is decreased by the presence of branches, pendant groups, or cross links.

Solubility parameters are useful for amorphous and some semicrystalline polymers. Crystallinity can yield behavior that appears anomalous. The Hildebrand solubility parameter is the square root of the cohesive energy density, which is the energy required to prevent 1 cm^3 of molecules from overcoming the intermolecular attractions between these molecules. When linear or branched polymers are exposed to solvents with solubility parameter values within ±1.8H, they dissolve. Solvents are absorbed by polymers having solubility parameters outside this range. Likewise, linear polar polymers such as starch dissolve in water, but strongly hydrogen-bonded cellulose is insoluble in water. When electrolytes are present, they are transported by the water and may react with the polymer. The reactions are essentially the same as those that

*Adapted from Raymond B. Seymour, Determination of Chemical Susceptibility, *Engineering Plastics,* Volume 2, *Engineered Materials Handbook,* ASM International, 1988, pages 571 to 574

occur with small molecules that are similar in chemical composition to the mer unit of the polymer. The rate of water transfer is decreased by the presence of the electrolytes, but the rate increases if a reaction occurs between the electrolyte and the polymer molecule. Cross linking decreases the rate of chemical attack.

Aggressive corrosives, such as the oxidizing nitric, chromic, and concentrated sulfuric acids, attack polymer surfaces. The permeability of the new surface to the additive corrosive may or may not be enhanced. For example, the attack of natural rubber by hydrochloric acid produces rubber hydrochloride, which is resistant to subsequent attack by hydrochloric acid and many other nonoxidizing corrosives. The ozonide formed by the reaction of ozone and natural rubber also resists further attack. However, the ozonide is a brittle, nonelastic polymer that cracks when stretched, permitting penetration by corrosives.

Additive Effects

As already stated, all components of a copolymer contribute to chemical susceptibility. In like turn, the chemical susceptibility of all additives must be taken into account. Because additives are not attached to the polymer molecule by chemical bonds, they are more readily attacked by chemicals present in the environment.

Plasticizers. Some naturally occurring polymers, such as rubber, are flexible and readily processed by appropriate machines. However, other naturally occurring polymers, such as gutta-percha, lignin, shellac, cellulose, and proteins, are relatively stiff. Both the cellulose present in paper and the proteins present in leather are readily softened but not dissolved by water, which is compatible with the polar polymers. In contrast, cellulose nitrate is intractable and is not softened by nonsolvents. However, flexible celluloid was produced in 1869 by adding camphor as a plasticizer to this manmade plastic (Ref 2). Plasticizers that are added to polymers such as polyvinyl chloride (PVC) lower the melt viscosity, elastic modulus, and T_g of intractable polymers (Ref 3).

Polyvinyl chloride was plasticized in 1933 by the addition of tricresyl phosphate or dibutyl phthalate (Ref 4). The plasticizer in PVC and other polymers is present as a cluster of molecules between the polymer chains. The volume of the polymer remains unchanged, but the movement of the polymer chain is less restricted, as indicated by an increase in the dielectric constant of the plasticized polymer. The thermodynamics of plasticizer-polymer interaction may be measured by determining the depression of vapor pressure and melting point by means of osmotic pressure measurements, light scattering, gas-liquid chromatography, solution viscosity, polymer swelling, and calorimetry.

The compatibility of a plasticizer with the polymer may also be predicted from the Hildebrand solubility parameter. The difference in solubility parameters between the polymer and the plasticizer should be less than 1.8H, or 6.3 $(J/cm^3)^{1/2}$ $[26.3(cal/cm^3)]^{1/2}$. More than one compatible additive may be used as a plasticizer as long as the mixture meets the solubility parameter requirements.

When added to an intractable polymer such as PVC, a plasticizer such as dioctyl phthalate (DOP) increases light stability, flammability, and susceptibility to attack by fungi and corrosives. Thus, in spite of moderately good resistance to attack by an oxidizing acid such as nitric acid, the resistance of PVC is lowered by the presence of plasticizers. The DOP plasticizer in PVC undergoes hydrolytic decomposition, the degree of which depends on the concentration of plasticizer present. Plasticizers with dielectric constants of 4 to 8 are compatible with PVC (Ref 5).

Processing Aids, Lubricants, and Impact Improvers. Lubricants and processing aids are added to intractable polymers to reduce both the sticking of the polymer to metal surfaces and the energy required for processing. Acrylic polymers are used both as processing aids and as impact improvers. Many molded or extruded PVC articles contain all three types of additives. If these additives are attacked by hostile environments or are extracted by solvents, they provide pathways for attack of the polymer by other chemicals.

Antioxidants and Ultraviolet (UV) Stabilizers. The resistance of natural rubber to corrosive environments is enhanced by vulcanization (cross linking by sulfur). In 1922, it was shown that the degradation of natural rubber in air was the result of oxygen absorption and the subsequent production of organic peroxides (Ref 6). These and other investigations showed that this deterioration of rubber and other unsaturated polymers could be prevented by the addition of hindered phenolic or secondary aromatic amine antioxidants.

Phenol consists of a benzene ring (C_6H_5) with a hydroxyl group (OH). When the carbon atoms adjacent to the hydroxyl-substituted carbon atom have bulky substituents in place of the hydrogen atoms, the compound is called a hindered phenol. Although this additive hinders oxidative degradation of rubber, it also provides a pathway for chemical attack, particularly by alkaline solutions.

Secondary amines (such as phenyl-β-naphthylamine [Ar_2NH], with Ar representing an aromatic hydrocarbon such as benzene) are also used as antioxidants in rubber. Because these additives react readily with acids, they provide a pathway for chemical attack, particularly by acids (Ref 7).

Organic polymers are also degraded by oxygen in the presence of ultraviolet radiation. The rate of this photo-oxidation may be decreased by the addition of pigments, such as carbon black or UV stabilizers (such as derivatives of hydroxy benzophenone). These additives may also provide a pathway for attack by aggressive chemicals (Ref 8).

Pigments. In the absence of strong interfacial bonds between pigments and polymers, corrosives are able to permeate polymer composites. However, this deleterious effect is decreased when coupling agents, such as organosilanes or organotitanates, are present (Ref 9). Pigments, such as carbon black, are inert to corrosives, but hydrophilic fillers, such as clay, will absorb water; carbonates, such as limestone, are decomposed by inorganic acids. The chemical resistance of metal-filled chemically resistant polymers is a function of the corrosion resistance of the metal filler.

Flame Retardants. When present as additives in polymers, chlorinated hydrocarbons may enhance the permeation of organic solvents. However, unless these additives are preferentially dissolved, they have little adverse effect on the permanence of polymers in corrosive environments. In contrast, the presence of hydrophilic flame retardants, such as ammonium polyphosphate (APP) or alumina trihydrate (ATH), enhances susceptibility of a polymer to polar environments.

Thermal Degradation

Similar to many other intrinsic properties, the thermal stability of polymers is a function of their structure (Ref 10) and that of any additives present. In the absence of an aggressive environment, thermal degradation occurs at elevated temperatures. The initial step in thermal degradation is the cleavage of the weakest bonds present in the polymer. The critical bond is the aliphatic CH, which permits a number of facile radical reactions that lead to cross linking or chain degradation. Their replacement (fluoropolymers or aromatic polymers) results in substantially improved stability.

The bond strength in polymers produced by step reactions, or condensation polymerization techniques, are usually stronger than those produced by chain or addition polymerization techniques. Thermal resistance is enhanced by the presence of double-stranded ladder polymer chains or heterocyclic rings and by cross linking (Ref 11).

In the absence of stabilizers, polymers with low ceiling temperatures, such as poly-α-methylstyrene, and polymers that can release small molecules, such as PVC, polyacrylonitrile (PAN), and polyisobutylene (PIB), undergo thermal degradation at relatively low temperatures via unzipping. When heated at 270 °C (520 °F), polymethylmethacrylate (PMMA) depolymerizes, producing methyl methacrylate. Polyoxymethylene (POM) also may be thermally depolymerized to produce formaldehyde, but this degradation is prevented in commercial polymers by capping the terminal hydroxyl groups or by producing a more stable copolymer of formaldehyde and ethylene oxide.

Hydrogen chloride is evolved when PVC is heated and the polymeric residue is a dark-colored conjugated diene, that is, a molecule with alternate ethylenic groups (–CH=CH–CH=CH–). Hydrogen cyanide is evolved when PAN is heated, and the unsaturated product then forms black graphitelike rings. Cellulose and starch also lose water when heated, and the final product is carbon (char). In all instances, the amorphous degradation processes are autocatalytic; that is, the degradation products accelerate the degradation reactions.

Thermal Oxidative Degradation

Thermal degradation also occurs in an autocatalytic process, at lower temperatures, when oxygen is present in the environment (Ref 12). Thus, oxygen is absorbed at 100 °C (212 °F) by the amorphous regions of polyethylene (PE). Regions that are initially inaccessible to oxidation become accessible as the tie chains break, but, as a result of realignment of the dangling segments, they again become inaccessible to oxygen. The initial oxidation of saturated polymeric hydrocarbons produces hydroperoxides by the insertion of oxygen atoms between the hydrogen and tertiary carbon atoms. Hydroperoxides are also produced by the oxidation of the alpha hydrogen atoms, that is, those next to the ethylenic double bonds in unsaturated polymeric hydrocarbons.

This oxidative degradation of polymers is catalyzed by heavy metals, such as copper, which undergo a reduction-oxidation reaction in the presence of oxygen. The rate of these degradation reactions may be followed by measuring molecular weight through gel-permeation chromatography or viscometry, measuring oxygen uptake, and monitoring the rate of formation of new groups, such as carbonyl (–C=O) groups, using Fourier transform infrared (FTIR) spectroscopy.

Photo-Oxidative Degradation

Because there have been many investigations of the outdoor oxidative degradation of natural rubber, considerable information is available on the photo-oxidative degradation of polymers. Because pure hydrocarbons absorb oxygen at wavelengths below 290 nm, it is believed that contaminants and adventitious photosensitizers, such as those with carbonyl and hydroperoxide groups, are responsible for the photo-oxidative degradation of polymers such as HDPE at wavelengths of 300 to 400 nm. In any case, the initial step is electronic excitation to produce an excited molecule that radiates excess energy before degradation. Fortunately, photo-oxidative degradation occurs predominately on the polymer surface.

Environmental Corrosion

The atmosphere is polluted by man-made smoke, industrial gases, and photochemical smog, as well as by the natural processes of volcanic eruptions, dust storms, and forest fires. The principal chemical pollutants are sulfur oxides, which, when oxidized, usually form sulfur trioxide or metallic sulfates. Nitrogen oxides, like sulfur oxides, are present in smog, hydrocarbons, and carbon monoxide. Industrial pollutants account for about 25% of the sulfur oxides and less than 10% of the nitrogen oxides in the atmosphere. The most significant source of air pollutants is the exhaust gas from the combustion of gasoline.

The damage by sulfur dioxide (SO_2) may be readily observed in deteriorated paper and leather goods. Sulfur dioxide also attacks coated metals at the metal-polymer interface in a process called filiform (thread-like) corrosion.

Nitroxy radicals ($\cdot NO_2$) abstract hydrogen atoms from saturated polymers, such as HDPE, to produce a macroradical (electron-deficient polymer) and nitrous acid (HNO_2). Nitroxy radicals also attack unsaturated polymers at a rate that is accelerated in the presence of UV radiation (Ref 13).

The atomic oxygen molecule (O) present in smog attacks polymers with tertiary hydrogen atoms, such as polypropylene (PP), as well as other polymers, such as POM (acetal). Singlet oxygen ($^1O_2$) reacts with unsaturated polymers, such as rubber, to produce hydroperoxy groups. Ozone (O_3) forms ozonides with rubber. These ozonides are inflexible, and hence cracking occurs when they are stretched.

Chemical Corrosion

In spite of the complexity of macromolecules, susceptibility to water-base chemicals is relatively simple and is comparable to that of small molecules with similar structures, that is, model compounds. Nobel laureate Flory showed that the rate of reaction of groups in polymers is independent of the size of the molecule, and thus the kinetics for reactions with readily available sites are similar to those for the model compounds (Ref 14).

Because these reactions are diffusion controlled, they occur more rapidly in amorphous than in crystalline regions. In general, fluorinated and chlorinated polymers and polymeric hydrocarbons, like their low-molecular-weight counterparts, have good resistance to chemical degradation. However, polyesters, polyamides, and polyacetals are readily hydrolyzed to produce derivatives of these reactants (Ref 15).

Polyolefins, such as PE and PP, are slowly attacked by oxidizing acids such as nitric acid and by nonoxidizing acids in the presence of oxidizing agents when carbonyl (C=O) and sulfate (SO_4) groups are present. These polymeric hydrocarbons are attacked by chromic acid at room temperature (Ref 16). Polyethers, such as POM, polyvinyl butyral, and acetals are readily hydrolyzed. All polyesters are hydrolyzable, but the aromatic polyesters are more resistant to this type of degradation than are the aliphatic polyesters.

Degradation Detection

Changes in color and texture, which may occur with the chemical attack of polymers, may be observed visually. These effects may be amplified by accelerated exposure conditions. Changes in physical properties, such as hardness, and in compressive strength and molecular weight may also be monitored during exposure. In addition, dynamic mechanical testing has been used to show the rate of change in physical properties when polymers are exposed to aggressive environments (Ref 17). Several reviews on the evaluation of such tests are available (Ref 18–20).

Environmental stress-cracking tests on bent, notched specimens that record the time lapse before breakage are described by ASTM D 1693. Other readily observable tests have been developed on the bleeding of color, migration, and volatility of plasticizers (ASTM D 1203), water vapor transmission (ASTM E 96), hardness, tensile strength, flexural strength, compressive strength, and impact resistance. A large number of tests have also been described for measuring the durability of polymeric coatings (Ref 21).

Tests have also been described for measuring the resistance of elastomers to liquid fuels (Ref 22). The solvent resistance of acrylonitrile elastomers is a function of the acrylonitrile content (Ref 23). It is also recognized that a trace of solvent can cause the mechanical failure of elastomers in hostile environments (Ref 24).

While visual and physical tests are considered significant and continue to be used, instrumental and spectroscopic tests provide more information on changes in molecular microstructure as a result of exposure of polymers to aggressive environments (Ref 25).

Sensitive pressure transducers have been used to measure oxygen absorption of polymers (Ref 26). Most of these tests are nondestructive, but a small, representative sample must be used when pyrolytic gas chromatographic procedures are used (Ref 27).

Ultraviolet-visible spectroscopy has been used and amplified by derivatization, such as the formation of hydrazones (Ref 28). Chemiluminescence, such as excimer fluorescence, has been used to monitor the degradation of polymers (Ref 29). Fourier transform infrared spectroscopy has been used to detect changes in structure during polymer exposure to hostile environments. The difficulties associated with overlapping in infrared absorption can be overcome by derivatization; for example, carboxylic groups may be converted to acid fluorides by treating the polymer with sulfur tetrafluoride (SF_4) (Ref 30). Internal reflectance spectroscopy has also been used to study changes on the polymer surface (Ref 31).

Nuclear magnetic resonance (NMR) spectroscopy, which is based on the interaction between nuclear dipole moments and a magnetic field, has been used to determine moisture content, modulus variation, degradation, aging, diffusion, and the degree of cure of polymers (Ref 32). The oxidation of HDPE and polymer degradations have been monitored by ^{13}C NMR (Ref 33). Electron spin resonance (ESR) has been used to monitor the production of macroradicals and low-molecular-weight radicals resulting from the cleavage of polymer chains (Ref 34).

The x-ray photoelectron analysis (ESCA) technique can be used to assign and monitor chemical peaks, such as peaks of CF_2 and CH_2, on the polymer surface and at various depths in the polymer sample. For example, ESCA has been used to show the disappearance of fluorine groups and the formation of carbon atoms with reactive sites when polytetrafluoroethylene is reacted with sodium metal in liquid ammonia (Ref 35).

Effect of Environment on Performance*

The mechanical properties of polymeric materials are often segregated into short-term and long-term properties. The category of short-term properties includes such things as tensile and impact strengths. Long-term properties include creep, stress relaxation, and creep (stress) rupture. Both categories of properties are affected by exposure to external chemical environments.

With any polymeric material, chemical exposure may have one or more different effects. Some chemicals act as plasticizers, changing the polymer from one that is hard, stiff, and brittle to one which is softer, more flexible, and tougher. Often these chemicals can dissolve the polymer if they are present in large enough quantity and if the polymer is not crosslinked. Other chemicals can induce environmental stress cracking (ESC), an effect in which brittle fracture of a polymer will occur at a level of stress well below that required to cause failure in the absence of the ESC reagent. Finally, there are some chemicals that cause actual degradation of the polymer, breaking the macromolecular chains, reducing molecular weight, and diminishing polymer properties as a result. Each of these effects is examined in subsequent paragraphs.

Plasticization, Solvation and Swelling

Certain interactions between liquid chemicals and polymers can be understood through the use of solubility parameters. The Hildebrand solubility parameter is the square root of the cohesive energy density, the latter being the energy required to vaporize one mol of a liquid (Ref 36). Solubility parameter values for both low-molecular-weight liquids and polymeric materials are tabulated in many references (Ref 37, 38). When linear or branched thermoplastic polymers are exposed to large enough quantities of solvents having solubility parameters within approximately ±2H of that of the polymer, dissolution of the polymer will occur. In smaller quantities, these solvents will be adsorbed by the polymer. With polymer-solvent combinations having solubility parameter differences outside this range, some adsorption of the solvent by the polymer may still occur. When large differences between solvent and polymer solubility parameters exist, the solvent will have no apparent effect. Amorphous polymers absorb chemicals more readily than crystalline polymers, and the rate varies inversely with the degree of crystallinity. Cross-linked polymers will not dissolve, but will swell significantly when exposed to chemicals having similar solubility-parameter values.

The impact of these interactions on the mechanical properties and failure of an affected polymer are many. One effect can be plasticization of the polymer by the adsorbed chemical. Plasticization of a polymer can result in the polymer being transformed from a rigid, glassy material to a soft flexible material (Ref 39, 40). Great advantage is taken of this effect in the polyvinyl chloride (PVC) industry. The PVC can be altered from the rigid material from which plastic pressure pipes are made to the extremely soft, flexible material from which medical tubing is made through the judicious use of plasticizers (Ref 41). However, the plasticization effect reduces both tensile strength and stiffness of the affected plastic and also accelerates the creep rate of the material if it is under stress. If an application of a particular plastic requires a certain minimum level of strength or stiffness, unintentional plasticization by exposure to a chemical that the plastic can adsorb could accelerate failure by reducing those properties (Ref 42).

Plasticizer loss from an intentionally plasticized polymer may also have an adverse effect on polymer performance. Plasticizer migration from PVC is a well-known phenomenon (Ref 43) that results not only in embrittlement and/or a loss of flexibility of the PVC part but also loss of the plasticizer chemical(s) into the environment. Plasticizer migration from flexible PVC products also results in some small amount of shrinkage of the products. This may be problematic if close dimensional tolerances are necessary.

The effects mentioned previously will occur in both amorphous and crystalline plastics, but they may not be as visibly evident in crystalline ones. Crystalline plastics often do not appear to be as affected by interactions with solvents, since diffusion of solvent into the crystalline regions is much more limited (Ref 44). Adsorption of solvents into the amorphous regions of a crystalline polymer will create the discussed effects within those regions of the polymer morphology; this can result in changes in polymer mechanical properties. For example, nylon plastics will absorb moisture from the air. An extremely dry nylon may be rather brittle, while that same nylon exposed to 50% relative humidity for several days can be quite tough. However, the short-term tensile strength and modulus of the hydrated nylon will be somewhat reduced, as will the long-term (creep rupture) strength of the material. Creep deformation of the hydrated nylon will proceed more rapidly than that in the dry material at the same level of stress. Hydrocarbon liquids will have similar effects on polyethylene.

In cross-linked rubber products, swelling will result when exposure to chemical solvents occurs (Ref 45). The swelling can be rather extreme when the solubility parameter difference between rubber and solvent is small. Highly swollen rubbers will exhibit a severe loss of strength and stiffness, a high creep rate, and a high rate of stress relaxation or compression set. This last phenomenon can have an extremely adverse effect on rubber gaskets and O-rings by allowing the compressive stress in such a seal to decrease to a level so low that leakage of the seal can occur. Even with polymer/solvent pairs for which the solubility-parameter difference is somewhat greater and swelling seems not as severe, these effects can occur. Rubber compounds are sometimes intentionally plasticized like PVC in order to achieve a desired level of compressibility, flexibility, and so forth while using a specific polymer as the base for the compound. Loss of these intentional plasticizers into the environment will have the same effects as with PVC.

With respect to identifying adsorbed chemicals in plastics, since the chemical is adsorbed, it can usually be extracted in some way and identified. In some cases, simply heating the plastic will drive off the chemical, which can be collected and fed into a gas chromatograph (GC) or a GC/mass spectrometry (GC/MS) analysis. Sometimes extraction with a second chemical solvent is necessary, one that is a better solvent for the plasticizer than the polymer is. The extract can then be tested for the presence of other chemicals to identify plasticizers. When the quantity of absorbed chemical is great, an infrared spectrum of the plasticized plastic can be obtained by FTIR. An FTIR spectrum of the unplasticized plastic, or a reference spectrum from a spectral library of such, can then be subtracted from the subject spectrum by the onboard computer of the instrument. The resultant subtraction spectrum will often be that of the absorbed chemical, which can then be identified by its infrared spectrum.

Environmental Stress Cracking

Environmental stress cracking (ESC) of plastics has been defined as "... the failure in sur-

*Adapted from Donald E. Duvall, Effect of Environment on the Performance of Plastics, *Failure Analysis and Prevention,* Volume 11, *ASM Handbook,* ASM International, 2002, p 796–799 face initiated brittle fracture of a specimen or part under polyaxial stress in contact with a medium in the absence of which fracture does not occur under the same conditions of stress" (Ref 46). Virtually all plastics are stress cracked by some chemical environments. The biggest problem with this is that each plastic has its own set of stress cracking reagents, and those chemicals that stress crack one type of plastic will have no effect on others. Thus, the potential stress cracking effect of a specific chemical on a specific plastic must be known from prior work or elucidated by direct experiment in order to know whether or not a problem exists.

Several physical characteristics are typical of environmental stress crack failures:

- Failure is always nonductile, even in plastics that would normally exhibit a ductile yielding failure mechanism.
- The brittle fracture is surface initiated.
- The surface at which cracking initiated was in contact with a chemical reagent.
- The plastic was mechanically stressed in some way; internal (residual) stresses or externally applied stresses both qualify.

Chemicals that induce ESC usually have no other apparent effect on the plastic in question, that is, no swelling (or dissolution in large quantities of the chemical) and no physical or chemical changes in the polymer that might be detected by analytical methods. In the absence of a mechanical stress, the ESC chemical has no discernible effect. Conversely, the magnitude of stress that will cause ESC will not cause fracture if imposed in the absence of the stress cracking reagent. Thus conventional chemical resistance tests run on plastics, in which unstressed tensile bars are soaked in a chemical and withdrawn periodically for testing, give no indication of the possibility of ESC for any polymer-reagent system. It is only in the presence of both mechanical stress and chemical environment that ESC occurs.

Failures from ESC may occur early or late in the life of a product. In some cases, ESC will occur as soon as a part is loaded, if the reagent is already present on the surface of a previously unstressed part. Stress cracking reagents also impact the creep rupture properties of plastics by shortening the time for brittle fracture to occur over that which exists in the absence of the reagent. Sometimes ESC reagents will create brittle fracture at a low stress level in a polymer such as polyethylene that normally fails in a highly ductile manner. These effects are summarized in the creep rupture curves of Fig. 1. High-density polyethylene exhibits ductile failure (elongations to break of several hundred percent) at stresses near to its reported yield stress. At lower stresses and longer failure times, a different molecular mechanism controls failure, and brittle fracture occurs at elongations of less than 5%. Presence of an ESC reagent on the surface of a plastic can dramatically shorten the time for failure to occur at a given stress level and change the failure mechanism from highly ductile to macroscopically brittle.

One condition in which the possibility of ESC should be considered is when there is an apparent stain or other deposit residue on the surface of a fractured part. Even though stress cracking chemicals are not adsorbed into the plastic to any significant extent, surface residues are often left behind that can be identified. Once the chemical that left the deposit is known, reference to prior work reported in the literature may tell whether or not it is an ESC agent for the particular plastic in question. If no previously identified problems with that chemical can be found, laboratory stress cracking tests of the polymer/chemical combination can be conducted to assess the likelihood of ESC.

Polymer Degradation by Chemical Reaction

Another effect that chemical environments can have on plastics is to actually degrade the polymer, that is, to break down the polymer chains into lower molecular weight compounds that no longer have the desirable strength or toughness properties of the original. Certain polymer types are more susceptible than others to specific degradation mechanisms, but all polymers can be degraded by at least one mechanism. The most common degradation mechanisms are discussed in the following paragraphs.

Hydrolysis. Polymers created by stepwise reactions, for example, polyesters and nylons, form water as a reaction product along with the polymer. Under certain circumstances of exposure to aqueous environments, the polymerization reaction can essentially be reversed and the polymer broken down. Normally, these hydrolytic degradation reactions occur at extremely slow rates, and certainly nylon and polyester fabrics can be repeatedly washed in water without adverse effect. However, at conditions of either low (>4) or high (<10) pH, the rate of hydrolysis may become perceptible and result in molecular weight reduction and mechanical property diminution.

These polymer types can also degrade during processing (i.e., extrusion or injection molding) if there is moisture in the material. Even at neutral pHs, the elevated temperatures used for polymer molding or extrusion (175 to 250 °C, or 350 to 500 °F, or more) will cause hydrolytic degradation if there is moisture in the resin. Because of this, polymer resin manufacturers advise drying of the material just prior to processing to reduce the moisture content to a low enough level that hydrolysis will not occur while the resin is heated in the manufacturing equipment. It is often the case with plastics that are susceptible to hydrolytic degradation that a reduced polymer molecular weight is found during the failure analysis. The challenge for the analyst then becomes deciding whether the degradation occurred during fabrication of the part or on exposure to an aqueous service environment.

Thermal Degradation. High-molecular-weight polymers will also break down upon exposure to elevated temperatures. Sufficient thermal energy can be input to a polymeric material to break the covalent chemical bonds that hold polymer molecules together. This bond breakage (chain scission), if it occurs in the polymer backbone, will reduce molecular weight. Chain scission of side-chain branches may also alter the polymer structure sufficiently to change appearance or mechanical properties enough to create a premature failure.

As with hydrolysis, thermal degradation can occur both in processing and in an end-use environment. In molding or extrusion operations, the molten plastic is exposed not only to elevated temperatures but also to mechanical shearing. The combination of the two may reduce molecular weight to the extent that performance properties will suffer. Thermally degraded plastics also tend to discolor, and sometimes a plastic product is deemed a failure because it no longer has the desired cosmetic appearance due to thermal degradation. These same changes may be observed in end-use environments, albeit at much slower rates due to the lower temperatures at which plastics are usually used.

Thermal degradation of polymers is a chain reaction that begins when an atom (usually a hydrogen atom) is abstracted from the polymer chain, leaving behind an unpaired electron from the broken covalent bond at an atom on the chain (Ref 47). The free radical thus formed may react in several different ways, one of which results in chain scission and molecular-weight reduction. Fortunately for plastics usage, there are chemical additives compounded into polymers that will react with the unpaired electron, interrupt the chain reaction, and postpone or at least greatly retard thermal decomposition. If thermal degradation is believed to be a contributing factor to failure, the type and amount of these additives should be checked to be certain that failure was not due to degradation in an unprotected polymer material.

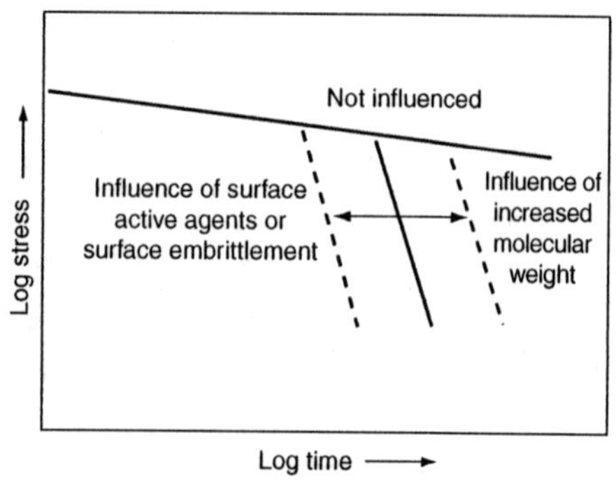

Fig. 1 Effect of environmental stress cracking agents on creep rupture performance

Oxidation. Many polymers, especially the olefins and others with long olefinic segments in the polymer-chain backbone, will oxidize when exposed to oxygen-containing environments. As with thermal degradation, oxidation usually commences by formation of a free radical on the polymer chain. An oxygen atom from the environment will then react with the unpaired electron to form a hydroperoxy radical. This will then degrade by one of several reactions, some of which result in chain scission and property loss (for details of oxidation chemistry, see Ref 47).

As with thermal degradation, chemical additive stabilizers and antioxidants can be added to the polymer that will break the chain reaction in a variety of ways, preserving polymer properties at least until the additives have been consumed. If oxidative degradation is a possible contributing factor to a premature failure, it becomes necessary to determine what allowed it to occur. It may be that the stabilizers were not present originally in the proper types or amounts. If the polymer resin did contain antioxidants, then what caused them to become ineffective must be determined. In some cases, stabilizers can bloom to the surface of a plastic part and be removed by ablation, dissolution, or evaporation into the environment (Ref 48). In other cases, the additives may simply have been consumed doing the job for which they were intended, and premature oxidation occurred because the service environment was at a higher temperature than the design engineer anticipated. All these can lead to an oxidized polymer with reduced mechanical properties, unacceptable appearance, or other deficiencies.

Photodegradation. With many polymeric materials, UV radiation can be the source of energy that will abstract an atom from the polymer backbone and start the degradation process. It is well known that prolonged outdoor exposure of plastics will initially cause color changes that may be undesirable. Oxidation initiated by UV radiation will result in eventual loss of properties as well. Once again, there are chemical additives that will retard these processes, but eventually they will be consumed and degradation will proceed. The plastics design engineer must be certain that the UV radiation stabilizers are present in the proper types and amounts to yield a product that will operate for its intended life without undergoing an inordinate amount of degradation from the exposure. There are both accelerated indoor and outdoor test methods that are used to assess the level of stability to UV exposure of a plastic material.

Surface Embrittlement

An adverse effect of polymer degradation on plastic part performance does not require changes in the bulk of the material in that part. In many cases, it is only necessary to cause degradation in a thin surface layer of the part in order for performance to be compromised. This phenomenon has been observed in both long-term and short-term properties of many polymeric materials.

In fact, polymer degradation is often limited, at least initially, to the surfaces of exposed plastic products. Oxidative degradation initiated by either purely thermal means or by UV radiation occurs initially at the surfaces, because that is where oxygen concentration is the greatest. In order for oxidation to occur deeper in a specimen, it must diffuse in. Because oxygen reacts very rapidly with free-radical species, oxidation below the surface of a polymer part is diffusion limited and occurs very slowly compared to surface oxidation. Hydrolytic degradation also occurs first and most rapidly at surfaces since that is where the concentration of water is the greatest. Since often all that is needed for premature failure to occur is to generate a sufficient level of degradation at the surface, even if the bulk material within the plastic part is unaffected, the fact that degradation is initially limited to the surfaces creates problems for product performance.

The impact of surface degradation on short-term properties has been demonstrated by many authors. Numerous studies have shown that considerable reductions in tensile strength, impact strength, and toughness have been observed for oxidation degradation extending only a short distance into a specimen (Ref 49, 50). Figure 2 illustrates this effect in polyethylene. Choi (Ref 52, 53) demonstrated that the creep rupture behavior of a polyethylene pipe resin could be compromised by a certain level of oxidative degradation occurring only in the first 50 μm below the surface of a 2.5 mm (0.1 in.) thick specimen. Figure 3 (Ref 53) illustrates how surface degradation of a plane strain tension specimen alters the ductile brittle transition in polyethylene creep rupture.

The existence of this surface embrittlement phenomenon requires that the evaluation of surface-initiated brittle fracture in an otherwise ductile polymer include characterization of material taken only from the surface. The results can be compared to similar testing of the core of the failed part or to testing of a control sample of the material to determine whether or not degradation of material at the surface is a contributing factor.

Temperature Effects

One final effect of environment on polymer performance is that of temperature. Polymeric materials will exhibit a transition between two very different types of mechanical behavior as the environmental temperature passes through T_g (Ref 54). Figure 4 shows how the modulus of an amorphous polymer changes as temperature is increased or decreased through this critical

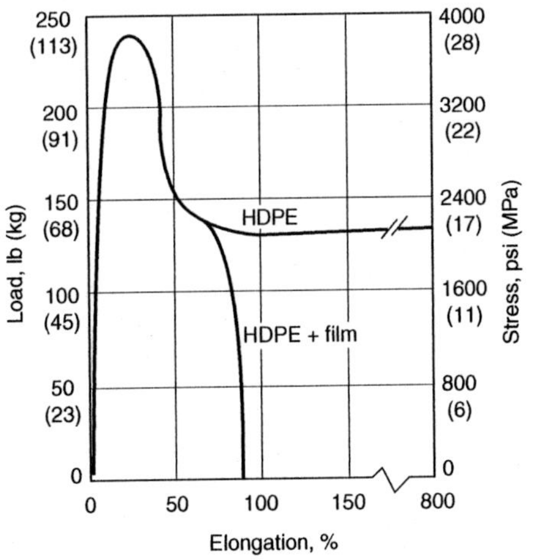

Fig. 2 Effect of thin brittle film on stress-strain behavior of high density polyethylene. Source: Ref 51

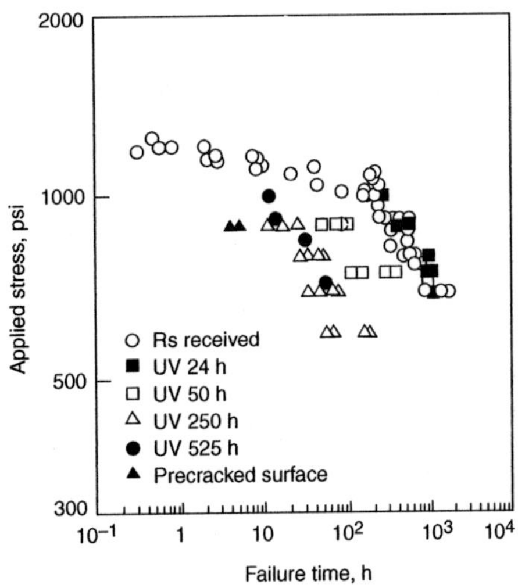

Fig. 3 Effect of surface embrittlement from varied UV exposure times on creep rupture behavior of polyethylene at 80°C (175 °F). Source: Ref 52

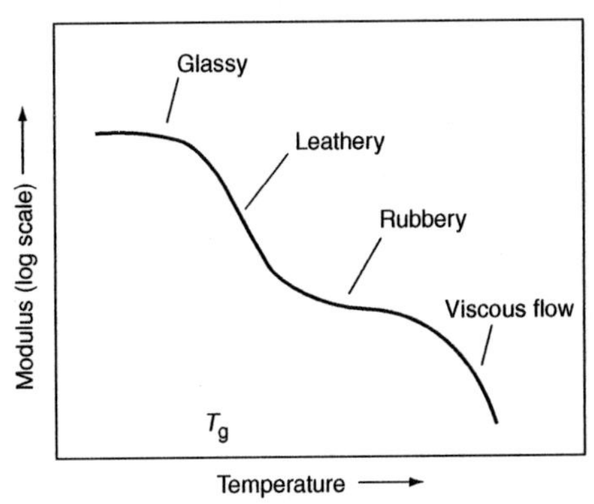

Fig. 4 Modulus versus temperature for a typical linear polymer. Source: Ref 54

region. At temperatures well below T_g, this polymer behaves like a glassy material, with a relatively high modulus and low energy to break. At temperatures well above T_g, the same polymer has two or three orders of magnitude lower modulus and will either flow like a very viscous liquid or fail in tension at high extension, depending on whether or not the polymer is cross linked. In the temperature range over which the glass transition is occurring, a mixed mechanical property behavior will occur, with reduced modulus and increased ductility versus the glassy state but not as extreme as the rubbery state.

This phenomenon also manifests itself in semicrystalline polymers, where the amorphous component of the material also exhibits a T_g. The T_g is always lower than the melting temperature (T_m) of a semicrystalline polymer. The extent to which mechanical properties are altered as temperature changes around T_g depends on the relative amounts of crystalline and amorphous material that exist in the polymer in question. Of course, when temperature is above both T_g and T_m, the material exhibits either rubbery or viscous fluid behavior depending on whether or not it is cross linked.

The actual temperature range over which the glass-transition phenomenon occurs will vary somewhat as the rate of deformation of the polymer changes. High-speed (high-strain-rate) deformation favors nonductile failure while low-speed (low-strain-rate) deformation favors more ductile failure. The actual numerical value of T_g will vary with the rate of testing used to make the measurement. Thus, when assessing temperature effects on failure mode, it is necessary to know how the environmental temperature compares to the polymer T_g. It is also necessary to factor in loading rate, especially when the environmental temperature is near T_g.

REFERENCES

1. J. Crank and G.S. Park, Ed., *Diffusion in Polymers,* Academic Press, 1968
2. J.W. Hyatt, U.S. Patent 105,338, 1869
3. R.B. Seymour and R.D. Deanin, Ed., *History of Polymeric Composites,* VNU Science Press, 1987
4. W.L. Semon, U.S. Patent 1,929,453, 1933
5. P.H. Foss and M.T. Shaw, *J. Vinyl Technol.,* No. 12, 1985, p 165
6. C. Moureau and C. Dufraisse, *Bull. Soc. Chim.,* Vol 31 (No. 4), 1922, p 1152
7. R. Gachter and H. Muller, *Plastics Additives Handbook,* Hanser, 1985
8. R.B. Seymour, Ed., *Additives for Plastics,* Vol I and II, Academic Press, 1978
9. R.B. Seymour and R.H. Steiner, *Plastics for Corrosion Resistance Applications,* Reinhold, 1955
10. R.B. Seymour and C.E. Carraher, *Structure Property Relationships in Polymers,* Plenum Press, 1984
11. R.B. Seymour and G.S. Kirshenbaum, Ed., *High Performance Polymers: Their Origin and Development,* Elsevier, 1986
12. R.K. Eby, Ed., *Durability of Macromolecular Materials,* ACS Symposium Series, American Chemical Society, 1979
13. B. Ranby and J.F. Rabek, chapter 17, *The Effects of Hostile Environments on Coatings and Plastics,* D.P. Garner and G.A. Stahl, Ed., ACS Symposium Series, 229, American Chemical Society, 1983
14. R.B. Seymour, *Plastics Versus Corrosives,* John Wiley & Sons, 1982
15. R.A. McCarthy, *Encyclopedia of Polymer Science and Engineering,* Vol 3, J.I. Kroschwitz, Ed., Wiley-Interscience, 1985
16. D.J. Carllson and D.M. Wiles, *Encyclopedia of Polymer Science and Engineering,* Vol 1, J.I. Kroschwitz, Ed., 1985
17. L. La Marre and C. de Tourreil, *Elastomerics,* Vol 119 (No. 11), 1986, p 17
18. M.S. Allen, Ed., *Degradation and Stabilization of Polyolefins,* Applied Science, 1983
19. M.C. Billingham, D.C. Bott, and A.S. Manke, *Dev. Polym. Deg.,* Vol 3 (No. 63), 1987
20. J.L. Koenig, *Adv. Polym. Sci.,* Vol 54 (No. 87), 1983
21. R. Lambourne, chapter 19, *Paint and Surface Coatings,* Halsted Press, 1987
22. J.R. Dunn and R.G. Vara, *Elastomerics,* Vol 119 (No. 5), 1986, p 29
23. H.A. Pfisterer and J.R. Dunn, *Rubber Chem. Technol.,* Vol 53 (No. 357), 1988
24. S. Ogintz, *Elastomerics,* Vol 119 (No. 11), 1987, p 21
25. D.W. Dwight and H.R.N. Lawrence, *Elastomerics,* Vol 119 (No. 7), 1987, p 20
26. D.W. Grattan, D.J. Carlsson, and D.M. Wiles, *Chem. Ind. London,* 1978, p 228
27. S.A. Liebman and E.J. Levy, chapter 35, *Polymer Characterization,* C.D. Cramer, Ed., ACS Advances in Chemistry Series, American Chemical Society, 1983
28. M.C. Billingham and P.D. Calvert, *Degradation and Stabilization of Polyolefins,* N.S. Allen, Ed., Applied Science, 1983
29. O.I. Soutar, a chapter in *Polymer Yearbook,* No. 2, R.A.P. Harwood, Ed., Academic Press, 1985
30. D.J. Carlsson and S.M. Minera, *J. Appl. Polym. Sci.,* Vol 27 (No. 1589), 1982
31. P. Blais, M. Day, and D.M. Wiles, *J. Appl. Polym. Sci.,* Vol 17 (No. 1895), 1975
32. G.A. Matzkanin, *Plast. Eng.,* Vol 43 (No. 5), 1987, p 37
33. F.A. Bovey, F.S. Schilling, and H.N. Cheng, *Adv. Chem. Ser.,* Vol 169 (No. 133), 1978
34. K.T. Suvi, *Adv. Polym. Sci.,* Vol 12 (No. 131), 1973
35. D.J. Clark, A. Dilks, and H.R. Thomas, *Dev. Polym. Deg.,* Vol 1 (No. 87), 1977
36. M.P. Stevens, *Polymer Chemistry: An Introduction,* 2nd ed., Oxford University Press, 1990, p 43
37. E.A. Grulke, Solubility Parameter Values, Chap. VII, *Polymer Handbook,* 4th ed., J. Brandup, E.H. Immergut, and E.A. Grulke, Ed., John Wiley & Sons, 1999
38. *Physical Properties of Polymers Handbook,* J.E. Mark, Ed., AIP Press, 1996
39. E. Miller, Properties Modification by Use of Additives, *Engineered Materials Handbook,* Vol 2, *Engineering Plastics,* ASM International, 1988, p 493–507
40. A. Kumar and R. Gupta, *Fundamentals of Polymers,* Section 2.3, McGraw-Hill, 1998.
41. J.K. Sears and J.R. Darby, Solvation and Plasticization, Chap. 9, *Encyclopedia of PVC,* Vol 1, 2nd ed., Marcel Dekker, 1986.
42. M. Ezrin, *Plastics Failure Guide: Cause and Prevention,* Hanser Publishers, 1996, p 22–23
43. J.-M. Vergnaud, *Polym.-Plast. Technol. Eng.,* Vol 20 (No. 1), 1983, p 1–22
44. F.W. Billmeyer, *Textbook of Polymer Science,* 3rd ed., Wiley-Interscience, 1984, p 344
45. S.L. Rosen, *Fundamental Principals of Polymeric Materials,* Section 3.4, 2nd ed., Wiley-Interscience, 1993
46. J.B. Howard, *SPE J.,* Vol 15, 1959, p 397
47. N. Grassie and G. Scott, *Polymer Degradation and Stabilization,* Cambridge University Press, 1985
48. U.W. Gedde et al., *Polym. Eng. Sci.,* Vol 34 (No. 24), 1994, p 1773
49. P.K. So and L.J. Broutman, *Polym. Eng. Sci.,* Vol 22 (No. 14), 1982, p 888
50. J.C.M. de Bruijn, "The Failure Behavior of High Density Polyethylene Products with an Embrittled Surface Layer Due to UV Exposure," Delft University of Technology, Delft, Netherlands, 1991.
51. L.J. Broutman, "Surface Embrittlement of Polyethylene," GRI-81-0030, Final Report to the Gas Research Institute, Chicago, Nov 1981
52. S.-W. Choi, "Surface Embrittlement of Polyethylene," Ph.D. dissertation, Illinois Institute of Technology, Chicago, 1992
53. S.W. Choi and L.J. Broutman, Proceedings of the 11th Plastic Fuel Gas Pipe Symposium (San Francisco), 3–5 Oct 1989, p 296–320
54. A. Eisenberg, The Glassy State and the Glass Transition, Chap 2, *Physical Properties of Polymers,* 2nd ed., American Chemical Society, 1992

Characterization of Weathering and Radiation Susceptibility*

ALL ENGINEERING PLASTICS are affected by outdoor weather. Weather and radiation factors that contribute to degradation in plastics include temperature variations, moisture, sunlight, oxidation, microbiologic attack, and other environmental elements.

The results of exposing plastics to these conditions can be discoloration, loss of mechanical strength, embrittlement, and loss of electrical insulation and resistance properties. The degree to which a particular material degrades depends on its susceptibility to each of the above factors. Outdoor degradation factors affect some materials more than others because of their chemical structure.

A plastic material may have excellent resistance to a particular weathering factor. However, the plastic material may be highly susceptible to degradation when exposed to other weathering factors or to a combination of factors. For example, elevated temperatures will increase the oxidation rate of a material as well as the rate of photochemical reactions.

Test methods designed to simulate natural weathering at an accelerated rate have been developed to help predict the reaction of a plastic material to weathering prior to its field use. These tests expose specimens to extreme conditions to accelerate the aging process so that long-term weathering effects can be estimated in a shorter, more useful time period. Artificial weathering tests do not necessarily perfectly forecast the response that a plastic material will exhibit in actual use. It is important to note that accelerated weather aging may accelerate the degradation of a material beyond a point where it will no longer represent the actual reaction that will occur over an extended time period.

This article presents a general overview of outdoor weather aging factors, their effects on plastic materials, and the accelerated test methods that can be used to estimate the reaction of a plastic component during actual use.

Degradation Factors

The following factors, present in outdoor weathering, will contribute to material degradation. Although the characteristics of exposure to each degradation factor are described individually, it is important to note that a material is typically exposed to more than one factor during outdoor use. Therefore, degradation is usually driven by one principal factor at a time. However, multiple factors may affect property performance.

Ultraviolet (UV) Radiation. Degradation due to UV radiation (sunlight) is the primary concern when plastics are meant for outdoor use. The photochemical effect of sunlight on a plastic material depends on the absorption properties and bond energies of the material. The wavelengths that have the most effect on plastics range from 290 to 400 nm (2900 to 4000 Å). Table 1 shows the wavelengths that have the greatest photochemical effect on various plastics.

The activation spectrum (a plot of specific degradation characteristics versus the incident wavelength) of a material indicates its sensitivity to the exposed wavelengths. Each activation spectrum is measured by observing a specific reaction to degradation. Ultraviolet radiation must be present for degradation by a photochemical process to occur. Therefore, the absorption properties of the plastic are important in determining the activation spectrum. Figure 1 shows the activation spectrum of polycarbonate, using yellowing as the measured reaction factor.

Stabilizers can be added to a polymer to influence wavelength sensitivity and radiation absorption. Other additives, such as pigments, also change the absorption characteristics of the material. The effect of adding a UV stabilizer to a polyester is illustrated in Fig. 2. The addition of pigment can act as a UV screen to varying degrees. Therefore, an unpigmented material would seem to be most susceptible to UV degradation. However, the use of certain pigments with specific polymers actually photosensitizes the material and can accelerate UV degradation.

The absorption of UV radiation alone may not necessarily cause the degradation of a plastic material. A wavelength whose photon energy corresponds to a particular bond energy in the polymer chain can break the bond (chain scission). For this reason, the observed degradation will vary with wavelength. Longer wavelengths tend to penetrate more deeply into a plastic material but have a moderate degradative effect because they are not easily absorbed. Shorter wavelengths tend to have a greater effect on the surface of the material because their total energy can be absorbed within a few micrometers of the surface.

Ultraviolet radiation absorption on the surface of a material can result in chalking, which is a surface film that breaks molecular bonds. Ultraviolet radiation also causes discoloration (yellowing and bleaching) and loss of physical and electrical properties.

Elevated or lowered temperatures can degrade a plastic material. There are generally three aspects to elevated-temperature exposure: elevated temperature over a long period of time, elevated temperature over a short period of time, or cyclic exposure to elevated and lowered temperatures, such as may occur during alternating day and night exposure.

Exposure to lowered temperature can cause a plastic to become brittle; the modulus of the material increases, while the elongation and impact resistance may decrease. Cracking and a propensity to fracture on impact can occur. Exposure to elevated temperature can result in loss of mechanical properties (embrittlement, loss of impact strength, flexibility, and elongation) and loss of electrical properties. Discoloration, cracking, chalking, loss of gloss, and flaking can occur. Cyclic exposure can result in mechanical fatigue failure and cracking due to alternating expansion and contraction of the material. It can also result in electrical failure due to the formation of minute cracks that may become contaminated with dirt and moisture, forming a conductive medium that promotes electrical tracking. Cyclic exposure is an important factor when considering the service environment of a plastic material. For example, the

*Adapted from Laura C. Delre and Robert W. Miller, Characterization of Weather Aging and Radiation Susceptibility, *Engineering Plastics,* Volume 2, *Engineered Materials Handbook,* ASM International, 1988, pages 575 to 580

temperature in Reno, NV, has a daily mean variation of 20 °C (37 °F) and can range annually from –30 to 40 °C (–19 to 106 °F).

Moisture is absorbed when a plastic is exposed to water or humidity. Outdoor exposure to moisture can include rain, snow, humidity, and condensation. The chemical effect, known as hydrolysis, may be a major factor in the degradation of condensation polymers such as polyesters, polyamides, and polycarbonates. Water can also attack the bonds between the polymer and an additive, such as a pigment, resulting in chalking. Rain can wash away any additives, such as flame retardants, that may have bloomed or migrated to the surface as a result of sunlight or another outdoor exposure factor.

The moisture absorbed by a plastic may affect its electrical insulation resistance, dielectric and mechanical strength properties, dimensions, and appearance. The type of water exposure a plastic encounters, as well as the physical shape of the plastic part, will affect the degree of change in properties that the material experiences.

Oxidation. Although most plastic materials react slowly with oxygen alone, elevated temperature and UV radiation will accelerate the oxidation process. Oxygen that is aided by heat (thermal oxidation) and UV radiation (photo-oxidation) will attack the bonds in a polymer chain. Depending on the material, oxygen may either form carbonyl groups or cross link.

Oxidation can also occur when certain materials are exposed to ozone, which is a by-product of high-voltage partial discharge (corona). The presence of ozone, combined with oxygen, will affect polymers by attacking covalent bonds, such as those present in natural rubber materials. Rubber is used to insulate high-voltage equipment, such as ignition cable, where a high degree of flexibility is essential. Because ozone is generated by and often surrounds high-voltage equipment, it is beneficial to use materials with good ozone resistance. Materials such as butyl rubber, synthetic rubbers, and some thermoplastic elastomers have few double bonds for the ozone to attack.

Microbiological Attack. The resin portion of a plastic material is generally not susceptible to attack by fungi or bacteria. It is the additives, such as plasticizers, lubricants, stabilizers, pigments, and their carrier systems, that can be susceptible to microbiologic attack. Additives that are not distributed evenly will provide areas of preferential growth for fungi or bacteria. The most favorable conditions for growth are high temperature and humidity. The growth of fungi and bacteria on a plastic material will result in discoloration, surface attack, and loss of optical transmission.

Changes in electrical properties are primarily caused by the surface growth of fungi and bacteria, as well as pH changes created by their excre-

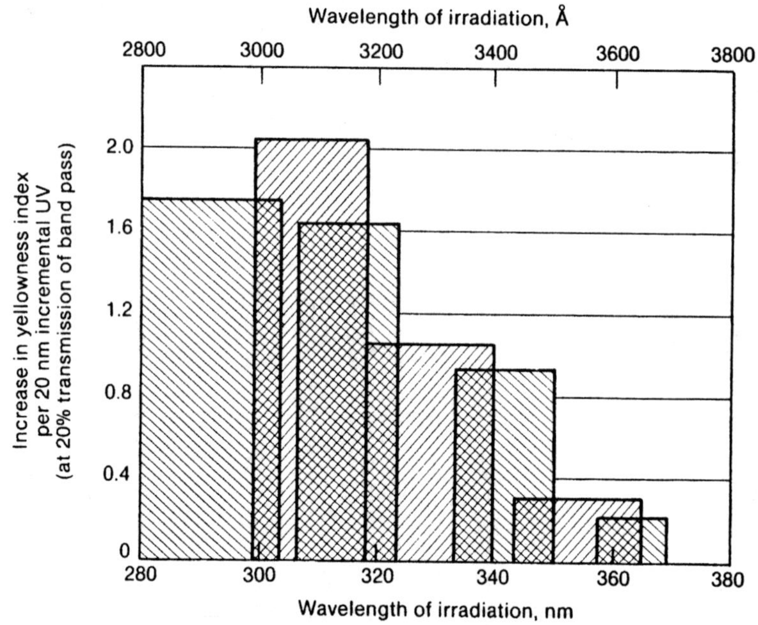

Fig. 1 Activation spectra of 760 μm (30 mil) polycarbonate source using 6000 W xenon weatherometer with borosilicate filters plus short-wavelength cutoff filters. Source: Ref 2

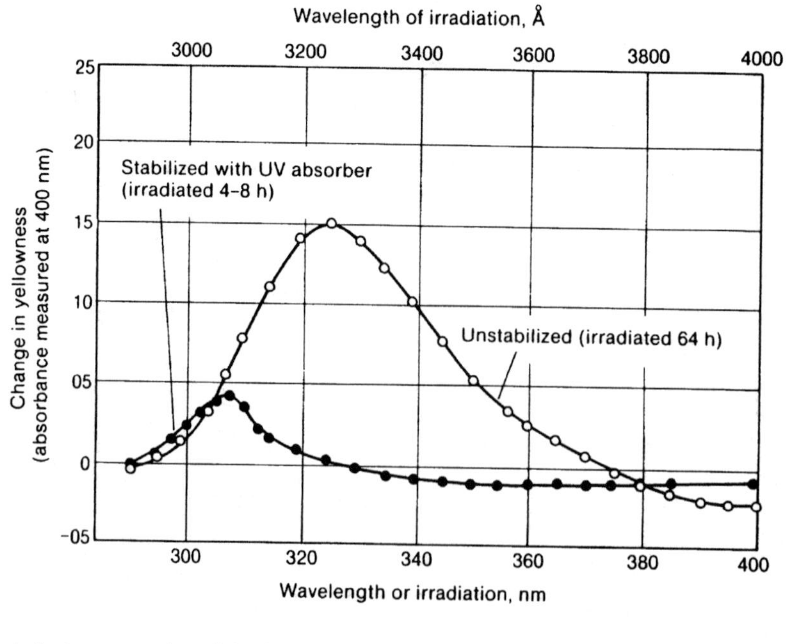

Fig. 2 Activation spectra of unstabilized and stabilized 3200 μm (125 mil) polyester using 1000 W xenon arc with borosilicate glass filter. Source: Ref 2

Table 1 Wavelength of maximum photochemical sensitivity

Polymer	Wavelength nm	Wavelength Å
Polyesters (various formulations)	325	3250
Polystyrene	318	3180
Polyethylene	300	3000
Polypropylene (non-heat stabilized)	370	3700
Polyvinyl chloride	320	3200
Polyvinyl chloride, copolymer with polyvinyl acetate	327–364	3270–3640
Polyvinyl acetate	280	2800
Polycarbonate	285–305, 330–360	2850–3050 3300–3600
Cellulose-acetate-butyrate	296	2960
Styrene-acrylonitrile	290, 325	2900, 3250

Source: Ref 1

ment and the presence of moisture. Removing the susceptible additives or adding a preventive material such as a fungistat may result in a change in properties. Removing an additive may increase modulus, decrease weight, or accelerate the deterioration of electrical properties, such as insulation resistance, dielectric constant, power factor, and dielectric strength.

Test Methods

This section describes the tests used to predict the behavior of a plastic material to outdoor exposure. The test methods can be used to characterize material performance when subjected to specific and well-defined factors. The results from the same method and conditions must be used to compare anticipated performance among materials. No one test can be used to evaluate completely the effects of weather aging on a material. Conducting a series of tests will promote the best estimate of future behavior. The tests chosen should closely represent the service environment.

The fadeometer was originally developed to test paints and dyes for colorfastness. It simulates the exposure to sunlight received by a material that is indoors. It is now commonly used to test plastic materials for color stability and degradation when exposed to sunlight through window glass. It can also be used to demonstrate how combinations of stabilizers, dyes, and pigments will react to UV radiation. For example, the fadeometer was used to help develop an acetate film for store windows that would absorb UV radiation; the film was ultimately used by merchants to protect merchandise from exposure to sunlight.

The light source for the fadeometer is an enclosed carbon arc, which was chosen for this application because its spectral energy distribution is close to that of natural sunlight. The light source consists of carbon electrodes enclosed in a borosilicate glass globe. This enclosure filters the light impinging the test specimens. The specimens are mounted on a cylindrical rack that rotates around the light source. The fadeometer can also be equipped with a xenon arc light source or with both a carbon arc and xenon arc that are used independently for tests involving either light source.

There are two ASTM International test methods that reference the use of a UV light source without water exposure. ASTM G 23-81 (Ref 3) cites the single enclosed carbon arc, and ASTM G 26-84 (Ref 4) cites the xenon arc light source.

It is important to note that fadeometer testing should not be considered representative of outdoor aging, because it does not include other important outdoor aging factors, such as the presence of moisture.

Weatherometers include open-flame carbon arc and twin enclosed carbon arc types.

The sunshine open-flame carbon arc weatherometer is a lamp that operates in free-flowing air. It can be used with light filters mounted on a stainless steel filter frame located between the light source and the test specimens. The test can also be conducted without a filter, increasing the amount of light below 350 nm (3500 Å) incident on the test specimens. Figure 3 shows the spectral distribution of this type of carbon arc with and without filters, compared to natural daylight in Miami, FL. Table 2 gives the increase in the shorter-wavelength light produced by the absence of filters. Table 2 also shows that the sunshine carbon arc has a spectral distribution closer to natural daylight for wavelengths above 400 nm (4000 Å) than the twin enclosed carbon arc light source.

Because shorter wavelengths are usually easily absorbed at the surface of a plastic material, the sunshine carbon arc can be used without filters to test the resistance of a plastic to chalking. This device is equipped with temperature and humidity controls, as well as a water spray, to simulate outdoor conditions. The use of the sunshine carbon arc weatherometer is discussed in ASTM G 23-81 (Ref 3).

The twin enclosed carbon arc weatherometer utilizes two enclosed violet carbon arc lamps. The lamps are positioned such that one is alongside but 76 mm (3.0 in.) lower than the other. This positioning allows for more uniform irradiance on the test specimens. The specimens are mounted on a cylindrical drum that rotates around the light source. The carbon arcs are each enclosed in a borosilicate glass globe. The borosilicate glass filters out light below 275 nm (2750 Å), which is not present in natural sunlight. The glass enclosure also protects the test specimens from possible contamination by the combustion products of the carbon arc. This weatherometer is equipped with temperature and humidity controls. There is also a water spray that, if used while the light source is on, provides a thermal shock to the test specimen. When the light source is off, the water spray exposes the specimen to 100% humidity conditions, similar to what a plastic would experience outdoors at night. This weatherometer also uses a black panel temperature control. This control consists of a temperature sensor mounted on a metal panel that is then coated with a black finish that absorbs light radiation. It is positioned next to the test specimen, where it measures the

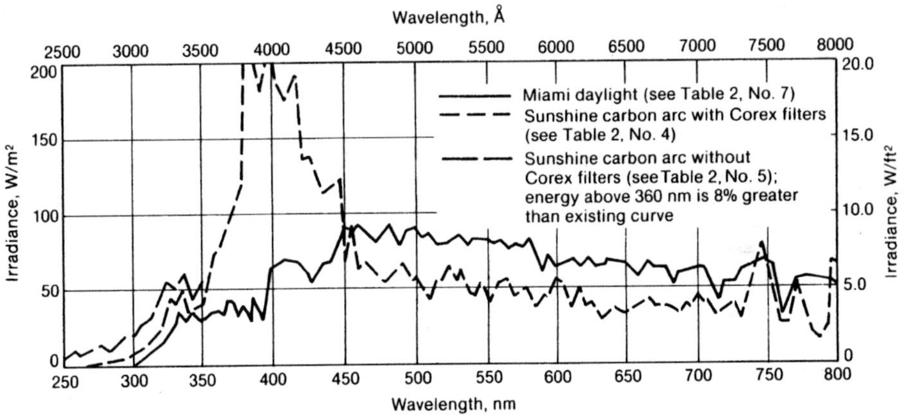

Fig. 3 Spectral power distribution of sunshine carbon arc lamp. Source: Ref 5

Table 2 Spectral power distribution

			Irradiance ranges					
			at 250–300 nm		at 300–400 nm		at 400–800 nm	
No.	Light source	Filter or condition	W/m²	W/ft²	W/m²	W/ft²	W/m²	W/ft²
1	Single enclosed carbon arc	Borosilicate	0	0	85	7.9	95	9
2	Twin enclosed carbon arc	Borosilicate	0	0	85	7.9	95	9
3	Sunshine carbon arc	Soda lime	0	0	90	8.4	250	23
4	Sunshine carbon arc	Corex	0.5	0.050	90	8.4	250	23
5	Sunshine carbon arc	None	3.8	0.35	110	10	275	25
6	Natural daylight	Horizontal plane (0°)(a)	2.0	0.20	66	6	580	55
7	Natural daylight, average optimum	26° south, cloudless, Miami, FL	0	0	30	3	315	30
8	Natural daylight, average optimum	45° south, cloudless, Miami, FL	0	0	30	3	280	26

(a) Calculated from Ref 6. Source: Ref 5

approximate temperature that the specimen encounters.

Reference 3 also cites the use of this type of carbon arc weatherometer and outlines four test methods:

- Method 1 is a continuous exposure to light with an intermittent water spray.
- Method 2 is an alternating exposure to light and darkness with an intermittent water spray.
- Method 3 is a continuous exposure to light without a water spray.
- Method 4 is an alternating exposure to light and darkness without a water spray.

Methods 1 and 2 attempt to simulate natural weathering in an accelerated manner. Methods 3 and 4 are typically used to predict color changes or fading of a material.

With the capability for temperature and humidity control, as well as a water spray, the twin enclosed carbon arc combines what are thought to be the primary causes of degradation in plastic materials: UV radiation, temperature, and moisture. Temperature and humidity conditions under the weatherometer can easily be accelerated by increasing these factors to a degree that is greater than what a plastic material would experience during actual outdoor weathering. Accelerating the UV radiation exposure is difficult because plastics react differently to different wavelengths in the UV spectrum. Therefore, it is important to have a light source producing a spectral distribution that is close to that of natural sunlight. This should achieve a representative artificial effect indicative of what a plastic will experience in actual use.

The spectral distribution for the twin enclosed carbon arc light source is shown in Fig. 4. The spectral distribution below 350 nm (3500 Å) is not highly representative of natural sunlight. This may be a significant consideration in choosing a light source because plastics are most affected by wavelengths in the range between 290 nm and 400 nm (2900 and 4000 Å), particularly those wavelengths below 350 nm (3500 Å). After specimens have been conditioned using this type of artificial weathering device, they can be tested for retention of mechanical or electrical properties and observed for changes in color or chalking.

Natural Environmental Testing. The original weathering test is outdoor exposure. Specimens are mounted vertically facing south and then tested for retention of properties after being exposed to natural weather elements for some length of time. Another type of outdoor test uses specimen racks inclined at 5 or 45° with a southern exposure. These tests are usually conducted in southern states to expose the specimens to as much sunlight as possible. Results obtained using accelerated outdoor weathering devices are often compared to the results obtained from specimens mounted on stationary outdoor racks facing south at 45°. It can then be determined if the accelerated test method being used is an

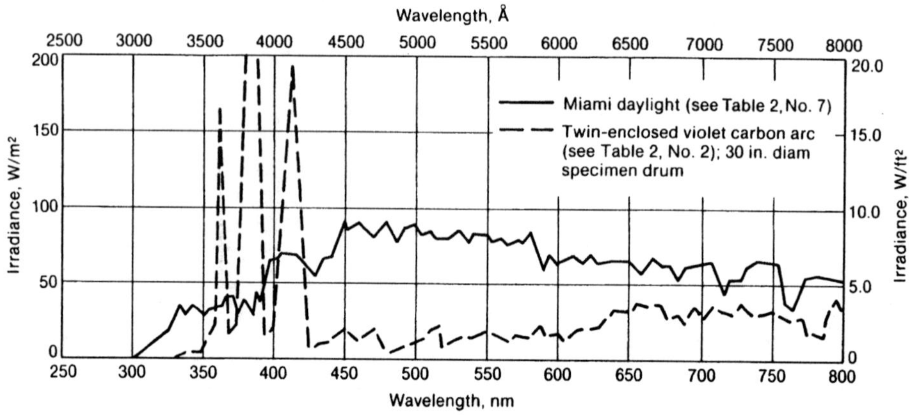

Fig. 4 Spectral power distribution of enclosed violet carbon arc lamp. Source: Ref 5

Table 3 Factors affecting irradiance levels

No.	Xenon arc: Inner filter glass	Xenon arc: Outer filter glass	Test conditions		Irradiance ranges(a): at 250–300 nm, W/m²	at 250–300 nm, W/ft²	at 300–400 nm, W/m²	at 300–400 nm, W/ft²	at 400–800 nm, W/m²	at 400–800 nm, W/ft²	at 340 nm, W/m²	at 340 nm, W/ft²	at 420 nm, W/m²	at 420 nm, W/ft²
1	Borosilicate	Borosilicate	Weathering tests at black panel temperatures above 50 °C (120 °F)	Min Max	0.1 0.2	0.009 0.020	25 65	2 6	240 665	20 60	0.2 0.55	0.020 0.050	0.45 1.30	0.040 0.120
2	Quartz	Borosilicate	Light fastness and weathering tests with somewhat more and shorter UV than is found in natural daylight	Min Max	0.4 1.0	0.037 0.09	30 75	3 7	260 685	25 65	0.25 0.65	0.025 0.060	0.55 1.5	0.050 0.14
3	Quartz	Quartz	Light fastness and weathering tests with considerably more and shorter UV than is found in natural daylight	Min Max	6.8 13.0	0.63 1.20	45 80	4 8	325 605	30 55	0.40 0.75	0.040 0.070	0.7 1.30	0.07 0.120
4	Borosilicate	Soda lime	Light fastness tests at black panel temperatures above 50 °C (120 °F)	Min Max	0 0	0 0	25 60	2 6	255 670	24 60	0.2 0.45	0.020 0.040	0.55 1.45	0.050 0.135
5	Infrared absorbing	Borosilicate	Light fastness tests at black panel temperatures from 38–50 °C (100–120 °F)	Min Max	0 0	0 0	20 50	2 5	165 455	15 40	0.15 0.35	0.015 0.035	0.45 1.25	0.040 0.116
6	Natural daylight		Horizontal plane (0°)(b)	Max	2.0	0.20	66	6	580	55	...	...	...	...
7	Natural daylight		45° south, cloudless, Miami, FL	Average optimum	0	0	30	3	280	26	0.3	0.030	0.7	0.07

(a) Irradiance measured in W/m² at a distance of 310 mm (12 in.). Variations of 10% may be experienced in typical operating conditions. (b) Calculated from Ref 6. Source: Ref 7 accurate way of predicting how a material will react to natural weathering.

A type of outdoor test that increases natural sunlight intensity on a specimen is an equatorial mount with mirrors for acceleration (EMMA) or an EMMA with water spray (EMMAQUA). The specimen mounts of this device automatically follow the sun to keep the rays of the sun normal to the specimen surface. Aluminum mirrors are used to increase the sunlight intensity on the specimens. The temperature around the specimens is controlled by blowers. Using an EMMA can accelerate aging by approximately eight times compared to the stationary 45° southern exposure rack test.

A xenon arc lamp is an alternative light source for the weatherometer and the fadeometer. The lamp is cooled by water that circulates around it. The water also filters out long-wavelength infrared energy. The xenon arc weatherometer is also equipped with a light-filtering system. Glass filters can be used in different combinations to achieve a desired spectral distribution. Several combinations are listed in Table 3. Because the level of irradiance decreases as the xenon arc burns, the device automatically compensates for irradiance changes, allowing test specimens to receive a constant level.

The total amount of irradiation exposure is predetermined by the operator. When the desired amount is achieved, the test is automatically ended. The amount of heat that the specimens receive from the xenon arc is controlled using black panel thermometers. Conditions of humidity, condensation, and rain are also controlled in the xenon arc weatherometer. Periods of darkness allow specimens to recover, just as they would during nighttime outdoor exposure. The specimens are mounted on a rotating drum surrounding the xenon arc lamp, allowing uniform irradiance of all specimens.

Of all the light sources discussed, the 6500-W xenon arc lamp has the closest spectral distribution to natural sunlight (Table 4). This is especially important in the UV range from 290 to 400 nm (2900 to 4000 Å), and particularly below 350 nm (3500 Å), where most of the degradation to plastics takes place. Figure 5 shows the close correlation between the spectral distribution of the xenon arc lamp and that of natural sunlight in Miami, FL. This close correlation to natural sunlight is important in artificial weathering tests because it is not known how exposure to higher levels of irradiant energy, at various wavelengths over the entire spectrum of natural sunlight, will influence test results on plastics.

The use of a xenon arc weathering device is cited in ASTM G 26-84 (Ref 4). There are four test methods outlined:

- Method 1 is a continuous exposure to light with an intermittent water spray.
- Method 2 is an alternating exposure to light and darkness with an intermittent water spray.
- Method 3 is a continuous exposure to light without a water spray.
- Method 4 is an alternating exposure to light and darkness without a water spray.

Methods 1 and 2 try to simulate natural weathering in an accelerated manner, while methods 3 and 4 are used to predict color changes or fading of a material. After exposure, specimens can be tested for retention of mechanical and electrical properties and can be observed for surface changes.

Fluorescent sunlamps are used in the UVCON test device, manufactured by the Atlas Electrical Devices Company, and in the QUV cyclic ultraviolet weathering tester, manufactured by the Q-Panel Company. These devices expose test specimens to alternating cycles of condensation and fluorescent UV light. Both devices have a light source consisting of eight fluorescent sunlamps mounted on two banks. Because the intensity of these lamps decreases with use, one lamp is replaced in each bank at regular intervals to maintain a consistent amount of irradiance on the specimens. The condensation system allows condensation to form on the specimen surfaces during periods of darkness. Because there is no water spray, the specimens are never subjected to thermal shock during light exposure. The specimens are mounted on two racks, with each rack facing a bank of four lamps. The fluorescent sunlamps emit a low amount of radiant heat, which aids in the temperature control of the system. These fluorescent sunlamp test devices are relatively inexpensive compared to carbon arc and xenon arc test devices.

The UV light emitted from fluorescent sunlamps ranges from 280 to 350 nm (2800 to 3500 Å). The peak intensity occurs at 310 nm (3100 Å). The intensity of these fluorescent sunlamps below this level is greater than that of natural sunlight, which is the primary cause of accelerated aging using this light source. Above 400 nm (4000 Å), the intensity of the fluorescent sunlamp is very low compared to natural sunlight. The distribution of irradiance for fluorescent sunlamps is also listed in Table 4. Careful consideration must be given to each material being tested with regard to wavelength sensitiv-

Table 4 Comparative distribution of irradiance

Band pass	Sunlight(a), %	6500-W xenon(b)(c), %	6500-W xenon(d) (modified dew cycle), %	Open-flame carbon arc(b), %	Fluorescent(b), %	EMMAQUA(a), %
Below 300 nm (3000 Å)	0.01	0.01	2.5	0.2	14.0	...
300–340 nm (3000–3400 Å)	1.6	1.5	3.0	2.0	70.0	...
340–400 nm (3400–4000 Å)	4.5	5.0	6.5	11.0	13.0	...
Total below 400 nm (4000 Å)	6.1	6.5	12.0	13.2	97.0	6.1
400–750 nm (4000–7500 Å)	48.0	51.5	49.0	32.0	3.0	...
Above 750 nm (7500 Å)	46.0	42.0	39.0	55.0	0.0	...
Total above 400 nm (4000 Å)	94.0	93.5	88.0	87.0	3.0	94.0

(a) Calculated from Ref 6. (b) Ref 8. (c) Borosilicate inner and outer filters, water cooled. (d) Quartz/borosilicate filter combination, water cooled. Source: Ref 9

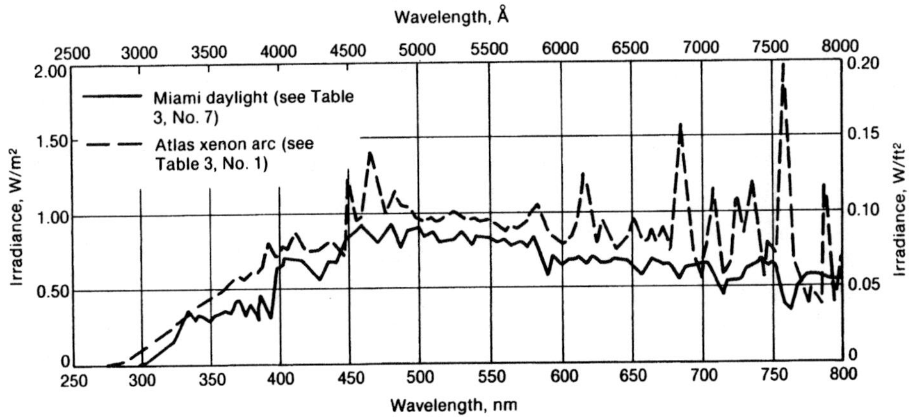

Fig. 5 Power distribution of xenon arc lamp compared to Miami, FL, daylight. Source: Ref 6

ity. A material that is sensitive to wavelengths above 400 nm (4000 Å) will not react the same way to fluorescent light as it will to natural sunlight. Fluorescent sunlamp exposures should therefore be used when the desired exposure is to be limited to irradiation below 350 nm (3500 Å). This may be useful when studying the effects of UV stabilizers in a plastic material.

The use of fluorescent sunlamp devices is cited in ASTM G 53-84 (Ref 10). The scope of the test method described in this standard is to simulate the deterioration caused by the UV energy in sunlight and by rain or dew. After exposure, specimens can be tested for retention of mechanical and electrical properties and can be observed for surface changes.

Fungal Resistance. The additives in plastic materials, such as plasticizers, lubricants, stabilizers, and colorants, are vulnerable to attack by fungi. Specification ASTM G 21-70 (Ref 11) is a method for testing the effect of fungi on plastics.

First, each specimen surface is inoculated with a spore suspension. The specimens are then allowed to incubate in an environment of high relative humidity (85% or higher) and at a temperature of approximately 30 °C (85 °F). The duration of the incubation period is at least 21 days, with fungal growth recorded every 7 days. The specimens can then be observed for visible effects, such as the amount of growth on the specimen or discoloration. Physical and electrical effects can be measured after the specimens have been cleaned with a mercuric chloride solution and allowed to dry thoroughly.

The standard suggests that other properties be tested because physical changes can occur on plastic films or coatings, which have more surface area for the fungi to attack. The resistance of a plastic to fungi may be affected by natural weathering due to the reactions of the additives to UV radiation, temperature, and moisture.

Bacterial Resistance. The additives in plastic materials are also vulnerable to attack by bacteria. Specification ASTM G 22-76 (Ref 12) is a method for testing the effects of bacteria on plastics.

First, test specimens are exposed to the bacteria. The specimens are then allowed to incubate at a temperature of approximately 37 °C (100 °F) and a relative humidity of at least 85%. This incubation period lasts for at least 21 days. The observation of bacterial growth is not as well defined as it is for fungal growth. It is therefore suggested in the ASTM standard that other properties be tested since physical changes can occur without much bacterial growth.

The specimens can be tested for retention of physical or electrical properties after being cleaned with a solution of mercuric chloride and allowed to dry thoroughly. The resistance of a plastic to bacteria may be affected by natural weathering due to the reaction of the additives to UV radiation, temperature, and moisture.

REFERENCES

1. R.C. Hirt and N.D. Searle, Energy Characteristics of Outdoor and Indoor Exposure Sources and Their Relation to the Weatherability of Plastics, *Applied Polymer Symposia,* No. 4, 1967, p 61–83
2. N.D. Searle, The Activation Spectrum and Its Significance to Weathering of Polymetric Materials, *Atlas Sun Spots,* Vol 14 (No. 13), 1984
3. "Standard Practice for Operating Light-Exposure Apparatus (Carbon-Arc Type) with and without Water for Exposure of Nonmetallic Materials," G 23-81, *Annual Book of ASTM Standards,* American Society for Testing and Materials
4. "Standard Practice for Operating Light-Exposure Apparatus (Xenon-Arc Type) with and without Water for Exposure of Nonmetallic Materials," G 26-84, *Annual Book of ASTM Standards,* American Society for Testing and Materials
5. "Series C Weather-Ometer Xenon and Carbon Arc Systems for Accelerated Lightfastness and Weathering Tests," Bulletin 1400, Atlas Electric Devices Co., 1983
6. Publication 20(TG-2.2), CIE, 1972
7. "Ci35 Controlled Irradiance Exposure System," Bulletin 1360, Atlas Electric Devices Co., 1984
8. R.A. Kinmonth and J.E. Norton, *JC-TAX,* Vol 49, 1977, p 663
9. J.L. Scott, Does Correlation Exist between Accelerated and Conventional Outdoor Exposures? Part II, *Atlas Sun Spots,* Vol 10 (No. 23), 1980
10. "Standard Practice for Operating Light-Exposure Apparatus (Fluorescent UV-Condensation Type) for Exposure of Nonmetallic Materials," G 53-84, *Annual Book of ASTM Standards,* American Society for Testing and Materials
11. "Standard Practice for Determining Resistance of Synthetic Polymeric Materials to Fungi," G 21-70, *Annual Book of ASTM Standards,* American Society for Testing and Materials
12. "Standard Practice for Determining Resistance of Plastics to Bacteria," G 22-76, *Annual Book of ASTM Standards,* American Society for Testing and Materials

SELECTED REFERENCES

- "Compact Series Fade-Ometer and Weather-Ometer for Accelerated Light-fastness and Weathering Tests," Bulletin 1380, Atlas Electric Devices Co., 1979
- D.R. Dreger, How Dependable Are Accelerated Weathering Tests for Plastics and Finishes?, *Mach. Des.,* 29 Nov 1973, p 61–67
- W.E. Driver, *Plastics Chemistry and Technology,* Van Nostrand Reinhold, 1979, p 144
- J.B. Dym, *Product Design with Plastics,* Industrial Press, 1983, p 7, 41
- "Fade-Ometer and Weather-Ometer For Accelerated Lightfastness and Weathering Tests," Bulletin 1300B, Atlas Electric Devices Co., 1974
- J.R. Fried, Polymer Technology—Part 8: Polymer Resins, Blends, and Composites, *Plast. Eng.,* Sept 1983, p 38–39
- G.W. Grossman, Correlation of Laboratory to Natural Weathering, *J. Coatings Technol.,* Vol 49 (No. 633), Oct 1977, p 45–54
- W.B. Hardy, Light Stabilization of Polymers, *Atlas Sun Spots,* Vol 13 (No. 31), 1983
- "IEEE General Principles for Temperature Limits in the Rating of Electric Equipment," Std 1-1969, The Institute of Electrical and Electronics Engineers, 1969
- M.R. Kamal and R. Saxon, Recent Developments in the Analysis and Prediction of the Weatherability of Plastics, *Applied Polymer Symposia,* No. 4, 1967, p 1–28
- R.A. Kinmonth, A Correlation Review—Published Results from 1967–1977 Part III, *Atlas Sun Spots,* Vol 7 (No. 18), 1978
- R.A. Kinmonth, When and Why to Use Fluorescent Sunlamps, *Atlas Sun Spots,* Vol 11 (No. 25), 1981
- R.A. Kinmonth, Wavelength Sensitivity, *Atlas Sun Spots,* Vol 12 (No. 28), 1982
- G.F. Kinney, *Engineering Properties and Applications of Plastics,* John Wiley & Sons, 1957, p 151, 156, 226
- *Modern Plastics Encyclopedia,* McGraw-Hill, 1986–1987, p 416–417
- J. Scott, Does Correlation Exist Between Accelerated and Conventional Outdoor Exposures?, *Atlas Sun Spots,* Vol 9 (No. 21), 1979–1980
- "UVCON A Laboratory Device for Screening Materials Sensitive to Ultra Violet Light and Condensation," Bulletin 1340, Atlas Electric Devices Co., 1976

Flammability Testing*

A MATERIAL IS FLAMMABLE if it is subject to easy ignition and rapid flaming combustion. Combustion is defined as an oxidation process that occurs at a rate fast enough to produce temperature rise and (usually) light, either as a glow or flame (Ref 1). Fire is defined as destructive burning, as manifested by light, flame, heat, and/or smoke. The combustible materials in most fatal fires are natural or synthetic polymers (Ref 2). A large amount of work has been done to improve the fire resistance of polymers. Along with this effort has come the development of flammability tests and codes and regulations that cite these tests.

Fire Resistance of Polymeric Materials

There are two basic approaches to improving the fire resistance of a polymeric material: modifying the basic polymer so that exposure to heat and oxygen will not produce combustion and using flame-retardant additives. The first approach does not improve the fire resistance of polymeric materials already being used, and polymers modified to improve heat resistance may also have processing, performance, and cost limitations. The second method is generally more cost effective. Flame-retardant additives include antimony trioxide, chlorine and bromine compounds, phosphorus compounds, and aluminum trihydrate (Ref 3).

Fire-resistant engineering plastics and elastomers are used in the building, transportation, appliance, and electronic industries (Ref 4). For any given application, three types of fire safety requirements must generally be met: customer requirements, government requirements, and insurance requirements (Ref 5). These requirements are not necessarily independent, and will vary for specific applications within each market. For example, aircraft have much more demanding fire safety requirements than motor vehicles because exit may not be possible in the case of a fire.

Overview of the Burning Process

The burning process can be considered on the microscale, where changes to individual polymer molecules are examined; on the macroscale, where the burning of a "unit mass" such as a 1 g quantity of a material is described; and on the mass scale, where the burning of a complete system such as a room or structure is characterized (Ref 5). The following discussion considers the burning process on a macroscale, where it comprises heating, decomposition, ignition, combustion, and propagation.

The rate of heating, or temperature rise, depends on the flow rate of applied heat; temperature differential; material properties such as specific heat, thermal conductivity, and latent heat of fusion; and vaporization or other changes that may occur during heating. While these material properties play an important role in the flammability of a polymeric material, they are not considered further in this article.

When the material reaches the decomposition temperature, one or more of the following types of products is evolved:

- Combustible gases (CH_4, C_2H_6, CO, acetone, etc.)
- Noncombustible gases (CO_2, HCl, HBr, H_2O)
- Liquids
- Solids (ash, char)
- Entrained solid particles or polymer fragments, which appear as smoke

Some of these products are more desirable than others. For example, solid residue helps preserve structural integrity, protects adjacent unit masses from decomposition, and impedes the mixing of air with combustible gases. Noncombustible gases dilute the combustible gas/oxygen mixture, and reduce the temperature of the flame (Ref 6).

Combustion begins when combustible gases ignite in the presence of sufficient oxygen or oxidizing agent. Ignition is affected by the temperature and composition of the gas mixture, and by several material characteristics: the flash-ignition temperature (the temperature at which decomposition gases can be ignited with a pilot flame), the self-ignition temperature (the temperature at which reactions within the material become self-sustaining to the point of ignition), and the limiting oxygen concentration (the minimum concentration of oxygen that will support combustion).

Once combustion has begun, the process has reached fully developed burning, and extinguishment becomes more important than inhibition. At this stage the most important material characteristic is the heat of combustion (the heat released by the combustion of a unit mass).

Propagation will occur if sufficient energy is available to bring an adjacent unit mass to the combustion stage. Energy may be supplied by the heat of combustion or external sources, and energy may be depleted by heat loss to surroundings.

The characteristics of the burning process vary with time. Figure 1 shows an example of temperature, smoke evolution, and concentrations of oxygen, CO, and CO_2 plotted as a function of time.

Flammability Test Methods

As noted previously, the burning process is a complicated one that is influenced by many factors (Ref 2). Energy transfer modes, oxygen availability, presence of flame-retardant additives, and the nature of the materials involved all influence the burning process. As a result, no two fires are alike, and it is difficult to develop meaningful laboratory simulations. Both the relatively simple tests performed on basic polymers and their compounds and the larger-scale tests performed on fabricated structures are criticized for their lack of ability to predict real-life performance (Ref 8). In spite of this, flammability tests continue to be used to measure and describe the response of materials or systems to heat or flame exposure under laboratory conditions.

Many organizations are involved in the characterization and specification of flammability properties. ASTM D 3814-91 "Standard Guide for Locating Combustion Test Methods for Plastics" (Ref 9), includes test methods promulgated by ASTM International, Canadian Standards Association (CSA), National Fire Protection Association (NFPA), Society of Automotive Engineers (SAE), Underwriters' Laboratories (UL), and government agencies. While 43 ASTM standards, 23 UL standards, 11 NFPA standards, and many others are cited in ASTM D 3814-91, the list is not exhaustive and

*Adapted from Rebecca Tuszynski, Flammability Testing, *Engineered Materials Handbook Desk Edition,* ASM International, 1995, pages 454 to 458

the user should assume that other tests exist for specific materials or applications.

There is movement toward increased uniformity in flammability testing of plastics. In the United States, ASTM and UL tests are most widely used, and efforts to make these tests more similar are progressing. Commonality among tests is also desirable on an international scale. Both the International Organization for Standardization (ISO) and the International Electrotechnical Commission (IEC) are developing standard tests similar to those of ASTM and UL. The American National Standards Institute (ANSI) is the official ISO representative in the United States. Although ANSI does not write standards, it does approve those issued by ASTM and similar organizations (Ref 7).

Several categorization strategies have been used for flammability tests, including tests for specific fire response characteristics, research tests versus acceptance tests, tests for different levels of severity, and tests for basis of origin. Test methods are classified in two ways in the following discussion: by fire response characteristics and by particular applications of polymeric materials.

Tests for Fire Response Characteristics

Ease of Ignition. Almost any polymeric material can be made to ignite given enough heat, oxygen, and time (Ref 5). Many types of energy sources are used for ignition tests, including heated air, flames generated by various types of small laboratory burners, hot surfaces, and high current or high voltage arcs. Simple "pass/fail" ignition tests provide fixed conditions of heat, oxygen, and time. The test specimen may or may not ignite within these conditions (see, for example, ASTM E 136, "Test Method for Behavior of Materials in a Vertical Tube Furnace at 750 °C"). Other tests, such as ASTM D 1929 and ASTM D 3713, provide quantitative results.

ASTM D 1929, "Standard Test Method for Ignition Properties of Plastics" (Ref 10), uses a hot-air ignition furnace. Both flash-ignition and self-ignition temperatures can be measured. Table 1 lists flash- and self-ignition temperature values for several types of polymers.

ASTM D 3713, "Standard Test Method for Measuring Response of Solid Plastics to Ignition by a Small Flame" (Ref 11), uses a small flame produced by a laboratory burner applied to the base of a sample held in a vertical position. The performance of the material is reported as an index of the time for which the material does not ignite. This method can be used in conjunction with ASTM D 3801, which is described below.

Flame spread, or propagation, can be defined as the rate of travel of a flame front under given conditions of burning (Ref 5). It can also be viewed as a series of progressive ignition events from a continuous flame front moving over a material (Ref 7), so the distinction between ignition and flame spread tests may be somewhat artificial. Sample geometry and direction of air-flow are extremely important in these tests. The following tests use a variety of sample sizes and configurations.

ASTM E 84, "Standard Test Method for Surface Burning Characteristics of Building Materials" (also known as the Steiner tunnel test or the 25 foot tunnel test) (Ref 12), was established in 1940 and is the oldest flame spread test (Ref 2). It provides a comparison of the surface burning characteristics of materials on a relatively large scale; sample dimensions are 7.62 by 0.496 m (300 by 19.5 in.). The specimen mounting simulates the underside of a ceiling exposed to a fairly severe flaming ignition source (Fig. 2). The combination of the horizontal sample orientation with the lower surface exposed and a concurrent airflow provides the most severe flame spread conditions. Performance is compared to that of red oak (which is given a rating of 100) and noncombustible asbestos board (which has a rating of 0). This test, which is intended for building and interior finish materials, does not adequately measure flame spread on materials that drip, melt, or disintegrate.

ASTM E 162, "Standard Test Method for Surface Flammability of Materials Using a Radiant Heat Energy Source" (Ref 14), uses a refractory panel maintained at about 670 °C (1238 °F). A pilot burner ignites the top of a 152 by 457 mm (6 by 18 in.) test specimen that is mounted at 30° from the vertical. The top of the test specimen is closer (121 mm, or 4.75 in.) to the radiant source. The progress of the flame is monitored as it travels downward. Figure 3 shows a diagram of the apparatus.

ASTM D 635, "Standard Test Method for Rate of Burning and/or Extent and Time of Burning of Self-Supporting Plastics in a Horizontal Position" (Ref 16), and ASTM D 3801, "Standard Test Method for Measuring the Comparative Extinguishing Characteristics of Solid Plastics in a Vertical Position" (Ref 17) are widely used. In ASTM D 635, a 125 by 12.5 mm (5 by 0.5 in.) specimen is held in a horizontal position and ignited at one end with the flame from a laboratory burner. The flame is applied for 30 s. An average burning rate is reported if the specimen burns tb a mark made 100 mm (4 in.) from the ignited end, or time and extent of burning are reported if the specimen does not burn to the mark. ASTM D 3801 uses a similar specimen, but the sample is held vertically. A standard test flame is applied for two 10 s applications. The flaming time before extinguishment is recorded after the first application, and the times of flaming and glowing extinguishment are recorded after the second application.

ASTM D 568 is frequently mentioned in the older literature as the appropriate test for evaluating flexible plastics in a vertical position. This test was discontinued in 1991.

Two other tests, ASTM D 4804, "Standard Test Methods for Determining the Flammability Characteristics of Nonrigid Solid Plastics" (Ref 18), and ASTM D 5048, "Standard Test Method for Measuring the Comparative Burning Characteristics and Resistance to Burn-Through of Solid Plastics Using 125-mm Flame" (Ref 19), are gaining in popularity. The former allows for the testing of flexible materials, which cannot be tested using D 635 or D 3801; the latter is a more demanding test because it uses a larger flame.

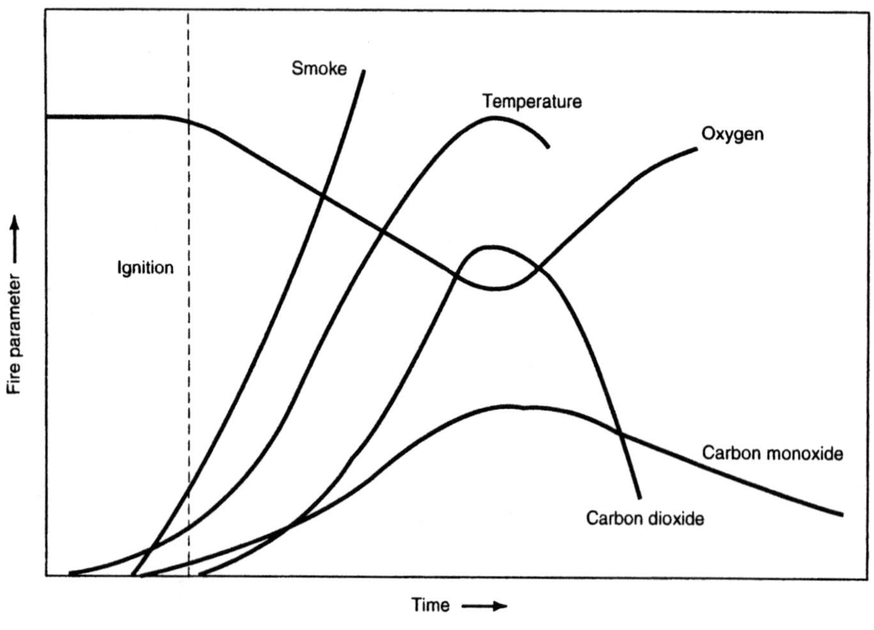

Fig. 1 Temperature, smoke evolution, and concentration of oxygen, CO, and CO_2 plotted as a function of time for the burning process. Source: Ref 7

ASTM D 4986, "Standard Test Method for Horizontal Burning Characteristics of Cellular Polymeric Materials" (Ref 20), describes a procedure for comparing the relative rate, extent, and time of burning of cellular (foamed) polymeric materials. In this test, a 50 by 150 mm (2 by 6 in.) test specimen is supported horizontally, and dry cotton is placed 175 mm (7 in.) under the test specimen. One end of the specimen is exposed to a specified flame for 60 s, and the burning characteristics, including any ignition of the dry cotton by flaming particles from the test specimen, are noted.

Heat release is caused by various exothermic chemical reactions that occur during combustion, especially the generation of CO and CO_2 and the consumption of O_2. The rate of heat release is the primary characteristic determining the size, growth, and suppression requirements of a fire environment (Ref 21). Many heat release tests have been suggested, but only a few have been developed to full scientific or regulatory status (Ref 7). These include the use of the cone and Ohio State University (OSU) calorimeters, which are described in ASTM E 1354, "Standard Test Method for Heat and Visible Smoke Release Rates for Materials and Products Using an Oxygen Consumption Calorimeter" (Ref 22), and ASTM E 906, "Standard Test Method for Heat and Visible Smoke Release Rates for Materials and Products" (Ref 23). Both test methods use radiant heat sources to generate heat fluxes as high as 100 kW/m^2. The specific heat flux(es) and whether an external ignition source is used are not specified in the test procedures, but must be determined separately for each material and/or application.

The OSU calorimeter uses four discrete silicon carbide heating elements. The heat release rate is measured by monitoring the temperature rise in the exhaust gas flow. The cone calorimeter uses a heater rod tightly wound into the shape of a truncated cone. The specimens are burned under ambient conditions while being exposed to the specified external heat flux. Oxygen concentration and exhaust gas flow rate are measured and used to calculate the heat release rate. The use of oxygen consumption measurements rather than temperature measurements to calculate heat release improves precision (Ref 22) because heat loss has a major effect on the latter. Mass loss rate, the time to sustained flaming, and smoke production can also be measured.

Ease of extinguishment can be evaluated using ASTM D 2863, "Standard Test Method for Measuring the Minimum Oxygen Concentration to Support Candle-like Combustion of Plastics (Oxygen Index)" (Ref 24), commonly known as the limiting oxygen index (LOI) test (Fig. 4). The limiting oxygen index is the minimum concentration of oxygen in an oxygen/nitrogen mixture that will support combustion. Table 2 gives representative values. While LOI is a fundamental property of the material being tested, it does not necessarily characterize burning behavior. The major limitation of this test is the absence of energy feedback to the specimen, since most of the energy is carried away by convection. A sample with a high oxygen index, which indicates a lower tendency toward burning, may still burn vigorously when it is preheated by another heat source (Ref 2).

Evolution of Smoke or Toxic Gases. While most of the destruction associated with fires occurs during flaming, most of the deaths are caused by smoke and toxic gases (Ref 2). Smoke evolution can be measured either optically or gravimetrically (Ref 5). Optical measurements can be either static or dynamic (Ref 2). There is often a significant difference between the amounts of smoke generated under smoldering (combustion of a solid without flame) or flaming

Table 1 Flash-ignition and self-ignition temperatures for selected polymers

Polymer	Flash-ignition temperature		Self-ignition temperature	
	°C	°F	°C	°F
Polyethylene (PE)	341–357	646–674	349	660
Polyvinyl chloride (PVC)	391	736	454	849
Polyvinylidene chloride (PVDC)	532	990	532	990
Polystyrene (PS)	345–360	653–680	488–496	910–925
Acrylonitrile-butadiene-styrene (ABS)	...	...	466	871
Polymethylmethacrylate (PMMA)	280–300	536–572	450–462	842–864
Polycarbonate (PC)	375–467	707–873	477–580	891–1076
Polyether-imide (PEI)	520	968	535	995
Polyether sulfone (PES)	560	1040	560	1040
Polytetrafluoroethylene	...	...	530	986
Cellulose nitrate	141	286	141	286
Cellulose acetate	305	581	475	887
Phenolic, glass fiber laminate	520–540	968–1004	571–580	1060–1076
Wool	200	392	...	...
Wood	220–264	428–507	260–416	500–781
Cotton	230–266	446–511	254	489

Source: Ref 5

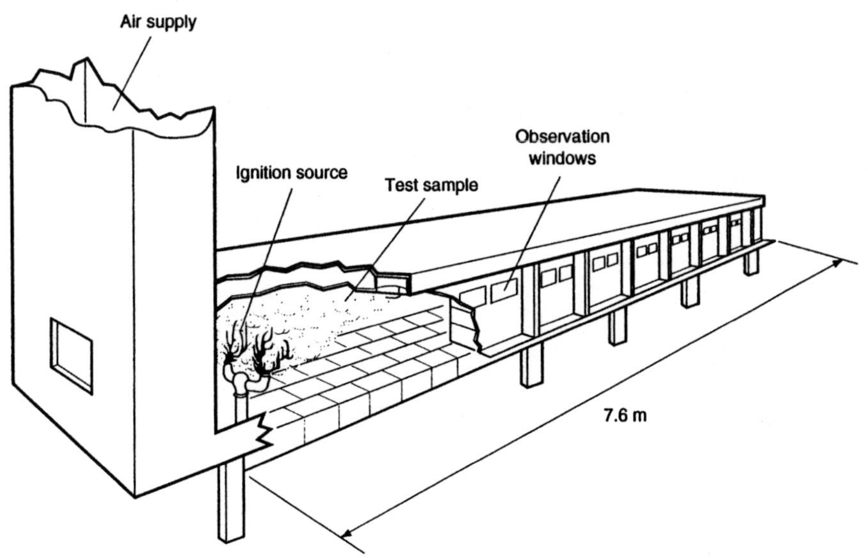

Fig. 2 The Steiner tunnel furnace used to evaluate the flame spread of materials in ASTM E 84. Source: Ref 13

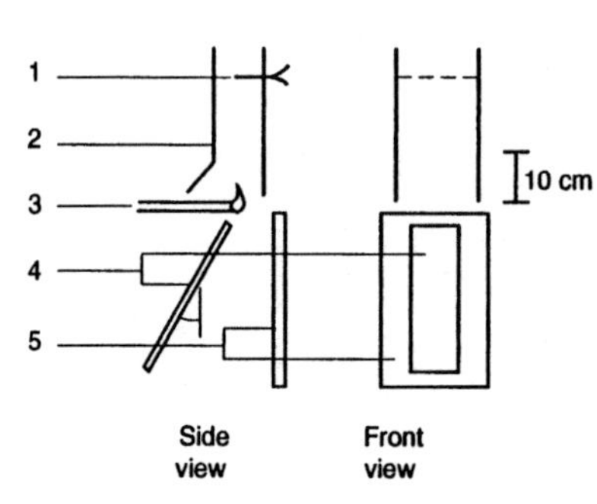

Fig. 3 Apparatus used in ASTM E 162. (1) Temperature sensor. (2) Exhaust stack. (3) Igniter. (4) Test specimen. (5) Radiant panel. Source: Ref 15

conditions (Ref 15). While smoke is generally undesirable, it does play an important role in the activation of fire-detection devices. There are two general approaches to tests for toxic gas evolution. The chemical compounds present in gaseous combustion products can be identified and analyzed, or the effects of gaseous combustion products on laboratory animals can be studied (Ref 5).

ASTM E 662, "Standard Test Method for Specific Optical Density of Smoke Generated by Solid Materials" (Ref 25), was the most widely accepted smoke evolution test in the United States as of 1989 (Ref 26). Also known as the NBS (National Bureau of Standards, now the National Institute of Standards and Technology, NIST) smoke test, it is a static test where the sample is held vertically while being exposed to a radiant flux of 25 kW/m^2. The smoke density is measured optically along a vertical path. ASTM D 2843, "Standard Test Method for Density of Smoke from the Burning or Decomposition of Plastics" (Ref 27), is also a static smoke chamber test, but the smoke density is measured across a horizontal path, and the specimen is subjected to flames from a gas jet.

ASTM E 84, E 1354, and E 906 (described above in sections on flame spread and heat release) have provisions for dynamic optical measurements of smoke density flowing past a specific location. ASTM D 4100, "Method for Gravimetric Determination of Smoke Particulates from Combustion of Plastic Materials" (also known as the Arapaho smoke test) is the best-known gravimetric method (Ref 5).

The chemical compounds evolved during combustion can be analyzed using many standard analytical techniques, including infrared analysis, gas chromatography, mass spectrometry, and ion-selective electrodes. Toxicological studies generally involve the exposure of rats or mice to the gaseous products of decomposition and/or combustion under controlled conditions. Test animals are monitored for incapacitation or lethality (death). Reference 5 has an excellent description of the appropriate test methods.

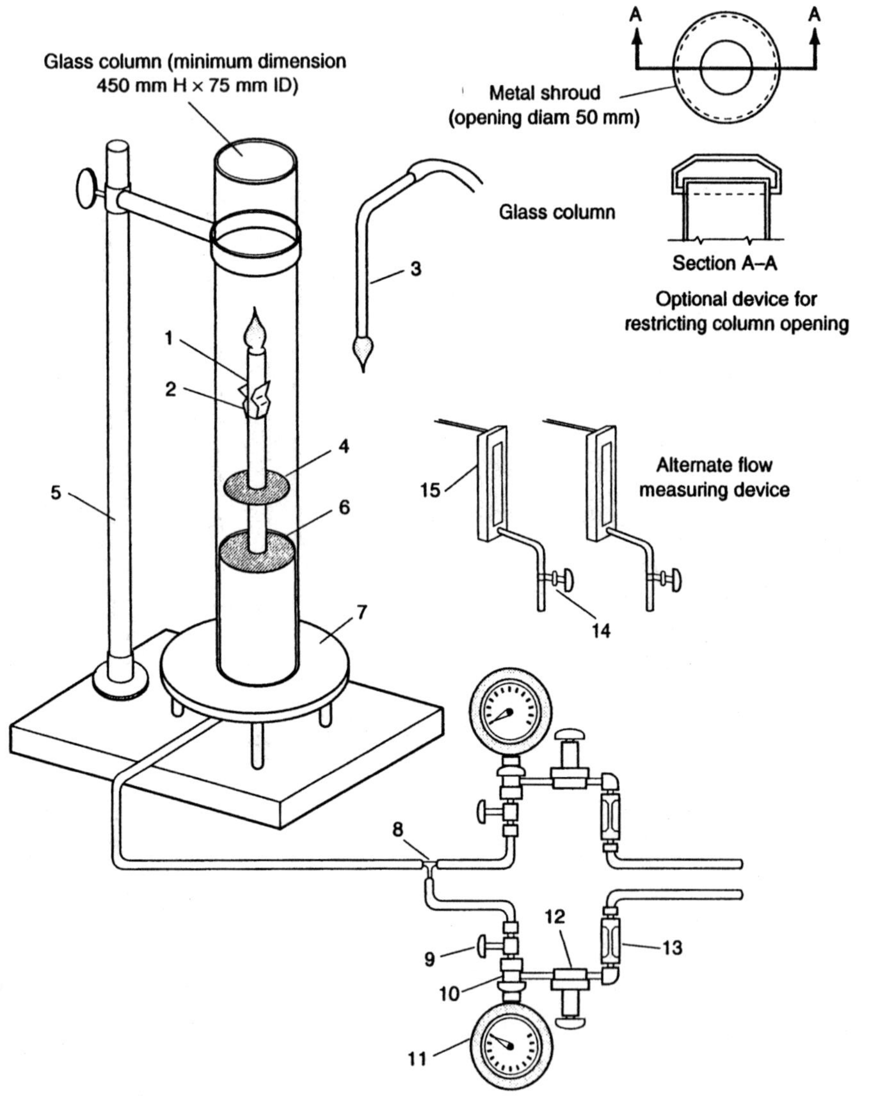

Fig. 4 Typical equipment used for the limiting oxygen index test (ASTM D 2863). (1) Burning specimen. (2) Clamp with rod support. (3) Igniter. (4) Wire screen. (5) Ring stand. (6) Glass beads in a bed. (7) Brass base. (8) Tee. (9) Cut-off-valve. (10) Orifice in holder. (11) Pressure gage. (12) Precision pressure regulator. (13) Filter. (14) Needle valve. (15) Rotameter. Source: Ref 24

Tests for Particular Applications of Polymeric Materials

Electrical Wire, Components, and Products. The most widely used tests for the flammability of plastics used in these applications are those found within UL94, *Test for Flammability of Plastic Materials for Parts in Devices and Appliances* (available from Underwriters' Laboratories, Inc., Northbrook, IL) and the ASTM and international counterparts to these tests (Table 3). These tests have a materials orientation and are typically performed on test specimens rather than parts.

ASTM D 876, "Methods of Testing Nonrigid Vinyl Chloride Polymer Tubing Used for Electrical Insulation," and ASTM D 2633, "Method of Testing Thermoplastic Insulations and Jackets for Wire and Cable," are specifically for plastics used in power or signal-carrying wires strung in air ducts or cable raceways (Ref 2).

Building Materials. The architects and engineers involved in building projects exercise some influence on materials choices, but the major influence is the building code involved. Three widely used model building codes—the Uniform Building Code (UBC), issued by the International Conference of Building Officials (ICBO), the Building Officials and Code Administrators International (BOCA) National

Table 2 Limiting oxygen index (LOI) values for unfilled polymers

Polymer	Limiting oxygen index
Polyacetal	15
Polymethylmethacrylate (PMMA)	17
Polypropylene (PP)	17
Polyethylene (PE)	17
Polybutylene terephthalate (PBT)	18
Polystyrene (PS)	18
Polycarbonate (PC)	26
Polyimide (PI)	32
Polyether sulfone (PES)	34–38
Polyvinyl chloride (PVC)	45
Polyvinylidene fluoride (PVDF)	44
Polyphenylene sulfide (PPS)	44–53
Polyvinylidene chloride (PVDC)	60
Polytetrafluoroethylene (PTFE)	95

Polymers burn with increasing difficulty as LOI increases. Source: Ref 8

Table 3 Flammability tests for plastics used in devices and appliances

Type of test	UL	ASTM	IEC In development(a)	IEC Current(a)	ISO	CSA
Horizontal						
Designation	UL94, Section 2	D 635	695-2-4/3	707	1210	C22.2 No. 017
Rating/classification	94HB	...	HB	FH 1,2,3	FH-1, -2, -3	HB
Vertical						
Designation	UL94, Section 3	D 3801	695-2-4/3	707	1210	C22.2 No. 017
Rating/classification	94V-0	V-0	V-0	FV 0	FV-0	V-0
	94V-1	V-1	V-1	FV 1	FV-1	V-1
	94V-2	V-2	V-2	FV 2	FV-2	V-2
Vertical specimens and horizontal plaques						
Designation	UL94, Section 4	D 5048	695-2-4/5	707	10351	C22.2 No. 017
Rating/classification	94-5VA	...	5VA	LF0	LFV-0	5VA
	94-5VB	...	5VB	LF1	LFV-1	5VB
Vertical, flexible materials						
Designation	UL94, Section 5A	D 4804	...	...	9773	...
Rating/classification	94VTM-0	...	...	...	...	...
	94VTM-1	...	...	...	...	...
	94VTM-2	...	...	...	...	...
Horizontal, foamed materials						
Designation	UL94, Appendix A	D 4986	...	...	9772	...
Rating/classification	94HBF	...	...	...	...	...
	94HF-1	...	...	...	...	...
	94HF-2	...	...	...	...	...

UL, Underwriters' Laboratories; ASTM, formerly American Society for Testing and Materials; IEC, International Electrotechnical Commission; ISO, International Organization for Standardization; CSA, Canadian Standards Association. (a) As of 1995

Building Code, and the Standard Building Code, issued by the Southern Building Code Congress International—classify plastics materials based on their flame spread rating, as determined by ASTM D 84. Additional information is provided in Ref 5.

Other tests also exist for fabrics and soft furnishings (upholstered furniture, mattresses, etc.) based on polymeric materials. These tests are not reviewed here, but detailed information can be found in Ref 2 and 5.

Specialized tests can be used to measure smolder susceptibility and flash-fire propensity. The former is particularly important for soft furnishings, where the classic ignition source is a lighted cigarette. Appropriate tests are described in Ref 5.

ACKNOWLEDGMENT

The author would like to thank D. Oates and A. Bertram of Underwriters' Laboratories for their help in providing information for this article.

REFERENCES

1. "Standard Terminology of Fire Standards," E 176-91d, *1992 Annual Book of ASTM Standards,* ASTM
2. Flammability, *Encyclopedia of Polymer Science and Engineering,* Vol 7, John Wiley & Sons, 1987
3. E. Miller, Properties Modification by Use of Additives, *Engineering Plastics,* Volume 2, *Engineered Materials Handbook,* ASM International, 1988, p 493–507
4. Additives, *Modern Plastics (Encyclopedia),* Vol 71 (No. 12), Mid-Nov 1994, p C-14
5. C.J. Hilado, *Flammability Handbook for Plastics,* 4th ed., Technomic Publishing Company, 1990
6. D.R. Schultz, Flame Retarding Materials, *Rubber World,* Vol 206 (No. 5), Aug 1992, p 23–26
7. R.P. Brown, *Handbook of Plastics Test Methods,* 3rd ed., Longman Scientific & Technical and John Wiley & Sons, 1988
8. J.A. Brydson, *Plastics Materials,* 4th ed., Butterworth, 1982
9. "Standard Guide for Locating Combustion Test Methods for Plastics," D 3814-91, *1995 Annual Book of ASTM Standards,* ASTM
10. "Standard Test Method for Ignition Properties of Plastics," D 1929-91a, *1995 Annual Book of ASTM Standards,* ASTM
11. "Standard Test Method for Measuring Response of Solid Plastics to Ignition by a Small Flame," D 3713-78 (1988) 1, *1995 Annual Book of ASTM Standards,* ASTM
12. "Standard Test Method for Surface Burning Characteristics of Building Materials," E 84-94, *1995 Annual Book of ASTM Standards,* ASTM
13. S. Kaufman, Flammability of Polymers: Test Methods, *Encyclopedia of Materials Science and Engineering,* Vol 3, M.B. Bever, Ed., Pergamon Press and the MIT Press, 1986, p 1797–1802
14. "Standard Test Method for Surface Flammability of Materials Using a Radiant Heat Energy Source," E 162-94, *1995 Annual Book of ASTM Standards,* ASTM
15. J.-M. Charrier, *Polymeric Materials and Processing,* Hansen Publishers, 1990
16. "Standard Test Method for Rate of Burning and/or Extent and Time of Burning of Self-Supporting Plastics in a Horizontal Position," D 635-91, *1995 Annual Book of ASTM Standards,* ASTM
17. "Standard Test Method for Measuring the Comparative Extinguishing Characteristics of Solid Plastics in a Vertical Position," D 3801-87, *1995 Annual Book of ASTM Standards,* ASTM
18. "Standard Test Methods for Determining the Flammability Characteristics of Nonrigid Solid Plastics," D 4804-91, *1992 Annual Book of ASTM Standards,* ASTM
19. "Standard Test Method for Measuring the Comparative Burning Characteristics and Resistance to Burn-Through of Solid Plastics Using 125-mm Flame," D 5048-90, *1992 Annual Book of ASTM Standards,* ASTM
20. "Standard Test Method for Horizontal Burning Characteristics of Cellular Polymeric Materials," D 4986-95, *1995 Annual Book of ASTM Standards,* ASTM
21. U. Sorathia, T. Dapp, and C. Beck, Fire Performance of Composites, *Mater. Eng.,* Vol 109 (No. 9), Sept 1992, p 10–12
22. "Standard Test Method for Heat and Visible Smoke Release Rates for Materials and Products Using an Oxygen Consumption Calorimeter," E 1354-92, *1995 Annual Book of ASTM Standards,* ASTM
23. "Standard Test Method for Heat and Visible Smoke Release Rates for Materials and Products," E 906-83, *1992 Annual Book of ASTM Standards,* ASTM
24. "Standard Test Method for Measuring the Minimum Oxygen Concentration to Support Candle-like Combustion of Plastics (Oxygen Index)," D 2863-91, *1995 Annual Book of ASTM Standards,* ASTM
25. "Standard Test Method for Specific Optical Density of Smoke Generated by Solid Materials," E 662-94, *1995 Annual Book of ASTM Standards,* ASTM
26. D.F. Lawson, Flammability of Elastomeric Materials, *Performance Properties of Plastics and Elastomers,* Vol 2, *Handbook of Polymer Science and Technology,* N.P. Cheremisinoff, Ed., Marcel Dekker, 1989
27. "Standard Test Method for Density of Smoke from the Burning or Decomposition of Plastics," D 2843-93, *1995 Annual Book of ASTM Standards,* ASTM

Electrical Testing and Characterization*

PLASTICS have become extremely popular among manufacturers of electrical products. Other materials, such as wood, glass, fabrics, mineral oil, and ceramics, were used prior to the application of plastics to provide mechanical support as well as shielding and insulation between electrically live components and the ground. As the electrical and electronics industry grew, so did the need for an alternative material that possessed the desired electrical properties and that could be produced economically. Many different formulations and methods of curing also resulted in plastic products that differ in their overall properties, and it is this versatility that makes plastics superior to other similar products (Ref 1, 2). Depending on the application, plastics may be formulated and processed to exhibit a single property or a designed combination of mechanical, electrical, chemical, thermal, optical, and aging properties. The Society of the Plastics Industry classifies plastic materials based on the aforementioned properties, which are derived from standards such as those of the American Society for Testing and Materials (ASTM), National Bureau of Standards, Manufacturing Chemists Association, federal government, and U.S. military.

This article first discusses electrical testing and recommended procedures for determining the electrical properties of insulating materials, with particular emphasis on plastics, followed by the electrical characteristics of various forms of plastics. The purpose of this article is to provide sufficient information to allow the reader to select the appropriate electrical test(s) for a particular application. Definitions of terms that are germane to this discipline are provided in the section "Terminology" in this article in this Volume.

Electrical Tests

The following tests are commonly used in the electrical and electronics industry to determine the electrical properties of insulating materials, particularly plastics.

Dielectric Breakdown Voltage and Dielectric Strength. The dielectric breakdown property is generally sought for materials used in applications in which an electrical field is present. In many cases, the dielectric strength of a material is the determining factor in the design of the apparatus in which it is to be used. The results obtained from this test provide part of the information needed for determining the suitability of a material for a given application and for detecting changes or deviations from normal characteristics due to processing variables, aging conditions, or other manufacturing or environmental situations. The test voltage is usually applied with simple test electrodes on opposite faces of the specimens. The specimens may be molded or cast, or cut from flat sheet or plate. Table 1 lists the typical electrode configurations used for various dielectric strength tests of insulating materials. The dielectric strength of materials can be determined by using either or both of the two commonly used test methods developed and published by the ASTM Electrical Insulating Materials Committee D-9 (Ref 3, 4). These two methods are described below.

Using Alternating Current (ac). One widely used test procedure for evaluating the dielectric breakdown characteristics of insulating materials is ASTM D 149 (Ref 3). In this method, alternating voltage, usually applied at a frequency of 60 Hz (or other specified frequencies), is increased from zero or from a level well below the breakdown voltage, in one of three prescribed methods, until dielectric breakdown failure of the test specimen occurs. The three methods of voltage applications are:

- Method A is a short-time test, in which the voltage is applied uniformly to the test electrodes from zero at one of the rates shown in Fig. 1(a) until breakdown occurs. Unless otherwise specified, the short-time test is normally used to determine the dielectric breakdown voltage
- Method B is a step-by-step test, in which the voltage is applied to the test electrodes at the preferred starting voltage in steps and durations as shown in Fig. 1(b) until breakdown occurs
- Method C is a slow rate-of-rise test, in which the voltage is applied to the test electrodes

Table 1 Typical electrodes for dielectric strength testing

These electrodes are those most commonly specified or referenced in ASTM standards. With the exception of type 5 electrodes, no attempt has been made to suggest electrode systems for other than flat surface material. Other electrodes may be used as specified in ASTM standards or as agreed upon between seller and purchaser where none of these electrodes in this table is suitable for proper evaluation of the material being tested.

Electrode type	Description of electrodes (a)(b)	Insulating materials
1	Opposing cylinders 51 mm (2 in.) in diameter, 25 mm (1 in.) thick with edges rounded to 6.4 mm (0.25 in.) radius	Flat sheets of paper, films, fabrics, rubber, molded plastics, laminates, boards, glass, mica, and ceramic
2	Opposing cylinders 25 mm (1 in.) in diameter, 25 mm (1 in.) thick with edges rounded to 3.2 mm (0.125 in.) radius	Same as for type 1, particularly for glass, mica, plastic, and ceramic
3	Opposing cylindrical rods 6.4 mm (0.25 in.) in diameter with edges rounded to 0.8 mm (0.0313 in.) radius(c)	Same as for type 1, particularly for varnish, plastic, and other thin film and tapes: where small specimens necessitate the use of smaller electrodes or where testing of a small area is desired
4	Flat plates 6.4 mm (0.25 in.) wide and 108 mm (4.25 in.) long with edges square and ends rounded to 3.2 mm (0.125 in.) radius	Same as for type 1, particularly for rubber tapes and other narrow widths of thin materials
5	Hemispherical electrodes 12.7 mm (0.5 in.) in diameter(d)	Filling and treating compounds, gels and semisolid compounds and greases, embedding, potting, and encapsulating materials
6	Opposing cylinders; the lower one 75 mm (3 in.) in diameter, 15 mm (0.60 in.) thick; the upper one 25 mm (1 in.) in diameter, 25 mm thick; with edges of both rounded to 3 mm (0.12 in.) radius(e)	Same as for types 1 and 2

(a) Electrodes are normally made from either brass or stainless steel. Reference should be made to the standard governing the material to be tested to determine which, if either, material is preferable. (b) The electrode surfaces should be polished and free from irregularities resulting from previous testing. (c) Refer to the appropriate standard for the load force applied by the upper electrode assembly. Unless otherwise specified the upper electrodes shall be 50 ± 2 g. (d) Refer to the appropriate standard for the proper gap settings. (e) The type 6 electrodes are those given in IEC Publication 243 for the testing of flat sheet materials. They are less critical as to the concentricity of the electrodes than types 1 and 2 electrodes. Source: Ref 3

*Adapted from Tony Ghaffari, Electrical Testing and Characterization, *Engineering Plastics,* Volume 2, *Engineered Materials Handbook,* ASM International, 1988, pages 581 to 583 from the starting voltage at the rate shown in Fig. 1(c) until breakdown occurs (Ref 3)

Using Direct Current (dc). This method covers the determination of dielectric breakdown voltage and dielectric strength of insulating materials under direct-voltage stress. One popular standard test procedure used by laboratories is ASTM D 3755 (Ref 4). This method is intended for use as a control and acceptance test for direct-voltage applications. It can also be used in the partial evaluation of materials for specific end-uses and as a means of detecting changes in a material that are due to specific deteriorating causes. Experience indicates that the breakdown value obtained from direct voltage will usually be approximately two to four times the rms (root mean square) value of the 60 Hz alternating voltage breakdown. Although other rates can be selected, there is usually one rate of voltage increase associated with this method, and it is a short-time method at a rate of 500 V/s (Ref 4).

For both test methods to produce results that are representative of the type of material being tested, sufficient breakdown values must be obtained and statistically analyzed. Maximum, minimum, average, standard deviation, coefficient of variation, and failure modes and types are among the variables that must be analyzed. The selection of direct or alternating voltage depends on the purpose for which the breakdown test is to be used and, to some extent, on the intended application of the material. It should be kept in mind that the results obtained from either of the test methods discussed can seldom be used to determine the dielectric behavior of a material in an actual application. In most cases, these results must be evaluated by comparison with results obtained from other functional tests or from tests on other materials, or both, in order to estimate their significance for a particular application. Specific information on dielectric breakdown tests is available in the appropriate ASTM standards (Ref 3, 4).

Dielectric Constant and Dissipation Factor. Insulating materials are generally used in two distinct ways: to support and insulate components of an electrical network from each other and from ground, and as the dielectric of a capacitor. For the first use, it is generally desirable to have the capacitance of the support as small as possible, consistent with acceptable mechanical, chemical, and heat-resisting properties. A low value of dielectric constant (relative permittivity) is therefore desirable. For the second use, it is desirable to have a high value of dielectric constant so that the capacitor dimensions can be as small as possible. With respect to ac losses (that is, dissipation factor, power factor, and so on), materials that are used to provide both insulation and capacitor dielectrics should have small losses to reduce the heating of the material and to minimize its effect on the rest of the network. In high-frequency applications, a low value of loss index is particularly desirable because for a given value of loss index the dielectric loss increases directly with frequency.

In comparisons of materials having approximately the same dielectric constant or when using any material under such conditions that its dielectric constant remains essentially constant, the quantity considered may also be the dissipation factor, power factor, phase angle, or loss angle (Ref 5). Table 2 identifies electrode systems for measuring permittivity and dissipation factor. Table 3 gives typical values for the dielectric constant of polar and nonpolar resins.

The comprehensive test method for determining dielectric constant, dissipation factor, loss index, power factor, phase angle, and loss angle of solid electrical insulating materials developed and published by ASTM Committee D-9 is ASTM D 150 (Ref 5). This standard test method is very popular among manufacturers and end-users of plastic materials. Several other methods that revert back to the discussions and theories contained in Ref 5 have been introduced by the same committee for specific materials such as polyethylene (Ref 6) and expanded cellular

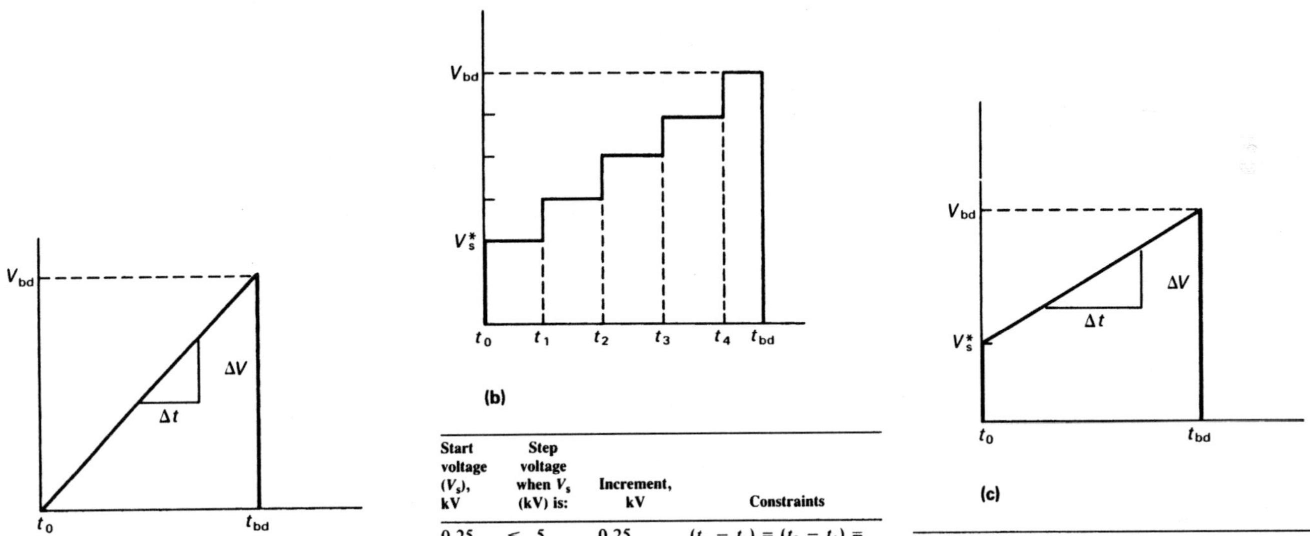

(a)

Rates, (V/s) ± 20%
100
200
500
1000
2000
5000

(b)

Start voltage (V_s), kV	Step voltage when V_s (kV) is:	Increment, kV	Constraints
0.25 . . .	≤ 5	0.25	$(t_1 - t_0) = (t_2 - t_1) = (t_3 - t_2) = 60 \pm 5$ s
0.50 . . .	> 5–10	0.50	. . .
1	> 10–25	1	20 } alternate
2	> 25–50	2	300 }
5	> 50–100	5	$120 \le t_{bd} \le 720$ s
10	> 100	10	. . .
20	. . .	. . .	(t_{bd} will usually exceed 720 s if 300 s time step is used.)
50	. . .	. . .	. . .
100 . . .	. . .	. . .	. . .

(c)

Rates, (V/s) ± 20%	Constraints
1	$t_{bd} > 120$ s
2	. . .
5	. . .
10	$V_{bd} = > 1.5\ V_s$
12.5	. . .
20	. . .
25	. . .
50	. . .
100	. . .

Fig. 1 Voltage profiles used in determining the dielectric strength of materials. (a) Short-time test, in which voltage is applied uniformly to test electrodes from zero using one of rates shown below figure until breakdown occurs. (b) Step-by-step test; use list below figure to select the initial voltage, which should be closest to 50% of the experimentally determined or expected breakdown voltage under short-time test. (c) Slow rate-of-rise test, in which voltage is applied to test electrodes from starting voltage and at rates shown below figure until breakdown occurs. Source: Ref 3

plastics (Ref 7), which are used for electrical insulation. Such variables as frequency, temperature, voltage, humidity, and weathering affect the dielectric constant and dissipation factor of a given material to varying degrees, depending on the level and duration of exposures.

Frequency. The changes in dielectric constant and loss index with frequency are produced by the dielectric polarizations that exist in the material. The two most important are dipole polarization due to polar molecules and interfacial polarization caused by inhomogeneities in the materials. Dielectric constant and loss index vary with frequency in the manner shown in Fig. 2.

Starting at the highest frequency where the dielectric constant is determined by electronic polarization, each succeeding polarization, either dipole or interfacial, contributes to the dielectric constant, and the result is that the dielectric constant has its maximum value at zero frequency. Each polarization furnishes a maximum of both loss index and dissipation factor. The frequency at which loss index is a maximum is called the relaxation frequency for that polarization. It is also the frequency at which the dielectric constant is increasing at the greatest rate and at which half its change for that polarization has occurred. A knowledge of the effects of these polarizations is often helpful in determining the frequencies at which measurements

Table 2 Electrode systems for measuring permittivity and dissipation factor, and associated calculations of vacuum capacitance and edge corrections

Type of electrode	Direct interelectrode capacitance in vacuum, pF	Correction for stray field at an edge, pF
Disk electrodes with guard ring 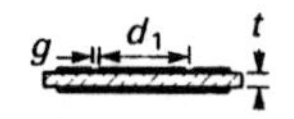	$C_v = \varepsilon_0 \frac{A}{t} = \frac{l}{\mu_0 c^2}\frac{A}{t}$ $= 0.0088542\frac{A}{t}$	$C_e = 0$
Disk electrodes without guard ring: diameter of the electrodes = diameter of the specimen 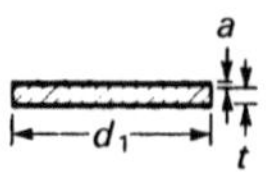	$A = \frac{\pi}{4}(d_1 + B * g)^2$ ...	where $a \ll t$, $C_e = (0.0087 - 0.00252 \ln t)P$
Equal electrodes smaller than the specimen 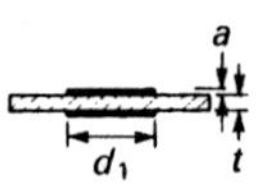	$C_v = 0.0069541\frac{d_1^2}{t}$	$C_e = (0.0019\ \kappa_x' - 0.00252 \ln t + 0.0068)P$ where: κ_x' = an approximate value of the specimen permittivity, and $a \ll t$
Unequal electrodes 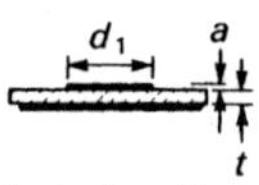	...	$C_e = (0.0041\ \kappa_x' - 0.00334 \ln t + 0.0122)P$ where: κ_x' = an approximate value of the specimen permittivity, and $a \ll t$ $C_e = 0$
Cylindrical electrodes with guard ring 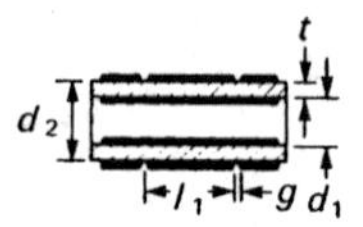	$C_v = \frac{0.055632\,(l_1 + B * g)}{\ln\frac{d_2}{d_1}}$	...
Cylindrical electrodes without guard ring 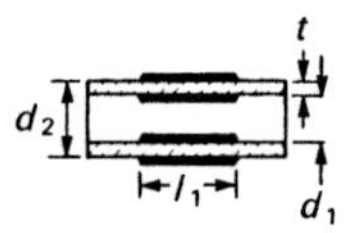	$C_v = \frac{0.055632\,l_1}{\ln\frac{d_2}{d_1}}$ (dimensions in millimeters)	If $\frac{t}{t + d_1} < \frac{1}{10}$ $C_e = (0.0038\ \kappa_x' - 0.00504 \ln t + 0.0136)P$ $P = \pi(d_1 + t)$ where κ_x' = an approximate value of the specimen permittivity

Calculation of capacitance—micrometer electrodes

Parallel capacitance	Definitions of symbols
$C_p = C' - C_r + C_{vr}$	C' is the calibration capacitance of the micrometer electrodes at the spacing to which the electrodes are reset C_{vr} is the vacuum capacitance for the area between the micrometer electrodes, which was occupied by the specimen, calculated as shown above C_r is the calibration capacitance of the micrometer electrodes at the spacing r r is the thickness of specimen and attached electrodes
The true thickness and area of the specimen must be used in calculating the permittivity. This double calculation of the vacuum capacitance can be avoided with only small error (0.2 to 0.5% due to fringing at the electrode edge), when the specimen has the same diameter as the electrodes, by using the following procedure and equation:	
$C_p = C' - C_v + C_{vt}$	C_v is the calibration capacitance of the micrometer electrodes at the spacing t C_{vt} is the vacuum capacitance of the specimen area t is the thickness of specimen

Source: Ref 5

Table 3 Typical values for the dielectric constant of polar and nonpolar resins

Polymer resin	Dielectric constant, κ
Nonpolar resins	
Polyethylene	2.3
Polystyrene	2.5–2.6
Polypropylene	2.2
Polytetrafluoroethylene	2.0
Polar resins	
Polyvinyl chloride (rigid)	3.2–3.6
Polyvinyl acetate	3.2
Polyvinyl fluoride	8.5
Nylon	4.0–4.6
Polyethylene terephthalate	3.25
Cellulose cotton fiber (dry)	5.4
Cellulose kraft fiber (dry)	5.9
Cellulose cellophane (dry)	6.6
Cellulose triacetate	4.7
Tricyanoethyl cellulose	15.2
Epoxy resins (unfilled)	3.0–4.5
Methylmethacrylate	3.6
Polyvinyl acetate	3.7–3.8
Polycarbonate	2.9–3.0
Phenolics (cellulose-filled)	4–15
Phenolics (glass-filled)	5–7
Phenolics (mica-filled)	4.7–7.5
Silicones (glass-filled)	3.1–4.5

Source: Ref 1

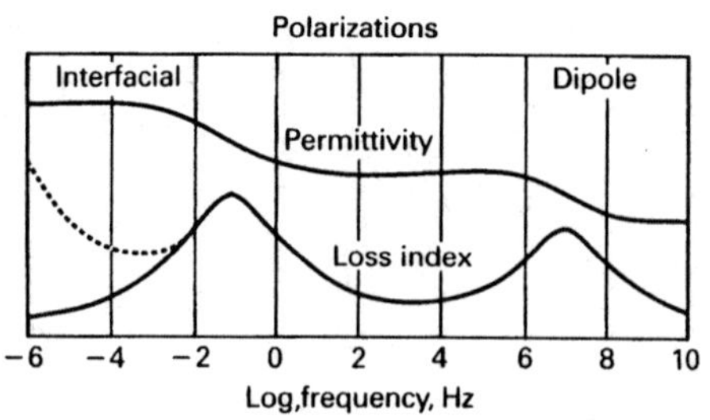

Fig. 2 Typical polarizations of insulating materials. Source: Ref 5

should be made. Any dc conductance in the dielectric caused by free ions or electrons, while having a direct effect on dielectric constant, will also produce a dissipation factor that varies inversely with frequency and becomes infinite at zero frequency, as shown by the dashed line in Fig. 2.

Temperature. The major electrical effect of increased temperature on an insulating material is an increase in the relaxation frequencies of its polarization. The temperature coefficient of dielectric constant at lower frequencies would always be positive, except for the fact that the temperature coefficient of permittivities resulting from many atomic and electronic polarizations is negative. The temperature coefficient will then be negative at high frequencies, zero at some intermediate frequency, and positive as the relaxation frequency of the dipole or interfacial polarization is approached. The temperature coefficient of loss index and dissipation factor may be either positive or negative, depending on the relationship of the measurement to the relaxation frequency. It will be positive for frequencies higher than the relaxation frequency and negative for lower frequencies (Ref 5).

Voltage. All dielectric polarizations, except interfacial, are nearly independent of the existing potential gradient until such a value is reached that ionization occurs in voids in the material or on its surface, or breakdown occurs (Ref 5).

Humidity. The major electrical effect of elevated humidity on an insulating material is to increase greatly the magnitude of its interfacial polarization, thus increasing conductance. These humidity effects are caused by the absorption of water into the volume of the material and by the formation of an ionized water film on its surface. The latter forms in a matter of minutes, while the former may require days and sometimes months to attain equilibrium, particularly for thick and relatively impervious materials (Ref 5).

Weathering, because it is a natural phenomenon, includes the effects of varying temperature and humidity, falling rain, severe winds, impurities in the atmosphere, and the ultraviolet light and heat of the sun. Under such conditions, the surface of an insulating material may be permanently changed, either physically, by roughening and cracking, and/or chemically, by the loss of relatively soluble components and by the reactions of the salts, acids, and other impurities deposited on the surface. Any water film formed on the surface will be thicker and more conducting, and water will penetrate more easily into the volume of the material (Ref 5).

When adequate correlating data are available, the dissipation factor or power factor can be used to indicate the characteristics of a material in other respects, such as dielectric breakdown, moisture content, degree of cure, and deterioration from any cause. However, deterioration due to thermal aging may not affect the dissipation factor unless the material is subsequently exposed to moisture. Although the initial value of the dissipation factor is important, the change in dissipation factor with aging may be much more significant (Ref 5).

In determining the dielectric constant, which is the real part of the relative complex permittivity, the equivalent parallel capacitance, C_p, of a given electrode configuration is measured with a sample material as a dielectric, and then the capacitance of vacuum (or air for most practical purposes), C_v, of the same electrode configuration is measured as the dielectric medium. The dielectric constant, κ', is then computed from Eq 1:

$$\kappa' = C_p/C_v \quad \text{(Eq 1)}$$

The dielectric constant of dry air at 23 °C (73 °F) and standard pressure at 101.3 kPa (14.7 psi) is 1.000536.

One method for measuring dissipation factor, *D,* is to measure the equivalent ac conductance, *G,* at the desired frequency and then compute the dissipation factor using Eq 2:

$$D = G/wC_p \quad \text{(Eq 2)}$$

where w is $2\pi f$, where f is the frequency at which the measurements were made.

In measuring the capacitance of a given material, the most important factors for minimizing the degree of uncertainty in the measurements are the fringing and stray capacitances. Several electrode systems, as proposed in Ref 5, take these factors into consideration. When a guarded electrode system (Fig. 3) is selected and used properly, it significantly reduces the errors caused by fringing and stray capacitors present around the edges. A three-terminal cell used for testing solid electrical insulating materials is shown in Fig. 4, and a guarded two-terminal micrometer electrode system is shown in Fig. 5 (Ref 1, 2). A fixed-plate, two-terminal, self-shielded test cell (Fig. 6), when used in accordance with ASTM D 1531, measures the dielectric constant and dissipation factor of polyethylene compounds by liquid-displacement procedures.

Other types of electrodes and their associated mathematical formulas for calculating the capacitance in vacuum and also the correction factors for the stray field are given in Table 2. A complete discussion of the electrode systems, specimen sizes and preconditioning, frequency ranges, and accompanying methods for measur-

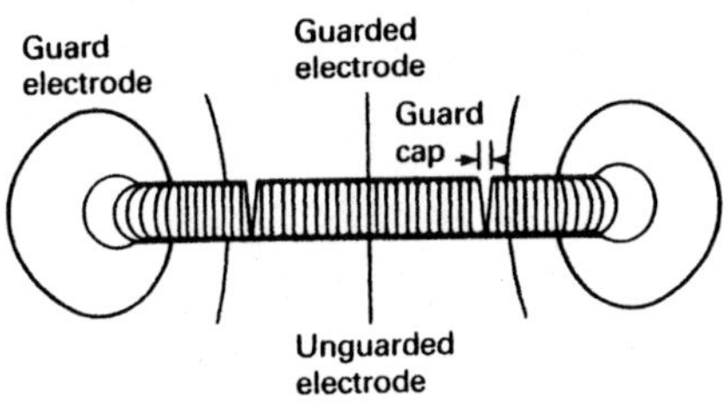

Fig. 3 Guarded three-terminal parallel-plate electrode system showing flux lines between electrodes. Source: Ref 5

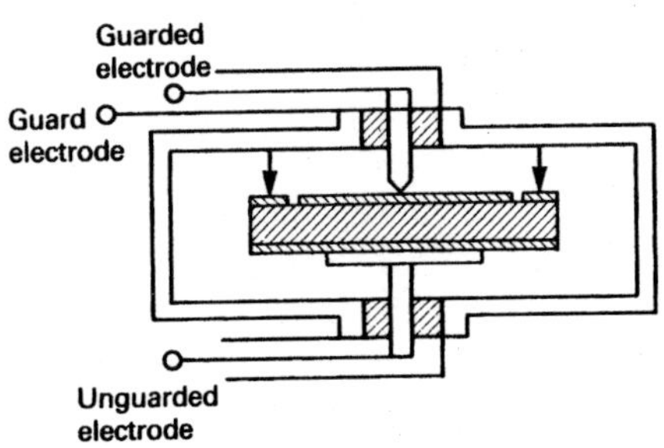

Fig. 4 Guarded three-terminal cell for testing solid materials. Source: Ref 5

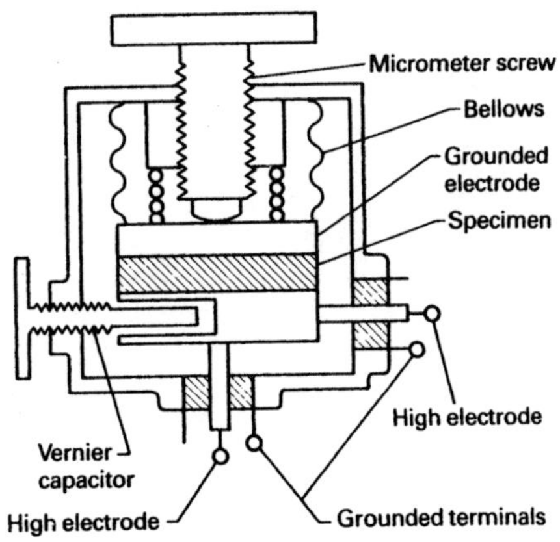

Fig. 5 Micrometer electrode system

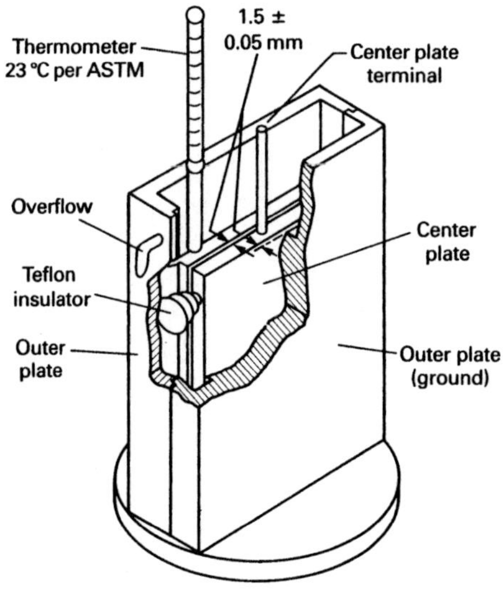

Fig. 6 Fixed-plate, two-terminal, self-shielded test cell for determining permittivity and dissipation factor of polyethylene compounds by the liquid-displacement method. Source: Ref 6

ing capacitance and ac loss is available in Ref 5. Typical values for the dielectric constant (permittivity) of some polar and nonpolar resins are given in Table 3.

Insulation Resistance, Volume, and Surface Resistivity or Conductivity. As mentioned previously, insulating materials are used to isolate the components of an electrical system from each other and from ground and to provide mechanical support for the components. Because the insulation resistance or conductance combines both volume and surface resistance or conductance, its measured value is most useful when the test specimen and electrodes have the same form as that required in actual use. Surface resistance or conductance changes rapidly with humidity, but volume resistance or conductance changes slowly, although the final change may eventually be greater.

Resistivity or conductivity can be used to predict, indirectly, the low-frequency dielectric breakdown and dissipation factor properties of some materials. Resistivity or conductivity is often used as an indirect measure of moisture content, degree of cure, mechanical continuity, and deterioration of various types. The usefulness of these indirect measurements is dependent on the degree of correlation established by supporting theoretical or experimental investigations. A decrease in surface resistance may result in either an increase in the dielectric breakdown voltage (because the electric field intensity is reduced) or a decrease in the dielectric breakdown voltage (because the area under stress is increased). In addition to the usual environmental variables, the dielectric resistance or conductance depends on the length of time of electrification and on the value of applied voltage. These parameters must be known to make the measured value of resistance or conductance meaningful (Ref 8).

Volume resistivity or conductivity can be used as an aid in designing an insulator for a specific application. The change in resistivity or conductivity with temperature and humidity may be great and must be known when designing for operating conditions. Volume resistivity or conductivity determinations are often used in checking the uniformity of an insulating material, either with regard to processing or to detect the conductive impurities that affect the quality of the material and that may not be readily detectable by other methods. Volume resistivities above 10^{19} $\Omega \cdot$ m obtained on specimens under usual laboratory conditions are of doubtful validity, considering the limitations of commonly used measuring equipment. Surface resistance or conductance cannot be measured accurately, but only approximated, because more or less volume resistance or conductance is nearly always involved in the measurement. The measured value is largely a property of the contamination that happens to be on the specimen at the time. However, the dielectric constant of the specimen influences the deposition of contaminants, and its surface characteristics affect the conductance of the contaminants. Surface resistivity or conductivity can be considered to be related to material properties when contamination is involved, but it is not a material property in the usual sense (Ref 8).

A commonly used test procedure for determining the insulation resistance, volume resistance or resistivity, and surface resistance or resistivity of electrical insulating materials, or the corresponding conductances and conductivities, is described in Ref 8. This method only covers the measurements made under the dc voltage application.

The resistance or conductance of a material specimen or a capacitor is determined from a measurement of either the current or the voltage drop under specified conditions. With the appropriate electrode systems, surface and volume resistance or conductance can be measured separately. The resistivity or conductivity can then be calculated when the required specimen and electrode dimensions are known.

In measuring the resistance or conductance of insulating materials, the electrodes should be of a type of material that is readily applied, allows intimate contact with the specimen surface, and introduces no appreciable error due to electrode resistance or contamination of the specimen. The electrode materials should be corrosion resistant under the conditions of the test (Ref 8). Several electrode configurations and methods of measurements are introduced and discussed in Ref 8. Three types of electrode systems are shown in Fig. 7 to 9.

The taper-pin electrodes shown in Fig. 7 are used for measuring the insulation resistance of materials that are in the form of plates, tubes, and rods. Figure 8 shows a circular guarded electrode system that is used in measuring the volume and surface resistance or conductance of flat specimens. Figure 9 depicts an electrode configuration used for measuring the volume and surface resistance and conductance of specimens in the form of tubes.

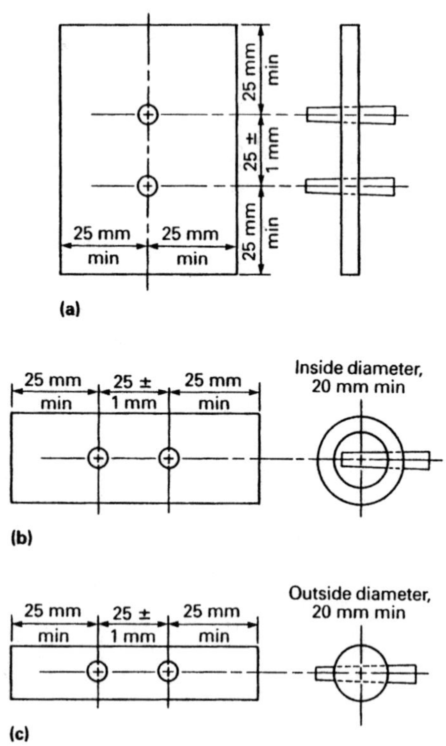

Fig. 7 Taper-pin electrodes for measuring the insulation resistance of (a) plate, (b) tube, and (c) rod specimens; min, minimum. Source: Ref 8

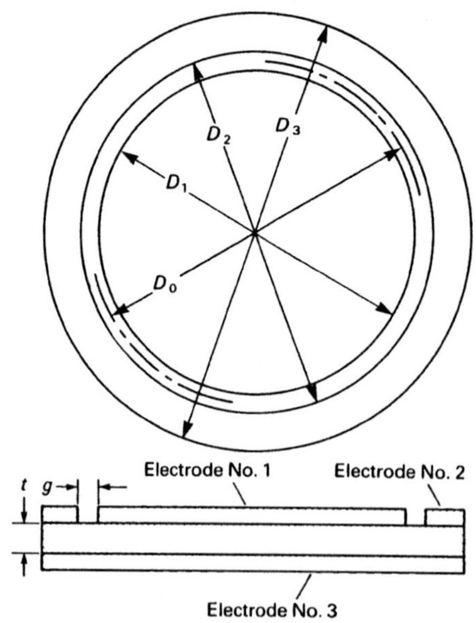

Fig. 8 Guarded three-terminal electrode system for measuring volume and surface resistance or conductance of flat specimens. $g \leq 2t$ volume resistivity, $g \geq 2t$ surface resistivity. Source: Ref 8

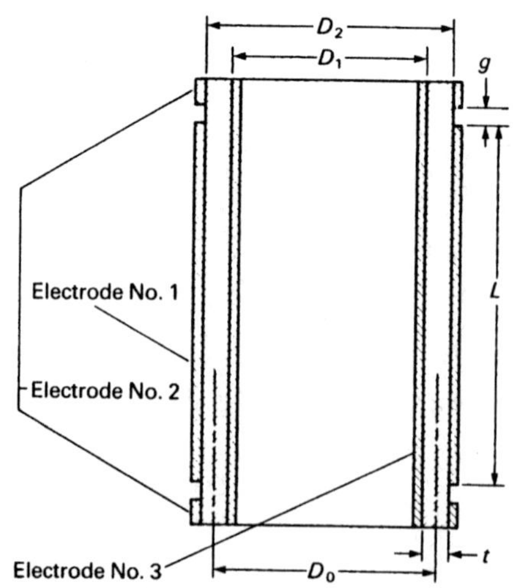

Fig. 9 Guarded three-terminal electrode assembly for measuring volume and surface resistance or conductance of tubular specimens. $D_0 = (D_1 + D_2)/2$; $L > 4t$; $g \leq 2t$ volume resistivity, $g \geq 2t$ surface resistivity. Source: Ref 8

Various methods for measuring insulation, volume, and surface resistances or conductances have been developed over the years:

- Voltmeter-ammeter method using a galvanometer
- Voltmeter-ammeter method using dc amplification or electrometer
- Voltage rate-of-change method
- Comparison method using a galvanometer or dc amplifier
- Comparison method using a Wheatstone bridge
- Direct-reading instruments

The electrical schematic of the voltmeter-ammeter method using a galvanometer is shown in Fig. 10. For a given electrode configuration, when all the electrical and dimensional measurements are made, the appropriate equations given in Table 4 can be used in calculating the volume or surface resistivity or conductivity of a sample material.

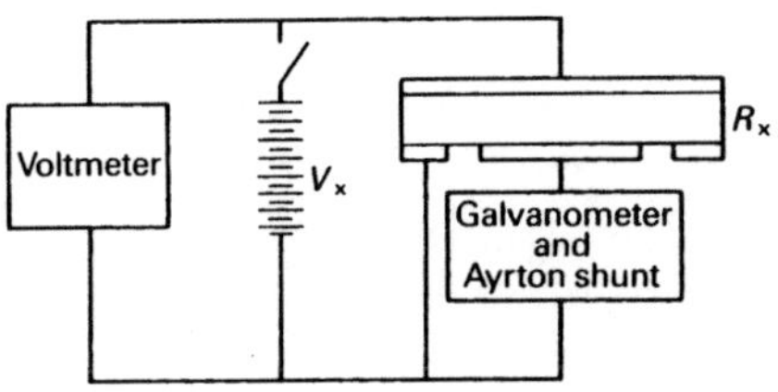

Fig. 10 Volume and surface resistivity or conductivity determination using a voltmeter-ammeter method utilizing a galvanometer. Source: Ref 8

Typically, a test report should contain the following information so that engineering decisions regarding manufacturing quality control or material acceptance or screening can be made quicker and, possibly, easier:

- Description and identification of the materials, such as name, grade, color, and manufacturer
- Shape and dimensions of the test specimens
- Type and dimensions of the electrodes
- Conditioning of the specimens, such as cleaning, predrying, hours at humidity, and temperature
- Test conditions such as specimen temperature and relative humidity at time of measurements
- Method of measurement
- Applied voltage
- Time of electrification of measurement
- Measured values of the appropriate resistances in ohms or conductances in siemens
- Computed values when required, for example, volume resistivity in Ω · m, volume conductivity in siemens per meter, surface resistivity in ohms (per square), or surface conductivity in siemens (per square)
- Statement as to whether the reported values are apparent or steady state

The precision and accuracy of this type of testing are inherently affected by the choice of method, apparatus, and specimen. Because of the variability of the resistance of a given specimen under similar test conditions and the nonuniformity of the same material from specimen to specimen, determinations are usually not reproducible to closer than 10% and are often even more widely divergent (a range of values of 10 to 1 may be obtained under apparently identical conditions).

Arc Tracking Resistance. High current as well as high-voltage, low-current arcing between conductors across the surface of insulating materials may carbonize the material and produce conducting tracks. Materials vary widely in their resistance to tracking, and there are a variety of dry and wet tests for this property. For example, ASTM D 495 (Ref 9) is intended to differentiate, in a preliminary fashion, among similar materials with respect to their resistance to the action of a high-voltage, low-current arc close to the surface of insulation. The arcing tends to form a conducting path or cause the material to become conducting because of the localized thermal and chemical decomposition and erosion. The usefulness of this method is severely limited by many restrictions and qualifications, some of which are described above.

Generally, this method is not used in the material specifications, and it will not permit conclusions to be drawn concerning the relative arc-resistance ranking of materials that may be subjected to other types of arcs, such as low-voltage arcs at low or high currents (caused by surges or by conducting contaminants). Because of its convenience and the short time required for testing, the dry arc resistance test is intended for the preliminary screening of materials, for detecting the effects of changes in formulation, and for quality control testing after correlation has been established with other types of simulated service arc tests and field experience. The test is usually conducted under clean, dry laboratory conditions that are rarely encountered in service; therefore, the prediction of the relative performance of a material in typical applications and in varying clean-to-dirty environments may be substantially altered (Ref 9).

The high-voltage, low-current dry arc resistance test is intended to simulate only approximately such service conditions as those existing in ac circuits operating at high voltage but at currents limited to tens of milliamperes. To distinguish more easily among materials that, by this test, have low arc resistance, the early stages of the test are mild, while later stages are successively more severe. The arc occurs intermittently between two electrodes resting on the surface of the specimen, in regular or inverted orientation. The severity is increased in the early stages by successively decreasing to 0 the time interval between flashes of uniform duration, and in later stages by increasing the current.

The arc resistance of a material is described by this method by measuring the total elapsed time of operation of the test until failure occurs. Four general types of failure have been observed (Ref 9):

Table 4 Calculations for volume and surface resistivity or conductivity for a given electrode assembly

Dimensions given in centimeters

Type of electrodes or specimen	Volume resistivity, Ω · cm(a)		Volume conductivity, S/cm
Circular	$\rho_v = \frac{A}{t} R_v$	$A = \frac{\pi(D_1 + g)^2}{4}$	$\gamma_v = \frac{t}{A} G_v$
Rectangular	...	$A = (a + g)(b + g)$	...
Square	...	$A = (a + g)^2$	...
Tubes	...	$A = \pi D_0(L + g)$	...
Cables	$\rho_r = \frac{2\pi L R_v}{\ln \frac{D_2}{D_1}}$	...	$\gamma_r = \frac{\ln \frac{D_2}{D_1}}{2\pi L R_v}$
	Surface resistivity, Ω/square	...	Surface conductivity, S/square
	$\rho_s = \frac{P}{g} R_s$		$\gamma_s = \frac{g}{P} G_s$
Circular	...	$P = \pi D_0$	...
Rectangular	...	$P = 2(a + b + 2g)$	...
Square	...	$P = 4(a + g)$	...
Tubes	...	$P = 2\pi D_2$	...

(a) A is the effective area of the measuring electrode for the particular arrangement employed; P is the effective perimeter of the guarded electrode for the particular arrangement employed; R_v is the measured volume resistance in ohms; G_v is the measured volume conductance in siemens; R_s is the measured surface resistance in ohms; G_s is the measured surface conductance in siemens; t is the average thickness of the specimen; D_0, D_1, D_2, g, L are dimensions indicated in Fig. 8 and 9 (see Appendix X2 in Ref 8 for correction to g); and a, b are lengths of the sides of rectangular electrodes. Source: Ref 8

- Many inorganic dielectrics become incandescent, at which point they are capable of conducting the current. Upon cooling, however, they return to their earlier insulating condition.
- Some organic compounds burst into flame without the formation of a visible conducting path in the substance.
- Some organic compounds fail by tracking; that is, a thin wiry line is formed between the electrodes.
- Some compounds experience carbonization of the surface until sufficient carbon is present to carry the current.

The sequence of time intervals and the associated current steps are given in Table 5. This test does not apply to materials that do not produce conductive paths under the action of an electric arc or materials that melt or form fluid residues that float conductive residues out of the active test area, thus preventing the formation of a conductive path. To overcome the limitations associated with the above test and to provide the optimal simulation of service conditions, ASTM Committee D-9 has developed standard test methods for insulating materials. Some of the tests are carried out in wet or high relative humidity and contaminated environments. The tests are discussed below.

ASTM D 2132 (Ref 10). This test is intended for insulating materials that may fail in service as a result of tracking, erosion, or both when the material is exposed to high humidity and contaminated environments. This test is particularly useful for organic insulations that are used in outdoor applications in which the surface of the insulation becomes contaminated with coatings of moisture and dirt, such as coal dust or salt spray. This method is an accelerated test that simulates extremely severe outdoor contamination. The synthetic dust used as a contaminant in this test has a composition, in parts by weight, of 85% 240-mesh flint, 9% 325-mesh clay, 3% technical grade salt, and 3% filter pulp paper. It is believed that the most severe conditions likely to be encountered in outdoor service in the United States will be relatively mild compared to the conditions specified in this method. Materials can be classified by this method as:

- Tracking resistant: Materials that fail well beyond 100 h of exposure
- Tracking affected: Materials that usually fail before 100 h
- Tracking susceptible: Materials that fail within 5 h

The dust and fog test chamber is shown in Fig. 11.

ASTM D 2303 (Ref 11). Several different test methods within this standard have been described. They differentiate among solid electrical insulating materials on the basis of their resistance to the action of voltage stresses along the surface of the solid when wet with an ionizable, electrically conductive liquid contaminant. Two tracking methods and one erosion test procedure, a variable-voltage method and a time-to-track method to evaluate resistance to tracking, and a method for the quantitative determination of erosion are discussed in this standard. Although a definite contaminant solution is specified, other concentrations or types of contaminants with suitable voltages can be used to simulate different service or environmental conditions.

In service, many types of contamination may cause tracking and erosion of different materials to different degrees. This standard recognizes the importance of such variability and suggests the use of special solutions to meet specific service needs. For example, an ionic contaminant containing a carbonaceous substance such as sugar can be used to cause tracking on very resistant materials such as polymethylmethacrylate (PMMA). Such contamination may be representative of some severe industrial environments. In this case, the time-to-track technique is used because time is needed to decompose the contaminant solution and to build up conducting residues on the sample surface. Very track resistant materials, such as PMMA, may erode rather than track under more usual contaminant conditions in service. Therefore, the use of this method for measuring erosion is important. For erosion studies, only tests as a function of time at constant voltage are useful (Ref 11).

In the field, the critical conditions and the resulting electrical discharges occur sporadically. Degradation, often in the form of a conducting track, develops very slowly until it ultimately bridges the space between conductors to cause complete electrical breakdown. In this method, the conducting liquid contaminant is continually supplied at an optimal rate to the surface of the test specimen in such a manner that essentially continuous electrical discharge can be maintained. By producing continuous surface discharge with controlled energy, it is possible to cause specimen failure within a few hours, which is similar to that occurring under long-time exposure to the erratic conditions of service. The test conditions, which are standardized and accelerated, do not reproduce all the conditions encountered in service. Therefore, caution is necessary when making inferences from the results of tracking tests concerning either direct or comparative service behavior (Ref 11).

ASTM D 3638 (Ref 12). This method evaluates, in a short period of time, the low-voltage (up to 600 V) track resistance or comparative tracking index of materials in the presence of aqueous contaminants (electrolytes). The surface of a specimen of electrical insulating material is subjected to a low-voltage alternating stress combined with a low current, which results from an aqueous contaminant that is dropped between two opposing electrodes every 30 s. The voltage applied across these electrodes is maintained until the current flow between them exceeds a predetermined value that constitutes failure. Additional specimens are tested at other voltages so that a relationship between applied voltage and number of drops to failure can be established through graphical means. The numerical value of the voltage that causes failure with the application of 50 drops of the elec-

Table 5 Sequence of 1 min current steps in the high-voltage, low-current, dry arc resistance test

Step	Current, mA	Time cycle(a)	Total time, s
⅛10	10	¼ s on, 1¾ s off	60
¼10	10	¼ s on, ¾ s off	120
½10	10	¼ s on, ¼ s off	180
10	10	Continuous	240
20	20	Continuous	300
30	30	Continuous	360
40	40	Continuous	420

(a) In the earlier steps, an interrupted arc is used to obtain a less severe condition than the continuous arc: a current of less than 10 mA produces an unsteady (flaring) arc. Source: Ref 9

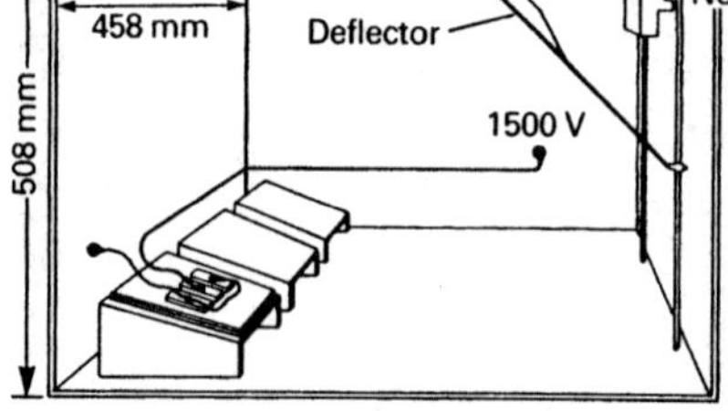

Fig. 11 Dust and fog test chamber. Minimum recommended dimensions are given. Source: Ref 10

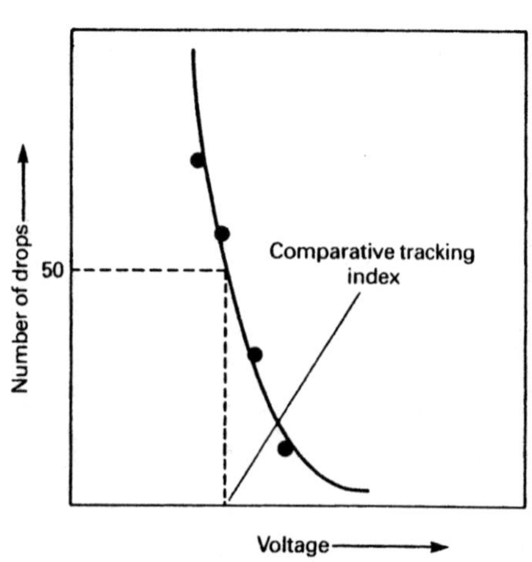

Fig. 12 Comparative tracking index and typical tracking voltage curve. Source: Ref 12

Table 6 Comparison of tracking resistance of various materials measured with seven test procedures

	Test procedure(a)						
Test method designation Units	ASTM D 495-61 Equivalent s/10	IEC 113; VDE Drops, 0.9 kV, Nekal	ASTM D 2132-62T Standard Dust-Fog, h, 1.5 kV	Linearly accelerated dust-fog, h, 1.5 kV	Differential Wet track W · min	Inclined plane I, V, kV	Inclined plane II, h, 2.5 kV
Polyvinyl chloride	0.5 Tr	...	0.5 Tr	0.5 Tr	...	...	...
Phenolic laminate, paper base	0.5 Tr	1	0.5 Tr	...	0.2 (d) Tr	1.5 Tr	...
Epoxy resin, unfilled	1.7 Tr	60 + No Tr	0.5 Tr	...	1.6 Tr	...	...
Polyamide resin	58 + Er	5 ... Tr	0.5 Tr	...	1.3 Int	...	...
Silicone resin, glass cloth	54 Tr	10 ... Tr	1.0 Tr	1.0 Tr	1.8 Tr	1.5 Tr	...
Melamine resin, glass cloth	47 Tr	6 Tr	3.5 Tr	2.5 Tr	2.3 Tr	2.3 Tr	0.2 Tr
Polyethylene	13 Tr	60 + No Tr	27 Tr	10 Er + Tr	...	...	...
Polyester, glass mat(b), 1	25 Tr	No Tr	50 Tr	12 Tr	3.7 (e) Tr	2 Tr	1.1 Tr
Polymethylmethacrylate	100 + Er	No Tr	90 Er	33 Tr	8.1 + Er	6 F	...
Polypropylene	310 Er	No Tr	180 Er	40 Tr	8.1 + Er	3.8 Tr	...
Epoxy resin(b)	100 + Er	No Tr	200 Er	...	...	...	...
Polyester, glass mat(b), 2	51 Tr	No Tr	350 Tr	90 Tr	6.4 Tr	3 Tr	11 Tr
Butyl rubber(b)	100 + Er	No Tr	450 Er	100 Er + Tr	8.1 + Er	6 F	...
Silicone rubber(c)	5 Tr		750 Er	120 Er + Tr		3.7 Tr	...
Polytetrafluoroethylene	310 + Er	No Tr	2700 Er	330 Tr	8.1 + Er	7 F	...

(a) Tr, tracked; No Tr, no tracking; Er, eroded; F, flame. (b) Hydrated, mineral-filled. (c) Nonhydrated, mineral-filled. (d) Failed 1.3 W (4.4 Btu/h), 1 s. (e) Failed 5.5 W (18.7 Btu/h), 18 s. Source: Ref 1

trolyte is arbitrarily called the comparative tracking index. This value provides an indication of the relative track resistance of the material. A typical tracking voltage curve is shown in Fig. 12. Table 6 indicates the difference between results obtained from seven test procedures on different materials and the correlation or lack of correlation between the tests.

Electrical Properties of Plastics and Their Characterizations

Plastics are the most widely used dielectric materials in the electrical and electronics industry. There are numerous plastic materials available with a wide variety of electrical, mechanical, and chemical properties. In terms of their electrical properties, plastics can be divided into thermosetting and thermoplastic materials, some of which are conductive or semiconductive. Elastomers, which are natural or synthetic rubberlike materials with outstanding elastic characteristics, are also used. Their designations and general electrical applications are given in Table 7, and Table 8 provides major electrical properties and a comparison with some popular rubber materials.

Thermosetting plastics are cured and hardened to a desired form at room temperature or higher. The chemical change in curing is permanent, and the material cannot be softened by reheating. Classification and general electrical applications of thermosetting plastics are given in Table 9, and major electrical properties are listed in Table 10.

Thermoplastics do not cure or set upon heating. They soften and can be shaped by molding into any desired form. Thermoplastics can be repeatedly resoftened by heating. Table 11 lists the electrical applications of several thermoplastics, and Table 12 shows their most important electrical properties.

Table 7 Designations and general electrical applications for elastomers

Elastomer designation			
ASTM D 1418	Trade name or common name	Chemical type	Major electrical applications
NR	Natural rubber	Natural polyisoprene	The best electrical grades are excellent in most electrical properties at room temperature.
IR	Synthetic natural	Synthetic polyisoprene	Same general electrical properties as natural rubber
CR	Neoprene	Chloroprene	Not as good electrically as NR or IR. However, good electrical properties for jacketing application. Coupled with all the other good properties, this elastomer has broad use for electrical wire and cable jackets.
SBR	GRS, Buna S	Styrene-butadiene	Electrical properties generally good but not specifically outstanding in any area
NBR	Buna N, nitrile	Acrylonitrile-butadiene	Electrical properties not outstanding; probably degraded by molecular polarity of acrylonitrile constituent
IIR	Butyl	Isobutylene-isoprene	Electrical properties generally good but not outstanding in any area
IIR	Chlorobutyl	Chloroisobutylene-isoprene	Same general properties as butyl
BR	Cis-4	Polybutadiene	Used principally as a blend in other rubbers
	Thiokol (PS)	Polysulfide	Widely used for potting of electrical connectors
R	EPR	Ethylene-propylene	Good general-purpose electrical properties
R	EPT	Ethyl-propylene terpolymer	Good general-purpose electrical properties
CSM	Hypalon (HYP)	Chlorosulfonated polyethylene	Not outstanding electrically
SIL	Silicone	Polysiloxane	Among the best electrical properties in the elastomer grouping; especially good stability of dielectric constant and dissipation factor at elevated temperatures
	Urethane (PUR)	Polyurethane diisocyanate	Good general-purpose electrical properties; some special, high-quality electrical grades available from formulator
ABR	Viton (FLU) acrylics	Fluorinated hydrocarbon polyacrylate	Not outstanding for or widely used in electrical applications

Source: Ref 2

Conductive or Semiconductive Plastics. Although plastics have traditionally been used as electrical insulators, there is a growing market for plastics with increased electrical conductivity. Insulating surfaces can generate and concentrate large electrostatic charges (30 to 40 kV) that discharge as an arc or spark when the material contacts a body of sufficiently different potential. Because electrostatic discharge (ESD) can damage or destroy sensitive electronic components and is capable of igniting highly flammable substances, conductive plastics are sought for use in the manufacture and assembly of microelectronics and explosives and in sensitive environments, such as hospital operating rooms. Insulating plastics are also transparent to electromagnetic radiation. Highly conductive plastics can be used to attenuate electromagnetic interference (EMI) from natural (lightning) and man-made (electronic devices and ESD) sources. Attenuating materials serve a dual purpose by protecting a device from incoming EMI and limiting EMI emissions from the device. This dual function has become more important because of recent legislation that is being enforced by the Federal Communication's Docket 20780, which limits the amount of EMI that a computing device using digital electronics can emit.

Table 8 Electrical properties of elastomers and comparison with rubbers

Material	ASTM D 1507 Dielectric constant(a)	ASTM D 1507 Power factor $\times 10^2$(a)	ASTM D 257 Volume resistivity, $\Omega \cdot m$	ASTM D 257 Surface resistivity, Ω	ASTM D 149 Dielectric strength MV/m	ASTM D 149 Dielectric strength V/mil
Natural rubber	2.7–5	0.05–0.2	10^{13}–10^{15}	10^{14}–10^{15}	18–24	450–600
Styrene-butadiene rubber	2.8–4.2	0.5–3.5	10^{12}–10^{14}	10^{13}–10^{14}	18–24	450–600
Acrylonitrile-butadiene rubber	3.9–10.0	3–5	10^{10}–10^{13}	10^{12}–10^{15}	16–24	400–600
Butyl rubber	2.1–4.0	0.3–8.0	10^{12}–10^{14}	10^{13}–10^{14}	16–32	400–800
Polychloroprene	7.5–14.0	1.0–6.0	10^{9}–10^{10}	10^{11}–10^{12}	4–20	100–500
Polysulfide polymer	7.0–9.5	0.1–0.5	10^{9}–10^{10}	...	10–13	250–325
Silicone	2.8–7.0	0.10–1.0	10^{11}–10^{15}	10^{13}	12–28	300–700
Chlorosulfonated polyethylene	5.0–11.0	2.0–9.0	10^{11}–10^{15}	10^{14}	16–24	400–600
Polyvinylidene fluoride copolymer, hexafluoropropylene	10.0–18.0	3.0–4.0	10^{11}	...	10–28	250–700
Polyurethane	5.0–8.0	3.0–6.0	10^{8}–10^{9}	...	18–20	450–500
Ethylene-propylene terpolymer	3.2–3.4	0.6–0.8	10^{13}–10^{15}	...	28–36	700–900

(a) At 1 MHz. Source: Ref 2

Table 9 Electrical application information for thermosetting plastics

Material	Major electrical application considerations	Common available forms
Alkyds	Excellent dielectric strength, arc resistance, and dry insulation resistance; low dielectric constant and dissipation factor	Compression moldings, transfer moldings
Aminos (melamine-formaldehyde and urea-formaldehyde)	Good general electrical properties, especially arc resistance	Compression moldings, extrusions, transfer moldings, laminates, film
Diallyl phthalates (DAP) (allylics)	Unsurpassed among thermosets in retention of properties in high-humidity environments; have among the highest volume and surface resistivities in thermosets; low dissipation factor	Compression moldings, extrusions, injection moldings, transfer moldings, laminates
Epoxies	Good electrical properties, useful over a wide range of environments	Castings, compression moldings, extrusions, injection moldings, transfer moldings, laminates, matched-die moldings, filament windings, foam
Phenolics	Among the least expensive, most widely used thermoset materials; excellent thermal stability to over 150 °C (300 °F) generally, and over 205 °C (400 °F) in special formulations	Castings, compression moldings, extrusions, injection moldings, transfer moldings, laminates, matched-die moldings, stock shapes, foam
Polyesters	Excellent electrical properties and low cost	Compression moldings, extrusions, injection moldings, transfer moldings, laminates, matched-die moldings, filament windings, stock shapes
Silicones (rigid)	Excellent electrical properties, especially low dielectric constant and dissipation factor, which change little up to 205 °C (400 °F) and over	Castings, compression moldings, transfer moldings, laminates
Urethanes (rigid foams)	Low-weight plastics; excellent electrical properties, which are basically variable as a function of density; easy to use for foam-in-place and embedding applications	Castings, coatings

Source: Ref 2

Conductive thermoplastics are actually composites that comprise electrically insulating plastic matrices and electrically conductive fillers. The conductive fillers may be particulates, plates, or fibers. Electrical conductivity is observed in the composite when the filler volume is sufficient to support a continuous electrical path through the composite. The critical filler volume needed to achieve conductivity depends on the resistivity, structure, and final dimensions of the filler in the melt-form composite. As the volume loading of filler is increased above the critical volume, the resistivity of the composite is decreased until a minimum is reached. A range of composite resistivities can be obtained by varying filler content.

Composites exhibiting 10^{-1} to 10^{2} Ω/square surface resistivity perform well as EMI/radiofrequency interference (RFI) shielding materials. Electrostatic discharge protection is provided by composites of 10^{2} to 10^{6} Ω/square resistivity. The resistivity requirement for antistatic composites that offer protection from low voltages is 10^{9} to 10^{13} Ω/square. The electrical test results of several thermoplastic composites that have been designed to shield EMI/RFI and to provide ESD protection are given in Ref 13. Data for electrical resistivity and shielding effectiveness were generated in accordance with ASTM test procedures, and the static decay rates were measured using the Federal Standard 101B, Method 4046.

Much confusion exists regarding the methods of testing the effective shielding of plastic materials. Distinctions must be made between an infinite homogeneous plane and real plane shields. Moreover, practical shields are housings or boxes with corners, joints, access holes, and so on. Plane shield measurements will yield the maximum available shielding effectiveness for the specified source distance. Several shielding effectiveness mechanisms exist, but in the frequency range of 30 to 1000 MHz, the electric conductivity due to electric conduction is most important. At 1000 MHz, the shielding effectiveness for the far field is approximated by:

$$\text{Shielding effectiveness (db)} = -20 \log R_0 + 45$$

where R_0 is the effective surface resistance. The use of metal fibers can cover the range of 0.1 to 10 Ω/square.

The simplest way to determine the shielding is to measure the dc surface resistivity. This does not guarantee good shielding, because inhomogeneities may cause aperture effects;

therefore, no shielding occurs at microwave frequencies. In a conductive plastic, the filler forms a mesh, and it is therefore difficult to make good contacts with it. The ASTM Committee D-9 proposes two test methods for two-dimensional configurations: the shielded-box method and the coaxial line test. The former uses larger samples, while special machining is necessary for the latter. Both methods require metallic contacts on the samples.

A new test method that evaluates the shielding effectiveness of materials from reflectivity measurements has been developed and compares well with results obtained from the shielded-box and the coaxial methods. This simple test, which is performed with a portable device, measures the reflection of a sample at 10 GHz and then compares it with the reflection of a metal plate at the same frequency. No contacts are needed. The result is then converted to the shielding effectiveness at 1000 MHz, taking into account the source-to-shield distance. Metal fibers can be used in a variety of ways for shielding purposes. Because of a high aspect ratio (length to diameter), only low weight percentages are needed. The fibers can also be used for conductive plastics. The shielding effectiveness of materials, particularly fiber-loaded materials, is easily determined with the new measuring technique.

Terminology

The terms used in connection with testing and specifying plastics for electrical applications are defined in this section (Ref 14). Complete definitions and related electrical terminologies are available in the Selected References in this article.

Arc Tracking. The process that produces surface tracks when arcs occur on or close to an insulating surface.

Capacitance. That property of a system of conductors and dielectrics that permits the storage of electrically separated charges when potential differences exist between the conductors. It is the ratio of a quantity, Q, of electricity to a potential difference, V. The units are farads when the charge is expressed in coulombs, and the potential is in volts: $C = Q/V$.

Conductance, Apparent dc. The ratio of the electrical current measured at the end of a specified electrification time to the steady, direct voltage applied to the specimen.

Conductance, Apparent dc Surface. The apparent dc conductance between two electrodes in contact with a specimen of insulating material when the current involved is limited to a thin film of moisture or other semiconducting material on the surface of the specimen.

Conductance, Apparent dc Volume. The apparent dc conductance of a specimen when the current measured is limited to the volume of the specimen.

Conductance, dc Insulation. The apparent dc conductance between two electrodes having a configuration such that both volume and surface conductance are included in an unknown ratio.

Conductivity, Apparent dc Volume. The apparent dc volume conductance multiplied by the function of specimen dimensions that transforms the conductance to that of a unit cube. The conductivity is usually expressed in the units of S/m, where S represents siemens.

Conductivity, dc Volume. The property of a material that permits the flow of electricity through its volume. It is numerically equal to the ratio of the steady-state current density to the steady, direct voltage gradient parallel with the current in the material.

Dielectric. A medium in which it is possible to maintain an electric field with little supply of energy from outside sources. The energy required to produce the electric field is recoverable in whole or in part. A vacuum, as well as any insulating material, is a dielectric.

Dielectric (Electric) Breakdown Voltage. The potential difference at which dielectric failure occurs under prescribed conditions, in an

Table 10 Electrical properties of thermosetting molding materials

		Diallyl phthalate			Epoxy		Melamine		
Property	ASTM method	Glass fiber filler	Mineral filler	Synthetic fiber filler	Glass fiber filler	Mineral filler	α-cellulose filler	Asbestos filler	Glass fiber filler
Volume resistivity, Ω · m	D 257	10^{14}	10^{11}	10^{14}	10^{12}	10^{12}	10^{12}	10^{10}	10^{9}
Dielectric strength, MV/m (V/mil)									
Short-time	D 149	18 (450)	17 (420)	16 (400)	16 (400)	16 (400)	16 (400)	17.2 (430)	12 (300)
Step-by-step	D 149	16 (400)	16 (400)	16.5 (410)	16 (400)	16 (400)	12 (300)	12.8 (320)	9.6 (240)
Dielectric constant, MV/m (V/mil)									
At 60 Hz	D 150	4.3	5.2	5.0	5.0	5.0	9.5	10.2	11.1
At 1 kHz	D 150	4.4	5.3	3.9	5.0	5.0	9.2	9.0	...
At 1 MHz	D 150	4.5	4.0	3.6	5.0	5.0	8.4	6.7	7.5
Dissipation factor									
At 60 Hz	D 150	0.01	0.03	0.026	0.01	0.01	0.030	0.07	0.14
At 1 kHz	D 150	0.004	0.03	0.004	0.01	0.01	0.015	0.07	...
At 1 MHz	D 150	0.009	0.02	0.012	0.01	0.01	0.027	0.041	0.013
Arc resistance, s	D 495	180	190	130	180	190	180	180	180

		Phenolic			Polyester		Silicone		
Property	ASTM method	Wood flour and cotton flock filler	Asbestos filler	Glass fiber filler	Glass fiber filler	Mineral filler	Glass fiber filler	Mineral filler	Ureaformaldehyde α-cellulose filler
Volume resistivity, Ω · m	D 257	10^{11}	10^{11}	10^{10}	10^{13}	10^{12}	10^{12}	10^{12}	10^{11}
Dielectric strength, MV/m (V/mil)									
Short-time	D 149	16 (400)	14 (350)	16 (400)	16.8 (420)	18 (450)	16 (400)	16 (400)	16 (400)
Step-by-step	D 149	15 (375)	12 (300)	10.8 (270)	15.6 (390)	14 (350)	12 (300)	15.2 (380)	12 (300)
Dielectric constant, MV/m (V/mil)									
At 60 Hz	D 150	13	50	7.1	7.3	7.5	5.2	3.6	9.5
At 1 kHz	D 150	9.0	30	6.9	4.68	6.2	5.0	...	7.5
At 1 MHz	D 150	6.0	10	6.6	6.4	5.5	4.7	6.3	6.8
Dissipation factor									
At 60 Hz	D 150	0.05	0.1	0.05	0.011	0.009	0.004	0.004	0.035
At 1 kHz	D 150	0.04	0.1	0.02	...	0.02	0.0035	...	0.025
At 1 MHz	D 150	0.03	0.4	0.012–0.026	0.008	0.015	0.002	0.002	0.25
Arc resistance, s	D 495	Tracks	120	120	180	150	250	420	150

Source: Ref 2

Table 11 Electrical application information for thermoplastics

Material	Major electrical application considerations	Common available forms
Acrylonitrile-butadiene-styrene	Good general electrical properties, but not outstanding for any specific electrical applications	Blow moldings, extrusions, injection moldings, thermoformed parts, laminates, stock shapes, foam
Acetals	Good electrical properties at most frequencies, which are little changed in humid environments up to 125 °C (257 °F)	Blow moldings, extrusions, injection moldings, stock shapes
Acrylics (PMMA)	Excellent resistance to arcing and electrical tracking	Blow moldings, castings, extrusions, injection moldings, thermoformed parts, stock shapes, film, fiber
Cellulosics	There are several materials in the cellulosic family, such as cellulose acetate, cellulose propionate, cellulose acetate butyrate, ethyl cellulose, and cellulose nitrate; widely used plastics in general, but not outstanding for electronic applications	Blow moldings, extrusions, injection moldings, thermoformed parts, film, fiber, stock shapes
Chlorinated polyethers	Good electrically	Extrusions, injection moldings, stock shapes, film
Ethylene-vinyl acetates	Not widely used in electronics	
Fluorocarbons (chlorotrifluoroethylene) (CTFE)	Excellent electrical properties; widely used in electronics but not quite so widely as TFE and FEP. Useful to about 205 °C (400 °F)	Extrusions, isostatic moldings, injection moldings, film, stock shapes
Fluorinated ethylene propylene (FEP)	Very similar properties to those of TFE, except useful temperature limited to about 205 °C (400 °F)	Extrusions, injection moldings, laminates, film
Polytetrafluoroethylene (PTFE)	Electrically one of the most outstanding thermoplastic materials; exhibits very low electrical losses and very high electrical resistivity; useful to over 260 °C (500 °F) and to below –185 °C (–300 °F); excellent high-frequency dielectric; among the best combinations of mechanical and electrical properties	Compression moldings, stock shapes, film
Polyvinylidine fluoride	Good electrically; useful to about 150 °C (300 °F); a major electronic application is wire jacketing	Extrusions, injection moldings, laminates, film
Nylons (polyamides)	Good general-purpose for electrical and nonelectrical applications; some nylons have limited use due to moisture-absorption properties	Blow moldings, extrusions, injection moldings, laminates, rotational moldings, stock shapes, film, fiber
Parylenes (polyparaxylylene)	Excellent dielectric properties; used primarily as thin films in capacitors and dielectric coatings; numerous polymer modifications exist	Film coatings
Phenoxies	Tough, rigid, high-impact plastic; useful for electronic applications below about 80 °C (175 °F)	Blow moldings, extrusions, injection moldings, film
Polyallomers	Thermoplastic polymers produced from two monomers; somewhat similar to polyethylene and polypropylene; electronic application areas similar to polyethylene and polypropylene; one of the lightest commercially available plastics	Blow moldings, extrusions, injection moldings, film
Polyamide-imides and polyimides	Among the highest-temperature thermoplastics available, having useful operating temperatures between 205 °C (400 °F) and about 370 °C (700 °F) or higher; excellent electrical properties, good rigidity, excellent thermal stability	Films, coatings, molded and/or machined parts, resin solutions
Polycarbonates	Good electrical properties for general electronic packaging application; available in transparent grades	Blow moldings, extrusions, injection moldings, thermoformed parts, stock shapes, film
Polyethylenes and polypropylenes (polyolefins or polyalkenes)	Excellent electrical properties, especially low electrical losses. There are three density grades of polyethylene: low (0.910–0.925 g/cm³), medium (0.926–0.940 g/cm³), and high (0.941–0.965 g/cm³).	Blow moldings, extrusions, injection molding, thermoformed parts, stock shapes, film, fiber, foam
Polyethylene terephthalates	Among the toughest of plastic films with outstanding dielectric strength properties; good humidity resistance; stable to 135–150 °C (275–300 °C)	Film, sheet, fiber
Polyphenylene oxides	Excellent electrical properties, especially loss properties to above 175 °C (350 °F) and over a wide frequency range	Extrusions, injection moldings, thermoformed parts, stock shapes, film
Polystyrenes	Excellent electrical properties, especially loss properties; conventional polystyrene is temperature limited, but high-temperature modifications exist that are widely used in electronics, especially for high-frequency applications	Blow moldings, extrusions, injection moldings, rotational moldings, thermoformed parts, foam
Polysulfones	Excellent electrical properties to above 150 °C (300 °F)	Blow moldings, extrusions, injection-molded thermoformed parts, stock shapes, film sheet
Vinyls	Good low-cost general-purpose thermoplastic materials but not specifically outstanding electrical properties; greatly influenced by plasticizers; many variations available, including flexible and rigid types; flexible vinyls, especially polyvinyl chloride, widely used for wire insulation and jacketing	Blow moldings, extrusions, injection moldings, rotational moldings, film sheet

Source: Ref 2

electrical insulating material located between two electrodes.

Dielectric Failure. An event that is evidenced by an increase in conductance in the dielectric under test and that limits the electric field that can be sustained.

Dissipation Factor (Loss Tangent), D. The ratio of the loss index to its relative permittivity, or $D = \kappa''/\kappa'$. It is also the tangent of its loss angle, δ, or the cotangent of its phase angle, θ, or $D = \tan\delta = \cot\theta = X_p/R_p = G/wC_p = 1/wC_pR_p$, where X_p is the parallel reactance, R_p is the equivalent ac parallel resistance, G is the equivalent ac conductance, C_p is the parallel capacitance, and $w = 2\pi$ times frequency.

Electrification Time. The time during which a steady, direct potential is applied to electrical insulating materials before the current is measured.

Erosion, Electrical. The progressive wearing away of electrical insulation by the action of electrical discharges.

Erosion Resistance, Electrical. The quantitative expression of the amount of electrical erosion under specific conditions.

Guard Electrode. One or more electrically conducting elements, arranged and connected in an electric instrument or measuring circuit so as to divert unwanted conduction or displacement currents from, or confine wanted currents to, the measurement device.

Ionization. The process by which electrons are lost from or transferred to neutral molecules or atoms to form positively or negatively charged particles.

Loss Angle, δ. The angle whose tangent is the dissipation factor or the arctan (κ''/κ'). It is also the difference between 90° and the phase angle.

Loss Index, κ″. The magnitude of the imaginary part of the relative complex permittivity. It is the product of relative permittivity and dissipation factor and it may be expressed as $\kappa'' = \kappa' D$.

Partial Discharge (Corona). An electrical discharge that only partially bridges the insulation between conductors. A transient, gaseous ionization occurs in an insulation system if the voltage stress exceeds a critical value, and this ionization produces partial discharges.

Partial Discharge (Corona) Level. The magnitude of the greatest recurring discharge during an observation of continuous discharges.

Power Factor, PF. The ratio of power in watts, W, dissipated in a material to the product of the effective sinusoidal voltage, V, and the current, I, in volt-ampere, VA. It may be expressed as the cosine of the phase angle or the sine of the loss angle ($PF = W/VI = \sin\delta = \cos\theta$). When the dissipation factor is less than 0.1, the power factor differs from the dissipation factor by less than 0.5%.

Quality Factor, Q. The reciprocal of the dissipation factor. This term has been referred to as storage factor.

Relative Complex Permittivity (Relative Complex Dielectric Constant), k.* The ratio of the admittance of a given configuration of the material to the admittance of the same configuration with vacuum as dielectric: $k^* = Y/Y_v = Y/jwC_v = \kappa' - j\kappa''$, where Y is admittance of the material and jwC_v is the admittance of vacuum.

Resistance, Apparent dc. The reciprocal of apparent dc conductance.

Resistance, Apparent dc Surface. The reciprocal of apparent dc surface conductance.

Resistance, Apparent dc Volume. The reciprocal of apparent dc volume conductance.

Resistivity, Apparent dc Volume. The reciprocal of apparent dc volume conductivity. Resistivity is usually expressed as $\Omega \cdot$ m.

Resistivity, dc Volume. The reciprocal of dc volume conductivity.

Scintillation. The multiple discharges or small arcs that originate in the more conductive areas of the insulation surface, and span less conductive area.

Track. A partially conducting path of localized deterioration on the surface of an insulating material.

Tracking. The process that produces tracks as a result of the action of the electric discharges on or close to the insulation surface.

Tracking, Contamination. Tracking caused by scintillations that result from the increased surface conduction due to contamination.

Tracking Resistance. The quantitative expression of the voltage and the time required to develop a track under specified conditions.

Table 12 Electrical properties of thermoplastic materials

Property	ASTM method	Acetal	ABS	Acrylic	Cellulose acetate	Cellulose acetate butyrate	Cellulose propionate	Chlorinated polyether	Chlorotrifluoroethylene	Nylon (polyamide)	Polycarbonate
Arc resistance	D 495	129	90	No track	200	...	180	...	>360	140	120
Dielectric constant											
At 60 Hz	D 150	3.8	3.0	4.0	7.5	6.4	4.0	3.1	2.8	5.5	3.2
At 1 MHz	...	3.8	3.0	3.5	7.0	6.3	4.0	3.0	2.7	4.9	3.0
At 1 GHz	...	3.8	3.0	3.2	7.0	6.2	3.6	2.9	2.5	4.7	3.0
Dissipation factor											
At 60 Hz	D 150	0.004	0.003	0.04	0.01	0.02	0.01	0.01	0.001	0.01	0.0009
At 1 GHz	...	0.004	0.005	0.02	0.01	0.05	0.01	0.01	0.09	0.03	0.01
Dielectric strength, step-by-step, MV/m (V/mil)	D 149	16 (400)	14 (350)	14 (350)	8 (200)	10 (250)	12 (300)	16 (400)	18 (450)	12.8 (320)	14.5 (364)
Volume resistivity, $\Omega \cdot$ m	D 257	10^{12}	10^{14}	10^{12}	10^{11}	10^{12}	10^{13}	10^{13}	10^{16}	10^{13}	10^{10}

Property	ASTM method	Polyethylene, low-density	Polyethylene, med-density	Polyethylene, high-density	Polypropylene	Polystyrene	Polysulfone	Polyphenylene oxide	Phenoxy	Polyvinyl chloride	Styrene-acrylonitrile	Tetrafluoroethylene
Arc resistance	D 495	140	200	200	185	100	122	75	...	80	150	<200
Dielectric constant												
At 60 Hz	D 150	2.4	2.4	2.4	2.6	3.4	3.1	2.6	4.1	3.6	3.4	2.1
At 1 MHz	...	2.4	2.4	2.4	2.6	3.2	3.1	2.6	4.1	3.3	2.5	2.1
At 1 GHz	...	2.4	2.4	2.4	2.6	3.1	3.1	2.6	3.8	3.4	3.1	2.1
Dissipation factor												
At 60 Hz	D 150	<0.0005	<0.0005	<0.0005	<0.0005	0.0004	0.0008	0.0004	0.001	0.007	0.004	<0.0002
At 1 MHz	...	<0.0005	<0.0005	<0.0005	<0.0005	0.0004	0.001	...	0.002	0.009	0.007	<0.0002
At 1 GHz	...	<0.0005	<0.0005	<0.0005	<0.0005	0.0004	0.005	0.0009	0.03	0.006	0.007	<0.0002
Dielectric strength, step-by-step, MV/m (V/mil)	D 149	16.8 (420)	20 (500)	22 (550)	18 (450)	12 (300)	16 (400)	16 (400)	16 (400)	15 (375)	12 (300)	17.2 (430)
Volume resistivity, $\Omega \cdot$ m	D 257	10^{14}	10^{14}	10^{14}	10^{14}	10^{14}	10^{15}	10^{11}	10^{11}	10^{14}	10^{14}	10^{16}

Source: Ref 2

REFERENCES

1. D.G. Fink and J.M. Carroll, *Standard Handbook for Electrical Engineers,* 10th ed., McGraw-Hill, 1969
2. D.G. Fink and D. Christiansen, *Electronics Engineers' Handbook,* 2nd ed., McGraw-Hill, 1982
3. "Test Methods for Dielectric Breakdown Voltage and Dielectric Strength of Solid Insulating Materials at Commercial Power Frequencies," D 149, *Annual Book of ASTM Standards,* American Society for Testing and Materials
4. "Test Method for Dielectric Breakdown Voltage and Dielectric Strength of Solid Electrical Insulating Materials under Direct-Voltage Stress," D 3755, *Annual Book of ASTM Standards,* American Society for Testing and Materials
5. "Test Methods for A-C Loss Characteristics and Permittivity (Dielectric Constant) of Solid Electrical Insulating Materials," D 150, *Annual Book of ASTM Standards,* American Society for Testing and Materials
6. "Test Method for Relative Permittivity (Dielectric Constant) and Dissipation Factor of Polyethylene by Liquid Displacement Procedure," D 1531, *Annual Book of ASTM Standards,* American Society for Testing and Materials
7. "Test Methods for Relative Permittivity and Dissipation Factor of Expanded Cellular Plastics Used for Electrical Insulation," D 1673, *Annual Book of ASTM Standards,* American Society for Testing and Materials
8. "Test Methods for D-C Resistance or Conductance of Insulating Materials," D 257, *Annual Book of ASTM Standards,* American Society for Testing and Materials
9. "Test Method for High-Voltage, Low-Current, Dry Arc Resistance of Solid Electrical Insulation," D 495, *Annual Book of ASTM Standards,* American Society for Testing and Materials
10. "Test Method for Dust-and-Fog Tracking and Erosion Resistance of Electrical Insulating Materials," D 2132, *Annual Book of ASTM Standards,* American Society for Testing and Materials
11. "Test Method for Liquid-Contaminant, Inclined Plane Tracking and Erosion of Insulating Materials," D 2303, *Annual Book of ASTM Standards,* American Society for Testing and Materials
12. "Test Method for Comparative Tracking Index of Electrical Insulating Materials," D 3638, *Annual Book of ASTM Standards,* American Society for Testing and Materials
13. J.M. Crosby and J.E. Travis, Conductive Thermoplastic Composites, *Rubber World,* Nov 1985
14. "Definitions of Terms Relating to Electrical Insulation," D 1711, *Annual Book of ASTM Standards,* American Society for Testing and Materials

SELECTED REFERENCES

- *Electrical and Electronics Terms,* ANSI/IEEE 100-1977, American National Standards Institute and the Institute of Electrical and Electronics Engineers, 1977
- *The Illustrated Dictionary of Electronics,* Tab Books, 1982

Optical Testing and Characterization*

OPTICAL TESTING of plastics includes characterization of materials and analysis of optical components. If a material is tested for transmission, haze, yellowness, and refractive index, knowledge of its optical properties is nearly complete. For optical components, surface irregularity, birefringence, and internal contamination must also be considered. These characteristics are a function of the material and fabrication method. Gloss and color also are affected by the base material and are measured as optical properties. Polymers differ from many other optical materials in the degree to which their properties change with wavelength, temperature, and moisture, and no optical testing of polymers or plastic optical components can be considered complete until these effects are evaluated.

Because most optical tests are performed as standard practice, industry standards have been developed and published by the American Society for Testing and Materials (ASTM). While virtually every optical characteristic can be tested in various ways, ASTM standards are used by the material manufacturers and are referred to throughout this article when applicable.

Transmission and Haze

Transmission is one of the most obvious characteristics of a transparent plastic. It is measured using a spectrophotometer or a hazemeter in accordance with ASTM D 1003. This method uses a collimated light source, an integrating sphere, and a detector. Measurements are taken on a flat, dust-free test specimen that has a diameter greater than 25 mm (1 in.) and that is thick enough for transmission losses to be significant (usually 3.2 mm, or ⅛ in., or greater). Any light scattered at greater than 2.5° is measured as haze. Haze results from the diffuse scattering from internal material inhomogeneities such as density differences, fillers, pigments, and voids. With the described setup, accuracies of 0.1 to 0.3% can be obtained. However, if haze is greater than 30%, ASTM E 167 should be used. Transmission is normally measured and plotted against wavelength.

If a material exhibits no internal absorption or internal haze in the visible or near-infrared range, as in the case of an optical quality acrylic material of reasonable thickness, the instrument will record 92% transmission (Fig. 1). Reflection accounts for losses of 8%, or 4% per surface.

Reflection loss is calculated as a close approximation by the formula:

$$R = \frac{(n^1 - n)^2}{(n^1 + n)^2}$$

where n and n^1 are the indexes of refraction of the two media involved (Ref 1). If a material has a refractive index of 1.49 (at 589 nm), such as acrylic, and the second medium is air (refractive index = 1), the reflection losses will be 3.9% per surface. If a material, such as polycarbonate (PC) or polystyrene (PS), has a refractive index of 1.59, the reflection losses will be 5.2% per surface. These numbers decrease with increasing wavelength and temperature. Thus, the maximum transmission of visible light for uncoated acrylic of any thickness is approximately 92%, and the maximum transmission of visible light for uncoated PC or PS is approximately 89 to 90%.

All plastic and glass materials have some degree of light scattering, which, in terms of transmission loss, is measured as haze (using ASTM D 1003). In addition, a material may have some degree of absorption that will also show up as transmission loss without an increase in haze. Because absorption and haze essentially increase linearly with an increase in thickness, they must be measured relative to the thickness of the test sample. The reflection loss must be subtracted before calculating transmission loss for a sample of a different thickness. Figure 2 provides curves of percent of transmission plotted against wavelength for typical plastics.

Yellowness

In their natural state, many plastic materials have a yellow or straw color. This can be seen as a falling-off in the blue region of the transmission curve around 400 nm, as is evident from curve 2 of Fig. 3. A blue toner is added to the material to make it appear "water clear," resulting in curve 3 of Fig. 3. It is the absorption of the green and red portions of the spectrum that make the part appear water clear. On thicker sections, this material would have a blue-gray appearance.

This same type of yellowness occurs when plastics are degraded, either during processing or from long-term exposure to heat or damaging radiation from ultraviolet or shorter wavelengths. Because it is so common, a standard test for yellowness (ASTM D 1925) was developed for transparent, nearly white, translucent, and opaque plastics. ASTM D 1925 was withdrawn in 1995, but it is briefly described in the following paragraphs.

Yellowness is defined in ASTM D 1925 as deviation in chroma from whiteness or water whiteness in the dominant wavelength range of 570 to 580 nm. The procedure is performed using a spectrophotometer; a yellowness index, based on a primary standard of magnesium oxide or an equivalent instrument standard, is calculated. Because the index is relative to magnesium oxide, the change in yellowness, rather than the absolute index, is often observed and reported. A positive change indicates increased yellowness, and a negative change indicates decreased yellowness or increased blueness. Laboratories can reproduce these results to approximately 0.3 units.

Because many companies do not have the necessary equipment for yellowness index measurements, test samples are often compared visually to a standard specimen under standard lighting conditions. In some cases, a color meter may be used, and its results can then be compared to a standard.

Refractive Index

The refractive index (n_D) of a material that is quoted in the literature is the index at 23 °C (73 °F) or 25 °C (77 °F) and at the specific wavelength of the D line of the sodium emission spectrum, which is 589.3 nm. The index is measured using the property of refraction in several ways.

The autocollimation method (Ref 2) uses an accurately made prism and measures the angle of refraction to calculate the index. With careful procedures, this method is capable of measurements to four or possibly five decimal places. However, more than four decimal places is extremely difficult to reproduce because of the

*Adapted from Donald L. Keyes, Optical Testing and Characterization, *Engineering Plastics,* Volume 2, *Engineered Materials Handbook,* ASM International, 1988, pages 594 to 598

difficulty of controlling temperature and moisture in the sample. It may be said here that the long-term consistency of the index of a production acrylic is probably approximately 2×10^{-4} (Ref 3); that of PC is about the same (Ref 4). Documentation of this information is not nearly as available as it is for glass optical materials, and usually the information is available to only two or three decimal places.

The standard for the plastics industry, ASTM D 542, calls for two methods of index measurement. One uses the Abbe refractometer, which measures a sample with one side polished to an optical flat and a perpendicular side that is also polished. This method is accurate to approximately three decimal places and possibly four. The second method is the microscopic method, which measures the apparent thickness of a flat sample with a microscope by focusing down through the sample and measuring the travel of the microscope stage. The number is compared to the actual thickness of the part. This method is only accurate to approximately two decimal places. It is unreliable for anisotropic materials, such as polyester or strained PC or PS, but both methods are suitable for many optical applications. A third method (Becké line method) uses the microscope with the plastic sample immersed in an oil of known refractive index. If the plastic is of a higher index than the oil and the immersed microscope objective lens is raised, the bright (Becké) line at the plastic/oil interface moves into the plastic. This method is accurate to four decimal places.

The change in refractive index with a change in wavelength is called the dispersion (Fig. 4).

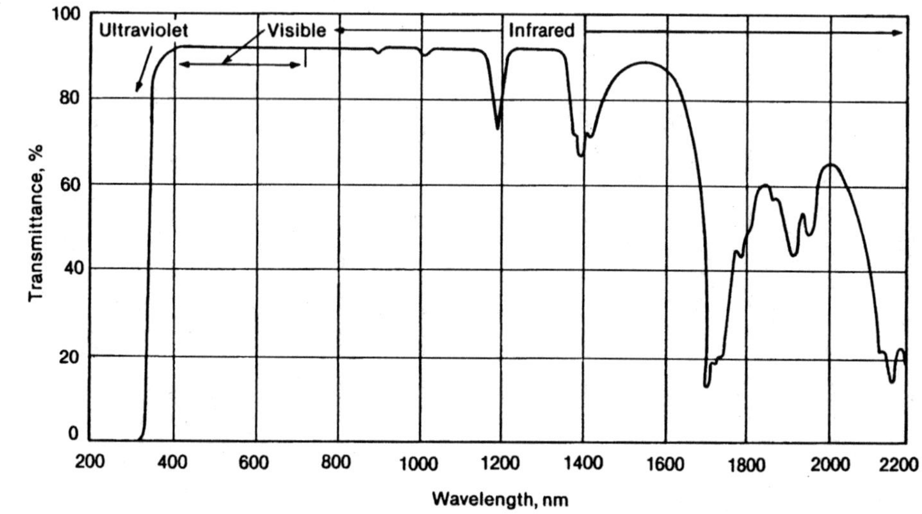

Fig. 1 Spectrophotometric transmission of acrylic

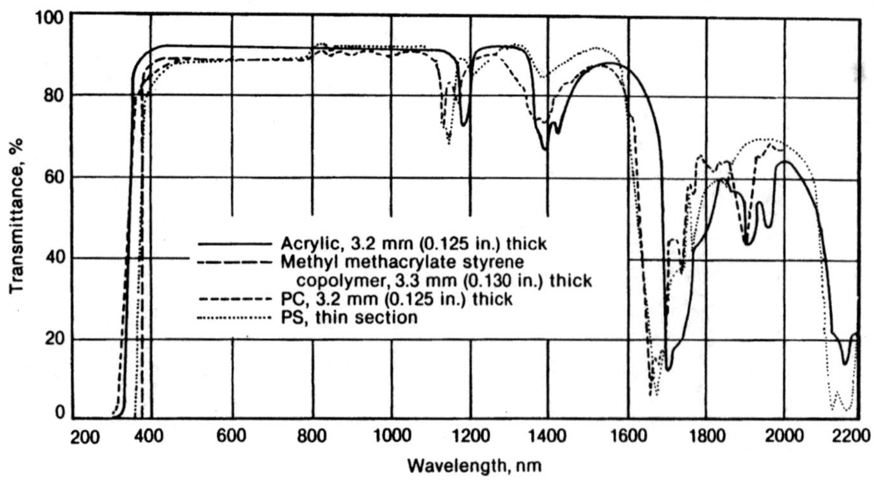

Fig. 2 Spectrophotometric transmission of principal optical plastics. Although the methyl methacrylate styrene copolymer curve stops at 800 nm, it is expected to be similar to the other polymers at higher wavelengths.

While the dispersion is different for different materials, the shape of the curves is similar. Therefore, the index is determined at three wavelengths, and the dispersion is calculated. The three wavelengths are n_C, 656.3 nm (hydrogen C line); n_D, 589.3 nm (sodium D line); and n_F, 486.1 nm (hydrogen F line). From these values, the reciprocal relative dispersion, called the Abbe *V* number, is calculated (Ref 1):

$$V = \frac{n_D - 1}{n_F - n_C}$$

The lower the *V* value, the greater the separation of white light into colors by a prism of that material.

The refractive index of plastic also changes with temperature and moisture. Because the change with temperature is small for glass and can therefore be neglected, it is also often ignored when using plastics. However, the number can be significant, as shown by the *dn*/*dt* in Table 1. In plastics, the index is related to density, and as the temperature rises and material expands, the index decreases. This relationship is nonlinear (Ref 5, 6). The *dn*/*dt* values in Table 1 are specified at or near room temperature and would increase with increasing temperature.

The refractive index change with moisture has been understood for some time, but the amount of the change is not well documented. As moisture is absorbed into a plastic surface, it creates some stress. The resultant compression creates a higher refractive index at the surface and an index gradient inward toward lower moisture levels. When the plastic is saturated, the index gradient disappears. Because this stress is uniform, it is not apparent in some photoelastic observations. Also, because the numbers are small, the optical characteristics are often overlooked. The index change is thought to be 1 or 2 in the third decimal place for acrylic. Other plastics have not been as carefully examined, but the effect appears to be present.

When testing a thick part for transmission, the light can be refracted toward the edge of the part as a result of the index gradient. When testing a part that reflects light internally, the index gradient can effectively cause the light ray to be shifted sideways (Fig. 5). Unless understood and accounted for, this effect can create significant confusion in certain geometries.

Birefringence

Photoelastic measurements and birefringence are subjects that are not well understood by many in the plastics industry. The subject, although somewhat complex, holds promise of lending great understanding to stress analysis and the optical characteristics of plastic materials.

A transparent material that has a refractive index (that is, relative velocity of transmission) that remains the same regardless of the direction or the polarization of transmitted light is said to be isotropic. A material for which the refractive index depends on the direction and polarization of light is anisotropic. An anisotropic material is said to be optically birefringent.

Some materials, plastic or otherwise, are naturally birefringent because of their crystalline structure. Others, such as amorphous plastics, are not birefringent unless strained (that is, deformed as a result of stress). The use of strain-induced birefringence (anisotropy) to study stress is called photoelasticity.

Light transmitted through a strained region splits into two perpendicularly polarized waves, each polarized in the direction of the principal strains, e_i. The difference in the refractive index, n, for the two polarized waves is:

$$n_1 - n_2 = k\,(e_1 - e_2)$$

where k is a material property called the strain-optical constant.

The retardation of one wave with respect to the other causes the waves to emerge from the material out of phase. This relative retardation, δ, is:

$$\delta = t\,k\,(e_1 - e_2)$$

where t is the thickness of the material.

When a strained part is held between two crossed polarizers and viewed in the direction of a light source (Fig. 6), a pattern of fringes can be seen, some of which may be colored, resulting from the birefringence. The object of photoelastic measurements is to measure the direction and amount of strain. Stress based on the material constants can then be calculated. Measurements are made using a light source, polarizers, quarter-wave plates, a compensator, and a filter in an appropriate assembly. This procedure is well described in ASTM D 4093, which also lists several worthwhile references. There is also a good introduction to the topic in Ref 7.

Surface Irregularity and Contamination

Surface irregularity is a characteristic that affects both transparent and opaque parts. If very small (difficult to measure) irregularities exist, they can be easily seen in the case of a shiny surface.

For the purpose of analysis, surface irregularities are separated into macroscopic irregularities, which occur over distances greater than 1 mm (0.040 in.), and microscopic irregularities, which occur over distances less than 1 mm (0.040 in.). Macroscopic irregularities tend to distort the visual perception of an object, while microscopic irregularities tend to increase light scattering and haze. Macroscopic irregularities on polished surfaces are visible to the unaided eye when they are approximately 0.0025 mm (0.0001 in.) in depth (or height) and are spread over an area approximately 3.2 mm (⅛ in.) in diameter. Thus, a 0.0025 mm (0.0001 in.) depression spread over 1.6 mm (1/16 in.) can be seen easily, but a 0.005 mm (0.0002 in.) depression spread over a 9.5 mm (⅜ in.) area is difficult to see.

The optical industry normally measures irregularity in wavelengths of light. Test instruments are often illuminated by helium-neon lasers, which operate at a wavelength of 0.633 μm (25 μin.). An irregularity of 0.0025 mm (0.0001 in.), then, is considered 4 wavelengths deep or high. Using this method, the irregularity of a surface can be deemed either better or worse, depending on the wavelength of light used.

Macroscopic surface irregularities are tested in several ways, depending on geometry and intended surface use. Curved surfaces of optical quality are tested using test glasses and interferometers, which are based on the reflection of monochromatic light from a master surface and the test surface. The reflected light creates an interference pattern that appears as a contour map of the test surface. The vertical distance between the lines on the contour map is a half wavelength of the source light, as shown in Fig. 7. Some curved surfaces are tested using very accurate coordinate measuring machines built for that purpose. Some of these machines are accurate to nearly 0.025 μm (1 μin.).

Flat surfaces are sometimes tested with test glasses, interferometers, or profilometers, the latter of which are widely used in the machine industry. Profilometers have accuracies in the range of 0.250 μm (10 μin.), which is adequate for most applications. A standard test (ASTM D 637) was developed for flat windows, where light is projected through the sample window and allows examination of the distortions of the image of a cross on a patterned screen. This test method does not work for other surface geometries, and the ASTM standard (D 637) was discontinued in 1995.

Microscopic surface irregularities are usually measured by a microscope or a profilometer. They tend to fall into three categories: scratches, digs (or pits), and a mottled surface called orange peel. Scratches and digs may result from tool marks or material handling. They are specified in various ways by "limit samples" used for quality control purposes. In the optics industry, there is a cosmetic specification for scratches and digs in accordance with military standard MIL-0-13830A. When a scratch/dig number is applied to a part, it specifies the maximum allowable length and number of scratches, as well as the maximum allowable dig size and accumulation of dig area. Scratches and digs are measured with a staged measuring

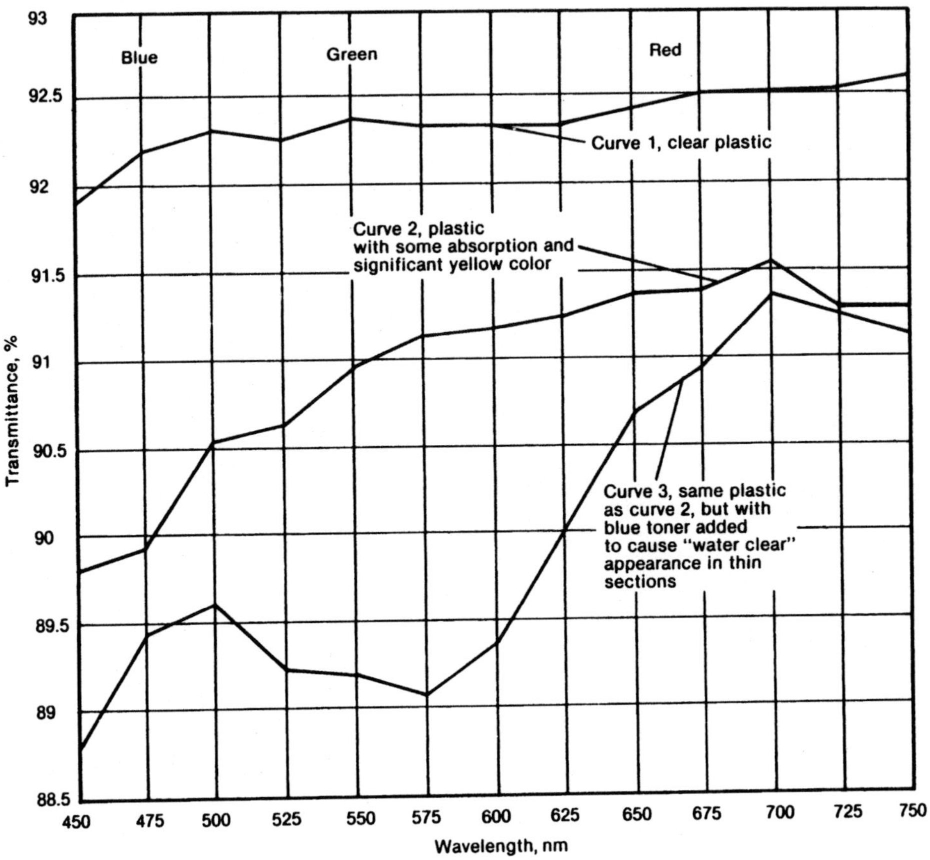

Fig. 3 Spectral transmission of three plastics

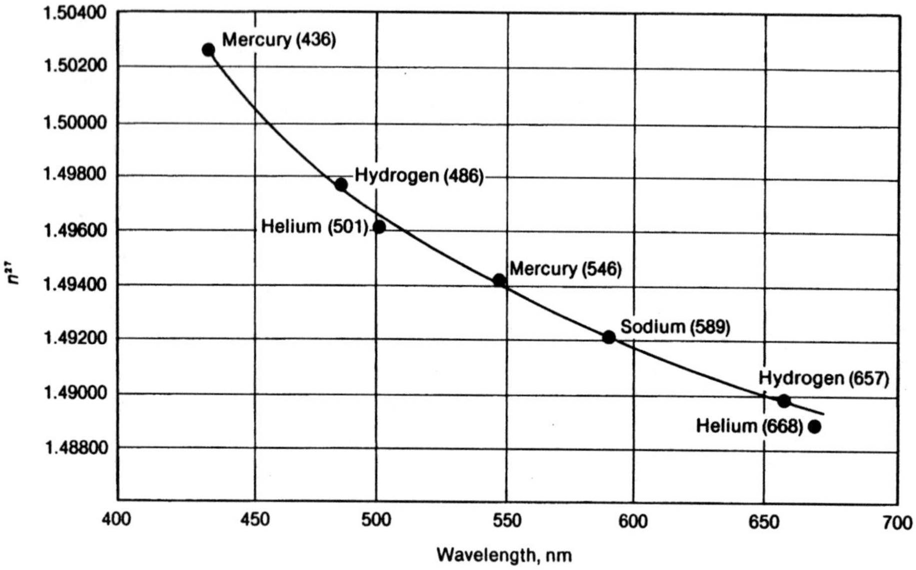

Fig. 4 Dispersion of acrylic polymer

microscope. The term orange peel describes a surface that usually results from the improper polishing of tooling or from the improper processing of parts. When it cannot be eliminated, it is measured either visually or by a haze measurement. Limit samples are then defined. For transparent windows or thin sheeting, ASTM D 1746 applies. This method determines the ratio of transmitted light to incident light and works when haze becomes significant because of microscopic surface irregularities.

The evaluation of contamination in a transparent or translucent material may be for either optical or cosmetic reasons. Cosmetic particulate contamination is usually specified and microscopically measured as anything greater than 0.1 mm (0.004 in.) in dimension. However, ophthalmic applications, compact disk technology, and a few other optical applications are creating new levels of product cleanness standards. Cosmetic standards for the ophthalmic industry are about 0.05 mm (0.002 in.), and standards for compact disks require the elimination of particles greater than about 1 μm (40 μin.). Some optical imaging applications demand lower levels of particulate contamination because the light-scattering particles create haze.

Table 1 Physical properties of principal optical plastics

Properties	Methyl methacrylate	Polystyrene	Polycarbonate	Methyl methacrylate styrene copolymer	ASTM method
Refractive index					D 542
n_D	1.491 (7)	1.590 (3)	1.586 (0)	1.564	
n_F	1.497 (8)	1.604 (1)	1.593 (4)	1.574 (4)	
n_C	1.489 (2)	1.584 (9)	1.576	1.558 (3)	
Abbe value	57.2	30.8	34.0	35	D 542
$dn/dt \times 10^{-5}$/K	−12.5	−12.0	−14.3	−14.0	. . .
Haze, %	<2	<3	<3.0	<3	D 1003
Luminous transmittance 3.2 mm (0.125 in.) thickness	92	87–90	85–91	90	D 1003
Critical angle, degrees	42.1	39.0	39.1	39.8	. . .
Deflection temperature at 2 °C/min (3.6 °F/min), °C (°F)					D 648
At 1.8 MPa (0.264 ksi)	198	180	280	. . .	
At 0.46 MPa (0.066 ksi)	214	230	270	212	
Coefficient of linear thermal expansion, 10^{-5}/K	3.6	3.5	3.8	3.6	D 696
Recommended maximum continuous service temperature for normal parts, °C (°F)	80 (175)	80 (175)	115 (240)	80 (175)	. . .
Water absorption, immersed 24 h at 23 °C (73 °F), %	0.3	0.2	0.15	0.15	D 570
Specific gravity	1.19	1.06	1.20	1.09	D 792
Hardness, 6.4 mm (0.25 in.) sample	M 97	M 90	M 70	M 75	D 785
Impact strength, Izod notch, J/m (ft · lbf/in.)	0.3–0.5	0.35	12–17	. . .	D 256
Dielectric strength, MV/m (V/mil)	19.7 (500)	19.7 (500)	15.7 (400)	17.7 (450)	D 149
Dielectric constant					D 150
At 60 Hz	3.7	2.6	2.90	3.40	
At 1 MHz	2.2	2.45	2.88	2.90	
Power factor					D 150
At 60 Hz	0.05	0.0002	0.0007	0.006	
At 1 MHz	0.03	0.0002–0.0004	0.0075	0.013	
Volume resistivity, Ω · m	10^{16}	$>10^{14}$	8×10^{14}	10^{13}	D 257

Note: This information is taken from available published data of raw-material manufacturers. Specific material formulation data should be confirmed prior to design and specification.

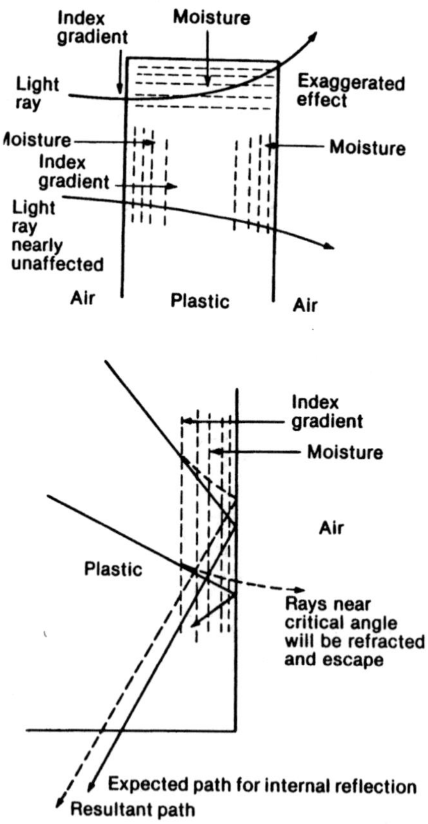

Fig. 5 Moisture-induced refractive index gradient

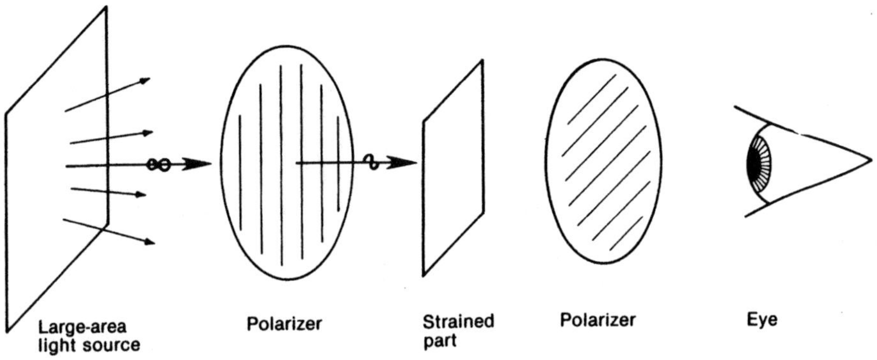

Fig. 6 Observation of birefringence

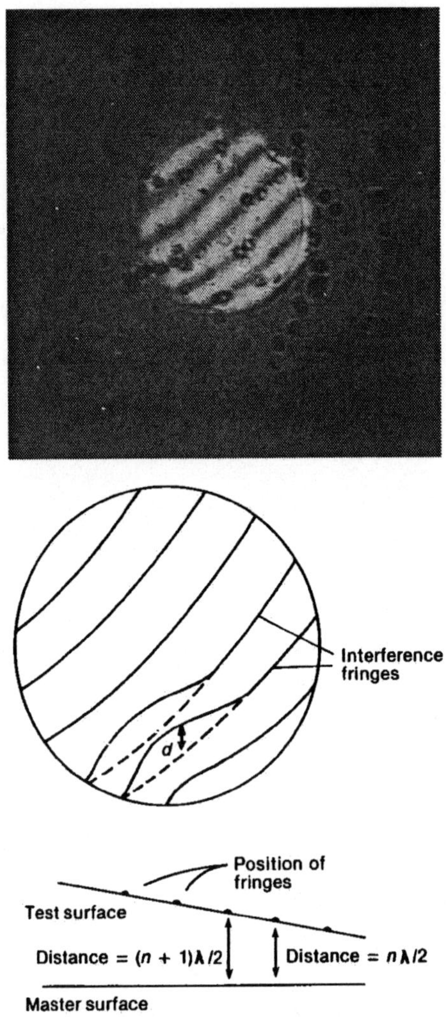

Fig. 7 Interference pattern of optical surface. Fringes occur as the distance between surfaces increases by one-half wavelength ($\lambda/2$); *d* represents dip or rise of $\lambda/2$ in the surface. At helium-neon wavelength, this represents 0.318 μm (12.5 μin.).

Surface Gloss and Color

Specular gloss, which is the luminous fractional reflectance of a specimen at the specular direction, is measured on opaque, flat parts. A gloss meter is used in accordance with ASTM D 523. Light is reflected from the surface into a detector at a 20°, 60°, or 85° angle, depending on the reflectance of the surface. The gloss meter is calibrated using polished black glass. Plastic films are measured for gloss in the same manner, in accordance with ASTM D 2457. Films are measured at 20°, 45°, and 60°.

Color is still difficult to measure. The perception of color is dependent on light spectral temperature, gloss, hue, background, and other factors. Color meters are available that can measure the consistency of a color more closely than the eye can see, and manufacturers continue to improve their measurement methods. The ASTM standard for color evaluation is D 1729, which calls for visual evaluation of color samples against a standard in a controlled environment. Evaluation is done using lights of various spectral temperatures to determine colors under different lighting conditions.

Ad Hoc Testing

For practical applications in which lenses, prisms, and light pipes are being used and tested, the necessary test instruments are often not available. The purpose of ad hoc tests is to determine the acceptability of a plastic part for its application, although the tests themselves can vary widely depending on part geometry and use. Typical examples of ad hoc testing follow.

A flat mirror or window needs to be checked for surface irregularity. The window may be held some distance from a grid pattern, and the straightness of the grid lines observed through the window. The test is somewhat subjective, but, in general, the greater the distance to the grid, the more sensitive the test. It is best to use distances similar to those required in actual use. If the surface of a window or mirror is held at arm's length from the observer and a distant (>6 m, or 20 ft) straight line, such as a long fluorescent tube, is reflected from the surface, the flatness of the surface may be inspected with accuracy. Deviations (from flat) of as little as three to four wavelengths of light may be observed by the practiced eye.

A lens should have a proper focal length, surface regularity, and the proper curvature. The focal length of a positively powered lens may be checked by using a distant light source other than the sun, such as a spotlight located across a large room. The lens should not be held toward the light. Instead, it should be held near a surface in a darkened area and its position adjusted for the smallest spot on the nearby surface. The distance from the back of the lens to the surface should be checked. This is called the back focal length (BFL) and can be compared to specifications. Holding the lens near its BFL over a good-quality straight line and viewing the line through the lens will usually give a reasonable indication of surface quality to the practiced eye.

REFERENCES

1. W.J. Smith, *Modern Optical Engineering,* McGraw-Hill
2. E.D. McAlister, J.J. Villa, and C.D. Salzbert, Rapid and Accurate Measurements of Refractive Index in the Infrared, *J. Opt. Soc. Am.,* Vol 46 (No. 7), July 1956
3. R.W. Jans, Acrylic Polymers for Optical Applications, *Soc. Photo-Opt. Instr. Eng. J.,* Vol 204, Physical Properties of Optical Materials, 1979
4. "Lexan Optical Properties," Bulletin CDC-683, Plastics Group, General Electric Company, June 1985
5. J.M. Cariou, J. Dugas, L. Martin, and P. Michel, Refractive Index Variations With Temperature of PMMA and Polycarbonate, *Appl. Opt.,* Vol 25 (No. 3), Feb 1986
6. R.M. Waxler, D. Horowitz, and A. Feldman, Optical and Physical Parameters of Plexiglas 55 and Lexan, *Appl. Opt.,* Vol 18 (No. 1), Jan 1979, p 101
7. R. Kingslake, *Applied Optics and Optical Engineering,* Vol 1, Academic Press, 1965

Mechanical Behavior and Wear

Mechanical Testing and Properties of Plastics: An Introduction*

PLASTICS ARE VISCOELASTIC MATERIALS, in which mechanical deformation can be very dependent on temperature as well as time. The time dependence can be broken down in terms of the rate at which a stress is applied, the duration of the applied stress, and the overall stress history. The exact temperature dependence will, in turn, be influenced by the thermal properties of the plastic itself, which are categorically different for amorphous thermoplastics, crystalline thermoplastics, and thermosets. The magnitude of the time dependence is very temperature dependent. At temperatures well below their glass-transition temperatures (T_g), glassy or semicrystalline polymers are weakly viscoelastic. For these polymers, test data based on a time-independent analysis will probably be adequate. As the temperature is increased, either by the environment or by heat given off during deformation, the time dependence of the mechanical response increases.

The deformation mechanisms of polymers also differs from that of metals. In most polymers, only about half of the work of plastic deformation is liberated as heat. The remainder is accounted for as damage due to cavitation and, in the case of noncavitation systems, changes in orientation of the long-chain molecules about their centers of gravity. In metals, the deformation is due to the changes in relative positions of the center of gravity of the metal molecules, while in polymers the entire molecule dominates the deformation process in terms of its conformation about its center of gravity. The large amount of stored internal energy during the plastic deformation of polymers can have many effects not observable with metals. One consequence of the plastic deformation of polymers is that if a plastically deformed sample is heated in the absence of external constraints, it will contract toward its original undeformed length. In many polymers, practically 100% recovery of the plastic deformation is possible without ever exceeding the melt temperature. If a plastically deformed sample is constrained at constant length while heated, very large increases in stress can be observed. Such experiments clearly illustrate the difference in the plasticity of metals and the apparent plasticity of polymers.

Typical mechanical properties of representative plastics are given in Table 1 (Ref 1), and this article briefly introduces some commonly used methods of the mechanical testing with further details in other articles in this Section. The testing of plastics includes a wide variety of mechanical tests (Table 2). The following sections briefly describe the test methods and comparative data for the mechanical property tests listed in Table 2. In addition, creep testing and dynamic mechanical analyses of viscoelastic plastics are also briefly described. For more detailed descriptions of these test methods and the other test methods listed in Table 2, readers are referred to Ref 2 and an extensive one-volume collection of International Organization for Standardization (ISO) and European standards for plastic testing (Ref 3).

Tensile Properties

The chemical composition and the long-chain nature of polymers lead to some important differences with metals. These differences include significantly lower stiffnesses, much higher elastic limits or recoverable strains, a wider range of Poisson's ratios, and time-dependent deformation from viscoelasticity. Thermoplastics also exhibit a unique variety of postyield phenomena. For example, Fig. 1 is a typical stress-strain plot for aluminum and polyethylene. The aluminum sample necks and extends to 50% strain. The polyethylene sample necks and extends to 350% strain as a consequence of the long-chain nature of polymers. The polyethylene also shows a stiffening due to chain alignment at the highest strains. This postyield stiffening involves shear deformation as described in Ref 3.

The ultimate tensile strengths of most unreinforced structural plastics range from 50 to 80 MPa (7 to 12 ksi) with elongation to final fracture much higher than metals (Fig. 2). It is very common to see large differences between metals and plastics in the amount of recoverable elastic strain. In metals, the amount of recoverable elastic strain is determined by the amount of strain that can be put into one of the metallic bonds before breaking. This amount of strain is typically less than 1%. In elastomers, the amount of recoverable strain can be 500% or more.

When recoverable strain is this large, the individual polymer chains must be prevented from flowing past each other during deformation. This is easily accomplished by cross links tying the chains together (Ref 6). Even in glassy polymers, in which internal energy effects are evident in the elasticity, the recoverable strain is limited by the strain required to break the weaker and longer-range van der Waals bonds. Because these bonds can be stretched farther than metallic bonds, it is possible to have a recoverable strain in glassy polymers of 5% or more. At such large strains, it is possible for the assumptions of small-strain elasticity to break down. Any standard test procedures based on small-strain elasticity may have to be modified to account for large elastic strains.

The mechanical behavior of polymers is also time dependent, or viscoelastic. Therefore, data based on short-term tests have the possibility of misrepresenting the tested polymer in a design application that involves long-term loading.

Under viscoelastic conditions, one method useful for obtaining long-term design data is the time-temperature superposition principle. This principle states that the mechanical response at long times at some particular temperature is equivalent to the mechanical response at short times but at some higher temperature (Ref 5). By determining shift factors, it is possible to determine which temperature to use in obtaining long-term data from short-term tests. This is essentially true for linear viscoelastic behavior in the absence of a phase change.

The short-term tensile test (ASTM D 638 and ISO 517) is one of the most widely used mechanical tests of plastics for determining mechanical properties such as tensile strength, yield strength, yield point, and elongation. The stress-strain curve from tension testing is also a convenient way to classify plastics (Fig. 3). A soft and weak material, such as polytetrafluo-

*Adapted from Mechanical Testing of Polymers and Ceramics, *Mechanical Testing and Evaluation*, Volume 8, *ASM Handbook*, ASM International, 2000, pages 26 to 41.

roethylene (PTFE), is characterized by low modulus, low yield stress, and moderate elongation at break point. A soft but tough material such as polyethylene (PE) shows low modulus and low yield stress but very high elongation at break. A hard and brittle material such as general-purpose phenolic is characterized by high modulus and low elongation. It may or may not yield before break. A hard and strong material such as polyacetal has high modulus, high yield stress, high ultimate strength (usually), and low elongation. A hard and tough material such as polycarbonate is characterized by high modulus, high yield stress, high elongation at break, and high ultimate strength.

Because of the diversity of mechanical behavior, the tension testing of plastics is subject to potential misapplication or misinterpretation of test results. This is particularly true for thermoplastics, which have some important differences with thermoset plastics. Compared to thermoset resins, thermoplastics exhibit more disruption or changes in the secondary bonding between the molecular chains during tension testing. This leads to a variety of postyield phenomena, such as the stiffening observed in polyethylene (Fig. 1). Another example is shown in Fig. 4. At the yield point the average axis of molecular orientation in thermoplastics may begin to conform increasingly with the direction of the stress. The term *draw* is sometimes used to describe this behavior. There is usually a break in the stress-strain curve as it begins to flatten out, and more strain is observed with a given increased stress. The result is that the giant molecules begin to align and team up in their resistance to the implied stress. Frequently, there is a final increase in the slope of the curve just before ultimate failure (Fig. 4). The extent to which this orientation takes place varies from one linear thermoplastic to the next, but the effect can be quite significant. Even the smallest amount of the teaming-up effect imparts greatly improved impact resistance and damage tolerance. In thermoplastics, there is much more area under the stress-strain curve than in conventional thermosets, which are more rigid networks with much less area under the stress-strain curve.

Because the deformation of thermoplastics is time-dependent, careful control of test duration and strain rate is important. A slower test (i.e., one at a low strain rate) allows more time for deformation and thus alters the stress-strain curve and lowers the tensile strength. This effect is shown in Fig. 5 for polycarbonate.

Short-term tensile properties are usually measured at a constant rate of 0.5 cm/min (0.2 in./min). It is recommended by the American Society for Testing and Materials (ASTM) that the speed of testing be such that rupture occurs in 0.5 to 5 min. Test coupons are either injection molded or compression molded and cut into a standard shape. In practice, injection-molded coupons are usually used.

The history of the plastic sample has some influence on tensile properties. A tensile bar prepared by injection molding with a high pressure tends to have higher tensile strength. A material that has been oriented in one direction tends to have a higher tensile strength and a lower elongation at break in the direction of orientation. In the direction perpendicular to the orientation, tensile strength is consistently lower. In a crystallizable material, stretching usually increases crystallinity.

Because the mechanical properties are sensitive to temperature and absorbed moisture, conditioning procedures for test specimens have been developed. These procedures are defined in ASTM D 618 and ISO 291.

Tensile Modulus. Because plastics are viscoelastic materials, stress-strain relationships are nonlinear and curved (usually convex upward). The curvature arises from two causes. First, the deflection axis is simultaneously a time axis, and during the test, molecular relaxation processes continuously reduce the stress required to maintain any particular strain. Second, as the strain increases, the molecular resistance to further deformation decreases; that is, the effective modulus falls.

The degree of curvature depends on the material and the test conditions. At high strain rates and/or low temperatures, the stress-strain relationship usually approximates to a straight line. However, if the curvature is pronounced, the stress-strain ratio must be either a tangent modulus or a secant modulus. The tangent modulus is the instantaneous slope at any point on the stress-strain curve, while the secant modulus is the slope of a line drawn from the origin to any point on a nonlinear stress-strain curve. These moduli may be conservative or nonconservative, relative to one another and depending on the location on the curve.

The accuracy of modulus data derivable from a stress-strain test may be limited, mainly because axiality of loading is difficult to achieve and because the specimen bends initially rather than stretches. In addition, the origin of the force-deflection curve is often ill defined, and the curvature there is erroneous, to the particular detriment of the accuracy of the tangent modulus at the origin and, to a lesser degree, that of the secant moduli. Under the very best experimental conditions, the coefficient of variation for the modulus data derivable from tensile tests can be 0.03 or lower, but more typically it is 0.10 (Ref 7). If the strain is derived from the relative movement of the clamps rather than from

Table 1 Typical room-temperature mechanical properties of plastics

Material	Tensile strength, MPa (ksi)	Elongation, %	Modulus of elasticity, GPa (10^6 psi)	Compressive strength, MPa (ksi)	Modulus of rupture, MPa (ksi)	Hardness
Thermosets						
EP, reinforced with glass cloth	350 (51)	...	175 (25)	410 (39)	485 (70)	
MF, alpha-cellulose filler	50–90 (7–13)	0.6–0.9	9 (1)	170–300 (25–44)	70–110 (10–16)	110–125 HRM
PF, no filler	50–55 (7–9)	1.0–1.5	5–7 (0.7–1)	70–200 (10–29)	80–100 (12–15)	124–128 HRM
PF, wood flour filler	45–60 (7–9)	0.4–0.8	6–8 (0.87– 1.16)	160–250 (23–36)	60–85 (9–12)	100–120 HRM
PF, macerated fabric filler	25–65 (4–9)	0.4–0.6	6–9 (0.87–1)	100–160 (15–24)	60–100 (9–15)	95–120 HRM
PF, cast, no filler	40–65 (6–9)	1.5–2.0	3 (0.43)	85–115 (12–17)	75–115 (11–17)	93–120 HRM
Polyester, glass-fiber filler	35–65 (5–9)	...	11–14 (1.6–2.0)	140–175 (20–25)	95–115 (14–17)	...
UF, alpha-cellulose filler	55–90 (8–13)	0.5–1.0	10 (1.5)	175–240 (25–35)	70–100 (10–15)	115–120 HRM
Thermoplastics						
ABS	35–45 (5–7)	15–60	1.7–2.2 (0.25–0.32)	25–50 (4–7)	...	95–105 HRR
CA	15–60 (2–9)	6–50	0.6–3.0 (0.1–0.4)	90–250 (13–36)	15–110 (2–16)	50–125 HRR
CN	50–55 (7–9)	40–45	1.3–15.0 (0.18–2)	150–240 (22–35)	60–75 (9–11)	95–115 HRR
PA	80 (12)	90	3.0 (0.43)	85 (12)	...	79 HRM, 118 HRR
PMMA	50–70 (7–10)	2–10	...	80–115 (12–17)	90–115 (13–17)	85–105 HRM
PS	35–60 (5–9)	1–4	3.0–4.0 (0.4–0.6)	80–110 (12–16)	55–110 (8–16)	65–90 HRM
PVC, rigid	40–60 (6–9)	5	2.4–2.7 (0.3–0.4)	60 (9)	...	110–120 HRR
PVCAc, rigid	50–60 (7–9)	...	2.0–3.0 (0.3–0.4)	70–80 (10–12)	85–100 (12–15)	...

ABS acrylonitrile-butadiene-styrene; CA, cellulose acetate; CN, cellulose nitrate; EP, epoxy; MF, melamine formaldehyde; PA, polyamide (nylon); PF, phenol formaldehyde; PMMA, polymethyl methacrylate; PS, polystyrene; PVC, polyvinyl chloride; PVCAc, polyvinyl chloride acetate; UF, urea formaldehyde; Source: Ref 1 an extensometer, the error in the calculated value of the tangent modulus at the origin can be 100% (Ref 6).

Yield stresses of plastics depend on a variety of molecular mechanisms, which vary among polymer classes and may not be strictly comparable. However, regardless of the underlying mechanisms, yield stress data have a low coefficient of variation, typically 0.03 (Ref 7). Brittle fracture strengths are much more variable, reflecting the distributions of defects that one might expect. The scatter due to the inherent defects in the materials is exacerbated when elongations at fracture are small because poor and variable alignment of the specimens induces apparently low strengths if the theoretical stresses are not corrected for the extraneous bending in the specimens (Ref 7).

Long-term uniaxial tensile creep testing of plastics is covered in ASTM D 2990 and ISO 899. ASTM D 2990 also addresses flexural and compressive creep testing. For the uniaxial tensile creep test in D 2990, the test specimen is either a standard type I or II bar, per ASTM D 638, that is preconditioned to ASTM D 618 specifications. The test apparatus is designed to ensure that the applied load does not vary with time and is uniaxial to the specimen. As with other tests, the test specimen must not slip in or creep from the grips. The load must be applied to the specimen in a smooth, rapid fashion in 1 to 5 s. If the test is run to specimen failure, the individual test cells must be isolated to eliminate shock loading from failure in adjacent test cells. Several types of tensile creep test systems are shown in Fig. 6.

Creep curves generally exhibit three distinct phases. First-stage creep deformation is characterized by a rapid deformation rate that decreases slowly to a constant value. The four-parameter model was proposed to describe long-term creep. In this model, the first-stage creep deformation was called retarded elastic strain. Second-stage creep deformation is characterized by a relative constant, low-deformation rate. In the four-parameter model, this was called equilibrium viscous flow. The final or third-stage creep deformation is creep rupture, fracture, or breakage.

The generalized uniaxial tensile creep behavior of plastics under constant load, isothermal temperature, and a given environment can be illustrated as ductile creep behavior (Fig. 7a) or as brittle creep behavior (Fig. 7b). At very low stress levels, both types of plastics exhibit similar first-stage and second-stage creep deformation. The onset of creep rupture may not occur within the service life of the product (let alone the test). As the stress level increases, first-stage and second-stage creep deformation rates remain relatively the same for these types, but the time of failure is of course considerably reduced. In addition, third-stage creep deforma-

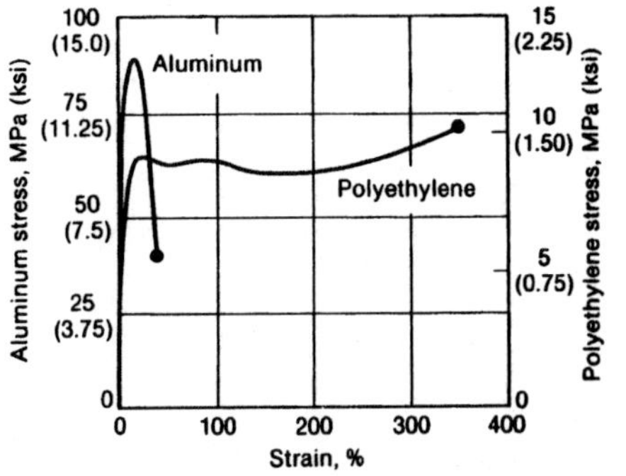

Fig. 1 Typical stress-strain curves for polycrystalline aluminum and semicrystalline polyethylene. Both materials neck. In polyethylene, chain alignment results in stiffening just before failure. Source: Ref 4

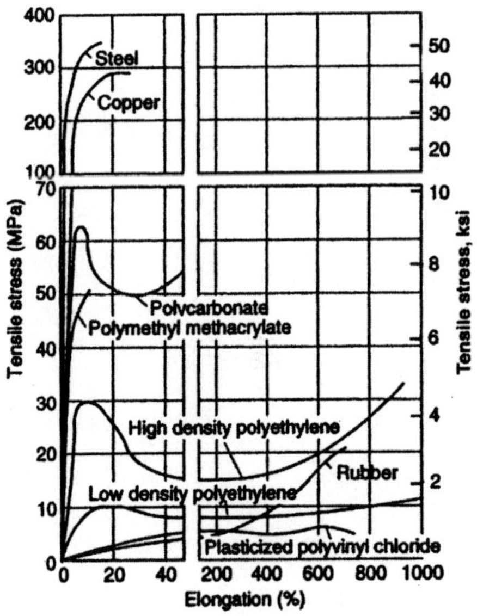

Fig. 2 Tensile stress-strain curves for copper, steel, and several thermoplastic resins. Source: Ref 5

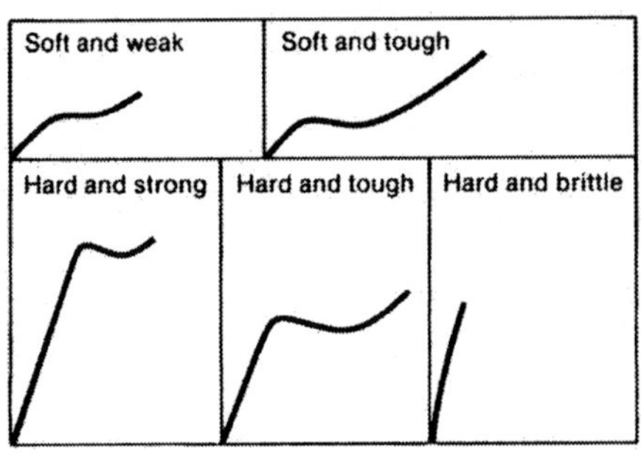

Fig. 3 Tensile stress-strain curves for several categories of plastic materials

Table 2 ASTM and ISO mechanical test standards for plastics

ASTM standard	ISO standard	Topic area of standard
Specimen preparation		
D 618	291	Methods of specimen conditioning
D 955	294-4	Measuring shrinkage from mold dimensions of molded thermoplastics
D 3419	10724	In-line screw-injection molding of test specimens from thermosetting compounds
D 3641	294-1,2,3	Injection molding test specimens of thermoplastic molding and extrusion materials
D4703	293	Compression molding thermoplastic materials into test specimens, plaques, or sheets
D 524	95	Compression molding test specimens of thermosetting molding compounds
D 6289	2577	Measuring shrinkage from mold dimensions of molded thermosetting plastics
Mechanical properties		
D 256	180	Determining the pendulum impact resistance of notched specimens of plastics
D 638	527-1,2	Tensile properties of plastics
D 695	604	Compressive properties of rigid plastics
D 785	2039-2	Rockwell hardness of plastics and electrical insulating materials
D 790	178	Flexural properties of unreinforced and reinforced plastics and insulating materials
D 882	527-3	Tensile properties of thin plastic sheeting
D 1043	458-1	Stiffness properties of plastics as a function of temperature by means of a torsion test
D 1044	9352	Resistance of transparent plastics to surface abrasion
D 1708	6239	Tensile properties of plastics by use of microtensile specimens
D 1822	8256	Tensile-impact energy to break plastics and electrical insulating materials
D 1894	6601	Static and kinetic coefficients of friction of plastic film and sheeting
D 1922	6383-2	Propagation tear resistance of plastic film and thin sheeting by pendulum method
D 1938	6383-1	Tear propagation resistance of plastic film and thin sheeting by a single tear method
D 2990	899-1,2	Tensile, compressive, and flexural creep and creep-rupture of plastics
D 3763	6603-2	High-speed puncture properties of plastics using load and displacement sensors
D 4065	6721-1	Determining and reporting dynamic mechanical properties of plastics
D 4092	6721	Dynamic mechanical measurements on plastics
D 4440	6721-10	Rheological measurement of polymer melts using dynamic mechanical procedures
D 5023	6721-3	Measuring the dynamic mechanical properties of plastics using three-point bending
D 5026	6721-5	Measuring the dynamic mechanical properties of plastics in tension
D 5045	572	Plane-strain fracture toughness and strain energy release rate of plastic materials
D 5083	3268	Tensile properties of reinforced thermosetting plastics using straight-sided specimens
D 5279	6721	Measuring the dynamic mechanical properties of plastics in torsion

Source: Ref 2

tion characteristics now differ considerably. The ductile plastic exhibits typical ductile yielding or irreversible plastic deformation prior to fracture. The brittle plastic, on the other hand, exhibits no observable gross plastic deformation and only abrupt failure.

Macroscopic yielding and fracture may not always be appropriate criteria for long-time duration material failure. For some plastics, stress crazing, stress cracking, or stress whitening may signal product failure and may therefore become a design limitation.

Creep strain is usually plotted against time on either semilog plots or log-log plots (Fig. 8). Extrapolation to times beyond the data can be difficult on the semilog plot (Fig. 8a). Replotting on log-log paper may allow easier extrapolation under one decade. Creep curves should not be extrapolated more than one decade, because some curvature still remains in the log-log plot. For small strains, the curves can be considered linear. These curves can usually be used to compare polymers at the same loading levels. Creep test data are also analyzed in various forms, as described further in the section "Creep Data Analysis" in this article.

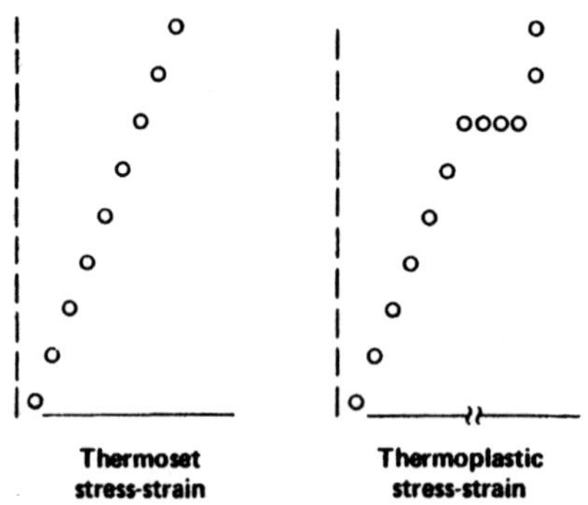

Fig. 4 Thermoset versus thermoplastic stress-strain behavior

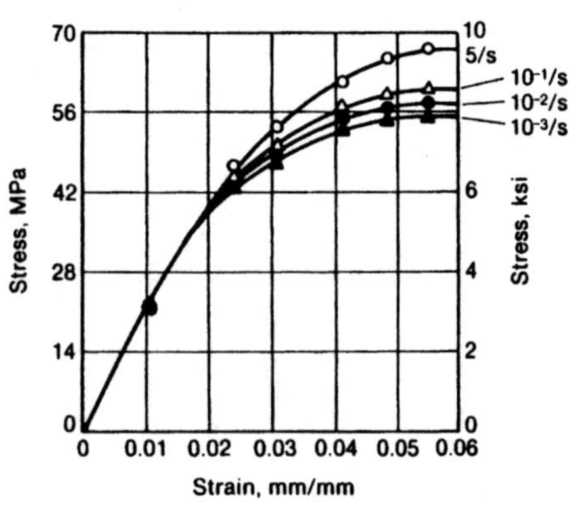

Fig. 5 Stress-strain curves for rubber-modified polycarbonate at room temperature as a function of strain rate

Other Strength/Modulus Tests

Compressive Strength Test (ASTM D 695 and ISO 604). Stress-strain properties are also measured for the behavior of a material under a uniform compressive load. The procedure and nomenclature for compression tests are similar to those for the tensile test. Universal testing machines can be used, and, as in tension testing, specimens should be preconditioned according to ASTM D 618 or ISO 291.

The standard test specimen in ASTM D 695 is a cylinder 12.7 mm (½ in.) in diameter and 25.4 mm (1 in.) in height. The force of the compressive tool is increased by the downward thrust of the tool at a rate of 1.3 mm/min (0.05 in./min). The compressive strength is calculated by dividing the maximum compressive load by the original cross section of the test specimen.

For plastics that do not fail by shattering fracture, the compressive strength is an arbitrary value and not a fundamental property of the material tested. When there is no brittle failure, compressive strength is reported at a particular deformation level such as 1 or 10%. Compressive strength of plastics may be useful in comparing materials, but it is especially significant in the evaluation of cellular or foamed plastics. Compression testing of cellular plastics is addressed in ISO Standards 1856 and 3386-1.

Typical compressive strengths for various plastics are compared in Fig. 9. Generally, the compressive modulus and strength are higher than the corresponding tensile values for a given material.

Compressive creep testing of plastic is addressed in ASTM D 2990. Normally creep information is given for tension loading.

Flexural Strength Test (ASTM D 790 and ISO 178). Flexural strength or cross-breaking strength is the maximum stress developed when a bar-shaped testpiece, acting as a simple beam, is subjected to a bending force. Two methods are used: three-point bending (Fig. 10) and four-point bending (Fig. 11). Four-point bending is useful in testing materials that do not fail at the point of maximum stress in three-point bending (Ref 9).

For three-point bending, an acceptable test specimen is one at least 3.2 mm (0.125 in.) thick, 12.7 mm (0.5 in.) wide, and long enough to overhang the supports (but with overhang less than 6.4 mm, or 0.25 in., on each end). The load should be applied at a specified crosshead rate, and the test should be terminated when the specimen bends or is deflected by 0.05 mm/min (0.002 in./min). The flexural stress (S) at the outer fibers at midspan in three-point bending is calculated from:

$$S = 3PL/2bd^2$$

in which P is the force at a given point on the deflection curve, L is the support span, b is the width of the bar, and d is the depth of the beam.

Because most plastics do not break from deflection, the flexural strength is measured when 5% strain occurs for most thermoplastics and elastomers. Fracture strength under flexural load may be more suitable for thermosets.

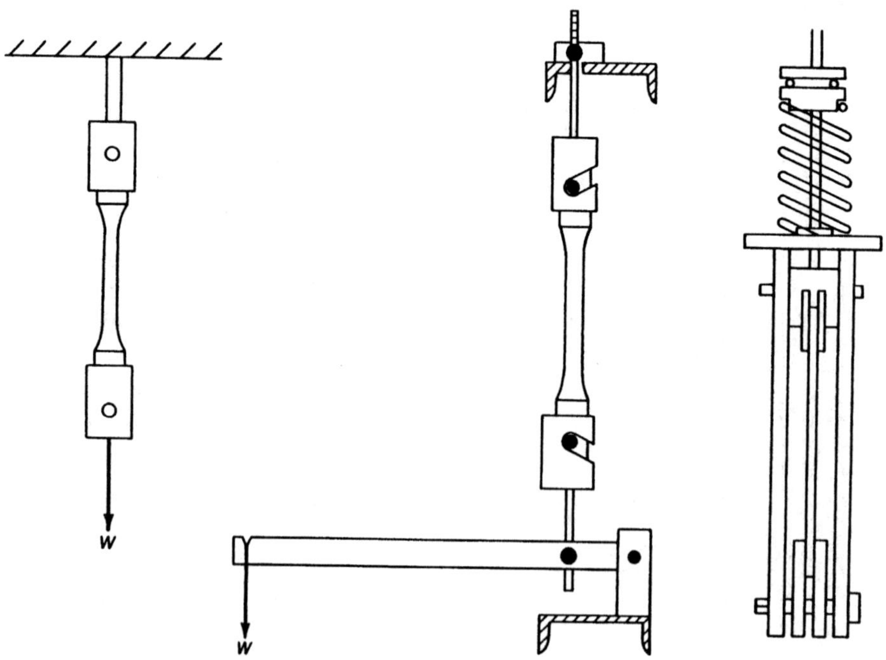

Fig. 6 Various equipment designs for the measurement of tensile creep in plastics

To obtain the strain, r, of the specimen under three-point test, apply:

$$r = 6Dd/L^2$$

in which D is the deflection to obtain the maximum strain (r) of the specimen under test. To obtain data for flexural modulus, which is a measure of stiffness, flexural stress is plotted versus strain, r, during the test; the slope of the curve obtained is the flexural modulus. Flexural moduli for various plastics are compared in Fig. 12.

Flexural creep tests (ISO 899-2) are done with standard flexural test methods where the deflection is measured as a function of time. The flexural creep modulus at time, t, (E_t) for three-point bending (Fig. 10) is calculated as:

$$E_t \frac{L^3 \cdot P}{4b \cdot d^3 \cdot s_t}$$

where s_t is the deflection at time, t.

Deflection Temperature under Load (ASTM D 648). Another measure of plastic rigidity under load is the deflection temperature under load (DTUL) test, also known as the heat-deflection temperature (HDT) test. In the standard ASTM test (D 648), the heat-deflection temperature is the temperature at which a 125 mm (5 in.) bar deflects 0.25 mm (0.010 in.) when a load is placed in the center. It is typically reported at both 460 and 1820 kPa (65 and 265 psi) stresses. The specimen is placed in an oil bath under a load of 460 or 1820 kPa (65 or 265 psi) in the apparatus shown in Fig. 13, and the temperature is raised at a rate of 2 °C/min (3.6 °F/min). The temperature is recorded when the specimen deflects by 0.25 mm (0.01 in.). Because crystalline polymers, such as nylon 6/6, have a low heat-deflection temperature value when measured under a load of 1820 kPa (265 psi), this test is often run at 460 kPa (65 psi).

The heat-deflection temperature is an often misused characteristic and must be used with caution. The established deflection is extremely small, and in some instances may be, at least in part, a measure of warpage or stress relief. The maximum resistance to continuous heat is an arbitrary value for useful temperatures, which are always below the DTUL value. The DTUL value is also influenced by glass reinforcement.

The heat-deflection temperature is more an indicator of general short-term temperature resistance. For long-term temperature resistance, one of the most common measures is the thermal index determined by the Underwriters' Laboratory (UL) (Ref 10). In this test, standard test specimens are exposed to different temperatures and tested at varying intervals. Failure is said to occur when property values drop to 50% of their initial value. The property criterion for determining the long-term use temperature depends on the application. Table 3 lists typical HDT values and the UL temperature index for various plastics.

Shear Strength Test (ASTM D 732). The specimen proscribed in ASTM D 732 is a disk or a plate with an 11 mm ($\frac{7}{16}$ in.) hole drilled through the center of the specimen. Testing can

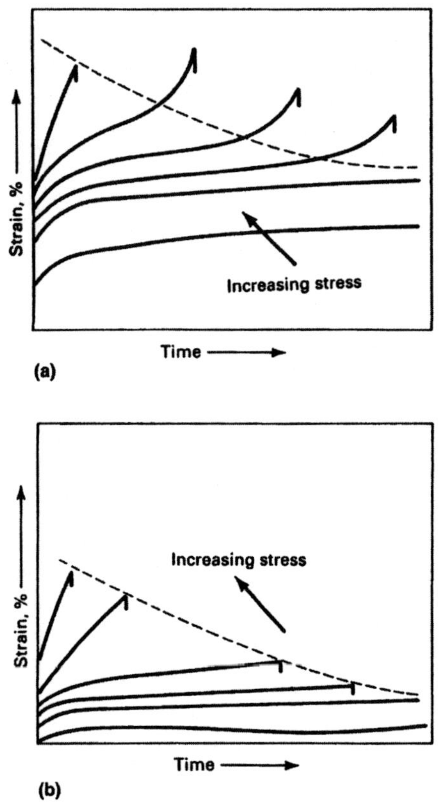

Fig. 7 Typical creep and creep rupture curves for polymers. (a) Ductile polymers. (b) Brittle polymers

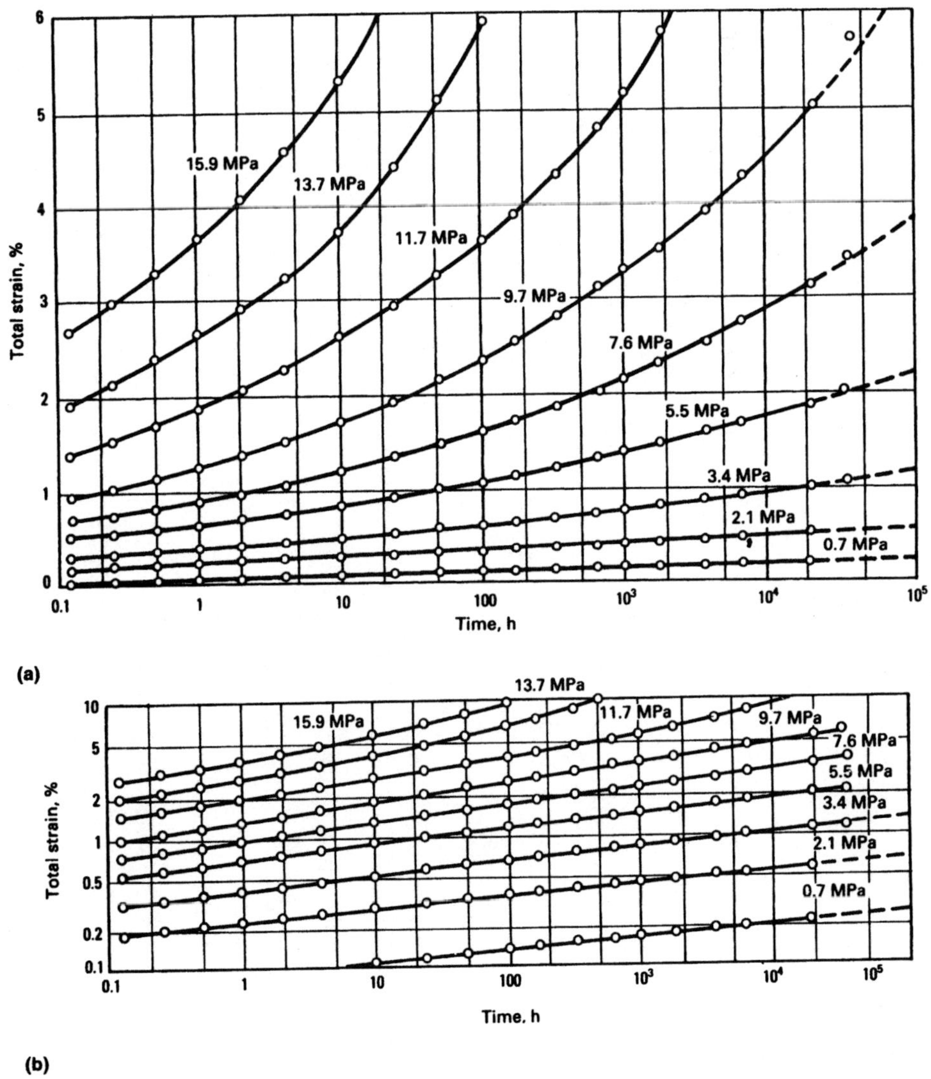

Fig. 8 Tensile creep strain of polypropylene copolymer. (a) Semilog plot. (b) Log-log plot

be done with a special fixture such as the one shown in Fig. 14. Shear strength is defined as the force for separation during loading divided by the area of the sheared edge. Shear strength is often estimated as one-half the tensile strength of a material. When a value for creep shear modulus is needed, it is reasonable to divide the creep tensile modulus by 2.8.

Creep Data Analysis

Mechanical tests under tensile, compressive, flexural, and shear loading can be performed as either short-term tests or long-term tests of creep deformation. Data for the long-term tests are typically recorded as time-dependent displacement values at various levels of constant stress (Fig. 15a). This type of data, however, can be displayed and analyzed in several forms as shown in Fig. 15. There is no universal method of graphically displaying tensile creep or, in fact, creep for compressive, shear, or flexural loading.

Creep Modulus. In addition to stress-strain plots versus time, creep behavior is also expressed as a creep modulus, $E(t)$, where:

$$E(t) = \sigma/\varepsilon(t)$$

where σ is the applied stress and $\varepsilon(t)$ is the creep strain as a function of time. The creep modulus is a measure of rigidity that can be applied for tensile, shear, compressive, or flexural load conditions. However, the creep modulus $E(t)$ is neither a design property nor a material constant. It is a time-dependent variable that is also a function of temperature and environment. The use of creep modulus data requires definition of intended design life and test conditions that accurately reflect the intended application.

Creep Rupture. Similar to creep modulus, creep rupture data depend strongly on temperature. Creep rupture, in many respects, is a more important parameter because it represents the ultimate lifetime of a given material. Two types of graphic representation can be constructed for the creep rupture envelope, as shown in Fig. 15(b) and (c). Figure 15(b) shows a semilog plot of creep rupture stress as a function of failure time. For most plastic candidates for long-term performance, the design life can be quite long—months or years. As a result, the log-log coordinate system (Fig. 15c) has greater utility. Furthermore, creep rupture data tend to display linearly on this coordinate scheme.

Creep strain data plots can be done in various forms. Figure 15 shows three methods of analyzing these data. Each method holds one variable (stress, strain, or time) to be constant. For constant stress, Fig. 15(d) and (e) apply. The data can be displayed either as a set of (usually near-linear) linear lines on log-log paper (Fig. 15d) or as curvilinear lines on semilog paper (Fig. 15e). The parallel straight lines on log-log coordinates are called a creep strain plot (Fig. 15d).

Isochronous Creep Data. If the time parameter is held constant, a set of isochronous (or constant time) stress-strain curves results (Fig. 15f). A linear coordinate system is used to display these results. The slopes of these isochronous creep curves produce the isochronous modulus graph (Fig. 15g). If the slopes of the semilog curves are replotted against time, a set of nearly linear lines on semilog paper results. This represents the time-dependent creep modulus plot (Fig. 15h). Most creep design data published in the United States are reported in this manner.

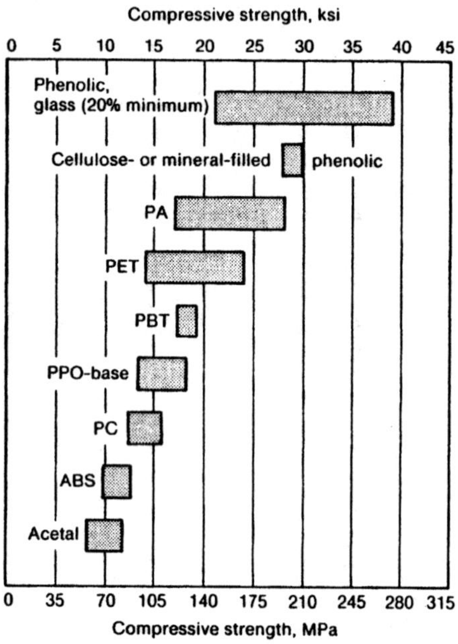

Fig. 9 Compressive strength of engineering plastics. PA, polyamide; PET, polyethylene terephthalate; PBT, polybutylene terephthalate; PPO, polyphenylene oxide; PC, polycarbonate; ABS, acrylonitrile-butadiene-styrene

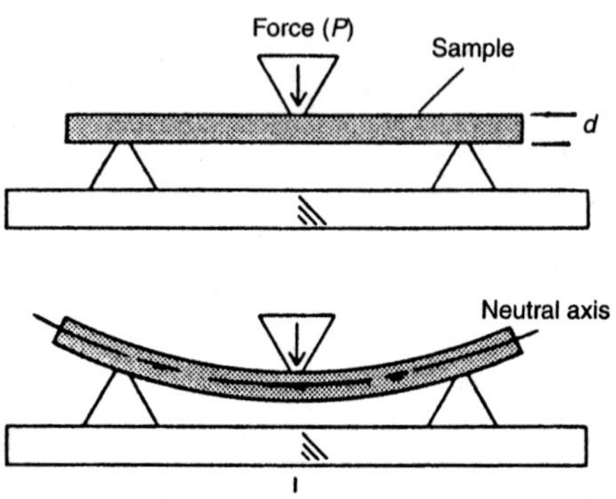

Fig. 10 Flexural test with three-point loading. Source: Ref 8

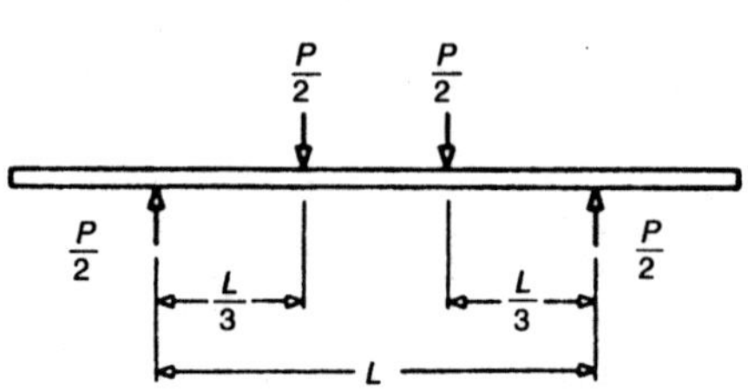

Fig. 11 Flexural test with four-point loading

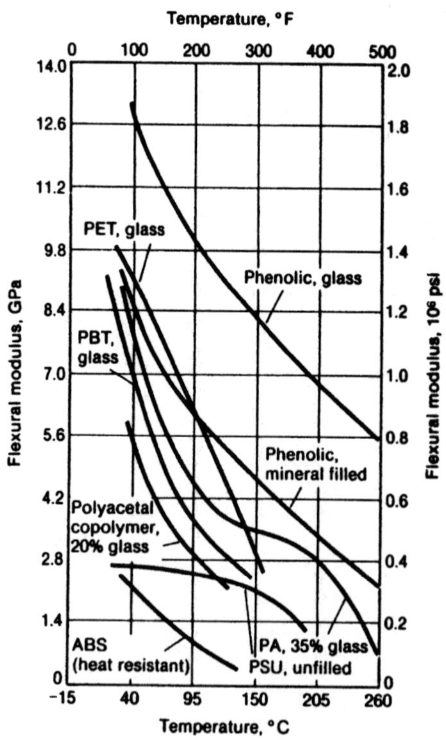

Fig. 12 Flexural modulus retention of engineering plastics at elevated temperatures. PET, polyethylene terephthalate; PBT, polybutylene terephthalate; ABS, acrylonitrile-butadiene-styrene; PA, polyamide; PSU, polysulfone

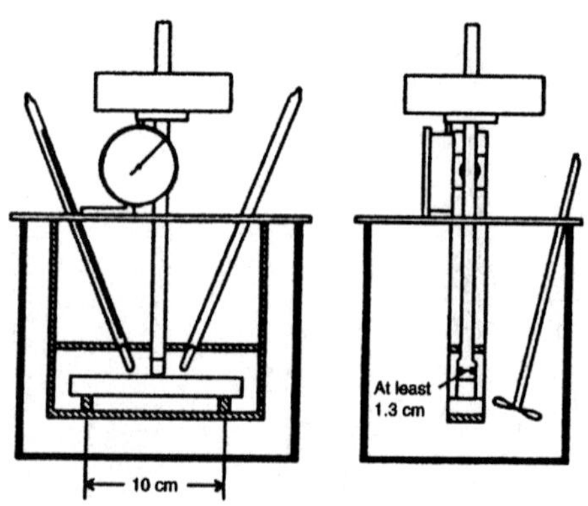

Fig. 13 Apparatus used in test for heat-deflection temperature under load (460 or 1820 kPa, or 65 or 265 psi)

Isometric Creep Data. If the strain is constant, isometric creep curves (Fig. 15i) result. The graph is usually semilog in time. Isometric creep data are used extensively in Europe. The isometric modulus data can also be extracted from these curves.

Dynamic Mechanical Properties

Dynamic mechanical tests measure the response of a material to a sinusoidal or other periodic stress. Because of the viscoelastic nature of plastics, the stress and strain are generally not in phase. Two quantities, a stress-to-strain ratio and a phase angle, are measured. Because dynamic properties are measured at the small deformation around the equilibrium position, they involve only relative displacement of polymer chains in the linear-response region. This measurement offers considerable information on structural property relationships, ranging from local interaction of segments to the macrostructure of polymer chains.

Dynamic mechanical tests give a wider range of information about a material than other short-term tests provide, because test parameters, such as temperature and frequency, can be varied over a wide range in a short time. Superposition of data from different temperatures is also possible. Dynamic data can be interpreted from the chemical structure and physical aggregation of the material. The results of dynamic measurements are generally expressed as complex modulus, which is defined as:

$$G^* = G' + iG''$$

or through the dissipation factor, tan δ, which is related to complex moduli by:

$$\tan \delta = \frac{G''}{G'}$$

Molecular weight, cross linking, crystallinity, and plasticization can affect the dynamic modulus. As a general rule, these factors affect G' the same way they affect complex modulus. In fact, one can convert shear modulus to complex modulus, and vice versa, at least from a theoretical point of view.

The dissipation loss factor is generally an indication of reduced dimensional stability. A material with a high loss factor, however, is useful for acoustical insulation. The dissipation factor shows a peak when there is a phase transition. Thus, it is a sensitive method for detecting the existence of transitions. Low-temperature transitions measured by this technique are related to high-impact properties for materials such as polycarbonate.

Impact Toughness

As would be expected, the impact toughness of plastics is affected by temperature. At temperatures below the glass-transition temperature, T_g, the material is brittle and impact strength is low. The brittleness temperature decreases with increasing molecular weight. This is the reverse of the effect of molecular weight on the T_g. When the temperature increases to near the T_g, the impact strength increases. With notch-sensitive materials such as some crystalline plastics, environmental factors can create surface microcracks and reduce impact strength considerably. Table 4 is a summary of fracture behavior of various plastics.

Although a number of standard impact tests are used to survey the performance of plastics exposed to different environmental and loading conditions, none of these tests provides real, geometry-independent material data that can be applied in design. Instead, they are only useful in application to quality control and initial material comparisons. Even in this latter role, different tests will often rank materials in a different order. As a result, proper test choice and interpretation require that the engineer have a very clear understanding of the test and its relationship to specific design requirements.

Because of differing engineering requirements, a wide variety of impact test methods have been developed. There is no one ideal method. The general classes of impact tests are shown in Fig. 16. However, this section briefly describes three of the most commonly used tests for impact performance: the Izod notched-beam test, the Charpy notched-beam test, and the dart penetration test.

Charpy Impact Test (ASTM D 256 and ISO 179). The Charpy geometry consists of a simply supported beam with a centrally applied load on the reverse side of the beam from the notch (Fig. 17a). The notch serves to create a stress concentration and to produce a constrained multiaxial state of tension a small distance below the bottom of the notch. The load is applied dynamically by a free-falling pendulum of known initial potential energy. The important dimensions of interest for these tests include the notch angle, the notch depth, the notch tip radius, the depth of the beam, and the width of the beam. All these quantities, as well as more detailed information specifying loading geometry and conditions, are described in ASTM D 256 and ISO 179.

Izod Impact Test (ASTM D 256 and ISO 180). Like the Charpy test, the Izod test involves a pendulum impact, but the Izod geometry consists of a cantilever beam with the notch located on the same side as the impact point (Fig. 17b). Because the pendulum hits the unnotched side of the sample in the Charpy test, Charpy values may be much higher impact

Table 3 Heat-deflection and Underwriters' Laboratories index temperatures for selected plastics

Material	Heat-deflection temperature at 1.82 MPa (0.264 ksi)		UL Index	
	°C	°F	°C	°F
Acrylonitrile-butadiene-styrene (ABS)	99	210	60	140
ABS-polycarbonate alloy (ABS-PC)	115	240	60	140
Diallyl phthalates (DAP)	285	545	130	265
Polyoxymethylene (POM)	136	275	85	185
Polymethyl methacrylate (PMMA)	92	200	90	195
Polyacrylate (PAR)	155	310	...	...
Liquid crystal polymer (LCP)	311	590	220	430
Melamine-formaldehyde (MF)	183	360	130	265
Nylon 6	65	150	75	165
Nylon 6/6	90	195	75	165
Amorphous nylon 12	140	285	65	150
Polyarylether (PAE)	160	320	160	320
Polybutylene terephthalate (PBT)	...	...	120	250
Polycarbonate (PC)	129	265	115	240
PBT-PC	129	265	105	220
Polyetheretherketone (PEEK)	...	...	250	480
Polyether-imide (PEI)	210	410	170	340
Polyether sulfone (PESV)	203	395	170	340
Polyethylene terephthalate (PET)	224	435	140	285
Phenol-formaldehyde (PF)	163	325	150	300
Unsaturated polyester (UP)	279	535	130	265
Modified polyphenylene oxide alloy (PPO) (mod)	100	212	80	175
Polyphenylene sulfide (PPS)	260	500	200	390
Polysulfone (PSU)	174	345	140	285
Styrene-malic anhydride terpolymer (SMA)	103	215	80	175

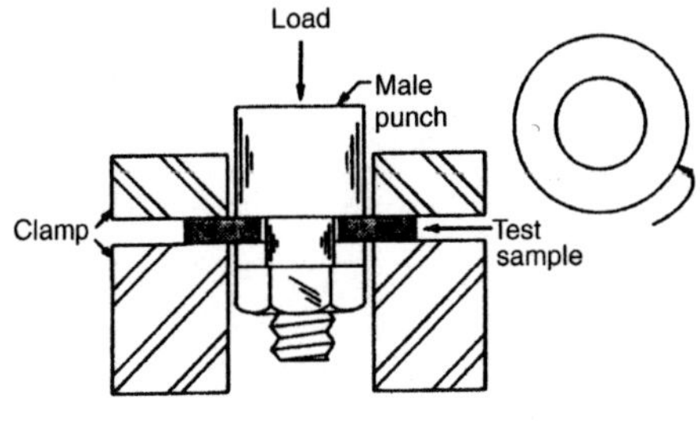

Fig. 14 Example of set for shear-strength testing of plastics

strength values than Izod test values. However, the two measurements can be correlated (Ref 12). The Izod test is usually done on 3.2 mm (⅛ in.) thick samples. Materials such as polycarbonate exhibit thickness-dependent impact properties. Below 6.4 mm (¼ in.), this material is ductile with a very high value, but above this thickness, the material has a much lower value.

Unnotched impact toughness tests in ASTM D 4812 and D 3029 (dart penetration test) have been replaced by ASTM D 5420. Impact values with unnotched samples are often considered a more definitive measure of impact strength, while the Izod test indicates notch sensitivity.

Dart Penetration (Puncture) Test. Another impact test that is often reported is the dart penetration (puncture) test. This test (Fig. 18) is different from the Izod and Charpy tests in a number of aspects. First, the stress state is two-dimensional in nature because the specimen is a plate rather than a beam. Second, the thin, plate-like specimen does not contain any notches or other stress concentrations.

The geometry and test conditions often applied using this specimen were described in ASTM D 3029 (now replaced by ASTM D 5420). The quantity most often quoted with respect to this test is the energy required for failure. Of course, these energy levels are very different from the notched-beam energies-to-failure, but they also do not represent any fundamental material property. A marked transition in mode of failure can also be observed with this specimen as the rate is increased or the temperature is decreased. However, this transition temperature is quite different from that measured in the notched-beam tests. Usually it displays a transition from ductile to brittle behavior at much lower temperatures than the notched specimens.

The dart penetration test is often performed with different specimens and indenter geometries. Linear elastic, small-displacement, thin-plate theory has occasionally been used to analyze test results in an effort to compare the performance of different materials tested with different specimen geometries. In all but the most brittle materials, this is an inappropriate simplification of the test. A number of very nonlinear events can take place during this test, including a growing indenter contact area, yielding, and large-displacement and large-strain deformation. References 13 to 15 provide more

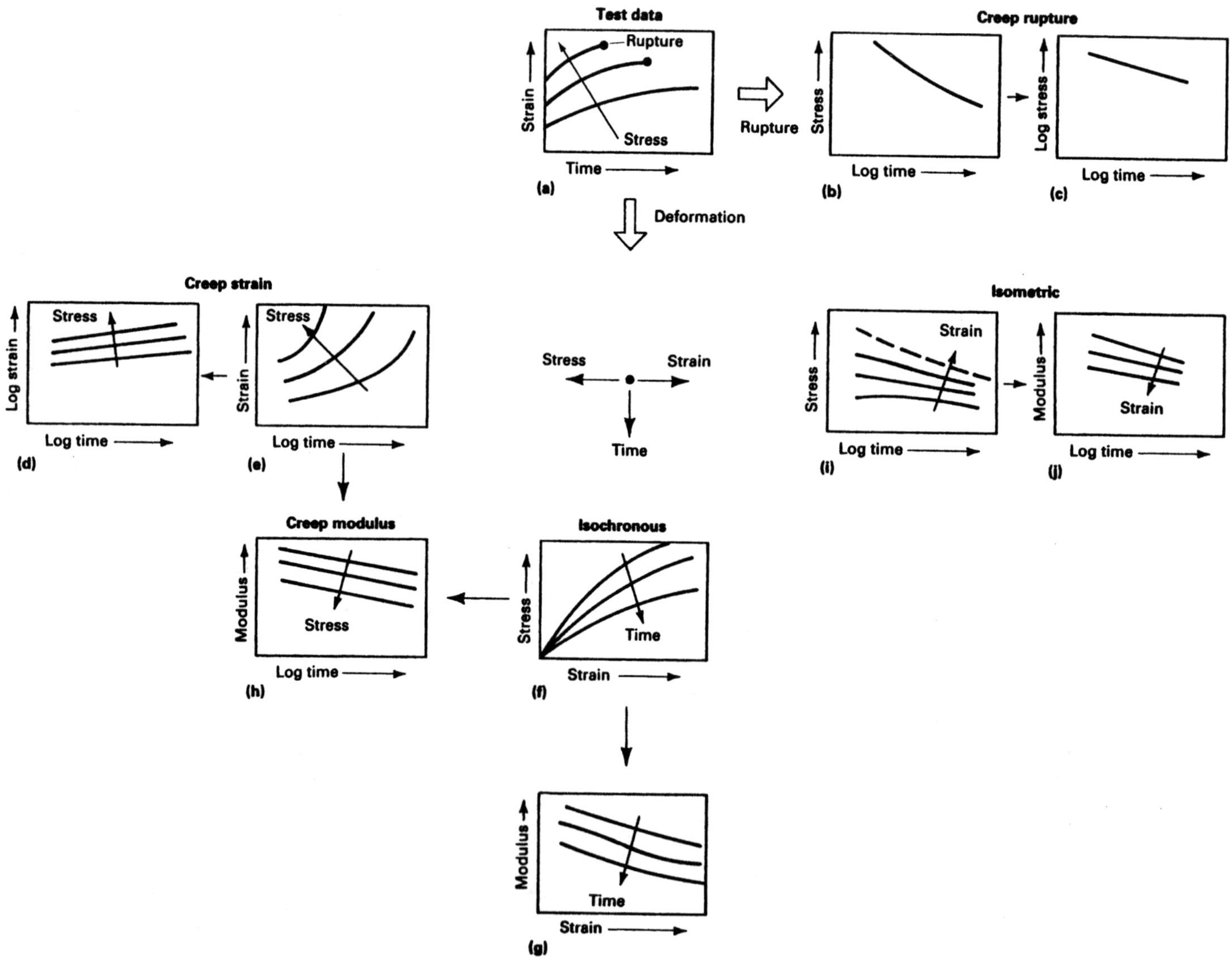

Fig. 15 Graphic representation of creep data showing various ways to plot time-dependent strain in response to time-dependent stress. See text for discussion.

details on these events and their effects on the test data.

Fracture Mechanics. Another way to evaluate the toughness of materials is by fracture toughness testing, where the value of the critical stress-intensity factor for a material can be measured by testing standard cracked specimens, such as the compact-tension specimen. Standard test methods and specimen geometries are defined for measuring the critical stress-intensity factors for metals (ASTM E 399), but similar standards have yet to be officially defined for plastics. It appears that many of the recommendations of the ASTM E 399 test procedure for metals are equally worthwhile for plastics, although the ductile nature and low yield strength of plastics pose problems of specimen size.

In fracture toughness testing, the sample size can be reduced as long as all dimensions of the laboratory specimen are much larger than the plastic zone size. In fracture testing, according to ASTM E 399, the thickness, B, must be:

$$B \geq 2.5\left(\frac{K_{IC}}{\sigma_y}\right)^2 \approx 16r_p$$

where K_{Ic} is the plane-strain fracture toughness, σ_y is the yield stress, and r_p is the radius of the plastic zone, which is given by:

$$r_p = \frac{1}{2\pi}\left(\frac{K_{Ic}}{\sigma_y}\right)^2$$

By ensuring that the thickness is much larger than the yield zone size (at least 16 times larger), the laboratory specimen will be in the state of plane strain. Because of the hydrostatic stresses that develop at crack tips under plane-strain conditions, yielding is suppressed, and a minimum value for fracture toughness is obtained. The plane-strain fracture toughness can be used with confidence in designing large components.

Similar arguments hold for polymer fracture testing. To design large polymer components or to design for polymer applications in which yielding is suppressed, it is important to measure fracture properties under conditions of plane strain. However, typical engineering plastics have fracture toughnesses in the range of 2 to 4 MPa$\sqrt{m}$ (1.8 to 3.6 ksi$\sqrt{\text{in.}}$) and yield strengths in the range of 50 to 80 MPa (7.3 to 11.6 ksi) (Ref 16). Plane-strain testing conditions would require sample thicknesses in the range of 1.6 to 16 mm (0.06 to 0.63 in.). The low-end range is a common size range, but the high end is more questionable. Engineering components designed with polymers almost never use polymers as thick as 16 mm (0.63 in.). Therefore, it is not clear that the plane-strain fracture toughness is the appropriate design data for engineering components in which the polymers will experience only plane-stress conditions. More importantly, fabricating thick polymer samples for plane-strain testing presents significant difficulties. Engineering plastics with fracture toughnesses

Table 4 Fracture behavior of selected plastics as a function of temperature

	Temperature, °C (°F)							
Plastics	**−20 (−4)**	**−10(14)**	**0(32)**	**10(50)**	**20(68)**	**30(85)**	**40(105)**	**50(120)**
Polystyrene	A	A	A	A	A	A	A	A
Polymethyl methacrylate	A	A	A	A	A	A	A	A
Glass-filled nylon (dry)	A	A	A	A	A	A	A	B
Polypropylene	A	A	A	A	B	B	B	B
Polyethylene terephthalate	B	B	B	B	B	B	B	B
Acetal	B	B	B	B	B	B	B	B
Nylon (dry)	B	B	B	B	B	B	B	B
Polysulfone	B	B	B	B	B	B	B	B
High-density polyethylene	B	B	B	B	B	B	B	B
Rigid polyvinyl chloride	B	B	B	B	B	B	C	C
Polyphenylene oxide	B	B	B	B	B	B	C	C
Acrylonitrile-butadiene-styrene	B	B	B	B	B	B	C	C
Polycarbonate	B	B	B	B	C	C	C	C
Nylon (wet)	B	B	B	C	C	C	C	C
Polytetrafluoroethylene	B	C	C	C	C	C	C	C
Low-density polyethylene	C	C	C	C	C	C	C	C

A, brittle even when unnotched; B, brittle, in the presence of n notch; C, tough

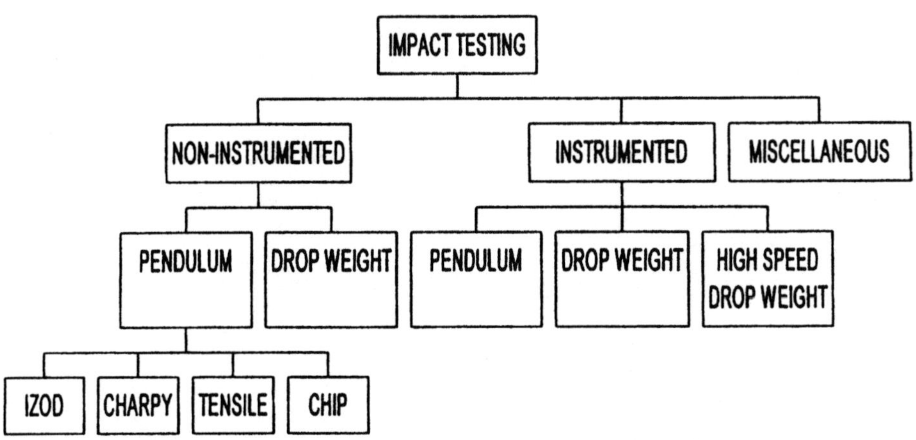

Fig. 16 Categories of impact test methods used in testing of plastics. Source: Ref 11

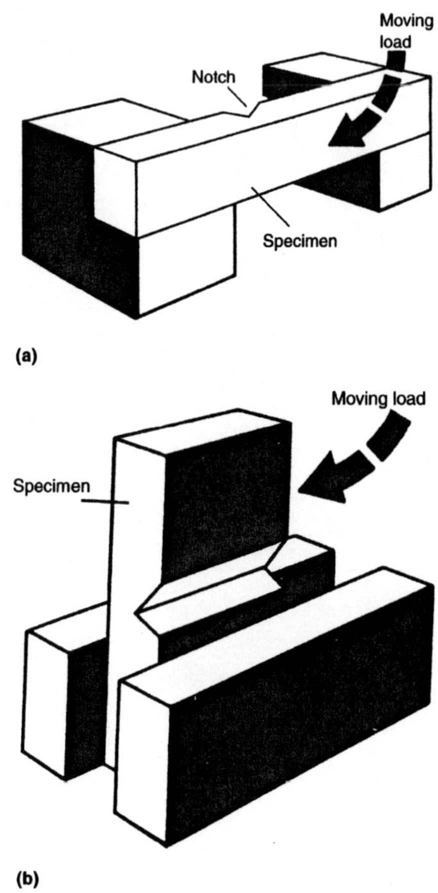

Fig. 17 Specimen types and test configurations for pendulum impact toughness tests. (a) Charpy method. (b) Izod method

in the range of 2 to 4 MPa$\sqrt{m}$ (1.8 to 3.6 ksi$\sqrt{in.}$) are not particularly tough. Rubber-toughened polymers can have much higher toughness. Also, because the yield strength of rubber-toughened polymers is usually lower, the thickness requirements for plane-strain fracture testing are such that potential laboratory specimens cannot be prepared. Some of these toughened polymers can be tested with *J*-integral techniques adapted from the *J*-integral metals standard (Ref 17, 18). Another technique known as the essential work of fracture technique has been considered. It has the potential to provide both plane-stress and plane-strain fracture toughness results for polymers. The essential work of fracture data can be obtained on thin polymers having thicknesses similar to those of typical polymer components (Ref 19).

Hardness Tests

Typical hardness values of common plastics are listed in Table 5. Rockwell testing and the durometer test method are the most common, although another type of hardness test for plastics is the Barcol method. A rough comparison of hardness scales for these methods is in Fig. 19, but it must be understood that any conversions from Fig. 19 are only rough estimates that vary depending on the materials. Hardness conversions are complicated by several material factors such as elastic recovery and, for plastics, the time-dependent effects from creep behavior. More information on the hardness testing of plastics is also given in the article "Selection and Industrial Applications of Hardness Tests" in *Mechanical Testing and Evaluation,* Volume 8 of the *ASM Handbook* (2000).

Rockwell hardness tests of plastics (ASTM D 785 and ISO 2039) are ball-indentation methods, where hardness is related to the net increase in the depth of an indentation after application of a minor load and a major load. The ball diameter and the loads are specified for each of the Rockwell scales, which are R, L, M, E, and K in order of increasing hardness. The Rockwell test is used for relatively hard plastics such as thermosets and structural thermoplastics such as nylons, polystyrene, acetals, and acrylics. Typical Rockwell values are shown in Fig. 20.

The durometer (or Shore hardness) method (ASTM D 2240 and ISO 868) registers the amount of indentation caused by a spring-loaded pointed indenter. This method is used for softer plastics and rubbers, and 100 is the highest hardness rating of this scale. Two types of durometers are used: type A and type D, as described further in the article "Selection and Industrial Applications of Hardness Tests" in *Mechanical Testing and Evaluation,* Volume 8 of the *ASM Handbook* (2000).

The Barcol hardness test (ASTM D 2583) is mainly used for measuring the hardness of reinforced and unreinforced rigid plastics. A hardness value is obtained by measuring the resistance to penetration of a sharp steel point under a spring load. The instrument, called the Barcol impressor, gives a direct reading on a 0 to 100 scale. The hardness value is often used as a measure of the degree of cure of a plastic.

International Rubber Hardness Degrees (IRHD) Testing. The IRHD hardness test is very similar to durometer testing with some important differences. Durometer testers apply a load to the sample using a calibrated spring and a pointed or blunt-shaped indenter. The load therefore will vary according to the depth of the indentation, because of the spring gradient. The IRHD tester uses a minor-major load system of constant load and a ball indenter to determine the hardness of the sample. This method is described further in the article "Miscellaneous Hardness Tests" in *Mechanical Testing and Evaluation,* Volume 8 of the *ASM Handbook* (2000).

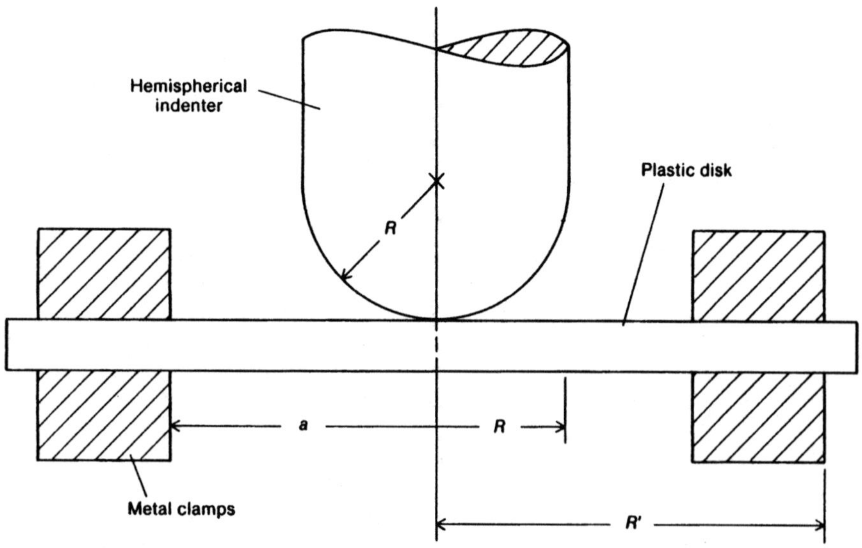

Fig. 18 Puncture test geometry

Fatigue Testing

Compared to testing of metals, the testing of plastics is a relatively recent pursuit. Because engineers and designers always use knowledge gained from previous experience, the methods used to test plastics in fatigue are largely based on methods developed for metals, with accommodations to account for the more obvious differences between the two materials.

For example, as previously noted in the section "Dynamic Mechanical Properties" in this article, the role of high hysteresis losses in the repeated stressing of plastics is very important. Unlike metals, plastics deform in a largely nonelastic manner, resulting in part of the mechanical energy being converted into heat within the material. The gradual buildup of heat may be sufficient to cause a loss in strength and rigidity. This effect is further aggravated by the low thermal conductivity of plastics and a general increase in hysteresis losses with an increase in temperature. Hysteresis losses are also a function of the loading rate (frequency), the type of load (bending, tension, or torsion), and the volume of material under stress. The hysteresis losses increase with loading rate and the volume of material under stress.

This also can be further extended to include the effects of different loading waveforms (sinusoid, saw tooth, or square) on the fatigue strength of viscoelastic materials. In addition, absorbed water and environmental variables also influence the fatigue strength of plastics.

Elastomers and Fibers

Polymers can exhibit a range of mechanical behaviors that characterize their various classifications as elastomers, plastics, and fibers (Fig. 21). This section briefly describe tension testing of elastomers and fibers.

Tension Testing of Elastomers

Elastomers have the ability to undergo high levels of reversible elongation that, in some cases, can reach up to 1000%. This high degree of reversible elongation allows stretching and recovery similar to that of a rubber band.

More than 20 different types of polymers can be used as bases for elastomeric compounds, and each type can have a significant number of contrasting subtypes within it. Properties of different polymers can be markedly different: for instance, urethanes seldom have tensile strengths

Table 5 Typical hardness values of selected plastics

Plastic material	Rockwell HRM	Rockwell HRR	Durometer, Shore D	Barcol
Thermoplastics				
Acrylonitrile-butadiene-styrene	...	75–115	...	...
Acetal	94	120	...	...
Acrylic	85–105	...	...	...
Cellulosics	...	30–125	...	...
Polyphenylene oxide	80	120	...	...
Nylon	...	108–120	...	...
Polycarbonate	72	118	...	...
High-density polyethylene	...	...	60–70	...
Low-density polyethylene	...	...	40–50	...
Polypropylene	...	...	75–85	...
Polystyrene	68–70	...	...	...
Polyvinyl chloride (PVC) (rigid)	...	115	...	...
Polysulfone	70	120	...	...
Thermosets				
Phenolic (with cellulose)	100–110	...	...	...
Phenolic (mineral filler)	105–115	...	...	...
Unsaturated polyester (clear cast)	...	...	...	34–40
Polyurethane (high-density integral skin foam)	...	...	36–63	...
Polyurethane (solid reaction injection-molded elastomer)	...	...	39–83	...
Epoxy (fiberglass reinforced)	106–108	...	...	...

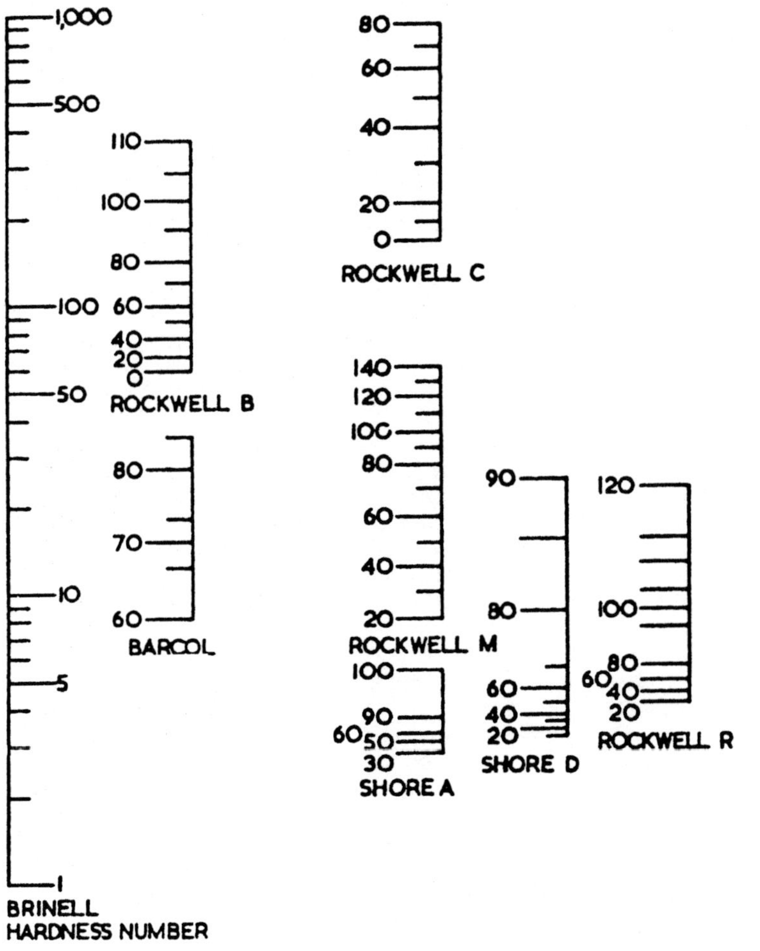

Fig. 19 Approximate relations among hardness scales for plastics

below 20.7 MPa (3.0 ksi), whereas silicones rarely exceed 8.3 MPa (1.2 ksi). Natural rubber is known for high elongation, 500 to 800%, whereas fluoroelastomers typically have elongation values ranging from 100 to 250%.

Literally hundreds of compounding ingredients are also available, including major classes such as powders (carbon black, clays, silicas), plasticizers (petroleum-base, vegetable, synthetic), and curatives (reactive chemicals that change the gummy mixture into a firm, stable elastomer). A rubber formulation can contain from four or five ingredients to 20 or more. The number, type, and level of ingredients can be used to change dramatically the properties of the resulting compound, even if the polymer base remains exactly the same.

ASTM D 412 is the U.S. Standard for tension testing of elastomers. It specifies two principal varieties of specimens: the more commonly used dumbbell-type, die cut from a standard test slab 150 mm by 150 mm by 20 mm (6 in. by 6 in. by 0.8 in.), and actual molded rings of rubber. The second type was standardized for use by the O-ring industry. For both varieties, several possible sizes are permitted, although, again, more tests are run on one of the dumbbell specimens (cut using the Die C shape) than on all other types combined. Straight specimens are also permitted, but their use is discouraged because of a pronounced tendency to break at the grip points, which makes the results less reliable.

The power-driven equipment used for testing is described, including details such as the jaws used to grip the specimen, temperature-controlled test chambers when needed, and the

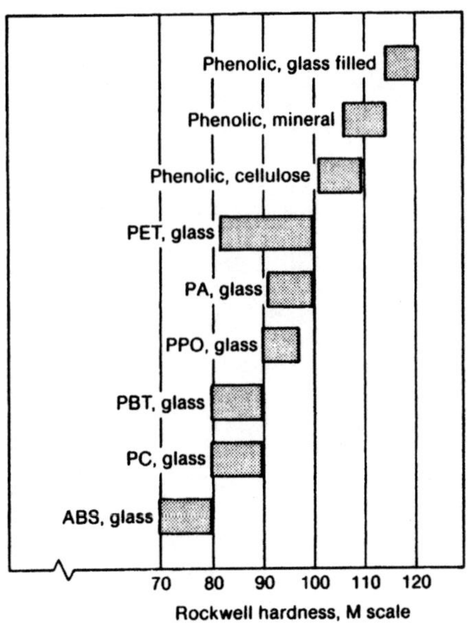

Fig. 20 Rockwell hardness of engineering plastics. PET, polyethylene terephthalate; PA, polyamide; PPO, polyphenylene oxide; PBT, polybutylene terephthalate; PC, polycarbonate; ABS, acrylonitrile-butadiene-styrene

crosshead speed of 500 mm/min (20 in./min). The testing machine must be capable of measuring the applied force within 2%, and a calibration procedure is described. Various other details, such as die-cutting procedures and descriptions of fixtures, are also provided.

The method for determining actual elongation can be visual, mechanical, or optical, but the method must be accurate within 10% increments. In the original visual technique, the machine operator simply held a scale behind or alongside the specimen as it was being stretched and noted the progressive change in the distance between two lines marked on the center length of the dogbone shape. The degree of precision that could be attained using a handheld ruler behind a piece of rubber being stretched at a rate of over 75 mm/s (3 in./s) was always open to question, with 10% being an optimistic estimate.

More recent technology employs extensometers, which comprise pairs of very light grips that are clamped onto the specimen and whose motion is then measured to determine actual material elongation. The newest technology involves optical methods, in which highly contrasting marks on the specimen are tracked by scanning devices, with the material elongation again being determined by the relative changes in the reference marks.

Normal procedure calls for three specimens to be tested from each compound, with the median figure being reported. Provision is also made for use of five specimens on some occasions, with the median again being used.

Techniques for calculating the tensile stress, tensile strength, and elongation are described for the different types of test specimens. The common practice of using the unstressed cross-sectional area for calculation of tensile strength is used for elastomers, as it is for many other materials. It is interesting to note that if the actual cross-sectional area at fracture is used to calculate true tensile strength of an elastomer, values that are higher by orders of magnitude are obtained.

In recent years, attention has been given to estimating the precision and reproducibility of the data generated in this type of testing. Interlaboratory test comparisons involving up to ten different facilities have been run, and the later versions of ASTM D 412 contain the information gathered.

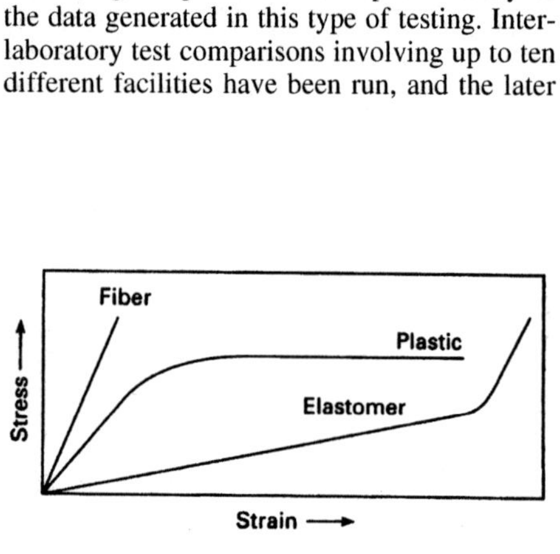

Fig. 21 Typical stress-strain curves for a fiber, a plastic, and an elastomer. Source: Ref 20

Variability of the data for any given compound is to some degree related to that particular formulation. When testing was performed on three different compounds of very divergent types and property levels, the pooled value for repeatability of tensile-strength determinations within labs was about 6%, whereas reproducibility between labs was much less precise, at about 18%. Comparable figures for ultimate elongation were approximately 9% (intralab) and 14% (interlab).

Similar comparisons of the 100% modulus (defined in "Modulus of the Compound" later in this section) have shown even less precision, with intralab variation of almost 20% and interlab variation of more than 31%. This runs counter to the premise that modulus should be more narrowly distributed than tensile strength, because tensile strength and ultimate elongation are failure properties and as such are profoundly affected by details of specimen preparation. Because the data do not support such a premise, some other factor must be at work. Possibly that factor is the lack of precision with which the 100% strain point is observed, but, in any case, it is important to determine the actual relationship between the precision levels of the different property measurements.

Significance and Use of Tensile-Testing Data for Elastomers. It is important to note that the tensile properties of elastomers are determined by a single application of progressive strain to a previously unstressed specimen to the point of rupture, which results in a stress-strain curve of some particular shape. The degree of nonlinearity and in fact complexity of that curve will vary substantially from compound to compound. Tensile properties of elastomers also have different significance than those of structural materials.

Tensile Strength of Elastomers. Because elastomers as a class of materials contain a substantial number of different polymers, the tensile strength of elastomers can range from as low as 3.5 MPa (500 psi) to as high as 55.2 MPa (8.0 ksi); however, the tensile strengths of the great majority of common elastomers tend to fall in the range from 6.9 to 20.7 MPa (1.0 to 3.0 ksi).

It should also be noted that successive strains to points just short of rupture for any given compound will yield a series of progressively different stress-strain curves; therefore, the tensile-strength rating of a compound would certainly change depending on how it was flexed prior to final fracture. Thus, the real meaning of elastomer tensile strength may be open to some question. However, some minimum level of tensile strength is often used as a criterion of basic compound quality, because the excessive use of inexpensive ingredients to fill out a formulation and lower the cost of the compound will dilute the polymer to the point that tensile strength decreases noticeably.

The meaning of tensile strength of elastomers must not be confused with the meaning of tensile strength of other materials, such as metals. Whereas tensile strength of a metal may be validly and directly used for a variety of design purposes, this is not true for tensile strength of elastomers. As stated early in ASTM D 412, "Tensile properties may or may not be directly related to the end-use performance of the product because of the wide range of performance requirements in actual use." In fact, very seldom if ever can a given high level of tensile strength of a compound be used as evidence that the compound is fit for some particular application.

Elongation of Elastomers. Ultimate elongation is the property that defines elastomeric materials. Any material that can be reversibly elongated to twice its unstressed length falls within the formal ASTM definition of an elastomer. The upper end of the range for rubber compounds is about 800%, and although the lower end is supposed to be 100% (a 100% increase of the unstressed reference dimension), some special compounds with limits that fall slightly below 100% elongation still are accepted as elastomers.

Just as with tensile strength, certain minimum levels of ultimate elongation are often called out in specifications for elastomers. The particular elongation required will relate to the type of polymer being used and the stiffness of the compound. For example, a comparatively hard (80 durometer) fluoroelastomer might have a requirement of only 125% elongation, whereas a soft (30 durometer) natural rubber might have a minimum required elongation of at least 400%.

However, ultimate elongation still does not provide a precise indication of serviceability, because service conditions normally do not require the rubber to stretch to any significant fraction of its ultimate elongative capacity. Nonetheless, elongation is a key material selection factor that is more applicable as an end-use criterion for elastomers than is tensile strength.

Modulus of the Compound. Another characteristic of interest is referred to in the rubber industry as the modulus of the compound. Specific designations such as 100% modulus or 300% modulus are used. This is due to the fact that the number generated is not an engineering modulus in the normal sense of the term, but, rather, is the stress required to obtain a given strain. Therefore, the 100% modulus, also referred to as M-100, is simply the stress required to elongate the rubber to twice its reference length.

Tensile modulus, better described as the stress required to achieve a defined strain, is a measurement of the stiffness of a compound. When the stress-strain curve of an elastomer is drawn, it can be seen that the tensile modulus is actually a secant modulus—that is, a line drawn from the origin of the graph straight to the point of the specific strain. However, an engineer needing to understand the forces that will be required to deform the elastomer in a small region about that strain would be better off drawing a line tangent to the curve at the specific level of strain and using the slope of that line to determine the approximate ratio of stress to strain in that region. This technique can be utilized in regard to actual elastomeric components as well as lab specimens.

Tension Set. A final characteristic that can be measured, but that is used less often than the other three is called tension set. Often, when an elastomer or rubber is stretched to final rupture, the recovery in length of the two sections resulting from the break is less than complete. It is possible to measure the total length of the original reference dimension and calculate how much longer the total length of the two separate sections is. This is expressed as a percentage. Some elastomers will exhibit almost total recovery, whereas others may display tension set as high as 10% or more. Tension set may also be measured on specimens stretched to less than breaking elongation.

The property of tension set is used as a rough measurement of the tolerance of high strain of the compound. This property is not tested very often, but, for some particular applications, such a test is considered useful. It could also be used as a quality control measure or compound development tool, but most of the types of changes it will detect in a compound will also show up in tests of tensile strength, elongation, and other properties, and so its use remains infrequent.

Tensile-Test Curves. Figure 22 is a plot of tensile-test curves from five very different compounds, covering a range of base polymer types and hardnesses. The contrasts in properties are clearly visible, such as the high elongation (>700%) of the soft natural rubber compound compared with the much lower (about 275%) elongation of a soft fluorosilicone compound. Tensile strengths as low as 2.4 MPa (350 psi) and as high as 15.5 MPa (2.25 ksi) are observed. Different shapes in the curves can be seen, most noticeably in the pronounced curvature of the natural rubber compound.

Figure 23 demonstrates that, even within a single elastomer type, contrasting tensile-property responses will exist. All four of the compounds tested were based on polychloroprene, covering a reasonably broad range of hardnesses, 40 to 70 Shore A durometer. Contrasts are again seen, but more in elongation levels than in final tensile strength. Two of the compounds are at the same durometer level and still display a noticeable difference between their respective stress-strain curves. This shows how the use of differing ingredients in similar formulas can result in some properties being the same or nearly the same, whereas others vary substantially.

Tests for Determining the Tensile Strength of Fibers

Mechanical properties of fibers are very dependent on test method. Two basic methods are the single-filament tension test and the two tensile test of a group or strand of fibers.

Single-filament tensile strength (ASTM D 3379) is determined using a random selection of single filaments made from the material to be tested. Filaments are centerline-mounted on special slotted tabs. The tabs are gripped so that the test specimen is aligned axially in the jaws of a constant-speed movable-crosshead test machine. The filaments are then stressed to failure at a constant strain rate. For this test method, filament cross-sectional areas are determined by planimeter measurements of a representative number of filament cross sections as displayed on highly magnified micrographs. Alternative methods of area determination use optical gages, an image-splitting microscope, a linear weight-density method, and others.

Tensile strength and Young's modulus of elasticity are calculated from the load elongation records and the cross-sectional area measurements.

The specimen setup is shown in Fig. 24. Note that a system compliance adjustment may be necessary for single-filament tensile modulus.

Tow Tensile Test (ASTM D 4018). The strength of fibers is rarely determined by testing single filaments and obtaining a numerical average of their strength values. Usually, a bundle or yarn of such fibers is impregnated with a polymer and loaded to failure. The average fiber strength is then defined by the maximum load

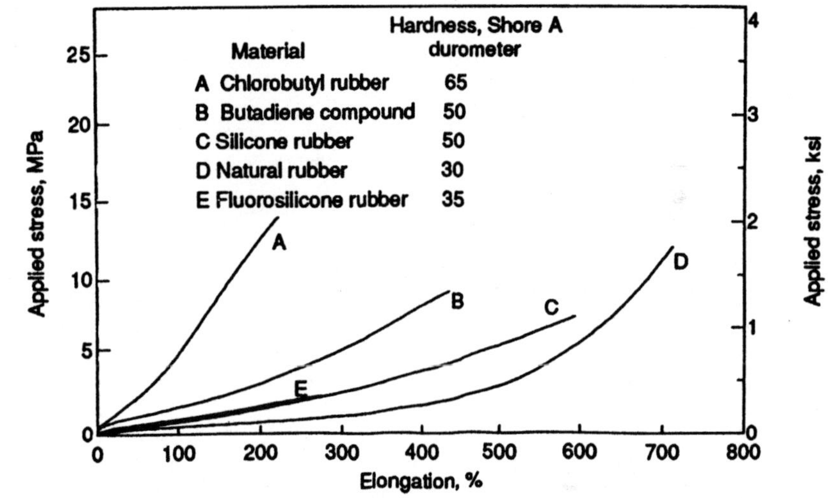

Fig. 22 Tensile test curves for five different elastomer compounds

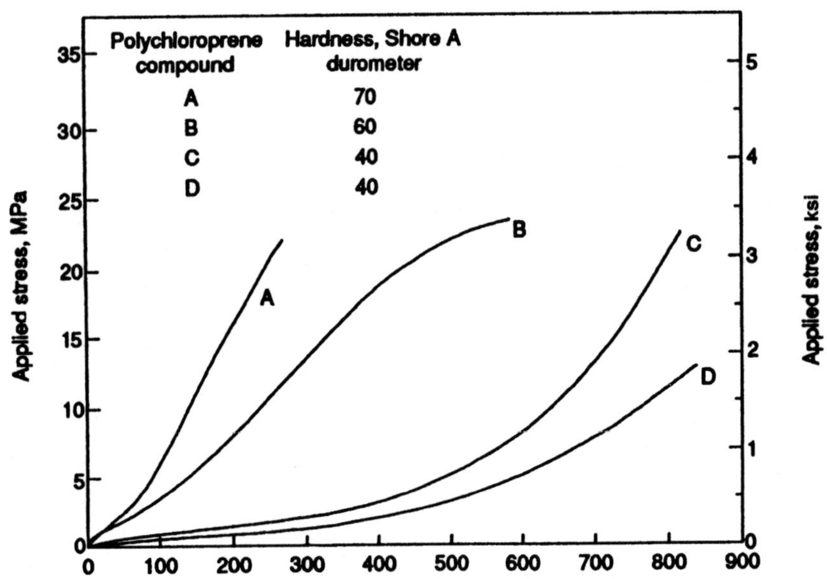

Fig. 23 Tensile test curves for four polychloroprene compounds

divided by the cross-sectional area of the fibers alone.

Using ASTM D 4018 or an equivalent is recommended. This is summarized as finding the tensile properties of continuous filament carbon and graphite yarns, strands, rovings, and tows by the tensile loading to failure of the resin-impregnated fiber forms. This technique loses accuracy as the filament count increases. Strain and Young's modulus are measured by extensometer.

The purpose of using impregnating resin is to provide the fiber forms, when cured, with enough mechanical strength to produce a rigid test specimen capable of sustaining uniform loading of the individual filaments in the specimen.

To minimize the effect of the impregnating resin on the tensile properties of the fiber forms, the resin should be compatible with the fiber, the resin content in the cured specimen should be limited to the minimum amount required to produce a useful test specimen, the individual filaments of the fiber forms should be well collimated, and the strain capability of the resin should be significantly greater than the strain capability of the filaments.

ASTM D 4018 Method I test specimens require a special cast-resin end tab and grip design to prevent grip slippage under high loads. Alternative methods of specimen mounting to end tabs are acceptable, provided that test specimens maintain axial alignment on the test machine centerline and that they do not slip in the grips at high loads. ASTM D 4018 Method II test specimens require no special gripping mechanisms. Standard rubber-faced jaws should be adequate.

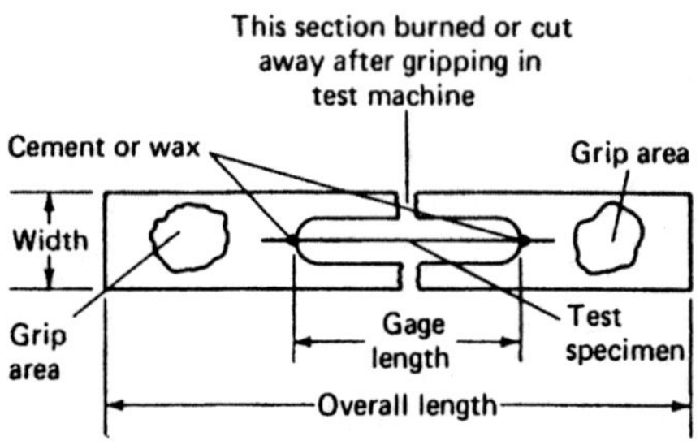

Fig. 24 Typical specimen-mounting method for the single-filament fiber tension test (ASTM D 3379)

REFERENCES

1. *Modern Plastics Encyclopedia,* McGraw Hill, 2000
2. V. Shah, *Handbook of Plastics Testing Technology,* 2nd ed., John Wiley & Sons, 1998
3. *ISO/IEC Selected Standards for Testing Plastics,* 2nd ed., ASTM, 1999
4. J. Nairn and R. Farris, Important Properties Divergences, *Engineering Plastics,* Vol 2, *Engineered Materials Handbook,* ASM International, 1988, p 655–658
5. T. Osswald, *Polymer Processing Fundamentals,* Hanser/Gardner Publications Inc., 1998, p 19–43
6. K.M. Ralls, T.H. Courtney, and J. Wulff, *Introduction to Materials Science and Engineering,* John Wiley & Sons, 1976
7. S. Turner, Mechanical Testing, *Engineering Plastics,* Vol 2, *Engineered Materials Handbook,* ASM International, 1988, p 547
8. R.B. Seymour, *Polymers for Engineering Applications,* ASM International, 1987, p 155
9. V. Shah, *Handbook of Plastics Testing Technology,* 2nd ed., John Wiley & Sons, 1998
10. Recognized Components Directories, Underwriters Laboratories
11. F.N. Kelly and F. Bueche, *J. Polym. Sci.,* Vol 50, 1961, p 549
12. R.D. Deanin, *Polymer Structure, Properties and Applications,* Cahners, 1982
13. R.P. Nimmer, Analysis of the Puncture of a Polycarbonate Disc, *Polym. Eng. Sci.,* Vol 23, 1983, p 155
14. R.P. Nimmer, An Analytical Study of Tensile and Puncture Test Behavior as a Function of Large-Strain Properties, Polym. Eng. Sci., Vol 27, 1987, p 263
15. L.M. Carapelucci, A.F. Yee, and R.P. Nimmer, Some Problems Associated with the Puncture Testing of Plastics, J. Polym. Eng., June 1987
16. J.G. Williams, Fracture Mechanics of Polymers, Ellis Horwood, 1984
17. D.D. Huang and J.G. Williams, J. Mater. Sci., Vol 22, 1987, p 2503
18. M.K.V. Chan and J.G. Williams, Int. J. Fract., Vol 19, 1983, p 145
19. Y.W. Mai and B. Cotterell, Int. J. Fract., Vol 32, 1986, p 105
20. R. Seymour, Overview of Polymer Chemistry, *Engineering Plastics,* Vol 2, *Engineered Materials Handbook,* ASM International, 1988, p 64

Creep, Stress Relaxation, and Yielding*

AS RELATED PHENOMENA, creep, stress relaxation, and the effect of strain rate on yielding are generally much more significant to polymers at room temperature than they are to metals. Two additional influences on the mechanical behavior of polymers are hydrostatic pressure (or the hydrostatic component of the stress) and the aging effect that polymers often undergo at room temperature, depending on their thermal and mechanical history.

This article describes the general aspects of creep, stress relaxation, and yielding for homogeneous polymers. The word homogeneous is used to exclude copolymers and blends that undergo microphase separation. However, the differences between crystalline and amorphous polymers are identified.

The application of this review to the practical problems of failure analysis are the same as for other solids; namely, excessive creep leads to intolerable dimensional changes in an engineering structure, stress relaxation causes a loosening of fasteners, and yielding produces a permanent change in shape that renders the structure inoperable. It is important to know how the applied stress, strain or strain rate, and temperature determine the rate of creep and stress relaxation and produce yielding. In addition, the effects of thermal and mechanical history must be considered.

Polymers deform by two mechanisms: shear flow and crazing (Ref 1). Shear flow is a bulk phenomenon, and the plastic deformation is often homogeneous, except for the shear banding that occurs at high strains. Also, the density change during shear flow is small. Crazing is a localized form of deformation that initiates at points of stress concentration. The crazes form as thin sheets that are approximately 0.1 to 10 μm (4 to 400 μin.) thick and spread as a planar zone. Within the craze, the density is on the order of 50% of the original density. The creep, stress relaxation, and yield behavior for crazes is quite different than it is for shear flow deformation. This article only emphasizes the various types of shear flow deformation. Sometimes, both shear flow and crazing occur in the same specimen. It is very important to determine, in each case, whether shear flow or crazing is the dominant mechanism. Crazes can be readily observed with the light microscope. Complete details on this failure mode are given in the article "Crazing and Fracture" in this Section of the book.

Creep Failure

Failure caused by creep is discussed first, because it is easy to show how the creep curve is directly related to stress relaxation and yielding. Initially, the effects of aging and the mechanical history are neglected in order to focus on the primary variables, stress, σ, and temperature, T. Generally, the strain, ε, is some function of σ, T, and time, t. As a first approximation, the strain is often separated into three functions:

$$\varepsilon = F_1(\sigma)F_2(T)F_3(t) \quad \text{(Eq 1)}$$

The various forms that have been proposed for these functions are given as follows.

The power-law representation for stress is:

$$F_1(\sigma) = A\left(\frac{\sigma}{\sigma_0}\right)^n \quad \text{(Eq 2)}$$

where A and n are constants and σ_0 is related to the yield point, which is defined in the section "Yield Failure" in this article. Equation 2 has essentially two material constants that must be determined experimentally for each polymer; n is generally greater than 1. Equation 2 is only an approximation, because n usually increases with stress, and the stress dependence of n must be determined by experiments. The magnitude of the stress relative to the yield point is an important factor made evident by Eq 2.

For small strains of less than approximately 1%, n is generally equal to 1. This region is called linear viscoelastic behavior, for which the important variable is the compliance, or the strain divided by the stress. Polymers such as polyethylene, however, show nonlinear behavior at strains less than 1%.

Another form that has been used for $F_1(\sigma)$ is:

$$F_1(\sigma) = B \sinh \sigma b \quad \text{(Eq 3)}$$

where B and b are material constants to be determined by experiments. Equation 3 reduces to $n = 1$ (Eq 2) at low stress, and at high stress, becomes:

$$F_1(\sigma) = c\, e^{\sigma b} \quad \text{(Eq 4)}$$

where c and b are experimental constants.

The exponential form of the stress function is often combined with the effect of temperature at temperatures below the glass transition temperature, T_g, where the creep rate is given by:

$$\dot{\varepsilon} = \dot{\varepsilon}_0\, e^{-(Q-\sigma v)/RT} \quad \text{(Eq 5)}$$

where $\dot{\varepsilon}_0$, Q, and v are material constants, and R is the gas constant.

The temperature function, $F_2(T)$, generally is represented in two forms. In the case of amorphous polymers at temperatures above the T_g, the Williams-Landel-Ferry (WLF) equation is considered to be the best representation:

$$\log \frac{\dot{\varepsilon}}{\dot{\varepsilon}_{T_g}} = \frac{17.4(T - T_g)}{51.6 + (T - T_g)} \quad \text{(Eq 6)}$$

for T is greater than T_g, where $\dot{\varepsilon}_{T_g}$ is the creep rate at T_g.

In some cases, Eq 6 does not completely describe the effect of temperature, because the quantity $\dot{\varepsilon}_{T_g}$ may vary with temperature. It is indeed important to note that above T_g, the effect of temperature on creep rate and stress relaxation is well described by the WLF equation for many amorphous polymers. (Reference 2, on viscoelasticity, is recommended for a more complete understanding of this phenomenon.)

Below T_g, the temperature effect is generally given by:

$$\dot{\varepsilon} = \dot{\varepsilon}_0\, e^{-Q/RT} \quad \text{(Eq 7)}$$

where $\dot{\varepsilon}_0$ and Q are experimental constants. However, Q depends on temperature. The value of Q that applies in a given temperature range is related to Q corresponding to the so-called α, β, and γ peaks that are observed by internal friction measurements for small stresses. Values of $\dot{\varepsilon}_0$ and Q, which are representative of polymers, are given in Ref 3. In some cases, the creep behavior is governed by a certain Q at the beginning of the creep test, but as time passes, another mechanism begins to operate, which is governed by a Q that corresponds to the next higher temperature range. One of the main problems in trying to predict the long-term creep behavior of plastics is that one type of kinetics may apply for the

*Adapted from the article by Norman Brown, "Creep, Stress Relaxation, and Yielding," in *Engineering Plastics*, Volume 2, *Engineered Materials Handbook*, ASM International, 1988, pages 728 to 733

initial part of the creep curve, and another mechanism of creep may come into play later on. It is possible that more than two mechanisms may operate.

The time function in Eq 1 is often represented by:

$$F_3(t) = t^m \quad \text{(Eq 8)}$$

where m can vary from approximately 0.06 to 0.6 for a number of polymers (Ref 1). Also, m can increase with stress and temperature (Ref 1).

However, if more than one mechanism occurs, the simple monotonic change in strain with time, as represented by Eq 8, may not occur. This more complex behavior (Fig. 1), in which the initial part of the creep curve has been associated with the β transition and the latter part with the α, has been demonstrated (Ref 4).

At low temperatures, the creep behavior may become logarithmic, as in the case of metals, so that:

$$\varepsilon = \varepsilon_0 \ln (t/t_0) \quad \text{(Eq 9)}$$

Also, at higher temperatures, polymers may exhibit the combination of transient creep, as represented by Eq 8, and then what appears to be steady-state creep, where:

$$\dot{\varepsilon} = \text{constant} \quad \text{(Eq 10)}$$

The previous equations represent situations in which the creep curves are well behaved. If one or a combination of the previous creep equations holds for the lifetime of the structure under stress, then the dimensional changes in the structure are predictable. However, there is also the possibility of a more catastrophic type of creep failure, in which there is a sudden onset of large-scale yielding, or even fracture.

First, the conditions for the onset of failure by large-scale yielding should be considered. For polyethylene (Ref 5, 6) and polymethyl methacrylate (Ref 7), when the creep strain becomes approximately 8 to 12%, large-scale yielding occurs. This result is consistent with the observation (Ref 8) that the shear yield strain of linear polymers is generally in the range of 5 to 15%, as measured by the ordinary stress-strain test, and depends somewhat on strain rate and temperature. This important point will be further discussed in the section on yield failure in this article.

It is important to note that if a craze forms in the specimen, and it is only a single craze, then fracture may occur before there is an appreciable homogeneous creep strain in the specimen as a whole. In the case of polyethylene at room temperature and above, both general shear creep and crack growth can occur simultaneously. Whether the specimen fails by general yielding or by brittle fracture depends on the stress; lower stresses favor brittle fracture. This ductile-brittle transition is analyzed in detail in Ref 9. At each temperature, there is a critical stress for the ductile-to-brittle failure mode, as shown in Fig. 2 (Ref 10). The transition stress from ductile-to-brittle behavior increases with decreasing temperature, primarily because the shear yield point increases with decreasing temperature. The ductile behavior is somewhat insensitive to the details of the morphological structure of the polyethylene (Ref 11) and depends primarily on the yield point. However, the susceptibility to the brittle fracture mode is strongly influenced by molecular weight, molecular weight distribution, and the degree and type of branching in the molecule. An increase in the molecular weight produces an increase in resistance to brittle fracture. The creep rate and yield strength are relatively insensitive to molecular weight in the range of molecular weights that exist in most engineering plastics. There are, however, processing limitations on the maximum useful molecular weight, because the viscosity of the melt increases rapidly with molecular weight.

One of the most important factors that causes brittle failure under a constant stress is the presence of points of stress concentration in the structural component. One source of stress concentrators is poor design of the component. Another is defects, such as dirt particles in the material, and yet another is the notches and scratches produced during component processing or by mishandling the finished component in the course of installation.

To summarize, the creep equation for a particular polymer is usually based on relatively short-term experimental creep curves relative to the lifetime of the engineering structure. In order to make predictions about the creep strain after prolonged periods, it is useful to insert an appropriate combination of Eq 2 to 10 into Eq 1, as dictated by the experimentally observed creep curves. If the creep strain is predicted to exceed 8 to 12%, the possibility of catastrophic failure by large-scale yielding can be anticipated. If there is evidence of crazing or incipient crack growth at a point of stress concentration in the structure, then there is the possibility of catastrophic failure by brittle fracture.

Stress Relaxation Failure

The kinetics of stress relaxation are intimately determined by the creep behavior of the

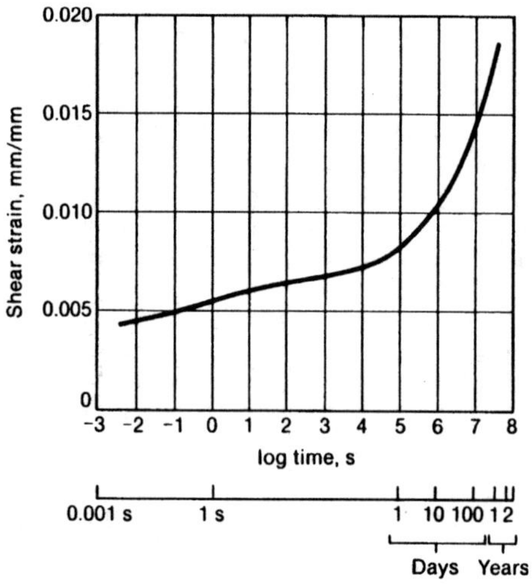

Fig. 1 Creep of polymethyl methacrylate under a torsional stress at 30 °C (85°F). Sources: Ref 4

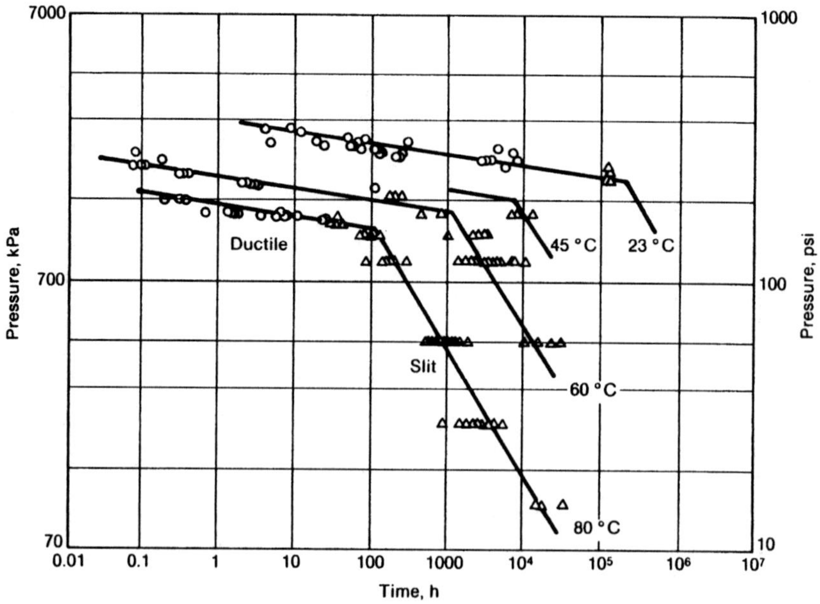

Fig. 2 Effect of internal pressure on time-to-failure of polyethylene gas pipe at various temperatures. Ductile regime indicates yield failure. Slit regime indicates brittle failure. Source: Ref 10

plastic. Whereas creep occurs under constant stress or load, stress relaxation occurs under constant strain or displacement and may be viewed as creep under a constantly decreasing stress. A simple model for the relationship between creep and stress relaxation is presented.

In stress relaxation, a specimen is generally rapidly strained to a certain strain or displacement, which is then held constant. The situation can be simply described by an equation in which the total strain ε_T is constant and consists of two parts: an elastic component, ε_E, and a creep component, ε_c. Thus:

$$\varepsilon_E + \varepsilon_c = \text{constant} \quad \text{(Eq 11)}$$

If Eq 11 is differentiated, then:

$$\dot{\varepsilon}_E + \dot{\varepsilon}_c = 0 \quad \text{(Eq 12)}$$

where $\dot{\varepsilon}_E$ and $\dot{\varepsilon}_c$ are the elastic and creep strain rates, respectively.

Now, the elastic strain is related to the stress by an elastic modulus, M, so that Eq 12 becomes:

$$\frac{\dot{\sigma}}{M} + \dot{\varepsilon}_c = 0 \quad \text{(Eq 13)}$$

where $\dot{\sigma}$ is the stress rate, and the specific form of $\dot{\varepsilon}_c$ is derived from the appropriate creep curve, as discussed in the previous section, "Creep Failure," in this article. Then, Eq 13 can be solved to obtain σ as a function of time, temperature, and the initial strain. Generally, M is assumed to be a constant, but it usually varies with stress, because polymers generally exhibit nonlinear stress-strain relationships. To exemplify how Eq 13 can be used, assume that the creep curve can be represented by:

$$\varepsilon_c = A\sigma^n e^{-Q/RT} t^m \quad \text{(Eq 14)}$$

where the material constants A, n, Q, and m have all been determined by the appropriate creep experiments. Then:

$$\dot{\varepsilon}_c = m A\sigma^n e^{-Q/RT} t^{m-1} \quad \text{(Eq 15)}$$

Inserting Eq 15 into Eq 13 and solving for σ, gives:

$$\frac{1}{\sigma^{(n-1)}} - \frac{1}{\sigma_0^{(n-1)}} = (n-1)\, M A\, e^{-Q/RT}\, t^m \quad \text{(Eq 16)}$$

where σ_0 is the stress at $t = 0$, and M was assumed to be a constant.

During stress relaxation, failure may occur when the stress relaxes to some fraction of the initial stress. Thus, it is important to know how long it will take for failure. If failure occurs when $\sigma = K\sigma_0$, where K is a fraction of σ_0, then the time for failure is given by:

$$t = \left(\frac{1}{\sigma_0^{(n-1)}} \frac{1 - K^{(n-1)}}{K^{(n-1)}} \frac{e^{Q/RT}}{(n-1)\, M A} \right)^{1/m} \quad \text{(Eq 17)}$$

Note that the time to failure increases as M decreases; M is related to the stiffness of the system. By analogy, if a nut is tightened on a bolt, then M can be effectively decreased by inserting a spring under the nut. A bent spring washer often serves this purpose. If A is made smaller, the time to failure is increased; A relates to the creep behavior of the material. Thus, the more creep resistant the plastic, the slower the rate of stress relaxation. The other important factor is temperature. As the temperature is decreased, the time for failure increases. If it is necessary to operate at an elevated temperature, then a material with a high Q, the activation energy for creep, is desirable.

Finally, in order to maintain over an extended time a high percentage of the stress that is applied initially, the initial stress should be kept as low as possible. If the device is overtightened initially, the fraction of stress relaxation will be greater than for a device that is initially tightened less, assuming the polymer is nonlinear.

Yield Failure

The stress-strain curves of polymers usually show a maximum stress beyond which permanent deformation (plastic strain) is produced. This maximum in the stress occurs under tensile, compressive, and shear loading and is common for most polymers (Fig. 3, 4). This maximum stress is called the yield point. Usually, when a structure or part of a structure is deformed beyond the yield point, it becomes so distorted that it is inoperable. It is important to understand the basic factors that produce yielding.

Note in Fig. 3 that the yield strain is approximately 12%. For most polymers, the yield strain lies in the range of 5 to 15%, especially if the temperature is not near T_g. For metals, the yield strains are generally 100 to 10,000 times smaller.

The yield stress depends primarily on the elastic modulus (Ref 8, 14–17). The effect of temperature and strain rate on the relationship between yield point and elastic modulus is discussed in the section "Effect of Crystallinity" in this article. At low temperatures, relative to the T_g for crystalline and amorphous polymers, the average relationship between yield point and modulus is given by:

$$\tau_y = CG \quad \text{(Eq 18)}$$

where τ_y is the shear yield point, G is shear modulus, and the ratio C ranges from approximately 0.03 to 0.13, with an average value of 0.076, as shown in Table 1. The ratio C is on the small side for crystalline polymers.

Thus, if the shear modulus is known at a given temperature, the shear yield point can be estimated from the previous equation. Often, only Young's modulus, E, is known. Young's modulus is related to G for linear behavior (small strains) by the equation:

$$G = \frac{E}{2(1 + \nu)} \quad \text{(Eq 19)}$$

where ν is Poisson's ratio. For most polymers, ν has a value of 0.36 to 0.38.

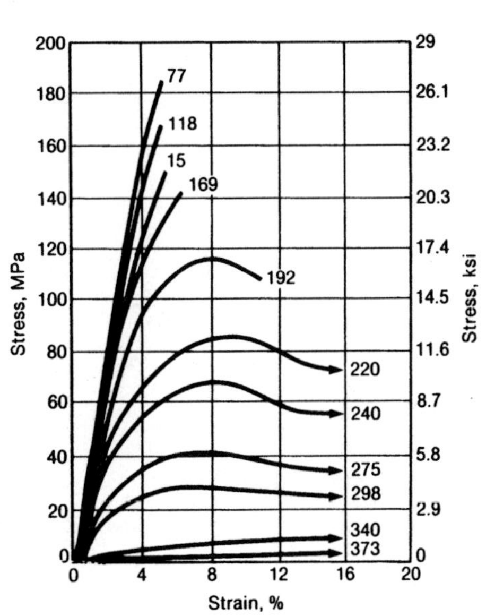

Fig. 3 Stress-strain curves in tension for quenched polychlorotrifluoroethylene at various temperatures, given in degree Kelvin. Sources: Ref 12

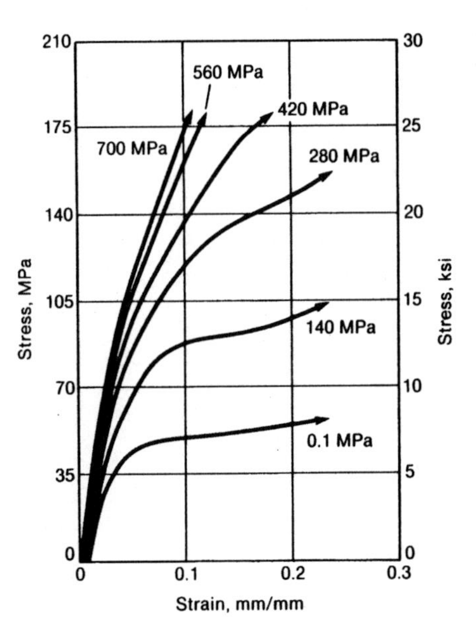

Fig. 4 Nominal compressive stress curves of polychlorotrifluoroethylene at various pressures. Sources: Ref 13

The shear yielding of polymers depends primarily on the shear component of the stress tensor but also on the hydrostatic component of the stress tensor. Figure 4 shows the effect of pressure on the yield point. Under a multiaxial stress field, a modification of either the Tresca or the von Mises criterion is applicable. The modified Tresca criterion is given by:

$$\frac{\sigma_1 - \sigma_2}{2} = \tau_y - \mu P \quad \text{(Eq 20)}$$

and the modified von Mises criterion is given by:

$$1/\sqrt{6}\,[(\sigma_1 - \sigma_2)^2 + (\sigma_2 - \sigma_3)^2 + (\sigma_3 - \sigma_1)^2]^{1/2} = \tau_y - \mu P \quad \text{(Eq 21)}$$

where σ_1, σ_2, and σ_3 are the principal stresses, τ_y is yield point in pure shear, $P = (\sigma_1 + \sigma_2 + \sigma_3)/3$, and μ (the change in yield point with respect to a change in pressure) is a material parameter. Generally, the von Mises criterion gives a somewhat better fit to the experimental data. However, from a design standpoint, either of these criteria differs by not more than approximately 15%, with the Tresca criterion giving a more conservative estimate for the probability of failure.

Table 1 Ratio of shear yield point to shear modulus at 0 K

Polymer	τ_y/G (experimental)
Polystyrene	0.069
Polymethyl methacrylate	0.133
Polycarbonate	0.050
Polyethylene terephthalate	0.091
Polychlorotrifluoroethylene	0.065
Polyethylene, high density	0.027
Polypropylene, isotactic-quenched	0.030
Average (excluding polyethylene)(a)	0.076 ± 0.03

τ_y, shear yield point; G, shear modulus. (a) Polyethylene was excluded because it is highly crystalline. Source: Ref 17

The yield point in tension is less than it is in compression. By the same token, the creep rate in tension is faster than the creep rate in compression for the same magnitude of applied stress. A relationship between the shear yield point and the uniaxial yield point based on the von Mises criterion is given by:

$$\tau_y = \frac{(1 \pm \mu)/\sqrt{3}}{\sqrt{3}}\,\sigma_y \quad \text{(Eq 22)}$$

where μ is the property of the material that determines the effect of pressure on the yield point. The positive sign in the previous equation is for tension, and the negative sign is for compression. The difference between the tensile and compressive yield point is generally approximately 10 to 20%. Table 2 gives the relative values of τ_y and σ_y for a group of polymers at a low temperature. The relative values of τ_y and σ_y are not expected to change appreciably with temperature. However, the absolute value of τ_y and σ_y are functions of temperature and strain rate.

Figure 3 shows a series of stress-strain curves over a range of temperatures. It may be noted that below a critical temperature, the polymer becomes brittle, because brittle fracture occurs prior to yielding. All polymers undergo a ductile-brittle transition temperature at some low temperature. It is important to know the ductile-brittle transition temperature before a polymer is placed in a low-temperature service environment. The ductile-brittle transition temperature is lower for compressive loading than for tensile loading.

Figure 4 shows that for an amorphous polymer, the yield point goes to zero when the temperature reaches T_g. For a crystalline polymer, the yield point approaches zero when the polymer begins to melt. It must be remembered that the usual so-called crystalline polymer is only partly crystalline and melts over a range of temperatures. To a first approximation, the yield point decreases linearly with increasing temperature. However, there are certain temperature ranges in which the change in yield point with temperature is somewhat more rapid. These ranges coincide with the temperature ranges in which the modulus also changes rapidly with temperature. A useful equation for the relation between yield point and temperature, T, for an amorphous polymer is given by:

$$\tau_y = \frac{CG}{T_g}(T_g - T) \quad \text{(Eq 23)}$$

where C is the ratio τ_y/G at 0 K; for typical polymers, values are given in Table 1, and the average value is 0.08 for amorphous polymers. The shear modulus, G, depends on temperature. The values of the shear and Young's modulus versus temperature for many polymers are given in Ref 3.

The yield point depends on the strain rate. The following equation is a good description of the most usual relationship:

$$\sigma_y = \sigma_0 + \beta \ln \dot{\varepsilon} \quad \text{(Eq 24)}$$

where σ_0 is the yield point at some reference strain rate, and β is a material parameter that must be determined by experiment. Because the yield point increases with strain rate, the ductile-brittle transition temperature also increases with strain rate. Thus, in the vicinity of the ductile-brittle transition temperature, a high strain rate may produce brittle fracture instead of a yield failure if the stress is sufficiently high. A brittle fracture always requires less energy than a yield failure.

Table 2 Yield points at very low temperatures

Polymer	Ref	Test	σ_y/G (experimental), GPa	σ_y/G (experimental), 10^6 psi	Yield criterion	$\mu(d\sigma_y/dP$	τ_y (calculated in Eq 22), GPa	τ_y (calculated in Eq 22), 10^6 psi
Polystyrene	18	Compression	0.30	0.045	Tresca	0.25	0.13	0.020
Polystyrene	19	Compression	0.27	0.040	Tresca	0.25	0.10	0.015
Polymethyl methacrylate	20	Compression	0.82	0.120	von Mises	0.16	0.43	0.060
Polymethyl methacrylate	21	Compression	0.87	0.130	von Mises	0.16	0.45	0.065
Polymethyl methacrylate	22	Compression	0.73	0.110	von Mises	0.16	0.38	0.055
Polycarbonate	23	Shear	0.18	0.025	...	...	0.18	0.025
Polycarbonate	24	Tension	0.26	0.040	von Mises	0.12	0.16	0.020
Polycarbonate	25	Tension and compression	0.30 (average)	0.045 (average)	...	...	...	...
Polyethylene terephthalate	24	Tension	0.24	0.035	von Mises	0.09	0.15	0.020
Polychlorotrifluoroethylene	12	Tension	0.25	0.036	von Mises	0.12	0.15	0.020
Polyethylene, high density	26	Tension	0.20	0.030	von Mises	Below 0.05	0.12	0.017
Polypropylene, isotactic quenched	27	Tension	0.11	0.016	von Mises	0.12	0.068	0.010

σ_y, uniaxial yield point; G, shear modulus; μ, material parameter; P, pressure; τ_y, shear yield point. Source: Ref 17

Effect of Crystallinity

In addition to a T_g, crystalline polymers have a melting point, T_m. Generally, T_g = ½ to ⅔ of T_m in degrees Kelvin. In order to understand the effect of percentage crystallinity on the yield point, it is important to consider whether the temperature of interest is above or below T_g. If the semicrystalline polymer is above T_g, then the yield point increases approximately linearly with percent crystallinity. Basically, the yield strength depends on the volume of the crystalline regions and does not depend greatly on the detailed morphology of the crystals (Ref 11).

Below T_g, the strength of the amorphous region approaches that of the crystalline region and may even exceed it. If the amorphous region has a yield strength equal to that of the crystalline region, the yield point will be independent of crystallinity. This is the case for various polyethylenes where, at approximately 160 K, the yield point is independent of crystallinity (Ref 26).

In the case of a synthetic fluorine-containing resin, as shown in Fig. 5, at approximately 200 K, the state of higher crystallinity has a higher yield point, but at 78 K, the least crystalline state has the higher yield point (Ref 28). This occurs because the amorphous regions have a yield point that is a high fraction of the modulus. However, the crystals do not have an intrinsically high yield strength, because they usually deform by the mechanism of dislocation. In the case of amorphous polymers, the dislocation mechanism of deformation is not expected. Considerations of free volume are more fruitful.

The Aging of Polymers

After a polymer is cooled to room temperature from above T_g, its structure slowly changes with time. The general effect is that the molecules slowly become more closely packed in the amorphous regions. This reconfiguration lowers the average energy of the system. The physical aging phenomenon has been associated with a decrease in free volume (Ref. 29), resulting in a population shift in molecular conformations that are less favorable for shear yielding. The aging process reduces the creep rate, compared to unaged materials. The aging process can be reversed by reheating the polymer close to or above T_g.

If the aged polymer is deformed to a sufficiently high strain on the order of the yield strain, there will be an unpacking of the molecules and reconfiguration of the molecular population in the direction of easier shear. As a result, the subsequent creep rate will be greater than for the aged and undeformed material. It should be noted that a large deformation leads to orientation strengthening.

The process is reversible in that after the deformed state is aged, it tends to approach the aged state that existed before the deformation. A review of the kinetics of aging and its relationship to the thermal and deformation history has been extensively investigated in Ref 29 to 31.

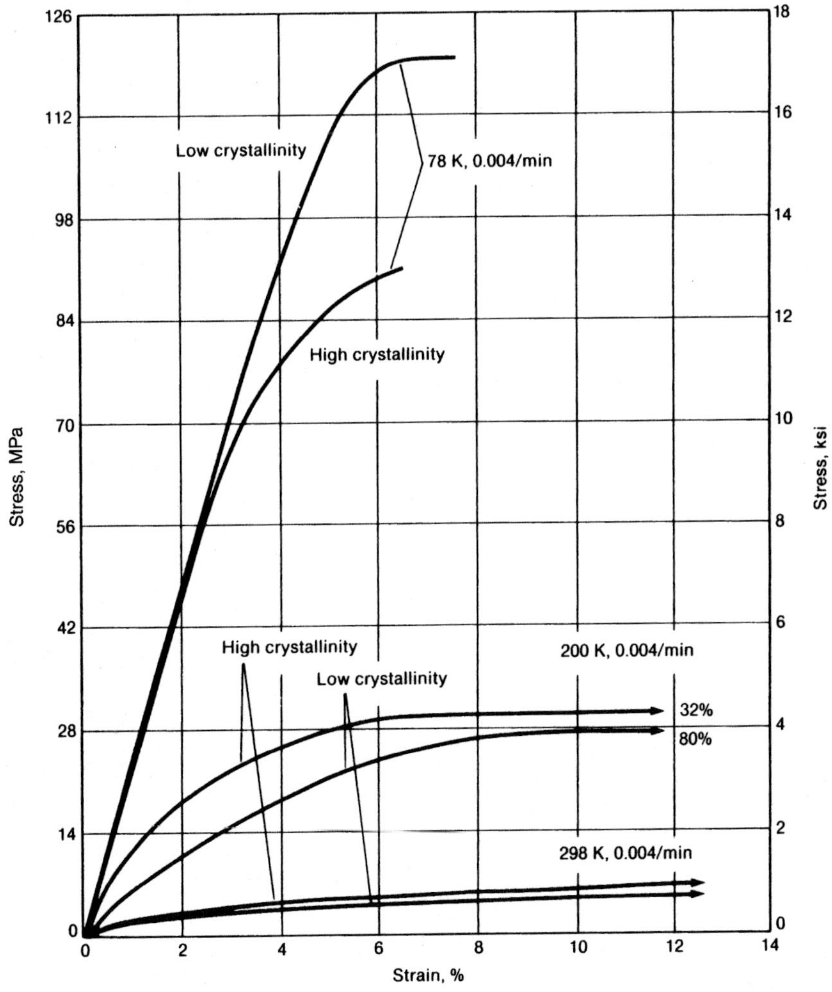

Fig. 5 Stress-strain behavior of a synthetic fluorine-containing resin for high end low crystallinity at various temperatures. Source: Ref 28

REFERENCES

1. R.P. Kambour and R.E. Robertson, Chap. 11, *Polymer Science,* A.D. Jenkins, Ed., Elsevier, 1972
2. J.D. Ferry, *Viscoelastic Properties of Polymers,* 3rd ed., John Wiley & Sons, 1980
3. N.G. McCrum, B.E. Read, and G. Williams, *Anelastic and Dielectric Effects in Polymeric Solids,* John Wiley & Sons, 1967
4. W. Lethersich, *Br. J. Appl. Phys.,* Vol 1, 1950, p 294
5. J.M. Crissman and L.J. Zapas, *Polym. Eng. Sci.,* Vol 19, 1979, p 99, 104
6. T.S. Hin and B.W. Cherry, *Polymer,* Vol 25, 1984, p 727
7. J.M. Crissman and G.B. McKenna, *J. Polym. Sci. Phys.,* Vol 25, 1987, p 1667
8. N. Brown, Yield Behavior of Polymers, *Failure of Plastics,* W. Brostow and R.D. Cornelussen, Ed., Hanser, 1986
9. N. Brown, J. Donofrio, and X. Lu, *Polymer,* Vol 28, 1987, p 1326
10. C.G. Bragaw, in *Eighth Plastic Fuel Pipe Symposium,* American Gas Association, 1983, p 40
11. R. Popli and L. Mandelkern, *J. Polym. Sci. Phys.,* Vol 25, 1987, p 441
12. Y. Imai and N. Brown, *Polymer,* Vol 18, 1977, p 298
13. A.A. Silano and K.D. Pae, in *Advanced Polymer Science and Engineering,* K.D. Pae, D.R. Murrow, and Y. Chen, Ed., Plenum Press, 1972
14. R. Buchdahl, *J. Polym. Sci. A, Polym. Chem.,* Vol 28, 1958, p 239
15. R.E. Robertson, Report 64-RL(3580C), General Electric Company, 1964
16. N. Brown, *Mater. Sci. Eng.,* Vol 8, 1972, p 69
17. N. Brown, *J. Mater. Sci.,* Vol 18, 1983, p 2241
18. P.B. Bowden and S. Raha, *Philos. Mag.,* Vol 22, 1970, p 463
19. J.P. Cavrot, J. Haussy, J.M. Lefebvre, and B. Escaig, *Mater. Sci. Eng.,* Vol 36, 1978, p 95
20. C. Bauwens-Crowet, *J. Mater. Sci.,* Vol 8, 1973, p 968
21. J. Haussy, J.P. Cavrot, B. Escaig, and J.M. Lefebvre, *J. Polym. Sci.,* Vol 18, 1980, p 311
22. P. Beardmore, *Philos. Mag.,* Vol 19, 1969, p 389
23. W. Wu and A.P.L. Turner, *J. Polym. Sci. Phys.,* Vol 13, 1975, p 19
24. J.R. Kastelic and E. Baer, *J. Macromol. Sci. Phys. B,* Vol 7 (No. 4), 1973, p 679
25. C. Bauwens-Crowet, J.C. Bauwens, and G. Holmes, *J. Mater. Sci.,* Vol 7, 1972, p 176
26. E. Kamei and N. Brown, *J. Polym. Sci. Phys.,* Vol 22, 1984, p 543
27. H.G. Olf and A. Peterlin, *J. Polym. Sci. Phys.,* Vol 12, 1974, p 2209
28. S. Fischer and N. Brown, *J. Appl. Phys.,* Vol 44, 1973, p 4322
29. L.C.E. Struik, Ed., *Physical Aging in Amorphous Polymers and Other Materials,* Elsevier, 1978
30. L.C.E. Struik, Ed., Chap. 11, *Physical Aging in Amorphous Polymers and Other Materials,* Elsevier, 1978
31. J. Bauwens, Chap. 12, *Physical Aging in Amorphous Polymers and Other Materials,* L.C.E. Struik, Ed., Elsevier, 1978

Crazing and Fracture*

HIGH-IMPACT PLASTICS, because of their light weight and toughness, are commonly used in structural components. Unfortunately, changes occurring in these materials during their service life can either severely limit their usefulness, requiring replacement, or cause catastrophic failure due to fracture on impact. This deterioration of physical properties is usually associated with a phenomenon known as crazing.

In some cases, the deterioration process can be quite complex. For example, longevity may be severely limited by the combined effect of an applied mechanical stress and a hostile environment. This phenomenon, commonly referred to as stress-corrosion cracking, leads to the transition from a ductile or high-impact material to a glassy or brittle plastic. Analytic criteria for the exposure-time dependence are nonexistent, which precludes conservative design in the use of these materials.

The failure process in stress-corrosion cracking can be viewed as three consecutive events that occur while the material is under a given load in a defined environment. If a critical strain is exceeded, a cavitational phenomenon called crazing occurs in the member (appearing as whitening or opacity in clear plastics). The morphology of crazes has been widely investigated and is reviewed in the literature (Ref 1–4). The crazes then degenerate into a microcrack through fibril breakage, which can easily be differentiated from a craze, because it has no morphology and will not support a load. The microcrack then grows until it reaches a critical size, at which point catastrophic failure takes place. It has been stated that "crack propagation in glassy polymers may be more exactly termed the formation and breaking of crazes" (Ref 5).

General Polymeric Behavior

Glassy thermoplastics are usually described as amorphous polymeric materials. The structure is commonly referred to as an array of random chains. The cohesive energy holding the material together is the result of intermolecular forces of attraction, such as van der Waals forces, hydrogen bonding, and so forth. The physical properties of these polymers are a consequence of their long chain lengths or high molecular weights.

If a rectangular test specimen is strained in tension and the resultant stress is determined, Young's modulus can be determined from the slope of the engineering stress-strain curve. For conventional structural materials such as steel, whose behavior is linear, the modulus can be calculated from the apparent stress and strain at any point as long as it is taken below the proportional limit of the material. Hence, Young's modulus, E, is calculated as:

$$E = \frac{\sigma}{\varepsilon}$$

where σ is the engineering stress, and ε is the tensile strain in the material. If the material is allowed to remain under the applied strain for a long period of time, the stress remains constant. As a result of this behavior, Young's modulus for steel is generally taken to be time independent.

On the other hand, if a glassy thermoplastic is strained, a quite different behavior is observed. The resultant stress decreases as a function of time. The apparent modulus, calculated as the ratio of the stress to the strain at a given point, also varies as a function of time. The time-dependent Young's modulus, $E(t)$, is found from the relationship:

$$E(t) = \frac{\sigma(t)}{\varepsilon_0}$$

where $\sigma(t)$ is the time-dependent stress, and ε_0 is the constant applied strain. This behavior, commonly referred to as stress relaxation, is a result of the polymer molecules sliding past each other, which allows the stress to decrease under the applied strain.

Figure 1 is a plot of the logarithm of the stress relaxation modulus, $E(t)$, as a function of log time (or temperature) for an amorphous thermoplastic material and shows four distinct types of behavior. The first region is the glassy plateau, in which the plastic shows glassy or brittle behavior. Next is the transition region, which is the range of times (or temperatures) over which the plastic exhibits tough or leathery behavior. In the third region, commonly referred to as the rubbery plateau, the material exhibits soft or elastomeric behavior. Last is the flow region, in which the plastic behaves like a liquid. The dashed line at the end of region 3 indicates the effect of lightly cross linking the material. Chemically cross linking the material prevents flow, which in turn obviates region 4.

The importance of Fig. 1 is that this type of mechanical response is general for all amorphous polymers and is often referred to as viscoelastic behavior. By changing the time scale or the temperature of the experiment, the apparent modulus of the material is observed to change. These changes are predictable using the time-temperature superposition principle, and the shift (factor) produced by changing temperature can be calculated using the Williams-Landel-Ferry equation (Ref 6). This allows plastics engineers to predict the modulus of the material, which in turn allows them to design against failure due to elastic instability (creep, stress relaxation, and so forth).

Ductile-Brittle Transitions

A well-known experiment in polymer science is to take a rubber ball and bounce it off the wall. The material in the ball exhibits the behavior shown in region 3 (rubbery) of Fig. 1. The temperature of the ball is then lowered by immersing it in liquid nitrogen. After the ball is allowed sufficient time to cool, it is thrown against a wall. In this case, the ball is observed

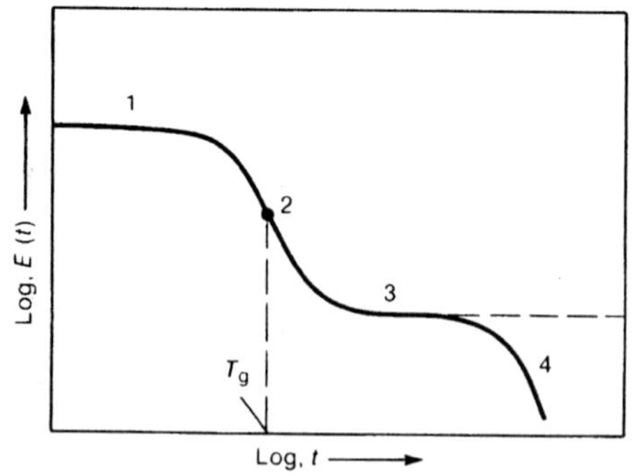

Fig. 1 Log relaxation modulus, $E(t)$, as a function of log time, t (or temperature), for a glassy thermoplastic material. 1, glassy plateau; 2, transition region; 3, rubbery plateau; 4, flow region. T_g, glass transition temperature

*Adapted from the article by Stephen P. Petrie, "Crazing and Fracture," in *Engineering Plastics*, Volume 2, *Engineered Materials Handbook*, ASM International, 1988, pages 734 to 740

to shatter or show region 1 (glassy) behavior. By changing the temperature of the experiment, one produces the same effect as observed by changing time. From the observed response, it appears that the material behavior curve slides along the x-axis from left to right when the temperature is lowered.

The material property of importance in this experiment is the glass transition temperature, T_g, of the plastic, which is shown in Fig. 1. If the ambient temperature of the material is above T_g, the behavior will be rubbery. On the other hand, if the ambient temperature is well below the T_g, the observed response will be brittleness. This relationship holds for all amorphous polymers. Some T_gs for common polymers are given in Table 1 (Ref 6).

In order to get a better understanding of ductile-brittle transitions, it is helpful to examine the stress-strain behavior of a ductile plastic from initial loading to fracture. Figure 2 shows the stress-strain behavior for a ductile plastic. As the strain is increased, the stress rises (nonlinearly) to a point at which there appears to be a change in behavior. Beyond a critical strain, the stress appears to decrease with increasing strain. This critical strain is referred to as the yield strain, ε_y, and the corresponding stress is the yield stress σ_y, as indicated in Fig. 2.

In a tensile test, at this critical strain, one would observe a "neck," or gross deformation in the specimen, produced by the localized flow of material. In a structural component, this would produce failure due to the permanent deformation produced in the member. Plastics engineers generally use a design stress that is well below the yield stress of the material in order to prevent failure from yielding or shear flow.

As the strain is increased, the material continues to draw or flow, and the stress again increases up to a point at which the specimen breaks or fractures, which is shown by "X" in the figure. The stress at the break, σ_b, and the strain at the break, ε_b, are frequently used as materials specifications. However, because most failures are not produced in this manner, these values are of little use in design applications.

Another parameter that can be obtained from the stress-strain behavior is the modulus of toughness, M_T, which is the area under the stress-strain curve. This can be calculated from the relationship:

$$M_T = \int \sigma \cdot d\varepsilon$$

which is integrated from zero strain to the strain at the break. The modulus of toughness represents the work, or energy, on a per volume basis to deform and fracture the material. Like the breaking strength, the modulus of toughness is useful in material comparisons but has little use in design.

Figure 3 illustrates the typical stress-strain relationship for a ductile plastic, showing the effect of temperature and strain rate. As the strain rate is increased (or the temperature is decreased), the properties of the material change quite drastically. These effects produce an increase in the yield stress of the material to the point at which the material no longer yields but becomes brittle. In other words, the behavior of the material has undergone a ductile-brittle transition, as evidenced by the loss of the yield point. Instead of the material failing by means of a shear flow mechanism, it fails because of brittle crack propagation.

In addition, the area under the stress-strain curve (the modulus of toughness) decreases drastically with increasing strain rate or decreasing temperature, indicating much more brittle behavior (or lower toughness values) under these conditions. These changes in the mechanical properties of the polymer are similarly associated with the viscoelastic nature of polymeric materials.

Because brittle fracture occurs more frequently than ductile failure, more research has been done in this area. Glassy thermoplastic materials, such as polystyrene (PS) and polymethyl methacrylate (PMMA), are well below their T_gs at room temperature. Under most test conditions, this leads to a brittle type of failure, which in turn facilitates the study of the brittle-fracture process. In addition to being brittle, both PS and PMMA are transparent and have good optical properties. This makes it possible to see the nucleation and growth of defects clearly. The observations made by scientists and researchers in the past have led to the current knowlege of crazing and its role in the fracture process.

Table 1 Glass transition temperatures for selected plastics

Material	Glass transition temperature °C	°F
PS	101	215
PMMA	107	225
PVC	78	170
PC	150	300

PS, polystyrene; PMMA, polymethyl methacrylate; PVC, polyvinyl chloride; PC, polycarbonate

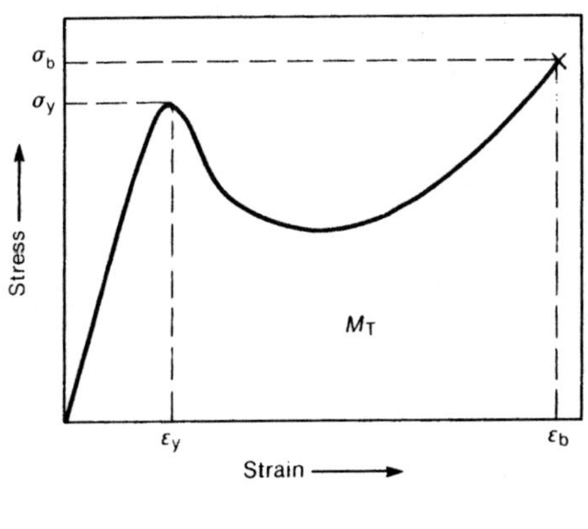

Fig. 2 Stress-strain behavior of a ductile plastic. M_T, modulus of toughness; σ_b, stress at the break; σ_y, yield stress; ε_y, yield strain; ε_b, strain at the break

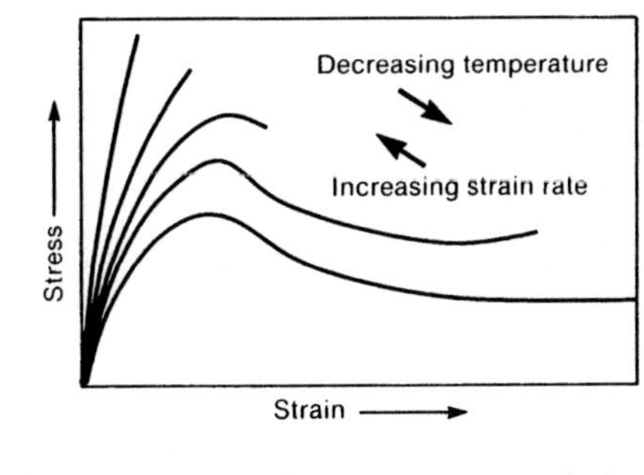

Fig. 3 Typical tensile stress-strain curves of a ductile plastic, showing the effect of strain rate and temperature

Crazing

Because crazes have a different refractive index and because they reflect light, they are easily visible, especially when viewed at the correct angle with the aid of a directed light source, such as a fiber-optic illuminator. When viewed with this type of light source, the craze appears to have a silvery appearance much like a very fine crack. The shape of a craze is depicted in Fig. 4.

Although crazing has been observed for many years in plastics, the phenomenon was often referred to as craze cracking, or simply, cracking. Sauer, Martin, and Hsaio were the first to distinguish between crazing and cracking (Ref 7). In their study with PS, the investigators observed that although the entire cross section of the specimen crazed, it did not fracture. As shown in Fig. 5(a), the specimen was still able to support a load, compared to the case in Fig. 5(b) in which a specimen cracked and could not support a load. From their observation, it became apparent that a craze contains polymeric material, while a crack does not. Distinct molecular orientation was also observed in the craze structure by means of x-ray analysis of the material. Sauer et al. concluded that crazing is a mode of

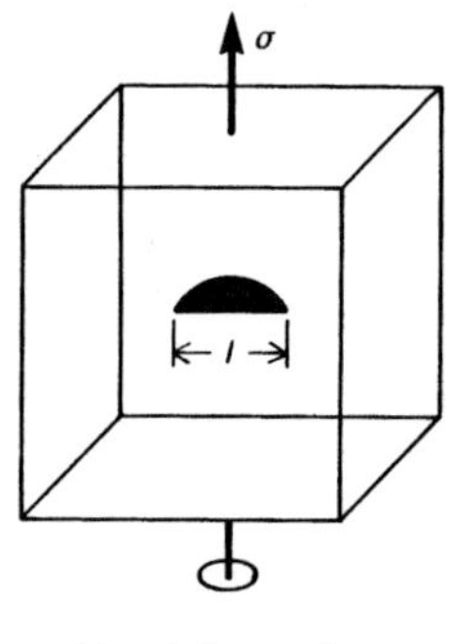

Fig. 4 Craze of length l grown in a transparent plastic under stress, σ

plastic deformation rather than mechanical cracking.

The deformation during crazing is constrained laterally; that is, it resists Poisson contraction. The density of the material decreases with increasing strain. The volume fraction of polymer in the craze, V_f, has been estimated (Ref 8) to be:

$$V_f = \frac{1}{1 + \varepsilon}$$

where ε is the tensile strain, which can be in the range of 40 to 60% (Ref 9).

In order to gain information about the structure or morphology of the craze, liquid sulfur has been used to craze poly (2,6-dimethyl 1,4-phenylene oxide) (PPO) in the bulk (Ref 10). The samples were quenched to room temperature under strain to prevent contraction of the craze structure. The sulfur does not affect the structure of the craze and allows it to be sectioned with minimal damage, due to its effect as a reinforcing agent. Transmission electron microscopy of the sections revealed an interconnected network of spheroids from 10 to 20 nm (100 to 200 Å) in diameter. Similarly, an iodine-sulphur eutectic has been successfully used to impregnate the craze structure in PS but was found to deteriorate the structure of crazes in polycarbonate (PC) through plasticization (Ref 11).

Researchers (Ref 12) studied the crazing of PS in bulk and thin-film forms without the aid of impregnants. The polymer fraction of the craze was observed to consist of fibrils (20 to 40 nm, or 200 to 400 Å in diameter) that extended across the craze and were oriented normal to the interface. The transition from polymer matrix to crazed structure was sharp and well defined (Ref 10, 12). Researchers examined the microstructure of solvent-induced crazes in thin films of PC and found similar results (Ref 13). From the previous results obtained by several researchers, the physical structures of crazes formed under different conditions were found to be quite similar. A physical model for the craze structure could be two sections of bulk polymer interconnected by a system of oriented polymer fibrils (Fig. 6).

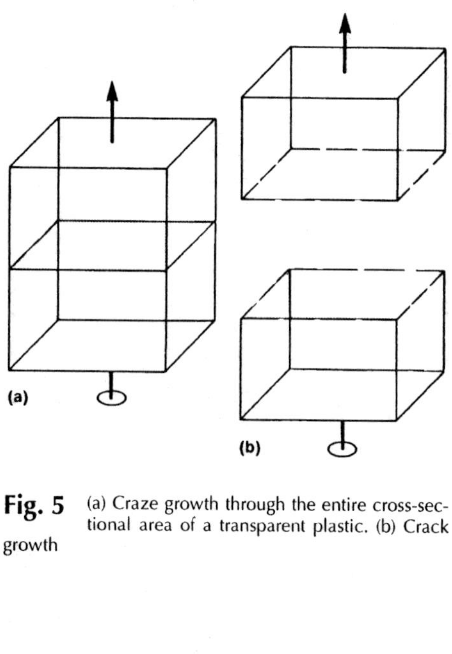

Fig. 5 (a) Craze growth through the entire cross-sectional area of a transparent plastic. (b) Crack growth

Fracture

The brittle fracture of glassy thermoplastics has been the subject of many studies. Several workers have investigated the growth of crazes to gain an insight into the mechanism of crack nucleation and growth. Evidence of void coalescence caused by fibril failure in PS is reported in Ref 12. Newton's ring formation in crazes ahead of a propagating crack in PMMA has been reported (Ref 14). The rings were thought to originate from preexisting flaws or inhomogeneities in the craze. The rings, or secondary fractures, grew radially until the main crack engulfed them. Researchers (Ref 15, 16) investigated the nucleation of cracks in preformed crazes in PS using visible light microscopy. As the craze was strained, small voids were observed to develop. As the craze was further strained, the voids began to coalesce and form a cavity within the craze. At a critical strain level, the cavity began to propagate and subsequently merged with other propagating cavities. Microscopic evidence of craze formation was found ahead of a propagating crack in PMMA (Ref 1).

Indirect evidence for the importance of crazing in the brittle fracture of glassy thermoplastics has been obtained by examining the fracture surfaces, using a variety of techniques. By using visible light microscopy, a highly oriented layer of polymeric material was found on the fracture surface of PMMA (Ref 17). It was found that the fracture surface of PMMA scattered small-angle x-rays in a similar manner as a craze in the bulk material (Ref 18). The refractive index of the material on the fracture surface was the same as that of the craze (Ref 19). These observations were subsequently extended to other glassy polymers, and it was concluded that the fracture surfaces are essentially the residual layers of crazed regions (Ref 5). From both direct and indirect results, crazing has been shown to play an important role in the brittle fracture of glassy thermoplastics in that it raises crack propagation energy by a factor of 10^3 or 10^4 versus a noncrazing glassy resin.

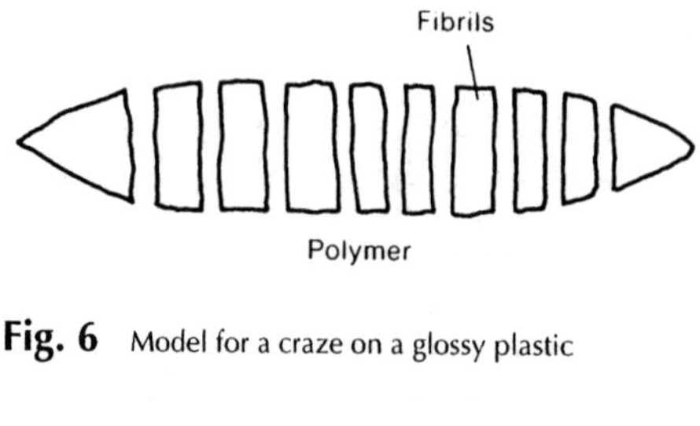

Fig. 6 Model for a craze on a glossy plastic

Environmental Effects

Although crazing was first reported more than 45 years ago, and many investigators have spent considerable effort working in this area, there is still much to be learned about the growth and structure of crazes, especially in hostile environments, such as organic solvents. It was reported that liquid nitrogen has an effect on the crazing of PC at 78 K, as opposed to an inert atmosphere, such as helium (Ref 20). Researchers have also shown that carbon dioxide has an effect on the crazing of PC (Ref 21).

There are two main theories to explain the ability of a fluid to cause crazing in a glassy thermoplastic. The two mechanisms are plasticization and wetting. The former theory, first proposed by Maxwell and Rahm (Ref 22), states that the role of the organic solvent is to plasticize and swell the polymer, thus reducing the T_g and resistance to flow. Because of the abundance of strong experimental evidence, this is the more accepted hypothesis today (Ref 2). The latter theory states that the liquid wets the surface of the polymer (Ref 23, 24), causing a reduction in the energy required for craze formation. The main evidence given for this mechanism is that crazing occurs even though only the slightest absorption of solvent can be detected in the polymer. However, a researcher reported that a small amount of absorption of dimethyl formamide could be detected in PPO when the polymer was stressed, and concluded that the observed crazing could be explained by the plasticization theory (Ref 2). On the other hand, it may be that the combined effect of an applied stress and solvent may be required when the liquid is not "hostile" enough to cause crazing by itself.

Initiation Criteria

In 1949, it was suggested that a critical strain of 0.75% must be obtained before crazing could occur in PS at room temperature (Ref 22). Because, for a given polymer, crazing is a function of the applied stress, strain, strain rate, time, temperature, and environment, it is not surprising that other investigators have suggested different criteria.

Several parametric descriptions of craze initiation have been given in the literature, using stress, strain, and dilation as the essential criteria (Ref 22, 25, 26). It has been suggested that the stress-intensity factor, rather than the applied stress, is the required criterion (Ref 27). It has also been established that crazes form in cyclic loading at stress amplitudes well below the static threshold level (Ref 28).

Recognizing that cavitation occurs during the crazing process, it has been suggested (Ref 26) that the dilative component of the stress must be involved in the initiation. Researchers (Ref 25) concluded from their crazing experiments with PMMA under biaxial stress conditions that the combination of a flow stress and a dilative stress is required for craze initiation. Using PS that was stressed in the tension-compression quadrant, researchers (Ref 29) obtained similar results, which they related to strain in the tensile direction.

Researchers (Ref 30) used hot, stretched specimens of PS and PMMA, which were cut at different angles to the draw direction ($\Theta = 0$). Their results showed that the ease of craze formation was inversely proportional to the orientation. They postulated that the dependence of the craze orientation on the angle, Θ, rules out the principal stress as the criterion, and that the crazes grow along resultant strain trajectories.

Researchers (Ref 31) used the inclusion of a steel ball in a PS matrix to effect a separation of the principal axes. A study of the resultant polar angle at which crazing occurred revealed that the principal strain gave the best fit. Other experiments with one or two rubber balls (softer inclusions) again showed the principal strain criterion to be the most plausible (Ref 32). The investigators concluded that the principal strain was the essential criterion for dry crazing.

Researchers (Ref 33) obtained similar results in their study of the crazing of PC in ethanol. A plate of PC containing a cylindrical inclusion of steel was tested in a tank containing the alcohol. It was found that the crazes nucleated at an angle that corresponded to the maximum in the principal strain. From the stress analysis of the composite and the experimental results, the investigators also concluded that the direction of craze growth was perpendicular to the direction of the principal strain.

Craze Growth

In transparent applications, such as canopies for aircraft, crazing can present severe problems. Because crazes reflect and scatter light, they can produce haze in the transparency. This forward scattering of light can decrease visibility to a point that renders the material no longer useful. Both the rate of nucleation and the rate of craze growth are important in determining the amount of haze that will result. It is important to note that crazing has caused the replacement of canopies, in most cases, before any cracking was observed. For this reason, there is much interest in craze kinetics as a tool to predicting the lifetime of the in-service transparent materials.

The kinetics of craze growth are not well cataloged and are poorly understood (Ref 2). In fact, with the exception of few studies, most investigators have measured only the growth of one dimension (length) with time, even though crazes are three dimensional. Light reflection was used to measure crazing (Ref 22, 34), but, unfortunately, it was not possible with this method to differentiate between craze initiation and growth. Researchers (Ref 35) simultaneously studied the areal growth and change in craze thickness with time. From the resulting craze profiles, a blunting of the craze tip with time was observed. Double-exposure holographic interferometry was used to investigate the craze opening displacement profile of solvent crazes grown in PS from preexisting cracks (Ref 36).

Most investigators have studied craze growth by measuring the length as a function of time. It was found that in PMMA, the craze length, l, was proportional to the logarithm of time if the induction time, t_0, for the craze to appear was taken into account (Ref 37). Hence:

$$l = \mathrm{K} \log\left(\frac{t}{t_0}\right)$$

where K is a proportionality constant. A linear relationship was also found between the craze growth rate and stress:

$$\frac{dl}{dt} = \frac{1}{\mathrm{m}}(\sigma - \sigma_0)$$

where σ is the applied stress, σ_0 is the threshold stress required for crazing to occur, and m is a constant (Ref 38).

In order to localize the craze growth in the material, several investigators have used a fracture mechanics approach to grow crazes from a notch or sharp crack. Researchers (Ref 39) employed a viscoelastic model to describe the craze growth rate from a sharp notch for a specimen of PC immersed in kerosene. The crazing of PMMA from a sharp notch in methanol was investigated (Ref 27). Researchers (Ref 40) measured the kinetics of craze growth from a fatigue crack in an ASTM E 399 (Ref 41) compact tension specimen. The data for the growth of crazes in PC in ethanol fit quite well with the model proposed in Ref 27. Below a level of critical stress-intensity factor, K_c, craze growth appeared to be controlled by end diffusion, and fracture was prevented by craze arrest. Above K_c, craze growth was controlled by either end flow or the combination of end flow and side flow. In both cases, the final result was fracture. As the specimen thickness was increased, side flow was precluded because of the transition from plane stress to plane strain.

Researchers (Ref 40) also found that the craze growth rate in thin specimens (3.2 mm, or ⅛ in.) was qualitatively predictable in terms of the state of stress. As the specimen thickness was increased, causing a transition from plane stress to plane strain, the state of stress had a diminishing effect on the craze growth rate, as predicted. Linearly elastic fracture mechanics was found to be a good tool for evaluating the growth of flaws.

Effect of Crazing on Toughness

The plane-strain fracture toughness, K_{Ic}, of PC was measured as a function of thickness and thermal history (Ref 40). A k-level statistical design was used to evaluate the effects and any possible interaction that might occur. The effect of thickness was found to be the most important of the variables. As the thickness was increased to 12.7 mm (½ in.) (plane strain), the effect of thermal history was found to be insignificant.

In subsequent experiments, the K_{Ic} was determined as a function of stress level and time of exposure in solvent. A k-level statistical design was again used to evaluate the significance of the effects. The effect of the interaction between the stress and solvent was the most significant, which indicates statistically that the mechanism is one of stress corrosion, rather than of stress or solvent alone.

Two opposing effects were noted in the K_{Ic} values. The solvent initially appeared to toughen the polymer, a result that can be explained by a plasticization effect, which eases flow. The K_{Ic} appears to have a maximum value as a function of time of exposure at a given K value. Beyond this time, t_{max}, the K_{Ic} decreases, and brittle failure eventually occurs.

Testing for Brittle Behavior

Design and materials engineers are often concerned with the selection of a plastic for a given application. This process can be quite complex, especially when designing a particular part against fracture. While it is possible to design against deformation and yielding using material properties such as Young's modulus and the yield strength, a design to prevent crack propagation is far more difficult to bring about, because no single material property exists on which to do the calculation.

As previously discussed, the glass transition temperature, T_g, is often used to predict whether or not a material will exhibit brittle behavior. A well-known exception is the case of PC. Although the T_g of PC is 145 °C (295 °F), which is well above room temperature, the material exhibits ductile behavior. The explanation given for this is that the shear yield stress is lower than the craze stress. A plot of the shear yield stress and the craze stress as a function of temperature

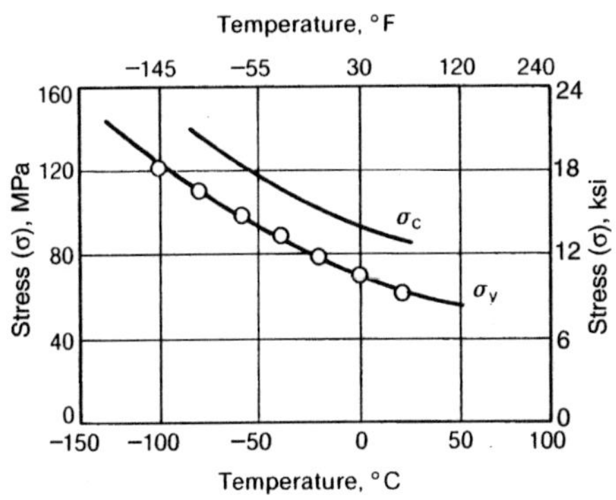

Fig. 7 Variation of yield and craze stress (σ_y and σ_c, respectively) with temperature. Source: Ref 42

(at a fixed strain rate) is given in Fig. 7. (The shear yield stress of PC at room temperature is approximately 60 MPa, or 8.7 ksi.) From the plot, it is apparent that as the material is strained, it reaches the yield stress first, and therefore, yield occurs before crazing. A similar plot for the behavior of PMMA would show that the craze stress is below that of the yield stress. PMMA shows brittle behavior at room temperature. From these observations, it appears that the stress (or strain) to shear and yield (in a given time scale) are better criteria to predict ductile or brittle behavior.

Test to Determine Stress-to-Craze Value. A simple test fixture (Ref 43), shown in Fig. 8, can be used to determine the minimum stress required to produce crazing. A rectangular test specimen is loaded in cantilever bending to produce a tensile stress across its top surface. The specimen can be allowed to creep for a given time and then be examined microscopically to determine whether crazing has occurred. Similarly, a specimen may be tested with a standard liquid on its surface to determine the craze resistance of a material to a given environmental factor for a given time. This method was used (Ref 44) to determine the lowest stress to craze Plexiglas-55 (ATOFINA Chemicals, Inc.) in a variety of test fluids for both dry and water-saturated acrylic. The results of this work are given in Table 2. It is interesting to note that no external stress was required to craze the water-saturated acrylic in isopropanol or *n*-octanol. The results of this type of test are useful to the engineer not only to determine craze resistance but also to determine environmental effects (moisture).

Test to Determine Strain-to-Craze Value. A simple bending fixture, which has proved to be quite useful in screening materials, is shown in Fig. 9 (Ref 45). By turning the screw of a given pitch, a given displacement, δ, is obtained. The maximum strain at the fulcrum, ε_{max}, is calculated as:

$$\varepsilon_{max} = \frac{6\delta t}{l^2}$$

where t and l are the thickness of the sample and length of the span, respectively. This apparatus has the advantage over commercially available elliptical fixtures in that the strain is variable rather than fixed.

Researchers used the fixture (Ref 45) to determine the critical strain to craze for PMMA and two other plastics in a variety of liquids having different solubility and hydrogen bonding parameters. When the investigators tabulated the results, they distinguished liquids that were solvents from those that were swelling agents, and they distinguished crazing from cracking. As previously discussed, this type of fixture allows stress relaxation in the material as a function of time.

Fracture Toughness Testing

For the brittle fracture mode, two other physical properties that are useful in the selection of materials are the impact strength and plane-strain fracture toughness of the plastic.

Impact Strength. The ASTM D 256 impact test (Ref 46) is the most widely used test for evaluating the toughness of plastics. In the test, a standardized machine, illustrated in Fig. 10, is used. Two geometries are possible, but the Izod type is almost universally used with plastics. In the test, a hammer of defined weight and fixed height falls and hits the sample at a given velocity. If the test progresses properly, the cantilever bending of the specimen starts a crack, which propagates through the thickness, causing the specimen to break or fracture. The broken piece of the sample is then tossed, and the hammer comes to a final rest. From the difference between the heights at start and final rest positions (or from a scale on the commercial testing instrument), the energy required to break the sample can be obtained. This energy is normalized on a thickness basis, giving the energy per

Table 2 Lowest stress to craze Plexiglas-55
Manufactured by ATOFINA Chemicals, Inc.

	Polymer condition			
	Water-saturated		Dry	
Test fluid	MPa	ksi	MPa	ksi
Water	32	4.6	79	11
Water/isopropanol 1/1	4	0.6	42	6.1
Isopropanol	0	0	25	3.6
n-octanol	0	0	34	4.9
Ethylene glycol	5	0.7	38	5.5
Iso-octane	26	3.8	59	8.6
Lubricating oil(a)	22	3.2	58	8.4

(a) Turbine engine lubricating oil. Jet oil II. MIL-L-23699B

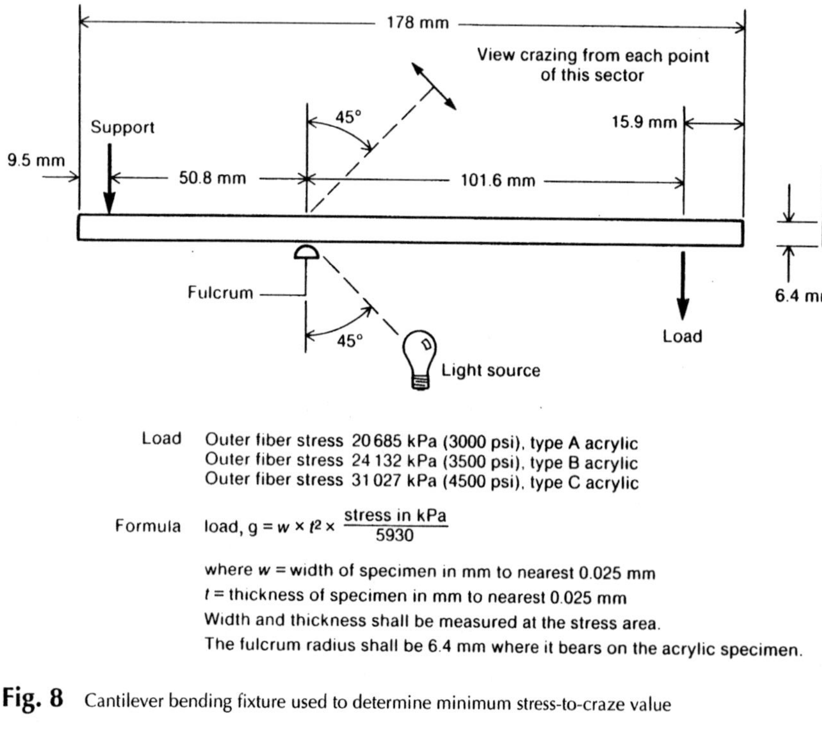

Fig. 8 Cantilever bending fixture used to determine minimum stress-to-craze value

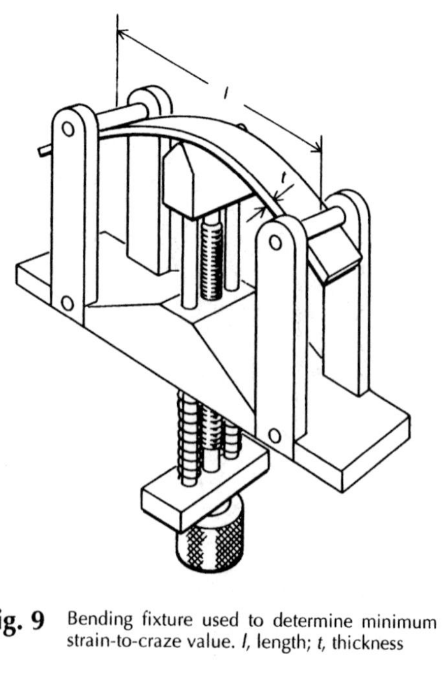

Fig. 9 Bending fixture used to determine minimum strain-to-craze value. *l*, length; *t*, thickness

thickness "of notch" in J/m (ft·lbf/in.). (If the dimensions of thickness are cancelled out, the origin of the term *impact strength* becomes obvious.) To produce a brittle failure in the material, a notch of defined geometry is machined into the specimen. This stress concentration leads to a hydrostatic tension at the notch, which in turn enhances cavitation and crazing (the start of brittle failure) instead of shear yielding.

Unfortunately, the results of this test are of a comparative nature and are not directly useful in design. However, the test results can be quite useful to the engineer in selecting materials. Impact strengths for some common plastics are given in Table 3 (Ref 46).

The impact test can also be used to evaluate the effects of thickness, temperature, orientation, and crystallinity (Ref 46). A very important practical aspect of this test is that it allows the engineer to evaluate the effect of molding a defect, such as a hole, into the material. By reversing the specimen in the vise, a comparison can be made between the notched and unnotched strengths. By machining notches of different radii, notch sensitivity can be determined. A plot of impact strength as a function of the radius of the tip of the notch for different plastics is given in Fig. 11. The slope of the resultant line is an indication of the notch sensitivity of the material. From the results shown in the figure, it appears that plastics differ widely in behavior.

Plane-strain fracture toughness testing is another test that indicates the fracture toughness, K_{Ic}, of a material. The test is based on fracture mechanics concepts and involves the use of standard specimens subjected to given rates of loading at a defined temperature. In a standard test (Ref 41), the specimen contains a very sharp notch (a fatigue crack) to enhance brittle failure. The K_{Ic} is a function of the fracture load and the length of the crack at failure. A critical stress-intensity factor, K_Q, is calculated from these parameters, which are determined in the test procedure. The K_Q must then be qualified as a valid K_{Ic}. If the K_Q does not meet the criteria, the thickness of the material is increased until the K_Q does meet the criteria. Unfortunately, with ductile plastics such as PC, this requires thicknesses in the order of 12.7 mm (½ in.) to meet the conditions given in the test procedure. For this reason, other test methods are used to obtain K values.

Some K_{Ic} values that have been reported are listed in Table 4. These values represent the resistance of the polymer to crack propagation (toughness) and are useful as a guide in design.

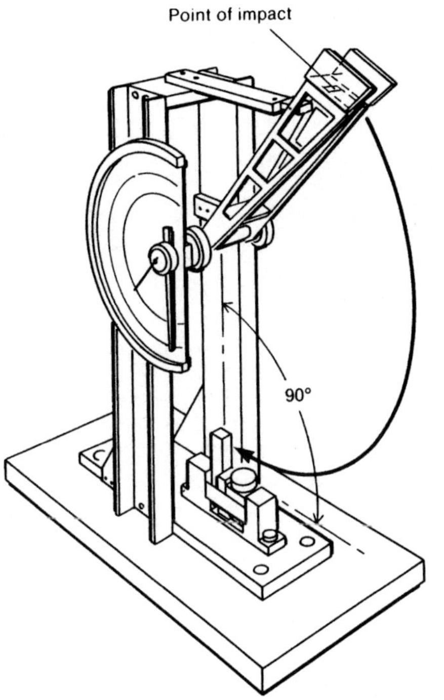

Fig. 10 Izod impact tester

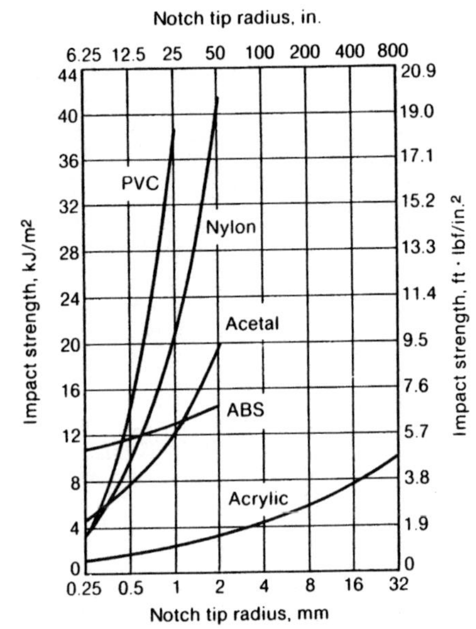

Fig. 11 Impact strength as a function of the radius of the tip of the notch for different polymers. PVC, polyvinyl chloride; ABS, acrylonitrile-butadiene-styrene

Table 3 Notched Izod impact strength of rigid plastics at 24 °C (75 °F)

Plastic	Impact strength, notched J/m	Impact strength, notched ft · lbf/in.
Polystyrene (PS)	13.3–31.3	0.25–0.40
Cellulose acetate	53.3–298	1.0–5.6
Cellulose nitrate	267–373	5.0–7.0
Ethyl cellulose	187–320	3.5–6.0
Nylon 6/6	53.3–160	1.0–3.0
Nylon 6	53.3–160	1.0–3.0
Polyoxymethylene (POM)	107–160	2–3
Polyethylene (PE), low density	>853	>16
PE, high density	27–1070	0.5–20
Polypropylene (PP)	27–107	0.5–2
Polyvinyl formal (PVF)	53.3–1070	1–30
Phenol-formaldehyde (PF), general purpose	13.3–18.7	0.25–0.35
PF, cloth filled	53.3–160	1–3
PF, glass fiber filled	530–1600	10–30
Polytetrafluoroethylene (PTFE)	107–215	2.0–4.0
Nylon 6/12	53.3–74.6	1.0–1.4
Nylon 1	96	1.8
Polyphenylene oxide (PPO), 25% glass fibers	74.6–80	1.4–1.5
Polyester, glass fiber filled	2–20	1.8–8.2
Epoxy, glass fiber filled	10–30	9.1–27.3
Polyimide (PI)	0.9	0.82

Table 4 Plane-strain fracture toughness (K_{Ic}) values at 20 °C (68 °F)

Material	K_{Ic}, MPa$\sqrt{m}$	K_{Ic}, Ksi$\sqrt{in.}$
Polymethyl methacrylate (PMMA)	0.7–1.6	0.6–1.5
Polystyrene (PS)	0.7–1.1	0.6–1.0
Polycarbonate (PC)	2.2	2.0
Polyether sulfone (PESV)	1.2	1.1
High-impact PS	1.0–2.0	0.91–1.8
Acrylonitrile-butadiene-styrene (ABS)	2.0	1.8
Polyvinyl chloride (PVC)	2.0–4.0	1.8–3.6
Polypropylene (PP)	3.0–4.5	2.7–4.1
Polyethylene (PE)	1.0–6.0	0.91–5.5
Polyoxymethylene (POM) (acetal)	≃4	≃3.6
Nylon	2.5–3.0	2.3–2.7
Epoxy	0.6	0.55
Polyester	0.6	0.55
Polyethylene terephthalate (PET)	≃5	≃4.6

REFERENCES

1. S. Rabinowitz and P. Beardmore, Craze Formation and Fracture in Glassy Polymers, *Critical Reviews in Macromolecular Science,* Vol 1, E. Baer, Ed., CRC Press, Chemical Rubber Company, 1972
2. R.P. Kambour, *J. Polym. Sci., Macromol. Rev.,* Vol 7, 1973, p 1
3. E.J. Kramer, Crazing in Polymers, *Advances in Polymer Science,* Vol 52/53, H.H. Kausch, Ed., Springer Verlag, 1983
4. R.P. Kambour, *Encyclopedia of Polymer Science and Engineering,* Vol 4, 2nd ed., John Wiley & Sons, 1986
5. R.P. Kambour, *J. Polym. Sci.,* Vol 4, Part A-2, 1966, p 17
6. J.J. Aklonis and W.J. MacKnight, *Introduction to Polymer Viscoelasticity,* 2nd ed., Wiley-Interscience, 1983
7. J.A. Sauer, I. Martin, and C.C. Hsiao, *J. Appl. Phys.,* Vol 20, 1949, p 507
8. R.P. Kambour and R.E. Robertson, The Mechanical Properties of Plastics, *Polymer Science, A Materials Science Handbook,* A.D. Jenkins, Ed., North Holland, 1972
9. R.P. Kambour, *J. Polym. Sci.,* Vol 2, Part A-2, 1964, p 4159
10. R.P. Kambour and A.S. Holik, *J. Polym. Sci.,* Vol 7, Part A-2, 1969, p 1393
11. R.P. Kambour and R.R. Russell, *Polymer,* Vol 12, 1971, p 237
12. P. Beahan, M. Bevis, and D. Hull, *Philos. Mag.,* Vol 24, 1971, p 1267
13. E.L. Thomas and S.J. Israel, *J. Mater. Sci.,* Vol 10, 1975, p 1603
14. R.P. Kambour, *J. Polym. Sci.,* Vol 8, Part A-2, 1970, p 583
15. J. Murray and D. Hull, *Polymer,* Vol 10, 1969, p 451
16. J. Murray and D. Hull, *J. Polym. Sci.,* Vol 8, Part A-2, 1970, p 583, 1521
17. J.P. Berry, *J. Polym. Sci.,* Vol 50, 1961, p 107
18. R.P. Kambour, *J. Polym. Sci.,* Part A, 1965, p 1713
19. R.P. Kambour, *J. Polym. Sci.,* Vol 2, Part A, 1964, p 4165
20. M.F. Parrish and N. Brown, *J. Macromol. Sci. Phys.,* Vol 138 (No. 3–4), 1973, p 655
21. W.V. Wang and E.J. Kramer, *Polymer,* Vol 23, 1982, p 1667
22. B. Maxwell and J.F. Rahm, *Ind. Eng. Chem.,* Vol 41, 1949, p 1988
23. H.A. Stuart, G. Markowski, and D. Jeschke, *Kunststoff,* Vol 54, 1964, p 618
24. R.L. Bergen, Jr., *SPE J.,* Vol 24 (No. 3), 1968, p 77
25. S.S. Sternstein and L. Ongchin, *Polym. Prepr.,* Vol 10 (No. 2), 1969, p 1117
26. R.P. Kambour, *Polymer,* Vol 5, 1964, p 143
27. G.P. Marshall, L.E. Culver, and J.G. Williams, *Proc. R. Soc. (London) A,* Vol 319, 1970, p 165
28. S. Rabinowitz, A.R. Krause, and P. Beardmore, *J. Mater. Sci.,* Vol 8, 1973, p 11
29. R.J. Oxborough and P.W. Bowden, *Philos. Mag.,* Vol 28, 1973, p 547
30. P. Beardmore and S. Rabinowitz, *J. Mater. Sci.,* Vol 10, 1975, p 1763
31. T.T. Wang, M. Matsuo, and T.K. Kwei, *J. Appl. Phys.,* Vol 42, 1971, p 4188
32. M. Matsuo, T.T. Wang, and T.K. Kwei, *J. Polym. Sci.,* Vol 10, Part A-2, 1972, p 1085
33. J. Miltz, A.T. DiBenedetto, and S. Petrie, *J. Mater. Sci.,* Vol 11, 1978, p 1427
34. Y. Sato, *Kobunshi Kagaku,* Vol 23, 1966, p 69
35. M.I. Bessanov and E.V. Kuvshinskii, *Sov. Phys. Solid State,* Vol 3, 1969, p 950
36. H.G. Krenz, D.G. Ast, and E.J. Kramer, *J. Mater. Sci.,* Vol 11, 1976, p 2198
37. U.R. Regel, *Tech. Phys. (USSR),* Vol 1, 1956–1957, p 353
38. J.A. Sauer and C.C. Hsiao, *Trans. ASME,* Vol 75, 1953, p 895
39. M. Kitagawa and K. Motomura, *J. Polym. Sci., Polym. Phys. Ed.,* Vol 12, 1974, p 1979
40. S. Petrie, A.T. DiBenedetto, and J. Miltz, *Polym. Eng. Sci.,* Vol 20, 1980, p 385
41. "Standard Test Method for Plane-Strain Fracture Toughness of Metallic Materials," E 399, *Annual Book of ASTM Standards,* American Society for Testing and Materials
42. J.G. Williams, *Fracture Mechanics of Polymers,* John Wiley & Sons, 1984
43. "Standard Test Method for Stress Crazing of Acrylic Plastics in Contact with Liquid or Semi-Liquid Compounds," F 484, *Annual Book of ASTM Standards,* American Society for Testing and Materials
44. P.J. Burchill and R.H. Stacewicz, *J. Mater. Sci. Lett.,* Vol 1, 1982, p 488
45. P.I. Vincent and S. Raha, *Polymer,* Vol 13, 1972, p 283
46. L.E. Nielsen, *Mechanical Properties of Polymers and Composites,* Vol 2, Marcel Dekker, 1974

Fracture Resistance Testing*

POLYMERIC MATERIALS are many and varied, ranging from pure glasses to blends to semicrystalline solids. Their mechanical properties range from pure elasticity with very high strains to fracture (rubbers or elastomers) to almost pure Hookian elasticity with low strains to fracture (glasses); the majority of polymers have properties somewhere between these two extremes. Virtually all polymeric materials show some form of inelastic behavior (Ref 1, 2). The elastomers show hysteresis, and the glasses show some form of yielding. The inelastic behavior is not restricted to the tip of a crack but is present in some form or another throughout the material. The inelasticity is a direct result of the time dependence of the motions of the polymer chains. With the exceptions of certain untoughened epoxy resins and related thermosets, inelasticity is the norm. Hence, the expectation of many theories of fracture mechanics that Hookian behavior can be assumed is not to be realized. Even theories that assume elastic-plastic criteria are inadequate, because they assume plastic behavior at the crack tip and elastic behavior throughout the remainder of the specimen, whereas in the real materials, there is viscoelastic deformation of some form or other occurring in the bulk of the specimen.

The presence of inelasticity in the entire specimen, as well as at the crack tip, results in additional energy being required for crack propagation. Hence, in any mechanical test, the energy measured to propagate a crack consists of the surface energy of the crack, energy of plastic deformation at the crack tip, and energy of inelastic deformation of the entire specimen (Ref 3). Because the latter two forms of energy absorption are a direct result of the time-dependent behavior of the polymer chains, the energy absorbed displays a strong dependence on the rate at which stress is applied. The crack opening displacements in polymeric materials can be quite large and, hence, the microstrain at a crack tip will be similarly large. In polymeric materials displaying minimal levels of plasticity and/or inelasticity, such as untoughened epoxies, the crack opening displacement is quite small. At the other extreme is the elastomer, or rubber, where the crack opening displacement is so large that the process is usually referred to as tearing. The crack opening displacement can reflect two extremes in deformation behavior: shear yielding or crazing (Ref 3). Both reflect large amounts of plastic deformation at the crack tip. In the case of some polymers, for example, polycarbonate, a large yield zone is observed. In others, the phenomenon is referred to as crazing, where the apparent crack is really a zone of fibrous material produced by the stress field ahead of the crack. This phenomenon can be present in glassy materials as well as semicrystalline materials, and it corresponds to microyielding to levels of several hundred percent strain. A similar phenomenon can also be observed in unnotched specimens where regions in the bulk of the specimen display what is usually described as stress whitening.

In addition to the behavior described previously, polymers are also sensitive to the environment, both gaseous and liquid (Ref 3). An example of the effects of gaseous environments is the effect of atmospheric ozone on crack propagation rates in natural rubber (Ref 4). In the case of a liquid, the behavior can be caused by several different effects (Ref 5, 6). First, there is always the possibility that the liquid may be a solvent and be absorbed by the polymer; the absorption process may occur more rapidly at the tip of a crack. In this case, the liquid will plasticize the polymer, lowering its glass transition temperature and thereby altering all of its fundamental properties. Second, the liquid may react chemically with the polymer, changing its fundamental structure and properties on a microscopic or macroscopic scale. Third, the liquid may simply wet the polymer, lowering the surface energy and making crack or craze propagation much easier. A well-known example of such behavior is the effect of carbon tetrachloride on polycarbonate.

Historical Development

Fracture in polymers was first studied intensively for rubber, and tests were developed logistically in the early 1900s (Ref 7, 8). Standard test methods included tensile testing with "dog-bone" specimens, where the breaking strength was obtained. By the 1920s, standard tests for tear strength, using "trouser-type" specimens, were in use. Such methods are still in common use, the tensile test to failure using a dog-bone specimen being one of the most popular for the characterization of all kinds of polymers (Ref 8). As new polymers are developed and testing is needed, tensile testing on sheets or thin films as a method of characterization still tends to be preferred over the standardized ASTM International tests for fracture strength. This may occur sometimes due to the amount of specimen available and at other times due to the simplicity of specimen preparation and characterization.

Fracture testing using standardized linear fracture mechanics approaches, such as plane-strain fracture toughness/strain energy release rate (K_{Ic}/G_{Ic}) methods, has been used for decades as a means of carrying out fracture testing (Ref 3, 9). However, because of the previously mentioned inelasticity problems, polymers have stress distributions at the tip of a crack that cannot be calculated or described adequately by the assumptions of classical elasticity theory. Such approaches clearly cannot describe adequately the behavior of even the most well-behaved systems. Early attempts at describing the fracture phenomenon in a more realistic manner recognized that the most important parameter describing the phenomenon was the energy absorbed by the fracture process (Ref 10). The energy balance approach was suggested very early by Griffith (Ref 11) but was used for rubber by Rivlin and Thomas (Ref 12) who used a $\mathfrak{J}$ to describe the total work needed to create a unit area of surface (or the tearing energy, in the case of rubber). Attempts at applying this approach were made successfully by several investigators (Ref 3, 13, 14). The beginning of a generalized theory of fracture mechanics, not requiring linear fracture assumptions, was developed (Ref 15). Study of fracture then concentrated for several years on the development and understanding of the mechanisms of craze formation, because, clearly, the formation of crazes ahead of the crack is the major contributor to the energy absorbed in fracture in most polymers (Ref 15). Indeed, because crazing is the precursor to fracture itself, it justifies attention on that ground alone.

A concurrent development, which now is used in the testing of polymers, is the *J*-integral method; it is essentially the equivalent of G_I for a nonlinear system. Discovered by Rice (Ref 16, 17) and developed independently by other investigators (Ref 18, 19), the *J*-integral method has been applied successfully to polymers (Ref 20–22). The disadvantage of the method is that it requires multiple specimens in its strict form,

*Adapted from the article by Kevin M. Kit and Paul J. Phillips, "Fracture Resistance Testing of Plastics," in *Mechanical Testing and Evaluation,* Volume 8, *ASM Handbook,* 1997, p 649–653

discouraging widespread use. A single specimen method was developed and used successfully on polypropylene (Ref 23), but it has not yet been converted into a standard ASTM method.

Fracture Test Methods for Polymers

Several methods have been developed specifically for determining the fracture toughness of polymeric materials. ASTM D 5045 (Ref 24) describes a method for determining the linear elastic fracture toughness (K_{Ic} and G_{Ic}) of polymers. This methodology is appropriate for highly cross-linked thermosets (e.g., epoxy) or glassy thermoplastics incapable of significant plastic deformation (e.g., polystyrene). ASTM D 6068 (Ref 25) describes a method for measuring *J-R* curves (a measure of elastic-plastic fracture toughness) for polymer specimens that are not large enough to experience conditions of plane strain during loading. However, methods originally developed to characterize the elastic-plastic fracture of ductile metallic materials are most commonly used (with slight modifications) to characterize ductile polymers. These methods are based on the concept of the *J*-integral to determine plane-strain fracture toughness values. To date, the most commonly used method is that of ASTM E 813 (Ref 26). This method was discontinued in 1989 and replaced by ASTM E 1737 (Ref 27). The differences between the two are minor, but the methods for data analysis and reporting described in ASTM E 1737 should now be followed.

J-Integral Testing

ASTM E 1737 is more general than ASTM E 813 and describes the method for determining either J_{Ic} or J_c under plane-stress conditions. J_{Ic} is the critical value of the *J*-integral at which onset of stable crack growth occurs. If stable crack growth is not observed, then J_c is defined as the value of the *J*-integral at which unstable crack growth (i.e., failure) occurs. The *J*-integral is a measure of the amount of energy absorbed (due to both elastic and plastic responses) during the growth of a crack through the material of interest.

Experimentally, J is determined as a function of crack extension, Δa, in a notched specimen loaded in tension. J is calculated according to (Ref 28):

$$J = \frac{2U}{bB} \quad \text{(Eq 1)}$$

where U is the area under the load-displacement curve, and B and b are the dimensions of the specimen in the plane of the crack. Testing is most commonly performed on single-edge notched bend or on compact tension specimens containing machined notches. ASTM E 1737 specifies that the specimen be fatigued so that a sharp precrack is formed at the base of the notch. However, this is not a viable technique for most thermoplastic polymers. The accepted method for creating a precrack in polymer samples is to tap a fresh, unused razor blade into the notch immediately preceding the test, as specified in ASTM D 6068 and ASTM D 5045.

To ensure the existence of plane-strain conditions at the crack tip, specimen thickness, B, and the original uncracked ligament, b_0 (i.e., the distance the crack would have to extend to separate the specimen into two pieces), must be greater than $25J_{Ic}/\sigma_y$, where J_{Ic} is the elastic-plastic fracture toughness, and σ_y is the yield strength. Because J_{Ic} is generally not known a priori, specimen dimensions must be based on an estimated value of J_{Ic} and then verified after testing. It has been shown (Ref 29) that the specimen size requirements specified by ASTM E 1737 can be relaxed for some polymers, such as low-density polyethylene and a polypropylene copolymer, to $B, b_0 > 17J_{Ic}/\sigma_y$.

In order to arrive at a value of J_{Ic}, *J*-integral values are plotted as a function of crack extension, Δa, to form a so-called *R*-curve. This data may be collected using single-specimen or multiple-specimen techniques. The multiple-specimen technique is widely accepted as a valid measure of the elastic-plastic fracture toughness of polymers and is commonly employed. However, results from the much simpler single-specimen technique have also been shown to be valid, and the implementation of this technique is increasing. These techniques differ only in the determination of the *R*-curve; specimen requirements and data analysis to determine J_{Ic} are identical. Both are summarized in the following sections.

Multiple-Specimen Technique. In both techniques, it is desirable to determine J at a minimum of ten equally spaced Δa points. In the multiple-specimen technique, each J-Δa point on the *R*-curve is generated with a different specimen. Each specimen is loaded to a level judged to produce a desired, stable crack growth extension, Δa, and is then unloaded. Polymer specimens are then removed from the test frame and fractured in liquid nitrogen. (This last step deviates from ASTM E 1737, which specifies that the specimens be fatigued first.) The precrack, stable crack growth and freeze-fracture regions of the fracture surface are usually easily identifiable (Ref 25), and an optical microscope is used to measure Δa (the length of the stable crack growth region) at nine points equally spaced across the thickness of the specimen. These nine values are averaged, as described by ASTM E 1737. J is then calculated according to:

$$J = J_{el} + J_{pl} \quad \text{(Eq 2)}$$

where J_{el} and J_{pl} are the elastic and plastic components of J, calculated as:

$$J_{el} = \frac{K^2(1 - \nu^2)}{E} \quad \text{(Eq 3)}$$

$$J_{pl} = \frac{\eta A_{pl}}{B_N b_o} \quad \text{(Eq 4)}$$

where K is a function of maximum load and specimen geometry, ν is Poisson's ratio, E is Young's modulus, A_{pl} is the area under the load-displacement curve for the entire loading-unloading cycle, and B_N is specimen thickness. For single-edge notch and compact tension specimen, $\eta = 2$, while for the disk-shaped compact tension specimen, η is a function of geometry. Equations for K for each specimen type are given in Annex 4 of ASTM E 1737.

Single-Specimen Technique. The single-specimen technique relies on the ability to determine the extent of crack growth, Δa, while the specimen is loaded in the test frame. If this can be done, then many J-Δa data pairs can be collected from one specimen. Crack growth is usually determined by an elastic compliance method or by an electrical resistance method. In the elastic compliance method, the specimen is unloaded periodically during the test. At each unloading point, Δa is calculated as a function of the slope of the unload line, Young's modulus, and specimen geometry. However, due to the viscoelastic behavior of polymers, accurate determination of crack lengths by this method is suspect (Ref 30, 31).

Another method determines the crack length by measuring the voltage drop across the uncracked ligament through which a constant direct current is passed. This method is also not generally applicable to polymers, because most are poor conductors. However, a method has been developed (Ref 23) that involves measuring crack extension directly with a video camera. A thin copper grid deposited on the surface of the specimen serves as a scale reference. Another *J*-integral technique that has been successfully applied to polymers is the normalization method (Ref 31). This method does not require specimen unloading or in situ measurements of crack growth. The crack length is calculated by separating total displacement into elastic and plastic components, each of which is a function of crack length. After a fitting procedure is used to establish a relationship between plastic displacement and crack length, the actual crack length can be calculated at any point on the load-displacement curve. Researchers (Ref 31) used this technique to determine J_{Ic} for two rubber-toughened nylons and found their results very close to values obtained by the standard multiple-specimen method.

Determination of J_{Ic}. Before the data can be analyzed, it must be checked to verify that it spans a sufficiently large range of Δa. This procedure to determine qualifying data is detailed in ASTM E 1737. Qualifying J data must also be less than the smaller of $b_0\sigma_y/20$ and $B\sigma_y/20$ to ensure that all data points are measured under plane-strain conditions. Qualified data are fit by the method of least squares to the curve described by:

$$\ln J = \ln C_1 + C_2 \ln\left(\frac{\Delta a}{k}\right) \quad \text{(Eq 5)}$$

where C_1 and C_2 are fitting parameters, and k = 1 mm (0.04 in.). A linear blunting line must also be constructed along the line defined by:

$$J = 2\sigma_y \Delta a \quad \text{(Eq 6)}$$

where σ_y is the average of the 0.2% offset yield strength and the ultimate tensile strength. The blunting line accounts for deflection that occurs due to plastic deformation near the crack tip prior to the onset of stable crack growth. ASTM E 1737 specifies that the J value at the intersection of the fit data and a line offset 0.2 mm (0.008 in.) from the blunting line defines an interim value, J_Q, which is used to verify the existence of plane-strain conditions. If both B and b_0 are indeed greater than $25J_{Ic}/\sigma_y$, and some additional data qualifications are met, then the value of J_Q is taken to be equal to J_{Ic}. Experimental and fit R-curves for an acrylonitrile-butadiene-styrene (ABS) copolymer are shown in Fig. 1, along with the blunting and 0.2 mm (0.008 in.) offset lines. The intersection of the fit R-curve and the 0.2 mm (0.008 in.) offset line indicates a J_{Ic} of 5.31 kJ/m^2.

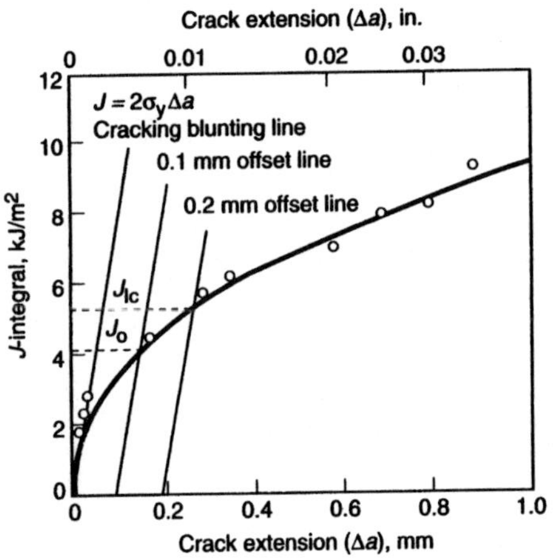

Fig. 1 Experimental R-curve for an acrylonitrile-butadiene-styrene copolymer showing power-law fit, blunting line, and 0.2 mm (0.008 in.) offset line. Source: Ref 32

Modifications for Polymeric Materials. Due to the unique properties of polymers, several modifications to the J-integral method have been proposed and used. Some of these modifications that affect the collection of J-Δa data have already been mentioned, and these are quite widely accepted as standard.

In some cases, crack tip blunting may not occur before or during stable crack growth in polymers. Crack tip blunting can be verified by direct microscopic observation or if J data follow the blunting line ($J = 2\sigma_y\Delta a$) for small amounts of crack growth. Some of the data in Fig. 1 lie on the blunting line, indicating that blunting does occur (Ref 32). If blunting is not known to occur, J_{Ic} should be determined by extrapolating a linear fit to the J-Δa data to zero crack growth ($\Delta a = 0$). It has been argued in Ref 33 that J-Δa data should, under conditions of plane strain, follow:

$$J = J_{Ic} + \frac{dJ}{d\Delta a}\Delta a \quad \text{(Eq 7)}$$

For small crack growth, J should vary linearly with Δa, and the value of J_{Ic} should be determined as previously explained. Optical microscopy (Ref 34, 35) has shown that crack blunting does not occur in certain grades of high-density polyethylene, toughened nylon 6/6, ABS, and toughened polycarbonate. As further evidence, the J data collected from the high-density polyethylene (Ref 35) do not follow the blunting line for small Δa, as shown in Fig. 2.

If crack tip blunting does occur, the procedure described will yield conservative values of J_{Ic}. If blunting is known to occur, then J_{Ic} should be determined by the methods of ASTM E 1737 or ASTM E 813. The determination of J_{Ic} by ASTM E 813 differs in that J_{Ic} is taken at the intersection of a linearly fit R-curve and the blunting line. This construction is shown in Fig. 3 for the same data used in Fig. 1. The intersection of the linear R fit and the blunting line indicates a J_{Ic} of 3.95 kJ/m^2 (compare to the ASTM E 1737 value of 5.31 kJ/m^2). The method in ASTM E 813 usually gives more conservative values than that in ASTM E 1737. Researchers (Ref 29, 31, 35, 36) have analyzed J data of high-impact polystyrene (HIPS), ABS, a polycarbonate (PC)/ABS blend, and a polycarbonate/polybutylene terephthalate (PBT) blend by three methods (ASTM E 1737, ASTM E 813, and the no-blunting method described previously). As can be seen in Table 1, the no-blunting method is the most conservative, while ASTM E 1737 is the least conservative. If no direct evidence of crack tip blunting exists, the most conservative method for calculating J_{Ic} should be used.

Several workers have shown that the plane-strain thickness requirements specified by ASTM E 813 and ASTM E 1737 are too conservative in certain cases, while not conservative enough in others. Researchers (Ref 37, 38) have shown that the requirement is too conservative for tough thermoplastics, ultrahigh-molecular-weight polyethylene (J_{Ic} = 95 kJ/m^2), and rubber-toughened nylon 6/6 (J_{Ic} = 30 kJ/m^2). Both studies found that size-independent values of J_{Ic} were obtained for specimen thicknesses greater than $6J_{Ic}/\sigma_y$, which is approximately 25% of the recommended minimum thickness. Conversely, investigators (Ref 39) found that size-independent values of J_{Ic} for a relatively brittle PC/ABS blend (J_{Ic} = 4 kJ/m^2) were not obtained until the thickness was greater than $64J_{Ic}/\sigma_y$, which is more than twice the recommended minimum thickness. In light of these results, it is recommended that J_{Ic} be determined for various thicknesses to ensure that the true plane-strain value is obtained.

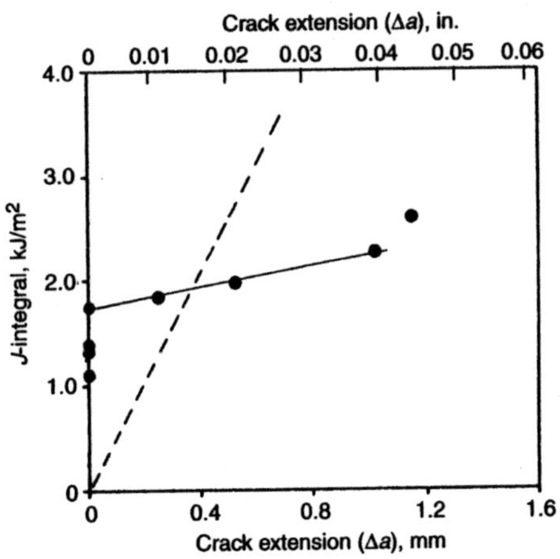

Fig. 2 Experimental R-curve for a high-density polyethylene showing the dashed blunting line and the absence of blunting behavior. Source: Ref 35

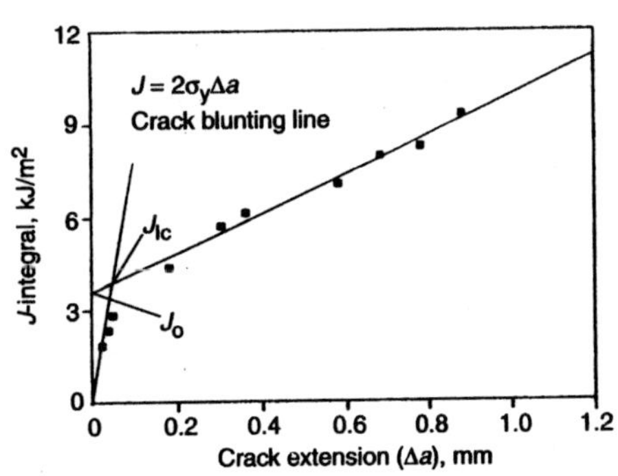

Fig. 3 Experimental R-curve for an acrylonitrile-butadiene-styrene copolymer showing linear fit and blunting line. Source: Ref 32

Linear Elastic Fracture Toughness

Other methods also exist to determine the plane-strain fracture toughness of polymers.

Table 1 Comparison of elastic-plastic fracture toughness (J_{Ic}) data for several polymers determined by different methods

	J_{Ic}, kJ/m^2			
Method	HIPS	ABS	PC/ABS	PC/PBT
No blunting	3.24	3.57	3.00	5.47
ASTM E 813	3.60	3.95	3.55	7.17
ASTM E 1737	4.30	5.31	7.85	13.41

HIPS, high-impact polystyrene; ABS, acrylonitrile-butadiene-styrene; PC/ABS, polycarbonate/acrylonitrile-butadiene-styrene; PC/PBT, polycarbonate/polybutylene terephthalate

ASTM D 5045 specifies a procedure for determining the critical strain energy release rate, G_{Ic}, of polymers. This parameter is equivalent to J_{Ic} for materials that exhibit linear (or nearly linear) elastic behavior (Ref 40). ASTM D 5045 specifies the use of single-edge notch bend or compact tension specimens. Precracks are created by tapping a fresh, unused razor blade into the machined notch immediately preceding the test. The samples are then loaded to a level that causes a 2.5% apparent crack extension. However, significant deviation from linear elastic behavior must not occur at this load level. The procedure for testing this requirement is detailed in ASTM D 5045. An interim value of the critical strain energy release rate, G_Q is determined by:

$$G_Q = \frac{U}{Bb\phi} \qquad \text{(Eq 8)}$$

where ϕ is a function of b and the original crack length, a. This interim value can be qualified as the plane-strain critical strain energy release rate if plane-strain conditions are verified. The standard specifies that B, b, and a must be greater than $2.5(K_{Ic}/\sigma_y)^2$, where K_{Ic} is the plane-strain fracture toughness and is related to G_{Ic} by:

$$K_{Ic} = \frac{E}{1 - \nu^2} G_{Ic} \qquad \text{(Eq 9)}$$

Using this relation, the size requirement for plane-strain conditions can be written as:

$$B, b, a > \frac{2.5E}{\sigma_y(1 - \nu^2)} \frac{G_{Ic}}{\sigma_y} \qquad \text{(Eq 10)}$$

Using typical values for E (1 GPa, or 145 ksi), σ_y (60 MPa, or 8.7 ksi), and ν (0.4), the size requirement is $B, b, a > 50G_{Ic}/\sigma_y$, which is twice the size requirement for determining plane-strain J_{Ic}.

Due to the viscoelastic properties of polymers, test temperature and strain rate should be well controlled and reported. The standard recommends 23 °C (73 °F) and a crosshead speed of 10 mm/min (0.4 in./min). The orientation of the specimen with respect to processing direction (e.g., extrusion direction and mold flow direction) should also be reported because of the strong dependence of mechanical properties on molecular orientation that often develops during processing.

Testing of Thin Sheets and Films

In order to ensure the existence of plane-strain state, the dimensions of the sample normal to the applied stress are usually required to be greater than $25J_{Ic}/\sigma_y$, where J_{Ic} is the elastic-plastic fracture toughness, and σ_y is the yield strength. Both J_{Ic} and σ_y are generally considerably lower than the corresponding values for metallic materials, but the ratio J_{Ic}/σ_y is usually much larger for polymeric materials. Therefore, the plane-strain size requirements for polymeric fracture specimens are often unrealistic (on the order of 5 cm, or 2 in.). In many applications, the properties of polymeric materials are strongly dependent on the level of molecular orientation and crystallinity. These levels, in turn, are strongly dependent on the thermal and mechanical histories experienced during processing. Specimens that are produced to fulfill the plane-strain condition are likely to have quite different thermal and mechanical histories than polymer materials processed into sheet or film. Therefore, the thicker test specimens do not reflect the actual properties of the polymer for the intended application. For these reasons, ASTM D 6068 is often a more desirable method than the plane-strain method of ASTM E 813 or ASTM E 1737. This method was developed specifically for the determination of R-curves from thin sheets or films. However, this is not a valid method for determining J_{Ic}, and results should not be reported as such. When using this method, specimen size and the values of C_1 and C_2 (which characterize the power-law fit of the R-curve) should be reported.

Other Methods

Alternative methods for determining the fracture toughness of polymer materials have recently been proposed. Most notable are the normalization and hysteresis methods, which are both single-specimen techniques. The normalization method does not require unloading cycles or on-line crack measurement and has been used successfully for metallic materials (Ref 31). The method is based on the assumption that the load, P, on the specimen can be represented by:

$$P = G(a)H(v_{pl}) \qquad \text{(Eq 11)}$$

where $G(a)$ is a known function of crack length and specimen geometry, and $H(v_{pl})$ is a function of plastic displacement, v_{pl}. After the form of $H(v_{pl})$ is fit to experimental data, values of a (and hence J) can be determined at any point on the load-displacement curve. J_{Ic} can then be determined from the R-curve using the methods described previously. Researchers (Ref 31) found that the results of this method are slightly less conservative than those determined by ASTM E 813 and more conservative than ASTM E 1737 for two rubber-toughened nylons (nylon 6/6 and an amorphous nylon).

The hysteresis method requires the application of multiple load-unload cycles to successively larger displacements (Ref 30, 32, 41), as shown in Fig. 4. The area between the loading

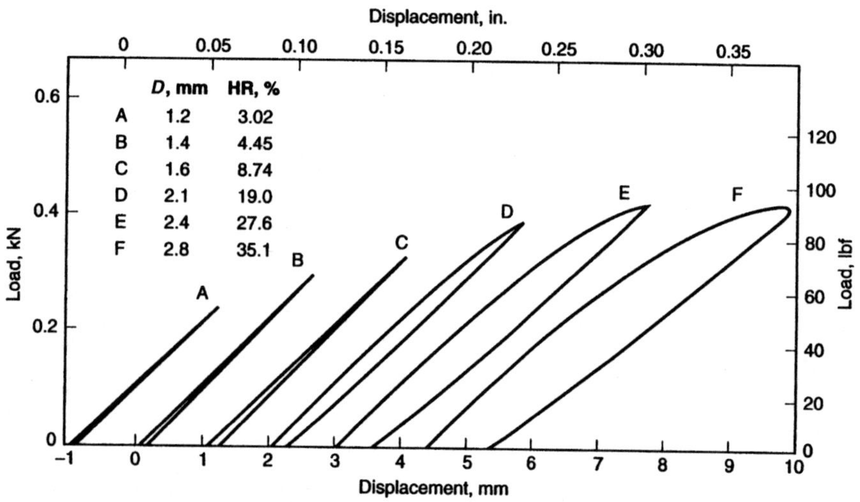

Fig. 4 Hysteresis loops for several loading-unloading cycles for a polycarbonate/polybutylene terephthalate blend. D, specimen displacement; HR, ratio of hysteresis energy to total strain energy. Source: Ref 41

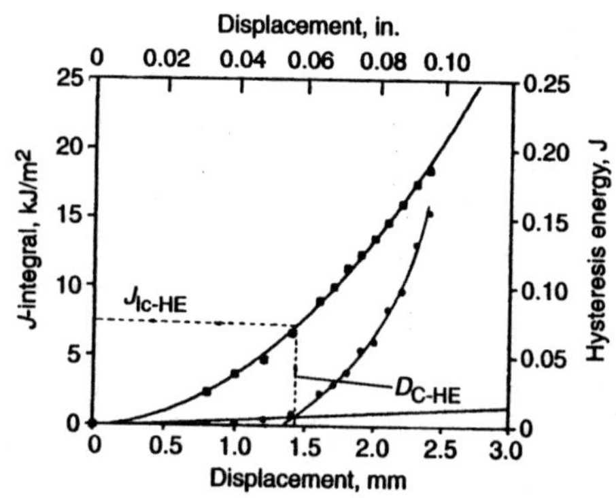

Fig. 5 J-Integral and hysteresis energy (HE) vs. displacement (D) for a polycarbonate/polybutylene terephthalate blend. Test rate, 2 mm/min (0.08 in./min) $J_{Ic\text{-}HE}$ and $D_{C\text{-}HE}$ are critical values of J and D for initation of crack propagation. Source: Ref 41

and unloading lines on the load-displacement curve is defined as the hysteresis energy, and this is plotted against maximum displacement for each loading cycle, as shown in Fig. 5. For small displacements, crack growth does not occur, and the hysteresis energy varies linearly with displacement. These data are fit with a linear blunting line. After crack growth commences, the hysteresis energy varies nonlinearly with displacement and can be fit with a power law. The displacement at which the linear blunting line intersects with the power-law curve is taken as the critical displacement to initiate crack growth, and the value of J at this displacement is taken as J_{Ic}. It has been found that the results of this method are slightly less conservative than those determined by ASTM E 813 and more conservative than ASTM E 1737 for several polymers (ABS, PC/ABS, HIPS, and PC/PBT) (Ref 32, 36, 41, 42).

REFERENCES

1. N.G. McCrum, B.E. Read, and G. Williams, *Anelastic and Dielectric Effects in Polymeric Solids,* Wiley, 1967
2. I.M. Ward, *Mechanical Properties of Solid Polymers,* Wiley, 1983, p 15
3. E.H. Andrews, Cracking and Crazing in Polymeric Glasses, *The Physics of Glassy Polymers,* R.N. Haward, Ed., Wiley, 1973, p 394
4. R. Natarajan and P.E. Reed, *J. Polym. Sci. A, Polym. Chem.,* Vol 2 (No. 10), 1972, p 585
5. G.A. Bernier and R.P. Kambour, *Macromolecules,* Vol 1, 1968, p 393
6. E.H. Andrews, G.M. Levy, and J. Willis, *J. Mater. Sci.,* Vol 8, 1973, p 1000
7. L.E. Weber, *The Chemistry of Rubber Manufacture,* Griffin, London, 1926, p 336
8. K. Memmler, *The Science of Rubber,* R.F. Dunbrook and V.N. Morris, Ed., Reinhold, 1934, p 523
9. G.R. Irwin, in *Encyclopaedia of Physics,* Vol 6, Springer Verlag, 1958
10. N.G. McCrum, C.P. Buckley, and C.B. Bucknall, *Principles of Polymer Engineering,* Oxford University Press, 1997, p 201
11. A.A. Griffith, *Philos. Trans. R. Soc. (London) A,* Vol 221, 1921, p 163
12. R.S. Rivlin and A.G. Thomas, *J. Polym. Sci.,* Vol 10, 1953, p 291
13. R.P. Kambour, *Applications of Polymers Symposium,* Vol 7, John Wiley & Sons, 1968, p 215
14. J.P. Berry, *J. Polym. Sci. A, Polym. Chem.,* Vol 2, 1964, p 4069
15. E.H. Andrews, *J. Mater. Sci.,* Vol 9, 1974, p 887
16. J.R. Rice, *J. Appl. Mech. (Trans. ASME),* Vol 35, 1968, p 379
17. J.R. Rice, *Fracture,* Vol 2, 1968, p 191
18. J. Begley and J.D. Landes, in *Fracture Toughness,* STP 514, ASTM, 1972, p 1
19. J.D. Landes and J. Begley, in *Fracture Toughness,* STP 514, ASTM, 1972, p 24
20. J.M. Hodgkinson and J.G. Williams, *J. Mater. Sci.,* Vol 16, 1981, p 50
21. S. Hashemi and J.D. Williams, *Polym. Eng. Sci.,* Vol 26, 1986, p 760
22. Y.W. Mai and P. Powell, *J. Polym. Sci. B. Polym. Phys.,* Vol 29, 1991, p 785
23. M. Ouederni and P.J. Phillips, *J. Polym. Sci. B., Polym. Phys.,* Vol 33, 1995, p 1313
24. "Standard Test Methods for Plane Strain Fracture Toughness and Strain Energy Release Rate of Plastic Materials," D 5045, *Annual Book of ASTM Standards,* Vol 08.03, ASTM, 1996
25. "Standard Test Method for Determining *J-R* Curves of Plastic Materials," D 6068, *Annual Book of ASTM Standards,* Vol 08.03, ASTM, 1996
26. "Standard Test Method for J_{Ic}, A Measure of Fracture Toughness," E 813, *Annual Book of ASTM Standards,* Vol 03.01, ASTM, 1989
27. "Standard Test Method for *J*-Integral Characterization of Fracture Toughness," E 1737, *Annual Book of ASTM Standards,* Vol 03.01, ASTM, 1996
28. J.D. Landes and J. Begley, in *Fracture Toughness,* STP 560, ASTM, 1974, p 170
29. S. Hashemi and J.G. Williams, *Plast. Rubber Process. Appl.,* Vol 6, 1986, p 363
30. M.-L. Lu and F.-C. Chang, *Polymer,* Vol 36, 1995, p 2541
31. Z. Zhou, J.D. Landes, and D.D. Huang, *Polym. Eng. Sci.,* Vol 34, 1994, p 128
32. M.-L. Lu, C.-B. Lee, and F.-C. Chang, *Polym. Eng. Sci.,* Vol 35, 1995, p 1433
33. J.W. Hutchinson and P.C. Paris, in *Elastic-Plastic Fracture,* STP 668, ASTM, 1979, p 37
34. I. Narisawa and M.T. Takemori, *Polym. Eng. Sci.,* Vol 29, 1989, p 671
35. H. Swei, B. Crist, and S.H. Carr, *Polymer,* Vol 32, 1991, p 1440
36. C.-B. Lee, M.-L. Lu, and F.-C. Chang, *J. Appl. Polym. Sci.,* Vol 47, 1993, p 1867
37. B.M. Rimnac, T.W. Wright, and R.W. Klein, *Polym. Eng. Sci.,* Vol 28, 1988, p 1586
38. B.D. Huang, in *Toughened Plastics I: Science and Engineering,* C.K. Riew and A.J. Kinloch, Ed., Vol 233, *ACS Advances in Chemistry Series,* American Chemical Society, 1993, p 39
39. M.-L. Lu, K.-C. Chiou, and F.-C. Chang, *Polymer,* Vol 37, 1996, p 4289
40. K.J. Pascoe, in *Failure of Plastics,* W. Brostow and R.D. Corneliussen, Ed., Hanser Publishers, 1989, p 119
41. M.-L. Lu and F.-C. Chang, *J. Appl. Polym. Sci.,* Vol 56, 1995, p 1065
42. M.-L. Lu, K.-C. Chiou, and F.-C. Chang, *Polym. Eng. Sci.,* Vol 36, 1996, p 2289

Impact Loading and Testing*

THE MATERIAL AND ENGINEERING issues associated with plastic components subjected to impact are discussed in this article. Impact resistance can be considered to be the relative susceptibility of a component to failure due to stresses applied at high rates. The two most significant concepts in this definition are those of failure due to mechanical stress and high rates of loading.

Failure of a plastic component can take many forms. There are situations in which excessive elastic deformation will constitute failure. A plastic automotive bumper is a good example of this class of failure. A bumper system is required to absorb specified levels of energy while simultaneously protecting the rest of the automobile from damage. If the plastic bumper withstands an impact without damage but undergoes such a large displacement that it dents the sheet metal of the automobile, then it has failed in its function. In other applications, the criteria defining failure may be linked directly to damage. If a plastic panel is to be used in an exterior automotive body application, the ability to withstand an impact without denting may be an appropriate measure of failure. In still other situations, visible damage may not constitute failure as long as the plastic component has not been punctured or penetrated by an impactor. Plastic containers or impact-resistant plastic window sheet may be subject to this type of failure criterion. Perhaps the most catastrophic failure associated with impact occurs when a plastic component shatters or at least fractures in a brittle manner. This event is usually intolerable, and every engineering precaution must be taken to avoid it.

In addition to failure definition, the other important factor that must be considered in treating impact events is the high rate at which the loads are applied. The effects of time and rate dependence enter into impact problems in two ways. In some problems, time must be considered simply because inertia effects must be considered in any engineering equilibrium relation if the loading rate is high enough. However, this is not related to any unique type of behavior associated with plastics and therefore is not discussed here. Normal transient dynamic relations are equally applicable for plastics and metals. The second manner in which time and rate influence impact problems is more closely related to differences in material behavior. Material properties can show rate dependence, the effects of which will definitely be visible in impact events. If the material is a plastic that yields, the stress at which it occurs is often a function of the rate of the loading event. Even more important is the fact that the mode of failure for a particular plastic may be very dependent on the rate of loading. A plastic that exhibits extreme ductility when loaded at slow rates may fracture in a brittle fashion when loaded more rapidly. This is one of the most significant issues associated with the concept of impact resistance—the interrelationship between the rate of loading imposed on a plastic component and its failure.

There are several other factors, in addition to the rate of loading, that can play major roles in determining the impact resistance of a plastic component. One of the most important parameters is temperature. Decreasing temperature has an effect on plastics that is quite similar to increasing rate. At lower temperatures, the yield stresses of plastics are generally higher. Furthermore, if the temperature is sufficiently decreased, the mode of failure can change from a ductile event to an event of extremely brittle nature. The temperature (or temperature range) over which this change in failure characteristics occurs for a given geometry and load is referred to as the ductile-to-brittle transition temperature. Unfortunately, transition temperatures are usually dependent on component geometry, load, and rate of deformation.

Engineers with a background in the use of other materials will recognize both similarities and differences in the behavior of plastics discussed here and the behavior of the materials with which they have had experience. Impact resistance has also been an issue for other engineering materials. In fact, the impact behavior for metals described in Ref 1 is similar to many of the phenomena observed in plastics. For example, the concept of a ductile-to-brittle transition temperature is also well known in metals, as is the fact that notched metal components are more prone to brittle failure than unnotched specimens. However, there are also major differences. The modulus of elasticity of polycarbonate (PC), for example, is 2.1 GPa (0.30×10^6 psi) as compared to 210 GPa (30.0×10^6 psi) for steel. Although ductile metals often undergo local necking during a tensile test, followed by failure in the neck, many ductile plastics exhibit the phenomenon of a propagating neck. Although the initial appearance of the neck is physically similar to its metal counterpart, the cross-sectional area in many necked plastics stabilizes and then propagates along the length of a tensile specimen under a constant cross-head load. The engineering effect of this characteristic difference in material behavior can also have an important influence on some aspects of impact resistance.

To address impact resistance issues, this article is divided into two major sections. The first section, which deals primarily with material behavior, covers the effects of loading rate, temperature, and state of stress on both deformation and mode of failure. Where possible, well-defined examples of impact events are given, as well as techniques for analyzing and predicting those events. Because puncture resistance is an important characteristic associated with impact behavior, the large-strain material properties of plastics and their relationship to puncture are discussed. This section also discusses standard impact tests, along with their associated results. The largest single disadvantage of these tests is that none of them provides true material properties. Fracture toughness, which is a material property derived from the use of linear elastic fracture mechanics, does have a role in understanding impact resistance. A brief discussion of the linear elastic fracture mechanics method is presented, along with an example of its effectiveness as a predictive tool for impact performance. Unfortunately, standardized test techniques for the accurate measurement of plane-strain fracture toughness do not exist for plastic materials, and the debate continues over the usefulness and applicability of plane-strain fracture toughness as a material property for plastics. Other issues with a bearing on impact performance, such as processing, chemical attack, and aging, are also briefly described.

The second section of this article describes the engineering calculations routinely used to predict the performance of thin plastic beams, plates, and shells. This particular class of thin structures is pertinent to plastic design, because most of the standard processes used to fabricate plastic components (injection molding, blow molding, and compression molding) lead to thin-wall parts. One of the basic assumptions underlying the theories used to predict the behavior of such parts is that of small displacements (rota-

*Adapted from the article by Ronald Nimmer, "Impact Loading," in *Engineering Plastics,* Volume 2, *Engineered Materials Handbook,* ASM International, 1988, pages 679 to 700 tions). As mentioned previously, the moduli of elasticity for engineering plastics are generally much lower than those associated with metals. This, in turn, can lead to large deformations. Therefore, the issue of assuming small displacements for the calculation of plastic structure performance is discussed and its limitations described. Another consequence of the very low modulus of elasticity associated with plastics is the potential importance of buckling and collapse of thin plastic structures. The loss of load-carrying ability associated with such an event can be a significant design issue for a plastic part, even though it was not a problem for the metal part it replaced. An example of such a situation and some plastic design solutions are offered.

Material Considerations in Impact Response

A number of fundamental material properties influence impact resistance and must be quantified if rational engineering design with plastics is to be feasible. Some of the critical material behavior has been studied in detail and quantified in the form of well-defined engineering data, while in other cases, the understanding is more qualitative in nature. In this section, well-understood behavior and engineering techniques are discussed first, followed by the less well-understood topics.

Because ductile plastics are often used in impact situations, yield stress is the first mechanical property to be discussed. For many engineering plastics, the yield stress (and in some cases, the elastic modulus) is dependent on strain rate and temperature. Under severe impact loads, such dependence will obviously influence the engineering response of a plastic component. This issue is perhaps one of the more well-understood areas of material response associated with impact, and extensive data exist for many plastics. Typical material properties are presented, as well as an example of the effectiveness of using these data to predict the engineering performance of a simple plastic component subjected to impact.

Understanding and characterizing some aspects of impact performance require much more complete knowledge of material properties than simply Young's modulus and yield stress. The puncture resistance of a plastic, for example, requires that its stress-strain behavior be quantified to very large strains (40 to 300%, depending on the plastic). The unique behavior of plastics in this strain range can have a significant effect on puncture resistance. Although material properties in this range are not as easy to measure or as well understood as elastic modulus and yield stress, some data do exist. There is also some initial understanding of how these properties affect puncture resistance. Consideration of this large-strain behavior is the second subject discussed with respect to material performance.

Strain rate and temperature, in addition to their effect on yield stress, can have a significant effect on the failure mode of a plastic component. Like other materials, plastics can exhibit a transition in failure mode as the rate of loading increases. A transition of this nature can cause catastrophic results. A material that exhibits great ductility when tested at slow rates may fracture abruptly without absorbing any significant energy at a higher rate of loading. Alternatively, if a plastic does not experience such a transition in failure mode as a function of rate at room temperature, it may become evident as the test temperature is reduced. Stress concentrators can also have a significant effect on such transitions, with the presence of a crack being the most critical. In any case, a ductile-to-brittle transition in behavior within the design envelope for such structures as motorcycle helmets or instrument panels would be unacceptable.

Qualitatively, a great deal is known about these transitions from ductile to brittle failure as a function of rate, temperature, and stress state. However, no fundamental material data are available (as there are in the case of yield stress, for example) that can be easily applied for engineering design. However, the basic concepts of rate and temperature effects on yield stress discussed earlier do provide a foundation of understanding for this critical issue, and the discussion presented in this section centers on this approach.

A number of tests used as standard indicators of impact resistance by the plastics industry are also briefly discussed in this section. None of the tests provides fundamental material properties that are quantitatively useful in a design analysis sense. However, the engineer should understand the behavior that they reflect qualitatively.

Unlike the results of other impact tests, plane-strain fracture toughness, the next subject to be discussed, is a material property. If it is properly measured within its range of applicability, it can be a useful engineering property relevant to impact resistance. Although the philosophy of fracture mechanics is often used for polymers, its level of development and application as an engineering tool for plastics is not as extensive as for other materials. A brief outline of its usefulness, as well as an example of its applicability in the area of impact for plastics, is presented.

Finally, there are a number of issues related to process and environment, such as weld lines, chemical attack, and aging, that will affect the impact resistance of plastic. Brief discussions of these issues are provided as additional information.

Loading Rate Effects on Polymer Deformation. It is well established that the strain rate imposed on a plastic has a distinct effect on its deformation. As for other solids, the relationship between stress and strain rate, $\dot{\varepsilon}$, takes the form (Ref 2):

$$\sigma_y = B_1 + B_2 \ln \dot{\varepsilon} \qquad \text{(Eq 1)}$$

where σ_y is yield stress, and B_1 and B_2 depend on the polymer and the temperature. In general, higher rates and lower temperatures lead to higher yield stresses, and lower rates and higher temperatures lead to lower yield stresses.

The temperature and strain-rate dependence of the yield stress for most plastics has been well investigated and reported in the literature. Reference 3, for example, provides such data for polycarbonate (PC), polyether-imide (PEI), and polybutylene terephthalate (PBT). Figure 1 illustrates stress-strain relationships for PC, an amorphous polymer, at room temperature and at a variety of strain rates, $\dot{\lambda}$. The authors report that permanent deformation is observed when the maxima in the stress-strain curves are reached and therefore use this as their definition of yield stress. As can be seen in Fig. 1, there is very little effect of strain rate evident in the lower-stress regions of the curves. However, there is a distinct increase in yield stress for increasing strain rate. Figure 2 illustrates the strain-rate dependence of the yield stress for different temperatures. As can be seen, the yield stress increases significantly as the temperature decreases.

Similar results describing the stress-strain behavior of PEI, another amorphous polymer, are shown in Fig. 3 and 4. One interesting variation in the behavior of PEI is that there appears to be an observable difference in the initial stress-strain data, normally characterized by Young's modulus, for very slow strain rates. In contrast to PC and PEI, PBT is a semicrystalline polymer. Nonetheless, it displays behavior very similar to that of PC and PEI with regard to rate and temperature dependence of stress-strain behavior, as shown in Fig. 5 and 6.

All the data shown in Fig. 1 through 6 are the results of tensile tests. Polymers exhibit a hydrostatic pressure dependence in yield stress, and, as a result, there are ordered differences in yield stress data measured in compression, shear, and tension (Ref 4). For PC, this dependence results in a yield stress difference between tensile and compressive data of only approximately 6 to 8% at room temperature and over a range of strain rates from 10^{-5} to 10^{-1}/s. The compressive yield stress for PC is higher than the tensile yield stress, with the yield stress measured in shear lying between the two.

If necessary, these well-known effects of temperature, rate, and stress state can be incorporated into component analyses, using either simple closed-form equations or the more precise numerical predictions often used in design. For example, with an elastic material model, the strain rate in an impacted component can be approximated. Using this approximation for strain rate and the temperature to which the component is exposed, data similar to those given in Ref 3 can be used to establish an approximate yield stress for the material in this application. Clearly, the strain rate in a component will be different at different locations. However, because the effect of rate on yield stress is only significant over orders of magnitude in rate, an approximation of yield stress based on maximum strain rate in a component is usually quite adequate in predicting the elasto-plastic response of the component.

As an example of this approximation, consider the boxlike structure shown in Fig. 7. This structure, molded from a rubber-modified PC, was subjected to a dynamic load applied with the bar shown at the top of the figure (Ref 5). Figure 8 illustrates the stress-strain response as a function of strain rate that was measured in standard tensile tests. Using a linear finite-element analysis, a relationship was established between the applied cross-head displacement rate and the maximum strain rate in the component. Next, elastoplastic analyses of the component were carried out using yield stresses associated with room temperature and the maximum strain rate in the component. Figures 9 and 10 illustrate the accuracy of this approximation in predicting the load-displacement behavior of the component and the maximum load sustained during the impact events as a function of strain rate. The maximum load sustained during these impact tests was associated with local regions of yielding and plasticity, expected from the tensile test results. The more detailed discussion of this comparison in Ref 5 demonstrates that predictions of strains on the component also agreed with strain gage data to within 5%. As can be seen, the correlation with experimental measurement is very good. In addition, for this material and component test, the rate effect on deformation was not extremely significant from a design standpoint. Standard low-speed tensile test data would probably have been sufficient for design at the strain rates considered here.

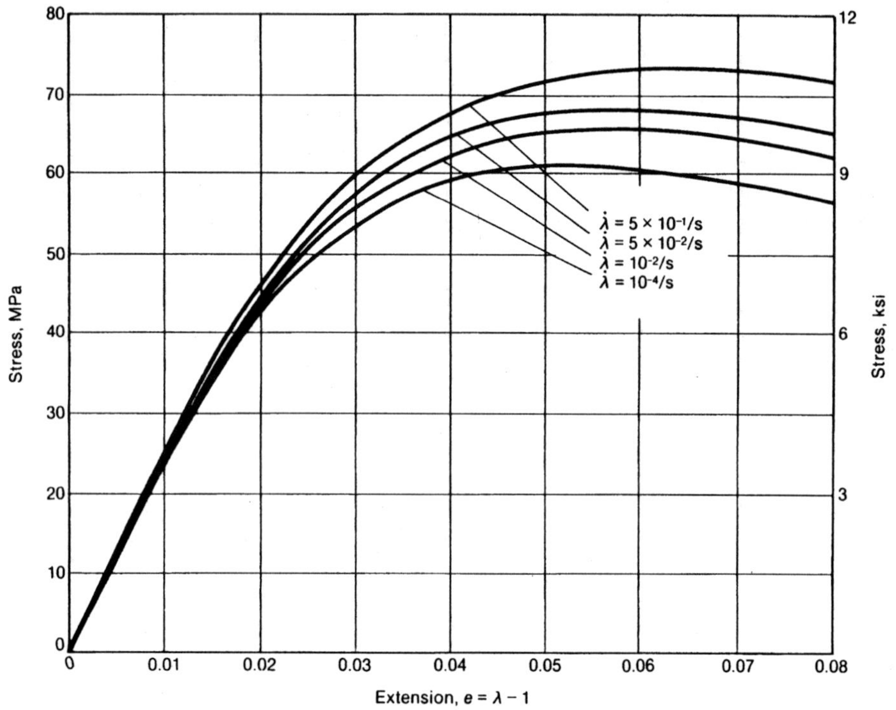

Fig. 1 Stress-strain behavior of polycarbonate as a function of strain rate, $\dot{\lambda}$, at 22.2 °C (72 °F). (Note: For small strains, extension, e, is approximately equal to engineering strain, ε.)

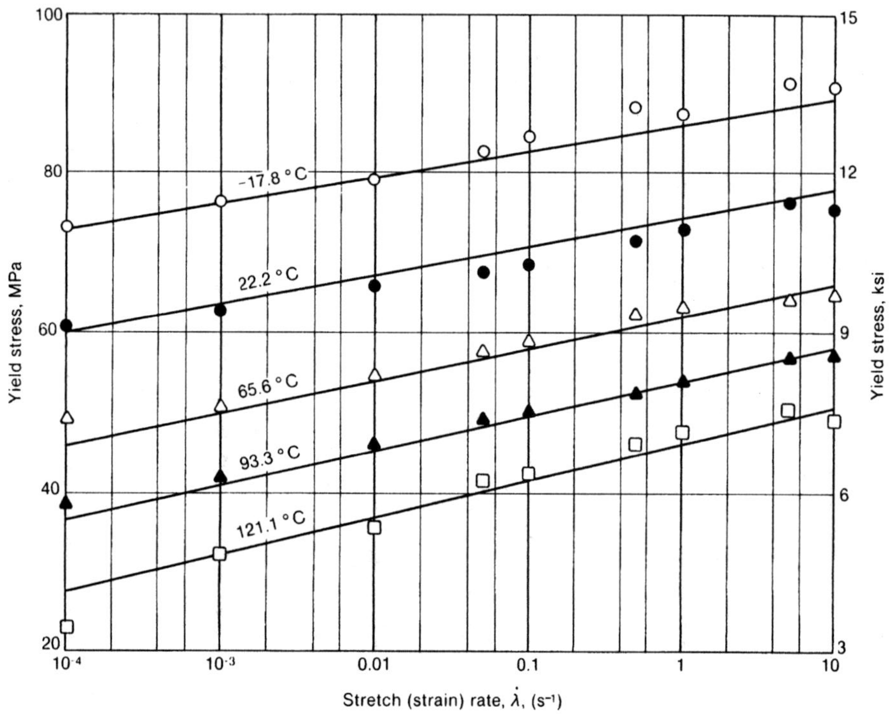

Fig. 2 Strain-rate and temperature dependence of yield stress for polycarbonate

Material Behavior in the Large-Strain Range. In the previous example of the impacted box section, knowledge of Young's modulus and yield stress, as well as their dependence on temperature and strain rate, was sufficient for an accurate engineering analysis of the component. Despite the fact that yielding, a maximum in the load-deflection curve, and permanent deformation all took place during this test, there are examples of much more severe deformation and damage to plastic materials due to impact. The puncture test, for example, subjects a plastic plate to strain and deformation that are much more severe than those experienced by the impacted box section. Compared to the strains of approximately 5 to 10% in the impacted box section, a punctured plate made of a ductile plastic will experience true strains of 40 to 300%, depending on the material.

Because puncture resistance is often a desirable characteristic for structures subject to impact, a discussion of the fundamental material properties that govern this behavior is appropriate. Figure 11 illustrates the axisymmetric geometry of a typical puncture test, and Fig. 12 illustrates the severe permanent deformation incurred by a PC plate subjected to this test at room temperature. The discussion that follows is a general overview because of the complexity of the behavior and the limited amount of pertinent material data that are available for characterizing performance in this regime. The discussion also is limited to ductile plastics, because they are often selected for their puncture resistance.

Material properties at strain levels well beyond the yield range have a substantial effect on observed puncture resistance in ductile plastics. Of special significance is the fact that for many plastics, substantial strain hardening occurs only at very large values of true strain. Proper measurement of stresses and strains is very difficult at such large deformation levels. As a result, considerably fewer fundamental material data are available in this range than for elastic and yield behavior. However, accurate experimental work has been done in this area (Ref 6–9), and it is very relevant to puncture resistance. As an introduction to the effects of large-strain behavior on puncture resistance, effects of large-strain behavior on phenomena observed during tensile tests are examined.

Figure 13 illustrates the true stress/true strain behavior of PC for true strains approaching 0.8. The experimental data are from the investigation reported in Ref 7. Significant strain hardening can be clearly observed as true strain levels exceed 0.4. This type of material behavior is not seen in standard metals and has some very interesting consequences. Although ductile metals will neck and fail locally at the reduced cross section, many plastics display the phenomenon of a propagating neck during tensile tests. This phenomenon has been described and explained from several points of view, including both mechanical and material aspects (Ref 10–14). In PC, for example, the neck appears almost immediately with the onset of yield. However, instead of remaining local in nature and progressing immediately to failure, the neck in PC reaches a limit in area reduction and then propagates along the length of the tensile specimen under constant cross-head load. Figure 14 shows a sequence that illustrates this event.

The necking process itself is associated with the range of behavior after yield in Fig. 13 and prior to the final hardening process. In this range, large strains are accumulated with very little change in true stress level. The termination of the initial necking process is associated with the onset of the material hardening that occurs at a true strain level of approximately 0.4 for PC in Fig. 13. If the modulus of the material characterizing this hardening for a given plastic is large enough, the cross-sectional thinning in the initial neck area will be terminated. As the original neck stabilizes, the cross-sectional thinning process is forced into adjacent locations, and the neck propagates. Mechanics and numerical procedures are available for treating this process, and it has been shown that with the appropriate true stress versus true strain data, modeled predictions of the physical process agree very well with experiments (Ref 14, 15).

The phenomenon of a propagating neck is a very important mechanism with regard to energy absorption. In contrast to the limited volume of material dissipating energy through plastic deformation in a localized neck, the characteristic of a propagating neck allows the plastic energy dissipation to occur over much larger volumes of material as the necked region is forced to grow. As one might expect, this fundamental material behavior also has a significant effect on the puncture resistance of ductile plastics.

Using a finite-element approach that incorporates both large-strain continuum mechanics and nonlinear material behavior (Ref 16), the effects of the large-strain material properties of plastics on puncture resistance can be investigated. The details of the modeling approach and assumptions are discussed in Ref 15, 17, and 18. Figure 15 illustrates the axisymmetric finite-element

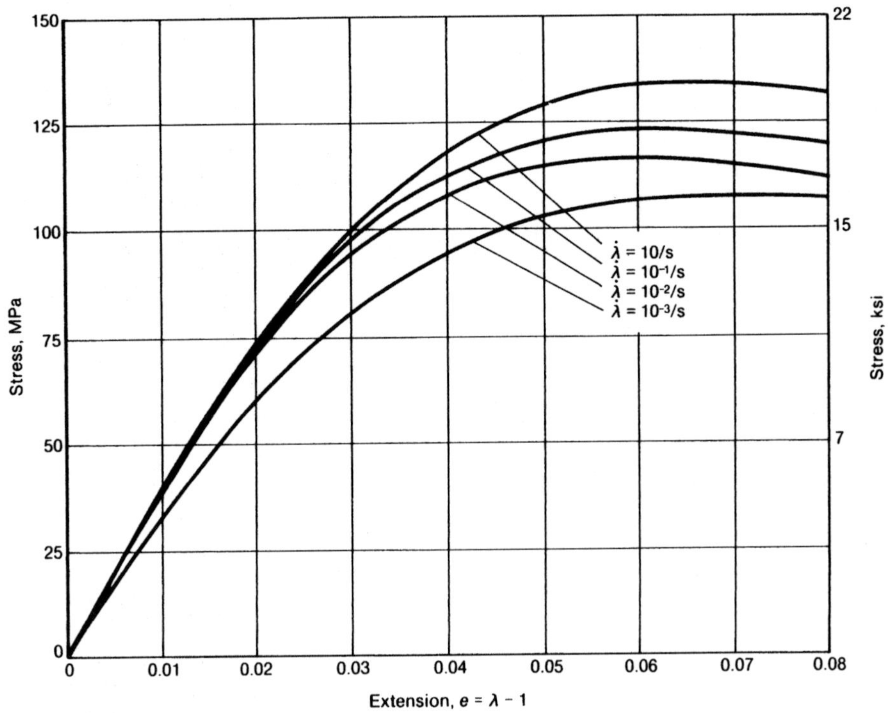

Fig. 3 Stress-strain behavior of polyether-imide as a function of strain rate, $\dot{\lambda}$, at 22.2 °C (72 °F). (Note: For small strains, extension, *e*, is approximately equal to engineering strain, ε.)

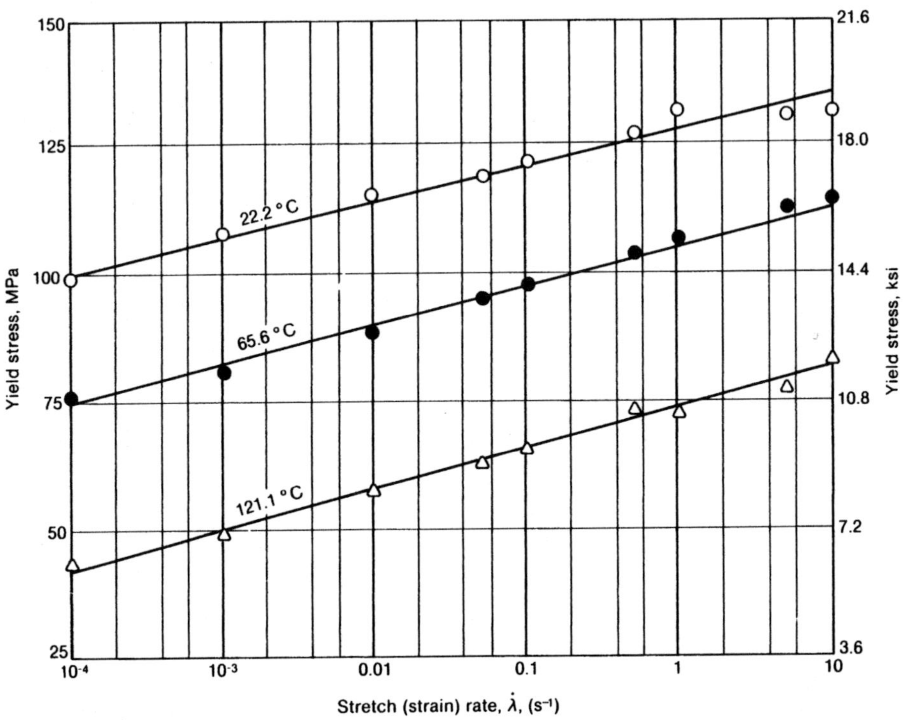

Fig. 4 Strain-rate and temperature dependence of yield stress for polyether-imide

model used in this investigation. With regard to the material model, a trilinear, elastic-plastic constitutive curve was used to approximate ductile plastic behavior. Such a curve is fitted to the PC true stress versus true strain data shown in Fig. 13. Using this material model, there are now two additional material parameters, ε_d and E_3, that characterize the large-strain behavior of the material.

With the finite-element representation shown in Fig. 15 and the trilinear material model, parameter studies showing the effects of ε_d and E_3 on the predicted load-displacement behavior during a puncture test can be carried out. In Fig. 16, the draw strain, ε_d, is held constant at 0.4, and the hardening modulus, E_3, is varied from 0 to 414 MPa (60 ksi). Young's modulus is 2.1 GPa (0.310×10^6 psi), and the yield stress is 69 MPa (10 ksi)—values representative of PC. As can be seen, the higher values of E_3 provide significant improvements in both maximum load and energy absorbed during the puncture test. Figure 17 illustrates the differences in the predicted deformations of the coupon at maximum load when E_3 is 0 and 276 MPa (40 ksi). In a manner similar to the propagating neck of the tensile test, a finite value for E_3 stabilizes the noticeable thinning that occurs immediately under the center of the indenter and forces this process to occur at larger distances from the centerline. This redistribution of load allows significantly more energy to be absorbed and makes the material with a larger E_3 significantly more

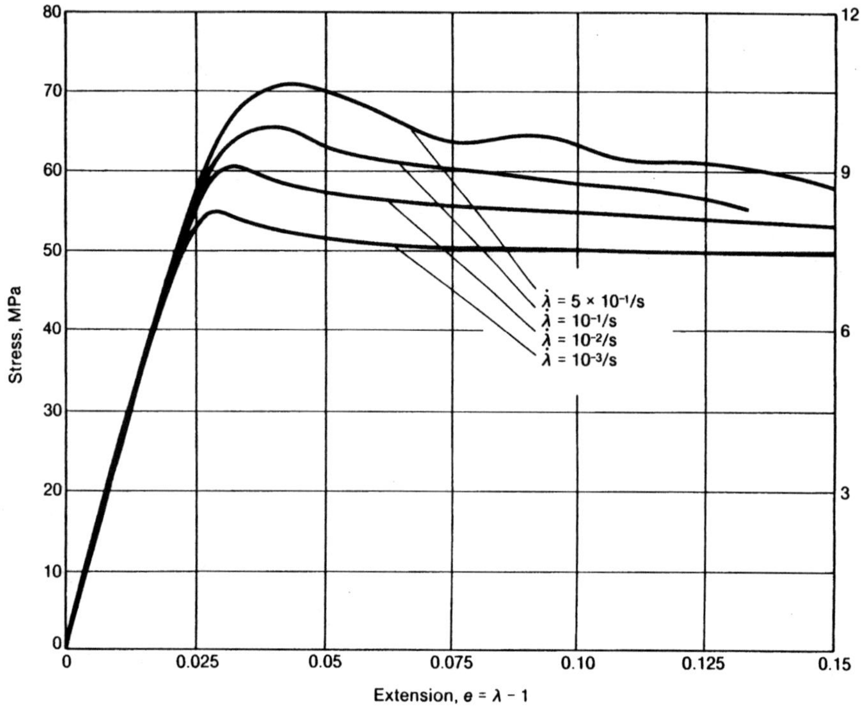

Fig. 5 Stress-strain behavior of polybutylene terephthalate as a function of strain rate, $\dot{\lambda}$, at 22.2 °C (72 °F). (Note: For small strains, extension, *e*, is approximately equal to engineering strain, ε.)

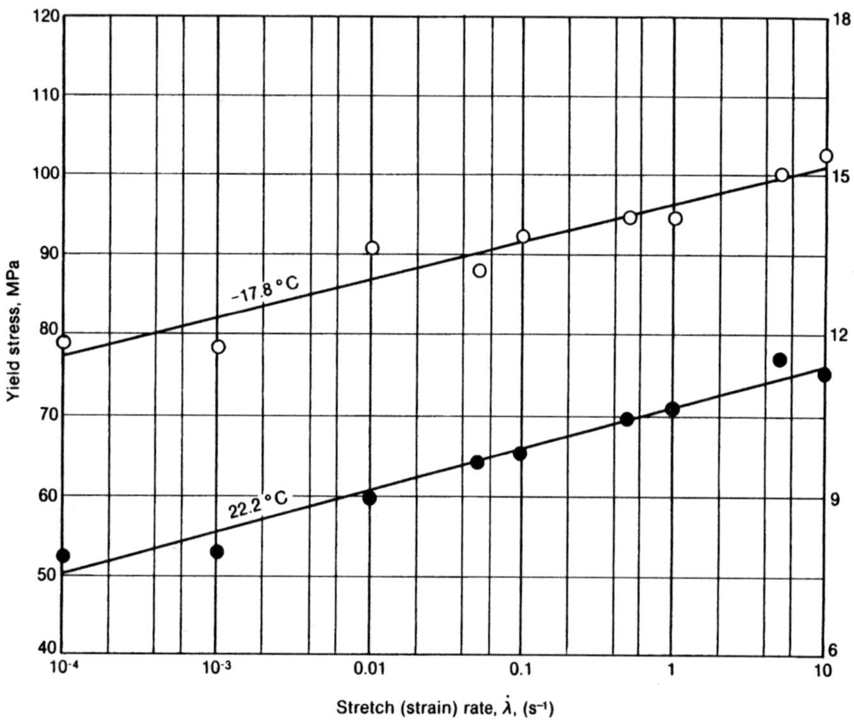

Fig. 6 Strain-rate and temperature dependence of yield stress for polybutylene terephthalate

Fig. 7 Impact test of a polycarbonate box section

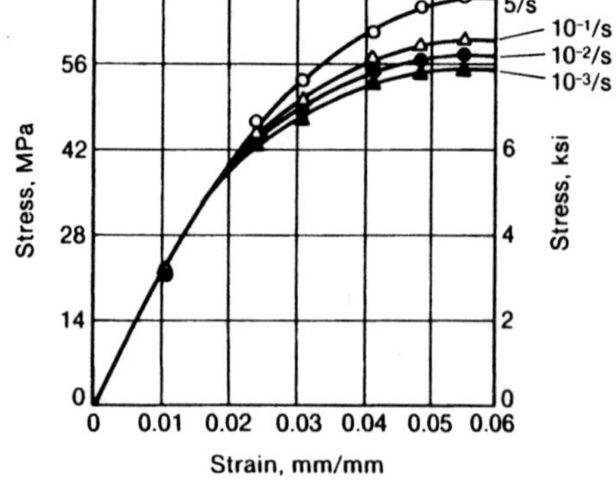

Fig. 8 Stress-strain curves for rubber-modified polycarbonate at room temperature as a function of strain rate

puncture resistant, if all other material properties are equal.

The experimental data shown in Fig. 16 represent a puncture test on PC at room temperature. At up to approximately 8 mm (0.30 in.) of indenter displacement, the predicted responses for all values of E_3 are identical and agree well with the PC test data. After this point, which is only approximately 35% of the indenter displacement at experimental failure, the predicted response is significantly affected by the large-strain hardening modulus, E_3, emphasizing the importance of this property on puncture resistance.

Similarly, a parameter study with respect to the draw strain, ε_d, is illustrated in Fig. 18. In this case, the hardening modulus, E_3, is held constant at 276 MPa (40 ksi), and the draw strain, ε_d, is varied from 0 to infinity. As can be seen, the draw strain value also plays an important role in the puncture resistance of plastic. If the draw strain becomes too large, it tends to offset the benefits of a high E_3 value.

As an indication of the accuracy of the modeling used for these parameter studies, comparisons with puncture experiments have been made at different temperatures and loading rates and for different indenter and disk geometries. True stress versus true strain data for PC taken from Ref 7 were used, and the effects of temperature and strain rate on Young's modulus and yield stress were accounted for. The comparisons of load-displacement predictions with experimental data are shown in Fig. 19. Given the simplicity of the trilinear material model and the severity of the deformation involved, the comparisons are very reasonable. The analytical predictions were terminated after maxima in the load-displacement curves, which were associated with rapid thinning of the disk material in the general vicinity of the indenter. It is not clear whether this thinning process is the same failure phenomenon that results in final rupture of the disk, but in the investigations reported in this article, the predicted maximum loads were always conservative. Additional material data relevant to final failure, as well as more detailed material modeling at large strain, would be necessary to assess the predictability of final rupture.

Puncture tests similar to those discussed in the previous paragraphs are often used as indicators of impact resistance for plastics. It is clear that the well-defined engineering data characterizing Young's modulus and yield strength as functions of temperature and strain rate are not sufficient to rank ductile plastics on the basis of energy absorbed during such a test. The analytic models described here illustrate the fact that the results of these tests are significantly affected by large-strain material properties that are not standardly measured (that is, true stress versus true strain data after yield). However, despite the substantial effects of these large-strain properties on the outcome, these tests do not measure fundamental material data useful to an engineer in the same sense as the Young's moduli and yield strengths measured in tensile tests.

The experimental behavior and the analysis discussed thus far have assumed ductile material response. Unfortunately, the type of response illustrated in Fig. 16 through 19 is not characteristic of all impact events, even for a very ductile plastic such as PC. For example, if the temperature of the plate puncture test discussed previously is reduced from 23 °C (73 °F) to below −90 °C (−130 °F), the load deflection takes on the much more linear form shown in Fig. 20. In addition, the associated energy absorbed during the test is substantially reduced, and the failure mode is brittle, as shown in Fig. 21. Such a change in the characteristic failure mode of material can be a disastrous surprise for an engineer who thought he was dealing with a very ductile material. The factors that contribute to this type of transition in behavior are discussed in the following section.

Effect of Strain Rate and Temperature on Failure Mode. As indicated in the introduction to this article, high loading rates are closely associated with impact events. Therefore, a significant consideration in the design of components that will be subjected to impact events is whether the mode of failure of a plastic component will be significantly affected by the strain rate imposed during the impact event. A plastic that is normally very ductile, with large strains to failure at slow strain rates, may fail in a completely brittle fashion at a relatively small strain if the rate of loading is sufficiently increased. This type of behavior can be observed in a simple tensile test. Nylon, for example, will yield, neck, and experience significant values of elongation at low strain rates. However, as the strain rates are increased, the amount of elongation to failure is significantly reduced until, at a sufficiently high rate, the nylon tensile specimen will break in a brittle manner with little or no plastic deformation. Such behavior obviously has significant implications in a component for which

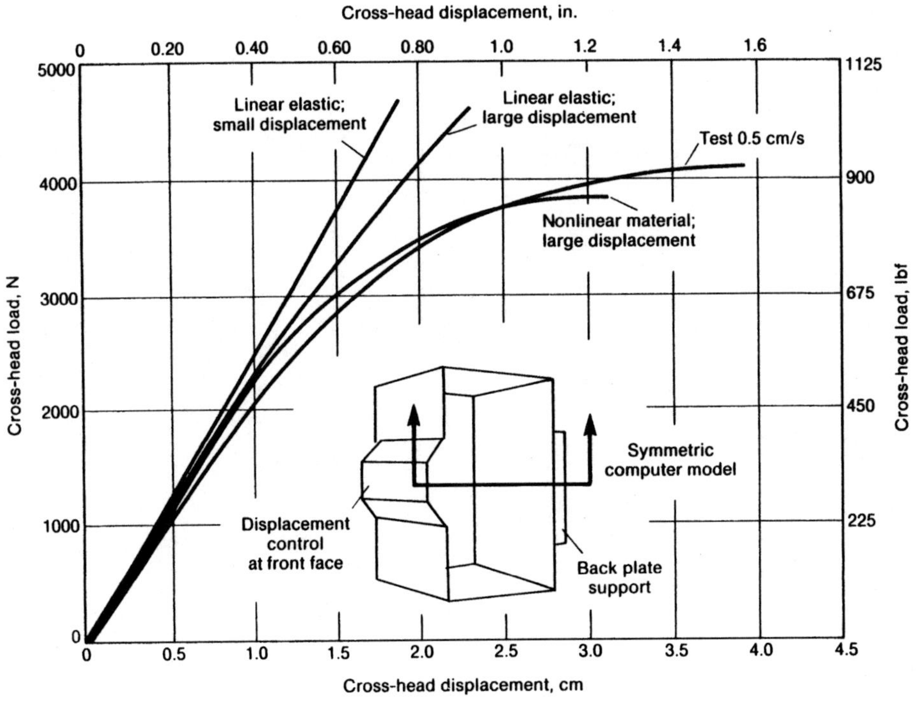

Fig. 9 Load-displacement behavior of an impacted rubber-toughened polycarbonate box

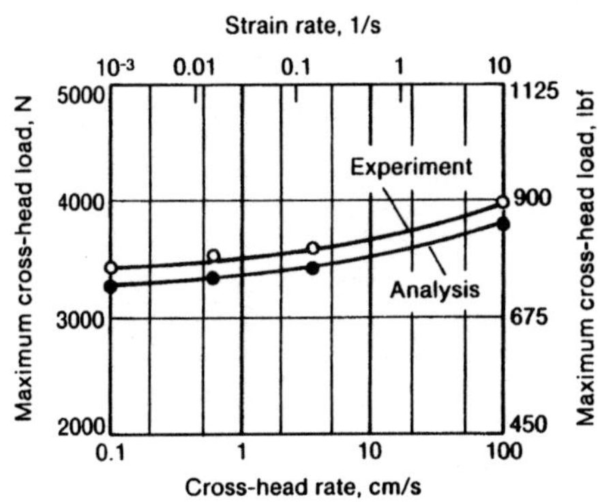

Fig. 10 Comparison of test values with predictions of the maximum load of an impacted polycarbonate box

impact is a design consideration. A plastic exhibiting this behavior could be tolerant of overloads at low strain rates because of its ability to redistribute load through the yielding process, but at the high rates, failure may be catastrophic, intolerant in nature, and associated with low energy absorption levels.

The rate at which impact loading takes place is not the only variable that influences the mode of failure of a plastic. As one would expect from the discussion of the rate dependence of yield stress, the temperature at which the load takes place will also significantly affect the failure process. In general, for a given rate of loading, a plastic will exhibit a transition in failure (from ductile to brittle in nature) as the temperature is reduced. Actually, the transition from ductile to brittle failure is most often discussed in terms of a transition temperature at a given strain rate.

Transitions from ductile to brittle failure as a function of temperature and strain rate are not unique to plastics. The technical history associated with this problem for metals is discussed in Ref 1. The work reported in Ref 1 suggests that it is useful to understand issues associated with this transition as observed for metals in terms of two distinct deformation and failure modes: one producing brittle fracture by separation and the other corresponding to yield through sliding. Similarly, these two competing modes of deformation are used in Ref 4 to discuss the ductile-to-brittle transition in the failure of plastics.

Using this concept of competing modes of deformation and failure, the interrelationship of rate and temperature with regard to ductile-to-brittle transitions can then be illustrated qualitatively. Figure 22 shows two sets of curves, one describing the brittle failure stress at two strain rates for a hypothetical plastic as a function of temperature, and the other describing the yield stress of the same material for the same strain rates. Although the brittle failure stress usually shows only a small dependence on rate, the yield stress is affected much more significantly (Ref 19). Furthermore, the yield stress usually also shows a stronger dependence on temperature than the brittle failure stress, as reflected in the steeper yield curves as a function of temperature in Fig. 22. The net effect illustrated in Fig. 22 is that as the strain rate is increased, the transition temperature defining the intersection of the ductile and brittle failure limits moves to higher temperatures.

A second factor that influences the ductile-to-brittle transition temperature is the state of stress imposed on the component. For multiaxial states of stress defined by the three principal stress components σ_1, σ_2, and σ_3, the character of the stress state can be described mechanically as having a dilatational (volumetric) component that can be quantified with the hydrostatic stress defined as:

$$\sigma_H = \tfrac{1}{3}[\sigma_1 + \sigma_2 + \sigma_3] \qquad \text{(Eq 2)}$$

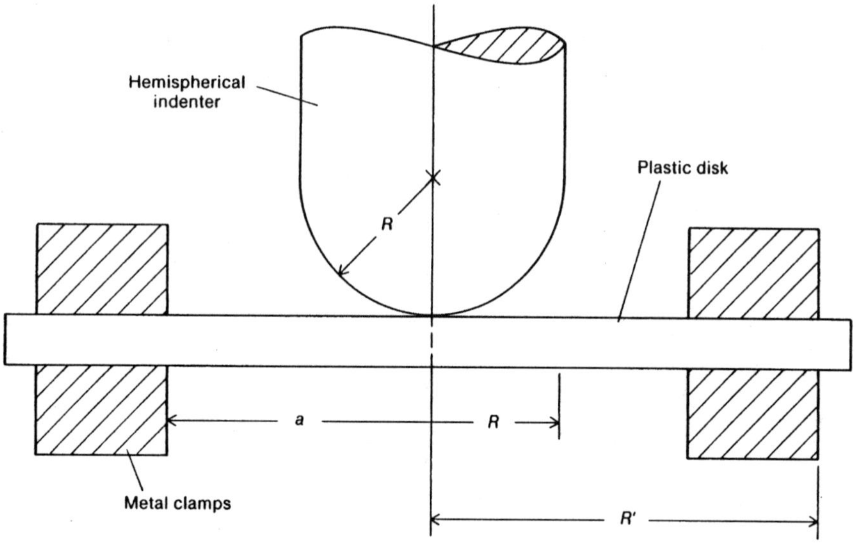

Fig. 11 Schematic puncture test geometry

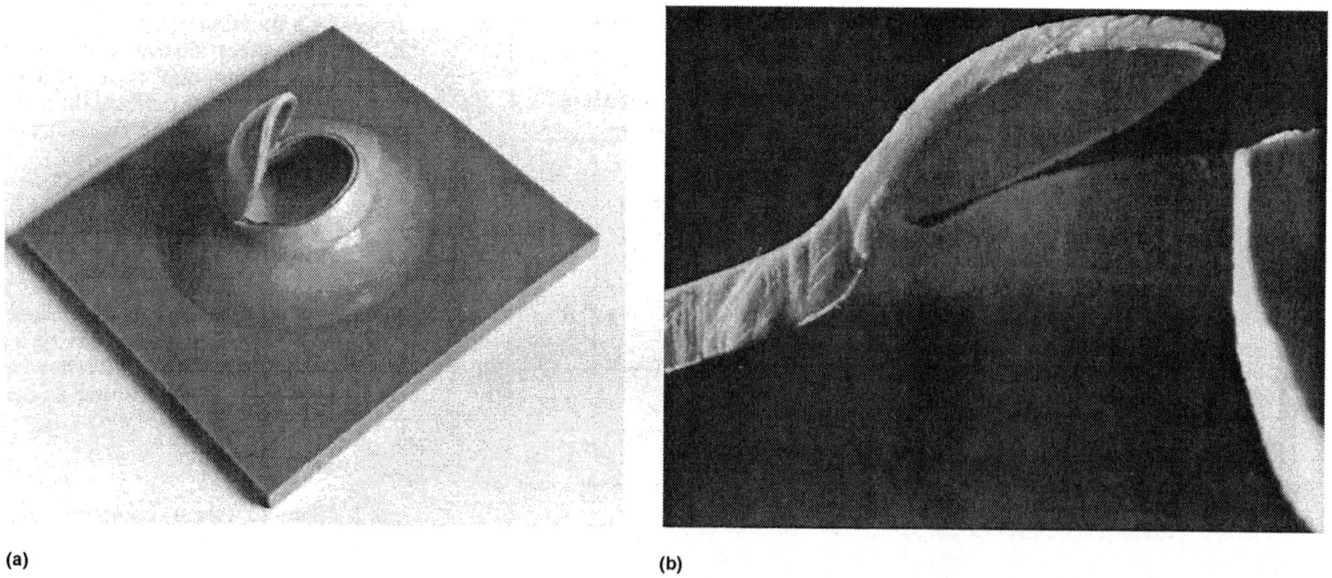

Fig. 12 Permanent deformation of flat polycarbonate plate due to puncture test. (a) View from specimen underside. (b) Cross section of puncture area showing thinning of section

and a deviatoric (shear) component quantified by the octahedral shear stress:

$$\sigma_{oct}^2 = \tfrac{1}{3}[(\sigma_1 - \sigma_2)^2 + (\sigma_2 - \sigma_3)^2 + (\sigma_3 - \sigma_1)^2] \quad \text{(Eq 3)}$$

States of stress in which the ratio of hydrostatic stress to octahedral shear stress (τ_0) is high are generally associated with higher transition temperatures than stress states dominated by shear deformation. This effect can be illustrated in several ways. Reference 20 shows that for a PC material and a strain rate of 8/s, the biaxial state of stress in the grooved tensile bar shown in Fig. 23(c) ($\sigma_H/\tau_0 \approx 0.8$) is associated with a higher transition temperature than that associated with a tensile test ($\sigma_H/\tau_0 = 0.408$). Still greater values of the ratio of hydrostatic to effective stress are achieved if a beam specimen with a notch is tested. For the case shown in Fig. 23(d), a three-dimensional tensile state of stress is imposed on the material in the vicinity of the notch, and the transition temperatures from ductile to brittle failure are still higher. The hydrostatic nature of the stress in a notched beam can be still further magnified as the beam is made thicker, thus increasing the through-thickness constraint and stress. As a result, the ductile-to-brittle transition temperature can be moved to still higher temperatures for a given notch radius and strain rate by making the beam thicker. The net practical result of these effects is that the likelihood of a brittle fracture of a plastic component during an impact event is much higher at higher strain rates, lower temperatures, and for highly constrained (thick) plastic components in the presence of notches and similar stress concentrators. Reference 4 summarizes work reported in this area with respect to plastics and notes some of the observed similarities in behavior with respect to other materials.

Although the qualitative effects of the parameters discussed previously are well understood, an accurate, quantitative approach to predicting failure loads and ductile-to-brittle transition temperatures for an arbitrary component based on fundamental material properties has not been demonstrated. Using tensile coupons, Menges (Ref 21, 22) has studied the effect of rate of loading on strain to failure and suggests that there is evidence of a minimum strain to failure as rates of loading are increased. He also suggests, along with other investigators (Ref 23, 24) that designing a structure to maintain strains below this level would produce a safe design. Unfortunately, the suggested minimum strain to failure is very low for the plastics he has considered and would severely restrict the designer into designing his part as if it were a brittle structure, even when the selected material was most likely chosen because of its ductility.

An alternative approach is the definition of transition temperature for a given material, below which brittle failure at the rate of loading associated with design would have to be expected. The definition of this temperature would allow the designer to choose a material consistent with his design requirements and expected range of temperature in application. Along these lines, a way to estimate the ductile-to-brittle transition temperature as a function of deformation rate based on the dynamic mechanical testing of plastics was suggested (Ref 22). Good agreement between estimates and experimental data from tensile tests for the materials considered was found. However, for such materials as PC, the transition temperatures shown are much lower than those found in the notched

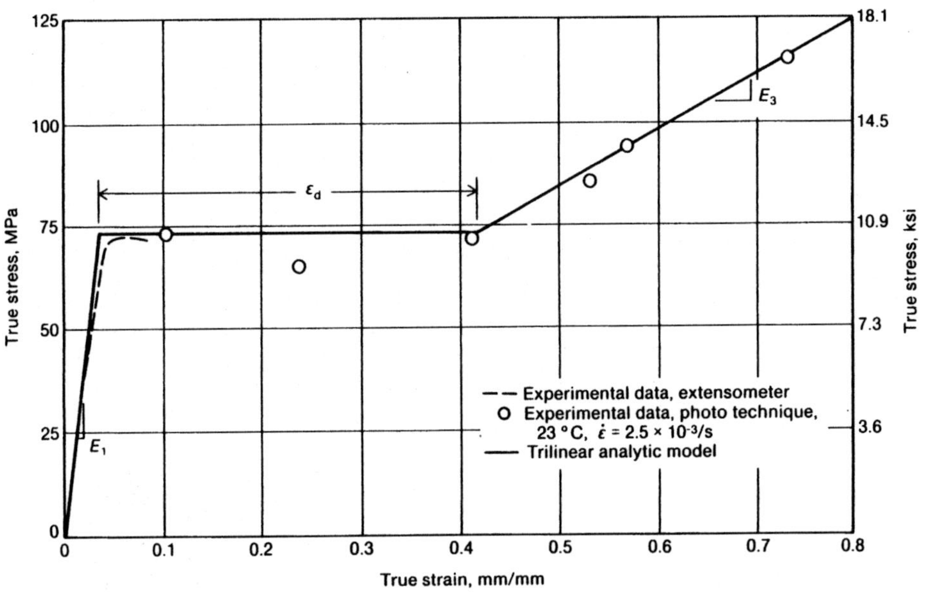

Fig. 13 True stress/true strain behavior of polycarbonate. E_1 = 2.06 GPa (0.30 ksi); yield stress, σ_y = 73 MPa (10.6 ksi); draw strain, ε_d = 0.375; hardening modulus, E_3 = 145 MPa (21 ksi)

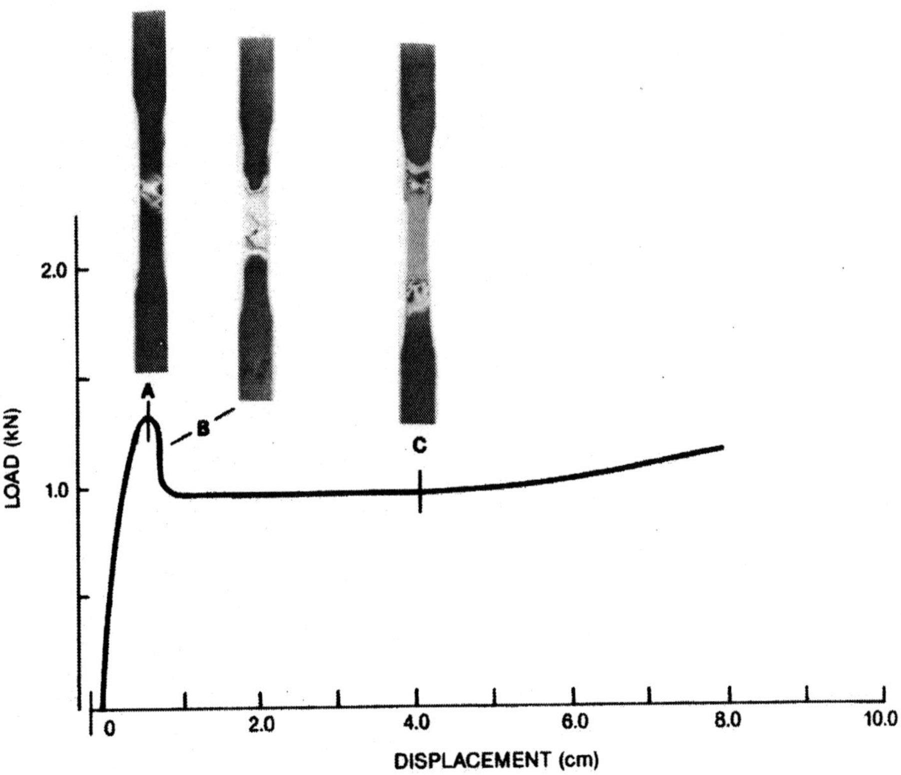

Fig. 14 Phenomenon of propagating neck in a polycarbonate tensile specimen. 1 kN = 0.11 tonf; 1 cm = 0.4 in.

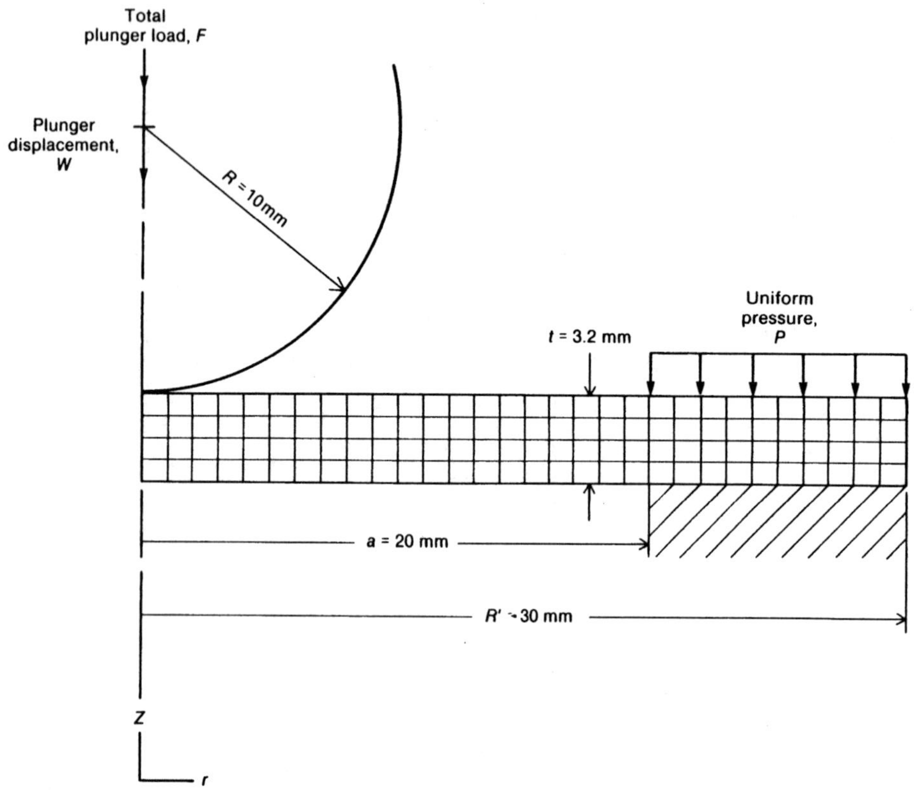

Fig. 15 Finite-element model of puncture test. t, thickness

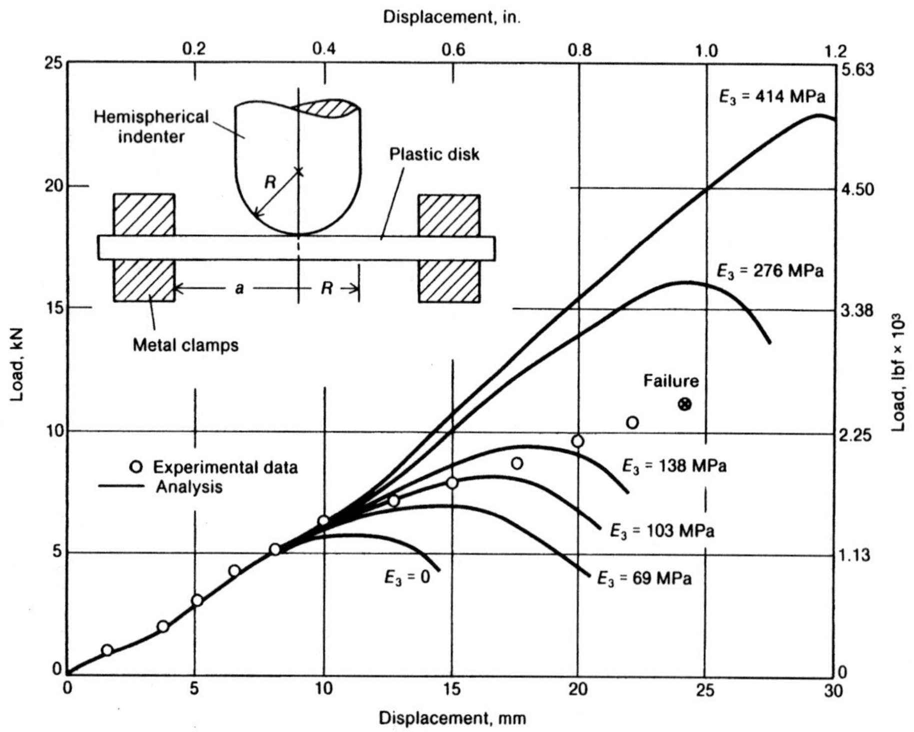

Fig. 16 Predicted puncture test response as a function of final material hardening modulus, E_3. Draw strain, $\varepsilon_d = 0.40$, yield stress, $\sigma_y = 69$ MPa (10 ksi)

impact tests for the same material reported by other investigators (Ref 25). This is, of course, further evidence of the effect of stress state on transition temperature. Brostow (Ref 26) presents a method that does attempt to account for the effects of notches on the observed transition temperature. However, his predictions in Ref 27 for the impact transition temperature of low-density polyethylene do not account for the effects of rate. Pragmatically, the definition of a ductile-to-brittle transition temperature would seem to be an extremely important engineering variable. However, there does not appear to be an agreed-upon method of measuring and reporting such a quantity in a manner that is easily applicable for the engineer.

Impact Tests. Although a number of standard impact tests are used to survey the performance of plastics exposed to different environmental and loading conditions, none of these tests provides real, geometry-independent material data that can be applied in design. Instead, they are only useful in application to quality control and initial material comparisons. Even in this latter role, different tests will often rank materials in a different order. As a result, proper test choice and interpretation require that the engineer have a very clear understanding of the test and its relationship to his own design requirements. Perhaps three of the most commonly used tests for impact performance are the Izod and Charpy notched beam tests and the dart penetration test. Each is briefly discussed as follows.

The Izod and Charpy tests are very similar in that they are both notched beam specimens subjected to bending moments. The notch serves to create a stress concentration and to produce a constrained multiaxial state of tension a small distance below the bottom of the notch. Both of these effects tend to make the test severe from the standpoint of early transition to brittle behavior as a function of both rate and temperature. The geometries for the two tests are shown in Fig. 24.

The Charpy geometry consists of a simply supported beam with a centrally applied load on the reverse side of the beam from the notch. The Izod geometry, on the other hand, consists of a cantilever beam with the notch located at the root of the beam. In both cases, the load is applied dynamically by a free-falling pendulum of known initial potential energy. The important dimensions of interest for these tests include the notch angle, the notch depth, the notch tip radius, the depth of the beam, and the width of the beam. All these quantities, as well as more detailed information specifying loading geometry and conditions, are listed in the standards of ASTM International, such as ASTM D 256 (Ref 28). As the pendulum falls in both of these tests, its original potential energy is converted to kinetic energy. Some of this energy is in turn used to break the specimen, which is encountered at the low point of the pendulum arc. The pendulum then continues its swing and comes to a halt at a height less than its starting location.

The energy expended to break the specimen can then be calculated as the difference between the initial and final potential energies of the pendulum. The value generally reported from these tests is the energy required to break the bar divided by the net cross-sectional area at the notch. The unit of this measure is Joules per square meter (J/m^2), and the parameter is referred to as impact strength.

It should be emphasized that the units of this measure are not those of stress, nor is impact strength, as defined by these tests, a material property. The measurement cannot be used to design a component. In fact, the values of impact strength are significantly affected by the parameters defining the specimen geometry, such as notch tip radius and depth. Even the identification of a transition temperature can be significantly affected by geometry, such as the width of the beam. Wider beams tend to provide more plane-strain constraint, and transition temperatures with wider beams often appear more distinct and at higher temperatures than results from narrower beams. Consequently, when comparing materials by using the value of impact strength, it is imperative that the test geometries be identical. In addition, the transition temperature identified by plotting the impact strength measured with an Izod or Charpy specimen as a function of temperature may or may not be appropriate for the component under consideration.

Another impact test that is often reported is the dart penetration (puncture) test. This test (Fig. 11) is different from the Izod and Charpy tests in a number of aspects. First, the stress state is two dimensional in nature, because the specimen is a plate rather than a beam. Second, the thin, platelike specimen does not contain any notches or other stress concentrations. The geometry and test conditions often applied using this specimen are described in ASTM D 3029 (Ref 29). The quantity most often quoted with respect to this test is the energy required for failure. Of course, these energy levels are very different from the notched beam energies to failure, but they also do not represent any fundamental material property. A marked transition in mode of failure can also be observed with this specimen as the rate is increased or the temperature is decreased. However, this transition temperature is quite different from that measured in the notched beam tests. Usually, it displays a transition from ductile to brittle behavior at much lower temperatures than the notched specimens.

The dart penetration test is often performed with different specimens and indenter geometries. Linear elastic, small-displacement, thin-plate theory has occasionally been used to analyze test results in an effort to compare the performance of different materials tested with different specimen geometries. In all but the most brittle materials, this is an inappropriate simplification of the test. A number of very nonlinear events can take place during this test, including a growing indenter contact area, yielding, and large-displacement and large-strain deformation. References 17, 18, and 20 provide more details on these events and their effects on the test data.

Obviously, the issues associated with designing for impact are complex, involving significant effects due to rate, temperature, and state of stress. As an aid to materials selection, plastics should be considered with respect to three categories (Ref 30):

- Brittle even when unnotched
- Brittle in the presence of a notch
- Tough; that is, specimens do not break completely even when sharply notched

As indicated previously, some materials will display more than one of these behavior types as a function of temperature. Table 1 lists the behavior of various plastics as a function of temperature. After assessing his own design requirements, an engineer could use Table 1 to guide his initial material choices.

Fracture Mechanics and Impact Failure. A final engineering concept relevant to the areas of impact and ductile-to-brittle transition is fracture mechanics, which is discussed in the preceding article, "Fracture Resistance Testing," in this book. The geometry of interest in this case is shown in Fig. 25. The stress state in the vicinity of the crack tip can be understood as the limiting case of an elliptical notch as the notch tip radius approaches 0. For such a geometry, the stresses perpendicular to the crack line approach infinity at the crack tip. Their variation as a function of distance from the crack tip is expressed as:

$$\sigma_{11} = \frac{K_I}{\sqrt{2\pi r}}$$

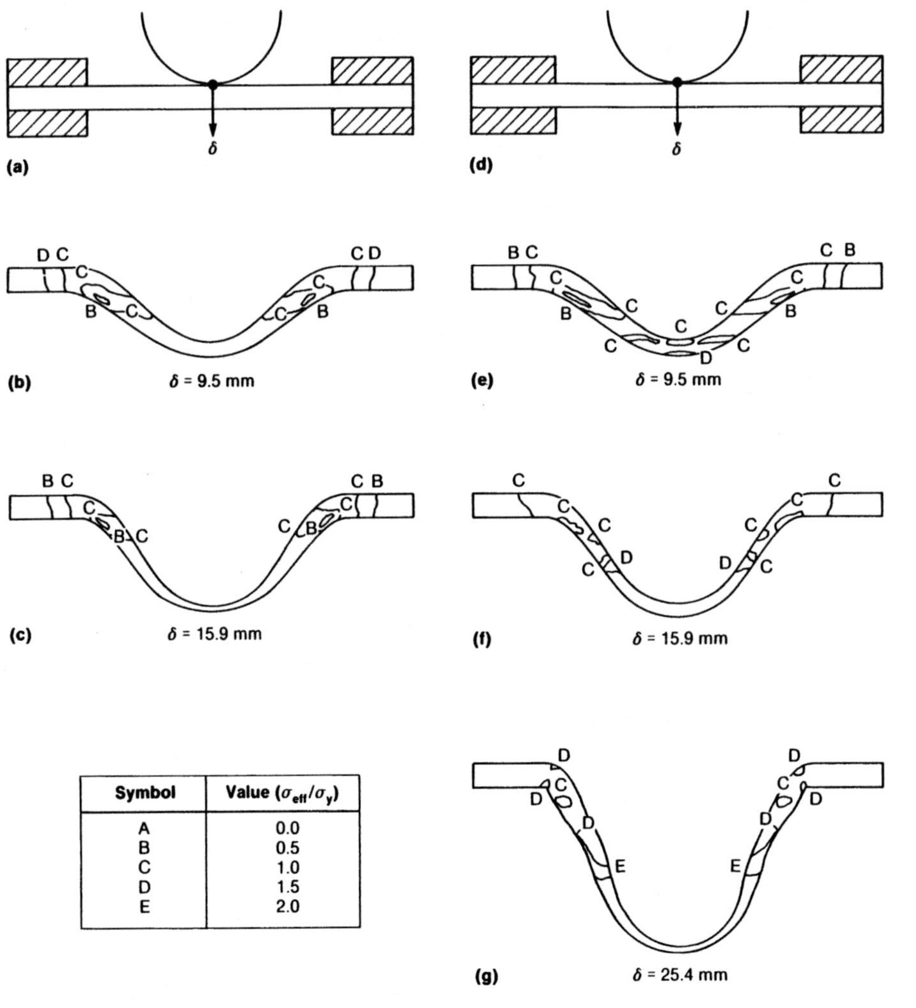

Symbol	Value (σ_{eff}/σ_y)
A	0.0
B	0.5
C	1.0
D	1.5
E	2.0

Fig. 17 Comparison of profiles at maximum load for simulated puncture tests, E_3, hardening modulus; σ_y, yield stress. (a) $E_3 = 0$. (b) $E_3 = 0$ and $\delta = 9.5$ mm (0.375 in.). (c) $E_3 = 0$ and $\delta = 15.9$ mm (0.625 in.). (d) $E_3 = 276$ MPa (40 ksi). (e) $E_3 = 276$ MPa (40 ksi) and $\delta = 9.5$ mm (0.375 in.). (f) $E_3 = 276$ MPa (40 ksi) and $\delta = 15.9$ mm (0.625 in.). (g) $E_3 = 276$ MPa (40 ksi) and $\delta = 25.4$ mm (1.0 in.)

where K_I is a function of the applied load and the component and crack geometries. Obviously, stress can no longer be used as a failure criterion at the crack tip, because the value of stress goes to infinity in a linear elastic material. However, under the proper circumstances, there is a true material property that can be used to predict failure in this case: the critical stress-intensity factor, K_{Ic}, which is also referred to as plane-strain fracture toughness.

The value of the critical stress-intensity factor for a material can be measured by testing standard cracked specimens, such as the compact-tension specimen illustrated in Fig. 26. Standard test methods and specimen geometries are defined for measuring the critical stress-intensity factors for metals as part of ASTM E 399 (Ref 31), but these standards have yet to be officially defined for plastics. Although the issue is still argued, it appears that many of the recommendations of the ASTM E 399 test procedure for metals are equally worthwhile for plastics (Ref 32). For the compact-tension test, the stress-intensity factor K_I can be related to the test load as:

$$K_I = P\,Y_{CT} \qquad \text{(Eq 4)}$$

where Y_{CT} is a function of the specimen geometry defined in Ref 31. Using Eq 4, a relationship between the load at failure, P_F, and the stress-intensity factor associated with that failure load, K_Q, can be established:

$$K_Q = P_F\,Y_{CT} \qquad \text{(Eq 5)}$$

If the fracture event meets the standards of applicability for linear elastic fracture mechanics as defined in Ref 31, then the stress intensity at failure, defined in Eq 5, is referred to as the plane-strain fracture toughness, or critical stress-intensity factor, K_{Ic}, and represents a material property.

As a material property, K_{Ic} can then be used to quantify the brittle failure of other more general components. These cracked components will also be characterized by a stress-intensity relationship, which can be expressed as:

$$K_I = P\,Y_{COM} \qquad \text{(Eq 6)}$$

where Y_{COM} is now the function of geometric variables relating the stress-intensity factor to the applied load. Because failure by fracture is defined to occur when the stress-intensity factor reaches K_{Ic}, the lower-bound failure load of the component structure can be defined as:

$$P_F = \frac{K_{Ic}}{Y_{COM}} \qquad \text{(Eq 7)}$$

The concept of fracture mechanics has proved to be appropriate for a wide range of materials, including glass, ceramics, and many metals. For the more brittle of these materials, this approach is valid over most of the range of temperatures of engineering interest. In other materials, there is a marked transition in behavior as a function of temperature very similar in nature to that noted for the notched impact tests. Below this transition temperature, linear elastic fracture mechanics is valid. However, at temperatures above this transition, plasticity in the vicinity of the crack tip becomes significant, the material requires much more energy to fail, and the approach of linear elastic fracture mechanics is no longer valid.

Plastics offer a similar range of behavior. Some brittle plastics show extensive ranges of applicability for linear elastic fracture mechanics. In other cases, a transition temperature exists. Figure 27 illustrates the dependence of the apparent fracture toughness, K_Q, of a rubber-toughened thermoplastic material as measured using compact-tension specimens. In the vicinity of –15 °C (5 °F), there is a very sudden change in the values of apparent fracture toughness. For the material and rate considered, linear elastic fracture mechanics procedures were found to be valid for temperatures below –15 °C (5 °F) (Ref 32). However, above this temperature, it can be shown that the test results do not meet the requirements for the definition of true fracture toughness properties, and the procedures of linear elastic fracture mechanics cannot be applied. These tests, using compact-tension specimens, were all conducted at a constant stress-intensity rate. The rate of loading can affect fracture behavior. Higher loading rates tend to move the transition temperature to higher values, thus promoting brittle fracture at higher temperatures.

The important point to be emphasized about fracture toughness measurements in contrast to Izod and Charpy data is that in the range of temperatures below transition, plane-strain fracture toughness is recognized to be a true material property that can be used to predict failure in general components as a function of crack length. Figure 28 illustrates a comparison between analytical predictions of failure and experimental measurements in a low-temperature impact event (Ref 32). The component tests considered were channel sections, also shown in Fig. 28, made of a rubber-toughened thermoplastic blend whose fracture behavior was illustrated in Fig. 27. The component tests were conducted at –50 °C (–60 °F), well below the transition temperature, and at the same stress-intensity rate as the compact-tension tests used to define K_{Ic} (100 MPa $\sqrt{m}$/s, or 90 ksi $\sqrt{in.}$/s). The solid lines in Fig. 28 represent the bounds on the analytical prediction as a result of uncertainty in boundary conditions.

For this material, significant sensitivity was noted in the compact-tension tests with respect

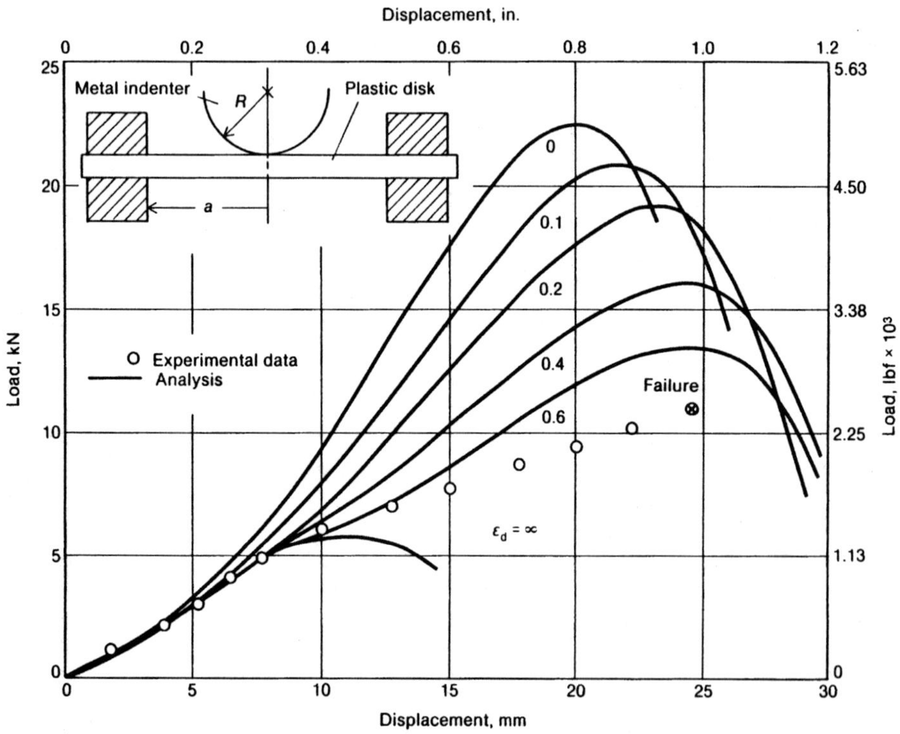

Fig. 18 Predicted puncture test response as a function of material draw strain, ε_d. Yield stress, σ_y = 69 MPa (10 ksi); hardening modulus, E_3 = 276 MPa (40 ksi)

to crack preparation. The cracks had to be introduced through fatigue at low loads, or they tended to become blunt. Whenever this blunting occurred in the compact-tension tests, higher values of apparent fracture toughness were observed. The cracks in the component were also introduced in fatigue from a razor notch. When sharp cracks were achieved (open circle, Fig. 28), the test results agreed well with the predictions. When blunt cracks appeared, along with stress whitening at the crack tip, higher fracture loads were observed (closed circle, Fig. 28). This is consistent with the concept of linear elastic fracture mechanics as a lower-bound failure prediction. The agreement between analysis and experiment is very good.

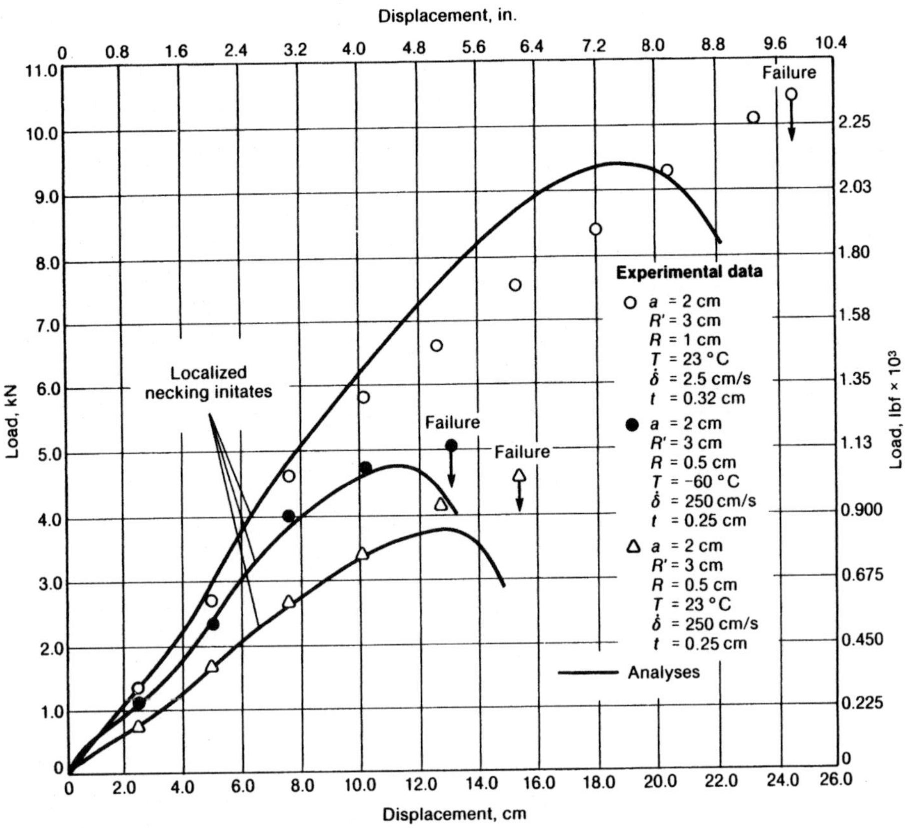

Fig. 19 Comparison of experimental results and predictions of puncture tests on polycarbonate. *T*, temperature; *t*, thickness

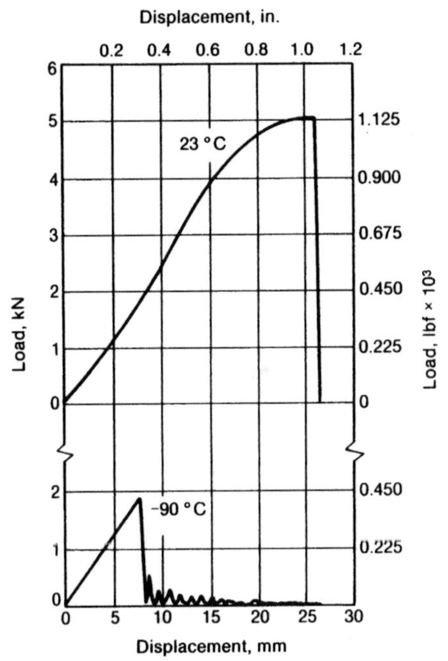

Fig. 20 Comparison of load-deflection behavior of polycarbonate disk at room temperature and at −90 °C (−130 °F)

(a) (b)

Fig. 21 Comparison of failed polycarbonate disk from puncture tests (a) at room temperature and (b) at −90 °C (−130 °F)

Figure 29 illustrates the severe effect that a crack can have during a low-temperature impact event. The component test considered in Fig. 29 is the same impacted box used in the previous discussion. The linear load-displacement curve is representative of a cracked box section, while the very nonlinear curve represents test results at the same rate and temperature for a part without a crack. Without the presence of a crack, the channel section experienced a maximum load due to yielding but no fracture. With the crack introduced, there was linear behavior to brittle failure. The difference in the amount of energy that can be absorbed during the two impact events is significant.

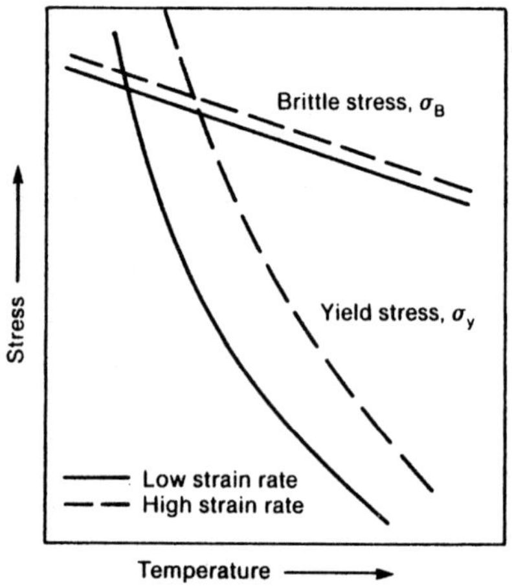

Fig. 22 The interrelationship of temperature and rate defining transition from ductile to brittle failure

Other engineering considerations also influence the impact resistance of plastics. For example, the processing conditions under which the component is made can have a significant effect on the impact strength of the material. Raising the temperature in the barrel of an injection molding machine can affect both the impact strength and the transition temperature from ductile to brittle fracture. Processing requirements vary for different plastics and must be well understood. The standardized impact tests discussed previously can be of use as quality-control tools to ensure that a material with a history of good impact performance does not suffer from the effects of poor processing.

In addition to processing temperature, the engineer must also consider the potential existence of weld lines in the molded part. A weld line is a locus of points in a molding at which two fronts of molten plastic meet during the molding process. The strength of these weld lines may be inferior to the plastic, as measured by standard tests on molded coupons without weld lines.

Plastics are susceptible to chemical attack from a variety of agents. A common result of this reaction is a severe embrittlement of the material. In choosing a plastic for a particular component application, the engineer should consider the chemical environment in which the component will function. He must then assess the effects that the chemical associated with the environment will have on the plastic.

Photooxidation of unsaturated rubber (a component of most rubber-toughened polymers) due to exposure to the ultraviolet component of sunlight (Ref 33) usually causes a severe reduction of impact properties at lower temperatures. There is usually a much smaller effect on impact strength at higher temperatures. Solutions to this problem (often referred to as aging) are discussed in Ref 33.

Design and Analysis Techniques for Thin Plastic Components

Injection-molded plastics are most often thin structures because of their manufacturing process. As a result, they are often analyzed using the simplified continuum theories of beams, plates, and shells. Two assumptions that

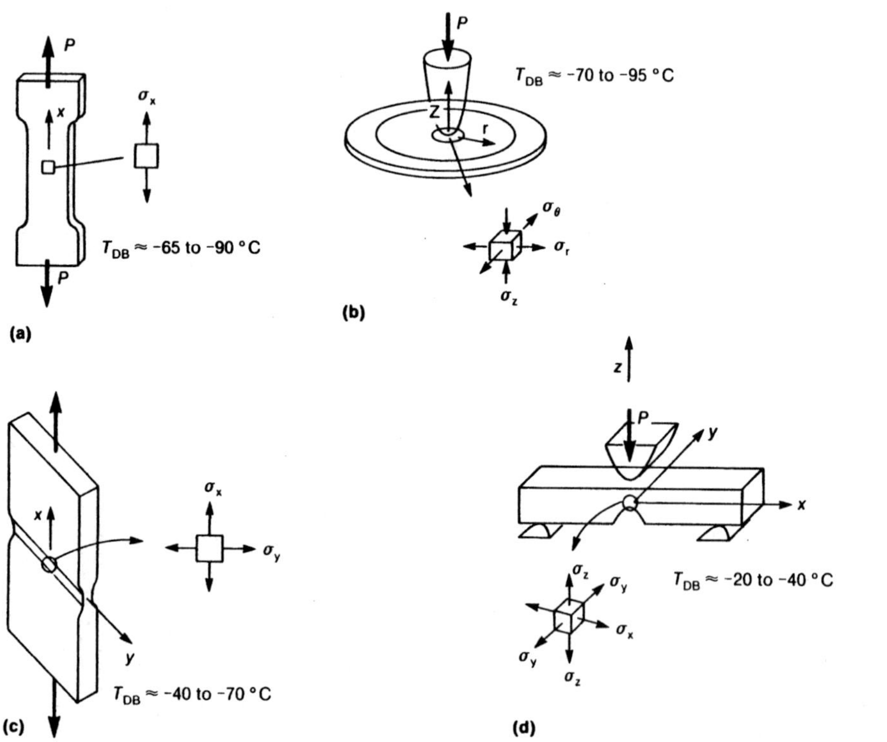

Fig. 23 Effect of stress state on the ductile-to-brittle transition temperature, T_{DB}, for polycarbonate. *P*, pressure; σ, stress. (a) Tensile test. (b) Puncture test. (c) Strip biaxial test. (d) Notched beam test

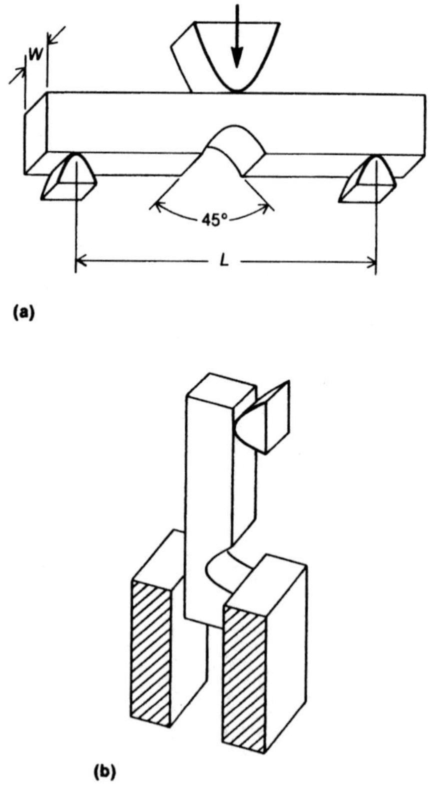

Fig. 24 Schematics of (a) Charpy and (b) Izod notched beam geometries. *W*, width; *L*, length

are regularly made when the customary engineering equations are applied to solve problems of this nature are that the rotations describing the continuum deformation are small (often referred to as the small-displacement assumption) and that the strains imposed on the structure are small. Because of the very different mechanical property values associated with plastics compared to metals, both of these assumptions need to be examined. The elastic moduli of polymers are routinely as much as two orders of magnitude less than those of metals. That can often result in larger rotations and displacements in thin plastic structures than an engineer may be accustomed to with metals. Although metals often have yield strains of the order of 0.001 to 0.005, many plastics are characterized by yield strains near 0.05. As long as the engineer limits his interest to strains in the vicinity of or less than the yield strain, the small-strain assumption will still be reasonably accurate. However, the same cannot be said for the applicability of the small-rotation (small-displacement) assumption when applied to thin plastic components.

To develop an understanding of the significance of moderately large rotations in thin plastic structures, the typical behavior of beams, plates, and shells is examined as the displacements and rotations imposed on these structures grow. For the examples that are discussed, the physical significance of the mathematical assumption of small rotations is that, as long as displacements (and rotations) in a thin structure are small enough, there will be an insignificant amount of stretching along the beam length or in the plate surface. In such cases, the deformation can be accurately described in terms of bending theories alone, and the boundary conditions for displacements in the plane of the structure will not have an effect on the stiffness of the beam or plate under an out-of-plane load. Such a simplification is not possible, however, if displacements and rotations become large.

At these larger displacement levels, the out-of-plane stiffness of the thin structure is affected by in-plane forces and boundary conditions. As a result of the in-plane forces generated at moderately large rotation levels, thin structures under lateral load may exhibit higher stiffness than would be expected from linear, small-displacement theories. The range of applicability of small-displacement linear solutions in terms of displacement and rotation size is discussed. For a beam, the presence or the absence of this additional, nonlinear stiffness depends on the in-plane boundary conditions at the beam ends. As is shown, there are boundary conditions for which no additional stiffness is generated and small-displacement theory will be accurate. However, in the case of a plate with lateral support on all four edges, this stiffening effect will be present for all possible in-plane boundary conditions. The exact conditions at the boundary will affect the amount of added stiffness, but it will always be present to some extent. Although this stiffening effect in thin plastic structures is a consequence of purely geometric considerations, its importance for plastic structures in comparison to those made of metal is directly related to the large yield strains associated with plastic materials.

The Small-Rotation Assumption and Beams. Consider the simply supported beam under a concentrated, centrally located load shown in Fig. 30. Using the standard Euler-Bernoulli beam theory, the strain in the x-direction of the beam is described in terms of the u (displacement in the x-direction) and w (displacement in the z-direction) displacements of the neutral axis of the beam. For the problem

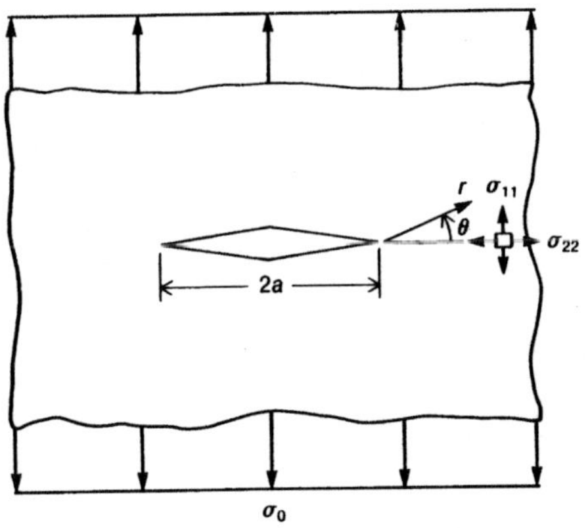

Fig. 25 Crack of length $2a$ in tensile stress field

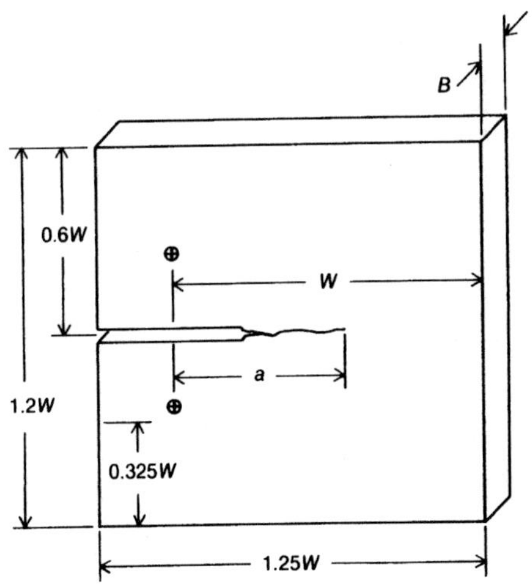

Fig. 26 Schematic of compact-tension specimen. W, width; a, crack length

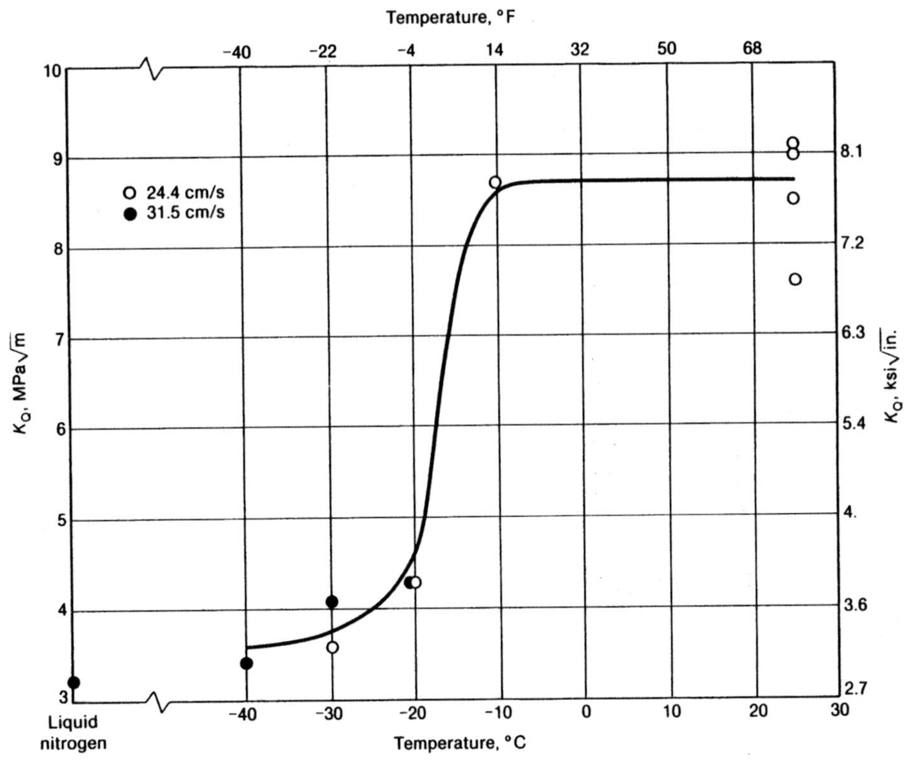

Fig. 27 Apparent fracture toughness as a function of temperature. 6.4 mm (0.25 in.) thick specimens

considered here, it is assumed that the cross section of the beam is a simple rectangle, as shown in Fig. 30, thus making the neutral axis pass through the center of the cross section. If the fully nonlinear, one-dimensional strain-displacement relations of continuum mechanics are used, along with the Euler-Bernoulli assumption that plane sections remain plane, then the strain in the x-direction at any point in the beam can be written as:

$$\varepsilon_x = \frac{du}{dx} + \frac{1}{2}\left(\frac{du}{dx}\right)^2 + \frac{1}{2}\left(\frac{dw}{dx}\right)^2 - zK \quad \text{(Eq 8)}$$

where K is the curvature at the neutral axis of the deformed beam and is defined in terms of the lateral displacement, w, as:

$$K = \frac{d^2w/dx^2}{[1 + (dw/dx)^2]^{3/2}} \quad \text{(Eq 9)}$$

Because $[dw/dx]^2$ is generally much less than 1, the curvature can usually be accurately approximated as:

$$K \cong \frac{d^2w}{dx^2} \quad \text{(Eq 10)}$$

Furthermore, because $du/dx \ll dw/dx$ in laterally loaded beams, Eq 8 can be simplified to:

$$\varepsilon_x \approx \frac{du}{dx} + \frac{1}{2}\left(\frac{dw}{dx}\right)^2 - zK \quad \text{(Eq 11)}$$

Equation 11 includes the square of the beam rotation, $[dw/dx]^2$, again. However, although neglecting it in Eq 9 only required that it be small in comparison to 1, the linearization of Eq 11 requires that $[dw/dx]^2 \ll |du/dx|$, which is a much more stringent requirement. Because there are two quantitatively different assumptions with regard to the size of rotations, the term *moderately large* is sometimes used to describe the situation where $[dw/dx]^2 \approx |du/dx|$. If this latter requirement on the size of the rotation squared ($[dw/dx]^2 \ll |du/dx|$) is met, then Eq 11 can be linearized to:

$$\varepsilon_x \cong \frac{du}{dx} - zK \quad \text{(Eq 12)}$$

which is the well-known strain-displacement relation for linear beam theory and allows uncoupled solutions for the u- and w-displacements.

Now consider the boundary conditions at the ends of the beam. If one or both ends of the beam are assumed to be free to move in the x-direction, then there can be no net force, F, in that direction. Using linear elasticity to relate stress and strain and then integrating the x component of stress over the cross section of the beam, the result is:

$$F = E\int_A \left[\frac{du}{dx} + \frac{1}{2}\left(\frac{dw}{dx}\right)^2 - z\frac{d^2w}{dx^2}\right] dy\, dz$$

$$= EA\left[\frac{du}{dx} + \frac{1}{2}\left(\frac{dw}{dx}\right)^2\right] \equiv 0 \quad \text{(Eq 13)}$$

where A is the cross-sectional area of the beam, E is the elastic modulus, and the rotations are allowed to become moderately large. Using Eq 13 along with Eq 11, the governing strain-displacement relation for the beam under a central, lateral load and the previously mentioned boundary conditions is:

$$\varepsilon_x \cong -zK \quad \text{(Eq 14)}$$

Therefore, under the boundary condition of free displacement in the x-direction at one or both ends of the beam, the full, nonlinear strain-displacement relation given in Eq 11 reduces to the linear equation given in Eq 14 without the requirement that $[dw/dx]^2 \ll |du/dx|$. Using Eq 14 and applying the equations of equilibrium results in the standard linear equation for beam theory.

The situation is different, however, if both ends of the beam in Fig. 30 are prevented from moving in the x-direction. Under this set of boundary conditions, it is no longer possible to equate the force in the x-direction of the beam to 0. As a result, Eq 13 is no longer available, and Eq 11 retains its full nonlinear form. In this case, unless the requirement that $[dw/dx]^2 \ll |du/dx|$ is applicable, the governing differential equations for equilibrium will be nonlinear. From a physical point of view, the bending and stretching deformation of a beam become increasingly coupled if the ends of the beam are restrained in the x-direction and the rotations become moderately large; that is, $[dw/dx]^2 \approx |du/dx|$.

Although an exact solution to these equations is not straightforward, an approximate solution can be obtained by using the theory of minimum total potential energy and approximations for the displacements that fulfill the displacement boundary conditions. Using simple trigonometric expressions, the first approximations for the lateral and in-plane displacements w and u are:

$$w = W_0 \sin\frac{\pi x}{L} \quad \text{(Eq 15)}$$

$$u = U_0 \sin\frac{2\pi x}{L} \quad \text{(Eq 16)}$$

The terms W_0 and U_0 are unknown coefficients that must be determined through minimizing the total potential energy $(U + V)$ of the system, expressed as:

$$U + V = \int_0^L \int_{-t/2}^{t/2} \frac{Eb}{2}\varepsilon_x^2\, dxdz - P\,W_0 \quad \text{(Eq 17)}$$

Differentiating the total potential energy expression (Eq 17) with respect to W_0 and U_0 and setting both expressions equal to 0 results in two nonlinear, algebraic equations for W_0 and U_0. Using these two equations, the load-displacement behavior of the fully nonlinear problem can be derived as:

$$P = \frac{\pi^4 EI}{2L^3} W_0 + \frac{\pi^4 EA}{4L^3} W_0^3 \quad \text{(Eq 18)}$$

Table 1 Plastics behavior as a function of temperature

Plastics	Temperature, °C (°F)							
	−20 (−4)	−10 (14)	0 (32)	10 (50)	20 (68)	30 (85)	40 (105)	50 (120)
Polystyrene	A	A	A	A	A	A	A	A
Polymethyl methacrylate	A	A	A	A	A	A	A	A
Glass-filled nylon (dry)	A	A	A	A	A	A	A	B
Polypropylene	A	A	A	A	B	B	B	B
Polyethylene terephthalate	B	B	B	B	B	B	B	B
Acetal	B	B	B	B	B	B	B	B
Nylon (dry)	B	B	B	B	B	B	B	B
Polysulfone	B	B	B	B	B	B	B	B
High-density polyethylene	B	B	B	B	B	B	B	B
Rigid polyvinyl chloride	B	B	B	B	B	B	C	C
Polyphenylene oxide	B	B	B	B	B	B	C	C
Acrylonitrile-butadiene-styrene	B	B	B	B	B	B	C	C
Polycarbonate	B	B	B	B	C	C	C	C
Nylon (wet)	B	B	B	C	C	C	C	C
Polytetrafluoroethylene	B	C	C	C	C	C	C	C
Low-density polyethylene	C	C	C	C	C	C	C	C

A, brittle even when unnotched; B, brittle, in the presence of a notch; C, tough where I is the moment of inertia. In Eq 18, the linear term in W_0 can be recognized as the approximate solution to the standard Euler-Bernoulli equation for linear beam theory. The cubic term provides additional stiffening to the system as the displacements become large. This stiffening is the physical consequence of stretching the length dimension of the beam as a result of lateral displacements perpendicular to the original length of the beam. As can be seen in Fig. 30, this nonlinear effect on the lateral stiffness of the beam becomes important after the displacement of the beam becomes significant with respect to the beam thickness ($W_0/t \approx 0.3$). This in turn leads to the description of these results as a large-displacement solution; that is, displacements significant with respect to the thickness of the beam can be accurately accommodated.

For most realistic beam structures, the boundary conditions at the ends are more likely to be approximated by the set of boundary conditions allowing free displacement in the x-direction. Effective fixity of the u-displacements at the ends is difficult to accomplish. As a result, such problems can usually be treated very accurately with normal, linearized beam theory even when the rotations $(dw/dx)^2$ become large with respect to $|du/dx|$. Reference 34 includes an example of the effect of very large rotations, $[dw/dx]^2 \approx 1.0$, on the prediction of lateral deflection for a beam with unrestrained ends. In such a case, large differences with the predictions based on small-rotation theory appear when displacements approach 10 to 20% of the beam length. However, using the fully linearized Eq 10 and 12 to predict the deformation and stress in plastic beams should generally be accurate for engineering problems, even though the low modulus of these plastic materials will produce large displacements in comparison to the thickness of a beam. The situation is not as simple for two-dimensional structures, such as flat plates.

The Small-Rotation Assumption and Plates. Although the assumptions associated with thin-plate theory are the same as those for beam theory, the conditions under which these assumptions are effective are very different because of the two-dimensional geometry of flat-plate structures. The expressions for strains in terms of displacements for flat plates are recognizable extensions of the beam expression given in Eq 11. There are now two direct strains that must be treated, as well as a shear strain, and they can be written as:

$$\varepsilon_x = \frac{du}{dx} + \frac{1}{2}\left(\frac{dw}{dx}\right)^2 - zK_{xx} \qquad \text{(Eq 19a)}$$

$$\varepsilon_y = \frac{du}{dy} + \frac{1}{2}\left(\frac{dw}{dy}\right)^2 - zK_{yy} \qquad \text{(Eq 19b)}$$

$$\gamma_{xy} = \frac{du}{dy} + \frac{dy}{dx} + \left(\frac{dw}{dx}\right)\left(\frac{dw}{dy}\right) - 2zK_{xy} \qquad \text{(Eq 19c)}$$

where K_{xx}, K_{yy}, and K_{xy} are curvature measures that can be approximated as:

$$K_{xx} \cong \frac{d^2w}{dx^2} \qquad \text{(Eq 20a)}$$

$$K_{yy} \cong \frac{d^2w}{dy^2} \qquad \text{(Eq 20b)}$$

$$K_{xy} \cong \frac{d^2w}{dxdy} \qquad \text{(Eq 20c)}$$

As for beams, the small-rotation assumption can be obtained by neglecting the quadratic rotation terms in Eq 19(a) to (c).

A significant difference in the range of problems for which small-rotation theory is effective now appears for plates in comparison to the beams previously discussed. Because of the two-dimensional nature of flat plates, the simple equilibrium equation (Eq 13) used for beams that have ends unrestrained in the in-plane direction is no longer available. As a result, the full, nonlinear equations of equilibrium in terms of the lateral displacement, w, and the two in-plane displacements, u and v, remain coupled, even if the edges of the flat plate are completely unrestrained in the in-plane directions. Therefore, although small-rotation beam theory is very accurate for moderately large rotations as long

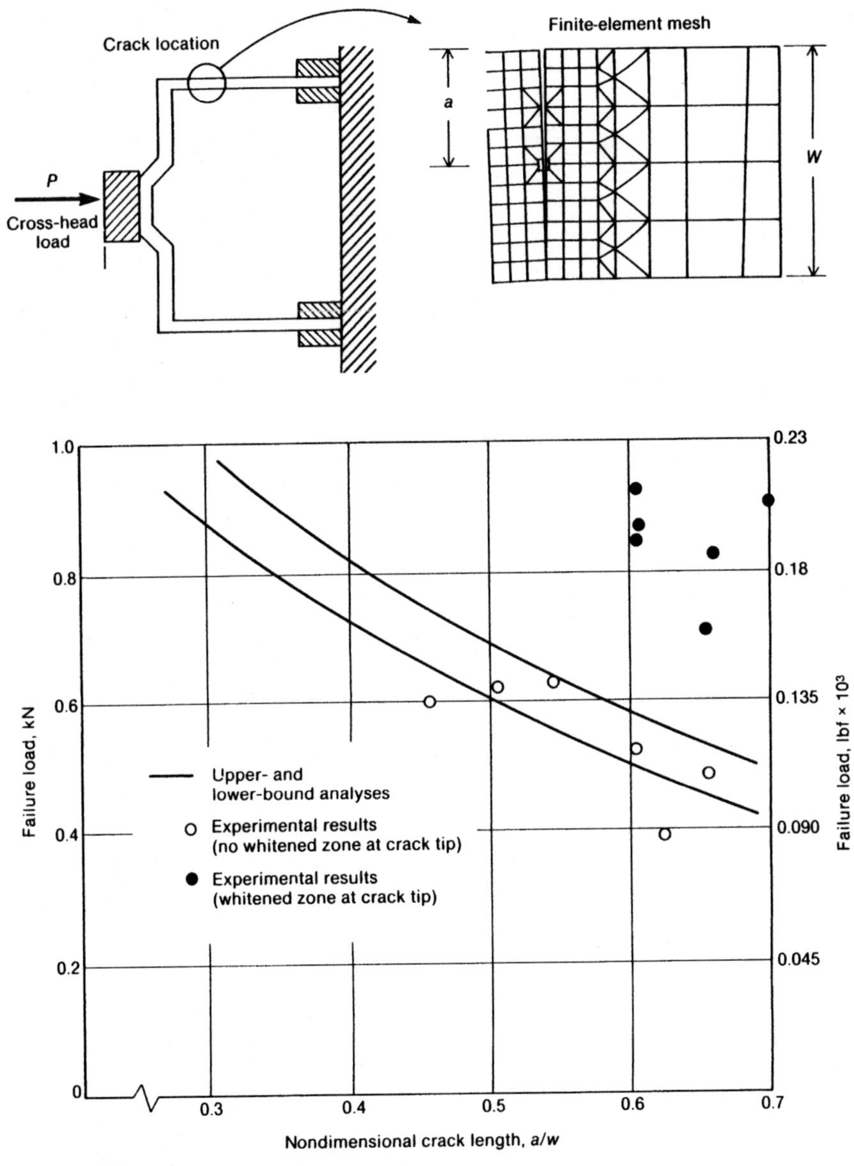

Fig. 28 Comparison of predicted and measured loads during the low-temperature impact of cracked specimens

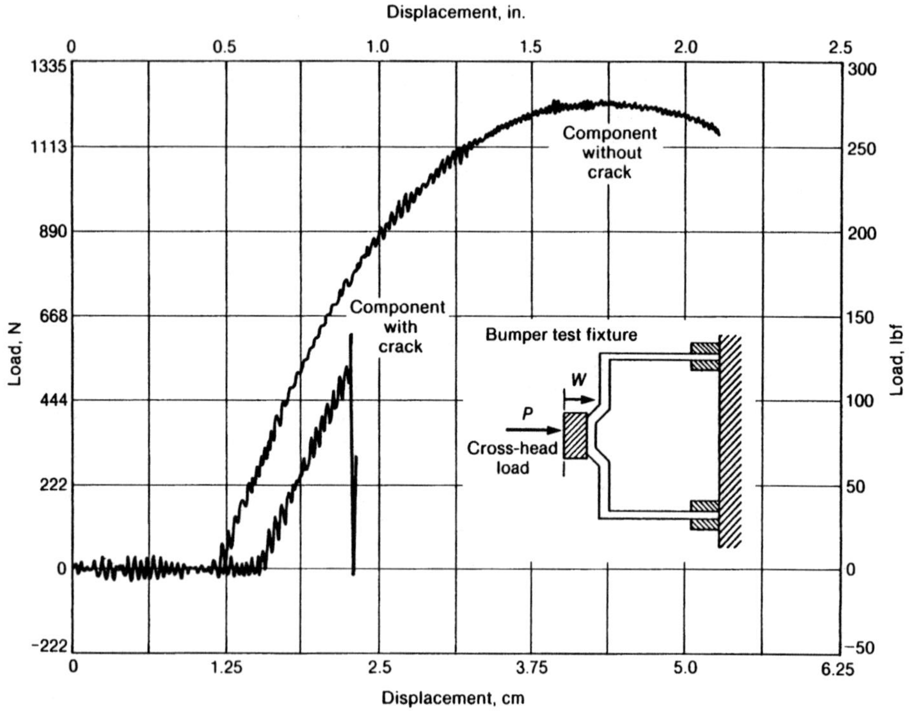

Fig. 29 Comparison of low-temperature impact performance in cracked and uncracked specimens; –50 °C (–60 °F). *P*, pressure; *W*, width

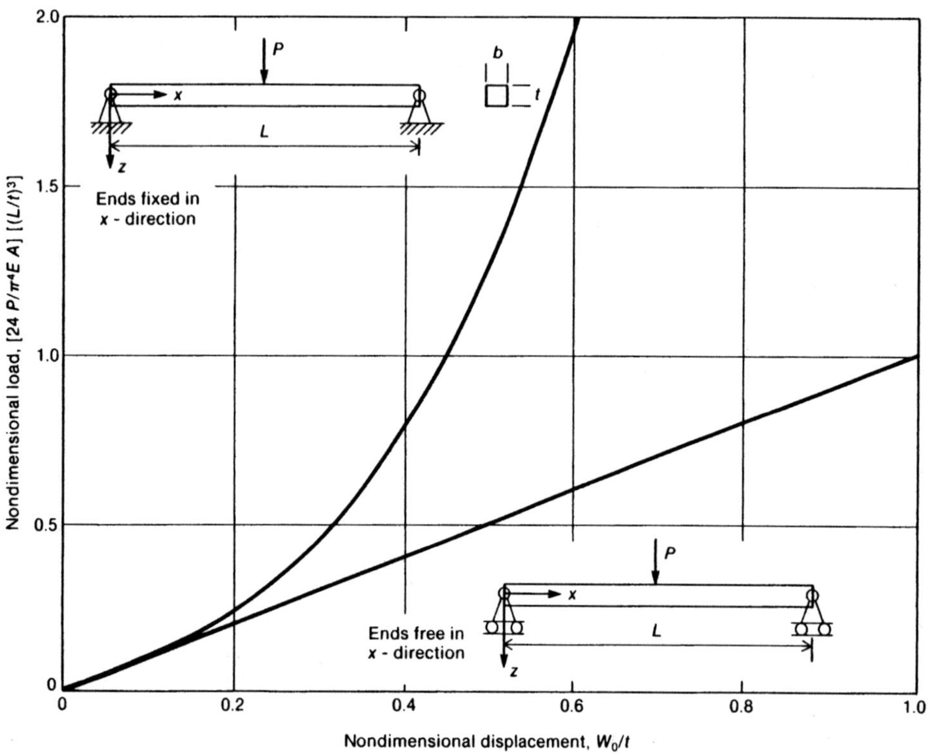

Fig. 30 Linear and nonlinear beam behavior as a function of end fixity. *P*, pressure; *L*, length; *t*, thickness; *A*, cross-sectional area

as the beam ends are unrestrained in the neutral axis direction, the same cannot be said for flat plates. Realistically sized flat plates supported against lateral displacements on their entire periphery show significant nonlinear behavior when lateral displacements become significant with respect to the plate thickness, even if the plate edges are completely unrestrained in the in-plane directions. If in-plane restraint is applied at the edges, then the range of displacements over which small-rotation theory is effective becomes even more restricted. Figure 31 shows two nonlinear load-deflection curves for a flat, simply supported plate subjected to a uniform pressure with two different in-plane edge-restraint conditions. For the lower curve, the edges are free of all in-plane restraint, and for the upper curve, the in-plane displacements, u and v, are constrained to be 0 everywhere along the edges. Both show noticeable nonlinear stiffening as the lateral displacement exceeds the value of the plate thickness.

Physically, the stiffening effect evident in Fig. 31, even for free edges, is a result of the fact that a flat surface cannot be transformed into a doubly curved surface without increasing its surface area, that is, without stretching the surface. In contrast, a flat surface can be transformed into a cylinder (singly curved surface) with only bending deformations. A simple experiment illustrating this point can be conducted with a piece of paper. Although the paper can be easily rolled up into a cylinder, it cannot be wrapped around a sphere without wrinkling or tearing it. Therefore, if all four edges of a plate are restrained in the lateral (w) direction, then small-rotation thin-plate theory will only be accurate as long as this surface-stretching effect is negligible. If, on the other hand, only one edge or two opposite edges are restrained laterally, then the physical situation is similar to that of a beam, because the plate can take on a cylindrical deformation without stretching its midsurface. In such a case, small-rotation theory again has a wider range of applicability.

With the beginnings of an understanding of thin-plate theory and its limitations, the question of why the nonlinear, moderately large rotation range of this theory should be more important for plastic materials than for metals can now be addressed. To explore this issue, the simply supported plate subjected to a uniform lateral pressure is examined. For this example, the edges of the plate to be completely unrestrained in the in-plane directions are considered. After this problem is solved (the finite-element technique is used), the maximum strain at the center of the plate can be calculated and is plotted in Fig. 32 as a function of the center-plate deflection nondimensionalized by the plate thickness. Both linear and nonlinear solutions to the problem are shown in this figure.

At this stage of the discussion, no assumption about the plate material has been mentioned. The nondimensionalized behavior shown in Fig. 32 is independent of material properties. First, let Fig. 32 be interpreted under the assumption that the plate is made of steel with a yield stress of 280 MPa (40 ksi) and a Young's modulus of 210 GPa (30×10^6 psi). If, in addition, the practical engineering constraint is imposed that requires strains to be less than the yield strain of the material, then for steel it can be seen that the linear portion of the curve in Fig. 32 associated with small-rotation theory is entirely adequate. If Fig. 32 is interpreted in terms of aluminum with a yield stress of 210 MPa (30 ksi) and a Young's modulus of 70 GPa (10×10^6 psi), then the nonlinear effects of large rotations do become evident before yield, but the linear theory still has a wide range of applicability. In contrast, however, if Fig. 32 is used to assess the behavior of a plastic material with a yield stress of 70 MPa (10 ksi) and a modulus of 2.1 GPa (0.30×10^6 psi), then it is strikingly clear that there is an extremely large range of structural behavior in which the material would be completely linear elastic but in which linear, small-rotation plate theory is completely inadequate. This comparison makes it clear that nonlinear, moderately large rotation solutions to thin-plate problems can be extremely important for efficient design with plastics. If linear small-rotation theory is used, then the engineer is likely to calculate an invalid displacement that is significantly too large for an applied lateral load, and a good application of plastic might be discarded for the wrong reasons.

Unfortunately, predicting plate deformation analytically by using nonlinear, moderately large rotation plate theory is not as straightforward as using the linear theory. Many calculations based on the moderately large rotation theory have been published (Ref 35 is one of the early examples), but they are not in a form that makes them as easy to use as the extensively tabulated results of small-rotation theory. In some cases, it may be efficient to use approximate solutions to these nonlinear problems generated for a few important geometries and boundary conditions by such techniques as the minimum total potential energy method illustrated earlier for the beam problem. These solutions are compatible with personal computers and provide for quick initial design studies. In situations in which the geometry is more complex and/or a higher degree of accuracy is required, it may be necessary to use a finite-element program with the capability to handle moderately large rotation plate problems. Programs capable of this task are widely available and have become much easier to use. If the lateral loading of flat plates is a routinely encountered problem, it is possible to create additional computer software programs, treating specific geometries and making the preparation of data and the interpretation of results from finite-element programs extremely efficient (Ref 36).

Instability and Collapse of Thin Plastic Components. The theory and methods applied in the previous section to illustrate the behavior of thin plates as they undergo displacements significant in comparison to their thickness were developed well before structural use of plastic became widespread. The importance of these methods for plastics arises due to the low moduli and large yield strains that are characteristic of many polymers. An additional subject area of impact-related structural mechanics with a similar relationship to plastics is the area of buckling and structural collapse.

Buckling and collapse constitute another class of mechanical behavior whose importance is accentuated by the low moduli of most plastic materials and the generally thin nature of the structures into which they are molded. This type of response may often be important in the design of plastic structures subjected to impact loads, even though it was not a factor in the metallic structure that it is replacing. Proper treatment of these issues in a design analysis sense does not require any characteristically new analysis techniques unique to plastics. However, it does require that an engineer be appropriately aware of this generic issue during his design process. In addition, it may be useful to have access to design engineering tools capable of addressing these events. A good example of such a situation is the application of a thermoplastic automotive bumper subjected to the standard impact events required by federal regulation.

One of the most important impact standards affecting the design of thermoplastic bumpers in the United States is the pendulum impact test. The details of this test are documented in a fed-

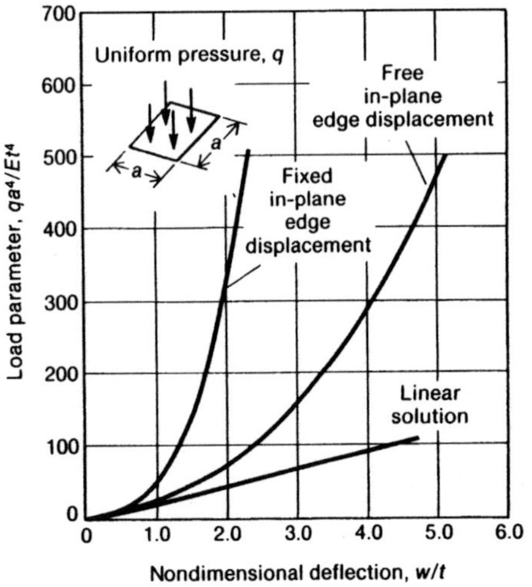

Fig. 31 Nonlinear plate behavior for two in-plane edge conditions and simple supports, *t*, thickness

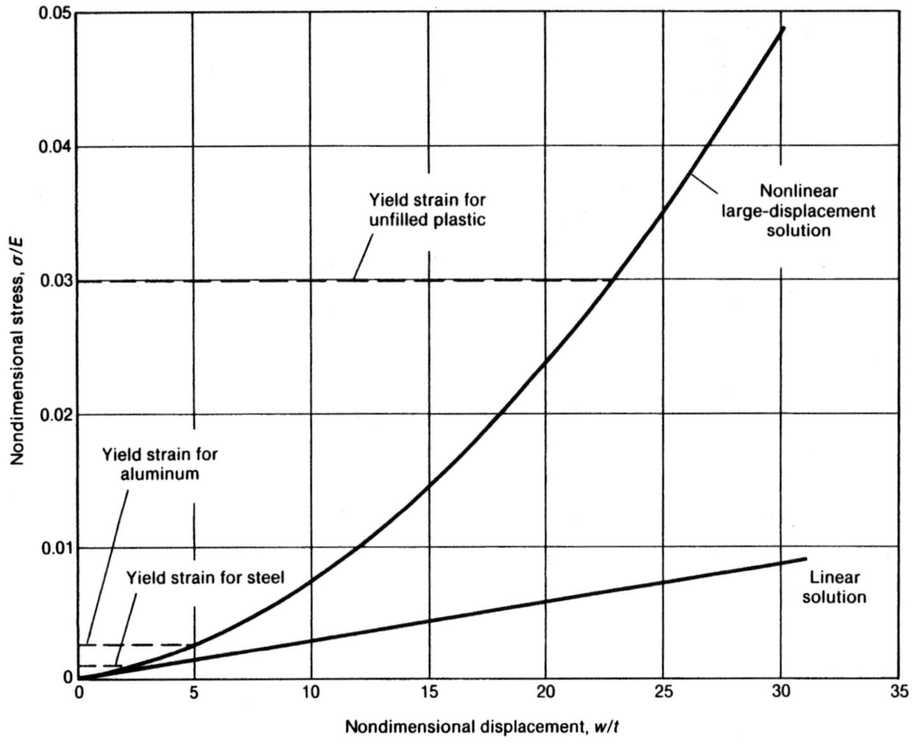

Fig. 32 Nonlinear regions for metal and plastic plates. σ, stress; *t*, thickness

eral standard (Ref 37). In this test, a standardized pendulum impacts the bumper with the automobile in neutral. The pendulum mass is equal to the mass of the car, and the velocity at impact is required to be at least 4 km/h (2.5 miles/h). Some manufacturers are designing to an 8 km/h (5 miles/h) standard. Of course, one of the requirements of a bumper is to manage the energy associated with this impact in such a manner that other parts of the car structure, such as sheet metal, are not damaged. Therefore, there is usually a maximum displacement that the bumper must not exceed during the impact event.

As indicated previously, plastics have Young's moduli that are often two orders of magnitude less than those of steel. For situations in which structural stiffness is a design issue, the shape of the plastic part must often be used to help offset this significant difference in material stiffness. Consequently, an attractive cross-sectional shape for a thermoplastic bumper has been a box section because of its effectiveness in producing bending stiffness. A generic, box-beam, thermoplastic bumper is examined in the subsequent example. Because of the complexity of some of the issues, finite-element analysis is used. It should be noted that this is not unusual in the area of plastic design. Several discussions are available in the literature on the use of finite-element analysis to assist in the design of plastic bumpers for impact using both linear (Ref 38) and nonlinear (Ref 39) techniques.

Figure 33 illustrates the geometry of the generic bumper beam considered in this study. The finite-element model also shown in Fig. 33 makes use of the symmetry of the central load condition and thus only represents the upper half of the bumper. The additional left-right condition of symmetry available to the analyst for this model was not imposed, because the same model was used to study other load cases without the same left-to-right symmetry conditions. The inset in Fig. 33 defines the cross-sectional geometry of the upper-beam half. The boundary conditions at the automobile rails are also defined.

Figure 33 also illustrates the central pendulum impact that was studied during this investigation. In this analysis, the pendulum is assumed to be rigid with respect to the bumper, and the impact event is modeled to account for the contact nature of the load application. Reference 39 contains detailed information on how this was accomplished.

In addition to the box-beam geometry shown in Fig. 33, current plastic bumper designs often include some internal web and stiffening structure. The primary purpose of the internal structure is to prevent crushing of the cross section of the beam under the central pendulum impact. There may also be molded box structures in the vicinity of the attachment locations whose primary function is to provide load path to the automobile frame while preventing crush of the box section when impact occurs over an attachment. Both of these generic structures are illustrated in Fig. 34.

Because these internal structures are important to the present discussion of buckling and collapse, an effort is made to include their effect in the following analysis. In order to accomplish this purpose, the internal bulkhead and box structures are simply modeled as contact elements. These elements will prevent local, cross-sectional collapse after the clearance between the front bumper face and the internal structure is closed. This behavior can be modeled using the "gap" elements available in many nonlinear finite-element codes. The effect of these structures in preventing section collapse can be quite dramatic, as is demonstrated.

Figure 35 compares load-displacement behaviors for several different bumper configurations as predicted by nonlinear finite-element analysis techniques, including large rotations,

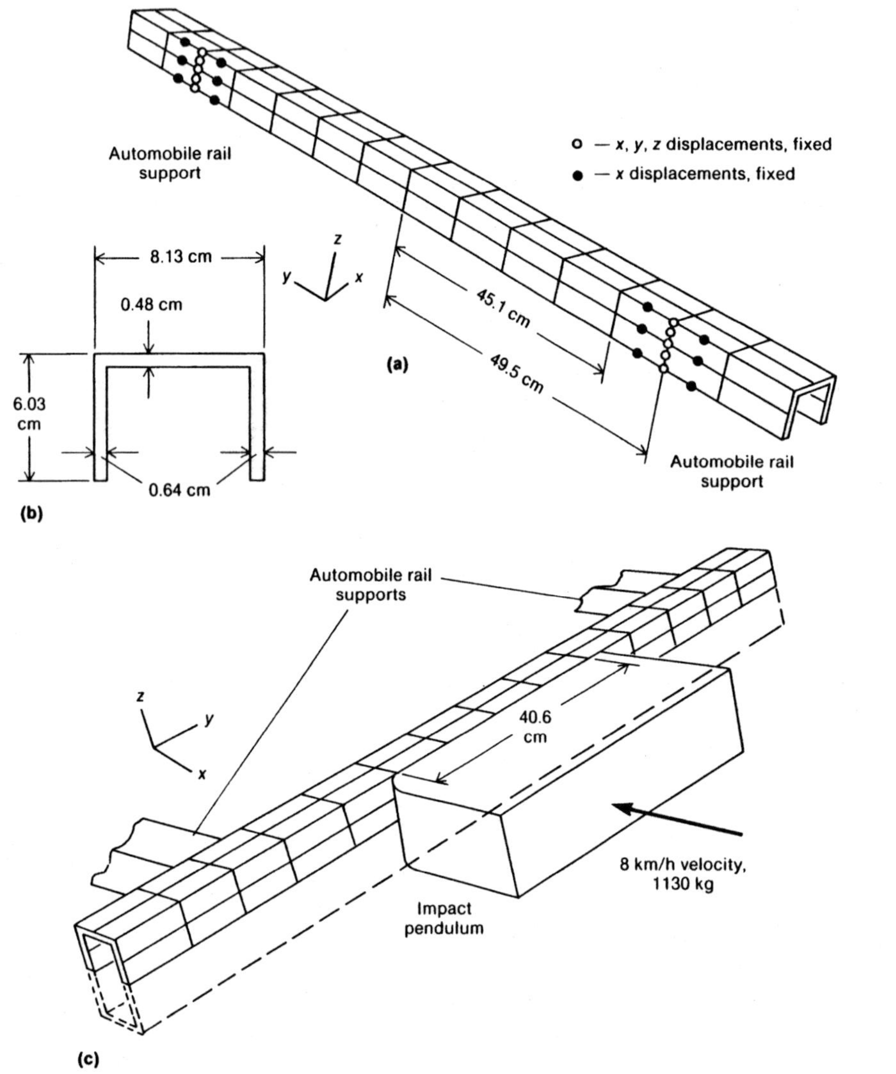

Fig. 33 Finite-element model of centerline pendulum impact. (a) Back view. (b) Cross-sectional view. (c) Front view

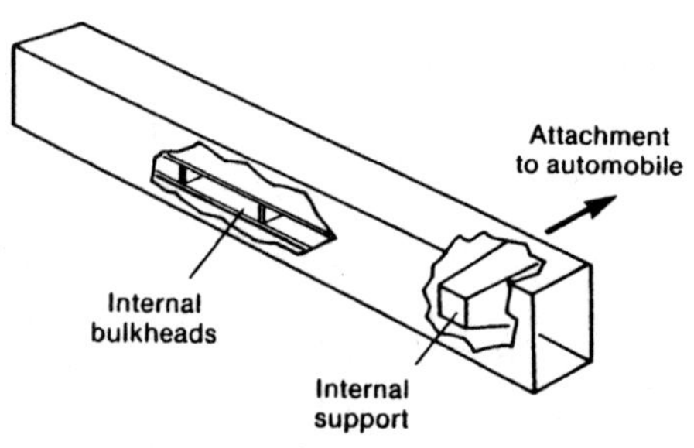

Fig. 34 Schematic of generic thermoplastic box-beam bumper

and illustrates the importance of nonlinear considerations. For reference, the linear solution to the pendulum impact of a thin-wall box-beam bumper is also plotted in Fig. 35. As can be seen, when the same box beam without bulkheads is analyzed using nonlinear analysis techniques, the predicted response is much more flexible than the linear results. In fact, as shown in Fig. 35, the nonlinear prediction reveals a maximum in the load-displacement curve. This maximum load is associated with the fact that the cross section of the bumper is collapsing in the vicinity of the outside pendulum edge (the point of load transfer), as illustrated in the cross-sectional view in Fig. 36.

To prevent this collapse, a designer might mold bulkhead structure into the cross section of the bumper. If such bulkheads are provided in the central 405 mm (16 in.) of the bumper (where the pendulum impacts for this problem), the load-displacement behavior is quite different, as can be seen from Fig. 35. With these bulkheads, the beam is much stiffer, collapse is prevented in the central regions of the bumper, and much higher loads and energy absorption levels are attained.

Although cross-sectional collapse is prevented, there is still evidence of another significant nonlinearity. As shown in Fig. 37, the front face of the box section has buckled because of the large compressive stresses induced there due to beam bending. Because this type of plate buckling is stable in nature, the front face does continue to carry increased stress (below yield stress), and additional load is still possible after buckling. However, a designer might still wish to avoid such behavior.

It may also be advantageous to use bulkheads in the area of the automobile rail attachments, because locally high loads and distortion are also experienced in this area, as can be seen in Fig. 37. If this design option is modeled, there is still another significant increase in the stiffness of the overall system, as can be seen in Fig. 35.

The most important point to be made regarding this example is that there are essential design considerations for thermoplastic bumpers (as well as other components) that require an engineering understanding of nonlinear events such as buckling and collapse. Numerous processes are being applied to the manufacture of these bumpers, including injection molding, blow molding, and thermoforming. In addition, design requirements vary from country to country and from company to company. If the cost penalties of inefficient build-and-test approaches are to be avoided, nonlinear analyses will be required for assessing the effectiveness of new designs that will evolve from these different manufacturing methods.

Summary

To design plastic structures for impact resistance, an engineer must deal with deformation and failure as a function of loading rate, stress state, and temperature. He must also understand the restrictions that limit the capability of his analytical techniques to predict the performance of plastic components in contrast to metal.

The effects of rate and temperature on the yield stress of most plastics is well investigated, and a substantial amount of fundamental material data describing plastics behavior is available. Furthermore, there is growing, positive experience with respect to the effective use of this engineering data to predict the response of plastic components.

If puncture resistance is an important aspect of the response of a component to impact, then, in addition to yield stress, material properties that characterize the large-strain response of plastics become very important. Although these

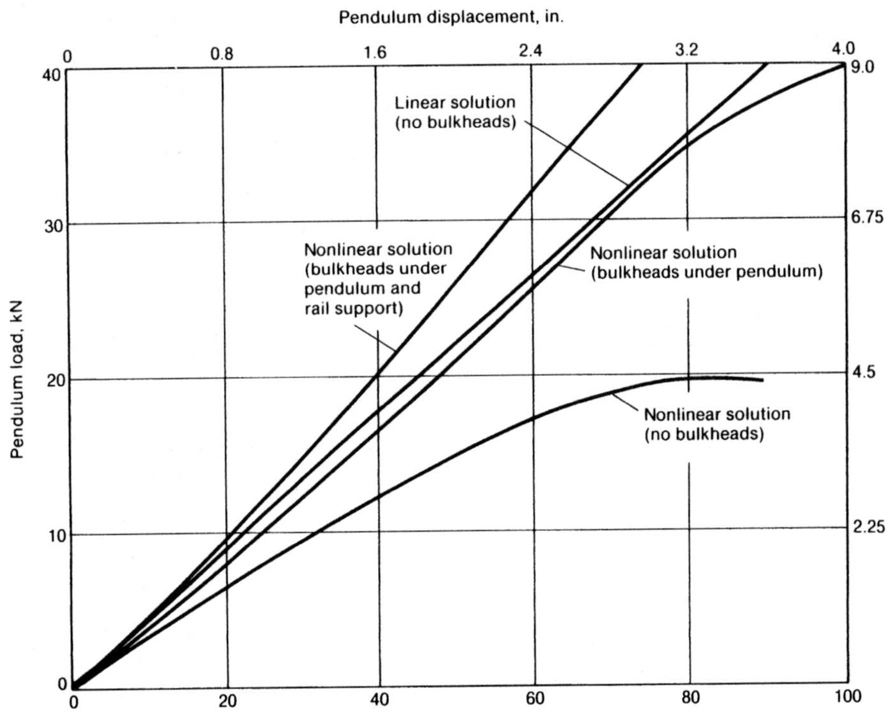

Fig. 35 Nonlinear effects of cross-sectional collapse on load-displacement behavior

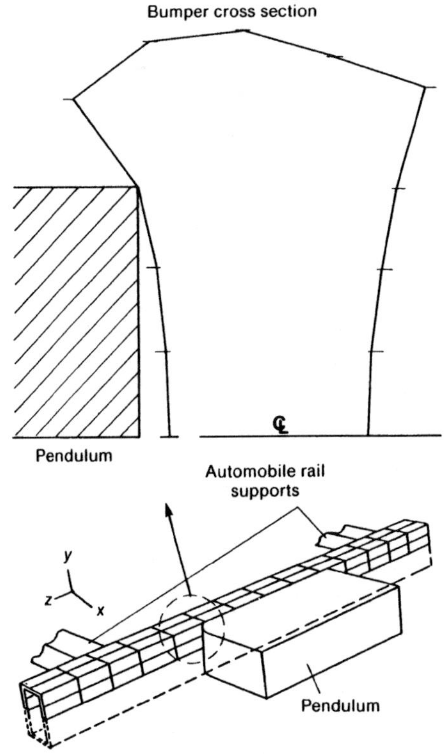

Fig. 36 Schematic of collapse of unreinforced cross section

properties are more difficult to measure and not as well investigated as yield strength, some fundamentally sound data are being reported. Furthermore, the quantitative effects that these properties have on puncture resistance are beginning to be understood.

As in other engineering materials, ductile plastics often exhibit a temperature range in which there is a transition in failure from ductile deformation (at higher temperatures) to brittle fracture (at lower temperatures). It has been observed that the rate of loading has a much more significant effect on the yield stress of many plastics than it does on the brittle failure stress. Because increasing the rate of loading on a plastic serves to increase the yield stress, it is possible to increase the rate of loading enough to make the stress required to yield a plastic material higher than the stress required to break it in a brittle manner. Thus, by increasing the rate of loading, a change in failure mode from ductile deformation to brittle failure may occur. The state of stress in the component also plays a major role in this phenomenon, and it has been observed that brittle failure is more likely for stress states characterized by larger components of triaxial tensile stresses.

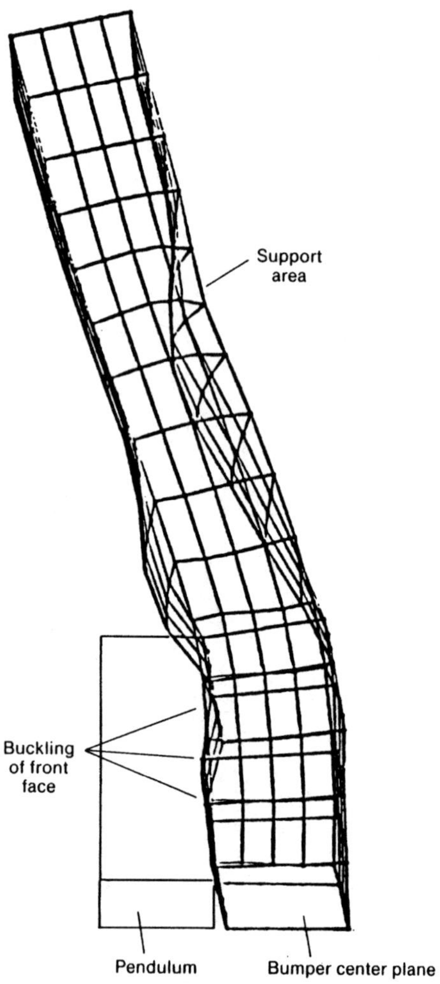

Fig. 37 Schematic of buckling of front face of bumper

A number of impact tests, including the Charpy, Izod, and dart penetration tests, have been developed, and they are used to provide qualitative information concerning many of the issues discussed previously. Unfortunately, none of these tests provides fundamental material properties that can be used in an engineering procedure to predict general part performance.

In contrast, proper measurement of plane-strain fracture toughness does provide a true material property that can be used to predict brittle failure loads as a function of crack length within the range of applicability of linear elastic fracture mechanics. Although debate continues over the issue, linear elastic fracture mechanics does seem to have a range of applicability for treating the brittle failure of ductile plastics at low temperatures. There remains a great need in the areas of standardized testing of plastics for fracture toughness as well as experience in the engineering application of these data.

When designing a plastic component for impact resistance, an engineer must consider a number of related issues that influence the effectiveness of the component. Processing temperatures that are too high for the plastic being used can adversely affect its impact resistance. A mold design that results in a knit line (that is, a line at which two fronts of polymer flow meet) can create a line of weakness susceptible to failure during impact. Plastics can be susceptible to attack by various chemicals that may cause embrittlement and poor impact performance. The environment of the component should be considered closely when choosing a plastic for any application. Exposure to ultraviolet rays can also reduce impact resistance at low temperatures through photooxidation of unsaturated rubber in rubber-toughened polymers.

With regard to engineering analysis, standard calculations can be used in many cases to guide the design of plastic components for impact. However, there are specific situations in which care and, perhaps, more sophisticated analyses are warranted. Because of their large strains to yield, plastic panel structures often violate the limitations of small-rotation (displacement) theory, which is at the basis of classical plate equations. In such situations, these equations can drastically overestimate the deflections incurred by a plastic plate subjected to lateral loading. The application of nonlinear strain-displacement equations can greatly improve this situation and lead to much more effective design of plastic panels. In addition, because of their low moduli, plastic components may also be susceptible to buckling or collapse and the load limitation associated with this phenomenon. Therefore, an engineer working with designs that use plastics should carefully consider whether such behavior may be an issue in his particular design. Again, nonlinear analysis methods, such as those available in many finite-element programs, can provide very useful design information on buckling and collapse.

REFERENCES

1. S.P. Timoshenko, *Strength of Materials,* Van Nostrand Reinhold, 1958, p 462–470
2. N. Brown, *Failure of Plastics,* W. Brostow and R.D. Corneliessen, Ed., Hanser Publishers, 1986, p 104
3. V.K. Stokes and H.F. Nied, Solid Phase Sheet Forming of Thermoplastics—Part I: Mechanical Behavior of Thermoplastics to Yield, *J. Eng. Mater. Technol.,* Vol 108, April 1986, p 107
4. I.M. Ward, *Mechanical Properties of Solid Polymers,* John Wiley & Sons, 1983, p 379, 424–432
5. R.P. Nimmer, H. Moran, and G.R. Tryson, Impact Response of a Polymeric Structure—Comparison of Analysis and Experiment, *Proceedings of the 1984 Society of Plastics Engineers Annual Technical Meeting* (New Orleans, LA), Society of Plastics Engineers, 1984, p 565
6. H.F. Nied and V.K. Stokes, Solid Phase Sheet Forming of Thermoplastics—Part II: Large Deformation Post-Yield Behavior of Plastics, *J. Eng. Mater. Technol.,* Vol 108, 1986, p 113
7. H.F. Nied, V.K. Stokes, and D.A. Ysseldyke, High Temperature, Large Strain Behavior of Polycarbonate, Polyetherimide and Poly(Butylene Terephthalate), *Polym. Eng. Sci.,* Vol 27 (No. 1), 1987, p 101
8. G.W. Halden and Y.C. Lo, The Solid-Phase Flow Behavior of Ductile Thermoplastics, *Proceedings of 1983 Annual Technical Meeting of the Society of Plastics Engineers* (Chicago, IL), Society of Plastics Engineers, 1983, p 366–367
9. D. Lee, Modeling of Polycarbonate Stress-Strain Behavior, *Polym. Eng. Sci.,* Vol 27 (No. 2), 1987
10. P.I. Vincent, The Necking and Cold Drawing of Rigid Plastics, *Polymer,* Vol 1, March 1960, p 7–19
11. I.M. Ward, *Mechanical Properties of Solid Polymers,* John Wiley & Sons, 1983, p 329
12. J.W. Hutchinson and K.W. Neale, Neck Propagation, *J. Mech. Phys. Solids,* Vol 31 (No. 5), 1983, p 405–426
13. R.P. Nimmer and L. Miller, Neck Propagation in Tensile Tests, *J. Appl. Mech.,* Vol 51, 1984, p 759
14. D.M. Parks, A.S. Argon, and B.S. Bagepalli, "Large Elastic-Plastic Deformation of Glassy Polymers, Part II: Numerical Analysis of Necking and Drawing," MIT Program in Polymer Science Report, Massachusetts Institute of Technology, March 1985

15. R.P. Nimmer, Predicting Large Strain Deformation of Polymers, *Polym. Eng. Sci.*, Vol 27, 1987, p 16
16. K.J. Bathe, E. Ramm, and E.L. Wilson, *Int. J. Num. Meth. Eng.*, Vol 9, 1975, p 353–386
17. R.P. Nimmer, Analysis of the Puncture of a Polycarbonate Disc, *Polym. Eng. Sci.*, Vol 23, 1983, p 155
18. R.P. Nimmer, An Analytical Study of Tensile and Puncture Test Behavior as a Function of Large-Strain Properties, *Polym. Eng. Sci.*, Vol 27, 1987, p 263
19. P.I. Vincent, *Polymer,* Vol 1, 1960, p 427
20. L.M. Carapelucci, A.F. Yee, and R.P. Nimmer, Some Problems Associated with the Puncture Testing of Plastics, *J. Polym. Eng. Sci.*, June 1987
21. G. Menges, *Werkstoffkunde der Kunststoffe,* Hanser Publishers, 1979
22. G. Menges, Deformation and Failure of Thermoplastics on Impact, *Failure of Plastics,* W. Brostow and R.G. Corneliessen, Ed., Hanser Publishers, 1986, p 171, 179
23. R.F. Boyer, *Polym. Eng. Sci.*, Vol 8, 1968, p 161
24. B. Kleinemeier and H.E. Boden, "Das Spannungs-Dehnungsveihalten von Thermoplasten unter Slopbeanpruchen," Final Report on Research Project AIF4094, IKV, 1980
25. E. Plati and J.G. Williams, *Polymer,* Vol 16, 1975, p 915–920
26. W. Brostow and R.G. Corneliessen, *J. Mater. Sci.*, Vol 16, 1981, p 1665
27. W. Brostow, Impact Strength: Determination and Prediction, *Failure of Plastics,* W. Brostow and R.G. Corneliessen, Ed., Hanser Publishers, 1986, p 203
28. "Impact Resistance of Plastics and Electrical Insulating Materials," D 256-84, *Annual Book of ASTM Standards,* Vol 08.01, American Society for Testing and Materials, 1987, p 80–102
29. "Standard Test Methods for Impact Resistance of a Rigid Plastic Sheeting or Part by Means of a Tup (Falling Weight)," D 3029-84, *Annual Book of ASTM Standards,* Vol 08.02, American Society for Testing and Materials, 1987, p 749–764
30. P.I. Vincent, *Impact Tests and Service Performance of Thermoplastics,* Plastics Institute, 1970
31. "Plane-Strain Fracture Toughness of Metallic Materials," E 399-83, *Annual Book of ASTM Standards,* Vol 03.01, American Society for Testing and Materials, 1987, p 680–715
32. R.P. Nimmer, K. Weiss, J. McGuire, M. Takemori, and T. Morelli, Engineering Application of Linear Elastic Fracture Mechanics for a Ductile Polymer Part, *Proceedings of the Society of Plastics Engineers Annual Technical Meeting* (Atlanta, GA), Society of Plastics Engineers, 1988
33. C.B. Bucknall, *Toughened Plastics,* Applied Science, 1977, p 301
34. J.G. Williams, *Stress Analysis of Polymers,* John Wiley & Sons, 1980, p 160
35. S. Levy, "Bending of Rectangular Plates with Large Deflections," Technical Note 846, National Advisory Committee for Aeronautics, 1942
36. M.D. Minnichelli, C.M. Mulcahy, and G.G. Trantina, Automated Structural Analysis of Plastic Sheet, *Adv. Polym. Technol.*, Vol 6 (No. 1), 1986, p 73–78
37. Federal Automotive Standards, Part 581—Bumper Standard (Docket 74-11; Notice 12, Docket 73–19; Notice 9), Sept 1978
38. P.M. Glance, "Computer-Aided Engineering Analysis of Automotive Bumpers," SAE Technical Paper 840222, presented at the International Congress and Exposition (Detroit, MI), Society of Automotive Engineers, Feb/March 1984
39. R.P. Nimmer, O.A. Bailey, and T.W. Paro, "Analysis Techniques for the Design of Thermoplastic Bumpers," SAE Technical Paper Series 870107, presented at the International Congress and Exposition (Detroit, MI), Society of Automotive Engineers, Feb 1987

Fatigue Testing and Behavior*

FATIGUE FRACTURE OF ENGINEERING PLASTICS due to cyclic loading conditions is of critical concern when designing polymeric components for structural employment. Like all engineering materials, failure often ensues in the plastic as a consequence of accumulated irreversible damage or growth of a fatigue flaw to a critical dimension. The fatigue life of a polymeric component is controlled by a number of factors; in general, polymers are more sensitive to the testing environment than metal or ceramic counterparts. These variables include the stress or strain amplitude of the loading cycle; the mean stress of the cycle; the presence of stress concentrations or initial defects in the component; the frequency, temperature, and environment of the test; and the molecular properties of the polymer. These factors are of considerable interest and practicality for the safe design of structural polymeric components subjected to repetitive loading.

When designing for the fatigue life of an engineering plastic, one of two distinct philosophies is generally practiced. The total-life approach is used with unnotched specimens that are assumed to be defect free, and this methodology is predicated on the notion that fatigue failure is a consequence of both crack nucleation and subsequent growth. Conversely, the defect-tolerant approach bases the fatigue life of a component on the number of loading cycles needed to propagate a crack of an initial size to a critical dimension. Over the last few decades, numerous researchers have provided detailed reviews (Ref 1–8) of fatigue behavior in polymers based on both total-life and fracture mechanics approaches.

This article provides a review of fatigue test methodologies and an overview of general fatigue behavior in engineering plastics. Many factors affect the fatigue performance of engineering materials, including molecular and mechanical variables as well as the design of the fatigue test. The appropriate test conditions should be used when evaluating the life of the polymer. If a structural component is likely to be free of defects and stress concentrations or if the component is likely to spend the majority of its lifetime in the initiation stage of crack growth, then the total-life philosophy is preferred. On the other hand, fracture mechanics should be used for safety-critical fatigue designs and in flawed structural components likely to sustain a high degree of stable crack growth prior to fracture.

Fatigue Crack Initiation

Stress-Based Loading. The traditional total-life philosophy for fatigue life prediction is based on an endurance limit established from stress-log cycle plots, also known as *S-N* curves. In these tests, uncracked specimens are subjected to a constant amplitude load cycle until failure occurs. Often, fatigue tests are performed on closed-loop servohydraulic universal test machines, which enable fatigue tests to be performed under a variety of waveforms over a range of test frequencies (including spectrum loadings). The stress amplitude can be applied through torsion, rotation, beam bending, or axial loads. Specimens (Fig. 1) are chosen according to the loading method employed. For example, the standard fatigue test for plastics in ASTM D 671 specifies repeat flexural stress as a standard fatigue test. In this test, a triangulated specimen geometry (Fig. 1d) provides a uniform flexural stress across the entire gage section.

In fatigue testing, the applied stress, σ_a, is typically described by the stress amplitude of the loading cycle:

$$\sigma_a = \frac{\sigma_{max} - \sigma_{min}}{2} \qquad \text{(Eq 1)}$$

where σ_{max} is the maximum stress, and σ_{min} is the minimum stress of the fatigue cycle. The stress amplitude, also denoted as *S*, is generally plotted against the number of cycles to failure, *N*, on a linear-log scale, which is called an *S-N* plot. A general trait of these plots is that the number of cycles to failure increases as the stress amplitude is reduced. In some polymers, there is a critical stress level, often referred to as the *endurance limit* of the material, below which the specimen does not fail in less than 10^7 cycles.

Figure 2 shows the *S-N* behavior of several commodity plastics. It should be noted that nylon and polyethylene terephthalate (PET) do not exhibit an endurance limit. Other plastics—including polyethylene (PE), polypropylene oxide (PPO), polystyrene (PS), polytetrafluoroethylene (PTFE), polypropylene (PP), polymethyl methacrylate (PMMA), and epoxy (EP)—exhibit a stress limit below which failure does not occur in less than 10^7 cycles for these testing conditions. Because plastics are sensitive to many factors, including frequency, temperature, mean stress, and molecular structure, the fatigue test conditions must closely mimic the service conditions of the polymeric component. Despite the simple nature of these experimental tests, the *S-N* approach is widely accepted in the engineering plastics community for design applications where stress concentrations are expected to be minimal or where the fatigue life of the component is likely to be dominated by the nucleation of a crack.

Displacement and Strain-Based Loading. While stress-based tests are appropriate for the evaluation of plastics chosen for load-controlled applications, these tests may not be suitable for circumstances where the structural component is likely to experience fluctuations in displacement or strain. In such instances, strain- or displacement-based tests may be more appropriate. In such tests, the configuration typically is based on a fixed cantilever subjected to repeated constant deflection. The initial stress range will typically decay under cyclic loading (much like a stress-relaxation experiment) and is caused by plastic deformation or softening of the polymer. Hysteretic heating from deformation can further result in an inaccurate prediction of the cyclic stress amplitude (Ref 9). However, because stresses generally decay in this type of test, thermal failures are rarely encountered (Ref 5).

Strain-based tests are often used for components with accumulated strain or blunt notches (Ref 5). The majority of strain-based fatigue tests are performed using fully reversed loading conditions, generally accompanied by a cyclic softening phenomenon in plastics (Ref 2, 9). Under cyclic strain conditions, the fatigue response is best characterized by the cyclic stress-strain curve. This curve is created by testing several specimens subjected to a range of controlled cyclic strain limits. Tests are continued for each specimen until the hysteresis loops become saturated. A curve is fit through the amplitude of these saturated hysteresis loops in order to establish a cyclic stress-strain curve.

*Adapted from the article by Lisa A. Pruitt, "Fatigue Testing and Behavior of Plastics," in *Mechanical Testing and Evaluation,* Volume 8, *ASM Handbook,* ASM International, 2000, pages 758 to 767

Figure 3 shows a comparison of the cyclic and monotonic stress-strain curves for several polymers. An interesting feature of these polymers is that they all soften and exhibit lower yield points under cyclic strain than under monotonic conditions (as opposed to metals, which can exhibit either cyclic softening or cyclic hardening).

The cyclic strain life data can also be portrayed in a manner analogous to the *S-N* approach. The total strain amplitude can be divided into elastic and plastic strain amplitude components. The strain amplitude of the fatigue cycle is plotted against the number of cycles or load reversals to failure, which provides an empirical relationship between the strain amplitude of the fatigue cycle and the number of cycles to specimen failure (Ref 6):

$$\varepsilon_a = \frac{\sigma_f'}{E}(2N_f)^b + \varepsilon_f'(2N_f)^c \qquad \text{(Eq 2)}$$

Here, ε_a is the strain amplitude, σ_f' is the strength coefficient, $2N_f$ is the number of load reversals to failure, ε_f' is the ductility coefficient, and b and c are material constants. The first term on the right side of Eq 2 is the elastic component of the strain amplitude, and the second term is the

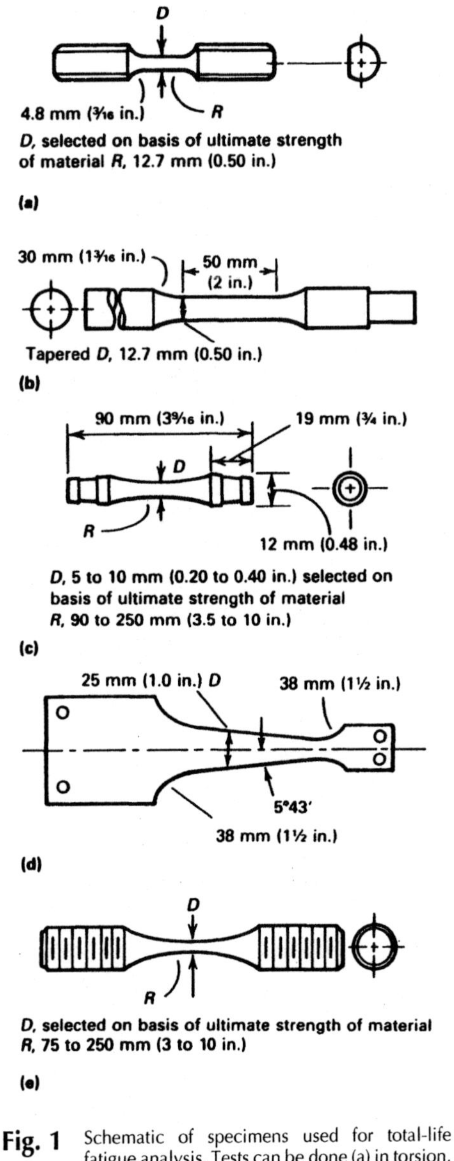

Fig. 1 Schematic of specimens used for total-life fatigue analysis. Tests can be done (a) in torsion, (b) with a rotating cantilever, (c) with a rotating beam, (d) with cantilever reverse bending, or (e) under axial loading

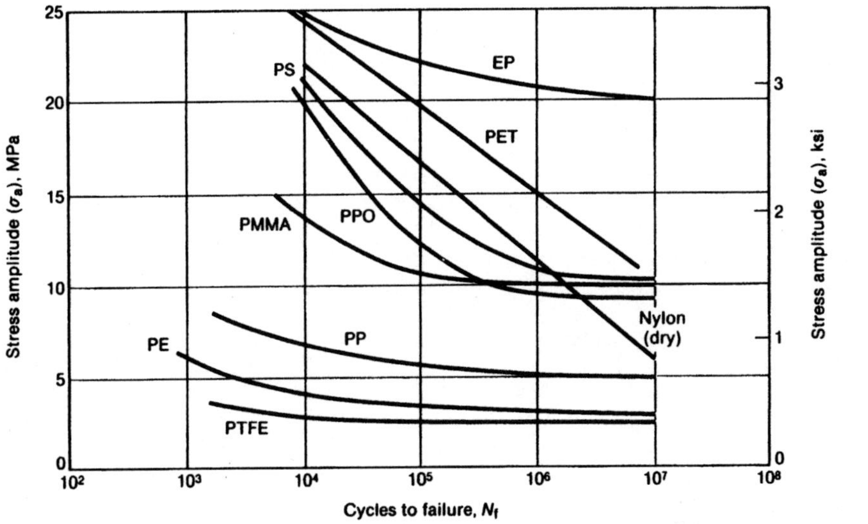

Fig. 2 Stress amplitude versus cycles to failure, or *S-N* behavior, of several commodity plastics. PS, polystyrene; EP, epoxy; PET, polyethylene terephthalate; PMMA, polymethyl methacrylate; PPO, polypropylene oxide; PE, polyethylene; PP, polypropylene; PTFE, polytetrafluoroethylene

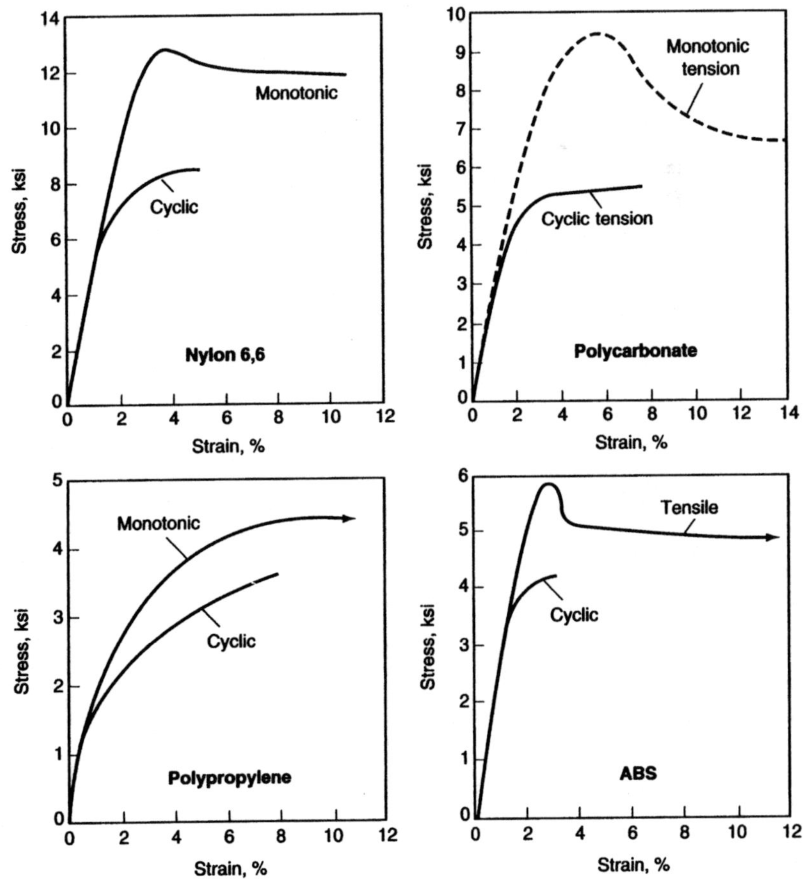

Fig. 3 Comparison of the cyclic and monotonic stress-strain curves for several polymers. ABS, acrylonitrile-butadiene-styrene. Source: Ref 2

plastic component of the strain amplitude. Tests dominated by the elastic component of strain are considered high-cycle fatigue with little plastic strain. Low-cycle fatigue tests are identified by the relatively small number of cycles or reversals to failure and the large degree of plastic strain.

An important concern in the testing of polymers is that the attributes of the fatigue test are crucial to the relative ranking of fatigue resistance among various polymers. For example, polymers with higher damping capacities can be less resistant to fatigue. These polymers experience thermal heating when tested under constant stress amplitude. Conversely, these same polymers can have enhanced fatigue resistance if tested under constant deflection conditions. Moreover, the relative placement of fatigue resistance in polymers correlates strongly to whether the tests are performed under adiabatic or isothermal conditions (Ref 5).

Thermal Fatigue and Hysteretic Heating. Due to the viscoelastic nature of polymeric solids, a portion of the strain energy dissipates under cyclic loading conditions. This heat generation results in an increase in specimen temperature until the heat generated per cycle is equal to the heat dissipated through conduction, convection, and radiation. The temperature of the specimen depends strongly on the frequency of the test, the amplitude of the applied stress or strain, and the damping properties of the polymer. In some instances, especially in unnotched specimens, the temperature of the polymer specimen can locally surpass the glass transition or flow temperature of the polymer. Research (Ref 9) has shown that the energy dissipated per second, $\dot{E}$, is given by:

$$\dot{E} = \pi \nu J''(\nu, T, \sigma)\sigma^2 \qquad \text{(Eq 3)}$$

where ν is the frequency, J'' is the loss compliance, T is temperature, and σ is the peak stress of the fatigue cycle. The rate of change of temperature for adiabatic heating conditions in which the heat generated is transferred into temperature rise is given as:

$$\frac{dT}{dt} = \frac{\dot{E}}{\rho C_p} \qquad \text{(Eq 4)}$$

where ρ is the mass density, and C_p is the heat capacity of the polymer. In general, the increase in temperature will scale with increase in frequency, stress amplitude, and internal friction of the polymer. Figure 4 shows the thermal fatigue behavior of polyacetal and its dependence on test frequency. The temperature rise in the specimen can be monitored with a thermal couple or an infrared sensor. In many polymer systems, the thermal work influenced by high damping and low thermal conductivity contributes to micromechanisms of permanent deformation, including craze formation, shear bands, voids, or even microcracks (Ref 8).

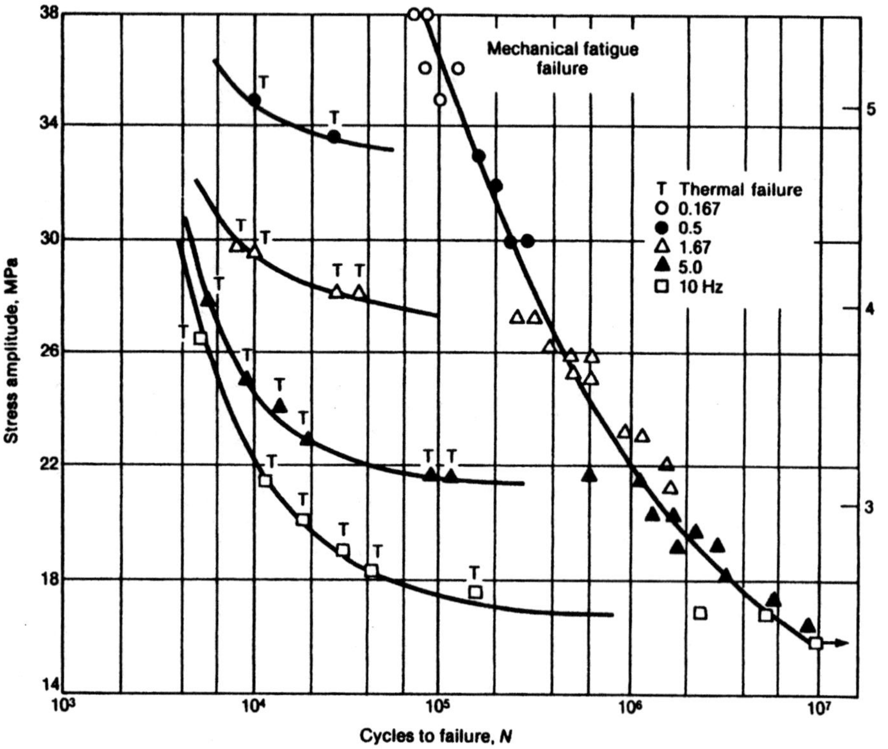

Fig. 4 Plot showing the effect of increasing test frequency and stress amplitude on the fatigue failure of polyacetal

The hysteresis loops observed under cyclic loading conditions can provide useful insight into micromechanisms of fatigue damage. Figure 5 shows the hysteresis loops after various numbers of fatigue cycles in both high-impact polystyrene (HIPS) and acrylonitrile-butadiene-styrene (ABS). An interesting observation is that the hysteresis loops are symmetric for ABS, while the hysteresis loops become much larger for the tensile portion of the fatigue cycle in HIPS as the fatigue test progresses. These events are understandable, considering ABS undergoes shear yielding mechanisms and the HIPS undergoes crazing that requires a tensile component of stress. Crazing results from fibrillation or polymeric drawing ahead of the fatigue flaw. The advancement of the craze zone is associated with damage accumulation in the leading fibrils; thus, the tensile portion of the hysteresis loop grows as damage accumulates in the specimen.

Fatigue Crack Propagation

Fracture Mechanics Concepts and the Defect-Tolerant Philosophy. The use of fracture mechanics for fatigue design is based on the tacit assumption that structural components are intrinsically flawed and capable of sustaining a considerable amount of stable crack growth before failure. The fatigue life of a component based on this defect-tolerant approach is dictated by the number of loading cycles needed to propagate a crack of an initial size to a critical dimension. Fracture mechanics is used widely in the characterization of fatigue crack propagation behavior of advanced engineering plastics capable of sustaining a substantial subcritical crack growth prior to fracture. Fracture mechanics is also used in safety-critical applications where defect-tolerant life estimates are essential. Characterizing crack growth behavior in polymers can be complicated by fatigue cracks known to propagate at different rates, depending on the near-tip damage micromechanisms, mean stress, frequency, or test environment. These factors are of significant interest and practicality for the safe design of structural polymeric components subjected to repetitive loading.

The stress-intensity factor, K, derived from linear elastic fracture mechanics, is the parameter used to describe the magnitude of the stresses, strains, and displacements in the region ahead of the crack tip. The linear elastic solution (Fig. 6) for the normal stress, σ_{yy}, in the opening mode of loading is written as a function of distance, r, and angle, θ, away from the crack tip (Ref 10):

$$\sigma_{yy} = \frac{K_I}{\sqrt{2\pi r}} \cdot \cos\frac{\theta}{2}\left(1 + \sin\frac{\theta}{2}\sin\frac{3\theta}{2}\right) \qquad \text{(Eq 5)}$$

where K_I is the mode I (opening mode) stress-intensity factor. This parameter incorporates the boundary conditions of the cracked body and is a function of loading, crack length, and geometry. The stress-intensity factor can be found for a wide range of specimen types and is used to scale the effect of the far-field load, crack length, and geometry of the flawed component. Standard specimens employed in fatigue crack propagation studies are the single-edge-notch specimen (Fig. 7a) and the compact-tension specimen (Fig. 7b). The form of the stress-intensity factor for the compact-tension geometry is given as (Ref 10):

$$K_I = \frac{P}{B\sqrt{W}} \cdot f(\alpha)$$

$$f(\alpha) = \frac{(2+\alpha)}{(1-\alpha)^{3/2}}[0.886 + 4.64\alpha - 13.32\alpha^2 + 14.72\alpha^3 - 5.6\alpha^4] \quad \text{(Eq 6)}$$

where P is the remote far-field load, B is the specimen thickness, W is the width, and α is the ratio a/W that increases as the fatigue crack, a, advances in length.

Regimes of Fatigue Crack Propagation. Fracture mechanics provides a design approach for predicting the life of a cracked structural component under cyclic loading conditions. While the micromechanisms of deformation differ for metals, polymers, and ceramics, the fatigue crack propagation behavior of these materials share many macroscopic similarities. As with crystalline materials, there are three distinct regimes of crack propagation for polymers under constant amplitude cyclic loading conditions. These regimes include the slow crack growth or threshold regime, the intermediate crack growth or Paris regime, and the rapid crack growth or fast fracture regime (Fig. 8). The velocity of an advancing fatigue crack subjected to a constant stress amplitude loading is determined from the change in crack length, a, as a function of the number of loading cycles, N. The fatigue crack propagation rate per cycle, da/dN, is found from experimentally generated curves, in which a is plotted as a function of N. For constant amplitude loading, the rate of crack growth increases as the crack grows longer. Researchers have suggested (Ref 11) that the stress-intensity factor range, $\Delta K = K_{max} - K_{min}$, which captures the far-field cyclic stress, crack length, and geometry, should be the characteristic driving parameter for fatigue crack propagation. This basis of the Paris relationship states that da/dN scales with ΔK through the power-law relationship:

$$\frac{da}{dN} = C \cdot \Delta K^m \quad \text{(Eq 7)}$$

where C and m are empirical constants. These constants can be strongly affected by polymer

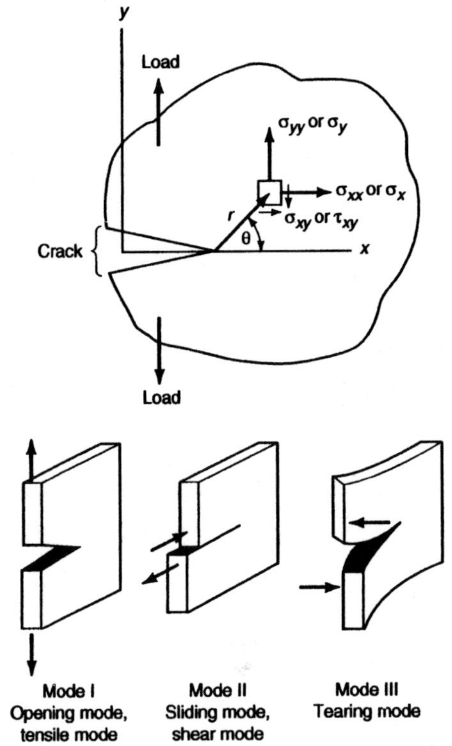

Fig. 6 Coordinate system for crack-tip stresses in model I loading (see Eq 5)

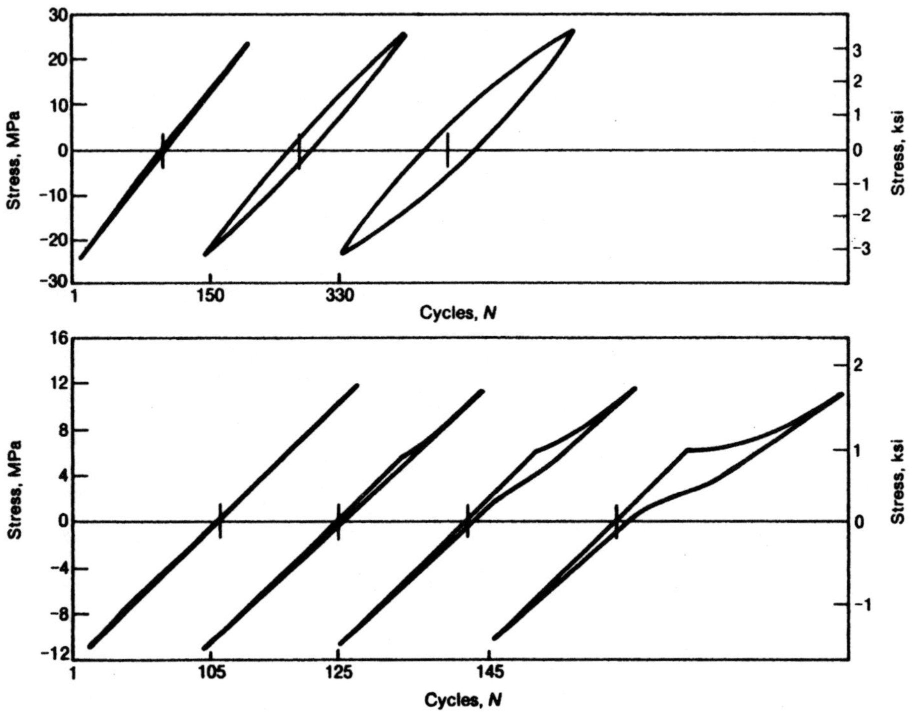

Fig. 5 Hysteresis loops after various numbers of fatigue cycles in both high-impact polystyrene (HIPS) (bottom) and acrylonitrile-butadiene-styrene (ABS) (top). Note the lack of symmetry in the HIPS due to crazing mechanisms. See text for discussion

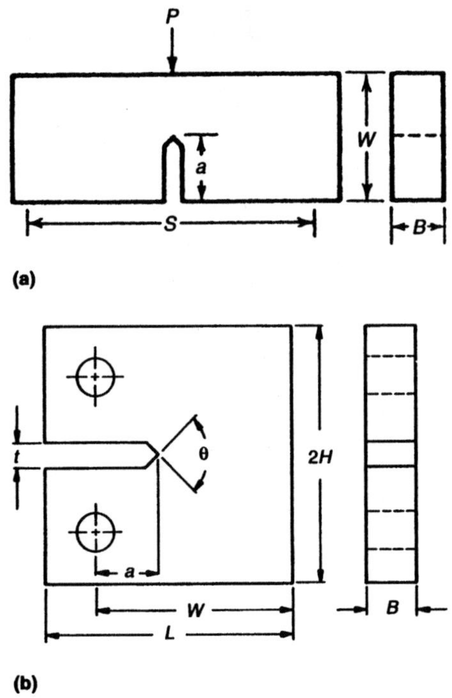

Fig. 7 Specimens employed in fatigue crack propagation studies. (a) Single-edge-notch specimen. (b) Compact-tension specimen

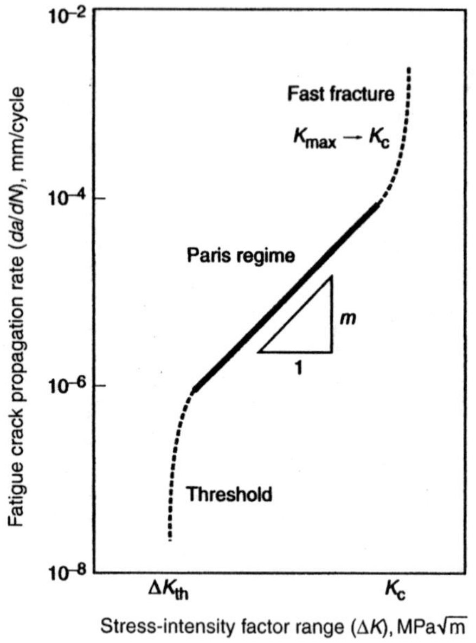

Fig. 8 Schematic illustration of the three distinct regimes of crack propagation rate observed in fatigue testing under constant amplitude loading conditions. For polymers, typical values of m range from 3 to 50, depending on the polymer system.

morphology, test frequency, stress ratio (the stress ratio, R, is defined as the ratio of the minimum stress to the maximum stress of the fatigue cycle) of the fatigue cycle, as well as by test temperature and environment. Figure 8 shows that the Paris equation is valid for intermediate ΔK levels spanning crack propagation rates from approximately 10^{-6} to 10^{-4} mm/cycle.

The Paris relationship is a useful tool for fatigue life prediction. It is implied in this defect-tolerant approach that all structural components are intrinsically flawed with an initial crack size, a_i. Assuming the fatigue loading is performed under constant stress amplitude conditions, the geometric factor, $f(\alpha)$, does not change within the limits of integration. Fracture occurs when the crack reaches a critical value, a_c. The Paris equation can be integrated to predict the fatigue life of the component:

$$N_f = \frac{2}{(m-2)Cf(\alpha)^m(\Delta\sigma)^m\pi m/2} \cdot \left[\frac{1}{a_i^{(m-2)/2}} - \frac{1}{a_c^{(m-2)/2}}\right] \text{ for } m \neq 2 \quad \text{(Eq 8)}$$

Crack Shielding Mechanisms in Polymers. The crack driving force near a fatigue crack tip, ΔK_{tip}, will be lower than the applied crack driving force, ΔK_a, when extrinsic toughening mechanisms are present. The presence of extrinsic toughening mechanisms shields the crack tip, thereby decreasing the crack driving force and the crack growth rate. A researcher (Ref 12) has expressed the extrinsic crack-tip shielding effect:

$$\Delta K_{tip} = \Delta K_a - K_s \quad \text{(Eq 9)}$$

where K_s is the stress-intensity factor due to shielding. Under cyclic loading conditions, there are three general types of shielding mechanisms: crack deflection, process zone shielding, and contact shielding (Fig. 9). Shielding due to crack path deflection results in improvements in the fatigue crack propagation behavior over all ranges of ΔK. By contrast, process zone shielding mechanisms operate more effectively at high ΔK levels, whereas contact shielding mechanisms are more effective at low ΔK levels.

The amount of shielding due to crack path deflection has been modeled (Ref 13). The author derived the effective fatigue crack driving force and subsequent crack growth rates by

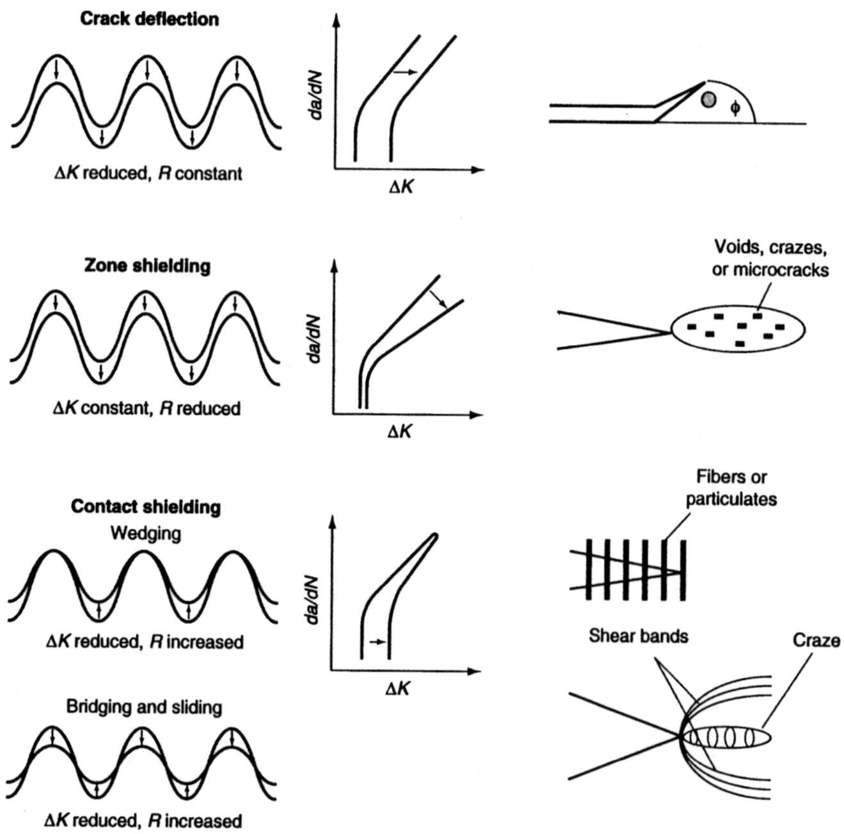

Fig. 9 Schematic illustration of the three types of shielding mechanisms: crack deflection, zone shielding, and contact shielding

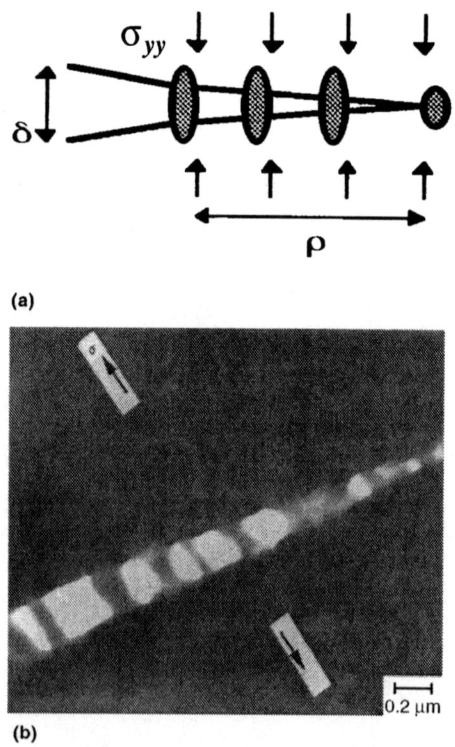

Fig. 10 Crazing. (a) Schematic of a craze zone preceding the crack. Note the craze consists of load-bearing fibrils and void space. (b) Transmission electron micrograph of a craze preceding a fatigue crack in polycarbonate

analyzing a small segment of the crack with an out-of-plane deflection:

$$\Delta K_{tip} = \frac{b\cos^2(\theta/2) + c}{b + c}\Delta K_a \qquad \text{(Eq 10)}$$

$$\frac{da}{dN} = \frac{b\cos\theta + c}{b + c} \cdot \left(\frac{da}{dN}\right)_n \qquad \text{(Eq 11)}$$

where θ is the deflection angle, b is the deflected distance, and c is the undeflected distance.

The amount of shielding caused by process zone mechanisms depends on the nature of the plastic deformation of the crack tip, such as massive crazing or shear banding (Ref 14–18). The yielding in front of the crack caused by far-field tensile loading results in the formation of a plastic or permanent deformation zone. For a crazeable polymer (Fig. 10), a Dugdale (Ref 13), or strip yield, approximation is used to estimate the size of this plastic zone, r_d:

$$r_d = \frac{\pi}{8}\left(\frac{K_I}{\sigma_y}\right)^2 \qquad \text{(Eq 12)}$$

where σ_y is the craze stress. For an elastic, perfectly plastic material behavior, the plane-stress plastic zone (Ref 13), r_p, can be estimated:

$$r_p \approx \frac{1}{\pi}\left(\frac{K_I}{\sigma_y}\right)^2 \qquad \text{(Eq 13)}$$

where σ_y is the yield stress.

Under cyclic loading, a reversed cyclic plastic zone will be generated within the monotonic plastic zone. For an elastic, perfectly plastic material, this region of residual tensile stress is one-fourth the size of the monotonic plastic zone described in Eq 13. Cyclic plastic zones have been observed in several amorphous polymer systems and are important in the inception of cracks under cyclic compression loading (Ref 13). Qualitatively, it is easy to see that the size of the plastic zone increases with ΔK; therefore, process zone shielding mechanisms are effective at high ΔK levels.

Contact shielding involves physical contact between mating crack surfaces because of the presence of asperities, second-phase particles, and/or fibers. Premature contact between the crack surfaces occurs during unloading at a stress-intensity level known as K_{cl}, which is the closure stress intensity. The degree of shielding caused by closure effects can be calculated:

$$\Delta K_{tip} = K_{max} - K_{cl} \qquad \text{(Eq 14)}$$

where K_{max} is the maximum stress intensity of the fatigue cycle. It should be noted from Eq 14 that the ΔK_{tip} is less than ΔK_a, thus effectively lowering the stress intensity felt at the crack tip. Mechanisms such as contact shielding and fiber bridging can contribute to this phenomenon. Contact shielding can arise from contact between asperities or from fiber bridging in reinforced or blended polymers. Fiber bridging has been shown to be a viable shielding mechanism in short fiber composites (Ref 19). In summary, extrinsic shielding mechanisms can be used to improve resistance to fatigue crack propagation in engineering polymers.

Factors Affecting Fatigue Performance of Polymers

Molecular Variables. Polymers are sensitive to a number of molecular variables, including molecular weight, molecular weight distribution, crystallinity, chain entanglement density, and cross-linking (Ref 3, 5, 20, 21). In general, as the molecular weight of the polymer is increased, the fatigue resistance of the polymer is enhanced. Some polymers (as mentioned earlier) are susceptible to craze nucleation that leads to subsequent crack growth and fatigue failure. Figure 10(a) schematically illustrates the load-bearing fibrils that comprise the craze zone. Figure 10(b) shows a transmission electron micrograph of a craze preceding a fatigue crack in polycarbonate (PC).

When a critical amount of damage has accumulated, the crack advances through the load-bearing fibrils of the craze zone. This advancement can occur by a void growth mechanism potentially enhanced by temperature, chemical environment, or rupture of the highly stressed fibrils. The craze often advances in a discontinuous manner and results in discontinuous crack growth in certain stress regimes (Ref 5). Research (Ref 22) has shown that craze stability depends on numerous factors, including the molecular weight and chain entanglement density of the polymer. The stability or strength of the craze can be improved by increasing the molecular weight of the polymer. Numerous studies indicate that increasing the molecular weight of the polymer increases craze strength, creep-rupture strength, and endurance limit under cyclic loading conditions (Ref 3, 5, 18, 22).

Semicrystalline polymers provide improved fatigue resistance over glassy amorphous polymers. One explanation is that the composite, two-phase structure offers enhanced toughness. Improved strength is provided by the more rigid crystalline phase, and ductility is provided by the more compliant amorphous phase. Semicrystalline polymers provide higher fracture energies and can accommodate both amorphous and crystalline modes of plasticity.

The arrangement of the crystallites within the amorphous phase or the polymer morphology is also important to the resistance of fatigue. For example, branched versions of PE offer decreased resistance, while very high-molecular-weight versions of PE with an enhanced level of tie molecules provide superior resistance to fatigue crack propagation in comparison to generic linear PE (Ref 3). In general, semicrystalline polymers, such as nylon, polyacetal, and PE, offer excellent resistance to fatigue crack propagation and provide high S-N endurance limits (Ref 5). In comparison, amorphous glassy polymers often suffer inferior fatigue strength due to a lack of shielding or toughening modes, because many amorphous polymers are used below the glass transition temperature and are incapable of large amounts of ductile or viscous deformation. Figure 11 shows a comparison of fatigue crack propagation behavior in the Paris regime for several amorphous and semicrystalline polymers, and it is evident that the semicrystalline polymers offer improved fatigue crack growth resistance.

Effect of Reinforcements. The addition of rubber particles to a ductile or brittle polymer provides a process zone shielding mechanism involving massive shear banding of the matrix, which leads to improved fatigue crack propagation resistance. The role of crack-tip shielding mechanisms on the crack growth rate regime has been modeled (Ref 12). According to Ref 12, the occurrence of a process zone shielding mechanism should change the slope (m) in the Paris regime but should not change the crack growth behavior at low crack growth or near-threshold regime. Experimental support of this model has been given (Ref 23).

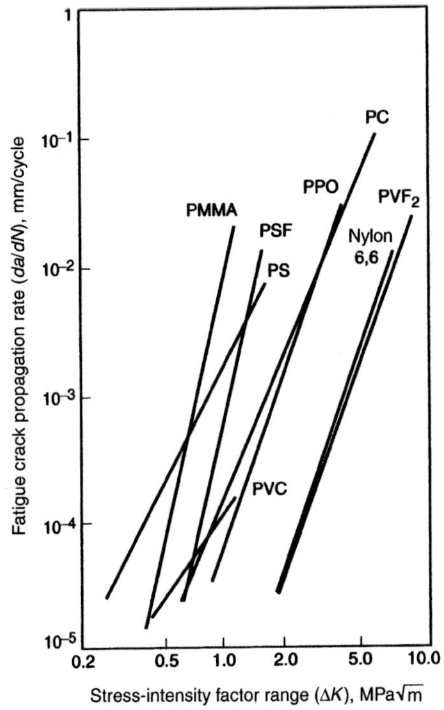

Fig. 11 Comparison of fatigue crack propagation behavior in the Paris regime for several amorphous and semicrystalline polymers. Note enhanced fatigue resistance of the semicrystalline polymers. PC, polycarbonate; PMMA, polymethyl methacrylate; PPO, polypropylene oxide; PVF, polyvinyl formal; PS, polystyrene; PVC; polyvinyl chloride; PSF, polysulfone. Source: Ref 5

Figure 12 shows that the addition of rubber decreases the slope, *m*, and retards crack growth at high crack growth rates due to toughening mechanisms. At low values of stress-intensity range (ΔK), however, the crack growth rates for the rubber-toughened epoxies are nearly identical to those of the unmodified (neat) resin. At low ΔK levels, the process zone in front of the crack tip is small; the rubber reinforcements are not highly stressed; hence, the crack grows with minimal plasticity in this regime. Conversely, at high ΔK levels, the process zone is much larger than the size of the particles, and the rubber additions within this region are highly stressed. The subsequent rubber particle cavitation causes significant additional plasticity in the matrix, and the crack propagation rate is reduced.

Blending rubber-toughened polymers with a small amount of inorganic filler can also improve fatigue crack propagation resistance. Researchers (Ref 23) have studied several glass-filled, rubber-toughened blends, and they believe the improved fatigue crack propagation resistance is the result of a synergistic interaction between the hollow glass filler at the crack tip and the plastic zone triggered by the rubber particles. The synergistic effect occurs by crack bridging via the glass phase and enhanced plasticity due to the presence of the rubber particles.

An interesting distinction must be made between fatigue crack initiation and propagation studies. The addition of rubber particles or reinforcements results in an increased resistance to crack propagation. However, this same material often exhibits a decreased resistance to fatigue crack initiation or flaw inception. The second-phase addition serves as a nucleation site for crazes, voids, or shear bands and results in a decreased threshold for crack inception. For example, HIPS studies (Ref 5) have shown an increased resistance to crack propagation but a degraded resistance to crack inception when compared to the neat polystyrene resin. Thus, designers need to have a clear understanding of the component design and loading environment when making their materials selection.

Mean Stress Effects. The fatigue response of a polymeric material is highly sensitive to the mean stress, σ_m, of the fatigue cycle:

$$\sigma_m \frac{\sigma_{max} + \sigma_{min}}{2} \quad \text{(Eq 15)}$$

Depending on the structure of the polymer and the micromechanisms of deformation, there are two distinct responses to an increase in mean stress. For a nominal stress-intensity range, some polymers exhibit an increase in crack propagation rate, while others show a decrease in crack growth rate. The published research on the effects of mean stress and *R*-ratio covers a broad range of polymer classes, including amorphous, semicrystalline, cross-linked, and rubber-modified polymers (Ref 24–34). Table 1 provides a summary of the effect of increased mean stress for several advanced polymer systems.

Increasing crack growth rates associated with an increased stress ratio or mean stress are observed in epoxy resins, PMMA, high-density polyethylene (HDPE) copolymers, PS, PVC, and nylon. A number of different explanations and relationships have been proposed to rationalize the effect of mean stress on fatigue crack propagation. Researchers (Ref 28) suggested that the fatigue crack growth rates could be scaled to the stress-intensity factor with the following relationship:

$$\frac{da}{dN} = \beta \cdot \lambda^n = \beta \cdot (K_{max}^2 - K_{min}^2)^n \quad \text{(Eq 16)}$$

where β is a coefficient that depends on the loading environment, frequency, and material properties, and *n* is a material constant.

A micromechanistic explanation for this response to an increase in stress ratio or mean stress is also possible. Polymers that become more prone to fracture with increasing mean stress are most likely affected by the monotonic fracture process associated with the maximum portion of the loading cycle as it approaches a critical stress-intensity level. In general, these polymers are susceptible to crazing, chain scission, or cross-link rupture. For these polymer types, an increase in mean stress results in faster crack propagation rates.

Remarkably, several polymer blends offer improved resistance to crack propagation as the mean stress is increased (Ref 5). These polymers include ABS, HIPS, PC, low-density or branched PE, low-molecular-weight PMMA, and rubber-toughened PMMA (Table 1). Hertzberg postulated that the strain energy normally available for crack extension is consumed through deformation or structural reorganization ahead of the crack tip. The use of strain energetics to describe fracture processes in polymers (Ref 1, 37) is formulated:

$$E = E_0 \cdot \left[\frac{C}{C - f(\psi)}\right] \quad \text{(Eq 17)}$$

Here, *E* is the total energy used by the solid to create a new unit area of surface through crack advance, E_0 is the energy expended for an ideal elastic solid, *C* is a material constant that depends on the strain state of the polymer, and ψ is the hysteresis ratio. The energy lost due to inelastic energy expenditure is captured by ψ. If the energy loss is large, the amount of energy

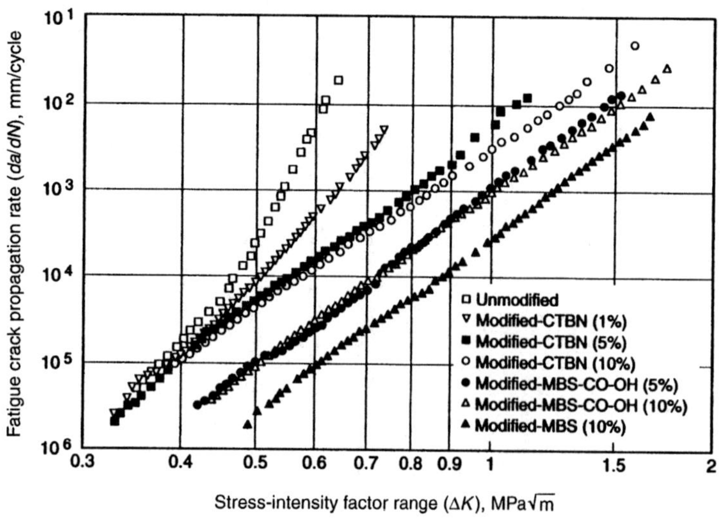

Fig. 12 Fatigue crack propagation behavior for a rubber-toughened epoxy. The addition of rubber decreases the slope, *m*, at high crack growth rates due to toughening mechanisms and retarded crack growth. CTBN, carboxyl-terminated polybutadiene acrylonitrile rubber; MBS, methacrylate-butadiene-styrene

Table 1 Effect of increasing mean stress on polymer fatigue crack propagation

Polymer	Reference
Increasing crack propagation rate with increasing mean stress	
High-density polyethylene	15
Nylon	5
High-molecular-weight PMMA	21, 25
Polystyrene	30, 35
Epoxy	24
Polyethylene copolymer	26
Decreasing crack propagation rate with increasing mean stress	
Low-density polyethylene	5
Polyvinyl chloride	34
Low-molecular-weight PMMA	21
Rubber-toughened PMMA	15
High-impact polystyrene	15, 35
Acrylonitrile-butadiene-styrene	15
Polycarbonate	24, 32, 33, 36
Toughened polycarbonate copolyester	27

PMMA, polymethyl methacrylate

needed to cause fracture increases; hence, it is expected that the crack growth rate will be reduced as ψ increases. Hertzberg and Manson (Ref 5) proposed that the effect of mean stress on the fatigue crack propagation resistance of polymeric materials is directly linked to the parameter Ψ. Thus, polymeric materials with a molecular structure susceptible to hysteretic losses or polymers capable of structural reorganization are likely to be more resistant to fatigue crack propagation as the mean stress is increased. These polymers have near-tip processes that dissipate elastic energy ahead of the crack tip: rubber toughening, orientation hardening, chain slip, and shear banding

Variable amplitude fatigue plays an important role in the design of polymeric components subjected to variations in the load cycle.

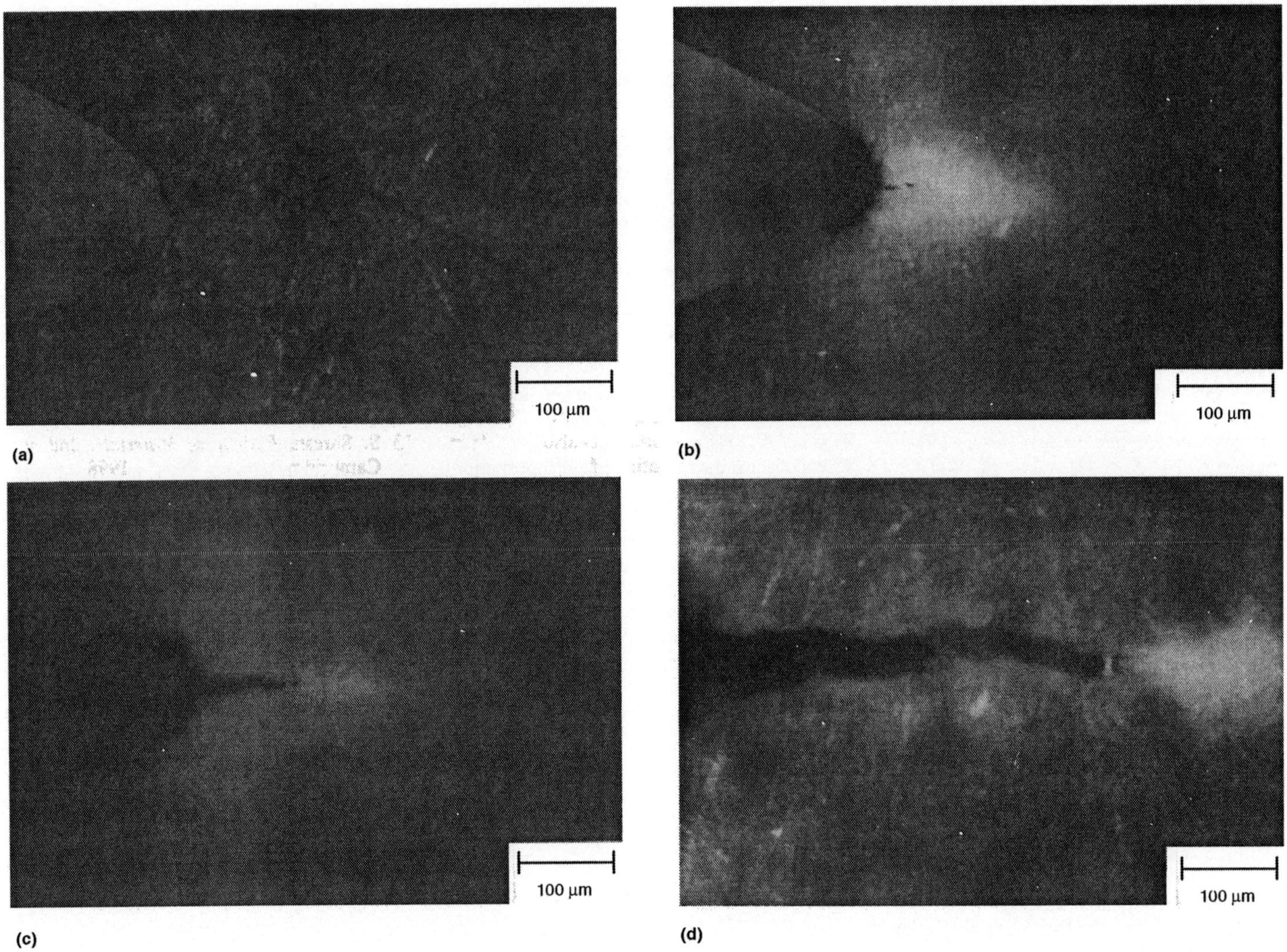

Fig. 13 Optical micrographs showing the nucleation and growth of a mode I fatigue crack in the plane of the notch as a result of cyclic compression loading in high-impact polystyrene. (a) Crazing before fatigue cycling. (b) Nucleation of fatigue crack after 15,000 cycles. (c) Crack growth after 20,000 cycles. (d) Crack growth after 50,000 cycles. Source: Ref 44

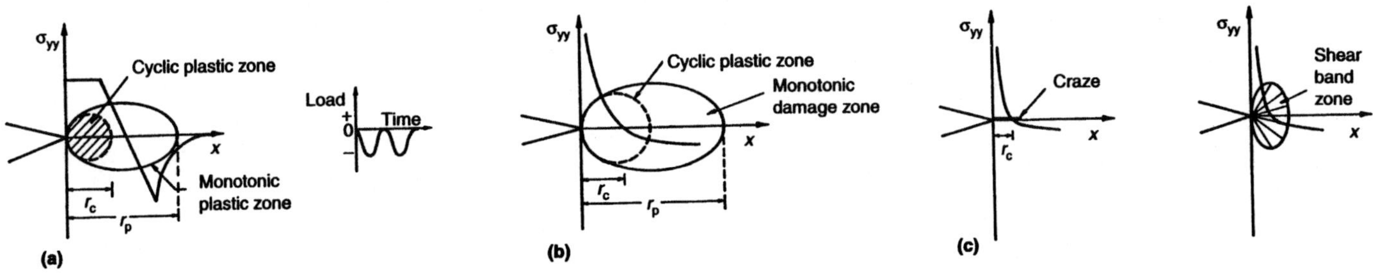

Fig. 14 Schematic illustrating the possible mechanisms of permanent deformation ahead of the notch tip. (a) Cyclic plastic zone typical of metals. (b) Cyclic damage zone typical of ceramics. (c) Craze of shear-band zones typical of polymer. Source: Ref 44

Further, variable amplitude fatigue is a concern for components likely to experience periodic or unanticipated tensile or compressive overloads. It is conventional to model the effect of variable amplitude loading using the concept of cumulative damage (e.g., Palmgren-Miner mean accumulation rule). While this concept has strength in crack initiation models, it does not capture the role of overload type or order in the loading sequence and the subsequent effect on a propagating crack.

Application of a single tensile overload can extend the life of a cracked component by retarding the rate of crack advance (Ref 13). This transient crack propagation behavior is often controlled by several mechanisms, including crack closure (Ref 38), residual compressive stresses on unloading (Ref 31, 39, 40), and crack-tip blunting (Ref 41). The closure concept justifies the retardation of crack velocity in terms of residual compressive stresses left in the plastically deformed wake of the advancing crack. The result is premature contact between the crack faces while the specimen is still in the tensile portion of the fatigue cycle, and it effectively reduces the stress-intensity range driving the crack advance. This crack closure mechanism has been proposed to describe crack retardation in PMMA (Ref 42) and for PC (Ref 43).

Blunting has been proposed to describe the reduced crack velocity following tensile overloads (Ref 44). Although crack-tip blunting can temporarily affect the crack velocity subsequent to the overload, it does not explain a prolonged regime of crack retardation. Researchers (Ref 36) have shown that the zone of residual compressive stresses sustained at the crack tip on unloading in amorphous polymers increases in size and magnitude as the far-field tensile load is increased. These residual compressive stresses sustained at the crack tip are believed to decrease the crack propagation rate following the tensile overload. The crack has to grow through this zone of residual compression before it can return to its initial crack propagation rate for the ΔK sustained prior to overload. This trend has been observed in numerous polymer systems (Ref 35, 36). Many current life prediction models are formulated on the basis of residual compressive stresses for the rationalization of crack retardation.

Compressive overloads can also be detrimental to the life of a structural component, because the overload can result in an enhanced rate of crack propagation. The application of fully compressive cyclic loads results in the inception and growth of fatigue cracks ahead of stress concentrations and notches in polymers (Ref 36, 40). Figure 13 shows the nucleation and growth of a mode I fatigue crack in the plane of the notch as a result of cyclic compression loading in HIPS. The source of this crack growth is the generation of a zone of residual tensile stresses on unloading from far-field compression. Permanent deformation ahead of the notch tip in polymers can be induced by crazing, shear flow, chain reorientation, or a combination thereof (Fig. 14).

In summary, the application of compressive overloads to polymers with stress concentrations can result in the generation of residual tensile stresses and concomitant enhancement of crack velocity, resulting in shortened component life.

Waveform and Frequency Effects. Many polymers, due to the viscoelastic nature, are highly sensitive to the waveform or frequency of a fatigue test. Some crazeable polymers, such as PS, PMMA, and HIPS, exhibit decreased crack propagation as the test frequency is increased, while materials such as PC, nylon, and polysulfone exhibit no sensitivity (Ref 5). In crazeable polymers, the increased test frequency can diminish chain disentanglement effects at the crack tip and result in a decreased rate of crack propagation. A researcher (Ref 45) proposed that the crack propagation in a polymer could be described as the sum of the elastic and viscoelastic contributions:

$$\frac{da}{dN} = C_1\left(\frac{\Delta K}{K_{Ic}}\right)^{n_1} + C_2\left(\frac{\Delta K}{K_{Ic}}\right)^{n_2} \cdot \frac{1}{\nu} \qquad \text{(Eq 18)}$$

where K_{Ic} is plane-strain fracture toughness, and the first term is the elastic contribution, which includes an elastic compliance term, C_1. The second term is the time-dependent contribution, which includes a creep compliance term, C_2, and the test frequency, ν.

Strain rate can also play a critical role in the fatigue response of time-dependent polymers. Researchers (Ref 46) found strong sensitivity to waveform in PVC, PS, PMMA, and especially vinyl urethane (VUR). The square wave provides a high strain rate in ramp-up, then it subjects the specimen to a longer period of peak load than a triangular waveform with the same stress amplitude. This difference in load function can cause major differences in fatigue crack propagation. For example, in VUR, the fatigue crack propagation rate is reduced by a factor of 6 when switching from a triangular to a square wave loading function (Ref 47). This behavior is attributed to the higher strain rate that dominates for very flexible polymers, such as VUR. Another important factor is the amount of creep sustained at the peak load of the fatigue cycle. Polymers, such as PMMA, that are susceptible to creep damage will generally perform poorly when tested under the square waveform loading due to creep at peak load (Ref 46).

Environmental factors can play a critical role in the fatigue performance of engineering polymers. Many amorphous polymers are known to be susceptible to chemically induced crazing (Ref 5). In such instances, the crack inception values can be substantially reduced in the presence of aggressive media (Ref 5). For example, PC is known to nucleate surface crazes in the presence of acetone vapor. Many rubbers are susceptible to oxidation-induced embrittlement (Ref 48). Many medical polymers, such as orthopedic-grade ultrahigh-molecular-weight polyethylene (UHMWPE) or bone cement (PMMA), degrade due to oxidation embrittlement and chain scission. These mechanisms are induced by ionizing modes of sterilization and subsequent aging (Ref 49–58). An example of the embrittling effect of gamma radiation sterilization on the fatigue crack propagation resistance of medical-grade UHMWPE used for total joint replacements is provided in Fig. 15. While not all aggressive environments and effects on polymers are discussed here, it is clear that care should be taken to conduct fatigue tests that mimic not only the mechanical loads but also the

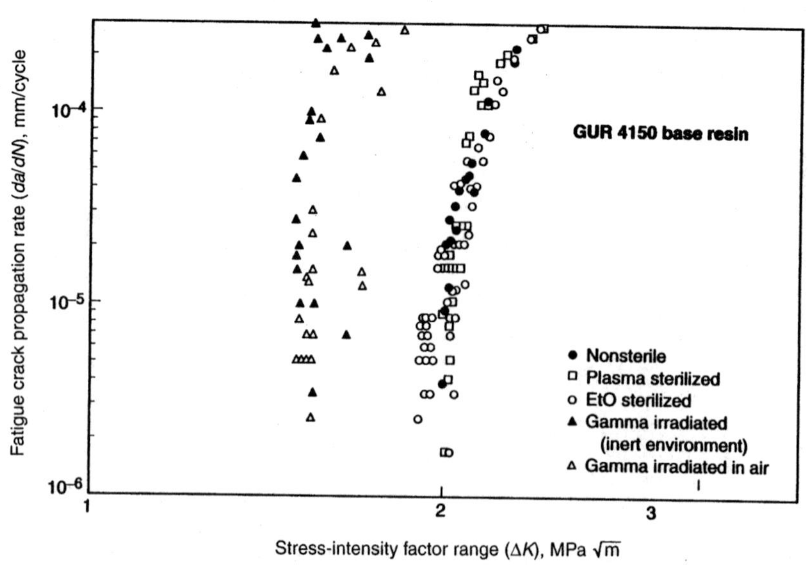

Fig. 15 Fatigue plot illustrating the devastating effect of gamma radiation sterilization on the fatigue resistance of orthopedic-grade ultrahigh-molecular-weight polyethylene used for total joint replacements

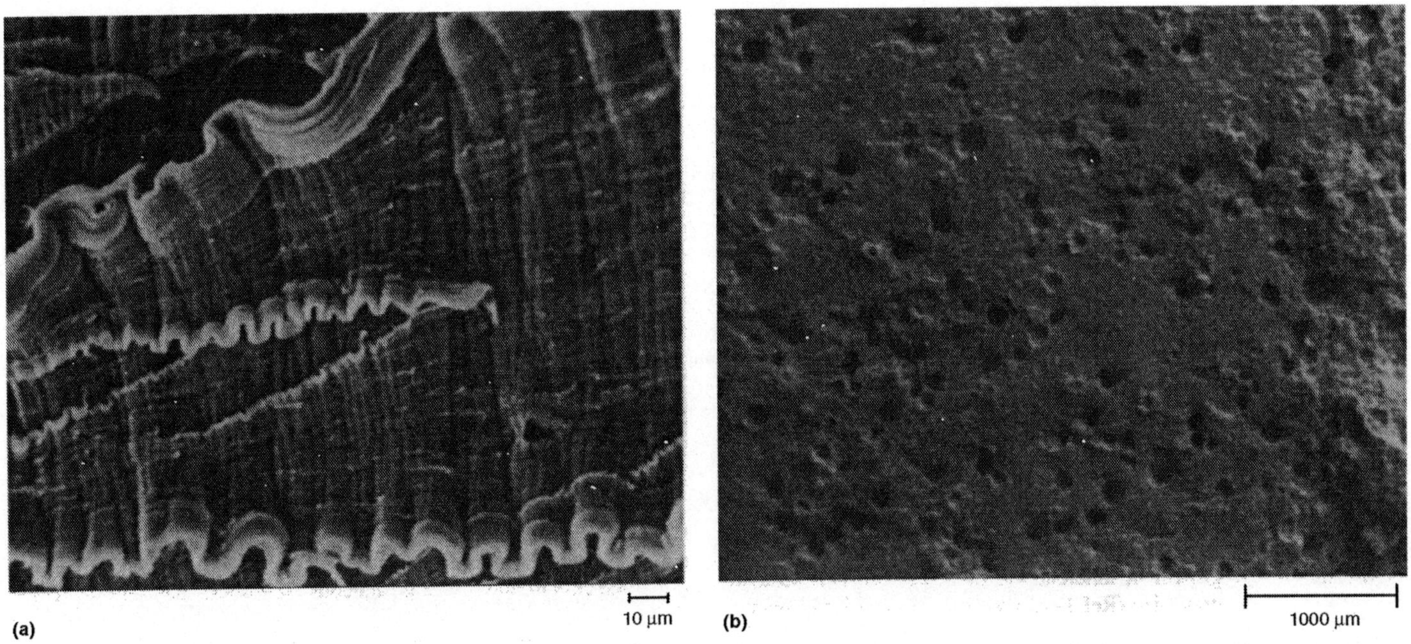

Fig. 16 Scanning electron micrographs depicting (a) the ductile mechanisms observed in pristine ultrahigh-molecular-weight polyethylene and (b) the brittle mechanisms found in acrylic bone cement

chemical and aging environments that are most likely to be encountered in the lifetime of the device.

Fractography

One of the most useful tools in failure analysis is fractography, the study of fracture surfaces. Scanning electron microscopy (SEM) can provide vast insight into failure mechanisms of polymers subjected to cyclic loading. Analysis of the surface with SEM provides the site of crack inception. Discontinuous growth bands are often encountered with polymers that undergo crazing. In such instances, the damage must accumulate before the leading fiber can fail and the crack can advance. This results in discontinuous growth bands or markings that are observed in fractography (Ref 5). Fractographic examination can also provide information on the formation of discontinuous or continuous crack growth bands. Figure 16(a) shows the typical ductile mechanisms observed in pristine UHMWPE, while Fig. 16(b) shows a typical brittle failure in acrylic-based bone cement. Fractography can provide useful insight into the nature of fracture processes acting at the crack tip and is a valid supplement for thorough fatigue characterization of engineering polymers (Ref 59).

REFERENCES

1. E.H. Andrews, *Testing of Polymers,* W. Brown, Ed., Wiley, 1969, p 237
2. P. Beardmore and S. Rabinowitz, *Treat. Mater. Sci. Technol.,* Vol 6, 1975, p 267
3. J.A. Sauer and G.C. Richardson, *Int. J. Fract.,* Vol 16, 1980, p 499
4. R.W. Hertzberg, M.D. Skibo, and J.A. Manson, *Fatigue Mechanisms,* STP 675, ASTM, 1979, p 471
5. R.W. Hertzberg and J.A. Manson, *Fatigue of Engineering Plastics,* Academic Press, 1980
6. R.W. Hertzberg and J.A. Manson, in *Encyclopedia of Polymer Science and Engineering,* Wiley, 1986, p 378
7. H.H. Kausch and J.G. Williams, in *Encyclopedia of Polymer Science and Engineering,* Wiley, 1986, p 341
8. J.A. Sauer and M. Hara, *Advances in Polymer Science 91/92,* H. H. Kausch, Ed., Springer-Verlag, 1990, p 71
9. J.D. Ferry, *Viscoelastic Properties of Polymers,* Wiley, 1961
10. G.C. Sih, *Handbook of Stress Intensity Factors,* Lehigh University Press, 1973
11. P.C. Paris, M.P. Gomez, and W.P. Anderson, *Trends Eng.,* Vol 13, 1961, p 9
12. R.O. Ritchie, *Proc. of the Fifth Int. Conf.,* M.G. Yan, S.H. Zhang, and Z.M. Zheng, Ed., Pergamon Press, 1988, p 5
13. S. Suresh, *Fatigue of Materials,* 2nd ed., Cambridge University Press, 1998
14. A.G. Evans, Z.B. Ahmad, D.G. Gilbert, and P.W.R. Beaumont, *Acta Metall.,* Vol 34, 1986, p 79
15. C.B. Bucknall and W.W. Stevens, in *Toughening of Plastics,* Plastics and Rubber Institute, London, 1978, p 24
16. R.W. Hertzberg, *Deformation and Fracture Mechanics of Engineering Materials,* 4th ed., Wiley, 1996
17. T.L. Anderson, *Fracture Mechanics: Fundamentals and Applications,* CRC Press, 1995
18. A.S. Argon, *Proc. of Seventh Int. Conf. on Advances in Fracture Research,* K. Samala, K. Ravi-Chander, D.M.R. Taplin, and P. Rama Rao, Ed., Pergamon Press, 1989
19. R.W. Lang, J.A. Manson, R.W. Hertzberg, and R. Schirrer, *Polym. Eng. Sci.,* Vol 24, 1984, p 833
20. A.J. Kinloch and F.J. Guild, in *Advances in Chemistry Series 252: Toughened Plastics II: Novel Approaches in Science and Engineering,* American Chemical Society, Washington, D.C., 1996, p 1
21. T.R. Clark, R.W. Hertzberg, and N. Nohammadi, in *Eighth Int. Conf. Deformation, Yield and Fracture of Polymers,* Plastics and Rubber Institute, London, 1991, p 31/1
22. E.J. Kramer and L.L. Berger, *Advances in Polymer Science 91/92,* H.H. Kausch, Ed., Springer-Verlag, 1990, p 1
23. H.R. Azimi, R.A. Pearson, and R.W. Hertzberg, *J. Mater. Sci.,* Vol 31, 1996, p 3777
24. S.A. Sutton, *Eng. Fract. Mech.,* Vol 6, 1974, p 587
25. B. Mukherjee and D.J. Burns, *Mech. Eng. Sci.,* Vol 11, 1971, p 433
26. Y.-Q. Zhou and N. Brown, *J. Poly. Sci. B, Polym. Phys.,* Vol 30, 1992, p 477
27. E.J. Moskala, in *Eighth Int. Conf. Deforma-*

tion, Yield and Fracture of Polymers, Plastics and Rubber Institute, London, 1991, p 51/1
28. S. Arad, J.C. Radon, and L.E. Culver, *J. Mech. Eng. Sci.,* Vol 13, 1971, p 75
29. A.S. Argon and R.E. Cohen, *Advances in Polymer Science 91/92,* H.H. Kausch, Ed., Springer-Verlag, 1990, p 300
30. J.A. Manson, R.W. Hertzberg, and P.E. Bretz, *Advances in Fracture Research,* D. Francois, Ed., Pergamon Press, 1981, p 443
31. L. Pruitt and S. Suresh, *Polymer,* Vol 35, 1994, p 3221
32. L. Pruitt and D. Rondinone, *Polym. Eng. Sci.,* Vol 36, 1996, p 1300
33. M.T. Takemori, *Polym. Eng. Sci.,* Vol 22, 1982, p 937
34. N.J. Mills and N. Walker, *Polymer,* Vol 17, 1976, p 335
35. L. Pruitt, Ph.D. dissertation, Brown University, 1993
36. L. Pruitt, R. Herman, and S. Suresh, *J. Mater. Sci.,* Vol 27, 1992, p 1608
37. S.M. Cadwell, R.A. Merrill, C.M. Sloman, and F.L. Yost, *Ind. Eng. Chem. Anal. Ed.,* Vol 12, 1940, p 19
38. W. Elber, *Eng. Fract. Mech.,* Vol 2, 1970, p 37
39. S. Suresh, *Eng. Fract. Mech.,* Vol 18, 1983, p 577
40. L. Pruitt and S. Suresh, *Philos. Mag. A,* Vol 67, 1993, p 1219
41. J.R. Rice, in *Fatigue Crack Propagation,* STP 415, ASTM, 1967, p 247
42. F.J. Pitoniak, A.F. Grandt, L.T. Montulli, and P.F. Packman, *Eng. Fract. Mech.,* Vol 6, 1974, p 663
43. R. Murakami, S. Noguchi, K. Akizono, and W.G. Ferguson, *J. Fract. Eng. Mater. Struct.,* Vol 6, 1987, p 461
44. D.H. Banasiak, A.F. Grandt, and L.T. Montulli, *J. Appl. Polym. Sci.,* Vol 21, 1977, p 1297
45. M.P. Wnuk, *J. Appl. Mech.,* Vol 41 (No. 1), 1974, p 234
46. R.W. Hertzberg, J.A. Manson, and M.D. Skibo, *Polym. Eng. Sci.,* Vol 15, 1975, p 252
47. J.S. Harris and I.M. Ward, *J. Mater. Sci.,* Vol 8, 1973, p 1655
48. K. Dawes and L.C. Glover, *Physical Properties of Polymers Handbook,* J.E. Mark, Ed., AIP Press, 1966, p 557
49. S.M. Kurtz, D.L. Bartel, and C.M. Rimnac, *Trans. 40th Annual Meeting of the Orthopedic Research Society* (San Francisco), Orthopedic Research Society, 1994, p 584
50. G.M. Connelly, C.M. Rimnac, T.M. Wright, R.W. Hertzberg, and J.A. Manson, *J. Orth. Res.,* Vol 2, 1984, p 119
51. T.M. Wright, C.M. Rimnac, S.D. Stulberg, L. Mintz, A.K. Tsao, R.W. Klein, and C. McCrae, *Clin. Orth.,* Vol 276, 1992, p 126
52. M. Goldman and L. Pruitt, *J. Biomed. Mater. Res.,* Vol 40 (No. 3), 1998, p 378–384
53. M. Goldman, R. Gronsky, and L. Pruitt, *J. Mater. Sci.: Mater. Med.,* Vol 9, 1998, p 207–212
54. C.M. Rimnac, T.M. Wright, and R.W. Klein, *Polym. Eng. Sci.,* Vol 28, 1988, p 1586
55. M. Goldman, R. Ranganathan, R. Gronsky, and L. Pruitt, *Polymer,* Vol 37 (No. 14), 1996, p 2909–2913
56. D. Baker, R. Hastings, and L. Pruitt, *Polymer,* 1999
57. L. Pruitt and R. Ranganthan, *Mater. Sci. Eng. C: Biomimetic Mater., Sens. Syst.,* Vol 3, 1995, p 91–93
58. L. Pruitt, J. Koo, C. Rimnac, S. Suresh, and T. Wright, *J. Orth. Res.,* Vol 13, 1995, p 143–146
59. L. Engel, H. Klingele, G.W. Ehrenstein, and H. Schaper, *An Atlas of Polymer Damage,* Wolfe Science Books, Vienna, 1981

Fatigue Failure Mechanisms*

FAILURE OF STRUCTURAL MATERIALS under cyclic application of stress or strain is not only a subject of technical interest but one of industrial importance as well. The understanding of fatigue mechanisms (damage) and the development of constitutive equations for damage evolution leading to crack initiation and propagation as a function of loading history represent a fundamental problem for scientists and engineers. Although loading conditions such as creep, stress relaxation, and continuous deformation lend considerable information to the study of fatigue behavior, fatigue does introduce additional factors, such as loading frequency, upper and lower loading limits, and loading waveform. These features appear to produce failure characteristics not otherwise encountered. Additionally, the dissipative nature of polymers results in high mechanical hysteresis. Because of their low thermal conductivity, a large portion of the mechanical work done is converted into heat, which complicates the analysis of fatigue data, particularly at high loading frequencies.

The traditional approach to fatigue lifetime prediction due to Wohler (Ref 1) involves using the endurance limit concept developed from the stress range versus number of cycles to final fatigue, or *S-N,* curve (Wohler diagram) as described in the preceding article, "Fatigue Testing and Behavior." In spite of its empirical nature, the *S-N* approach is a commonly accepted design criterion for fatigue resistance in engineering plastics. For example, ASTM D 671 for plastics specifies repeat flexural stress (fatigue) as a standard test. Loading methods and testpiece configurations have been agreed on, as discussed by many investigators (Ref 2–4). Loads may be applied by bending, by torsion, or axially, and testpieces may be in the form of a plate or rod, with or without an artificially introduced notch or crack.

Relationships between stress amplitude and cycles to failure for different plastic materials (Ref 5) are shown in Fig. 1, which also shows that dry nylon and polyethylene terephthalate (PET) do not appear to possess a stress limit below which failure does not occur after a large number of cycles. The *S-N* relationship for these two materials is essentially linear, with no indication of an endurance limit. Moreover, polymer fatigue behavior is generally sensitive to temperature, frequency, and environment (Ref 6), as well as molecular weight (MW), molecular weight density (MWD), and aging. *S-N* curves that do not account for these effects should not be used exclusively without looking at test conditions.

Experiments conducted to construct a *S-N* curve are time-consuming; several samples at each stress are required to account for the statistical nature of the data obtained. It is common for fatigue lifetime data from well-controlled samples to spread over a few orders of magnitude. This, in fact, reflects the complex nature of the fracture processes involved. In this regard, it is useful to consider a unique experiment conducted in the mid-1950s (Ref 7) and later compiled by other researchers (Ref 8). In this experiment (Fig. 2), 400 identical specimens of EN-24 steel were tested near their endurance limit. Although the lifetime was measured as the number of cycles to failure, mostly clustered within one decade, the scatter spreads over three decades. On the other hand, the error in the fatigue stress limit falls within a reasonable range of less than ±5%.

Heterogeneities inherent in the microstructure of most materials result in a random field of defects whose geometry, size, and orientation are also random. Such a random field of defects, influenced by the imposed stress, gives rise to a complex process of growth and interaction of defects, which ultimately leads to the initiation of macroscopic cracks. A crack propagates first in a stable manner to a stage at which it undergoes a transition to unstable (uncontrolled) propagation. The lifetime of a structure is accordingly composed of two stages, namely, crack initiation and crack propagation. Depending on the severity of defects, crack initiation may comprise 20 to 80% of the total lifetime. Hence, sound lifetime prediction relies on knowledge of the law of crack initiation and that of slow crack propagation.

Mechanisms of Fatigue Failure

Depending on the stress amplitude and the frequency of load application, fatigue failure of some polymers has been observed to occur by one of two general mechanisms. The first involves thermal softening (or yielding), which precedes crack propagation, leading to ultimate failure. This mechanism dominates in certain materials at large stress amplitudes within a particular range of frequency of load applications (Ref 9–11). At a lower stress amplitude, on the other hand, a conventional form of fatigue crack propagation (FCP) mechanism is generally observed. Low frequency is also found to cause fatigue fracture by conventional crack propagation at high stress amplitude. The interrelation of the two mechanisms, stress amplitude and frequency, for polyoxymethylene (acetal) is shown in Fig. 3 (Ref 10).

The high damping and low thermal conductivity of polymers cause a strong dependency of temperature rise on the rate of load application (frequency) and on the deformation level (stress or strain amplitude). From a thermodynamic point of view (Ref 12), part of the mechanical work done during cyclic loading is spent on irreversible molecular processes (Ref 13), leading to microscopic deformations such as crazes, shear bands, voids, and microcracks. The other part of the mechanical work evolves as heat. Both processes are obviously interdependent and relate to the specific nature of the relaxation time spectrum of the macromolecules considered.

The total work done is measurable from the hysteresis loop encountered in fatigue testing. Attempts were thus made to characterize fatigue damage from the evolution of the stress-strain relationship reflected in the hysteresis loop (Ref 9, 14, 15). Fatigue data on unnotched samples of acrylonitrile-butadiene-styrene (ABS) and high-impact polystyrene (HIPS), at $^1/_{30}$ Hz and under a tension-compression square waveform, were obtained (Ref 15). Hysteresis loops recorded after various numbers of cycles are shown in Fig. 4. The loops of ABS tend to be symmetrical, which is attributed to shear yielding being the dominant form of damage in this polymer. In HIPS, the hysteresis loop area is much larger for the tension half-cycle than for the compression half-cycle. This area remains almost constant for a number of cycles and then increases significantly, as demonstrated clearly in Fig. 4. This effect is attributed to the production and growth of crazes after some induction period. For ABS, an increase of stress amplitude causes the energy to dissipate per half-cycle and the speci-

*Adapted from the article by Abdelsamie Moet and Heshmat Aglan, "Fatigue Failure," in *Engineering Plastics,* Volume 2, *Engineered Materials Handbook,* ASM International, 1988, pages 741 to 750 men temperature to rise significantly, while the secant modulus decreases. However, this type of mechanistic analysis is particularly useful for uncracked specimens with a single dominant fatigue damage mechanism.

An FCP mechanism frequently involves damage formation, which precedes crack initiation and propagation. On the other hand, the micromechanism underlying yielding (softening or thermal failure) remains unclear. Fatigue experiments on polycarbonate indicate that cyclic softening is caused by profuse crazing prior to fracture (Ref 14). Microscopic examination of the same polymer (Ref 16, 17) shows that fatigue crazes are terminated by, and interact through, pairs of shear bands. A crack ultimately initiates and propagates within one of the crazes. Related studies on polystyrene (PS) and polymethyl methacrylate (PMMA) show that long fatigue life at low stress amplitude is associated with more profuse crazing along the gage length (Ref 18).

Current evidence suggests that large deformation or softening can precede crack initiation under certain loading conditions. In cases such as this, subsequent crack propagation may occur in a localized region of the transformed material. Although the nature of such deformation depends on the molecular structure and its interaction with the stress field, specimen separation remains to occur by means of FCP. A similar situation is encountered in creep fracture of polyethylene in which brittle fracture is observed at low load and ductile fracture is observed at high load (Ref 19). The latter is nothing but a brittle crack propagation through a large yielded zone.

Thus, fatigue failure of polymers can occur by two means: thermal fatigue failure and mechanical fatigue failure. Although mechanical failure behavior of polymers is the main consideration in this article, thermal fatigue is discussed first.

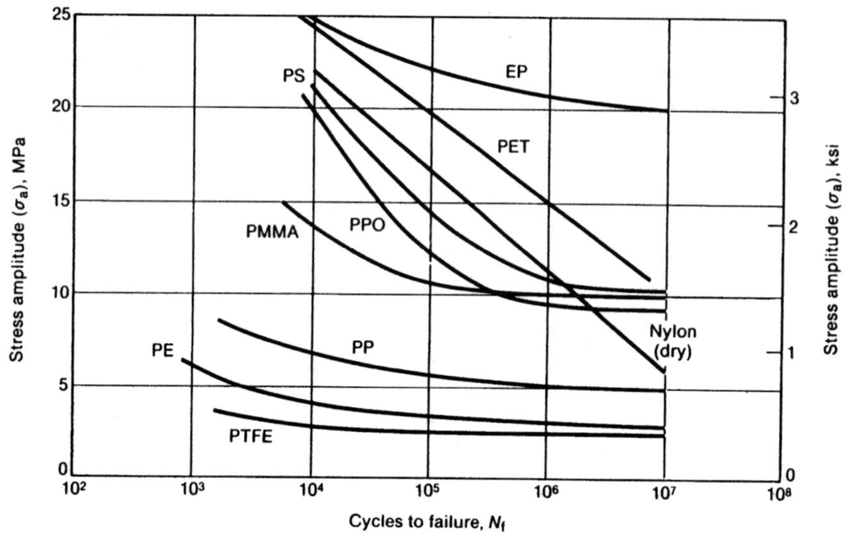

Fig. 1 Stress amplitude versus cycles-to-failure curves for several polymers tested at a frequency of 30 Hz. PS, polystyrene; EP, epoxy; PET, polyethylene terephthalate; PMMA, polymethyl methacrylate; PPO, polyphenylene oxide; PE, polyethylene; PP, polypropylene; PTFE, polytetrafluoroethylene. Source: Ref 5

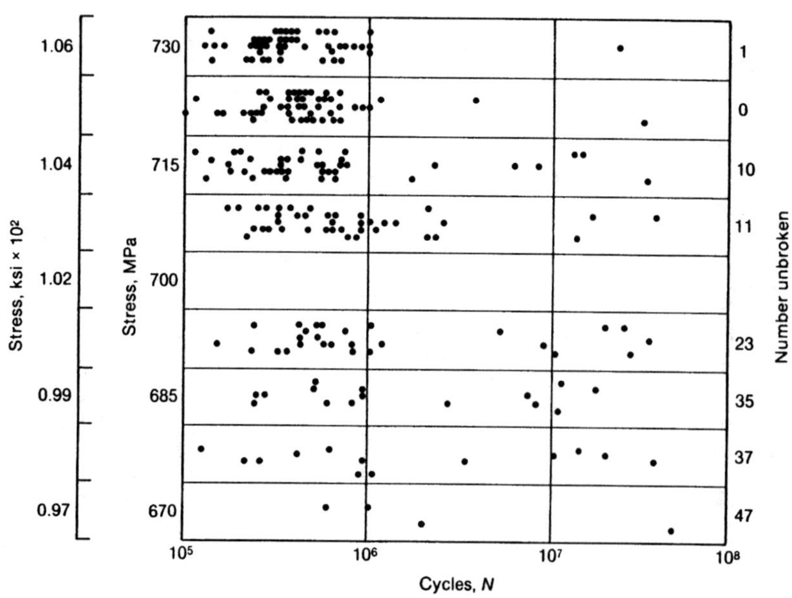

Fig. 2 Stress-number of cycles to fatigue (*S-N*) behavior of 400 specimens of EN-24 steel tested near the endurance limit. Source: Ref 5

Thermal Fatigue Failure

Because polymers are viscoelastic materials, plastic flow is commonly observed when they are fatigued at higher levels of strain. At moderate strain, however, they exhibit mechanical hysteresis. This is due to the energy-dissipative nature of these materials when they undergo cyclic testing. Thus, some of this inelastic deformation energy is transferred into heat. With each cycle, heat buildup raises the temperature of the specimen. A plastic material heating up in fatigue will display either thermal stability or instability.

Thermal stability exists when the heat-transfer rate to the surroundings, by means of conduction, convection, or radiation, equals the rate of heat generated. The temperature of the specimen stabilizes, and the material is able to withstand the fatigue load, but at a reduced stress level because of the reduction in the strength and stiffness of the material at that particular elevated temperature.

Thermal instability occurs when the heat-transfer rate to the surroundings by conventional heat-transfer mechanisms is less than the rate of heat generated by successive fatigue cycles. In this case, the temperature of the material increases until its properties decline to a point at which it can no longer withstand the load. This is called a thermal failure. The difference between the rate of heat generated and the rate of heat dissipated to the surroundings, which raises the temperature of the material, depends on stress amplitude, frequency of loading, specimen geometry, test environment, and the internal friction, thermal conductivity, and heat capacity of the material.

The cumulative effect of heat generated per unit time under continual cyclic load may be described by (Ref 20):

$$\dot{U}g = \pi f E'' \sigma_{max}^2 \quad \text{(Eq 1)}$$

where f is the applied frequency, E'' is the loss compliance at the temperature and frequency of the test, and σ_{max} is the maximum applied stress.

Because of the viscoelastic behavior of polymers, stress and strain are not in phase during cyclic loading. The magnitude of the phase angle difference varies considerably with the plastic. According to Ref 21, the phase angle, δ, is very large for certain unfilled semirigid thermoplastics, such as polytetrafluoroethylene (PTFE).

The loss compliance, E'', in Eq 1 is related to the phase angle, δ_c, by:

$$\tan \delta_c = \frac{E''}{E'} \quad \text{(Eq 2)}$$

where E' is the storage compliance associated with the elastic stiffness of the material. The loss compliance, E'', is associated with the loss of energy as heat.

Thermal effects associated with cyclic loading of different polymers have been studied by many investigators (Ref 10, 22–31). For example, researchers (Ref 22) have stated that PS, which possesses a very low internal friction, can withstand fatigue tests at approximately 30 Hz and a σ_{max} of approximately 15 MPa (2.2 ksi), with the accompanying temperature rise being less than 2 K (Ref 24). Under the same conditions, however, polyethylene (PE) samples would rapidly melt, and PMMA would fail by thermal rupture. A temperature rise of 80 K has been reported (Ref 30, 31) for PMMA tested at 50 Hz.

Research (Ref 26) showed the *S-N* curve for unfilled PTFE (Fig. 5), along with temperature rise curves for each of the individual tests on which the *S-N* curve was based. The test configuration for these data was cantilever bending at constant load. It can be seen from Fig. 5 that, at the highest stress, temperature rise is rapid, and failure occurs in a short time. Temperature rise decreases and failure occurs at longer times until a stress is reached at which no failure occurs. This is called runout and defines the stress level at which heat generated within the specimen is in equilibrium with heat transferred to the surroundings (thermal stability). This stress level is significant for design and material comparison purposes. It is the stress below which the part or specimen cycles for a long time without thermal rupture.

Other fatigue variables that affect temperature rise are the frequency of applied load, the thickness of the specimen, and the loss compliance, each of which is described subsequently.

Structural metals are relatively insensitive to load frequency over a fairly large range. However, as indicated by Eq 1, this is an important variable for plastics, because it contributes to heat dissipation. In addition, the loss compliance increases with increasing frequency. Thick specimens tend to generate more heat, which leads to a greater temperature rise with a given set of conditions. This is due to the smaller surface area/volume ratio of the longer specimens.

Loss compliance is a fundamental material variable that controls energy dissipation and therefore temperature rise of plastics under cyclic load. It increases with frequency and material temperature. It also tends to increase rapidly through transitions, such as the glass transition.

At room temperature, the loss compliance can provide a basis for a general classification of plastics by failure mechanism (Ref 21), and its value can be measured using dynamic tests. These classifications are:

- *Group 1:* Materials with low ambient loss compliance, less than 0.1×10^{-10} m²/N (6.9×10^{-7} in.²/lbf), fail primarily by crack propagation and/or thermal stability. This group includes rigid polyvinyl chloride (PVC), polyphenylene oxide (PPO), polysulfone (PSU), urea, diallyl orthophthalate (DAP), phenolic, and epoxy (EP).
- *Group 2:* Materials with intermediate loss compliance, 0.1 to 0.5×10^{-10} m²/N (6.9 to 34.5×10^{-7} in.²/lbf), tend to fail by temperature rise and crack propagation occurring simultaneously. This group includes PMMA, acetal, PET, alkyd, and polycarbonate (PC).
- *Group 3:* Materials with high loss compliance, 0.5 to 5×10^{-10} m²/N (34.5 to 345×10^{-7} in.²/lbf), tend to fail exclusively by thermal failure. This group of materials includes fluoroplastics, polypropylene (PP), PE, and nylon.

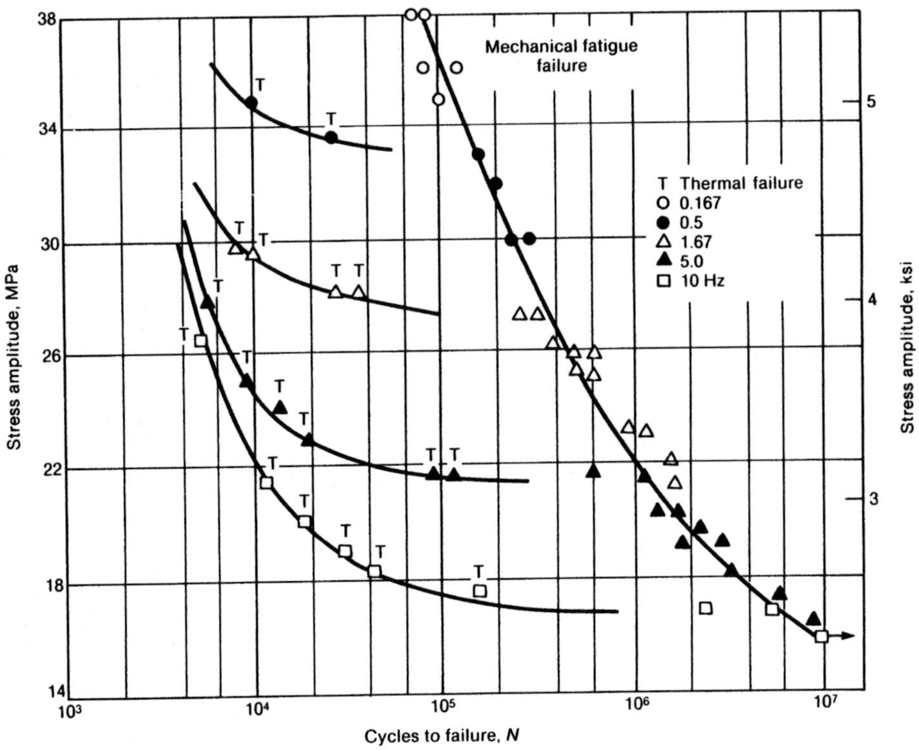

Fig. 3 Thermal fatigue failure and conventional fatigue crack propagation fracture during reversed load cycling of acetal. Source: Ref 10

Mechanical Fatigue Failure

The other main failure mechanism, mechanical fatigue, involves the initiation of a crack and its subsequent propagation. This is discussed as follows in terms of fatigue crack initiation and FCP.

Fatigue Crack Initiation. The initiation of macroscopic cracks on the order of 10^{-3} under fatigue loading is studied by means of two complementary approaches. A fracture mechanics approach is used to characterize fatigue crack initiation (FCI) by a threshold value of the stress-intensity factor, K_{th}, or its range, ΔK_{th}. Below this threshold value, macroscopic cracks remain dormant. This type of study involves the use of fracture mechanics concepts. Generally, fatigue load is applied to a notched specimen, and the first measurable crack or notch extension, Δa, is recorded. Fatigue threshold signifies that not every precrack will extend, so that a certain condition must be met for $\Delta a/\Delta N$ to exist. Accordingly, the threshold value ΔK_{th} is interpreted as that minimum of stress-intensity factor range, ΔK, that is required to make the precrack grow. The hypothesis is that the crack growth is linearly related to the crack opening displacement (COD) (Ref 32). If the COD during loading exceeds a threshold value, a permanent step (Δa) of the crack is assumed to remain open on unloading. In other words, fatigue threshold describes the first crack jump. The corresponding ΔK, that is, ΔK_{th}, is thought of as a material property characterizing the resistance to crack initiation. However, this quantity is probably dependent on a number of factors, including temperature, frequency, environment, MW, and MWD. Alternatively, the related energy release rate, ΔG_{th} ($=\Delta K^2/E$), has also been considered (Ref 33).

On the other hand, micromechanistic investigations of initially uncracked and initially cracked polymer specimens emphasize the role of crazing in FCI in glassy and semicrystalline polymers. For example, in a 6 mm (0.25 in.) thick extruded, unnotched PC specimen exposed to high strain fatigue, the formation of microcrazes terminated by shear bands precedes crack initiation (Ref 16). The crazing density appears to reach a critical level at which the main fatigue crack initiates within one of the crazes. Once initiated, subcritical crack propagation occurs through a craze surrounded by a pair of shear bands (Fig. 6), forming what is known as an epsilon crack (Ref 17). Similar FCI behavior is observed in tension-compression fatigue of unnotched PC sheet (Ref 9).

With HIPS and ABS, analysis of hysteresis loops reveals that FCI occurs because of crazing and shear banding, respectively (Ref 15). The magnitude of crazing developed prior to crack initiation depends on the stress level and test frequency. From the review of optical-interference measurements (Ref 34), it is inferred that a crack initiates and propagates in glassy polymers under certain conditions through a single craze. Optical micrography, on the other hand, shows that a few (Ref 2) or a myriad (Ref 35) of crazes precede FCI. What appears to be common to all of these observations is that a critical level of damage ought to be reached to cause initiation. This critical level of damage seems to correspond to the sudden crack jump characterized by ΔK_{th}. Efforts are underway to develop techniques for quantitative damage analysis (Ref 35). Presently, fatigue damage on the microscale leading to crack initiation as well as crack propagation in polymers can be measured quantitatively. Knowledge of submicroscopic events, such as diffusion of chain molecules, disentanglement, fibrillation, and chain scission, which constitute the underlying phenomena of damage formation, remains qualitative in nature, yet significantly important. A thermodynamic approach (Ref 36) treats the phenomenon as local instability and proposes a framework to establish the law of crack initiation. However, a quantitative measure of the initiation time from a smooth bar specimen is still not possible at this time.

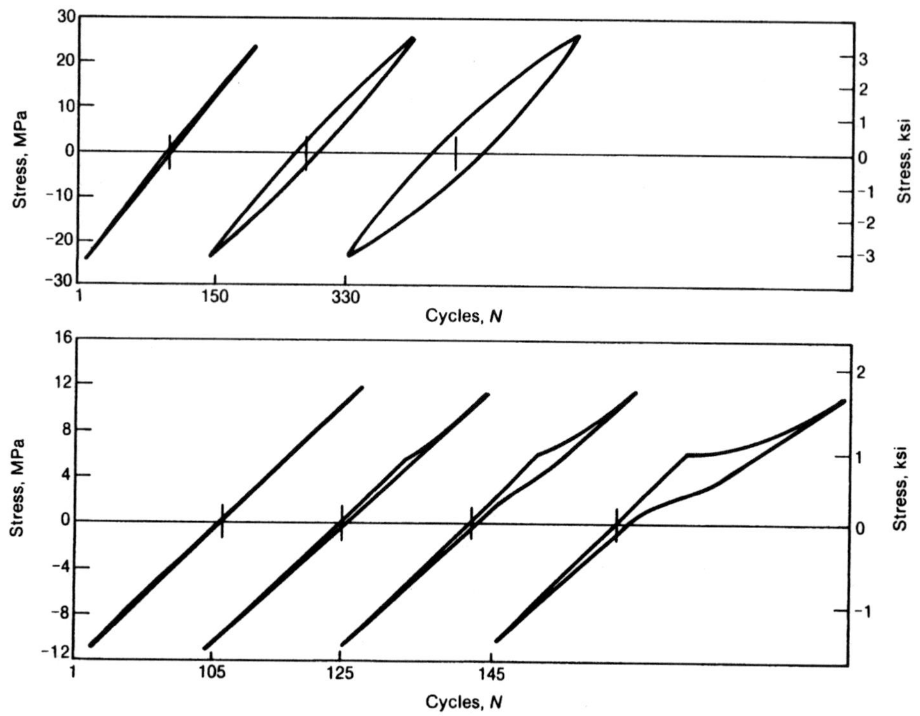

Fig. 4 Hysteresis loops after various cycles in acrylonitrile-butadiene-styrene tested at stress amplitude (σ_α) = 25.4 MPa (3.68 ksi) and in high-impact polystyrene tested at σ_α = 11.6 MPa (1.68 ksi)

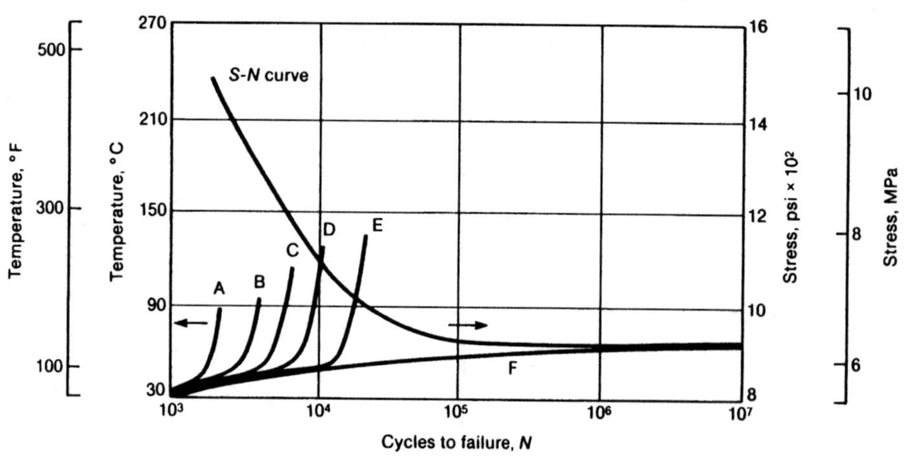

Fig. 5 Stress-number of cycles to failure (S-N) curve and corresponding temperature rise curves for individual test specimens of unfilled polytetrafluoroethylene, showing thermal failure. A: 10.3 MPa (1.5 ksi), 2×10^3 cycles, 100 °C (212 °F); B: 9.0 MPa (1.3 ksi), 4×10^3 cycles, 115 °C (240 °F); C: 8.3 MPa (1.2 ksi), 6.1×10^3 cycles, 125 °C (255 °F); D: 7.6 MPa (1.1 ksi), 9.5×10^3 cycles, 130 °C (265 °F); E: 6.9 MPa (1.0 ksi), 19×10^3 cycles, 141 °C (285 °F); F: 6.3 MPa (0.91 ksi), 10^7 cycles, 60 °C (140 °F). Source: Ref 26

Fatigue Crack Propagation. Advances in fracture mechanics in the past inspired tremendous interest in FCP, which evolved as an independent discipline. Attempts to formulate the law of subcritical (slow, stable, or quasistatic) crack propagation under intermittent load application play a central role in the effort. In spite of the mechanistic differences between metals and polymers in FCP, the formal approach remains the same, because it is founded on the ideas of fracture mechanics.

An FCP experiment usually involves measurements of the average incremental crack length, Δa, from a sharp notch of a known depth, a_0, in a specimen of a defined geometry. The average crack speed is given by ($\Delta a/\Delta N$), where ΔN is the number of cycles corresponding to a crack extension, Δa. Commonly used geometries include single-edge notched (SEN) and compact-tension (CT) specimens. A double cantilever geometry is better suited for the studies of FCP in adhesive bond lines. Although a variety of loading cycles may be applied, it is common to study FCP under tension loading programs of different waveforms, such as sinusoidal, triangular, or rectangular. The majority of FCP experiments, however, are conducted under tensile sinusoidal loads. The frequency of load applications, the load amplitude, and the stress level determined by its maximum or mean values represent the basic loading variables (Ref 2). The load amplitude is usually expressed as the load ratio, which is the ratio of minimum stress to its maximum, that is, $R = \sigma_{min}/\sigma_{max}$.

In the fracture mechanics approach, the stress-intensity factor, which is a measure of the stress singularity at the crack tip, characterizes the stress field associated with a sharp crack in an elastic continuum. Three geometric configurations are used to model the crack. Mode I refers to the crack opening with displacement normal to the fracture surface. Mode II refers to shear or antisymmetric crack surface separation. Mode III refers to tearing in which, again, the crack is antisymmetrically opened. Generally, fracture can be characterized by some combination of these modes, with mode I being the most common configuration.

For crack propagation by opening (mode I) in a SEN specimen, the stress-intensity factor, K_I, is given by:

$$K_I = \sigma\sqrt{\pi a}\, f\left(\frac{a}{W}\right) \quad \text{(Eq 3)}$$

where W is the width of the specimen, and σ is the stress applied remotely, that is, at the grips. The function f(a/W) is a geometric correction factor whose solutions can be obtained from the boundary value problem. Solutions for various geometries can be found in stress analysis handbooks (Ref 37). In fatigue, a maximum and minimum of the stress-intensity factor corresponds to the stress limits, that is, σ_{max} and σ_{min}. Thus, a stress-intensity factor range ($\Delta K = K_{max} - K_{min}$) is usually considered.

The ideal sharp planar crack, which presumably separates two adjacent rows of atoms, ought to be compared with a real crack-tip geometry (Fig. 7). Clearly, the difference is great. It is therefore instructive to consider the quantities calculated from linear fracture mechanics in view of such differences. Parameters such as K or ΔK are useful as correlative tools, particularly because they possess an invariant nature.

Crack growth equations have been used to describe FCP in polymers as well as in metals. The rate of FCP is correlated with experimental conditions, such as applied stress, temperature, and frequency, and material parameters, such as molecular composition or microstructure. The equation proposed by Paris and Erdogan (Ref 38) has gained the widest acceptance. It states:

$$\frac{da}{dN} = C_1(\Delta K)^{m_1} \quad \text{(Eq 4)}$$

where da/dN is the cyclic crack growth rate, ΔK is the stress-intensity factor range, and C_1 and m_1 are material- and loading-dependent constants. The following equation (Ref 39), based on the fracture toughness, K_c, measured at high propagation rates, is a modified Paris equation of the form:

$$\frac{da}{dN} = \frac{C_2\,(\Delta K)^{m_2}}{K_c(1 - R) - \Delta K} \quad \text{(Eq 5)}$$

where C_2 and m_2 are constants, and R is the stress ratio, $\sigma_{min}/\sigma_{max}$. Equation 5 was further modified (Ref 40) by replacing the fracture toughness term with the plane-strain fracture toughness parameter, K_{Ic}, to give:

$$\frac{da}{dN} = \frac{C_3\,(\Delta K)^{m_3}}{\sqrt{K_{Ic}\,(1 - R) - \Delta K}} \quad \text{(Eq 6)}$$

where C_3 and m_3 are constants.

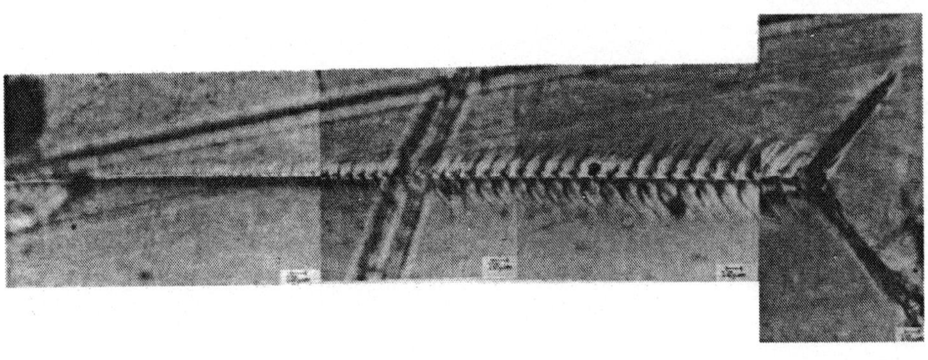

Fig. 6 Crack propagation through a craze surrounded by a pair of shear bands (an epsilon crack) in polycarbonate. Source: Ref 17

Researchers (Ref 41) carried out extensive fatigue experiments on PMMA. They expressed the rate of FCP as a function of the mean stress-intensity factor, K_m, and the frequency, f, by:

$$\frac{da}{dN} = C_4 K_m^{m_4}(\Delta K)^{m_5} f^{m_6} \quad \text{(Eq 7)}$$

where C_4, m_4, m_5, and m_6 are material constants.

To account for mean stress-intensity effects, researchers (Ref 42, 43) postulated an equation of the form:

$$\frac{da}{dN} = C_5\,\lambda^{m_7}$$

$$\lambda = (K_{max}^2 - K_{min}^2) = 2\,K_m\,(\Delta K) \quad \text{(Eq 8)}$$

where K_{max} and K_{min} are the maximum and minimum cyclic stress intensities, and C_5 and m_7 are constants. Researchers (Ref 44) further extended Eq 8 in order to incorporate both the shear modulus, G, and Poisson's ratio, ν, using the relation:

$$\frac{da}{dN} = C_6 \frac{[(1 - \nu^2)\,\lambda]^{m_8}}{[2G(1 + \lambda)]} \quad \text{(Eq 9)}$$

where C_6 and m_8 are constants.

Other investigators (Ref 45) adapted the formulation in Eq 8 to describe the effect of mean stress. They hoped that the equation would predict FCP for the entire range of the loading spectrum from the threshold value ΔK_{th} to K_c. This equation is in the form:

$$\frac{da}{dN} = C_7\phi^{m_9} \quad \text{(Eq 10)}$$

Here, ϕ is defined as:

$$\phi = \frac{2K_m(\Delta K - \Delta K_{th})}{K_c^2 = K_{max}^2}$$

and C_7 and m_9 are constants.

Another researcher (Ref 46) proposed a more generalized law based on Eq 8 in which the FCP rate is expressed as a function of K_{max}, K_{min}, and K_m. This equation is in the form:

$$\frac{da}{dN} = C_8\,(K_{max}^2 - K_{min}^2)^{m_{10}}\,(K_m^2)^{m_{11}} \quad \text{(Eq 11)}$$

where the parameters C_8, m_{10}, and m_{11} are functions of frequency, environment, loading conditions, and material properties.

The inadequacy of the Paris equation to predict FCP rates at both low and high levels of ΔK has led to the development of the other fatigue models. This equation suggests that the rate of FCP is a logarithmically linear function of ΔK. In fact, typical FCP behavior, as illustrated in Fig. 8, falls into three distinct regions. Region I starts with a threshold value of the stress-intensity factor range, ΔK_{th}, below which propagation of the crack is not observed. The value of K_{th} has been attributed to the attainment of a sufficient level of activity in the notch tip region to cause its propagation (Ref 47). The initial slope of region I is usually very steep. As the crack becomes longer, that is, as ΔK becomes larger, reduced crack acceleration occurs, leading to region II. The FCP curve is effectively linear in region II in the majority of cases. The rate of FCP approaches its asymptotic value at $K = K_c$, where a transition from a stable condition to crack propagation resembling an avalanche occurs.

The commonly observed linearity of the FCP rate within region II promoted the general acceptance of Eq 4 to describe the phenomenon. A lack of linearity in some polymers is immediately obvious when the test is conducted over a wide range of ΔK. Nevertheless, Paris plots can still be used to evaluate the relative resistance of materials to FCP (Fig. 9) (Ref 48). This is achieved by examining the rate of FCP at a particular value of ΔK. The higher the *da*/*dN*, the lower the FCP resistance. Alternatively, the higher the ΔK for a particular *da*/*dN*, the more resistant the material is supposed to be.

Careful examination of the results in Fig. 9 indicates that region II is not necessarily observed within the same ΔK span (see PS and PMMA). Hence, the comparison could be misleading, because curve crossover is observed. Had the entire FCP been recorded, a more certain assessment of the resistance to FCP would have been possible. Therefore, it is helpful to examine the FCP behavior of the two PVC composites (Ref 49) shown in Fig. 10. The energy release rate, J_I, is more appropriately correlated with the rate of crack propagation from geometric and thermodynamic viewpoints. Comparison of the two curves addresses the resistance to FCP in terms of two questions: How long does it last, and how strong is it? The large, reduced crack acceleration observed in the case of 10% glass fiber (low gradient of region II) results in a higher fracture toughness as measured from the respective critical energy release rate, J_{Ic}. The lifetime of the FCP, on the other hand, is evaluated from the speed at which reduced crack acceleration occurs. Thus, the 30% glass-fiber composite lasts longer under the same fatigue conditions, although it displays lower J_{Ic}.

The importance of more complete characterization of FCP is further dramatized by the reported fatigue crack deceleration (Ref 35, 50, 51). This behavior is shown in Fig. 11 (Ref 51). Solid lines represent the FCP previously reported. The data points representing the rate of FCP in the same material examined over a wide range of ΔK qualitatively deviate from our conviction based on the Paris equation and related power models. A decrease in (*da*/*dN*) is observed with increasing ΔK.

However, the Paris equation can be useful in some cases for comparing the resistance of materials to crack propagation and their endurance limit. The comparison can be made either between two different materials at the same testing conditions or for one material at different testing conditions. For example, FCP data for PMMA at 1 Hz and at different testing temperatures were obtained (Ref 52). The data were then statistically fit (Ref 53) to the Paris equation (Fig. 12). In spite of the intersections at low temperature range, a comparison of the PMMA resistance to FCP at high temperature range can easily be made. Thus, the resistance of the PMMA to FCP decreases with the increase of the environment temperature. Using a thermodynamic approach, a generalized model that describes FCP over the entire range of temperature and stress was developed (Ref 54, 55). A modified form of this model is presented here.

Crack Layer (CL) Model. The resistance of a material to crack propagation depends on the energy expended on irreversible deformation (damage) in the vicinity of the crack tip. The objective of crack propagation studies is to identify and determine the material parameters responsible for the resistance of the material to crack propagation, that is, to determine fracture toughness. It is thus hoped to establish predictive relationships to aid in the assessment of the lifetime of load-bearing structural components and thereby to guide the development of crack-resistant materials.

Recently, the CL theory has been developed and successfully applied to several materials (Ref 12, 35, 49, 54–64). The crack is always preceded by a zone of transformed (damaged)

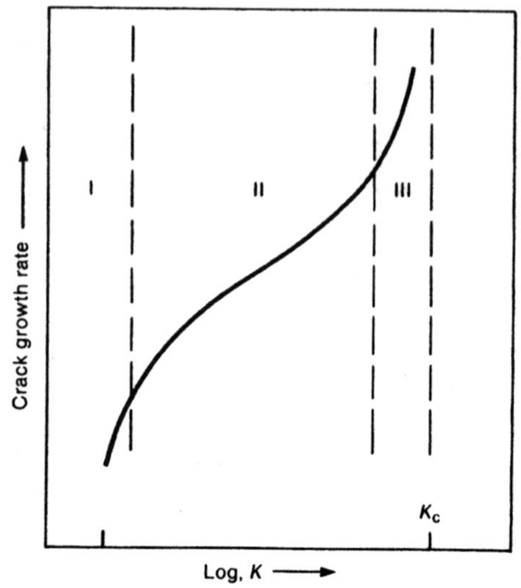

Fig. 8 An S-shaped fatigue crack propagation. *K*, stress-intensity factor; K_c, fracture toughness curve indicating its three characteristic regions.

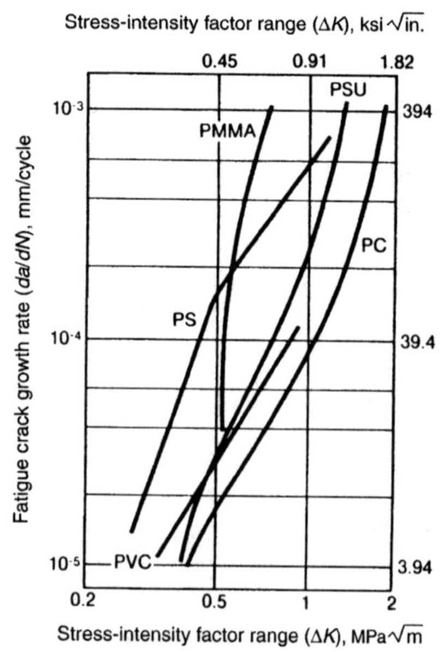

Fig. 9 Fatigue crack propagation behavior of various polymers. PSU, polysulfone; PMMA, polymethyl methacrylate; PC, polycarbonate; PS, polystyrene; PVC, polyvinyl chloride. Source: Ref 48

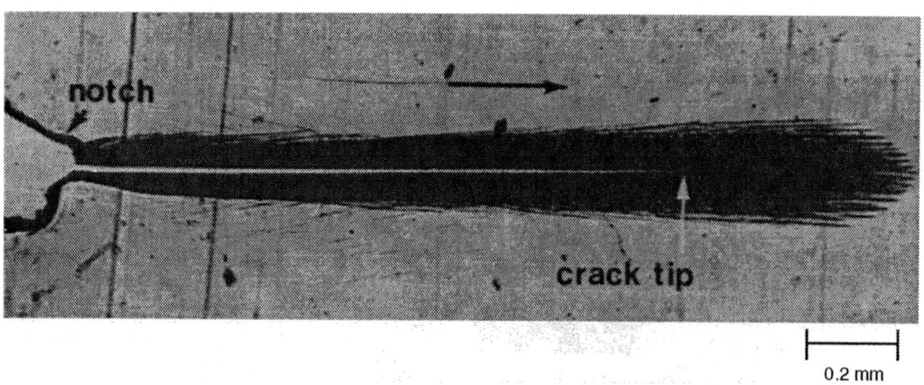

Fig. 7 Side view of a crack associated with a "crowd" of crazes in a fatigued single-edge notch of 0.25 mm (0.10 in.) thick polystyrene

material, as illustrated in Fig. 13. Stress concentration that is due to the crack induces irreversible deformation processes in the active zone. Depending on the material and loading conditions, the active zone may or may not be detectable during a crack propagation experiment, but it exists nonetheless. The crack and the preceding and surrounding damage are considered a single thermodynamic entity. Because fracture is envisioned as motion of the active zone, the crack growth resistance of a material is a measure of the resistance to such motion. As the crack propagates, its active zone evolves. Active zone evolution is an irreversible process that is adequately described by the thermodynamics of irreversible processes (Ref 12). Accordingly, the driving force for crack extension is defined as the derivative of Gibbs potential with respect to the crack length (flux of the process). The thermodynamic force for crack propagation was derived as the difference between the energy release rate, J_I, and the energy required for crack advance. The latter is expressed as the specific energy of damage, γ^*, multiplied by the amount of damage associated with crack advance, R_I. Thus, evolution of the energy barrier ($\gamma^* R_I - J_I$) guides the fracture process. The rate of crack propagation is accordingly expressed as:

$$\frac{da}{dN} = \frac{dD/dN}{\gamma^* R_I - J_I} \qquad \text{(Eq 12)}$$

where dD/dN is the cyclic rate of energy dissipated on submicroscopic processes, leading to damage formation and growth within the active zone. The physical meaning of D and the way it may be evaluated are considered next.

Part of the irreversible work associated with crack propagation, W_i, is expended on submicroscopic processes, leading to damage accu-

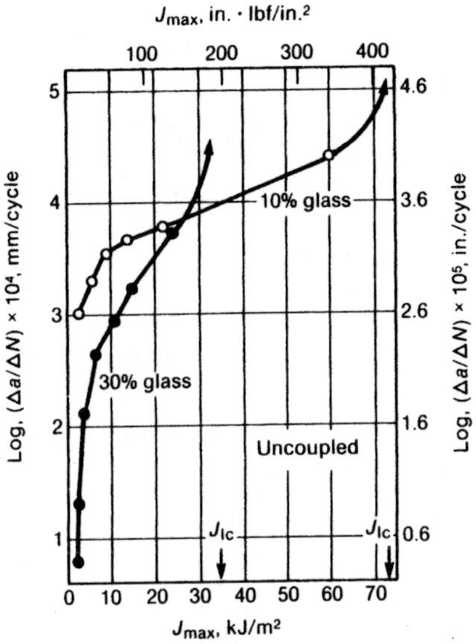

Fig. 10 The rate of fatigue crack propagation of injection-molded glass-reinforced polyvinyl chloride composites containing 10 and 30% glass as a function of the energy release rate, J_I. Arrows indicate the critical energy release rate, J_{Ic}, for each.

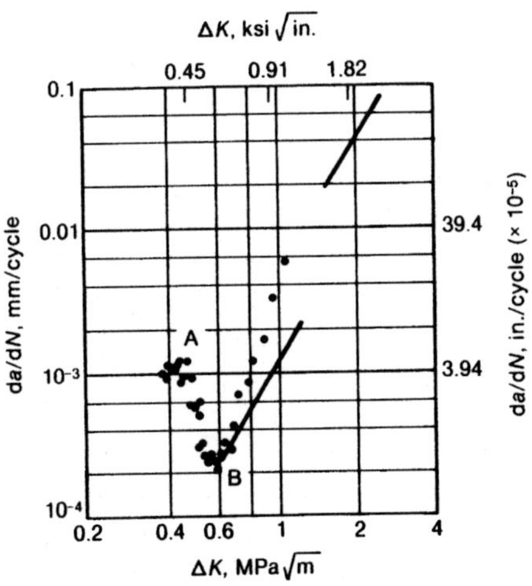

Fig. 11 Fatigue crack propagation rates (*da/dN*) at 10 Hz as a function of stress-intensity factor range (ΔK) in low-density polyethylene. *da/dN* decreases with increasing ΔK. Source: Ref 51

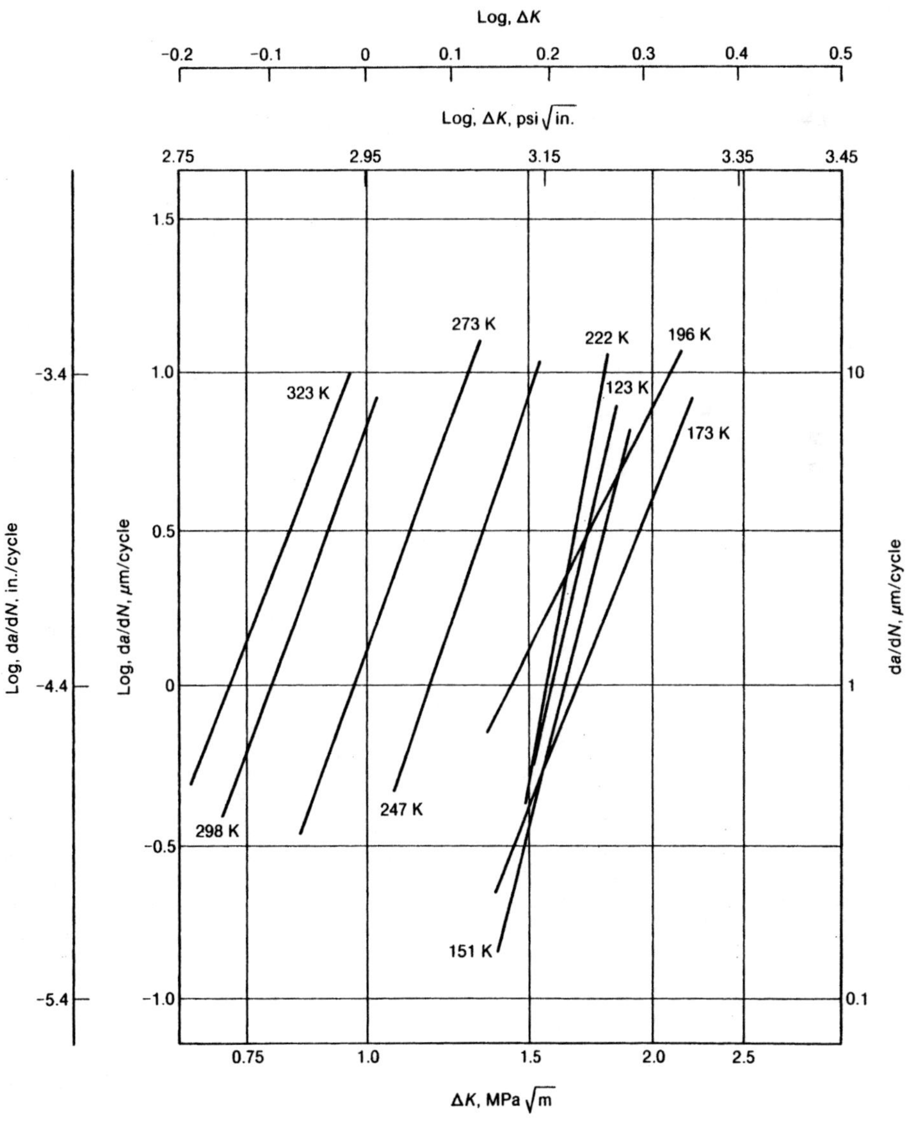

Fig. 12 Fatigue behavior of polymethyl methacrylate at 1 Hz for the Paris model. Temperature range is 123 to 323 K. *da/dN*, fatigue crack growth propagation; ΔK, stress-intensity factor range. Source: Ref 53

mulation within the active zone. The other part evolves as heat, Q. Thus:

$$D = W_i - Q \qquad \text{(Eq 13)}$$

In highly dissipative materials, W_i can be evaluated from load-displacement relationships. In principle, Q can also be measured, for example, by calorimetric techniques. Nevertheless, the rate of energy dissipation may be expressed as:

$$\frac{dD}{dN} = \beta' \frac{dW_i}{dN} \qquad \text{(Eq 14)}$$

where β' represents the portion of dW_i/dN expended on damage accumulation within the active zone. Experimentally, dW_i/dN can be extracted from the area within the hysteresis loop associated with each loading-unloading cycle recorded during a fatigue experiment.

In brittle materials, the extent of irreversible work is too small to be measured. The rate of dissipation has been shown to be proportional to the active zone length times the energy release rate (Ref 64). Following the Dugdale-Barenblatt model (Ref 65, 66), the active zone length is found to be proportional to the energy release rate. Thus:

$$\frac{dD}{dN} = \beta J_I^2 \qquad \text{(Eq 15)}$$

where β is the coefficient of energy dissipation.

Techniques for evaluating the amount of damage accumulation within the active zone in various materials have been outlined in various publications. Crack propagation, however, is often preceded by an active zone whose magnitude cannot, at present, be evaluated accurately from direct optical observations. Means of approximating its relative magnitude should therefore be devised. This is approached as follows.

At uncontrolled (critical) crack propagation, the denominator of Eq 12 approaches zero, that is:

$$J_{Ic} = \gamma^* R_{Ic} \qquad \text{(Eq 16)}$$

The subscript "c" indicates the transition from subcritical to critical crack propagation. Substituting γ^* from Eq 16 into Eq 12 gives:

$$\frac{da}{dN} = \frac{\beta J_I^2}{\mu J_{Ic} - J_I} \qquad \text{(Eq 17)}$$

where $\mu = R_I/R_{Ic}$ is a damage evolution coefficient. Equation 17 obviously calls for accurate measurement of the critical energy release rate to compute μ.

Plots of the rate of crack propagation in terms of Eq 17 provide a direct means for evaluating the coefficient of energy dissipation, β, and damage evolution coefficient, μ, to elucidate the resistance of the material to crack propagation. Figure 14 displays the applicability of the CL

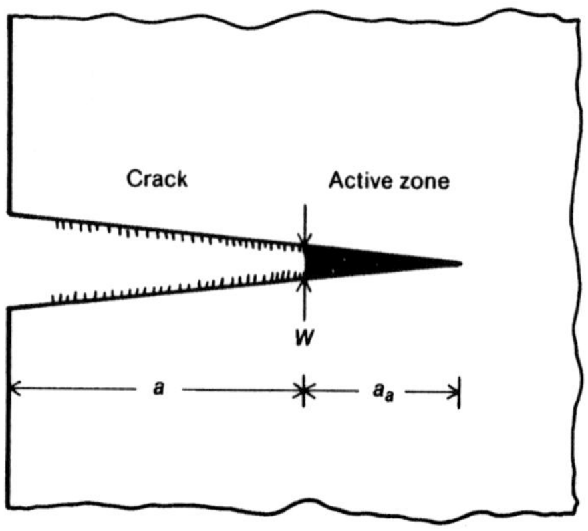

Fig. 13 A crack, a, preceded by an active zone, a_a. W, width

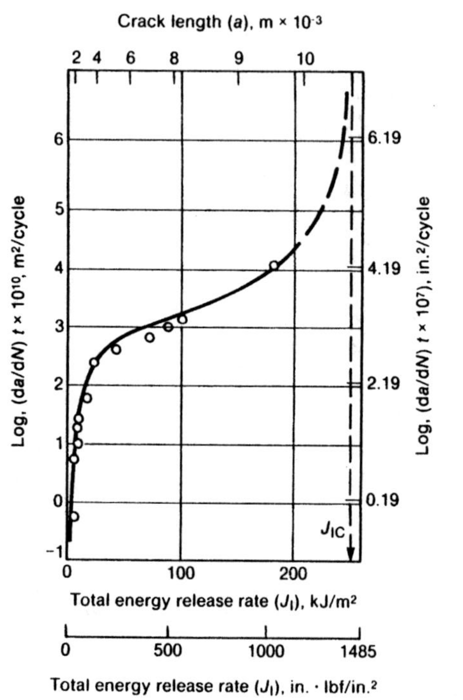

Fig. 14 Crack growth rate (da/dN) as a function of the energy release rate, J_I, for a single-edge notched polycarbonate specimen with 0.33 mm (0.013 in.) thickness

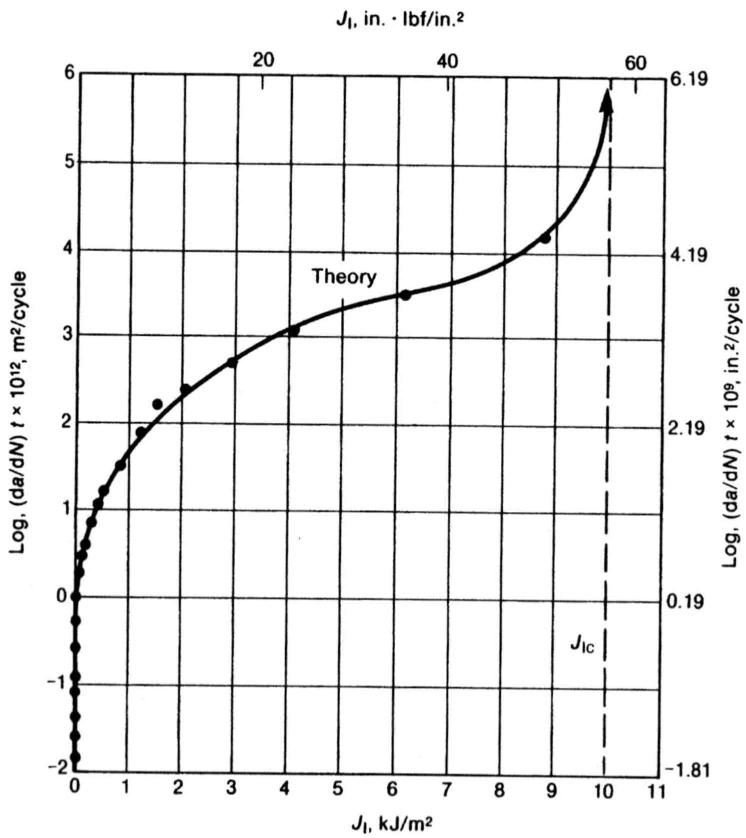

Fig. 15 Crack growth rate (da/dN) as a function of the energy release rate, J_I, (tearing energy) for a rubber compound. J_{Ic}, critical energy release rate

model to FCP in PC (Ref 61). Experimental data of a natural rubber vulcanizate (Ref 67) have been analyzed using the CL model; Fig. 15 displays the applicability of this model to FCP in this rubber compound. The CL model obviously provides a good description of the entire range of crack propagation for both PC and the rubber compound.

Fatigue crack propagation data for PC and PMMA were obtained at 1 Hz over a temperature range of 100 to 373 K (Ref 52) and examined by the CL model. The data were then statistically fit to all the fatigue models discussed previously (Ref 53). These investigators concluded that the CL model describes the fatigue behavior of these polymers over the entire range of temperature and stress.

According to the CL theory, the denominator of Eq 17 represents the energy barrier for crack advance. It is the evolution of this barrier that controls crack propagation, where J_I is the amount of accumulated potential energy that is released on crack advance, and μJ_{Ic} represents the energy required for CL translation. The evolution of the normalized energy release rate, J_I/J_{Ic}, for EP (Ref 68), PC, and the rubber compound is shown in Fig. 16. The energy barriers evolve differently, illustrating the distinctive resistance of each material to FCP.

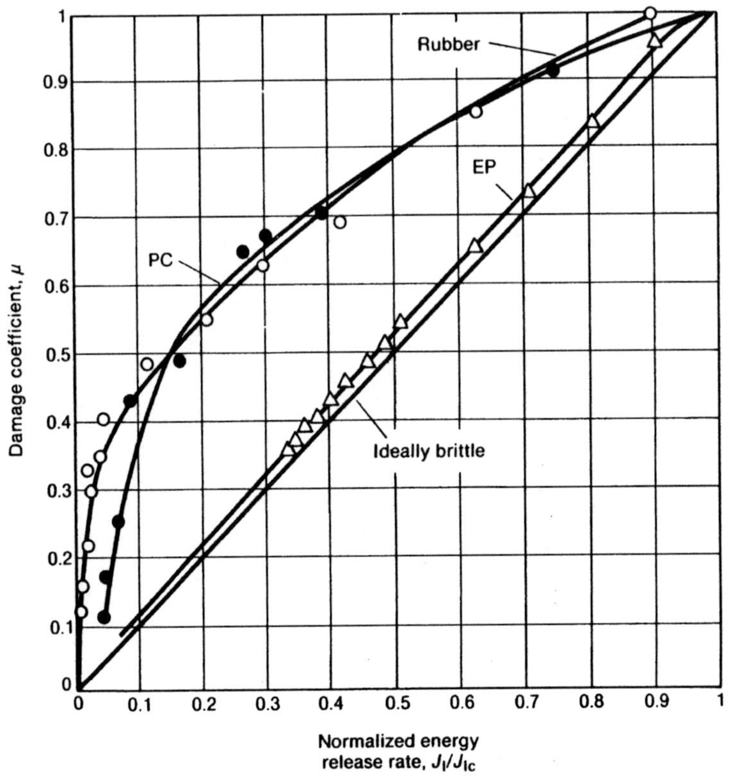

Fig. 16 Evolution of the damage coefficient, μ, with respect to the normalized energy release rate (J_I/J_{Ic}) in epoxy (EP), polycarbonate (PC), and the rubber compound

REFERENCES

1. A. Wohler, *English Abstract in Engineering,* Vol II, 1871, p 199
2. R.W. Hertzberg and J.A. Manson, *Fatigue of Engineering Plastics,* Academic Press, 1980
3. G.J. Lake and P.B. Lindley, in *Physical Basis of Yield and Fracture, Conference Proceedings,* Institute of Physics, 1966, p 176
4. E.H. Andrews, in *Testing of Polymers IV,* Interscience, p 237
5. M.N. Riddell, *Plast. Eng.,* Vol 30 (No. 4), 1974, p 71
6. J. Burke and V. Weiss, *Fatigue: Environment and Temperature Effects,* Plenum Press, 1983
7. J. Clyton-Cave, R.J. Taylor, and E. Ineson, *J. Iron Steel Inst.,* Vol 180, 1955, p 161
8. F.A.M. McClintock and A.S. Argon, Ed., *Mechanical Behavior of Materials,* Addison-Wesley, 1966
9. S. Rabinowitz and P. Beardmore, *J. Mater. Sci.,* Vol 9, 1974, p 81
10. R.J. Crawford and P.P. Benham, *Polymer,* Vol 16, 1975, p 908
11. J.F. Mandell, K.L. Smith, and D.D. Huang, *Polym. Eng. Sci.,* Vol 21, 1981, p 1173
12. A. Chudnovsky and A. Moet, *J. Mater. Sci.,* Vol 20, 1985, p 630
13. H.H. Kausch, Intersegmental Interactions and Chain Scission, *Failure of Plastics,* Hanser Publishers, 1986
14. M.E. Mackay, T.G. Teng, and J.M. Schultz, *J. Mater. Sci.,* Vol 14, 1979, p 221
15. C.B. Bucknall and W.W. Stevens, *J. Mater. Sci.,* Vol 15, 1980, p 2950
16. N.J. Mills and N. Walker, *J. Mater. Sci.,* Vol 15, 1980
17. M.T. Takemori and R.P. Kambour, *J. Mater. Sci.,* Vol 16, 1981, p 1108
18. J.A. Sauer, private communication, 1985
19. H.H. Kausch, *Polymer Fracture,* Springer-Verlag, 1978
20. J.D. Ferry, *Viscoelastic Properties of Polymers,* John Wiley & Sons, 1961, p 14, 609
21. *Modern Plastics Encyclopedia,* McGraw-Hill, 1986–1987, p 426, 427
22. J.A. Sauer and G.C. Richardson, *Int. J. Fract.,* Vol 16, 1980, p 499
23. T.R. Tauchert and S.M. Afzal, *J. Appl. Phys.,* Vol 38, 1967, p 4568
24. E. Foden, D.R. Morrow, and J.N. Sauer, *J. Appl. Polym. Sci.,* Vol 16, 1972, p 519
25. J.A. Sauer, E. Foden, and D.R. Morrow, *Polym. Eng. Sci.,* Vol 17, 1977, p 246
26. M.N. Riddell, G.P. Koo, and J.L. O'Toole, *Polym. Eng. Sci.,* Vol 6, 1966, p 363
27. G.P. Koo, M.N. Riddell, and J.L. O'Toole, *Polym. Eng. Sci.,* Vol 7, 1967, p 182
28. R.J. Crawford and P.P. Benham, *J. Mater. Sci.,* Vol 9, 1974, p 18
29. I. Constable, J.G. Williams, and D.J. Barns, *J. Mech. Eng. Sci.,* Vol 12, 1970, p 20
30. P.P. Oldyrev and V.M. Parfeer, *Polym. Mech.,* Vol 10, 1974, p 148
31. P.P. Oldyrev and V.M. Parfeer, *Polym. Mech.,* Vol 11, 1975, p 682
32. K. Hellan, *Introduction to Fracture Mechanics,* McGraw-Hill, 1984
33. S. Mostovoy and E.J. Ripling, *Polym. Sci. Tech.,* Vol 9B, 1975, p 513
34. W. Doll, *Adv. Polym. Sci.,* Vol 52–53, 1983, p 105
35. A. Chudnovsky, A. Moet, N.J. Bankart, and M.T. Takemori, *J. Appl. Phys.,* Vol 54, 1983, p 5562
36. A. Chudnovsky, NASA Contractor Report 174634, National Aeronautics and Space Administration, March 1984
37. H. Tada, P.C. Paris, and G. Irwin, *The Stress Analysis of Cracks Handbook,* Del Research Corporation, 1973
38. P. Paris and F. Erdogan, *Trans. ASME,* Dec 1963, p 528
39. R.G. Forman, V.E. Kearney, and R.M. Engle, *J. Basic Eng.,* Vol 89, 1967, p 459
40. S. Pearson, *Eng. Fract. Mech.,* Vol 4, 1972, p 9
41. B. Mukhergee and D.J. Burns, *J. Exp. Mech.,* Vol 11, 1971, p 433
42. S. Arad, J.C. Radon, and L.E. Culver, *Polym. Eng. Sci.,* Vol 12, 1972, p 193
43. S. Arad, J.C. Radon, and L.E. Culver, *J. Mech. Eng. Sci.,* Vol 13, 1971, p 75

44. J.C. Radon, S. Arad, and L.E. Culver, *Eng. Fract. Mech.,* Vol 6, 1974, p 195
45. C.A.M. Branco, J.C. Radon, and L.E. Culver, *J. Test, Eval.,* Vol 3, 1975, p 195
46. H.M. El-Hakeem, Ph.D. thesis, University of London, 1975
47. J.F. Knott, *Fundamentals of Fracture Mechanics,* Butterworth, 1973
48. M.D. Skibo, R.W. Hertzberg, J.A. Manson, and S.L. Kim, *J. Mater. Sci.,* Vol 12, 1977, p 531
49. P.X. Nguyen and A. Moet, *J. Vinyl Technol.,* Vol 7, 1985, p 160
50. Y.W. Mai, *J. Mater. Sci.,* Vol 9, 1974, p 1896
51. P.E. Bretz, R.W. Hertzberg, and J.A. Manson, *Polymer,* Vol 22, 1981, p 575
52. G.C. Martin, Ph.D. thesis, University of Minnesota, 1976
53. E.P. Tam and G.C. Martin, *J. Macromol. Sci. Phys.,* Vol B23, 1985, p 415
54. A. Chudnovsky and A. Moet, *Org. Coat. Plast. Chem. Prep.,* Vol 45, 1981, p 607
55. A. Moet and A. Chudnovsky, *Org. Coat. Plast. Chem. Prep.,* Vol 45, 1981, p 616
56. A. Chudnovsky and A. Moet, *J. Elastomers Plast.,* Vol 18, 1985, p 50
57. J. Botsis, A. Moet, and A. Chudnovsky, in *Proceedings of the 29th Annual Technical Conference* (ANTEC), Society of Plastics Engineers, 1983, p 444
58. K. Sehanobish, E. Baer, A. Chudnovsky, and A. Moet, *J. Mater. Sci.,* Vol 20, 1985, p 1934
59. J. Botsis, Ph.D. thesis, Case Western Reserve University, 1984
60. N. Haddaoui, A. Chudnovsky, and A. Moet, *Polym. Mater. Sci. Eng.,* Vol 49, 1983, p 117
61. N. Haddaoui, A. Chudnovsky, and A. Moet, *Polymer,* Vol 27, 1986
62. L. Koenczoel and K. Sehanobish, *J. Macromol. Sci. Phys.,* Vol B26 (No. 3), 1987, p 307
63. K. Sehanobish, A. Moet, and A. Chudnovsky, *Polymer,* Vol 28, 1987
64. J. Botsis, A. Chudnovsky, and A. Moet, *Int. J. Fract.,* Vol 33, 1987, p 263, 277
65. D.S. Dugdale, *J. Mech. Phys. Solids,* Vol 8, 1960, p 100
66. G.I. Barenblatt, *Adv. Appl. Mech.,* Vol 7, 1962, p 55
67. J. Lake and P. Lindley, *Rubber J.,* Part 1, Oct 1964
68. X. Wang, K. Sehanobish, and A. Moet, *Polym. Compos.,* Vol 9 (No. 3), 1988

Friction and Wear Testing*

TRIBOLOGY comes from the Greek word *tribos,* to rub (Ref 1); *friction* is derived from the Latin verb *fricare,* which has the same meaning (Ref 2). Tribology is the science and technology of interacting surfaces in relative motion (Ref 3), or, more simply expressed, the study of friction, wear, and lubrication. This article focuses on friction and wear as they relate to polymeric materials.

The study and evaluation of friction are driven by the need to control it (Ref 4). In applications such as bearings and gears, low friction is desirable, whereas high friction is required in materials used in brakes, clutches, and road surfaces. In each case, constant, reproducible, and predictable friction behavior is essential.

Wear, along with corrosion and obsolescence, is one of the most common life-determining processes for consumer goods and machinery (Ref 5). While some types of material removal are beneficial (cutting, grinding, and polishing), wear of a single component (e.g., a bearing in the main rotor of a helicopter) can lead to catastrophic failure (Ref 6). The nearly imperceptible wear of many identical components can lead to the generation of large quantities of waste, such as mountains of used automobile tires.

Either external or internal (self) lubrication can be used to reduce both friction and wear. A detailed discussion of lubrication, especially as it relates to friction and wear, is provided in Ref 7 and in *Friction, Lubrication, and Wear Technology,* Volume 18 of the *ASM Handbook.*

Friction, Wear, and Lubrication

Friction and wear are inevitable when two surfaces undergo sliding or rolling under load (Ref 7). The control of friction and wear is essential for both performance and economic reasons. An understanding of friction and wear processes aids in the evaluation and selection of materials used in friction and wear applications.

Friction

Friction (or friction force) is the resisting force tangential to the common boundary between two bodies when, under the action of an external force, one body moves or tends to move relative to the surface of the other (Ref 8). The dimensionless ratio of the friction force (F) to the normal force (N) pressing the two bodies together is the coefficient of friction, μ (or f):

$$F/N = \mu \qquad \text{(Eq 1)}$$

The friction force required to set a body in motion is typically greater than the force needed to sustain the motion. The respective coefficients of friction are the static coefficient of friction, μ_s, and the kinetic (or dynamic) coefficient of friction, μ_k.

Typical values for kinetic coefficient of friction are 0.03 for a well-lubricated bearing, 0.5 to 0.7 for dry sliding, and 5 or more for clean metal surfaces in a vacuum (Ref 2). A coefficient of friction of 0.2 to 0.3 allows for comfortable walking, but if ice is one of the mating surfaces, the coefficient of friction can be 0.05 or less.

Along with the observation that, within a relatively wide range of conditions, friction force is proportional to normal force, it has been noted that the coefficient of friction is independent of both the apparent area of contact and the velocity between the contacting surfaces. These observations form the classical laws of friction; however, many deviations are found (Ref 9).

Friction in polymers is caused by many of the same mechanisms that cause friction in metals: adhesion of the contacting surfaces, asperity contact leading to plastic deformation and plowing, elastic deformation energy, and interference and local deformation caused by third bodies (Ref 4). In addition, differences in mechanical properties, such as viscoelasticity, strain-rate sensitivity, and thermal conductivity, may lead to additional friction-generating mechanisms. The two most important friction-generating mechanisms for polymeric materials are surface adhesion and mechanical deformation.

Figure 1 shows a polymer surface in contact with a hard asperity. Two friction dissipation zones are shown. The interfacial shear zone is a very thin layer (approximately 100 nm) at the surface. Slip may occur at the interface for some polymeric materials, but it more commonly occurs within the polymer itself, because the adhesive bond between the asperity and the polymer is stronger than the cohesive strength of the polymer. In the latter case, a layer of polymer, called a transfer layer or transfer film, is formed on the facing material. Once this transfer layer has been formed, subsequent traversals result in lower friction and wear, because the contact becomes polymer-on-polymer (Ref 7).

Within the deformation zone, and for cases where there is no surface adhesion, friction is caused by energy dissipation in the material below the indenter (Ref 4). The type of recovery depends on the particular polymeric material. For glassy polymers, much of the energy dissipation manifests as microcracking, while grooves may form in ductile polymers. Other polymeric materials may exhibit no permanent deformation, because they are viscoelastic and recover the original strain; energy is dissipated in the hysteresis of the deformation.

The preceding discussion of friction has focused on sliding surfaces. Rolling friction is an equally complex phenomenon with many contributing mechanisms, including varying amounts of sliding or slip (Ref 7). Coefficients of rolling friction are typically much smaller (5×10^{-3} to 10^{-5}) than coefficients of sliding friction.

Wear

Material can be removed from a solid surface by melting or sublimation, by chemical dissolution, or by physical separation of atoms from the surface (Ref 10). The last can be accomplished either by a single high-strain event or by a series of strains. Wear, the process of material loss or displacement from one or both of two solid surfaces in relative motion, is an example of the latter case (Ref 11, 12).

Wear of Polymers. Five major types of wear processes have been identified (Ref 1, 5):

- *Abrasive wear* occurs when a hard, solid particle or asperity comes in contact with a softer surface. The softer material experiences both material loss and deformation of the remaining portion.
- *Adhesive wear* occurs when surfaces in contact bond together through local welding of asperities or cohesive bonding. If the bonded junctions are stronger than one of the solids,

*Adapted from the article by Rebecca Tuszynski, "Friction and Wear Testing," in *Engineered Materials Handbook Desk Edition,* ASM International, 1995, pages 459 to 466 wear arises from a shearing process within the solid.

- *Fatigue wear* is the result of periodic stress variations between wearing surfaces. Even though the forces may be less than that required to permanently deform the material, local fatigue produces cracks that eventually lead to the removal of relatively large pieces of material in the form of pitting, spalling, or flaking.
- *Chemical or corrosive wear* is found when chemical reactions occur along with mechanical wear. This form of wear may lead to a weight gain rather than a weight loss. This type of wear is less common for many polymeric materials (because of their general chemical stability) than for metals.
- *Fretting wear* occurs when two surfaces have oscillatory relative motion of small amplitude. Small debris particles are produced at a relatively slow rate; the buildup of this debris is a notable feature of fretting wear. Corrosion can occur with fretting if the appropriate chemical species are present.

Lubrication

Lubrication reduces friction and wear between surfaces in relative motion by the application of a solid, liquid, or gaseous substance (lubricant) (Ref 1). A lubricant is therefore any substance that is used to reduce friction and wear between moving surfaces. Lubricants are often externally applied, but solid materials may also be internally (self) lubricated.

There are several basic lubrication regimes, ranging from hydrodynamic lubrication, where there is no contact between the surfaces, to boundary lubrication, where there is considerable asperity interaction between the contacting surfaces (Ref 7). In the hydrodynamic regime, friction is due only to viscous dissipation within the lubricant and has little or nothing to do with the nature of the contacting materials (Ref 4). The characteristics of the contact surfaces begin to play a significant role in friction and wear once boundary lubrication conditions are reached.

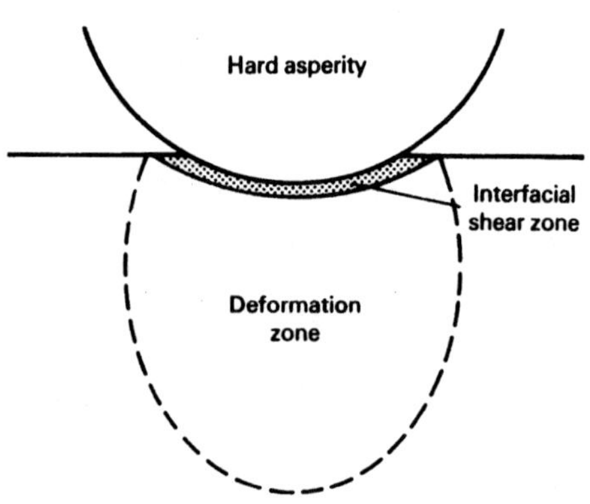

Fig. 1 Schematic of a polymer surface in contact with a hard asperity. Two friction dissipation zones are shown: the interfacial shear zone and the deformation zone. Source: Ref 4

Friction and Wear Applications for Polymeric Materials

Table 1 shows polymers and composites that are used in representative friction and wear applications. Several of these polymers (polytetrafluoroethylene, or PTFE, polyamides, and polyethylene) are self-lubricating; that is, they form transfer films that reduce friction (Ref 7). Other plastics are formulated with lubricating additives such as molybdenum disulfide (MoS_2), graphite, or PTFE.

Plastics are widely used for bearings. Self-lubricated plastics are recommended for continuous service (Ref 14), but unlubricated plastics may be used for parts that see only intermittent use. Unlike metal bearings, plastic bearings rarely seize in case of loss of lubrication. They also have good tolerance to high stresses from excessive loading due to shaft misalignment.

Plastics, especially self-lubricating polyamides and acetals, are also used in gears, rolling elements, cams, and many other machine components. Polyamides have a tendency to absorb moisture and swell, which can create problems in high-humidity or water-immersion applications (Ref 14). Acetals are very popular for friction and wear applications because of their combination of very good mechanical properties and moderate cost. Their good mechanical properties make it unnecessary to use glass or inorganic fiber reinforcements to improve strength. This is advantageous, because these additives can lead to abrasive wear in some applications.

Elastomers such as natural and synthetic rubbers and fluoroelastomers can be used as soft-lined plain bearings, flexible thrust-pad bearings, seals, seal rings, and abrasion-resistant parts (Ref 7). Elastomers typically have a high tolerance to abrasive particles, high resiliency, and low wear. However, they can swell on contact with certain liquids, and they may harden with decreasing temperature.

Friction and Wear Test Methods

Laboratory-scale friction and wear testing is usually performed either to rank the performance of candidate materials for an application or to investigate a particular wear process (Ref 13). Friction and wear testing generally uses one of two basic strategies: In one case, the test is made as representative of the application conditions as possible; in the other, the test is accelerated by means of increased temperatures, loads, and so on. This has the advantage of saving

Table 1 Representative friction and wear applications of polymers and composites

Material(a)	Seals	Gears	Compressor rings	Pivot bearings	Slideways	Abrasive service	Water immersion	High-temperature service
Unfilled thermoplastics								
PTFE	X	...	...	X	X	...	...	...
Acetal	X	X	...	...	...	...	X	...
Polyamide	...	X	...	...	...	X	...	...
UHMWPE	X	X	...	...	X	X	X	...
Filled thermoplastics								
Polyamide + MoS_2	...	X	...	X	...	...	...	...
Acetal + oil	...	X	...	X	...	...	...	...
Polyamide + oil	...	X	...	X	...	...	...	...
Polyurethane + fillers	...	...	...	...	...	X	X	...
High-temperature polymers								
Polyimide (filled)	X	X	X	...	...	...	...	X
Polyamide-imide	...	X	...	...	...	...	...	X
Filled PTFE								
PTFE/glass fibers	...	...	X	X	...	...	...	...
PTFE/graphite	X	...	X	...	...	...	...	...
PTFE/bronze	...	...	X	...	X	...	...	...
PTFE/glass/MoS_2	X	...	X	...	...	...	...	...
Reinforced thermosets								
Polyester laminate	...	X	...	X	X	...	X	...
Asbestos/phenolic	...	X	...	X	X	...	X	...
Cotton/phenolic	...	X	...	...	X	...	X	...

(a) PTFE, polytetrafluoroethylene; UHMWPE, ultrahigh-molecular-weight polyethylene. Adapted from Ref 13 time, but wear mechanisms not present in the actual application may be introduced, and the ranking of the tested materials may not represent their performance in service.

ASTM C 808, "Standard Guideline for Reporting Friction and Wear Test Results of Manufactured Carbon and Graphite Bearing and Seal Materials" (Ref 15), was developed for a specific class of materials but offers a general reporting format that may be useful to anyone concerned with friction and wear testing. The suggested reporting format includes a description of the test device and test techniques, a description of the test specimen and the mating surface, and a report of friction and wear results.

ASTM G 118-93, "Standard Guide for Recommended Data Format of Sliding Wear Test Data Suitable for Databases" (Ref 16), offers suggestions for the organization of test data that will be stored in a computerized database. ASTM G 115-93, "Standard Guide for Measuring and Reporting Friction Coefficients" (Ref 17), tabulates current ASTM International friction test standards, points out the factors that must be considered when determining coefficients of friction, and suggests a standard reporting format for friction data.

Friction Tests

An inclined plane test is often used to measure the static coefficient of friction. A body at rest on a flat surface will begin to move when the surface is tilted to a certain angle, θ (Fig. 2). The static coefficient of friction is given by:

$$\mu_s = F/N = \tan\theta \qquad \text{(Eq 2)}$$

While this test is a simple means of measuring static coefficient of friction, it is more typical to use force measurements to determine both static and kinetic coefficients of friction (Ref 2).

Many committees within ASTM have developed tests for measuring coefficients of friction (Ref 18). Most of these are directed toward a particular application or material. For example, Committee D-7 on wood has developed D 2394, "Simulated Service Testing of Wood and Woodbase Finish Flooring," for testing floor finishes against shoe sole leather. Committee D-20 on plastics has developed two tests to measure coefficients of friction: D 1894 and D 3028. Both are also American National Standards.

ASTM D 1894-90, "Standard Test Method for Static and Kinetic Coefficients of Friction of Plastic Film and Sheeting" (Ref 19), describes the determination of μ_s and μ_k of plastic film and sheeting using a variety of test assemblies, as shown in Fig. 3 (where sled A and plane B are the materials of interest). In each case, the force required to move a sled across a plane is measured. A test speed of 150 ± 30 mm/min (0.5 ± 0.1 ft/min) is specified. Both the force required to initiate motion and the average force required to sustain motion are recorded and used to calculate μ_s and μ_k. While this test is written for the evaluation of plastic film, it can be used to evaluate other materials, such as coated metals and paper.

ASTM D 3028-90, "Standard Test Method for Kinetic Coefficient of Friction of Plastic Solids" (Ref 20), uses a variable-speed frictionometer to measure kinetic coefficients of friction. Three types of test specimens can be used: 20.0 ± 0.1 mm diameter rigid fixed specimens that weigh 5.0 ± 0.1 g, 100.0 ± 0.1 mm diameter rigid moving specimens, or film or sheeting mounted on a 100 mm diameter mounting wheel. Two procedures are described. In procedure A, coefficients of friction are measured at velocities of 0.25, 0.50, 1.0, 2.0, and 3.0 m/s. The test is performed as rapidly as possible to minimize wear effects. Testing performed by procedure B is intended to show the effects of time, velocity, and wear on coefficient of friction. A velocity is selected (1.0 m/s is suggested as a default), and readings are taken every 30 s until they reach a constant value.

The relationship between wear and friction is an important consideration when selecting a friction test. Several ways in which wear may affect friction are illustrated in Fig. 4. Figure 4(a) shows how friction force varies with time when a system experiences no wear. The friction is essentially constant. Figure 4(b) shows a system where the friction force increases with time and finally reaches a steady state. This type of behavior may be seen in a system where both surfaces experience heavy wear: The coefficient of friction is low for the original surfaces, but it rises and stabilizes as new surfaces are exposed. Figure 4(c) shows a system that experiences a variety of wear events. Friction changes as the wear processes change. This type of behavior may be observed in a system where retained wear debris can either increase or decrease friction.

If a system will wear, and the friction of worn surfaces is of interest, friction should be measured in a test that produces wear. ASTM D 3028-90, described previously, has that capability, as do several of the wear tests described subsequently.

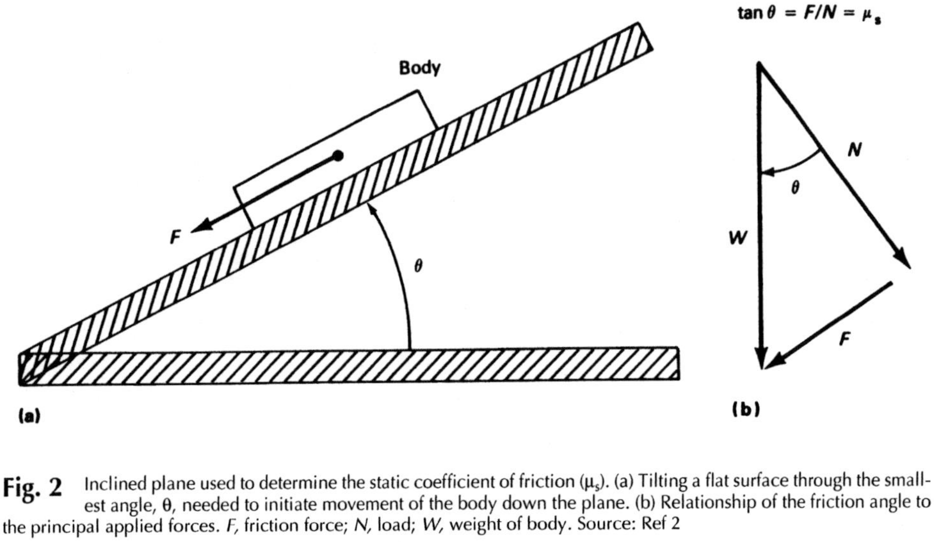

Fig. 2 Inclined plane used to determine the static coefficient of friction (μ_s). (a) Tilting a flat surface through the smallest angle, θ, needed to initiate movement of the body down the plane. (b) Relationship of the friction angle to the principal applied forces. *F*, friction force; *N*, load; *W*, weight of body. Source: Ref 2

Wear Tests

Wear processes (comprised of wear by abrasion, adhesion, and fatigue) are complex. Many different test apparatuses and methods have been developed to simulate particular wear mechanisms, but there is no single general-purpose wear test that establishes a unique wear parameter or rating (Ref 12).

A concern with all wear tests, regardless of the test apparatus used, is the actual measurement of wear. Common wear measurements include weight loss, volume loss, displacement scar width or depth, and indirect measures such as the time required to wear through a coating or the load required to cause a change in reflectance (Ref 12). Weight loss is straightforward, but it may not account for material displacement, and it should not be used to compare materials with different densities. Volume loss can be calculated from weight loss or estimated based on wear geometry. Displacement scar width and depth are related to volume and can be easily measured, but the results of different types of tests are not comparable. Indirect measures are typically limited in scope and applicability and do not easily provide fundamental wear parameters. Ideally, the wear measurement method should reflect the actual service performance of the system. It should be repeatable and as objective as possible.

Specific wear rate can be used to compare the performance of materials under the same operating conditions. For polymers and composites, there are conditions of low pressure and ambient temperature where the wear rate is essentially independent of these parameters (Ref 13). The

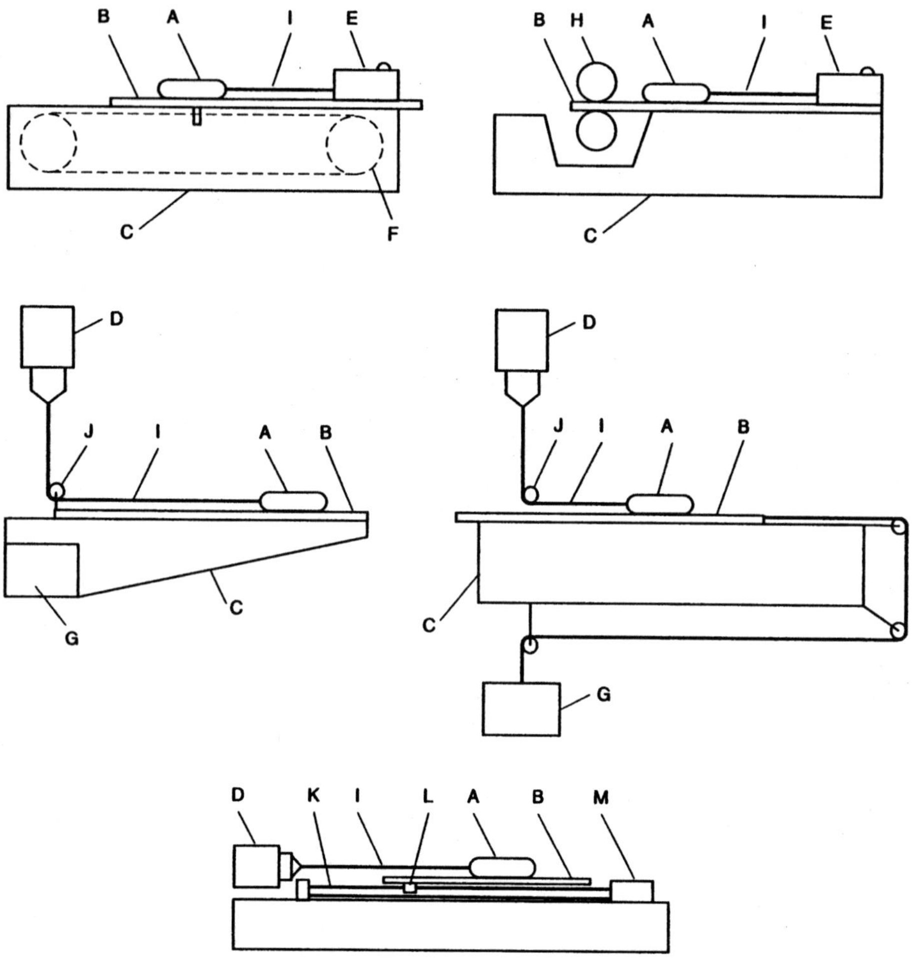

Fig. 3 Different assemblies used for the determination of coefficients of friction by ASTM D 1894. A, sled; B, plane; C, supporting base; D, gage; E, spring gage; F, constant-speed chain drive; G, constant-speed tensile tester crosshead; H, constant-speed drive rolls; I, nylon monofilament; J, low-friction pulley; K, worm screw; L, half nut; M, hysteresis, synchronous motor. Source: Ref 19

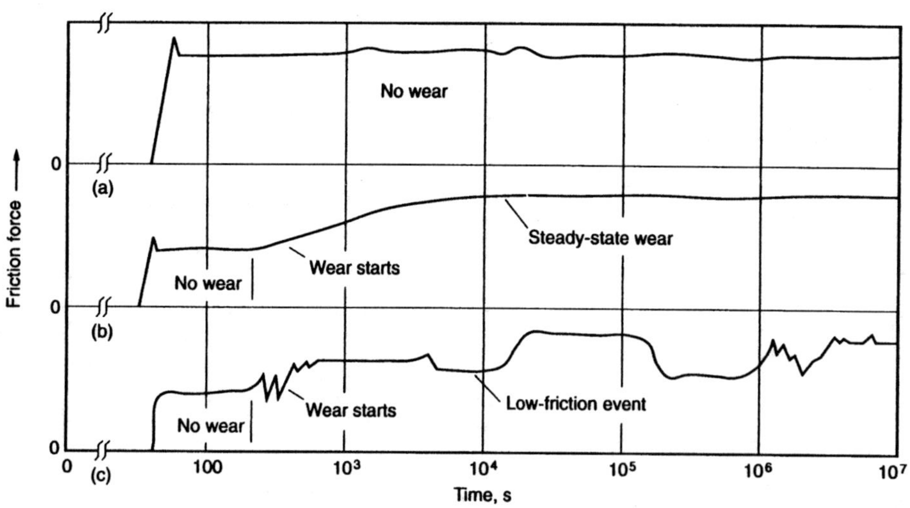

Fig. 4 The effect of system wear on friction force. (a) System that does not experience any wear. (b) System where friction force increases with time until reaching a steady-state condition. (c) System where friction force varies with each event in the wear process. Source: Ref 18

specific wear rate under these conditions (volume of material worn per unit applied load per unit distance of sliding) is designated k_0; however, specific wear rates obtained using another wear test may differ. Comparison of k_0 values may be useful for both materials selection and component design. Several specific wear rates (determined by the thrust washer test with a mild steel counterface, described subsequently) are shown in Fig. 5. Many different geometries are used in commercial wear testing devices, and Fig. 6 shows several examples.

Wear Tests with Abrasive. Many wear tests include the use of an abrasive material, either supplied as a third body or bonded to the counterface.

ASTM D 1044-90, "Standard Test Method for Resistance of Transparent Plastics to Surface Abrasion" (Ref 21), uses the Taber abraser to evaluate the abrasion resistance of transparent plastics. The amount of light diffused by the abraded track is measured according to the procedure outlined in ASTM D 1003 (Ref 9). The Taber abraser was better known as a test for determining the weight loss of a specimen traversed by either a hard or resilient abrading wheel under a specified load for a particular number of revolutions (Ref 9, 11, 22, 23). However, the current version of ASTM D 1044 contains the statement that "recent attempts to employ the Taber abraser for volume loss determinations of various plastics, like earlier ones, have been unsuccessful because of excessively large coefficients of variation attributed to the data. Insufficient agreement among the participating laboratories has rendered the use of volume loss procedure inadvisable as an ASTM test method." The standard recommends the use of ASTM D 1242 for the evaluation of abrasion resistance of plastics by volume loss.

ASTM D 1242-87, "Standard Test Methods for Resistance of Plastic Materials to Abrasion" (Ref 24), outlines two tests that are suitable for flat plastic surfaces. Method A calls for loose abrasive (No. 80 TP aluminum oxide grit is suggested) that is applied to the test surface under controlled conditions. The results are expressed as volume loss, calculated from test specimen weight loss and density. Method B calls for a "bonded abrasive abrading machine" that is capable of testing multiple specimens. The grade of abrasive material, load, and length of time can all be varied as needed. A standard zinc calibration specimen is included with each run of test specimens. Volume loss is calculated and reported, along with the weight loss of the zinc standard run at the same time. This test is also an American National Standard.

ASTM D 673-88, "Standard Test Method for Mar Resistance of Plastics" (Ref 25), quantifies the abrasion resistance of glossy plastics by measuring the loss of gloss caused by impacting carborundum grit (Ref 9). A flat sample is held at 45° and struck with increasing amounts of No. 80 silicon carbide dropped from a height of 635 mm. The relatively mild airborne abrasive action is similar to that encountered by many items in actual use, and correlation with field experience has been demonstrated for this test.

ASTM G 65-91, "Standard Test Method for Measuring Abrasion Using the Dry Sand/Rubber Wheel Apparatus" (Ref 26), offers four procedures for the determination of scratching abrasion resistance of metallic materials. It can also be used to test a wide variety of plastics (Ref 23). A wheel faced with a chlorobutyl rubber tire revolves against a stationary test specimen while a flow of sand is forced between the wheel and the specimen. The severity of the test is adjusted by varying test duration and the force with which the specimen is applied to the wheel. This test can be used to rank the performance of materials in an abrasive environment, but it should not be used to predict the exact resistance of a given material in a specific environment.

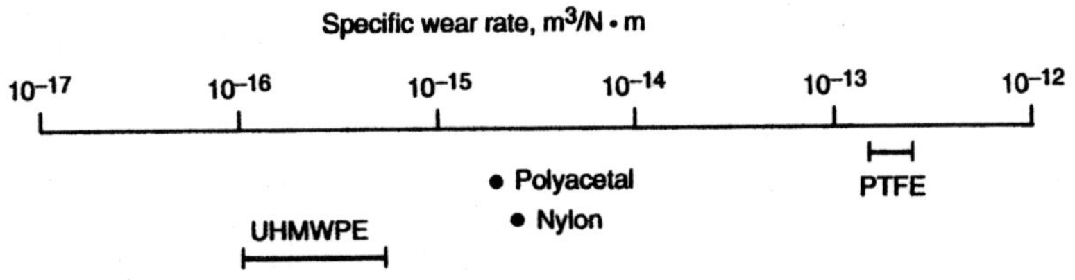

Fig. 5 Specific wear rates for selected polymeric materials. UHMWPE, ultrahigh-molecular-weight polyethylene; PTFE, polytetra fluoroethylene. Source: Ref 13

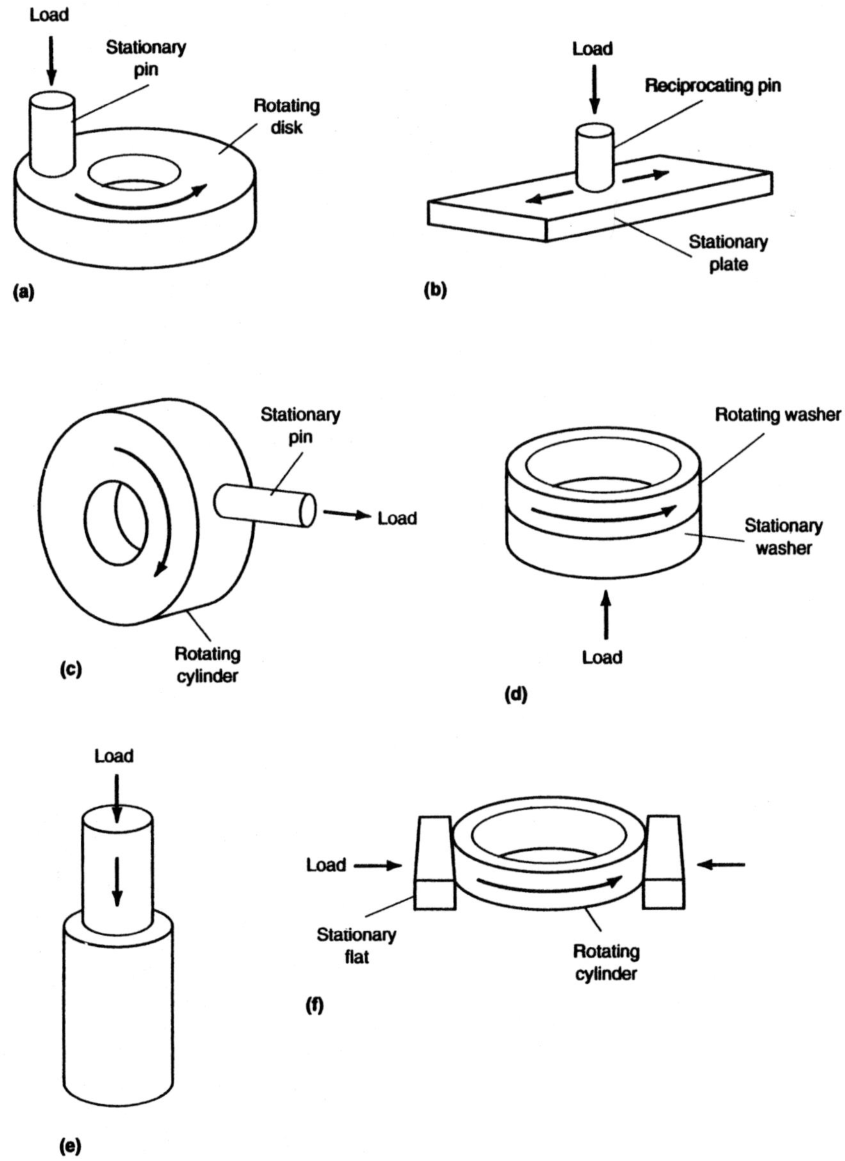

Fig. 6 Examples of test geometries that may be used for sliding friction and wear tests. (a) Pin-on-disk. (b) Pin-on-flat (reciprocating). (c) Pin-on-cylinder. (d) Thrust washer. (e) Pin-into-bushing. (f) Rectangular flats on rotating cylinder. Source: Ref 7

ASTM G 75-89, "Test Method for Determination of Slurry Abrasivity (Miller Number) and Slurry Abrasion Response of Materials (SAR Number)" (Ref 27), covers a laboratory procedure that allows for the determination of either the relative abrasivity of any slurry or the response of different materials to different slurries. The Miller number ranks the abrasivity of slurries in terms of the wear of a standard reference material (27% chromium iron). The SAR number is an index of the relative abrasion response of materials as tested in a particular slurry. A major use of the SAR number is in the ranking of construction materials for use in pumping a particular slurry.

Wear Tests for Elastomers. ASTM D 1630-94, "Standard Test Method for Rubber Property—Abrasion Resistance (Footwear Abrader)" (Ref 28), gives a quantitative measure of scuffing abrasion resistance of soft rubber and polyurethane specimens (Ref 23). A rotating abrasive medium is attached to a drum and rubbed against stationary specimens. The number of revolutions required to abrade 2.5 mm (0.1 in.) of the test specimen (R_1) is determined and compared to the number of revolutions required to abrade a reference material to the same degree (R_2). The abrasive index is calculated:

$$\text{Abrasive index} = (R_1/R_2) \times 100 \qquad \text{(Eq 3)}$$

Another test for evaluating the abrasion resistance of elastomers is ASTM D 2228-88, "Standard Test Method for Rubber Property—Abrasion Resistance (Pico Abrader)" (Ref 29). This method compares the abrasion resistance of soft vulcanized rubber compounds and similar materials to that of a reference standard material. A pair of tungsten carbide knives is used to abrade the surface. This test may be used to estimate the relative abrasion resistance of elastomers, but "no correlation between this accelerated test and service performance is given or implied."

ISO 4649 is an abrasion test for elastomers that is also used for many plastics (Ref 9). The test specimen is held in a chuck and traversed over a rotating drum that is covered with a sheet of the abradant. This test is easy to run and requires a relatively small specimen, but it lacks the versatility of some of the other abrasion tests.

Wear Tests without Abrasive. ASTM D 3702-90, "Standard Test Method for Wear Rate of Materials in Self-Lubricated Rubbing Contact Using a Thrust Washer Testing Machine" (Ref 30), is commonly used to rank the scuffing and sliding wear resistance of polymers, especially the harder, noncompressible plastics and composites (Ref 23). A disc-shaped specimen with a contact area of 1.29 cm^2 (0.20 $in.^2$) is rotated under load against a stationary steel washer. Test duration is selected to give at least 0.1 mm (0.004 in.) of wear; typical test durations are in the range of 50 to 4000 h. The wear rate is calculated from change in thickness. This test is an American National Standard. The ASTM method notes that the test machine may also be used to measure coefficient of friction, but the procedure is not described.

ASTM G 77-91, "Standard Test Method for Ranking Resistance of Materials to Sliding Wear Using Block-on-Ring Wear Test" (Ref 31), is a standard but single-purpose test (Ref 23). A block-on-ring friction and wear machine (described in ASTM D 2714) is used to rank pairs of materials according to their sliding wear characteristics under various conditions. A test block is loaded against a test ring that rotates at a given speed for a given number of revolutions. Volume loss is calculated from block scar width and ring weight loss. Friction is continuously monitored using a load cell.

ASTM F 732-82 (Reapproved 1991), "Standard Practice for Reciprocating Pin-on-Flat Evaluation of Friction and Wear Properties of Polymeric Materials for Use in Total Joint Prostheses" (Ref 32), ranks materials with regard to friction levels and wear rates under simulated physiological conditions. A reciprocating pin-on-flat wear machine is used to compare the wear rates of candidate materials. Ultrahigh-molecular-weight polyethylene (UHMWPE), which may have wear rates as low as 100 μg per million cycles, is recommended as a reference standard.

PV Limit

The concept of *PV* limit (where *P* is contact pressure and *V* is velocity) is important for plastics used in sliding applications (Ref 5). It has been shown experimentally that if *PV* does not exceed a limiting value for a given system, the operation is basically satisfactory, and only acceptable amounts of wear will occur (Ref 11). The *PV* limit may be established using any wear test apparatus that is capable of changing load pressure and velocity, although different apparatuses will provide different *PV* limits to some extent. Also, the relevance of the *PV* limit to a given application depends on how closely the test apparatus resembles the application.

Reference 7 describes two generally accepted methods for establishing the *PV* limit. In the first method, the velocity is held constant while the load is increased incrementally. Friction force and/or the temperature of the interface is monitored. At some load, friction force and/or temperature no longer stabilize, and the *PV* limit is established for this test velocity. Tests performed at several velocities allow limiting *PV* curves as a function of velocity; the *PV* limit generally decreases with an increase in sliding velocity. This test procedure should be used only to determine a short-term *PV* rating.

The second method is based on the theory that the wear rate, *k*, is essentially constant below the *PV* limit. A plot of wear rate versus load at constant velocity shows limiting *PV* as that point where *k* is no longer constant. This method is appropriate for determining the *PV* limit for a sustained operation.

Figure 7 compares the results of the two methods for the same system. The *PV* limit defined by the second method is lower than that defined by the first.

The *PV* limit value depends on the equipment and temperature used (Ref 33). Comparing individual test values is difficult, especially if no information is given about test conditions. However, *PV* limits have been collected for various polymeric materials under dry conditions (Table 2). In general, acetal, polyethylene, and polyamides are used for low-*PV* service, PTFE is used for moderate-*PV* service, and polyphenylene sulfide, polyamide-imide, and polyimide are used for high-temperature and/or high-*PV* service.

Friction and Wear Test Data for Polymeric Materials

In addition to listing *PV* limits, Table 2 compares the friction and wear performance of many polymeric materials. Of these materials, PTFE is capable of the lowest coefficient of friction (approximately 0.04 to 0.1) because of the formation of transfer films. PTFE can be used as a filler in other polymeric materials to improve lubricity (see data for acetal, polycarbonate, and others in Table 2). Graphite can also be used to lower coefficient of friction. Fillers and reinforcing fibers, such as glass or carbon fibers, are added to polymeric materials to improve their strength (Ref 6), but these additives are not always beneficial to friction and wear characteristics.

Additional coefficient of friction data for polymeric materials and other materials may be found in Ref 34. Because coefficient of friction data depend on both the materials involved and

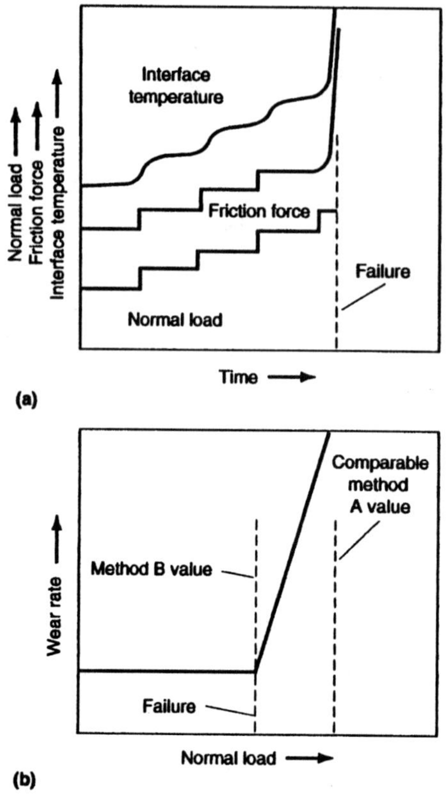

Fig. 7 Schematic representation of friction, interface temperature, and wear rate changes during the determination of contact pressure and velocity (*PV*) limit by (a) constant velocity and incremental load increases or (b) wear rate vs. load at constant velocity. Source: Ref 7

Table 2 Contact pressure and velocity (*PV*) limits and coefficients of friction for various unfilled and filled polymeric materials under dry conditions

Material (filler)(a)	*PV* limit at 22 °C (72 °F), MPa · m/s (at velocity, V, m/s)	Coefficient of friction
PTFE	0.06 (0.5)	0.04–0.1
PTFE (glass fiber)	0.35 (0.05–5.0)	0.1–0.25
PTFE (graphite fiber)	1.05 (5.0)	0.1
Acetal	0.14 (0.5)	0.2–0.3
Acetal (PTFE)	0.19 (0.5)	0.15–0.27
Polyamide	0.14 (0.5)	0.2–0.4
Polyamide (graphite)	0.14 (0.5)	0.1–0.25
Polycarbonate	0.01 (0.5)	0.35
Polycarbonate (PTFE)	0.06 (0.5)	0.15
Polycarbonate (PTFE, glass fiber)	1.05 (0.5)	0.2
Polyphenylene sulfide	3.50 (0.5)	0.15–0.3
Polyphenylene sulfide (PTFE, carbon fiber)	3.50 (0.5)	0.1–0.3
Polyamide-imide	3.50 (0.5)	0.15–0.3
Polyamide-imide (PTFE, graphite)	1.75 (0.5)	0.08–0.3
Phenolic	0.17 (0.05)	0.9–1.1
Phenolic (PTFE)	1.38 (0.5)	0.1–0.45

Note: Coefficients of friction measured for sliding on steel. These are approximate values taken from various publications. (a) PTFE; polytetrafluoroethylene. Source: Ref 7 the method of measurement, it is important to know which test was used to generate the data. The data reported in Ref 34 were obtained under a variety of conditions and should be used only as approximate guides.

UHMWPE has the highest abrasion resistance and highest impact strength of any plastic (Ref 35). Because of its high abrasion resistance, UHMWPE is used in bearings, gears, pump parts, and prosthetic joints. Table 3 shows comparative abrasion resistance (reported as volume loss relative to UHMWPE) and coefficients of friction for several materials, including polyamide, PTFE, and acetal. The effect of lubrication (with either water or oil) is also shown.

The thrust washer test (ASTM D 3702, described previously) can provide information about wear rate and static and kinetic coefficients of friction. Reference 36 describes wear and friction testing of various composites at several temperatures. Selected data from this reference are shown in Table 4. In this case, wear data are presented in the form of a wear factor, K, which is calculated as follows:

$$K = W/FVT \quad \text{(Eq 4)}$$

where W is wear volume in cubic millimeters, F is force in newtons, V is velocity in meters per second, and T is time in seconds. This gives a wear factor with units of $mm^3/N \cdot m$. The data in Table 4 show how increasing the amount of lubricating filler improves the wear factor for this particular resin at higher temperatures (260 °C, or 500 °F).

REFERENCES

1. M.J. Furey, Tribology, *Encyclopedia of Materials Science and Engineering,* Vol 7, M.B. Bever, Ed., Pergamon Press and MIT Press, 1986, p 5145–5157
2. J. Larsen-Basse, Introduction to Friction, *Friction, Lubrication, and Wear Technology,* Vol 18, *ASM Handbook,* P.J. Blau, Ed., ASM International, 1992, p 25–26
3. H. Czichos, Introduction to Friction and Wear, *Friction and Wear of Polymer Composites,* K. Friedrich, Ed., Elsevier, 1986, p 1–23
4. J. Larsen-Basse, Basic Theory of Solid Friction, *Friction, Lubrication, and Wear Technology,* Vol 18, *ASM Handbook,* P.J. Blau, Ed., ASM International, 1992, p 27–38
5. A.W. Ruff and K.C. Ludema, Wear, *Encyclopedia of Materials Science and Engineering,* Vol 7, M.B. Bever, Ed., Pergamon Press and MIT Press, 1986, p 5273–5278
6. J.K. Lancaster, Abrasion and Wear, *Encyclopedia of Polymer Science and Engineering,* Vol 1, 2nd ed., John Wiley & Sons, 1985
7. B. Bhushan and B.K. Gupta, *Handbook of Tribology,* McGraw-Hill, Inc., 1991
8. J.R. Davis, Ed., *ASM Materials Engineering Dictionary,* ASM International, 1992
9. Friction and Wear, *Handbook of Plastics Test Methods,* 3rd ed., R.P. Brown, Ed., Longman Scientific & Technical, 1988, p 174–184
10. K.C. Ludema, Introduction to Wear, *Friction, Lubrication, and Wear Technology,* Vol 18, *ASM Handbook,* P.J. Blau, Ed., ASM International, 1992, p 175
11. J.-M. Charrier, *Polymeric Materials and Processing,* Hansen Publishers, 1990
12. R.G. Bayer, Wear Testing, *Mechanical Testing,* Vol 8, *Metals Handbook,* 9th ed., American Society for Metals, 1985, p 601–608
13. J.C. Anderson, The Wear and Friction of Commercial Polymers and Composites, *Friction and Wear of Polymer Composites,* K. Friedrich, Ed., Elsevier, 1986, p 329–362
14. K. Budinski, *Engineering Materials Properties and Selection,* Reston Publishing Co., Inc., 1983
15. "Standard Guideline for Reporting Friction and Wear Test Results of Manufactured Carbon and Graphite Bearing and Seal Materials," C 808-75 (Reapproved 1990), *1992 Annual Book of ASTM Standards,* ASTM
16. "Standard Guide for Recommended Data Format of Sliding Wear Test Data Suitable for Databases," G 118-93, *1995 Annual Book of ASTM Standards,* ASTM
17. "Standard Guide for Measuring and Reporting Friction Coefficients," G 115-93, *1995 Annual Book of ASTM Standards,* ASTM
18. K.G. Budinski, Laboratory Testing Methods for Solid Friction, *Friction, Lubrication, and Wear Technology,* Vol 18, *ASM*

Table 3 Kinetic coefficients of friction (dry and lubricated) and relative abrasion resistance for selected polymeric materials

Resin(a)	Kinetic coefficient of friction: Dry	Water	Oil	Relative abrasion resistance
UHMWPE	0.10–0.22	0.05–0.10	0.05–0.08	100
Polyamide	0.15–0.40	0.14–0.19	0.02–0.11	150
Polyamide/MoS_2	0.12–0.20	0.10–0.12	0.08–0.10	...
PTFE	0.04–0.25	0.04–0.08	0.04–0.05	530
Acetal	0.15–0.35	0.10–0.20	0.05–0.10	700

Note: Test method for coefficient of friction not specified. Relative abrasion resistance is reported as abrasion resistance relative to UHMWPE = 100; test method not specified. (a) UHMWPE, ultrahigh-molecular-weight polyethylene; PTFE, polytetrafluoroethylene. Adapted from Ref 35

Table 4 Wear factors and coefficients of friction for various polyetheretherketone (PEEK) composites at different temperatures using the thrust washer test

Composite(a)	Temperature °C	Temperature °F	Load kPa	Load psi	Wear factor (K), $mm^3/N \cdot m$: Plastic	Steel	Coefficient of friction: Static	Kinetic
PEEK + 15% carbon fiber + 10% PTFE	23	73	280	40	13	0.3	0.08	0.17
	150	300	280	40	34	0.1	0.07	0.15
	260	500	280	40	120	0.03	0.08	0.17
PEEK + 15% carbon fiber + 15% PTFE	23	73	280	40	20	1.8	0.09	0.15
	260	500	280	40	50	1.0	0.17	0.23
PEEK + 15% carbon fiber + 10% graphite	23	73	350	50	18	1.3	0.05	0.15
	260	500	280	40	70	0.2	0.10	0.13
PEEK + 15% carbon fiber + 10% graphite + 10% PTFE	23	73	280	40	10	0.4	0.07	0.20
	260	500	280	40	40	0.6	0.13	0.22

Note: All tests performed at 0.25 m/s (50 ft/min). (a) PTFE, polytetrafluoroethylene. Adapted from Ref 36

Handbook, P.J. Blau, Ed., ASM International, 1992, p 45–58

19. "Standard Test Method for Static and Kinetic Coefficients of Friction of Plastic Film and Sheeting," D 1894-90, *1992 Annual Book of ASTM Standards,* ASTM
20. "Standard Test Method for Kinetic Coefficient of Friction of Plastic Solids," D 3028-90, *1992 Annual Book of ASTM Standards,* ASTM
21. "Standard Test Method for Resistance of Transparent Plastics to Surface Abrasion," D 1044-90, *1992 Annual Book of ASTM Standards,* ASTM
22. V. Shah, *Handbook of Plastics Testing Technology,* John Wiley & Sons, Inc., 1984
23. Test Screens Wear-Resistant Materials, *Adv. Mater. Process.,* Vol 104 (No. 2), Aug 1991, p 44–46
24. "Standard Test Methods for Resistance of Plastic Materials to Abrasion," D 1242-87, *1992 Annual Book of ASTM Standards,* ASTM
25. "Standard Test Method for Mar Resistance of Plastics," D 673-88, *1992 Annual Book of ASTM Standards,* ASTM
26. "Standard Test Method for Measuring Abrasion Using the Dry Sand/Rubber Wheel Apparatus," G 65-91, *1992 Annual Book of ASTM Standards,* ASTM
27. "Test Method for Determination of Slurry Abrasivity (Miller Number) and Slurry Abrasion Response of Materials (SAR Number)," G 75-89, *1995 Annual Book of ASTM Standards,* ASTM
28. "Standard Test Method for Rubber Property—Abrasion Resistance (NBS Abrader)," D 1630-83, *1992 Annual Book of ASTM Standards,* ASTM
29. "Standard Test Method for Rubber Property—Abrasion Resistance (Pico Abrader)," D 2228-88, *1992 Annual Book of ASTM Standards,* ASTM
30. "Standard Test Method for Wear Rate of Materials in Self-Lubricated Rubbing Contact Using a Thrust Washer Testing Machine," D 3702-90, *1992 Annual Book of ASTM Standards,* ASTM
31. "Standard Test Method for Ranking Resistance of Materials to Sliding Wear Using Block-on-Ring Wear Test," G 77-91, *1992 Annual Book of ASTM Standards,* ASTM
32. "Standard Practice for Reciprocating Pin-on-Flat Evaluation of Friction and Wear Properties of Polymeric Materials for Use in Total Joint Prostheses," F 732-82 (Reapproved 1991), *1992 Annual Book of ASTM Standards,* ASTM
33. H. Winkler, Selecting Materials for Wear Applications, *Plast. Des. Forum,* Oct 1994, p 39
34. P.J. Blau, Appendix: Static and Kinetic Friction Coefficients for Selected Materials, *Friction, Lubrication, and Wear Technology,* Vol 18, *ASM Handbook,* P.J. Blau, Ed., ASM International, 1992, p 70–75
35. H.L. Stein, Ultrahigh Molecular Weight Polyethylenes (UHMWPE), *Engineering Plastics,* Vol 2, *Engineered Materials Handbook,* ASM International, 1988, p 167–171
36. Friction and Wear of Thermoplastic Composites, *Friction, Lubrication, and Wear Technology,* Vol 18, *ASM Handbook,* P.J. Blau, Ed., ASM International, 1992, p 820–826

SELECTED REFERENCES

- E.R. Booser, Ed., *Handbook of Lubrication—Theory and Practice of Tribology,* Vol 1–3, CRC Press, 1983–1994
- L.-H. Lee, Ed., *Polymer Wear and Its Control,* No. 287, *ACS Symposium Series,* American Chemical Society, 1985
- A.W. Ruff and R.G. Bayer, Ed., *Tribology: Wear Test Selection for Design and Application,* STP 1199, ASTM, 1993
- "Standard Terminology Relating to Wear and Erosion," G 40-92, *1992 Annual Book of ASTM Standards,* ASTM
- Y. Yamaguchi, *Tribology of Plastic Materials,* Elsevier, 1990

Wear Failures of Plastics*

PLASTICS (or polymers**) are used in a variety of engineering and nonengineering applications where they are subjected to surface damage and wear. Examples of the tribological (involving sliding between two surfaces) use of plastics include gears and cams of various machines, tires, break pads, conveyors, hoppers, automobile body parts, aircraft, spacecrafts, hip/knee joint replacement, roller-skating wheels, and household appliances (washing machine, tubs, etc.). Wear of material parts is a very common cause of failure or low working life of machines, leading to financial loss and life hazards. Therefore, it is important to understand how polymers and other materials wear. Similar to the wear of metal, polymer wear is affected by several factors that may be broadly divided into three groups: mechanical, environmental, and thermal. These three groups of factors largely decide the mechanism of wear of a polymer surface when it comes in contact with another surface.

Historically, polymer wear has been studied based on the prevailing wear mechanisms at the contact zone (between the polymer surface and a hard counterface), which led to several methods of classification. The classification of polymer wear mechanisms that has often been followed in the literature is based on three methodologies of defining types of wear (Ref 1). The first classification is based on the two-term model that divides wear mechanisms into two types—interfacial and bulk. The second classification is more phenomenological and is based on the perceived wear mechanism. This classification includes fatigue wear, chemical wear, delamination wear, fretting, erosion, abrasion, and transfer wear. The third classification is specific to polymers and draws the distinction based on mechanical properties of polymers. In the third classification, wear study is separated as elastomers, thermosets, glassy thermoplastics, and semicrystalline thermoplastics. These classifications provide a useful basis for understanding wear failures in polymers. More often than not, wear of a polymer is a complex phenomenon that involves several of the wear mechanisms listed previously in any one wear process. For the purpose of this article, details on several of the aforementioned classifications are expanded, using wear data and micrographs from published works. The primary goals are to present the mechanisms of polymer wear and to quantify wear in terms of wear rate (rate of removal of the material). This analysis is restricted mostly to base polymers (with no fillers). Normally, polymers used in tribological applications are subjected to sliding against hard surfaces such as metals. A polymer-polymer sliding pair, except in few instances, usually produces undesirable high friction and high wear conditions due to enhanced adhesion between the polymer. Also, poor conductivity of the polymers results in elevated temperature at the polymer/polymer interface, leading to melting and rapid wear. Therefore, the focus of this article is on the wear of polymers when slid against metallic surfaces.

Interfacial Wear

The notion of interfacial wear arises from the popular two-term model of frictional energy dissipation (Ref 2). This model states that in any frictional phenomenon, where frictional energy is released at the contact points between two sliding surfaces, there can be two types of energy dissipation—interfacial and bulk. Although subjectively defined, the interface may be considered the region of the material very close (a few microns) to the contact point. This region of the material is almost instantly affected by the stress and thermal conditions arising at the contact points due to sliding. The interfacial wear is defined as the removal of the material due to interfacial friction energy dissipation between asperities, leading to events such as material softening, transfer wear, and chemical wear. A schematic of the processes involved in the interfacial wear is shown in Fig. 1.

A distinction within the interfacial wear process may be made based on whether or not the frictional heat dissipation is isothermal or quasi-adiabatic. Isothermal heat dissipation can change the mechanical property of the interface zone as opposed to the quasi-adiabatic, which affects only the transfer layer normally present at the true interface. The chemical-wear mechanism is initiated if the frictional heat can chemically affect the polymer surface, resulting in the production of degraded polymer molecules. The other important parameter to consider in interfacial wear is the roughness of the counterface. For rough and hard counterfaces, the wear mode is generally that of bulk or cohesive wear. Interfacial wear is initiated only when the counterface is smooth enough to form interfacial junctions between the polymer and the counterface. An excellent example of interfacial wear with isothermal condition is that of polytetrafluoroethylene (PTFE) sliding against a metal surface. When PTFE is slid against a smooth metal surface, friction is high in the beginning but drops to a lower value after some sliding. Because of the presence of frictional stress and heat, the PTFE molecular chains are oriented in the direction of sliding, and a transfer film is deposited onto the counterface. The molecular orientation in PTFE is responsible for the drop in friction coefficient. Although the friction coefficient is low for PTFE, wear is generally high because of the thermal softening of the interface zone and the easy removal of the material. This is one of the reasons why PTFE has not been used very widely for tribological applications. Figure 2 shows micrographs of oriented PTFE molecules deposited on the counterface after wear.

The quasi-adiabatic interfacial wear involves glassy thermoplastics (not cross linked) and cross-linked polymer systems such as elastomers and thermosets. These polymers show a range of wear behavior. For example, thermosets, which do not soften due to thermal energy, undergo chemical degradation at the interface. These degraded products detach themselves from the main body of the polymer and form transfer film and debris at the interface. The wear rate can be very high if the prevailing interface temperature is high. An important application of thermosets in a tribological context is in brake pads, where the base polymer is mixed with several additives for optimal friction, wear, and mechanical strength.

Although friction models are available for interfacial sliding, theoretical wear quantification is difficult. This is because wear depends on a number of parameters other than the mechanical and physical properties of the material.

*Adapted from the article by Sujeet K. Sinha, "Wear Failure of Plastics," in *Failure Analysis and Prevention,* Volume 11, *ASM Handbook,* ASM International, 2002, pages 1019 to 1027

**The terms *plastic* and *polymers* have some distinctions. However, in this article, the two terms mean engineering plastics. Engineering plastics are polymers that contain a very small percentage of additives, such as plasticizers and antioxidants, in order to enhance their physical and mechanical properties.

These parameters include temperature, sliding speed, normal pressure, counterface roughness, and the rheological properties of transfer film. The exact influence of each parameter on wear is rarely known. Few attempts have been made to obtain wear laws using empirical means. In one such example involving PTFE, the effects of temperature and normal pressure in relating linear wear (thickness removed per unit sliding distance) with sliding speed have been rationalized (Ref 5). According to the work, if linear wear, x (length per unit sliding distance), is assumed to be directly proportional to the sliding speed, v, at a constant temperature, T_0, and pressure, p_0, then linear wear can be expressed by:

$$x = \frac{k_0\,(a_T v)\,(p/p_0)^n}{b_s} \qquad \text{(Eq 1)}$$

where n is a constant greater than unity, and k_0 is a proportionality constant. a_T and b_s are shift factors that depend on the temperature. a_T and b_s are obtained through experimentation by shifting the data on the speed axis and wear rate axis (on a wear rate/sliding speed plot), respectively, such that they coincide with similar data obtained at a temperature of 29 °C (84 °F). The authors claim that the relation can be applied to other polymer systems, too.

Cohesive Wear

Cohesive wear is defined as subsurface or bulk wear when the interacting surfaces produce damage to the material far deeper into the material than only at the interface. This type of wear is also referred to in the literature as plowing or abrasive wear. Subsurface damage in material can be caused by surface sliding in two ways. First, if a polymer is sliding against a rough and hard surface, the asperities of the hard surface can plow into the bulk of the polymer, removing debris. These debris materials generally get transferred to the counterface, forming a transfer film (also known as the third body), which eventually makes the counterface appear smoother. The formation of a stable film at the counterface leads to a change in the wear rate of the polymer. The second cause of subsurface damage is through subsurface fatigue cracks, which can lead to the removal of material when these cracks grow to the surface of the polymer. Fatigue wear removes the material in chunks or flakes.

Considerable attention has been given by researchers to the creation of a model for cohesive or abrasive wear of polymers. The most notable model for wear involving bulk properties of the polymer was given by Ratner-Lancaster (Ref 6). The relation is given as:

$$V = \frac{K\mu W v}{HS\varepsilon} \qquad \text{(Eq 2)}$$

where V is the wear volume, K is a proportionality constant also termed the wear rate, v is the sliding speed, μ is the coefficient of friction, H is the indentation hardness, S is the ultimate tensile strength, W is wear rate, and ε is the elon-

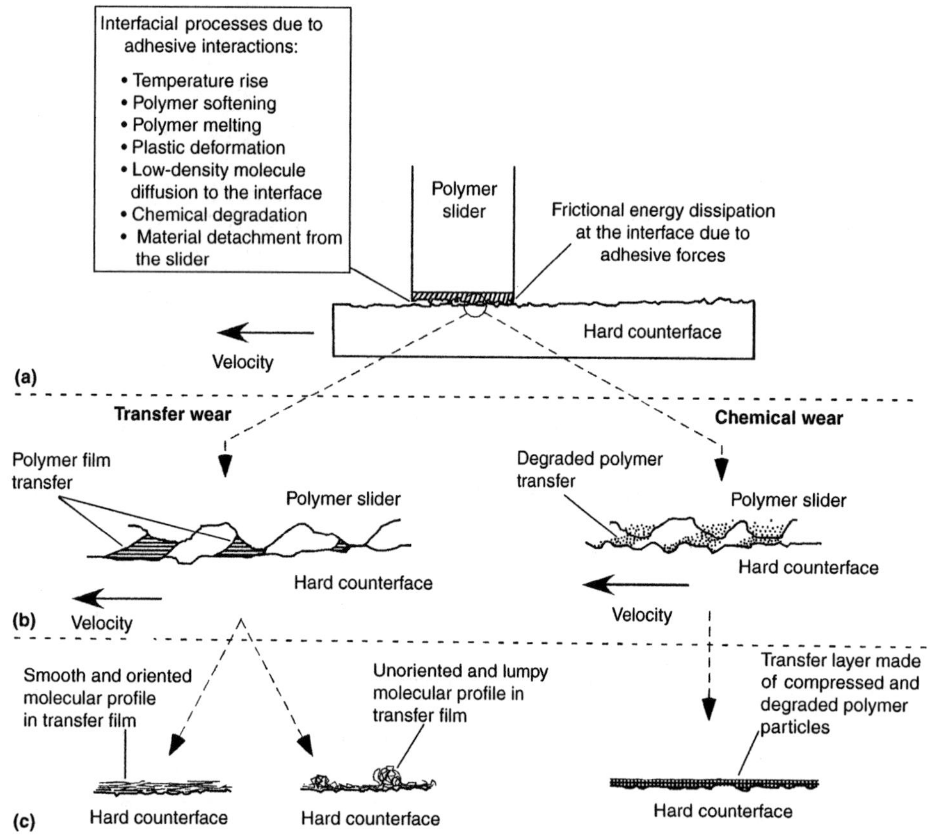

Fig. 1 Interfacial wear processes. (a) Initial contact of the two surfaces. (b) Running-in process where the soft polymer molecules are gradually transferred to the hard counterface as third body. (c) Steady-state wear process where the wear and friction phenomena are influenced mainly by the shear and adhesive properties of the transferred film. Reprinted with permission from Ref 1

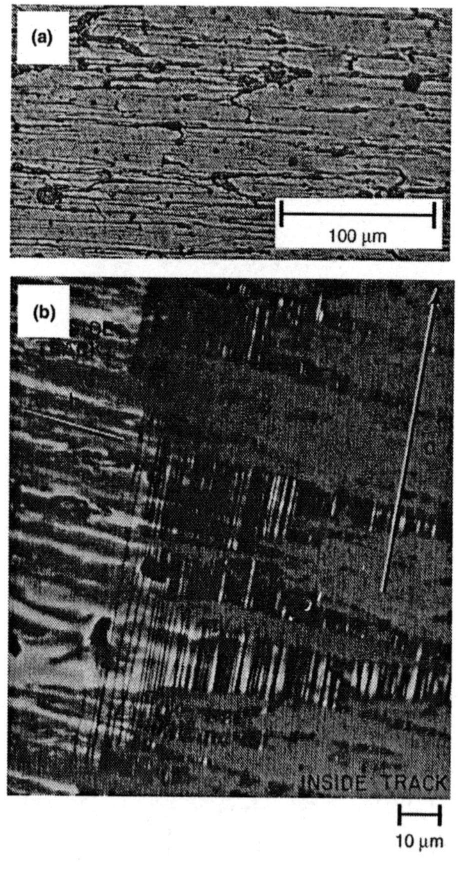

Fig. 2 Micrographs of oriented polytetrafluoroethylene (PTFE) films on the counterface. (a) PTFE transfer film on a glass slide. The film thickness varies between 50 and 500 nm, and sometimes it can show a lumpy feature when the sliding test is carried out at high loads. The film is highly birefringent, indicating that the molecules are oriented parallel to the sliding direction. Reprinted with permission from Ref 3. (b) PTFE transfer film when a PTFE pin is slid over a metallic surface. PTFE covers the counterface, making fibers and layers over one another. The orientation of the fibers in the transfer film can easily change if the sliding direction is changed. Reprinted with permission from Ref 4

gation to break of the polymer. Some evidence of the usefulness of the Ratner-Lancaster relation may be found in the work by another researcher (Ref 7). In this work, the wear rate ($mm^3 \cdot mm^{-1} \cdot kg^{-1}$) was plotted against the reciprocal of the product of S and ε, which furnished, as predicted by Eq 2, a straight line. In contrast to Eq 2, another author has followed a different approach, where wear is thought to be nonlinearly proportional to pressure, sliding velocity, and temperature (Ref 8). He proposed an empirical relation of the type:

$$\Delta w = K p^a v^b T^c$$

where Δw is the weight loss of the polymer, and a, b, and c are material-dependent variables. Yet another wear model was proposed (Ref 9) in which the authors arrived at an empirical relation using the principles of dimensional analysis. Using data obtained for polyoxymethylene (POM) and PTFE-filled POM, they obtained a relation given as:

$$V = \frac{1.5K\gamma^{1.775}p^{1.47}Z^{1.25}}{E^{3.225}} \quad \text{(Eq 3)}$$

where γ is the surface energy, Z is the sliding distance, and E the modulus of elasticity of the polymer. Another variation of Eq 3 may be found in Ref 10.

Figure 3 presents specific wear rate (wear volume per unit sliding distance per unit normal load) for a number of polymer systems under abrasive or nonabrasive sliding conditions (Ref 5, 11–18). The data are shown for a variety of experimental conditions as reported in the literature. Although the experimental conditions used in these tests were different, some trends may be noticed. Polybenzimidazole (PBI) and ultrahigh-molecular-weight polyethylene (UHMWPE) show, among all polymers, very high wear resistance. Extremely poor wear resistance is demonstrated by polymethyl methacrylate (PMMA), polystyrene (PS), and phenolic resin. Figure 4 shows worn surfaces of polyetheretherketone (PEEK) (Ref 19) and UHMWPE (Ref 20). These polymer surfaces show scars of wear by plowing and plastic deformation.

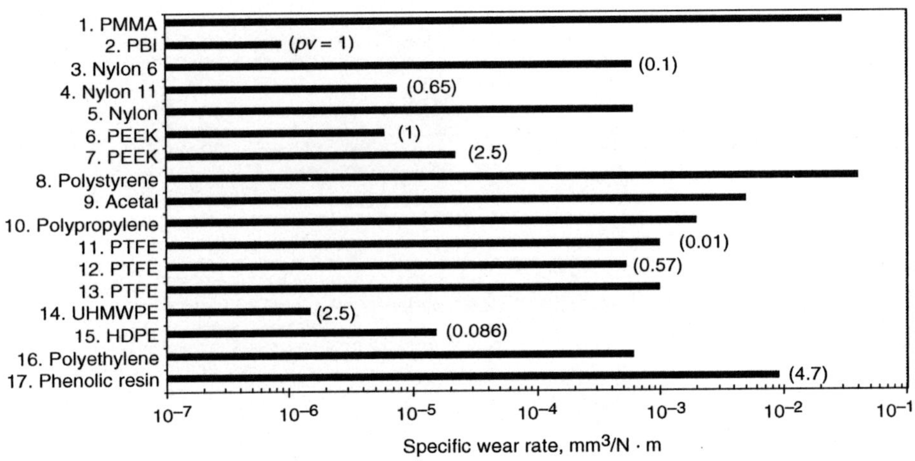

Specimen	Material	Counterface roughness (R_a), µm	Sliding speed (v), m/s	1/$S\varepsilon$(a)	Normal pressure (p) MPa	Normal pressure (p) ksi	Temperature °C	Temperature °F	Ref
1	PMMA	1.2	...	0.09	...	...	...	...	11
2	PBI	...	1	...	1	0.15	20	68	17
3	Nylon 6	...	5×10^{-3}	...	20	2.9	...	...	15
4	Nylon 11	0.11	1	...	0.65	0.09	...	...	13
5	Nylon	1.2	...	0.1	...	...	...	...	11
6	PEEK	...	1	...	1	0.15	20	68	17
7	PEEK	0.05	0.5	...	5	0.73	...	...	12
8	Polystyrene	1.2	...	5	...	...	...	...	11
9	Acetal	1.2	...	0.5	...	...	...	...	11
10	Polypropylene	1.2	...	0.1	...	...	...	...	11
11	PTFE	...	0.2	...	0.05	0.007	...	...	16
12	PTFE	...	0.1	...	5.66	0.82	29	84	5
13	PTFE	1.2	...	0.2	...	...	...	...	11
14	UHMWPE	0.05	0.5	...	5	0.73	...	...	12
15	HDPE	0.9	0.03	...	2.8	0.41	...	...	14
16	Polyethylene	1.2	...	0.09	...	...	...	...	11
17	Phenolic resin	0.05	5.6	...	0.84	0.12	...	...	18

PMMA, polymethyl methacrylate; PBI, polybenzimidazole; PEEK, polyetheretherketone; PTFE, polytetrafluoroethylene; UHMWPE, ultrahigh-molecular-weight polyethylene; HDPE, high-density polyethylene. (a) S, tensile strength; ε, elongation to break

Fig. 3 Specific wear rate for a number of polymers, as reported in the literature. The experimental conditions as reported in the literature are given in the table. *pv*, pressure × velocity

Elastomers

The study of wear of elastomers has evolved primarily from the interest in the friction and wear of automobile tires and industrial seals. Extensive studies were carried out on relatively softer rubbers, such as polyisoprene, butyl rubber, and natural rubber (Ref 21, 22). Through extensive experimentation on the sliding of rubber against hard surfaces, it was found that the process of sliding for rubber takes place through a series of detachments at the contact points, giving it the look of a wave (Fig. 5). These waves initiate at the front edge of the slider, due to excessive buckling of rubber in the front, and run to the rear of the slider. When a slider in contact with an elastomer is pushed forward, the adhesive force (between the slider and the elastomer) generates compressive tensile stress at the front edge, leading to buckling and folding of the elastomer in the form of a wave. The detached part further relaxes the material, thus facilitating the movement of the slider. A later study of the wear of rubbers and tires (Ref 23) concluded that for elastomeric materials, there are two ways in which the frictional energy is dissipated, leading to wear. The flow chart the investigator produced is redrawn in Fig. 6. In order to model abrasive action of asperities on elastomers, several tests using sharp needles have also been carried out in the past. The process of wear by a sharp needle or an asperity is schematically shown in Fig. 7. Although there are a number of models available that quantify the frictional work done during sliding on rubber, a wear model for elastomers is still unavailable presently, except for the abrasive case where the Ratner-Lancaster relation can be applied.

Thermosets

Thermosets have found applications mainly in automobile brakes, gears, cams, and clutch parts, where they are subjected to sliding. Brake pads are one area where thermosets such as phenolic and epoxy resins have been used and studied extensively. These polymers do not soften when the temperature rises at the interface, and thus, they prevent the component from yielding or failing in a catastrophic manner during service. However, thermal energy dissipated due to frictional work can induce chemical degradation and wear at the sliding surface. Thermosets have generally been filled with fibers and particles as additives in order to increase the strength and wear resistance of the material (Ref 24–27). Fillers include glass fiber, aramid fiber, and metal oxide particles of various kinds.

The role of aramid fibers, in the context of brake pads, caught special attention from tribologists when there was an effort to replace asbestos used in brake pads with aramid fibers (Ref 18, 28–30). Figure 8 compares the specific wear rate of a few formulations of thermoset composites. The relevant micrograph is given in Fig. 9. It is seen from these results that the wear resistance of phenolic resin increases by almost 2 orders of magnitude when fillers such as carbon and aramid fibers are added to the phenolic resin matrix. The micrograph (Fig. 9) shows that a transfer layer is formed on the polymer surface in addition to the transfer layer found on the counterface. These strong and highly adhesive transfer layers help improve the wear resistance of the polymer composite.

Glassy Thermoplastics

Traditionally, glassy thermoplastics have not been used as typical tribological materials. This is because they show mechanical instability at the glass transition temperature. However, they are often subjected to sliding, scratching, or abrasion in various working environments. For example, a window pan or automobile body part made of glassy polymer may be subjected to water, dust, and occasional scratching, or a bathtub may have water plasticization coupled with sliding and compression. Some glassy thermoplastics filled with fibers or particulate fillers have been used for tribological applications. The problem encountered with such polymers is their tendency to fail in a catastrophic manner when the glass transition temperature is reached. Examples of this class of polymer are PMMA, PS, and polycarbonate. Cross-linked polymer PEEK also behaves in a way similar to glassy polymers (Ref 31). See Fig. 10 for the changes in deformation behavior in sliding of PEEK when the operating temperature is close to the glass transition temperature for PEEK.

The study of glassy thermoplastic surfaces has mainly focused on understanding the damage processes under a variety of experimental and ambient conditions (Ref 32, 33). For example, damage modes can be studied using the concept of wear maps. Figure 11 gives such a map of PMMA for different normal load and imposed strain conditions. A range of studies have been carried out (Ref 34, 35) in order to understand the role of the third body in fretting wear of PMMA. These studies with PMMA concluded that the formation of the third body and the wear rate depend on the kinematics of sliding. In linear sliding, as opposed to torsional sliding, the wear rate is low. The worn area showed debris material in rolled and compacted forms. The authors concluded that the energy dissipation in the linear sliding case occurred mainly by the rolling and shearing actions on the rolled debris, which reduced the frictional work required for sliding. Therefore, wear in the linear sliding case was low.

Semicrystalline Thermoplastics

The most versatile use of polymers in tribological application has been for the semicrystalline group of polymers. Semicrystalline thermoplastics include PTFE, polyethylene (PE), UHMWPE, and nylon. These polymers, in homogeneous or heterogeneous forms, have found applications in gears, bearings, automobile piston seals, knee/hip joint replacement, and so forth. Semicrystalline thermoplastics do soften in the presence of thermal energy; however, the way thermal energy is transmitted from the interface to the bulk depends on the thermal properties of the individual polymer. Based on this behavior, the mode of wear for semicrystalline polymers can be divided into two groups: adiabatic and isothermal. Furthermore, the isothermal type, a common case, is subdivided into three categories based on the way polymer transfer film is deposited onto the hard counterface. Figure 12 delineates these groups of wear processes for semicrystalline thermoplastics in isothermal heat-transfer conditions.

Early studies on the friction and wear of thermoplastics was motivated by the prospect of finding an ultralow-friction polymer material

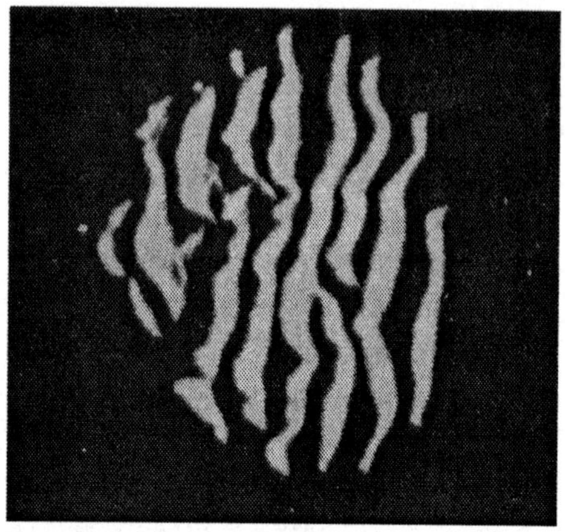

Fig. 5 Waves of detachment when an elastomer is slid against a hard and smooth surface. The rubber moves forward in the form of ripples of wave on its contact surface with a smooth and hard counterface. These so-called waves of detachment can produce wear in the form of rolls of detached material or the third body. Reprinted with permission from Ref 22

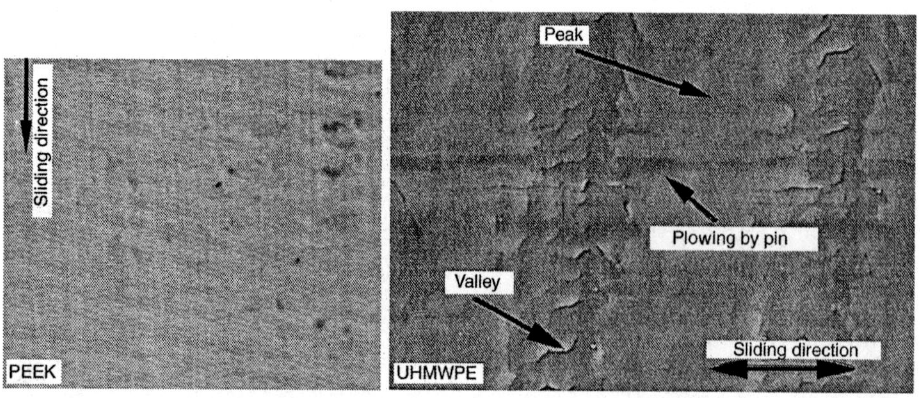

Fig. 4 Micrographs showing surfaces of worn polymers when they were slid against abrasive surfaces. Polyetheretherketone (PEEK) (left) reprinted with permission from Ref 19. Ultrahigh-molecular-weight polyethylene (UHMWPE) (right) (reprinted with permission from Ref 20) surfaces show scars of abrasive and plowing actions of hard counterfaces.

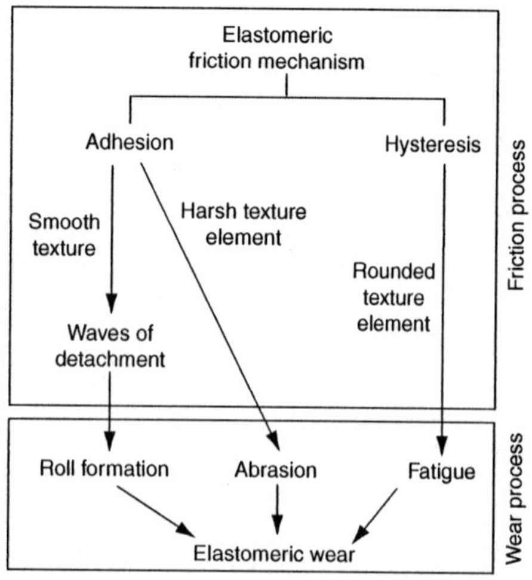

Fig. 6 Classification of the processes of friction leading to wear for elastomers (adapted from Ref 23). The diagram clarifies the role of friction in determining the wear mechanism for elastomeric polymers.

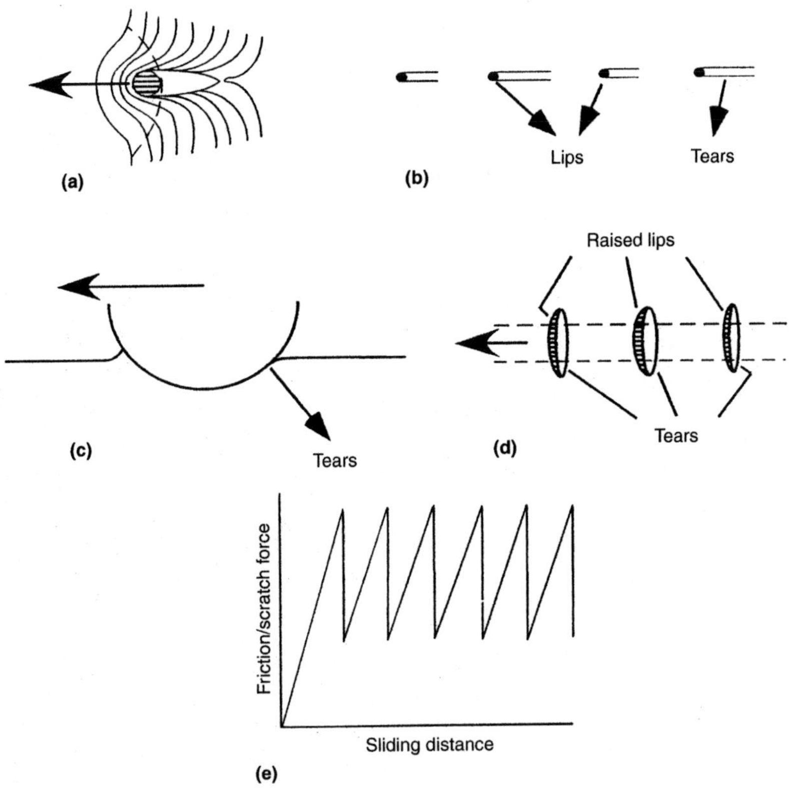

Fig. 7 Damage created on the surface of an elastomer by isolated stress concentration. (a) Surface deformation pattern when a sharp needle or conical indentor with acute angle is slid on the surface of an elastomer. The elastomer surface is pulled in the direction of motion and fails in tension behind the contact at $\pi/2$ to the tensile field. (b) After the needle jumps forward, the surface relaxes, and tensile tears are evident on the surface but are now in the direction of motion. (c) Tearing of an elastomer due to tractive stress with a large unlubricated indentor. The tear is generated at the rear of the contact region and is almost at right angles to the sliding motion. (d) A raised lip of elastomer is formed, but no material is actually removed. (e) A typical friction/scratching force profile when a slider is passed over an elastomer. Reprinted with permission from Ref 1

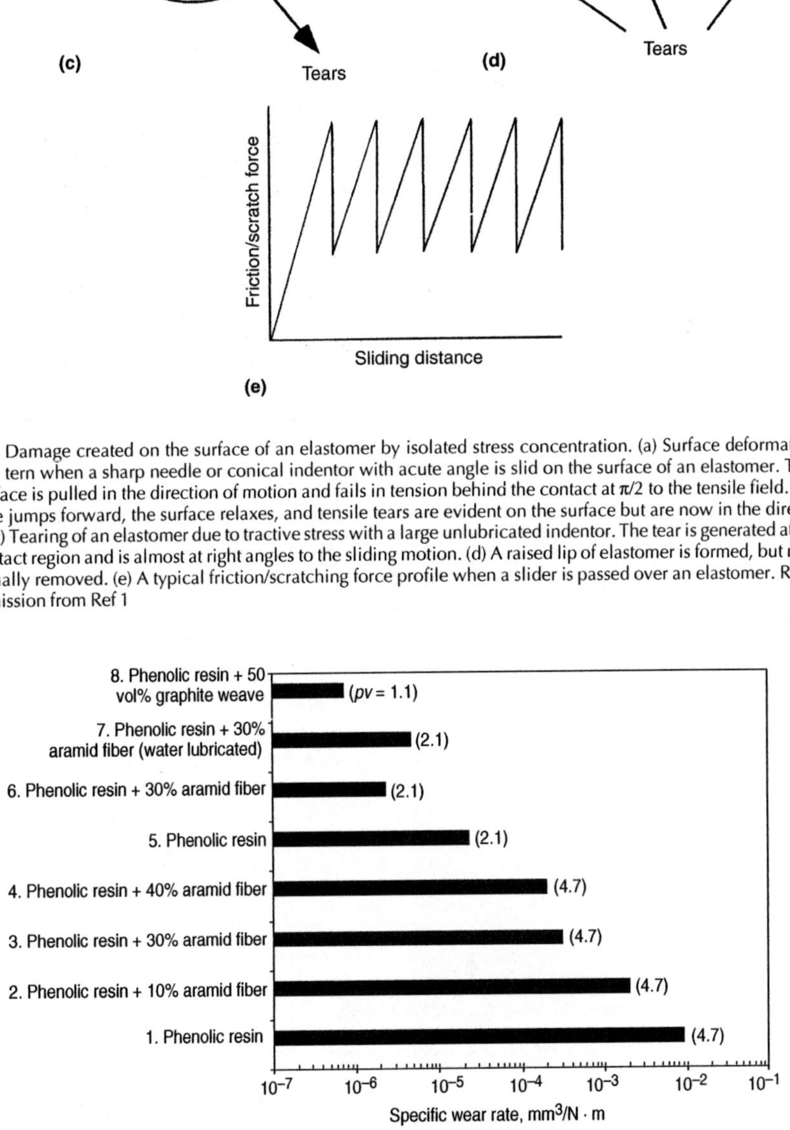

Specimen	Sliding speed (v), m/s	Normal pressure (p) MPa	Normal pressure (p) ksi	Counterface roughness (R_a), μm	Ref
1–4	5.6	0.84	0.12	0.5	18
5–7	0.5	4.25	0.62	0.05–0.1	29
8(a)	1.6	0.69	0.10	0.05	27

(a) N_2 atmosphere at room temperature

Fig. 8 Specific wear rates for phenolic resin and its composites. The data are reported for various experimental conditions and *pv* (pressure × velocity) factors, as reported in the literature.

(Ref 3, 36). Polytetrafluoroethylene provided a very low friction coefficient (~0.06), although the corresponding wear rate was high. The reason for low friction was found to be highly oriented PTFE molecules that were transferred to the counterface during sliding (Ref 3). The interface of the polymer also showed highly oriented molecules that extended out of the samples showing fibers. In order to reduce the wear rate and use the excellent low-friction property of PTFE, this polymer has often been used with fillers to form composites. Polytetrafluoroethylene itself has also been used as filler for other polymeric systems, such as PE. Figure 13 gives the wear rate of PTFE and some of its composites when slid against hard metallic surfaces. For surface-treated PTFE (such as γ-irradiation), the situation may be different. Evidence shows that for such a system there may be an increase in the crystallinity of the polymer at the surface and consequently, a decrease in the wear rate (Ref 37).

The wear process for a semicrystalline thermoplastic polymer may seem to depend very much on the transfer film and its rheological properties, although evidence is also available showing that the loading condition can also change the wear mechanism. For UHMWPE, Wang and others (Ref 38) found that the microscopic surface wear depends on the tensile and elongation properties of the polymer. However, under intense and nonconformal loading conditions, the wear mechanism could change to macroscopic subsurface wear due to fatigue. Thus, the wear mechanism can change if the loading condition is changed. The authors provided a model for the wear of semicrystalline thermoplastics that resembles the Ratner-Lancaster model for abrasive wear of polymers; for microscopic surface wear:

$$V \propto \frac{L^{3/2} R_a^{3/2}}{S^{3/2} \varepsilon}$$

for macroscopic subsurface wear:

$$V \propto \frac{1}{N} = (\Delta\varepsilon_p/\varepsilon)^{(1/\alpha)}$$

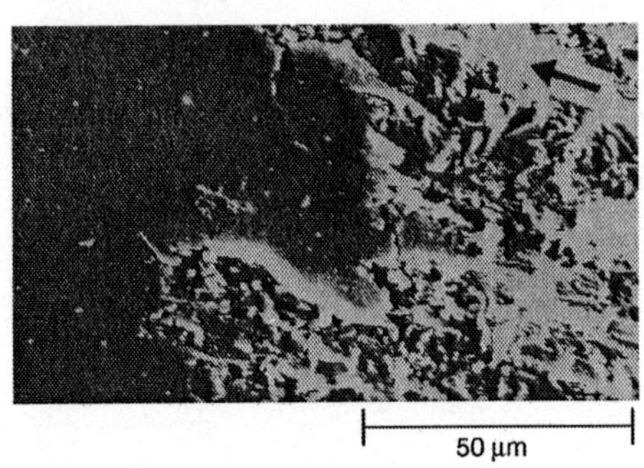

Fig. 9 Micrograph of the worn surface for a phenolic resin/aramid fiber composite (Ref 29) showing partial coverage of the polymer pin by transfer film

where V is the wear volume, L is the normal load, R_a is the counterface roughness, S is the ultimate tensile strength of the polymer, ε is the elongation at break, N is the cyclic fatigue life of the polymer, $\Delta\varepsilon_p$ is the inelastic strain amplitude, and α is a material constant obtained from the low-cycle fatigue test using the Coffin-Manson equation (Ref 39).

Environmental and Lubricant Effects on the Wear Failures of Polymers

Except for elastomers, polymers in general are not used in lubricated conditions. However, polymers are often subjected to environmental conditions that affect their friction and wear performances. For example, polymers used in marine applications get exposed to seawater, and a machine component such as a gear or brake pad may come in direct contact with leaking oil or water. For elastomers, their applications in seal rings and automobile tires regularly expose the material to lubricants, chemicals, and water. For industrial seals, the presence of lubricant protects it from dry contact with metal parts and the consequent severe wear. This kind of wear not only lowers the life of the seal but also affects the metal part. It has been observed that soft elastomer can wear the metal part it comes in contact with (Ref 40). In an effort to increase the life of seals, a number of studies have been carried out to estimate the film thickness of the lubricant for elastomer pressed against a metal (Ref 41–43). The other example of the use of polymers in a lubricating environment is that of the knee/hip joint replacement using UHMWPE (Ref 44, 45). UHMWPE is widely used in making acetabular sockets for hip joints that normally slide against a ceramic ball. The presence of synovial body fluid ensures low friction by lubricating the surfaces. This fluid does not seem to chemically affect the polymer, although it does affect the way transfer film is formed at the counterface. The main problems in the application of UHMWPE for knee/hip joint replacement are the production of wear particles, which tend to become points of bacterial infection growth for the patient, and the wear of the metallic or ceramic counterface, leading to increased wear of the polymer.

Environmental fluids and humidity have been found to affect many polymers in two ways. The first is the change in the adhesive and flow properties of the transfer film, and the second effect is that of changing the mechanical properties of the bulk of the polymer due to plasticization. In the presence of a liquid, the adhesion of the transfer film is normally decreased, leading to high wear of the polymer (see Fig. 8 for wear data on a water-lubricated sliding case). This is because the deposited polymer on the counterface is constantly removed during sliding, requiring further wear of the bulk of the polymer. The effect of liquid on the mechanical properties of the bulk polymer largely depends on the polarities of the polymer and the liquid, as well as on the surface tension of the liquid (thus, the surface energy of the polymer) (Ref 46). Many polymers plasticize in the presence of water and some chemical liquids, because liquid molecules can easily migrate into the bulk of the polymer. Plasticization of a polymer drastically reduces its mechanical strength and hardness, which gives rise to a substantial reduction in the wear resistance. The effect of lubricants on PE has been studied (Ref 47). The authors found that when oleamide and stearamide are applied to the surface of PE, the lubricants interact with polymer molecules and form a chemically bonded monolayer on the outer surface of the polymer. This can drastically reduce the coefficient of friction when the polymer slides against a hard surface.

Summary and Case Study

Wear of polymers is an important aspect of their failure analysis and lifetime prediction. Wear failure of polymers is controlled by a number of factors, which include mechanical properties of polymers, such as ultimate tensile strength, elongation to break and hardness, sliding speed, normal load, coefficient of friction, counterface roughness, rheology and adhesive property of the transfer film, and thermal properties of polymers. The adhesive strength of the transfer layer to the counterface has strong influence on the wear rate. Strong adherent transfer film normally gives low wear rate. Abrasive action of the asperities, adhesive force, thermal softening, chemical degradation, and subsurface fatigue are some of the factors that initiate mate-

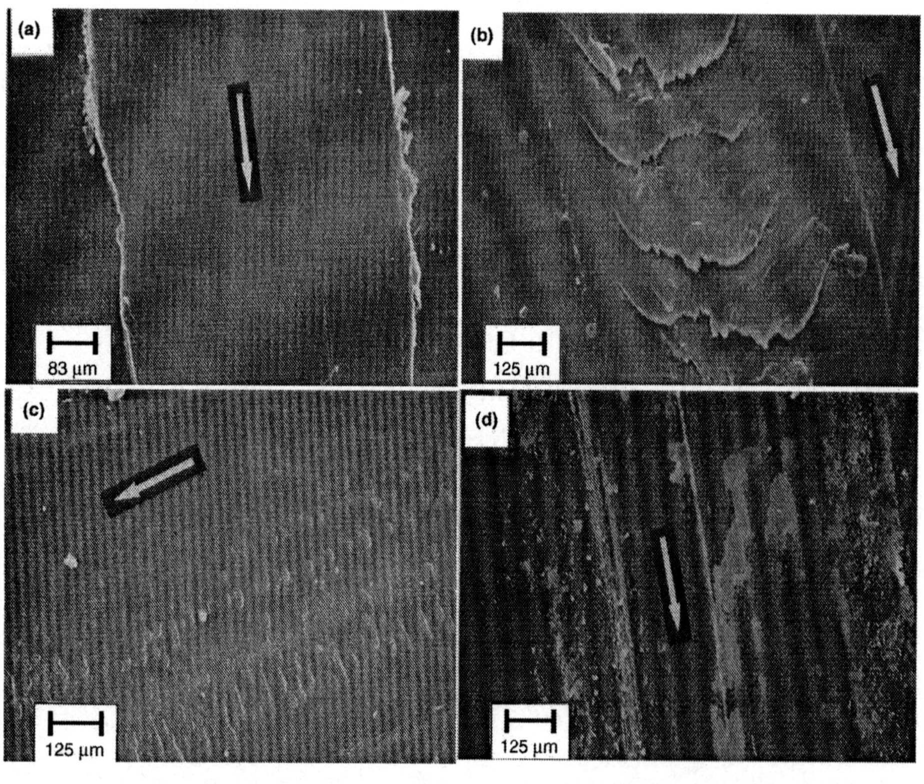

Fig. 10 Micrographs of worn polyetheretherketone (PEEK) surfaces at various operating temperatures. These pictures highlight the changes in the surface deformation behavior of the polymer with temperature. (a) 90 °C (194 °F). (b) 152 °C (306 °F). (c) 180 °C (356 °F). (d) 225 °C (437 °F). Arrows indicate the sliding direction. Glass transition temperature for PEEK used in the experiment was 148 °C (300 °F). Reprinted with permission from Ref 29

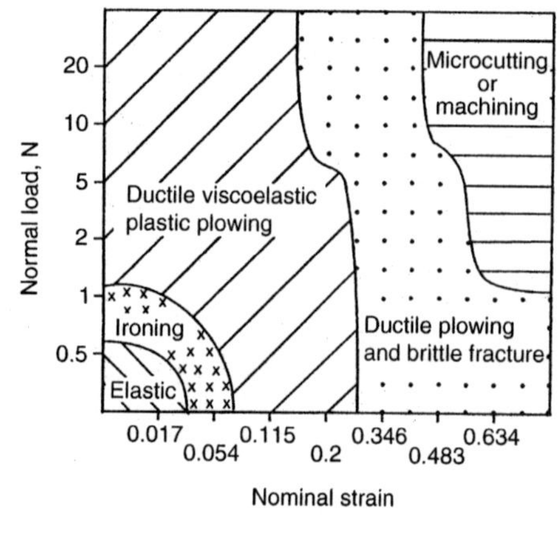

Fig. 11 Scratching damage maps for polymethyl methacrylate. Scratching velocity = 0.004 mm/s, and nominal strain is defined as 0.2 × tan θ; 2θ being the included angle of the indenter.

rial removal during the process of polymer wear. The effect of lubricants depends on how lubricant molecules attach themselves to the polymer molecules, making bonds between the two molecular entities. Many polymers, in the presence of water or lubricant molecules, plasticize, which reduces friction, but wear can be high because of the decrease in the mechanical strength of the polymer due to plasticization. Lubricants, in general, reduce the adhesion of the transfer layer to the counterface, leading to easy removal of the transfer layer and a high wear situation for the polymer.

A Case Study: Nylon as a Tribological Material. First synthesized in 1935 by Carothers (Ref 48), nylon is among few very important semicrystalline industrial thermoplastics. Nylon is the commercial name for those aliphatic polyamides that are made exclusively from ω-amino acids (Ref 49). There are several forms of nylon, generally denoted by nylon-*n* or nylon-*m,n,* where *m* and *n* stand for the number of main chain carbon atoms in constituent monomer(s). Among all varieties of nylons, nylon 6 and nylon 6/6 are the most widely produced and used materials because of their excellent mechanical properties and low cost. Nylon 11 and nylon 12, which show better performance in terms of low moisture absorption when compared to other nylons, are also used extensively; however, they are expensive.

Historically, nylons have been very popular materials for many tribological applications, such as sliding fittings, bearings, and gears. Possibly the greatest advantage of using nylon as tribological material over metals is that no external lubricant is needed, and the vibration noise is far less for nylon than for metals. Nylon parts can be extrusion molded with superior strength properties and low overall production cost.

Nylon sliding against nylon is a poor tribological pair due to high friction and high thermal effects (Ref 50). Even pure nylon sliding against metal surfaces does not perform well. However, nylon is an excellent low-friction and wear-resistant material if used in the form of plastic composite sliding against metal surfaces. This can be observed from the few studies that are available in the literature on nylon. Table 1 provides friction and wear results on a few types of nylon and its composites.

Similar to the case of many other plastics, the tribological performance of nylon greatly depends on its ability to form adherent and stable transfer film on the hard metal counterface. Several studies have shown that if pure nylon is used in sliding, the transfer film is weak and patchy. This kind of transfer film can be easily removed from the counterface due to the dynamic actions of sliding. Interfacial temperature also plays its role in making the transfer layer soft and weak. With certain types of fillers in nylon, it has been found that the composite makes a very thin but adherent transfer layer. This transfer layer protects the bulk of the polymer from further wear. Common fillers with advantageous effects on the wear resistance of nylon are glass fiber (Ref 50), CuS, CuO, CuF_2 (Ref 51), and PTFE (Ref 13). In one study (Ref 50), aramid and carbon fibers were also used as fillers for nylon. However, the investigators found high friction for these two fillers and concluded that interfacial heating due to high friction could damage the nylon matrix, leading to accelerated wear, especially in the high load and speed conditions.

The main disadvantage with the use of nylons is their water-absorbent characteristics. Mechanical properties, such as elastic modulus and hardness, as well as physical properties, such as glass transition temperature of nylon, drastically reduce with the increase in the absorbed water content in nylon. In this respect, nylon 11 and nylon 12 are superior to nylon 6 and nylon 6/6. The percentage water absorption at saturation and 20 °C (70 °F) temperature for nylon 11 and

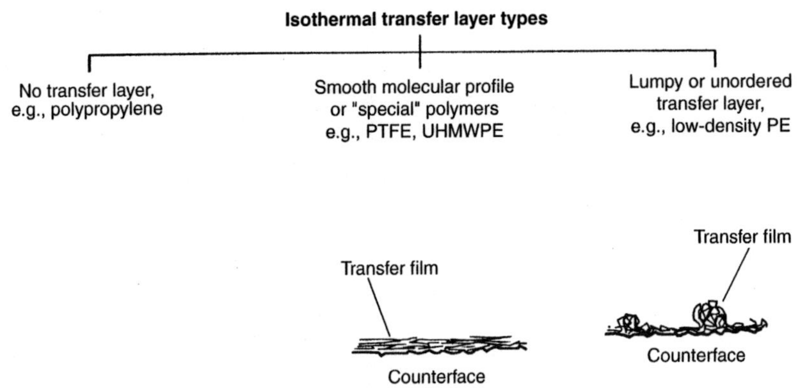

Fig. 12 Generic types of transfer wear behavior when semicrystalline polymers are slid on a hard, smooth surface. In most of the cases, there is a formation of transfer layer on the counterface, although the shear and adhesive properties of the transfer films will vary depending on the mechanical properties of the polymer and the surface topography of the counterface. PTFE, polytetrafluoroethylene; UHMWPE, ultrahigh-molecular-weight polyethylene; PE, polyethylene. Reprinted with permission from Ref 1

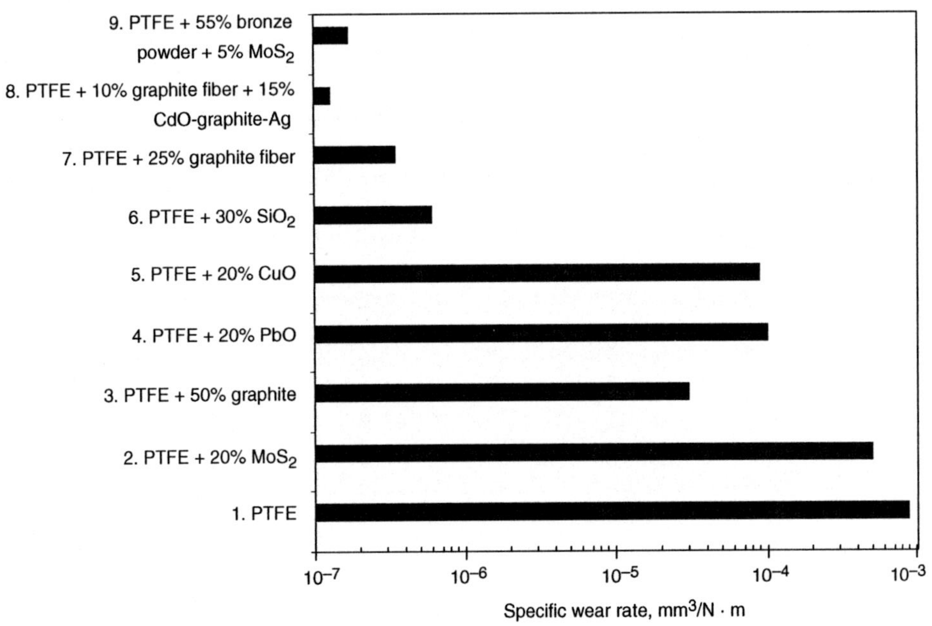

Fig. 13 Wear rate of polytetrafluoroethylene (PTFE) and its composites under different experimental conditions. For specimens 1 to 4: sliding speed (v) = 0.2 m/s; normal pressure (p) = 0.05 MPa (0.007 ksi). Source: Ref 16. For specimens 7 to 9: sliding speed (v) = 1.6 m/s; normal pressure (p) = 0.69 MPa (0.10 ksi); counterface roughness (R_a) = 0.025 μm. Source: Ref 27

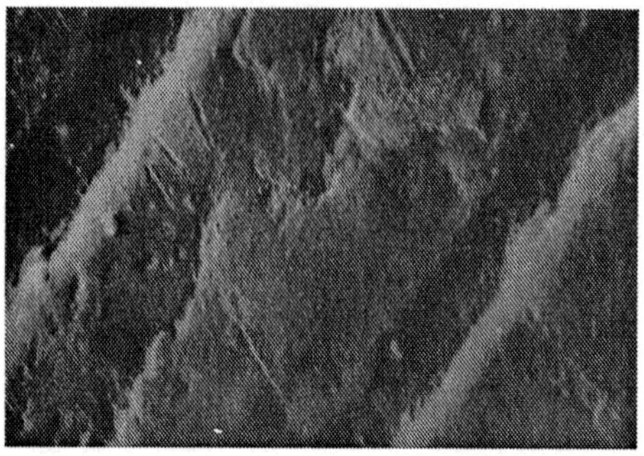

Fig. 14 Wear marks on the surface of a nylon/polyethylene antifriction bearing. The bearing was in contact with a rotating steel shaft. 417×. Source: Ref 53

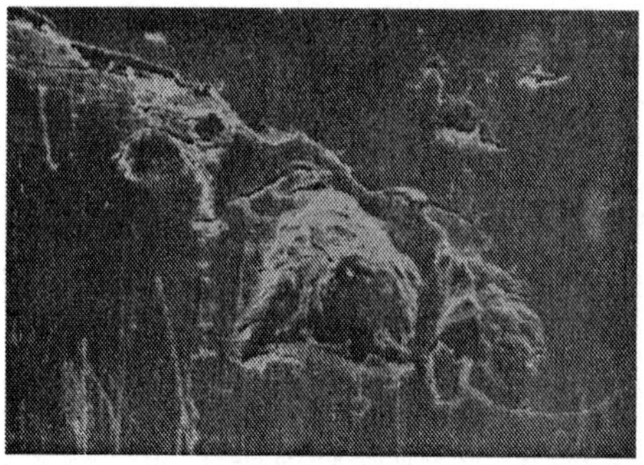

Fig. 15 Pitting and surface microcracks on the tooth flank of an oil-lubricated nylon driving gear. 37×. Source: Ref 53

Fig. 16 Failed polyoxymethylene gear wheel that had been in operation in a boiler-room environment. 305×. Source: Ref 53

Table 1 Friction and wear for nylons

Nylon type	Friction coefficient	Specific wear rate, $\times 10^{-6}$ mm^3/N · m	Test conditions	Ref
Nylon 11	0.31	7.48	Normal pressure = 0.65 MPa; sliding speed = 1 m/s; quench-hardened AISI steel counterface (R_a = 0.11 μm)	13, 51
Nylon 11 + 35% CuS	0.42	1.8	Normal pressure = 0.65 MPa; sliding speed = 1 m/s; quench-hardened AISI steel counterface (R_a = 0.11 μm)	13, 51
Nylon 11 + 5.6% glass fiber	0.38–0.5	2.97	Normal pressure = 0.65 MPa; sliding speed = 1 m/s; quench-hardened AISI steel counterface (R_a = 0.11 μm)	13
Nylon 11 + 20.7% glass fiber	0.38–0.5	1.66	Normal pressure = 0.65 MPa; sliding speed = 1 m/s; quench-hardened AISI steel counterface (R_a = 0.11 μm)	13
Nylon 6/6	0.62	...	Normal load = 200 N; sliding against nylon 6/6	50
Nylon 6/6 + 30% glass fiber	0.1–0.3	...	Normal load = 200 N; sliding against nylon 6/6	50
Nylon 6/6 + 30% glass fiber + 15% PTFE	0.05–0.1	...	Normal load = 200 N; sliding against nylon 6/6	...
Nylon 6	0.3	589	Normal load = 825 kN (pressure = 20 MPa); sliding velocity = 5 mm/s; steel counterface ($R_a \approx$ 5 μm), extremely high pressure	15

PTFE, polytetrafluoroethylene; R_a, counterface roughness

nylon 12 is 1.6% each (Ref 48), while this value for nylon 6 is 10.9% (Ref 48). The percentage of water absorption depends on the amount of crystallinity in the polymer—the higher the crystallinity, the lower the water absorption. A loss of mechanical strength for nylon results in increased wear rate.

One can conclude from this case study that for nylon, the wear resistance characteristics can be enhanced if low-water-absorbing forms (such as nylon 11 or nylon 12) of nylon reinforced with fillers, such as glass fiber, CuS, CuO, or PTFE, are used. To the author's knowledge, so far there is no available published work on the friction and wear characteristics of nylon 12.

Failure Examples (Ref 52)

Example 1: Wear Failure of an Antifriction Bearing. Shown in Fig. 14 is the worn surface of an antifriction bearing made from a nylon/PE blend. The bearing was worn in contact with a steel shaft. Movement of the shaft against the bearing caused abrasive marks (Fig. 14). Fine iron oxide particles acted as an abrasive, producing the failure mechanism observed.

Example 2: Failure of a Nylon Driving Gear. Figure 15 shows pitting on the tooth flank of a nylon oil-lubricated driving gear. The pitting produced numerous surface microcracks in association with large-scale fragmentation (frictional wear). The stress-cracking effect of the lubricating oil is believed to have played a role in initiating the observed microcracks.

Example 3: Failure of a Polyoxymethylene Gear Wheel. A polyoxymethylene gear wheel (Fig. 16) exhibits a different failure mechanism. This component had been in operation in a boiler room and is believed to have failed because of considerable shrinkage. The oriented crystalline superstructures and the microporosity are reported to be due to postcrystallization. The porosity is attributed to the difference in densities between the amorphous (1.05 g/cm^3, or 0.04 lb/in.^{3}) and the semicrystalline (1.45 g/cm^3, or 0.05 lb/in.^{3}) states (Ref 53). Breakdown along the crystalline superstructure started mainly at the mechanically stressed tooth flanks. In addition, oil vapors, humidity, and other degradative agents could also have contributed to the observed failure.

REFERENCES

1. B.J. Briscoe and S.K. Sinha, Wear of Polymers, *Proc. Inst. Mech. Eng. J., J. Eng. Tribol.,* Vol 216, 2002
2. B.J. Briscoe and D. Tabor, Friction and Wear of Polymers, *Polymer Surfaces,* D.T. Clark and J. Feast, Ed., John Wiley & Sons, 1978
3. C.M. Pooley and D. Tabor, Friction and Molecular Structure: The Behaviour of Some Thermoplastics, *Proc. R. Soc. (London) A,* Vol 329, 1972, p 251–274
4. R.P. Steijn, The Sliding Surface of Polytetrafluoroethylene: An Investigation with the Electron Microscope, *Wear,* Vol 12, 1968, p 193–212
5. Y. Uchiyama and K. Tanaka, Wear Law for Polytetrafluoroethylene, *Wear,* Vol 58, 1980, p 223–235
6. J.K. Lancaster, Friction and Wear, *Polymer Science, A Materials Science Handbook,* A.D. Jenkins, Ed., North-Holland Publishing, 1972
7. B.J. Briscoe, Wear of Polymers: An Essay on Fundamental Aspects, *Tribol. Int.,* Aug 1981, p 231–243
8. S.K. Rhee, *Wear,* Vol 16, 1970, p 431

9. M.K. Kar and S. Bahadur, The Wear Equation for Unfilled and Filled Polyoxymethylene, *Wear,* Vol 30, 1974, p 337–348
10. N. Viswanath and D.G. Bellow, Development of an Equation for the Wear of Polymers, *Wear,* Vol 181–183, 1995, p 42–49
11. D.G. Evans and J.K. Lancaster, The Wear of Polymers, *Wear,* Vol 13, *Materials Science and Technology,* D. Scott, Ed., Academic Press, 1979, p 85–139
12. J.P. Davim and N. Marques, Evaluation of Tribological Behaviour of Polymeric Materials for Hip Prostheses Applications, *Tribol. Lett.,* Vol 11 (No. 2), 2001, p 91–94
13. S. Bahadur and V.K. Polineni, Tribological Studies of Glass Fabric Reinforced Polyamide Composites Filled with CuO and PTFE, *Wear,* Vol 200, 1996, p 95–104
14. C.H. da Silva, D.K. Tanaka, and A. Sinatora, The Effect of Load and Relative Humidity on Friction Coefficient Between High Density Polyethylene on Galvanized Steel—Preliminary Results, *Wear,* Vol 225–229, 1999, p 339–342
15. F. Van De Velde and P. De Baets, The Friction and Wear Behaviour of Polyamide 6 Sliding Against Steel at Low Velocity Under Very High Contact Pressure, *Wear,* Vol 209, 1997, p 106–114
16. S. Bahadur and D. Gong, The Action of Fillers in the Modification of the Tribological Behaviour of Polymers, *Wear,* Vol 158, 1992, p 41–59
17. K. Friedrich, Z. Lu, and A.M. Hager, Recent Advances in Polymer Composites' Tribology, *Wear,* Vol 190, 1995, p 139–144
18. T. Kato and A. Magario, The Wear of Aramid Fibre Reinforced Brake Pads: The Role of Aramid Fibres, *STLE Tribol. Trans.,* Vol 37 (No. 3), 1994, p 559–565
19. J.V. Voort and S. Bahadur, The Growth and Bonding of Transfer Film and the Role of CuS and PTFE in the Tribological Behavior of PEEK, *Wear,* Vol 181–183, 1995, p 212–221
20. J. Song, P. Liu, M. Cremens, and P. Bonutti, Effects of Machining on Tribological Effects of Ultra High Molecular Weight Polyethylene (UHMWPE) Under Dry Reciprocating Sliding, *Wear,* Vol 225–229, 1999, p 716–723
21. A. Schallamach, Abrasion of Rubber by a Needle, *J. Polym. Sci.,* Vol 9 (No. 5), 1952, p 385–404
22. A. Schallamach, How Does Rubber Slide? *Wear,* Vol 17, 1971, p 301–312
23. D.F. Moore, Friction and Wear in Rubbers and Tyres, *Wear,* Vol 61, 1980, p 273–282
24. T. Liu and S.K. Rhee, High Temperature Wear of Semi-Metallic Disk Brake Pads, *Wear of Materials 1977,* American Society of Mechanical Engineers, 1977, p 552–554
25. S.K. Rhee and P.A. Thesier, "Effects of Surface Roughness of Brake Drums on Coefficient of Friction and Lining Wear," paper 720449, Society of Automotive Engineers, 1971
26. J.K. Lancaster and J.P. Giltrow, The Role of the Counterface Roughness in the Friction and Wear of Carbon Fibre Reinforced Thermosetting Resins, *Wear,* Vol 16, 1970, p 357–374
27. B. Bhushan and D.F. Wilcock, Wear Behaviour of Polymeric Compositions in Dry Reciprocating Sliding, *Wear,* Vol 75, 1982, p 41–70
28. B.J. Briscoe, I. Ramirez, and P.J. Tweedle, *Proc. Int. Conf. Disc Brakes for Commercial Vehicles* (London), 1–2 Nov 1988, Mechanical Engineering Publications Ltd., 1988, p 5
29. S.K. Sinha and S.K. Biswas, Friction and Wear Behaviour of Continuous Fibre as Cast Kevlar-Phenolic Resin Composite, *J. Mater. Sci.,* Vol 27, 1992, p 3085–3091
30. S.K. Sinha and S.K. Biswas, Effect of Sliding Speed on Friction and Wear of Uni-Directional Aramid Fibre-Phenolic Resin Composite, *J. Mater. Sci.,* Vol 30, 1995, p 2430–2437
31. J. Hanchi and N.S. Eiss, Jr., Tribological Behaviour of Polyetheretherketone, a Thermotropic Liquid Crystalline Polymer and in situ Composites Based on Their Blends Under Dry Sliding Conditions at Elevated Temperatures, *Wear,* Vol 200, 1996, p 105–121
32. B.J. Briscoe, P.D. Evans, E. Pelillo, and S.K. Sinha, Scratching Maps for Polymers, *Wear,* Vol 200, 1996, p 137–147
33. B.J. Briscoe, E. Pelillo, and S.K. Sinha, Characterisation of the Scratch Deformation Mechanisms for Poly(methylmethacrylate) Using Surface Optical Reflectivity, *Polym. Int.,* Vol 43, 1997, p 359–367
34. B.J. Briscoe, A. Chateauminois, T.C. Lindley, and D. Parsonage, Fretting Wear Behaviour of Polymethylmethacrylate Under Linear Motions and Torsional Contact Conditions, *Tribol. Int.,* Vol 31 (No. 11), 1998, p 701–711
35. B.J. Briscoe, A. Chateauminois, T.C. Lindley, and D. Parsonage, Contact Damage of Poly(methylmethacrylate) During Complex Microdisplacements, *Wear,* Vol 240, 2000, p 27–39
36. K. Tanaka and T. Miyata, Studies on the Friction and Transfer of Semi-Crystalline Polymers, *Wear,* Vol 41, 1977, p 383–398
37. B.J. Briscoe, P.D. Evans, and J.K. Lancaster, Single Point Deformation and Abrasion of γ-Irradiated Poly(tetrafluoroethylene), *J. Phys. D, Appl. Phys.,* Vol 20, 1987, p 346–353
38. A. Wang, D.C. Sun, C. Stark, and J.H. Dumbleton, Wear Mechanism of UHMWPE in Total Joint Replacement, *Wear,* Vol 181–183, 1995, p 241–249
39. R.W. Hertzberg and J.A. Manson, *Fatigue Engineering Plastics,* Academic Press, 1980, p 54–61
40. A.N. Gent and C.T.R. Pulford, Wear of Steel by Rubber, *Wear,* Vol 49, 1978, p 135–139
41. D. Dowson and P.D. Swales, An Elastohydrodynamic Approach to the Problem of the Reciprocating Seal, paper F3, *Proc. Third Int. Conf. Fluid Sealing,* British Hydromechanics Research Association
42. G.J. Field and B.S. Nau, Theoretical Study of EHL of Reciprocating Rubber Seals, *ASLE Trans.,* Vol 18, 1974, p 48–54
43. S.C. Richards and A.D. Roberts, Boundary Lubrication of Rubber by Aqueous Surfactant, *J. Phys. D, Appl. Phys.,* Vol 25, 1992, p A76–A80
44. A. Wang, A. Essner, V.K. Polineni, D.C. Sun, C. Stark, and J.H. Dumbleton, Lubrication and Wear of Ultra-High Molecular Weight Polyethylene in Total Joint Replacements, *New Directions in Tribology,* I. Hutchings, Ed., Mechanical Engineering Publications Ltd., London, 1997, p 443–458
45. A. Unsworth, The Effects of Lubrication in Hip Joint Prostheses, *Phys. Med. Biol.,* Vol 23, 1978, p 253–268
46. J.M. Senior and G.H. West, Interaction Between Lubricants and Plastic Bearing Surfaces, *Wear,* Vol 18, 1971, p 311–323
47. B.J. Briscoe, V. Mustafaev, and D. Tabor, Lubrication of Polyethylene by Oleamide and Stearamide, *Wear,* Vol 19, 1972, p 399–414
48. W.H. Carothers, U.S. Patents 2,130,947 and 2,130,948
49. S.M. Aharoni, *n-Nylons: Their Synthesis, Structure, and Properties,* John Wiley & Sons, 1997
50. C.J. Hooke, S.N. Kukureka, P. Liao, M. Rao, and Y.K. Chen, The Friction and Wear of Polymers in Non-Conformal Contacts, *Wear,* Vol 200, 1996, p 83–94
51. S. Bahadur, D. Gong, and J.W. Anderegg, The Role of Copper Compounds as Fillers in Transfer Film Formation and Wear of Nylon, *Wear,* Vol 154, 1992, p 207–223
52. A. Moet, Failure Analysis of Polymers, *Failure Analysis and Prevention,* Vol 11, *ASM Handbook,* American Society for Metals, 1986
53. L. Engel, H. Klingele, G.W. Ehrenstein, and H. Schaper, *An Atlas of Polymer Damage,* Prentice Hall, 1981

Wear Failures of Reinforced Polymers*

REINFORCED POLYMERS are used extensively in applications where resistance to adhesive and abrasive wear failure is important for materials selection. Polymers form a special class of materials because of their self-lubricity, which allows them to function without external conventional liquid lubrication. However, polymers also have some inherent tribological limitations, such as significantly low thermal conductivity, dissipativity, and diffusivity, as compared with metals. Frictional heat generated at the sliding contacts cannot be dissipated properly, and hence, flash temperatures at sliding contacts remain high. Their poor thermal stability also makes them more vulnerable due to loss of mechanical strength with an increase in the surface temperature. The thermal expansion coefficients of polymers are ten times greater than those of metals, posing problems related to dimensional clearances. In addition to the creeping tendency, polymers have low dimensional stability and rigidity. They have poor compressive strengths (approximately 30 times less) compared with those of other classes of tribomaterials. These inherent limitations restrict the utility of the polymers under severe operating conditions, such as high loads, speeds, and temperatures. Therefore, reinforcements (fibrous or particulate) generally are used to increase the load-carrying capacity, strength, resistance to creep, and wear. Limitations of strength and thermal conductivity can be overcome efficiently by the right selection of reinforcements and fillers in the appropriate amount, combination, and processing technology.

The tribological performance of reinforced polymers is governed by the type of base matrix, the nature of the filler (type, amount, size, shape, aspect ratio, distribution, orientation, combination with fillers, and the quality of bonding with the matrix), and the operating conditions. Fibers are far more wear resistant than the matrix and hence control the wear of the composite. Continuous fiber-reinforced composites with a thermoset-polymer matrix (such as phenolics, epoxy, etc.) may have low wear rates and higher strength than those with thermoplastics, because a higher ratio of fiber is achievable with a thermoset matrix. Incorporation of fillers also can modify wear resistance of polymers up to the order of 4. Solid lubricants mostly reduce wear and, rather essentially, friction coefficient. Similarly, reinforcement generally reduces wear but not always the coefficient of friction.

Generally, reinforced polymers have multiple or multifunctional fillers that can have synergistic and/or antagonistic interactions with respect to wear performance. Therefore, the tribological behavior of reinforced polymers is an empirical evaluation depending on specific material conditions, operating parameters, and environment. It is also ill advised to compare tribological properties of materials evaluated in different laboratories, configurations, environments, and so on. Their ranges or performance rankings are important for materials selection in a particular wear situation. Table 1 indicates tribological regimes for various kinds of reinforced polymers. Several review articles also provide background on the tribology of polymers and composites (Ref 1–11).

This article briefly reviews the abrasive and adhesive wear failure of:

- Particulate-filled polymers
- Short-fiber-reinforced polymers (SFRP)
- Polymers with continuous fibers
- Mixed reinforcements (either with fiber and filler, or with two types of fillers known as hybrid composites) and fabrics

Each section discusses various aspects, such as friction and wear performance of the composites, correlation of performance with various materials properties, and studies on wear of failure mechanisms by scanning electron microscopy (SEM).

Apart from the anisotropy of reinforced polymers (especially fiber-reinforced polymers, or FRP), synergistic and/or antagonistic effects in the case of a combination of two fibers or fillers is one of the most important aspects of composite tribology. This topic is briefly discussed, with some emphasis on various mathematical models, starting with the simple rule of mixtures. More complex methods developed for describing the wear performance of each type of composite are also available, but such models, along with the data on wear mechanisms and friction and wear performance, serve more for tailoring future composites. This subject is mentioned only briefly because it is beyond the main purpose of characterizing wear failures from the perspective of failure analysis and prevention.

Abrasive Wear Failure of Reinforced Polymers

Polymer composites are extensively used for sliding components in earth-moving equipment, rock and ore crushers, dies in powder metallurgy, extruders and chutes, and so on, where the major wear failure mechanism is either two-body or third-body abrasion. The abrasion is a

Table 1 Tribopotential of polymers and composites for a variety of applications

Composite material	Tribological applications	Maximum tribological regime
Neat and short-fiber-reinforced composition (SFRP)	Seals, goats, slideways bearings, and abrasive wear application	$PV < 15$ MPa · m/s $V < 5$ m/s, $\mu > 0.03$ $T < 250$ °C $W_S > 10^{-16}$ m^3/Nm
Continuous-fiber-reinforced composites (UD)	Underwater or high-temperature applications, netrospace scals and bearings	$PV < 100$ MPa · m/s $V < 5$ m/s $T < 320$ °C, $\mu > 0.09$ $W_S > 10^{-17}$ m^3/Nm
Thin-layer composites with metallic supports	Pivot bearings, high-pressure applications	$PV < 300$ MPa · m/s $V < 1$ m/s $T < 320$ °C, $\mu > 0.06$ $W_S > 10^{-18}$ m^3/Nm

UD, unidirectional; P, pressure; V, sliding speed; W_S, specific wear rate; T, temperature; μ, coefficient of friction. Source: Ref. 1

*Adapted from the article by J. Bijwe, "Wear Failures of Reinforced Polymers," in *Failure Analysis and Prevention,* Volume 11, *ASM Handbook,* ASM International, 2002, p 1028–1044

net result of microscopic interactions of the surface and the abradant, as shown in Fig. 1 (Ref 12). Microplowing and microcutting are the dominant processes in the abrasion of ductile materials, while microcracking is important in brittle materials.

As previously noted, the wear of reinforced polymers is influenced by the properties of the matrix, the filler, and their bonding; the ratio of grooving depth to the filler size; the shape and size; orientation distribution; hardness of the grit and filler; and the operating parameters. In the case of abrasive wear of reinforced polymers, the role of the filler is very much different from the role of the filler in adhesive wear mode. Abrasive wear behavior generally is evaluated by abrading it under load against hard, rough surfaces, such as paper or wheel impregnated with silicon carbide (SiC), flint, alumina (Al_2O_3), and so on, or a rough metallic disc.

Abrasive wear failure of FRPs generally occurs because of matrix shearing and fiber debonding followed by fiber cracking and cutting. Generally, fibers in the normal (N) direction are more wear resistant than those in the parallel (P) direction, which, in turn, are better than those in the antiparallel (AP) direction. If the size of the filler is smaller than the abrading grit, then the filler particle is easily dislodged and dug out. Fillers of medium size are most effective in this case. With increase in volume fraction (V_f) of the filler or fiber, wear may increase, decrease, or show minima at some concentration, generally approximately 20 to 30%. Abrasive wear may increase with a decrease in modulus of elasticity of fiber or matrix as a result of higher debonding of fibers. If the filler hardness is higher than the abrading grit, wear decreases. If the filler or matrix is brittle, wear increases due to cracking and flaking (Ref 2).

Abrasive Wear of Particulate-Reinforced Polymers. Extensive work has been done in this area (Ref 13). Figure 2 shows relative abrasive wear loss of quartz- and glass-filled polymethyl methacrylate (PMMA) slid against SiC, SiO_2, and $CaCO_3$ abrasives as a function of filler volume fraction. The performance clearly depends on the ratio of hardness of the abrasives to the filler, the interfacial adhesion of the filler with the matrix PMMA, and the volume fraction.

An upper bound on abrasive wear resistance is modeled by the following equation, known as the inverse rule of mixtures (IROM), where overall wear behavior is assumed to be a function of the individual contribution from each phase. Wear resistance (W^{-1}) (as modeled by Khruschov) (Ref 14) is:

$$W^{-1} = \sum V_i W_i^{-1} \quad \text{(Eq 1)}$$

where W is the total wear volume, as a linear function of volume fraction of the phase present, V_i is the volume fraction of the ith phase, and W_i is the wear volume due to ith phase. Equation 1 provides an upper bound to the wear resistance. However, the basic assumption that each phase shows a wear rate proportional to the applied load is not always correct, especially when hard phases and ceramic fillers are involved. IROM cannot be applied in such a case, because of nonlinearity in wear load relation. Hence, another model, known as the linear rule of mixtures (LROM) (Eq 2), was developed (Ref 15) and is suitable for composites containing a combination of similar phases:

$$W = \sum V_i W_i \quad \text{(Eq 2)}$$

These models, based on volume fraction of the reinforcing phases, showed considerable deviation from the experimental values, especially when the size of the abrasives was large and the load was high. As seen in Fig. 3 (Ref 16), the deviation clearly increased with the size of the filler and applied load. In reality, the abrasive wear of a multiphase system is the macroscopic sum of all the microscopic events generated by the abradants and hence depends on the size of the grain. When grain size equals or exceeds the microstructure, filler pullout is inevitable, leading to large wear to the extent depending on the quality of the interface. The deviation depends on the size of the grain, microstructure, load, and volume fraction. The

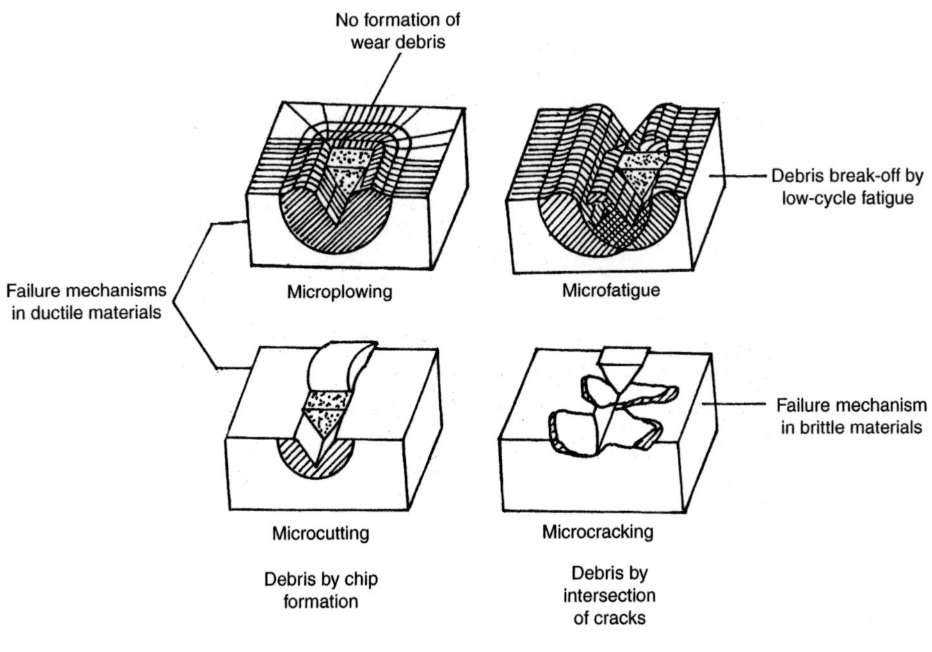

Fig. 1 Schematic of different interactions during sliding of abrasive particles against the surface of material. Source: Ref 12

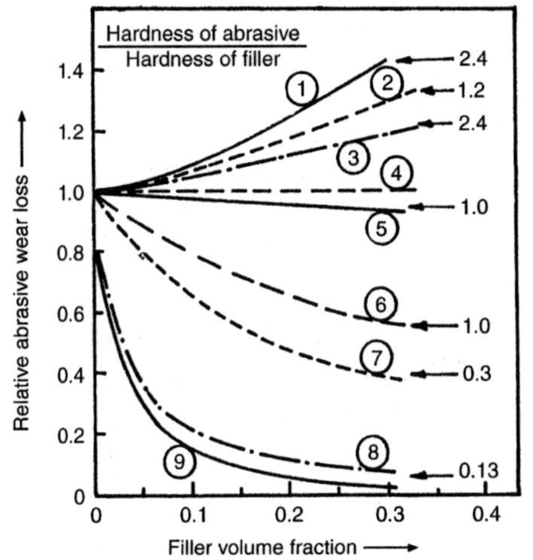

Fig. 2 Relative abrasive wear loss of polymethyl methacrylate (PMMA) and composites filled with quartz and glass against abrasives SiC (45 μm), SiO_2 (10 μm), and $CaCO_3$ (3 μm) as a function of filler volume fraction, V_f. WIB, weak interfacial bond; SIB, strong interfacial bond. 1, WIB-quartz filler against SiC; 2, WIB-glass filler against SiO_2; 3, SIB-quartz filler against SiC; 4, unfilled PMMA; 5, WIB-quartz filler against SiO_2; 6, SIB-quartz filler against SiO_2; 7, WIB-glass filler against $CaCO_3$; 8, WIB-quartz filler against $CaCO_3$; 9, SIB-quartz filler against $CaCO_3$. Source: Ref 13

rules of mixture are not useful in the quantitative prediction of industrial wear rates but for the development of wear-resistant materials in the laboratory.

Researchers (Ref 17) developed a theoretical model for the optimal filler loading based on the random packing model of particles in a polymer. Using the computer program based on this model and the data on filler particle size distribution with the density of filler and polymer and the distance between the particles of the filler in the critical packing state, the filler proportion in the polyamide (PA) 11 was calculated. The maximum mechanical strength and highest abrasive resistance of the composites with optimal filler contents calculated with the equation agreed well with the experimental data.

Abrasive Wear of SFRPs. Figure 4 indicates the effect of size, orientation, hardness, modulus, and brittleness of the fiber on the abrasive wear performance of the composites (Ref 2). The performance of the composites generally deteriorates in the case of SFRPs. An increase in wear performance was reported for seven composites and deterioration for six in the case of thirteen polymers reinforced with 30% short carbon fibers (CF) (Ref 18). Generally, a reduction in *Se* factor or *HSe* factor (where *S* is ultimate tensile strength, *e* is the elongation to break, and *H* is hardness) is observed to be responsible for the deterioration in performance, and Ratner-Lancaster plots show good correlation (Ref 8, 18, 19). Table 2 shows several properties correlated with the wear behavior of such composites by various researchers.

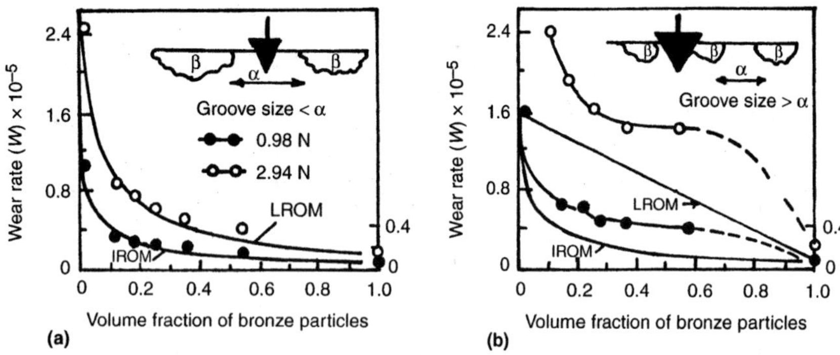

Fig. 3 Dimensionless wear rate, *W*, at two loads as a function of volume fraction of bronze particles in epoxy-Cu-Al system (Cu-Al particle diameter ≈100 μm). Abrading surface, silicon carbide. (a) Fine grade; (b) coarse grade. α, space between two particles, β. LROM; linear rule of mixtures; IROM, inverse rule of mixtures. Source: Ref 16

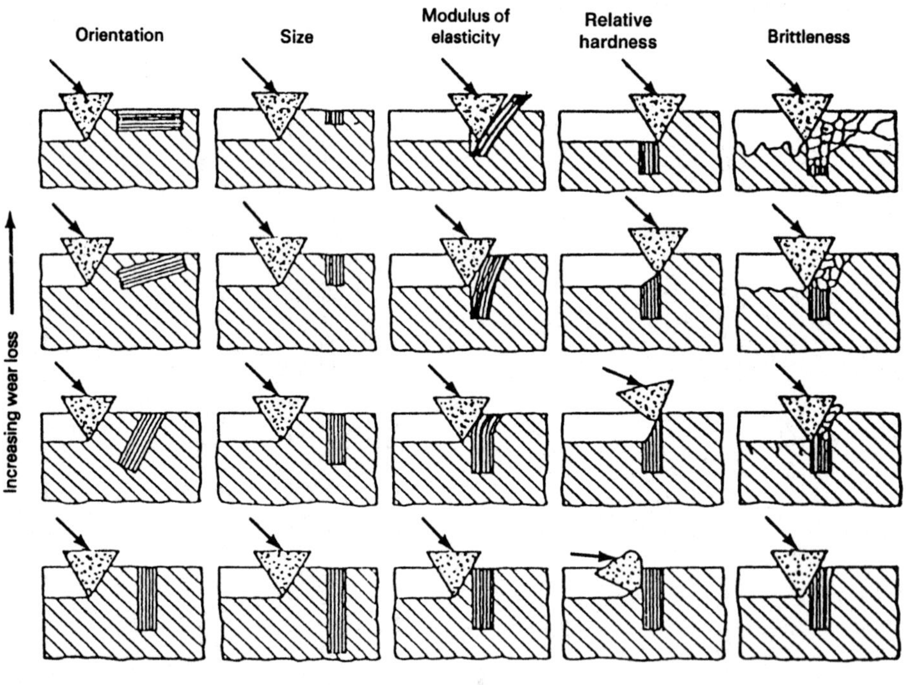

Fig. 4 Influence of various properties of reinforcing phase on abrasive wear of composite. Source: Ref 2

The maxima in the specific wear rate/volume fraction relation also depends on the size of abrasives. As seen in Fig. 5 (Ref 29), the 10% glass fiber (GF) loading in polyether-imide (PEI) showed maximum wear for the abrasive grain size of ≅118 μm, while for 175 μm size grits, loading was at 20%. The wear minima in both cases, however, was at 30% loading of short GF. The difference in the severity of the fiber damage in the 10% PEI composite due to these different grit sizes can be seen in Fig. 6(a) and (b). The various stages of fiber cracking, cutting, and pulverization in 30% GF composite are shown in Fig. 6(c). Generally, the wear rate of SFRPs shows increase with fiber volume fraction. The performance, however, deteriorates disproportionately if the combination of filler, solid lubricants, and fibers is included in the composite (Ref 8).

A wear equation was developed for SFRPs based on the crack propagation theory for describing the effect of load (F_N) on specific wear rate (W_S) (Ref 30):

$$W_S = K_6 \frac{V_S V_C}{EH\varepsilon_f \mu_\alpha F_N^\delta} \quad \text{(Eq 3)}$$

where K_6 is a parametric constant, V_S is sliding speed, E is elastic modulus, H is hardness, and ε_f is wear failure strain. The equation is valid only when thermal activation is insignificant, as in the case of low sliding speed. μ_α is the number between unity and zero; V_C, the crack growth velocity, is related to fiber aspect ratio, elastic modulus of fiber, shear modulus of the matrix, stress of friction on damaged matrix-fiber interface, fiber fraction, and many other complex characteristic properties. If F_N is large, thermal effect becomes significant, and the exponential term in the following equation controls the wear rate:

$$W_S = K_9 \frac{V_S F_N^{2/(\beta-5)}}{EH\varepsilon_f \mu_\alpha} \exp \frac{-\gamma_2}{F_N} \quad \text{(Eq 4)}$$

Table 2 Details of the literature on (abrasive) wear property correlation of polymers and composites

Resin(a)	Fiber/filler(b), wt%	Wear rate due to filler	Correlation of abrasive wear with(c)	Ref
PA 5, PA 66, PVC, PTFE, PP, EP, PMMA, polyester, PC, phenolic, PE, acetal copolymer	CF/30	Increased for six polymers and decreased for seven polymers	$(Se)^{-1}$	18
PA 66	CF/10, 20, 30, and 40 PTFE/15	Increased	$(Se)^{-1}$	19
PA, 6, PMMA, PE, POM, PA 66, PP, PTFE, PC, PTFCB, polyphenylene oxide, and PVC	...	...	Cohesive energies	20
PA 6, PTFE, PP, POM, HDPE, PVC, and PMMA	...	...	Plowing component of friction, hardness, tensile strength, and elongation to failure of polymers	21
PET	GF/30 Glass/30 Sphere	Increased	The hardness, macrofracture energy, and the probability factor for microcracking	22
PA 6	GF/15 and 20 PTFE/3 Bronze and copper powder/6	Increased	$(HSe)^{-1}$, fracture toughness, fracture energy, and durability factor	23
PTFE	...	...	$(S^2e)^{-1}$	24
PTFE	...	...	$(S_{max}S_y)$ and disperity strain (r/R)	25
PA 66, PPS, PEEK, PC, PES, and PEI	CF/30 CF/30 and 40	...	Roughness of abrasive paper, $(Se)^{-1}$ and (ESe^{-1})	26
EP and PEEK	CF, GF, and AF/60	...	Asperity size	27
ABS, PA, PE, PP, PS, and POM	...	...	(Depth of wear groove $\times f_{ab}$ value)$^{-1}$	2
PES and PMMA	...	...	K_{Ic}	28
PEI	GF/16, 20, and 25 PTFE/15 Graphite and MeS_2/15	Increased	$(Se)^{-1}$	8
PI	Graphite/15 and 40 PTFE/15 MoS_2/15	Increased	$(Se)^{-1}$	8

(a) PA, polyamide; PVC, polyvinyl chloride; PTFE, polytetrafluoroethylene; EP, epoxy; PMMA, polymethyl methacrylate; PC, polycarbonate; PE, polyethylene; POM, polyoxymethylene; PP, polypropylene; PTFCE, polytrifluorochloroethylene; HDPE, high-density polyethylene; PET, polyethylene terephthalate; PPS, polyphenylene sulfide; PEEK, polyetheretherketone; PES, polyether sulfone; PEI, polyether-imide; ABS, acrylonitrile-butadiene-styrene; PS, polystyrene; PI, polyimide. (b) CF, carbon fiber; GF, glass fiber; AF, aramid fiber. (c) S, ultimate tensile strength; e, elongation to break; H, hardness; E, elastic modulus; F_{ab}, the ratio of volume of material removed as wear debris to the volume of the wear groove; K_{Ic}, fracture toughness

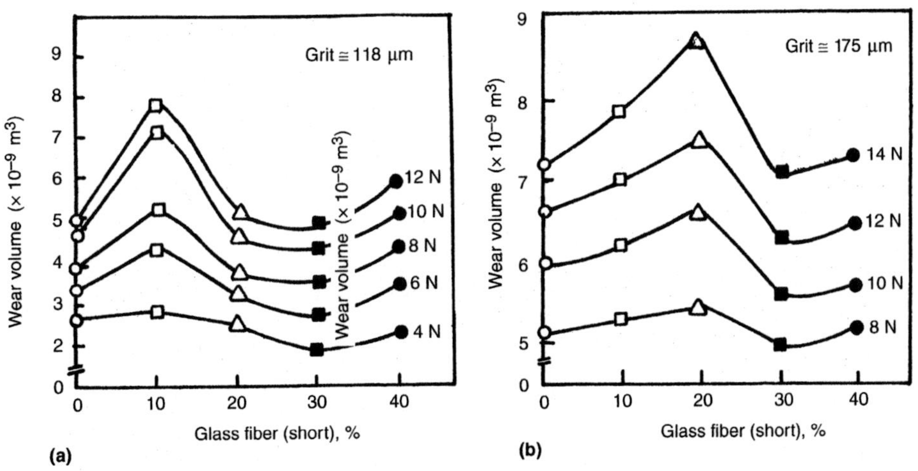

Fig. 5 Abrasive wear volume at various loads and SiC abrasive papers as a function of volume fraction of short glass fibers (GF) in polyether-imide. Speed, 5 cm/s in single-pass condition; distance slid, 3.26 m. (a) 120 grade, grit size ≃118 μm. (b) 80 grade, grit size ≃175 μm. Source: Ref 29

where K_9 is a parametric constant, and γ_2 and β are dimensionless constants. When V_S is high, material softens, causing crack propagation to be easier than in the case of low V_S. Hence, depending on the magnitude of V_S, wear rate increases, decreases, or remains constant. From the F_N^{δ} term, W_S decreases with F_N when $\delta > 2/\beta$, because $V_C \propto F_N^{2/\beta}$. When V_C becomes thermally activated, then, at a certain point, the effect of V_C can override the F_N^{δ} term, and thus, W_S increases with F_N.

A wear model was developed for correlating wear behavior of various types of composites with the materials properties (Ref 31). As seen in Fig. 7, surface deformation and wear mechanisms are functions of hardness and fracture energy of the matrix. When pressure, P, increases a certain critical value, P_{crit}, the contribution due to brittle fracture of the material to the wear increases during the transition in wear mechanism at certain volume fraction. The fracture toughness of a material (K_{Ic}) can be correlated to H as:

$$P_{crit} \propto \frac{K_{Ic}^2}{H} \quad \text{(Eq 5)}$$

If load is high, P_{crit} is approached earlier. For a given apparent contact area (A), effective pressure (P_{eff}) is related to effective contact area (A_{eff}) as:

$$P_{eff} \propto \frac{PA}{A_{eff}} \quad \text{(Eq 6)}$$

If the material has high hardness, low fracture toughness, sharp grains, or rough counterface, the probability of P_{eff} reaching P_{crit} is high (Fig. 7a). Wear rate (W) as a function of hardness shows a change in wear mechanism, as shown in Fig. 7(b). Transition II shows a disproportionately higher wear rate with an increase in P (for a material of hardness H). Transition I represents reduction in P_{crit} ($P < P_{crit}$ to $P > P_{crit}$) due to increase in H at constant pressure, P, leading to additional microcracking events. When load increases and H also increases due to higher V_f, the same roughness of the abradant becomes more detrimental, and wear dominated by microcracking becomes more prominent. Taking into account various factors, such as probability factor (Ω^*) of microcracking, hardness and size of abradant, and particle density/area, specific wear rate, W_S, was correlated with H of the composite, modified wear coefficient (Ω), and fracture energy, G_{Ic} (Ref 31).

Abrasive Wear of Continuous Unidirectional FRPs. In-depth studies on continuous steel fiber-reinforced epoxy (EP) polymer and polymethyl methacrylate (PMMA) (Ref 2) focused on various aspects such as area fraction, volume fraction, mean free path between the fibers (λ), and the ratio λ/D, where D was the size of abrasives. Based on these studies, the general trends in abrasive wear of FRP composites and the properties and alignment of fibers were summarized (Fig. 4) (Ref 2). Studies were done on abrasive wear performance of FRP composites of EP (thermoset polymer) and polyetherether-

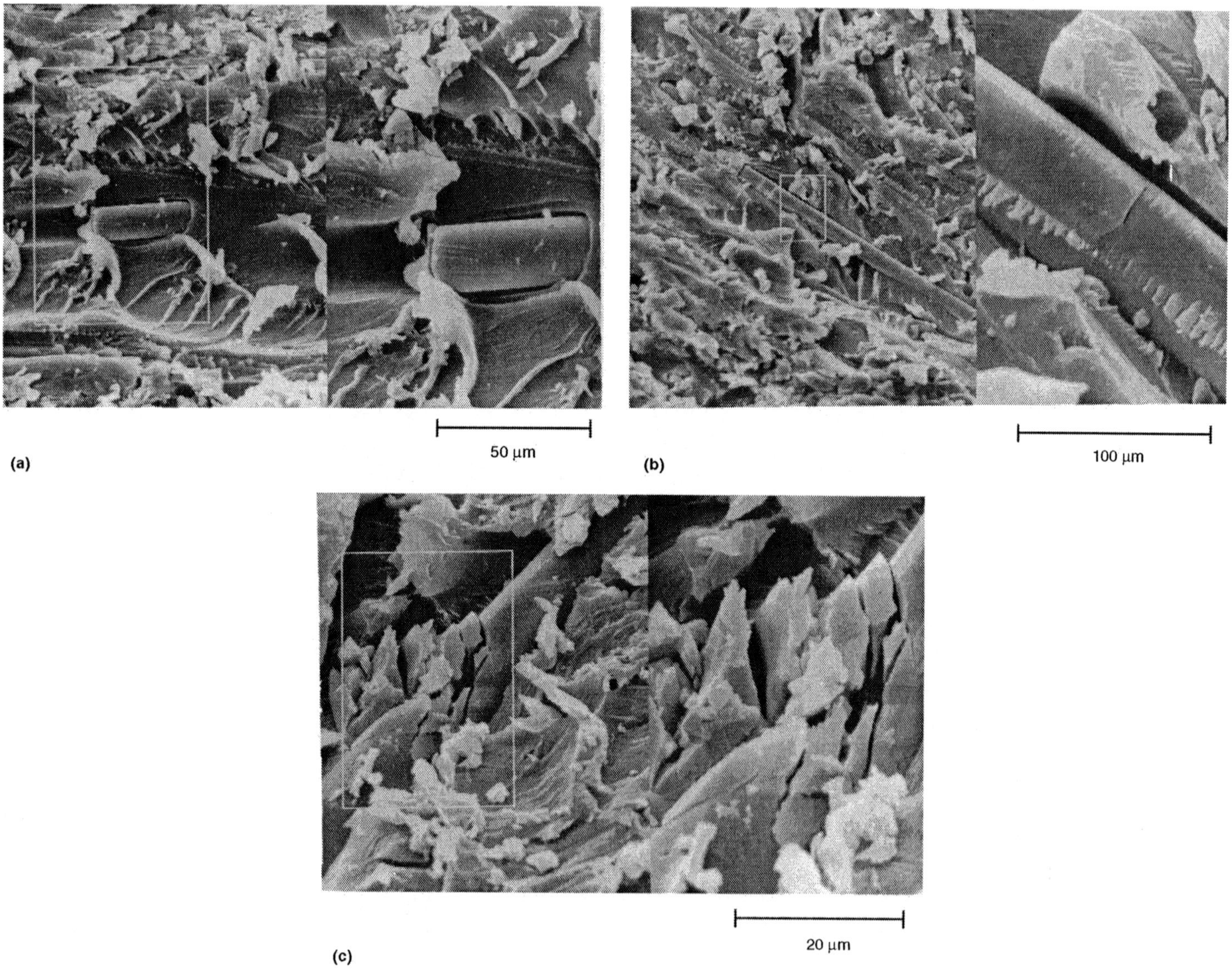

Fig. 6 Scanning electron micrographs of abraded surfaces of composites against 80-grade SiC paper and under 14 N load. (a) Polyether-imide (PEI) + 10% glass fiber (GF) showing extensive damage to matrix and fiber; cavities left after fiber consumption. PEI + 30% GF. (b) Fiber on the stage of microcracking. (c) Initiation of fiber pulverization. Source: Ref 29

ketone (PEEK) (thermoplastic polymer) reinforced with various fibers, such as AS4 carbon fiber (CF) (62 vol%), glass fiber (GF) (58 vol%) and K49, aramid fibers (AF) in epoxy and AS4CF (55 vol%) and K49, and AF (60 vol%) in PEEK (Ref 32, 33).

A summary of the wear behavior (Table 3) indicates clearly that unidirectional (UD) fiber reinforcement in the antiparallel (AP) orientation was detrimental in all the cases, while normal orientation (O_N) was always beneficial. Aramid fiber was most effective in improving the resistance when it was in O_N. Interestingly, parallel orientation of AF was worse than the AP orientation in the case of PEEK. The various wear mechanisms suggested in different orientations are schematically shown in Fig. 8(a), while Fig. 8(b) shows the ideal composite with very good abrasive wear resistance based on these studies. It should contain a thermoplastic polymer such as PEEK, and continuous AFs in the O_N and CFs in the parallel orientation (O_P) (Ref 32). A cyclic wear model based on volume fraction of fiber, wear resistance, and elastic moduli of its constituents shows a very good matching of experimental data with the calculated one (Ref 34). The wear model developed for FRPs (Ref 31) also shows good correlation for CF-reinforced UD composite of epoxy matrix.

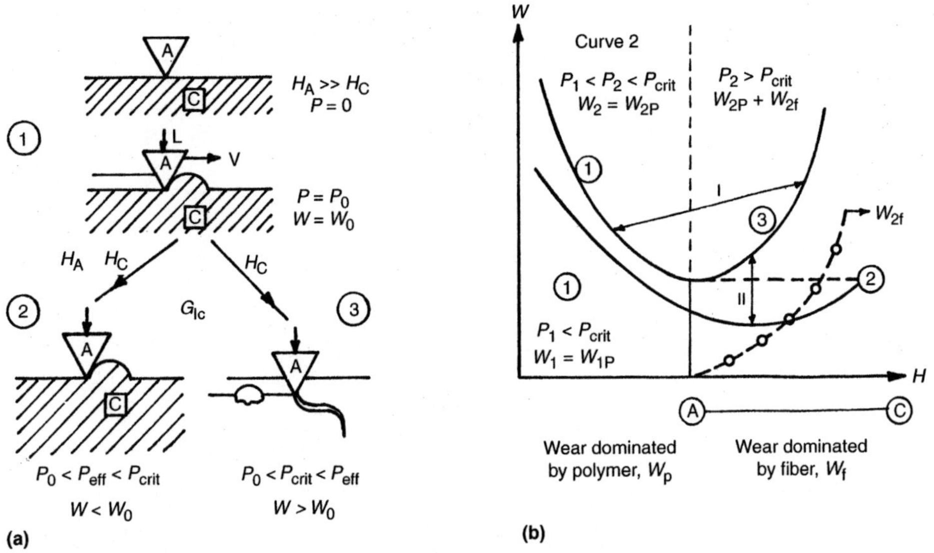

Fig. 7 (a) Abrasive wear mechanisms and surface deformation as a function of pressure, *P;* material hardness, *H;* and fracture energy, G_{Ic}. *L,* normal load; *V,* velocity. (b) Curves 1 to 3 correspond to the schematic in (a), possible schematic of the wear rate, *W,* as a function of hardness, *H,* of wearing material. Curve 1 is a normal curve showing a reduction in wear with increase in hardness, while curve 2 reflects changes in trends when microcracking. W_f plays a role at higher pressure. Source: Ref 31

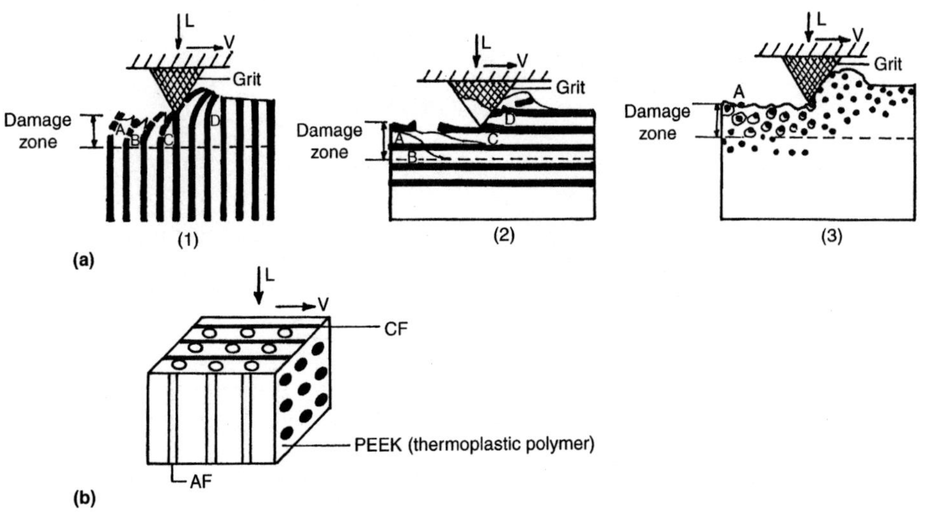

Fig. 8 (a) Schematic of basic wear failure mechanisms observed in parallel, P (a1) (a2), and antiparallel, AP (a3), orientations. (a1) A, fiber slicing; B, fiber-matrix debonding; C, fiber cracking; and D, fiber bending (especially in the case of aramid fiber, AF, or carbon fiber, CF). (a2) A, interlaminar crack propagation; B, fiber cracking; C, fiber-matrix debonding; and D, fiber fracturing. (a3) A, fiber fracturing. (b) Ideal composite for high abrasive wear resistance. *L,* normal load; *V,* velocity; PEEK, polyetheretherketone. Source: Ref 32

Abrasive Wear of Fabric-Reinforced Polymer Composites. Very limited research work is done on the abrasive wear behavior of bidirectionally (BD) reinforced composites. The influence of various fabrics and their orientations on the abrasive wear behavior of composites of thermoplastic PEI is covered in Ref 35. The weave of glass fabric and load were also influencing factors. As compared with short glass fibers (Ref 29), fabric proved to be significantly beneficial for enhancing the abrasive wear resistance of PEI. Worn surfaces of PEI_{AF} (O_P); PEI_{AF} (O_N); PEI hybrid (PEI_{HY}) (O_N), PEI_{CF} (O_P); and PEI_{CF} (O_N) are shown in Fig. 9 (Ref 36). The extensive softening of AF due to high contact pressure, especially in O_N followed by fibrillation, was the most distinct feature observed on PEI_{AF} and PEI_{HY} surfaces. The fiber cracking breakage, pulverization, removal, and plastic deformation of the matrix were minimal in the case of AF composites. The presence of AF hindered the removal of PEI matrix also. In fact, the smooth topography looked more like adhesive wear case rather than the abrasive wear against SiC paper of 175 μm size. In other cases, excessive breakage and removal of fibers, which led to higher wear, were observed. The influence of amount and type of fiber fraction in BD composites and fiber fraction orientation on abrasive wear behavior by sliding the EP composites against 70 μm Al_2O_3 paper has been reported (Ref 31).

Table 3 Influence of fiber orientation on the abrasive wear behavior of continuous fiber-reinforced polymer composite

	UD reinforcement and wear resistance (normalized) (W_S^{-1}) composite/(W_S^{-1}) epoxy			
Polymer	CF	AF	GF	Neat polymer
EP	...	...	...	1.0
N	1.8	7.9	2.3	...
P	1.7	1.1	1.4	...
AP	0.9	1.1	0.8	...
PEEK	...	...	...	1.8
N	1.7	15.2	2.4	...
P	1.6	0.9	1.8	...
AP	1.0	1.3	1.2	...

UD, unidirectional; W_S wear rate; CF, carbon fiber; AF, aramid fiber; GF, glass fiber; EP, epoxy; N, normal; P, parallel; AP, antiparallel; PEEK, polyetheretherketone. Source: Ref 33

Sliding (Adhesive) Wear Failure of Polymer Composites

The polymer composites that are used for sliding wear applications, such as bush bearings, bearing cages, slides, gear seals, and so forth, in industries such as textile, food, paper, pharmaceutical and such are particulate-filled, fiber-reinforced, or mixed composites. In the case of particulate fillers, the particle size is very important for achieving desired performance. The fillers such as ZrO_2, SiC, and so forth are known for their hardness and beneficial effects on abrasive wear resistance of a composite. However, they have a detrimental effect on adhesive wear of polymer composites. Interestingly, the same fillers have proved to be very beneficial when the size is in nanometers (Ref 37–40).

The performance of FRP composites depends on the type of fiber and matrix, volume fraction, distribution, aspect ratio, alignment, and adhesion to the matrix. In accordance with Eq 7, the higher the aspect ratio (l/r, where l and r are the length and radius of fiber, respectively), the greater is the contact load transferred from the matrix to the fiber and the greater the wear resistance, W_R (inverse of wear rate) (Ref 41):

$$\sigma_f = 2\tau l r^{-1} + \sigma_m \qquad \text{(Eq 7)}$$

where σ_f is the contact stress, σ_m is the compressive stress of the matrix in the composite loaded against counterface under a load, L, and τ is the tangential stress produced because of the difference in the moduli of matrix and fiber. It is, however, not true that with increase in the concentration of fibers, wear resistance increases continuously. In fact, either it deteriorates or becomes constant beyond a typical optimal concentration in the case of short fibers.

Generally, short fibers (0.1 mm to 3 mm, or 0.004 to 0.118 in.) in approximately the 20 to 30% concentration range are used for reinforcement and reducing wear in thermoplastics.

Figure 10 (Ref 2) highlights some trends generally observed in the wearing of composites against a smooth metal. Increase in load and speed results in higher wear of FRP through different mechanisms. High load results in more fiber cracking and pulverization, leading to deterioration in load-carrying capacity, while high speed accelerates the debonding of fibers/fillers. This results in easy peeling off or pulling out of the reinforcing phase. High-modulus fibers are more effective in wear reduction than the high-strength fibers. Moreover, the higher the modu-

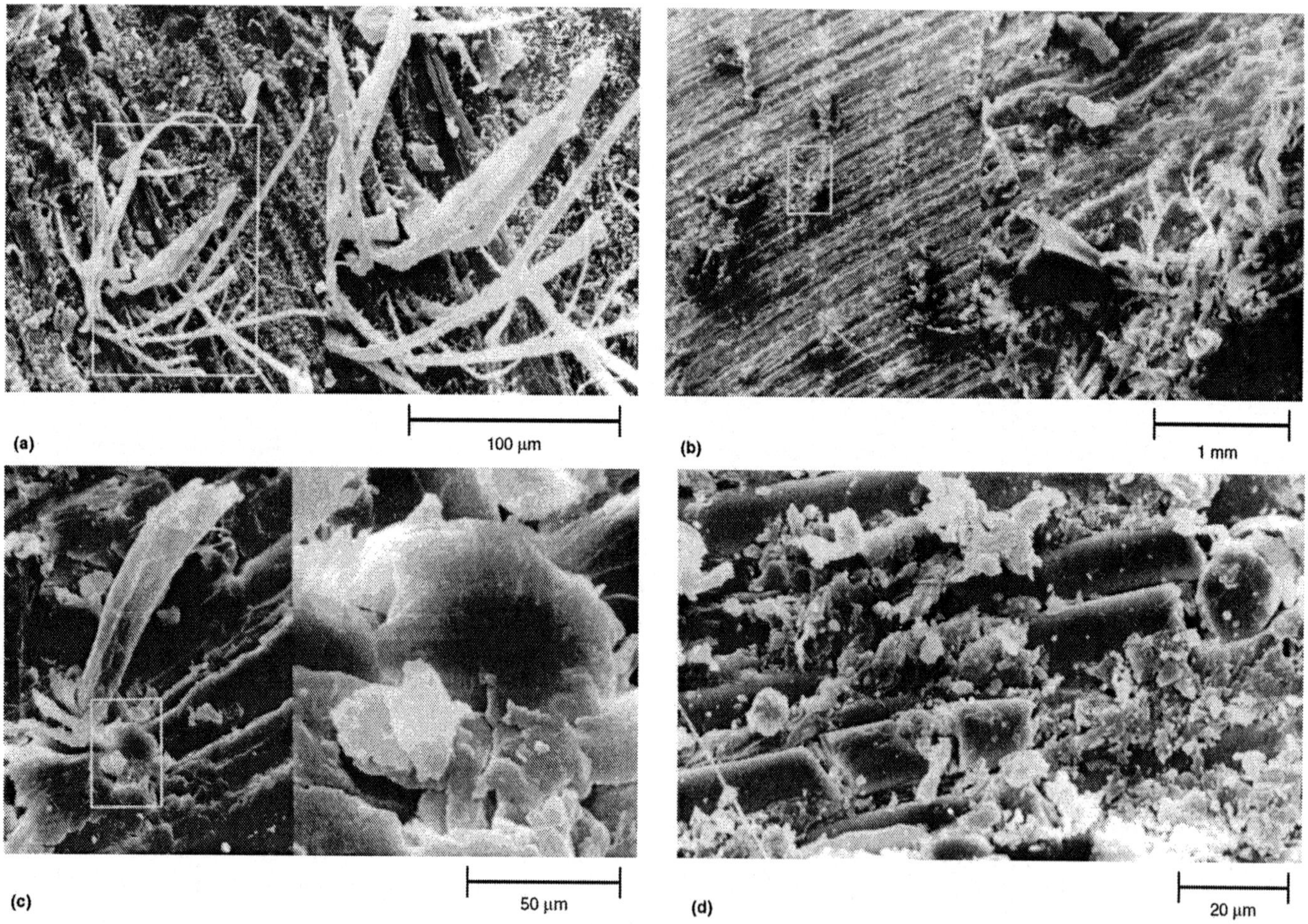

Fig. 9 Scanning electron microscope micrographs of abraded polyether-imide (PEI) composites reinforced by various fabrics; normal load, 12 N; SiC paper, 80 grade (grit size, 175 μm); distance slid, 10 m (33 ft). O_P, fabric parallel to the sliding plane; O_N, fabric normal to the sliding plane; AF, aramid fiber; CF, carbon fiber; HY, hybrid; GF, glass fiber. (a) PEI_{AF} (O_P) showing extensive elongation and fibrillation of ductile and soft AF during abrasion (b and c) for PEI_{AF} (O_N). (b) Smooth surface topography due to molten AF (high contact pressure, half the fibers being in normal direction). (c) Enlarged view of AF tip indicating extensive elongation and melting. (d and e) $PEI_{CF+AF(HY)}$. (d) PEI_{HY} (O_P) abraded from CF side showing multiple microcutting in CF. (e) PEI_{HY} (O_N) showing excessive melting of AF and third-body abrasion due to loose grit (middle portion) on the softened matrix. (f) PEI_{CF} (O_P) excessive breakage of an array of CF in both directions, resulting in high wear. (g) PEI_{CF} (O_N) CF tips showing less fiber damage (and hence less wear); cavities due to fiber pull-out. (h) PEI_{GF} (O_P) excessive damage to GF in both directions due to microcutting. Source: Ref 36

lus of fiber or composite, the less is the wear. The higher the aspect ratio of the reinforcement, the lower is the wear. For a typical volume fraction of a filler, there exists an optimal value of the mean free path for the minimum wear; the same is true for the filler size. High strength and high elastic modulus of the matrix enhance the support to the fillers. The higher the brittleness of the matrix, the higher the crack propagation tendency and the higher the wear of a composite.

Apart from these factors, the most important wear-controlling factor in the case of a polymer composite is the efficiency of interaction of the polymer and filler with the counterface. If the interaction reinforces the thin-film transfer efficiency of the polymer, the friction coefficient (μ) and the extent of frictional heating reduce, leading to less damage to the polymer, filler, and their adhesion. This generally results in less wear. The nature of transferred film on the counterface plays a key role in controlling wear performance of a composite. If this adheres to the counterface firmly, wear of the composite decreases. If the fillers are capable of enhancing this adhesion by forming chemical bonds through chemical or physical interaction with the counterface during sliding, the film is more firmly attached, resulting in significant reduction in wear.

Adhesive Wear of Particulate-Filled Composites. The influence of nanometer-sized particulate fillers such as SiC, ZrO_2, and Si_3N_4 has been studied in PEEK (Ref 37–40). At particular concentrations and sizes of the filler, minimum wear rate (K_0) and minimum μ were observed. However, minima of K_0 and μ did not always match. For example, these matched in the case of ZrO_2 at 7.5% (K_0 decreased by 1.8 times and μ by 1.3 times). For Si_3N_4, K_0 decreased continuously with increased filler amount. The μ, however, was minimal at 8 vol% loading (1.5 times decrease in μ and 7 times decrease in K_0). In a further work on simultaneous addition of SiC filler (3.3 vol%, which proved to be the best for maximum reduction in the wear rate and μ) and solid lubricant polytetrafluoroethylene (PTFE) in increasing amounts up to 40 vol%, a synergistic effect was observed.

Figure 11 (Ref 40) shows the influence of PTFE on μ and K_0 of PEEK and PEEK + SiC (3.3 vol%) composite. It was interesting to observe that at up to 5% PTFE loading in a PEEK-SiC composite, μ rose to a high value, showing a negative contribution of PTFE toward μ. Without PTFE, μ was quite low (Fig. 11b). Beyond 10% loading, a combination of PTFE and SiC showed synergism, and μ lower than that with the individual fillers was observed. Thus, this combination showed antagonism and synergism in particular ranges. K_0, however, was lowest without PTFE in the PEEK-SiC composite. A combination of PTFE with SiC thus showed a negative effect for wear behavior. Interestingly, inclusion of the single filler PTFE proved beneficial. These studies

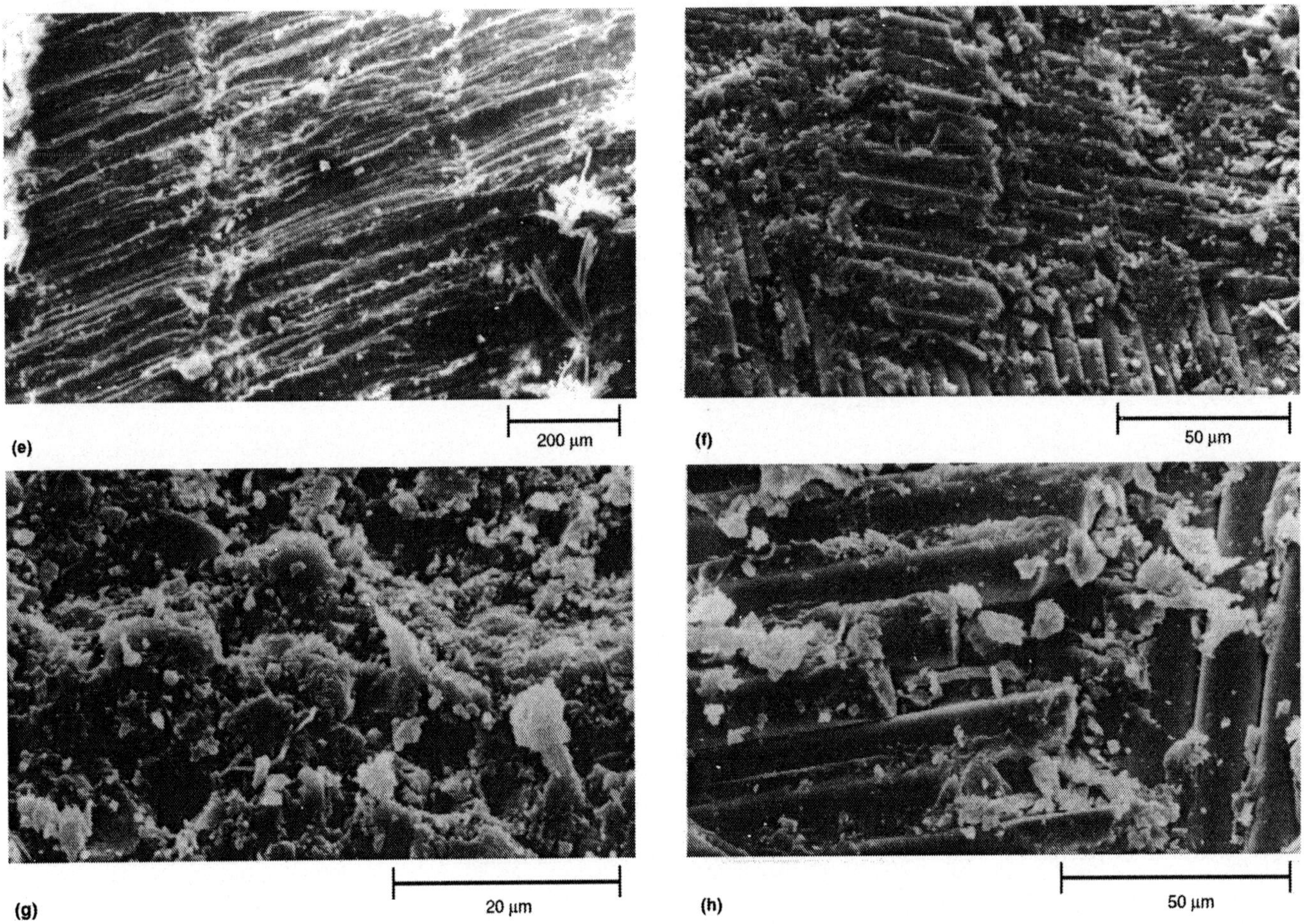

Fig. 9 (continued) (e) PEI_{HY} (O_N) showing excessive melting of AF and third-body abrasion due to loose grit (middle portion) on the softened matrix. (f) PEI_{CF} (O_P) excessive breakage of an array of CF in both directions, resulting in high wear. (g) PEI_{CF} (O_N) CF tips showing less fiber damage (and hence less wear); cavities due to fiber pullout. (h) PEI_{GF} (O_P) excessive damage to GF in both directions due to microcutting. Source: Ref 36 brought out a very important aspect of the influence of solid lubricant. It worsened the performance of a composite when combined with other potential fillers. This detrimental effect was due to the formulation of SiF_x, confirmed during x-ray photoelectron spectroscopic analysis, due to chemical reaction between SiC and PTFE. This SiF_x was responsible for raising the μ of a composite. When PTFE contents exceeded the amount required for the chemical reaction with 3.3 vol% SiC, the excess unreacted PTFE started showing a beneficial effect, and μ started decreasing. The film transfer efficiency of the filler was poor when PTFE and SiC were in combination, and this led to deterioration in wear performance.

The influence of increasing the amount of PTFE in PEEK (Ref 42, 43) (Fig. 12a) showed that the PTFE significantly benefited both μ and the specific wear rate of PEEK. However, with an increase in PTFE, although μ decreased continuously, K_0 showed excessive increase beyond 80% loading of PTFE. K_0 was minimal at 5% PTFE contents, while μ was minimal for 100% PTFE. A 12 to 18% loading range of PTFE was found to be optimal for the friction and wear combination.

Wear rates (W_C) of PEEK-PTFE composite have been described (Ref 42, 43) as follows:

$$W_C = (1 - V_{fL})W_M + V_{fL}W_L \quad \text{(Eq 8)}$$

where V_{fL} is the volume fraction of lubricant PTFE, and W_M and W_L are the wear rates of the PEEK matrix and lubricant PTFE. Figure 12(b) shows the synergistic effect of the lubricant on wear. This could not be explained with the help of linear correlation. Such synergism could be described as:

$$\frac{1}{W_C} = (1 - V_{fL})\frac{1}{W_M^*} + V_{fL}\frac{1}{W_L^*} \quad \text{(Eq 9)}$$

where W_M^* indicates wear rate in the presence of PTFE:

$$W_M^* = W_M \cdot f \quad \text{(Eq 10)}$$

where f is the lubricating efficiency factor, and W_M^* is the effective wear rate of the matrix (Ref 4).

Adhesive Wear of SFRP or Mixed (SFRP + Particulate-Filled) Composites. Short fibers

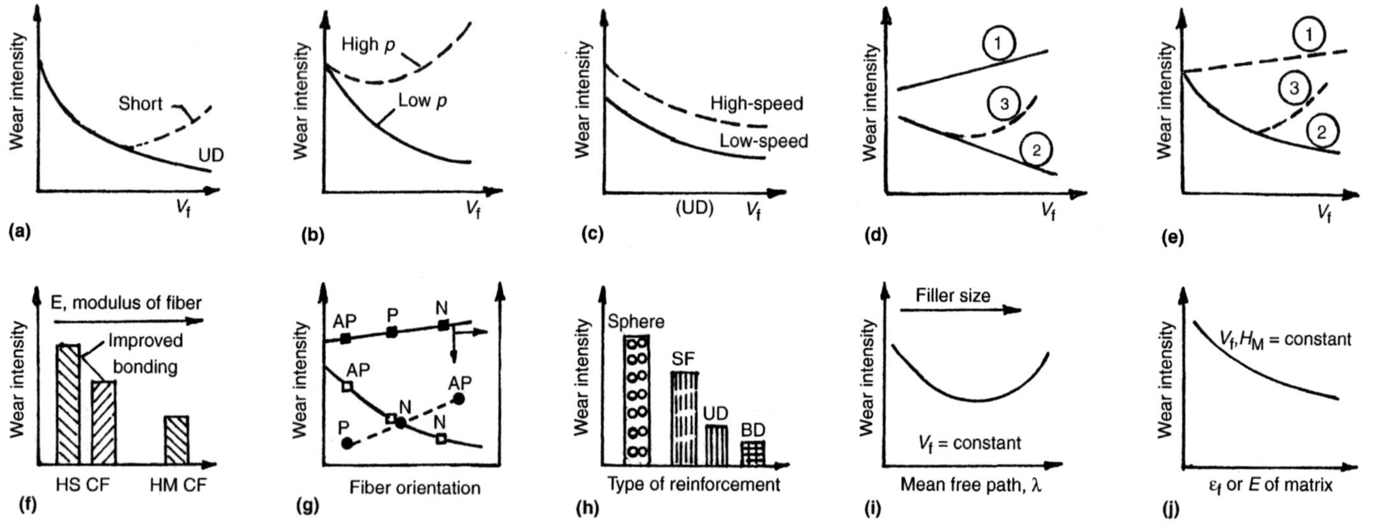

Fig. 10 General trends indicating effect of microstructure of a composite and the properties of fillers on adhesive wear of composites. V_f, volume fraction; UD, unidirectional; p, applied pressure; H_M, hardness of matrix. AP, P, and N refer to orientations of fibers with respect to sliding direction: AP, antiparallel; P, parallel; and N, normal. HS CF and HM CF, high-strength and high-modulus carbon fibers; SF, short fiber; BD, bidirectional; ε_f, fracture strain; and E, Young's modulus. 1, abrasive of larger size; 2, nonabrasive filler/solid lubricant/abrasive filler in nanometer size/long fibers or fabric; 3, short fibers. Source: Ref 2

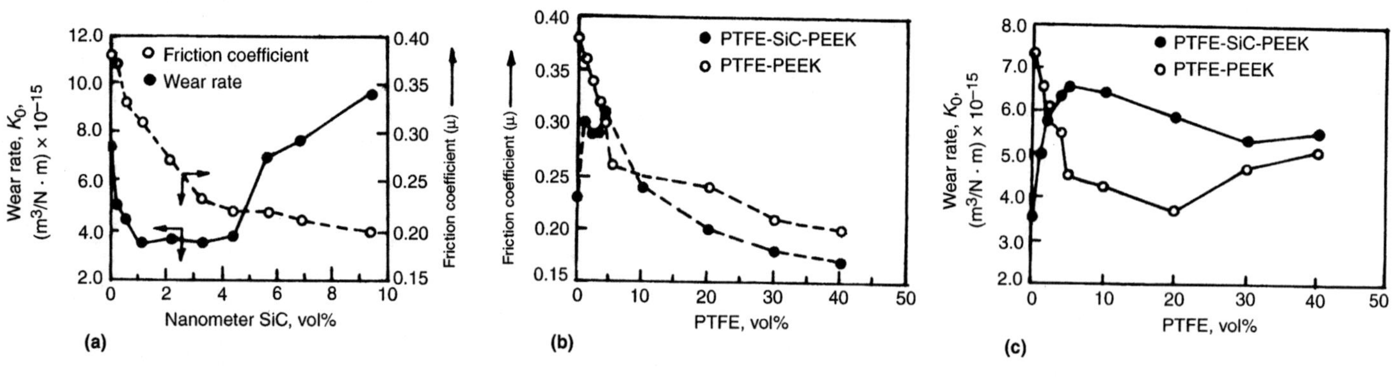

Fig. 11 Influence of fillers on friction and wear behavior of polyetheretherketone (PEEK) composites; L, normal load, 196 N; speed, 0.445 m/s; counterface, plain carbon steel ring. (a) Nanometer-sized SiC in PEEK. (b) and (c) Polytetrafluoroethylene (PTFE) in PEEK and PEEK + SiC (3.3 vol% constant) composites. Source: Ref 40

are very effective in modifying the wear performance and friction behavior (except in the case of glass fibers) of the composite. The combination of fibers and lubricants usually shows synergism. Inclusion of short glass fibers benefited the polyphenylene sulfide (PPS) maximum and the PEEK minimum (Ref 4). Beyond typical fiber loading, extent of improvement slowed down for polyethernitrile (PEN) and polyether sulfone (PES) but not for PPS. Similarly, the type of carbon fiber and its volume fraction in PEN influenced the wear behavior significantly. Pitch-based carbon fiber proved more beneficial than the polyacrylonitrile (PAN)-based. Beyond 15% loading, the extent of improvement, however, was marginal (Ref 4). Results of a study on the short-fiber and solid-lubricated composites are shown in Fig. 13(a), while Fig. 13(b) shows the influence on pressure × velocity (*PV*) factor and high temperature on the wear rate of composites. Figure 13(b) indicates that unreinforced polybenzimidozole (PBI) is far superior to the reinforced and lubricated PEEK. Systematic and step-by-step inclusion of the lubricant and reinforcement (short glass fibers) in PEI could improve friction and wear behavior very significantly, as seen in Fig. 14 (Ref 8). Polyetherimide is a hard and ductile polymer. It did not transfer any film on the counterface but did transfer a molten material in the severe *PV* condition. The thin coherent film of PTFE transferred on the mild steel counterface (Fig. 15a) in the case of $PEI_{PTFE15\%}$ composite was responsible for the lowest μ in the series of composites. For the composite containing three lubricants and short glass fibers, the film transfer was not as coherent and thin (Fig. 15b) as in the earlier case, and μ was a little higher. Various wear failure mechanisms observed on the worn surfaces of composites during SEM studies are shown in Fig. 16 (Ref 44–48).

A wear model to describe the adhesive wear behavior of SFRPs based on the microscopic observations on the worn surfaces of the FRPs has been developed (Ref 49). Because fiber cracking and fiber-matrix debonding occur sequentially, a combined process can be considered. The fiber debris removed from the matrix can act as third-body abrasives and also needs to be included in models. Hence, the sliding wear rate of composite ($W_{s,c}$) is the sum of wear rates that account for the sliding process ($W_{s,s}$) and others, which account for the additional wear mechanism ($W_{s.fci}$), the postsliding wear process; that is, wear due to fiber fracture, fiber-matrix interfacial debonding pulverization, fibrillation, and so forth. Correlation was observed between the experimental and calculated data (Ref 49).

Adhesive Wear of Unidirectional (UD) FRP Composites. Pioneering research in this area included investigation of various factors influencing the wear performance of FRP (Ref 18, 41, 50–52), while others developed the wear model based on in-depth studies on UDFRPs (Ref 53–56). The tribostudies were later extended to various composites containing reinforcement with short and continuous fibers and fabrics (Ref 1, 4, 5, 22, 27, 31–33, 42, 49, 57, 58). Figure 17 summarizes some results of selected UD composites (Ref 54).

Wear behavior of FRP depends on the properties of fibers, their orientations, and bonding with the matrix and the counterface material, including operating conditions. In the case of a brittle matrix with EP, generated cracks propagate right through the fiber if the bonding between fiber and matrix is strong. In the case of

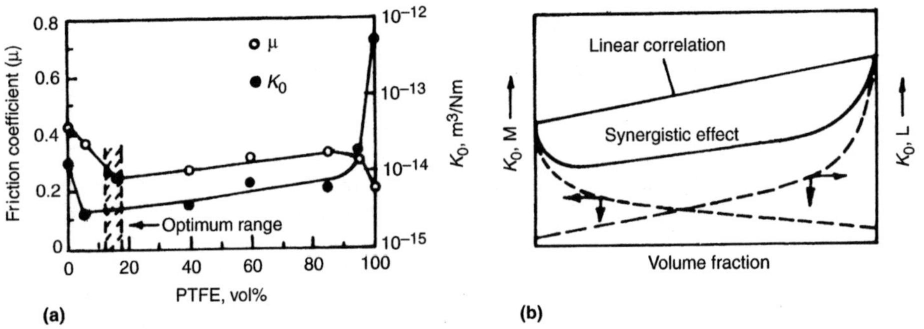

Fig. 12 (a) Influence of polytetrafluoroethylene (PTFE) on friction and wear performance of polyetheretherketone composites, and the optimal range of PTFE amount for best combination of friction coefficient (μ) and wear rate (K_0). (b) Linear correlation and synergistic effect as a result of two opposite trends. K_0, M and K_0, L represent specific wear rates of matrix and lubricant (PTFE). Source: Ref 42, 43

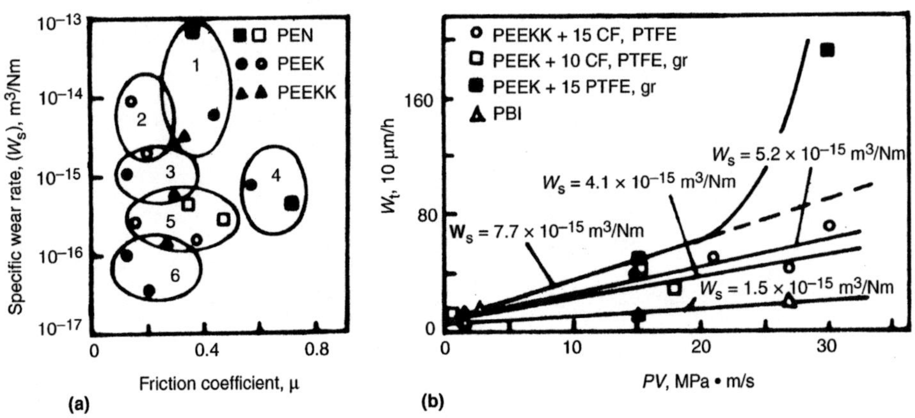

Fig. 13 (a) Indicative trends in influence of reinforcement and solid lubrication on friction and wear of high-temperature polymers. Pressure (*P*) = 1 MPa; velocity (*V*) = 1 m/s. PEN, polyethernitriale; PEEK, polyetherether ketone; PEEKK, polyetheretherketoneketone; PTFE, polytetrafluoroethylene; PBI, polybenzimidazole; CF, carbon fibers; gr, graphite. 1, neat polymers; 2, polymers + PTFE; 3, polymers + graphite/PTFE; 4, polymers + glass fibers (GF); 5, polymers + carbon fibers (CF); and 6, polymers + CF/GF + PTFE. (b) Influence of pressure × velocity (*PV*) factor on wear rate of fiber-reinforced plastics (*T*, 220 °C; *V*, 3 m/s). Source: Ref 4

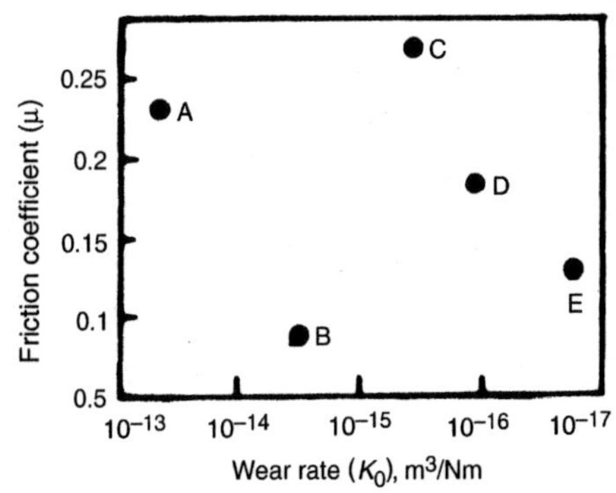

Fig. 14 Influence of inclusion of fillers (individually and simultaneously) on friction and wear performance of polyether-imide (PEI) composites against mild steel (normal load, 43 N; speed, 2.1 m/s; for a 25 N counterface mild steel). A, PEI; B, ($PEI_{PTFE15\%}$); C, ($PEI_{GF20\%}$); D, ($PEI_{GF16\%+graphite\ 20\%}$); and E, ($PEI_{GF25\%+PTFE15\%+(MoS_2+graphite)15\%}$). PTFE, polytetrafluoroethylene; GF, glass fiber

a highly ductile matrix, such as polyurethane (PU), the cracks cannot propagate through the matrix and fiber. The fiber bends with the matrix under the asperity contact, and the wear rate is controlled by the wear rate of the fiber (Ref 6). Analysis of worn surfaces of UD graphite-fiber-reinforced PU indicated that the fiber tips (fibers perpendicular to sliding plane), which were originally circular, became elliptical and bent during sliding.

The following mechanisms for wear failure of FRPs sliding against smooth metals under pressure, *p*, have been proposed (Ref 53):

- Wear thinning of the fiber due to continuous sliding for a distance, *D*, under load, *L*
- Subsequent breakdown of the fiber due to strain, $\mu p/E$ (*E*, modulus of elasticity), of FRP caused by the frictional force, load, and sliding distance
- Peeling off of the fibers from the matrix because of strain, $\mu p/E$, exceeding interlaminar shear strength

In-depth studies on UD composites (Ref 5) focused on investigating the influence of type and orientation of reinforcements in the selected matrices on friction and wear behavior. The various failure mechanisms operative in wearing FRP are shown in Fig. 18 (Ref 57). Various finite-elemental micromodels have been developed for explaining the failure mechanisms in different fiber orientations based on the evaluation of contact and stress conditions produced by a sliding of hemispherical steel asperity. Various deformed shapes of microstuctures in three orientations have been discussed based on deformation of fibers mainly by compression and bending/shear-type loadings (Ref 57).

Various wear failure mechanisms evident from SEM studies on the worn surfaces of UD composites of PA 66 are shown in Fig. 19 (Ref 5). CF/PA 66 (P) composite shows wear failure of CFs parallel to the sliding direction by various mechanisms, leading to fiber thinning, cracking, pulverization, and debonding from the matrix. Wearing of AF in O_N led to a smooth surface with little fiber-matrix debonding (Fig. 19b). The surface also showed microcracking (middle portion parallel to the width of the micrograph), delamination, and microcracking of the fibers at the edge. When the AF was in the O_P direction, it showed a tendency to be peeled from the surface due to poor wetting to the matrix (Fig. 19c).

Adhesive Wear of Fabric-Reinforced Composites. The friction and wear performances of GF-PEEK (UD) composite and graphite fabric (five-harness satin weave) PEEK (BD) composite were compared (Ref 58), and it was concluded that the wear rate of BD composite was lower than that of the UD composite by an order of 1. Friction behavior was also better for the BD rather than the UD composite. The temperature sensitivity of the former was remarkably lower than that of the latter.

Adhesive Wear of Hybrid Composites. The tribology of composites reinforced with continuous fibers of two types in different proportions in EP matrix was investigated (Ref 53). The wear behavior of hybrid UD composites containing fibers of glass and carbon in EP composite was also studied (Ref 59). Figure 20 shows that the wear rates of these composites were lower than the values expected from the LROM equation but higher than the minimum values indicated by the IROM. The dotted curve shown in Fig. 20 is for the calculated values in accordance with an equation (developed in Ref 59), which fit reasonably well. However, the model did not consider the possibility of mutual interaction of the constituents causing a deviation in the wear resistance of a hybrid composite based on the rule of mixture calculation. The practical application of such composite was justified on the basis of performance-to-cost ratio.

Various hybrid composites based on three matrices (namely, amorphous polyamide, or PA, PA 66, and EP) containing three reinforcements, as shown in Table 5, were tailored (Ref 5). Among various investigations on these composites, behavior of just one composite (AF-CF-PA 66) is shown in Fig. 21 (Ref 5). The stacking sequence for the sandwich hybrids (namely, the composite with CF placed in the surface layer and AF in the core) was an important influencing factor and proved to be superior to CF in the core and AF in the surface layer. Thus, the positive hybrid effect was found in the former case. In the latter, wear behavior was in accordance with a linear correlation between two limits. Figure 22 highlights schematics of various wear failure mechanisms operative in the wearing of

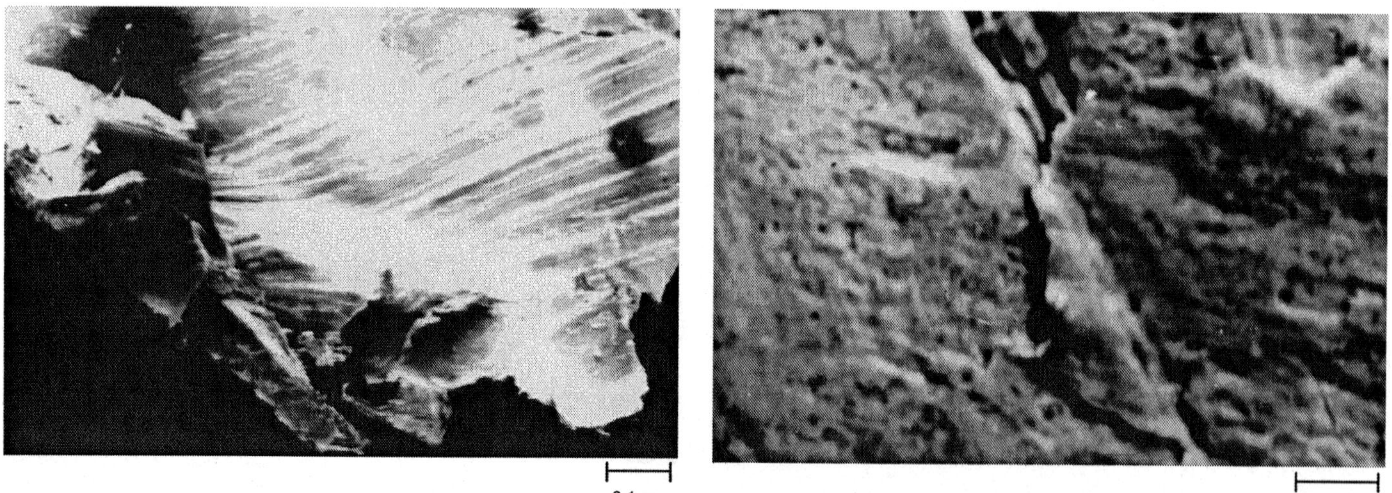

(a) (b)

Fig. 15 Scanning electron micrographs of worn surfaces of polyether-imide (PEI) composites indicating (a) transfer of thin and coherent film of polytetrafluoroethylene (PTFE) on the steel disc responsible for lowest friction coefficient (μ) exhibited by ($PEI_{PTFE15\%}$). (b) Film transfer (less coherent and thin) in the case of ($PEI_{GF25\%+PTFE15\%+(MoS_2+graphite)15\%}$) responsible for slightly higher μ than the $PEI_{PTFE15\%}$. GF, glass fiber. Source Fig. 15(a): Ref 44. Source Fig. 15(b): Ref 45

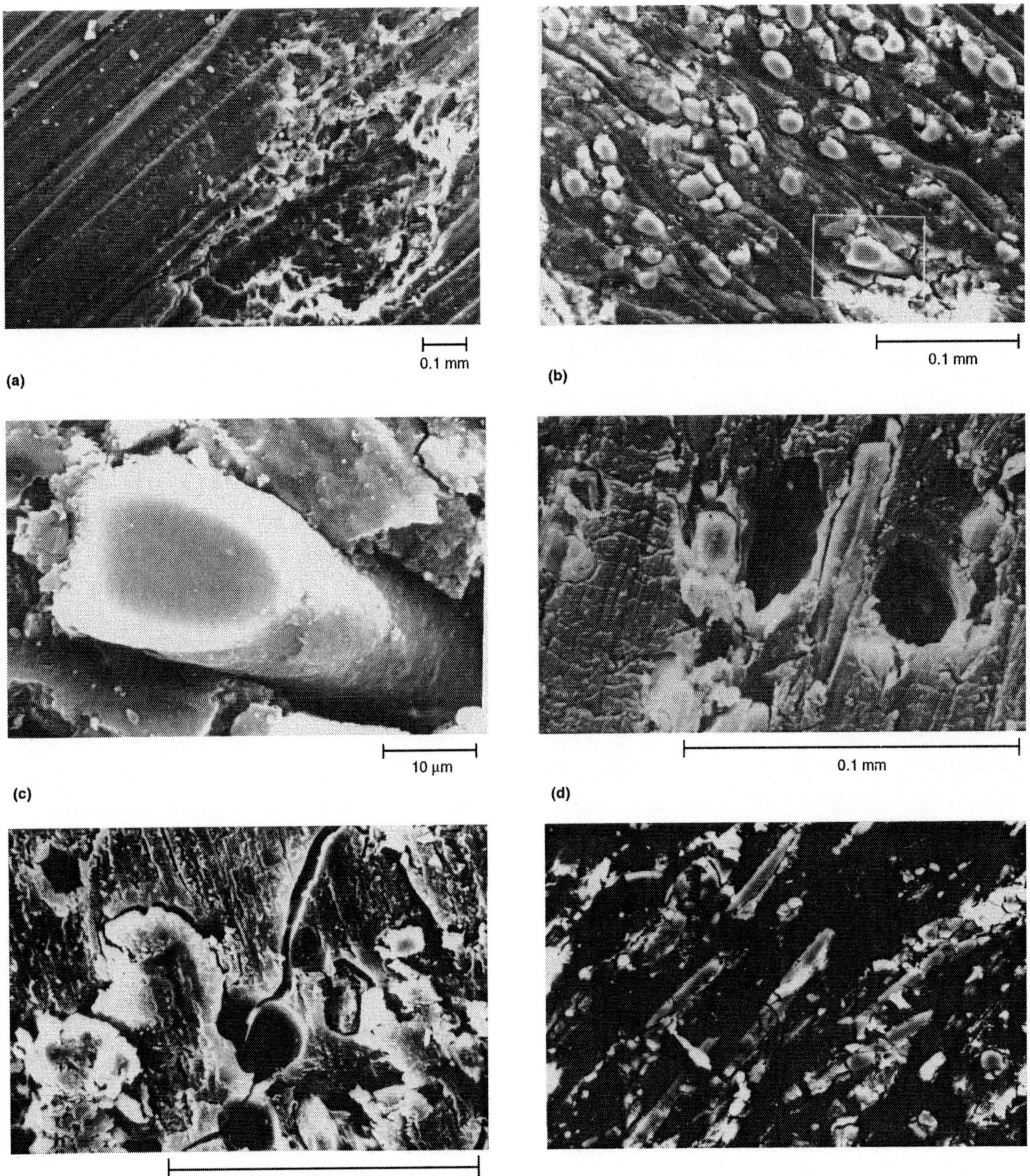

Fig. 16 Wear failure of polyether-imide (PEI) and composites. (a) Failed surface of PEI while sliding against very smooth (R_a, 0.06 μm) aluminum surface, resulting in high friction coefficient (normal load, *L*, 28 N; velocity, *V*, 2.1 m/s), Left part shows severe melt flow of PEI; middle portion shows crater with chipped-off molten material (Ref 46). (b–e) Worn surface of PEI_{GF+gr} (*L*, 112 N; *V*, 2.1 m/s). (b) Severe melt flow of polymer in sliding direction, with maximum fibers normal to the surface, cracks generated in sliding direction, and a pulled-out fiber. (c) Magnified view of pulled-out fiber from the matrix, with worn elliptical and polished tip with excessive fiber-matrix debonding aggravating wear of composite. (d) Multiple parallel microcracks perpendicular to the sliding direction indicating fatigue with cavities due to fiber consumption, deterioration in fiber-matrix adhesion, and wear thinning of longitudinal fiber. (e) Deep cracks initiating and propagating from fiber to fiber with pits formed due to graphite extraction and fiber consumption, back-transfer of molten polymer from the disc to the pin surface (patches in the left portion of the micrographs) (*L*, 132 N; *V*, 1 m/s). (f–h) Worn surfaces of $PEI_{GF+gr+PTFE+MoS_2}$ (with fibers parallel to the sliding surface), worn under *L*, 72 N, and *V*, 2.1 m/s, showing (f) microcracking of fibers, (g) deterioration in the fiber-matrix adhesion and peeled-off fiber, and (h) wear thinning of fibers with still more deterioration in fiber-matrix adhesion (*V*, 2.1 m/s; *L*, 132 N). GF, glass fiber; gr, graphite, PTFE, polytetrafluoroethylene. Source: Ref. 44–48

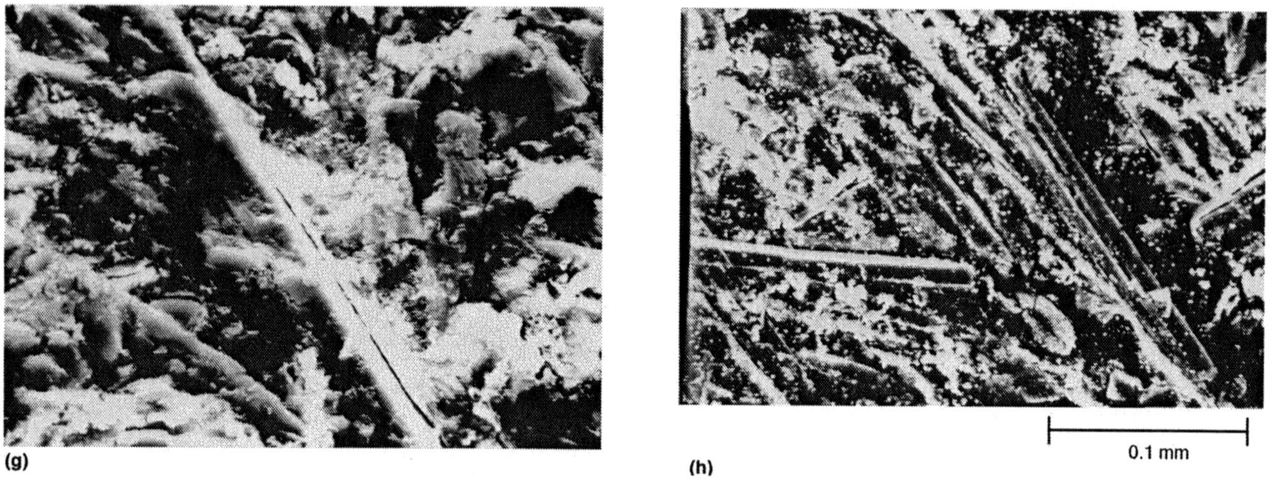

Fig. 16 (continued) (g) deterioration in the fiber-matrix adhesion and peeled-off fiber, and (h) wear thinning of fibers with still more deterioration in fiber-matrix adhesion (*V*, 2.1 m/s; *L*, 132 N). GF, glass fiber; gr, graphite, PTFE, polytetrafluoroethylene. Source: Ref. 44–48

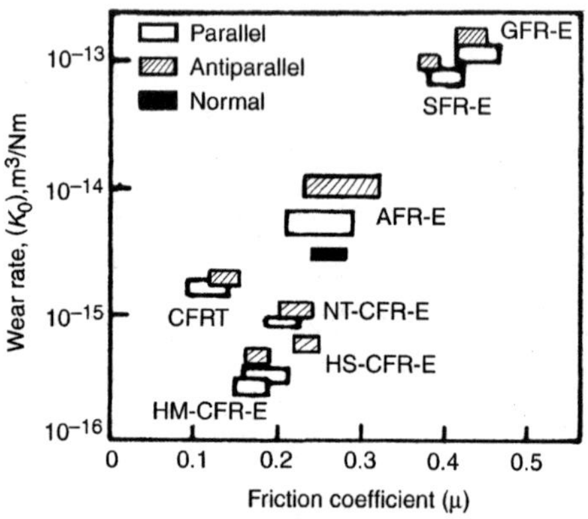

Fig. 17 Specific wear rate and friction coefficient of unidirectional composites (see Table 4) in three orientations (pressure, 1.5 N/mm^2; velocity, 0.83 m/s; distance slid, 16 km)

Table 4 Details of unidirectional (UD) composites studied in adhesive wear mode

No.	Reinforcement (UD)	Resin and volume fraction: Epoxy	Polyester	PTFE/Teflon
1	High-strength carbon fiber (HS-CFR)	HS-CFR-E (42, 52, 59, 65)	HS-CFR-EST (42, 50, 57, 65)	HS-CFRT (42, 67)
2	High-moduling carbon fiber (HM-CFR)	HM-CFR-E (65)	...	...
3	High-strength carbon fiber NT-CFR	NT-CFR-B (65)	...	...
4	E-glass fiber (GFR)	GFR-E (60, 68, 76)	GFR-EST (52, 58, 70)	...
5	Stainless steel fiber (SFR)	SFR-E (56, 62, 69, 75)	SFR-EST (48, 54, 70, 76)	SFRT (42, 67)
6	Aramid fiber (Kevlar fiber) (AFR)	AFR-E (40, 50, 60, 70)	AFR-EST (70)	AFRT (42, 67)

PTFE, polytetrafluoroethylene; NT, nontreated. Source: Ref. 54

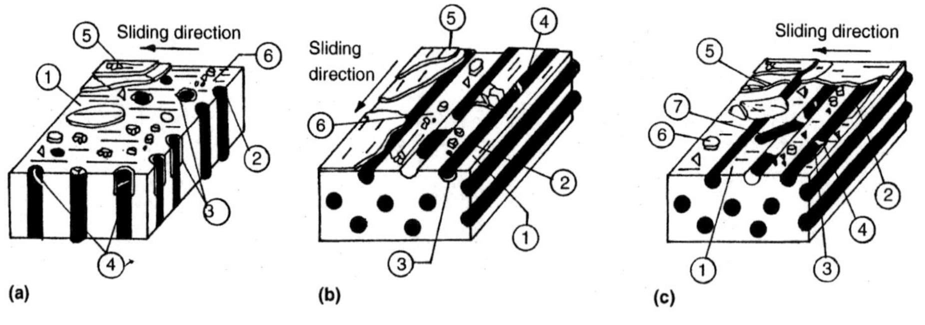

Fig. 18 Failure wear mechanisms in fiber-reinforced polymers sliding with fibers in different orientations. (a) Normal orientation; (b) parallel orientation; (c) antiparallel orientation. 1, wear failure of matrix by microplowing, microcracking, and microcutting; microplowing; 2, sliding and wear thinning of fibers; 3, interfacial separation of fiber and matrix; 4, fiber cracking; 5, back-transferred polymer or organic fibers (film and layered wear debris) showing delamination and cracking; 6, metallic and wear debris transferred from the counterface; 7, pulled-out or peeled-off fiber pieces

such BD and UD composites (Fig. 22a, b). Studies by SEM on the worn surfaces of BD hybrid composites (sandwich structure) are shown in Fig. 23. The following favorable interactions could be seen on the surfaces of the hybrid composites:

- The probability of termination of cracks by the strong barrier of CFs was high when cracks were generated in the core of AF due to poor wetting.
- Stiffer CFs were better bonded to the matrix and prevented edge delamination and fibrillation of AFs.
- Third-body formation due to transferred material consisting of fiber debris separates the counterface and contributes to the load-carrying capacity. Thus, debris of AF O_N in the core, especially in the vicinity of the CF/AF interfacial region, stayed there temporarily. This third body decelerated the wear process further.

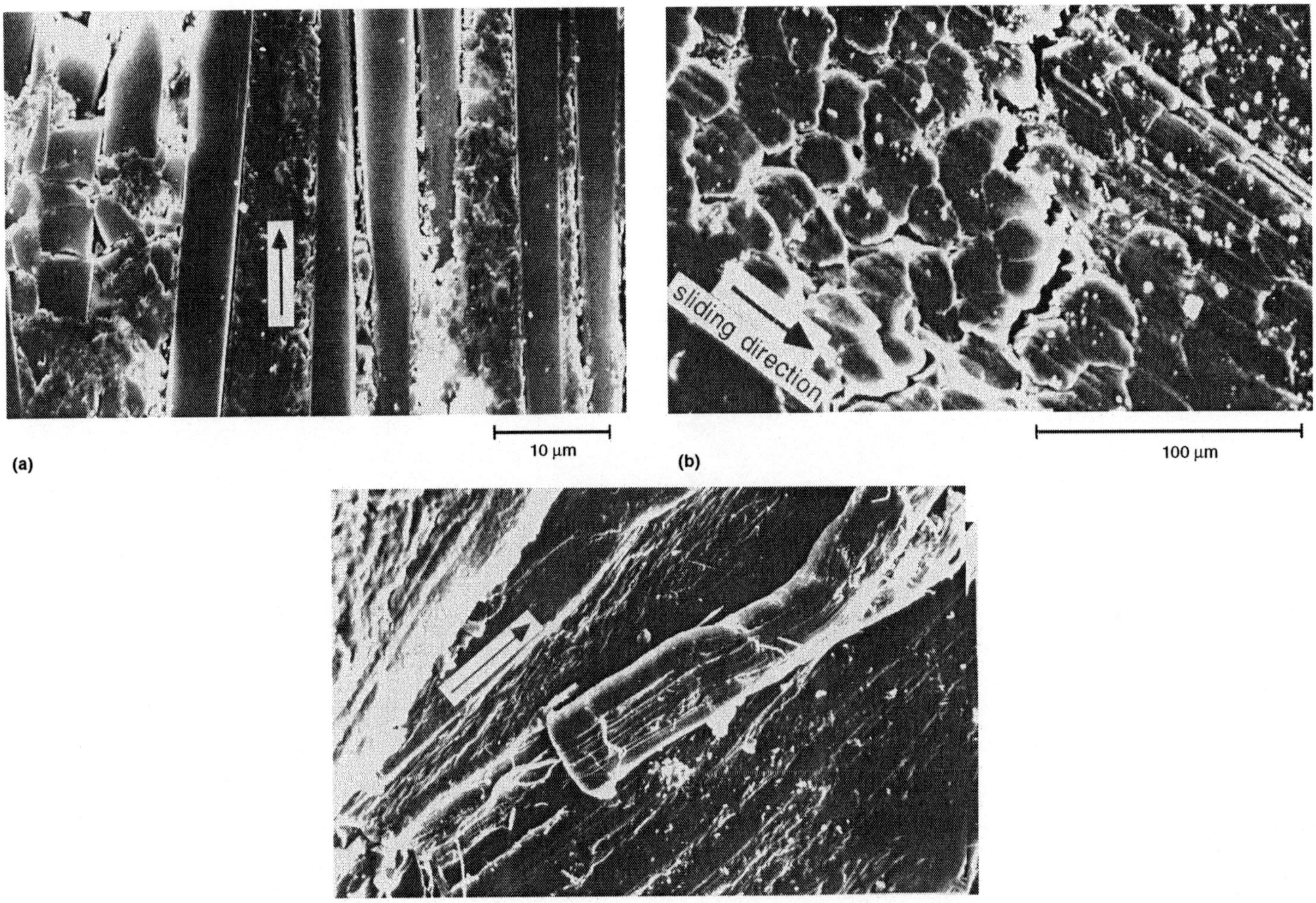

(a) (b) (c)

Fig. 19 Scanning electron microscope micrographs of worn surfaces of PA 66 unidirectional composites. (a) Carbon fiber (parallel, P) showing fiber thinning, fiber fracture, fiber pulverization (left portion), and fiber-matrix debonding (middle portion). (b) Aramid fiber (AF) in the normal orientation showing fiber cracking (edge); (c) AF(P) showing pullout of aramid fiber. Source: Ref 5

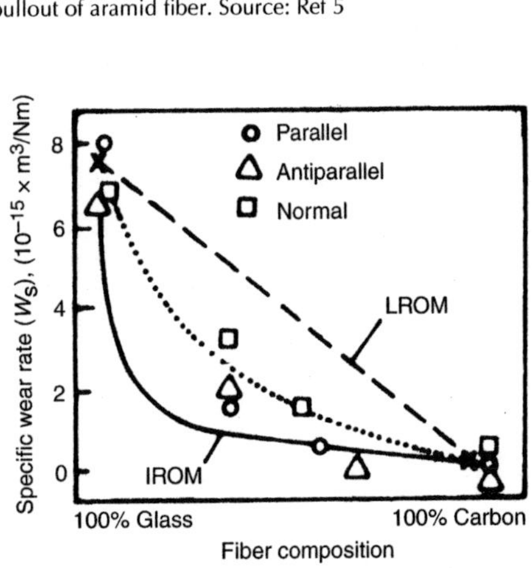

Fig. 20 Specific wear rate as a function of fiber composition in hybrid composite (normal load, 93 N; velocity, 0.5 m/s; nominal volume fraction, 0.57), with dotted curve for calculated values in accordance with equation in Ref 59. IROM, inverse rule of mixture; LROM, linear rule of mixture. Source: Ref 59

Table 5 Details of hybrid composite

Matrix	No.	Designation	Total volume fraction (V_f) $V_{f1} + V_{f2}$	f_1/f_2%	Hybrid type	Sliding direction	Hybrid efficiency factor	Specific wear rate (W_s) reduction at f_1/f_2 = 50/50(a)
Am PA	1	CF(0)-AF(90)-CF(0)	59–65	V	S	P-N-P	>0	21%
	2	AF(0)-CF(90)-AF(0)	61	50/50	S	N-P-N	0	. . .
	3(a)	CF(0)-AF(0)-CF(0)	61	50/50	S	P-P-P	<0	22%
	3(b)	. . .	. . .	. . .	. . .	N-N-N	>0	27%
PA 66	4	CF(0)-AF(90)-CF(0)	20–40	V	S	P-N-P	>0	45%
	5	CF(0)-AF(90)-CF(0)	20–40	V	L	P-N-P	>0	64%
	6(a)	CF(0)-GF(90)-CF(0)	25	50/50	S	N-P-N	0	. . .
	7(a)	GF(0)-AF(90)-CF(0)	35	50/50	S	P-N-P	<0	135%
	7(b)	. . .	. . .	. . .	. . .	AP-N-AP	. . .	127%
EP_2	8(a)	CF(0)-AF(0)-CF(0)	62	50/50	S	P-P-P	<0	63%
	8(b)	. . .	. . .	. . .	. . .	N-N-N	>0	36%

Am, amorphous; PA, polyamide; EP, epoxy; CF, carbon fiber; AF, aramid fiber; GF, glass fiber; V, variable; S, sandwich; L, layer structure; P, parallel; N, normal; AP, antiparallel. (a) $W_{S\,red}$ expressed in % of rule of mixtures value $W_{s\,f1/f2}$. Source: Ref 5

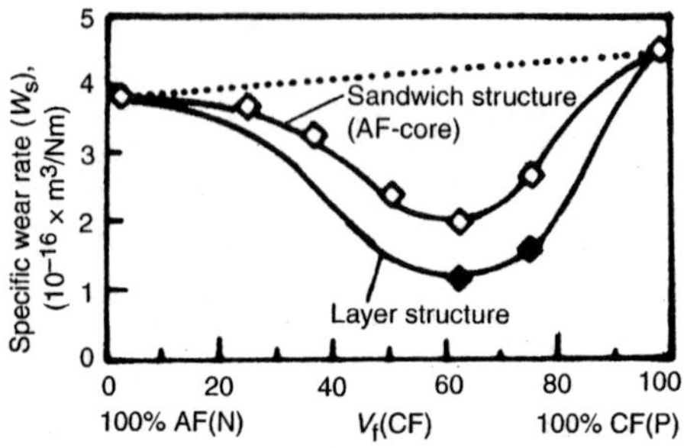

Fig. 21 Specific wear rates of hybrid composites formulated by two structures, sandwich and layer, (composite aramid fiber/carbon fiber polyamide amorphous). AF, aramid fiber; CF, carbon fiber; N, normal; V_f, volume fraction; P, parallel. Source: Ref 5

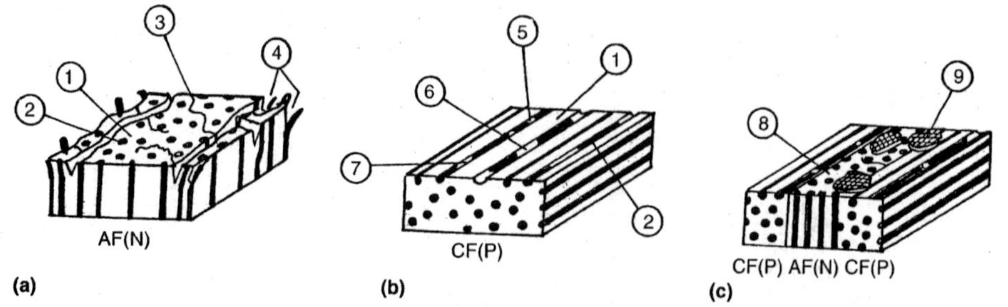

Fig. 22 Failure wear mechanisms of unidirectional fiber-reinforced polymer composites with different orientations of fibers with respect to sliding direction against a smooth metal surface. (a) Normal (N) aramid fibers (AF). (b) Parallel (P) carbon fibers (CF). (c) Wear reduction mechanism due to hybridization. 1, wear of matrix by plowing, cracking, cutting as a result of plastic deformation; 2, wear thinning of fiber or tip resulting in elliptical, well-polished tip; 3, matrix cracking; 4, edge delamination and fiber fibrillation; 5, fiber cracking; 6, pulverized fiber wear debris; 7, deterioration in fiber-matrix adhesion, fiber pulverization, pullout, and peeling off followed by removal; 8, inhibition to matrix cracking; and 9, layer of back-transferred film or wear debris. Source: Ref 2

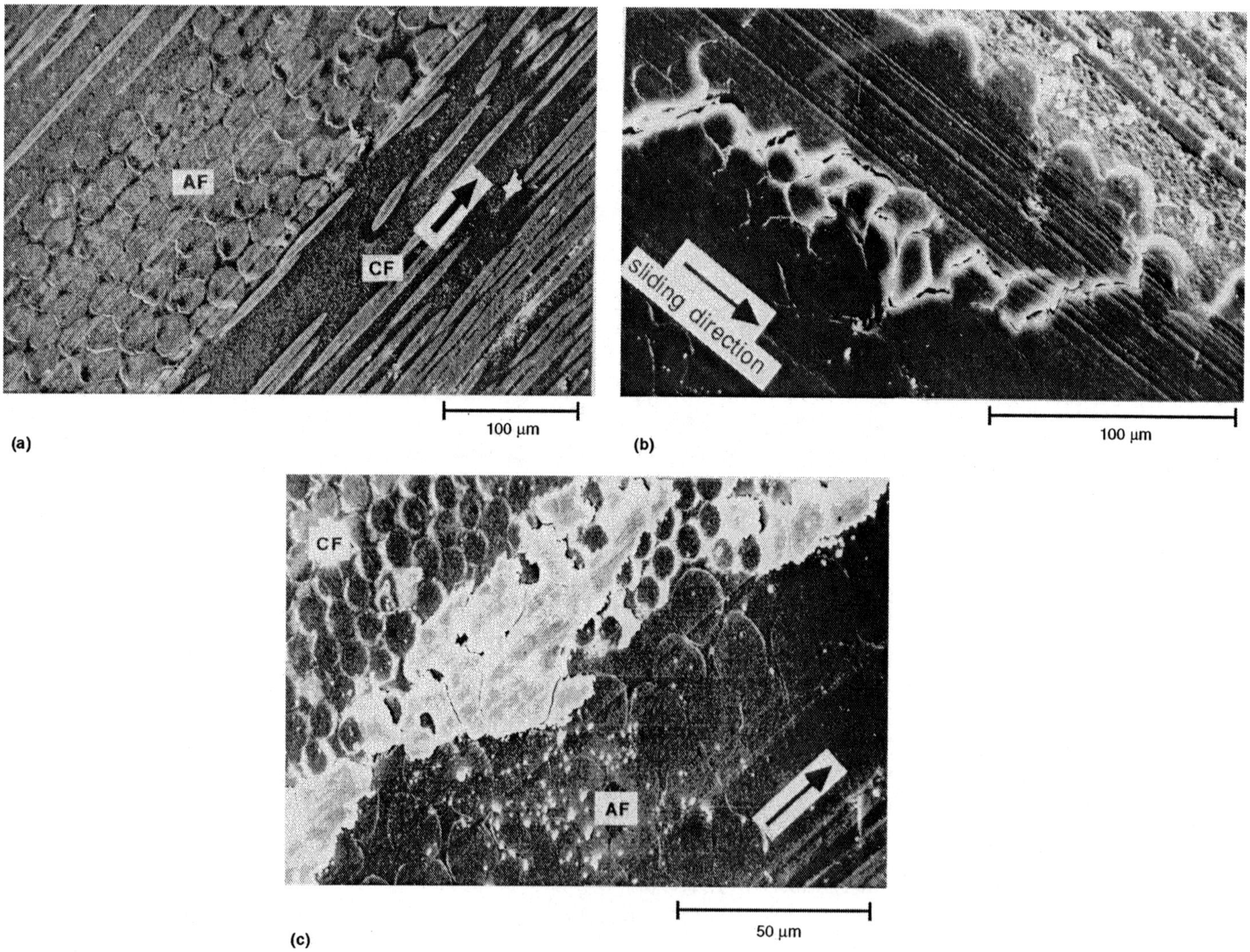

Fig. 23 Scanning electron microscope micrographs of worn surfaces of PA 66 hybrid composites. AF, aramid fiber; CF, carbon fiber; N, normal; P, parallel. (a) Hybrid (layer) composite, AF(N)/CF(P). (b) Hybrid (sandwich) composite, AF(N)/CF(P), stopping crack responsible for less wear. (c) AF(N)/CF(P) composite, accumulation of protective patch work (back-transferred layer) (middle portion). Source: Ref 5

ACKNOWLEDGMENTS

The original article acknowledges Professor K. Friedrich, Research Director, Institute for Composite Materials Ltd., University of Kaiserslautern, Germany, for the kind consent to present his work and the provision of SEMs describing failure mechanisms of some important composites. J. John Rajesh and Nidhi Dhingra are thanked for help in research and typing.

REFERENCES

1. K. Friedrich and R. Walter, Microstructure and Tribological Properties of Short Fiber/Thermoplastic Composites, *Tribology of Composite Materials Conf. Proc.*, P.K. Rohatgi, P.J. Blau, and C.S. Yust, Ed., ASM International, 1990, p 217–226
2. K.-H. Zum Gahr, *Microstructure and Wear of Materials, Tribology Series 10,* Elsevier, Amsterdam, 1987
3. J. Bijwe and M. Fahim, Tribology of High Performance Polymers—State of Art, *Handbook of Advanced Functional Molecules and Polymers,* H.S. Nalwa, Ed., Gordon and Breach Publishers, Amsterdam, 2001, p 265–321
4. K. Friedrich, Z. Lu, and A.M. Hager, Recent Advances in Polymer Composite Tribology, *Wear,* Vol 190, 1995, p 139–144
5. K. Friedrich, Wear Models for Multiphase Materials and Synergistic Effects in Polymeric Hybrid Composites, *Advances in Composite Tribology, Composite Materials Series 8,* K. Friedrich, Ed., Elsevier, 1993, p 209–273
6. N.P. Suh, *Tribophysics,* Prentice-Hall Inc., 1986
7. K. Friedrich, Ed., *Friction and Wear of Polymer Composites, Composite Materials Series 1,* Elsevier, 1986
8. U.S. Tewari and J. Bijwe, Tribological Behavior of Polyimides, *Polyimides: Fundamentals and Applications,* M.K. Ghosh and K.L. Mittal, Ed., Marcel Dekker Inc., 1996, p 533–586
9. S.W. Zhang, State-of-the-Art of Polymer Tribology, *Tribol. Int.,* Vol 31 (No. 1–3), 1998, p 46–60
10. H. Czichos, *Tribology: A System Approach to the Science and Technology of Friction, Lubrication and Wear, Tribology Series 1,* Elsevier, 1978
11. E. Santner and H. Czichos, Tribology of Polymers, *Tribol. Int.,* Vol 22, 1989, p 103–109
12. K.-H. Zum Gahr, Wear by Hard Particles, *Tribol. Int.,* Vol 31, 1998, p 587–596
13. S.V. Prasad and P.D. Calvert, Abrasive Wear of Particle Filled Polymers, *J. Mater. Sci.,* Vol 15, 1980, p 1746–1754
14. M.M. Khruschov, Principles of Abrasive Wear, *Wear,* Vol 28, 1974, p 69–88
15. K.-H. Zum Gahr, *Proc. International Conf. on Wear of Materials* (Vancouver) American Society of Mechanical Engineers, 14–18 April 1985, p 45–58
16. W. Simm and S. Freti, Abrasive Wear of Multiphase Materials, *Wear,* Vol 129, 1989, p 105–121
17. S. Bahadur and D. Gong, Formulation of the Model for Optimal Proportion of Filler in Polymer for Abrasive Wear Resistance, *Wear of Materials,* American Society of Mechanical Engineers, 1991, p 177–185
18. J.K. Lancaster, Friction and Wear, *Polymer Science: A Material Science Handbook,* A.D. Jenkins, Ed., North Holland, 1972
19. U.S. Tewari, J. Bijwe, J.N. Mathur, and I. Sharma, Studies on Abrasive Wear of Carbon Fiber (Short) Reinforced Polyamide Composites, *Tribol. Int.,* Vol 25, 1992, p 53–60
20. J.P. Giltrow, A Relationship Between Abrasive Wear and the Cohesive Energy of Materials, *Wear,* Vol 15, 1970, p 71–78
21. M. Vaziri, R.T. Spurr, and F.H. Stott, An Investigation of the Wear of Polymeric Materials, *Wear,* Vol 122, 1988, p 329–342
22. K. Friedrich and M. Cyffka, On the Wear of Reinforced Thermoplastics by Different Abrasive Papers, *Wear,* Vol 103, 1985, p 333–344
23. J.J. Rajesh, J. Bijwe, and U.S. Tewari, Influence of Fillers on Abrasive Wear of Short Glass Fiber Reinforced Polyamide Composites, *J. Mater. Sci.,* Vol 36, 2001, p 351–356
24. B.J. Briscoe, P.D. Evans, and J.K. Lancaster, The Influence of Debris Inclusion on Abrasive Wear Relationships of PTFE, *Wear,* Vol 124, 1988, p 177–194
25. B.J. Briscoe, and P.D. Evans, The Influence of Asperity Deformation Condition on the Abrasive Wear of γ-Irradiated Polytetrafluoroethylenes, *Wear,* Vol 133, 1989, p 47–64
26. C. Lhymn, K.E. Templemeyer, and P.K. Davis, Abrasive Wear of Short Fiber Composites, *Composites,* Vol 16 (No. 2), 1985, p 127–136
27. M. Cirino, K. Friedrich, and R.B. Pipes, Evaluation of Polymer Composites for Sliding and Abrasive Wear Applications, *Composites,* Vol 19 (No. 5), 1988, p 383–392
28. M.K. Omar, A.G. Atkins, and J.K. Lancaster, The Role of Crack Resistance Parameters in Polymer Wear, *Physica D: Appl. Phys.,* Vol 19, 1986, p 177–195
29. J. Bijwe, J. Indumathi, and A.K. Ghosh, Evaluation of Engineering Polymeric Composites for Abrasive Wear Performance, *J. Reinf. Plast. Compos.,* Vol 18 (No. 17), 1999, p 1573–1591
30. C. Lhymn, Effect of Normal Load on Specific Wear Rate of Fibrous Composites, *Wear,* Vol 120, 1987, p 1–27
31. K. Friedrich, Wear of Reinforced Polymers by Different Abrasive Counterparts, *Friction and Wear of Polymeric Composites, Composite Materials Series 1,* K. Friedrich, Ed., Elsevier, 1986, p 233–287
32. M. Cirino, R.B. Pipes, and K. Friedrich, The Abrasive Wear of Continuous Fiber Reinforced Polymer Composites, *J. Mater. Sci.,* Vol 22, 1987, p 2481–2492
33. M. Cirino, K. Friedrich, and R.B. Pipes, The Effect of Fiber Orientation on the Abrasive Wear Behavior of Polymer Composite Materials, *Wear,* Vol 121, 1988, p 127–141
34. B.K. Yen and C.K.H. Dharan, A Model for Abrasive Wear of Fiber Reinforced Polymer Composites, *Wear,* Vol 195, 1996, p 123–127
35. J. Bijwe, J. Indumathi, and A.K. Ghosh, On the Abrasive Wear Behavior of Fabric Reinforced Polyetherimide Composites, *Wear,* 2003
36. J. Indumathi, "Friction and Wear Studies on Polyetherimide and Composites," Ph.D. dissertation, Indian Institute of Technology Delhi, New Delhi, 2000
37. Q.H. Wang, Q.J. Xue, H. Liu, and J. Xu, The Effect of Particle Size on Nanometer ZrO_2 on Tribo-Behavior of PEEK, *Wear,* Vol 198, 1996, p 216–219
38. Q.H. Wang, J. Xu, W.C. Shen, and Q. Liu, An Investigation of the Friction and Wear Properties of Nanometer Si_3N_4 Filled PEEK, *Wear,* Vol 196, 1996, p 82–86
39. Q.H. Wang, W.C. Shen, J.F. Xu, and Q.J. Xue, The Effect of Nanometer SiC Filler on the Tribological Behavior of PEEK, *Wear,* Vol 209, 1997, p 316–321
40. Q.H. Wang, J. Xue, W.M. Liu, and J.M. Chen, Friction and Wear Characteristics of Nanometer SiC and Poly(tetraflouroethylene) Filled Poly(etheretherketone), *Wear,* Vol 243, 2000, p 140–146
41. J.K. Lancaster, Composites for Aerospace Dry Bearing Application, *Friction and Wear of Polymer Composites, Composite Material Series 1,* K. Friedrich, Ed., Elsevier, 1986
42. P. Lu and K. Friedrich, On Sliding of Friction and Wear of PEEK and its Composite, *Wear,* Vol 181–183, 1995, p 624–631
43. A.M. Hager and M. Davis, Short Fiber Reinforced, High Temperature Resistant Polymer for a Wide Field of Tribological Applications, *Advances in Composite Tribology, Composite Materials Series 8,* K. Friedrich, Ed., Elsevier, 1993, p 107–157
44. J. Bijwe, U.S. Tewari, and P. Vasudevan, Friction and Wear Studies of Internally Lubricated Polyetherimide Composite, *J. Synth. Lub.,* Vol 6, 1989, p 179–202
45. J. Bijwe, U.S. Tewari, and P. Vasudevan, Friction and Wear Studies of Polyetherimide Composite, *Wear,* Vol 138, 1990, p 61–76
46. J. Bijwe, U.S. Tewari, and P. Vasudevan, Friction and Wear Studies of Bulk Polyetherimide, *J. Mater. Sci.,* Vol 25, 1990, p 548–556
47. J. Bijwe, U.S. Tewari, and P. Vasudevan, Friction and Wear Studies of a Short Glass

Fiber Reinforced Polyetherimide Composite, *Wear,* Vol 132, 1989, p 247–264
48. U.S. Tewari and J. Bijwe, Tribological Investigations of Polyetherimide Composite, *J. Mater. Sci.,* Vol 27, 1992, p 328–334
49. H. Voss and K. Friedrich, Wear Performance of a Bulk Liquid Crystal Polymer and its Short Fiber Composites, *Tribol. Int.,* Vol 19, 1986, p 145–156
50. J.P. Giltrow and J.K. Lancaster, Carbon Fiber Reinforced Polymer as Self Lubricating Materials, *Proc. Inst. Mech. Eng.,* Vol 182 (3N), 1967–1968, p 147
51. J.P. Giltrow and J.K. Lancaster, The Role of the Counterface in the Friction and Wear of Carbon Fiber Reinforced Thermosetting Resin, *Wear,* Vol 16, 1970, p 359–374
52. J.K. Lancaster, Estimation of the Limiting *PV* Relationship for Thermoplastic Bearing Materials, *Tribology,* Vol 4, 1971, p 82–86
53. T. Tsukizoe and N. Ohmae, Friction and Wear Performance of Unidirectionally Oriented Glass, Carbon, Aramid and Stainless Steel Fiber-Reinforced Plastics, *Friction and Wear of Polymer Composites,* K. Friedrich, Ed., Elsevier, 1986, p 203–231
54. T. Tsukizoe and N. Ohmae, Wear Mechanism of Unidirectionally Oriented Fiber Reinforced Plastics, *J. Lubr. Technol. (Trans. ASME),* Vol 99 F(4), 1977, p 401–407
55. T. Tsukizoe and N. Ohmae, Friction and Wear of Partially Oriented Fiber Reinforced Plastics—Tribological Assessment for CFRP, GFRP and SFRP, *J. Jpn. Soc. Lubr. Eng.,* Vol 2 (No. 5), 1976, p 330 (in Japanese)
56. T. Tsukizoe and N. Ohmae, Friction Properties of Composite Materials, *Trans. Jpn. Soc. Mech. Eng.,* Vol 43 (No. 367), 1977, p 115 (in Japanese)
57. T. Goda, K. Varadi, and K. Friedrich, Finite Element Micro-Models to Study Contact States, Stress and Failure Mechanisms in a Polymer Composite Subjected to a Sliding Steel Asperity, *Research News,* Budapest University of Technology and Economics, (3), 2000, p 8–14
58. P.B. Mody, T.W. Chou, and K. Friedrich, Effect of Testing Conditions and Microstructure on the Sliding Wear of Graphite Fiber/PEEK Matrix Composites, *J. Mater. Sci.,* Vol 23, 1988, p 4319–4330
59. H.M. Hawthorne, Wear in Hybrid Carbon/Glass Fiber Epoxy Composite Materials, *Proc. International Conf. on Wear of Materials* (Reston, VA), K.C. Ludema, Ed., American Society of Mechanical Engineers, 11–14 April 1983, p 576–582

Environmental Effects

Thermal Stresses and Physical Aging*

ENGINEERING PLASTICS, as a general class of materials, are prone to the development of internal stresses that arise during processing or during service when parts are exposed to environments that impose deformation and/or temperature extremes. These stresses, which manifest themselves in the properties of the plastic, may affect mechanical, optical, and electrical properties and may influence dimensional stability, permeability, and resistance to hostile environments. The effects may be beneficial or detrimental. It is possible, for example, to delay brittle fracture in plastic members by introducing residual compressive stresses on the surfaces of Izod impact specimens (Ref 1). On the other hand, the buildup of stresses during processing can result in voids or cracks that diminish mechanical properties.

Thermal stresses are largely a consequence of high coefficients of thermal expansion and low thermal diffusivities. These effects, which are exacerbated when there is a large difference between the glass transition temperature, T_g, and the ambient temperature, represent a problem in high-performance (high-temperature) thermoplastics such as polysulfone (PSU) or polyetherketone (PEK) because they develop significant thermal stresses on cooling. It is this coefficient of thermal expansion (CTE) mismatch between polymers and fillers, especially anisotropic composites, that can lead to diminished mechanical properties. However, by a judicious choice of fillers, it is possible to lower the CTE of the composite to such an extent that it can be used in conjunction with metal parts, even at cryogenic temperatures. Although time-consuming techniques can be used to analyze thermal stresses, several useful qualitative tests are described in this article.

Flow-induced orientation effects are also discussed. Orientation commonly occurs in processing by such techniques as extrusion, injection molding, pultrusion, and calendering. The anisotropy of physical properties that accompanies orientation must be considered. Of paramount concern are the dimensional instabilities that arise from anisotropic CTEs.

Physical aging, which is also examined, is the process by which plastics cooled below the T_g gradually approach thermodynamic equilibrium. The approach to equilibrium can lead to drastic or sometimes subtle changes in physical properties. Although aging is often characterized by monitoring changes in excess enthalpy and entropy, these measurements do not necessarily directly correlate with changes in physical properties. Aging temperature ranges have been determined for a number of plastics. It is accepted that the aging range extends from the T_g down to the highest secondary transition associated with small-scale molecular motion. Secondary transitions that are not well separated from the T_g or that involve cooperative motion are possible exceptions to this rule. At any rate, the densification and possible configurational changes that accompany aging alter mechanical properties, especially ductile-brittle behavior.

Cooling stresses, orientation, and the nonequilibrium thermodynamic state of glasses often simultaneously accompany processing. Orientation is removed by annealing above the T_g. The effect of thermal stresses can be somewhat isolated from aging effects by the layer removal technique. However, aging may also be affected by temperature gradients, because material near the core of the sample is annealed for longer times. By being aware of these phenomena, an engineer should be able to select plastics and processing techniques more efficiently and to broaden product applications.

In an attempt to explain the stresses encountered in the plastics industry, the first section of this article defines the different types of internal stresses. Then, each type of thermal stress is discussed in detail, with reference to the mechanism of generation and the effect on engineering properties. Methods of detecting and measuring internal stresses are presented. Next, orientation effects are described. Finally, numerous aspects of physical aging are discussed.

Classification of Stress

Internal stress phenomena have been extensively studied in metals and inorganic glasses. The increased use of engineering plastics in structural materials has necessitated increasing the body of knowledge on this subject. Some excellent reviews in this area are available in the literature (Ref 2–4). There are three types of internal stresses in amorphous polymers (this classification scheme can be extended to include semicrystalline polymers and composites), as follows (Ref 4).

The first type, thermal or cooling stresses, results from rapid, inhomogeneous cooling through the T_g range in amorphous polymers or through the solidification range in semicrystalline polymers. When cooling proceeds from the outer layer inward, large thermal gradients are formed, and thermal stresses are frozen in. Thermal stresses also arise from the thermal mismatch between materials in a composite that have different thermal expansion properties.

The second type of internal stress consists of orientation and orientation stresses. Processing involves deformation at elevated temperatures. This deformation produces molecular orientation, which is accompanied by internal stresses. If cooling is rapid enough, these stresses are frozen in, along with orientation. This results in anisotropic physical properties and a propensity for dimensional instability.

The third type of internal stress includes quenching stresses and physical aging quenching stresses. These arise when an amorphous polymer is cooled to below the T_g at a rate that is too rapid for the molecules to attain a true glassy equilibrium state. Molecular motion below T_g is extremely slow, and the approach to the equilibrium state is a result of this slow motion. The approach to equilibrium gives rise to what is called physical aging. The net result is a slow contraction of material. This contraction relieves the internal stresses. The dimensional changes here are not as pronounced as those that result from the relaxation of thermal or orientation stresses. Nonetheless, mechanical properties are significantly affected by the aging process. Deformation by swelling can generate similar internal stresses.

Categorically, these three types of stresses are defined as separate entities. It is emphasized, however, that it is often difficult to isolate and quantify internal stresses in plastic parts. This is because the processing conditions may generate more than one type of internal stress; that is, orientation stresses are likely to be accompanied by cooling stresses.

*Adapted from the article by Julie P. Harmon and Charles L. Beatty, "Thermal Stresses and Physical Aging," in *Engineering Plastics*, Volume 2, *Engineered Materials Handbook*, ASM International, 1988, pages 751 to 760

Thermal Stresses

The buildup of cooling stresses in plastic parts is a result of the effects of low thermal diffusivity, high thermal expansion properties, and the variation of mechanical properties with temperature. An understanding of these properties and the way in which they compare with those of other classes of materials (inorganic glasses and metals) clarifies the fact that plastics are prone to the development of thermal stresses.

Thermal expansion can be defined in a number of ways (Ref 5):

- Specific thermal expansivity, in which $e = \partial v/\partial T)_p$ in units of $cm^3/g \cdot °C$
- Volume CTE, in which $\alpha = 1/v(\partial v/\partial T)_p$ in units of 1/°C
- Linear CTE, in which $\beta = 1/L(\partial L/\partial T)_p$ in units of 1/°C

In the preceding cases, v is the specific volume, T is the temperature, and L is the length.

Thermal diffusivity is related to thermal conductivity (Ref 5):

- Thermal conductivity, $\lambda, = C_p \cdot \rho\mu \cdot L$ in units of $J/s \cdot m \cdot °C$
- Thermal diffusivity, $D, = \lambda/C_p \cdot \rho$ in units of cm^2/s

In these cases, C_p is the specific heat capacity at constant pressure, ρ is the density, and μ is the velocity of sound.

Table 1 lists room-temperature linear CTEs for various materials. It is apparent that plastics have large CTEs compared to those of metals or inorganic glass. However, the room-temperature thermal diffusivity, as given in Table 2, is much lower for plastics than for metals and is slightly lower than that of inorganic glass. It should be emphasized that the values listed in Tables 1 and 2 are room-temperature values. The CTEs and thermal diffusivities of plastics vary with temperature. In general, the variation in CTE with temperature above and below the T_g is not significant when compared with the large increase in the CTE that is encountered at T_g (Ref 5). The value of CTE of a glassy polymer is approximately one-half that of a liquid polymer (Ref 6). For crystalline polymers, the sharp increase in CTE occurs at the melting point. Values for glassy and crystalline components are approximately equal.

The thermal conductivity of polymers goes through a broad maximum at T_g. For crystalline polymers, thermal conductivity increases with the percentage of crystalline and amorphous phases, ρ_c and ρ_a (Ref 5):

$$\frac{\lambda_c}{\lambda_a} \sim \left(\frac{\rho_c}{\rho_a}\right)^6 \qquad \text{(Eq 1)}$$

For the purpose of thermal stress determination, it is convenient to consider the thermal conductivity constant as the temperature varies.

Mechanical Properties Versus Temperature. Mechanical properties, specifically the Young's modulus, also vary with temperature. The glassy modulus of an amorphous polymer shows a large decrease as the polymer changes from the glassy to the rubbery state (Ref 4). The modulus of atactic polystyrene (PS) (M_n = 217,000) decreases 5000-fold in the transition region (Ref 9). There is less of a decrease at the glass transition in crystalline polymers. Crystalline isotactic PS showed a decrease in modulus by 400 times in the range from below T_g to near the melting temperature (Ref 9).

Solidification. Thermal stresses are due to inhomogeneous cooling during solidification. Solidification arises at such a temperature when the polymer matrix becomes stiff enough to support stress. This temperature occurs at the glass transition in amorphous thermoplastic and somewhat below the melting temperature in semicrystalline polymers (Ref 10).

The situation is somewhat different in cross-linked or thermoset processing. In this case, solidification takes place at the cure temperature, because cross links enable the matrix to support stress (Ref 10). Unsaturated polyester and epoxy resins are included in this category. Stress support is a consequence of an increase in modulus that occurs as cooling proceeds through the T_g, below the melt temperature, T_m, or at the cure temperature. When thermoplastics or semicrystalline polymers are cooled from the outside in, the solidification temperature is reached rapidly at the surface of the mold. As cooling proceeds, the solidification boundary moves inward to the core. Compressive forces are generated at the surface of the mold by the solidifying internal layers. Internal layers are in a state of tension. The amount of contraction that takes place from the solidification temperature to the ambient temperature is determined by the CTE. In crystalline polymers, an additional volume decrease accompanies the crystallization process, because crystalline fractions are denser than amorphous fractions.

The percentage of volume decrease that accompanies cooling from the solidification temperature to room temperature was measured for several polymers (Ref 10). The results are shown in Table 3. Potential volume changes are generally much higher for crystalline polymers than for amorphous thermoplastics. Inhomogeneous cooling inhibits the volume contraction in the inner layers of material when the solidification temperature is reached at the mold surface. The result is that the thermal stresses generated during cooling persist (become residual stresses).

In addition, crystallization against a cool mold surface produces an inhomogeneous semicrystalline morphology. The cool mold surface results in higher nucleation and growth rates, which produce a columnar spherulitic regime called the transcrystalline layer. The low thermal conductivity of the polymer melt allows steep temperature gradients to develop. The lower crystallization temperature at the surface, compared to the higher crystallization temperature of the interior, results in thinner lamellae, smaller-diameter spherulites, and a higher population of

Table 1 Linear coefficients of thermal expansion

Material	10^{-6}/K
Polymethyl methacrylate(a)	50–90
Polyacrylonitrile)(a)	66
Cellulose acetate(a)	100–150
Nylon 6(a)	80–83
Nylon 11(a)	100
Polycarbonate(a)	68
Polyethylene terephthalate(a)	65
Polyphenylene sulfide(a)	49
Polyethylene, branched(a)	100–220
Polypropylene(a)	81–100
Polystyrene(a)	50–83
Polyvinyl chloride(a)	50–100
Glass(b)	8.3–9.7
Gold(b)	14.3
Cast iron(b)	10.6
Carbon graphite(b)	7.9
Hardened stainless steel(b)	17.3

(a) Source: Ref 6. (b) Source: Ref 7

Table 2 Thermal diffusivity at room temperature

Material	Diffusivity	
	$cm^2/s \times 10^4$	$in.^2/s \times 10^4$
Magnesium oxide(a)	1400	217
Iron(a)	2000	310
Glass(a)	20–60	3–9
Polypropylene(b)	9.5	1.5
Polystyrene(b)	11.1	1.7
Polymethyl methacrylate(b)	11.8	1.8
Polyvinyl chloride(b)	12.5	1.9
Polyethylene terephthalate	14.3	2.2

(a) Source: Ref 8. (b) Source: Ref 5

Table 3 Percentage of volume decrease on cooling

Material	Solidification temperature		Volume contraction, $\Delta V/V_0$, from solidification temperature, %
	°C	°F	
Polyethylene	120	250	22.0
Polyethylene terephthalate	200	390	13.6
Polycarbonate	160	320	3.2
Polysulfonate	185	365	3.2
Epoxy BP907	177	350	2.5

Source: Ref 10

tie molecules. Therefore, the tensile strength and modulus of the transcrystalline layer are significantly higher than those of the interior.

Transcrystalline layer thicknesses of 1 to 20 μm (40 to 790 μin.) are not uncommon. In addition to this layered composite semicrystalline structure, the more rapid crystallization that occurs in the transcrystalline layer is subject to continued partial slow crystallization after removal from the mold. Therefore, the stress distribution in a molded semicrystalline polymer can be complex in terms of depth from the mold surface, the morphology obtained, and the time and temperature (that is, aging storage conditions) after the molding process.

Thermal Stress Distribution. Thermal expansion, thermal diffusivity, and the variation of mechanical properties with temperature are factors that engineers need to be familiar with in working thermal stresses. A researcher (Ref 4) formulated a model for amorphous polymers encompassing these parameters. He modeled inhomogeneous cooling in flat sheets using thermoelastic theory. The main conclusions of this theory are:

- Stresses are proportional to A, which is defined as $A = \alpha E/(1 - \nu)$, where α is the expansion coefficient, E is the Young's modulus of the glassy polymer, and ν is Poisson's ratio.
- Stresses are determined by the temperature difference $T_0 - T_\infty$ and $T_g - T_\infty$, where T_0 is the initial temperature above T_g, and T_∞ is the final temperature below T_g.
- Stresses are further determined by $m = HL$/(thermal diffusivity), where m is the dimensionless Biot number, H is a convective heat-transfer coefficient, and $2L$ is the sample thickness.

In an effort to simplify the model, the researcher neglected time-temperature effects on mechanical properties, volume relaxation effects, and nonlinear stress effects. Figure 1 shows a plot of

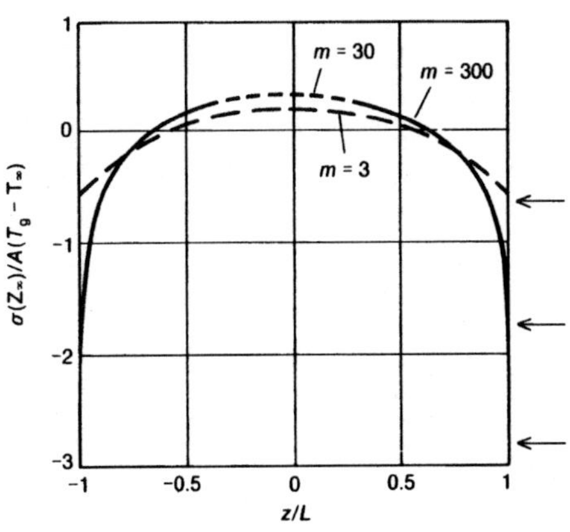

Fig. 1 Residual stress distributions for various values of the Biot number, m, for the case that the initial temperature, T_0, lies far above the glass transition temperature, T_g. T_∞, final temperature below T_g; L, sample thickness. Source: Ref 4

residual-stress distributions generated from this model. As the Biot number decreases, residual stresses decrease. This model makes apparent the factors that can be optimized in reducing residual thermal stresses:

- Select materials with lower glass moduli, or lower the CTEs.
- Decrease convective heat transfer, that is, air cool instead of ice cool.
- Increase thermal diffusivity. This results in more homogeneous cooling. It has been reported, for example, that thermal stresses are reduced by the incorporation of small amounts of conductive metal filler into the polymer matrix (Ref 8).

Thermal Mismatch. In composite structures, thermal stresses arise both from inhomogeneous cooling and as a result of thermal mismatch due to differences in CTEs between the filler and matrix polymer. Ideally, in terms of processing, one should attempt to minimize thermal stress buildup by minimizing the differences in CTEs. This is not always practical, because most fillers are inorganic materials with low CTEs.

Isotropic Effects. When spherical or particulate fillers are used, thermal contraction of the matrix exerts a compressive stress on the particle surface. This may actually be beneficial in that it can minimize decreases in the modulus due to poor filler-matrix adhesion (Ref 11). In extreme cases, however, cracking occurs, and the strength of the composite is diminished. The use of particulate fillers, especially spherical ones, offers the advantage of a lack of anisotropic shrinkage often found with fibrous fillers. In instances where polymers are joined with metal parts, it is desirable to decrease the CTE by the addition of filler to the polymer matrix. If the part is then exposed to extremes in temperature, the thermal mismatch between the filled polymer and metal surfaces is minimized, and thermal stresses are less likely to induce failure. In recent years, there has been an increased use of filled polymers in cryogenic applications due to the ability to decrease CTEs by using fillers (Ref 12). For spherical fillers, the CTE of the composite is expressed as (Ref 11):

$$\alpha = \alpha_1 \phi_1 + \alpha_2 \phi_2 - (\alpha_1 - \alpha_2)\,\phi_1 \phi_2 \left[\frac{1/B_1 - 1/B_2}{\phi_1/B_1 + \phi_2/B_2 + 3/4G_1}\right] \quad \text{(Eq 2)}$$

where the subscripts 1 and 2 refer to polymer and composite, respectively; ϕ is the volume fraction; α is the CTE; B is the bulk modulus; and G is the shear modulus.

Powder-filled epoxy resin systems have been developed for use as spacers between superconductive cables in synchrotron accelerators (Ref 13). In this system, the part must withstand thermal cycling between room temperature and liquid helium temperature. The experimenters adjusted the thermal contraction of the epoxy matrix to match that of the cable by adding appropriate amounts of calcium carbonate, talc, or asbestos fillers. They used a model for predicting thermal contraction in polymer matrices filled with spheres (Ref 14). Thermal contraction is defined as:

$$\alpha = \frac{L_1 - L_2}{L_1} \quad \text{(Eq 3)}$$

where L_1 is the sample length at room temperature, and L_2 is the sample length at liquid helium temperature.

Unfilled epoxy systems have a thermal contraction four times greater than that of the cable material. The experimenters succeeded in reducing the thermal contraction to match that of the metal without sacrificing the mechanical properties of the composite.

Anisotropic Effects. In the previous example, filler-matrix contraction is isotropic. In fiber-reinforced composites, the CTEs of the fibers are anisotropic, and they are often layered anisotropically. As a result of this anisotropy, the effect of thermal stress may be more severe than that of powder-filled systems.

Two researchers studied the thermal stress buildup in unidirectional graphite- and aramid-filled composites (Ref 10). The longitudinal CTE for graphite is -0.36×10^{-6}/K; for aramid, -2×10^{-6}/K. The radial CTE for graphite is 18×10^{-6}/K; for aramid, 59×10^{-6}/K. The lower limit of thermal stress in the longitudinal direction was approximated as:

$$\sigma_\parallel = \frac{\Delta_\alpha E_m E_L V_f}{E_m V_m + E_L V_f}(T_0 - T) \quad \text{(Eq 4)}$$

where Δ_α is the difference in CTEs of the matrix and the fiber in the longitudinal direction; T is the temperature; T_0 is the solidification temperature; E_L and E_m are the longitudinal modulus of the fiber and the modulus of the matrix, respectively; and V_f and V_m are the volume fractions of fiber and matrix, respectively.

The researchers measured the average difference in principal stresses, $\langle \sigma_\parallel - \sigma_\perp \rangle$, by photoelasticity, and from this they estimated the transverse stress, $\sigma_\perp$. The values are given in Table 4. Thermal stresses are obviously higher in the PSU composite than in the epoxy (EP) composite. This is because the volume contrac-

Table 4 Thermal stresses

Principal stresses	Graphite/polysulfone, $V_f = 0.33$	Graphite/epoxy BP907, $V_f = 0.35$
$\langle \sigma_\parallel - \sigma_\perp \rangle$, MPa (ksi)	31.4 (4.5)	20.0 (2.9)
$\langle \sigma_\parallel \rangle$, MPa (ksi)	25.7 (3.7)	15.2 (2.2)
$\langle \sigma_\perp \rangle$, MPa (ksi)	−5.7 (−0.8)	−4.8 (−0.7)

V_f, volume fraction. Source: Ref 10

tion of PSU is higher than that of EP (Table 3). The laminates are molded at pressures from 0.34 to 3.4 MPa (0.05 to 0.5 ksi). In high-pressure molding (100 to 500 MPa, or 14.5 to 72.5 ksi), the volume contraction of the resin is decreased, and thermal stresses are likely to be less severe.

The situation becomes more complex when one considers anisotropic fiber arrays. Laminates are often formed from layers of unidirectional plies, the direction of the plies being rotated to increase strength. Two types of residual thermal stresses develop: microstresses around each fiber and macrostresses between the plies.

One researcher used strain gages to measure directional expansion coefficients in unidirectional- and angle-ply laminates (Ref 15). Residual strains were measured from differences in the coefficients between angle-ply and single-ply laminates for particular directions in relation to the fibers. Residual stresses were calculated directly from these residual strains. In graphite-epoxy and aramid-epoxy laminates, residual stresses at room temperature exceeded the transverse tensile strength of the unidirectional composite. In addition, these stresses did not relax with time. Residual stresses and the tendency toward cracking were strongly dependent on ply lay-up. Transverse stresses increased from zero to a maximum as the angle between the two plies varied from 0 to 90°. Although ply rotation can be optimized to yield a zero net expansion coefficient, this may result in a considerable buildup of thermal stresses.

Thermal Stress Measurement. Thermal stresses can be determined quantitatively by complex, time-consuming methods. However, it is often more practical for the engineer to use qualitative methods to estimate the severity of the problem.

Methods developed for determining thermal stresses in metals have been adapted for use with polymers. One researcher proposed a method of estimating the average internal stress in a cross section of metal by stress relaxation (Ref 16). In stress relaxation tests, strain is kept constant, and stress decay is monitored as a function of time. Other researchers suggested a method of analyzing stress relaxation data to obtain the average internal stress, σ_i (Ref 8, 17, 18). The maximum slope of a stress log time (t) plot, F, is determined by:

$$F = -\left(\frac{d\sigma}{d \log t}\right)_{max} \quad \text{(Eq 5)}$$

A plot of F versus initial stress, σ_0, yields a straight line, intersecting the σ_0 axis at a value equivalent to the internal stress. This method is time-consuming and yields only an average stress rather than a stress profile (Ref 2, 19).

The layer removal technique is useful in measuring residual-stress distributions. It is applicable in a practical sense only to flat bars, plates, and pipes (Ref 2). Thin layers are removed from one face of the sample with high-speed milling machines. The face becomes unbalanced, and the sample takes on a curvature, ρ. A plot of curvature versus depth of removed material can be converted into a stress-versus-depth profile. Formulas for stress distributions using the layer removal technique are reviewed in Ref 2, and applications of this technique are abundant in the literature (Ref 1, 19, 20, 21).

For the case of a rectangular bar with no directional effects in the plane of the specimen (Ref 22):

$$\sigma_x(Z) = \sigma_y(Z) = \frac{-E}{6(1-\nu)}\left[(Z_0 + Z_1)^2 \frac{d\rho x(Z_1)}{dZ_1} + 4(Z_0 + Z_1)\rho x(Z_1) - 2\int_{Z_4}^{Z_0} \rho x(Z)\, dZ\right] \quad \text{(Eq 6)}$$

where E is the elastic modulus, ρx is the curvature parallel to the x direction, $Z = \pm Z_0$ are the original upper and lower surfaces, and Z_1 is the upper surface after layer removal.

As a word of caution, layer removal techniques assume that no gradient in modulus exists throughout the specimen thickness. However, density and modulus increase in the direction from the sample edge to the core (Ref 22, 23). This effect is more complex in oriented specimens, as discussed in the following section.

Unbalanced forces may result in curvature in unbalanced cross-ply laminates or in coatings cured on metal substrates (Ref 24, 25). Thermal stresses and strains have been related to the curvature in such systems. In processing simulated laminates, plies are separated by a release ply that is removed after cooling. Internal stresses are analyzed in terms of the deformation that occurs on separation (Ref 26).

Thermal Stress Evaluation. The engineer is often faced with the need to use less rigorous qualitative techniques. Microscopy may be valuable in revealing the presence of voids or cracks induced by thermal stresses and possible skin-core or crystalline morphology changes in molded parts (Ref 2, 19).

A number of qualitative techniques are given in Ref 8. Surface hardness, for example, decreases because of internal tensile stresses and increases for compressive stresses. The researchers (Ref 8) suggest a method for assessing the effect of internal stress on cracking tendencies. Here, a hole is drilled in the sample, and the sample is exposed to certain liquids. When such a sample of PS was placed in contact with n-hexane, samples with lower internal stresses showed less cracking. Also, they correlated shrinkage to thermal stresses: $\varepsilon_s = (\alpha_m - \alpha_s)/\alpha_m$, where ε_s is the mold shrinkage, α_m is the length dimension of the mold cavity, and α_s is the sample length. They noted a decrease in mold shrinkage in metal-filled samples. Modulus, tensile strength, and elongation-to-rupture usually show a weak dependence on internal stress level. (Of course, only average values of these parameters are measured when a tensile specimen is pulled if the gradient in properties in the thickness direction is not taken into account.) The previous methods for evaluating thermal stresses are, again, complicated by concurrent aging and orientation effects in processed parts.

Orientation Effects

As mentioned earlier, processing at elevated temperatures often results in residual orientation on cooling to below T_g. By using a model for rubber elasticity and the theory of the kinetic origin of rubber elasticity, a researcher (Ref 4) has shown that molecular orientation is accompanied by residual entropy stresses. Orientation and entropy stresses affect anisotropy effects. For example, the linear CTE depends on the direction of orientation, while the volume CTE for oriented and unoriented material is the same:

$$\alpha = \beta_{\parallel} + 2\beta_{\perp} \quad \text{(Eq 7)}$$

where $\beta_{\parallel}$ and $\beta_{\perp}$ are linear CTEs parallel and perpendicular to molecular alignment, respectively (Ref 5). Orientation-induced anisotropy is also evident in small-strain mechanical properties, creep, and yield behavior. The characterization of orientation and the effects of orientation on physical properties are discussed in Ref 27 to 31.

Of interest here are the more complicated, combined effects of thermal stresses and orientation that result from processing conditions. One of the most straightforward ways of separating these effects is to measure residual stresses in a processed part that has orientation and residual thermal stresses and to compare these results with identical specimens that are heated above the T_g to remove orientation and then quenched. The latter gives information on the thermal stress profile only.

The residual thermal stress profile for flat sheets (Fig. 1) is parabolic, with the compressive stresses on the surface and the core in tension. A team of researchers characterized residual stresses in injection-molded parts where flow-induced stresses and orientation accompany thermal stresses (Ref 23, 32, 33). Flow-induced tensile stresses maximize at the mold surface. When combined with thermal stresses, compression stresses on the surface are reduced. Extreme conditions inducing orientation result in tensile stresses at the surface.

The researchers investigated the effect of melt temperature, injection pressure, injection time, and mold temperature on residual stresses as determined by the layer removal technique. Curvature measurements were taken in the direction corresponding to stresses parallel to the flow direction. Polyphenylene oxide (PPO), PSU, and an amorphous polyamide (PA) were studied.

The conclusions of this work depict the effect of processing-induced orientation and residual stresses on mechanical properties, especially tensile strength, ultimate elongation, and elastic modulus. When melt and mold temperatures were optimized for each polymer, PPO and PSU showed an increase in modulus parallel to the injection direction as the injection rate increased (Ref 32, 33).

The ultimate properties of PSU were measured. Increasing the injection rate increased the ultimate strength and decreased the elongation-to-break. The decrease in the latter is associated with an increase in orientation parallel to the deformation axis. Increasing the injection pressure had similar effects on the ultimate elongation of PSU.

Polyamide samples were studied in more detail; that is, tensile properties were measured after successive layer removals on both sides. A distinct gradient in mechanical properties was noted. The elastic modulus increased with increasing injection pressure, and as the remaining sample thickness increased at low temperatures, ultimate properties increased toward the center of the sample; the opposite occurred at higher pressures.

It is interesting to note that in thermally treated and quenched PPO samples, gradients in the tensile properties increased on approach to the center of the sample (Ref 22, 23). Density increases paralleled this effect. This may be due to the need to consider aging effects. Aging effects are more pronounced in the inner layers of the sample, which cool at slower rates. Apparently, orientation and flow-induced stresses reverse this effect.

The failure properties of anisotropic molded parts are also anisotropic. It has been shown that the fracture energy of nylon 11 varies with respect to the flow direction in injection-molded samples (Ref 34). Notched Charpy impact tests at room temperature at a constant set hammer speed indicated that fracture occurs in stable and unstable propagation stages, each associated with a strain energy release rate. The strain energy release rates for both stable and unstable crack propagation are higher (by a factor of as much as 1.2) perpendicular to the flow direction. This is attributed to anisotropic craze resistance. This resistance is higher in the flow, or orientation, direction.

Processing-induced anisotropy also increases susceptibility to hostile environments. For example, extruded in-line drawn PS samples sorb hydrocarbons at accelerated rates transverse to the orientation direction. Such oriented samples also show increased dissolution rates (Ref 35).

This indicates that processing induces complex orientation and residual-stress effects. In analyzing these effects, sample orientation with respect to the flow directions and gradients in successive sample layers must be taken into account. Residual stress calculations by the layer removal technique must consider variations in the modulus, which occur in both annealed-quenched specimens and as-is injection-molded specimens.

Physical Aging

Physical aging at temperatures below the T_g has been extensively studied in linear amorphous polymers (Ref 3, 4, 36–39). This work has been extended to include cross-linked (Ref 37, 39) and filled (Ref 37, 40) polymers. It is speculated that aging occurs above the T_g in crystalline polymers because of the pinning action of crystalline components, which retard mobility (Ref 19). Aging is also known to occur in inorganic glasses and polycrystalline metals (Ref 3, 37). Of primary concern to the engineer is the fact that the onset of aging occurs when the polymer is cooled to the T_g and may continue for long periods of time, during which there is a simultaneous change in mechanical properties. Especially critical is that aging may have a profound effect on failure properties.

Thermodynamic Equilibrium. When cooled through the T_g, polymers behave like undercooled liquids; that is, they are not at thermodynamic equilibrium. As a result, the structure of the glassy polymer is continually changing in the course of the approach to the thermodynamic equilibrium state. This is best understood qualitatively from the concept of free volume (Ref 19). The nonequilibrium state is associated with free volume, which is accessible for molecular rearrangements. The approach to equilibrium is accompanied by a decrease of free volume, which then limits mobility and, therefore, rearrangement. The process of aging, at least in part, is associated with gradual densification of the material. This structural packing in turn causes the material to become more brittle (Ref 41).

The nonequilibrium state results in excess enthalpy, entropy, and volume. Physical aging has been characterized by decreases in excess volume and enthalpy (Ref 36, 38, 42–45). Although both quantities qualitatively parallel changes associated with aging, no quantitative relation to the free energy of the system has been found (Ref 46).

In volume relaxations, the typical increase in density is of the order of 0.5% (Ref 47). The most convenient method of recording changes in volume of this magnitude is volume dilatometry. Volume changes determined by this method are reported to have an accuracy of 1 to 5×10^{-5} cm^3/g (0.6 to 3×10^{-3} ft^3/lb). The use of the dilatometer is discussed in Ref 37 and 48.

One researcher measured the isobaric volume recovery in PS samples quenched from a temperature above T_g to temperatures below T_g (Ref 23). He developed a phenomenological theory to account for behavior such as this based on a distribution of retardation times (Ref 42). The model encompassed material constants and retardation spectra. It accurately predicted the response of glassy polymers to thermal treatments.

Another researcher proposed a mechanism for volume (V) relaxation based on defect annihilation (Ref 49). If positive and negative density defects or excess volumes combine during annealing, volume recovery follows second-order kinetics. By assuming that $V - V_\infty$ is proportional to the concentration of defects at time, t, he observed second-order kinetics at long aging times.

In another experiment, a researcher monitored enthalpy relaxations by differential scanning calorimetry (Ref 38, 43, 44). Briefly, enthalpy relaxation is accompanied by an absorption of energy or an endotherm in the region of the glass transition. Aging also shifts the position of the glass transition to higher temperatures. The energy absorbed increases with annealing time and approaches a maximum characteristic of each annealing temperature. Figure 2 shows typical endotherms resulting from annealing EP at times ranging from 0 to 52,000 min at 23 °C (73 °F). Measurements were taken with a differential scanning calorimeter at a heating rate of 10 °C/min (18 °F/min). The area under the curve represents the difference in enthalpy between initial and final temperatures. The area difference between that of the quenched sample (0 time) and that for a particular aging time gives a quantitative value for the enthalpy relaxation (Ref 39).

Aging and Physical Properties. The engineer is concerned with determining the effect of aging on mechanical properties and the susceptibility to solvent or corrosive environments. Changes in thermodynamic properties induced by aging can only be related qualitatively to changes in mechanical properties. The measurement of volume relaxation, for example, will not provide information that can quantitatively predict the effect of aging on yield stress or elongation-to-break. Aging is commonly interpreted in terms of a shift in relaxation times. Because rearrangements on the molecular scale involve more than one mode of motion, a spectrum or

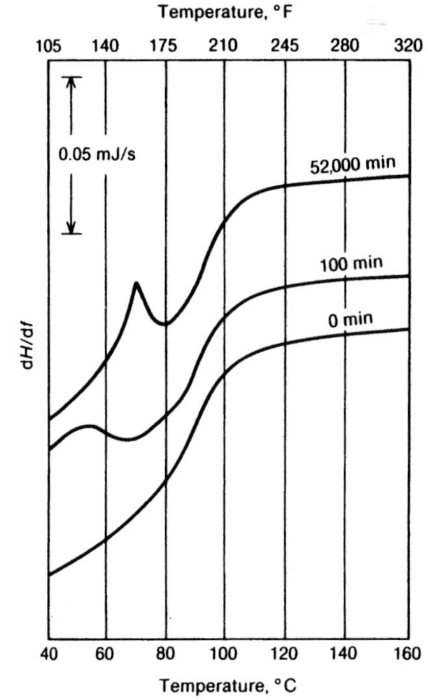

Fig. 2 Annealing time effects on differential scanning calorimetry traces of epoxy 828-0-0. Annealed at 23 °C (73 °F). H, convective heat-transfer coefficient. Source: Ref 39

distribution of relaxation times is used to model aging phenomenon. A primary effect of aging is a shift of the intrinsic relaxation time distribution to longer times. Enthalpy and volume recovery are best characterized by a broad distribution of relaxation times. Such broad distributions, however, predict that the effect of aging on mechanical properties is sluggish compared to experimental results (Ref 50).

General Effects of Aging on Mechanical Properties. The effects of aging on mechanical properties, especially creep behavior, have been extensively studied (Ref 3, 37). Figure 3 shows a series of tensile creep curves for polyvinyl chloride (PVC) quenched from 90 to 40 °C (195 to 105 °F). The creep curve shifts by 4.5 decades in aging from 0 to 1000 days (Ref 3). For a fixed time, the magnitude of the compliance changes by almost 50%. Such creep curves can be superimposed by a horizontal shift, indicating that aging is related to relaxation time changes. This horizontal shift, log *a*, varies linearly with log t_e, where t_e is the aging time. This is expressed as an aging shift rate:

$$\mu = \frac{d \log a}{d \log t_e} \qquad \text{(Eq 8)}$$

Above T_g, where no aging occurs, μ is 0. At the T_g and throughout the aging range, μ is unity. This is the case for most polymers with relatively flexible chains. Polymers such as cellulose-acetate-butyrate (CAB) have rigid backbone structures. CAB has a maximum μ value of 0.75 (Ref 51). This is attributed to the inability of the rigid structure to age as well as more flexible structures.

Another phenomenon associated with aging is thermoreversibility. In an experiment similar to that shown in Fig. 2, a sample was reheated after 1000 day aging. It was then quenched and aged for 1 day. The creep curve for this sample superimposed the creep curve for the original 1 day aging curve (Ref 3).

This extensive work in the area of aging has led to a characterization of basic aging aspects, some of which are (Ref 3):

- Aging affects the long-term behavior of plastics. Aging time is the main parameter that affects small-strain properties.
- In the aging range, all polymers age the same. The small-strain behavior of PS, for example, is similar to that of polycarbonate (PC) and other glassy structures.
- Aging is a general phenomenon; it has been observed in all glassy structures, such as bitumen, shellac, amorphous sugar, and molded dry cheese powder.
- Aging persists for long periods of time. If t_∞ is the time required to reach equilibrium glass structure at a temperature T below T_g, t_∞ increases by approximately a factor of 10 per 3 °C (5 °F) in $T_g - T$.
- Aging occurs at temperatures from T_g down to the temperature of the highest secondary transition associated with localized rather than segmental motion. Below this temperature, aging appears to cease.

Ductile-Brittle Behavior. In reference to this last point, the magnitude of the temperature range from the local mode transition to the glass transition is also an indication of the temperature range over which a polymer exhibits ductile versus brittle behavior. Polymers, such as PC, with broad aging temperature ranges will exhibit ductility over a broad range in temperatures. The converse is true of PS and polymethyl methacrylate (PMMA).

The work on aging has shed new light on ductile versus brittle behavior. It should be mentioned that such behavior is characterized by a ductile-to-brittle transition temperature. This temperature is not a material property; instead, it depends on the strain rate and the imposed stress configuration. As the strain rate increases, the ductile-to-brittle transition shifts to higher temperatures. At this transition temperature, there is competition between the ductile mode of failure (shear banding) and the brittle mode of failure (crazing) (Ref 52). At a particular strain rate, then, polymers exhibit ductile behavior over a range of temperatures, $T - T_g$, the breadth of which increases as the aging temperature range increases.

In the past, it was thought that secondary transitions were responsible for enhancing ductility in polymers. In some cases, there were direct correlations between ductile behavior and secondary transitions, but this observation does not extend to all polymers. This topic is discussed in Ref 53. More recent studies indicate that aging has a profound influence on ductile-versus-brittle behavior. Both aging and ductility are believed to require some segmental mobility (Ref 3). Below the secondary transition, there is no segmental mobility, and both ductile behavior and aging cease.

Aging and Transition Behavior. Dielectric and dynamic mechanical spectroscopies reveal the effects of aging in relation to primary and secondary transitions. Quenched polymers exhibit higher damping than slow-cooled or annealed polymers. The T_g is also lower in quenched specimens. There is evidence that quenching decreases the modulus (Ref 54). Of interest is the fact that quenching broadens the low-temperature end of the damping peak for the glass transition. Because aging is associated with main-chain or segmented motion, it is important to know whether or not the broadening of the glass transition damping peak extends into regions encompassing or affecting the secondary transition region (Ref 3).

Two researchers monitored the effect of cooling rate on primary and secondary transitions in amorphous methacrylate polymers by dielectric relaxation (Ref 55). The polymers investigated were PMMA, polyethyl methacrylate (PEMA), polybutyl methacrylate (PBMA), and polyisobutyl methacrylate (P-iso-BMA). They found, in general, that annealing lowered damping in regions below T_g and above the secondary transition. Previously, they reported that in some cases, quenching resulted in the appearance of a new damping peak between these two transitions (Ref 56).

In PEMA, PBMA, and P-iso-BMA, the secondary transition is well separated from the glass transition. They found that the region below T_g affected by quenching was constant, irrespective of frequency (frequencies were varied from 60 Hz to 50 kHz), even though the β peak moved to higher temperatures according to the activation energy. This indicates the volume contraction effects, mainly segmental motion. Effects are not discernible in polymers where the damping peaks of the T_g and secondary transition overlap. This is the case with PMMA. Secondary transitions have a much lower activation energy than the glass transition. Conse-

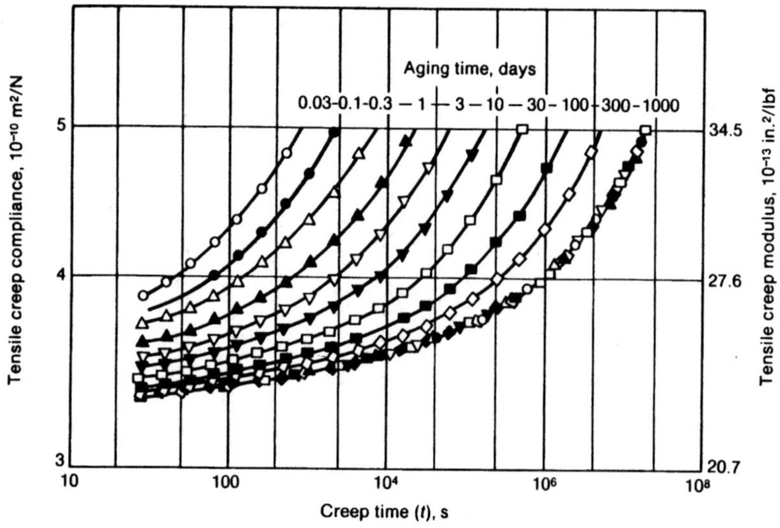

Fig. 3 Polyvinyl chloride quenched from 90 to 40 °C (195 to 105 °F). Accurate to ±2%. Source: Ref 37

quently, the secondary peak shifts with frequency at a more rapid rate, merging with T_g at higher frequencies. At 60 Hz, annealing effects were apparent in the relaxation spectrum of PMMA. At higher frequencies, when the damping peaks merged, no annealing effects were apparent. As mentioned earlier, PMMA has a much smaller aging temperature range than other amorphous polymers, such as PC.

Aging and High-Strain Behavior. Aging effects are apparent in small-strain creep experiments and in mechanical and dielectric measurements. In considering the failure of plastic parts, it is important to extend aging studies to include effects due to higher deformations.

Creep rates have been examined at stresses that induce nonlinear deformation (Ref 37). At higher deformations, the aging effect is erased. Nonlinear creep curves were shifted to the right as aging time increased. The shifting appeared to be horizontal in the samples examined. The shift was still characterized by the double logarithmic shift rate, μ, which decreases with increasing stress. Nevertheless, the shift was the same for the polymers investigated. Both shear and tensile deformations were examined. The shift, again, was independent of the type of deformation.

In a study of nonlinear deformations and physical aging in PMMA, two researchers measured the torque and normal force in stress relaxation tests for different aging times (Ref 57). They found that from 40 to 60 °C (105 to 140 °F), the aging curves for linear, small-strain deformations could not be superimposed by horizontal and vertical shifts. At higher temperatures or at deformations in the nonlinear range, the curves were superimposed by a horizontal shift. Shifts decreased as the strain increased. In addition, shift rates were significantly lower for torque than for normal force measurements. It has been suggested that torque measurements are less sensitive than normal force measurements to aging-induced structural rearrangements. Large deformations may also affect the structure in such a way that torque and normal force measurements respond differently. The inability to superimpose data by a horizontal shift factor is due to possible contributions from the secondary relaxation process. Again, PMMA has a short aging range. It is worthwhile to investigate a broader range of polymers by this method of testing.

In recent years, mechanical testing has evolved to such an extent that large-strain behavior can be probed by very sensitive techniques. This is done by the superposition of small stresses or strains onto large stresses. These techniques depict the erasing of aging that follows large deformations. A group of researchers examined the behavior of PMMA by such a technique (Ref 58). They used torsional microcreep to monitor the effect of aging time on creep behavior. On aged specimens, they applied increasing longitudinal stresses and simultaneously measured microcreep. Microcreep alone clearly depicted the aging. In the early stages of microcreep (200 to 1200 s) at 90 °C (195 °F), microcreep was logarithmic:

$$\gamma_1 = A\tau \ln \alpha t \quad \text{(Eq 9)}$$

where γ_1 is the strain, A is the constant that characterizes strain rate, τ is the stress, $\alpha = 10$ and is a constant to account for time of stress application, and t is time.

After this stage, microcreep reached a stable strain rate characterized by:

$$\dot{\gamma}_s = B\tau \quad \text{(Eq 10)}$$

During logarithmic creep, the effect of aging time, t_e, on the constant, A, followed the form:

$$10^{11} A = 1.02 \log t_e + 4.92 \quad \text{(Eq 11)}$$

The parameter B appeared to decrease with t_e in a logarithmic fashion. Both parameters characterize strain rate. The logarithmic decrease with t_e is thought to be analogous to the horizontal shift. When a series of increasing longitudinal stresses was applied during microcreep on one aged sample, the parameters A and B decreased with stress increase in an opposite fashion to those in the aging study. This effect was noted only at longitudinal strains greater than 1%. This indicates that the structure that evolves during high-strain deformations is similar to that in quenched, unaged samples.

High-strain deformations are also found to influence aging in yield stress measurements. It has long been known that tensile, flexural, torsional, and compressive yield stresses increase with aging (Ref 37, 38, 59, 60). Plots of yield stress versus logarithm of strain rate for various aging times show behavior similar to that of small-strain creep behavior (Fig. 3). The magnitude of the shift is, again, much smaller than that in small-strain experiments (Ref 37). Also of importance is whether or not aging shifts the mode of failure from yield to brittle fracture. An excellent example of this is shown in Fig. 4. Aged and unaged amorphous polyethylene terephthalate (PET) was pulled in tension at a strain rate of 10%/min. Aging for only 90 min at 50 °C (120 °F) resulted in brittle behavior. Aging at room temperature for only 4 days produced the same effect (Ref 44).

The effect of aging on a yield or embrittlement is not always only an effect of a decrease in the volume due to aging. Intramolecular and intermolecular conformations influence the effect of aging on deformation. A team of researchers studied the effect of PC structure on high-deformation behavior and aging (Ref 47). The polymers used were bisphenol A polycarbonate, polyester carbonate with varying ratios of terephthalate and isophthalate esters, bisphenol A/phenophthalein random copolycarbonate, and PSU.

The effect of aging on density was monitored by dynamic mechanical spectroscopy. This technique revealed a secondary transition in PCs at approximately 70 °C (160 °F). This transition is believed to be due to a cooperative motion of three to four monomer units and is therefore not highly localized.

Stress-strain curves for the samples showed a postyield stress drop. This stress drop decreased with increasing damping peaks associated with the secondary transition. Aging increased the postyield stress drop and decreased the height of the damping peak. Increasing T units increased peak height and decreased the stress drop. The magnitude of this stress drop is an indication of the amount of conformational change and packing that is unfavorable for ductile deformation. Damping peak heights also correlated with time-to-embrittlement due to aging for a particular strain rate. The higher the peak height, the greater the resistance to embrittlement. Densities showed a decrease with aging, but these data did not correlate with embrittlement data. This is an indication that both free volume and conformational changes take place during aging-induced deformation effects.

The effect of aging on the degree of ductility is further depicted in shear band studies. Shear banding is a form of inhomogeneous deformation observed in compression studies. It has been well characterized in polymers such as PMMA, PS, PET, and PC (Ref 43, 61–65). Shear yield induces the formation of two types of slip bands: coarse and fine. Temperature, strain rate, and aging influence the type of band formed. Higher temperatures, slower strain rates, and decreased aging favor the formation of the more diffuse fine bands. Fine bands induce ductile fracture after large deformations, while coarse bands induce brittle fracture after propagating through the specimens. Annealing increases the tendency to form coarse bands (Ref 61). This is yet another demonstration of the effect of aging on embrittlement.

As expected, aging influences other high-strain properties as well. Aging induces a decrease in strain-to-break for rubber-modified and pure EP systems (Ref 39). Similarly, in graphite EP complexes, decreases in tensile strength, toughness, and strain-to-break accompanied increases in aging times (Ref 40).

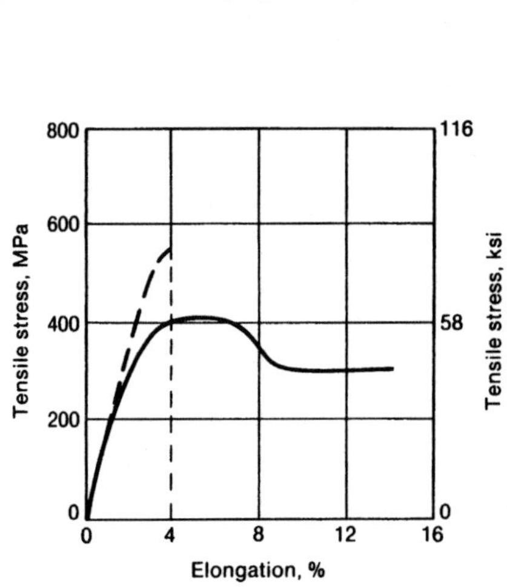

Fig. 4 Tensile stress-strain curves for amorphous polyethylene terephthalate film unannealed (solid line) and annealed at 51 °C (125 °F) for 90 min (dashed line). Source: Ref 44

Other effects of aging on materials extend beyond the realm of density increases and mechanical properties. Of particular importance to the engineer is the influence of aging on the susceptibility to solvent or swelling environments. For example, it has been shown that annealing decreases the diffusion coefficient of methane and propane in glassy polymers (Ref 66). Here, the vapor concentration was low enough so that swelling did not occur. Another researcher studied the effect of annealing on the sorption of *n*-hexane in glassy polymers (Ref 67). Annealing decreased sorption rates by factors as high as 100. Equilibrium solubilities were unaffected. It is important, then, to consider the thermal history of polymer parts that are exposed to such environments.

Use of High-Modulus Graphite Fibers in Amorphous Polymers

High-modulus graphite fibers impart strength to composites. However, thermal stresses build up during processing in the temperature range from solidification to ambient temperature. If the structures survive in this temperature range, the question arises as to how far these structures can be cooled before thermal stresses cause failure.

In this study, six amorphous polymers are examined for suitability in graphite composites. The longitudinal stress due to the presence of the fiber is calculated for different temperature ranges, ΔT, from the solidification temperature, T_0, to T_f, where T_f is the room temperature, liquid nitrogen temperature, or liquid helium temperature. The longitudinal tensile stress is significantly higher than the transverse stress and is therefore likely to be the stress controlling the failure. The compressive strength of amorphous polymers is also greater than the tensile stress. The stress, $\sigma_{\parallel}$, is calculated as follows (Ref 10):

$$\sigma_{\parallel} = \frac{\Delta_{\alpha} E_m E_L V_f}{E_m V_m + E_L V_f}(T_0 - T) \qquad \text{(Eq 12)}$$

where the symbols are as defined for Eq 4, and E_m = 220 GPa (32 × 10^6 psi). If $E_L \gg E_m$ (as is the case with graphite fibers as compared to amorphous polymers), then Eq 12 can be approximated as (Ref 10):

$$\sigma_{\parallel} = \int_{T_f}^{T_0} \Delta\sigma\, E_m\, dT \qquad \text{(Eq 13)}$$

Because $\sigma_{\parallel}$ is actually a lower limit of residual stress (Ref 10), a safety factor should be used when considering failure. For the purpose of comparison, the safety factor is eliminated.

The relevant properties of the polymers are shown in Table 5. Table 6 shows calculated values of $\sigma_{\parallel}$ versus the tensile strength of the polymer. In these calculations, a modulus invariant with temperature is assumed. The calculations are valid for V_f at 10% and greater. Above 10%, $\sigma_{\parallel}$ is constant (Ref 10).

The $\sigma_{\parallel}$ values for cooldown from solidification temperatures to room temperature indicate that the polymer matrix is likely to remain intact. On cooling to liquid nitrogen or liquid helium temperature, it is doubtful that any structures would hold up. The EP compound is the closest to being able to withstand these temperatures. Advances in EP chemistry have resulted

Table 5 Properties of polymers

Material	Linear coefficient of thermal expansion, 10^{-6}/K	Tensile modulus, GPa	Tensile modulus, 10^5 psi	Tensile strength, MPa	Tensile strength, ksi	Temperature solidification, °C	Temperature solidification, °F
Polymethyl methacrylate	70	2.76	4.00	65.6	9.5	97	207
Polyacrylonitrile	66	3.76	5.45	62.0	9.0	95	203
Polycarbonate	68	2.38	3.45	65.6	9.5	150	302
Polystyrene	63	2.77	4.02	42.8	6.2	102	216
Polyvinyl chloride	75	3.27	4.75	46.9	6.8	80	176
Cast epoxy	55	2.41	3.50	58.7	8.5	170	338

Source: Ref 6

Table 6 Calculated values of longitudinal stress, $\sigma_{\parallel}$, versus tensile strength of various polymers

Material	Tensile strength, MPa	Tensile strength, ksi	$\sigma_{\parallel}$ Cool to room temperature, MPa	ksi	$\sigma_{\parallel}$ Cool to liquid nitrogen, MPa	ksi	$\sigma_{\parallel}$ Cool to liquid helium, MPa	ksi
Polymethyl methacrylate	66	9.5	15	2.1	57	8.3	71	10.3
Polyacrylonitrile	62	9.0	18	2.6	73	10.6	90	13.1
Polycarbonate	66	9.5	21	3.0	57	8.2	68	9.9
Polystyrene	43	6.2	14	2.0	52	7.6	66	9.5
Polyvinyl chloride	47	6.8	14	2.0	44	6.4	86	12.5
Epoxy	59	8.5	19	2.8	35	5.1	59	8.5

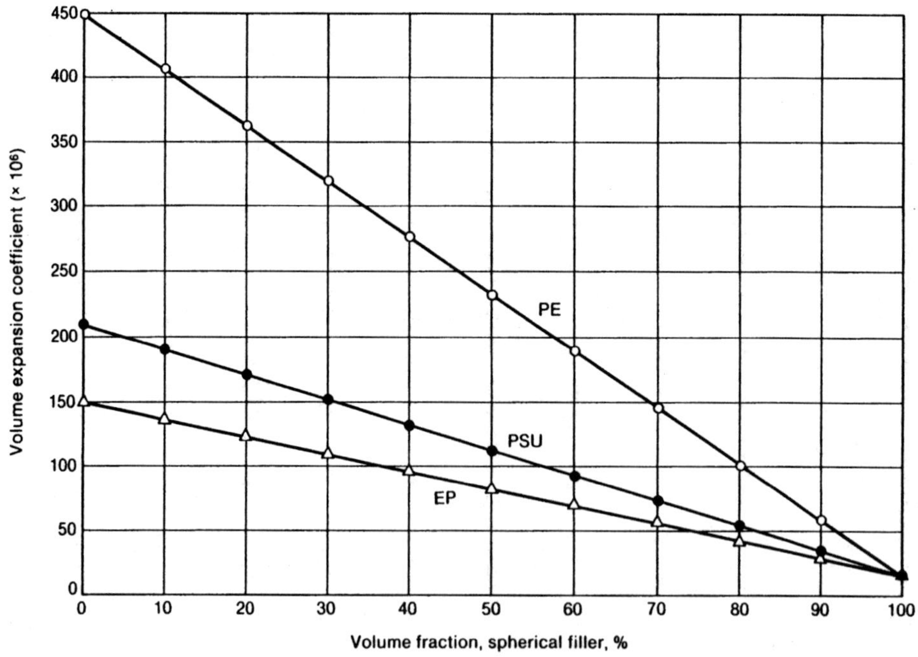

Fig. 5 Expansion coefficients, per linear rule of mixtures. PE, polyethylene; PSU, polysulfone; EP, epoxy

in stronger EP compounds. This is one approach to the problem of expanding the useful temperature range for graphite EPs. In addition, it is possible to use a combination of two fillers: a spherical filler and a fiber. The spherical filler reduces the expansion coefficient with less residual-stress effects. This reduction in σ_c is calculated from the linear rule of mixtures (Ref 11):

$$\sigma_c = \phi_1\sigma_1 + \phi_2\sigma_2 \qquad \text{(Eq 14)}$$

where ϕ is the volume fraction, and the subscripts 1 and 2 refer to matrix and fillers. Equation 14 is a close approximation. Figure 5 shows a plot of σ_c versus volume fraction of filler for three representative polymers. For EP, a 30% level of filler significantly reduces σ_c.

Residual-stress modeling in three-component systems is complicated. Nonetheless, calculations such as those in Table 6 give the engineer an approximate direction in which to proceed.

REFERENCES

1. P. So and L.J. Broutman, Residual Stresses in Polymers and Their Effect on Mechanical Behavior, *Polym. Eng. Sci.,* Vol 6 (No. 12), Dec 1976, p 785
2. J.R. White, Origins and Measurement of Internal Stress in Plastics, *Measurement Techniques for Polymer Solids,* R.P. Brown and B.E. Read, Ed., Elsevier, 1984, p 165
3. L.C.E. Struik, Physical Aging in Plastics and Other Glassy Materials, *Polym. Eng. Sci.,* Vol 17 (No. 3), March 1977, p 165
4. L.C.E. Struik, Orientation Effects and Cooling Stresses in Amorphous Polymers, *Polym. Eng. Sci.,* Vol 18 (No. 12), Aug 1978, p 799
5. D.W. Van Krevelen, *Properties of Polymers,* Elsevier, 1976, p 67, 395
6. *Modern Plastics Encyclopedia,* Vol 59 (No. 10A), Plastics Catalogue Corporation, Oct 1982, p 466
7. *Handbook of Chemistry and Physics,* 4th ed., Chemical Rubber Publishing Company, 1958, p 2239
8. J. Kubát and M. Rigdahl, Reduction of Internal Stresses in Injection Molded Parts by Metallic Fillers, *Polym. Eng. Sci.,* Vol 16 (No. 12), Dec 1976, p 792
9. J.V. Schmitz, Ed., *Testing of Polymers,* Vol 2, Interscience, 1966, p 208
10. J.A. Nairn and P. Zoller, Matrix Solidification and the Resulting Residual Thermal Stresses in Composites, *J. Mater. Sci.,* Vol 20, 1985, p 355
11. L.E. Nielsen, *Mechanical Properties of Polymers and Composites,* Vol 2, Marcel Dekker, 1974, p 434
12. G. Claudit, F. Disdier, and M. Locatelli, Interesting Low Temperature Thermal and Mechanical Properties of a Particular Powder-Filled Polyimide, *Nonmetallic Materials and Composites at Low Temperatures,* A.F. Clark, R.P. Reed, and G. Hartwig, Ed., Plenum Press, 1978, p 131
13. K. Ishibashi, W. Wake, M. Kobayashi, and A. Katase, Powder-Filled Epoxy Resin Composites of Adjustable Thermal Contraction, *Nonmetallic Materials and Composites at Low Temperatures,* Plenum Press, 1978, p 291
14. G. Hartwig and W. Weiss, Die Thermische Ausdehnung von Pulvergefüllten Epoxidharzen, *Mater. Sci. Eng.,* Vol 22, 1976, p 261
15. I.M. Daniel, Thermal Deformations and Stresses in Composite Materials, *Thermal Stresses in Severe Environments,* Plenum Press, 1980, p 607
16. J.C.M. Li, Dislocation Dynamics in Deformation and Recovery, *Can. J. Phys.,* Vol 45, 1967, p 493
17. J. Kubát, J. Peterman, and M. Rigdahl, Internal Stresses in Polyethylene as Related to Its Structure, *Mater. Sci. Eng.,* Vol 19, 1975, p 185
18. J. Kubát and M. Rigdahl, A Simple Model for Stress Relaxation in Injection Molded Plastics with an Internal Stress Distribution, *Mater. Sci. Eng.,* Vol 21, 1975, p 63
19. L.D. Coxon and J.R. White, Residual Stresses and Aging in Injection Molded Polypropylene, *Polym. Eng. Sci.,* Vol 20 (No. 3), 1980, p 230
20. A. Bhatnagar and L.J. Broutman, Effect of Annealing and Heat Fusion on Residual Stresses in Polyethylene Pipe, *Proceedings of the 43rd Annual Technical Conference,* Society of Plastics Engineers, 1985, p 545
21. J.G. Williams and J.M. Hodgkinson, The Determination of Residual Stresses in Plastic Pipe and Their Role in Fracture, *Polym. Eng. Sci.,* Vol 21 (No. 13), 1981, p 822
22. A. Siegmann, A. Buchman, and S. Kenig, Comments on the Layer Removal Method for Measurements of Residual Stresses in Plastics, *J. Mater. Sci. Lett.,* Vol 16, 1981, p 3514
23. A. Siegmann, S. Kenig, and A. Buchman, Residual Stresses in Polymers II: Their Effect on Mechanical Behavior, *Polym. Eng. Sci.,* Vol 21 (No. 15), 1981, p 997
24. F.R. Jones and M. Mulheron, Generation of Thermal Strains in GRP, *J. Mater. Sci. Eng.,* Vol 18, 1983, p 1533
25. M. Shimbo, M. Ochi, and K. Arai, Effect of Solvent and Solvent Concentration on the Internal Stress of Epoxide Resin Coatings, *J. Coatings Technol.,* Vol 57 (No. 728), 1985, p 93
26. J.A. Manson and J.C. Seferis, Internal Stress Determination by Process Simulated Laminates, *Proceedings of the 45th Annual Technical Conference,* Society of Plastics Engineers, 1987, p 1446
27. I.M. Ward, Ed., *Structure and Properties of Oriented Polymers,* John Wiley & Sons, 1975
28. B.E. Read, J.C. Duncan, and D.E. Meyer, Birefringence Techniques for the Assessment of Orientation, *Measurement Techniques for Polymer Solids,* R.P. Brown and B.E. Read, Ed., Elsevier, 1984, p 143
29. S. Raha and P.B. Bowden, Birefringence of Plastically Deformed Poly-(Methyl Methacrylate), *Polymer,* Vol 31, 1972, p 174
30. E. Saiz, E. Riande, and J.E. Mark, Birefringence in Poly(Methyl Acrylate) Networks in Elongation, *Macromolecules,* Vol 17, 1984, p 899
31. I.M. Ward, Review: The Yield Behavior of Polymers, *J. Mater. Sci.,* Vol 6, 1971, p 1397
32. A. Siegmann, S. Kenig, and A. Buchman, Residual Stresses in Injection-Molded Amorphous Polymers, *Polym. Eng. Sci.,* Vol 27 (No. 14), 1987, p 1069
33. A. Siegmann, A. Buchman, and S. Kenig, Residual Stresses in Polymers III: The Influence of Injection-Molding Process Conditions, *Polym. Eng. Sci.,* Vol 22 (No. 9), 1982, p 560
34. T. Vu-Khanh and F.X. De Charentenay, Mechanics and Mechanism of Impact Fracture in Semi-Ductile Polymers, *Polym. Eng. Sci.,* Vol 25 (No. 13), 1985, p 841
35. L. Nicolais, E. Driolo, H.B. Hopfenberg, and A. Apicella, Effects of Orientation and the Penetration Crazing and Dissolution of Polystyrene by N-Hexane, *Polymer,* Vol 20, 1979, p 459
36. A.J. Kovacs, La Contraction Isotherme du Volume des Polymères Amorphes, *J. Polym. Sci.,* Vol XXX, 1958, p 131
37. L.C.E. Struik, *Physical Aging in Amorphous Polymers and Other Materials,* Elsevier, 1978
38. S.E.B. Petrie, Thermodynamic Equilibrium in Glassy Polymers, *Polymeric Materials,* American Society for Metals, 1973, p 55
39. Z.H. Ophir, J.A. Emerson, and G.L. Wilkes, Sub-T_g Annealing Studies of Rubber Modified and Unmodified Systems, *J. Appl. Phys.,* Vol 49 (No. 10), 1978, p 5032
40. E.S.W. Kong, S.M. Lee, and H.G. Nelson, Physical Aging in Graphite Epoxy Blends, *Polym. Compos.,* Vol 3 (No. 1), 1982, p 29
41. S. Matsuoka, Thermodynamic Aspects of Brittleness in Glassy Polymers, *Toughness and Brittleness of Plastics,* R. Deanin and A.M. Crugnola, Ed., Advances in Chemistry Series 154, American Chemical Society, 1972, p 3
42. A.J. Kovacs, A Multiparameter Approach for Structural Recovery of Glasses and Its Implication for Their Physical Properties, *Annals of the New York Academy of Sciences,* Vol 371, J.M. O'Reilly and M. Goldstein, Ed., New York Academy of Sciences, 1981, p 38
43. S.E.B. Petrie, Thermal Behavior of Annealed Organic Glasses, Part A-2, *J. Polym. Sci.,* Vol 10, 1972, p 1255
44. S.E.B. Petrie, The Effect of Excess Thermodynamic Properties Versus Structure Formation on the Physical Properties of Glassy Polymers, *J. Macromol. Sci. Phys.,* Vol B12 (No. 2), 1976, p 225
45. R.E. Robertson, The Aging of Glassy Poly-

mers as Determined by Scanning Calorimeter Measurements, *J. Appl. Phys.*, Vol 49 (No. 10), 1979, p 5048
46. S.S. Sternstein, Mechanical Properties of Glassy Polymers, *Properties of Solid Materials, Part B*, J.M. Schultz, Ed., Academic Press, 1977, p 541
47. R.A. Bubeck and S.E. Bales, Changes in Yield and Deformation of Polycarbonates Caused by Physical Aging, *Polym. Eng. Sci.*, Vol 24 (No. 10), 1984, p 1142
48. L.H. Sperling, Chapter 6, in *Introduction to Physical Polymer Science*, John Wiley & Sons, 1986, p 239
49. J.C.M. Li, Physical Chemistry of Some Microstructural Phenomena, *Metall. Trans. A*, Vol 9 (No. 10), 1978, p 1353
50. Y.P. Chen and J.J. Aklonis, Multiordering Parameter Models of Volume and Enthalpy Recovery Generalized to Treat Physical Aging: A Quantitative Investigation, *Polym. Eng. Sci.*, Vol 27 (No. 17), 1987, p 1275
51. G. Levitar and L.C.E. Struik, Physical Aging in Rigid Chain Molecules, *Polymer*, Vol 24, 1983, p 1071
52. R.P. Kambour, D. Faulkner, E.E. Kampf, S. Miller, G.E. Niznik, and A.R. Schultz, Toughness Enhancement by Introduction of Silicone Blocks into Polycarbonates of Bisphenol Acetone and Bisphenol Fluorenone, *Toughness and Brittleness of Plastics*, R.D. Deanin and A.M. Grugnola, Ed., Advances in Chemistry Series 154, American Chemical Society, 1976, p 312
53. R.F. Boyer, Dependence of Mechanical Properties on Molecular Motion in Polymers, *Polym. Eng. Sci.*, Vol 8 (No. 3), 1968, p 161
54. L.E. Nielsen, *Mechanical Properties of Polymers*, Vol 2, Marcel Dekker, 1974, p 165
55. J.L. Gómez and R. Díaz, Effect of the Cooling Rate in the Formation of Glass on the α and β Relaxations of Some Amorphous Polymers, *Polym. Eng. Sci.*, Vol 24 (No. 15), 1984, p 1202
56. R. Díaz and J.L. Gómez, The Influence of Thermal History of Dielectric Properties of Poly (Vinyl Chloride), *Polym. Eng. Sci.*, Vol 22, 1982, p 845
57. G.B. McKenna and A.J. Kovacs, Physical Aging of Poly (Methyl Methacrylate) in the Nonlinear Range: Torque and Normal Force Measurements, *Polym. Eng. Sci.*, Vol 24 (No. 10), 1984, p 1138
58. N. Gowri Shankar, Y.A. Bertin, and J.L. Gacougnolle, Analysis of the Evolution of Microcreep During Physical Aging and Mechanical Deformation in Poly(Methyl Methacrylate) Using A Microstructural Model, *Polym. Eng. Sci.*, Vol 24 (No. 11), 1984, p 921
59. C. Bauwens-Crowet and J.C. Bauwens, The Relationship Between the Effect of Thermal Pre-Treatment and the Viscoelastic Behavior of Polycarbonate in the Glassy State, *J. Mater. Sci.*, Vol 14, 1979, p 1817
60. T.E. Brady and G.S. Yeh, Yielding Behavior of Glassy Amorphous Polymers, *J. Appl. Phys.*, Vol 42 (No. 12), 1971, p 4622
61. J.C.M. Li and J.B.C. Wu, Pressure and Normal Stress Effects in Shear Yielding, *J. Mater. Sci.*, Vol 11, 1976, p 445
62. J.C.M. Li, Behavior and Properties of Shear Bands, *Polym. Eng. Sci.*, Vol 24 (No. 10), 1984, p 750
63. L. Camwell and D. Hull, Crazing and Fracture Associated with Interaction of Shear Bands in Polystyrene, *Philos. Mag.*, Vol 27, 1973, p 1135
64. P.B. Bowden and S. Raha, The Formation of Micro Shear Bands in Polystyrene and Polymethylmethacrylate, *Philos. Mag.*, Vol 22, 1970, p 463
65. W. Wu and A.P.L. Turner, Shear Bands in Polycarbonate, *J. Polym. Sci.*, Vol 11, 1973, p 2199
66. S.P. Chen, Effect of Volume Relaxation, Temperature and Chemical Structure on Diffusion of Gaseous Hydrocarbons Through Glassy Polymers, *Polym. Prepr.*, Vol 51 (No. 1), 1974, p 71
67. H.B. Hopfenberg, V.T. Stannett, and G.M. Folk, Sorption Kinetics and Equilibria in Annealed Glassy Polyblends, *Polym. Eng. Sci.*, Vol 15 (No. 4), 1975, p 261

Environmental Stress Crazing*

CRAZING can be described as the formation of regions of plastic deformation normal to the local tensile strain. In glassy thermoplastics, crazes appear as whitened areas that are visually indistinguishable from cracks. Figure 1 illustrates the phenomenon on a sample of polycarbonate. Viewed microscopically, however, it is evident (Fig. 2) that fibrous material bridges these regions, which means that crazes can be load bearing.

Although crazing is generally the precursor to cracking, the distinction between a craze and a crack is far from an academic issue. The concept of rubber toughening of thermoplastic materials is based on using the crazing phenomenon to advantage so that fracture energy can be absorbed over a wide area of the structure rather than localized at a single area of weakness. In addition, the appearance of a craze at the tip of a preexisting crack can have significant implications, from a fracture mechanics standpoint, in predicting crack growth rates under stable crack growth conditions, and thus is critical in determining the service life of a given thermoplastic system.

As engineering plastics find their way into new and more demanding applications, their resistance to failure in specific chemical environments becomes a critical consideration. Often, environmental stress crazing (ESC) is the life-limiting mode of failure. For example, polyethylene, which could hardly be considered an engineering plastic because of its wide acceptance in items such as milk containers, is now finding its way into more demanding areas, such as in-ground liners for solid and hazardous waste disposal. These liners are placed under landfills to prevent groundwater contamination, an application that requires a service life of at least 30 years. Resistance to cracking due to the combined effects of stress and chemicals leaching from the waste essentially determines whether the service life objective is met.

The engineer who wishes to work with thermoplastics in a given environment needs to consider particular questions and problems:

- Why certain environments promote crazing in polymers under stress
- How to identify environments that promote crazing in specific polymer systems
- What, if anything, can be done to optimize materials to improve resistance to environmentally induced crazing
- How to identify appropriate tests to determine the susceptibility of polymers to this mode of failure in specific environments

The phenomenon of ESC in glassy amorphous thermoplastics has been recognized for almost 40 years. Direct evidence of crazing by ESC of semicrystalline polytetrafluoroethylene was observed as early as 1973 and then later in polyethylene and nylon (Ref 2–5); thus, craze growth and breakdown in these materials also are described in this article.

Molecular Mechanism

For glassy thermoplastics, the crazing phenomenon is manifest as linear regions of local plastic deformation forming perpendicular to the applied stress. The craze itself is a highly voided, spongy structure of material oriented across its width.

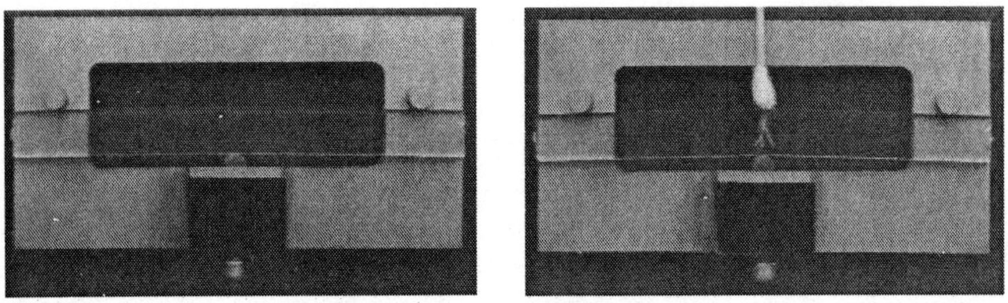

(a) (b)

Fig. 1 Environmental stress crazing in a sample of polycarbonate under three-point bending. (a) Sample before exposure to acetone. (b) Sample after exposure to acetone (on cotton swab)

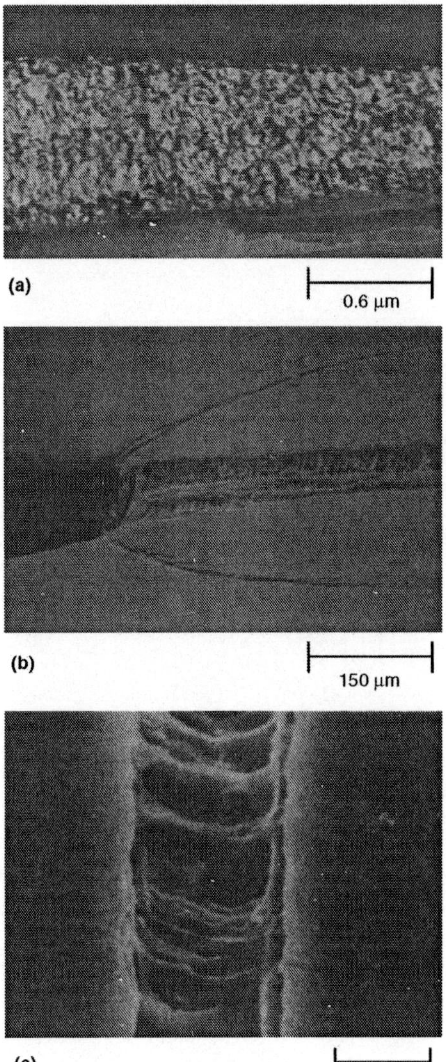

(a) 0.6 μm

(b) 150 μm

(c) 10 μm

Fig. 2 Electron microscope views of crazes. (a) In polyphenylene oxide. Source: Ref 1. (b) In polyethylene. (c) In nylon. Source: Ref 2

*Adapted from the article by Arnold Lustiger, "Environmental Stress Crazing," in *Engineering Plastics,* Volume 2, *Engineered Materials Handbook,* ASM International, 1988, pages 796 to 804

For a craze to form from a previously undisturbed glassy matrix, considerable molecular mobility must somehow be introduced into a structure in which the polymer chains are essentially stiff. A possible scenario for introducing this mobility has been suggested in Ref 6: During the process of intrinsic crazing, which is crazing in the absence of an accelerating environmental effect, increased free volume can open up in local regions of the polymer under stress. This occurs because the intermolecular forces between adjacent polymer chains are low relative to the yield point of the material. As the local free volume in the vicinity of these chains increases, molecular mobility likewise increases. Subsequent cavitation and drawing take place within this softened material. Although the same intermolecular forces are overcome during yielding, the important point about crazes is that they initiate at defects where the stress is concentrated.

When a specific environment acts as a crazing agent, it tends to weaken these intermolecular forces even further. The environment thus acts as a solvent or plasticizer, essentially lubricating the polymer chains so they can move past each other. The overall effect of a plasticizer is to lower the glass transition temperature, T_g, of the polymer. When the T_g decreases to below room temperature in the region of the craze, the material ceases to behave as a glass, and hence, significant flow processes can occur as material is drawn across the craze width. As a result, environmentally induced crazes tend to be much longer and more extensive than intrinsic crazes (Ref 1).

Polyethylenes. Although ESC of polyethylene takes place in environments in which the plasticization effects are not as obvious, the same mechanism as that mentioned previously generally becomes pertinent. However, because polyethylene is semicrystalline, the action of the environment is limited to the amorphous regions.

Figures 3 to 5 are simplified views of what is generally believed to occur during the ESC of polyethylene. If a tensile load is applied normal to the face of the semicrystalline lamellae, it can be seen that the tie molecules in the amorphous regions that connect adjacent lamellae will stretch. At a certain point, however, they can be stretched no farther. Under a long-term low-level load, tie molecules begin to disentangle under the effect of a plasticizing environment, and interlamellar failure begins to take place (Ref 7, 9–13). However, as adjacent regions in the polymer undergo interlamellar failure (thereby acting as voids), material that is in-between experiences much higher stresses (Fig. 4). Ductile deformation can occur at these high stresses, resulting in fiber formation as the lamellae break up into smaller units (Fig. 5). Aligned fibers therefore form across the craze (Fig. 2b), although not nearly to the elongations typical of ductile deformation. It should be mentioned, however, that under much longer-term lower-level loads, researchers found a totally fiber-free fracture surface (Ref 13). They sug-

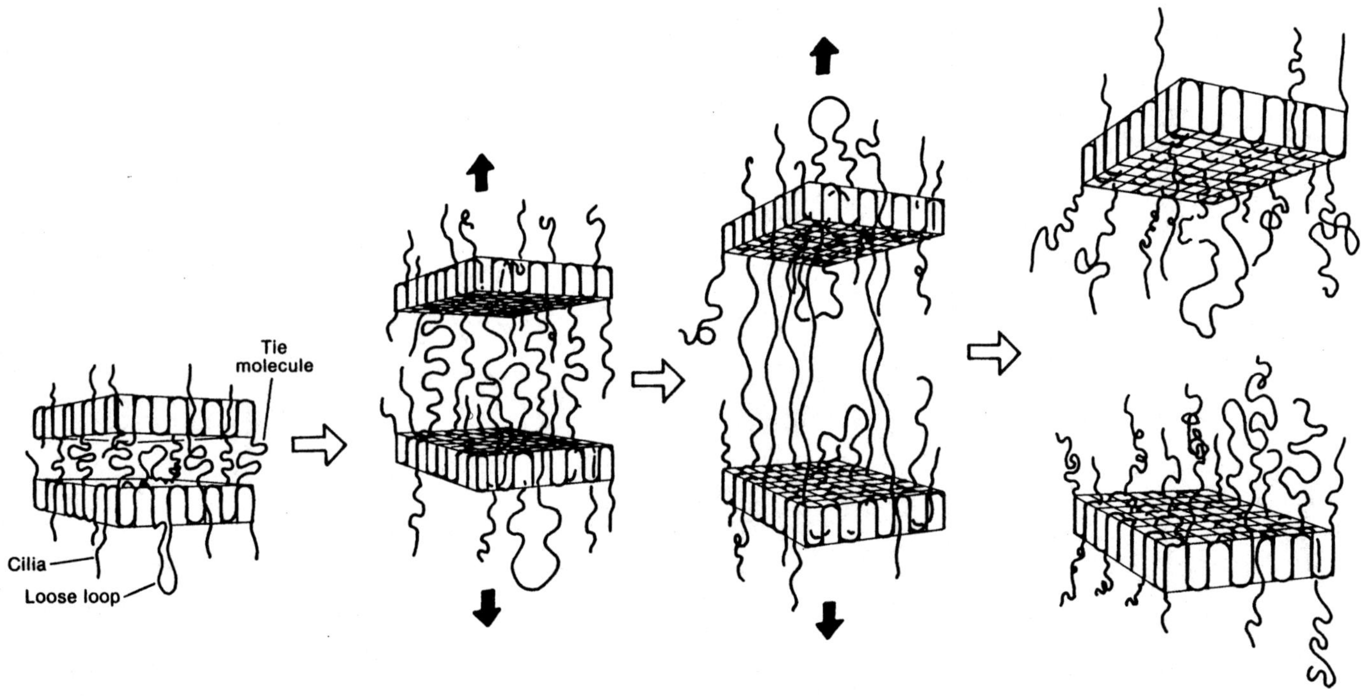

Fig. 3 Steps in the interlamellar failure of polyethylene. Source: Ref 7

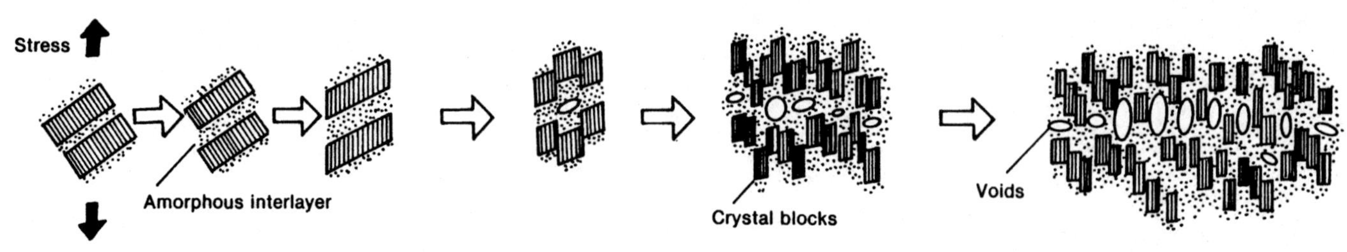

Fig. 4 Void formation due to interlamellar failure. Source: Ref 8

gested that under these conditions, interlamellar failure occurs exclusively, without the ductile deformation in between. In such a situation, crazes presumably do not form; only cracking takes place.

Crazing in nylons is found to occur in the presence of inorganic salts of various metals, such as zinc and cobalt chloride. It has been well demonstrated (Ref 14, 15) that cracking occurs because of the disruption of hydrogen bonding in the plastic as the environment becomes attracted to the dipolar amide groups. The amide N–H protons then bond with either water in the environment or with hydrated metal halide molecules. Other types of metal halides, such as lithium and magnesium chlorides, form proton-donating solvated constituents that act as solvents for the plastic. In the latter case, the typical mechanism associated with glassy plastics, described previously, becomes pertinent.

Environmental Criteria

Glassy Thermoplastics. As explained earlier, ESC agents tend to weaken intermolecular forces between polymer chains. A measure of the strength of these forces is given by the cohesive energy density (CED) (Ref 16):

$$\text{CED} + \frac{\Delta E_v}{V_1}$$

where ΔE_v is the molar energy of vaporization, and V_1 is the molar volume of the liquid. The square root of the CED parameter, called the solubility parameter, δ, is particularly useful. A liquid with a solubility parameter, δ_0, close to that of a given polymer, δ_p, generally dissolves the polymer. Similarly, such environments in contact with polymers under stress result in craze formation. Figure 6 clearly shows such a correlation between the critical strain to craze and the solubility parameter of polyphenylene oxide. Similar correlations exist for polysulfone and polystyrene (Ref 18, 19).

Although correlating δ_0 with δ_p provides an excellent rule of thumb for defining possible crazing agents, its sole use is frequently insufficient. Figure 7(a) shows that no simple correlation between δ_0 and δ_p exists for various aliphatic hydrocarbons in polycarbonate (Ref 20), as was shown previously for polyphenylene oxide. The δ_p for polycarbonate is approximately 42 $(\text{J/cm}^3)^{1/2}$ (10 $(\text{cal/cm}^3)^{1/2}$).

One of the parameters that must be taken into account, in addition to δ, is the molar volume of the solvent, V_0 (Ref 20). The larger the molar volume, the more difficult it is to enter between adjacent polymer chains, despite increasingly compatible solubility parameters. Figure 7(b) displays the same data as Fig. 6, except the data are normalized for differences in molar volume, displaying excellent correlation with critical strain to craze, ε_c.

Other complications arise when the environment is a polar liquid. When this is the case, it has been found that by separating the solubility parameter of the liquids into polar, δ_a, and nonpolar, δ_d, components, two-dimensional ESC maps can be developed to adequately describe ε_c (Ref 20). It should be emphasized, however, that this approach has been applied only to polycarbonates, although conceptually there is no reason to expect that it would not apply to other glassy thermoplastics. As shown in Fig. 8, plastics with both polar and nonpolar solubility parameters near those of the solvent (that is, with data points near the origin of the plot) tend to have low values of critical strain to craze. Solubility parameters, as well as their various components, are listed in Ref 21 for a variety of plastics and solvents.

Finally, there is a significant body of literature (Ref 22–26) that describes the crazing behavior of various plastics in contact with liquid nitrogen, argon, oxygen, and carbon dioxide. This phenomenon can be partially explained by invoking the plasticization mechanism (as detailed previously), which is that the gas is absorbed at the tip of an incidental flaw or defect, easing the polymer flow processes involved in the nucleation and growth of a craze. However, another mechanism is also described in the literature. It is proposed that, in addition to absorption and resulting plasticization, the gas is adsorbed onto the polymer, reducing its surface energy and thereby facilitating the creation of new surfaces in the holes and voids of the craze. An equation that separates these two perceived effects has been developed (Ref 23):

$$\sigma_c = 3\left[\left(\frac{2\gamma\beta_s}{r}\right) + (4.3\,\sigma_y\beta_p)\right]$$

where σ_c represents the critical stress to craze, γ represents surface energy, r represents the radius of submicroscopic voids, σ_y represents the yield point for shear deformation, β_s represents the factor by which surface energy is reduced by the environment, and β_p represents the factor by which yield point is reduced by the environment.

Although two of the parameters in the preceding equation are not available experimentally, the equation is useful in conceptualizing the two separate effects of plasticization and surface energy reduction. Subsequent literature clearly suggests, however, that the plasticization effect is dominant.

Polyethylenes. It has been suggested (Ref 27, 28) that ESC of polyethylenes involves the same environmental criteria as do glassy polymers. The solubility parameter of polyethylene is 35 $(\text{J/cm}^3)^{1/2}$ (8 $(\text{cal/cm}^3)^{1/2}$), and that of its most widely used ESC agent, nonylphenoxypoly(ethyleneoxy)ethanol, a surfactant better known by its trade name Igepal CO-630, is 40.8 $(\text{J/cm}^3)^{1/2}$ (9.75 $(\text{cal/cm}^3)^{1/2}$). Igepal does not swell the polymer to any appreciable extent because of its large molar volume. However, under stress, enough free volume can open up in the amorphous regions of the polymer so that the relatively large Igepal molecule can be accommodated. This process is known as stress-induced plasticization. In addition to its occurrence in various surfactants, ESC has commonly been reported in various alcohols (Ref 29) and silicone

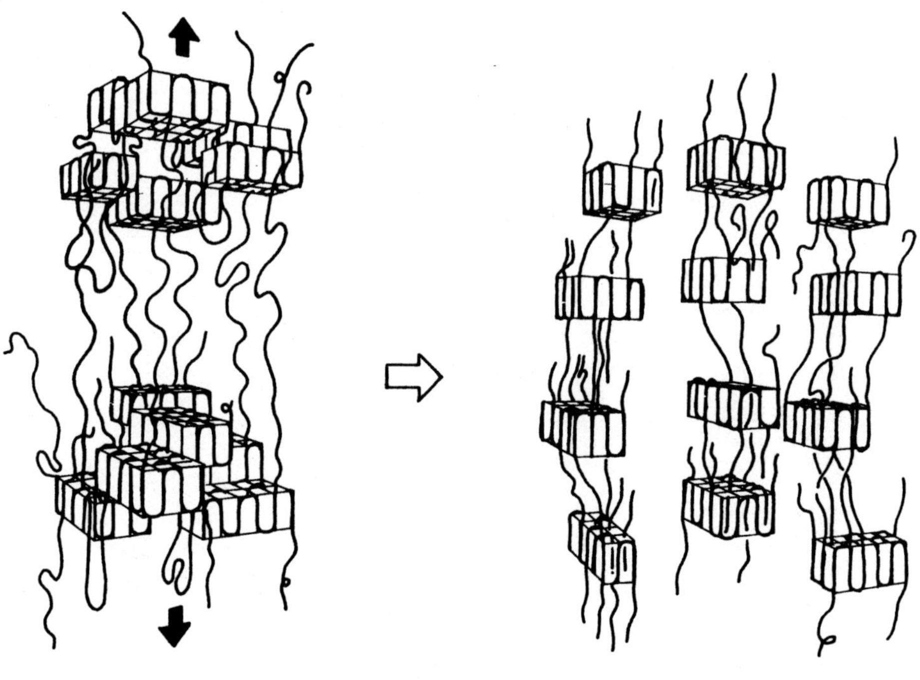

Fig. 5 Fiber formation within craze due to interlamellar deformation between adjacent voids. Source: Ref 7

fluids (Ref 30), presumably due to the same mechanism.

Nylons. The mechanism of disruption of hydrogen bonds, as involved in the ESC of nylon by certain metal salts, and the mechanism of solvation are both difficult to predict a priori. Table 1 gives a number of such environments and their ESC activity in nylons.

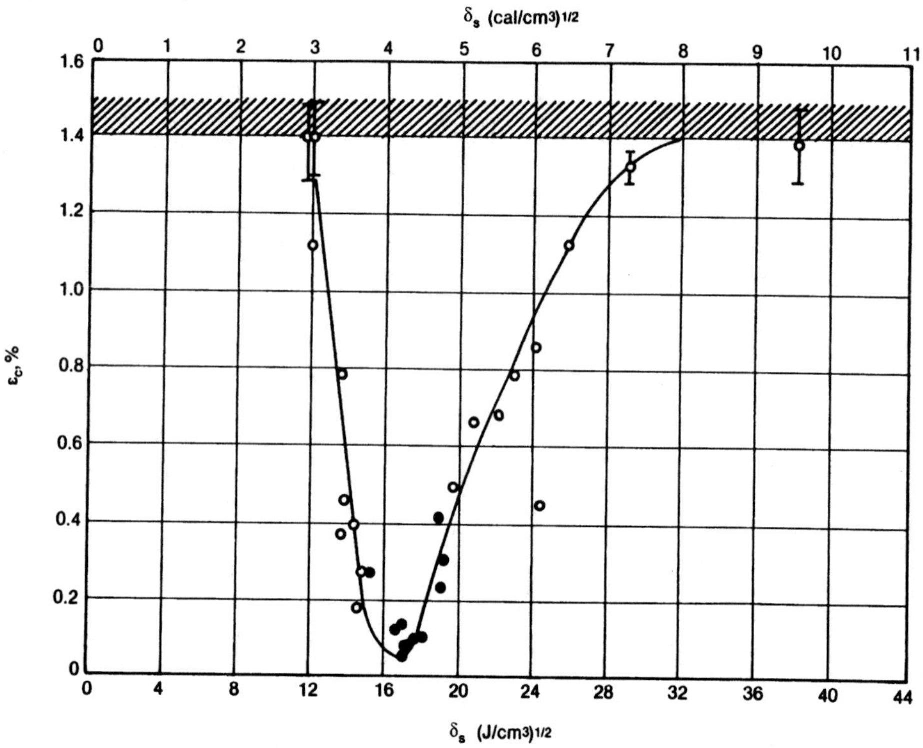

Fig. 6 Critical strain for environmental craze initiation, ε_c, in polyphenylene oxide versus solubility parameter of the solvent, δ_s. Filled data points are cracking agents with no apparent crazes at the crack tip. Source: Ref 17

Material Optimization

Glassy Thermoplastics. Polymer orientation is the major material modification that can significantly improve craze resistance (Ref 1). If the system can be designed so that the applied stress is parallel to the orientation direction of the polymer, ESC resistance can be increased by a factor of 2 to 4, as has been reported for polymethyl methacrylate. On the other hand, if the direction of applied stress is perpendicular to the direction of orientation, the opposite effect can occur. This orientation effect can be readily understood, because chain segments preoriented in the stress direction require higher stress to be further oriented during crazing. By contrast, increasing molecular weight has a negligible effect on craze resistance. Significant improvements are often evident on blending a second ESC-resistant phase. These improvements can become quite dramatic at the point of phase inversion, that is, when the second phase becomes continuous and the first, discontinuous. Incorporation of glass fibers can also improve ESC resistance. Under these conditions, the fibers support the applied load and stop the growing cracks (Ref 32).

Polyethylenes. It follows from the discussion earlier that polyethylene materials containing relatively few tie molecules are more susceptible to ESC. Conversely, materials with more tie molecules are more resistant to this type of failure. However, it should be added that if the proportion of tie molecules to crystalline molecules is too high, the material will display high ductility but very low stiffness.

Visualizing the mechanism of brittle failure in terms of this model can help identify molecular parameters of importance in order to opti-

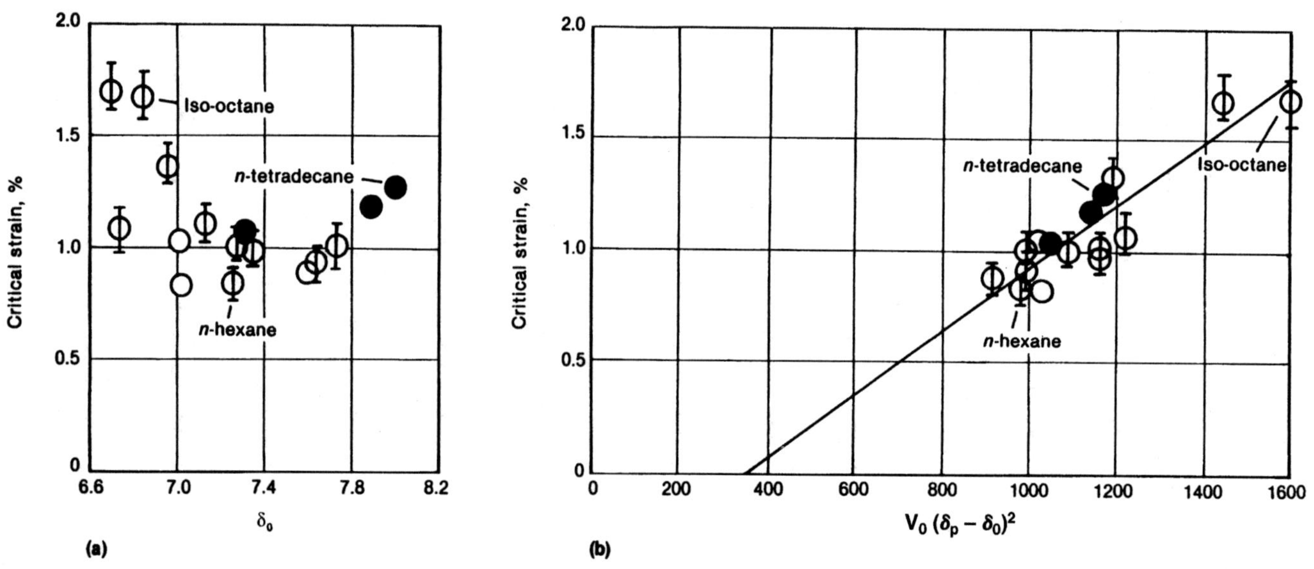

Fig. 7 Critical strain for environmental craze initiation in polycarbonate. (a) Versus solubility parameter of the solvent, δ_0. (b) Versus molar volume, V_0, times the square of the difference in solubility parameters between polymer, δ_p, and solvent, δ_0. Source: Ref 20

mize polyethylene resistance to ESC. Some of these parameters are discussed as follows.

Molecular Weight. The higher the molecular weight, the greater the resistance to ESC (Ref 33, 34). Figures 3 to 5 illustrate that the longer the polymer chains as a result of increased molecular weight, the greater the tie molecule concentration. Because commercial polymers are polydisperse, the entire molecular weight distribution is a critical factor (Ref 35). Because melt index is inversely proportional to molecular weight, it is desirable to work with material that has a low melt index to attain optimal ESC resistance.

However, the decision to use a polyethylene with a low melt index (that is, high melt viscosity) constitutes one of the classic engineering trade-offs relative to the use of this material. Although toughness and failure resistance are improved with increased molecular weight, the difficulty of processing a material with high melt viscosity must be considered. In addition, many in-service uses of the material necessitate melt fusion, specifically for pipe and liner applications, which are made considerably more difficult with materials of high melt viscosity.

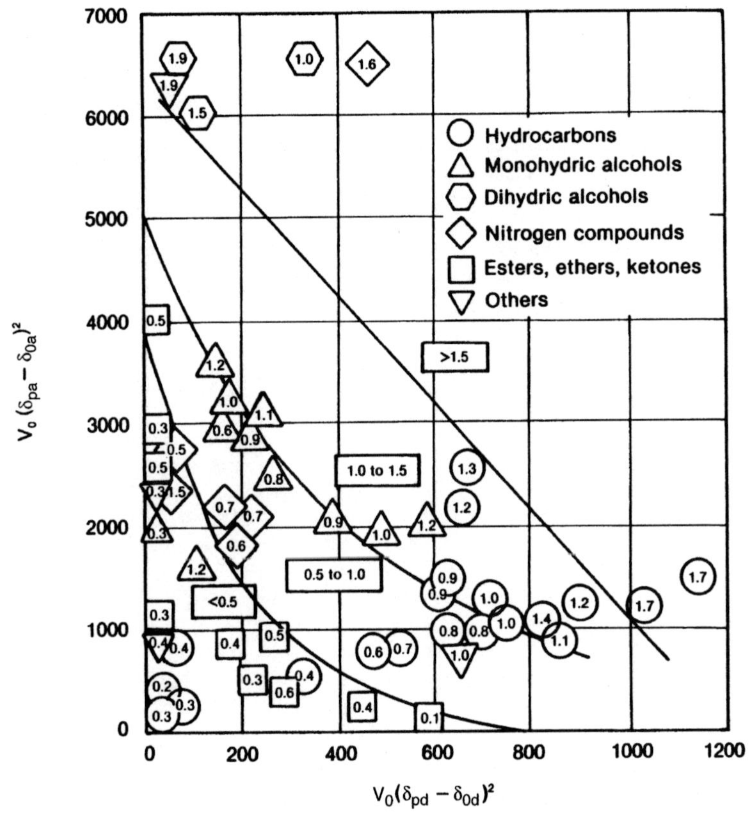

Fig. 8 Solubility parameter map of critical strain to craze in polycarbonate, taking into account molar volumes, V_0, and polar contributions to the solubility parameter. The numbered symbols represent critical strain to craze. δ_{pa}, solubility parameter for a polar polymer; δ_{0a}, solubility parameter for a polar liquid; δ_{pd}, solubility parameter for a nonpolar polymer; δ_{0d}, solubility parameter for a nonpolar liquid. Source: Ref 20

Comonomer Content. ESC resistance can be dramatically improved with the placement of a small amount of comonomer on the polyethylene chains to inhibit crystallinity in medium and linear low-density polyethylenes. Higher comonomer concentrations and longer comonomer chain branches (that is, 1-hexene or longer) probably do not enter the tightly packed lamellar lattice and therefore produce additional intercrystalline tie molecules (Ref 36), as shown in Fig. 9.

Density/Degree of Crystallinity. The more crystalline the material, the lower its ESC resistance (Ref 33). This is because of the fewer number of tie molecules that hold it together. As a result, quenched material has better ESC resistance than material that is cooled slowly after processing from the melt (Ref 37, 38).

However, the use of lower-density material also constitutes a trade-off in engineering properties: Failure resistance and toughness improve with lower crystallinity, but stiffness and yield point are reduced. In many applications, these properties must be considered when designing a structure that must resist deformation from a variety of in-service loads.

Because resistance of polyethylene to ESC is so sensitive to these parameters, optimizing the material to resist this failure mode has been a high priority among material producers. Improvements, specifically in the optimal use of comonomer, have been very dramatic in recent years. Crack resistance has improved by an order of magnitude or more in many cases. This is particularly true in the relatively recent development of linear low-density polyethylene, which incorporates the longer comonomers into its backbone chain in relatively high quantities.

Table 1 Activity of metal halides and thiocyanates in the crazing of nylon

Solvent	Metal ion	Activity(a)			
		Thiocyanate	Chloride	Bromide	Iodide
Water	Zinc	+++	+++	+++	+++
Water	$Cobalt_{II}$	+++	++	++	0
Water	Calcium	–	–	0	0
Water	Barium	++	–	–	–
Water	Lithium	+++	+	+++	–
Water	$Iron_{III}$	++	+	0	0
Water	Ammonium	++	–	–	–
Methanol	Zinc	–	+++	+++	+++
Methanol	$Cobalt_{II}$	–	++	0	0
Methanol	Calcium	++	++	0	0
Methanol	Barium	–	–	++	++
Methanol	Lithium	+++	++	+++	++
Methanol	$Iron_{III}$	–	++	0	0
Methanol	Ammonium	–	–	–	–

(a) +++, highly active; ++, active; +, weakly active; –, inactive; 0, not tested. Source: Ref 31

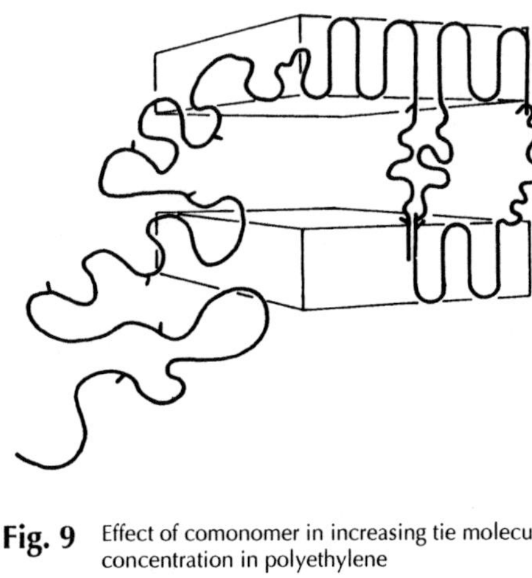

Fig. 9 Effect of comonomer in increasing tie molecule concentration in polyethylene

Nylons. Just as in glassy plastics, orientation significantly improves the ESC resistance of nylons in the direction of stress (Ref 39). However, it has been found that slight orientation with subsequent relaxation reduces failure times, presumably because of the expansion and coalescence of preexisting microcracks. Although average molecular weight does not appear to make a significant difference, removing the lowest molecular weight "tail" of the molecular weight distribution by water extraction significantly improves ESC resistance. In contrast to polyethylene, it was found that slow cooling also improves ESC resistance.

Testing

There are basically two types of tests used to determine relative susceptibility to ESC: those based on a constant load, and those based on a constant strain. There is an important conceptual limitation using either approach that must be addressed before discussing the various testing options available. In addition, it is important to normalize data initially for differences in the yield point when comparing different materials in a given test.

Constant-Load Versus Constant-Strain Testing (Ref 40). A present limitation of ESC testing is the inability to isolate the yield stress property as a parameter independent of the failure resistance of the plastic. Thus, constant-strain tests have been criticized because of stiffness variations between specimens. These variations give rise to an ambiguity when interpreting the results of these tests: Do differences between times to failure mirror a real difference in ESC resistance, or do these differences merely reflect the higher stress levels in the stiffer specimens? A similar objection can be directed to constant tensile load testing. Although a load is constant in the test, the response to it varies among materials. Therefore, specimen stiffness again becomes a complicating material parameter that obscures ESC resistance as an independent property. Hence, the question arises as to whether a material fails quickly in this test as a result of its low ESC resistance or because its low stiffness allows more deformation under the constant load.

A prime example of the confusion created by this situation was demonstrated by testing high- and low-density polyethylene in both constant-strain and constant-load tests and comparing the data. The constant-strain test, in this case, was the bent-strip test (Ref 41), which involves notching polyethylene samples longitudinally, bending them in a channel, and placing them in a solution of Igepal CO-630 at 50 °C (120 °F). A schematic of the test is shown in Fig. 10. The constant-load test involved subjecting a double-edge-notched specimen of the same dimensions to various loads in a detergent solution until failure. The data are shown in Table 2.

As is readily evident, high-density polyethylene fails faster than low-density polyethylene in the constant-strain bent-strip ESC test. On the other hand, the same samples exhibit the opposite effects in the constant tensile load ESC test.

The reason for the apparent contradiction in failure trends becomes clear when one considers the influence of mechanical properties on the response of a material to load (Fig. 11). Because of its relative stiffness, high-density polyethylene is stressed close to or beyond the yield point in a constant-strain test, and cracking takes place in the portion of the bend at which the material is just below the yield strain. Conversely, low-density polyethylene is more susceptible to failure than the high-density material in the constant tensile load test for the same reason that the high-density material failed faster than the low-density material in the constant-strain test; that is, the yield point was more closely reached by the less stiff, low-density material in the constant tensile load test. In contrast, the yield point was not even approached in the stiffer, high-density material under the same loading conditions.

Therefore, unless the yield points between two specimens are very close, neither a constant

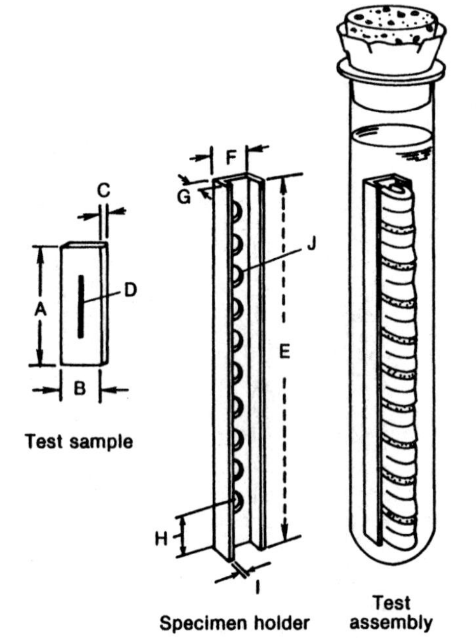

Fig. 10 The bent-strip test for polyethylene. Appropriate dimensions are given in Ref 41.

Table 2 Constant-strain versus constant-load testing of high- and low-density polyethylene

	Bent strip, failure time, h	Constant-load test, failure time, h: At 3.51 MPa (0.50 ksi)	Constant-load test, failure time, h: At 9.0 MPa (1.3 ksi)
High-density polyethylene	<1	4.7	0.6
Low-density polyethylene	20	1.9	Yield

Source: Ref 40

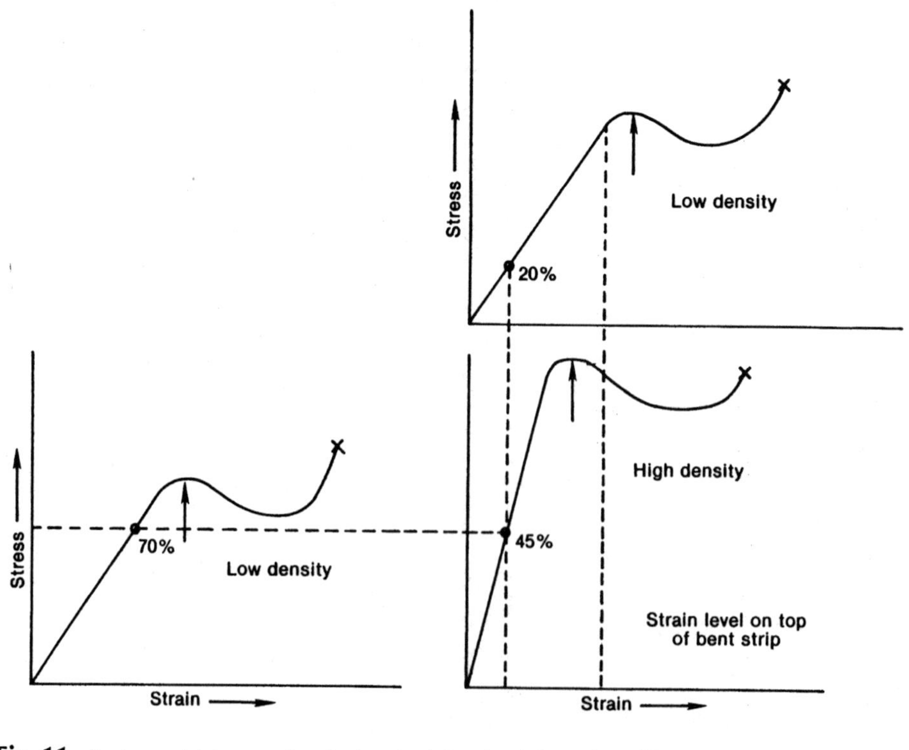

Fig. 11 Environmental stress-crack testing in polyethylene in relation to the yield point. Source: Ref 40

stress nor a constant strain provides good criteria for discerning ESC resistance. The parameter deserving closer examination as the ordinate of an ESC plot in a constant-load situation is the percentage of yield stress or reduced stress. For a more realistic comparison of materials, this percentage should be kept constant, although the actual stress may vary widely between specimens.

Constant Tensile Load Testing. To illustrate the utility of the reduced stress parameter, the constant tensile load test was implemented directly on polyethylene pipe used for natural gas distribution. In this test, a ring 12.7 mm (0.50 in.) wide is cut from the pipe, which is then axially notched on both the inside and the outside, each notch depth being 25% of the minimum wall thickness. The ring is then placed in a split-ring fixture and subjected to a constant load in the presence of a 1% solution of Igepal CO-630 (Fig. 12).

Figure 13(a) displays initial stress versus failure time data for seven polyethylene piping materials. The typical curve displays a shallowly sloped region followed by a steeply sloped region, although for two materials no such slope change is evident. Generally, ductile-type failure, showing large deformation and necking, occurs in the shallowly sloped region of the curve. In the region of lower stress and steeper slope, ESC occurs, characterized by little deformation at the point of failure. The point at which the slope changes has been termed a ductile-brittle transition, although prior use of this concept has been limited to impact fracture. The location of this transition gives a relative indication of ESC resistance: The later the transition, the better the resistance.

Comparing polyethylene materials with differing yield points results in a wide band of scattered data in the ductile portion of the curve when the ordinate is labeled nominal stress. However, when the data are plotted in terms of reduced stress, as in Fig. 13(b), they tend to fall very close to the same straight line in the shallowly sloped region of the curve.

Constant-Strain Testing. In practice, it is considerably more difficult to normalize constant-strain data than load data, because it is generally impractical to vary the strain between specimens to obtain the same percentage of yield strain. In the case of polyethylene, a given strain is induced in a given test regardless of the type of polyethylene being tested. The most common test is the bent-strip test, mentioned previously.

For piping materials, a closely related test is the so-called compressed-ring test, in which a 12.7 mm (0.50 in.) wide ring of pipe is notched in the same way as in the bent-strip test and compressed between two plates (Ref 42). The specimen configuration is the same as in the bent-strip test but allows the properties of the extruded product to be measured directly, rather than remolding the pipe into flat plaques.

Although effective material comparison is precluded in these constant-strain tests because of the difficulty of normalizing the data for yield point differences, the test is, in fact, very effective for quality-control purposes. Virtually any change in the product that is due to either basic resin or process variations results in different failure times. These changes include differences

Fig. 12 Constant tensile load test setup for polyethylene pipe. At upper left is close-up view of specimen in the fixture. The specimen is placed under load in a given environment in the metal can, which is fastened to the base of the assembly. Source: Ref 40

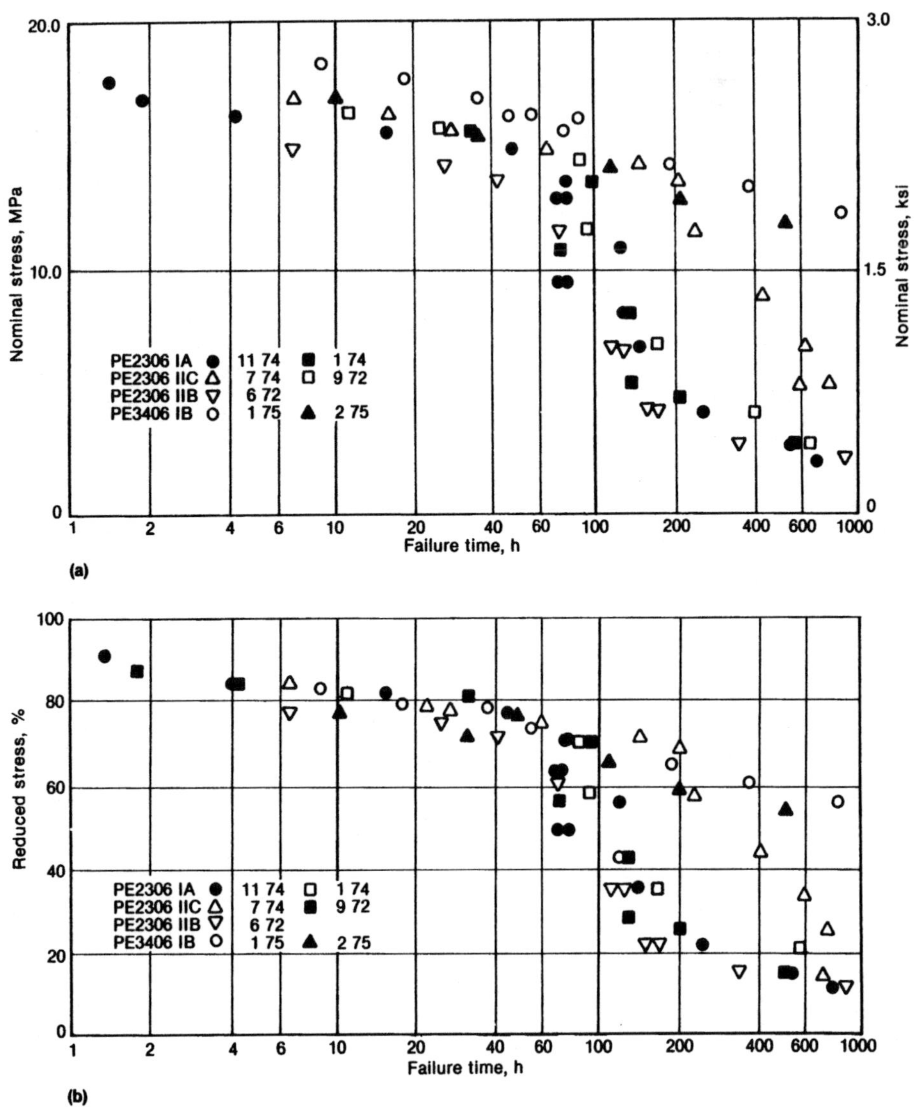

Fig. 13 Failure time for seven polyethylene piping materials in Igepal. (a) Plotted against nominal (initial) stress. (b) Plotted against reduced stress. Source: Ref 7

in molecular weight, degree of crystallinity, and comonomer content, as already mentioned, but can also include additives and surface features (Ref 32).

ESC Testing of Glassy Plastics and Nylon. It has been suggested in the literature that crazes will initiate in a plastic when a critical limit is reached in stress, strain, dilation, distortion strain energy, or stress bias. Researchers (Ref 43) evaluated all these criteria and concluded that critical strain is the most consistent. However, from a practical standpoint, a designer may wish to test either under constant strain or constant load, based on in-service conditions.

A variety of constant-strain and constant-load tests appear in the literature. Generally, the constant-load tests use the same principle as the test described previously. However, instead of razor notches, holes or a condition of no stress concentration at all is imposed on the specimen. Constant-strain tests can involve an imposed curvature in which a specimen is bent to conform to a given radius, or they can involve free bending similar to the bent-strip test. Alternatively, they can impart strain through three-point bending. Two reviews of testing methods appear in Ref 44 and 45. Two investigators used constant-strain tests, coupled with strain gages and force transducers, to determine stress relaxation in the specimens as the tests proceeded (Ref 46, 47). For nontransparent specimens, stress relaxation must be the measure of craze initiation, because the crazes cannot be detected visually. Other tests on more complicated specimen configurations involve imposing biaxial stress or inserting steel or metal balls into the specimen. Alternatively, there are a number of end-use tests available that require immersion of a plastic product in the stress-cracking agent to determine whether residual molding stresses are sufficient to craze it (Ref 45).

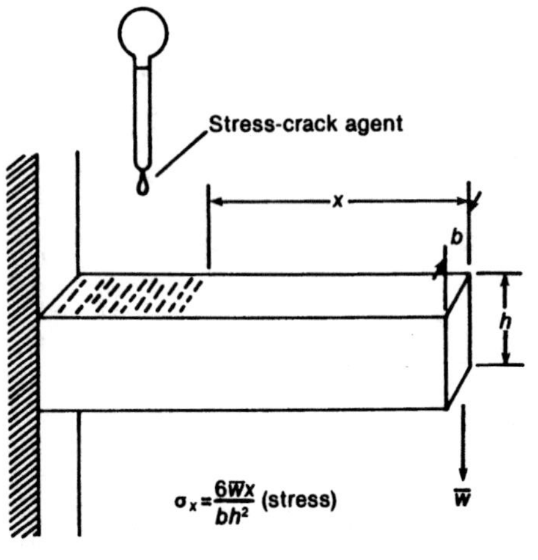

Fig. 14 Constant-load test for glassy plastics. Source: Ref 48

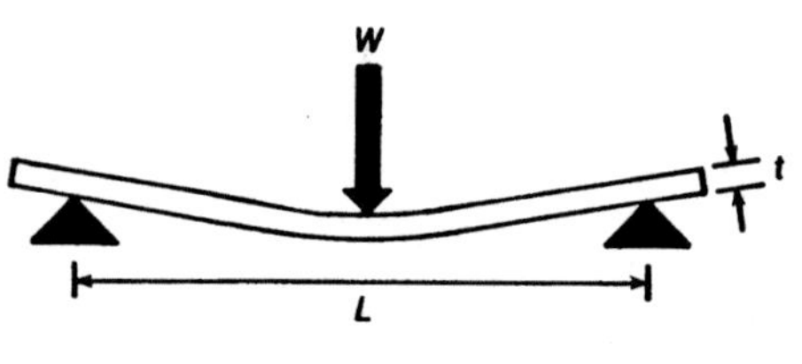

Fig. 15 Constant-strain (three-point bending) test for glassy plastics. *W*, applied load; *L*, span; t, thickness. Source: Ref 47

A simple, representative constant-load test for glassy plastics is shown in Fig. 14. In this test, a cantilever beam made from the specimen, with one end fixed, is placed under load, and drops of the crazing environment are placed on the top side of the specimen. Using the equation shown in Fig. 14, a crazing stress can be calculated based on the distance (from the fixed end) that the crazes are visible. The crazing behavior of polymethyl methacrylate was investigated in this way (Ref 48).

A constant-strain test is shown in Fig. 15, in which a specimen is placed under three-point bending. The maximum strain on the outer surface of the specimen occurs opposite the center loading pin and is given by the equation:

$$\varepsilon = \frac{2WL}{2EDt^2}$$

where W is applied load, L is span, E is Young's modulus, D is sample width, and t is sample thickness. The crazing behavior of polycarbonate in the presence of various gasoline components was determined using this test (Ref 47).

The literature on nylon stress crazing discusses various testing procedures. Researchers (Ref 14, 15) used either a stressed film oriented biaxially or a film stressed by suction over a circular orifice. Other investigators used a simple constant tensile load (Ref 2) or a tensile machine (Ref 39) to study the phenomenon.

REFERENCES

1. R.P. Kambour, A Review of Crazing and Fracture in Thermoplastics, *J. Polym. Sci. D, Rev.,* Vol 7, 1973, p 1
2. R.P. Burford and D.R.G. Williams, The Morphology and Mechanism of Crack Propagation in the Presence of Inorganic Salts, *J. Mater. Sci.,* Vol 14, 1979, p 2872
3. A. Lustiger and R.D. Corneliussen, Environmental Stress Crack Growth in High Density Polyethylene, *J. Polym. Sci. C, Polym. Lett.,* Vol 17, 1979, p 269
4. A. Lustiger and R.D. Corneliussen, The Role of Crazes in the Crack Growth of Polyethylene, *J. Mater. Sci.,* Vol 22, 1987, p 2470
5. S. Bandopadhyay and H.R. Brown, Direct Evidence for the Existence of a Craze at the Crack Tip in Environmental Stress Cracking of Polyethylene, *Polym. Eng. Sci.,* Vol 20, 1980, p 720
6. A.N. Gent, Hypothetical Mechanism of Crazing in Glassy Plastics, *J. Mater. Sci.,* Vol 5, 1970, p 925
7. A. Lustiger and R.L. Markham, The Importance of Tie Molecules in Preventing Polyethylene Fracture Under Long Term Loading Conditions, *Polymer,* Vol 24, 1983, p 1647
8. K. Friedrich, Crazes and Shear Bands in Semi-Crystalline Thermoplastics, *Advances in Polymer Science 1952–1953, Crazing in Polymers,* Springer-Verlag, 1983
9. P.D. Frayer, P.P.L. Tong, and W.W. Dreher, The Role of Intercrystalline Links in the Environmental Stress Cracking of High Density Polyethylene, *Polym. Eng. Sci.,* Vol 17, 1977, p 27
10. S. Bandopadhyay and H.R. Brown, Evidence of Interlamellar Failure in Environmental Stress Cracking of Polyethylene, *J. Mater. Sci.,* Vol 12, 1977, p 2131
11. S. Bandopadhyay and H.R. Brown, Environmental Stress Cracking and Morphology of Polyethylene, *Polymer,* Vol 19, 1978, p 589
12. T.W. Haas and P.H. MacRae, Microscopic Observation of Fracture in Spherulitic Films of Linear PE Under Biaxial Stress, *SPE J.,* Vol 24, 1968, p 27
13. M.K.V. Chan and J.G. Williams, Slow Stable Crack Growth in High Density Polyethylene, *Polymer,* Vol 24, 1983, p 234
14. P. Dunn and G.F. Sansom, The Stress Cracking of Polyamides by Metal Salts, Part 1: Metal Halides, *J. Appl. Polym. Sci.,* Vol 13, 1969, p 1641
15. P. Dunn and G.F. Sansom, The Stress Cracking of Polyamides by Metal Salts, Part 2: Mechanism of Cracking, *J. Appl. Polym. Sci.,* Vol 13, 1969, p 1657
16. F. Rodriguez, *Principles of Polymer Systems,* McGraw-Hill, 1970
17. R.P. Kambour, Crazing, *Encyclopedia of Polymer Science and Technology,* Vol 4, John Wiley & Sons, 1986
18. R.P. Kambour, E.E. Romagosa, and C.L. Gruner, Swelling, Crazing and Cracking of an Aromatic Copolyether-Sulfone in Organic Media, *Macromolecules,* Vol 5, 1972, p 335
19. R.P. Kambour, C.L. Gruner, and E.E. Romagosa, Solvent Crazing of "Dry" Polystyrene and "Dry" Crazing of Plasticized Polystyrene, *J. Polym. Sci. B, Polym. Phys.,* Vol 11, 1973, p 1879
20. C.H.M. Jacques and M.G. Wyzgoski, Prediction of Environmental Stress Cracking of Polycarbonate from Solubility Considerations, *J. Appl. Polym. Sci.,* Vol 23, 1979, p 1153
21. J. Brandrup and E.H. Immergut, Ed., *Polymer Handbook,* John Wiley & Sons, 1975
22. A. Peterlin and H.G. Olf, Environmental Effects on Low Temperature Crazing of

Crystalline Polymers, *J. Polym. Sci., Polym. Symp.*, Vol 50, 1975, p 243

23. N. Brown, A Theory for Environmental Craze Yielding of Polymers at Low Temperatures, *J. Polym. Sci. B, Polym. Phys.*, Vol 14, 1973, p 2099
24. N. Brown and S. Fischer, Nucleation and Growth of Crazes in Amorphous Polychlorotrifluoroethylene in Liquid Nitrogen, *J. Polym. Sci. B, Polym. Phys.*, Vol 13, 1975, p 1315
25. N. Brown and Y. Imai, Craze Yielding of Polycarbonate in N_2, Ar, and O_2 at Low Pressures and Temperatures, *J. Appl. Phys.*, Vol 46, 1975, p 4130
26. E. Kamei and N. Brown, Crazing in Polyethylene, *J. Polym. Sci. B, Polym. Phys.*, Vol 22, 1984, p 543
27. H.R. Brown, A Theory of the Environmental Stress Cracking of Polyethylene, *Polymer,* Vol 19, 1978, p 1186
28. K. Tonyali, C.E. Rogers, and H.R. Brown, Stress Cracking of Polyethylene in Organic Liquids, *Polymer,* Vol 28, 1987, p 1472
29. C.J. Singleton, E. Roche, and P.H. Geil, Environmental Stress Cracking of Polyethylene, *J. Appl. Polym. Sci.*, Vol 21, 1977, p 2319
30. M.E.R. Shanahan and J. Schultz, A Kinetic Effect in the Environmental Stress Cracking of Polyethylene due to Liquid Viscosity, *J. Polym. Sci. B, Polym. Phys.*, Vol 14, 1976, p 1567
31. P. Dunn and G.F. Sansom, The Stress Cracking of Polyamides by Metal Salts, Part III: Metal Thiocyanates, *J. Appl. Polym. Sci.*, Vol 13, 1969, p 1673
32. L.M. Robeson, Overcoming Stress Rupture of Amorphous Thermoplastics in Organic Environments, *Proceedings of Symposium on Problem Solving in Plastics,* National Association of Corrosion Engineers, 1971, p 87
33. J.B. Howard, Stress-Cracking, *Crystalline Olefin Polymers,* R.A.V. Raff and K.W. Doak, Ed., John Wiley & Sons, 1964
34. W.A. Dukes, The Endurance of Polyethylene Under Constant Tension While Immersed in Igepal, *Br. Plast.*, 1961, p 123
35. J.N. Herman and J. Biesenberger, Molecular Weight Distribution and Environmental Stress Cracking of Linear Polyethylene, *Polym. Eng. Sci.*, Vol 6, 1966, p 341
36. M.J. Hannon, Microscopic Aspects of Brittle Failure of Polyethylene Below the Yield Stress, *J. Appl. Polym. Sci.*, Vol 18, 1974, p 3761
37. J.B. Howard and W.M. Martin, Effects of Thermal History on Some Properties of Polyethylene, *SPE J.*, Vol 16, 1960, p 407
38. M.E.R. Shanahan, C. Chen-Fargheon, and J. Schultz, The Influence of Spherulitic Size on the Environmental Stress Cracking of Low Density Polyethylene, *Makromol. Chem.*, Vol 181, 1980, p 1121–1126
39. A.C. Reimschuessel and Y.J. Kim, Stress Cracking of Nylons Induced by Zinc Chloride Solutions, *J. Mater. Sci.*, Vol 13, 1978, p 243
40. A. Lustiger, Environmental Stress Cracking, the Phenomenon and Its Utility, *Failure of Plastics,* W. Brostow and R.D. Corneliussen, Ed., Hanser-Verlag, 1986
41. "Standard Test Method for Environmental Stress-Cracking of Ethylene Plastics," D 1693, *Annual Book of ASTM Standards,* American Society for Testing and Materials
42. A. Lustiger, R.L. Markham, and M.M. Epstein, Environmental Stress Crack Growth in Medium-Density Polyethylene Pipe, *J. Appl. Polym. Sci.*, Vol 26, 1981, p 1049
43. T.T. Wang, M. Matsuo, and T.K. Kwei, Criteria of Craze Initiation in Glassy Polymers, *J. Appl. Phys.*, Vol 42, 1971, p 4188
44. H.J. Orthmann, Environmental Stress Cracking of Thermoplastics, *Ger. Plast.*, Vol 73, 1987, p 17
45. W.V. Titow, A Review of Methods for the Testing and Study of Environmental Stress Failure in Thermoplastics, *Plast. Polym.*, Vol 43, 1975, p 98
46. L.F. Henry, Prediction and Evaluation of the Susceptibilities of Glassy Thermoplastics to Environmental Stress Cracking, *Polym. Eng. Sci.*, Vol 14, 1974, p 167
47. M.G. Wyzgoski and C.H.M. Jacques, Stress Cracking of Plastics by Gasoline and Gasoline Components, *Polym. Eng. Sci.*, Vol 17, 1977, p 854
48. D.M. Bigg, R.I. Leininger, and C.S. Lee, Stress-Cracking Behavior of Poly-(Methyl Methacrylate) and a Poly-(Methyl Methacrylate)-Ethyl Acrylate Copolymer, *Polymer,* Vol 22, 1981, p 539

Moisture-Related Failure*

ONE ASPECT of plastics corrosion resistance is the adverse effects of moisture on the capability of a plastic to perform its task in a structural application. Although plastics are often regarded as being much more corrosion resistant than metals are, this view has pitfalls. There are important differences in how these two materials resist chemical attack. Generally, metals are attacked at their interfaces by electrochemical processes. Either they are dissolved (in aqueous solutions) or surface deposits build up, which may or may not limit the corrosive process. In either case, structural performance is adversely affected. Corrosion by diffusion of attacking species into metal is rare, although a well-known example that is directly related to failure mechanisms is hydrogen embrittlement of steel.

On the other hand, diffusion of species into plastics is common. The amount of such diffusion and its effect on properties range from none to catastrophic, such as the complete dissolution of certain plastics in appropriate solvents. The amount of diffusant absorbed may not be directly correlated to its effect on properties, especially in comparisons of different plastics. Adverse effects arising from penetrants are not necessarily chemical in nature; that is, no chemical bonds need be altered for deleterious effects to occur. This type of damage has been called physical corrosion (Ref 1, 2). Furthermore, this type of behavior characterizes most aspects of the interaction of water with structural plastics.

Chemical reactions that may occur in plastics cannot be studied electrochemically because they are generally nonconducting. Such reactions may encompass the range of organic reactions into which the polymer backbone and the various attacking species may enter. While attack by aqueous solutions of acids, alkalies, or oxidants is common, chemical attack of structural plastics by water itself is somewhat rare. Well-known exceptions are the hot-water degradation of polycarbonate (PC) and the thermosetting polyesters.

It is significant that in cases involving no attack of a plastic by an active ion, the presence of dissolved ion actually may act to diminish the amount (and the effect) of water absorption. Water absorption may be viewed as an osmotic phenomenon; that is, the amount of water absorption is governed by the thermodynamic activity of water either in solution or in the vapor phase (Ref 1–3). Examples of this behavior are given in Fig. 1, which compares the effect of water absorption in vinyl ester/styrene copolymers in distilled water and in saturated sodium chloride (NaCl) solution. This figure also compares the effect of a decreased polarity of the polymer, in this case by varying the amount of styrene coreactant.

Organic polymeric materials generally absorb moisture to some measurable degree when immersed directly in water or when exposed to atmospheric moisture. The amount of moisture absorbed depends on the chemical nature of the material, whereas the rate of absorption of water by plastics often follows well-developed mathematical models. This absorption can have both reversible and irreversible effects on properties and performance. Sorbed moisture has been shown to act as a plasticizer, reducing the glass transition temperature, T_g, and the strength properties of a plastic. This effect on properties has been shown to be essentially reversible. However, as already indicated, sorbed water can also induce physical corrosion, or irreversible mechanical damage, by means of microcracking or crazing, as well as irreversible chemical degradation of the polymer structure.

Table 1 lists the water absorption values for selected plastics as determined by ASTM D 570 after a 24 h immersion at 25 °C (77 °F). Generally, the more polar the nature of the plastic, the greater its affinity for water. As Table 1 shows, materials of low polarity, such as polyethylene (PE) and polypropylene (PP), absorb less than 0.01 wt% water. PC, polysulfone (PSU), and cellulose acetate, which have increasingly greater polarities that are greater than those of polyolefins, display increasing affinities for water. Equilibrium value for water absorption will be significantly higher for many plastics, as will water absorption values obtained at elevated temperatures. No simple correlation between the number of polar groups and the solubility of water in a plastic exists because of such factors as the accessibility of the polar groups, the relative strength of the water-water versus the water-plastic bonds, and the degree of crystallinity (Ref 4).

Mechanisms of Moisture-Induced Damage

Bulk effects, such as a loss of stiffness, a lowering of the T_g, and an increase in creep and stress relaxation, are the most commonly observed problems that can lead to premature failure. These essentially stem from the same cause, which is related to the penetration of moisture into the bulk of the polymer, thereby

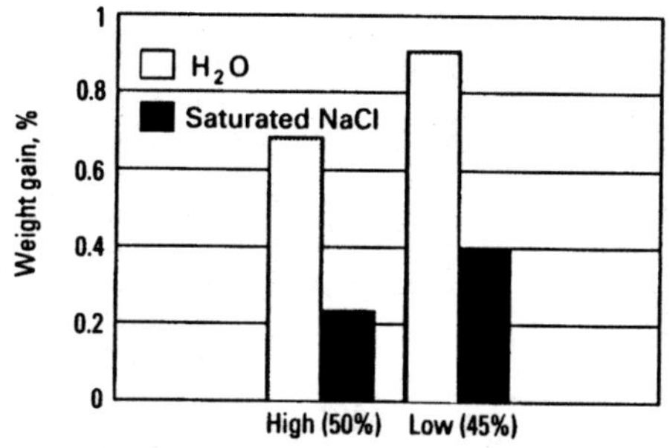

Fig. 1 Percent gains in weight of ASTM C 581 laminates of a high- and low-styrene-containing vinyl ester exposed at 66 °C (150 °F) to a distilled and saturated sodium chloride solution for 120 days

Table 1 Water absorption values for selected polymers

Plastic	Water absorption, wt%
PTFE	0.00
PE, high density	<0.01
PP	<0.01
PVC	0.03
PS	0.05
PC	0.15
PSU	0.22
POM	0.25
Nylon 11	0.25
Polyvinyl butyral	1.0
Nylon 6	1.3
Cellulose acetate	1.7

*Adapted from the article by R. Charles Allen and Ronald S. Bauer, "Moisture-Related Failure," in *Engineering Plastics*, Volume 2, *Engineered Materials Handbook*, ASM International, 1988, pages 761 to 769

plasticizing it and inducing the undesirable effects. In general, the magnitude of the plasticization effects depends on the polarity of the polymer; these effects do not appreciably occur in nonpolar polymers, but rather, depend on the nature of the individual polymer. Because plastics differ widely in their susceptibility to these effects, the characteristics of important structural plastics are discussed as follows. The effects of moisture on modulus, creep, and stress relaxation cannot be discussed without noting the very important aggravating role played by temperature and stress.

The Effect of Moisture on T_g. Lowering the T_g is probably the most widely studied phenomenon related to failure of plastics, but it is not itself a true failure mechanism. For a structural plastic, the T_g is, simply put, an approximate upper limit of its useful temperature range. The plastic, which is hard and glassy below this temperature, becomes rubbery above it. However, the T_g is also a measure of the onset of long-range molecular movement in the plastic. Because the transition from glass to rubber is not a thermodynamic transition but a manifestation of viscoelasticity, the exact value of the T_g depends on the method used to measure it and the rate at which the temperature is changed during the measurement. For this reason, these parameters must be specified when reporting T_g measurements and when comparing data of different plastics. Semicrystalline plastics, in some cases, are exempt from the concern of exceeding the T_g, because their crystalline melting point, always above their T_g, represents their temperature limit.

The effect of absorbed moisture on the T_g is invariably to lower it. This is consistent with the role of water as a plasticizer. As a rule of thumb, the more water absorbed, the lower the T_g. For this reason, a nonpolar plastic such as polystyrene (PS) is less affected than, for example, polymethyl methacrylate (PMMA). The lowering of T_g is sometimes quantitatively discussed in terms of several mixing formulas (Ref 5, 6). Currently, the most often used expression is:

$$T_g = \frac{X_1C_{p1}T_{g1} + X_2C_{p2}T_{g2}}{X_1C_{p1} + X_{p2}C_{p2}} \quad \text{(Eq 1)}$$

In this expression, T_g, T_{g1}, and T_{g2} are the glass transition temperatures of the polymer mixture, polymer 1, and diluent 2, respectively. The expressions C_{p1} and C_{p2} are the discontinuities in the heat capacities at the glass transitions of the components. This expression was first derived (Ref 7) for polymer blends and was based on the Gibbs-DiMarzio entropy theory (Ref 8). Couchman provided an alternative derivation, based on a purely thermodynamic exposition (Ref 9). The extension of the Couchman approach to plastic-diluent systems, especially epoxy-water systems, has been carried out by researchers (Ref 10–12). Couchman's derivation (but not the result) has been criticized (Ref 13). The discussion is important to an understanding of the glassy state.

These relationships can be quite useful for predicting the loss of properties due to moisture. Such a relationship is shown in Fig. 2. Because the modulus falls precipitously at the glass transition, these types of data give an absolute upper temperature limit. Most conservative design requires all application temperatures to be remote from the glass transition region. The relationship between T_g and the amount of absorbed water can be affected by many factors, such as additives, thermal pretreatments, presence or absence of fillers or reinforcements, and, in thermosets, amount and type of curative, degree of cure, and so forth.

Measuring the T_gs of moisture-containing resins is not accomplished without a good deal of care, because water is often lost during the measurement. This can be especially serious for high-T_g polymers, such as those used in high-technology applications. Examples of this are given in Table 2, with water loss being measured by thermogravimetric analysis. Essentially all the absorbed water may be lost unless proper precautions are taken. Measurements of T_g are often carried out by differential scanning calorimetry (DSC) (as was done by researchers in Ref 10–12); DSC techniques are particularly adaptable to preventing loss of moisture during T_g measurements.

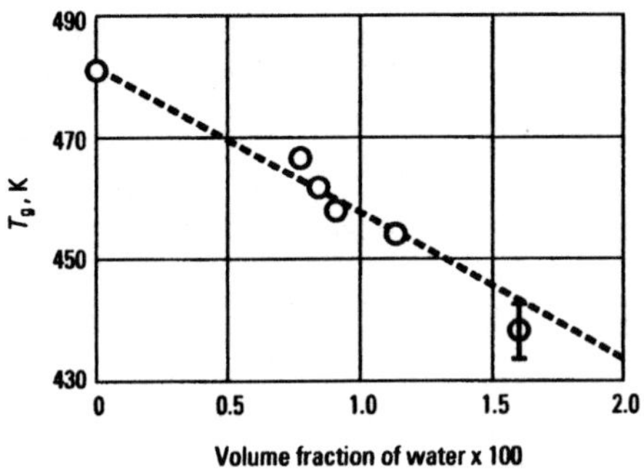

Fig. 2 Glass transition depression data (calculated). Curve as predicted by Eq 1. Source: Ref 10

A method that satisfactorily measures the T_g in water-saturated thermoplastics and thermosets that do not have an excessively high cross-link density is to seal the plastic and a small amount of water in a high-pressure DSC pan and then measure in a normal manner (Ref 14). The pan contains three phases (liquid and gaseous forms of water, and polymer) and thus has but one degree of freedom by Gibb's phase rule, namely temperature, because this is necessarily the independent variable in a measurement of T_g. Results of using this method are given in Table 3, where they are contrasted with measurements in standard DSC analysis. The plastic, in principle, remains fully saturated with moisture during the run. A T_g measured by this technique is thus a worst case value of a plastic that is fully moisture saturated. A major drawback is that no values of T_g of intermediate saturations can be obtained. Other DSC techniques seem to be satisfactory as well. Simply sealing a moisture-containing plastic into a DSC high-pressure pan may be adequate, especially if the

Table 2 Water losses during temperature scans (thermogravimetric tests)

Resin/curing agent or plastic	Beginning water content, wt%	Water loss, wt%
EPON Resin 826(a)/diaminodiphenyl sulfone	2.28	–0.91 at 40 °C/min (70 °F/min) –2.02 at 10 °C/min (18 °F/min)
PC	0.32	–0.2 at 40 °C/min (70 °F/min) –0.39 at 10 °C/min (18 °F/min)
PSU	0.57	–0.25 at 40 °C/min (70 °F/min) –0.52 at 10 °C/min (18 °F/min)

(a) Trade name of Shell Chemical Company

Table 3 Differential scanning calorimetry comparison of glass transition temperature (T_g) results from sealed and unsealed pans

	T_g, unsealed pan				T_g, sealed pan, resin and water	
	Dry		Wet			
Resin/curing agent or plastic	°C	°F	°C	°F	°C	°F
EPON Resin 826(a)/EPON curing agent Y(b)	167	333	134 (1st scan)	275	125	257
EPON Resin 826/methylenedianiline	165	330	...	...	122	252
EPON Resin 826/Jeffamine D-230(b)	92	200	88	190	62	145
EPON Resin 826/Jeffamine D-400(b)	50	120	...	...	30	85
PC	148	300	139	280	132	270
PSU	184	365	184	365	158	315

(a) Tradename of Shell Chemical Company. (b) Tradename of Texaco Chemical Company plastic sample is large in comparison to the vapor space. A drawback of the DSC method is that it generally fails to give measurable T_gs for resins having a very high cross-link density, in particular, some of the aerospace epoxy resins.

Perhaps the most popular method of T_g measurement is dynamic mechanical analysis (DMA). Not only is the glass transition clearly distinguished, but other relaxations as well, and the shear or tensile complex moduli are measured and have a clear connection to the moduli of interest for engineering design. In Fig. 3, representative dynamic mechanical data are given, with some of the commonly used measures of T_g pointed out. To date, it is not possible to run experiments in an autoclave to prevent loss of moisture. Nonetheless, there is much to recommend this technique as a routine screening method for the T_g of moisture-containing plastics and composites.

In one common mode of operation, the temperature is increased in jumps of 5 to 10 °C (9 to 18 °F) and held for 2 min, after which the dynamic mechanical parameters are measured. This technique tends to give conservative values for the T_g of dry plastics but is not a good technique to use when determining the T_g for moisture-containing ones, particularly if the T_g is above 100 °C (212 °F). In many instances, the plastic specimen is completely dried out by this technique. A better technique is to ramp the temperature, as in the DSC and DMA techniques, at 10 °C/min (18 °F/min), for example. A further improvement (Ref 15) is to enclose the specimen in a polytetrafluoroethylene (PTFE) bag containing oil saturated with water. Thermomechanical analysis is also a recognized method for measuring T_g (Ref 16).

Another method of assessing T_g of composite materials by modulus measurement is used in the aerospace industry. A number of specimens of a size suitable for measuring flexural modulus are placed in a humidity chamber until they reach saturation. Then, the flexural modulus is determined for individual specimens at increasing temperatures in oil baths. The modulus-versus-temperature curve is plotted, and the T_g is identified by a rapid drop of modulus on the curve. Although rather tedious and time-consuming, this method has proved to be very satisfactory for advanced composite structures, for which performance is critical. Shear modulus (G_{12}) also can be determined, using ±45° tension tests.

It is likely that more work has been done on the effect of moisture on the T_g of epoxy-resin systems than on any other plastic system. Of particular interest is the system based on tetraglycidyl methylenedianiline (TGMDA)/diaminodiphenyl sulfone (DDS). The chemical structures of these materials, which are the principal epoxy-matrix resin systems currently used in advanced composite aircraft/aerospace applications, are given in Fig. 4. As can be seen from Table 4, the TGMDA/DDS can absorb as much as 6.5 wt% water. This absorbed water results in a dramatic drop in T_g (Ref 17–19). The reduction of the T_g resulting from the absorbed moisture is also given in Table 4 and corresponds with the 13 to 15 °C/wt% (25 to 30 °F/wt%) water content, as predicted by researchers (Ref 11). Also, the amount and rate of moisture absorption of a typical TGMDA/DDS laminate were found to increase with periodic exposure to thermal spikes (Ref 20), such as those experienced on a supersonic aircraft. The absorptivity coefficient of a graphite-epoxy laminate was shown to double with such an exposure.

Not all epoxy-resin systems absorb as much water as the TGMDA/DDS system, because the amount of water absorbed by an epoxy resin depends on the polarity of the epoxy-resin system. Figure 5 gives the equilibrium water uptake values obtained by researchers (Ref 21) for two epoxy resins of different polarities, with three different curing agents also having varying degrees of polarity. The values are of the expected order, because the TGMDA has more polarity than the diglycidyl ether of bisphenol A (DGEBA), and the polarity of the curing agents follows the order SO_2 is greater than CH_2, which is greater than O (Ref 21). The effect of moisture on the T_g depends on the amount of moisture absorbed, which in turn depends on the chemical structure of the cured resin.

Effect of Moisture on Creep and Stress Relaxation. Creep and stress relaxation are more important considerations in plastic materials than they are in metals. Unlike the T_g, these can be true failure mechanisms. For a plastic used above its T_g, these phenomena can be quite

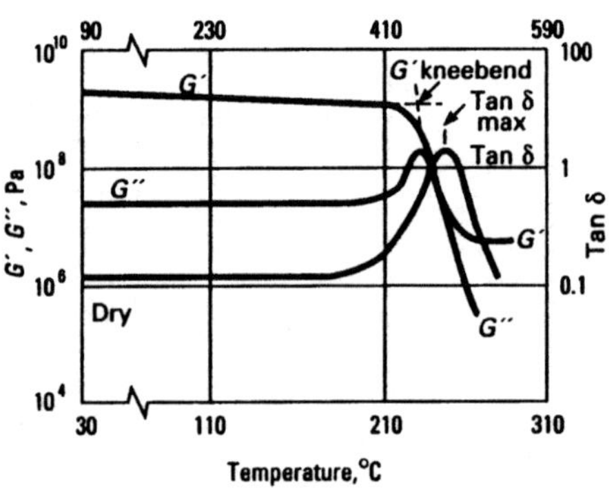

Fig. 3 Typical dynamic mechanical spectrum of high-temperature epoxy-resin system. G', storage modulus. G'', loss modulus

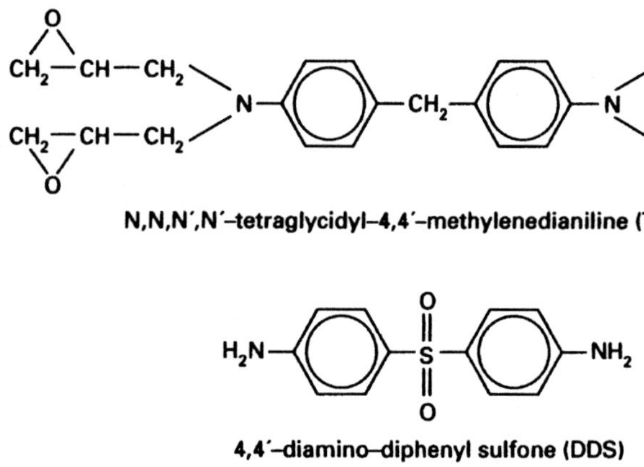

Fig. 4 Chemical structures of TGMDA and DDS

Table 4 Effect of water on the glass transition temperature of tetraglycidyl methylenedianiline/diaminodiphenyl sulfone (TGMDA/DDS) systems

	System I(a)	System II(b)	System III(c)
Moisture gain, wt%	6.5(d)	5.5(d)	5.0(e)
Glass transition temperature			
Dry, °C (°F)	246 (475)	175 (350)	200 (390)
Wet, °C (°F)	144 (290)	112 (235)	140 (285)
°C/wt% water absorbed	15.7	11.5	12.0

(a) NARMCO 5208 (Ref 20). (b) TGMDA/32 phr DDS/$BF_3 \cdot H_2NCH_3$ (Ref 18). (c) TGMDA/50 phr DDS (Ref 17). (d) Immersion in water at 71 °C (160 °F). (e) Immersion in water at 60 °C (140 °F) important, particularly in amorphous plastics with no microcrystalline phase and therefore no microcrystallites to act as "anchors." Most plastics used for structural applications are used at temperatures below their T_gs. However, creep and stress relaxation can still represent significant factors that a designer must consider.

Creep is the name for the increase in deformation that occurs under a constant load, in addition to the initial elastic deformation, and stress relaxation is the decrease in stress with time after stressing to a constant deformation. A designer considering the use of a plastic composite leaf spring in the rear suspension of an automobile would want to know how much the rear of the car would sag after a given period of time. This sag is an example of creep. When designing a plastic oil pan in which the bolts are tightened to a certain stress, a designer must ensure that the bolts will not become loose in service. The loosening of the bolts would be an example of stress relaxation. In this case, the bolts are tightened against an initial modulus of the oil pan, and the plastic is deflected a definite, although unmeasured, amount.

The cause of both creep and stress relaxation is the relaxation of molecular segments under stress on a time scale greater than that of loading. Higher temperatures accelerate the relaxational processes, of course, making creep and stress relaxation more important. An excellent and practical review of creep and stress relaxation is found in Ref 22 and 23. A fundamental approach is given in Ref 24. Creep and stress relaxation are not only important in a practical sense as possible failure mechanisms, but they are also used in laboratory studies of the viscoelastic nature of plastics.

It is very useful to know that, in many cases, the creep and stress relaxation curves at several temperatures can be laid side by side and shifted horizontally to form a single curve (Fig. 6). If one curve is not shifted and the other curves are shifted to the left and right of it, the temperature of the unshifted curve is called the reference temperature. Any one of the available curves may be chosen to represent the reference temperature. The superposed curve resulting from the shifting is called a master curve, on which the horizontal scaling of the reference curve (usually logarithmic) is simply extended to the left and right. The amount that each curve is shifted, identified as a_T, is measured and recorded.

The construction of the master curve makes it possible to predict the creep or stress relaxation at the reference temperature (or other temperatures) for very long periods of time, even years, using experimental measurements that extend only over hours, weeks, or months. This remarkable property, called the time-temperature equivalency, occurs in most of the commonly encountered plastics. Examples of creep and stress-relaxation curves are shown in Fig. 6, 7, and 8. Figure 7 is a master curve of the tensile creep of a commercial grade of PC (Ref 22), while Fig. 6 is previously unpublished data on the flexural creep of a composite material. The constituents of the composite are an aromatic-amine-cured epoxy resin and uniaxial 67 wt% glass roving reinforcement. In the creep tests, the reinforcement was parallel to the long axis of the specimens.

There are precautions for using time-temperature superposition for the prediction of long-term behavior. The time-temperature superposition is valid, strictly speaking, only if linear viscoelastic equations are applicable; this is called Boltzmann linearity. Thus, in practice, the application should be limited to regions in which an increase in stress from S to kS produces a change (increase) in deformation from $x(t)$ to $kx(t)$. For instance, in the case of the epoxy composite mentioned previously, stress levels up to at least 345 MPa (50 ksi) were tolerated at temperatures below 93 °C (200 °F). However, above this temperature, only lower stresses gave linear behavior. Although a 345 MPa (50 ksi) stress level translates to approximately 20 MPa (3 ksi) stress in the resin (which is a high load for viscoelastic behavior, even for a glassy polymer), linearity is apparently maintained because deflection is limited by the reinforcement. Another possible pitfall is that a new failure mechanism that has not been taken into

Resin structure	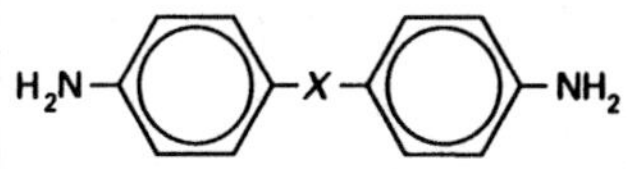		
	$X = -O-$	$X = -CH_2-$	$X = -SO_2-$
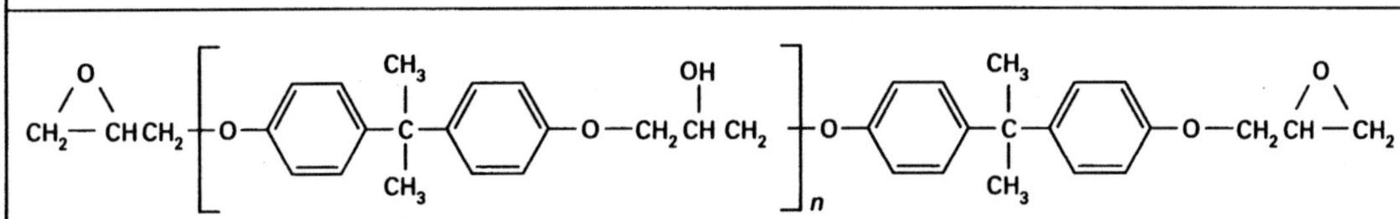 Bisphenol A epoxy resin molecule	1.9	2.1	2.8
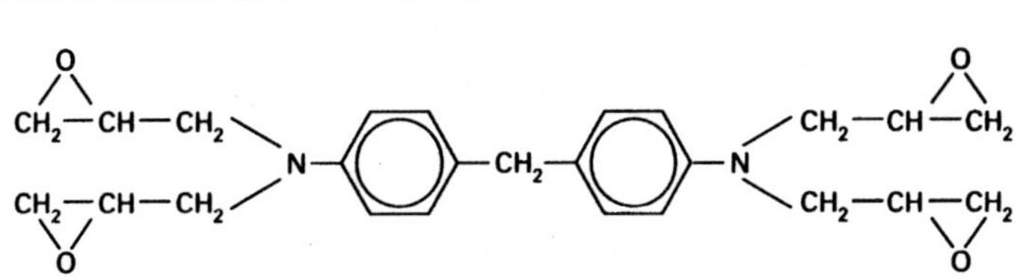 TGMDA	2.8	4.3	5.1

Fig. 5 Comparison of water absorption of epoxy-resin systems of differing polarities. TGMDA, tetraglycidyl methylenedianiline

account may appear after a long period. Chemical degradation could be one such mechanism, and the effect of water, which is discussed later, is yet another.

Relaxation processes are usually accelerated by moisture; as noted previously, the severity of this effect depends on the nature of the plastic, the amount of water absorbed, the temperature, and the severity of the loading to which the part is subjected. In the case of the flexural creep of the epoxy-resin composite discussed previously, the behavior of specimens soaked for long times in water at 66 °C (151 °F) and then tested for creep, also at 66 °C (151 °F), was essentially the same as for the dry specimens tested similarly; thus, no additional shift need be applied to the creep curves. However, in experiments that were similar but were conducted at 93 °C (200 °F), a shift factor due to water (–log a_w) equal to 3.94 had to be applied to the curve for the water-soaked specimen to bring it into reasonable agreement with the master curve. This demonstrated that a wet composite will creep at 93 °C (200 °F) to the same extent as a dry one at approximately 113 °C (235 °F). Furthermore, nonlinearity in the Boltzmann sense was introduced at lower stress because of the increase in compliance. In addition, other failure mechanisms peculiar to the effect of moisture on composites were sometimes observed. A very careful and detailed study of the effect of moisture and temperature on the creep of polyester resins was conducted (Ref 25). It was found that the effect of moisture and temperature on the shift of the data was not independent but interactive.

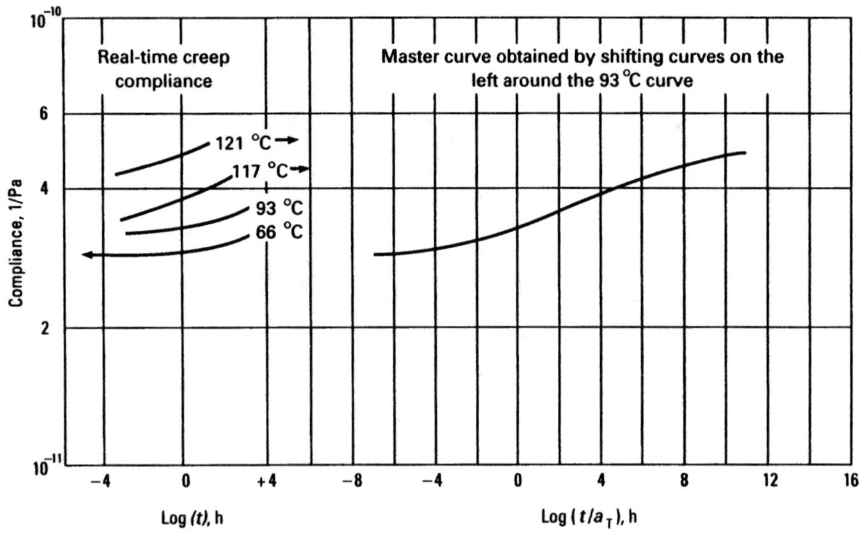

Fig. 6 Flexural creep compliance of parallel glass-fiber-reinforced aromatic-amine-cured epoxy resin (EPON Resin 826). *t*, time; a_T, amount of curve shift

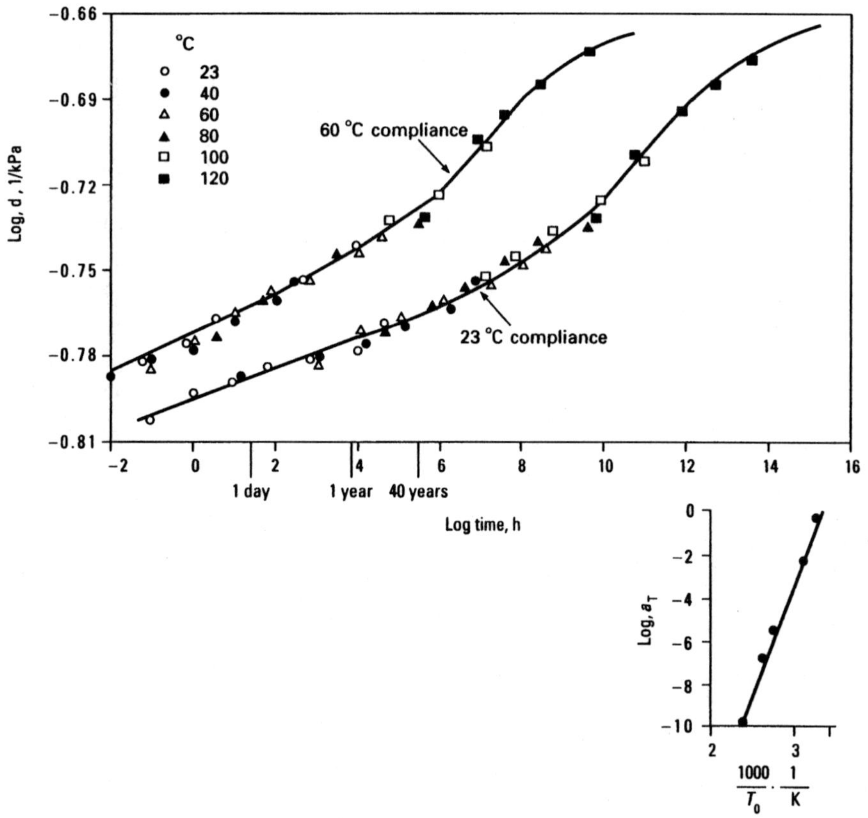

Fig. 7 Polycarbonate creep compliance at 23 °C (73.4 °F) and 60 °C (140 °F) with Arrhenius plot of shift factor. a_T, amount of curve shift. Source: Ref 22

Moisture-Induced Fatigue Failure. Resistance to fatigue failure is an important criterion in the design of many plastic parts, and it is fair to say that many plastic parts actually fail by this mechanism. However, the effect of moisture on fatigue of plastics has not been widely studied. In the few studies reported, the effect of absorbed moisture is not uniformly bad but in fact sometimes acts to improve performance. Manson and Hertzberg (Ref 26) found that fatigue crack propagation in PC was slower above a certain stress level than in dry nitrogen, but the converse was true below this stress level. More complete studies have since been made on plastic materials having an important application as marine ropes. In these studies, polyester materials were affected very little by water or seawater; nylon 6/6 was more affected. The presence of ions, either in seawater or added to solution, did not promote stress cracking in these fibers and ropes (Ref 27–29).

Moisture-Induced Failure in Composites. This review would not be complete without a discussion of moisture-induced failure mechanisms in composite materials. The discussion focuses mainly on reinforcing glass and carbon fibers and is limited, for clarity, to composites reinforced with continuous, uniaxially oriented fibers. Damage mechanisms may take several forms. The plastic matrix is subject not only to the damage mechanisms discussed previously, but also to interfacial and stress-cracking mechanisms. An example of interfacial failure is the much-discussed loss of compressive strength in carbon-fiber-reinforced epoxy composites under hot, wet conditions. Fiber buckling may occur in this type of failure, allowed by the creep of the more compliant matrix. Another type of moisture-induced damage often seen in composite materials is delamination, which also may contribute to the compressive failure described previously. These damage mechanisms are seen in shear and compression but not in tensile loadings. Stress cracking, however, is a tensile failure and is more commonly associated with glass fibers than with carbon. Water alone is known to cause this effect, but the situation is greatly aggravated in fibers exposed to aqueous solutions of acids (Ref 30, 31). These failures, when they occur, may be sudden and, in critical applications, can be catastrophic.

Blistering is a delamination failure that occurs in some composites when exposed to moisture. According to Ref 3, this phenomenon is promoted by osmotic pressures that build up when diffusing water increasingly dilutes ionic species at the fiber-matrix interface. This damage mechanism is fairly common in polyester composites as well as in vinyl ester resins.

Effect of Moisture on Mechanical Properties

Thermoset resins, specifically, epoxy and polyester, are described as follows.

Epoxy Resins. Absorbed water not only results in a depression of the T_g of plastic materials but also causes a loss in other performance properties. For example, researchers (Ref 32) have shown that the epoxy-resin system based on the DGEBA cured with tetraethylenetriamine (TETA), and conditioned in water at 20 and 50 °C (68 and 120 °F), absorbed 2.96 and 3.22 wt% water, respectively, resulting in a decrease of their elastic moduli by 5.9 and 6.6 MPa (60 and 67 kgf/cm^2) (Ref 32). Other investigators noticed the same behavior in an epoxy-resin adhesive based on DGEBA cured with di(1-aminopropyl-3-ethoxy) ether (Ref 33). After a 24 h immersion in water, the adhesive became highly ductile, resulting in reduced yield strength, tensile strength, and modulus.

The TGMDA/DDS epoxy-resin system, used for over 15 years, suffers from high moisture absorption and a rather dramatic loss of performance in hot/wet conditions. Table 5 summarizes the data obtained on two TGMDA/DDS systems cured with 54 and 100% of the stoichiometric quantities of curing agent (Ref 34). As can be seen from the table, the two systems absorbed 4.7 and 5.8 wt% water, which resulted in a decrease of the moduli under hot/wet conditions of 19 and 36%, respectively, when measured while still immersed in water at 93 °C (200 °F) after 2 weeks of immersion in water at this same temperature. The effect of high levels of moisture on the performance above 93 °C (200 °F) of a TGMDA/DDS system is apparent from the data summarized in Table 6.

Degradation of amine-cured epoxy-resin matrix properties by water, because of the formation of microcavities, is discussed in Ref 35 to 37. It was shown that sorbed moisture induces irreversible damage in the resin, the amount depending on the temperature and humidity levels to which the material is exposed. As shown in Table 7, on exposure to water vapor at 65% relative humidity, the failure mode of the epoxy-resin system being studied changed from ductile to brittle. It was proposed that this loss of properties is associated with the formation of microcavities after long-term exposure of the epoxy resin to high humidity. It has been shown that water-soluble inclusions, such as undissolved salts, in a glassy epoxy-resin matrix can nucleate crack formation when the resin is immersed in water (Ref 38).

Polyester Resins. The most widely used class of thermoset resins for fiber-reinforced applica-

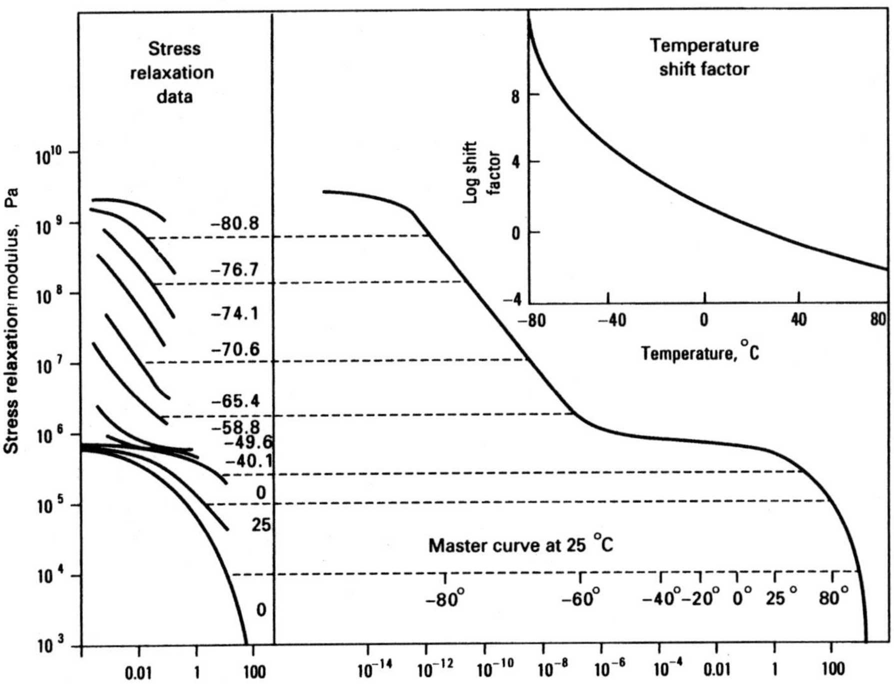

Fig. 8 Time-temperature superposition principle illustrated with polyisobutylene data. Reference temperature of the master curve is 25 °C (77 °F). Insert graph shows the amount of curve shifting required at different temperatures. Source: Ref 5

Table 5 Effect of absorbed moisture on the physical properties of a tetraglycidyl methylenedianiline/diaminodiphenyl sulfone (TGMDA/DDS) system

Components	A	B
TGMDA, parts	100	100
BPA epoxy novolac(a), pph	8.2	8.2
DDS, pph	28	51.9
Glass transition temperature		
Dry, °C (°F)	242 (470)	262 (505)
Wet, °C (°F)	174 (345)	170 (340)
Flexural properties, RT/dry		
Strength, MPa (ksi)	117 (17)	131 (19)
Modulus, GPa (10⁶ psi)	4.0 (0.580)	3.8 (0.548)
Flexural properties, wet(b)		
Strength, MPa (ksi)	83 (12)	76 (11)
Modulus, GPa (10⁶ psi)	3.2 (0.471)	2.4 (0.352)
Modulus, wet/dry(c) % retention	81	64
Moisture gain, %	4.7	5.8

BPA, bisphenol A; RT, room temperature. pph, parts per hundred. (a) EPI-REZ SU-8 (Interez, Inc.). (b) Tested in water at 93 °C (200 °F) after 2 weeks immersion at 93 °C (200 °F). (c) Tested in water at 25 °C (77 °F) after 2 weeks immersion at 25 °C (77 °F)

Table 6 Hot/wet neat resin properties of a tetraglycidyl methylenedianiline/diaminodiphenyl sulfone (TGMDA/DDS) system

System: 90 parts TGMDA, 10 parts EPI-REZ SU-8

Flexural properties	Dry	Wet
Moisture content, %	...	3.6(a)
At 25 °C (77 °F)		
Strength, MPa (ksi)	140 (20)	90 (13)
Modulus, GPa (10⁶ psi)	3.8 (0.550)	3.5 (0.500)
Elongation, %	3.9	2.5
At 150 °C (300 °F)		
Strength, MPa (ksi)	76 (11)	40 (6)
Modulus, GPa (10⁶ psi)	2.6 (0.370)	1.5 (0.215)
Elongation, %	3.2	3.5
At 175 °C (350 °F)		
Strength, MPa (ksi)	83 (12)	35 (5)
Modulus, GPa (10⁶ psi)	2.3 (0.330)	1.1 (0.155)
Elongation, %	4.2	4.0

EPI-REZ SU-8 (Interez, Inc.). (a) After 48 h immersion in boiling water tions is the polyester-based resins. The structures of the three most common types—the orthophthalic esters, the isophthalic esters, and the bisphenol A/fumarate resins—are given in Fig. 9. The effect on mechanical properties of exposure to 100 °C (212 °F) water on glass laminates based on these three types of polyesters is given in Fig. 10. Despite the fact that the orthophthalic polyester shows a dramatic loss of properties at 100 °C (212 °F), it is the most widely used matrix-resin system in fiberglass-reinforced boats. The rapid decline in properties results from exposure to water at temperatures well above its T_g and well above general-use temperatures. Researchers report that glass-reinforced orthophthalate ester was untestable after 10 months of exposure at 100% relative humidity at 93 °C (200 °F) and significantly deteriorated at 82 °C (180 °F) (Ref 39). A glass-reinforced isophthalate polyester test specimen was relatively unaffected under the same conditions. Unfortunately, no tests were carried out at lower temperatures.

The mechanism of the moisture-induced failure of polyesters is quite complex. For example, researchers have shown that the action of water on polyesters is a combination of the result of leaching of low-molecular-weight components initially present in the resin, chemical attack on the ester linkages, and plasticization by the sorbed water (Ref 40). Also, investigators have concluded that in neat polyester resin, degradation begins with water uptake, swelling, and leaching of nonbound substances (Ref 41). However, these workers have found that hydrolysis is the principal irreversible process, accelerated by osmotic-induced cracking. This osmotic process was ascribed to traces of water-soluble, phase-separated materials, particularly glycols.

It was reported in Ref 3 that interfacial bonding between polyester and clean glass fibers was rapidly destroyed by diffused water, with the initiating mechanism being dependent on the glass composition. Bond fracture with E-glass and C-glass reinforcing fibers was shown to be due to osmotic pressure generated at the interface by water-soluble materials leached from the glass fibers. However, when coupling agents were used, debonding was important only with hot-water immersion. What effect this moisture-induced disbonding had on the mechanical properties of the glass-reinforced systems was, however, not examined. A method for predicting laminate properties of isophthalate esters after long-term immersion in water has been developed (Ref 42). These expressions were used to predict tensile property changes at 15 °C (60 °F) over a period of 15 years. Similar methods were used to predict property changes at 30 °C (85 °F) and were experimentally confirmed for absorption times of nearly 3 years.

Thermoplastics described subsequently include polyester, PA, PC, PSU, polyoxymethylene (POM), and polyolefin.

Polyester. In an early, long-term aging study at 0 and 100% relative humidity on poly(1,4-butylene terephthalate) (PBT), it was found that the dry samples aged 18 months showed little degradation (Ref 43). However, the results indicated that those aged at 100% relative humidity at 45 °C (115 °F) would lose up to half the initial value of mechanical properties in 4 to 10 years. Three grades of PBT were aged up to 3 years at 100, 75, 50, and 11% relative humidity and temperatures of 66 to 93 °C (150 to 200 °F) (Ref 44). The decrease in mechanical properties caused by hydrolysis occurs rapidly at higher temperatures and relative humidities. The hydrolytic degradation of polyesters results from

Table 7 Effect of moisture on an epoxy-resin adhesive system

Mechanical properties	Dry	24 h immersion at 100 °C (212 °F)	Dried 48 h after 65 °C (150 °F)(a)	3 months at 65% RH at 25 °C (77 °F)
Yield strength, MPa (ksi)	48 (7.0)	24 (3.5)	58 (8.4)	...
Tensile strength, MPa (ksi)	41 (5.9)	24 (3.5)	52 (7.5)	54 (7.8)
Elongation, %	7.1	37	6.8	5.1
Modulus, GPa (10^6 psi)	1.70 (0.25)	1.02 (0.15)	1.56 (0.25)	1.77 (0.26)
Failure mode	Ductile	Ductile	Ductile	Brittle

RH, relative humidity. (a) Dried 24 h in a vacuum oven at 65 °C (150 °F) after 24 h immersion at 100 °C (212 °F)

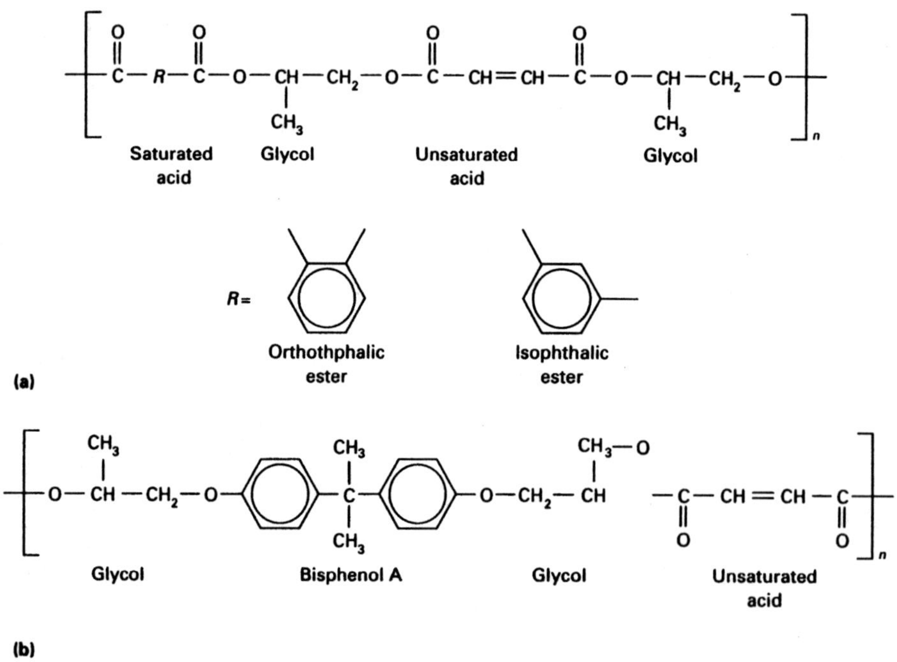

Fig. 9 Structure of unsaturated polyester resins. (a) Phthalate esters. (b) Bisphenol A/fumarate resins

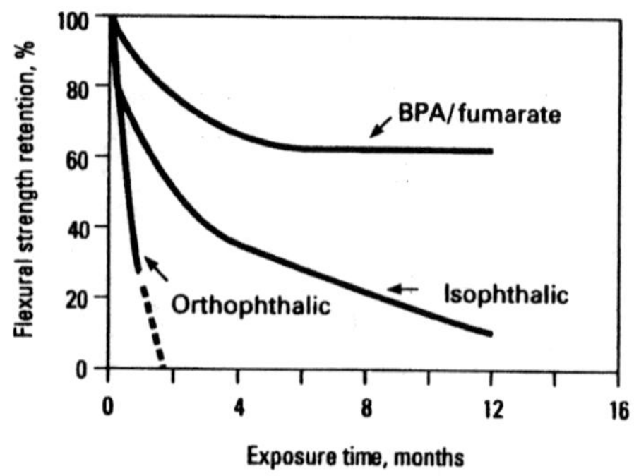

Fig. 10 Degradation of glass laminates in water at 100 °C (212 °F) for different polyester-resin matrices. BPA, bisphenol A

the scission of the polymer chain at the ester linkage, which leads to a progressive reduction of molecular weight. Loss of tensile properties occurs rapidly after aging at higher temperatures and relative humidities. Elongation shows similar trends, although the loss of elongation is even more rapid. For example, the elongation of the material under almost all aging conditions fell essentially to zero in less than 32 weeks, whereas, at 32 weeks, greater than 50% retention of tensile strength was noted under several different aging conditions. Thus, toughness, or impact strength, is lost long before tensile strength half-life is reached.

Researchers (Ref 44) have derived equations from Arrhenius plots for making life-cycle predictions at any temperature and humidity combination for both unfilled and filled PBT. The combined effect of temperature and humidity on unfilled PBT is written:

$$\ln_{t_{1/2}} = \frac{12,680}{T} - 1.36\ln(R) - 31.73 \quad \text{(Eq 2)}$$

where $t_{1/2}$ is the tensile strength half-life (days), T is the temperature (K), and R is the (fractional) relative humidity. A similar equation for glass-reinforced PBT is written:

$$\ln_{t_{1/2}} = \frac{13,134}{T} - 1.33\ln(R) - 33.41 \quad \text{(Eq 3)}$$

Gardner and Martin (Ref 44) point out, however, that these equations may be less reliable at low relative humidity, because other reactions, such as esterification and oxidation, become more important than hydrolysis at low relative humidities. The researchers did not study other polyesters, but similar trends could be expected in polyethylene terephthalate (PET), which would be expected to undergo the same hydrolysis, oxidation, and esterification reactions as PBT.

Polyamide. The effects of water on both unfilled and filled nylon 6, nylon 6/6, and nylon 6/10 have been studied (Ref 45). The water absorption values after immersion at 23 °C (73 °F) for 160 days for unfilled samples of these three nylons were found to be approximately 10.2, 9.2, and 3.0 wt%. This would be expected, based on their polarity. Glass reinforcement reduces the rate of moisture pickup, with the most significant reduction being in nylon 6. For example, after 160 days under the previously mentioned conditions, the moisture absorption of nylon 6 containing 40% glass reinforcement is only approximately 4.6 wt%, and that of nylon 6/10, containing the same amount of glass reinforcement, is approximately 2.0 wt%.

These researchers found that absorbed moisture affected the various physical properties differently. It plasticized the nylons; increasing amounts of absorbed water in unfilled nylons resulted in increasing Izod impact strengths. These increases in impact strength were particularly dramatic for nylon 6 and nylon 6/6. The glass-reinforced systems showed only slight increases in impact strength at increasing levels of moisture. Tensile and flexural strength of both the unfilled and filled nylons were found to decrease steadily with increasing levels of moisture. However, glass-reinforced systems at all moisture levels were generally found to maintain a higher level of their properties at saturation than were the dry, unreinforced systems.

This study was done at 23 °C (73 °F). However, nylon, like polyester, is susceptible to hydrolysis and oxidation. Other investigators (Ref 46) found that unreinforced nylon 6/6, for example, is not suitable for long-term exposure to 100% relative humidity at temperatures of 66 °C (151 °F) or above. However, the strength of unreinforced nylon 6/6 was substantially reduced by long-term aging at 93 °C (200 °F), even at 0% relative humidity, indicating oxidative degradation of the plastic. Adding glass and stabilizers substantially improves the performance of nylons in hot/humid environments. Also, nylons that absorb less moisture retain their properties longer under hot/humid conditions. Nylon 12 loses approximately one-third of its strength because of absorbed water; however, further decreases, even at 66 °C (151 °F), were found to be quite small. At 93 °C (200 °F), glass-reinforced nylon 12 should be serviceable after 10 months at 100% relative humidity.

Polycarbonate. When aged under hot/humid conditions, PCs were found to have a rapid drop in weight-average molecular weight (Ref 47). For example, the weight-average molecular weight of a commercial-grade PC dropped to approximately 65% of its initial value after 40 weeks at 100% relative humidity and 65 °C (149 °F), and to 12% of its initial value at 93 °C (200 °F) and 100% relative humidity after the same length of time. Because the rate of hydrolysis is a function of both the temperature and the concentration of water in the plastic, the increased degradation at 93 °C (200 °F) is a result of both the increased temperature and increased water concentration. Water concentration at 65 °C (149 °F) and 100% relative humidity was found to be 0.416% in this study, and 0.470% at 93 °C (200 °F) and the same humidity level. At 75% relative humidity, the corresponding values were 0.337 and 0.365%. Thus, at constant temperature, the degradation rate over most of the temperature range is approximately halved when the relative humidity is reduced from 100 to 75%.

As indicated in the discussion on thermoplastic polyesters, the hydrolytic degradation leads to a progressive reduction in molecular weight and eventually to a loss of mechanical properties. In the study, the tensile strength was found to drop rapidly below a critical weight-average molecular weight of 33,800, and a transition from ductile to brittle failure was also observed at this point. Thus, brittle fracture occurred even in low-speed tensile tests of an injection-molding grade of PC after exposure to 100% relative humidity for more than 12 days at 93 °C (200 °F). Extrapolations of Arrhenius plots based on 18 month tests indicate that the ductile-brittle transition at 38 °C (100 °F) would be reached after 5 years at 100% relative humidity. Reducing the relative humidity from 100 to 75% would reduce the hydrolysis rate by half at 82 °C (180 °F) and 93 °C (200 °F).

Polysulfone. As seen in Table 1, PSU, like PC, absorbs a relatively low amount of water, but unlike PC, PSU is not susceptible to hydrolytic degradation. Researchers (Ref 39) found that humid aging did accelerate an annealing effect in a commercial grade of PSU. They found the elongation after just a few months of humidity aging dropped from 90% to 6 to 7% elongation. No further change occurred after that time. Elongation in dry specimens, aged at 93 °C (200 °F), also dropped to the same level but only after 18 months. During the 18 months of both dry and humid aging, all specimens had an increase of tensile strength of 6 to 16%; however, the effect appeared to occur faster in the humidity-aged specimens.

Polyoxymethylene. Unreinforced commercial grades of both acetal homopolymer and copolymer were found to be unaffected after 18 months of dry aging at 83 °C (181 °F) or at 66 °C (151 °F) and 100% relative humidity (Ref 39). In fact, the homopolymer showed only approximately a 6% drop in tensile strength after 1 month at 99 °C (210 °F) and 100% relative humidity, and a 3% drop at 82 °C (180 °F) and 100% relative humidity after 19 weeks. However, the tensile strength had dropped to such a point after 10 months at 82 °C (180 °F) and 100% relative humidity that the material was untestable. Under the same conditions, the tensile strength of the copolymer was unchanged. After a total of 18 months exposure at these conditions, the copolymer had lost approximately half of its original strength.

Because a strong odor of formaldehyde was observed in every 100% relative humidity test jar containing the acetal test specimens, the researchers suggested that under hot/humid conditions, these acetal polymers degraded by an unzipping mechanism.

Polyolefin. Because polyolefins such as PE, PP, and polybutylene absorb relatively little water (Table 1) and contain no chemical bonds that are easily hydrolyzable, they are essentially unaffected by aging in water. Martin and Gardner (Ref 39) found, in a commercial 30% glass-reinforced PP, for example, that the long-term influence of humidity is rather small. After an initial drop in both tensile strength and elongation, the properties stabilized and did not change further.

Polyolefins are, however, rapidly oxidized at elevated temperatures, particularly those containing a tertiary hydrogen on the backbone, as in PP and polybutylene. These materials are protected with stabilizer packages containing antioxidants and ultraviolet-light stabilizers. It has been shown (Ref 48) that hot-water deterioration of polyolefin consists of two separate reactions: antioxidant depletion by water (through leaching or hydrolysis) and polymer oxidation. Thus, the service life of a polyolefin in hot water is a function of the specific plastic and stabilizer package, the temperature, and the humidity (or water pressure, in the case of hot-water pipe).

The effect of water on other polymers with all-carbon atom backbones, such as polystyrene, polyvinyl chloride, polyvinylidene chloride, PTFE, and polyisobutylene would be expected to behave somewhat similarly.

REFERENCES

1. R.C. Allen, *Polym. Eng. Sci.,* Vol 19 (No. 5), 1979, p 329
2. R.C. Allen, Paper 6-D, presented at the 33rd RP/C Technical Conference, Society of the Plastics Industry, 1978
3. K.H.G. Ashbee and R.C. Wyatt, *Proc. R. Soc. (London) A,* A.312, 1969, p 553–564
4. D.W. van Krevelin, *Properties of Polymers,* Elsevier, 1980, p 423
5. L.E. Nielson, *Mechanical Properties of Polymers,* Van Nostrand Reinhold, 1962
6. F.N. Kelly and F. Bueche, *J. Polym. Sci.,* Vol 50, 1961, p 549
7. J.M. Gordon, G.B. Rouse, J.H. Gibbs, and W.M. Rosen, *J. Chem. Phys.,* 1977, p 4971
8. J.H. Gibbs and E.A. DiMarzio, *J. Chem. Phys.,* Vol 20, 1958, p 373
9. P.R. Couchman and F.E. Karasz, *Macromolecules,* Vol 11, 1978, p 117
10. P. Moy and F.E. Karasz, The Interactions of Water with Epoxy Resin, *Water Interactions in Polymers,* S.P. Rowland, Ed., Symposium Series 127, American Chemical Society, 1980
11. T. Ellis and F.E. Karasz, *Polymer,* Vol 25, 1984, p 664
12. G. Brinke, F.E. Karasz, and T. Ellis, *Macromolecules,* Vol 16, 1983, p 244
13. M. Goldstein, *Macromolecules,* Vol 18, 1985, p 277
14. R.C. Allen, in *Proceedings of the 18th SAMPE Technical Conference,* Society for the Advancement of Material and Process Engineering, 1986, p 583
15. D.J. Boll, W.D. Bascom, and B. Motiee, *Compos. Sci. Technol.,* Vol 24, 1985, p 253
16. E.L. McKague, J.D. Reynolds, and J.E. Halkias, Thermomechanical Testing of Plastics for Environmental Resistance, *J. Test. Eval.,* Vol 1 (No. 6), Nov 1973, p 468–471
17. C.E. Browning, *Polym. Eng. Sci.,* Vol 18, 1978, p 16–24
18. A. Apicella, L. Nicolais, G. Astarita, and E. Drioli, *Polymer,* Vol 20 (No. 9), 1979, p 1143–1148
19. E.L. McKague, J.D. Reynolds, and J.E. Halkias, *J. Polym. Sci.,* Vol 22, 1978, p 1643–1654
20. E.L. McKague, J.E. Halkias, and J.D. Reynolds, *J. Compos. Mater.,* Vol 9, 1975, p 2–9
21. E. Morel, V. Bellenger, and J. Verdu, *Polymer,* Vol 26, 1985, p 1719–1724
22. G.S. Brockway, *Plast. Des. Forum,* Jan/Feb 1982
23. G.S. Brockway, *Plast. Des. Forum,* March/April 1982
24. Ferry, *Viscoelastic Properties of Polymers,* 3rd ed., John Wiley & Sons, 1980
25. R.D. Maksimov, E.A. Sokolov, and V.B. Machalov, *Mekh. Polimorov,* Vol 3, May/June 1975, p 393 (in English)
26. J.A. Manson and R.W. Hertzberg, *CRC Crit. Rev. Macromol. Sci.,* Vol 1 (No. 4), 1973, p 433
27. P.E. Bretz, R.W. Hertzberg, and J.A. Manson, *J. Mater. Sci.,* Vol 14 (No. 10), 1979, p 2482
28. M.C. Kenney and J.F. McGarry, *J. Mater. Sci.,* Vol 20, 1985, p 2060
29. J.F. Mandell, M.G. Steckel, S.S. Chung, and M.C. Kenney, *Polym. Sci.,* Vol 27, 1987, p 1121
30. P.J. Hogg, *Composites,* Vol 14 (No. 3), 1983, p 262
31. A. Bledzki, R. Spaude, and G.W. Ehrenstein, *Compos. Sci. Technol.,* Vol 23, 1985, p 263
32. A. Apicella and L. Nicolais, *Ind. Eng. Chem. Prod. Res. Dev.,* (No. 23), 1984, p 88
33. D.M. Brewis, J. Comyn, and R.J.A. Shalash, *Polym. Commun.,* Vol 24, 1983, p 67–70
34. R.S. Bauer, *Polym. Prepr.,* Vol 28 (No. 1), 1986, p 33–34
35. A. Apicella, L. Nicolais, G. Astarita, and E. Drioli, *Polymer,* Vol 22, 1981, p 1064
36. A. Apicella and L. Nicolais, *Ind. Eng. Chem. Prod. Res. Dev.,* Vol 20, 1981, p 17
37. A. Apicella, L. Nicolais, G. Astarita, and E. Drioli, *Polym. Eng. Sci.,* Vol 21, 1981, p 17
38. R.F. Fedors, *Polymer,* Vol 21, 1980, p 713–715
39. J.R. Martin and R.J. Gardner, *Polym. Eng. Sci.,* Vol 21 (No. 9), 1981, p 557–565
40. A. Apicella, C. Migiaresi, L. Nicodemo, L. Nicolais, L. Iaccarino, and S. Roccotelli, *Composites,* Vol 13 (No. 4), 1982, p 406–410
41. H.P. Abeysinghe, W. Edwards, G. Pritchard, and G.J. Swampillai, *Polymer,* Vol 23 (No. 12), 1982, p 1785–1790
42. G. Pritchard and S.D. Speake, *Composites,* Vol 18 (No. 3), 1987, p 227–232
43. J.R. Martin, *Managing Corrosion Problems with Plastics,* National Association of Corrosion Engineers, Vol 60, 1975
44. R.J. Gardner and J.R. Martin, *J. Appl. Polym. Sci.,* Vol 25, 1980, p 2352–2361
45. G.W. Woodham and D.R. Pinkston, *SPE J.,* Vol 20, 1970, p 44–47
46. J.R. Martin and R.J. Gardner, *Polymer,* Vol 21 (No. 9), 1981, p 557–565
47. R.J. Gardner and J.R. Martin, *J. Appl. Polym. Sci.,* Vol 24, 1979
48. D.I. Lusk, *SPE ANTEC,* Vol 29, 1983, p 430–433

Organic Chemical Related Failure*

THE SUSCEPTIBILITY OF PLASTICS to environmental failure, when exposed to organic chemicals, limits their use in many applications. Organic liquids, such as cleaning fluids, detergents, gasoline, lubricants, and sealants, may seriously reduce the mechanical properties of plastics. It is therefore important for design engineers to consider the effects of environment on plastics.

The most serious problems arise when a material is exposed to aggressive fluids under stress. The nature of the failure is usually brittle, compared to that occurring in air. The applied stresses on the components may be as low as less than one-tenth of the yield (or failure) stress of the material in air. An understanding of the failure phenomenon associated with aggressive agents is important because of the many cases in which sufficiently high stresses are introduced into plastics in the form of residual stresses during processing, as in shrinkage, or in the form of external stresses incurred during the service life.

A combination of chemical and physical factors, along with stress, usually leads to a serious deterioration in properties. A stress or chemical environment alone does not appreciably weaken a material. The failure mechanism in a particular plastics-chemical environment can be quite complex and, in many cases, is not yet determined. Examples of individual cases are examined in a number of reviews on the subject (Ref 1–6).

Environmental factors can be classified into two categories: chemical and physical effects. Chemical attack occurs when chemical reactions result from the interaction between the environment and polymer molecules. This type of interaction may involve chain scission, which is an irreversible effect. Crazing and microcracking, which are also irreversible processes, may be more noticeable. Physical effects, which are reversible, include fluid absorption or swelling, that is, plasticization. With reversible effects, the material regains its original properties once the environment is desorbed. Both chemical and physical effects are discussed as follows.

Chemical Interactions

Aggressive environments may interact with plastics, causing changes in the chemical character and structure of individual molecules. These changes may involve a decrease in the molecular weight by chain scission or the incorporation of a new chemical group onto the polymer chain. Chain scission may cause a reduction in mechanical properties such as tensile strength, elastic modulus, and fracture toughness. In some cases, chain scission may be followed by depolymerization (unzipping) of the chains, which releases gaseous fragments that cause bubble formation, crazes, and crack formation. For example, polyoxymethylene can be depolymerized to formaldehyde in highly acidic or alkaline environments (Ref 6). The incorporation of new chemical groups onto the polymer chain through chemical reaction may also induce hardening, decrease toughness, and lower the resistance to aggressive environments (Ref 5).

Condensation polymers, such as acetals, polyamides, polyesters, and thioesters, are susceptible to hydrolysis (Ref 1–6). Polycarbonate (PC) and polyphenylene sulfide are attacked by formic acid and amines. Formic acid can decrease the tensile strength of polyphenylene sulfide by 25% (Ref 7).

The tensile properties of polyester-based polyurethane (PUR) samples have been studied as a function of the time of exposure to water, methanol, and a water-methanol mixture (Ref 8). It has been shown that water induces hydrolysis of the ester group, while methanol causes a transesterification reaction. In both situations, these reactions result in losses in tensile properties via molecular weight reduction through random chain scission. It has been observed that the tensile strength decreases as a function of increasing exposure time (Fig. 1). The methanol is believed to swell the PUR, thus reducing the mechanical properties by a plasticization mechanism and causing transesterification. In this system, the overall effect is a physiochemical process, because both a chemical reaction and swelling are observed.

Polycarbonate fails in a sodium hydroxide-ethanol mixture as a result of main-chain scission through hydrolysis under low stresses (Ref 3). It has been proposed that the fibrils in a very short craze, which forms at the crack tip, are exposed to the chemically aggressive environment. The hydrolytic cleavage of the exposed macromolecules causes failure of the fibrils, followed by crack propagation.

The failure mechanism in chemically aggressive agents is far more complex than a simple chain scission mechanism. It is often difficult to pinpoint the controlling factors in a failure process. Generally, sequential processes are thought to occur during crack growth. In the ozone cracking of rubbers, for example, it has been shown that the diffusion of ozone to the crack tip is the rate-controlling step, although ozone induces chain scission (Ref 4). Very few studies have been conducted in the area of chemically induced polymer cracking that

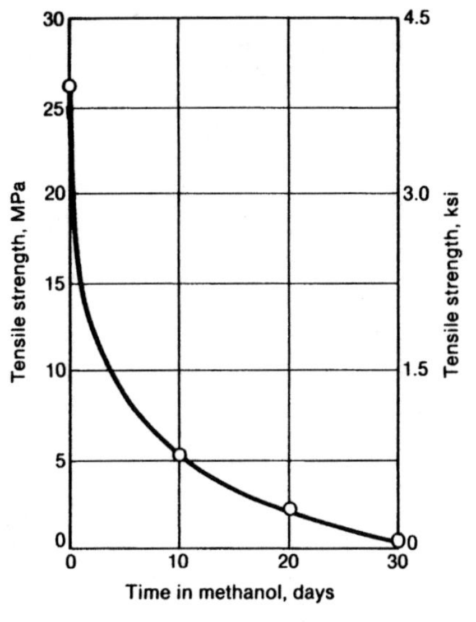

Fig. 1 Tensile strength of polyurethane aged in methanol at 60 °C (140 °F) as a function of exposure time. Source: Ref 8

*Adapted from the article by Koksal Tonyali, "Organic Chemical Related Failure," in *Engineering Plastics*, Volume 2, *Engineered Materials Handbook*, ASM International, 1988, pages 770 to 775

involve crack propagation tests. It is thought that the overall failure mechanism occurs through chain scission by the chemical reaction.

Physical Interactions

Generally, plastics are permeable to organic chemicals to varying degrees. This is a disadvantage, because the presence of environmental liquids in a plastic material has a profound effect on its mechanical properties. The action of sorption may induce plasticization, swelling, dissolution, recrystallization, and leaching of additives in solids, all of which reduce mechanical properties.

Characterization of the diffusion or sorption of penetrants into plastics may promote an understanding of environmental effects on failure properties. After a plastic component is exposed to an organic chemical, aggressive molecules may diffuse into the component, leading to plasticization. Swelling of the material results in high stresses, which can cause crazing or cracking. Fracture has been observed in many glassy plastics, such as polymethyl methacrylate (PMMA), polystyrene (PS), and PC, because of anisotropic swelling (Ref 9–14).

Swelling Kinetics. The diffusion of organic penetrants into plastics may occur either by a Fickian diffusion (case I) or a nonFickian process (case II), in which the rate of the diffusing front in the plastic follows the square root of time, for case I, or the first power of time, for case II (Ref 12–14). A sharp, moving front that propagates at a constant velocity through the material develops in the case II process, which is the one that usually occurs in glassy polymers. Although there is little evidence, it is also believed to occur in crystalline polymers, such as polyethylene(PE)-hydrocarbon systems (Ref 9). As a result of this sharp boundary between the swollen and unswollen material, high swelling stresses are introduced (Ref 10–16). These stresses are large enough to cause crazing or cracking.

The sorption of n-heptane in PS induces extensive bulk crazing in the swollen regions (Ref 17, 18). The formation of crazed networks is usually associated with case II swelling behavior. A model involving the case II swelling of PS-alkane systems to explain the propagation of a crazed front (that is, the velocity of the sharp boundary) is suggested in Ref 17 and 18, such that:

$$\dot{\lambda} = K(\alpha\pi - \sigma_c) \qquad \text{(Eq 1)}$$

where $\dot{\lambda}$ is the propagation rate of the swollen boundary (crazed front), K is a temperature-dependent constant, α is a constant, σ_c is the critical stress for craze propagation in dry PS, and π is the osmotic pressure induced by swelling. The theory uses an experimental relationship for the craze propagation of PS in air and replaces the external applied stress with $\alpha\pi$. However, the theory is applicable only to PS.

A theory to explain case II diffusion kinetics has been proposed for glassy polymers. It couples the diffusive processes ahead of the diffusive front with the mechanical resistance of the glassy polymer to swelling (Ref 16, 19–24). The growth rate of the front, $\dot{d}$, is:

$$\dot{d} = \sqrt{\frac{D(\phi_m)}{\phi_m}\left[\frac{\partial \phi_m}{\partial t}\right]_{\phi_m}} \qquad \text{(Eq 2)}$$

where ϕ_m is the solvent volume fraction at the position of maximum osmotic pressure, D is the diffusion coefficient, and $\partial\phi_m/\partial t$ is the mechanical resistance of the polymer. Attempts have been made to apply this model to the environmental failure of plastics (Ref 25, 26). Assuming that the external stress, σ, is added to the osmotic pressure, P, $\partial\phi_m/\partial t$ can be written as:

$$\frac{\partial \phi}{\partial t} = f(P + \sigma) \qquad \text{(Eq 3)}$$

Considering the simple viscous flow, Eq 3 can be given by:

$$\frac{\partial \phi}{\partial t} = \frac{P + \sigma}{\eta} \qquad \text{(Eq 4)}$$

where η is the viscosity of the swollen material under stress, which can be described by (Ref 21, 23, 24):

$$\eta = \eta_0 \exp(-m\phi) \exp[-n(P + \sigma)] \qquad \text{(Eq 5)}$$

where η_0 is the viscosity of the unswollen polymer, and m and n are constants. The diffusion constant may also be dependent on ϕ and $P + \sigma$ exponentially. Therefore, Eq 2 can be rewritten as:

$$\dot{d} = \left[\frac{D_0}{\eta_0}\,\frac{(P + \sigma)\exp[l(P + \sigma)]}{\phi \exp(k\phi)}\right]^{1/2} \qquad \text{(Eq 6)}$$

where D_0 is the diffusion constant in a glassy polymer, and k and l are constants. Equation 6 can be used only at low solvent activities in plastics (Ref 23, 24).

It is apparent from Eq 6 that the external stress increases the rate of advance of the propagating front. Indeed, it has been shown experimentally that mechanical deformation induces considerable increases in the propagation rate in the PMMA-methanol system (Ref 20–22). Therefore, as the total stress ($P + \sigma$) approaches the yield stress of the material, the failure rate should also increase. The propagation rate of the swollen front in the PMMA-methanol system ($\sim 1 \times 10^{-9}$ m/s, or 3.3×10^{-9} ft/s) without external stress at room temperature (Ref 19) and that of craze growth under low stresses (Ref 27) have been observed to be in the same range. Therefore, the craze growth rates and the swollen front propagation velocities are comparable.

Dissolution and Swelling. An understanding of the solution (or swelling) and dissolution of polymers in solvents is needed to postulate some explanations for environmental failure. Qualitatively, it is convenient to use the Flory-Huggins relationship. The basic idea is that like dissolves like; that is, if a solvent has characteristics similar to those of the plastic, it may dissolve the plastic (Ref 6). For a plastic-solvent system, the activity, a_1, is given by:

$$\ln a_1 = \ln \phi_1 + \phi_2 + \chi(\phi_2)^2 \qquad \text{(Eq 7)}$$

where ϕ_1 and ϕ_2 are the volume fraction of the solvent and of the plastic in a swollen plastic, respectively, and χ is the interaction parameter between the plastic and solvent molecules, which can be estimated by (Ref 28):

$$\chi = a + \frac{V_1}{RT}(\delta_p - \delta_s)^2 \qquad \text{(Eq 8)}$$

where a is a constant, V_1 is the molar volume of the solvent, R is the gas constant, T is the absolute temperature, and δ_p and δ_s are the solubility parameters (that is, square roots of the cohesive energy densities) of the plastic and solvent, respectively.

Equation 7 shows that at equilibrium swelling ($a_1 = 1$), $\ln \phi_1$ is proportional to $(\delta_p - \delta_s)^2$. As an approximation, this suggests that when the difference between the solubility parameters approaches 0, the solvent will be the most effective for dissolving the plastic. In the case of linear polymers, a value of $\chi < 0.5$ leads to full solubility, while $\chi > 0.5$ indicates partial solubility or swelling rather than dissolution. Partial solubility may arise either from limited compatibility or from the strain energy of a swollen network that resists further expansion (Ref 4, 29). The solvent uptake by the plastic induces swelling. The swollen material is plasticized; that is, its mechanical properties are below those of an unswollen solid, but the elongation at break increases.

Fracture processes may not occur at the equilibrium swelling. Figure 2 shows that the absorbed amount of alcohol present in PMMA can substantially reduce tensile yield stress (Ref 30).

Swelling causes plasticization, thus reducing the glass transition temperature, T_g, of the plastic. The T_gs of a swollen aromatic copolyethersulfone in various organic chemicals were determined as a function of the sorbed volume (Ref 31). It was found that the T_g decreases with increasing sorbed volume.

The role of solvent absorption in the crazing and cracking of plastics has been demonstrated for various systems (Ref 1, 2, 4, 30–35). The critical strain to induce crazing in polysulfone (PSU) as a function of the T_g of the solvent-equilibrated films is shown in Fig. 3. The T_g of the equilibrated plastic depends on the solubility parameter and the equilibrium swelling of the plastic, and the reduction in T_g decreases the critical strain for crazing due to the plasticization efficiency of the liquid.

Critical stresses (or strains) for the crazing of PS, which is internally plasticized with dichlorobenzene to varying degrees, were measured as a function of the T_g of the plasticized polymer (Ref 2). It was observed that a similar critical strain dependence on T_g is obtained when the samples are swollen to equilibrium in the environmental liquids. This observation supports the plasticization mechanism for environmental failure.

The critical strain or stress to obtain the crazing (or cracking) of plastics was measured in organic media, and it was observed that the behavior is determined approximately by the difference between the solubility parameters of the plastic and the organic agent (Ref 1, 2). Figure 4 shows the critical strain to induce crazing or cracking of poly(2,6-dimethyl-1,4-phenylene oxide) versus the solubility parameters of the aggressive environments (Ref 32). The liquids include alkanes, aliphatic alcohols, amides, ketones, esters, and halogenated alkanes. The plastic has a solubility parameter of 18.2 $\sqrt{J/cm^3}$ ($8.9\sqrt{cal/cm^3}$). Figure 4 shows that plasticization plays a major role in causing the failure of plastics exposed to aggressive agents; that is, the environment becomes the most effective when the difference between the solubility parameters approaches 0.

In strong polar or hydrogen-bonding liquids, the relationship between the failure properties and solubility parameters is not well correlated (Ref 2, 36). The effect of hydrogen bonding has been taken into account for PMMA, polyvinyl chloride, and PSU, and it has been shown that the solubility effect is similar to that of nonpolar liquids (Ref 36).

The molar volumes, V_m, of the environmental liquids are also found to be important in determining the environmental cracking behavior (Ref 37). The fracture of PC in linear aliphatic hydrocarbons is well described when the critical strain is plotted as a function of $V_m(\delta_p - \delta_s)^2$.

In strong swelling agents, T_g of a plastic is greatly reduced. The fibrils in a craze obtained in such an environment are highly plasticized and therefore cannot withstand external stresses. In this case, cracks are formed rapidly, followed by instantaneous fracture of the plastic (Ref 2, 35). However, in relatively weak swelling agents, the extent of plasticization is limited, and crazing is more pronounced than the formation of cracks. For example, the fatigue failure of PC was studied in various liquid environments. It was found that the craze growth rate at the crack tip decreased and that crack growth and dissolution became more important as the difference between the solubility parameters of the plastic and the solvent approached 0 (Ref 35).

Structural components are generally subjected to external loading during their service lives. The applied stress may affect the sorption kinetics of the environments and the equilibrium swelling (Ref 9, 16, 38, 39). The rate of diffusion for the stressed samples is enhanced by the applied stress due to the defects induced by deformation; therefore, the diffusion rate increases exponentially with stress (Ref 40). The effect of the applied stress on the equilibrium solubility is also considered (Ref 38, 39). The tensile stress increases the equilibrium solubility, which decreases the resistance of the material to crazing and cracking. If a stressed sample with microcracks (or defects) is considered, the stresses are highly concentrated at the crack tips, where the aggressive environment is sorbed more. Inhomogeneous swelling leads to a higher plasticization efficiency at the highly swollen regions, resulting in a reduced flow stress of the material.

It has been suggested that the fracture mechanics approach can describe the environmental crack growth behavior and that a unique relationship exists between the stress-intensity factor, K_I, and the crack speed, $\dot{c}$ (Ref 3, 27, 41). Such K_I and $\dot{c}$ plots consist of three regimes:

- Region I is controlled by the relaxation processes at the crack tip at low K_I values.

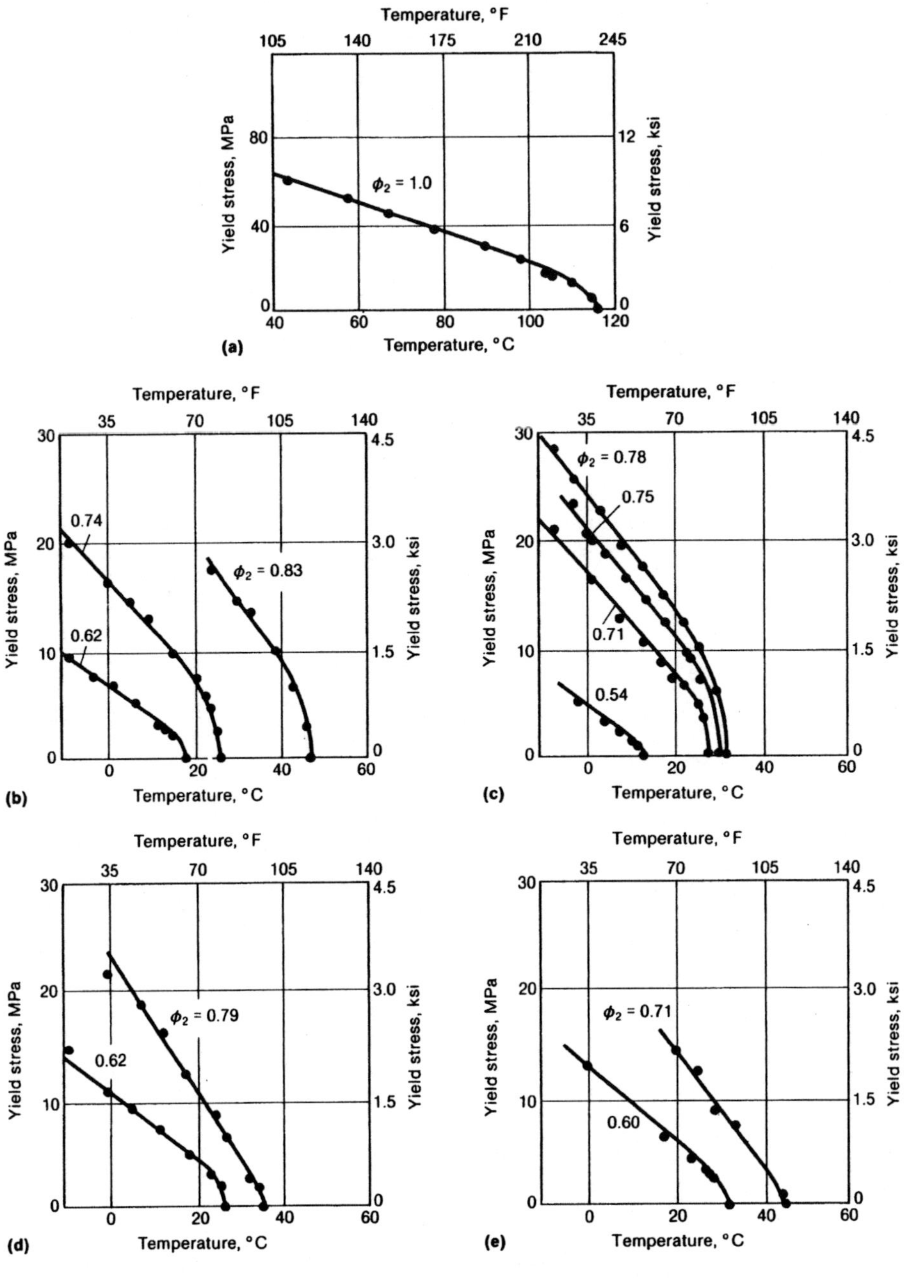

Fig. 2 Yield stress of swollen polymethyl methacrylate samples as a function of the polymer volume fraction, ϕ_2, and temperature. (a) Air. (b) Methanol. (c) Ethanol. (d) *n*-propanol. (e) *n*-butanol. Source: Ref 30

- Region II is determined by the hydrodynamic transport properties of the liquid at moderate K_Is, where the crack speed is inversely proportional to the viscosity of the environment and is usually constant.
- In region III, crack propagation occurs as in air.

The model has been used to interpret the kinetics of the environmental crazing/cracking behavior of polymers (Ref 3).

With organic agents, which sorb into polymers very little, failures of plastics are still observed under low stresses (Ref 1–4). In the absence of an applied stress, no apparent chemical or physical change is observed in plastics properties. The fracture surfaces show evidence that the failure is relatively brittle compared to that obtained in air. The environmental cracking of polyolefins in detergents and alcohols is an example of such a failure process (Ref 27, 41–46). It is generally agreed that the cause of the problem is some form of plasticization due to stress-induced swelling at the defect points (Ref 39). Using infrared spectroscopic techniques, it has been shown that the absorption of low-molecular-weight ethylene oxide adducts of the detergent and nonyl phenol occurs in PE (Ref 47). Furthermore, it has been observed that a small amount of dissolution of PE occurs in detergents. The absorption of alcohols is also observed (Ref 41, 42).

It has been argued that the environmental cracking of PE can be described by the three-region crack growth model (Ref 27, 41). The constant crack speed region (region II) especially has been attributed to the hydrodynamic flow-controlled behavior. At the moderate stress levels, the existence of a dry craze zone at the crack tip was reported (Ref 41). However, recent findings suggest that the constant crack speed region is not flow controlled (Ref 45, 46). The crack growth rates have been determined in the detergent solutions containing various detergent concentrations. It has been found that the constant crack speed increases with increasing detergent concentration. It is well known that the viscosity of a detergent solution is an increasing function of the detergent concentration. Therefore, the constant crack growth rate increases with increasing solution viscosity, which is in contrast to the flow-controlled model. However, in the case of a plasticization mechanism, the condition of the crack tip is irrelevant if it is filled fully or partially (dry craze zone). That is, as soon as some of the load-bearing fibrils are wetted at the crack tip, followed by swelling or dissolution, crack growth should occur (Ref 45), although the exact reasoning for dry craze zone formation is not completely understood.

The solution composition of the environmental media is found to be important in stress cracking (Ref 45, 46). It has been reported that the same amount of a detergent in alcohol is less aggressive than that in water, although the detergent and alcohol are more aggressive separately. The greater degree of aggressiveness in the water solution is attributed to micelle formation by the detergent molecules in water as opposed to the alcohol solution. In the water solution, a micelle contains highly aggressive detergent molecules that are held together, while in the alcohol solution, the aggressive molecules are individually dispersed. Therefore, as soon as a micelle reaches the crack tip, it can induce better plasticization locally, because the detergent activity is high. Furthermore, it has been shown that the environmental solution becomes more aggressive if the detergent concentration is beyond its critical micelle concentration. This is probably because the detergent molecules are not aggregated below the critical micelle concentration.

The addition of swelling agents to the water solution (such as xylene, which locates itself in the micelles only, being insoluble in water) induces a higher cracking efficiency (Ref 45, 46). The micelles act as carriers for aggressive molecules. This information is important, because very small amounts of aggressive agents, or impurities, may be present in cleaning solutions. Nonyl phenol is an example of such an agent found in nonionic detergents.

Surface Energy Effects. Organic liquids usually have low surface tensions and can be readily spread on plastics surfaces. This process has been considered for some time to reduce the surface energy of plastics to accelerate crazing and cracking. It is generally agreed that surface energy reduction appears to be of secondary importance in environmental failure (Ref 1, 2, 30, 48).

The surface energy effect for the PS-methanol system has been measured (Ref 48). It has been shown that crazing is primarily induced by reduction in the flow stress of the swollen material due to plasticization. This is also supported by the results for other glassy polymers (Ref 1, 2, 30).

Destruction of Hydrogen Bonding. Some organic acids can disrupt hydrogen bonding between the macromolecular chains in bulk polymers (Ref 1, 2). Solvent molecules can form a new hydrogen bond between the solvent and polymer molecules. This causes a dissolution process in the material. Polyamides such as nylons can be included in this class of materials, because formic acid or phenols can promote stress cracking (Ref 2).

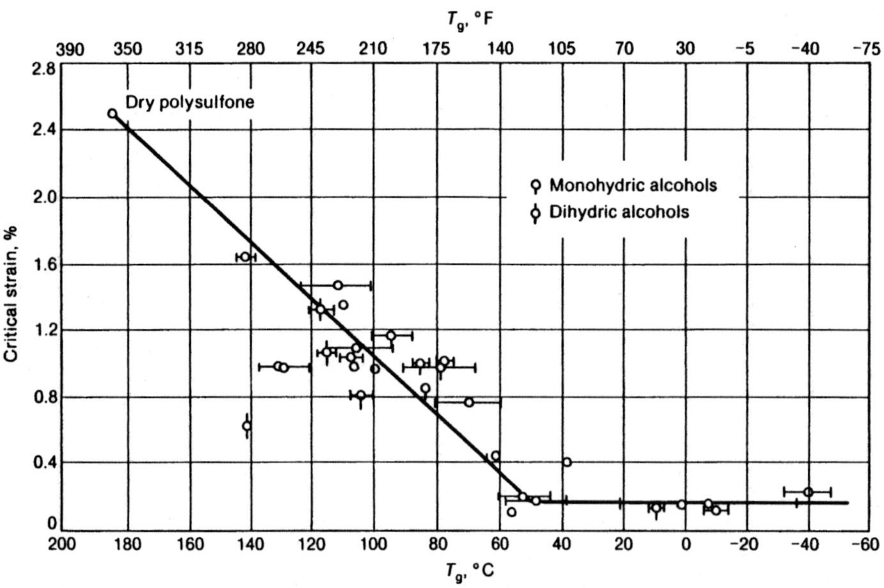

Fig. 3 Critical strain for the crazing or cracking of swollen polysulfone as a function of the glass transition temperature, T_g, of solvent-equilibrated films. Source: Ref 31

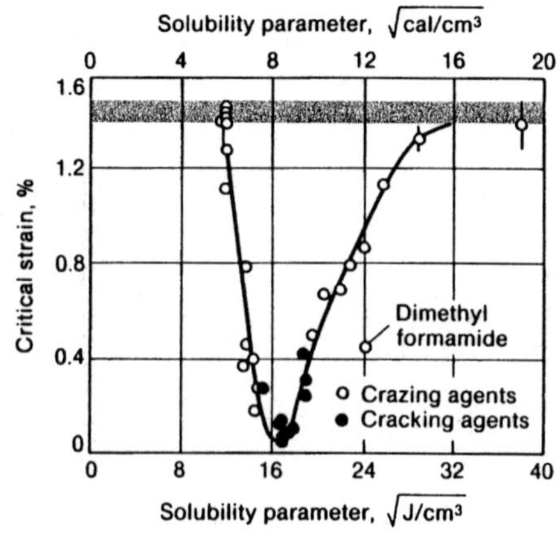

Fig. 4 Critical strain for the crazing or cracking of polyphenylene oxide as a function of solubility parameter. Crosshatched area shows range of critical strain values in air. Source: Ref 32

Solvent Recrystallization. The crazing of some glassy polymers is attributed to recrystallization of the polymer during swelling (Ref 2, 49). The diffusion of acetone into PC causes opacity to develop in the polymer as a result of an increase in crystallinity with the concurrent formation of macroscopic voids (Ref 50). It is proposed that the swelling agent reduces the T_g of the polymer sufficiently to allow the mobile polymer chains to crystallize (Ref 50). A stress-cracking environment should be an effective swelling agent to induce crystallization (Ref 49). As a result of chain ordering, the formation of crystallites introduces high shearing stresses that are sufficient to propagate crazes or cracks.

Incorporating a miscible polyester into a PC improves stress-cracking resistance in strong swelling agents (Ref 51). Polyester crystallizes much more rapidly than PC; therefore, further swelling is restricted because of the recrystallization that stabilizes the craze fibrils. The solvent recrystallization effects remain open to debate until extensive studies have been conducted.

Solvent Leaching of Additives. Additives such as plasticizers, fillers, stabilizers, and colorants are introduced into plastics to improve their physical properties. Leaching of these additives may create serious problems in the working life of plastics components (Ref 52). The chemical resistance of plasticized plastics to organic liquids is usually less than that of the unplasticized plastics, such as polyvinyl chloride (Ref 6). The interaction between the additives and the organic chemicals determines the resistance of the system in terms of solubility parameters.

Adding a plasticizer increases the mobility of the polymer chains, which enhances the effective diffusion coefficient of liquids (Ref 9, 13). Organic additives can be extracted from plastics even if they are not greatly soluble in the solvent. The diffusion and migration of additives from the material induce losses in physical properties because of the development of a somewhat porous structure in the solid (Ref 53). Such defects reduce mechanical properties for practical use. Plasticizer migration, or deplasticization, leads to embrittlement of the compound. On the other hand, the regions of additives may swell anisotropically, thus causing differential expansion or cluster formation, which results in crazing and cracking of the structure (Ref 9). In the case of a stabilizer, plastics durability, or resistance to oxidative degradation, is reduced because the stabilizer is leached from the plastic. Therefore, the interaction of organic liquids with additives as well as the plastic itself must be considered for design purposes.

REFERENCES

1. R.P. Kambour, Environmental Stress Cracking of Thermoplastics, *Corrosion Fatigue,* O.F. Deveraux and R.W. Staehle, Ed., National Association of Corrosion Engineers, 1972, p 681
2. R.P. Kambour, A Review of Crazing and Fracture in Thermoplastics, *J. Polym. Sci., D, Macromol. Rev.,* Vol 7, 1973, p 1
3. E.J. Kramer, Environmental Cracking of Polymers, *Developments in Polymer Fracture—1,* E.H. Andrews, Ed., Applied Science, 1979, p 55
4. E.H. Andrews, The Short and Long Term Performance of Polymers in Different Environments, *Br. Polym. J.,* Vol 10, 1978, p 39
5. V. Shah, *Handbook of Plastics Testing Technology,* John Wiley & Sons, 1984
6. R.B. Seymour, *Plastics vs. Corrosives,* John Wiley & Sons, 1982
7. V.C. Vives, J.S. Dix, and D.G. Brady, Polyphenylene Sulphide in Harsh Environments, *The Effects of Hostile Environments on Coatings and Plastics,* D.P. Garner and G.A. Stahl, Ed., ACS Symposium Series 229, American Chemical Society, 1983, p 65
8. D.L. Faulkner, M.G. Wyzgoski, and M.E. Myers, Jr., Polyurethane Aging in Water and Methanol Environments, *The Effects of Hostile Environments on Coatings and Plastics,* D.P. Garner and G.A. Stahl, Ed., ACS Symposium Series 229, American Chemical Society, 1983, p 173
9. C.E. Rogers, Polymer Films as Coatings, *Surfaces and Coatings Related to Paper and Wood,* R.H. Marchessault and C. Skaar, Ed., Syracuse University Press, 1967, p 463
10. T. Alfrey, E.F. Gurnee, and W.G. Lloyd, Diffusion in Glassy Polymers, *J. Polym. Sci.,* Vol C12, 1966, p 249
11. B. Rosen, Time Dependent Tensile Properties. Part III: Microfracture and Non-Fickian Vapor Diffusion in Organic Glasses, *J. Polym. Sci.,* Vol 49, 1961, p 177
12. H.B. Hopfenberg and V. Stannett, The Diffusion and Sorption of Gases and Vapors in Glassy Polymers, *The Physics of Glassy Polymers,* R.N. Havard, Ed., Applied Science, 1973, p 504
13. R.M. Felder and G.S. Huvard, Permeation, Diffusion and Sorption of Gases and Vapors, *Methods of Experimental Physics—Polymers,* Vol 16C, R.A. Fava, Ed., Academic Press, 1980, p 315
14. J. Comyn, Ed., *Polymer Permeability,* Elsevier, 1985
15. C.E. Rogers, Solubility and Diffusivity, *Physics and Chemistry of the Organic Solid State,* Vol II, D. Fox, M. Labes, and A. Weissberger, Ed., Interscience, 1965, p 150
16. A.H. Windle, Case II Sorption, *Polymer Permeability,* J. Comyn, Ed., Elsevier, 1985, p 75
17. G.C. Sarti, Solvent Osmotic Stresses and the Prediction of Case II Transport Kinetics, *Polymer,* Vol 20, 1979, p 827
18. G.C. Sarti and A. Apicella, Non-Equilibrium Glassy Properties and Their Relevance in Case II Transport Kinetics, *Polymer,* Vol 21, 1980, p 1031
19. N.L. Thomas and A.H. Windle, A Deformation Model for Case II Diffusion, *Polymer,* Vol 21, 1980, p 613
20. N.L. Thomas and A.H. Windle, Diffusion Mechanics of the System PMMA-Methanol, *Polymer,* Vol 22, 1981, p 627
21. N.L. Thomas and A.H. Windle, A Theory of Case II Diffusion, *Polymer,* Vol 23, 1982, p 529
22. A.H. Windle, The Influence of Thermal and Mechanical Histories on Case II Sorption of Methanol by PMMA, *J. Membrane Sci.,* Vol 18, 1984, p 87
23. C.Y. Hui, K.C. Wu, R.N. Lasky, and E.J. Kramer, Case II Diffusion in Polymers. I. Transient Swelling, *J. Appl. Phys.,* Vol 61, 1987, p 5129
24. C.Y. Hui, K.C. Wu, R.N. Lasky, and E.J. Kramer, Case II Diffusion in Polymers. II. Steady-State Front Motion, *J. Appl. Phys.,* Vol 61, 1987, p 5137
25. K. Tonyali, "Stress Cracking of Polyethylene in Organic Liquids," Ph.D. dissertation, Case Western Reserve University, 1986
26. H.R. Brown, A Model of Environmental Craze Growth in Polymers, J. Polym. Sci. B, Polym. Phys., Vol 27 (No. 6), May 1989, p 1273–1288
27. J.G. Williams, Applications of Linear Fracture Mechanics, *Adv. Polym. Sci.,* Vol 27, 1978, p 67
28. R.F. Blanks and J.M. Prausnitz, Thermodynamics of Polymer Solubility in Polar and Nonpolar Systems, *Ind. Eng. Chem., Fundam.,* Vol 3, 1964, p 1
29. L.R.G. Treloar, *The Physics of Rubber Elasticity,* 3rd ed., Clarendon, 1975
30. E.H. Andrews, G.M. Levy, and J. Willis, Environmental Crazing in a Glassy Polymer: The Role of Solvent Absorption, *J. Mater. Sci.,* Vol 8, 1973, p 1000
31. R.P. Kambour, E.E. Ramagosa, and C.L. Gruner, Swelling, Crazing and Cracking of an Aromatic Copolyether-Sulphone in Organic Media, *Macromolecules,* Vol 5, 1972, p 335
32. G.A. Bernier and R.P. Kambour, The Role of Organic Agents in the Stress Crazing and Cracking of Poly(2, 6-dimethyl-1, 4-phenylene oxide), *Macromolecules,* Vol 1, 1968, p 393
33. E.H. Andrews and L. Bevan, Mechanics and Mechanism of Environmental Crazing in a Polymeric Glass, *Polymer,* Vol 13, 1972, p 337
34. E.H. Andrews and G.M. Levy, Solvent Stress Crazing in PMMA: I. Geometrical Effects, *Polymer,* Vol 15, 1974, p 599
35. J. Miltz, A.T. DiBenedetto, and S. Petrie, The Effect of Environment on the Stress Crazing of Polycarbonate, *J. Mater. Sci.,* Vol 13, 1978, p 2037
36. P.I. Vincent and S. Raha, Influence of Hydrogen Bonding on Crazing and Cracking of Amorphous Thermoplastics, *Polymer,* Vol 13, 1972, p 283
37. C.H.M. Jacques and M.G. Wyzgoski, Pre-

diction of Environmental Stress Cracking of Polycarbonate from Solubility Considerations, *J. Appl. Polym. Sci.,* Vol 23, 1979, p 1153

38. A.N. Gent, Hypothetical Mechanism of Crazing in Glassy Plastics, *J. Mater. Sci.,* Vol 5, 1970, p 925
39. H.R. Brown, A Theory of the Environmental Stress Cracking of Polyethylene, *Polymer,* Vol 19, 1978, p 1186
40. J. Comyn, Kinetics and Mechanism of Environmental Attack, *Durability of Structural Adhesives,* A.J. Kinloch, Ed., Applied Science, 1983, p 85
41. R.A. Bubeck, Kinetics of Environmental Stress Cracking in High Density Polyethylene, *Polymer,* Vol 22, 1981, p 682
42. J.B. Howard, Stress Cracking, *Engineering Design for Plastics,* E. Baer, Ed., Reinhold, 1964, p 742
43. R.A. Isaksen, S. Newman, and R.J. Clark, Mechanism of Environmental Stress Cracking in Linear Polyethylene, *J. Appl. Polym. Sci.,* Vol 7, 1963, p 515
44. C.J. Singleton, E. Roche, and P.H. Geil, Environmental Stress Cracking of Polyethylene, *J. Appl. Polym. Sci.,* Vol 21, 1977, p 2340
45. K. Tonyali, C.E. Rogers, and H.R. Brown, Stress-Cracking of Polyethylene in Organic Liquids, *Polymer,* Vol 28, 1987, p 1472
46. K. Tonyali and H.R. Brown, Effects of Detergent Concentration and Ethylene Oxide Chain Length of the Detergent Molecule on Stress-Cracking of Low Density Polyethylene, *J. Mater. Sci.,* Vol 22, 1987, p 3287
47. J. Belcher, "Environmental Stress Cracking of Polyethylene," M.Sc. thesis, Monash University, 1981
48. H.R. Brown and E.J. Kramer, Effect of Surface Tension on the Stress in Environmental Crazes, *Polymer,* Vol 22, 1981, p 687
49. G.W. Miller, S.A.D. Visser, and A.S. Morecroft, On the Solvent Stress Cracking of Polycarbonate, *Polym. Eng. Sci.,* Vol 11, 1971, p 73
50. R.P. Kambour, F.E. Karasz, and J.H. Daane, Kinetic and Equilibrium Phenomena in the System: Acetone Vapor and Polycarbonate Film, *J. Polym. Sci.,* Vol A2 (No. 4), 1966, p 327
51. R.P. Kambour, C. Chu, and R.W. Avakian, Crystallizing Crazes: Probable Source of Solvent Stress Cracking Resistance in Polyester/Polycarbonate Blend, Part B, *J. Polym. Sci.,* Vol 24, 1986, p 2135
52. R.D. Deanin, *Polymer Structure, Properties and Applications,* Cahners, 1972
53. R. Ranby and J.F. Rabek, Environmental Corrosion of Polymers, *The Effects of Hostile Environments on Coatings and Plastics,* D.P. Garner and G.A. Stahl, Ed., ACS Symposium Series 229, American Chemical Society, 1983, p 291

Photolytic Degradation*

ENGINEERING PLASTICS of various types are currently used outdoors or are soon expected to be used outdoors (Ref 1, 2). In addition, new materials are continually being developed. An organic material used outdoors is exposed to a very hostile environment. Sunlight, oxygen, heat, humidity, atmospheric pollutants, and physical stresses all combine to produce changes in the chemical composition of the material. These changes may take the form of polymer molecular weight reduction due to main-chain cleavage, the formation of cross links, or the formation of oxidized and other functional groups. As the chemical composition of the material changes, its mechanical properties and physical appearance change. At some point, the changes in chemical composition become sufficiently extensive to render the material unfit for its design objectives, and the material fails.

Typically, a new material is evaluated for outdoor weatherability by placing it in a location known to have a harsh environment, such as Florida, and waiting for physical failure to occur (Ref 3). Typical types of failure include yellowing, chalking, surface embrittlement, loss of tensile or impact strength, and cracking. Chemical degradation usually proceeds from the top layer; this weakened surface can serve as a site for crack initiation. Once formed, cracks can propagate rapidly into the undegraded material below, causing failure.

Although a great deal of information has been gathered by following the physical performance of materials during outdoor exposure, the information tends to be empirical and therefore not easily extrapolated to new plastics systems. For example, it is not clear whether the individual components of new materials, such as plastics alloys/blends, will weather independently, allowing the weakest component ultimately to determine the rate at which physical properties are lost, or whether synergistic or antagonistic interactions between components are to be expected. Most of the weatherability data on plastics are restricted to simple systems.

Only recently, with advances in spectroscopic techniques, have relationships begun to develop between the chemistry that takes place in materials during outdoor exposure and the loss of mechanical properties. The absence of such relationships is surprising, because the photochemistry of the simpler plastics, as well as model compounds representing every type of functionality found in plastics, has been studied in great detail (Ref 4, 5). However, the photochemistry of plastics in the outdoors is very difficult to follow systematically. First, the chemistry itself is slow, with many years between the onset of exposure and failure. The amount of chemistry necessary to cause failure can actually be very small. The loss of a single bond can halve the molecular weight of a polymer chain. Finally, the chemistry occurs in the near absence of the usual laboratory controls. Light intensity and wavelength, temperature, humidity, and physical stresses, representing the more obvious variables, can span enormous ranges over the course of an experiment. Compounding this issue, failure events also tend to be stochastic and require multiple exposures to accumulate statistically reliable data. Several samples of the same plastic exposed to the same weathering will have a distribution of time-to-failure, and this must be taken into account in analyzing failure data.

Because there are few truly photostable plastics and because there is a lack of specific long-term data on many engineering plastics, it is generally assumed that these plastics require some protection in the form of stabilizers or an external coating. Although the use of coatings can eliminate concerns regarding plastic durability, coatings have their own set of durability issues and can add significantly to the cost of the final product. The photolytic instability and degradation of plastics have a direct analog in the corrosion of metals, in that function is slowly degraded in the external environment by chemical reactions and must be treated as limiting the true usefulness of the materials. In the future, the full and successful outdoor use of engineering plastics will undoubtedly depend on a clear understanding of the chemical changes induced by weathering and their relationship to physical properties.

This article provides a basic review of polymer photochemistry as it relates to the weatherability of engineering plastics. The present work considers only one aspect of weatherability chemistry, namely, the chemistry induced by exposure to sunlight in the open air. Mechanochemical, biochemical, hydrolytic, and air pollution induced degradation are not discussed. It is recognized that these other environmental factors can influence the rate of photochemistry (Ref 6). Elementary aspects that are discussed include the light wavelengths responsible for polymer photochemistry, problems with artificial light sources, general photooxidation and specific photochemical reactions important in plastics, and factors influencing the rate of degradation. The approaches used to stabilize plastics against photochemical damage, including ultraviolet light absorbers, oxidation inhibitors, and the use of protective coatings, are also considered.

Sunlight

Ultraviolet Light. When light is absorbed by a plastic, the energy is used to promote an electron in the absorbing chromophore to an excited state. Photochemistry begins when the stored light energy is used to drive a chemical reaction. No photochemistry occurs when the light is dissipated harmlessly as heat. The rate at which reaction occurs depends on the energy content and intensity of the light absorbed, the chemical nature of the chromophore excited, and its environment.

The energy content of sunlight at ground level, as well as its intensity as a function of wavelength, is modified considerably by the presence of ozone in the earth's atmosphere. Ozone absorbs sunlight at wavelengths shorter than 290 to 300 nm (2900 to 3000 Å). The ozone cutoff has often been ignored in the design of apparatuses used to accelerate the degradation of polymers for test purposes (specifically, fluorescent sunlamp devices). This is one of the reasons why accelerated weathering results do not always correlate with outdoor exposure results. Ozone limits the energy of photons reaching ground level to a maximum of 410 to 400 kJ/mol (98 to 95 kcal/mol). In contrast, many studies of polymer photochemistry have used mercury arc light sources. This source has light at wavelengths as low as 254 nm (2540 Å) and an energy level of 470 kJ/mol (112 kcal/mol), and it may induce chemistry that does not occur outdoors.

*Adapted from the article by John L. Gerlock and David R. Bauer, "Photolytic Degradation," in *Engineering Plastics,* Volume 2, *Engineered Materials Handbook,* ASM International, 1988, pages 776 to 782

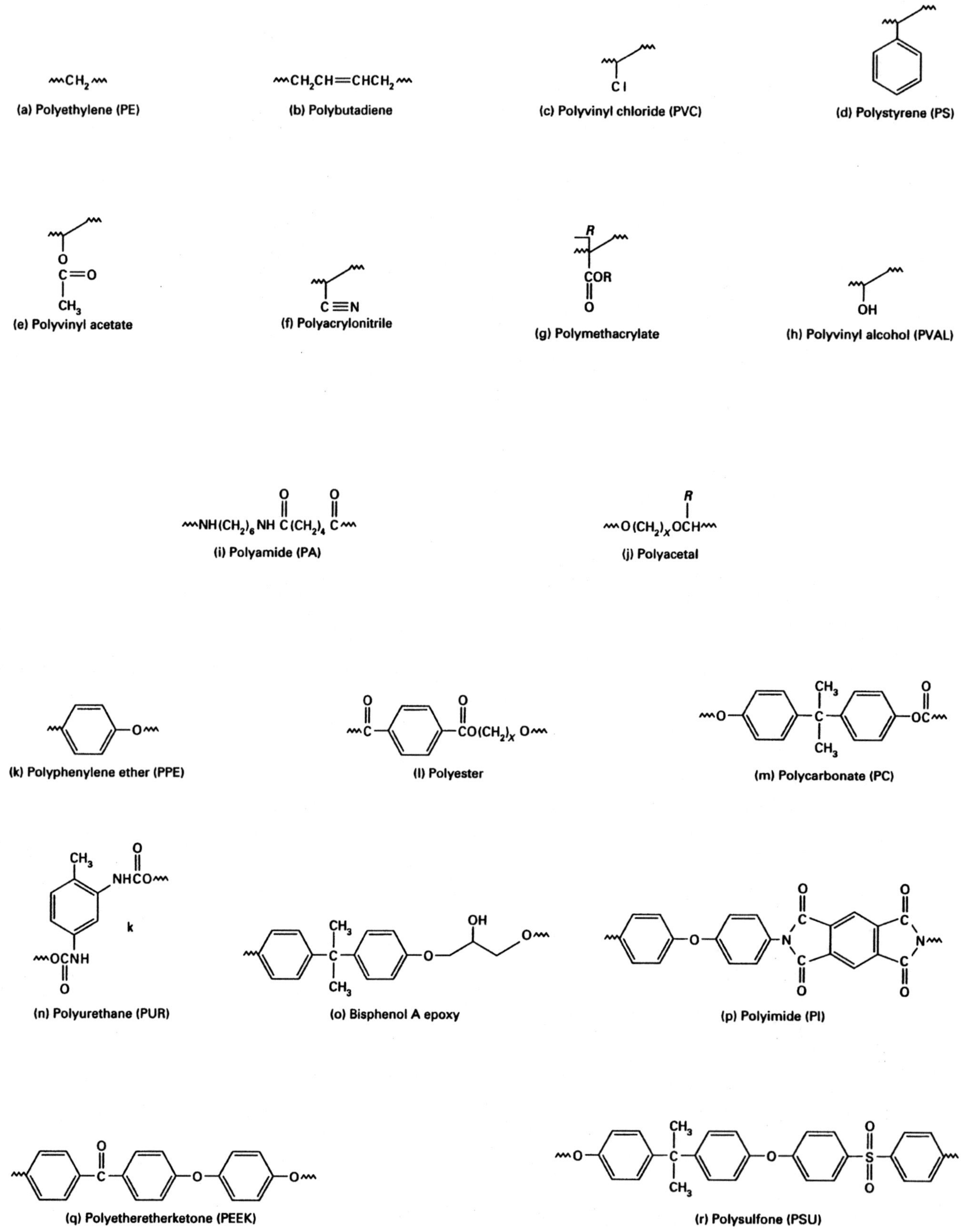

Fig. 1 Monomer units of common polymers. (a) through (j) are not sunlight absorbing; (k) through (r) are sunlight absorbing.

The activation energy of most photochemical reactions in the gas phase usually lies no more than 5 to 6% above the dissociation energy of the bond being broken. Typical bond dissociation energies in plastics range between 420 and 290 kJ/mol (100 and 70 kcal/mol). Therefore, it is not surprising that ultraviolet light at wavelengths shorter than 300 nm (3000 Å) is sufficient to break bonds and to initiate degradation. Fortunately, the intensity of the light in the 290 to 300 nm (2900 to 3000 Å) region is very low; if this were not the case, few present-day plastics would be of use outdoors. Light in the 290 to 320 nm (2900 to 3200 Å) range, with 410 to 370 kJ/mol (98 to 89 kcal/mol), is estimated to account for 0.5%, at most, of the radiant energy of the sunlight at noon in southern regions. Light in the 300 to 360 nm (3000 to 3600 Å) range, with 370 to 330 kJ/mol (89 to 79 kcal/mol), is more abundant, but it is less energetic and accounts for up to 2.5% of the total radiant energy of sunlight. Light in the 360 to 400 nm (3600 to 4000 Å) range, with 330 to 300 kJ/mol (79 to 71 kcal/mol), is sufficiently energetic to break only the weakest polymer bonds, and it accounts for over half of the ultraviolet component of sunlight.

Absorption of Ultraviolet Light. Figure 1 identifies some polymers that are commonly used in engineering plastics. These polymers can be divided into two broad categories based on whether or not the monomer unit contains a chromophore that absorbs the ultraviolet component of sunlight. The division is artificial in that nonsunlight-absorbing polymers invariably contain traces of sunlight-absorbing impurities. All these polymers photochemically degrade during outdoor exposure. Transparency alone is not a good measure of the utility of a polymer in the outdoors. Much depends on the criteria assigned to constitute failure. The surface of a photodegraded material consisting of a sunlight-absorbing polymer may yellow, while its interior remains physically sound and chemically unchanged because it is screened. Conversely, absorption of ultraviolet light and subsequent photodegradation may occur throughout the bulk of a nonsunlight-absorbing polymer. However, the division is useful in categorizing the type of chemistry likely to dominate degradation.

The photodegradation of nonsunlight-absorbing polymers is dominated by free-radical chain oxidation initiated by the photolysis of unwitting chromophoric impurities or chemical defects. In the photodegradation of sunlight-absorbing polymers, the direct photochemistry of specific functional groups leads to destruction of the polymer chain, with subsequent free-radical chain oxidation chemistry, causing further damage. Strategies for extending the useful life of plastics in the outdoors are keyed to these differences in degradation mode.

Polymer Photochemistry

Reactions in the Excited State. The first step in any photochemical reaction is the absorption of photon energy ($h\nu$) by a chromophore to create an excited state, denoted by A^* (Ref 7):

$$A + h\nu \rightarrow A^* \quad \text{(Eq 1)}$$

The excited state can either relax back to the ground state, A (through emission of a photon or heat), without any changes in chemistry or undergo chemical reaction. Two types of reactions dominate the chemistry of excited states in polymers: dissociation and hydrogen atom abstraction. Both of these processes form a pair of radicals in close proximity. The initially formed radical pair is thought to reside in a cage, surrounded by its polymer host. A number of reaction possibilities exist for radicals within the cage. They may escape the cage to become free radicals or may simply recombine. Recombination is favored in a highly viscous medium, such as a polymer, and accounts for the fact that the quantum yield for the formation of free radicals in polymers is usually very low (<1%) relative to model compounds in solution. Recombination within the cage may result either in the regeneration of the original chromophore or in a rearrangement, as illustrated by the well-known photo-Fries reaction observed in aromatic polycarbonates (PCs) (Fig. 2).

The product of this rearrangement is a phenyl salicylate ultraviolet absorber. This moiety acts to protect lower layers of PC from sunlight. In part, this accounts for the relatively good durability of the material. The PC fraction at or near the surface does yellow. In applications in which optical performance is important (for example, headlamps), this can be unacceptable, and coatings or ultraviolet absorbers are required.

The escape of cage radicals to form free radicals is illustrated in Fig. 3 by the Norrish I photocleavage of a terephthalate ester. This reaction results in the cleavage of the polymer chain and opens the possibility for free-radical chain oxidation, as is discussed shortly.

Dissociation is the predominant path for free-radical formation in most of the functional groups shown in Fig. 1.

Intermolecular hydrogen atom abstraction is illustrated in Fig. 4 by the reaction of excited benzophenone, an analog of polyetheretherketone (PEEK), with the hydrogen atom donor, PH. Hydrogen atom abstraction can result in the

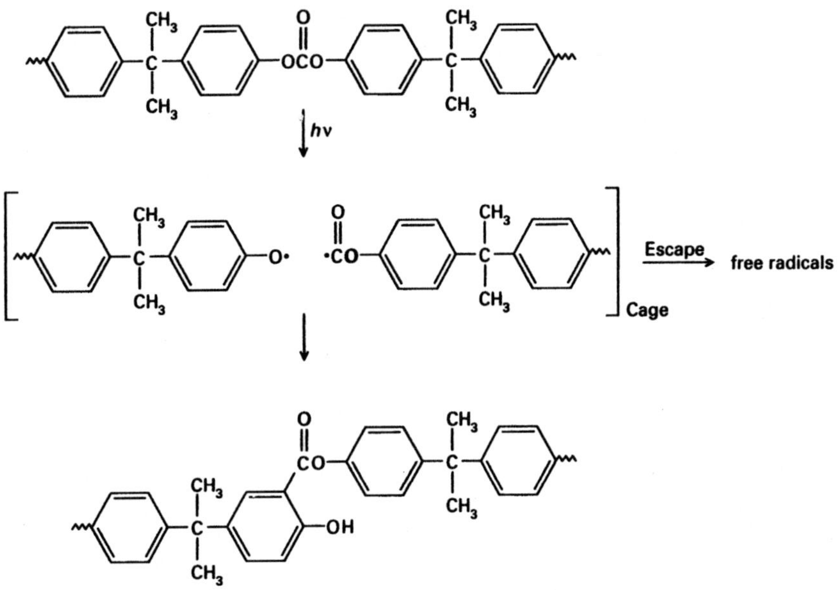

Fig. 2 Photo-Fries reaction in aromatic polycarbonate; h, Planck's constant; ν, photon frequency

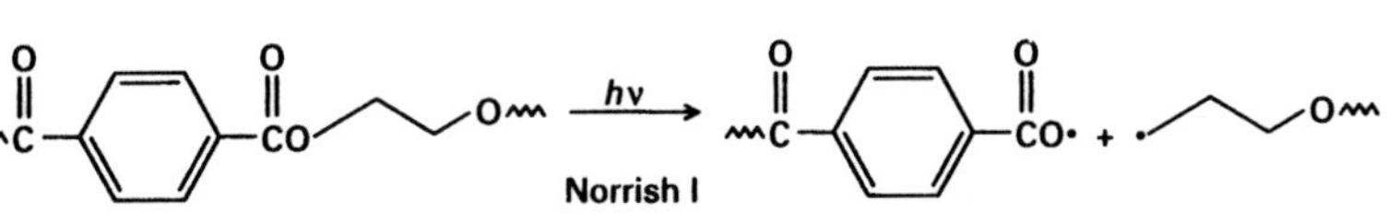

Fig. 3 Norrish I photocleavage of terephthalate ester

formation of polymer-polymer cross links, which will cause changes in polymer properties, especially embrittlement.

The Norrish II photocleavage of a terephthalate ester illustrates an intramolecular hydrogen atom abstraction, shown in Fig. 5, in which free radicals are not formed. This reaction results in the cleavage of a polymer chain.

Free-Radical-Induced Oxidation. The reactions illustrated in Fig. 2 through 5 also pertain to trace chromophores in nonsunlight-absorbing polymers. The nature of the chromophore-absorbing light is usually unknown. Typically, the chromophores are impurities, such as end groups, initiator and inhibitor fragments, and processing-related oxidation products. Specific chromophores include aliphatic and aromatic ketones, phenols, hydroperoxides, and transition metal compounds. On absorption of light, the excited states of these groups dissociate, or hydrogen atom abstract (for example, the reactions shown in Fig. 3 and 4), to yield free radicals. Some species may also transfer the excited-state energy to another species, which undergoes these reactions.

Because the specific chromophore(s) is not usually known, the photoinitiation reaction is usually generalized as polymer + light → 2Y·, as shown in Fig. 6. The photolysis of a noncovalently bound chromophoric impurity causes no direct damage to the polymer. A chromophore-based free radical, Y·, goes on to abstract a hydrogen atom from its polymer host to produce a polymer radical. This reaction destroys the identity of the initially formed free radical, and subsequent reactions reflect the free-radical chemistry of the polymer. The polymer radical may undergo a variety of transformations in which radical character is preserved. A reactive polymer radical may abstract a hydrogen atom from a neighbor to produce a less reactive polymer radical. In Fig. 7, for example, the tertiary benzylic radical is more stable than the secondary aliphatic radical.

If the resultant radical is sufficiently nonreactive, then reaction ceases. Intramolecular hydrogen atom abstraction results in the movement of a radical site down a polymer chain. Repeated intramolecular or intermolecular hydrogen atom abstraction between nearly equivalent radical sites is thought to be one of the ways in which radical sites move in polymers, for which translational motion is restricted. The formation of polymer radicals can lead to depolymerization, although this reaction is most often encountered when polymers are photolyzed at elevated temperatures (>100 °C, or 212 °F) and may not be important in the outdoors. In depolymerization, the radical site moves down a polymer chain in a stepwise fashion as monomer units are eliminated.

When oxygen is plentiful, the most likely reaction for a radical (either the primary event radical, Y·, or polymer radical, P·) is the reaction with oxygen. The oxygenated radicals, YOO· or POO·, abstract hydrogens from the polymer matrix to form a hydroperoxide and a new polymer radical. The newly generated polymer radical reacts with oxygen to complete an oxidation cycle, as illustrated in Fig. 6. This is termed propagation. A free radical will cycle or propagate through the loop, generating oxidized products, until it is terminated by reaction with another radical. Under oxygen-starved conditions, two polymer radicals can terminate either by disproportionation:

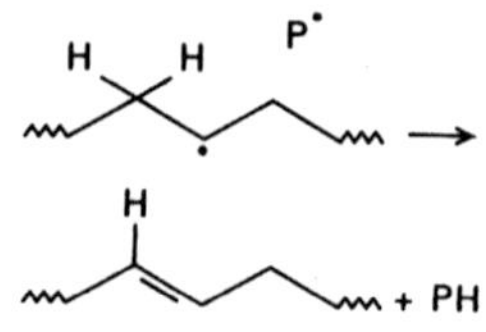

or by recombination:

$$P^{\cdot} + P^{\cdot} \rightarrow P\text{–}P$$

Recombination results in the formation of a polymer-polymer cross link. When oxygen is plentiful, most termination reactions involve oxygenated radicals:

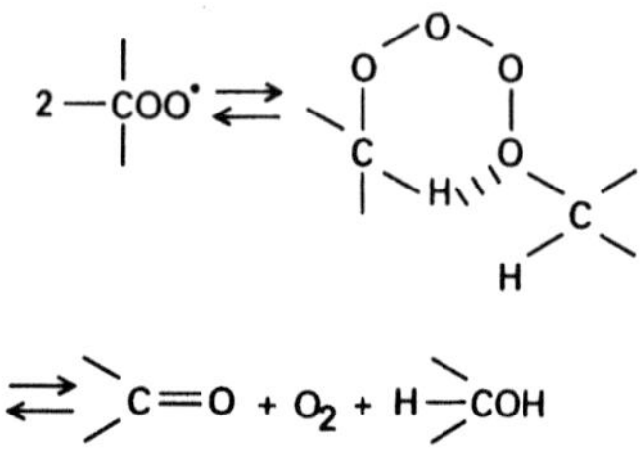

Subsequent reactions of peroxy radicals and hydroperoxides result in both chain scission and cross-link formation, shown in Fig. 8 and 9, respectively.

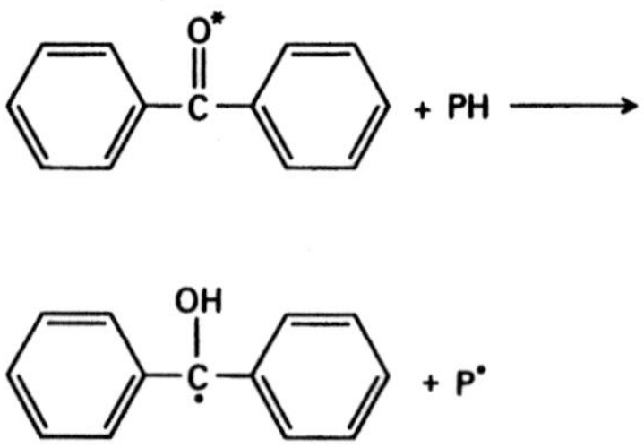

Fig. 4 Intermolecular hydrogen atom abstraction. P·, polymer radical

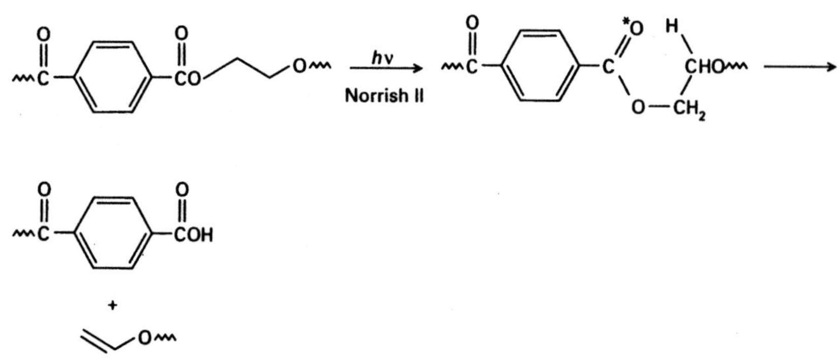

Fig. 5 Intramolecular hydrogen atom abstraction

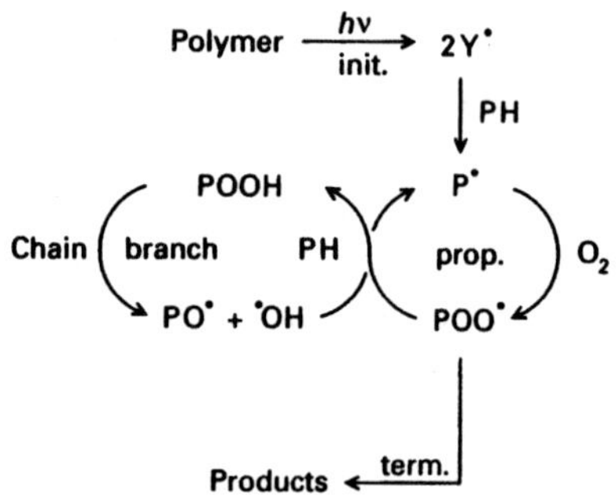

Fig. 6 Schematic of photooxidation cycle. Y·, chromophore-based free radical; P·, polymer radical

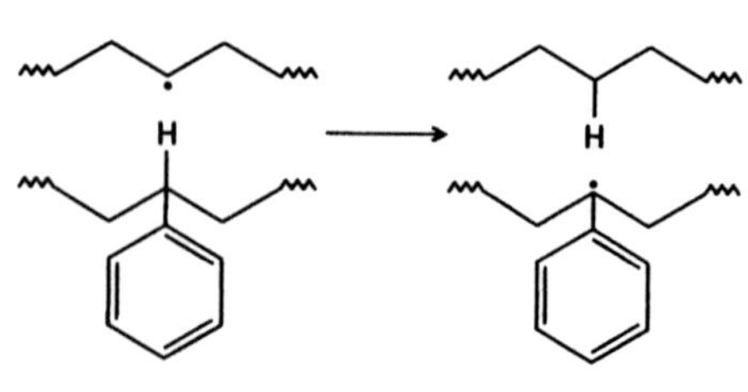

Fig. 7 Tertiary benzylic and secondary aliphatic radicals

The balance between chain scission and cross linking depends on the nature of the polymer. Polymers such as polybutadienes, polyacrylates, and polystyrenes tend to form cross links on degradation, while polymers such as polymethacrylates tend to degrade by chain scission. Oxidative-induced chain scission and cross linking occur in addition to the direct photoinduced chain scission that occurs in sunlight-absorbing polymers.

The reactions shown in Fig. 6 are a gross simplification of the reactions necessary to describe the photooxidation chemistry of even the simplest polymer. In an actual system, a variety of P· and POO· radicals will coexist. Alkoxy radicals, PO·, are also formed. The rate at which oxidation proceeds (Eq 2) is determined by the photoinitiation rate, W_i, the propagation rate constant, k_p (the rate constant for the hydrogen atom abstraction of PH by POO·), and the termination rate constant, k_t. Equation 2 is derived using the steady-state approximation for the reactive radicals shown in Fig. 6 (Ref 8):

$$\text{Photooxidation rate} = \frac{-d[\text{PH}]}{dt} = \frac{k_p\,[\text{PH}]\,(W_i)^{1/2}}{k_t^{1/2}} \quad \text{(Eq 2)}$$

The photooxidative chain length is the ratio of the photooxidation rate to the photoinitiation rate. The photoinitiation rate is proportional to the light intensity at the wavelength necessary to excite chromophoric impurities, as well as their concentration. As oxidation proceeds, the photochemistry of oxidation products contributes both to reaction complexity and to rate. The initial chromophores may be consumed while other chromophores are produced during the photooxidative cycle. A chromophore that is particularly important in the photooxidation of polyolefins is hydroperoxide. In polyolefins, the chromophore concentration, and therefore the photoinitiation rate, is initially very low. This leads to large photooxidative chain lengths (>100) with a slow buildup of hydroperoxides. Oxidation is relatively slow during this photooxidation stage, termed the induction period. Hydroperoxides decompose either thermally or by reaction with light to form alkoxy and hydroxyl radicals. This chain branching (Fig. 6) leads to an autocatalytic increase in the photoinitiation rate and the photooxidation rate.

Hydroperoxide-driven autooxidation is less important in polymers in which the photooxidation chain length is relatively small. This is likely to be the case with sunlight-absorbing polymers. Hydroperoxide photochemistry was found to be of little importance in a cross-linked acrylic coating in which the oxidative chain length was relatively short (<10). Hydroperoxide buildup was not observed, and the rate of photooxidation was relatively constant with time. The rate of oxidation in these coatings was found to be very sensitive to the concentration of ketone end groups formed during polymer synthesis. The introduction of unwitting chromophores may play a similar role in determining the photooxidative stability of engineering plastics. Here, processing usually involves both high mechanochemical and thermal stresses, conditions that are ideal for oxidative degradation.

Factors Controlling Photodegradation Rates. In addition to chromophore concentration, several other polymer variables can affect photoinitiation and photooxidation rates (Ref 9). The photooxidation rate depends on the ratio of $k_p/(k_t)^{1/2}$. The propagation rate constant, k_p, is determined by the ease of hydrogen atom abstraction from the host polymer by peroxy radicals. The ease of hydrogen atom abstraction from aliphatic alkanes is as follows: tertiary > secondary > primary. Aromatic hydrogens are much more difficult to abstract than aliphatic hydrogens. Hydrogens that are α to an ether oxygen are relatively easy to abstract. Amide hydrogens are also easily abstracted. Fluorine atoms are nearly impossible to abstract. Hydrogen atom abstractability can explain, in part, the different photodegradation rates of different polymers. For example, polyethylene (PE) has better photostability than polypropylene (PP), because PP has a large concentration of easily abstractable tertiary hydrogens, while PE has only secondary hydrogens. This leads to a higher ratio of $k_p/(k_t)^{1/2}$ in PP and to more rapid photooxidation.

Another very important variable in determining photooxidation rate is the rigidity of the polymer chain. The more rigid the chain, the less likely that photooxidation will occur on the chain. In semicrystalline polymers, photooxidation occurs almost exclusively in the mobile amorphous phase. Photooxidation can be very slow in the rigid crystalline phase. Therefore, increasing the crystallinity generally improves the photostability of a polymer. In amorphous polymers, the rate of photooxidation is strongly influenced by the glass transition temperature, T_g, of the polymer. For example, in a series of cross-linked acrylic copolymer coatings, the photoinitiation rate and the photooxidation rate decreased as the T_g of the acrylic copolymer increased (Ref 10). This effect can be explained as the influence of cage rigidity on free-radical escape efficiency. Polymer rigidity will also affect the propagation and termination rate constants. Photoinitiation and photooxidation are also affected by service temperature. Increasing the temperature increases polymer mobility, which increases cage escape efficiency and leads to more rapid photooxidation.

Ultimate mechanical failure depends on the rate of photodegradation and on the amount of chemical damage that a particular plastic can sustain before failure. The amount of chemical damage necessary to cause failure depends on a number of factors. One factor is the type of chemical reactions that occur. For thermoplastic polymers, chain scission, which results in a decrease in polymer molecular weight, may be a more important reaction than the oxidation of a side chain, which leaves the main polymer chain intact. The basic structure of the polymer is also important. A cross-linked polymer (thermoset) can tolerate a higher level of chain scission while maintaining its structural integrity. Finally, the level of stress that the polymer is subjected to will also affect how much chemistry will cause failure. Higher levels of stress will cause the plastic to fail at lower levels of chemical change. In addition, there is evidence that higher levels of stress can actually increase the rate of photodegradation.

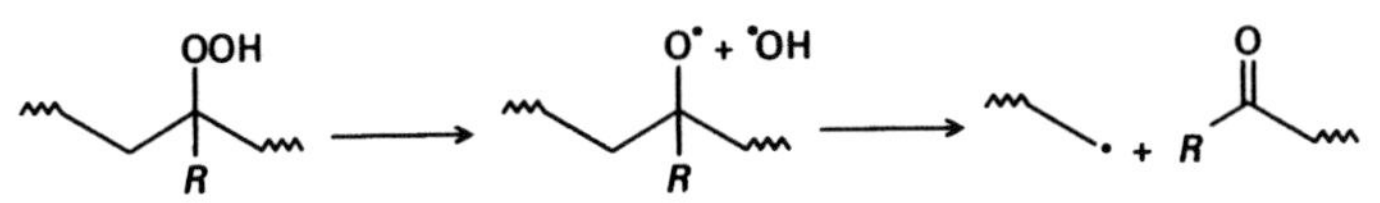

Fig. 8 Chain scission formation

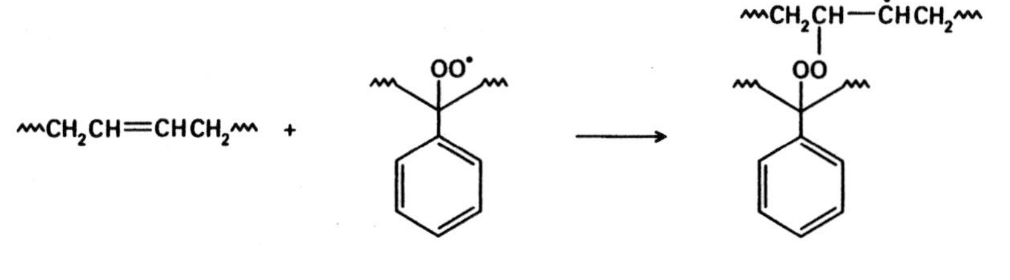

Fig. 9 Cross-link formation

Protection of Plastics from Sunlight

Ultraviolet Absorbers and Excited-State Quenchers. From the previous discussion, it should be clear that nearly all plastics require protection from sunlight in order to perform outdoors for long periods of time (Ref 11, 12). There are basically two strategies for stabilizing plastics against photodegradation. The first involves slowing the rate of initial photochemistry, while the second involves interfering with the propagation cycle of photooxidation. Generally, it is more difficult to inhibit degradation in sunlight-absorbing plastics than in nonsunlight-absorbing plastics. Stabilizers that interfere with the propagation cycle are not as effective in sunlight-absorbing plastics, because they cannot prevent the primary photochemical reactions.

The initiation of photochemistry is usually controlled by lowering the amount of ultraviolet light available in the plastic. This can be done by adding ultraviolet-absorbing or -scattering pigments, such as carbon black or titanium dioxide. This prevents ultraviolet light from reaching very far into the plastic. Further reductions in ultraviolet light intensity are obtained by using ultraviolet absorbers. There are a number of classes of commercially available ultraviolet absorbers, including phenyl salicylates, *o*-hydroxybenzophenones, and *o*-hydroxyphenylbenzotriazoles. The benzotriazoles are probably the most effective ultraviolet absorbers currently available. At the 1 wt% concentration level, benzotriazole effectively reduces the intensity of sunlight below 370 nm (3700 Å) by 99%, at a depth of 40 to 50 μm (1.6 to 2.0 mils). Transmission above 400 nm (4000 Å) is high, minimizing the effects on color. The performance of an ultraviolet absorber depends on its ability to dissipate the energy absorbed without degradation. Benzotriazoles and *o*-hydroxybenzophenones rapidly dissipate excited-state energy through internal hydrogen bond transfer, as shown in Fig. 10.

For example, the lifetime of the benzotriazole excited state is less than 100×10^{-12} s, thus minimizing its excited-state photochemistry. In addition to its light-absorbing capability, the performance of an ultraviolet absorber depends on its compatibility with the polymer matrix and its long-term permanence. If the ultraviolet absorber is incompatible with the polymer, it will tend to bloom out of the polymer and be ineffective.

Another approach to reducing the initiation rate is to add materials that quench excited states. This reduces the lifetime of the excited state, thus lowering the quantum yield. The effectiveness of quenchers depends strongly on the nature of the chromophore to be quenched. Some ultraviolet absorbers can also act as excited-state quenchers. For example, *o*-hydroxybenzophenones and benzotriazoles can be effective quenchers of aromatic excited states. Nickel chelation compounds are also used as quenchers in polymers, mainly polyolefins.

Free-Radical Scavengers. Ultraviolet absorbers can reduce the rate of the specific photochemical reactions, as well as the rate of free-radical oxidation (by reducing the rate of initiation of radicals). One limitation of ultraviolet absorbers is that they cannot be effective at the surface of the polymer. The second basic approach to stabilization is to inhibit the photooxidation cycle through the use of antioxidants. One class of commonly used antioxidants is that of the hindered phenols. Hindered phenols react with peroxy radicals to lower the steady-state concentration of polymer-based radicals and to shorten the photooxidative chain length, as shown in Fig. 11.

Hindered phenols are primarily used to minimize thermal oxidation during processing and end-use. This limits the formation of chromophores produced by thermal oxidation. Hindered phenols are not very effective as light stabilizers, because they and their radical scavenger products can absorb sunlight and initiate free-radical oxidation. Hindered phenols are

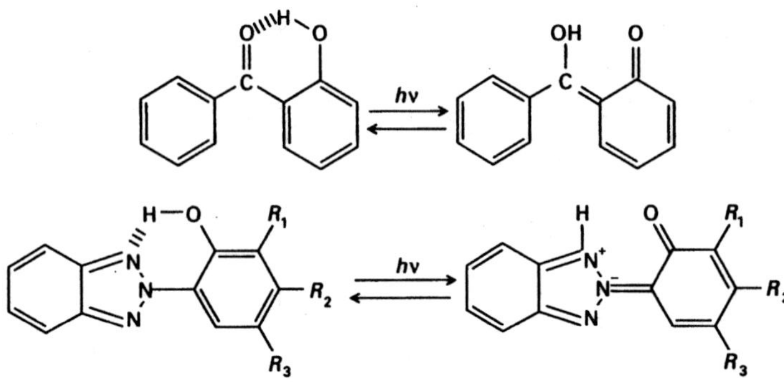

Fig. 10 Excited-state energy dissipation through internal hydrogen bond transfer

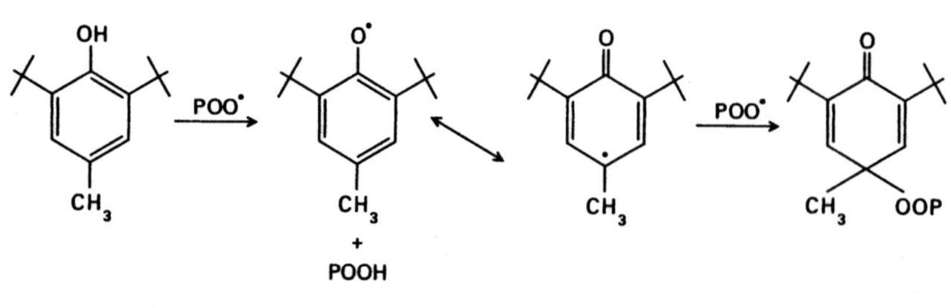

Fig. 11 Hindered phenol reaction with peroxy radicals

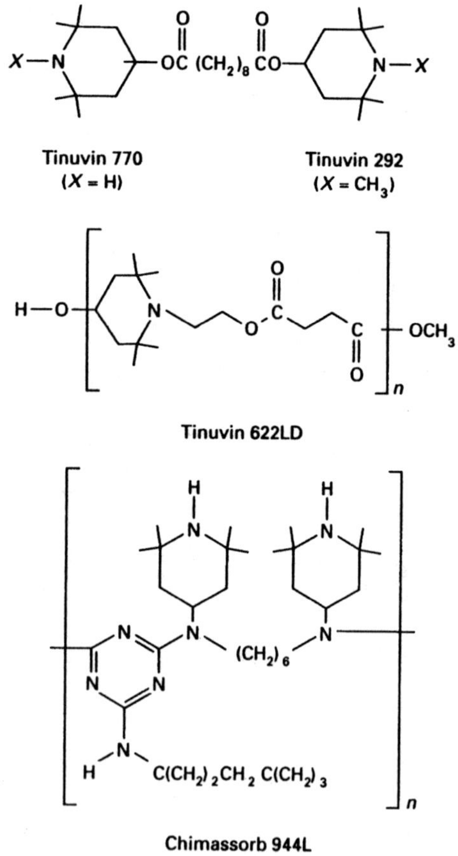

Fig. 12 Common hindered amine light stabilizers, including low-molecular-weight and polymer additives

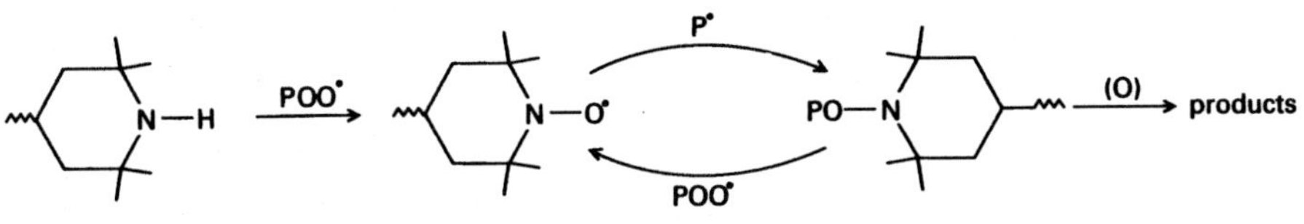

Fig. 13 Hindered amine converted to nitroxide by reaction with peroxy radicals

generally not photostable; they do not last very long on exposure to sunlight.

Another widely used class of stabilizer that inhibits photooxidation is the hindered amines. Typical hindered amines are shown in Fig. 12. The functional group that is important in preventing oxidation is the amine group in the tetramethyl piperidine ring. The amine is converted to a nitroxide by reaction with peroxy radicals, as shown in Fig. 13.

Nitroxides are efficient scavengers of alkyl and other radicals to form amino ethers (>NOP). Amino ethers can also react with radicals to regenerate nitroxides. In each case, a polymer free radical is removed from the oxidation cycle. A key advantage of the hindered amines is that one hindered amine can ultimately scavenge many radicals through the nitroxide/amino ether cycle. Another advantage is that neither the hindered amine nor the amino ether reaction products absorb sunlight. Thus, they do not initiate photochemistry.

Hindered amines may also act to decompose hydroperoxides to non-free-radical products, thus limiting the possibility of chain branching. Additives containing sulfur and phosphorus are also used to decompose hydroperoxides.

Hindered amines, although excellent photostabilizers, are generally not effective as stabilizers for thermal oxidation, because amino ethers (>NOP) are not thermally stable, although some hindered amines are effective in moderate-temperature (<120 °C, or 250 °F) oven aging but not during processing. Hindered amines can greatly reduce the photooxidative chain length. They are most effective in nonsunlight-absorbing polymers in which the oxidative chain length is long. Hindered amines, together with ultraviolet absorbers to reduce the initiation rate, provide the most effective overall stabilization for many polymers, particularly polyolefins. As was the case for ultraviolet absorbers, the effectiveness of hindered amine light stabilizers is, in large part, determined by their compatibility with the polymer host.

External Coatings. Although photostabilizers are added to plastics to prevent photodegradation, there are some disadvantages to their use. As noted previously, some photostabilizers may not be compatible with the plastic matrix or with the processing requirements. For example, it is not possible to use hindered amines in PCs, because they catalyze depolymerization during processing. In addition, photodegradation is usually limited to the first 100 to 200 μm (4 to 8 mils), even in unpigmented, nonsunlight-absorbing polymers. The stabilizers are present throughout the plastic. Therefore, most of the stabilizer is wasted (or, at best, serves as a reservoir) in the bulk of the plastic. Because stabilizing additives are generally more expensive than the host plastic, there is a cost penalty associated with their use. For this reason, coatings are often used to protect plastics from sunlight. There are numerous advantages to the use of protective coatings. They offer the potential for improved appearance and decoration. Coatings with ultraviolet absorbers protect the surface of the plastic as well as the bulk, provided the coating is not broken or scratched. They can be designed specifically for durability and can protect the plastic from attack by other environmental agents.

The design of coatings for plastics is rarely straightforward. Although a complete discussion of coatings is beyond the scope of this article, a few key issues should be mentioned. First, coatings often require baking for solvent removal and cure. The cure temperature must be below the heat-deflection temperature of the plastic to prevent shape changes. Some plastics, such as PCs, can craze or otherwise degrade in the presence of some solvents. The coating must not contain species that can attack the plastic. The choice of solvent in the coating formulation is critical for avoiding damage to the plastic and for obtaining good initial adhesion. Removal of contaminants, such as mold-release agents, is also important. The coating must be sufficiently durable to protect the plastic surface for the entire service life of the part. Finally, the viscoelastic properties of the coating must be matched to the plastic. For example, if a brittle coating is used over a ductile plastic, brittle failure of the plastic can occur on impact. Basically, the coating can act as a crack-initiating site. The crack then propagates into the plastic, causing premature failure. Despite these potential problems, coatings are widely and successfully used to protect plastics from outdoor exposure.

REFERENCES

1. V. Wigotsky, Alloys and Blends Gain Market Momentum, *Eng. Plast.*, Vol 62 (No. 7), July 1986, p 19
2. V. Wigotsky, Engineering Resins Widen Performance Boundaries, *Eng. Plast.*, Vol 43 (No. 1), Jan 1987, p 15
3. M.L. Ellinger, Weathering Tests, *Prog. Org. Coatings,* Vol 5, 1977, p 21
4. J.G. Calvert and J.N. Pitts, Jr., Photochemistry of the Polyatomic Molecules, *Photochemistry,* John Wiley & Sons, 1966, p 366–486
5. B. Ranby and J.F. Rabek, Photodegradation and Photo-Oxidation of Particular Polymers, *Photodegradation, Photo-Oxidation and Photostabilization of Polymers,* Wiley-Interscience, 1975, p 120–275
6. A.G.H. Dietz, H.S. Schwartz, and D.V. Rosato, Reinforced Plastic Composites, *Environmental Effects on Polymeric Materials,* D. Rosato and R.T. Schwartz, Ed., Interscience, 1968
7. N.J. Turro, *Modern Molecular Photochemistry,* Benjamin/Cummings, 1987
8. N.M. Emanual, E.T. Denisov, and Z.K. Maizus, The Chain Mechanism of Free Radical Oxidation, *Liquid Phase Oxidation of Hydrocarbons,* B.J. Hazzard, Trans., Plenum Press, 1967, p 1–18
9. I. Mita, Effect of Structure on Degradation and Stability of Polymers, *Aspects of Degradation and Stabilization of Polymers,* H.H.G. Jellinek, Ed., Elsevier, 1978, p 248–294
10. J.L. Gerlock, D.R. Bauer, and L.M. Briggs, Electron Spin Resonance Determination of Nitroxide Kinetics in Acrylic/Melamine Coatings: Relationship to Photodegradation and Stabilization Kinetics, *Polymer Stabilization and Degradation,* ACS Symposium Series 280, American Chemical Society, 1985, p 119
11. W. Schnabel and J. Kiwi, Photodegradation, *Aspects of Degradation and Stabilization of Polymers,* H.H.G. Jellinek, Ed., Elsevier, 1978, p 195–246
12. G. Scott, Ed., *Developments in Polymer Stabilization,* Vol 1 to 5, Applied Science, 1979–1984

Microbial Degradation*

MOST WIDELY USED PLASTIC materials developed in the past 50 years are characteristically durable and inert in the presence of microbes. This stability is important to long-term performance. However, in many applications, only short-term performance is required before the material is discarded, as in the fast food and packaging industries. It is considered advantageous for the discarded plastic to degrade when exposed to microbes. Other applications of plastics actually require degradability. Thus, an interesting challenge is to develop or modify plastics that possess the properties required for their service life but which have the capability of degrading in a timely and safe manner. For example, agricultural plastics applications, such as mulch films, seedling pots, and binding twine, have increased significantly, making biodegradation a desirable function of the plastic, to minimize disposal and soil pollution problems.

Biodegradation Mechanisms

Plastics remain relatively immune to microbial attack as long as their molecular weight remains high. A review of the biodegradability of plastics or lack thereof has been conducted (Ref 1). Many plastics, such as polyethylene (PE), polypropylene (PP), and polystyrene (PS), do not support microbial growth. Low-molecular-weight hydrocarbons, however, can be degraded by microbes. They are taken up by microbial cells, activated by attachment to coenzyme-A, and converted to cellular metabolites within the microbial cell, as shown in Fig. 1. However, these processes do not function well (if at all) in an extracellular environment, and the plastic molecules are too large to enter the cell. This problem does not arise with natural molecules, such as starch and cellulose, because conversion to low-molecular-weight components by enzyme reaction occurs outside the microbial cell. Photochemical or chemical reactions may decrease molecular weight to the point that microbial attack can proceed.

Upper limits of molecular weight, beyond which uptake and intracellular degradation do not occur, have not been established for all alkane-derived materials. Very slow degradation of paraffins, PE glycols, and linear alkyl benzene sulfonates occurs when the length of the polymer chain exceeds 24 to 30 carbon atoms (Ref 2–4). It could be concluded from these amply documented results that alkane-based plastics with molecular weights exceeding 400 to 500 daltons (that is, greater than 30 carbon atoms) must be degraded to smaller molecules by photochemical, chemical, or other biological means before biodegradation as defined here can proceed. For comparative purposes, low-density PE with a molecular weight average of $\overline{M}_w$ = 150,000 contains approximately 11,000 carbon atoms. Decreasing molecules of this size to biologically acceptable dimensions requires extensive destruction of the PE matrix. As is discussed later, this destruction can be partly accomplished in blends of PE and biodegradable natural polymers by the action of macroorganisms, such as arthropods, millipedes, crickets, slugs, roaches, and snails.

Efforts to construct plastics that fall within the logistical limits discussed previously were described by a number of groups during a 1987 symposium on degradable plastics (Ref 5). The ethylene/carbon monoxide copolymer (E/CO) described in Ref 5 provides a good example of the difficulties involved in attempting to make large molecules, such as those of PE, biodegradable. Insertion of carbon monoxide in the chain permits chain scission by a Norrish-type reaction in a photochemical process. It was found that E/CO polymers with 2.5% CO linkages lost approximately 98% of their original elongation after 40 h of sunlamp exposure. However, after 650 h of exposure, the samples that originally

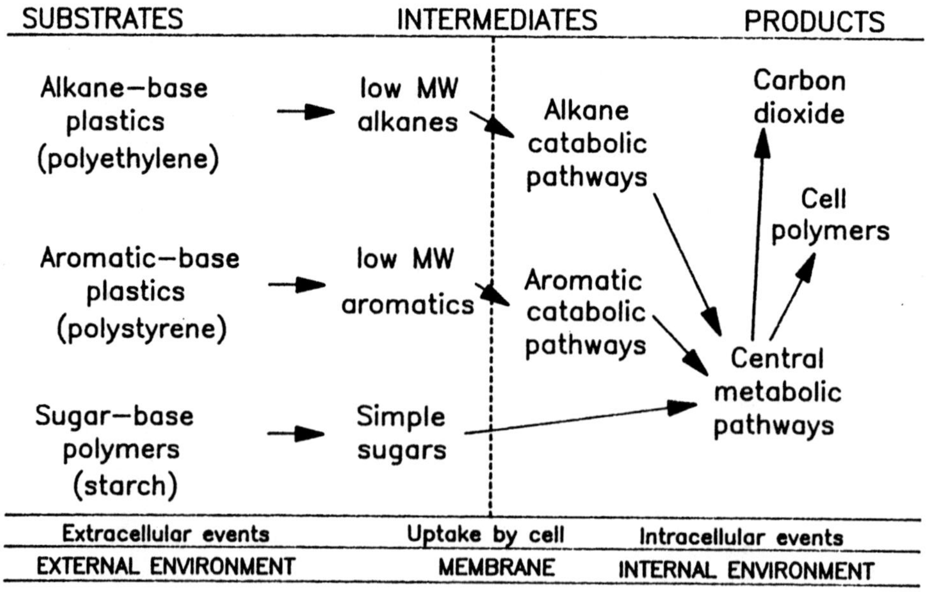

Fig. 1 Pathways for polymer biodegradation. The existence of extracellular enzymes that depolymerize polyethylene or polystyrene has not been documented. MW, molecular weight

*Adapted from the article by Richard P. Wool and Michael A. Cole, "Microbial Degradation," in *Engineering Plastics,* Volume 2, *Engineered Materials Handbook,* ASM International, 1988, pages 783 to 787

had a $\overline{M}_w$ = 618,700 and number-average molecular weight ($\overline{M}_n$) = 45,000 had photolytic products with $\overline{M}_w$ = 15,000 and $\overline{M}_n$ = 7300. It is likely that the photolysis products described by this group will degrade very slowly unless some other means of further reducing molecular weight can be found. Other authors in the same symposium reported similar photodecay rates and specifically stated (in Ref 5) that E/CO polymers were not biodegradable after ultraviolet light exposure and soil burial.

The most widely used plastics, that is, PE, PP, PS, polyurethane (PUR), polyvinyl chloride (PVC), and polyacrylic, are structurally substituted macroalkanes. These materials are homologues of smaller alkanes, such as motor fuels, motor oils, and surfactants. In discussing biodegradation of plastics, it is worthwhile to consider the definitions and means of measurement of biodegradation and biodeterioration when these terms are applied to smaller alkanes and derivatives such as surfactants. Because these definitions are central to the development of biodegradable plastics (or composite materials containing plastics), they are discussed as follows.

Biodeterioration and Biodegradation Definitions

Biodegradation has been discussed in depth from the perspective of surfactants (Ref 4). It was proposed that primary degradation occurs when the molecule has been oxidized or otherwise altered by bacterial (microbial) action, such that characteristic properties are no longer evident, or such that the molecule no longer responds to analytical procedures more or less specific to the detection of the original surfactant. This definition carries two conditions: first, that the molecule itself be chemically altered to an unspecified degree, and, second, that the alterations result in the loss of the utilitarian characteristics of the material.

Applying these concepts to plastics, biodeterioration can be defined as the biologically mediated loss of utilitarian or physical characteristics of a plastic or a hybrid material containing plastics or plastics as a major component. Thus, this definition includes physical disintegration or cosmetic disfigurement of a material but does not require extensive changes in the chemical structure of the components. This definition intentionally avoids criteria based on percentage loss of characteristic properties, or on the time required to attain a specified percentage loss, because the specific application of the material establishes the boundaries of acceptable or unacceptable deterioration rates. This definition describes the end result of many tests conducted according to ASTM G 21 (Ref 6) and reported as evidence of biodegradability of plastic materials.

Biodegradation is defined, for small molecules, as the complete conversion of the surfactant molecules to carbon dioxide, water, inorganic salts, and products associated with the normal metabolic processes of the bacteria (Ref 4). When applied to plastics, the definition of biodegradation may be stated as the conversion of all constituents of a plastic or hybrid material containing plastics to carbon dioxide, inorganic salts, microbial cellular components, and miscellaneous by-products characteristically formed from natural materials.

With this definition, biodegradable plastics are materials whose physical integrity and useful material properties are lost on contact with microbial and/or invertebrate activity in a natural environment for a limited period of time. The natural environment, whether in laboratory or field tests, with soil or aquatic systems, must be conducive to degradation of readily decayed natural materials, such as carbohydrates, lipids, and proteins. Deterioration is therefore an inevitable consequence of biodegradation, because the very chemical structures that composed the original material no longer exist.

Biodeterioration and Biodegradation Measurements

Biodeterioration involves cosmetic blemishing and/or loss of useful properties, which means that the ASTM procedures for determining the resistance of synthetic polymers to fungi and plastics to bacteria (Ref 6, 7) may be applicable in evaluating these effects. Method 2 of Ref 6, in which the material is buried in agar, is preferable to method 1, in which nonporous materials are used. With method 1, in which the film is placed on the agar surface, diffusion of nutrients from the agar through a nonporous film is likely to restrict microbial growth on the top of the film. Water availability on the usually hydrophobic film surface is also an impediment to surface growth.

A procedure based on microscope slide trapping techniques (Ref 8) seems preferable to the agar plate method. Because the organisms used in the ASTM procedure are a mere fraction of the microorganisms likely to make contact with films under use conditions, techniques that allow exposure to soil and/or natural waters are more likely to provide a more realistic assessment of biodeterioration potential. Trapping is a relatively simple and rapid visual technique that does not require extensive microbiological experience. The test is conducted by placing moist soil in a covered container that allows moisture control and gas exchange with the atmosphere. The test film is placed on the surface of the moist soil and covered with a glass microscope slide. The entire assembly is incubated at an appropriate temperature, and microbial colonization can be assessed as described in ANSI/ASTM D 3274 (Ref 9).

This procedure, developed for assessing paint film blemishing, provides good illustrations of graded degrees of blemishing. In their experiments, the authors (Ref 8) use filter paper (cellulose) as a positive control to ensure incubation conditions that are conducive to microbial activity; they also use low-density PE film as a test material on which microbial colonization and disfigurement are minimal. Blemishing of filter paper occurs within a couple of weeks under good conditions, that is, moist soil and temperatures of 20 to 30 °C (70 to 85 °F).

In most cases, polymer films and associated microorganisms adhere to the slides and can be recovered and stained with conventional bacteriological dyes. Incubation in aqueous systems is possible by attaching the short edges of a test film to a microscope slide with an appropriate waterproof adhesive. Growth on the adhesive components may occur, but its origin from the adhesive application site will be obvious.

It should be noted that loss of film integrity with this technique cannot be considered to be persuasive evidence of biodegradability, because smaller products can be formed that are neither visible nor subject to degradation. This situation is analogous to some extended photodegradation experiments in which polymer disintegration by ultraviolet radiation yields products that are smaller than the parent materials but are still chemically identifiable as polymer components.

Measurements of Biodegradation. Biodegradation involves loss of structural characteristics and mass (total carbon) of a material, as carbon dioxide and water-soluble components are formed. These changes are frequently accompanied by discoloration resulting from extensive growth of pigmented microorganisms. However, discoloration and visible microorganism growth may occur on films in the absence of any changes in the plastic material. This growth is supported by plasticizers and other low-molecular-weight components that may be mixed with the plastic itself. This type of growth is explicitly discussed in Ref 6, but many investigators may not recognize the difference between biodisfigurement on the surface of a plastic and actual decay of that plastic by biodegradation mechanisms throughout the material.

Biodegradation involves the conversion process and the chemical mechanisms whereby carbon from a biodegrading material is distributed among the products. Authors (Ref 10) reported evolution of $[^{14}C]$–CO_2 from ^{14}C-labeled, photodegraded PS added to soil. Release of $[^{14}C]$–CO_2 was measured for 8 weeks, at which time less than 0.2% of total ^{14}C had been evolved as CO_2. The authors conclude that complete biodegradation would eventually occur. However, the amount of ^{14}C evolved is much less than the theoretical maximum, and one cannot be assured that the process would go to completion. If the polymer were fully degraded, release of $[^{14}C]$–CO_2 would be approximately 50 to 60% (although the percentage is time dependent) of the total ^{14}C, with approximately 40 to 50% of carbon as microbial cell material under aerobic conditions. These percentages are based on the release of carbon from natural materials, such as simple sugars, organic acids, and numerous biopolymers (Ref 11–13).

Although the authors of Ref 10 concluded that release of any ^{14}C as $[^{14}C]–CO_2$ was evidence of biodegradability, the authors of this article suggest from analysis of the data that only 1% or less of the photodegradation products was biodegradable.

Lack of release of the expected amount of carbon as CO_2 may be attributed to one of two major causes: Either the material is not being degraded, or it is being incompletely degraded to form nondegradable compounds of lower molecular weight than the parent material. In the latter case, the material may lose some structural integrity and could be judged biodegradable when assessed by methods such as weight loss or depolymerization but would not necessarily be judged biodegradable if assessed quantitatively by CO_2 evolution, O_2 uptake, or microbial growth.

The decay of natural rubber (Ref 14) provides a good example of full decay. With natural rubber, a 100% weight loss of starting material and formation of cell protein equivalent to 27% of the initial rubber weight was found. Because the protein is approximately 50% of the cell dry weight, the actual cell mass was approximately two times the protein values, or approximately 54% of the substrate mass. With cells containing 50% C, approximately 27% of the substrate mass was recovered as cells. Adding 27%, as cells, to 10%, as $CHCl_3$-soluble products, yields 37% of the initial mass. This value is reasonably close to the cell yield expected on full degradation (the remaining carbon presumably having been lost as CO_2) and justifies the conclusion that the natural rubber under investigation was biodegradable.

Reference 15 provides an extensive review on biodegradable polymers, with an emphasis on degradation studies with pure enzymes or pure microbial cultures. The discussion implies that ASTM International procedures involving the formation of microbial cell mass, CO_2 production, O_2 consumption, material weight loss, and changes in structural characteristics may be used interchangeably to assess degradation. Frequently, in situations other than pure culture, most of these techniques are technically difficult and may yield inconclusive or misleading results. It should be emphasized that most of these techniques involve quantitative procedures with predictable end results. Data obtained with these methods can be used to compare test plastics with other materials that are known to be biodegradable.

A special point needs to be made in regard to the quantitative nature of testing because of the potentially misleading perception that any evidence of a microbial process implies material biodegradability. For example, when interpreting data obtained by Ref 6, the growth is measured as the percentage area covered by fungi without considering whether 100% fungal coverage represents a loss of any of the initial material mass. Although Ref 6 explicitly makes this point, it can be overlooked in practice. For a variety of reasons, the authors of this article propose reexamination of materials if degradability was assessed only by visual procedures (for example, fungal growth on surface) or if the mass balance of starting material and products was not established. On the other hand, it is highly likely that materials that have been interpreted as being nondegradable by ASTM methods are indeed nondegradable, unless their degradation is catalyzed by bacteria or environments not prescribed by the ASTM tests.

Experimental Example

Fungal Attack in Cellophane and Amylose Films. Other authors (Ref 16, 17) compared the fungal attack in cellophane (0.025 mm, or 1 mil) and amylose films (0.10 mm, or 4 mil). Fungal colonies were isolated from soil by using a selective solid medium. The samples were incubated at 23 °C (75 °F) on an agar medium for the prescribed length of time and then sterilized in ethylene oxide gas to stop the degradation process. The degraded material was analyzed by x-ray diffraction, infrared spectroscopy, and mechanical property tests. The degree of crystallinity, as measured by x-ray diffraction, was found to increase during the degradation by fungi. This is consistent with the preferential growth of the fungus through the amorphous regions of these semicrystalline polymers. Attack on amylase films by cell-free enzymes, either alpha or beta amylase, resulted in degradation of both the amorphous and the crystalline regions, with preferential degradation of the amorphous regions.

The ultimate and yield tensile stress of the amylose films declined during the degradation process, and the tensile modulus increased as the material became embrittled. Changes in dynamic mechanical properties of cellophane films were measured as a function of fungal degradation. After 16 days of degradation, the storage tensile modulus declined by approximately 40% over most of the temperature range studied, and considerable broadening occurred in the beta and gamma tan δ peaks. These results were interpreted in terms of the damage caused by cellulase enzymes that are released from a fungal overgrowth and become localized in the noncrystalline regions.

Starch-based PE films were formulated (Ref 18, 19) and consisted of up to 40% starch, urea, ammonia, and various portions of low-density PE (LDPE) and poly(ethylene coacrylic acid) (EAA). The EAA acted as a compatibilizer, forming a complex between the starch and the PE in the presence of ammonia. The resulting blend could be cast or blown into films and had physical properties approaching those of LDPE.

The ASTM G 21 method (Ref 6) was used to determine fungi resistance. The mixed fungal spore suspension consisted of *Aspergillus niger, Penicillium funiculosum, Trichoderma* sp., and *Pullularia pullulans.* Films containing 30% or more starch could develop a complete mold coverage. Samples were also placed on outdoor soil with their ends buried in the soil.

This experiment was designed to evaluate starch-based plastics as agricultural mulch film. Samples with 90% by weight starch were badly torn within 1 to 3 days after water soaking, primarily because of embrittlement and shrinkage on drying. Heavy mold growth occurred within 3 to 5 days on the buried portion of these samples. Similar deterioration was observed for samples containing 50 and 70% starch but not until approximately 5 to 7 days of exposure. Samples with 30 and 40% starch remained flexible and intact for at least 30 days. Deterioration of the latter films was primarily attributed to microbial attack, as evidenced by heavy mold growth, rather than to embrittlement and shrinkage. The rate of biodegradation of these films could be controlled by varying the amount of starch and by using a fungicide such as paraformaldehyde. Plasticizers could also be added, causing a significant increase in both the percentage elongation-at-break and the rate of fungal attack. These films were also found to be photodegradable, such that the use of carbon black ultraviolet stabilizer was recommended in cases in which prolonged exposure to sunshine is anticipated.

Films with Modified Starch Additives. The St. Lawrence Starch Company produces a master batch resin that can be mixed with PEs to produce films consisting of 6 to 15% starch (Ref 5). Degradation most likely proceeds by two interactive mechanisms. The starch is present in the polymer as phase-separated granules and is attacked by fungi and bacteria. This biodisfigurement process weakens the polymer matrix and increases the surface area of the plastic. It is hypothesized that unsaturated oils in the blend act as an autoxidant to form peroxides by reacting with metal salts in the soil or seawater, and the peroxides decompose to form free radicals that attack the polymer chain. The resulting breakdown of the polymer chain weakens the material and reduces the molecular weight to further enhance microbial attack. Related studies by the originator of this technique (Ref 5) demonstrate that the ultimate tensile strength of a modified LDPE film in a compost heap at 70 °C (160 °F) decreases appreciably after approximately 5 days. Other companies, such as Archer-Daniels-Midland Company, are using this technology to produce plastic films with a biodegradable component.

PHBV-Biodegradable Plastic. In the 1980s, ICI Americas Inc. worked on the concept of a natural thermoplastic polymer derived from bacteria known as poly(3-hydroxybutyrate-valerate), or PHBV (Ref 5). The naturally occurring soil bacterium *Alcaligenes eutrophus* is grown aseptically in conventional fermenters on a sugar feed-stock, usually glucose derived from corn or wheat starch. By controlling the culture conditions, the bacterium can be made to accumulate up to 80% of its dry matter as poly(3-hydroxybutyrate), or PHB. The polymer exists as discrete granules within the cell. The composition of the PHB polyester can be influenced by means of the nutrition of the organism, leading to the incorporation of 3-hydroxyvalerate into the polymer. The resulting material consists of the random copolymer PHBV. The homopolymer PHB is a highly crystalline, somewhat brittle material that melts at 176 °C (350 °F) and has properties similar to those of PP. PHBV containing 25% valerate melts at 138 °C (280 °F) and is much more ductile.

ICI Americas Inc. has demonstrated that this material can be blow molded, injection molded, and processed like conventional plastics. Because it is biologically synthesized in the bacterial cell, PHBV is completely biodegradable. Fungi, in particular, colonize the surface of PHBV in soils and secrete enzymes that digest the plastic. The products of this digestion are absorbed and metabolized, yielding carbon dioxide and water. The rate of biodegradation of a typical PHBV device in garden soil is given as 3 μm (0.12 mil) of thickness per week. Applications of this material are being considered in surgical and orthopedic devices, slow-release systems involving drugs and herbicides, personal hygiene products, and packaging applications.

Biodisintegration and Biodegradation Studies of Plastic-Starch Blends. The degradability of cornstarch-based PEs developed by Otey et al. (Ref 18, 19) was investigated (Ref 20, 21), as described in the section "Starch-Based PE Films" in this article. Such materials are also being researched and developed by Agri-Tech Industries. The blown-film material (75 μm, or 3 mils) consisted of 40% starch, 10% urea, 25% LDPE, and 25% EAA. Biodegradation of the starch component was studied by monitoring carbon dioxide evolution in loam clay (relatively high in organic matter and very fertile) and sandy soil (low in clay and organic matter). Pure LDPE films were used as controls in similar soils. Although differences in the rate of carbon dioxide evolution from these soils were found, all the starch was removed within approximately 90 days; the fastest rate (20 days) was in the fine, sandy soil. Typical results for one soil are shown in Fig. 2.

Invertebrate attack on films is more selective than microbial decay. Films made by the Otey formulation are consumed by a range of invertebrates, but these organisms would not consume the pure LDPE films or films containing approximately 6% starch made by U.S. and Canadian companies. When films are consumed, the remaining plastic material is highly subdivided and is likely to be more susceptible to chemical or microbial attack than intact films would be.

The accessibility of the starch in these plastic-starch blends was determined by enzyme digestion experiments using alpha and beta amylase and amyloglucosidase. Polymer strips were immersed in an appropriately pH-buffered solution for each enzyme. A single enzyme type was added to each tube and incubated at 20 °C (70 °F). Merthiolate was added to the tubes to prevent degradation of the products by microbial contaminants. The strips were transferred to fresh buffer and enzyme solutions at different times, and the solutions were analyzed for starch breakdown products (reducing sugars). Hydrolysis of starch in the blends ranged from 11% to greater than 90%, depending on the enzyme used and the blending technique. It was found that only the alpha amylase was able to degrade the starch in the blend. There was neither a release of reducing sugar with the beta amylase or the amyloglucosidase, nor a loss of starch into the solution in the absence of amylase. The total amount of reducing sugar released after 48 h of incubation represents more than 90% of the theoretical yield from the starch in the polymer film.

Changes in the physical characteristics of the preceding microbial- and enzyme-degraded films have been analyzed using percolation models (Ref 21, 22). Accessibility of biodegradable components, such as starch, can be treated as a conductivity (scalar) percolation problem (Ref 23). The most important consequence of this approach is that the biodegradable component must exceed the percolation threshold (~30% by volume) to permit continuous invasion of the fungi or bacteria throughout the bulk of the material. Percolation of fungi on a square lattice is shown in Fig. 3. The observed reduction of modulus, fracture stress, and elongation-to-break of the plastic-starch blends during degradation can be modelled using elasticity (vector) percolation (Ref 24, 25). Plastic materials containing starch additives with a volume fraction appreciably lower than 30% are not expected to undergo biodisintegration by fungal action alone. Such materials may, however, develop a surface growth of fungi, which is a biodisfigurement process that should not be mistaken for biodegradability.

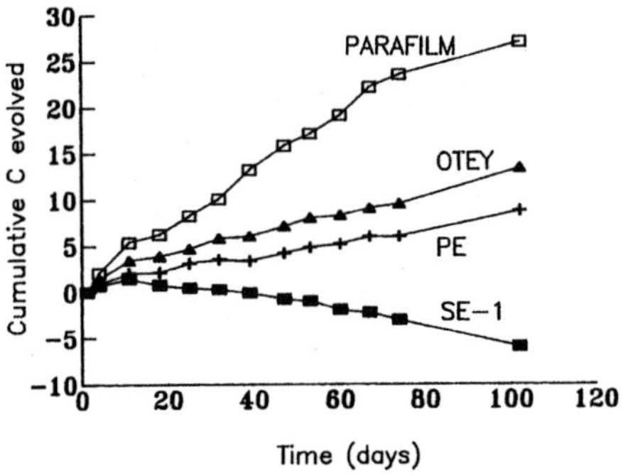

Fig. 2 Release of carbon dioxide from soil-buried films. The materials are SE-1, commercial starch-extended film with 6% starch, 94% polyethylene; Otey, 40% starch, 25% polyethylene, 25% ethylene acrylic acid; PE; commercial polyethylene film; and Parafilm (Royal Purple, Ltd.). Moist soil (100 g) and film (50 mg) were placed in Erlenmeyer flasks with a central vial containing KOH as the CO_2 trapping agent. The flask assemblies are described in Ref 22 and are suitable for nonradioactive materials when they are added in small quantities to soil. Carbon dioxide evaluation was determined by titration of residual alkali. Each point is the mean value of three replicate flasks per soil; standard deviations were ±10% of the mean value for each data point. *y*-axis; summation of milligram carbon evolved over the experimental period

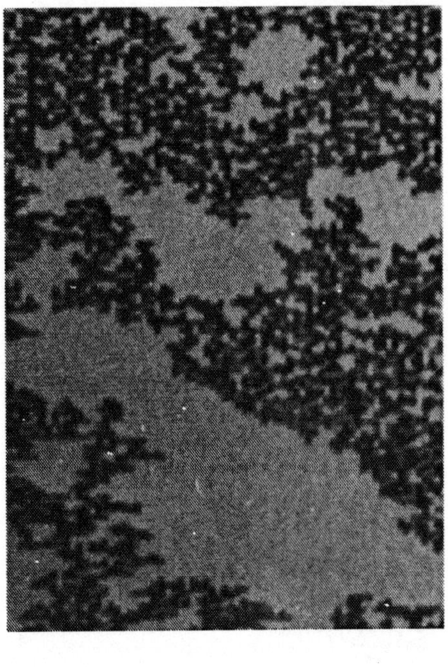

Fig. 3 Computer simulation of fungal invasion in two-dimensional matrix containing randomly distributed biodegradable fraction at the percolation threshold

REFERENCES

1. J.E. Potts, *The Encyclopedia of Chemical Technology,* John Wiley & Sons, 1982
2. J.D. Walker, J.F. Austin, and R.R. Colwell, Utilization of Mixed Hydrocarbon Substrate by Petroleum-Degrading Microorganisms, *J. Gen. Appl. Microbiol.,* Vol 21, 1975, p 27–39
3. D.F. Dwyer and J.M. Tiegje, Metabolism of Polyethylene Glycol by Two Anaerobic Bacteria, *Desulfovibrio de-sulfurican* and a *Bacteroides* sp., *Appl. Environ. Microbiol.,* Vol 52, 1986, p 852–856
4. R.D. Swisher, *Surfactant Biodegradation,* Marcel Dekker, 1987, p 415–457
5. *Proceedings of Society of the Plastics Industry Symposium on Degradable Plastics* (Washington, D.C.), June 1987
6. "Standard Practice for Determining Resistance of Synthetic Polymeric Materials to Fungi," G 21, *Annual Book of ASTM Standards,* American Society for Testing and Materials
7. "Standard Practice for Determining Resistance of Plastics to Bacteria," G 22, *Annual Book of ASTM Standards,* American Society for Testing and Materials
8. T.R.G. Gray and T.F. Bell, The Decomposition of Chitin in an Acid Soil, *Soil Organisms,* J. Doeksen and J. van der Drift, Ed., North-Holland, 1963, p 222–230

9. "Standard Method of Evaluating Degree of Surface Disfigurement of Paint Films by Fungal Growth or Soil and Dirt Accumulation," American National Standards Institute and D 3274, *Annual Book of ASTM Standards,* American Society for Testing and Materials
10. J.E. Guillet, T.W. Regulski, and T.B. MacAneney, Biodegradability of Photo-degraded Polymers, II, Tracer Studies of Biooxidation of Ecolyte PS Polystyrene, *Environ. Sci. Technol.,* Vol 8, 1974, p 923–925
11. W.J. Payne, Energy Yields and Growth of Heterotrophs, *Annu. Rev. Microbiol.,* Vol 24, 1970, p 17–52
12. C.M. Gilmour and J.T. Gilmour, Assimilation of Carbon by the Soil Biomass, *Plant Soil,* Vol 86, 1985, p 101–112
13. E.A. Paul and R.P. Voroney, Field Interpretation of Microbial Biomass Activity Measurements, *Microbial Ecology—Current Perspectives,* American Society for Microbiology, Washington, D.C., 1984, p 509–514
14. A. Tsuchii, T. Suzuki, and K. Takeda, Microbial Degradation of Natural Rubber Vulcanizates, *Appl. Environ. Microbiol.,* Vol 50, 1985, p 965–970
15. S.J. Huang, Biodegradable Polymers, *Encyclopedia of Polymer Science and Engineering,* Vol 2, J.L. Kroschwitz, Ed., John Wiley & Sons, 1985, p 220–243
16. S.A. Bradley and S.H. Carr, Effect of Fungal Degradation on Mechanical Properties of Cellophane, *J. Polym. Sci., Polym. Phys. Ed.,* Vol 14, 1976, p 125–133
17. S.A. Bradley, P. Engler, and S.H. Carr, Microbial Degradation of Polymer Solids, II, A Comparison of Fungal Attack in Cellophane and Amylose Films, *Applied Polymer Symposium, No. 22,* John Wiley & Sons, 1973, p 269–286
18. F.H. Otey and W.M. Doane, A Starch-Based Degradable Plastic Film, *Proceedings of Society of the Plastics Industry Symposium on Degradable Plastics* (Washington, D.C.), June 1987, p 39–40
19. F.H. Otey, R.P. Westhoff, and W.M. Doane, Starch-Based Blown Films, *Ind. Eng. Chem., Prod. Res. Dev.,* Vol 19 (No. 4), 1980, p 592–595
20. R.P. Wool, M.A. Cole, and J.S. Peanasky, Biodegradation Mechanisms in Polymers, *Bull. Am. Phys. Soc.,* Vol 33 (No. 3), March 1988
21. R.P. Wool, M.A. Cole, J.S. Peanasky, and J.L. Willett, "Biodegradation and Biodisintegration of Polymer-Starch Blends," paper presented to the American Chemical Society (Toronto), June 1988
22. M.A. Loos, A.K. Kontson, and P.C. Kearney, Inexpensive Soil Flask for ^{14}C-Pesticide Degradation Studies, *Soil Biol. and Biochem.,* Vol 12, 1980, p 583–585
23. S. Kirkpatrick, Percolation and Conduction, *Rev. Mod. Phys.,* Vol 45 (No. 4), 1973, p 574
24. Y. Kantor and I. Webman, Elastic Properties of Random Percolating Systems, *Phys. Rev. Lett.,* Vol 52 (No. 21), 1984, p 1891
25. S. Feng and P.N. Sen, Percolation on Elastic Networks: New Exponent and Threshold, *Phys. Rev. Lett.,* Vol 52 (No. 5), 1984, p 216

Failure Analysis of Plastics

Analysis of Structure*

FAILURE OF polymeric materials is a complex process. This article introduces procedures an engineer or materials scientist can use to investigate failures. It also gives a brief survey of polymer systems and key properties that need to be measured during failure analysis. More detailed examples and discussion of methods are given in the next article, "Characterization of Plastics in Failure Analysis," in this book.

Figure 1 identifies the interrelationships among composition, structure properties, and morphology. These can affect processing, which in turn can strongly affect the ultimate properties or end-use performance of a polymer. Although the compounding step introduces additional variables, they are often essential to allow processing and/or to develop specific properties that are not inherent in the virgin resin. It should be noted that the physical design of a part can be as important as selecting the proper resin or considering the effects of processing variables on performance and failure, but these aspects are not considered here.

The main intrinsic properties that characterize all polymer molecules are their size, weight, and weight distribution; chain stiffness or chain rigidity; the particular state, whether amorphous, semicrystalline, or crystalline; and the nature of any network structure. These elements, by which all polymers can be classified, are illustrated in Fig. 2 and described with examples in Table 1. Clearly, there should be no network structure or cross linking with engineering thermoplastics. However, in practice, small amounts can occur as a result of thermal and oxidative degradation, or, in specific cases, can be introduced by design in the processing operation.

A brief scheme of structure analysis as it relates to material failure is presented in Table 2. The examples described illustrate the type of information that can be obtained when using a particular analytical technique.

Problem Solving

A typical problem a material engineer must face is a piece of failed pipe. Identifying the pipe material can be accomplished by infrared (IR) spectroscopy or spectroscopic methods in general. With the chemical structure of the pipe established, the next property to address is the glass transition temperature, T_g, and the melt temperature, T_m, because they characterize the useful working temperature range of the material. Either differential scanning calorimetry (DSC) or thermomechanical analysis (TMA) can be used to determine both T_g and T_m, as well as other thermal events. Repeat analysis can be used on a single sample to reveal the effects of thermal or process history on the transition temperatures. The next step is to determine the molecular weight (MW) and/or molecular weight distribution (MWD) of the polymer. A primary method, such as light scattering, or secondary methods, such as dilute solution viscosity, gel permeation chromatography (GPC), or melt rheology, can be used to characterize the MW and MWD. Differences in MW and MWD can have a profound effect on ultimate properties. Wide-angle x-ray diffraction, as well as electron microscopy, can determine the morphology, or structure (amorphous or crystalline), of the material. This problem-solving approach can be extended from failed pipe to a polymeric structural member that failed under load, or a plastic gear that cracked under use conditions, or any other failure problem.

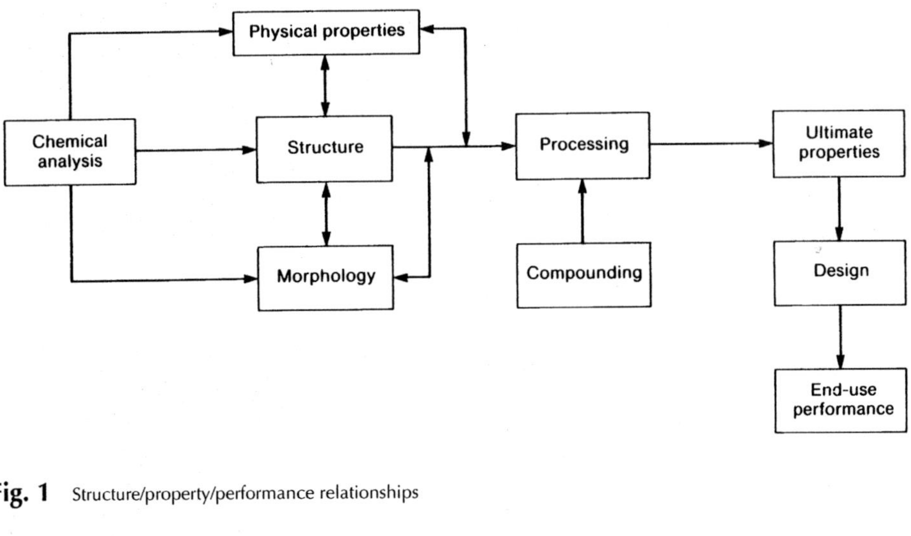

Fig. 1 Structure/property/performance relationships

Molecular Spectroscopy

Spectroscopic methods are widely used in the analysis of polymers (Ref 1–10). These methods include IR spectroscopy, which itself includes Fourier transform infrared (FTIR) detection, diffuse FTIR, micro-FTIR, attenuated total reflectance, and combined technologies such as gas chromatography FTIR and thermogravimetric analysis (TGA)/FTIR; solution and solid-state nuclear magnetic resonance (NMR) spectroscopy; ultraviolet-visible spectroscopy; mass spectroscopy; and Raman spectroscopy. Only the characterization of plastics by IR and NMR spectroscopy are reviewed here.

IR or FTIR Spectroscopy. The characteristic IR bands are fingerprints of the functional

*Adapted from the article by Alan T. Riga and Edward A. Collins, "Analysis of Structure," in *Engineering Plastics,* Volume 2, *Engineered Materials Handbook,* ASM International, 1988, pages 824 to 837

groups that make up polymers, as shown in Fig. 3 to 10, and are important both for identification and for correlation to performance and failure. Spectra of polymers are quite often simple, considering the complexity of the macromolecule.

Most polymer constituents absorb electromagnetic radiation in the IR region. Vibrational states of varying energy levels exist in the molecules. Transition from one vibrational state to another is related to absorption or emission of electromagnetic radiation. The frequency of the radiation that excites the molecule is related to the difference in energy states, E, by the following equation:

$$\text{Frequency} = \frac{E}{h} = \frac{c}{\text{wavelength}}$$

where h is Planck's constant, and c is the velocity of light. Organic molecules have absorption frequencies and corresponding wavelengths in the 1 to 50 μm IR region. The vibrational energy of a group of atoms is associated with a given frequency, which, in most cases, is independent of polymer chain length.

Some of the absorption frequencies are almost the same as those observed for the monomers. The IR spectra of synthetic polymers are illustrated in Fig. 3 to 6, 9 and 10. An alternate way to view polymer spectra is to evaluate the IR band assignments and absorption frequencies characteristic of various functional groups (Fig. 7, 8). The wavelength is usually given in microns or wave numbers (1/cm). The microstructure of macromolecules has been related to specific absorptions (Ref 7–9). Therefore, the IR spectra of a polymer can be used for identification and structural characterization. Computer-based IR spectrophotometers can store a vast number of polymer spectra that are commercially available, as well as those spectra generated in-house. Identifying a polymer based on a computer search, or a comparison to known collections of polymer spectra, is not a difficult task, to date.

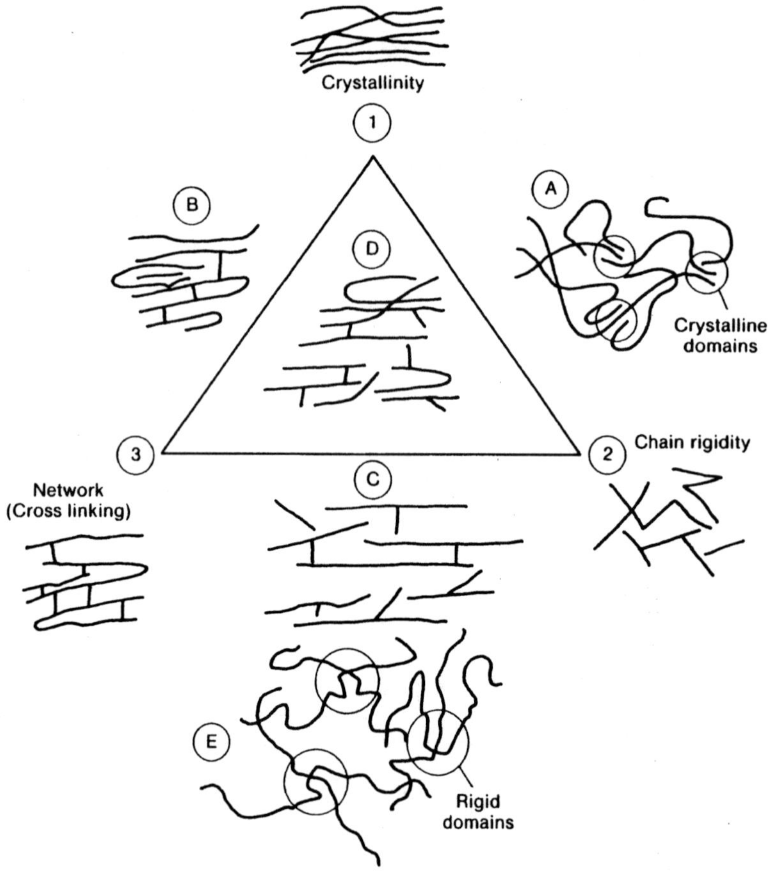

Fig. 2 Basic elements of engineering polymers. See Table 1 for explanation.

Table 1 Basic elements of engineering polymers

Location	Characteristics	Examples
1	Flexible and crystallizable chains	PE PP PVC PA
2	Cross-linked amorphous networks of flexible chains	Phenol-formaldehyde cured rubber Styrenated polyester
3	Rigid chains	PIs (ladder molecules)
A	Crystalline domains in a viscous network	PET Terylene (Dacron) Cellulose acetate
B	Moderate cross linking, with some crystallinity	Chloroprene rubber Polyisoprene
C	Rigid chains, partly cross linked	Heat-resistant materials
D	Crystalline domains with rigid chains between them and cross-linking chains	High-strength and temperature-resistant materials
E	Rigid-chain domains in a flexible-chain matrix	Styrene-butadiene-styrene, triblock polymer Thermoplastic elastomer

Note: See Fig. 2. PE, polyethylene; PP, polypropylene; PVC, polyvinyl chloride; PA, polyamide; PI, polyimide; PET, polyethylene terephthalate. Dacron, E.I. DuPont de Nemours & Co.

NMR spectroscopy is second only to IR spectroscopy in its importance as an analytical tool available to the polymer analyst. Nuclear magnetic resonance spectroscopy reveals the number of types of hydrogen or carbon atom and their surrounding electronic environments. The structure of an unknown polymer can be determined from IR functional group analysis. Nuclear magnetic resonance analysis provides the information that reveals the particular molecular cross linking (Ref 10–12). It also provides conformation of functional groups. Determining the sequence distribution can be achieved by NMR techniques for copolymer, block, and graft polymers only. This method requires that the sample be soluble in a suitable solvent.

The basis of NMR spectroscopy is the measure of absorbed energy required for nuclei to change their magnetic spin orientation while aligned in an applied magnetic field. Proton NMR is widely used, and typical field strengths are 1.41 and 51.48 T (14,100 and 51,400 gauss), with corresponding absorption of radiation at frequencies of 60 and 220 MHz. The ^{13}C nucleus, also of great interest to the polymer chemist, can be examined with a field strength of 1.0 T (10,000 gauss) at a frequency of 10.7 MHz.

Background. A universal reference compound, tetramethyl silane (TMS), is placed in the solution of the chemical to be measured, and the resonance frequency of each proton in the

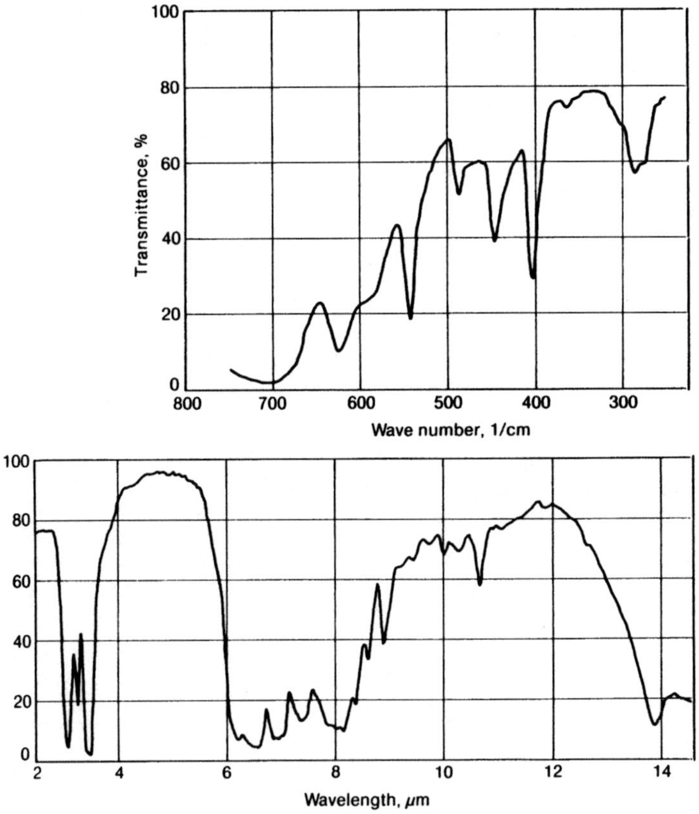

Fig. 3 Infrared spectra of nylon 11. Composition: nylon 11, polyundecal lactam. Used for plastics of all kinds. Preparation: multiple internal reflection, pressed film; far infrared, film from formic acid solution

unknown chemical is measured relative to the resonance frequency of the protons of the TMS. The frequency shift in hertz from TMS for a given proton depends on the field strength of the applied magnetic field. Therefore, the chemical shift, δ, a field-independent measure, is defined as (shift in hertz)/(NMR frequency in megahertz). The chemical shift describes the amount by which a proton resonance is shifted from TMS in parts per million based on the operating frequency of the spectrometer.

Solid-state NMR can be used to determine the structure of solid polymers (Ref 10–12). It differs from solution NMR in instrumentation and techniques because, with solids, problems that are associated with spectral resolution and sensitivity can arise. Solid-state NMR techniques require a high-power magnetic decoupler, various probes, magic-angle spinning, and fast spinning rates. The purpose of the solid-state instrumentation is to render the chemical shift to a single, interpretable peak, that is, to improve resolution, which is easily obtained in solution NMR, because Brownian motion assists in the generation of a single chemical shift.

Styrene-acrylonitrile (SAN) has a unique NMR spectrum associated with the structural sequences in this copolymer (Fig. 11, Table 3). The NMR spectra of isotactic and syndiotactic polypropylene (PP) clearly differentiate the stereoregularity of this polymer (Fig. 12, Ref 10). In addition to polyolefins, polystyrene (PS) and polymethyl methacrylate (PMMA) can also

Table 2 Practical information derived from polymer analysis methods

Test method	Property	Practical information provided
Gel permeation chromatography (GPC)	Weight-average molecular weight, $\overline{M}_w$; molecular weight distribution, MWD	Polymer MWD with soluble polymers
Low-angle light scattering	$\overline{M}_w$, MWD	$\overline{M}_w$ for homopolymers, interactions polymer/solvent
Osmometry, membrane	Number-average molecular weight, $\overline{M}_n$	$\overline{M}_n = 10^3$ to 10^6
Osmometry, vapor pressure	$\overline{M}_n$	$\overline{M}_n$ = 300 to 30,000, interactions polymer/solvent
Dilute solution viscosity	Inherent viscosity, reduced viscosity, intrinsic viscosity	Viscosity determination, viscosity average molecular weight
Quasi-elastic light scattering (QELS)	Macromolecular particle diameter, diffusion coefficient	Particle size in dilute dispersions, molecular aggregation
Differential thermal analysis (DTA)	Glass transition temperatures, T_g; melt/crystallization temperatures, T_m	Phase changes, T_g and T_m
Differential scanning calorimetry (DSC)	Heat of polymerization, fusion, T_g, T_m	Phase changes, reaction kinetics degree of cross linking, degradation inhibitor content and effectiveness
Thermogravimetric analysis (TGA)	Composition, weight loss with time or temperature	Thermal and oxidative stability, volatilization kinetics
Dynamic mechanical analysis (DMA)	Elastic modulus, loss modulus, tan delta	Mechanical properties, phase transitions, damping, softening cross linking
Thermal-mechanical analysis (TMA)	Penetration temperature, expansion coefficient	Phase changes, T_g, T_m, dimensional stability, modulus, compliance, deflection temperature under load, Vicat temperature
Mechanical spectroscopy	Viscosity, normal stress difference, shear elastic and loss modulus	Rheological properties, flow behavior, melt or solution elasticity, yield stress
Infrared (IR) spectroscopy, Fourier transform infrared spectroscopy (FTIR)	Chemical functional groups	Molecular information, chemical analysis
Attenuated total reflectance (ATR)	Surface functional groups	Molecular information
Nuclear magnetic resonance (NMR) spectroscopy	Chemical shift of nuclei	Molecular structure, functional group determination, sequence distribution
Mass spectroscopy	Mass/charge of ions produced	Molecular structure
X-ray diffraction analysis (XRD)	Crystalline polymer component	Crystalline structure, short- and long-range ordering, percent crystallinity, polymer blend/copolymer
Small-angle x-ray diffraction	X-ray scattering at low angle	Crystallite size and shape, long-range periodicity
Scanning electron microscopy (SEM)	Surface and particle morphology	Particle size and shape, surface features
Transmission electron microscopy	Polymer morphology	Polymer features and defects
X-ray photoelectron spectroscopy (XPS)	Elemental concentrations, oxidation states	Chemical composition of surfaces, determination of surface species
Auger electron spectroscopy (AES)	Elemental concentrations	Chemical compositions of surfaces
Wavelength dispersive x-ray analysis	Elements present on polymer surfaces	Identification of contaminants on polymer surfaces

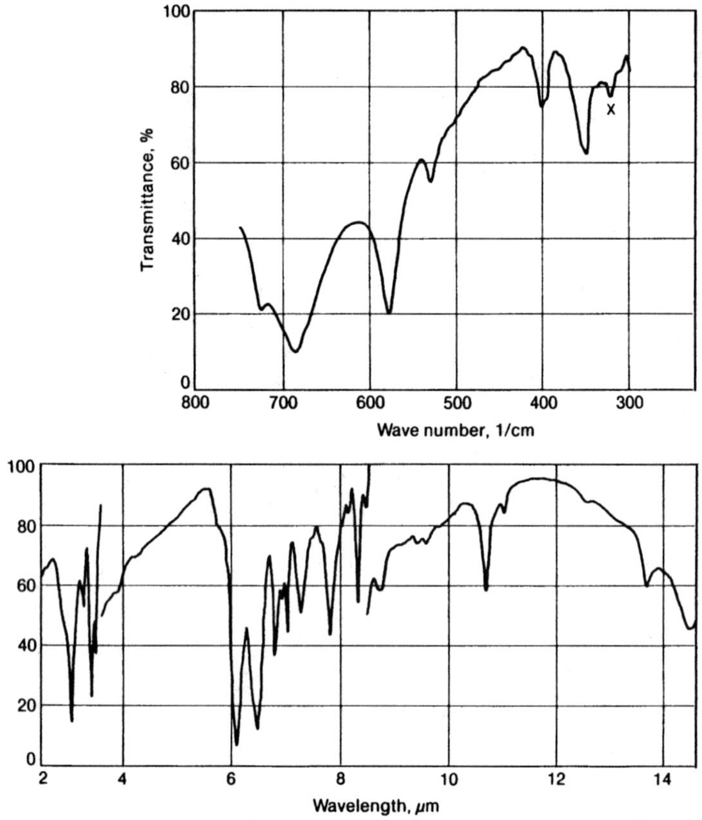

Fig. 4 Infrared spectra of nylon 6/6. Used for plastics of all kinds. Preparation: film from formic acid solution

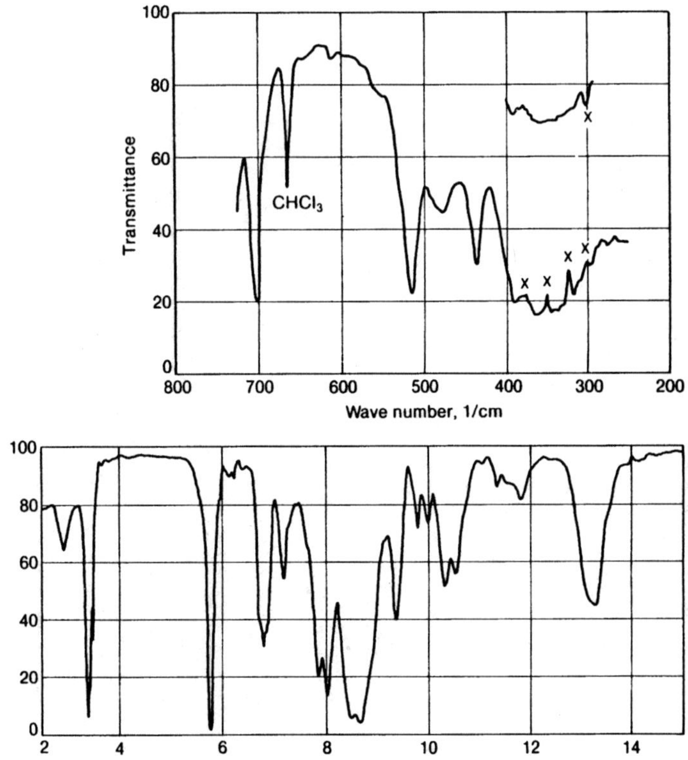

Fig. 5 Infrared spectra of poly(*n*-butyl methacrylate). Used for transparent plastic. Preparation: multiple internal reflection, KBr dispersion; far infrared, film from $CHCl_3$ solution (~0.12 mm, or 0.005 in., thickness)

be synthesized in an isotactic configuration and identified by NMR studies.

Molecular Weight

The MW of an engineering plastic is a very important, if not the most important, property, because a minimum MW is most often needed to achieve desired properties. Most physical and mechanical properties are a linear function of MW. Molecular weight should be carefully monitored and can be determined directly or indirectly in a number of ways. The primary methods include membrane osmometry, light scattering, sedimentation, and centrifugation (Ref 1–6, 13–16). Secondary but more commonly used methods are dilute solution viscosity (intrinsic viscosity), hydrodynamic chromatography, and bulk melt viscosity (melt flow) (Ref 13–16). With the exception of the latter, all of these primary and secondary methods are dilute-solution methods that can only be used on uncured polymers capable of being dissolved in an appropriate solvent.

A polymer consists of a collection of like molecules with a distribution of molecular weights. A typical distribution is given in Fig. 13. The distribution can be described by average molecular weights:

$$\text{Number average, } \overline{M}_n = \frac{\Sigma n_i M_i}{\Sigma n_i}$$

$$\text{Weight average, } \overline{M}_w = \frac{\Sigma n_i M_i^2}{n_i M_i}$$

$$\text{Z-average, } \overline{M}_z = \frac{\Sigma n_i M_i^3}{n_i M_i}$$

where n_i is the number of polymer molecules of molecular weight M_i. As a first order approximation, the ratio $\overline{M}_w/\overline{M}_n$ (known as the polydispersity index) is often used as an index of the MWD. However, there is no substitution for the complete distribution, because higher molecular moments are especially important for correlations with ultimate properties. For most common polymers, the range of the $\overline{M}_w/\overline{M}_n$ index is 1.0 to 80. Fractionation of polymers into less heterogeneous components by high-speed GPC gives more insight into polymer distribution than simply MWD, especially in the presence of additives (Fig. 14). The weight average and higher moments of the MWD are the important criteria when correlating with mechanical properties. In general, most properties increase with increasing MW. However, as MW increases, processing becomes more difficult.

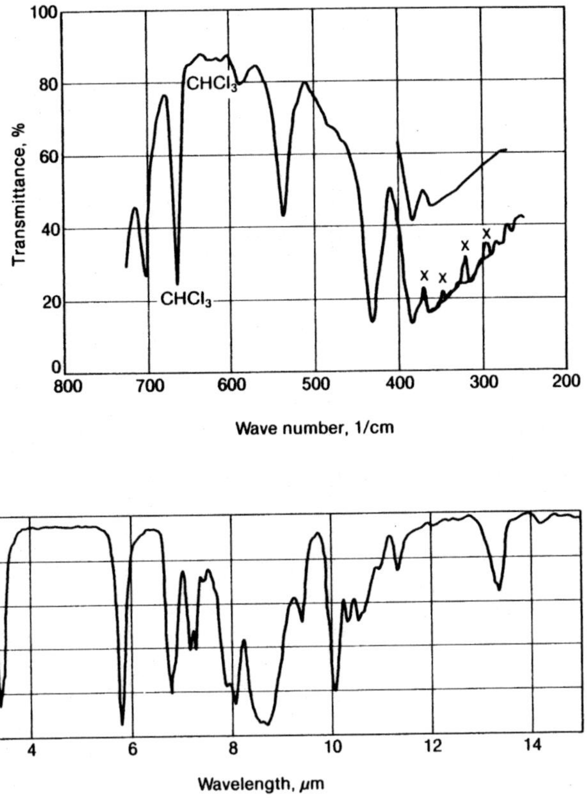

Fig. 6 Infrared spectra of poly(isobutyl methacrylate). Used for transparent plastic. Preparation: film from $CHCl_3$ solution

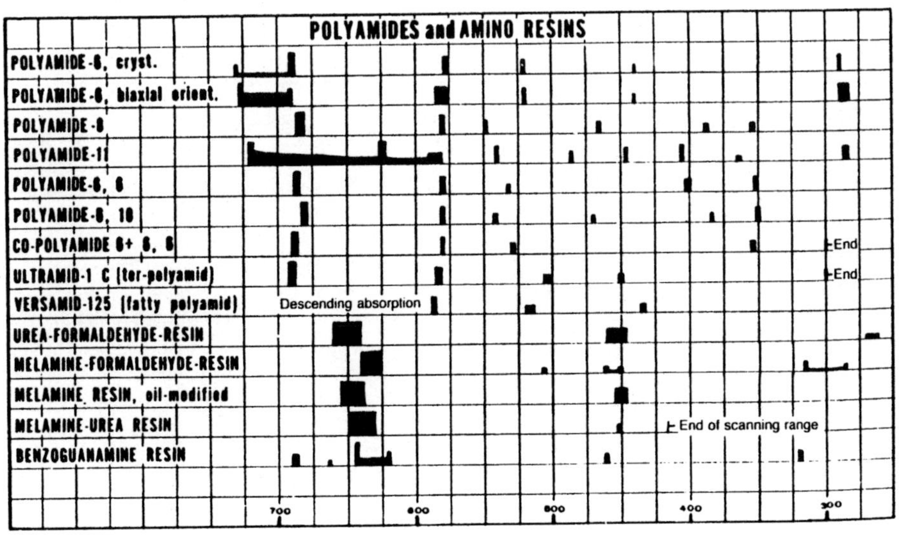

Fig. 7 Absorption frequencies of polyamides and amino resins. Source: Ref 8

Methods of Thermal Analysis

DSC and Differential Thermal Analysis (DTA). Engineering thermoplastics have been characterized by DSC and DTA (Ref 1–6, 18–20). With these methods, the following physical properties have been determined: T_g, T_c (the temperature at which crystallization occurs at a maximum rate), exothermic heat of polymerization or cure, T_m, heat of fusion, exothermic heat of stress relaxation, specific heat as a function of temperature, thermal and oxidative stability, and heat of volatilization of residual solvents (Fig. 15) (Ref 21–23). This information provides the engineer with differences between a potentially successful and a potentially inadequate sample, differences resulting from thermal or processing histories, the presence of undesirable contaminants, errors in formulation, and other causes, all of which can be directly responsible for failure.

Thermal analysis describes the techniques used in characterizing materials by measuring a physical or mechanical property as a function of temperature or time at a constant temperature. The more common methods used in thermal analysis are DSC, DTA, TGA, TMA, and dynamic mechanical analysis (DMA).

Differential scanning calorimetry measures the thermal energy absorbed by the sample (endothermic process) or given off (exothermic process). In the DSC method, the sample and reference are placed in thin metal (aluminum) pans, with the thermocouple sensors below the pans. Differential scanning calorimetry measurements can be made in two ways: by measuring the electrical energy provided to heaters below the pans necessary to maintain the two pans at the same temperature (power compensation), or by measuring the heat flow (differential temperature) as a function of sample temperature (heat flux). In the DTA method, the sensor thermocouple is placed either directly in the sample or close to the sample. The endothermic or exothermic heats of transition can be quantitatively measured by DSC, not by the DTA method. In short, DSC measures heat flow, while DTA measures temperature differentials.

A general relationship exists among MW, T_g, T_m, and polymer properties (Fig. 16). In general, T_m is greater than T_g over a wide MW range. The T_g and T_m reach a plateau at high MWs over 100,000. The T_g of a thermoset polymer has been reported to increase with cure time and temperature (Fig. 17). A higher MW is achieved with a shorter cure time (20 min) at the higher polymerization temperature (175 °C, or 350 °F) than at the lower temperature (140 °C, or 285 °F).

The T_gs of acrylonitrile-butadiene-styrene (ABS), observed as sigmoidal-shaped curves at −85 °C (−120 °F) for the butadiene (BD) phase and at 103 °C (220 °F) for the SAN phase, can be easily detected by DSC (Ref 4, 6, 19, 23). The SAN phase renders the material hard, while the BD phase gives the ABS resiliency. This material is a hard, tough polymer.

Polycarbonate (PC) is another example of a hard, tough polymer with a T_g at 141 to 150 °C (285 to 300 °F) and a low temperature transition at approximately –80 °C (–110 °F). A lower T_g for PC can indicate embrittlement and a latent or observed failure in the part (Ref 23).

The thermal history of a stress-cracked polyamide (PA) gear was determined by DSC. The T_g, T_m, heat of fusion, and amide functionality (determined by IR spectroscopy) characterized the polymer as a PA. The gear that failed had an exothermic stress-relaxation process prior to the melting endotherm. The stressed part did not exhibit an exotherm on reheating the sample in the DSC (Fig. 18). The stress-cracked gear was nylon 6/6 with no fiber-glass reinforcement, while a pass gear was nylon 6 with 30 wt% glass reinforcement, as determined by TGA.

As final examples of the utility of DSC, it can determine the effect of a plasticizer on the melting point of nylon 11 (Fig. 19), the amount of polyethylene (PE) in impact PC (Fig. 20), and the crystallinity of polyolefins (Fig. 21). All of these effects can be directly responsible for failure.

Thermomechanical analysis measures the dimensional change of a plastic as a function of time or temperature. The thermomechanical properties that have been measured are the T_g, softening point, coefficient of linear thermal expansion, heat-deflection temperatures (HDT), creep moduli, creep relaxation, degree of cure, viscoelastic behavior, and dilatometric properties (Ref 18, 20, 23–25).

Incomplete cure of a polymer is indicated by a low thermomechanical analysis T_g or by an increasing T_g with heat cycling (Fig. 22). Cracking of a plastic can occur with a partially cured part, because the ultimate strength has yet to be achieved with the proper MW and T_g. ASTM International has developed thermomechanical tests that approximate the strength and T_g of plastics, for example, the Vicat softening temperature and HDT under load (DTUL) test method. Vicat softening (ASTM D 1155) and HDTs (ASTM D 648) of plastics have been determined by TMA at the high stresses of 10.3 and 1.82 MPa (1.5 and 0.264 ksi), respectively. The TMA Vicat softening temperatures increased with MW for PMMA and PS (Fig. 23). The thermomechanical analysis DTUL varied linearly with the ASTM D 648 DTUL (Fig. 24).

Generalized tensile stress-strain curves for plastics are related to polymer properties (Fig. 25). Based on this generalization and the room temperature TMA creep modulus, as well as the percent of creep recovery, a scheme has been developed for ranking commercial polymers (Fig. 26, Ref 24). The purpose of the TMA creep test was to evaluate small pieces of a larger plastic part or limited amounts of a plastic from a

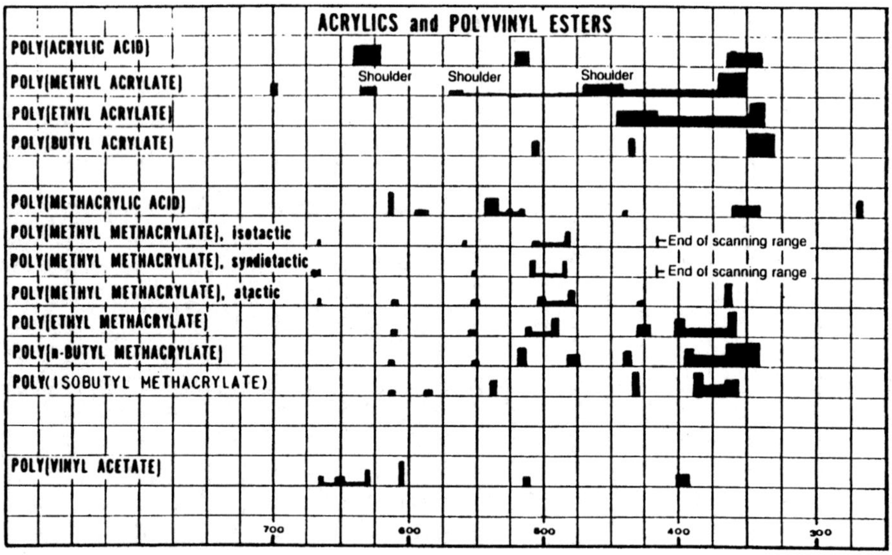

Fig. 8 Absorption frequencies of acrylics and polyvinyl esters. Source: Ref 8

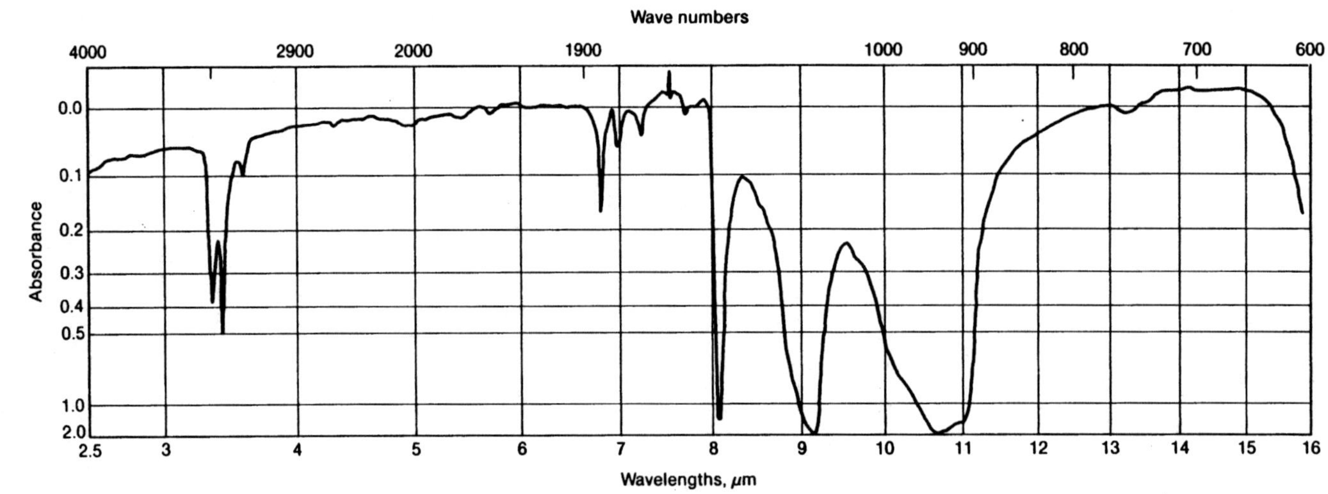

Fig. 9 Infrared spectra of polyformaldehyde

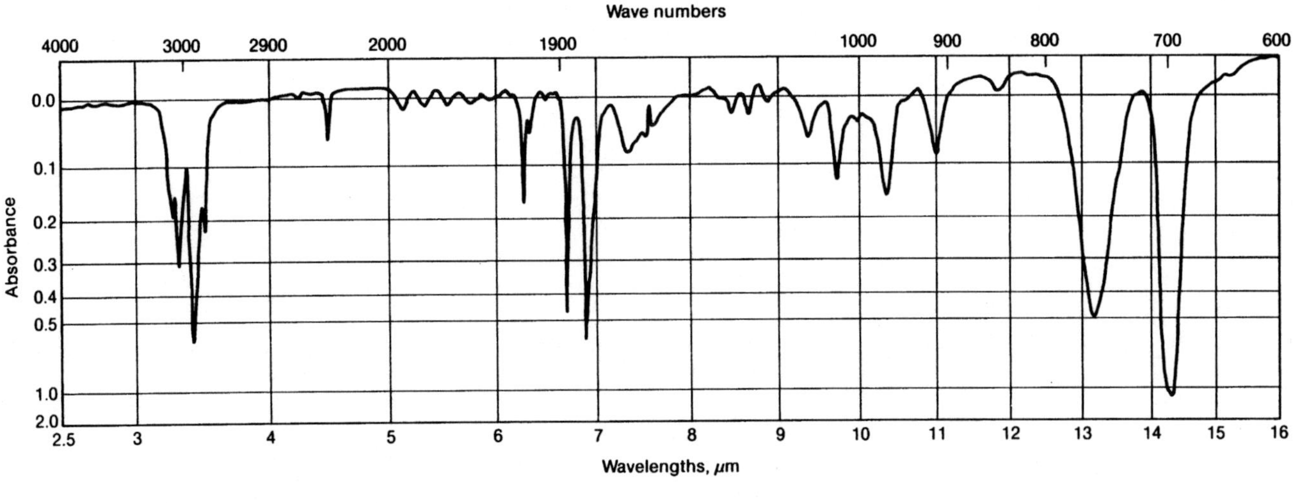

Fig. 10 Infrared spectra of acrylonitrile-butadiene-styrene

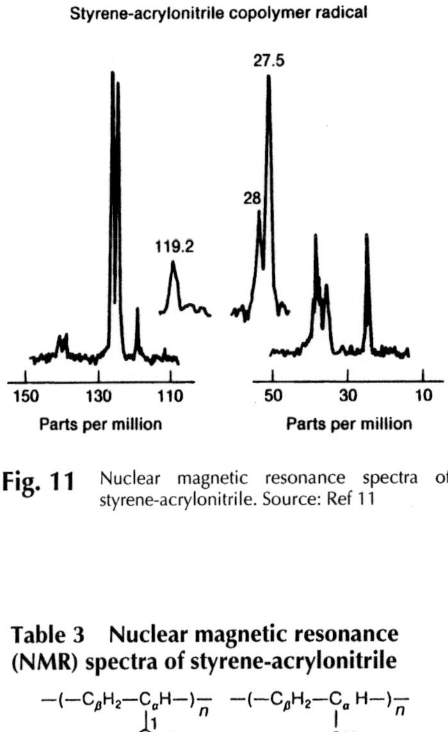

Fig. 11 Nuclear magnetic resonance spectra of styrene-acrylonitrile. Source: Ref 11

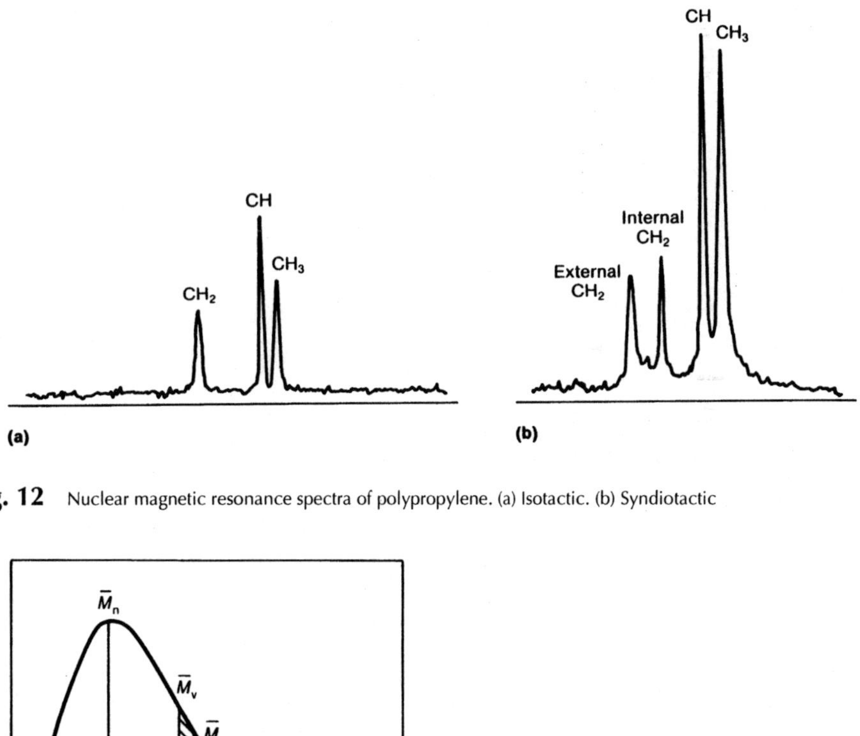

Fig. 12 Nuclear magnetic resonance spectra of polypropylene. (a) Isotactic. (b) Syndiotactic

Table 3 Nuclear magnetic resonance (NMR) spectra of styrene-acrylonitrile

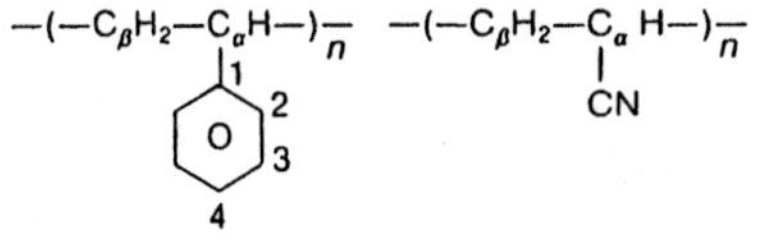

Styrene (S) Acrylonitrile (A)

Chemical shift (δ), ppm	Assignment
140.5	C_1 (S) of S*S*A
139	C_1 (S) of A*S*A
129.5	C_2 (S)
128.5	$(C_3 + C_4)$ (S)
119.2	–CN (A) of SAS
(44–33)	DMSO + C_α (S) + C_β (SS)
28	C_α (A) of A*A*S
27.5	C_α (A) of A*A*A

Analysis conditions: Nucleus: ^{13}C. Frequency: 25.2 MHz. Spectrometer: Varian XL100. Detection technique: FT-7000 pulses. Flip angle: 30°. Repetition time: 0.4 s. Solvent: dimethyl sulfoxide $(DMSO)d_6$. Temperature: 100 °C (212 °F). Lock: DMSO-d_6
Note: Composition (^{1}H NMR): S = 65 mol%. "A" units are isolated.

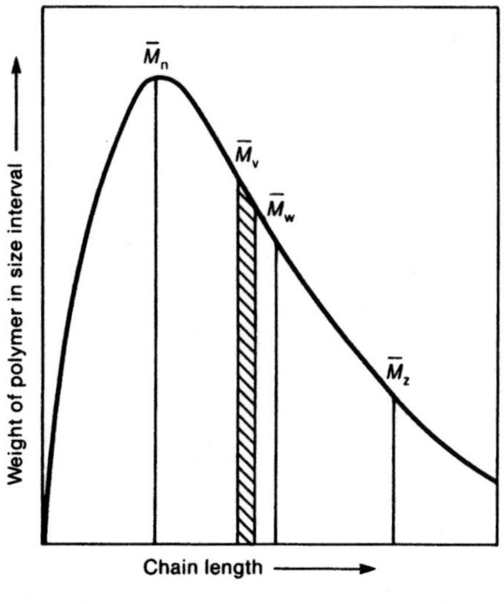

Fig. 13 Typical molecular weight distribution curve. $\overline{M}_n$, number-average molecular weight; $\overline{M}_v$, viscosity average molecular weight; $\overline{M}_w$, weight-average molecular weight; $\overline{M}_z$, Z-average molecular weight. Source: Ref 14

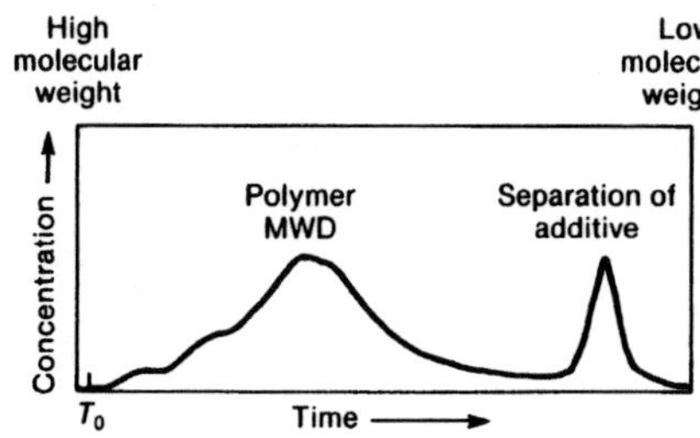

Fig. 14 Gel permeation chromatogram from a high-performance liquid chromatograph. MWD, molecular weight distribution. Source: Ref 17

failed part. The polymers are categorized by their mechanical properties: hard tough, hard brittle, soft tough, and soft weak. There is a good correlation between the TMA properties and the known tensile properties of these commercial polymers.

To establish a relationship between various thermal properties, seven engineering plastics from the Society of Plastics Engineers resin kit

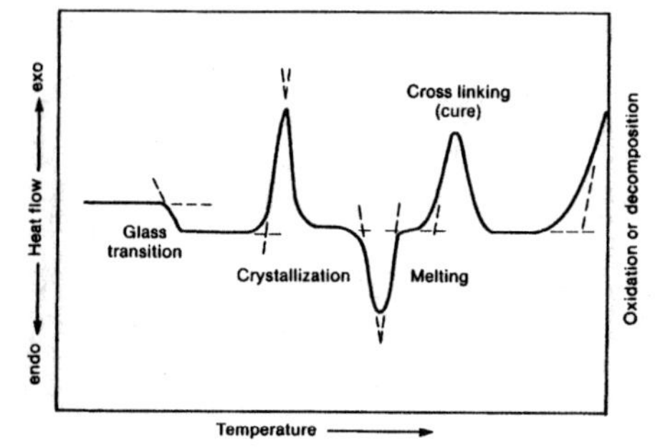

Fig. 15 Schematic differential scanning calorimetry thermogram

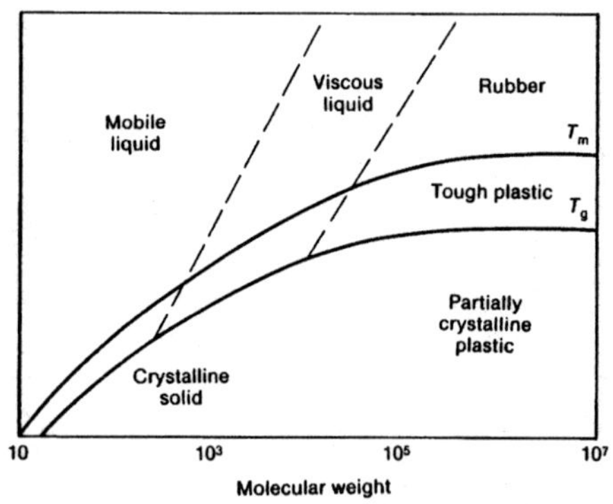

Fig. 16 Relationships among glass transition temperature (T_g), melt temperature (T_m), molecular weight, and polymer properties. Source: Ref 13

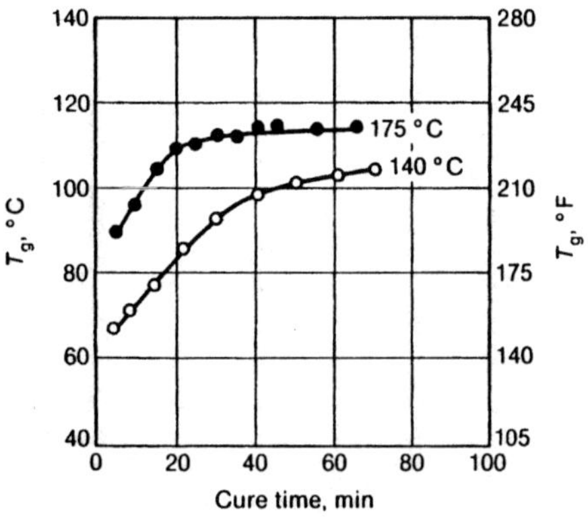

Fig. 17 Variation of glass transition temperature (T_g) with cure time and temperature. Source: Ref 23

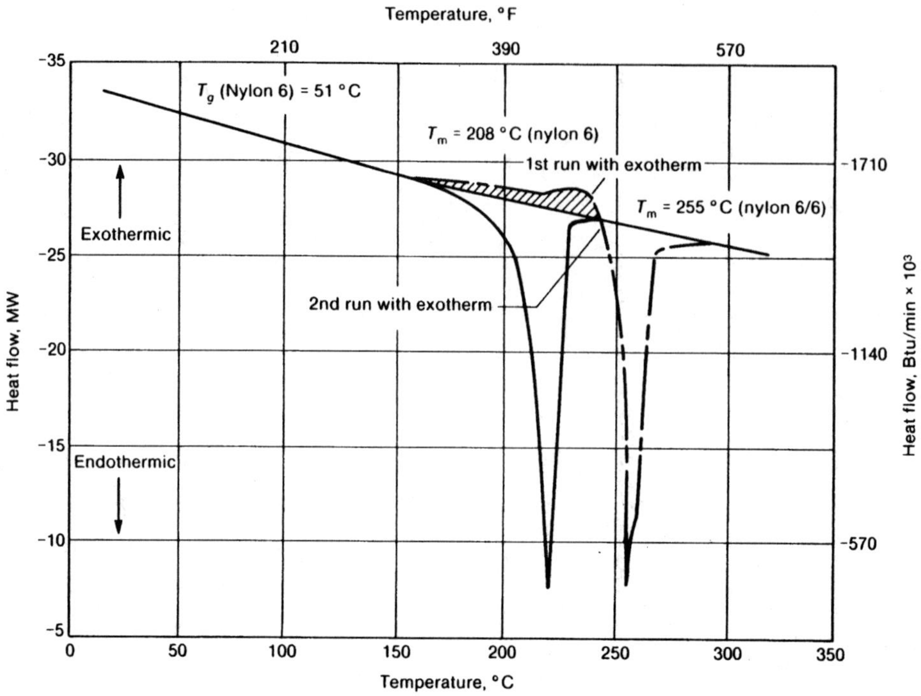

Fig. 18 Differential scanning calorimetry of nylon gears. MW, molecular weight; T_g, glass transition temperature; T_m, melt temperature

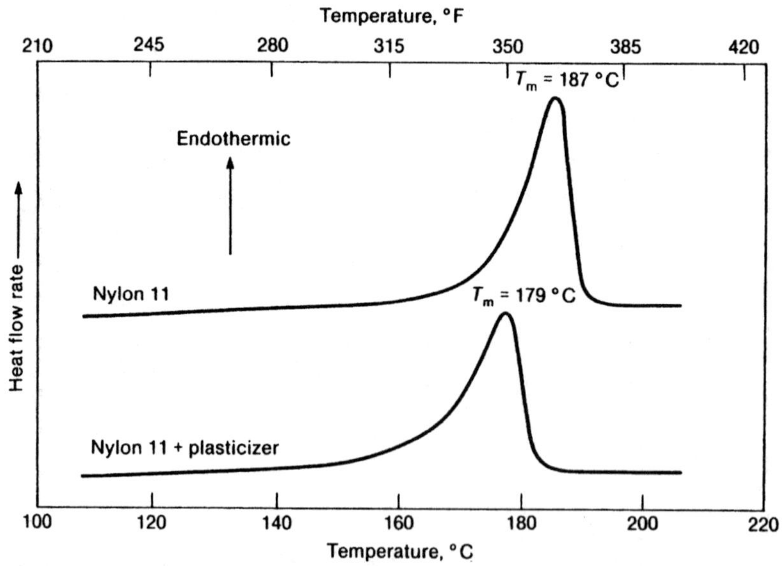

Fig. 19 Differential scanning calorimetry determination of the effect of a plasticizer on melting temperature (T_m) of nylon 11. Range, 0.0024 W (10 mcal/s); heating rate, 20 °C/min (36 °F/min); weight, 6.8 mg (0.105 gr), both samples. Source: Ref 19

(Ref 27) were evaluated by DSC, TMA, and TGA (Table 4, Fig. 27). There is a good correlation between DSC and TMA transition temperatures. Although the TGA onset temperature, a pyrolysis temperature (T_p), was used to rank the polymers, it is not related to the thermal or mechanical properties as determined by DSC, TMA, or Izod impact testing, under ASTM D 256.

Dynamic mechanical analysis can be used to evaluate structure-related processing and performance characteristics. This technique measures the viscoelastic response of a polymer when subjected to a sinusoidal stress. Because the applied strain is low, the measurements fall in the linear viscoelastic region. Dynamic mechanical analysis detects both the elastic and viscous components of the complex modulus

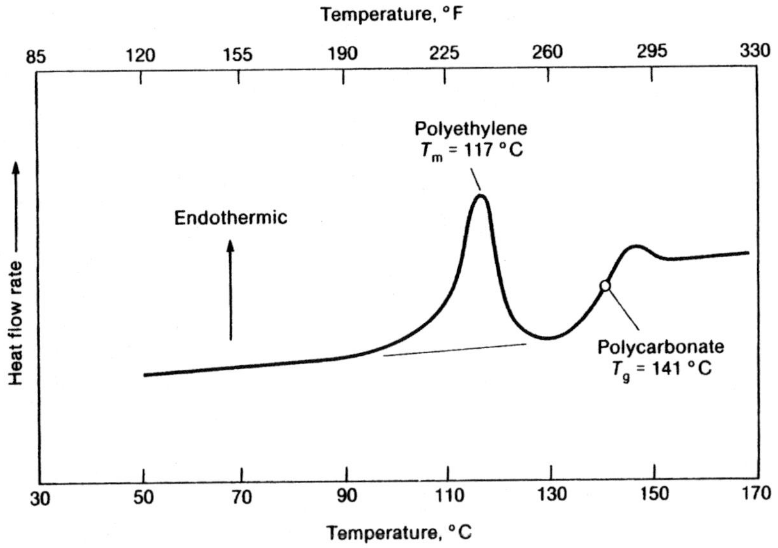

Fig. 20 Differential scanning calorimetry determination of polyethylene in impact polycarbonate. Range, 0.00048 W (2 mcal/s); heating rate, 20 °C/min (36 °F/min); weight, 23 mg (0.355 gr). T_m, melting temperature; T_g, glass transition temperature. Source: Ref 19

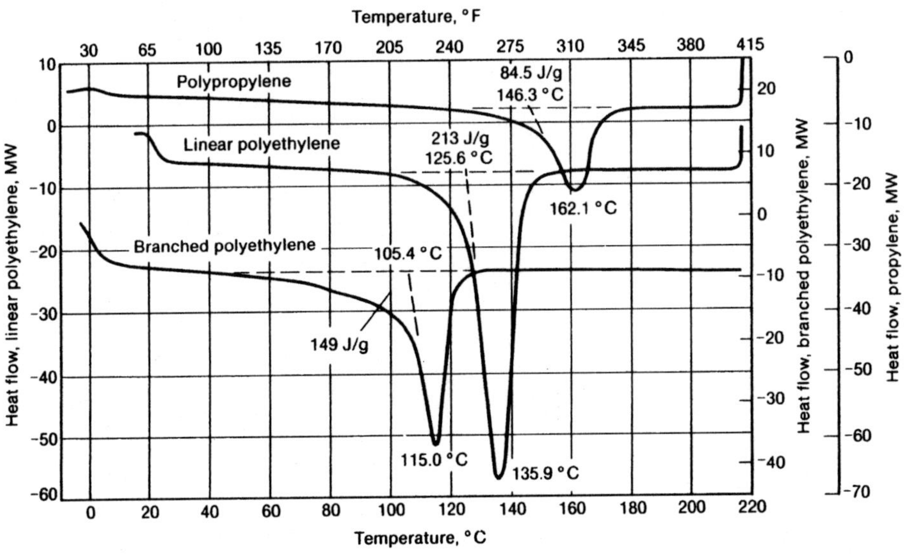

Fig. 21 Polyolefin melting profiles. MW, molecular weight. Source: Ref 23

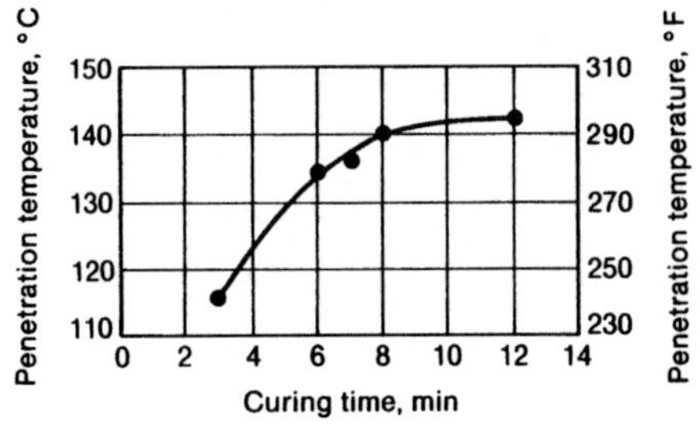

Fig. 22 Thermomechanical analysis evaluation of degree of cure by penetration. Source: Ref 23

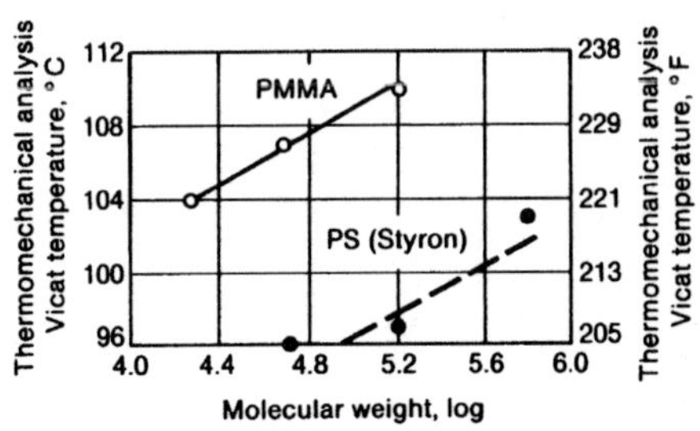

Fig. 23 Thermomechanical analysis, Vicat softening temperatures, under 10.3 MPa (1.5 ksi). Source: Ref 24

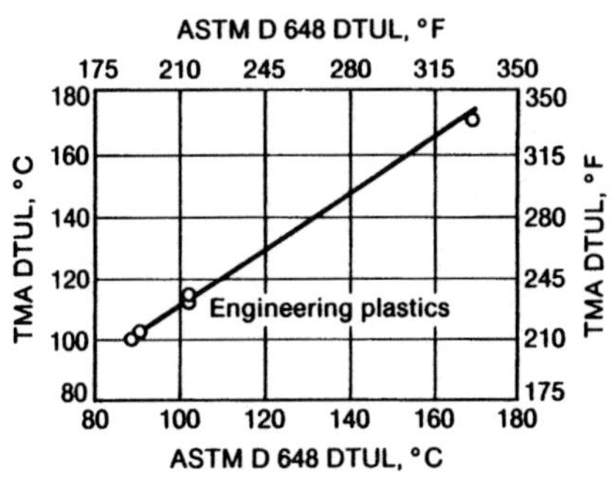

Fig. 24 Thermomechanical analysis (TMA) heat-deflection temperature under load (DTUL) at 1.82 MPa (0.264 ksi). Source: Ref 24

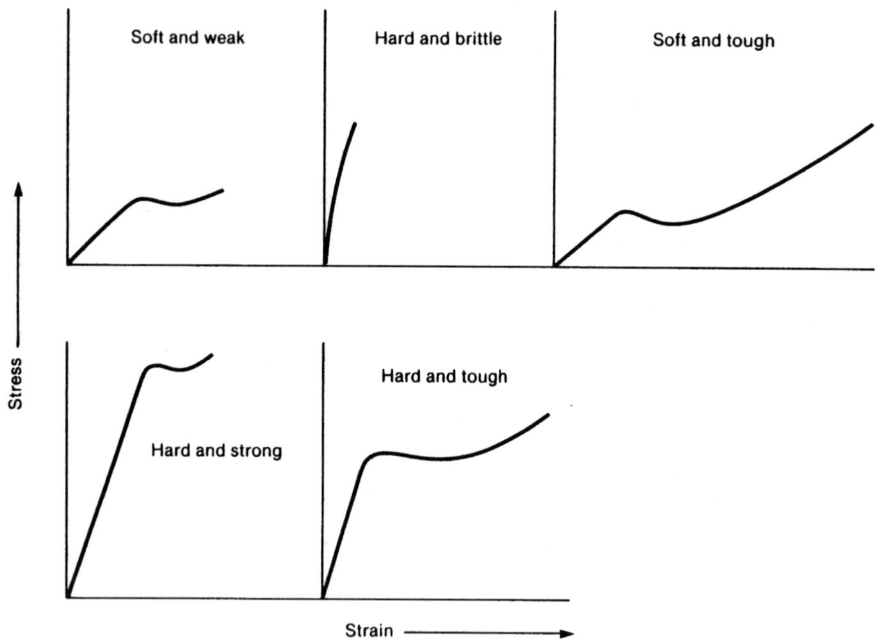

Fig. 25 Tensile stress-strain curve for several types of polymeric materials. Source: Ref 26

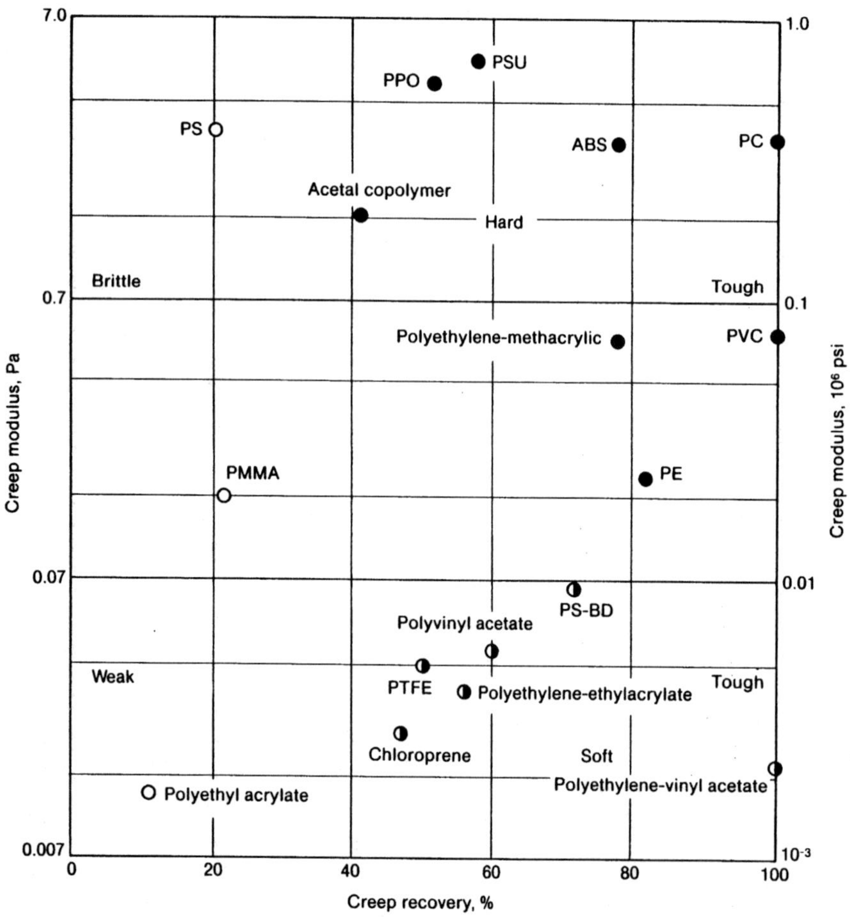

Fig. 26 Thermomechanical analysis properties of commercial polymers. PSU, polysulfone; PPO, polyphenylene oxide; PVC, polyvinyl chloride; PTFE, polytetrafluoroethylene. Source: Ref 24

when polymers are studied in the oscillatory mode. The in-phase response to a sinusoidally varying strain is the elastic, or storage, modulus (G') and the out-of-phase response is the viscous, or loss, modulus (G''). The relationship between the two moduli, G''/G', is the loss tangent, or dissipation factor, commonly known as tan δ (Ref 28, 29), and is related to the ability of the polymer to dissipate the energy of the applied stress.

The temperature can be varied in DMA, generating frequency (modulus) or damping (tan δ) spectra of an engineering plastic. As reported in Ref 28, "The dynamic parameters have been used to determine the glass transition region, relaxation spectra, degree of crystallinity, molecular orientation, cross linking, phase separation, structural or morphological changes resulting from processing, and chemical composition in polymer blends, graft polymers, and copolymers." Polymer composites have also been studied by DMA.

The tan δ spectra of impact-modified PP related well with impact resistance values from the drop weight index (DWI), ASTM D 3029 (Fig. 28).

The comparative modulus of nylon 6/6 clearly delineates the effect of added fiberglass reinforcement (Fig. 29) and added moisture (Fig. 30). In the first case, the modulus below the T_g was approximately 30% greater for the reinforced nylon 6/6, while at 30 °C (85 °F), moisture significantly lowered the modulus of this material (Ref 23).

Thermogravimetric analysis measures a change in sample weight with time or temperature. Typically, a polymer sample is examined from room temperature to above its decomposition or pyrolysis temperature in nitrogen (thermal stability). Often, a platinum pan or quartz boat is used for high-temperature studies to 1000 °C (1830 °F). Sample size may vary from 1.0 to 100 mg (0.0154 to 1.54 gr). The oxidative stability of polymers in air or oxygen can also be determined by TGA.

The relative thermal stability of polymers measured by TGA is illustrated in Fig. 31. Based on the onset temperature of thermal degradation, the polymers are ranked in order of stability: polyimide (PI) stability is greater than that of polytetrafluoroethylene (PTFE), which is greater than that of high-density polyethylene (HDPE), which is greater than that of PMMA, which is greater than that of polyvinyl chloride (PVC).

The composition of silica- and carbon-filled PTFE was determined by TGA (Fig. 32). PTFE is decomposed and volatilized in nitrogen, while the carbon filler is volatilized in air at 600 °C (1110 °F). The inorganic residue is silica.

A summary of polymer thermal properties as determined by thermal analysis and limited oxygen index (LOI) (ASTM D 2863) is given in Table 5 (Ref 31). According to this standard, LOI is "the minimum concentration of oxygen, expressed as volume percent, in a mixture of oxygen and nitrogen that will just support flaming combustion of a polymer initially at room temperature." The logarithm of the heat of combustion varied linearly with LOI for the polymers studied. The T_g, T_m, T_p, and combustion temperature (T_c) did not relate to the LOI.

A relatively new technique of combining TGA and FTIR spectroscopy has been developed in which the polymer degradation pattern (derivative TGA) can be superimposed on the IR spectral data. The TGA and reconstruction of the spectral data of PVC are given in Fig. 33 and 34 (Ref 32).

X-Ray Diffraction (XRD) Analysis

X-ray diffraction is used for analyzing crystalline phases in solid materials, determining the extent of crystallinity, and identifying crystalline structure. The crystalline state of a thermoplastic material affects the ultimate strength, the thermomechanical properties, and the end-use performance. The extent of crystallinity can be altered by the processing or thermal history of the plastic.

Crystalline polymers can be characterized by their XRD patterns (Ref 33–37). X-ray diffraction analysis is needed when the percent crystallinity may be related to field problems. The crystalline portion of a polymer will diffract when exposed to x-rays, for example, Cu k-alpha at 1.542 Å. Crystal diffraction follows Bragg's law:

$$n\lambda = 2d \text{ sine } \theta$$

where n is a constant (usually 1), λ is the wavelength of the x-ray, d is the interplanar spacing of the crystalline material, and sine θ is the experimental diffraction angle.

A powder camera or diffractometer is used when the diffraction angle can be varied and the resulting diffraction intensity measured (counts per second). The x-ray diagram of unoriented PE at 100 and 120 °C (212 and 250 °F) is cited in Fig. 35 (Ref 33). The three-dimensional crystalline order of PE can be seen as sharp peaks on the diffuse x-ray bands at 20 to 25° 2θ at 100 °C (212 °F), below the T_m. The crystallinity is lost at 120 °C (250 °F), because the sample temperature is above the T_m.

Some polymers have XRD patterns with multiple diffuse bands that are interpreted as two-dimensional or short-range ordering (Fig. 36). Typical half-band widths are 3 to 7° (2θ), while crystalline diffraction has band widths of 0.3 to 0.6° (2θ). Polymer blends can be distinguished from copolymers by XRD.

The effect of environmental conditioning of molded nylon 6 by water saturation was to vary the crystal structure (Fig. 37).

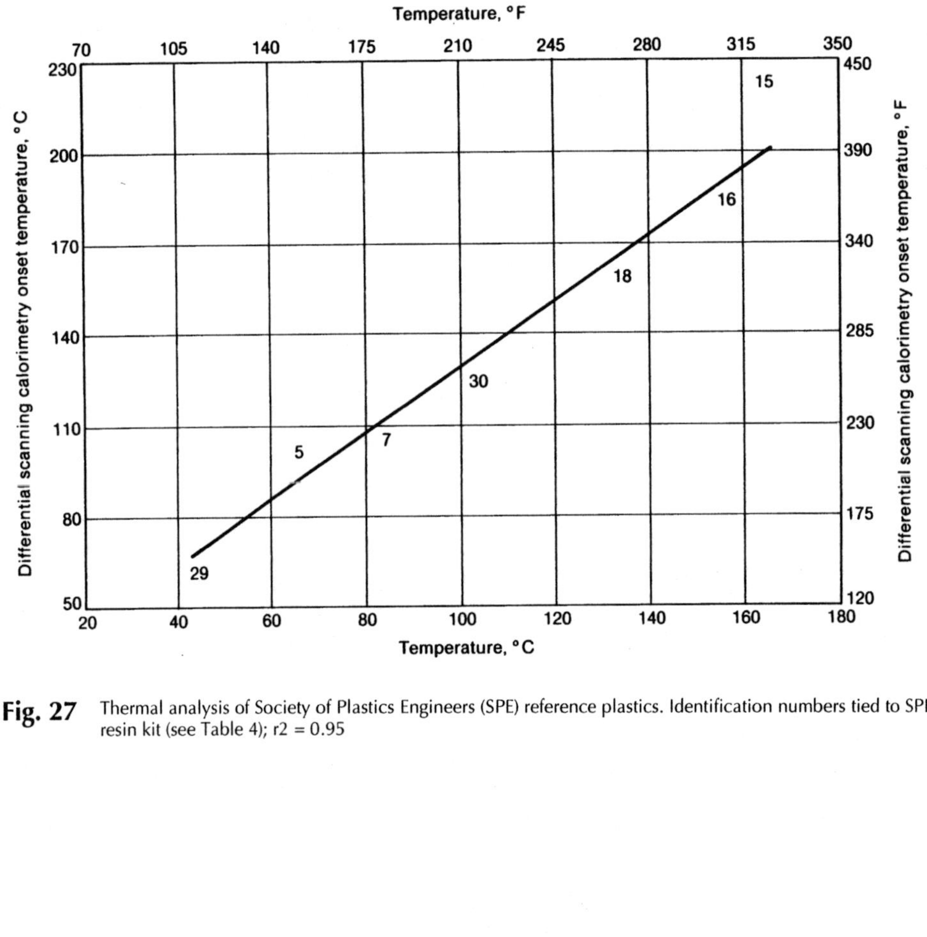

Fig. 27 Thermal analysis of Society of Plastics Engineers (SPE) reference plastics. Identification numbers tied to SPE resin kit (see Table 4); r2 = 0.95

Table 4 Thermal characterization of Society of Plastics Engineers (SPE) reference plastics

Polymer	SPE identification number	Onset temperature: DSC °C	DSC °F	TMA °C	TMA °F	Source of DSC or TMA transition	ASTM D 256 Izod impact: J/m	ft · lbf/in.	TGA: Extrapolated onset temperature °C	°F	wt% at 600 °C (1110 °F)
PVC, flexible	29	65	150	41	105	T_g	270	5.0	274	525	5.7
PVC, rigid	30	120	250	102	215	T_g	20	0.4	278	530	7.9
ABS, transparent	5	99	210	59	140	T_g, SAN	130	2.5	407	765	0
ABS, high impact	7	108	225	85	185	T_g, SAN	430	8.0	422	790	4
Nylon 6	16	183	360	155	310	T_m	160	3.0	439	820	1.2
Nylon 6/6	15	218	425	165	330	T_m	110	2.1	433	810	2.4
PET	18	151	305	130	265	T_g	40	0.7	517	960	28.3

Experimental conditions: Heating rate = 10 °C/min (18 °F/min); N_2 flow = 50 cm^3/min (3 $in.^3$/min) in DSC, TGA, and TMA; weight = 14–21 mg (0.21–0.32 gr) in DSC and 27–36 mg (0.42–0.55 gr) in TGA; in TMA, 5.0 g (0.18 oz) applied load, and height = 1.3–1.7 mm (0.05–0.07 in.)

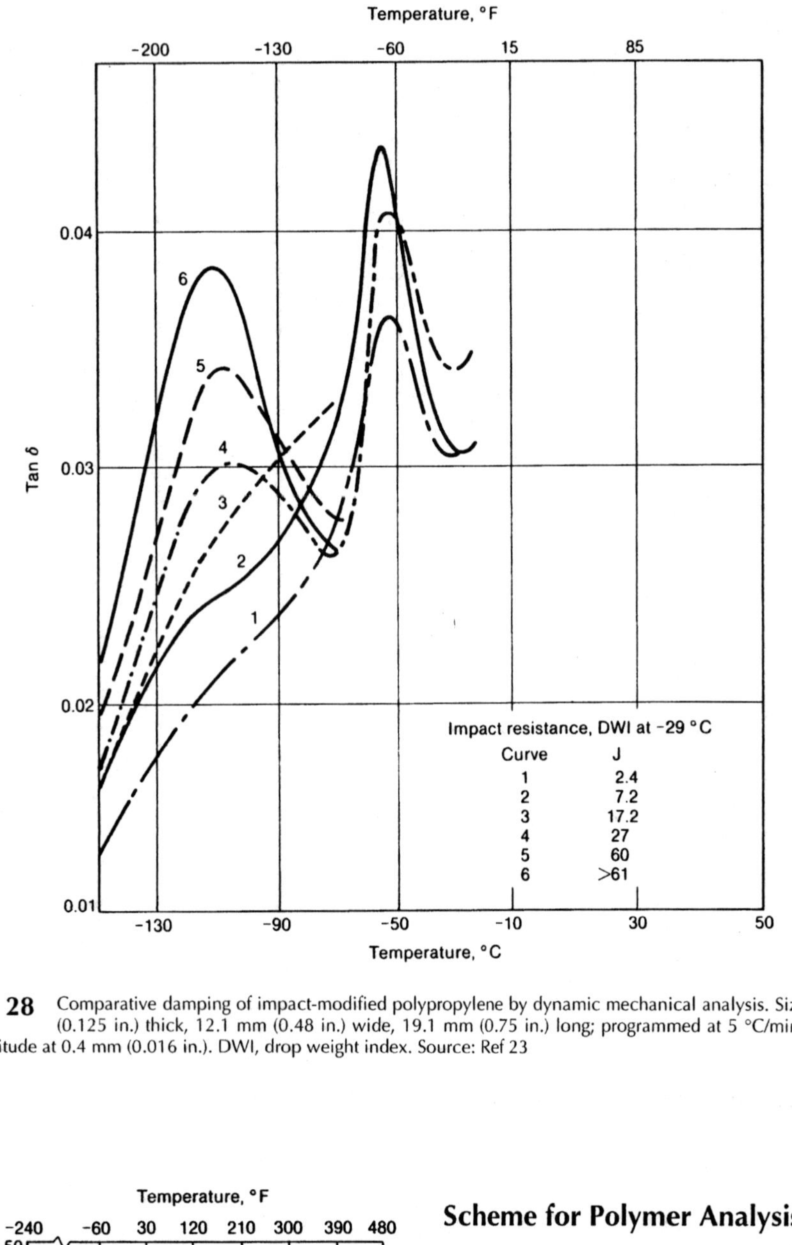

Fig. 28 Comparative damping of impact-modified polypropylene by dynamic mechanical analysis. Size, 3.18 mm (0.125 in.) thick, 12.1 mm (0.48 in.) wide, 19.1 mm (0.75 in.) long; programmed at 5 °C/min (9 °F/min), amplitude at 0.4 mm (0.016 in.). DWI, drop weight index. Source: Ref 23

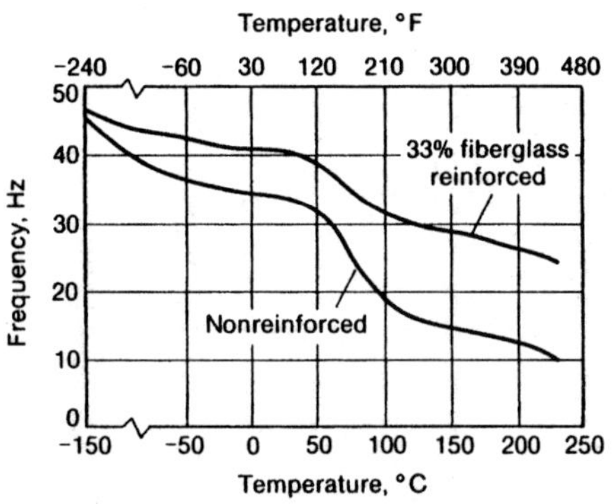

Fig. 29 Comparative modulus of nylon 6/6 measured by dynamic mechanical analysis. Size, 1.3 mm (0.05 in.) thick, 15.5 mm (0.6 in.) wide, 6.5 mm (0.25 in.) long; programmed at 5 °C/min (9 °F/min)

Scheme for Polymer Analysis

A minimal scheme for polymer analysis and characterization is set forth here to assist the design engineer. While a selection of tests can be based on cost, there are analyses that are needed for specific information and do not offer the engineer a choice, such as sequence distribution by NMR spectroscopy. The initial screening should include determination of the chemical composition by IR spectroscopy; determination of the MWD by GPC or of an indirect MW by dilute solution viscosity; determination of thermal history, T_g, T_m, crystallization temperature, and decomposition temperature by thermal analysis; study of solubility in selected solvents; and investigation of stress-strain characteristics, such as elongation, modulus, or stiffness.

A secondary round of tests should include study of polymer percent crystallinity by XRD, IR spectroscopy, or DSC; determination of the bulk viscosity from a flow curve; and determination of dynamic mechanical properties such as elasticity, ultimate strength, impact toughness, fatigue, and HDTs using ASTM procedures. The polymer structure should be determined by IR and/or NMR spectroscopy.

A third round of testing related to specific applications is the next step in characterizing a polymer or polymer system. Environmental stability and stress cracking should be checked, either in outdoor tests or in simulated performance tests. Swelling and solubility of the polymer in specific fluids related to performance conditions should be examined. Dimensional stability, coefficient of linear or volume expansion, and creep behavior should be evaluated by TMA or by an industry-accepted method. Finally, flammability and other regulatory specifications of the engineering plastics need to be established.

Procedure for Analyzing Milligram Quantities of Polymer Sample

This is a thermal-analytical scheme that can be expanded to include IR and NMR spectroscopy methods, which can also deal with milligram quantities of polymeric materials:

- Weigh a sample of 10 to 30 mg (0.15 to 0.45 gr) in a DSC pan and examine it at 100× magnification with a reflecting polarizing microscope. Note the color and morphology of the polymer.
- Evaluate the sample, in nitrogen, by DSC (or TMA) from –100 to 120 °C (–150 to 250 °F). Determine the T_g and any other transitions associated with the thermal history of the sample, such as moisture evolution or stress-relaxation exotherms.
- Reweigh the sample, and reexamine it under the microscope. Note any changes in polymer form, and determine the moisture content.
- Examine the sample in nitrogen in the DSC from 80 to 350 °C (175 to 660 °F). Determine melting range, processing temperatures, and exothermic cross-linking temperatures and heats. Calculate degree of cure by comparison to heats associated with complete cure.
- Program-cool DSC from 300 to –100 °C (570 to –150 °F) in nitrogen. Observe recrystallization heat and temperature for semicrystalline or crystalline polymers. Note T_g on cooling.
- Heat the DSC sample in nitrogen from –100 to 300 °C (–150 to 570 °F), and deter-

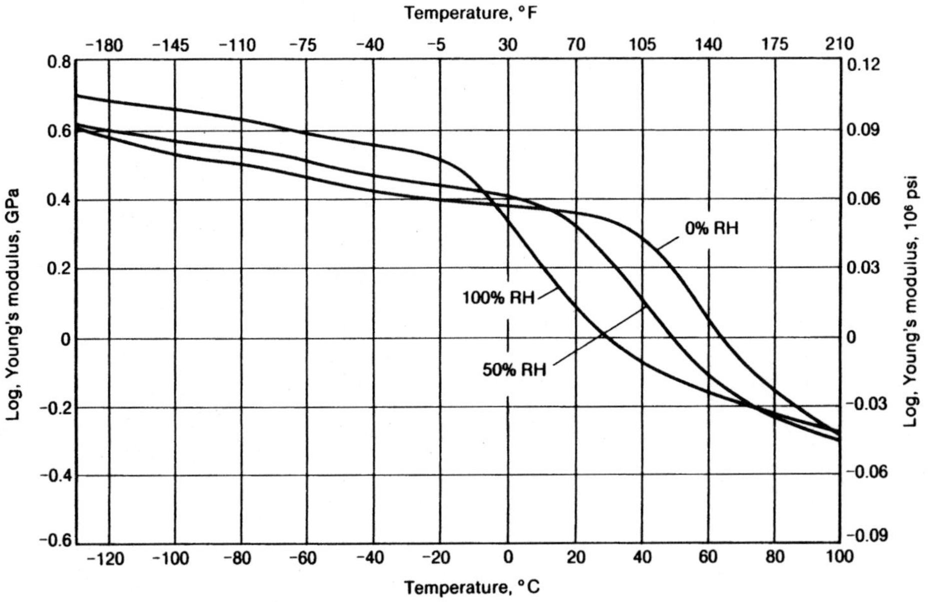

Fig. 30 Effects of moisture on nylon 6/6 measured by dynamic mechanical analysis. Size, 3 mm (0.12 in.) thick, 13 mm (0.5 in.) wide, 19 mm (0.75 in.) long; programmed at 5 °C/min (9 °F/min), in nitrogen. RH, relative humidity. Source: Ref 30

mine the process free-melt temperature and T_g.

- Transfer the DSC sample, after cooling to room temperature, to the TGA platinum boat. Examine the sample, in nitrogen, from 300 to 950 °C (570 to 1740 °F) to determine the thermal degradation characteristics. Examine the effluents from the TGA, trapped in a liquid nitrogen cold finger, by FTIR spectroscopy.

The following properties have been determined by this thermal-analytical exercise: T_g, T_m, heat of fusion and polymerization, cure or polymerization temperature, moisture evolution temperature and amount, polymer morphology and the effect of temperature, and effects of thermal history and processing. Finally, the TGA thermal degradation spectra, with accompanying IR curves, identify the functional chemical groups in the polymer. Using this scheme, a comparison of failed and virgin plastic parts can lead to a quality-assurance test and a reason for a performance failure.

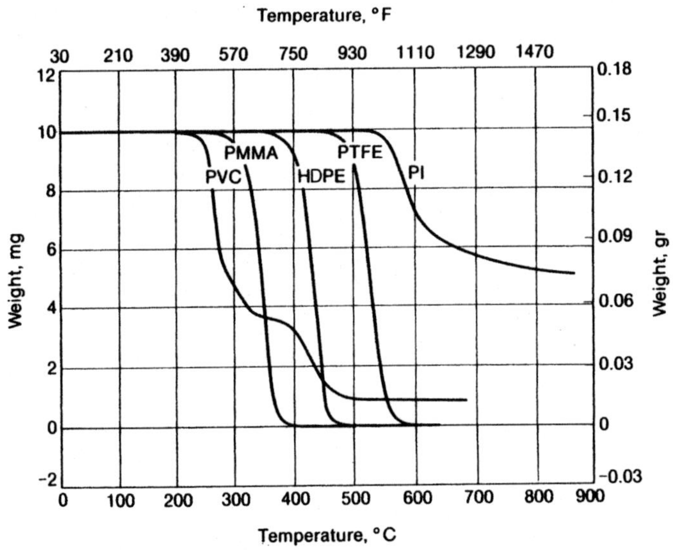

Fig. 31 Relative thermal stability of polymers by thermogravimetric analysis; 10 mg (0.15 gr) at 5 °C/min (9 °F/min), in nitrogen; HDPE, high-density polyethylene

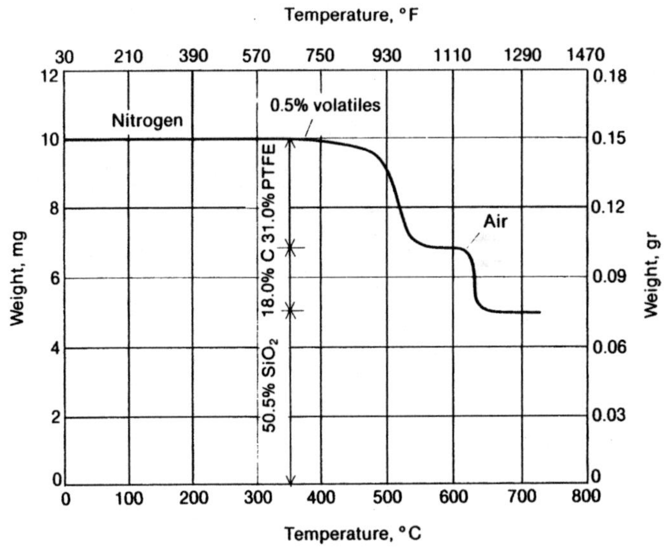

Fig. 32 Thermogravimetric analysis of silica- and carbon-filled PTFE; 10 mg (0.15 gr) at 5 °C/min (9 °F/min). Source: Ref 23

Table 5 Polymer thermal and oxidative properties

Polymer	T_g (softens)		T_m (melts)		T_p (pyrolysis)		T_c (combustion)		ΔH (change in enthalpy)		Limiting oxygen index
	°C	°F	°C	°F	°C	°F	°C	°F	kJ/g	10^3 Btu/lb	
Nylon 6	50	120	215	420	431	810	450	840	39	16.8	20.8
Nylon 6/6	50	120	265	510	403	755	530	990	32	13.8	20.8
Polyester	85	185	255	490	433	810	480	900	24	10.3	20.5
Acrylic	100	212	>220	>430	290	555	>250	>480	32	13.8	18.2
PP	−20	−4	165	330	469	875	550	1020	44	18.9	18.6
Modacrylic	<80	<175	>240	>465	273	525	690	1275	...	...	29.5
PVC	<80	<175	>180	>355	>180	>355	450	840	21	9.0	38
Polyvinylidene chloride	−17	1	195	385	>220	>430	535	995	11	4.7	60
PTFE	126	260	>327	>620	400	750	560	1040	4	1.7	95
Aramid honeycomb core	275	525	375	705	410	770	>500	>930	30	12.9	29.4
Aramid	340	645	560	1040	>590	>1095	>550	>1020	...	...	29
Polybenzimidazole	>400	>750	...	...	>500	>930	>500	>930	...	...	41

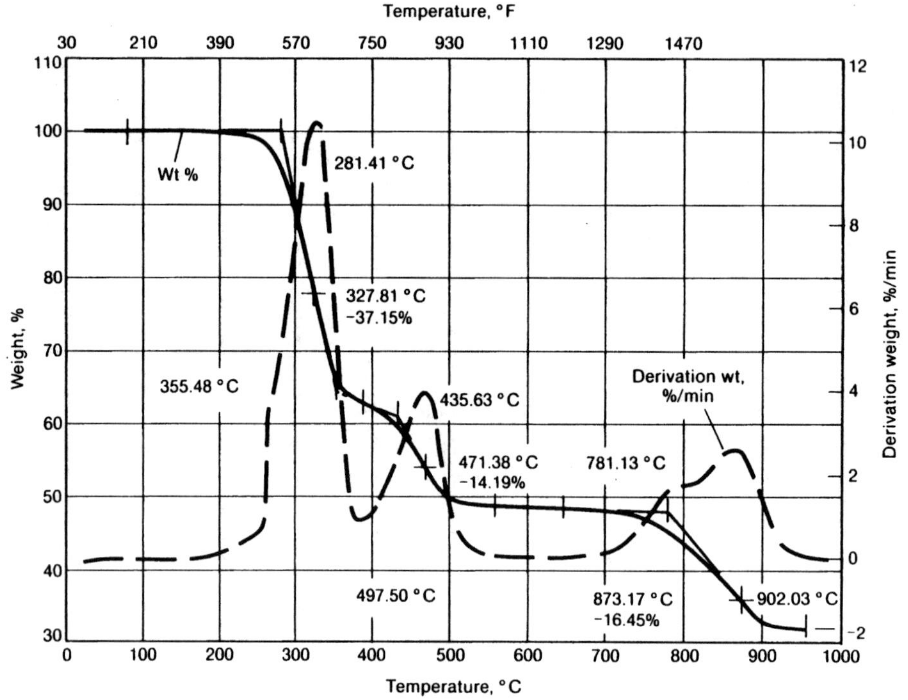

Fig. 33 Thermogravimetric analysis of polyvinyl chloride, 21.41 mg (0.33 gr), 20 °C/min (36 °F/min), to 950 °C (1740 °F), in nitrogen

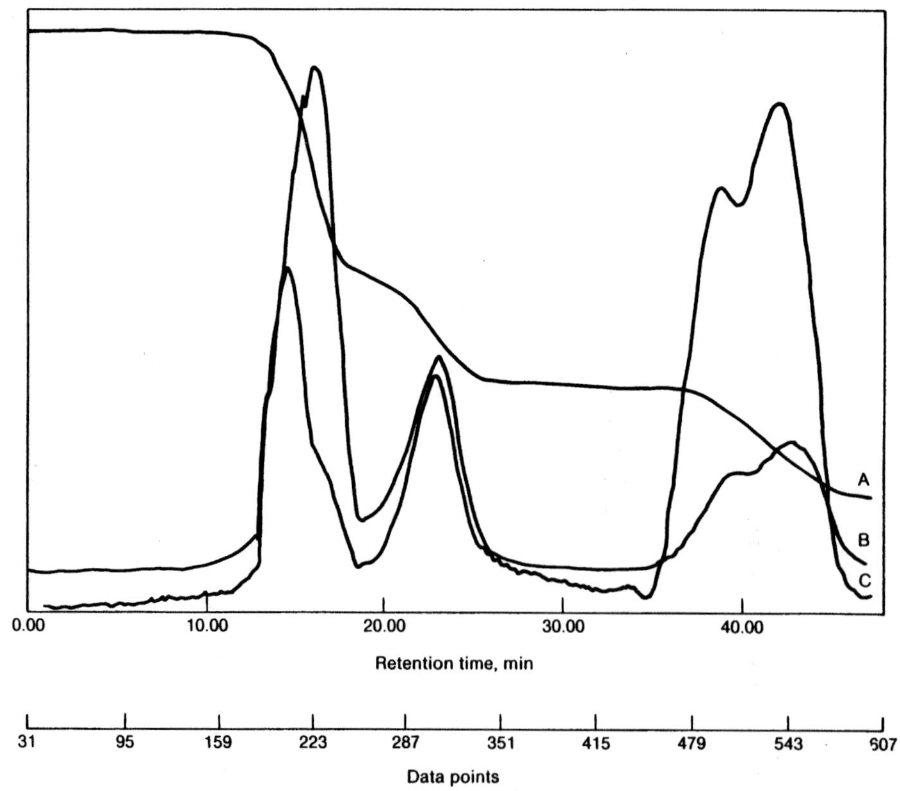

Fig. 34 Thermogravimetric analysis-Fourier transform infrared spectroscopy of polyvinyl chloride

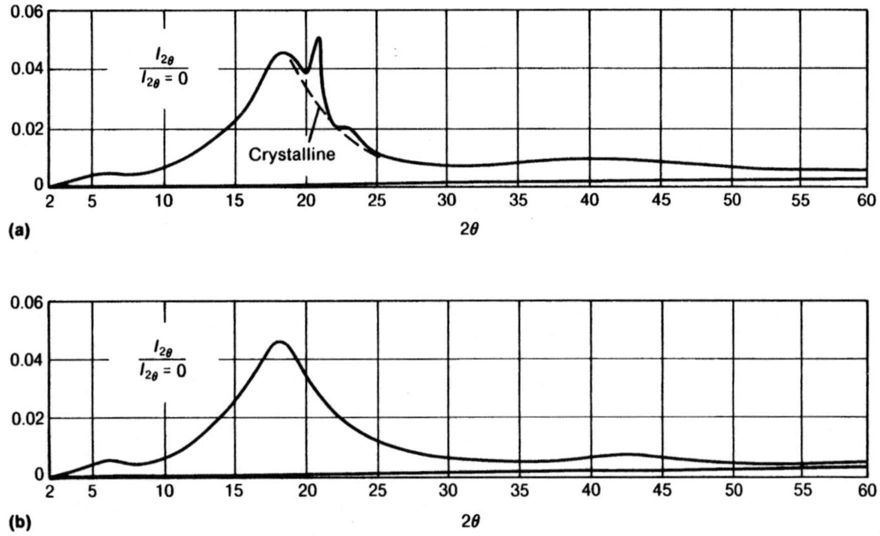

Fig. 35 X-ray diffraction curve of unoriented polyethylene. (a) At 100 °C (212 °F). (b) At 120 °C (250 °F)

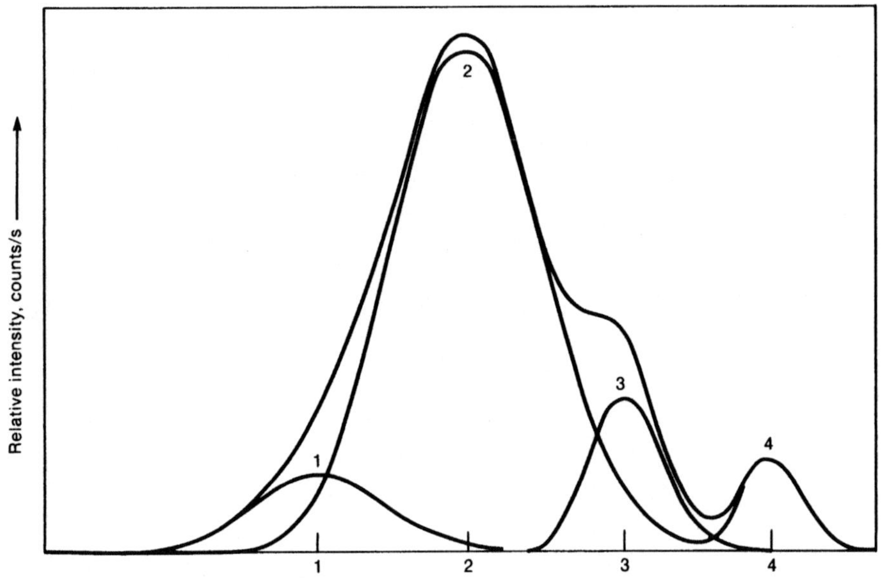

	1	2	3	4
Peak diffraction angle (2θ)	24.4°	19.2°	13.1°	7.7°
Half-band width, deg, (2θ)	5.4	6.4	3.2	3.1
Relative area, %	10	72	12	6

Fig. 36 X-ray diffraction curve of two-dimensional ordering in a polymer, short-range ordering. Source: Ref 38

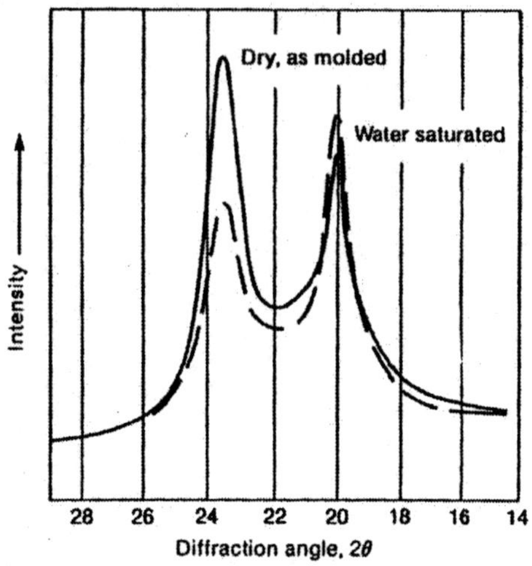

Fig. 37 Diffraction curves for nylon 6. Source: Ref 39

REFERENCES

1. J.E. Mark, A. Eisenberg, W. Graessley, L. Mandeikern, and J. Koenig, "Physical Properties of Polymers," paper presented to the American Chemical Society (Washington, D.C.), 1984
2. T. Smith, Physical Properties of Polymers—An Introductory Discussion, *Polym. Eng. Sci.,* Vol 13 (No. 3), 1973, p 161
3. J. Haslam and H.A Willis, *Identification and Analysis of Plastics,* Van Nostrand, 1967
4. F. Billmeyer, *Textbook of Polymer Science,* 2nd ed., Wiley-Interscience, 1981
5. R.J. Young, *Introduction to Polymers,* Chapman-Hall, 1981
6. W. Greive and A.T. Riga, *Instrumental Analysis of Plastics,* American Society for Testing and Materials, Nov 1986; also, Oct 1987
7. J.L. Koenig, Spectroscopic Characterization of Polymers, *Anal. Chem.,* Vol 59 (No. 19), 1987, p 1141A
8. D.O. Hummel, *Infrared Spectra of Polymers,* Vol 14, Interscience, 1966, p 193–194
9. W. Nyquist, "Infrared Spectra of Plastics and Resins," Dow Chemical Company, May 1961
10. L.W. Jelinski, NMR of Plastics, *Chem-tech,* Vol 16 (No. 3), 1986, p 186; also, Vol 16 (No. 5), 1986, p 312
11. Q.T. Pham, R. Petiand, and H. Waton, *Proton and Carbon NMR Spectra of Polymers,* John Wiley & Sons, 1985
12. F. Mikuis, V. Bartvska, and G. Maciel, Cross-Polarization Carbon-13 NMR with Magic-Angle Spinning, *Am. Lab.,* Nov 1979
13. E.A. Collins, J. Bares, and F.W. Billmeyer, *Experiments in Polymer Science,* John Wiley & Sons, 1973
14. W.R. Moore and B. Tidswell, Instrumentation of Molecular Weight Measurements, *Chem. Ind.,* Jan 1967
15. M. Ohama and T. Ozawa, Molecular Weight Determination of Polyamides by Vapor Pressure Osmometry, *J. Polym. Sci.,* A-2, Vol 4, 1966, p 817
16. A. Dondos and D. Patterson, Intrinsic Viscosity of Homopolymers and Graft Copolymers in Solvent Mixtures, *J. Polym. Sci.,* A-2, Vol 5, 1967, p 230
17. F. Peterson, Ed., Chromatography, *Lubrication,* Texaco Inc., Vol 65 (No. 2), 1979, p 24
18. W.P. Brennan, What is a T_g? A Review of the Scanning Calorimetry of the Glass Transition, *Perkin Elmer,* No. 7, March 1973
19. W.P. Brennan, Thermal Analysis: Useful Tool for Quality Control in a Complex Era, *Mod. Plast.,* Vol 56 (No. 1), 1979, p 98
20. P. Levy, Thermal Analysis—An Overview, *Am. Lab.,* Jan 1970
21. A.T. Riga, Inhibitor Selection for Vinyl Monomers by DSC, *Polym. Eng. Sci.,* Vol 18 (No. 12), 1976, p 836
22. A.T. Riga, Thermal Analysis as an Aid to Monomer Plant Design, *Polym. Eng. Sci.,* Vol 15 (No. 5), 1975, p 349
23. "Instrument Systems," E.I. Du Pont de Nemours & Company, Inc., 1987
24. A.T. Riga, Heat Distortion and Mechanical Properties of Polymers by Thermal-Mechanical Analysis, *Polym. Eng. Sci.,* Vol 14 (No. 11), 1974, p 764
25. A.P. Gray, Some Applications of the Model TMS-1 Precision Thermomechanical Analyzer, *Instrum. News,* Vol 20 (No. 1), 1974
26. C.C. Winding and G.D. Hiatt, *Polymeric Materials,* McGraw-Hill, 1961
27. "Resinkit," Society of Plastics Engineers, 1981
28. E. Galli, Properties Testing: Dynamical Mechanical Testing, *Plast. Compo.,* July/Aug 1984
29. R.P. Chartoff and B. Maxwell, Dynamic Mechanical Properties of Polymer Melts, *Polym. Eng. Sci.,* Vol 8 (No. 2), 1968, p 126
30. J. Chiu, Applications of Thermogravimetry to the Study of High Polymers, *Appl. Polym. Symp.,* (No. 2), 1966, p 25; also, Reprint RL-22, E.I. Du Pont de Nemours & Company, Inc.
31. D. Price, A. Horrocks, and M. Tunc, Textile Flammability, *Chem. Brit.,* Vol 23 (No. 3), 1987, p 235
32. R. Wieboldt, G. Adams, S. Lowry, and R. Rosenthal, *Analysis of a Vinyl Chloride Polymer by TGA-FTIR,* Nicolet, 1987
33. Walter and Reding, X-Ray Structure of Polyethylene, *J. Polym. Sci.,* Vol 21, 1956, p 561
34. S. Peiser, H. Rooksby, and A. Wilson, *X-Ray Diffraction by Polycrystalline Materials,* Reinhold, 1960
35. L. Alexander, *X-Ray Diffraction Methods in Polymer Science,* Wiley-Interscience, 1969
36. W.O. Statton, "Meaning of Crystallinity in Polymers," paper presented to the American Chemical Society Symposium (Phoenix), 1966
37. J. Turley, "X-Ray Diffraction of Polymers," Dow Chemical Company, Midland, MI, 1965
38. A.T. Riga, Distinguishing Amorphous Polymer Blends from Copolymers by Wide Angle X-Ray Diffraction, *Polym. Eng. Sci.,* Vol 18 (No. 15), 1978, p 1114
39. G. Campbell, The Effect of Water Sorption on Bulk Nylon 6 as Determined by X-Ray Crystallinity, *Polym. Lett.,* Vol 7, 1969, p 629

Characterization of Plastics in Failure Analysis*

THE ULTIMATE OBJECTIVE of a failure analysis is to ascertain the mode and the cause of the failure, regardless of the material from which the part was fabricated. The investigation is performed in generally the same manner, whether the failed component was produced from metal or plastic or a combination of these materials. Thus, the general steps required to conduct a comprehensive failure investigation are the same, and these are outlined in Fig. 1. In general, the failure analysis process is analogous to putting together a jigsaw puzzle. A failure analysis requires assembling bits of information into a coherent and accurate portrayal of how and why the part failed. Reaching the objectives of the plastic failure analysis, namely, the determination of the mode and cause of the failure, or expressed alternatively, evaluating how the part failed and why it failed, requires a scientific approach and a broad knowledge of polymeric materials.

Plastic components can fail via many different modes, including catastrophic mechanisms, such as brittle fracture, ductile overload, creep rupture, environmental stress cracking, molecular degradation, and fatigue. In the case of failure involving fracture, the determination of the failure mode involves identifying how the crack initiated and how it subsequently extended. This is usually ascertained using a number of visual-based techniques, such as stereomicroscopy, scanning electron microscopy (SEM), and the preparation of mounted cross sections. Noncatastrophic failure modes are also relevant, and these include discoloration, distortion, and contamination. Assessing the mode of the failure is often not as difficult as establishing why the part failed. Evaluating why the part failed usually requires analytical testing beyond the visual-based techniques. In many cases, a single cause cannot be identified, because multiple integrated factors may have contributed to the failure. All of the factors that affect the performance of a plastic component can be classified into one of four categories: material, design, processing, and service conditions (Ref 1). These factors do not act independently on the component but instead act in concert to determine the performance properties of a plastic component. This is represented graphically in Fig. 2 (Ref 1).

The principal differences between how failure analyses are performed on metal and plastic materials center on the techniques used to evaluate the composition and structure of the material. Unlike metals, polymers have a molecular structure that includes characteristics such as molecular weight, crystallinity, and orientation, and this has a significant impact on the properties of the molded article. Additionally, plastic resins usually contain additives, such as reinforcing fillers, plasticizers, colorants, antidegradants, and process aids. It is this combination of molecular structure and complex formulation that requires specialized testing (Ref 2). While the chemical composition of a failed metal component can often be evaluated using a single spectroscopic technique, a similar determination requires multiple analytical approaches for a part produced from a plastic resin.

This article reviews those analytical techniques most commonly used in plastic component failure analysis. The description of the techniques is not designed to be a comprehensive review and tutorial but instead is intended to make the reader familiar with the general principles and benefits of the methodologies. The descriptions of the analytical techniques are supplemented by a series of case studies. The technique descriptions refer to the case studies, and the two are written in a complimentary manner to illustrate the significance of the method. The case studies also include pertinent visual examination results and the corresponding images that aided in the characterization of the failures.

Fourier Transform Infrared Spectroscopy

Fourier transform infrared spectroscopy (FTIR) is a nondestructive microanalytical spectroscopic technique that involves the study of molecular vibrations (Ref 2). The analysis results provide principally qualitative, but also limited quantitative, information regarding the composition and state of the material evaluated. Fourier transform infrared spectroscopy uses infrared energy to produce vibrations within the molecular bonds that constitute the material evaluated. Vibrational states of varying energy levels exist in molecules. Transition from one vibrational state to another is related to absorption or emission of electromagnetic radiation (Ref 3). These vibrations occur at characteristic frequencies, revealing the structure of the sample. Fourier transform infrared spectroscopy produces a unique spectrum, which is comparable to the fingerprint of the material. It is the principal analytical technique used to qualitatively identify polymeric materials.

Method

Several different sampling techniques, all involving either transmission or reflection of the infrared energy, can be used to analyze the sample material. This allows the evaluation of materials in all forms, including hard solids, powders, liquids, and gases. Depending on the spectrometer and the corresponding accessories, most samples can be analyzed without significant preparation or alteration. In the analysis of polymeric materials, transmittance, reflectance, and attenuated total reflectance are the most common sampling techniques. Additionally, a microscope can be interfaced with the spectrometer to focus the infrared beam and allow the analysis of samples down to 10 μm. Regardless of the sampling technique, the beam of infrared energy is passed through or reflected off of the sample and directed to a detector. The obtained spectrum shows those frequencies that the material has absorbed and those that have been transmitted, as illustrated in Fig. 3. The spectrum can be interpreted manually or, more commonly, compared with voluminous library references with the aid of a computer.

*Adapted from the article by Jeffrey A. Jansen, "Characterization of Plastics in Failure Analysis," in *Failure Analysis and Prevention,* Volume 11, *ASM Handbook,* ASM International, 2002, pages 437 to 459

Results

The results generated through FTIR analysis are referred to as an infrared spectrum. The spectrum graphically illustrates the relative intensity of the energy absorbed on the *y*-axis versus the frequency of the energy on the *x*-axis. The frequency of the energy can be represented directly in microns (μm) or, more popularly, as reciprocal centimeters (cm^{-1}) referred to as wavenumbers. The discrete spectral features present with a FTIR spectrum are known as absorption bands.

Uses of FTIR in Failure Analysis

Material Identification. Possibly the most important use of FTIR in failure analysis is the identification of the base polymer used to produce the sample. The determination of the composition of the failed component is an essential part of the investigation. Because different polymers have a wide variation in their physical, mechanical, chemical resistance, and aging properties, the use of the wrong resin can yield detrimental results in many applications. Fourier transform infrared spectroscopy is well suited for the identification of polymers having different molecular structures, and this is illustrated in Fig. 4. Confirming that the failed article was produced from the specified material is the primary consideration of the failure analyst in assessing the cause of the failure. Thus, FTIR is often the first analytical test performed during a plastic failure analysis. The use of FTIR in characterizing the composition of the plastic-resin base polymer is illustrated in examples 1, 4, 7, and 9 in this article. One area where FTIR is inadequate is in differentiating between polymers having similar molecular structures, such as the members of the nylon family, and polyethylene terephthalate and polybutylene terephthalate. In these cases, other techniques, such as differential scanning calorimetry, must be used to augment the FTIR results.

Aside from the determination of the base polymer, FTIR is used to characterize other formulation constituents. Fourier transform infrared spectroscopic analysis can provide information regarding the presence of additives and filler materials. Due to the nonlinearity of infrared absorptivity of different molecular bonds, it is not possible to accurately state minimum concentration detection limits. However, it is generally considered that materials present within a compounded plastic resin at concentrations below 1% may be below the detection limits of the spectrometer. Given this restriction, it is likely that most major formulation additives, such as plasticizers, can be characterized, while low-level additives, including antioxidants, may go undetected. Given that FTIR is principally used for the analysis of organic materials, its use

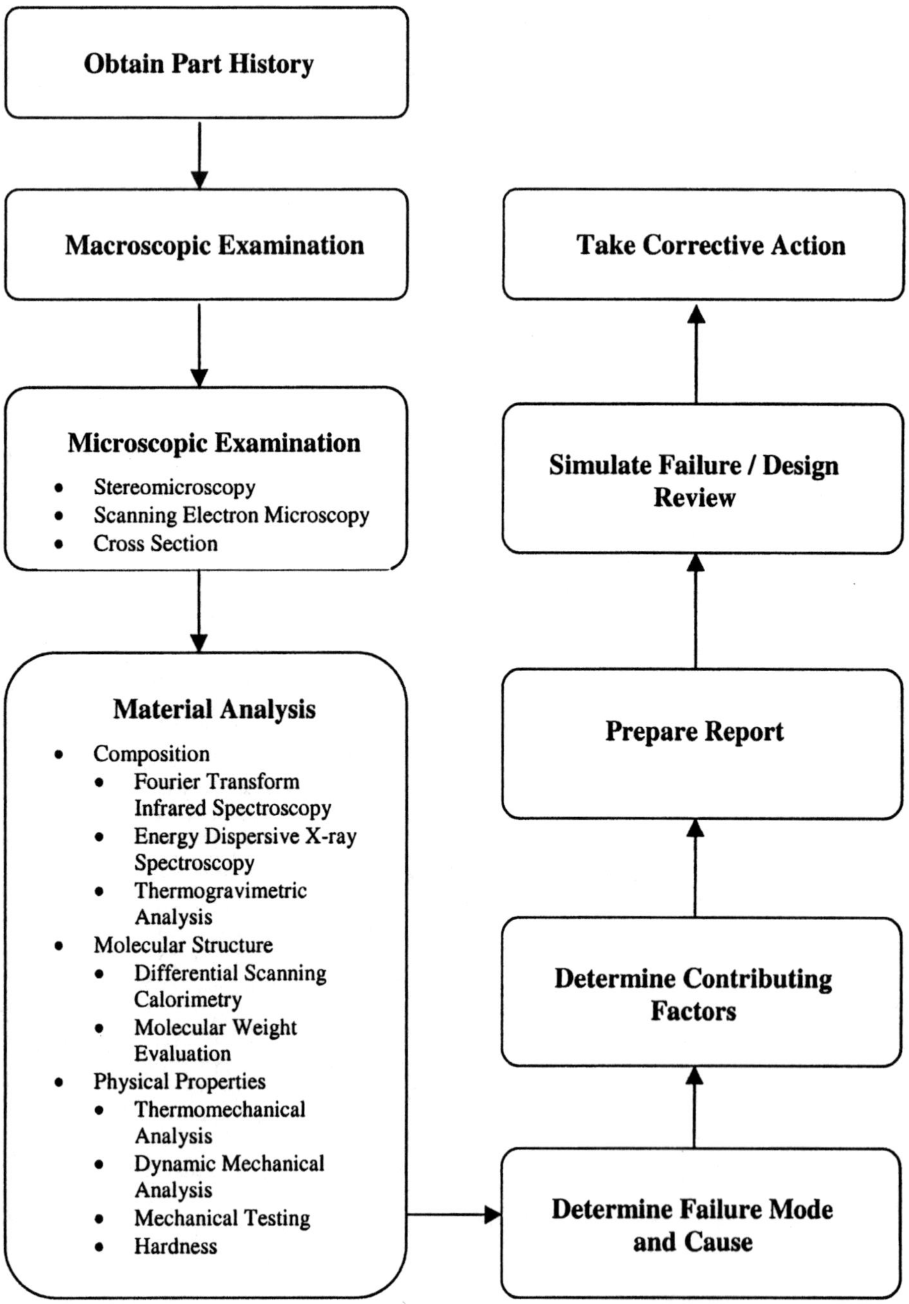

Fig. 1 Steps for performing failure analysis. The steps are the same regardless of the material.

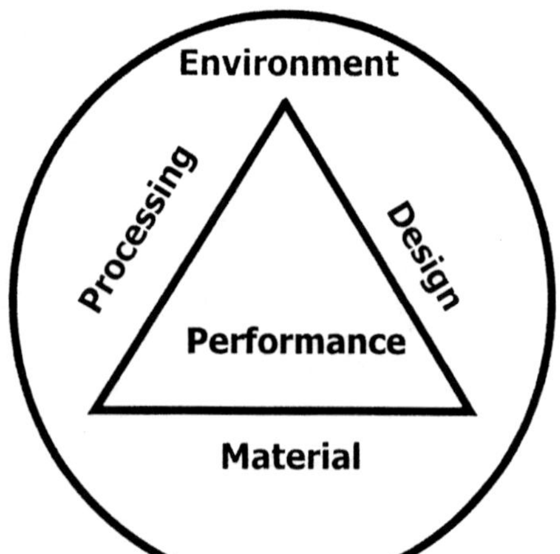

Fig. 2 Graphical representation of the four factors influencing plastic part performance

in the evaluation of inorganic filler materials is somewhat limited. However, some commonly used fillers, such as calcium carbonate, barium sulfate, and talc, produce unique, identifiable absorption spectra. Example 6 in this article shows the analysis of plastic-resin formulation constituents.

Contamination. Similar to its ability to identify the plastic formulation constituents, FTIR is extremely useful in the determination of contaminant materials within the failed part material. While contamination is never an intended part of a plastic compound, its presence certainly can have a tremendous impact on the properties of the molded component. Through the electronic manipulation of the obtained FTIR results, including spectral subtraction, extraneous absorption bands not attributed to the base resin can be used to characterize contaminant materials. Fourier transform infrared spectroscopy is useful in the identification of contaminant material, whether it is mixed homogeneously into the resin or present as a discrete inclusion. The role of FTIR in the identification of contaminants is discussed in examples 3 and 8 in this article.

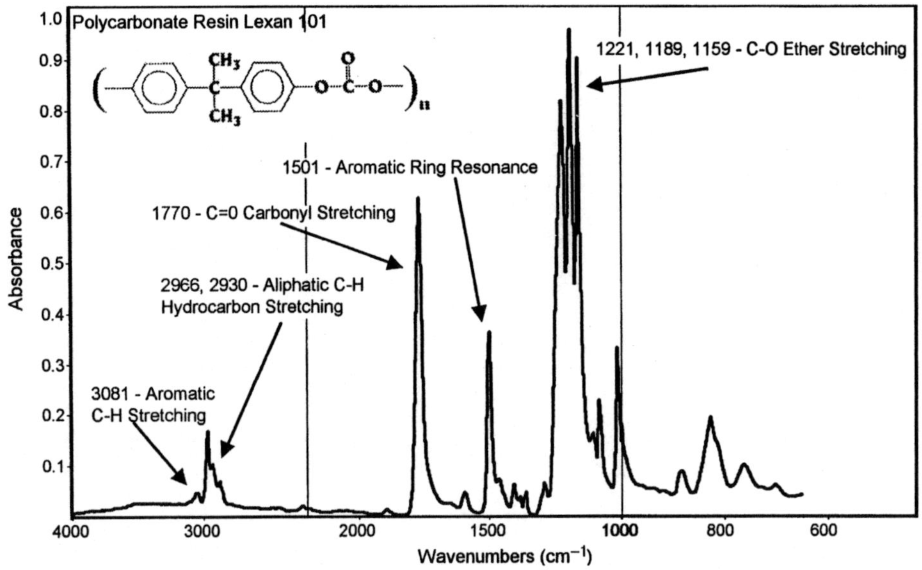

Fig. 3 A typical Fourier transform infrared spectroscopy spectrum illustrating the correlation between structure and absorption bands. Lexan, G.E. Plastics

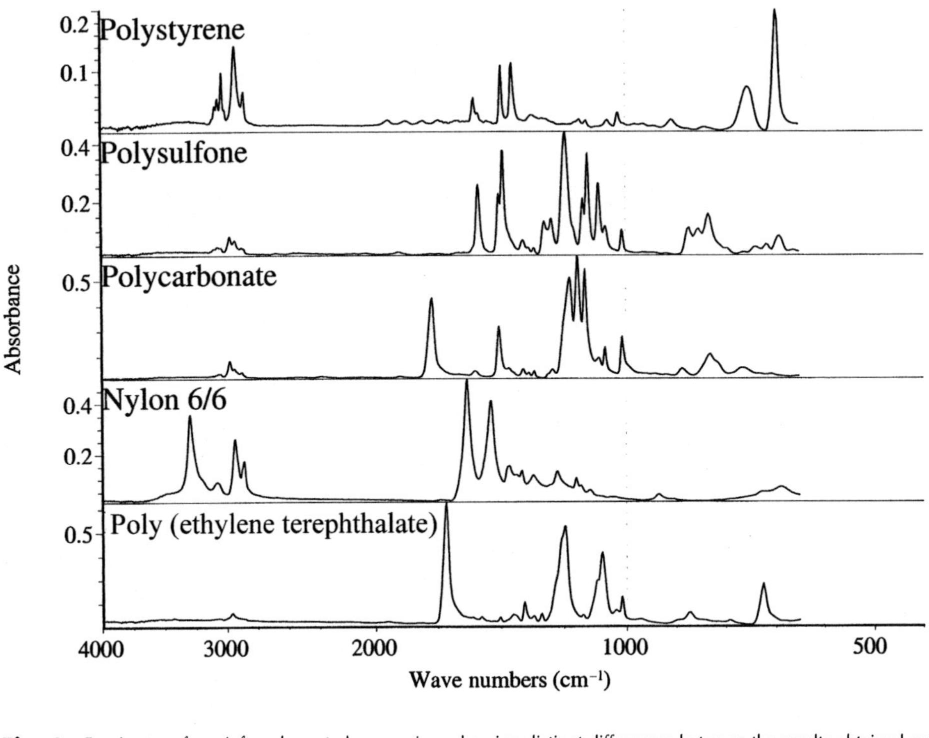

Fig. 4 Fourier transform infrared spectral comparison showing distinct differences between the results obtained on various plastic materials

Degradation. Fourier transform infrared spectroscopy is a valuable tool in assessing a failed component material for degradation, such as oxidation and hydrolysis. Molecular degradation, often involving molecular weight reduction, has a significant detrimental impact on the mechanical and physical properties of a plastic material. This degradation can result from several stages in the product life, including resin compounding, molding, and service. As a polymeric material is degraded on a molecular level, the bonds comprising the material are altered. Fourier transform infrared spectroscopy detects these changes in the molecular structure. While FTIR cannot readily quantify the level of degradation, it is useful in assessing whether the material has been degraded and determining the mechanism of the degradation. Specifically, several spectral bands and the corresponding molecular structure can be ascertained, including carbonyl band formation representing oxidation, vinylidene group formation as an indication of thermal oxidation, vinyl, vinylene functionality for photooxidation, and hydroxyl group formation indicating hydrolysis (Ref 4). Case studies showing the effectiveness of FTIR in assessing molecular degradation are presented in examples 1, 13, and 15 in this article.

Chemical Contact. Parallel to the application of FTIR in addressing polymeric degradation, the technique is also useful in evaluating the failed sample material for chemical contact. Plastic materials can be affected in several ways through contact with chemical agents. Depending on the polymer/chemical combination, solvation, plasticization, chemical attack, or environmental stress cracking can occur. In the case of property alteration through solvation or plasticization, FTIR can be helpful in identifying the absorbed chemicals. Because these chemicals are present within the failed plastic material, the likelihood of distinguishing the agents is high. Based on the observed spectral changes, mechanisms and chemical agents producing chemical attack, including nitration, sulfonation, hydrolysis, and aminolysis, can be detected (Ref 4). Environmental stress cracking, the synergistic effect of tensile stress while in contact with a chemical agent, is one of the leading causes of plastic failure. The chemical agent responsible for the cracking may be identified using FTIR.

However, given that such materials are often volatile organic solvents, the chemical may not be present within the sample at the time of the analysis. Examples 2, 9, and 14 in this article illustrate the identification of chemicals that had been in contact with a failed plastic component.

Differential Scanning Calorimetry

Differential scanning calorimetry (DSC) is a thermoanalytical technique in which heat flow is measured as a function of temperature and/or time. The obtained measurements provide quantitative and qualitative information regarding physical and chemical changes involving exothermic and endothermic processes, or changes in the heat capacity in the sample material (Ref 2). Differential scanning calorimetry monitors the difference in heat flow between a sample and a reference as the material is heated or cooled (Ref 5). The technique is used to evaluate thermal transitions within a material. Such transitions include melting, evaporation, crystallization, solidification, cross linking, chemical reactions, and decomposition. A typical DSC result is presented in Fig. 5. Differential scanning calorimetry uses the temperature difference between a sample material and a reference as the raw data. In the application, the instrumentation converts the temperature difference into a measurement of the energy per unit mass associated with the transition that caused the temperature change. Any transition in a material that involves a change in the heat content of the material can be evaluated by DSC (Ref 5), the limitation being that commercially available equipment may not be able to detect transitions within materials that are present at concentrations below 5% (Ref 4).

Method

Sample preparation for DSC analysis includes placing the specimen within a metal pan. The pan can be open, crimped, or sealed hermetically, depending on the experiment. A reference, either in the form of an empty pan of the same type or an inert material having the same weight as the sample, is used. The most commonly used metal pan material is aluminum; however, pans made of copper and gold are used for special applications. The sample and reference pans rest on thermoelectric disc platforms, with thermocouples used to measure the differential heat flow (Ref 5). Specimen size typically ranges between 1 and 10 mg, although this can vary depending on the nature of the sample and the experiment. The normal operating temperature range for DSC testing is –180 to 700 °C (–290 to 1290 °F), with a standard heating rate of 10 °C/min (18 °F/min). A dynamic purge gas is used to flush the sample chamber. Nitrogen is the most commonly used purge gas, but helium, argon, air, and oxygen can also be used for specific purposes. Often, two consecutive heating runs are performed to evaluate a sample. A controlled cooling run is performed after the initial analysis in order to eliminate the heat history of the sample. The first heating run assesses the sample in the as-molded condition, while the second run evaluates the inherent properties of the material.

Results

The plotted results obtained during a DSC analysis are referred to as a thermogram. The thermogram shows the heat flow in energy units or energy per mass units on the *y*-axis as a function of either temperature or time on the *x*-axis. The transitions that the sample material undergoes appear as exothermic and endothermic changes in the heat flow. Endothermic transitions require heat to proceed, while exothermic transitions give off heat.

Uses of DSC in Failure Analysis

Melting Point and Crystallinity. The primary use of DSC in polymer analysis is the detection and quantification of the crystalline melting process. Because the crystalline state of a polymeric material is greatly affected by properties including stereoregularity of the chain and the molecular weight distribution as well as by processing and subsequent environmental exposure, this property is of considerable importance (Ref 5). The melting point (T_m) of a semicrystalline polymer is measured as the peak of the melting endotherm. A composite thermogram showing the melting transitions of several common plastic materials is presented in Fig. 6. The T_m is used as a means of identification, particularly when other techniques, such as FTIR, cannot distinguish between materials having similar structures. This can be useful in identifying both the main resin and any contaminant materials. The material identification aspects of DSC are illustrated in examples 4, 5, 7, 8, 10–12, and 15 in this article.

The heat of fusion represents the energy required to melt the material and is calculated as the area under the melting endotherm. The level of crystallinity is determined by comparing the actual as-molded heat of fusion with that of a 100% crystalline sample. The level of crystallinity that a material has reached during the molding process can be practically assessed by comparing the heat of fusion obtained during an initial analysis of the sample with the results generated during the second run, after slow cooling. The level of crystallinity is important, because it impacts the mechanical, physical, and chemical resistance properties of the molded article. In general, rapid or quench cooling results in a lower crystalline state. This is the result of the formation of frozen-in amorphous regions within the preferentially crystalline structure. Examples 11 and 12 in this article show applications involving DSC as a means of assessing crystallinity.

Recrystallization, or the solidification of the polymer, is represented by the corresponding exothermic transition as the sample cools. The recrystallization temperature (T_c) is taken as the peak of the exotherm, and the heat of recrystal-

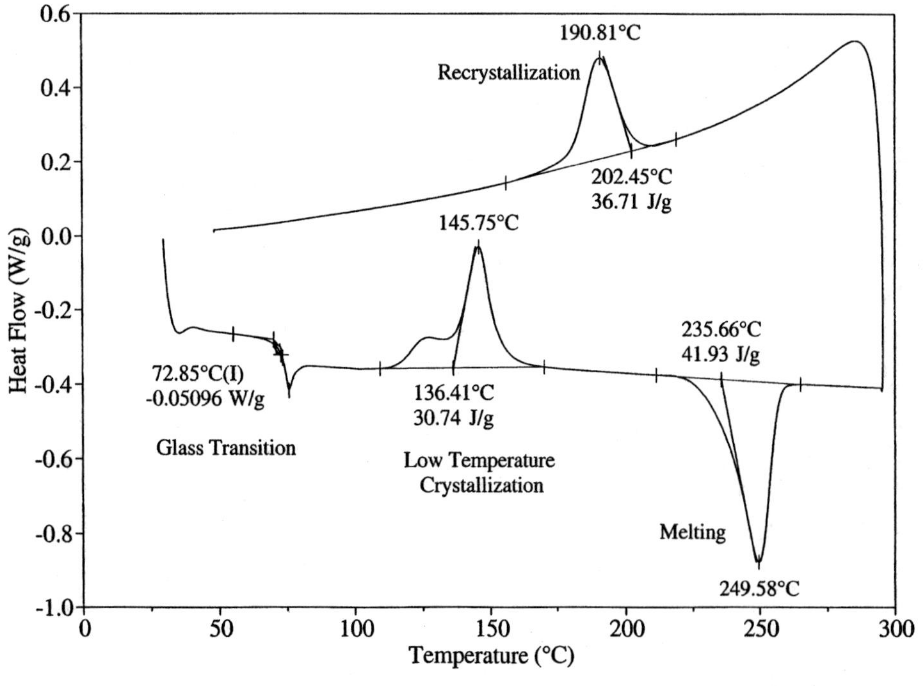

Fig. 5 Differential scanning calorimetry thermogram showing various transitions associated with polymeric materials. The (I) indicates that the numerical temperature was determined as the inflection point on the curve.

lization is the area under the exotherm. Some slow-crystallizing materials, such as polyethylene terephthalate and polyphthalamide, undergo low-temperature crystallization, representing the spontaneous rearrangement of amorphous segments within the polymer structure into a more orderly crystalline structure. Such exothermic transitions indicate that the as-molded material had been cooled relatively rapidly. Example 9 in this article shows how low-temperature crystallization was detected via DSC.

Glass Transition in Amorphous Plastics. Polymers that do not crystallize and semicrystalline materials having a significant level of amorphous segments undergo a phase change referred to as a glass transition. The glass transition represents the reversible change from/to a viscous or rubbery condition to/from a hard and relatively brittle one (Ref 6). The glass transition is observed as a change in the heat capacity of the material. The glass transition temperature (T_g) can be defined in several ways but is most often taken as the inflection point of the step transition. A composite thermogram showing the glass transitions of several common plastic materials is presented in Fig. 7. The T_g of an amorphous resin has an important impact on the mechanical properties of the molded article, because it represents softening of the material to the point that it loses load-bearing capabilities.

Aging, Degradation, and Thermal History. As noted by Sepe (Ref 5), "DSC techniques can be useful in detecting the chemical and morphological changes that accompany aging and degradation." Semicrystalline polymers may exhibit solid-state crystallization associated with aging that takes place at elevated temperatures. In some polymers, this may be evident as a second T_m at a reduced temperature. This second T_m represents the approximate temperature of the aging exposure. Other semicrystalline materials may show an increase in the heat of fusion and an increase in the T_m. The thermal aging of both the resin and the failed molded part is illustrated in example 10 in this article.

Amorphous resins exhibit changes in the glass transition as a result of aging. In particular, physical aging, which occurs through the progression toward thermodynamic equilibrium below the T_g, produces an apparent endothermic transition on completion of the glass transition.

Degradation and other nonreversible changes to the molecular structure of semicrystalline polymers can be detected as reduced values for the T_m, T_c, or heat of fusion. Instances of degradation detected by DSC are presented in examples 7, 14, and 15 in this article. Similarly, degradation in amorphous resins can be observed as a reduction in the T_g or in the magnitude of the corresponding change in heat capacity.

Further, the resistance of a polymer to oxidation can be evaluated via DSC by standard methods or experiments involving high-pressure oxygen or air exposure. Such evaluations usually measure the oxidative induction time or the temperature at which oxidation initiates under the experimental conditions. This can be used to compare two similar materials or to determine whether a plastic resin has undergone partial oxidation.

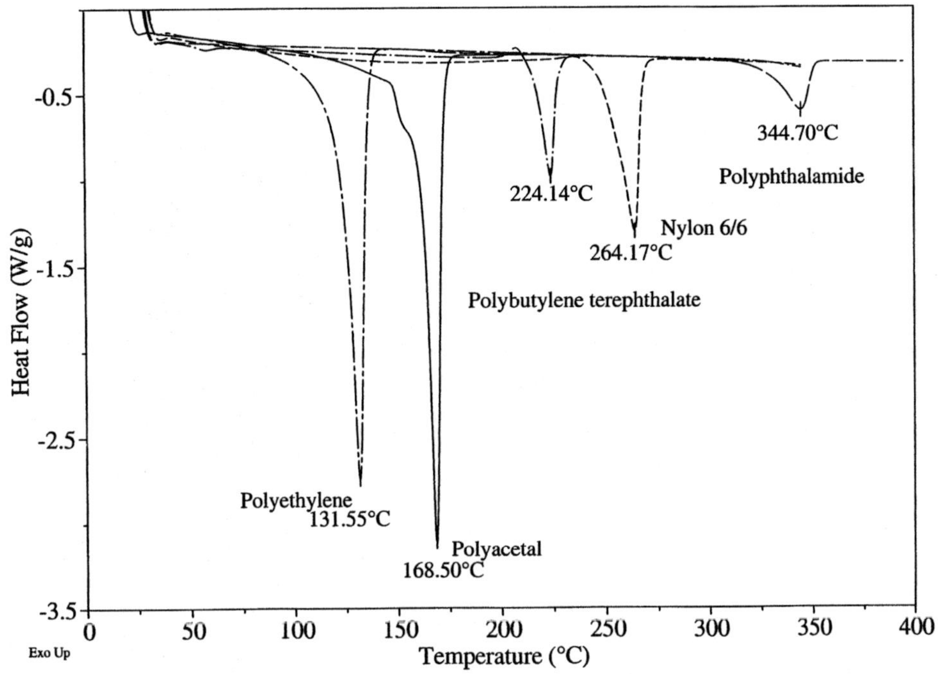

Fig. 6 Differential scanning calorimetry used to identify polymeric materials by determination of their melting point

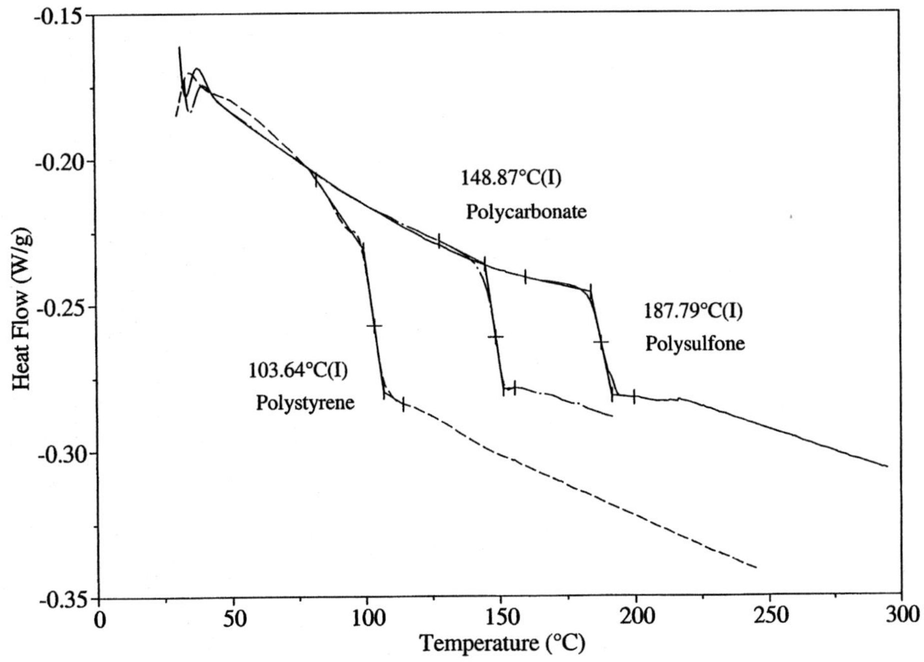

Fig. 7 Differential scanning calorimetry used to detect glass transitions within amorphous thermoplastic resins. The (I) indicates that the numerical temperature was determined as the inflection point on the curve.

Thermogravimetric Analysis

Thermogravimetric analysis (TGA) is a thermal analysis technique that measures the amount and rate of change in the weight of a material as a function of temperature or time in a controlled atmosphere. The weight of the evaluated material can decrease due to volatilization or decomposition or increase because of gas absorption or chemical reaction. Thermogravimetric analysis can provide valuable information regarding the composition and thermal stability of polymeric materials. The obtained data can include the volatiles content, inorganic filler content, carbon black content, the onset of thermal decomposition, and the volatility of additives such as antioxidants (Ref 4).

Method

Thermogravimetric analysis instruments consist of two primary components: a microbalance and a furnace. The sample is suspended from the balance while heated in conjunction with a thermal program. A ceramic or, more often, a platinum sample boat is used for the evaluation. As part of the TGA evaluation, the sample is usually heated from ambient room temperature to 1000 °C (1830 °F) in a dynamic gas purge of nitrogen, air, or a consecutive switch program. The composition of the purge gas can have a significant effect on the TGA results and, as such, must be properly controlled. The size of the sample evaluated usually ranges between 5 and 100 mg, with samples as large as 1000 mg possible. Minimal sample preparation is required for TGA experiments.

Results

The results obtained as part of TGA evaluation are known as a thermogram. The TGA thermogram illustrates the sample weight, usually in percent of original weight, on the *y*-axis as a function of time or, more commonly, temperature on the *x*-axis. The weight-change transitions are often highlighted by plotting the corresponding derivative on an alternate *y*-axis.

Uses of TGA in Failure Analysis

Composition. Thermogravimetric analysis is a key analytical technique used in the assessment of the composition of polymeric-based materials. The quantitative results obtained during a TGA evaluation directly complement the qualitative information produced by FTIR analysis. The relative loadings of various constituents within a plastic material, including polymers, plasticizers, additives, carbon black, mineral fillers, and glass reinforcement, can be assessed. The assessment of a plastic resin composition is illustrated in Fig. 8. These data are important as part of a failure analysis in order to determine if the component was produced from the correct material. The weight-loss profile of the material is evaluated, and, ideally, the TGA results obtained on the material exhibit distinct, separate weight-loss steps. These steps are measured and associated with transitions within the evaluated material. A thorough knowledge of the decomposition and chemical reactions is required to properly interpret the obtained results. In most situations, however, distinct weight-loss steps are not obtained, and, in these cases, the results are complemented by the corresponding derivative curve. Noncombustible material remaining at the conclusion of the TGA evaluation is often associated with inorganic fillers. Such residue is often further analyzed using energy-dispersive x-ray spectroscopy (EDS) in order to evaluate its composition. The use of TGA in characterizing plastic composition is presented in examples 8, 10, 12, and 15 in this article. Additionally, example 11 illustrates the quantification of an absorbed chemical within a failed plastic component.

Thermal Stability. Thermogravimetric analysis data can also be used to compare the thermal and oxidative stability of polymeric materials. The relative stability of polymeric materials can be evaluated by assessing the onset temperature of decomposition of the polymer. Quantitatively, these onset temperatures are not useful for comparing the long-term stability of fabricated products, because the materials are generally molten at the beginning of decomposition (Ref 5). However, a comparison of the obtained TGA thermograms can provide insight into possible degradation of the failed component material. Example 6 in this article illustrates a comparison of the thermal stability of two polymeric materials, while example 13 shows the effects of molecular degradation. Degradation experiments involving polymeric materials can also provide information regarding the kinetics of decomposition. Such studies provide information regarding the projected lifetime of the material. Such measurements, however, provide little information pertinent to a failure analysis.

Evolved Gas Analysis. Thermogravimetric analysis evaluations can also be performed whereby the evolved gaseous constituents are further analyzed using a hyphenated technique, such as FTIR or mass spectroscopy (MS). Such TGA-FTIR or TGA-MS experiments are referred to as evolved gas analysis.

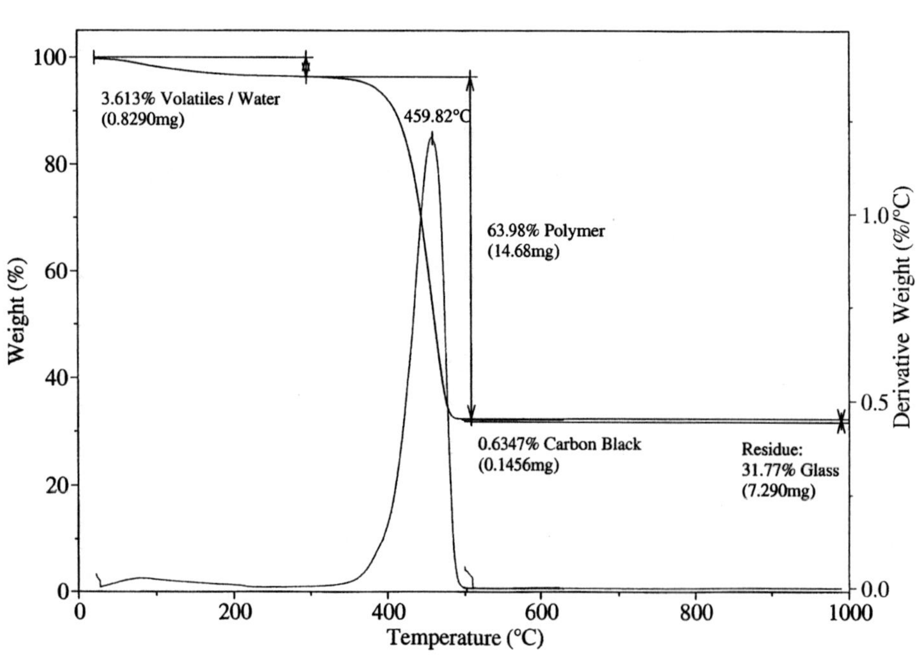

Fig. 8 Thermogravimetric analysis thermogram showing the weight-loss profile for a typical plastic resin

Thermomechanical Analysis

Thermomechanical analysis (TMA) is a thermal analysis method in which linear or volumetric dimensional changes are measured as a function of temperature, time, or force (Ref 2). Thermomechanical analysis is used to study the structure of a polymeric material by evaluating the implications of the material dimensional changes.

Method

Standard solid samples evaluated via TMA should be of regular shape, having two flat, parallel sides. Additionally, fiber and film samples can also be tested with minimal preparation. Experiments conducted to evaluate expansion and contraction of solid materials are performed on a quartz stage. The sample is placed on the stage, with a quartz probe resting on the opposing end. Thermomechanical analysis data can be acquired in compression modes, including expansion, penetration, dilatometry, rheometry, and flexure or tension mode (Ref 2). The analysis of film and fiber samples requires special fixturing, similar in principle to a universal mechanical tester. For all analysis configurations, the stage assembly is surrounded by a furnace and a cooling device. The normal operating range for TMA experiments is −180 to 1000 °C (−290 to 1830 °F), with a 5 °C/min (9 °F/min) heating rate commonly used. A compressive force is normally applied to the probe configuration throughout the evaluation for purposes of preload and stability.

Results

Plotted results generated through a TMA analysis, similar to the printed data obtained from all of the thermal analysis techniques, are referred to as a thermogram. The thermogram presents the sample dimension, either as length or a percentage of original length, on the *y*-axis, and as a function of temperature, time, or force on the *x*-axis. Temperature is the standard independent variable. Changes in the sample are presented as expansion or contraction.

Uses of TMA in Failure Analysis

Coefficient of Thermal Expansion. The coefficient of thermal expansion (CTE) is the change in the length of a material as a response to a change in temperature. The derivative of the slope of the line showing the dimensional changes with respect to temperature represents the CTE. This is a significant property when plastic materials are used under highly constrained conditions. This is commonly the case when plastic parts are used in conjunction with components produced from other materials, such as metals and ceramics. In general, the CTEs of polymeric materials are substantially greater than those of metals and ceramics. Thus, comparative testing of mating materials can produce data used to illustrate and even calculate the potential interference stresses on the materials in a multimaterial design. The evaluation of the CTEs of mating plastic and metal components is illustrated in examples 10 and 14 in this article.

Material Transitions. According to Sepe (Ref 5), "The CTE is an important property in itself; however, it is of particular value in polymers, because sudden changes in CTE can signal important transitions in the material structure." Within semicrystalline polymers, the T_g, signaling the conversion from a hard, brittle material to a rubbery condition, is accompanied by an increase in the CTE. A thermogram representing a typical semicrystalline resin is shown in Fig. 9. The physical properties of the material can be expected to be significantly different across this transition. Amorphous resins soften at the T_g, and because of this, samples undergo compression under the inherent load of the testing conditions. A thermogram showing the glass transition in an amorphous resin is shown in Fig. 10. The evaluation of the glass transition is presented in example 14 in this article. Thermomechanical analysis is generally accepted as a more accurate method for assessing the T_g of polymeric materials, relative to DSC. "By using the prescribed attachments and the appropriate force, TMA can be used to determine two commonly measured properties of plastic materials: the heat-deflection temperature and the Vicat softening temperature." (Ref 5).

Molded-In Stress. Internal molded-in stress is an important source of the total stress on a plastic component and is often sufficient to result in the failure of plastic materials. Such stresses are particularly important in amorphous resins, which are prone to environmental stress cracking. Molded-in stresses are commonly imparted through the forming process, especially injection molding and thermoforming. Molded-in stress is observed in amorphous resins as a marked expansion in the sample dimension at temperatures approaching the T_g, as illustrated in Fig. 11. This expansion is associated with rapid expansion as the internal stresses are relieved. This stress relief is due to molecular reorientation on attaining sufficient thermal freedom. In the absence of molded-in stress, the sample would compress due to the loss of load-bearing capabilities as the material undergoes glass transition.

Chemical Compatibility. The chemical compatibility of a plastic material with a particular chemical agent can be assessed using TMA. In particular, the volume swell of a polymeric material by a chemical can be tested. The sample material is constrained in a quartz vessel, and the chemical agent is added. Dilatometry is used to measure the volume expansion of the material over time.

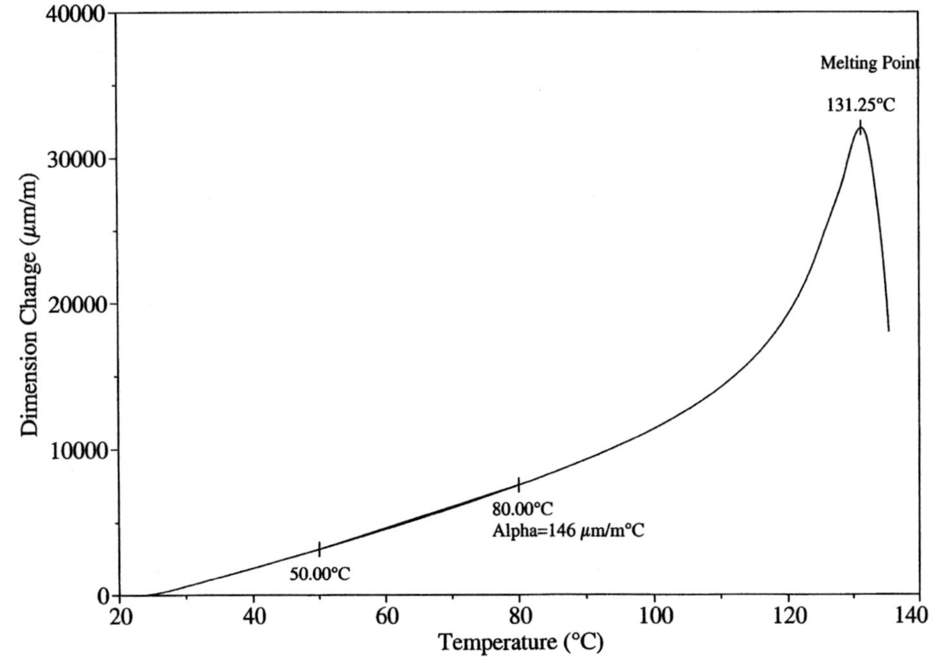

Fig. 9 Thermogravimetric analysis thermogram representing a typical semicrystalline plastic resin

Dynamic Mechanical Analysis

Dynamic mechanical analysis (DMA) is a thermoanalytical technique that assesses the viscoelastic properties of materials. Dynamical mechanical analysis evaluates the stiffness, as measured by modulus, as a function of temperature or time. Polymeric materials display both elastic and viscous behavior simultaneously, and the balance between the elastic recovery and viscous flow changes with temperature and time (Ref 5). Measurements can be made in several modes, including tension, shear, compression, torsion, and flexure. The results obtained as part of a DMA experiment provide the storage modulus, loss modulus, and the tangent of the phase-angle delta (tan delta). Dynamic mechanical analysis is not routinely used as a failure analysis technique, but it can provide valuable material information.

Method

Dynamic mechanical analysis experiments can be performed using one of several configurations. The analysis can be conducted to apply stress in tension, flexure, compression, shear, or torsion. The mode of the analysis determines which type of modulus is evaluated. The measurement of modulus across a temperature range is referred to as temperature sweep. Dynamic mechanical analysis offers an advantage over traditional tensile or flexural testing in that the obtained modulus is continuous over the temperature range of interest. In addition, special DMAs can also be conducted to evaluate creep through the application of constant stress or stress relaxation by using a constant strain. Dynamic mechanical analysis studies can be performed from –150 to 600 °C (–240 to 1110 °F), usually employing a 2 °C/min (4 °F/min) heating rate.

Results

The results obtained as part of a DMA evaluation are plotted to illustrate the elastic or storage modulus (E') and the viscous or loss modulus (E'') on the y-axes and as a function of temperature on the x-axis. Less frequently, time is used, depending on the type of experiment. Additionally, the tangent of the phase-angle delta (E''/E') is also calculated. A typical DMA thermogram is presented in Fig. 12.

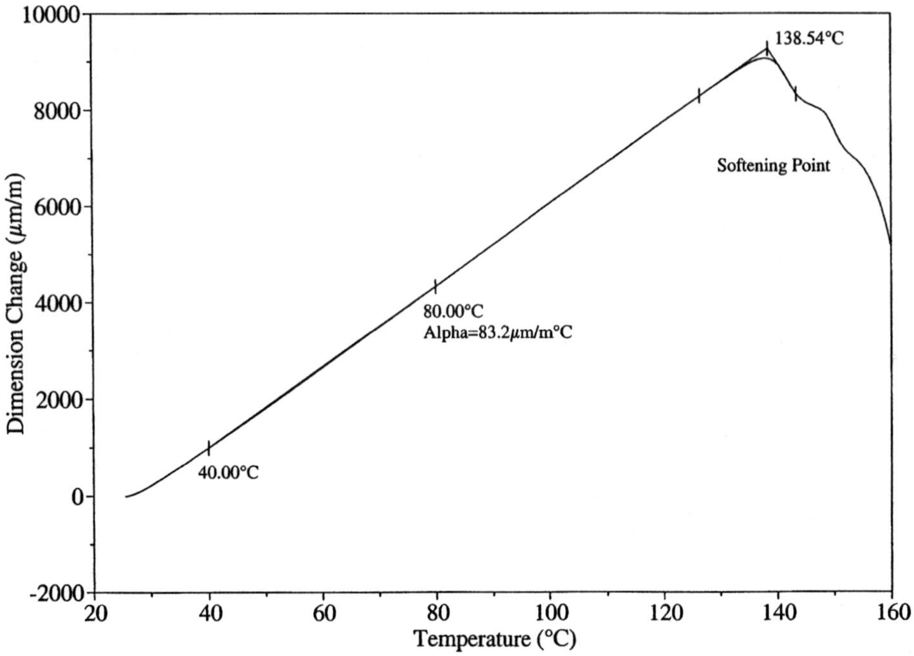

Fig. 10 Thermogravimetric analysis thermogram representing a typical amorphous plastic resin

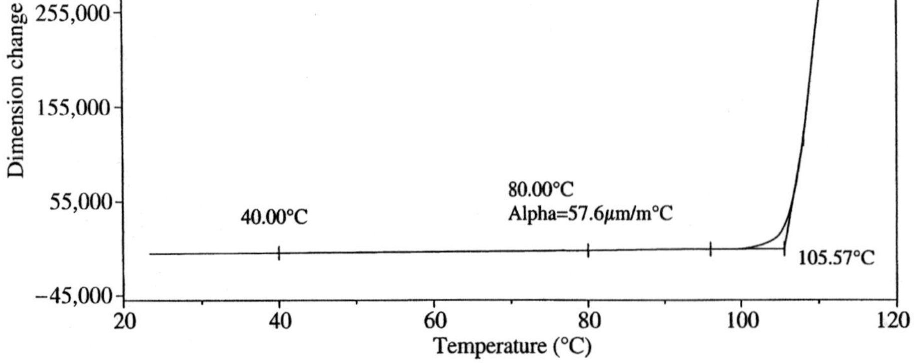

Fig. 11 Thermogravimetric analysis thermogram showing a high level of residual stress in an amorphous plastic resin

Uses of DMA in Failure Analysis

Temperature-Dependent Behavior. The temperature-dependent behavior of polymeric materials is one of the most important applications of DMA. In a standard temperature-sweep evaluation, the results show the storage modulus, loss modulus, and the tan delta as a function of temperature. The storage modulus indicates the ability of the material to accommodate stress over a temperature range. The loss modulus and tan delta provide data on temperatures where molecular changes produce property changes, such as the glass transition and other secondary transitions not detectable by other thermal analysis techniques. The superiority of DMA over DSC and TMA for assessing the glass transition is well documented (Ref 5). Secondary transitions of lesser magnitude are also important, because they can relate to material properties such as impact resistance. The ability of a plastic molded component to retain its properties over the service temperature range is essential and is well predicted by DMA.

Aging and Degradation. Changes in the mechanical properties of plastic resins that arise from molecular degradation or aging can be evaluated via DMA. Such changes can significantly alter the ability of the plastic material to withstand service stresses. While the cause and type of degradation cannot be determined, DMA can assess the magnitude of the changes. This can provide insight into potential failure causes.

Solid and Liquid Interactions. Sepe (Ref 5) notes that "DMA is sensitive to structural changes that can arise when a solid polymer absorbs a liquid material." This effect can arise from the absorption of water or organic-based solvents. Dynamic mechanical analysis experiments can assess changes in the physical properties of a plastic material that can result from such absorption, including loss of strength and stiffness. Example 11 in this article shows the changes in mechanical properties of a plastic resin associated with chemical absorption. The experiments can also evaluate the recovery after the removal or evaporation of the absorbed liquid.

Methods for Molecular Weight Assessment

The aspect of molecular structure, and specifically, molecular weight, makes polymeric materials unique among materials commonly used in engineering applications, including metals and ceramics. Molecular weight and molecular weight distribution are probably the most important properties for characterizing plastics

(Ref 4). These parameters have a significant impact on the entirety of characteristics of a plastic resin, including mechanical, physical, and chemical resistance properties. Molecular weight assessment can be used to evaluate the characteristics of a base resin or to assess the effects of compounding, molding, or service on the material. Changes in molecular weight can occur throughout the material life cycle and can significantly impact the performance of the molded part. Changes can result in molecular weight decreases through such mechanisms as chain scission, oxidation, and hydrolysis, or as increases through destructive cross linking. Because of this, the characterization of molecular weight is an important aspect of a thorough failure analysis.

Gel permeation chromatography (GPC), which is also referred to as size exclusion chromatography, is an analytical method used to characterize the molecular weight distribution of a polymeric material. Similar to all chromatographic techniques, GPC uses a packed column to segregate various constituents. One or multiple columns used in conjunction are used to separate the polymeric and oligomeric materials within the plastic resin. The polymer is further separated by molecular weight, producing essentially a histogram representing the molecular weight distribution of the material. From these results, a numerical average molecular weight can be calculated. Detectors, based on refractive index or ultraviolet detection, are used to identify the changes in molecular weight.

Gel permeation chromatography offers the advantage, unlike melt viscosity and solution viscosity techniques, of producing results that directly represent the actual molecular weight and molecular weight distribution of the plastic resin. Another advantage is that GPC requires a relatively small sample size, 30 to 120 μg, for a complete evaluation (Ref 7). The technique, however, is often complicated to perform, using sophisticated instrumentation, and difficult to interpret. Example 10 in this article reviews the use of GPC in a failure investigation.

Melt Flow Index. The melt flow index or melt flow rate (MFR) describes the viscosity of a plastic material in the molten state. The sample material is heated through the melting or softening point and extruded through a die having a standard-sized orifice. Different materials use various test conditions, including temperature and load. The method for determining the MFR is described in ASTM D 1238. Melt flow rate is the simplest technique for assessing the molecular weight of a plastic material and is inversely proportional to the molecular weight of the polymer (Ref 4). Melt flow rate is widely used to describe the molecular weight of a plastic resin and is commonly cited by suppliers on a material data sheet. The units used to indicate MFR are grams per 10 min. Examples 7, 11, 12, and 14 in this article describe the use of melt flow in assessing molecular weight in a failure analysis.

While MFR is relatively easily determined and is commonly used to describe molecular weight, the technique has several negative aspects. Melt flow rate does not measure the molecular weight distribution of the analyzed material and represents only the average molecular weight of the material. Because of this, the blending of polymers having different molecular weight distributions and average molecular weights can result in equal determinations between very different materials having distinct properties.

Solution Viscosity. The traditional approach for determining only the molecular weight of a resin, but not the molecular weight distribution, involves dissolving the polymer in a suitable solvent. However, the more structurally complicated macromolecules require the use of hostile solvents, tedious sample preparations, and costly time delays to obtain limited, single datapoint values. For example, the solution viscosity determination of polyvinyl chloride (PVC), according to ASTM D 1243, requires either a 1 or 4% concentration in cyclohexanone or dinitrobenzene, while polyamides, or nylons, require formic acid. Other engineering polymers might require tetrahydrofuran, dimethylformamide, dimethylsulfoxide, or other equally hostile solvents (Ref 8). The obtained solution viscosity values are only indications of molecular weight and do not reflect the absolute weight values (Ref 8). Example 9 in this article illustrates the use of solution viscosity in a failure investigation.

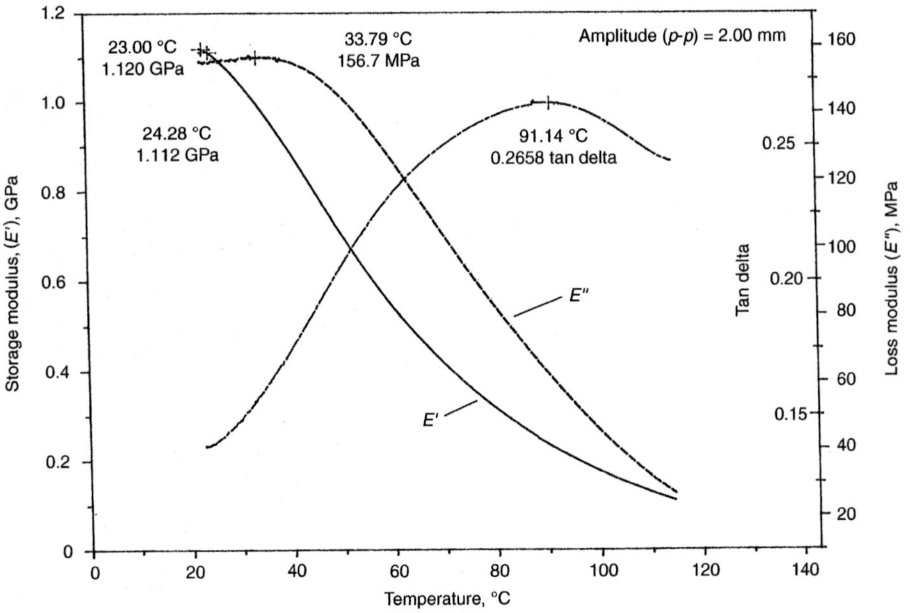

Fig. 12 Dynamic mechanical analysis thermogram showing the results obtained on a typical plastic resin. Tan delta is ratio of the loss modulus to the storage modulus.

Mechanical Testing

Because a wide range of mechanical tests are available to evaluate plastics and polymers, they initially do not seem to constitute a rational set. The totality of mechanical tests can be partitioned into logical groups in several distinct ways (Ref 9). One very useful way to classify the various mechanical test methods is to distinguish between tests that evaluate long-term properties, as opposed to those that evaluate short-term properties. Short-term tests include those that assess what are generally considered to be material properties. These include tensile tests, flexural tests, and the evaluation of impact resistance. Short-term tests, while generally easy to conduct and interpret, lack the ability to predict or assess the long-range performance properties of a material. As such, short-term tests are frequently listed on material data sheets.

Tests for Short-Term Properties. The most commonly performed mechanical test used to evaluate plastic material properties is the tensile test. This testing is performed on a dumbbell-shaped specimen and is outlined in ASTM D 638. Tensile testing provides data regarding the yield point in the form of yield strength and elongation at yield, the break properties as tensile strength at break and elongation at break, and the stiffness of the material as elastic modulus. Additionally, the tensile test generates information regarding the proportional limit.

A second short-term mechanical method that is used to evaluate plastic materials is flexural testing. Flexural testing simulates bending of the test sample. The test specimen is evaluated on a universal mechanical tester, and the tests can be performed using a three- or four-point bend configuration. Flexural testing provides two pieces of data: flexural modulus and break strength.

This testing is performed in accordance with ASTM D 790.

Several different types of tests are used to evaluate the impact properties of a plastic material. These include pendulum-based tests, such as Izod and Charpy tests, and falling weight tests, such as the dart penetration configuration. Unlike tensile and flexural testing, the results obtained from impact testing do not provide fundamental material properties. Instead, impact testing results are more performance-based. Given these different methodologies of assessing the impact properties of a plastic material, the falling weight or dart impact tests are generally considered to be superior to the pendulum configurations. Falling weight tests evaluate the sample material in two dimensions and not one, because the specimen is a plate rather than a beam. The data obtained during an instrumented falling weight impact test include the energy to maximum load, representing the energy required to initiate cracking, and the total energy to failure. The ratio of these two is an indication of the ductility of the material. Additionally, an examination of the test specimens is used to classify the failure mode from brittle to ductile. Falling weight impact testing is described as part of ASTM D 3029.

Tests for Long-Term Properties. Fatigue testing of plastic materials exposes the samples to cyclic stresses in an attempt to evaluate the samples in a manner that would produce fatigue failure while in service. Testing procedures are used to simulate flexural fatigue and tensile fatigue. The analyses are normally conducted in a way that does not excessively heat the specimen, thus altering the failure mode. The results of a fatigue test are shown in the form of a stress-number of cycles curve.

A second long-term test methodology assesses the creep resistance of the material. Creep testing exposes the sample to a constant stress over a prolonged period of time. This is done to simulate the effects of static stresses on the performance of a material in service. The extension or strain of the sample over time is measured. Traditional creep testing can take an extended period of time. Similar results can, however, be obtained through a DMA creep study, which can be performed in the course of a few days.

Mechanical Testing as Part of a Failure Analysis. The use of mechanical testing in a failure analysis is limited. The preparation of specimens from the failed component may not be possible. Further, published standard mechanical data, including yield strength, elastic modulus, and flexural modulus, are very dependent on the specimen configuration and testing conditions. Given that most published data are generated on specially molded test specimens, the testing of samples excised from molded articles may not provide an adequate comparison. In some cases, it is not apparent whether observed differences are the result of material deficiencies or variations in test specimen configuration. Instead, mechanical testing is most useful in comparing a known good or control sample with a failed part. Many times, this is best accomplished through some sort of proof load testing. Proof load testing involves measuring the strength and dimensional changes as a function of an applied load. In most cases, this testing involves producing a catastrophic failure within the test sample. The use of proof load testing as part of a failure analysis is illustrated in example 12 in this article.

Considerations in the Selection and Use of Test Methods

Through the application of analytical testing and a systematic engineering approach, it is possible to successfully ascertain the nature and cause of a plastic component failure. The testing, however, must be performed in a sound manner, with the obtained data only being as good as the analysis method. Further, the data presented by the analytical methods are often complicated and, in many cases, require an experienced analyst to be properly interpreted.

The aforementioned analytical tests are not meant to be an all-encompassing list of the methods used to evaluate failed plastic components. Certainly, there are numerous testing methodologies that provide data pertinent to a plastic component failure analysis. Other analysis techniques, including EDS and SEM, are important tools in a plastic component failure analysis. More specialized chromatographic methods, including gas chromatography and gas chromatography-mass spectroscopy, are extremely useful in assessing low-concentration additives within a plastic resin. Nuclear magnetic resonance spectroscopy is useful in polymeric analysis, providing information related to composition beyond FTIR. Nuclear magnetic resonance can provide data regarding stereoregularity, carbon content, chemical composition, and copolymer structure (Ref 10). Additionally, surface analysis spectroscopic techniques, such as secondary ion mass spectroscopy, x-ray photoelectron spectroscopy, and electron spectroscopy for chemical analysis, are specifically used to characterize very shallow surface layers. These techniques can be used to analyze material composition but are particularly suited for the analysis of surface contaminants (Ref 10). While these analytical techniques can provide valuable data as part of a plastics failure analysis, the tests described in this article are considered to be the most important in the majority of cases. Given the charge that this article be treated in a practical manner, test methods used less often were omitted. A summary showing both the treated and omitted analysis methods and the corresponding information gained is included in Table 1 (Ref 3).

Case Studies

Example 1: Embrittlement of a Polycarbonate Bracket. A plastic bracket exhibited relatively brittle material properties, which ulti-

Table 1 Practical information derived from polymer analysis methods

Test method	Properties measured	Use in failure analysis
Fourier transform infrared spectroscopy (FTIR)	Molecular bond structure	Material identification, contamination, degradation, chemical contact
Differential scanning calorimetry (DSC)	Heat of fusion, melting point, glass transition temperature, heat capacity	Material identification, level of crystallinity, aging/degradation, thermal history
Thermogravimetric analysis (TGA)	Weight loss over temperature or time	Composition, thermal stability, evolved gas analysis
Thermomechanical analysis (TMA)	Dimensional changes over temperature	Coefficient of thermal expansion, material transitions, molded-in stress, chemical compatibility
Dynamic mechanical analysis (DMA)	Elastic modulus, viscous modulus, tan delta	Temperature-dependent behavior, aging/degradation, solid-liquid interactions
Gel permeation chromatography (GPC)	Weight-average molecular weight, molecular weight distribution	Degradation, suitability of material for use
Melt flow rate (MFR)	Melt viscosity	Degradation, compliance with material specification
Solution viscosity	Intrinsic viscosity	Degradation
Mechanical testing	Strength and elongation properties, modulus	Compliance with material specification, mechanical properties
Scanning electron microscopy (SEM)	Surface and particle morphology	Fracture mode
Energy-dispersive x-ray spectroscopy (EDS)	Elemental concentrations	Material composition, fillers, additives
Nuclear magnetic resonance (NMR)	Molecular bond structure	Material identification
Mass spectroscopy (MS)	Molecular structure	Material identification, additives
X-ray photoelectron spectroscopy (XPS)	Elemental concentrations	Chemical composition of surfaces
Auger electron spectroscopy (AES)	Elemental concentrations	Chemical composition of surfaces

Source: Ref 3

mately led to catastrophic failure. The part had been injection molded from a medium-viscosity polycarbonate resin and had been in service for a short duration prior to the failure.

Tests and Results. A visual examination of the bracket revealed a series of surface anomalies, and it was suspected that the presence of the defects was related to the premature failure. The component base material was analyzed using micro-FTIR in the attenuated total reflectance (ATR) mode. The obtained spectrum exhibited absorption bands characteristic of polycarbonate, as shown in Fig. 13. No evidence of material contamination was found. A similar analysis was performed on the part surface in an area that showed the anomalous surface condition. The spectrum representing the surface was generally similar to the results obtained on the base material. However, the surface spectrum showed a relative increase in the intensity of a spectral band between 3600 and 3350 cm^{-1}, indicative of hydroxyl functionality. Additionally, the spectrum also showed changes in the relative intensities of several bands, as compared to the results representing the base material. A spectral subtraction was performed, and the results produced a good match with a library reference of diphenyl carbonate. This is illustrated in the spectral comparison presented in Fig. 14. Diphenyl carbonate is a common breakdown product produced during the decomposition of polycarbonate.

Conclusions. Overall, the obtained results suggested that the anomalous surface condition observed on the bracket represented molecular degradation of the polycarbonate. This is consistent with the brittle properties exhibited by the component. The most likely cause of the molecular degradation was improper drying and/or exposure to excessive heat during the injection molding process.

Example 2: Chemical Attack of Acrylonitrile-Butadiene-Styrene Grips. A set of plastic grips from an electric consumer product failed while in service. The grips had been injection molded from a general-purpose grade of an acrylonitrile-butadiene-styrene (ABS) resin. The parts had cracked while in use after apparent embrittlement of the material.

Tests and Results. An examination of the grips confirmed a severe level of cracking, covering the majority of the grip surface. Handling of the parts revealed that the grip material exhibited very little integrity, unlike the usual ductility associated with ABS resins. A white discoloration was also observed on the otherwise red grips. The surface of the grips was evaluated using SEM, revealing isolated areas that showed significant degradation in the form of material loss, as shown in Fig. 15. The observed morphology suggested selective degradation of the polybutadiene domains within the ABS resin.

Micro-FTIR in the ATR mode was used to analyze the base material and the surfaces of the grips. The results obtained on the base material were characteristic of an ABS resin. Analysis of the surface of the part produced a somewhat different result. The spectrum representing the grip surface contained absorption bands associated with ABS; however, the results contained additional bands of significant intensity. A spectral subtraction was performed, thereby removing the bands associated with the ABS resin. The obtained subtraction spectrum produced a very good match with glyceride derivatives of fats and oils. This identification is illustrated in Fig. 16.

Conclusions. It was the conclusion of the analysis that the grips failed via brittle fracture associated with severe chemical attack of the ABS resin. A significant level of glyceride derivatives of fatty acids, known to degrade ABS resins, was found on the part surface. The glyceride derivatives selectively attacked the polybutadiene domains within the molded ABS part, leading to apparent embrittlement and subsequent failure.

Example 3: Inclusion within an ABS Handle. The handle from a consumer product exhibited an apparent surface defect. The handle had been injection molded from a medium-viscosity-grade ABS resin. The anomalous appearance was objectionable to the assembler of the final product and resulted in a production lot being placed on quality-control hold.

Tests and Results. The surface of the part was examined using an optical stereomicroscope. The defect appeared as a localized area of lightened color, and the zone immediately surrounding the anomaly was slightly recessed. A mounted and polished cross section was prepared through the part, revealing distinct included material within the base molding resin. The inclusion, as shown in Fig. 17, did not appear to contain a significant level of the blue pigment, as present in the base material.

The preparation of the cross section not only allowed a thorough inspection of the defect but also served to facilitate further analysis of the material. The sample was initially analyzed using EDS. The results obtained on the included material showed exclusively carbon and oxygen, precluding an inorganic contaminant. The base resin and the included material were further analyzed using FTIR in the reflectance mode. The spectrum representing the base material contained absorption bands indicative of an ABS resin. Analysis of the included material produced distinctly different results. The spectrum obtained on the included material was characteristic of polybutadiene, the rubber-modifying agent present in ABS. This identification is presented in Fig. 18.

Conclusions. It was the conclusion of the evaluation that the handle contained an inclusion, which produced the apparent surface anomaly. The included material was identified as polybutadiene. The most likely source of the included polybutadiene was an undispersed gel particle formed during the production of the molding resin.

Example 4: Relaxation of Nylon Wire Clips. A production lot of plastic wire clips was failing after limited service. The failures were

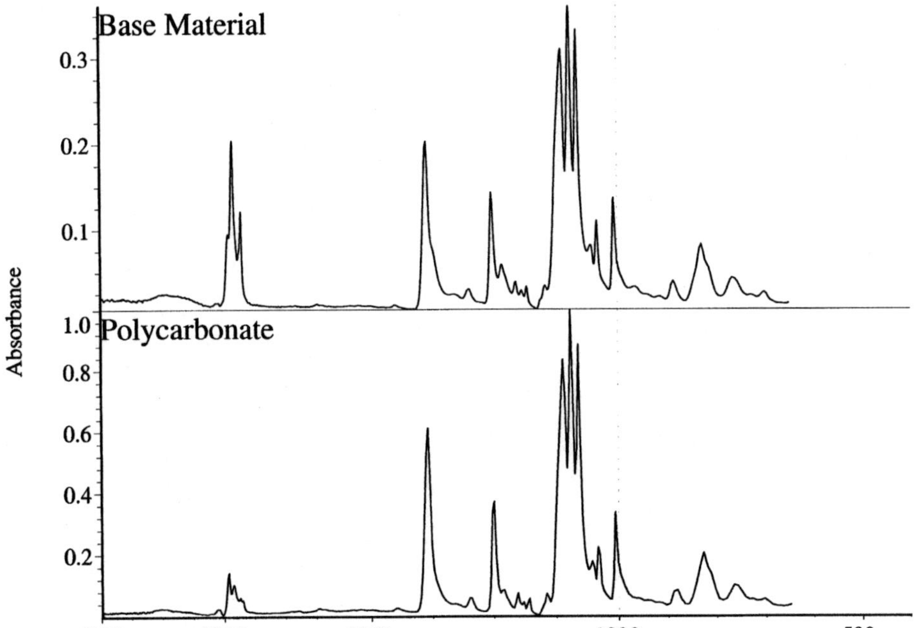

Fig. 13 The Fourier transform infrared spectroscopy spectrum obtained on the bracket base material, exhibiting absorption bands characteristic of polycarbonate

characterized by excessive relaxation of the clips, such that the corresponding wires were no longer adequately secured in the parts. No catastrophic failures had been encountered. Parts representing an older lot, which exhibited satisfactory performance properties, were also available for reference purposes. The clips were specified to be injection molded from an impact-modified grade of nylon 6/6. However, the part drawing did not indicate a specific resin.

Tests and Results. A visual examination of the clips showed that the failed parts were off-white in color, while the control parts had a pure white appearance. An analysis of both sets of parts was performed using micro-FTIR in the ATR mode. A direct comparison of the results produced a good match, with both sets of spectra exhibiting absorption bands that were characteristic of a nylon resin. The comparison, however, revealed subtle differences between the two sets of clips. The spectrum representing the reference parts showed a relatively higher level of a hydrocarbon-based impact modifier, while the results obtained on the failed parts showed the presence of an acrylic-based modifier. The differences in the spectra suggested that the two sets of clips were produced from resins having different formulations, particularly regarding the impact modifier.

The clip materials were further analyzed using DSC. The thermogram representing the reference part material, as shown in Fig. 19, exhibited an endothermic transition at 264 °C (507 °F), characteristic of the melting point of a nylon 6/6 resin. Additionally, the results contained a second melting point, of lesser magnitude, at 95 °C (203 °F). This transition was indicative of a hydrocarbon-based impact modifier, as indicated by the FTIR results. The thermogram obtained on the failed clip material also showed a melting point characteristic of a nylon 6/6 resin. However, no evidence was found to indicate a transition corresponding to the hydrocarbon-based modifier found in the control clip material.

Conclusions. It was the conclusion of the analysis that the control and failed clips had been produced from two distinctly different resins. While both materials satisfied the requirements of an impact-modified nylon 6/6 resin, differences in the impact modifiers resulted in the observed performance variation. From the results and the observed performance, it appeared that the material used to produce the failed clips had different viscoelastic properties, which produced a greater predisposition for stress relaxation.

Example 5: Embrittlement of Nylon Couplings. Molded plastic couplings used in an industrial application exhibited abnormally brittle properties, as compared to previously produced components. The couplings were specified to be molded from a custom-compounded glass-filled nylon 6/12 resin. An inspection of the molding resin used to produce the discrepant parts revealed differences in the material appearance, relative to a retained resin lot. Specifically, physical sorting resulted in two distinct sets of molding resin pellets from the lot that had generated the brittle parts. Both of these sets of pellets had a coloration that varied from that of the retained reference resin pellets. A sample of retained molding resin, which had produced parts exhibiting satisfactory performance, was available for comparative analysis.

Tests and Results. Micro-FTIR in the ATR mode was used to analyze the molding resin samples. The results obtained on the three molding resin samples were generally similar, and all of the spectra exhibited absorption bands characteristic of a nylon resin. Further analysis of the resin samples using DSC indicated that the control material results exhibited a single endothermic transition at 218 °C (424 °F), consistent with the melting point of a nylon 6/12 resin, as specified.

The DSC thermograms obtained on the two resin samples that produced brittle parts also exhibited melting point transitions associated with nylon 6/12. However, additional transitions were also apparent in the results, indicative of the presence of contaminant materials. The results obtained on one of the resin samples, as presented in Fig. 20, showed a secondary melting point at 165 °C (330 °F), indicative of polypropylene. The thermogram representing the second resin sample, as included in Fig. 21, displayed a second melting transition at 260 °C (500 °F), characteristic of a nylon 6/6 resin.

Conclusions. It was the conclusion of the analysis that the molding resin used to produce the brittle couplings contained a significant level of contamination, which compromised the mechanical properties of the molded components. Two distinct contaminants were found mixed into the molding pellets. The contaminant materials were identified as polypropylene and nylon 6/6. The source of the polypropylene was likely the purging compound used to clean the compounding extruder. The origin of the nylon 6/6 resin was unknown but may represent a previously compounded resin.

Example 6: Failure of Plasticized Polyvinyl Chloride Tubing. A section of clear

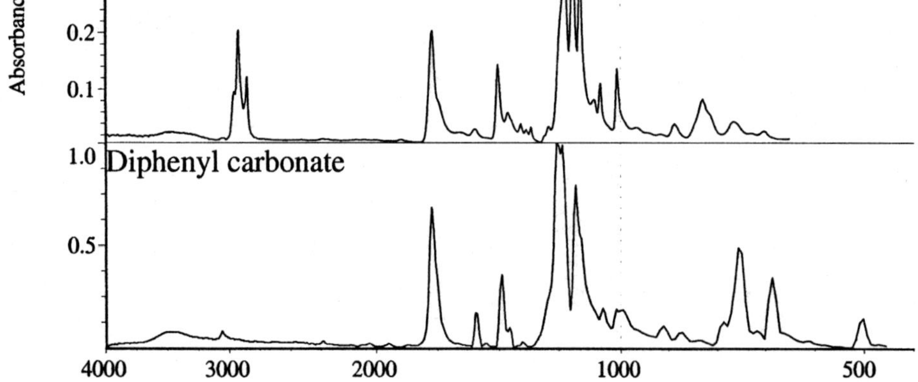

Fig. 14 Spectral comparison showing differences between the base material and surface spectra, attributed to diphenyl carbonate

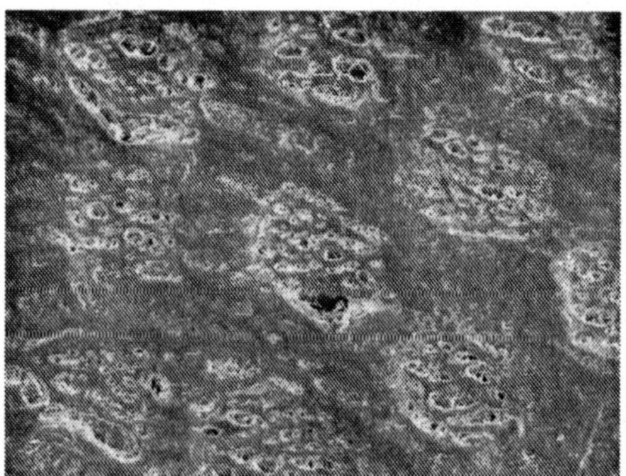

Fig. 15 Scanning electron image showing isolated degradation of the grip material. 30×

polymeric tubing failed while in service. The failed sample had been used in a chemical transport application. The tubing had also been exposed to periods of elevated temperature as part of the operation. The tubing was specified to be a polyvinyl chloride (PVC) resin plasticized with trioctyl trimellitate (TOTM). A reference sample of the tubing, which had performed well in service, was also available for testing.

Tests and Results. The failed and reference tubing samples were analyzed using micro-FTIR in the ATR mode, and the results representing the reference tubing material were consistent with the stated description: a PVC resin containing a trimellitate-based plasticizer. However, the spectrum representing the failed tubing material was noticeably different. While the obtained spectrum contained absorption bands characteristic of PVC, the results indicated that the material had been plasticized with an adipate-based material, such as dioctyl adipate. This identification is shown in Fig. 22.

In order to assess their relative thermal stability, the two tubing materials were analyzed via thermogravimetric analysis (TGA). Both sets of results were consistent with those expected for plasticized PVC resins. The thermograms representing the reference and failed sample materials showed comparable plasticizer contents of 28 and 25%, respectively. The results also showed that the reference material, containing the trimellitate-based plasticizer, exhibited superior thermal resistance relative to the failed material, containing the adipate-based material. This was indicated by the elevated temperature of weight-loss onset exhibited by the reference tubing material.

Conclusions. It was the conclusion of the evaluation that the failed tubing had been produced from a formulation that did not comply with the specified material. The failed tubing was identified as a PVC resin with an adipate-based plasticizer, not TOTM. The obtained TGA results confirmed that the failed tubing material was not as thermally stable as the reference material because of this formulation difference, and that this was responsible for the observed failure.

Example 7: Cracking of Polybutylene Terephthalate Automotive Sleeves. A number of plastic sleeves used in an automotive application cracked after assembly but prior to installation into the mating components. The sleeves were specified to be injection molded from a 20% glass-fiber-reinforced polybutylene terephthalate (PBT) resin. After molding, electronic components are inserted into the sleeves, and the assembly is filled with a potting compound. A retained lot of parts, which had not cracked, were available for reference purposes.

Tests and Results. The reference and failed parts were analyzed using micro-FTIR in the ATR mode. The spectra obtained on both sets of parts contained absorption bands characteristic of a thermoplastic polyester, such as PBT or polyethylene terephthalate (PET). Different types of polyester resins cannot be distinguished spectrally, because of the similar nature of their structures. However, subtle but distinct differences were apparent in the results, suggestive of degradation of the failed part material.

Differential scanning calorimetry was performed on the sleeve materials using a heat/cool/heat methodology. Testing of the reference material produced initial heating results indicative of a PBT resin, as illustrated by the melting point at 224 °C (435 °F). Analysis of the failed sleeve samples produced a melting transition at a significantly reduced temperature, 219 °C (426 °F). Additionally, the failed material transition was broader, and overall, the results suggested molecular degradation of the failed sleeve material. A comparison of the initial heating thermograms is presented in Fig. 23.

The identification of degradation was supported by the second heating DSC results, obtained after slow cooling. The second heating thermogram representing the failed sleeve material showed additional differences relative to the results obtained on the reference material. The failed material did not produce the bimodal melting endothermic transition normally associated with PBT after slow cooling. This was thought to be the result of molecular degradation, which produced shorter polymer chain lengths, therefore reducing steric hindrance. A comparison of the second heating thermograms is included in Fig. 24.

The sleeve materials were further analyzed using TGA. The thermograms obtained on the reference and failed samples were generally consistent, including equivalent glass contents. Additionally, the results were in agreement with those expected for a PBT resin.

The melt flow rates (MFRs) of the reference and failed sleeve materials were determined. Because no molding resin was available for comparison purposes, the nominal range from the specification sheet, 14 to 34 g/10 min, was used. The testing showed that the failed sleeve material had been severely degraded, producing

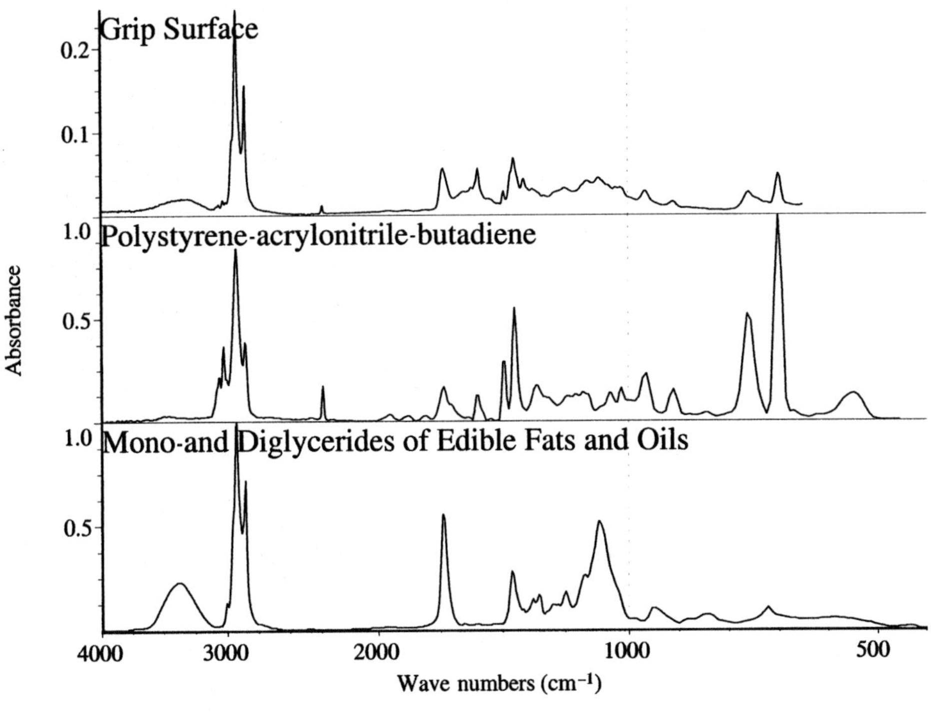

Fig. 16 The Fourier transform infrared spectroscopy spectrum obtained on the grip surface. The spectrum contains absorption bands indicative of glyceride derivatives of fats and oils in addition to bands associated with the base acrylonitrile-butadiene-styrene resin.

Fig. 17 Micrograph showing the included material within the handle. 24×

a MFR of 128 g/10 min. This was in agreement with the DSC data and indicated severe molecular degradation of the PBT resin. A review of the results generated by the reference parts also showed significant molecular degradation. While the extent of the degradation was less, the obtained MFR, 50 g/10 min, still demonstrated a substantial reduction in the average molecular weight.

Conclusions. It was the conclusion of the evaluation that the failed sleeves had cracked due to embrittlement associated with severe degradation and the corresponding molecular weight reduction. The degradation was clearly illustrated by the reduced melting point and uncharacteristic nature of the associated endothermic melting transitions as well as the substantial increase in the MFR of the molded parts. The reduction in molecular weight significantly reduced the mechanical properties of the sleeves. The cause of the degradation was not evident, but the likely source appears to be the molding operation and exposure to elevated temperature for an extended period of time. It is significant to note that the reference parts also showed a moderate level of molecular degradation, rendering them susceptible to failure over a longer duration.

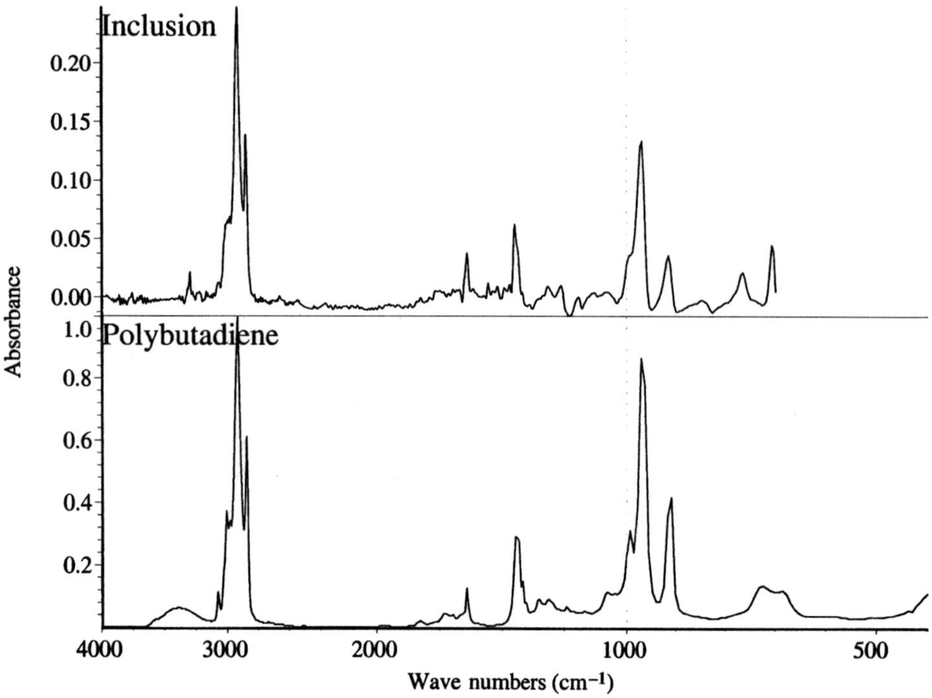

Fig. 18 The Fourier transform infrared spectroscopy spectrum obtained on the included particle, characteristic of polybutadiene

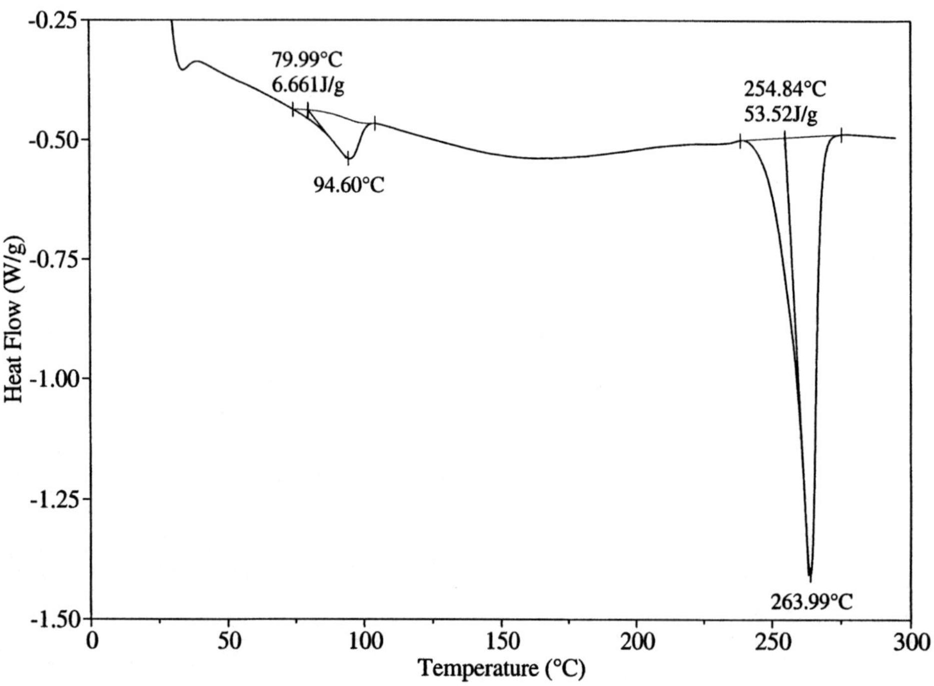

Fig. 19 The differential scanning calorimetry thermogram representing the reference clip material, exhibiting an endothermic transition characteristic of the melting of a nylon 6/6 resin. The results also showed a second melting transition attributed to a hydrocarbon-based impact modifier.

Example 8: Cracking of ABS Protective Covers. Numerous protective covers, used in conjunction with an electrical appliance, failed during assembly with the mating components. The failures were traced to a particular production lot of the covers and occurred during insertion of the screws into the corresponding bosses. The parts had been injection molded from an ABS resin to which regrind was routinely added. Retained parts, which exhibited normal behavior during assembly, were available for comparative analysis.

Tests and Results. A visual examination of the failed parts revealed relatively brittle fracture features, without significant ductility, as would be apparent as stress whitening or permanent deformation. Core material taken from the reference and failed parts was analyzed using micro-FTIR in the ATR mode. Both obtained spectra exhibited absorption bands associated with an ABS resin. However, the spectrum representing the failed part showed additional absorption bands. A spectral subtraction was performed, thereby removing the absorbances attributed to the ABS resin from the spectrum obtained on the failed part. The spectral subtraction results were consistent with a thermoplastic polyester, such as PET or PBT. However, these two materials cannot be distinguished spectrally, because of similarities in their structures. As such, the melting point is usually used to differentiate between these materials. The FTIR results indicated the presence of contaminant material exclusively within the ABS resin used to mold the failed covers.

In order to further identify the contaminant material, a sample taken from the failed part was analyzed via DSC. The obtained DSC thermogram, as presented in Fig. 25, showed a glass transition at approximately 101 °C (214 °F), consistent with the expected results for an ABS resin. The results also showed an additional endothermic transition at 222 °C (432 °F), indicative of a PBT resin. The failed cover material was also analyzed using TGA in order to assess the level of the contamination. The TGA analysis was performed using high-resolution temperature programming, and the results revealed adequate separation of the ABS and PBT resins. Based on the results, the contamination was estimated to account for approximately 23% of the failed cover material.

Conclusions. It was the conclusion of the evaluation that the appliance covers failed via brittle fracture associated with stress overload. The failures, which occurred under normal assembly conditions, were attributed to embrittlement of the molded parts, due to contamination of the ABS resin with a high level of PBT. The source of the PBT resin was not positively identified, but a likely source appeared to be the use of improper regrind.

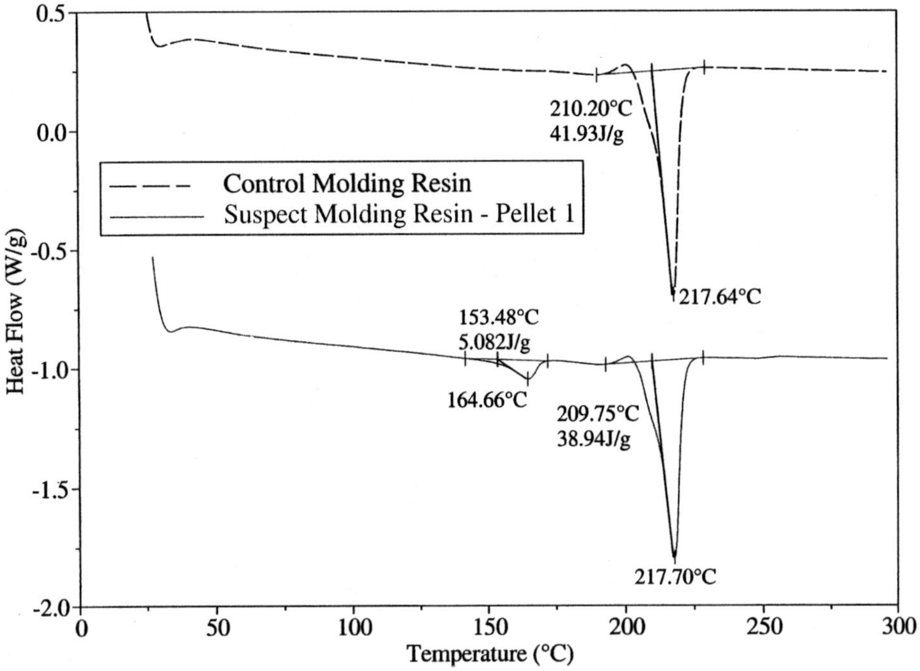

Fig. 20 The differential scanning calorimetry thermogram representing a molding resin pellet that had produced brittle parts. The thermogram shows a major melting transition associated with nylon 6/12 and a weaker transition attributed to polypropylene.

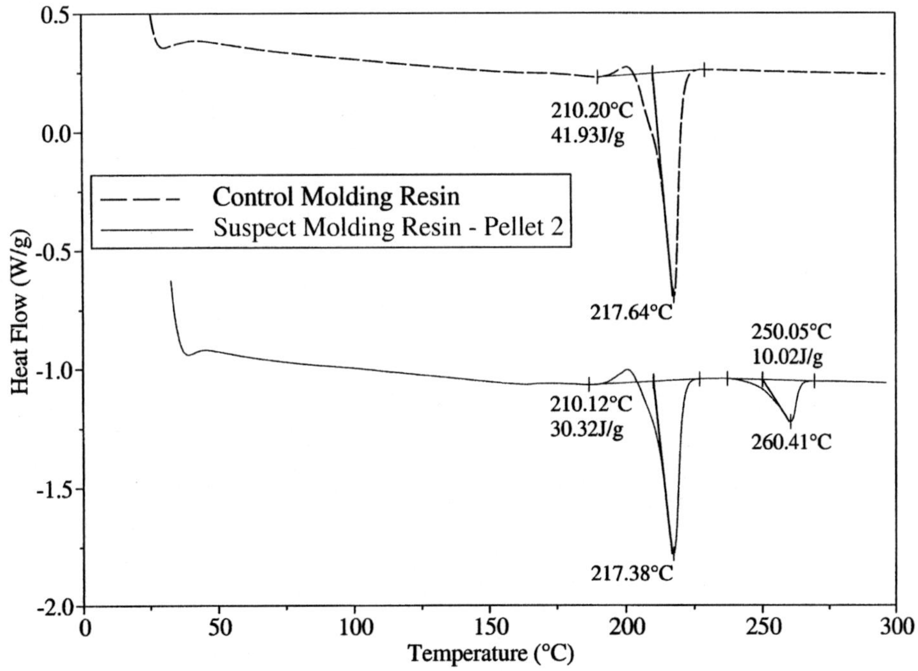

Fig. 21 The differential scanning calorimetry thermogram representing a second molding resin pellet that had produced brittle parts. The thermogram shows a major melting transition associated with nylon 6/12 and a weaker transition attributed to nylon 6/6.

Example 9: Failure of Polycarbonate/PET Appliance Housings. Housings from an electrical appliance failed during an engineering evaluation. The housings had been injection molded from a commercial polycarbonate/PET (PC/PET) blend. The parts were being tested as part of a material conversion. Parts produced from the previous material, a nylon 6/6 resin, had consistently passed the testing regimen. The housing assembly included a spring clip, which applied a static force on a molded-in boss extending from the main body of the housing. Grease was applied liberally within the housing assembly during production.

Tests and Results. A visual inspection of the tested parts showed catastrophic failure within the molded-in boss. The failures were consistent across all of the parts and were located at an area where the spring clip contacted the housing boss. While the final fracture zone exhibited limited features associated with ductility in the form of stress whitening, no such characteristics were apparent at the locations corresponding to the crack origins. The fracture surfaces were further examined via SEM. The SEM inspection showed the presence of multiple crack initiation sites along the side of the boss that had mated with the spring clip. No evidence of significant ductility was found with the crack initiation locations, as represented in Fig. 26. The overall features observed on the fracture surface were indicative of environmental stress cracking.

Micro-FTIR in the ATR mode was performed on the housing material, and the resulting spectrum was in agreement with the stated resin description, a blend of PC and polyester. No signs of material contamination were found. The housing material was further evaluated using DSC. The thermogram obtained during the initial heating run, as shown in Fig. 27, exhibited an endothermic transition at 253 °C (487 °F), characteristic of the melting point of a PET resin. The initial heating run results also showed a low-temperature exothermic transition associated with the crystallization of the PET resin. These results indicated that the material had not been fully crystallized during the molding process. The results generated during the second heating run, after slow cooling, did not show the low-temperature crystallization. The glass transition associated with the PC resin was observed in the second heating run.

In order to assess the molecular weight of the housing material, the intrinsic viscosity of the resin was measured. A comparison of the results with historical data revealed a substantial reduction in the viscosity of the failed part material. This indicated that the housing material had undergone significant molecular degradation during the injection molding process.

The grease present within the housing assembly was analyzed using micro-FTIR. The FTIR test results indicated that the grease was composed of a relatively complex mixture. The lubricant contained a hydrocarbon-based oil, a phthalate-based oil, lithium stearate, and an amide-based additive. These results were significant, because phthalate esters are known to be incompatible with PC resins.

Conclusions. It was the conclusion of the analysis that the appliance housings failed through environmental stress cracking. The required chemical agent was identified as a phthalate-based oil present within the grease used to lubricate the assembly. Specifically, the phthalate oil was not compatible with the PC portion of the resin blend. The source of the stress responsible for the cracking appears to be the interference related to the spring clip. While the previous parts, produced from the nylon 6/6 resin, were also under similar stresses, this resin was not prone to stress cracking in conjunction with the lubricant. Thus, the resin conversion was the root cause of the failures. Additionally, the test results also showed that the injection molding process left the material susceptible to failure. Specifically, the molded parts had been under-crystallized, reducing the mechanical strength of the molded articles, and, more importantly, the resin had been degraded, producing a reduction in the molecular weight and reducing both the mechanical integrity and chemical-resistance properties of the parts.

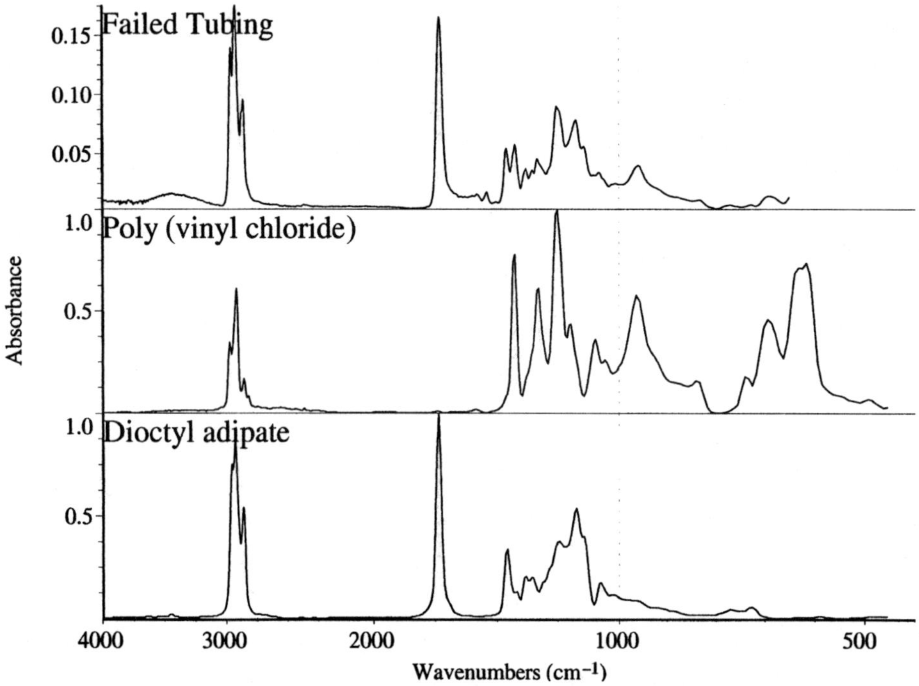

Fig. 22 The Fourier transform infrared spectroscopy spectrum obtained on the failed tubing material. The spectrum exhibits absorption bands indicative of a polyvinyl chloride resin containing an adipate-based plasticizer.

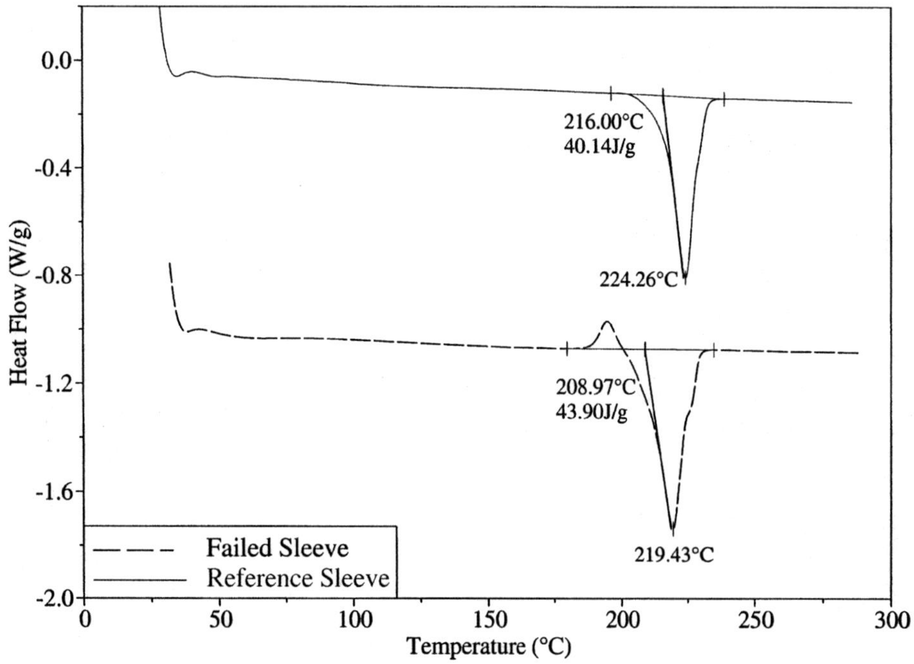

Fig. 23 A comparison of the initial heating run results, suggesting degradation of the failed sleeve material

Example 10: Failure of PET Assemblies. Several assemblies used in a transportation application failed during an engineering testing regimen. The testing involved cyclic thermal shock, immediately after which cracking was observed on the parts. The cracking occurred within the plastic jacket, which had been injection molded from an impact-modified, 15% glass-fiber-reinforced PET resin. The plastic jacket had been molded over an underlying metal coil component. Additionally, a metal sleeve was used to house the entire assembly. Prior to molding, the resin had reportedly been dried at 135 °C (275 °F). The drying process usually lasted 6 h, but occasionally, the material was dried overnight. The thermal shock testing included exposing the parts to alternating temperatures of –40 and 180 °C (–40 and 360 °F). The failures were apparent after 100 cycles. Molding resin and nonfailed parts were also available for analysis.

Tests and Results. The failed assemblies were visually and microscopically examined. The inspection showed several different areas within the overmolded jacket that exhibited cracking. The cracked areas were located immediately adjacent to both the underlying metal coil and the outer metal housing. The appearance of the cracks was consistent with brittle fracture, without significant signs of ductility. The examination also revealed design features, including relatively sharp corners and nonuniform wall thicknesses, that appeared to have likely induced molded-in stress within the plastic jacket.

The fracture surfaces were further inspected using SEM, and the examination revealed features generally associated with brittle fracture, as shown in Fig. 28. No evidence of microductility, such as stretched fibrils, was found. The fracture surface features indicated that the cracking had initiated along the outer jacket wall and subsequently extended through the wall and circumferentially around the wall. Throughout the examination, no indication of postmolding molecular degradation was found.

Micro-FTIR was performed in the ATR mode on a core specimen of the jacket material. The resulting spectrum was consistent with a thermoplastic polyester resin. Such materials, including PET and PBT, cannot be distinguished spectrally, and a melting point determination is usually used to distinguish these materials. The failed jacket material and reference molding resin were analyzed using TGA, and the results obtained on the two samples were generally consistent. This included relatively comparable levels of volatiles, polymer, carbon black, and glass reinforcement. Further, the results were in excellent agreement with those expected for the stated PET material.

The failed jacket and reference materials were evaluated via DSC. Analysis of the failed jacket material produced results that indicated a melting transition at 251 °C (484 °F), consistent with a PET resin. However, a second endothermic transition was also present. This transition, at 215 °C (420 °F), suggested the melting of annealed crystals, indicating that the part had been exposed to a temperature approaching 215 °C (420 °F). The thermal shock testing appeared to be the only possible source of this thermal exposure.

Analysis of the molding resin also produced results consistent with a PET resin. The results also exhibited a second melting endotherm at 174 °C (345 °F). Again, this transition was associated with melting of annealed crystals for material exposed to this temperature. The apparent source of the exposure was the drying process. This was well in excess of the stated drying temperature.

Further analysis of the assembly materials using thermomechanical analysis (TMA) produced significantly different results for the PET jacket and the steel housing material. Determination of the coefficients of thermal expansion (CTEs) showed approximately an order of magnitude difference between the two mating materials.

An assessment of the molecular weight of the failed jacket samples as well as a nonfailed part and the molding resin samples was performed using several techniques. A combination of MFR, intrinsic viscosity, and finally, gel permeation chromatography (GPC) was used, because of conflicting results. The MFR determinations showed that the drying process produced a considerable increase in the MFR of the resin, corresponding to molecular degradation in the form of chain scission. This was contrasted by the results generated by the intrinsic viscosity test-

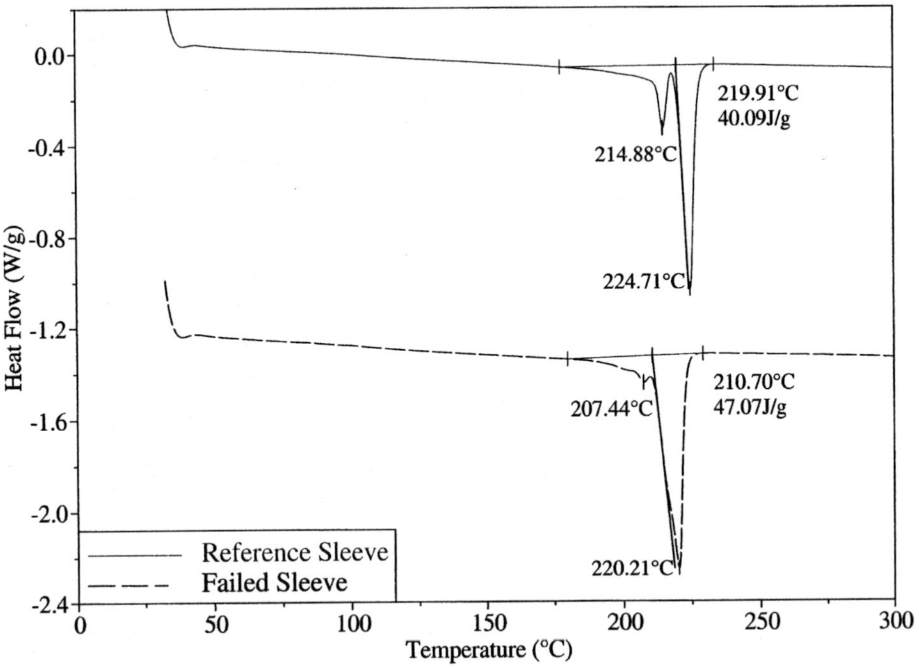

Fig. 24 A comparison of the second heating run results, further suggesting degradation of the failed sleeve material

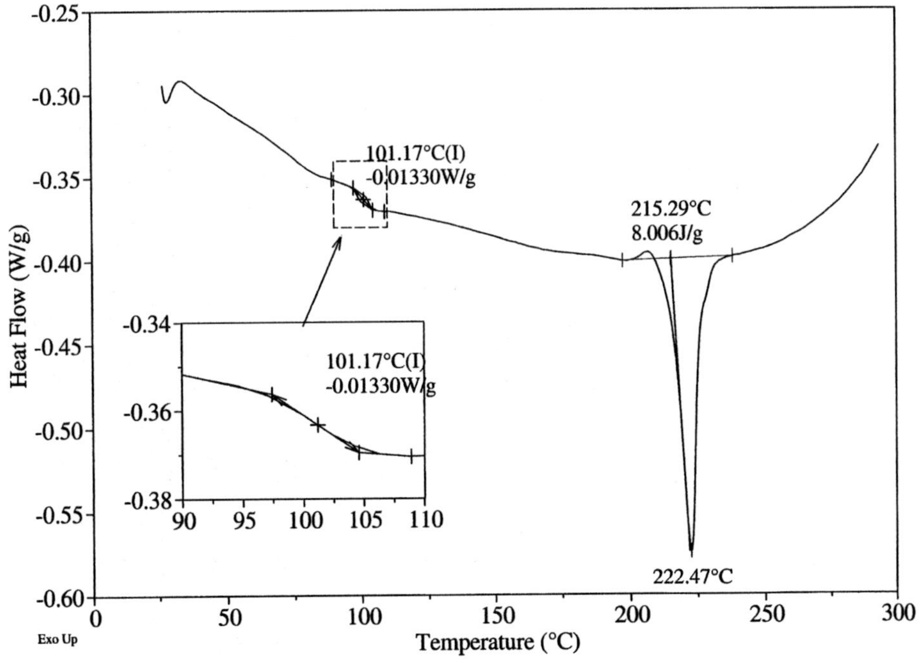

Fig. 25 The differential scanning calorimetry thermogram obtained on the failed cover material. The thermogram shows an endothermic transition associated with polybutylene terephthalate. The (I) indicates that the numerical temperature was determined as the inflection point on the curve.

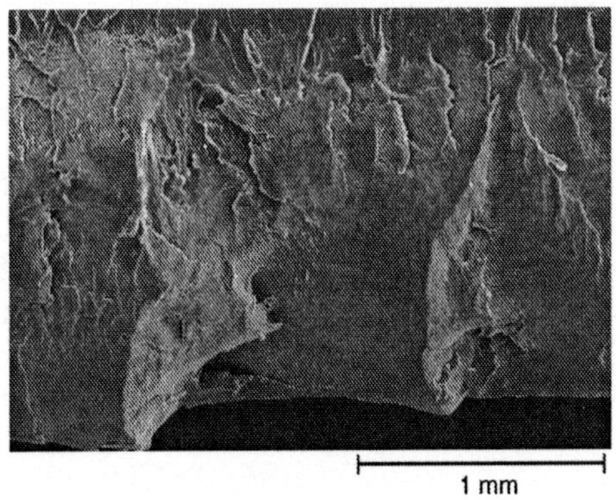

Fig. 26 Scanning electron image showing brittle fracture features at the crack initiation site, characteristic of environmental stress cracking. 24×

ing. These results showed an increase in the viscosity of the dried resin relative to the virgin resin. This increase was suggestive of an increase in molecular weight, possibly through partial cross linking. Testing of the resin samples and the molded parts via GPC produced results that reconciled the discrepancy. The GPC results showed that the drying process produced competing reactions of chain scission and cross linking. The net result was severe degradation of the dried resin, which predisposed the molding material to produce jackets having poor mechanical properties. The GPC testing showed that the molded jackets were further degraded during the injection molding process.

Conclusions. It was the conclusion of the investigation that the assemblies failed via brittle fracture associated with the exertion of stresses that exceeded the strength of the resin as-molded. The stresses were induced by the thermal cycling and the dimensional interference caused by the disparity in the CTEs of the PET jacket and the mating steel sleeve. However, several factors were significant in the failures. It was determined that the resin drying process had exposed the resin to relatively high temperatures, which caused substantial molecular degradation. The drying temperature was found to be approximately 173 °C (344 °F), well in excess of the recommendation for the PET resin. Further degradation was attributed to the molding process itself, leaving the molded jacket in a severely degraded state. This degradation limited the ability of the part to withstand the applied stresses. Additionally, the testing itself exposed the parts to temperatures above the recognized limits for PET, and this may have significantly lowered the mechanical properties of the part.

Example 11: Cracking of a Polyethylene Chemical Storage Vessel. A chemical storage vessel failed while in service. The failure occurred as cracking through the vessel wall, resulting in leakage of the fluid. The tank had been molded from a high-density polyethylene (HDPE) resin. The material held within the vessel was an aromatic hydrocarbon-based solvent.

Tests and Results. A stereomicroscopic examination of the failed vessel revealed brittle fracture surface features. This was indicated by the lack of stress whitening and permanent deformation. Limited ductility, in the form of stretching, was found exclusively within the final fracture zones. On cutting the vessel, significant stress relief, in the form of distortion, was evident. This indicated a high level of molded-in stress within the part. The fracture surface was further inspected using SEM. The observed features included a relatively smooth morphology within the crack origin location, which was indicative of slow crack initiation. This area is shown in Fig. 29. Features associated with more rapid crack extension, including hackle marks and river markings, were found at the midfracture and final fracture areas, as represented in Fig. 30. The entirety of the fracture surface features indicated that the cracking had initiated along the exterior wall of the vessel. The cracking extended transversely through the wall initially, and subsequently, circumferentially around the wall. Throughout the examination, no signs of postmolding molecular degradation or chemical attack were found.

The failed vessel material was analyzed using micro-FTIR in the ATR mode. The obtained spectrum exhibited absorption bands characteristic of a polyethylene resin. No evidence was found to indicate contamination or degradation of the material.

Material excised from the failed vessel was analyzed using DSC. The results showed a single endothermic transition associated with the melting point of the material at 133 °C (271 °F). The results were consistent with those expected for a HDPE resin. The results also showed that the HDPE resin had a relatively high level of crystallinity, as indicated by the elevated heat of fusion.

Thermogravimetric analysis was performed to further evaluate the failed vessel material.

Fig. 28 Scanning electron image showing brittle fracture features on the failed jacket crack surface. 20×

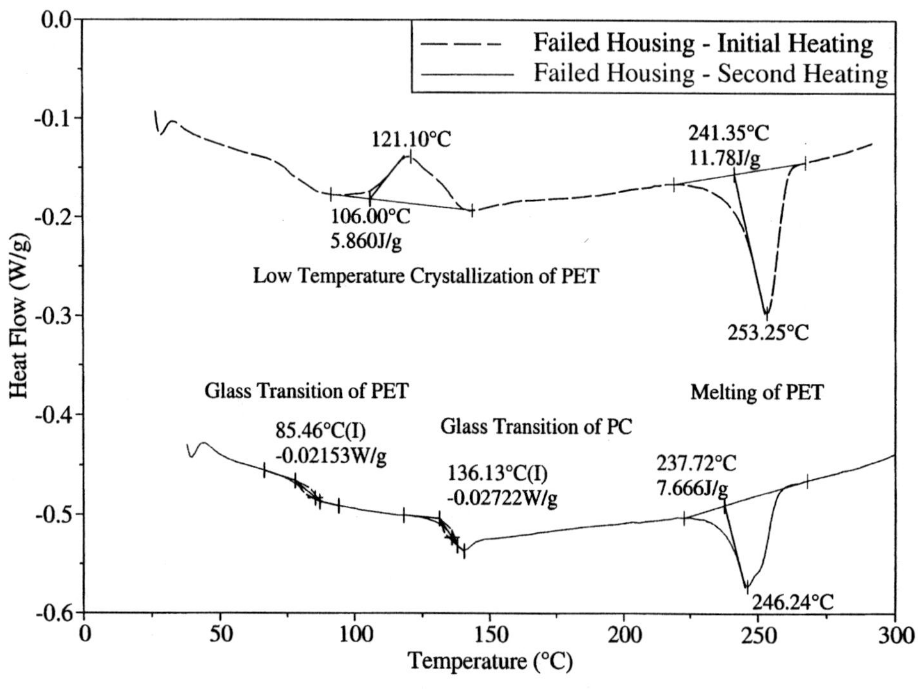

Fig. 27 The initial heating differential scanning calorimetry thermogram, exhibiting a melting transition consistent with a PET resin. A low-temperature crystallization exothermic transition was also apparent. The (I) indicates that the numerical temperature was determined as the inflection point on the curve.

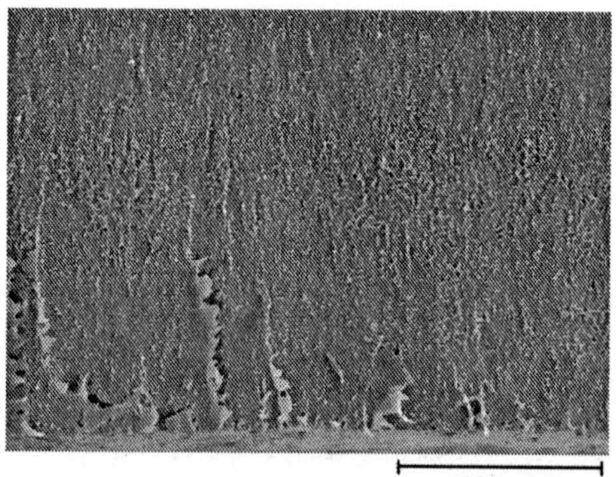

Fig. 29 Scanning electron image showing features associated with brittle fracture and slow crack growth within the crack initiation site. 100×

The TGA testing showed that the HDPE absorbed approximately 6.3% of its weight in the aromatic hydrocarbon-based solvent. Overall, the TGA results were consistent with those expected for a HDPE resin.

The MFR of the vessel material was evaluated, and the testing produced an average result of 3.8 g/10 min. This is excellent agreement with the nominal value indicated on the material data sheet, 4.0 g/10 min. As such, it was apparent that the vessel material had not undergone molecular degradation. The specific gravity of the resin was measured. The material produced a result of 0.965. This indicated that the material had a relatively high level of crystallinity, as suggested by the DSC results.

In order to assess the effects of the hydrocarbon-based solvent on the HDPE vessel, the material was evaluated using dynamic mechanical analysis (DMA). The vessel material was analyzed in two conditions. Material samples representing the vessel material in the as-molded condition as well as material from the failed vessel were evaluated. A comparison of the DMA results showed that in the saturated, equilibrium state, the HDPE resin lost over 60% of its elastic modulus at room temperature, because of the plasticizing effects of the solvent. A comparison of the DMA results, indicating the reduction in mechanical properties, is shown in Fig. 31.

Conclusions. It was the conclusion of the investigation that the chemical storage vessel failed via a creep mechanism associated with the exertion of relatively low stresses. Given the lack of apparent ductility, the stresses responsible for the failure appear to have been below the yield strength of the material. The source of the stress was thought to be molded-in residual stresses associated with uneven shrinkage. This was suggested by the obvious distortion evident on cutting the vessel. The relatively high specific gravity and the elevated heat of fusion are indicative that the material has a high level of crystallinity. In general, increased levels of crystallinity result in higher levels of molded-in stress and the corresponding warpage. The significant reduction in the modulus of the HDPE material, which accompanied the saturation of the resin with the aromatic hydrocarbon-based solvent, substantially decreased the creep resistance of the material and accelerated the failure. The dramatic effects of the solvent had not been anticipated prior to use.

Example 12: Failure of Polyacetal Latch Assemblies. Components of a latch assembly used in a consumer device exhibited a relatively high failure rate. The latches are used as a safety restraint, and failure in the field could result in severe injury. The failures were occurring after installation but prior to actual field use. Specifically, the failures occurred as cracking within retaining tabs used to secure a metal slide. The cracking was limited to an older design, with newer components showing no signs of failure. The latch assembly components were injection molded from an unfilled commercial grade of a polyacetal copolymer. As part of the evaluation, both failed parts representative of the older design and newer components were available for testing.

Tests and Results. A visual examination of the failed parts confirmed cracking within the retaining tab adjacent to the metal slide. The failures were present at consistent locations on all of the parts. The crack surfaces showed evidence of macroductility in the form of stress whitening within the final fracture zone exclusively. Throughout the visual examination, it was also apparent that the parts exhibited a very sharp corner formed by the retaining tab and the main body of the latch assembly body. Sharp corners are considered a poor design feature in plastic components, because they can result in severe stress concentration and can produce areas of localized poor fusion.

The fracture surface was further evaluated using SEM. The SEM examination showed a clear crack origin at the corner formed by the retaining tab. The crack origin areas exhibited brittle fracture features without signs of significant microductility. Secondary crazing was also apparent at the crack origin location. A typical crack initiation site is shown in Fig. 32. The overall features were suggestive of cracking caused by a relatively high strain rate event and/or very high stress concentration. The mid-fracture surface showed an increase in the apparent ductility, as evidenced by an overlapping morphological structure. The final fracture zone showed significant deformation and stretching, indicative of ductile overload. A laboratory failure was created by overloading the tab from a nonfailed part in a manner consistent with the insertion of the corresponding metal slide. The laboratory fractures exhibited surface features that were in excellent agreement with those exhibited by the failed parts, as apparent in Fig. 32.

The failed latch assembly material was analyzed using micro-FTIR in the ATR mode, and the obtained spectrum exhibited absorption

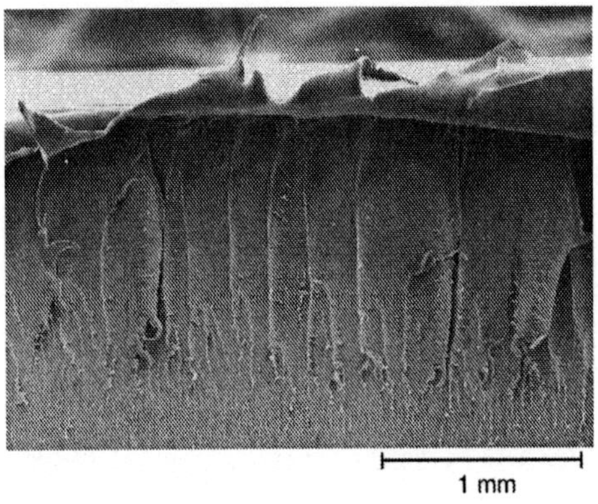

Fig. 30 Scanning electron image showing features indicative of rapid crack extension within the final fracture zone. 20×

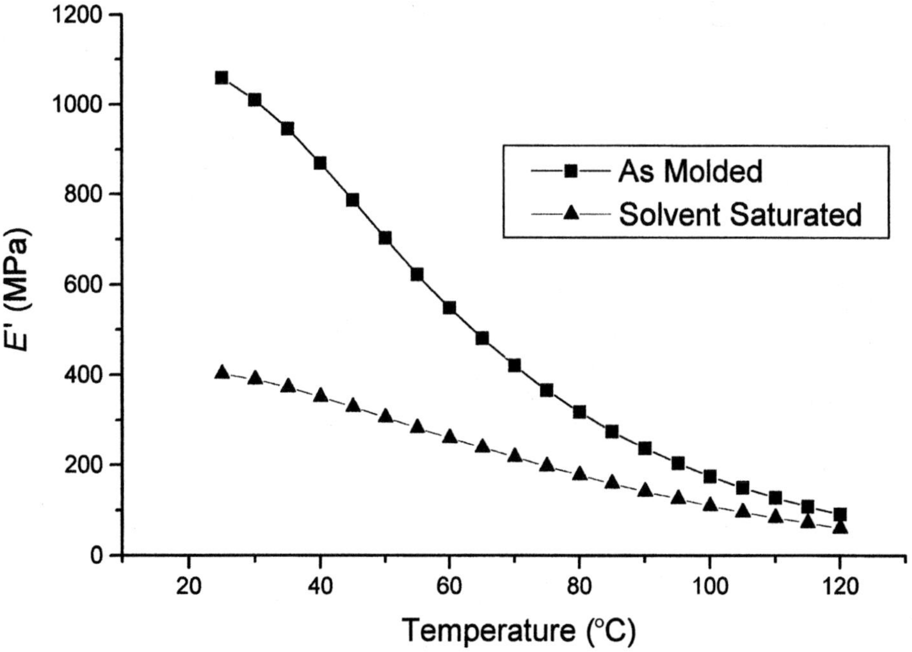

Fig. 31 A comparison of the dynamic mechanical analysis results, showing a loss of over 60% in the elastic modulus, E', as a result of the effects of the solvent

bands characteristic of a polyacetal resin. It is significant to note that polyacetal copolymers and homopolymers cannot be differentiated spectrally, and a melting point determination is often used to distinguish between these materials.

Differential scanning calorimetry was used to analyze the latch material. The obtained results showed that the material underwent a single endothermic transition at approximately 165 °C (330 °F), characteristic of the melting point of a polyacetal copolymer. The results also showed that the part was somewhat undercrystallized. This was evident through a significant increase in the heat of fusion between the initial heating run and the second heating run, after slow cooling. Undercrystallization can reduce the mechanical strength of the molded article and is usually the result of molding in a relatively cold tool. The level of undercrystallization found in the failed parts, however, was moderate in nature and not thought to be a major factor in the failures. Thermogravimetric analysis was also performed on the latch material, and the obtained results were consistent with those expected for an unfilled polyacetal copolymer.

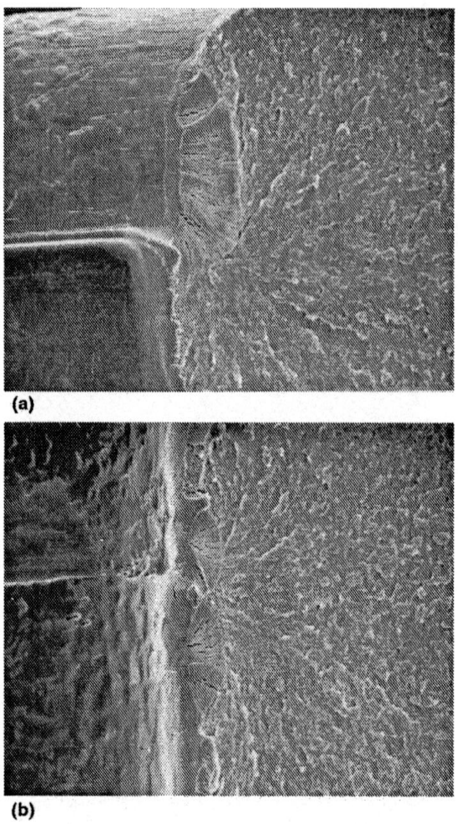

Fig. 32 Scanning electron images showing excellent agreement between the features present within the crack initiation sites of (a) the failed latch assembly and (b) the laboratory fracture. Both surfaces showed relatively brittle fracture features. 59×

The latch material was also analyzed to determine its MFR. Parts representing the older, failed components and the newer, current design were evaluated. Both sets of molded parts produced results ranging from 10.7 to 11.0 g/10 min. This was in good agreement with the nominal MFR for the molding resin, 9.0 g/10 min. Throughout the analytical testing of the failed latch material, no evidence was found to indicate contamination or degradation of the molded parts.

Mechanical testing was performed in order to assess the effect of the recent design change. Because of the configuration of the parts, standard mechanical testing could not be performed. Instead, a proof load test was devised to directly assess the stress required to produce failure within the tab, at an area consistent with the failure latch assembly. A direct comparison was made between the two sets of parts. The parts representing the older design, with the sharp corner at the retaining tab, produced an average value of 78.7 N (17.7 lbf) at failure. More importantly, the parts within this group produced an average tab extension of 0.76 mm (0.03 in.) at failure. The evaluation of the part representing the new design generated significantly different results. Specifically, the failures occurred at a higher load, 92.1 N (20.7 lbf), and a greater tab extension, 2.5 mm (0.10 in.). A comparison of the mechanical test results is shown in Fig. 33. This mechanical evaluation clearly illustrated the advantage afforded by the design change, effectively increasing the tab radius.

Conclusions. It was the conclusion of the evaluation of the failed latch assemblies that the parts failed via brittle fracture associated with stress overload. The stress overload was accompanied by severe apparent embrittlement resulting from a relatively high strain rate event and/or significant stress concentration. The relatively sharp corner formed by the retaining tab was shown to be a primary cause of the failures, with the newer, redesigned parts producing superior mechanical test results.

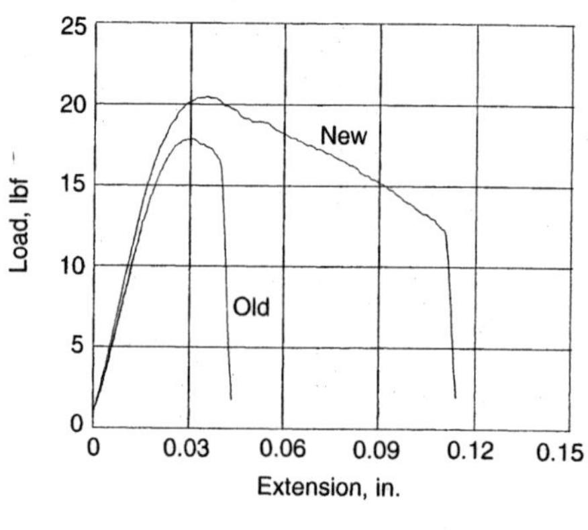

Fig. 33 A comparison of the mechanical test results, showing a significant improvement in the parts produced from the new design

Example 13: Failure of a Nylon Filtration Unit. A component of a water filtration unit failed while being used in service for approximately eight months. The filter system had been installed in a commercial laboratory, where it was stated to have been used exclusively in conjunction with deionized water. The failed part had been injection molded from a 30% glass-fiber- and mineral-reinforced nylon 12 resin.

Tests and Results. A visual examination of the filter component revealed significant cracking on the inner surface. The cracking ran along the longitudinal axis of the part and exhibited an irregular pattern. The surfaces of the part presented a flaky texture, without substantial integrity, and displayed significant discoloration. The irregular crack pattern, flaky texture, and discoloration were apparent on all surfaces of the part that had been in contact with the fluid passing through the component. Several of the crack surfaces were further examined using SEM. The fracture surface exhibited a coarse morphology, as illustrated in Fig. 34. The reinforcing glass fibers protruded unbounded from the surrounding polymeric matrix. The fracture surface also showed a network of secondary cracking. Overall, the observations made during the visual and SEM inspections were consistent with molecular degradation associated with chemical attack of the filter component material.

To allow further assessment of the failure, a mounted cross section was prepared through one of the cracks. The cross section, as presented in Fig. 35, showed a clear zone of degradation along the surface of the part that had contacted the fluid passing through the filter. The degradation zone extended into the cracks, which indicated massive chemical attack. The prepared cross section was analyzed using energy-dispersive x-ray spectroscopy, and the results obtained on the base material showed relatively high concentrations of silicon, calcium, and aluminum, with lesser amounts of sulfur and sodium in addition to carbon and oxygen. The results were consistent with a mineral- and glass-filled nylon resin. Analysis of the surface material, which exhibited obvious degradation, showed a generally similar elemental profile. However, significant levels of silver and chlorine were also found. This was important, because aqueous solutions of metallic chlorides are known to cause cracking and degradation within nylon resins.

The filter component material was further analyzed using micro-FTIR in the ATR mode. Analysis of the base material produced results characteristic of a glass- and mineral-filled nylon resin. However, analysis of the surface material showed additional absorption bands characteristic of substantial oxidation and hydrolysis of the nylon. A spectral comparison showing this is presented in Fig. 36. The presence of these bands is consistent with the high level of molecular degradation noted during the visual and SEM examinations.

Comparative TGA of the base material and the surface material also showed a significant difference. In particular, the results obtained on the surface material showed a lower temperature corresponding to the onset of polymer decomposition. This is illustrated in Fig. 37.

Conclusions. It was the conclusion of the evaluation that the filter component failed as a result of molecular degradation caused by the service conditions. Specifically, the part material had undergone severe chemical attack, including oxidation and hydrolysis, through contact with silver chloride. The source of the silver chloride was not established, but one potential source was photographic silver recovery.

Example 14: Failure of a PC Switch Housing. A housing used in conjunction with an electrical switch failed shortly after being placed into service. A relatively high failure rate had been encountered, corresponding to a recent production lot of the housings, and the failed part was representative of the problem. The housing had been injection molded from a commercially available, medium-viscosity grade of PC, formulated with an ultraviolet stabilizer. In addition to the PC housing, the design of the switch included an external protective zinc component installed with a snap-fit and two retained copper press-fit contact inserts. Control parts representing an earlier production lot were available for reference purposes.

Tests and Results. A visual examination of the submitted housing revealed massive cracking within the base of the part, including the retaining tabs securing the contacts. The fractures were primarily located adjacent to the copper contacts. Gray streaks, commonly referred to as splay, were also apparent on the PC housing, as shown in Fig. 38. Splay is often associated with molecular degradation, caused by insufficient drying or exposure to excessive heat, from the molding process. The visual examination also revealed that the contacts corresponding to the failed housing retaining tabs extended significantly, relative to contacts in nonfailed areas. This suggested a high level of interference stress between the contact and the tab. The fracture surface was further inspected using an optical stereomicroscope. The fracture surface showed no evidence of ductility, as would be evident in the form of stress whitening or permanent deformation. An oily residue was evident covering the crack surface.

The crack surface was further examined via SEM. The fracture surface exhibited multiple apparent crack origins and classic brittle fracture features, including hackle marks, river markings, and Wallner lines. A representative area on the fracture surface is shown in Fig. 39. No evidence of ductility, which would be apparent as stretched fibrils, was found. Overall, the observed features were indicative of brittle fracture associated with the exertion of stresses below the yield point of the material, over an extended period of time, by a creep mechanism.

The housing base material was analyzed using micro-FTIR in the ATR mode, and the resulting spectrum contained absorption bands characteristic of PC. The results produced an excellent match with a spectrum obtained on a reference part, without evidence of contamination. The oily residue found on the part, including the fracture surface, was also analyzed. The obtained spectrum was characteristic of an aliphatic hydrocarbon-based oil, with no signs of aromatic hydrocarbons or other chemicals known to produce stress cracking in PC resins.

The housing material was also analyzed using DSC. The DSC thermogram showed a single transition at 141 °C (286 °F), associated with the glass transition temperature (T_g) of the material. This temperature was somewhat lower than expected for a PC resin, which usually undergoes this transition closer to 150 °C (300 °F). This difference was thought to be an indication of potential molecular degradation.

Thermomechanical analysis was used to evaluate the failed retaining tab material, using an expansion probe. The TMA results confirmed

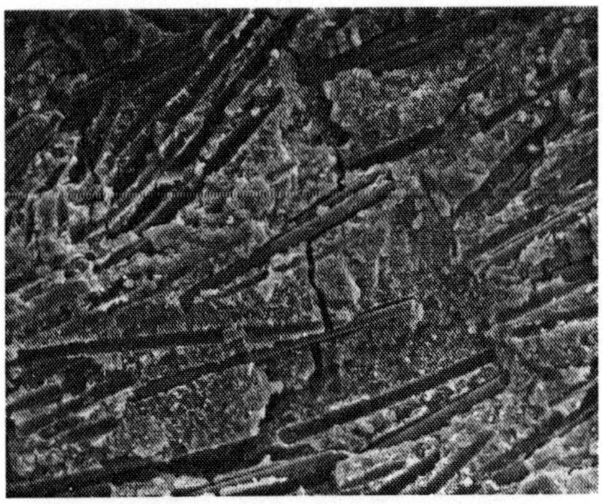

Fig. 34 Scanning electron image showing features characteristic of severe degradation of the filter material. 118×

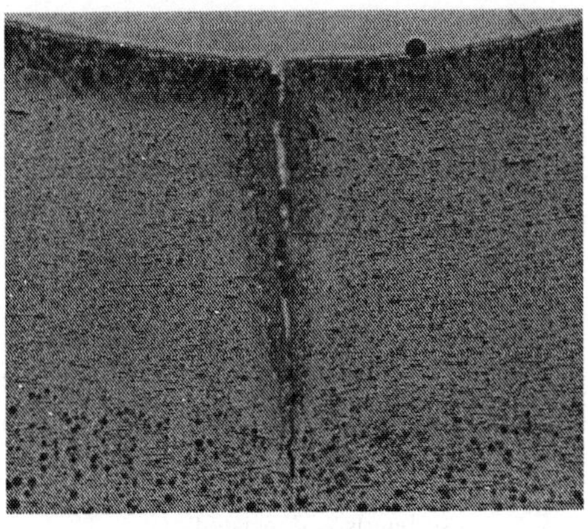

Fig. 35 Micrograph showing the cross section prepared through the filter component. 9×

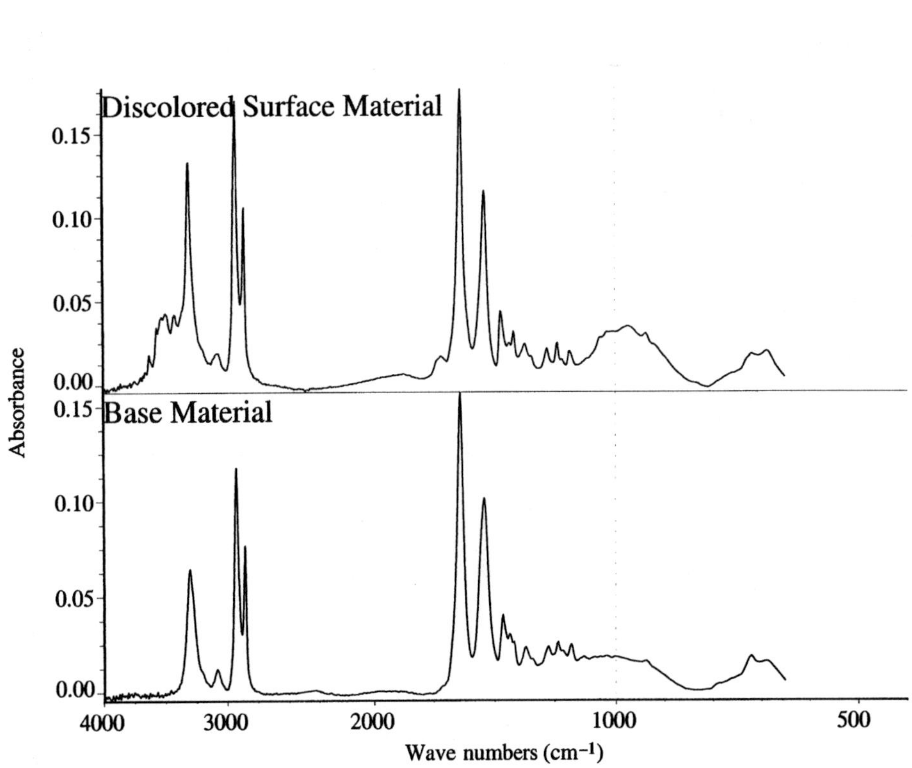

Fig. 36 Fourier transform infrared spectral comparison showing absorption bands associated with hydrolysis at 3350 cm^{-1} and oxidation at 1720 cm^{-1} in the results obtained on the discolored surface

the relatively low T_g, in particular, with a comparison to a reference part. A comparison showing this is presented in Fig. 40. No evidence was found in the results to indicate molded-in residual stress.

Melt flow testing of the housing samples showed the submitted reference part to have a MFR of 39.7 g/10 min, compared with 78.1 g/10 min for the failed components. The nominal value for the resin used to produce the housing was 9 to 12 g/10 min. This indicated not only severe molecular degradation within the failed housing material but also within the reference parts. The most likely source of the degradation was the molding process. This degradation was consistent with the presence of splay observed on the part as well as the reduced T_g.

Conclusions. It was the conclusion of the evaluation that the switch housings failed via brittle fracture, likely through a creep mechanism. The failure was caused by severe embrittlement of the housing resin associated with massive molecular degradation produced during the molding process. A potential contributing factor was the design of the part, which produced significant interference stresses between the contact and the mating retaining tab.

Example 15: Failure of Nylon Hinges. A production lot of mechanical hinges had failed during incoming quality-control testing. The hinges were used in an automotive application and had cracked during routine actuation testing. Similar parts had been through complete prototype evaluations without failure. However, a change in part supplier had taken place between the approval of the prototype parts and the receipt of the first lot of production parts. The mechanical hinges were specified to be injection molded from an impact-modified, 13% glass-fiber-reinforced nylon 6/6 resin. A resin substitution was suspected, corresponding to the supplier change. Samples representing the failed components and the original prototype parts were available for the failure investigation.

Tests and Results. A visual examination of the failed parts confirmed catastrophic cracking within the mechanical hinge in an area that would be under the highest level of stress during actuation. The failures did not show signs of macroductility, which would be apparent in the form of stress whitening and permanent deformation. The fracture surfaces of the failed parts were further inspected via SEM. While the presence of glass-reinforcing fibers can render a plastic resin inherently more brittle, a certain level of ductility is still expected at the 13% glass level. This ductility is often only apparent at high magnification and only between the individual glass fibers. However, the failed hinge components did not exhibit any signs of ductility even at high magnification, with the fracture surface showing only brittle features. A laboratory failure was created on one of the prototype parts by overloading the component. Examination of the fracture surface using SEM showed the normally anticipated level of ductility, as indicated by the overlapping, rose-petal morphology. The crack surfaces of both the failed part and the laboratory fracture are shown in Fig. 41.

Analysis of the failed components and the corresponding molding resin via micro-FTIR produced results characteristic of a nylon resin. The molding resin and failed parts generated generally similar results. However, a distinct difference was apparent in that the spectra obtained on the failed parts showed an additional absorption band at approximately 1740 cm^{-1}, indicative of partial oxidative degradation of the resin. A spectral comparison illustrating this is presented in Fig. 42. Because the parts had not yet been in service, this degradation was thought to have occurred during the molding process.

The failed parts were further tested using DSC. The obtained DSC results showed a melting point of 263 °C (505 °F), consistent with a nylon 6/6 resin. The molding resin was also analyzed via DSC, and a comparison of the results further indicated degradation of the molded nylon resin. This was apparent by a noted reduction in the heat of fusion in the results representing the failed parts.

The failed parts and the prototype parts were also analyzed using conventional thermogravimetric analysis (TGA), and both analyses produced results indicative of a nylon resin containing approximately 13% glass-fiber reinforcement. Further testing was performed using TGA in the high-resolution mode. This analysis

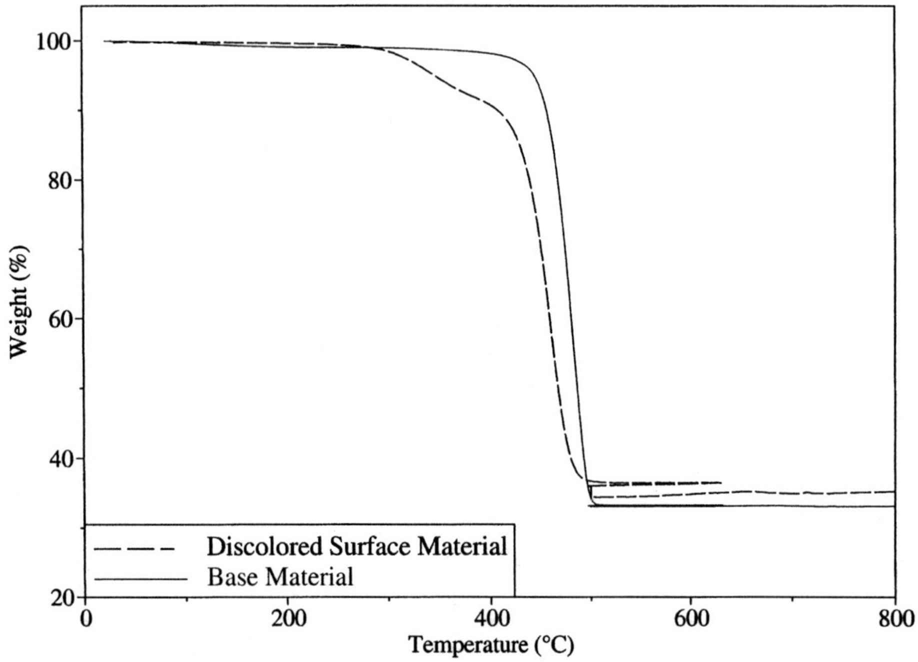

Fig. 37 Thermogravimetric analysis weight-loss profile comparison showing a reduction in the thermal stability of the discolored surface material relative to the base material

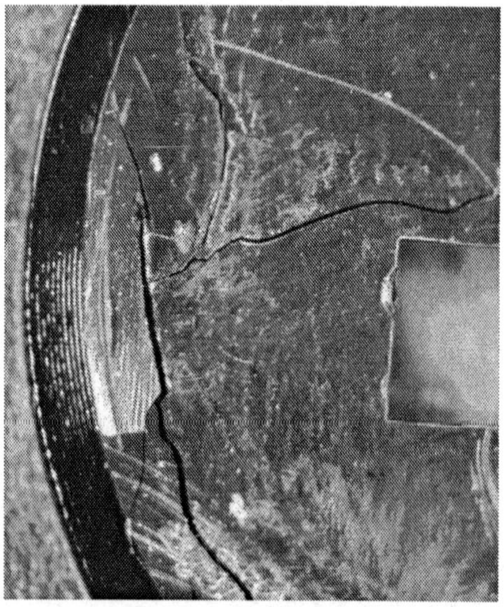

Fig. 38 A view of the housing showing gray streaks characteristic of splay

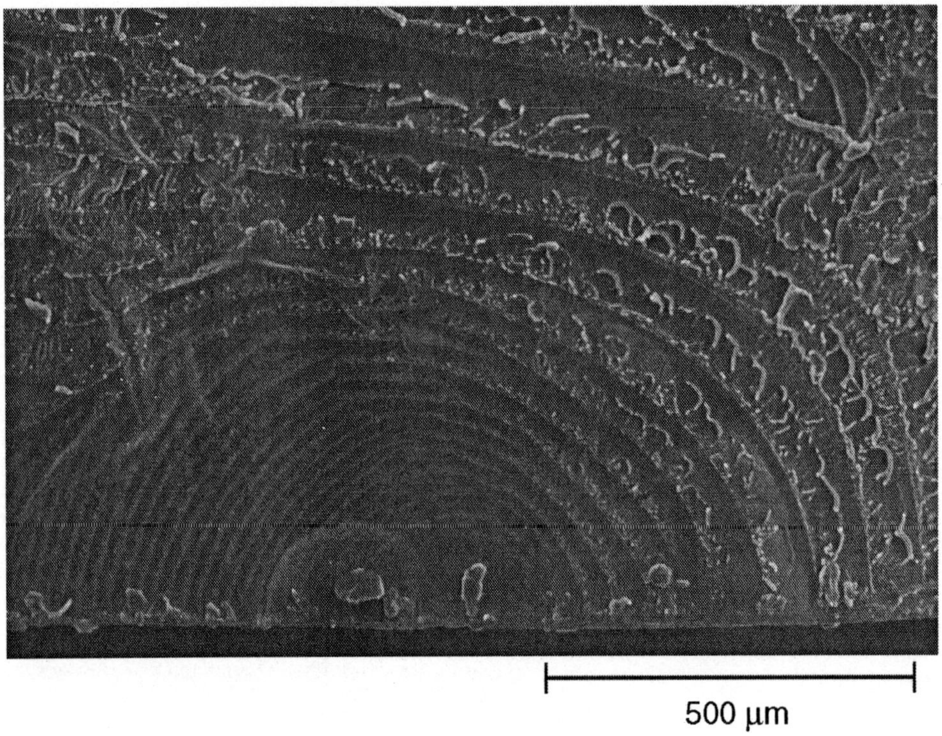

Fig. 39 Scanning electron image showing characteristic brittle fracture features on the housing crack surface. 100×

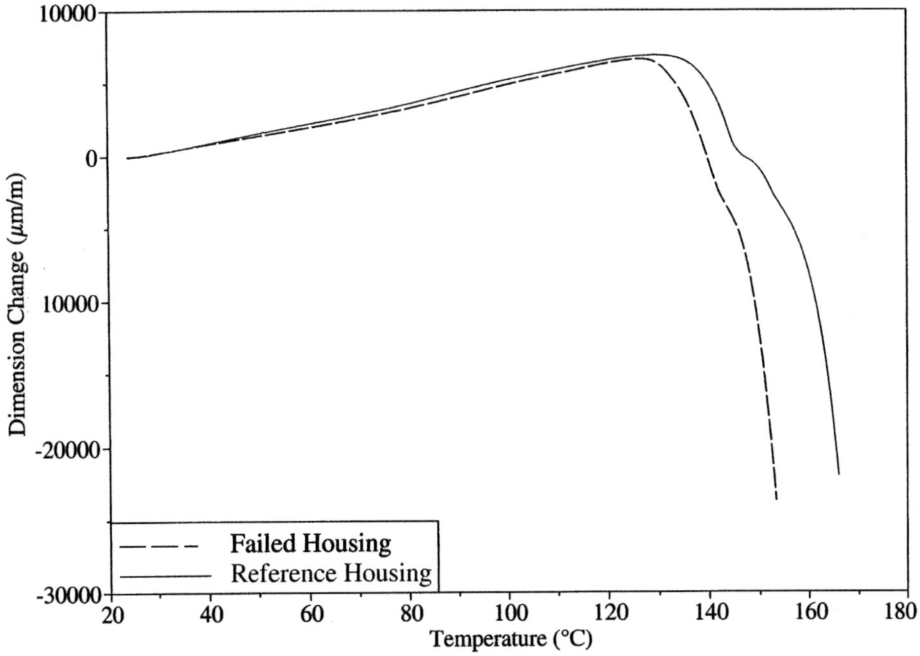

Fig. 40 The thermomechanical analysis results obtained on the failed and reference parts. The results exhibit differences corresponding to a reduction in the glass transition of the failed material.

was conducted in order to assess the level of impact-modifying rubber resin. The weight loss associated with the rubber was observed as a shoulder on the high-temperature side of the weight loss representing the nylon resin. This weight loss was particularly evident in the derivative curve. Because the weight losses could not be totally resolved, an absolute level of rubber could not be determined. However, a comparison of the results allowed a determination of the relative level of the impact modifiers in the two materials. This comparison showed a distinctly higher level of impact modifier in the prototype part material, relative to the failed part material.

Conclusions. It was the conclusion of this evaluation that the hinge assemblies failed through brittle fracture associated with stress overload during the actuation of the parts. The failed part material was found to be degraded, as indicated by both the FTIR and DSC analysis results. This degradation likely occurred either during the compounding of the resin or during the actual molding of the parts. A significant factor in the hinge failures is the conver-

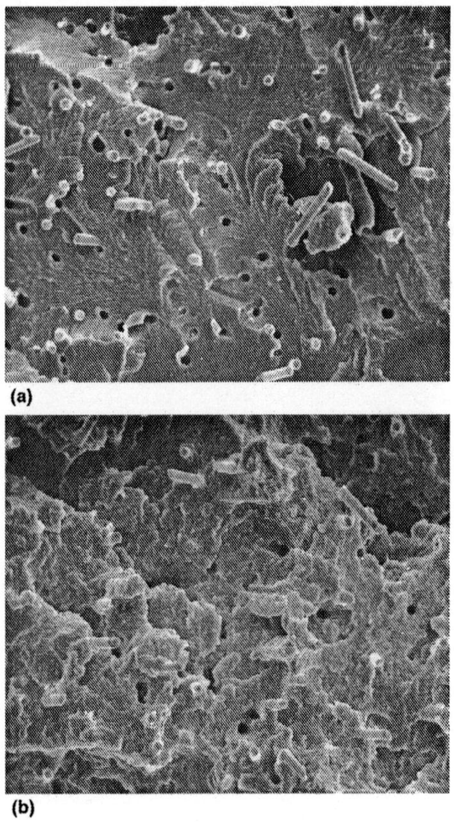

Fig. 41 Scanning electron images showing (a) brittle fracture features on the failed hinge and (b) ductile fracture features on the laboratory fracture. 118×

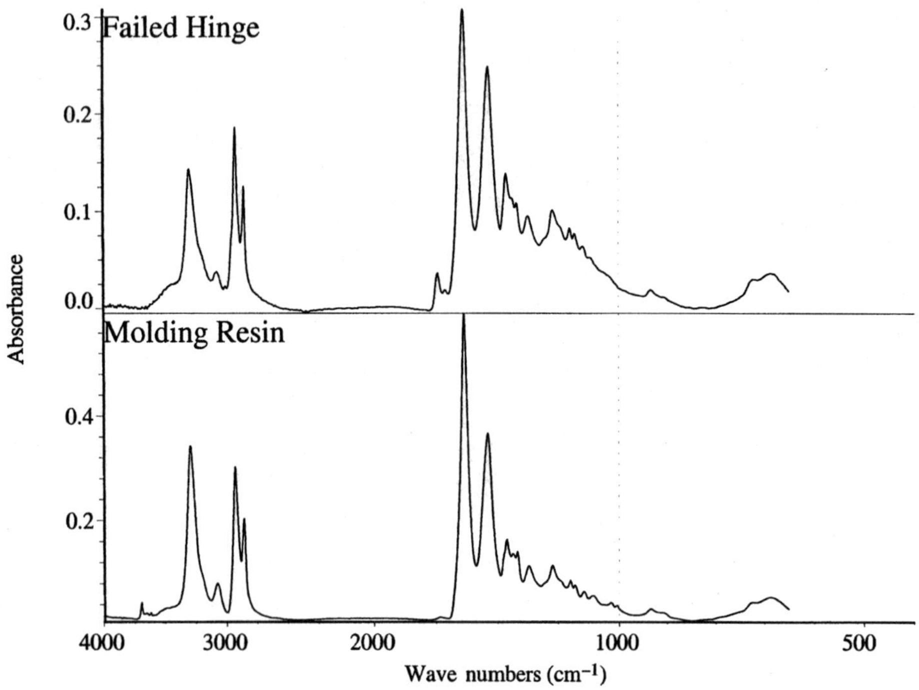

Fig. 42 Fourier transform infrared spectral comparison showing absorption bands at 1740 cm^{-1}, characteristic of oxidation within the results obtained on the failed parts

sion to a different grade of resin to produce the failed production parts as compared to the prototype parts. While both resins produced results characteristic of a 13% glass-fiber-reinforced, impact-modified nylon 6/6, the failed part material contained a significantly lower level of rubber. This decrease in rubber content rendered the parts less impact resistant and subsequently lowered the ductility of the molded hinge assemblies.

REFERENCES

1. J.A. Jansen, Conducting a Plastic Component Failure Investigation: Examples from the Appliance Industry, *International Appliance Technology Conference,* March 2002, p 2
2. J.A. Jansen, Plastic Component Failure Analysis, *Adv. Mater. Process.,* May 2001, p 56, 58, 59
3. A.T. Riga and E.A. Collins, Analysis of Structure, *Engineering Plastics,* Vol 2, *Engineered Materials Handbook,* ASM International, 1988, p 825, 826
4. J. Scheirs, *Compositional and Failure Analysis of Polymers,* John Wiley & Sons, 2000, p 109, 138, 153, 393, 415
5. M.P. Sepe, *Thermal Analysis of Polymers,* RAPRA Technology, Shawbury, U.K., 1997, p 3, 4, 8, 17, 19, 22, 24, 33
6. L.C. Roy Oberholtzer, General Design Considerations, *Engineering Plastics,* Vol 2, *Engineered Materials Handbook,* ASM International, 1988, p 21
7. "Polymer Characterization: Laboratory Techniques and Analysis," Noyes Publications, 1996, p 15
8. S.B. Driscoll, Physical, Chemical, and Thermal Analysis of Thermoplastic Resins, *Engineering Plastics,* Vol 2, *Engineered Materials Handbook,* ASM International, 1988, p 533
9. S. Turner, Mechanical Testing, *Engineering Plastics,* Vol 2, *Engineered Materials Handbook,* ASM International, 1988, p 545
10. M. Ezrin, Plastics Analysis: The Engineer's Resource for Troubleshooting Product and Process Problems and for Competitive Analysis, *Plast. Eng.,* Feb 2002, p 45, 46

SELECTED REFERENCES

- W. Brostow and R.D. Corneliussen, *Failure of Plastics,* Hanser Publishers, 1986
- T.R. Crompton, *Practical Polymer Analysis,* Plenum Press, 1993
- T.R. Crompton, *Manual of Plastics Analysis,* Plenum Press, 1998
- M. Ezrin, *Plastics Failure Guide,* Hanser Publishers, 1996
- *G.E. Engineering Thermoplastics Design Guide,* G.E. Plastics, 1997
- J.W. Gooch, *Analysis and Deformulation of Polymeric Materials,* Plenum Press, 1997
- J. Moalli, Ed., *Plastics Failure: Analysis and Prevention,* Plastics Design Library, 2001
- T.A. Osswald and G. Menges, *Materials Science of Polymers for Engineers,* Hanser Publishers, 1995
- R.C. Portney, Ed., *Medical Plastics: Degradation Resistance and Failure Analysis,* Plastic Design Library, 1998
- B.C. Smith, *Fundamentals of Fourier Transform Infrared Spectroscopy,* CRC Press, 1996
- E.A. Turi, Ed., *Thermal Characterization of Polymeric Materials,* Academic Press, Inc., 1981
- D. Wright, *Failure of Plastics and Rubber Products,* RAPRA Technology, Shawbury, U.K., 2001

Surface Analysis

MANY ANALYTICAL TECHNIQUES are available for the study and characterization of surfaces. These techniques provide data about the physical topography, physical properties, chemical composition, and chemical structure of the surfaces under study. Most of these techniques are based on bombarding the surface with photons, x-rays, ions, neutrons, or electrons and analyzing the radiation emitted and/or reflected from the surface. Other techniques use other interactions, such as physical probing of the surface.

Analyzing the chemistry and topography of failure surfaces is an important part of failure analysis. Many polymer materials depend on special treatment of surfaces. Surface analysis techniques can identify inadvertent contaminants introduced during manufacturing, storage, shipping, or handling. The deleterious effects of errors in the initial composition of ingredients, or of upsets in the manufacturing process, are often concentrated at surfaces and interfaces. Thus, even minute differences in the bulk can be magnified and detected easily at the surface. Similarly, environmental degradation often has its most pronounced effects at surfaces.

The workhorse instrument in surface analysis is a scanning electron microscope (SEM), which typically includes x-ray instrumentation for chemical characterization by energy-dispersive spectroscopy (EDS). In addition, other analytical techniques are available, either through an in-house laboratory or from an outside service laboratory. The most common analytical methods for chemical characterization of surfaces are shown in Table 1. The techniques to be applied to a particular failure depend on the type and size of the sample, the depth of analysis, the type of information sought, the ease of performing the analysis, the allowable destruction of the sample in either preparation or analysis, and the cost/time required. The information required about the surfaces in a failure analysis varies from failure to failure. No one technique can fully characterize a surface, but a full characterization is seldom required to solve a particular problem. Understanding the various analytical techniques allows an analyst to select the most appropriate method(s) to obtain the data needed for each failure. In many cases, a combination of analytical techniques may be required to evaluate the physical and chemical nature of the surface under study.

This article covers common techniques for surface characterization, including the modern SEM and methods for the chemical characterization of surfaces by Auger electron spectroscopy (AES), x-ray photoelectron spectroscopy (XPS), and time-of-flight secondary ion mass spectrometry (TOF-SIMS). Here, XPS is emphasized because of its preponderance in use for polymer analysis. Chemical characterization of surfaces by EDS instrumentation, which is commonly a module integrated with modern SEMs, is discussed in the section "Scanning Electron Microscopy" in this article. This article also highlights some principles of surface analysis and applications in polymer failure studies. Here, XPS is emphasized because of its preponderance in studies to date on polymer materials. Detailed physics of beam/specimen interactions and the electronics of instruments are not covered here. Instead, the focus is on qualitative and semiquantitative interpretation of spectra and those aspects of experimental technique that are important to practical failure analysis. Instrumentation and physics of these methods are described in more detail in *Materials Characterization,* Volume 10 of ASM Handbook.

Table 1 Evaluation techniques for chemical characterization of surfaces

Technique	Information	Analysis depth	Analysis area	Detection limit	Ease of use
EDS	Elemental	<5 μm	<1 μm	<1 at.%	Easy
WDS	Elemental	<5 μm	>1 μm	<0.1 at.%	Easy
AES	Elemental	<5 nm	>100 nm	<0.5 at.%	Moderate
XPS	Elemental, chemical structure	<5 cm	>10 μm	<0.1 at.%	Moderate
TOF-SIMS	Elemental, molecular structure	<2 nm	<5 μm	<1 ppm	Moderate
FTIR	Chemical structure	<5 μm	>10 μm	<100 ppm	Moderate
Raman	Chemical structure	>1 μm	>1 μm	<0.1 at.%	Difficult

Note: EDS, energy-dispersive spectroscopy; WDS, wavelength-dispersive spectroscopy; AES, Auger electron spectroscopy; XPS, x-ray photoelectron spectroscopy; TOF-SIMS, time-of-flight secondary ion mass spectrometry; FTIR, Fourier transform infrared (spectroscopy)

Scanning Electron Microscopy*

The SEM is one of the most versatile instruments for investigating the topology and chemistry of surfaces. Compared to the optical microscope, it expands the resolution range by more than 1 order of magnitude to approximately 10 nm in routine instruments, with ultimate values below 3 nm. Useful magnification thus extends beyond 10,000× up to 100,000×, closing the gap between the optical and the transmission electron microscope. Compared to optical microscopy, the depth of focus, ranging from 1 μm at 10,000× to 2 mm (0.08 in.) at 10×, is larger by more than 2 orders of magnitude. Scanning electron microscopes have been found particularly useful in failure analysis investigations, particularly because the SEM has the ability to image large, nonflat samples from low to high magnifications (approximately 10× to greater than 100,000×). The ability to obtain in-focus images of rough samples over a large change in vertical height is termed depth of field, and it is this trait that gives SEM images their characteristic three-dimensional appearance. The large depth of field is made possible by the relationship between the small size of the electron probe used as related to the size of the imaging pixel determined by the operating magnification (Ref 1).

While all SEMs exhibit similar characteristics, a number of variations exist, depending on the specific application and/or environment in which they are used. By changing the type of electron source used, the signals to be processed, and the type of vacuum systems, a SEM can be designed to enhance particular capabilities. For example, proper choice of the electron source can be used to enhance imaging, while changing the types of x-ray detectors can pro-

*Adapted from the article by L.S. Chumbley and L.D. Hanke, "Scanning Electron Microscopy," in *Failure Analysis and Prevention,* Volume 11, *ASM Handbook,* ASM International, 2002, pages 516 to 526 vide better chemical analysis. Modifying the vacuum system can allow examination of a wide range of sample types, including nonconductive or wet samples. Thus, a number of types of SEMs now exist with slightly different capabilities.

A conventional SEM typically uses a heated filament to produce electrons. These thermionic sources use tungsten or lanthanum hexaboride (LaB_6) as the filament material, tungsten being less expensive and more robust, while LaB_6 is brighter and lasts longer. Scanning electron microscopes that use a high electric field to remove electrons from a tungsten filament are termed cold-field-emission SEMs or high-resolution SEMs and have the highest resolution capabilities of all SEMs. A compromise is reached in the thermal-field-emission (or Schottky gun) SEM, where a combination of heat and electric field are used to produce electrons. These machines have better resolution than conventional SEMs, while being easier to use than the high-resolution machines.

A conventional SEM also requires that the sample be electrically conductive to prevent a charge buildup in the sample that affects the incoming primary and emitted secondary electrons, resulting in a poor, distorted image that is constantly changing in contrast and location. Because most polymers are not electrically conductive, they must be coated with a thin conductive layer, such as carbon or a metal, to allow examination using a standard SEM. The coating is applied by a method such as vacuum evaporation or sputtering. Coating must be performed carefully to avoid damaging the fracture surface (e.g., by excessive heating) or introducing artifacts associated with the coating process (Fig. 1, Ref 2).

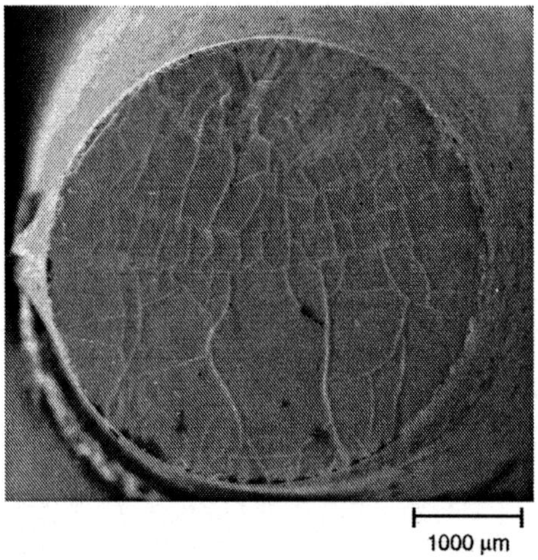

Fig. 1 Artifacts generated by improper platinum sputter coating of a 4.6 mm (0.18 in.) diameter polycarbonate rotating beam fatigue specimen. This SEM view shows a pattern in the coating reminiscent of "mudcracking." Source: Ref 2

Restrictions on nonconductive samples can be overcome by changing the design of the vacuum system and the choice of detection system used to image the sample. Microscopes adapted in this way are termed low- or variable-pressure microscopes. In this type of microscope, the pressure in the sample chamber is raised to a value on the order of 0.1 to 1 torr. Interaction of the electron beam with gas molecules in the region where the beam strikes the sample effectively creates a positively charged cloud of ions above the surface of the sample. This positive cloud offsets the negative charge buildup that occurs in insulating or poorly conductive samples and allows images of these types of samples to be obtained using backscattered electrons.

Modern microscopes can be purchased that allow the user to operate in either the high-vacuum, conventional imaging mode or in the low- or variable-pressure mode, making them extremely flexible for all types of investigations. The biggest advantage of low-pressure microscopes is that they allow the imaging of nonconductive surfaces, eliminating the coating that would be required for clear imaging in a conventional SEM. This is illustrated by Fig. 2, which shows a section of polyvinyl chloride tube that failed due to fatigue. The image obtained shows excellent contrast without necessitating that the sample be coated in any way. The chamber pressure for this particular example was 0.3 torr. Selecting a suitable chamber pressure is an important factor in using low-pressure SEMs.

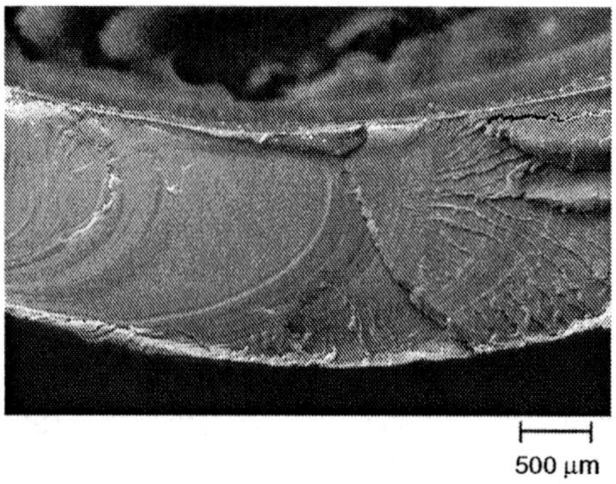

Fig. 2 Fatigue failure of a nonconductive polyvinyl chloride pipe imaged in the uncoated state using a low-pressure microscope. Source: Ref 1

If even greater capability to image nonconductive samples is desired, an environmental SEM can be purchased. This type of SEM operates at vacuum levels of 20 torr and allows both backscattered and secondary images to be obtained from the sample surface. This is made possible by the use of an electron detector that operates on the principle of induced current. In this instance, the interaction of the electron signals from the sample surface with the gas molecules above the sample induces a current in the detector, which consists of two parallel conductive plates, one placed above the sample and one that acts as the sample holder. By proper selection of where the current is measured, a secondary or backscattered image can be obtained.

Signals Generated by the Electron Beam. When an electron beam strikes a solid surface, electrons and x-rays are emitted from the surface. The energy distribution of these signals is shown qualitatively in Fig. 3. Electromagnetic radiation with energies lower than x-rays is also emitted. This is termed cathodoluminescence. There are three general types of instruments: the SEM, the scanning transmission electron microscope, and the scanning Auger microscope. These instruments all have in common the feature of obtaining information from the surface (or volume, for the scanning transmission electron microscope) of the sample by scanning an electron beam over a raster and analyzing the various signals generated. The scanning transmission electron microscope is closely related to the conventional transmission electron microscope (discussed in the article "Analytical Transmission Electron Microscopy" in *Materials Characterization,* Volume 10 of *ASM Handbook*). The scanning Auger microscope is designed to optimize information obtained from the Auger electron signal (as discussed in more detail in this article).

An image of the scanned surface region can be generated from any signal generated by the

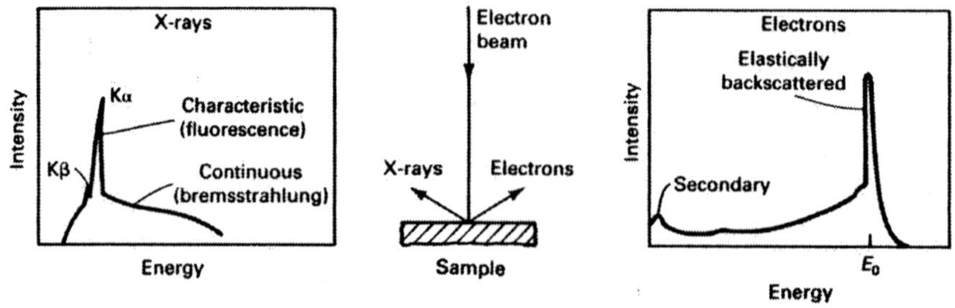

Fig. 3 Energy distribution of signals generated by the electron beam

electron beam. The electrons generated by the electron beam can be partitioned into three types: secondary, Auger, and backscattered (Fig. 4). The intensity scale shown in Fig. 4 has been increased to reveal the details not apparent in Fig. 3. High-resolution surface images from a SEM are typically generated by the detection of the secondary electrons that emanate from the surface. The secondary electron detector can also be used to detect backscattered electrons, although specialized backscatter detectors are available at relatively low cost. Generally, scanning transmission electron and scanning Auger microscopes also use a secondary electron detector; therefore, they can also operate in the SEM mode.

Secondary electrons generally display a peak intensity at approximately only 5 eV (Fig. 4). These electrons are the signal used to generate high-resolution images in the SEM. They can be generated by the primary electron beam or any scattered electron that passes near the surface. Comparison of Fig. 4(a) and (b) illustrates that the energies of the secondary and Auger electrons are fixed, but the backscattered electrons shift their energy values as the primary beam energy is changed.

In addition to the secondary electron detector, most SEMs are equipped with an x-ray detector to determine the energy of the emitted characteristic x-rays. Because each element in the periodic table has a different characteristic energy, the x-ray analyzer enables determination of the chemical analysis from point to point on the sample surface. After obtaining a scanned image of the surface with the secondary electron detector, the beam can be positioned over a particle or region of interest to obtain the x-ray spectrum from that point. Two types of x-ray detectors are used: wavelength-dispersive spectrometers and energy-dispersive spectrometers. Most SEMs are currently being equipped with energy-dispersive detectors. In contrast, the wavelength-dispersive detectors are large and slow. Because of their size, they are not compatible with scanning transmission electron microscopes. The energy-dispersive detector can be limited in detection of light elements, depending on the detector. Energy-dispersive spectroscopy is usually limited to elements with atomic numbers (Z) higher than beryllium; some detectors may be limited to analysis of elements above sodium ($Z = 11$).

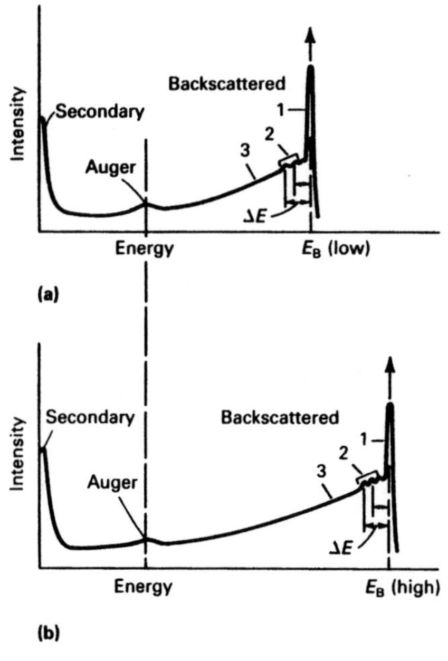

Fig. 4 The energy distribution of emitted electrons at (a) low beam energy (approximately 1 keV) and (b) a higher beam energy (approximately 5 keV)

Auger analysis has some advantages for light-element analysis. As with the characteristic x-ray emission, the energy of the Auger electrons is different for each element; therefore, analysis of Auger energies yields information on chemical identity. Characteristic x-rays and Auger electrons are generated as a result of the incoming electrons knocking out intershell electrons (K, L, and M, depending on atomic number) from atoms near the surface. These knock-out events occur within a scatter-volume near the surface. After a K electron is knocked out, the surface atom emits either a characteristic x-ray or an Auger electron. The probability for Auger emission exceeds that for x-ray emission as atomic number decreases. This is one of the reasons why Auger analysis has some advantages for light-element analysis. In addition, as with energy-loss electrons, the Auger energy levels sometimes shift when an atom becomes oxidized, nitrided, and so on; therefore, information on the chemical state of the surface atoms may sometimes be obtained from Auger analysis.

Scanning electron microscopes often contain a backscattered electron detector. These detectors generally measure an energy average of the three types of backscattered electrons:

- *Type 1:* elastically scattered
- *Type 2:* plasmon and interband transition scattered
- *Type 3:* inelastically scattered

Type 1 is referred to as primary backscattered electrons (BSEs), which are incident electrons that have undergone elastic Rutherford scattering from the nuclei of atoms in the sample. Imaging is somewhat different for primary (Type 1) BSEs. For this signal, as the size of the nuclei gets larger (i.e., higher atomic number), the BSE emission increases in a nearly linear fashion. Thus, for a flat, polished sample containing numerous phases, where little or no contrast may be seen using secondary electrons, an image of high contrast may still be produced using the BSE signal.

Unlike the secondary electrons that can be attracted to a detector, the high-energy BSE signal can only be collected by placing a detector in a position where it is likely to be struck by the emitted electrons. The most advantageous position is directly over the top of the sample. If the detector is designed such that different signals can be collected from different sides of the detector (usually divided into halves or quadrants), enhanced imaging, called topographical imaging, is possible by skillful use of the signals. By manipulating the signals received by different halves or quadrants of the BSE detector, small surface features that are indistinct in a secondary electron image can be imaged very well with BSEs. Topographical imaging works best on single-phase samples that are nearly flat and possess only slight surface roughness, such as pits or ledges.

Topographical imaging is possible due to the highly directional nature of the elastically scattered BSE signal. This characteristic, while useful in many investigations, also can cause confusion when using the BSE signal to observe rough samples. In this instance, it should be remembered that the image obtained is not directly related to the degree of surface roughness (as is the case for the secondary image) but is a function of how that roughness is angled with respect to the detectors rather than the incident beam. When imaging a rough sample using BSE, the image can be altered radically by changing how the detected BSE signals are manipulated. The signals from all detectors should be added together when viewing BSE images from a rough sample. This should be remembered when using low-pressure or variable-pressure SEMs, where only a BSE signal is available for imaging in the low-pressure mode. This is especially true if the sample is not single phase, because the difference in BSE emission due to varying atomic number adds an additional complication. Having noted these potential problems, BSE imaging of rough surfaces often can still provide a rapid identification of areas that may deserve closer scrutiny.

Figure 5(a) is an example of a satisfactory BSE image of a nonconductive material in a low-pressure microscope. Good contrast and resolution are obtained with 20 Pa gas pressure. The quality of the image contrast is degraded, however, if the pressure used for imaging becomes too great, as shown in Fig. 5(b).

Sample Volume Contributing to the Various Signals. To interpret correctly the physical significance of the various signals, it is important to understand the regions below the surface from which the signal is originating. This is shown schematically in Fig. 6. The Auger electrons are collected from sample depths of 0.5 to 3 nm below the surface, depending on their energy. In contrast, primary (characteristic) x-rays and BSEs emanate from deeper regions. It is important to realize that the x-ray sample volume and shape vary with the electron beam voltage and the sample atomic number. In general, higher voltages, lower density, and lower-atomic-number elements produce larger volumes, which tend to balloon out below the beam. Thin-foil samples can be used to reduce the x-ray generation volume. This is performed using the scanning transmission electron microscope; minimum sampling diameters of 30 nm can generally be achieved, with 5 nm possible in specialized instruments. Similar results can be obtained in a conventional SEM simply by using thin-foil samples, but the lower voltages of the SEM will produce larger volumes than for the scanning transmission electron microscope due to the volume shape change with voltage.

In summary, Tables 2 and 3 provide an overview of the three types of scanning electron beam instruments and summarize the source of the signals used. Table 2 compares the scanning transmission electron microscope and scanning Auger microscope with regard to top surface analysis. The scanning transmission electron microscope enables direct probing through the sample. With the scanning Auger microscope, surface films can be probed through by ion sputtering. The Auger signal in the scanning Auger microscope is much lower, and it is difficult to achieve resolutions exceeding 100 nm in the scanning Auger microscope using the Auger signal, because the lower Auger signal intensity requires the use of larger beam diameters. The scanning Auger microscope is extremely useful, however, because when equipped with secondary electron and x-ray detectors, it becomes a SEM, an electron probe microanalyzer, a scanning Auger microscope, and a scanning electron loss microscope.

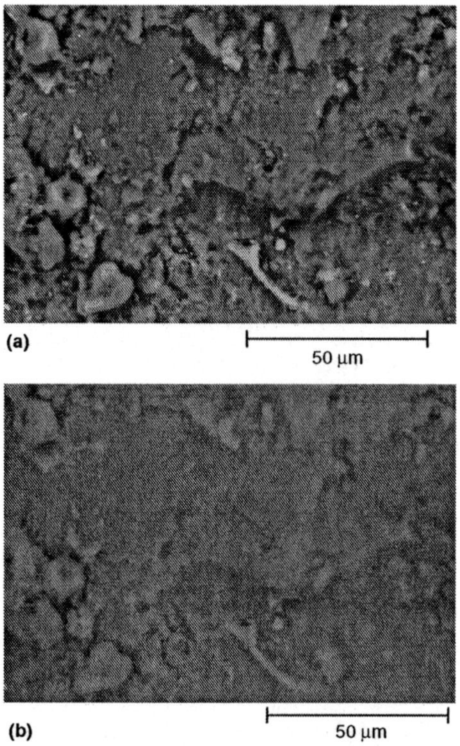

Fig. 5 Nonconductive material backscattered electron image using low-pressure imaging with (a) 20 Pa gas pressure and (b) 270 Pa gas pressure

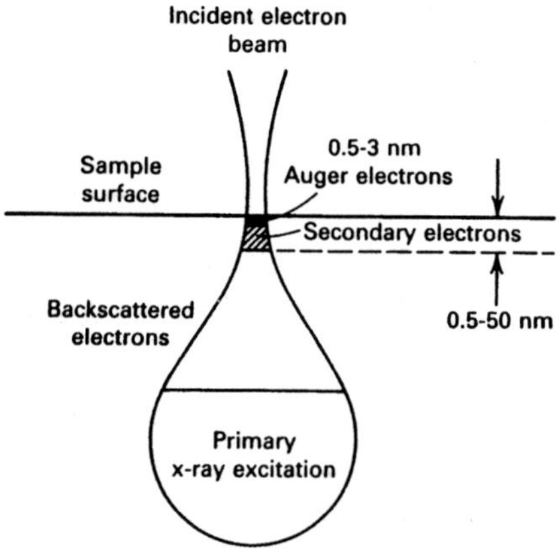

Fig. 6 Comparison of Auger electron escape depths with emission depths of backscattered electrons and x-rays

Chemical Characterization of Surfaces*

In many cases, a combination of analytical techniques (Table 1) may be required to evaluate the physical and chemical nature of the surface under study. This section briefly reviews the chemical characterization of surfaces by AES, XPS, and TOF-SIMS. Fourier transform infrared spectroscopy also is used extensively in the analysis of polymers (see the article "Characterization of Plastics in Failure Analysis" in this book).

Table 4 is a summary chart of techniques discussed in this section. In this brief review, it is not possible to develop all the capabilities of each technique, but some of the more important attributes are highlighted for a preliminary insight into the strength and usefulness of these techniques for chemical characterization of surfaces.

Overview of Surface Analysis

What Is Surface Analysis? The surface of a sample can mean different things to different people. To those performing analytical tests at the air/specimen interface of samples, the word *surface* can be anywhere from the top monolayer to as deep as several micrometers into the sample, depending on the depth of analysis of the technique that is being applied. Although there are scores of different techniques currently being used to study the surfaces of materials, only a fraction of these can be classified as true surface analysis tools that derive the majority of their analytical signal from the top few atomic layers. Three such techniques include AES, XPS—also known as electron spectroscopy for chemical analysis (ESCA)—and TOF-SIMS. The electron spectroscopy techniques of AES and XPS have depths of analysis, on average, of approximately 5 nm, while TOF-SIMS is even more surface sensitive, deriving the majority of its signal from the top 2 nm. Because these techniques are so surface sensitive, they are often employed in failure analysis or general surface characterization. Typical applications include identification of thin layers of contaminants on surfaces or at interfaces, evaluation of cleaning processes, and the identification of stains and discolorations.

Successful failure analysis can often be a matter of piecing together the different bits of information that different techniques can provide. Each of these three surface analysis techniques provides a different view of the sample surface and thus a different piece of the puzzle. This section describes the basic theory behind each of the different techniques, the types of data produced from each, and some typical applications. Table 4 summarizes the different features of these techniques to allow for at-a-glance comparisons. Also discussed are the different types of samples that can be analyzed and the special sample-handling procedures that must be implemented when preparing to do failure analysis using these surface-sensitive techniques.

Analytical Considerations. Proper design of experiment and the choice of samples for analysis are important factors in the likelihood of success when determining the cause of a failure. Failures can occur in the manufacturing of a device or some time after the device has been in use. In the ideal situation, good versus bad samples can be compared. This allows for differences in elemental concentrations and chemistries to be observed and related to their effects on device failure. Attempts are then made to link these differences to a known step in the manufacturing process or suspected contaminant in the end use of the product. The cause of failure may be obvious from the contaminants observed, or it may be extremely subtle. If a contaminant is

*Adapted from the article by John G. Newman, "Chemical Characterization of Surfaces," in *Failure Analysis and Prevention*, Volume 11, *ASM Handbook*, ASM International, 2002, p 527–537

Table 2 Comparison summary of scanning electron beam instruments equipped with secondary electron and x-ray detectors

Instrument	Features optimized	Minimum area: Surface pictures	Minimum area: Microchemical analysis	Comments
Scanning electron microscope	Surface pictures: above ~500× on polished and etched samples; at all magnifications on high depth-of-field surfaces; accuracy and sensitivity of microchemical analysis	4–5 nm (conventional scanning electron microscope) 2–3 nm (in-lens)	1–3 μm (EDS and/or WDS)	Can be equipped with a WDS x-ray detector that maximizes sensitivity and light-element analysis. WDS: elements with $Z > 4$. EDS: elements with $Z > 10$
Scanning transmission electron microscope	Small area microanalysis of thin films; small area diffraction	2–3 nm (SEM mode, in-lens)	5–30 nm (EDS)	Samples must be thinned; generally also functions as a transmission electron microscope; allows chemical analysis of particles characterized by transmission electron microscope observation
Scanning Auger microscope	Chemical analysis of (1) monolayers on surfaces made by in situ fracturing, and (2) low-Z elements on surfaces cleaned by in situ ion etching	~100 nm (Auger) 10 nm (SEM mode)	~100 nm (Auger) 1–3 μm (EDS)	Requires ultrahigh vacuum and careful surface preparation; can also detect electron loss signal

Table 3 Comparison summary of signals used in scanning electron beam instruments

Signal type	Type	Energy	Source	Use
X-ray	Characteristic (fluorescent)	Discrete values; different for each element: Cu Kα ~ 8000 eV; Si Kα ~ 1800 eV	Interband transitions: L→ K = Kα, M → K = Kβ; Kα: (1) lose K electron, (2) L → K, (3) photon ejects	Chemical analysis from micro areas in SEM, STEM, and SAM
	Continuous	Continuous	Deceleration electron	None (background noise)
Electron	Auger	Discrete values; different for each element; range: 100–1500 eV; Si LMM ~ 100 eV; Cu LMM ~900 eV	Interband transitions; LMM: (1) lose L electron, (2) M → L, (3) M electron ejects	Monolayer surface analysis in SAM
	Backscattered (elastic)	Essentially same as beam energy	Beam electron scattered back after elastic collision	Atomic number contrast, channeling contrast, channeling patterns, and magnetic contrast in SEM
	Backscattered (inelastic)	Energies less than beam energy	Beam electron scattered back after inelastic collision	
	Backscattered (plasmon and interband transition interactions)	1–1000 eV less than beam energy	Beam electron scattered back after collision, producing plasmon oscillations or interband transition	Surface analysis in SAM; light-element analysis in STEM where scattering is in forward direction
	Secondary	~5 eV	Loosely bound electrons scattered from surface	Main signal for image formation in SEM

SEM, scanning electron microscope; STEM, scanning transmission electron microscope; SAM, scanning Auger microscope

Table 4 General features of AES, XPS, and TOF-SIMS

Feature	Technique: AES	XPS	TOF-SIMS
Probe beam	Electrons	X-ray photons	Ions
Analyzed beam	Electrons	Electrons	Ions
Average sampling depth	5 nm	5 nm	2 nm
Detection limits	10^{-3}	10^{-4}	10^{-6}
Spatial resolution	10 nm	5–10 μm	150 nm
Information	Mostly elemental, SEM photos	Elemental, chemical	Elemental, molecular
Strengths	Ultimate small area analysis, imaging	Ease of use, quantification	Chemical and molecular analysis, imaging
Limitations	Semiconductive	Very few	Quantification difficult
Major applications	Semiconductors, electronics	All Industries	Polymers, contamination, trace metal analysis

found, but its root source is not known, then the next step, if possible, is to analyze samples and materials extracted at various steps in the manufacturing cycle or end use of the product.

In other instances, a direct comparison of a failed material to a good sample is difficult or even impossible, such as in interface delamination problems, where the bonding failure does not occur on good samples. In these cases, data from the failed area should be compared to reference materials to infer what should or should not be present. Reference materials can include starting substrate materials (before and after cleaning processes); pure compounds and coatings used in the manufacturing of the device; dried solvents, oils, greases, and cleaning solution residues used in processing; and any material that may have come in contact with the device.

In all cases, the surface analyst must rely on the knowledge of the process engineer to provide as much detail as possible regarding the nature of the failure and the history of the sample of interest. Doing analysis on samples where no processing and handling history are provided can result in the incorrect interpretation of the data and results that are very misleading.

Sample-Handling Issues. In no other class of analytical techniques is sample handling as critical an issue. Because AES, XPS, and TOF-SIMS provide information from the top few atomic layers, extreme care must be taken to keep the surface of interest from being covered up with extraneous contamination. Physically touching a sample with ungloved and even gloved hands is a common source of handling contamination. With ungloved hands, finger oils, salts, hand lotions, and other contaminants are easily transferred to touched surfaces. Many hand lotions contain siloxane, commonly referred to as silicone, as one of their main ingredients. Siloxanes are known to easily spread across surfaces. Thus, touching any part of a sample with siloxane-contaminated hands can lead to large areas of the sample becoming contaminated. Many assume that if sample handling is performed with gloved hands, their samples are safe from contamination. Unfortunately, the surfaces of many gloves are also laden with oils, salts, and/or lubricants that can be transferred to the sample during handling (Ref 3). Even so-called clean-room gloves are not necessarily clean of such contaminants, because they are manufactured more to be particle-free than contaminant-free. Common surface contaminants on powder-free gloves include silicon, chlorine, sodium, and zinc. Touching the area to be investigated with unclean tools can also introduce impurities onto the area of interest. In general, nothing should ever come in physical contact with the area of interest. Even the air surrounding the sample should be kept free of smoke and particles. Caution should also be taken to make sure that samples do not come in contact with common plastic bags during storage or shipping, because additives in the plastic can transfer to the sample and be detected with surface-sensitive tools. Clean, plastic petri dishes (polystyrene or polypropylene are common), glass containers, and even plain white typing paper are all acceptable ways of storing or shipping samples that will later be analyzed by one or more surface analysis methods. If tape must be used to secure samples during shipping, avoid using tapes that are laden with silicones, such as many conducting carbon tapes. Many clear cellophane tapes are fairly clean of silicones, but make sure to avoid placing the tape in the vicinity of the area of interest on the sample.

Sample Types. Most solid materials, if they are vacuum compatible and are of the proper geometric size to be accepted into the surface analysis instrument, can be analyzed by one or more of these surface analysis tools. This includes, but is not limited to, wafers, sheets, films, coatings, foils, chunks, powders, wires, tubes, printed circuit boards, computer chips, and other whole or partial pieces of devices. While surface analysis tools are used to analyze solid surfaces, dried residues from liquids and even low-vapor-pressure oils and greases can also be looked at. A minute drop or thin layer of the liquid or oil of interest can be placed on a clean substrate (silicon wafer, aluminum foil, etc.) and allowed to dry in air prior to its analysis.

One consideration, however, is the electrical property of the sample—is it conducting or insulating? In general, AES is better adapted to running conducting and semiconducting materials, while both conducting and insulating materials are readily analyzed by XPS and TOF-SIMS.

Auger Electron Spectroscopy

Auger electron spectroscopy is a surface analysis technique used to determine the elemental composition of the top few atomic layers of a surface or exposed interface in a solid material. Auger electron spectroscopy can detect all elements except hydrogen and helium and can provide semiquantitative information, with detection limits of 0.1 to 1 at.% for most elements. The most attractive attribute of AES is its ability to analyze extremely small features. Modern AES instruments with field-emission electron sources can produce secondary electron micrographs with spatial resolutions as small as 10 nm and can characterize sample features as small as 25 nm.

Auger electron spectroscopy is performed under ultrahigh vacuum conditions, using an electron beam typically in the 3 to 25 keV range. The AES process begins with electron bombardment of the sample material, causing an atom to eject an inner-shell electron, thus forming a vacancy. A second electron from a higher shell fills this inner-shell vacancy in an energy gain process. This energy can then cause the ejection of a third electron, referred to as an Auger electron. The energy of the escaping Auger electron is analyzed by an electron spectrometer. Because each element has a unique set of electron energies surrounding the nucleus, it also has a unique set of Auger peaks, and the resulting spectrum provides a fingerprint of the probed surface. The kinetic energy of the Auger electrons is typically between 40 and 2500 eV. In this energy regime, electrons can travel only a short distance before interacting with other atoms and losing energy. This short distance is referred to as the escape depth of the electron. Escape depths range from 0.5 to 10 nm, depending on the kinetic energy of the emitted electron and the material being analyzed. It is this small range of escape depths that gives AES its surface sensitivity. Those electrons that are close enough to the surface to escape without loss of energy are detected as Auger electron peaks. Those electrons that lose energy before leaving the sample surface add to the background of the spectrum.

As previously noted, scanning Auger microscopy is accomplished by scanning an electron beam across the surface of a sample while measuring resultant electron signals. This scanning process generates secondary electron microscopy images, BSE images, and Auger maps. Secondary electron microscopy images, which provide a topographic view of the sample by detecting low-energy electrons emitted from the surface, are used to locate specific areas for more detailed study. Backscattered electron images, involving higher-energy electrons that have undergone scattering processes before escaping from the sample, reveal atomic number contrast and crystallographic information. Auger maps, obtained by measuring the emitted Auger electron intensity while scanning the electron beam, reveal the lateral distribution of elements across the sample surface.

Because of the uncompensated loss of secondary electrons during electron bombardment, AES typically has a difficult time with insulating materials and, therefore, is primarily used for analyzing conducting and semiconducting solids. However, AES analysis of some inorganic insulating materials is also possible. The high charge density of the AES beam makes it impossible to use the technique on many insulating specimens. Nevertheless, in many suspected polymer failure cases, Auger-ion milling depth profiles can be obtained quickly from both failure surfaces, giving enough initial information to make preliminary hypotheses and to decide on the next steps to better understanding the cause of failure.

X-Ray Photoelectron Spectroscopy

X-ray photoelectron spectroscopy, also known as ESCA, is a surface analysis technique that can be used to determine both the elemental and chemical composition of the outermost atomic layers of a solid material. With the exception of hydrogen and helium, all elements can be detected. Detection limits are, on average, approximately 0.1 at.%; however, many of the heavier elements can be detected down to approximately 0.01 at.% (100 ppm). By monitoring subtle shifts in the atomic binding energies of the emitted photoelectrons, chemical speciation can be obtained from both organic and inorganic materials. Similar to AES, the average depth of analysis is approximately 5 nm.

X-ray photoelectron spectroscopy is accomplished by flooding the sample with x-rays of a known energy (typically Mg $K\alpha$ at 1253.6 eV or monochromated Al $K\alpha$ at 1486.7 eV). Absorption of these x-rays by the sample atoms causes photoelectrons to be emitted. The kinetic energy of the emitted photoelectrons is measured with an electron spectrometer. To a first approximation, the kinetic energy is determined from the following equation:

$$KE = h\nu - BE$$

where KE is the measured kinetic energy of the emitted photoelectron, $h\nu$ is the energy of the x-rays being used, and BE is the atomic binding energy associated with the emitted photoelectron. To maintain consistency in the spectral energy scale when using different x-ray energies, the data are displayed on a binding-energy, rather than kinetic-energy, scale. The binding energy associated with a peak is then used to establish its elemental identity and chemical state. The incoming x-rays penetrate microns into the surface of the sample. However, the escape depths for these low-kinetic-energy (less than 1500 eV) photoelectrons are less than 10 nm, thus making XPS an extremely surface-sensitive technique.

Quantification is possible with the use of elemental sensitivity factors (Ref 4) that have been determined empirically and found to be in agreement with the current theoretical models for quantification of XPS data.

If the sample is insulating, x-ray bombardment can create a positive-charge buildup on the surface of the sample, causing all of the spectral peaks to shift to apparently higher binding energies. To minimize this charging phenomenon, low-energy electrons or a combination of electron and ion floods (Ref 5) may be added to the sample surface. All peaks are then referenced to a peak of known energy, such as carbon, because hydrocarbons are present on most surfaces. This charge correction method allows one to obtain useful chemical-state information from insulating samples.

Common materials analyzed by XPS include metals, lubricants, semiconductors, metal oxides, glasses, ceramics, catalysts, plastics/polymers, coatings/thin films, and paper.

Types of XPS Data

Because the technique is so flexible in its sample-handling capabilities, the industries served by XPS are quite varied. Application areas include bond pads, paint and other thin-film adhesion problems, cleanliness concerns, corrosion analysis, identification of debris and discolorations, characterization of polymer surface functionalization, lubricant thickness measurements, optical and other thin-film profiling, and determination of oxidation state and oxide thickness of alloys. Types of XPS data, as briefly described subsequently, include:

- Survey spectra
- High-resolution spectra
- Depth profiles
- Angle-dependent analysis
- Maps and line scans

See the selected references at the end of this article for recommended reading on XPS.

Survey Spectra. Similar to AES, the first piece of analytical data obtained in an XPS experiment is typically a survey spectrum. A low-resolution XPS spectrum of an ethylene-chlorotrifluoroethylene copolymer is shown in Fig. 7. It is worth noting that the abscissa decreases in binding energy because the measured quantity, kinetic energy, is increasing. Each peak is labelled according to atom and orbital. Contamination is immediately apparent from the foreign atom, oxygen (1s orbital ~ 530 eV). Two different photoelectrons appear for chlorine (2s and 2p orbitals) and fluorine (1s and 2s orbitals). Also, there is a set of three fluorine Auger electron peaks between 600 and 650 eV. These electrons arise from a relaxation process that occurs immediately after photoionization. A diagram of orbital energy levels is shown in Fig. 8, with ionization of an inner (1s) orbital, or core level, depicted on the left. An Auger relaxation (Fig. 8b) fills the core hole with an electron from a higher level, with simultaneous emission of another electron from an equivalent or even higher-energy orbital. The multitude of possible combinations for Auger processes gives rise to a broad band of emission, often with much discrete structure. Auger peaks have not played a significant role in polymer failure analysis, but the difference in energy between principal photoemission and Auger electron peaks (the Auger parameter) has been used for refined chemical-state identification.

High-Resolution Spectra. Once it is determined what species are present on a surface, high-resolution scans are typically acquired on specific elements of interest to monitor binding-energy shifts that can provide chemical information. These scans are obtained using much higher energy-resolution conditions on the spectrometer compared to the survey spectra. Because the peak shapes obtained in this mode are more accurate, quantification is also better. It is on these high-resolution scans where curve-fits and other mathematical routines are performed to extract chemical-state information and quantification.

More information is contained in high-resolution XPS spectra that are collected for the most intense core level for each element. Figure 9 shows such a spectrum of the carbon 1s level from the specimen of Fig. 7, illustrating the chemical shift effect. The chemical structure of this copolymer contains C–H, C–Cl, and C–F bonds. Higher binding energies are due to the electron withdrawing (oxidizing) power of the substituent on carbon. Thus, many organic functional groups can be identified, and valencies in inorganic compounds can be determined.

Also, the peak areas are very reproducible and can be used to compute element ratios, thus providing a level of quantitative analysis. From first principles, the peak intensity, I, can be converted to atomic concentration, N:

$$N = \frac{I}{FK\sigma\lambda}$$

where F is the x-ray flux (kept constant), σ is the photoemission cross section (probability), λ is the electron mean free path, and K is a spectrometer function. It is possible to obtain reliable values for the conversion factor with the use of a few calibration standards. Commercial instruments have software packages that convert relative peak intensities into atomic concentration.

Other useful features in the XPS spectra are related to the bonding electrons in the valence band. The spectra of molecular orbitals between the Fermi level (BE = 0) and BE = 40 eV do provide "fingerprint" characteristics, but sensitivity and resolution are low on typical failure specimens. For cases in which resolution is good, additional levels of information on structure and bonding are available. Finally, a small peak that is 6 to 7 eV lower in kinetic energy than the main carbon peak appears in the spectrum of polymers with unsaturation (for example, polystyrene and polybutadiene). This so-called shake-up satellite can be useful in distinguishing among the polymers that, although comprising only carbon and hydrogen, have various amounts of carbon-carbon double bonds and conjugation.

Another example of the type of localized chemical bonding information that can be obtained with XPS is shown in the high-resolution carbon spectrum of polyethylene terephthalate (PET) (Fig. 10). Three relatively large peaks and one smaller peak are observed. The largest peak in the spectrum, located at a binding energy of 284.8 eV, is assigned to hydrocarbon-type carbon (C–C or C–H) where carbon is bound only to other carbon or hydrogen. In PET, this peak is associated with the nonoxidized carbons in the aromatic ring. At approximately

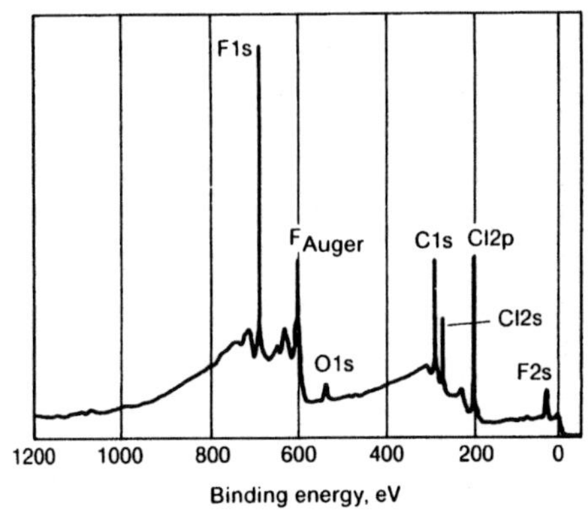

Fig. 7 Low-resolution x-ray photoelectron spectroscopy spectrum of an ethylene-chlorotrifluoroethylene copolymer

286.3 eV, a C–O peak is observed that is characteristic of the ethylene carbons in PET that are singly bound to oxygen. Located at 288.7 eV is a peak associated with the carboxylate carbons (ester O=C–O) that are bound to two oxygens. A small satellite peak is also observed at approximately 291.5 eV. This peak is indicative of the aromaticity within the PET molecule.

Depth Profiles. Elemental and, in some cases, chemical depth profiling is possible with the use of an inert gas sputter ion gun. The ion gun is used to slowly remove material, thus exposing a new surface to be analyzed. By sequentially sputtering and taking XPS data, a compositional depth profile can be generated. If the chemical matrix of the sample is not severely damaged by the sputter ion beam, chemical information may also be obtained as a function of depth. However, many oxides and especially organics are very susceptible to chemical modification during ion bombardment. Similar to AES, exact sputter rates are often difficult to determine. Therefore, sputter rates are often specified relative to a known sputter rate on a reference material, such as thermally grown SiO_2 on single-crystal silicon.

Angle-Dependent Analysis. Angle-dependent XPS is a method for nondestructively analyzing the in-depth chemical gradients in the top surface layers (<10 nm) of a smooth sample. By changing the sample tilt with respect to the energy analyzer (takeoff angle), the effective depth of analysis can easily be changed. This technique can be useful in cases where ion bombardment is known to modify the chemical makeup of the material being probed. It has been greatly used for studying such things as the depth of modification of plasma- and corona-treated polymers, the oxidation of metals and semiconductors, polymer-metal adhesion, additive surface migration, and the lubrication of various materials.

Maps and Line Scans. Many newer-generation XPS instruments are capable of obtaining spatially resolved information from a sample surface by acquiring photoelectron maps and line scans. A photoelectron map is a two-dimensional display of photoelectron intensity for a specific element from a given area on the sample surface. High intensities (bright areas) on the map indicate that more of that particular element or chemistry is present at that point than at lower-intensity (darker) points. Line scans are similar to maps, except that the data are obtained in one direction only. Spatial resolutions for state-of-the-art instruments are on the order of 5 to 10 μm.

XPS Application

The XPS technique is limited to the top few atomic layers of a specimen because of the short mean free path, λ, of electrons. Inelastic scattering causes an exponential decrease with depth (x) in the number of electrons detected with the initial photoemission kinetic energy, I_0:

$$I = I_0 \, e^{-x/\lambda}$$

In the range of 100 to 1000 eV, λ is approximately related to the electron kinetic energy through a power law with an exponent of approximately 0.75. Therefore, the depth analyzed does vary from atom to atom (that is, orbital binding energy). Analysis depth is specified by some multiple of the mean free path; that is, 63% of the signal emerges from 1λ deep, 90% from 2λ, and so on. Roughly speaking, when aluminum or magnesium anodes are used, XPS samples a depth of from 0.5 to 10 nm (5 to 100 Å).

Gradients of composition versus depth can be identified by analyzing the variation in peak intensity ratios either with changes in the angle between the specimen plane and the direction to the lens or analyzer slit, or with variation in photoelectron kinetic energy. The latter approach can be used when an element emits two photoelectrons with enough difference in energy to change the analysis depth appreciably. Also, there are multiple anodes for XPS that change the analysis depth because of variable incoming x-ray energy. For planar, relatively smooth specimens, a commonly used method is the so-called angular-dependent depth profile. As illustrated in Fig. 11, the analysis depth decreases in proportion to the sine of the angle between the plane of the specimen surface and the photoemission direction. Most modern spectrometers have the capability to manipulate that angle with the specimen in place. For failure analysis, two photoelectron exit angles—normal (90°) and grazing (10 to 30°)—are usually enough to indicate whether there are gradients within the top 2 to 10 atomic layers.

Experimental Techniques. Naturally, one must be concerned about inadvertent contamination of surfaces after the failure event, because the analysis is limited to the first few atomic layers. Often, it is possible to ignore or compensate for adventitious carbon, but great care must be taken during sample handling and transportation to avoid additional contamination. With a single fingerprint, an analysis can be hopelessly complicated. One effective transmit-

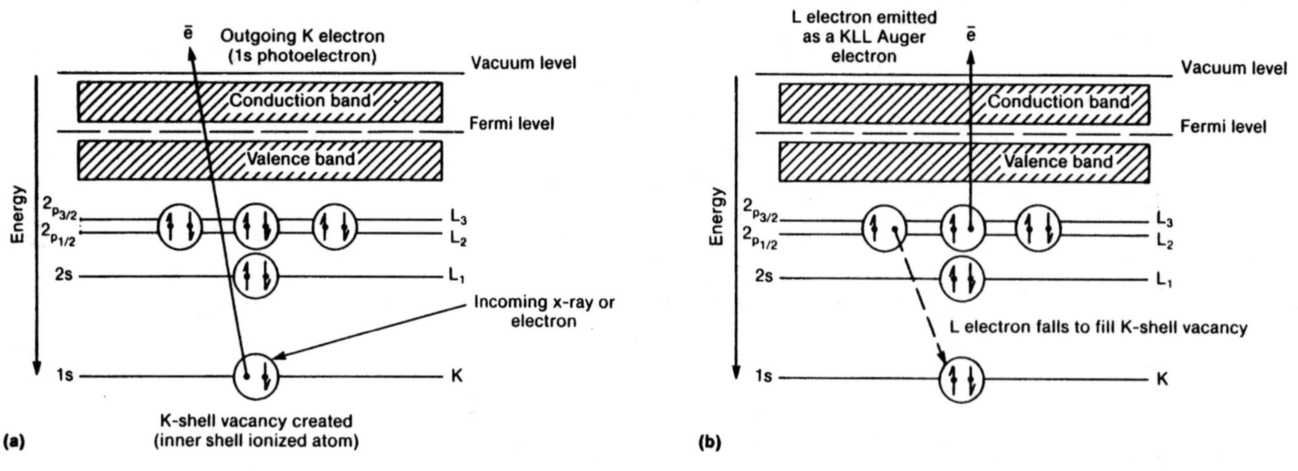

Fig. 8 Orbital energy level diagrams. (a) Photoelectron emission. (b) Auger relaxation

tal approach is to fasten two identical specimen surfaces in contact with each other, then cut them to size, and leave them fastened until just prior to analysis. Cotton gloves and tweezers must be used during handling, transfer, and mounting.

The components of a typical XPS spectrometer are shown in Fig. 12. High-vacuum pumps maintain the sample chamber and hemispherical electron energy analyzer at a pressure of 10^{-7} Pa (10^{-9} torr) or lower. This is necessary to minimize collisions of photoelectrons with gas molecules and to avoid additional carbonaceous contamination of the specimen during x-ray exposure. Specially designed x-ray sources are separated from the sample chamber by a thin sheet of aluminum foil. Some instruments are equipped with crystal diffraction x-ray monochrometers, which permit better resolution and signal-to-noise ratio. Photoelectrons emitted from the specimen are transferred by a lens system to a hemispheric capacitor electron energy analyzer. Single channeltron electron multiplier tubes are used as the collector, although the trend in new instruments is toward position-sensitive detectors. The spectrum is obtained by varying lens and analyzer potentials and stepping a narrow kinetic-energy window across the range of interest.

The specimen must be transferred into the analysis chamber through an intermediate vacuum lock chamber, with a single specimen fixed to the end of an insertion rod. Newer instruments can accept multiple specimens on a large carousel. Cotton gloves are always worn during this stage to minimize contamination. Rough vacuum pumping of the prechamber is followed by opening the valve to the analysis chamber and advancing the specimen holder into position in front of the x-ray window and the lens and analyzer. Data collection times range from minutes to hours, depending on the number of elements involved and the level of detail required. With a multiple-specimen carousel operating around the clock under computer control, it is possible to obtain XPS analysis of two dozen or more specimens per day. In the practice of failure analysis, however, it is more likely that the analyses will be conducted by investigators who can select succeeding steps based on interpretation of the preceding spectra. One to two hours per specimen is typical.

It is common for the XPS analysis chamber to be equipped with a gun that ionizes a gas (usually argon) and accelerates the ions onto the specimen surface. Ion impact causes clusters of atoms or fragments to be sputtered off the specimen. One purpose behind this is to remove contaminating overlayers, while another is to mill away layers of the specimen and determine changes in composition with increasing time periods of ion beam sputtering. Figure 13, an idealized sketch of a specimen cross section during ion bombardment, illustrates the principle behind this method of obtaining a so-called chemical depth profile.

Time-of-Flight Secondary Ion Mass Spectrometry

Time-of-flight secondary ion mass spectrometry is an analytical technique that uses a high-energy, primary ion beam to probe the surface of a solid material. The instrument is typically operated in the static mode for obtaining elemental and molecular-chemical information from both organics and inorganics. In this mode of operation, the sample integrity and chemistry are preserved by applying extremely low primary ion doses (less than 1×10^{12} ions/cm^2) during the entire experiment. This ensures that roughly less than 0.1% of the surface atoms or molecules are ever struck and damaged by the primary ion beam. Time-of-flight secondary ion mass spectrometry is the most surface sensitive of the surface analytical techniques, with a depth of analysis of only approximately 2 nm. The technique has extremely good detection sensitivities, with detection limits for most elements in the parts-per-thousand to parts-per-million range. By using a finely focused ion beam (typically, gallium or indium), it is also possible to record the lateral distribution of chemical species across a surface with micron to submicron spatial resolution. Because secondary ion

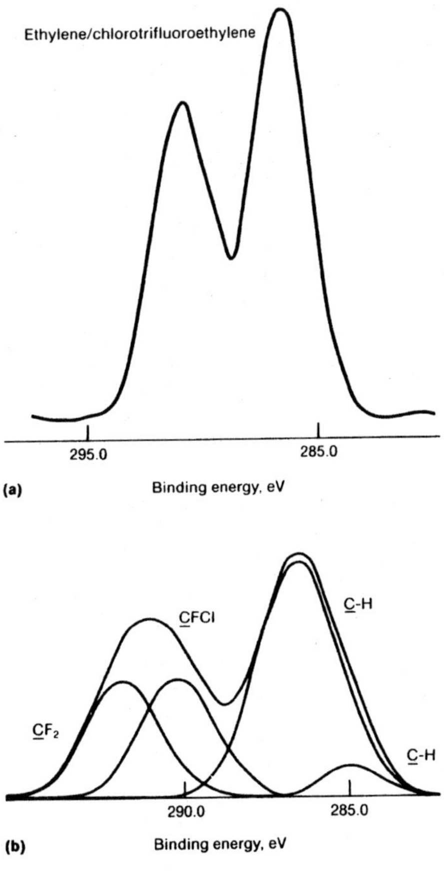

Fig. 9 High-resolution x-ray photoelectron spectroscopy spectrum of the carbon 1s region from Fig. 7. (a) Raw data. (b) Computer curve-fit, showing four individual components

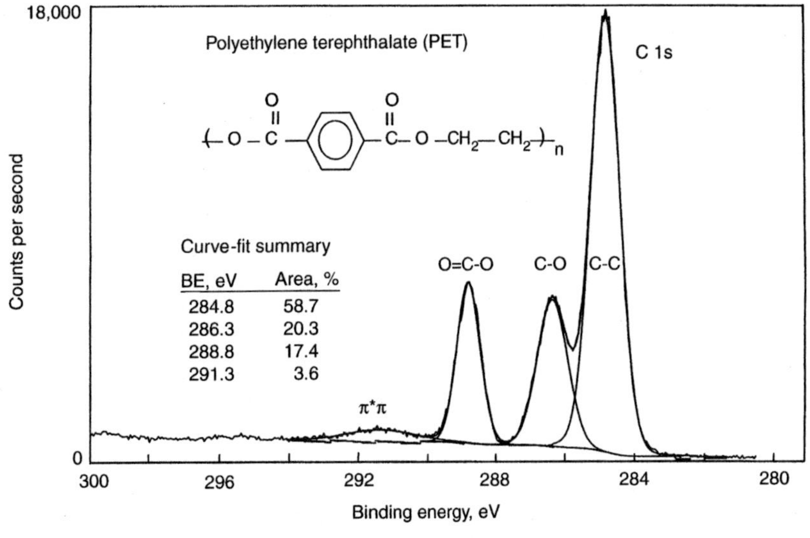

Fig. 10 X-ray photoelectron spectroscopy high-resolution spectrum of polyethylene terephthalate

intensities vary dramatically from element to element and are highly dependent on the matrix from which they are sputtered, quantification with TOF-SIMS can be extremely difficult.

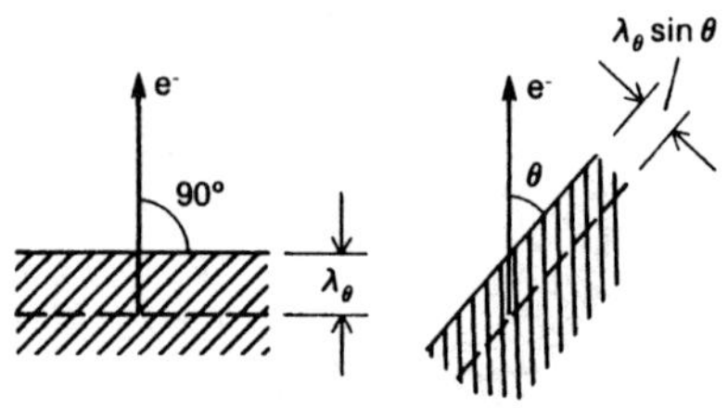

Fig. 11 Angular-dependent method for determining compositional gradients with x-ray photoelectron spectroscopy. Depth analyzed is proportional to sin θ.

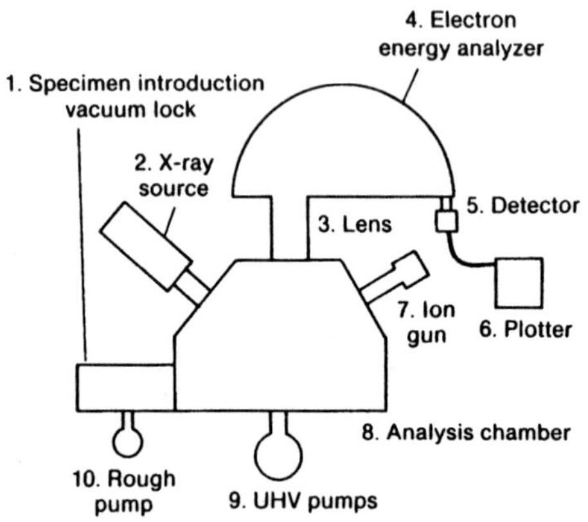

Fig. 12 Block diagram of a typical x-ray photoelectron spectroscopy spectrometer. UHV, ultra-high vacuum

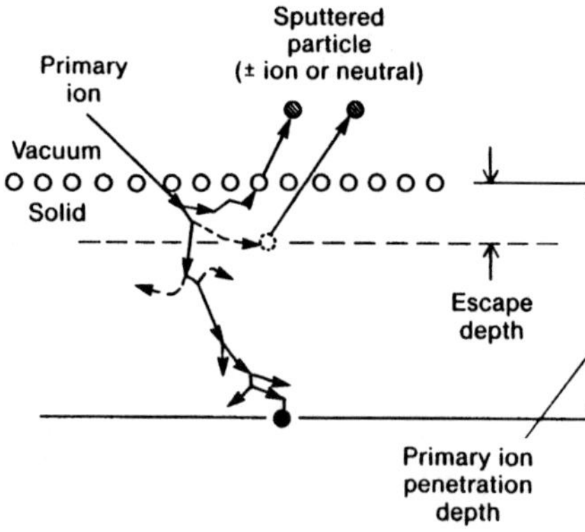

Fig. 13 Ion impact removal of atoms or clusters from solid surfaces. Mass analysis of the sputtered particles is the basis of the static SIMS technique. Simultaneous Auger electron spectroscopy analysis of the bottom of the etch crater produces chemical depth profiles.

Thus, in most cases, the technique is used more for qualitative purposes than for quantitative analyses. However, in select cases where appropriate standards are available (e.g., metals on silicon), accurate quantification can be performed (Ref 6).

When elemental depth profiles are required, most TOF-SIMS systems can also be operated as a dynamic SIMS instrument. In the dynamic mode, very high primary ion doses are used in order to obtain in-depth information. However, in this mode of operation, very little chemical specificity can be gleaned, due to the destructive nature of the primary ion beam. Therefore, the following discussion only focuses on the static mode of operation.

In a TOF-SIMS experiment, the desorption/ejection of secondary ions from the surface of a material is initiated by a short pulse (~1 ns) of primary ions that impinges on the surface at high angles of incidence. The momentum transfer from the primary beam to the solid initiates a collision cascade within the solid, much like a microscopic billiard game. A portion of this momentum is redirected back toward the surface, resulting in the ejection of atomic and molecular ions. Greater than 90% of these secondary ions originate from the outermost one to two layers of the solid, thus defining TOF-SIMS as an extremely surface-sensitive technique. By applying a potential between the sample surface and the mass analyzer, the desorbed secondary ions are extracted into a TOF mass spectrometer where their masses are separated in flight time, with very high accuracy, based on their mass-to-charge ratio (m/z). The resulting mass spectrum, typically in the 1 to 2000 dalton range, allows for the unambiguous identification of chemical moieties that can often be correlated to the original surface structure. An example of the type of chemical/molecular information that can be obtained from secondary ion mass spectra is shown in the positive ion spectrum of PET (Fig. 14). The spectrum shows a variety of large molecular cluster ions that are identified as fragments of the PET polymer chain. This provides information on the long-range molecular makeup of the sample as well as providing a unique fingerprint for this material.

On insulating materials, the loss of secondary electrons during ion bombardment can lead to a positive charge buildup and loss of signal from the sample surface. Therefore, auxiliary electron sources are often used to supply electrons to the sample surface to help neutralize the excess charge.

Because TOF-SIMS can easily handle insulating as well as conducting materials, the types of samples analyzed by TOF are often similar to those analyzed with XPS. However, because of its much smaller analytical probe size, TOF-SIMS systems can analyze features as small as a few micrometers in width and are commonly used for mapping distributions of elements and molecules across surfaces.

Application areas for TOF-SIMS include organic and inorganic contaminant identification, cleaning studies, surface segregation and modifications, corrosion, discolorations, drug distribution, diffusion studies, and chemical characterization. Two types of secondary ion data are simultaneously obtained in a TOF-SIMS experiment:

- A total-area mass spectrum
- A total secondary ion image

As the focused primary ion beam is digitally rastered across the sample surface, complete mass spectra of ion intensities versus mass-to-charge ratio are obtained at every pixel (256 × 256) within the raster. The summation of these spectra produces a total area mass spectrum (Fig. 15). From the same acquisition, a total ion image is also obtained that shows the lateral distribution of secondary ion signals from across the area of analysis. High intensities (bright areas) on the image indicate that more secondary ion signal is present at those points than at lower-intensity (darker) points. The variation in secondary ion intensity can be due to topographical effects or from differences in chemical composition. The total ion image is used to selectively analyze the mass spectra (chemistries) from areas that show differing amounts of brightness. Conversely, the total ion mass spectrum can be used to select specific elements or molecules for display in secondary ion maps that show the relative localized abundance of these species. Because both positive and negative secondary ions are created during ion bombardment, two separate experiments must be performed for complete characterization of a given sample.

Application Examples*

Modern instrumental surface analysis techniques, especially XPS (also known as ESCA), AES, and secondary ion mass spectroscopy (SIMS), provide a wealth of information about the chemistry of the top few atomic layers of solids. To varying degrees, small lateral dimensions (microphases) also can be resolved. This section highlights some applications in polymer failure studies. X-ray photoelectron spectroscopy is emphasized because of its prepon-

*Adapted from the article by David W. Dwight, "Surface Analysis," in *Engineering Plastics*, Volume 2, *Engineered Materials Handbook*, ASM International, 1988, pages 811 to 823 derance in studies to date on polymer materials. Because the microelectronics industry offers a variety of interfaces during manufacturing, case histories from that industry are described. Many other studies relevant to different areas, such as anticorrosion coatings, plasma or corona surface modification, and biocompatibility, are identified in the selected references in this article.

Ideally, the failure analyst would like to have available to him definitive information about the surface chemistry in respect to composition and bonding as well as its variation at the nanometer level, laterally and with depth into the bulk. The results of one analysis of a particular failure often cannot be interpreted unambiguously. It is fortunate that several techniques are available. Difficult problems require two or more types of analysis, after which a self-consistent picture must be derived from the combined results.

Features of the three techniques are compared in Table 5. X-ray photoelectron spectroscopy is relatively nondestructive and provides information on oxidation state, as well as quantitative elemental analysis. Information on composition versus depth can be obtained on smooth specimens, but rough specimens present serious limitations. Lateral resolution also is limited, because even in specially designed apparatus, the smallest spot is approximately 100 μm (3940 μin.). However, it should be noted that small spot XPS becomes more viable as more high-resolution instruments become available.

Of course, it is important to understand the topography of failure surfaces. The combination of XPS and SEM often provides enough information to generate a hypothesis about failure mechanisms. In a reasonable percentage of practical cases, the first analysis does point to some corrective action at the production level. Also, the results of preliminary failure analysis often lead to quick laboratory experiments to substantiate hypotheses. Surface analysis of accelerated-failure test specimens can be compared with analysis of field failures to clarify any differences in failure mechanisms.

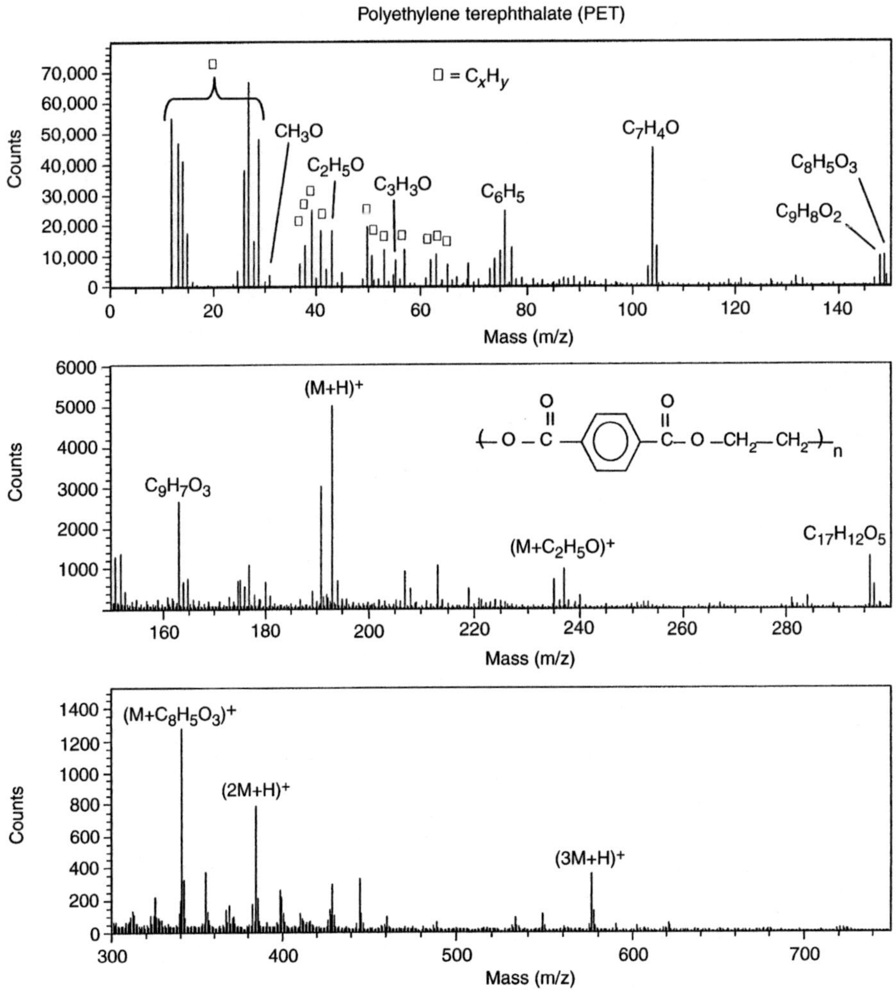

Fig. 14 Time-of-flight secondary ion mass spectroscopy mass spectrum of polyethylene terephthalate

Example 1: Delamination of Polyester Insulation from Brass Cable Connectors

A specialized connector designed for rapid, automatic clamping was punched from a laminate consisting of a tin-plated brass conductor adhesively bonded to a polyester film. After a couple of years of trouble-free use, a significant amount of the product began to delaminate during installation, or even in storage prior to use. The initial approaches to the problem included SEM analysis of failure surfaces as well as variations in the tin-plating step, the adhesives, and the polyester film. The failure surfaces looked smooth in the SEM, and the preliminary conclusion was that either the adhesive or the polyester film was the cause of the delaminations.

Analysis of Failures in the Field. Samples that delaminated in the field (designated *B*) were clamped lightly together for transmittal to the laboratory, thereby avoiding contamination of the failure surfaces. For comparison, a strongly bonded laminate (designated *A*) was peeled apart (requiring considerable force) and used for control surfaces. For both samples, there were two failure surfaces: the polyester side and the brass side. The first steps in the analysis were AES survey spectra of the brass side of both specimens. An important new insight was gained immediately from a comparison of the spectra shown in Fig. 16. Zinc is a constituent of the tin oxide surface at the top of the tin plate on the brass in sample *B* (Fig. 16b) but not sample *A* (Fig. 16a). The Auger spectrometer was then set to collect the spectral peak intensities for carbon, tin, oxygen, and zinc, while the surface was progressively etched away by argon ion bombardment. The results (composition versus depth) are shown in Fig. 17. Sample *A* has sharply decreasing carbon and oxygen constituents in the outer 3 to 4 nm (30 to 40 Å) and a corresponding sharply increasing tin component. These results are expected for the top of tin-plated brass: a thin tin oxide layer, containing some carbonaceous material (perhaps residual adhesive), or else a contaminant deriving from the organic brighteners or grain-size control agents used in the plating solutions. However, in sample *B*, it is clear that zinc and oxygen dominated the top surface and decreased only gradually throughout the profile in the low-strength case. Undoubtedly, the zinc diffused from the brass substrate throughout the tin plate.

This finding had an immediate impact on the problem-solving process: A suspected correlation of the onset of delamination problems with a change in the manufacturing process was corroborated. The change had involved a delay of days to weeks between the plating operation and the adhesive lamination of the polyester sheet. Formerly, the lamination had been done right after plating. Now, it all made sense in light of the AES and ion-milling depth profiles, which identified zinc oxide on the surface of the delaminated sample: The process change gave much more time for zinc to diffuse through the tin plate.

Comparison of the carbon, oxygen, and zinc compositions obtained with the angular-dependent XPS (Table 6) confirms the AES profile results. That is, in the low-strength case, the top 1.0 to 2.0 nm (10 to 20 Å) have more zinc and oxygen than the high-strength sample. Therefore, it is concluded that neither ion nor electron beam changed the atom concentration significantly during Auger profile measurement, because similar results were obtained with x-rays only. The difference between the grazing-exit and normal-exit angle results shows a carbon gradient from 95 to 82% in the strong-bond case, while in the weak-bond case, corresponding values of 79 and 59% confirm the depletion of the organic phase in surface and subsurface.

The subtitle of Table 6 summarizes the information obtained on oxidation states by analysis of the binding energy shifts in the high-resolution XPS spectra. First, tin in the top surface layer had a binding energy corresponding to tin oxide. After ion etching, the XPS peak position indicated only tin metal. The XPS result for zinc before and after sputtering showed only high binding energy (oxide) and no shift in peak position. This corresponds to the parallel zinc and oxygen compositions in the Auger profile. Moreover, zinc oxide was found on both sides of the low-strength fracture surface.

The third piece of information obtained from XPS is critical to the interpretation of the source of carbon. The question is whether the carbon on the fracture surfaces is polymer or components from the plating bath. The polymer spectra have high-binding-energy "shoulders," or peaks, corresponding to carbon bonded to oxygen and nitrogen. However, only a single, symmetric carbon peak appeared, and no nitrogen, on the brass side of the failure surface.

Laboratory Simulation of the Problem. Following the leads developed from analysis of the field samples, two laboratory specimens were studied to confirm the initial hypotheses and to develop an accelerated test for zinc diffusion. A thin tin coat was electroplated on a brass sheet, and one piece was immediately stored in a nitrogen atmosphere. A second piece was aged at 90 °C (195 °F) and high humidity for 1½ h and stored under nitrogen before analysis.

The tin plate on the laboratory samples was much thinner than on the production pieces. The AES results showed that zinc oxide formed in the surface region. In the accelerated-aging test (mild conditions), a significant increase in zinc oxide was shown, and carbon in the surface region was reduced by a factor of 2. These results confirm the diffusion of zinc through tin plate, forming an oxide interphase region with a concomitant decrease in carbon components in the surface region. These results indicate a potential synergistic effect of additive residues in the interphase, apparently acting as coupling agents to enhance bonding to polymers. The results are summarized as follows:

Observations

- High zinc and oxygen, low carbon on low-strength field and laboratory aged
- More ZnO and SnO on surface only; SnO in subsurface
- ZnO on polyester only in low-strength field sample

Conclusions

- Zinc diffuses rapidly through tin plate; process is accelerated by H_2O and temperature; forms surface oxide/hydroxide; displaces carbon
- Low strength yields brittle fracture in $Zn(OH)_2$
- Immediate lamination retards diffusion
- Carbon in tin plate acts as a coupling agent

On the low-strength field sample and the laboratory-aged sample, there were high concentrations of zinc and oxygen and low concentrations of carbon relative to the high-strength field sam-

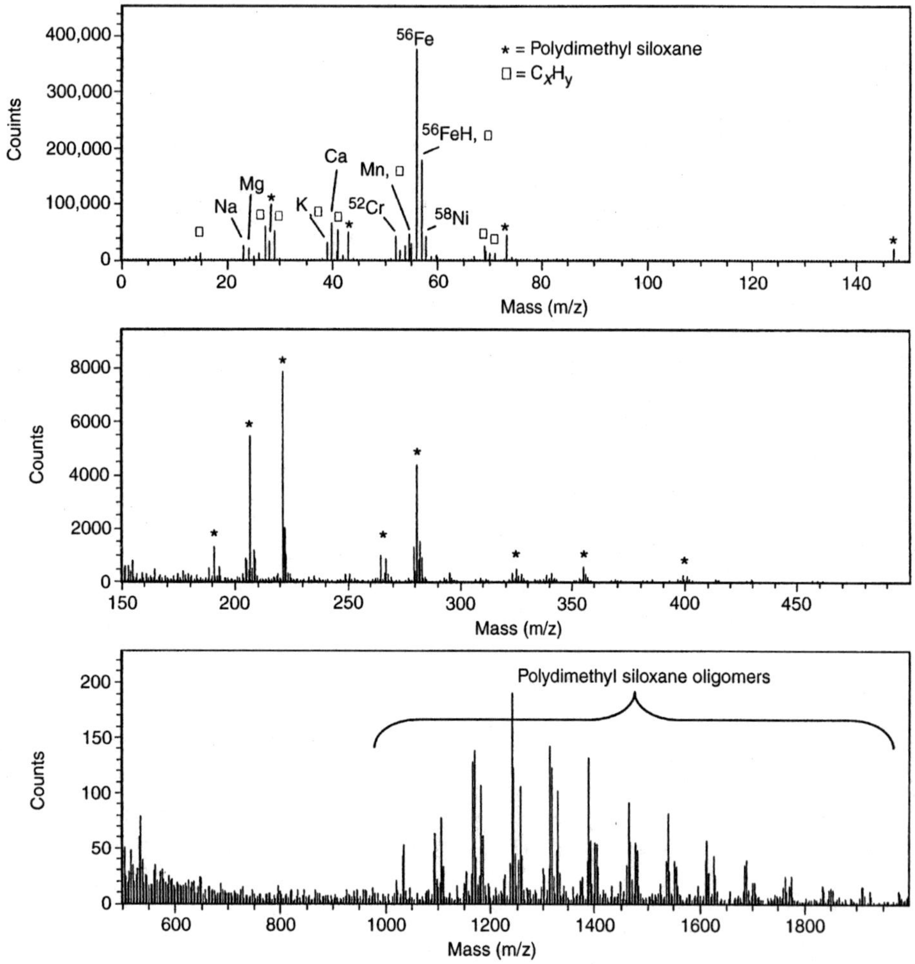

Fig. 15 Time-of-flight secondary ion mass spectrometry positive ion spectrum of stainless steel surface

ple and the fresh laboratory sample. Zinc oxide and tin oxide inhabited the top surface; tin metal was present in the subsurface, but zinc oxide persisted throughout. Zinc oxide was found on the polymer only in the low-strength field sample. It was concluded from these data that zinc diffuses rapidly through tin plate via the type of diffusion that is accelerated by moisture and temperature. These conditions also promote hydration of zinc oxide.

It should be possible to use surface analysis to monitor the kinetics of zinc diffusion through tin plate with variable process parameters. Zinc diffusion displaces organic carbon originally entrained in the tin plate. Low strength derives from two factors: brittle fracture occurs in a zinc hydroxide phase, and the depletion of carbon constituents in the tin plate surface region may prevent formation of strong bonds in the first place. Figure 18 illustrates these conclusions.

In Fig. 18(a), representing high strength, fractures proceed very close to the interface between the polyester and the surface of the tin plate. In Fig. 18(b), after aging, the fracture proceeds directly through a thick, friable surface of zinc hydroxide.

Possible remedial actions could involve: bonding immediately after tin plating, thicker tin plate, application of other barrier films, or storage of the tin-plated brass under dry, low-temperature conditions before bonding. In fact, thin nickel films that were electrodeposited before the tin-plating step did stop zinc migration, and the delamination problem was history.

Example 2: Printed Circuit Boards

Multiwire printed circuit boards (PCBs), illustrated in Fig. 19, are basically customized patterns of insulated wire placed onto adhesive-coated substrates, using ultrasonic energy to embed the wire into the thin adhesive. Interconnections are made by drilling holes though the wires and electrolessly plating the through-holes. RC-205, a multiwire adhesive developed by Kollmorgan Corporation, is composed of phenolic and epoxy constituents that provide rigidity, and solvent-swelled rubber components that provided wireable properties at room temperature. First, RC-205 is hot-roll laminated to the surface of etched copper. Then, wire is fed from a dispenser to a foot that is also an ultrasonic transducer for melting the swelled-rubber components in RC-205, a foot which embeds the wire into the adhesive. The table is computer controlled and moves directionally for wire

Table 5 Comparison of selected surface analysis techniques

Technique	Probe radiation	Analyzed emission	Advantages	Limitations
X-ray photoelectron spectroscopy (XPS or ESCA)	X-rays	Photoelectrons	Nondestructive Depth profiles Oxidation states Quantitative	Spot size ≥100 μm (3940 μin.) Detection limit, 0.1 to 1% No molecular information
Auger electron spectroscopy (AES)	Electron beam	Auger electrons	Very rapid Small spot (<50 nm, or 500 Å) Imaging Simultaneous ion milling depth profile	Charging in insulators Atomic information only Detection limit, 0.1 to 1%
Static secondary ion mass spectroscopy (SIMS)	Ion (or atom) beam	Secondary ions	Molecular information High sensitivity Small spot Imaging	Complex spectra Semiquantitative

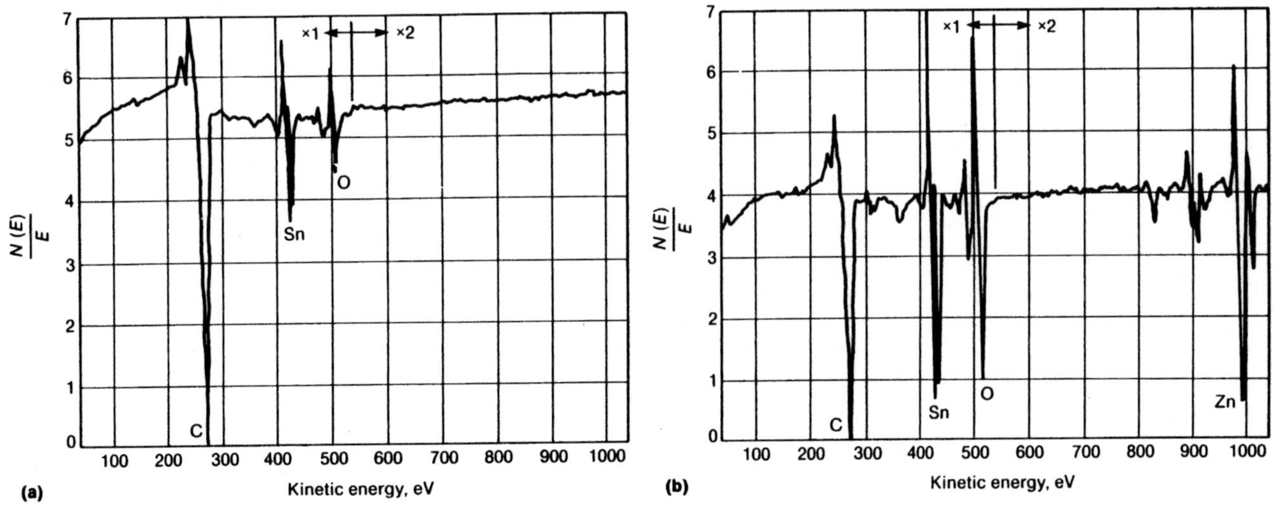

Fig. 16 Auger electron spectroscopy survey spectra comparing the metal sides of (a) high- and (b) low-peel-strength polyester-adhesive-metal laminates

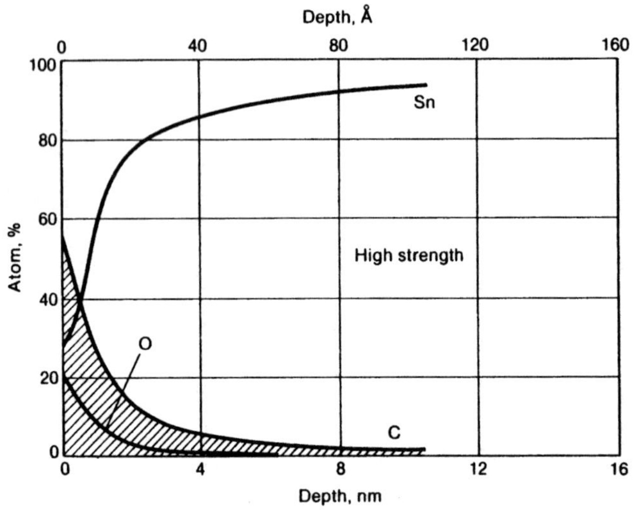

Fig. 17 Auger electron spectroscopy depth profiles of the specimens in Fig. 16

Table 6 Angular-dependent XPS results
Oxidation states: A, Tin outer layers are SnO on top of Sn^0. B, Zinc is all ZnO or $Zn(OH)_2$ (no Zn^0); gives brittle fracture. C, Carbon is all aliphatic, therefore not polymeric.

Specimen		Atom percentages Grazing	Normal
Strong bond	Carbon	95	82
	Oxygen	3	10
	Tin	2	8
Weak bond	Carbon	79	59
	Oxygen	14	24
	Tin	4	11
	Zinc	3	7

placement. After wiring, a glass-epoxy prepreg is laminated over the surface, and the epoxy and the adhesive are cured to set the wire position.

Structurally, RC-205 comes in the form of a resin-impregnated glass cloth sandwiched between two layers of adhesive. The adhesive coating on one side is thicker for bonding wires to the board; the opposite side bonds to the etched-copper format. The wire side is covered by polyester film, and the board side is protected by a polypropylene release sheet. These cover sheets are easy to distinguish to ensure that the multiwire adhesive is placed on the board right side up. Also, they protect the RC-205 layers from physical damage and help retain the solvent content of the film. The polypropylene film is removed when RC-205 is hot-roll laminated to the board, but the protective polyester sheet remains in place until the board is ready for wiring.

Many proprietary materials are involved, and they often exhibit lot-to-lot variations. Without the aid of advanced analytical techniques to identify cause-effect relationships, production problems typically appear and disappear mysteriously, year after year, without resolution. The eight basic steps in PCB fabrication are identified as follows:

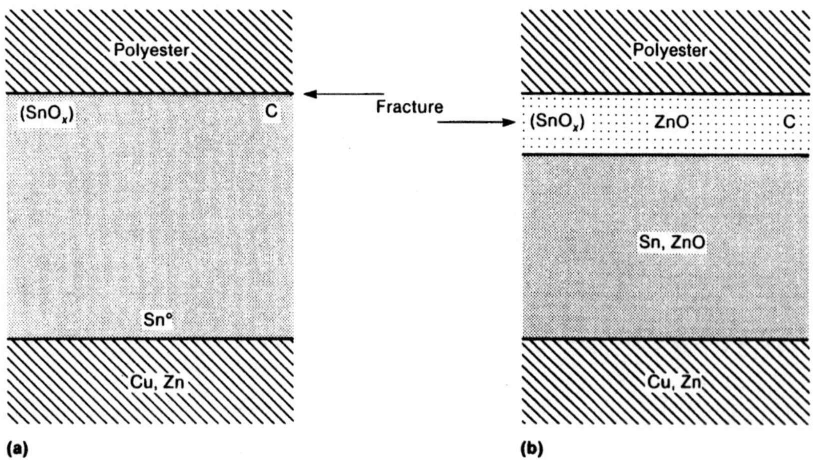

Fig. 18 Schematic models derived from x-ray photoelectron spectroscopy and Auger electron spectroscopy analysis of (a) high-strength and (b) low-strength polyester-adhesive-brass laminates

Start with copperclad catalytic-based material

- Shear laminate
- Stabilize laminate at 160 °C (320 °F)/8 h

Make format board

- Print and etch format by acid treatment
- Bake at 150 °C (300 °F)/2 h

Laminate with RC-205 adhesive

- Apply RC-205 adhesive to levels 1 and 2 (hot-roll lamination)

Do wiring

- Wire levels 1 and 2
- Postwire bake at 95 °C (200 °F)/1 h
- Flush press

Encapsulate wires

- Apply prepreg (platen press)
- Cure prepreg at 175 °C (350 °F)/0.75 h

Drill holes

- Apply Polyspotstik
- Drill
- Apply first water blast at 2.1 MPa (0.3 ksi), 2 m/min (7 ft/min) to remove debris
- Shrink-back bake at 140 °C (280 °F)/0.5 h
- Apply second water blast to wet holes before chemical hole-cleaning cycle
- Clean hole with chemicals
- Apply third water blast to remove residual hole-cleaning salts

Deposit electroless copper

- Plate with electroless copper plating
- Strip Polyspotstik
- Postcure at 160 °C (320 °F)/1 h

Make finishing touches

Many of these steps include significant mechanical, thermal, and chemical stresses. Delamination at any interface can cause dimensional instability and localized changes in the impedance characteristics of the board. A blister can result in either an open circuit, via tension on the wire-to-hole interconnections, or short circuits, by providing conductive (moisture) paths between pins.

The following sections describe several problems encountered in PCB manufacturing technology. X-ray photoelectron spectroscopy was selected as the first analytical tool because it generally provides the greatest information content and unambiguous interpretation. In the cases cited, few additional analyses were required to suggest a cause of failure that proved to be correct in terms of fixing the problem. Of course, analysis by additional techniques would enhance the quantitative and scientific certainty of the conclusions.

Delamination of Multiwire Adhesive from Copper Format. The most common problem encountered in the manufacture of multiwire PCBs is delamination of the wire adhesive from the copper format. This condition often occurs at the end of the PCB manufacturing process, after significant value has been added. The adhesive RC-205 and surface-treated copper foil form the interface where delamination occurs. Suppliers of copper foil treat the surface in various ways meant to roughen the surface and to provide a chemically inert barrier layer. Therefore, in the absence of any chemical reaction between the RC-205 multiwire adhesive and the foil surface, adhesion should primarily be mechanical, attributed mostly to the size and shape of asperities and porosity of the surface treatment layer, thereby anchoring the initially compliant polymer film.

The severity of the delamination problem also seemed to vary with different lots of RC-205 adhesive. The formulation of RC-205 lists 14 ingredients, which include two synthetic rub-

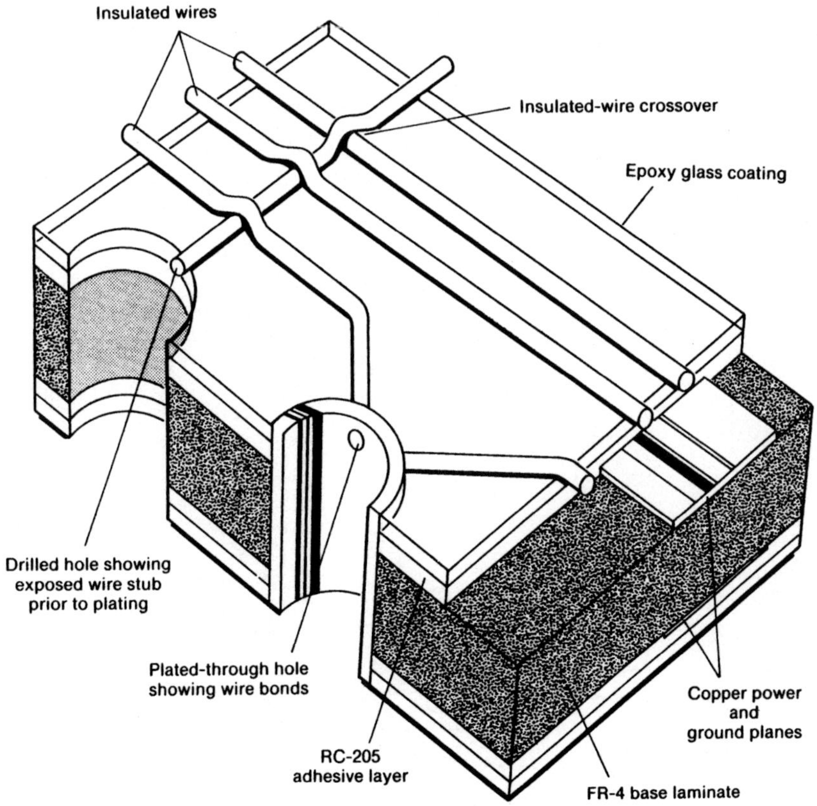

Fig. 19 Cross section of a multiwire circuit board

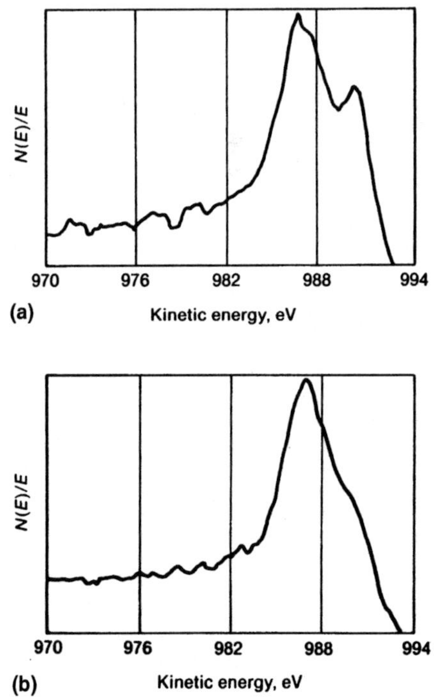

Fig. 20 Comparison of zinc LMM Auger peak after 5 nm (50 Å) of sputter etching. (a) *X* laminate. (b) *Y* laminate

bers (60% of dry weight composition), three solvents, silica and zirconium silicate fillers, a pigment, a leveling agent, a plating catalyst, two types of epoxy resins, a cross linker, and a phenolic resin. At first examination, none of those ingredients could be singled out as leading to chemical reactions at an interface with either zinc oxide (ZnO) or metallic zinc. However, during the mechanical grinding of the major rubber component of RC-205, a powdered polyvinyl chloride (PVC) copolymer product called VYHH is added to prevent reclumping. The resulting VYHH content of this rubber averages approximately 7%, but amounts as high as 15% are possible. Furthermore, in this case, the ground rubber was packaged in 25 kg (50 lb) boxes, and loosely adherent PVC powder segregated to the bottom. Thus, rubber taken from the bottom was even richer in PVC.

Surface Analysis of Copper Laminates. Surface analyses compared two copper foils (*X* produced delamination, while *Y* did not). Sputter rates were calibrated with a standard silicon dioxide (SiO_2) reference sample. Samples of as-received *X* and *Y* laminates were examined using XPS and AES to determine the cause of delamination. X-ray photoelectron spectroscopy results were essentially identical: surface compositions consisting of zinc, oxygen, carbon, and chromium. High-resolution XPS indicated that the zinc species present was ZnO in both the *X* and *Y* samples.

A second set of XPS spectra were obtained after sputter etching 5 nm (50 Å) from the surfaces. These results showed a distinct difference between the two materials. For the *Y* material,

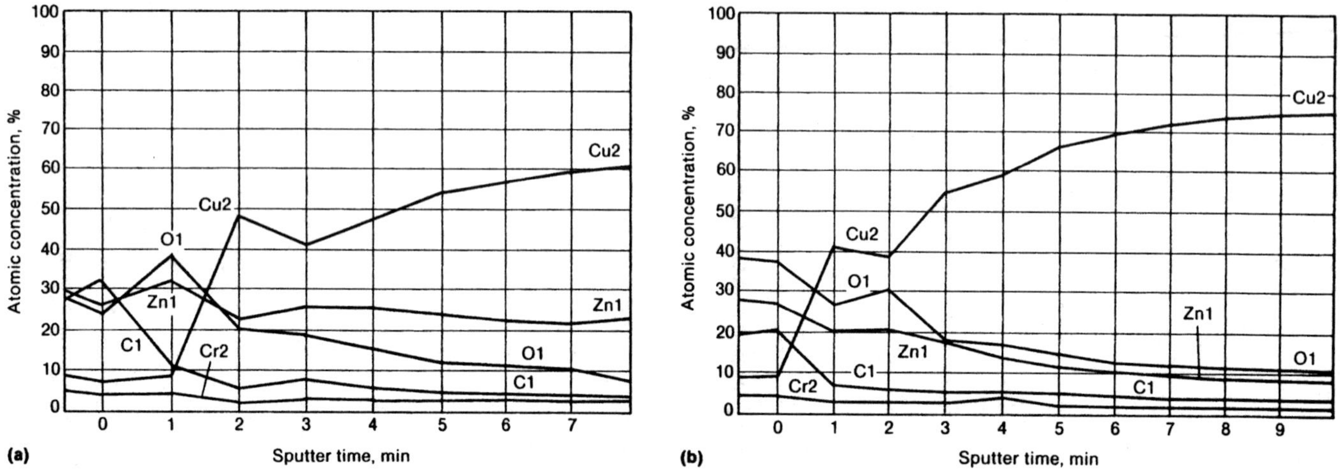

Fig. 21 Auger electron spectroscopy-ion milling depth profiles comparing laminates. (a) *X* laminate, 5 nm (50 Å)/min. (b) *Y* laminate, 10 nm (100 Å)/min

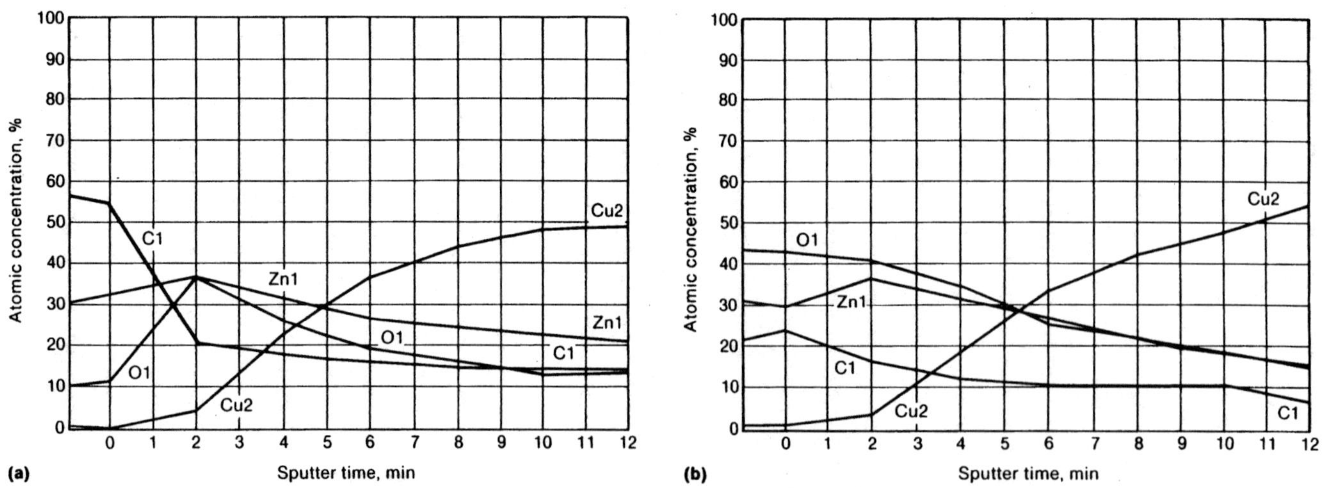

Fig. 22 X-ray photoelectron spectroscopy-ion milling depth profiles comparing laminates. (a) *X* laminate, 5 nm (50 Å)/min. (b) *Y* laminate, 5 nm (50 Å)/min

the XPS data showed that the surface contained predominantly ZnO. This was apparent from both the Zn/O ratio and the Auger parameter (2010.0 eV). Both values are indicative of ZnO. The results obtained for the *X* laminate after sputtering 5 nm (50 Å) showed metallic zinc in the zinc LMM Auger line in Fig. 20. The most intense peak is characteristic of ZnO, while the lower energy peak correlates with the peak position of metallic zinc.

Auger electron spectroscopy and x-ray photoelectron spectroscopy depth profiles obtained for both materials, using argon ion milling, are shown in Fig. 21 and 22, respectively. The most obvious difference to be seen in the comparison of the AES profile data is in the relative amounts of zinc and oxygen present in the *X* and *Y* materials. The results obtained for the *X* material (Fig. 21a) indicated a steady decline in the oxygen content with sputtering time, which would indicate the presence of zinc metal in the near-surface layers. Conversely, the profile results for the *Y* laminate (Fig. 21b) showed a 1 to 1 relationship between zinc and oxygen throughout the profile. This suggested that ZnO was present in the bulk of the material. The XPS profile results corroborated the results.

The surface topography of the *X* and *Y* laminates is compared in the electron micrographs in Fig. 23. The *Y* foil has a nodular surface with uniform particle size and shape, whereas the copper foil on the *X* laminate exhibits a wide distribution of particle size and shapes. Higher magnification shows that the nodules on the *Y* surface are covered with whiskerlike growth, while the nodules on the *X* surface are featureless.

Proposed Failure Mechanism. Polyvinyl chloride degrades under high temperature and/or high shear conditions by a dehydrochlorination reaction, which evolves hydrogen chloride (HCl). Release of HCl at the surface of copper can cause corrosion and eventual loss of adhesion. Generation of HCl in the presence of water in proximity to metallic zinc or zinc oxides causes hydrogen gas generation at the interface. As the temperature cycles during PCB processing, the gas creates high, localized pressure at the copper/RC-205 interface, and delamination results. The presence of both PVC in the RC-205 formulation and metallic zinc on the copper surface promotes a high incidence of failure.

Production Evaluation. The validity of the proposed failure mechanism was tested in a trial production run on 380 × 430 mm (15 × 17 in.) multiwire boards with approximately 12,000 holes, using unetched *X* substrates to simulate the worst case for adhesion. In this run, the standard multiwire process was modified by inserting a 10% acid dip, rinse, and bake cycle into the process before hot-roll lamination of RC-205. During acid treatment, gas evolution (metallic zinc reacting with HCl) was observed on the *X* material. Similar acid treatment of the *Y* laminate produced no gas. Finished boards were subjected to 260 °C (500 °F) hot air leveling and wave soldering to evaluate delamination. The test showed that the acid treatment was very effective in reducing delamination: It occurred only at very small spots around one or two holes per board, compared to a large number of isolated delamination patches observed on untreated *X* laminate. The acid treatment was then incorporated in two separate production runs totaling 76 multiwire boards. Only two boards had any RC-205 delamination, whereas untreated boards had a 100% reject rate.

Addition of an acid cleaning step to remove metallic zinc from the copper substrate surface permits the use of a wider range of materials. The best way to minimize delamination is to remove the reactive zinc from the substrate and reduce the acid-generating source (PVC) in RC-205 formulations.

The next iteration in the RC-205 delamination problem involved modifying the PVC content of the formulation. This was accomplished with the cooperation of the sole compounder of RC-205 as well as the converter of the liquid adhesive into a rolled film ready to bond to PCB formats. Low and high PVC content and a control lot of RC-205 were manufactured. The best adhesion results were obtained for acid-treated

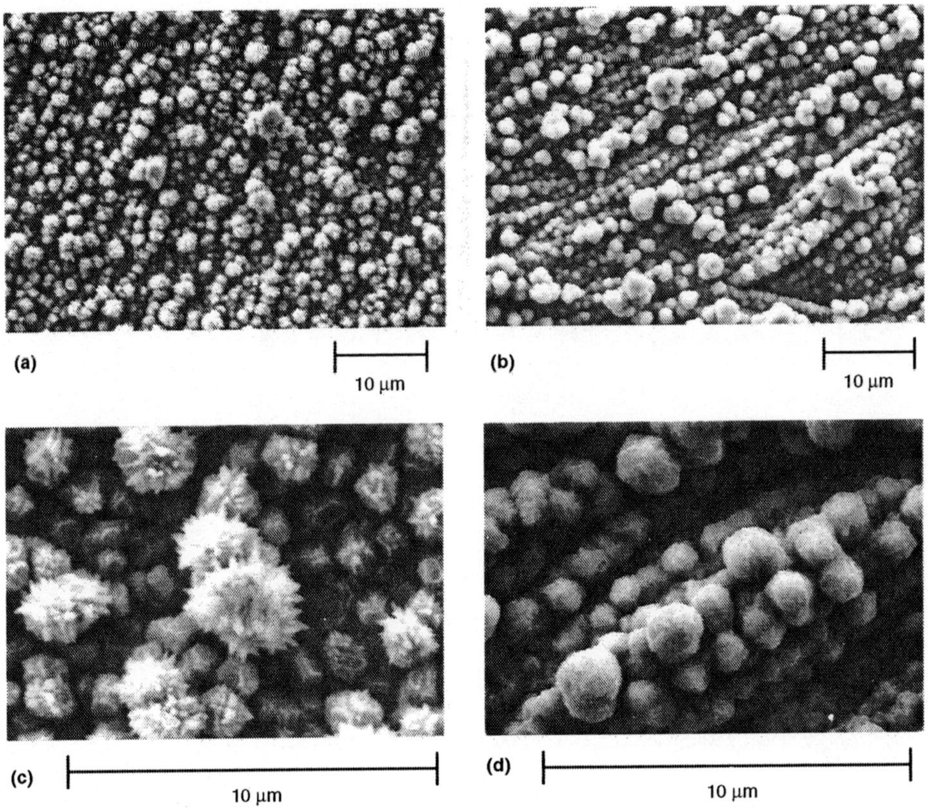

Fig. 23 Scanning electron micrographs comparing *Y* laminate (left) and *X* laminate (right). (a) 1125×. (b) 1120×. (c) and (d) 4480×

Table 7 Epoxy prepreg delamination

	Atomic concentration, %				
Element	Polyester release sheet	Polypropylene release sheet	RC-205 delamination surface	Epoxy prepreg delamination surface	RC-205
Carbon	62.8	98.9	76.2	76.2	74.2
Oxygen	27.8	1.1	19.2	20.0	20.5
Nitrogen	0.9	...	1.7	0.6	...
Chromium	3.9	...	1.3	1.5	...
Silicon	4.6	...	0.8	0.9	5.3
Chlorine	...	...	0.8	0.8	...

X laminate with the low-vinyl formulation. Untreated foil delaminated when both the high- and the low-vinyl-content RC-205 were used, demonstrating that the absence of metallic zinc from the surface is more critical to adhesion than is the vinyl content of RC-205.

Five production boards for each experimental RC-205 formulation were then manufactured using *Y* foil. On all boards, RC-205 had good adhesion, except for one board with the high-vinyl-content RC-205. This board had gross delamination in a wire repair area subjected to additional thermal stress from a soldering iron.

Epoxy Prepreg/RC-205 Delamination. Normally, the adhesion of epoxy to epoxy or epoxy to RC-205 is not a problem. However, a localized prepreg delamination appeared sporadically as white spots on the board surface because of debonding of the epoxy from the wired RC-205 layer.

The surface analyses of the epoxy prepreg and the RC-205 at the delamination interface are summarized in Table 7. One unique feature of the elemental composition of both surfaces was the presence of approximately 1.4% Cr. Because chromium-containing compounds were not found in either the epoxy or the RC-205, the polypropylene and polyester release sheets (Fig. 24) were also analyzed by XPS. The surface composition of polypropylene showed only 98.9% C and 1.1% O. The polyester sheet had 3.9% Cr and 4.6% Si, indicative of a release agent applied to impart antistick properties to this cover sheet. Subsequent investigation revealed the coating was an organic chrome complex sold as a release agent. The prepreg delamination problem was traced to the transfer of the release agent from the polyester cover sheet to the RC-205 due to inadequate polymerization of the coating. Because the release agent could not be removed from the multiwire adhesive surface, the only reasonable solution to this problem was a material change.

Prepreg from a different source resulted in a PCB exhibiting extensive prepreg delamination. The prepreg was peeled off, and both surfaces constituting the interface between the epoxy prepreg and the RC-205 were examined with XPS. Results comparing the delamination interface, polyester cover sheet, and epoxy prepreg (good bonding) are summarized in Table 8. Figure 25 shows that the transfer of silicon from the polyester cover sheet to the RC-205 surface was

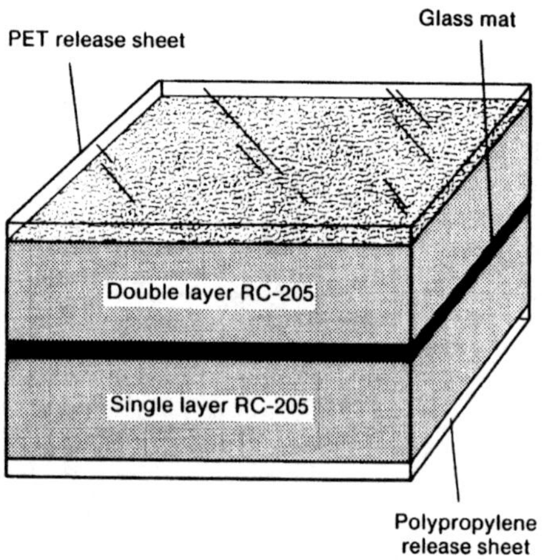

Fig. 24 Multilayer construction of RC-205 adhesive material

Table 8 Quantitative XPS data

Surface	Carbon, %	Oxygen, %	Silicon, %
Epoxy prepreg delamination interface	71.9	21.6	6.5
RC-205 delamination interface	57.7	26.3	16.0
Release silicon, polyester cover sheet on RC-205	58.1	25.2	16.7
Epoxy prepreg (good bonding)	78.7	13.8	...

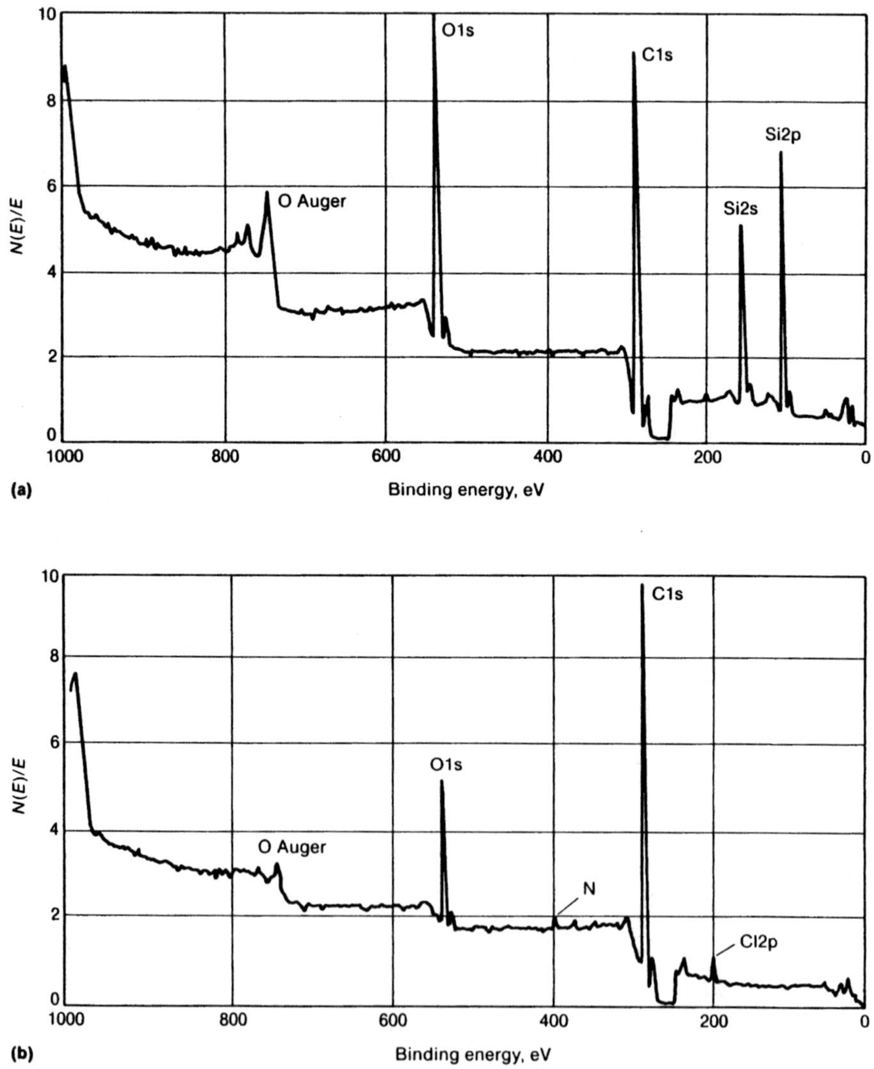

Fig. 25 X-ray photoelectron spectroscopy survey spectra of the opposite sides of the RC-205 material after removal of release sheets. (a) Polyester side. (b) Polypropylene side

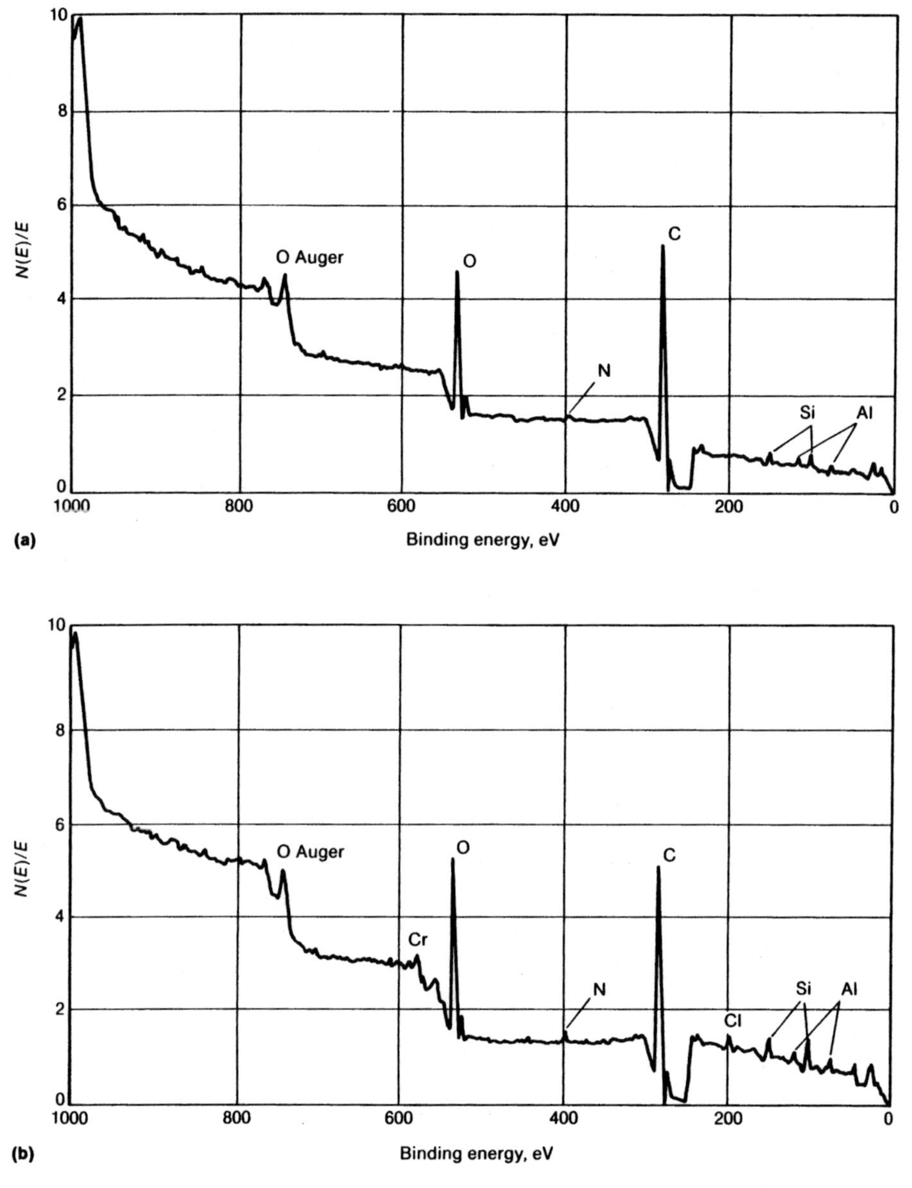

Fig. 26 X-ray photoelectron spectroscopy survey spectra of the failure surfaces from white, spotty delaminations. (a) Board surface. (b) Prepreg surface

Table 9 Epoxy prepreg delamination, XPS atomic concentration results

Surface	Carbon, %	Oxygen, %	Chromium, %	Nitrogen, %	Chlorine, %	Silicon, %	Aluminum, %
Epoxy delamination	73.2	19.8	1.6	2.1	0.2	2.1	1.0
Board delamination	80.2	17.1	0.0	0.9	0.2	1.0	0.6
As-received epoxy prepreg	78.7	13.8	...	7.5	...	...	...

the cause of that prepreg delamination. It is noteworthy that the composition of the release agent on polyester, compared to that on the RC-205, was nearly identical to that on the polyester release sheet, indicating release agent transfer to the RC-205 surface.

There is a third type of prepreg delamination that is similar in appearance to those described previously, but it occurs in smaller spots. The XPS spectra shown in Fig. 26, plus the quantitative data in Table 9, indicate the presence of bare glass fibers in the epoxy prepreg surface. Inorganic aluminum and silicon are from fiberglass, and chromium is present in the sizing agent on the glass fibers. Therefore, this type of prepreg delamination is due to prepreg resin starvation, that is, lack of complete wetting of the fiberglass by the epoxy resin during prepreg manufacture.

White Haze on PCBs Soldered by Vapor Phase Reflow. In the batch vapor phase reflow system shown in Fig. 27, assemblies to be soldered are coated with flux, placed on an elevator, and lowered into a high-temperature, saturated vapor, which condenses on the PCBs. As it condenses, it heats the PCB to 215 °C (419 °F), the boiling point of the FC-70 perfluorinated fluid. Batch vapor phase reflow machines have two vapor zones and condensing coils. The lower zone contains the perfluorinated fluid and is the vapor zone in which solder is reflowed. The primary condensing coil used to contain this fluid is cooled by flowing water, with an inlet water temperature of 40 to 50 °C (100 to 120 °F) (above the boiling point of Freon). The upper zone contains a vapor blanket of Freon TF, which functions to prevent evaporation of the more expensive primary fluid. The Freon is condensed on a secondary coil at a cooling temperature of 7 to 18 °C (45 to 65 °F). It then drips into a trough below the coil, is fed through a water bath (acid-stripper) to strip off acidic impurities, and then through molecular sieves to dry before being recycled back to the tank. The primary perfluorinated fluid in the sump becomes cloudy because of contamination by flux residues, which, if not removed, are deposited on the heating coils and carbonize. Boards soldered by vapor phase reflow should be bright and shiny. Intermittently, solder-plated back planes exhibit a white, hazy deposit after vapor phase reflow. The white haze is a deposit that cannot be removed by normal cleaning procedures. The time of appearance of this white haze on back planes correlates with the buildup of deposits on the walls between the primary and secondary cooling coils of the vapor phase unit.

X-ray photoelectron spectroscopy analysis of a vapor phase reflowed coupon that was not fluxed showed 3.8% Cl, 1.0% F, and a 69/31 tin/lead composition. A white haze spot on the same board showed 8.8% Cl and 3.2% F. It was concluded that the problem was corrosive attack on the solder by HCl and hydrofluoric vapors from breakdown products of Freon TF, flux residues, and Fluorinert fluid. A water leak in the vapor phase reflow cooling coil was fixed, the machine walls were extensively cleaned, a carbon filter was installed on the Fluorinert

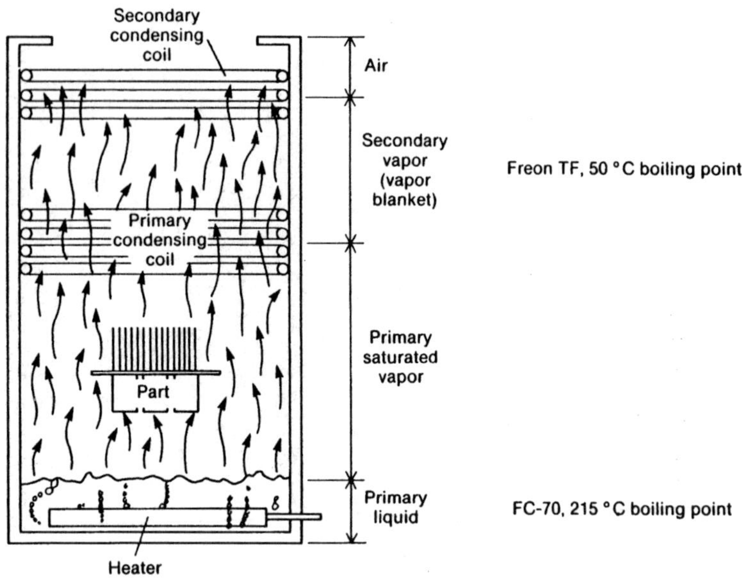

Fig. 27 Vapor phase reflow equipment

fluid, and the quality, as determined by visual inspection, improved dramatically.

However, a white haze problem still appeared periodically. X-ray photoelectron spectroscopy analysis of these stained areas showed none of the corrosive salts described previously but contained 1.6% Ca and smaller concentrations of sodium, magnesium, and chlorine. Vapor phase reflowed boards are cleaned with a mixed solvent alkaline spray to make the flux residues solvent, followed by tapwater rinsing. Thus, this type of white haze on the surface of the solder might be due to hard water, which resulted in a buildup of calcium and magnesium deposits in the spray nozzles. The reduced water flow in rinsing and the drying of hard water salts on the board could produce this type of white stain.

A quick check showed that the tapwater rinse had a pH of 8.3, which was traced to a lime treatment used to reduce the possibility of acidic corrosion of the local water company pipes. Use of softened water for flux removal eliminated calcium and magnesium deposits on solder.

Table 10 Quantitative XPS data

	Carbon, %	Oxygen, %	Phosphorus, %	Sulfur, %	Silicon, %
Modified polyphenylene oxide part delamination interface	77.6	19.9	0.9	0.1	1.6
Paint delamination interface	56.3	27.3	...	...	16.4
Modified polyphenylene oxide part surface, as-received	87.9	10.5	0.4	0.1	1.0
Paint, top side	72.2	26.7	0.3	...	0.6
Tape control	99.2	0.5	...	...	0.1
Tape peel from molded part	95.0	3.2	...	...	1.6

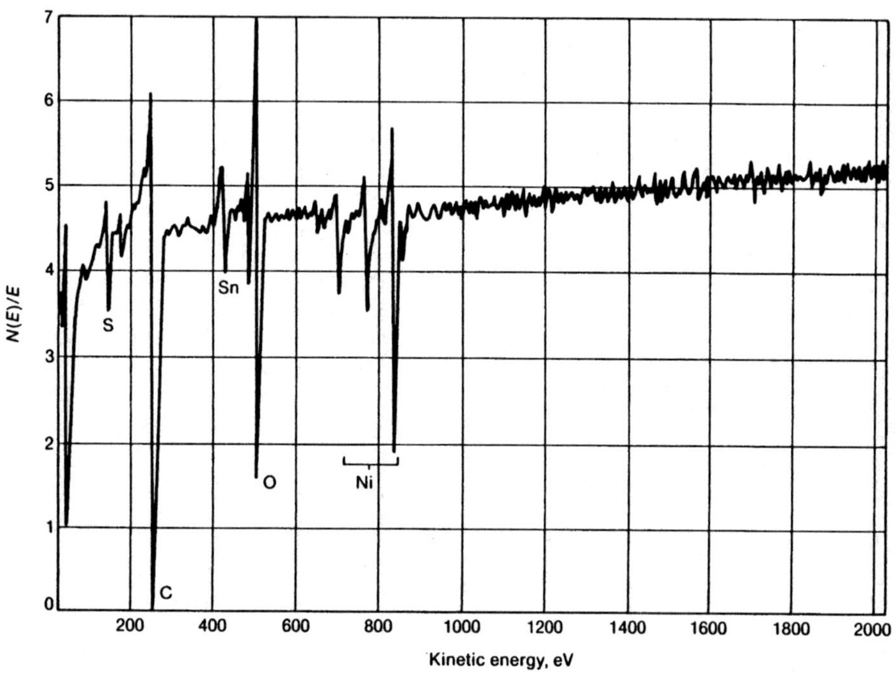

Fig. 28 Auger electron spectroscopy survey spectrum from integrated circuit chip solder pad failure surface

Example 3: Paint Delamination from a Molded Cabinet

The inside of a computer monitor cabinet (an injection-molded modified polyphenylene oxide foam material) was painted with a conductive nickel acrylic paint to provide electromagnetic interference shielding. Any delamination of this conductive coating will cause the cabinet to lose its shielding effectiveness, and flakes of this conductive paint could become an electrical shorting hazard.

The unique feature of the XPS results shown in Table 10 is the 16.4% Si detected on the paint delamination interface. This is an order of magnitude higher than the level detected on the outer molded surface or on the top side of the conductive paint. The amount of organic silicon detected at the paint interface points to a mold-release spray used when injection molding the part. Periodic spraying of the mold would also explain the sporadic nature of this problem; that is, paint delaminates on only one out of every five to ten parts. A simple quality-control test was developed with an adhesive tape pressed onto molded parts and then peeled to transfer any silicon release agent to the tape. X-ray photoelectron spectroscopy data collected from the peeled tape showed 1.6% Si.

Example 4: Delamination of a Surface-Mounted Integrated Circuit (IC) from a Solder Pad

An AES survey scan of the chip (Fig. 28) showed the presence of nickel at the failure interface. The PCB was nickel plated and then soldered using a solder paste for surface mounting the IC. However, nickel always forms a thick oxide that is solderable only after the oxide is removed. Solder pastes do not have flux that is active enough to remove Ni_3O_4. This problem was solved by substituting solder plating (tin/lead) for nickel plating.

REFERENCES

1. L.S. Chumbley and L.D. Hanke, Scanning Electron Microscopy, *Failure Analysis and Prevention,* Vol 11, *ASM Handbook,* ASM International, 2002, p 516–526
2. R. Lund and S. Sheybany, Fatigue Fracture Appearances, *Failure Analysis and Prevention,* Vol 11, *ASM Handbook,* ASM International, 2002, p 524
3. "XPS Analysis of Disposable Gloves," XPS application brief, Charles Evans & Associates, Sunnyvale, CA, 1999
4. J. Moulder et al., *Handbook of X-Ray Photoelectron Spectroscopy,* J. Chastain and R.C. King, Jr., Ed., Physical Electronics, Inc., 1995
5. P.E. Larson and M.A. Kelly, Surface Charge Neutralization of Insulating Samples in X-Ray Photoemission Spectroscopy, *J. Vac. Sci. Technol. A,* Vol 16 (No. 6), 1998, p 3483
6. M.A. Douglas and P.J. Chen, Quantitative Trace Metal Analysis of Silicon Surfaces by TOF-SIMS, *Surf. Interface Anal.,* Vol 26, 1998, p 984–994

SELECTED REFERENCES

General

- D.M. Brewis and D. Briggs, *Industrial Adhesion Problems,* Wiley-Interscience, 1985
- C.R. Brundle et al., *Encyclopedia of Materials Characterization,* L.E. Fitzpatrick, Ed., Butterworth-Heinemann, 1992, p 282– 299, 310–323, 549–558
- A.W. Czanderna, *Methods of Surface Analysis,* Elsevier, 1975
- D.W. Dwight, T.J. Fabish, and H.R. Thomas, *Photon, Electron and Ion Probes of Polymer Structure and Properties,* American Chemical Society, 1981
- W.J. Feast and H.S. Munro, *Polymer Surfaces and Interfaces,* John Wiley and Sons, 1987
- P.F. Kane and G.B. Larrabee, Ed., *Characterization of Solid Surfaces,* Plenum Press, 1974
- L.H. Lee, Ed., *Characterization of Metal and Polymer Surfaces,* Academic Press, 1977
- E.A. Leone and A.J. Signorelli, Surface Analysis, *A Guide to Materials Characterization and Chemical Analysis,* 2nd ed., J.P. Sibilia, Ed., VCH Publishers, Inc., 1996, p 221–259
- N.S. McIntyre, Ed., *Quantitative Surface Analysis of Materials,* STP 643, American Society for Testing and Materials, 1978
- K. Mittal, Ed., *Surface Contamination,* Plenum Press, 1979
- F. Settle, Ed., *Handbook of Instrumental Techniques for Analytical Chemistry,* Prentice Hall PTR, 1997, p 793–808, 811–827, 831–847

AES, XPS, and TOF-SIMS

- D. Briggs, Ed., *Handbook of X-Ray and Ultraviolet Photoelectron Spectroscopy,* Heyden & Sons, 1978
- D. Briggs and M.P. Seah, Ed., *Practical Surface Analysis by Auger and X-Ray Photoelectron Spectroscopy,* John Wiley & Sons, 1983
- T.A. Carlson, *Photoelectron and Auger Spectroscopy,* Dowden, Hutchinson, and Ross, 1978
- K. Childs et al., *Handbook of Auger Electron Spectroscopy,* 3rd ed., C.L. Hedberg, Ed., Physical Electronics, Inc., 1995
- L.E. Davis, Ed., *Handbook of Auger Electron Spectroscopy,* Perkin Elmer Corporation, 1972
- L. Fiermans, J. Vennik, and W. Dekeyser, *Electron and Ion Spectroscopy of Solids,* Plenum Press, 1978
- J. Moulder et al., *Handbook of X-Ray Photoelectron Spectroscopy,* J. Chastain and R.C. King, Jr., Ed., Physical Electronics, Inc., 1995
- G.E. Murlenberg, Ed., *Handbook of X-Ray Photoelectron Spectroscopy,* Perkin Elmer Corporation, 1979
- J.C. Vickerman and D. Briggs, *The Wiley Static SIMS Library,* John Wiley & Sons, 1996

Fracture and Fractography

THERE ARE MANY CAUSES AND FORMS of fracture, and careful analysis of fractured parts requires an understanding of the component design, service loading, environments, and structure-property relationships, the application of sound laboratory techniques in materials, and the examination and interpretation of fracture surfaces (i.e., fractography).

The purpose of this article is to introduce the subject of fractography and how it is used in failure analysis. Fractography is the science of revealing loading conditions and environment that caused the fracture by a three-dimensional interpretation of the appearance of a broken component. If the specimen is well preserved and if the analyst is knowledgeable, the fracture appearance reveals details of the loading events that culminated in fracture.

Fractographic analysis often reveals important clues about the cause of fracture. However, the overall examination is not necessarily confined to the fracture surface alone. The cause of the failure may sometimes be apparent from the examination alone, but in most cases, the fracture surface examination is intended to reveal the location of the fracture origin, providing valuable information about the local service environment as well as the state of stress responsible for the crack initiation and growth that eventually led to fracture. The subsequent stress analysis of the failed part can be considerably simplified, because attention can be focused at the location of crack initiation.

Structure and Behavior

Fractographic features are related not only to the geometry, loading conditions, and service environment, but also to the inherent properties controlled by the structure of the material. Polymers are typically amorphous or partially crystalline, where crystallinity decreases with complexity of the pendant atom groups (steric hindrance), chain branching, as well as with increasing molecular weight. Qualitatively, strength and modulus are increased as crystallinity increases, while ductility is usually reduced. In contrast to modeling of metallic material behavior, it is not common to describe behavior of polymeric material in terms of dislocation models and/or microscale slip and twinning processes. Behavior of polymeric materials also depends strongly on whether the service temperature is above or below the glass transition temperature where the molecule dramatically stiffens.

Under monotonic loading, the competing processes of ductile fracture by deformation and brittle fracture are influenced by structure, loading conditions, and temperature. Deformation occurs by viscoplastic flow processes in noncrystalline material such as thermoplastic polymers. Thermoset plastics are brittle. Viscoplastic deformation depends on temperature and strain rate. As the temperature is decreased, the material undergoes a glass transition, and if the pendant group is not too complex, the material may partially crystallize. Qualitatively, a decrease in temperature results in an increase in stiffness and strength with a decrease in ductility. Tensile curves shown in Fig. 1 illustrate the change in behavior as the temperature is decreased and illustrate the important observation that there is a transition from rubbery behavior to cold drawing behavior to glassy behavior as the temperature is decreased. A cold drawing nonoriented polymeric strain hardens very little, creating a strip necking zone. Continued loading may result in craze formation in this zone (Fig. 2). Crazes are not cracks but rather crack precursors.

Ductile polymeric materials develop a neck, as do metallic materials, but the mechanisms for neck formation are somewhat different. Neck formation in a metal occurs at the maximum load. In cold drawing polymeric materials, there is extensive chain straightening and alignment of the backbone chain in the direction of the applied load. In a cold drawing polymer, neck formation initiates at yield, and the slope of a load-elongation curve at maximum load is often still positive. After the instability at yield, the polymer reorients. During the initial stages of plastic flow, there is little hardening, but hardening increases as the chains become aligned (therefore often described as orientation hardening).

Plastic flow causes breakage of bonds, shear sliding, chain straightening, and chain alignment parallel to the applied load. There are also strain-rate effects. There is a visible nonuniformity of strain distribution at the neck/bulk material interface as the opaque oriented neck grows along the length of the specimen (Ref 2). Chain scission and fracture occurs rapidly soon after most of the chains are aligned. Analytical descriptions of the behavior after yield via a Considére type analysis are totally analogous to those for sharp necking at yield in strain aging low-carbon steels.

Cavitation and microvoid formation in polymers differs from that in metals. Cavities form in the necked region of metals after the onset of necking, because the geometry changes induce local triaxial stress components from the tensile load. The cavities grow and, through microvoid coalescence, result in a crack that soon causes fracture. Such cavitation processes in metals typically occur after the onset of necking (or from debonding at inclusions before necking), and fracture typically occurs soon thereafter. In contrast, cavitation processes in polymers can dominate the plastic deformation. These voids do not coalesce into a crack but instead become stabilized by fibrils containing oriented polymeric material. The resulting region consisting of voids and fibrils is known as a craze (Fig. 3). The long-chain nature of polymers is responsible for crazing. Without the long chains, it would not be possible to form the fibrils that span the craze and prevent the conversion of the craze into a crack. Although crazing is usually associated with the deformation of amorphous polymers, it has also been observed in semicrystalline polymers and in thermosetting resins.

In polymers, macroscopic yielding and fracture may not always be appropriate criteria for long-time duration material failure, because cavitation and localized damage can occur prior to necking (which often occurs at yielding rather than at maximum load for a metal). The designer must keep in mind that by their very nature, polymers cannot be fabricated at their maximum packing density. Molecular scale voids, known collectively as free volume, exist within the polymeric macromolecular structure. Applied stress serves to straighten the polymer chains and perhaps to redistribute the free volume so that sizable microvoids are formed. Depending on the nature of the polymer, these microvoids continue to grow into localized areas of stress whitening, stress crazing, stress cracking, or brittle failure. Therefore, design limitation and product failure for some plastics may be associated with stress crazing, stress cracking, or

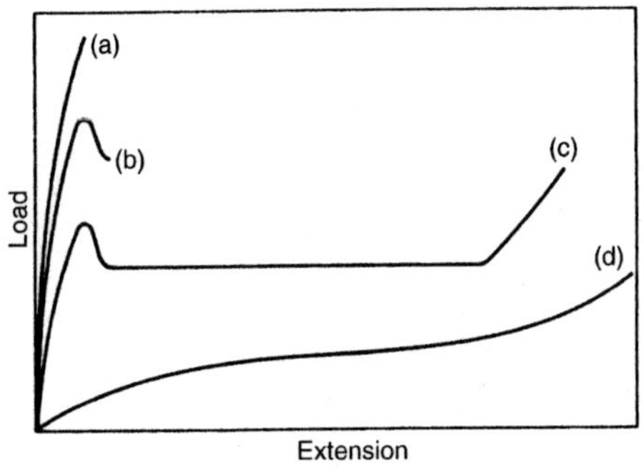

Fig. 1 Change in behavior of a polymeric material with increasing strain rate and/or decreasing temperature. (a) Brittle behavior. (b) Limited ductility behavior. (c) Cold drawing behavior. (d) Rubbery behavior. Curve (a) could represent testing below the glass transition temperature. Source: Ref 1

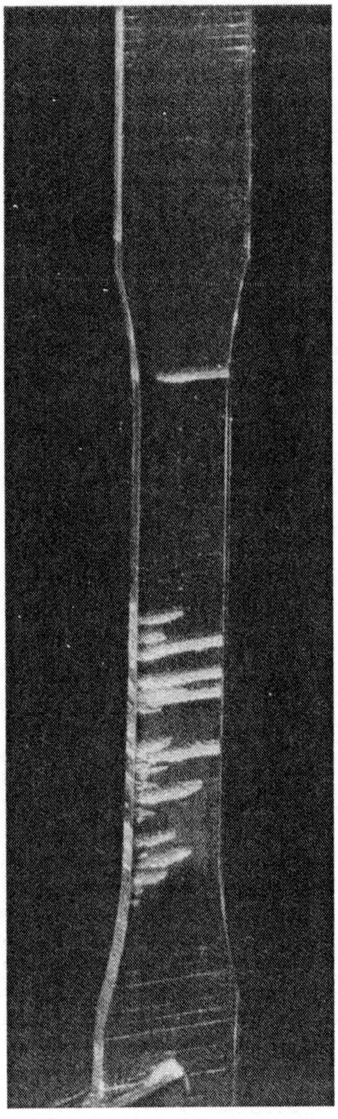

Fig. 2 Craze formation in a polycarbonate polymer in tension under alcohol. Source: Ref 2

stress whitening. This is shown in Fig. 4 by a set of time-dependent stress curves for unplasticized polyvinyl chloride (PVC). The upper line indicates ductile failure due to necking. The region around the dotted line denotes stress-whitening failure. The dashed line, representing craze initiation, is curved in a shape similar to the ductile failure curve, but it is apparent that craze initiation occurs at stress levels substantially below those of ductile failure. To prevent failure initiation, the strain level must be below 1%. This means that if the product has a lifetime of 10^8 s, or 3.2 yr, the applied stress cannot exceed 17 MPa (2.5 ksi).

The deformation and fracture behavior is generally a complex phenomenon that varies with material composition, microstructure, stress environment, time, and temperature (Ref 3). Such behavior for polymers, like metals, depends on external loading conditions, component geometry, and imperfection geometry. One way of displaying the effect of temperature on creep is by temperature-dependent isochronous creep moduli (typically for 10^3 h). In Fig. 5, the creep moduli for eight unfilled plastics are shown for three temperatures. Fillers, reinforcements, plasticizers, and other adducts can significantly influence creep data. Consider the temperature effect on the isochronous creep plot for polycarbonate (PC) (Fig. 6). As expected, the overall creep effect increases with increasing temperature. The crazing boundary is also influenced by temperature (Fig. 6).

Stress cracking implies the localized failure that occurs when localized stresses produce excessive localized strain. This localized failure results in the formation of microcracks that spread rapidly throughout the local area. Brittle materials are more prone to stress cracking than to stress whitening.

Stress whitening is a generic term describing many different microscopic phenomena that produce a cloudy, foggy, or whitened appearance in transparent or translucent polymers in stress. The cloudy appearance is the result of a localized change in polymer refractive index. Thus, transmitted light is scattered. Microvoid clusters of dimension equal to or greater than the wavelength of light are thought to be the primary cause of stress whitening. The microvoids can be caused by the delamination of fillers or fibers, or they can be localized failure around occlusions, such as rubber particles or other impact modifiers. Although stress whitening results in visually apparent changes, the load-bearing capabilities of a specimen may not be substantially reduced during stress whitening.

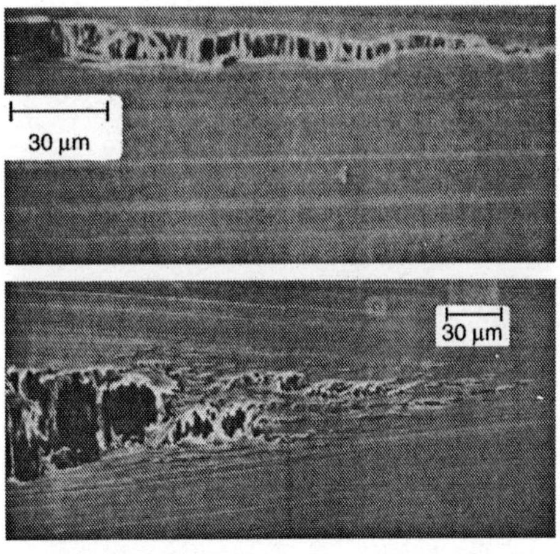

Fig. 3 Crazing fibrils in linear polyethylene (density, 0.964 g/cm³)

Shear bands (Fig. 7, Ref 4) are also microscopic localized deformation zones that propagate ideally along shear planes. Like crazes, shear-deformation bands, or slip lines, are traditionally thought to be the mechanism of irreversible tensile deformation in ductile amorphous polymers. Almost invariably, a compressive-stress state will cause shear deformation in polymers. Under monotonic tensile loading, PC is reported to deform by shear banding.

Stress Crazing. As previously noted, a craze is a microcrack that is spanned by plastic microfibrils, typically oriented in the direction of applied stress. The width of a craze is of the order of 1 to 2 μm, and it may grow to several millimeters in length, depending on its interaction with other heterogeneities. Being dilatational, crazes grow normal to the applied tensile component of the stress field. Craze fibrils are load-bearing elements whose strength and density depend to some extent on the molecular weight of the polymer. Because the fibrils are load-bearing, the microcrack usually does not open substantially before parallel microcracks form.

Fiber-forming plastics such as nylon, polyethylene (PE), and polypropylene (PP) readily stress craze, but crazing also occurs in non-fiber-forming plastics such as PC and polymethyl methacrylate (PMMA). Crazes occur in type I (normally brittle in tension) amorphous polymers, such as polystyrene (PS) and PMMA, under monotonic tensile loading. Crazes, however, have been observed in semicrystalline polymers, such as PP, and in type II (normally ductile in tension) amorphous polymers, such as PC (Ref 5), under tensile-fatigue loading. Higher-molecular-weight polymers develop fewer, longer, and stronger (i.e., stable) crazes. Although crazing is commonly associated with brittleness, higher-molecular-weight polymers have higher resistance to fracture. Implications of this phenomenon have been exploited commercially in the development of toughened polymers.

Because crazes have a different refractive index and because they reflect light, they are easily visible, especially when viewed at the correct angle with the aid of a directed light source, such as a fiber-optic illuminator. When viewed with this type of light source, the craze appears to have a silvery appearance much like that of a very fine crack. The formation of crazes is sensitive to many variables, such as polymer molecular weight, loading conditions, temperature, pressure, and the presence of solvents. Crazes will initiate in a plastic when a critical limit is reached in stress, strain, dilation, or distortion strain energy. Higher-molecular-weight plastics generally have greater resistance to crazing, while more crystalline plastics have lower resistance to crazing. This is because of the fewer number of tie molecules that hold it together.

The effect that solvents have on crazing and, therefore, on polymer deformation is another contrast to metals. When solvents initiate crazing in polymers, they generally lower the stress at which crazes form. This solvent-crazing phenomenon can drastically affect failure characteristics. Particular, large effects are observed in polymers that normally deform by shear yielding but can be induced to craze in the presence of solvents. When an active solvent causes the crazing stress to fall below the shear yielding stress, a polymer that normally exhibits ductile failures under tensile loading can show brittle failures precipitated by the crazing. Such solvent effects can be observed in failures of PC and polysulfone (Ref 6).

Solvent effects can be a source of failure in metals (such as in stress-corrosion cracking), but solvent effects in polymers are more numerous and more likely to be precursors to failure. Solvent-induced failures are a serious engineering problem (Ref 7) that remains a concern. Example 1 in this article is just one case of this. Nonsolvents that possess specific physiochemical affinity leading to wetting of the polymer are also known to cause premature failure (Ref 8). Environmentally enhanced failure generally involves crazing or cracking as the underlying mechanism.

Degradation by Heat and Light (Ref 4). Another serious consideration in failure analysis of polymers is the fact that their mechanical properties may severely deteriorate by exposure to heat and/or light. Thus, structural components serving at elevated temperatures or in outdoor applications may undergo degradation if the polymers employed are not properly stabilized. Mechanical stress is known to enhance degradative effects. Degradation usually involves molecular-weight reduction by one of several mechanisms (Ref 9). Surface embrittlement and consequent microcracking occur. This promotes crack initiation and possibly assists crack propagation. Figure 8 illustrates surface microcracking induced in a polyoxymethylene specimen exposed to ultraviolet light in the laboratory for 1000 h. It should be noted, however, that some polymers are intrinsically more resistant to degradation than others. The following common polymers are ranked with respect to their relative resistance to the deterioration of mechanical properties due to photodegradative effects (Ref 4):

Polymer	Relative resistance
Polymethyl methacrylate	n
Polyacrylonitrile	n
Polyoxymethylene	m
Polyethylene	m
Polyvinyl chloride	n
Polystyrene	w
Nylon 6	m
Polypropylene	vs
Polycarbonate	s
Polyurethane	m
Polysulfone	s
Polyphenylene oxide	s

n, not significant deterioration; m, moderate; w, weak; s, severe; vs, very severe

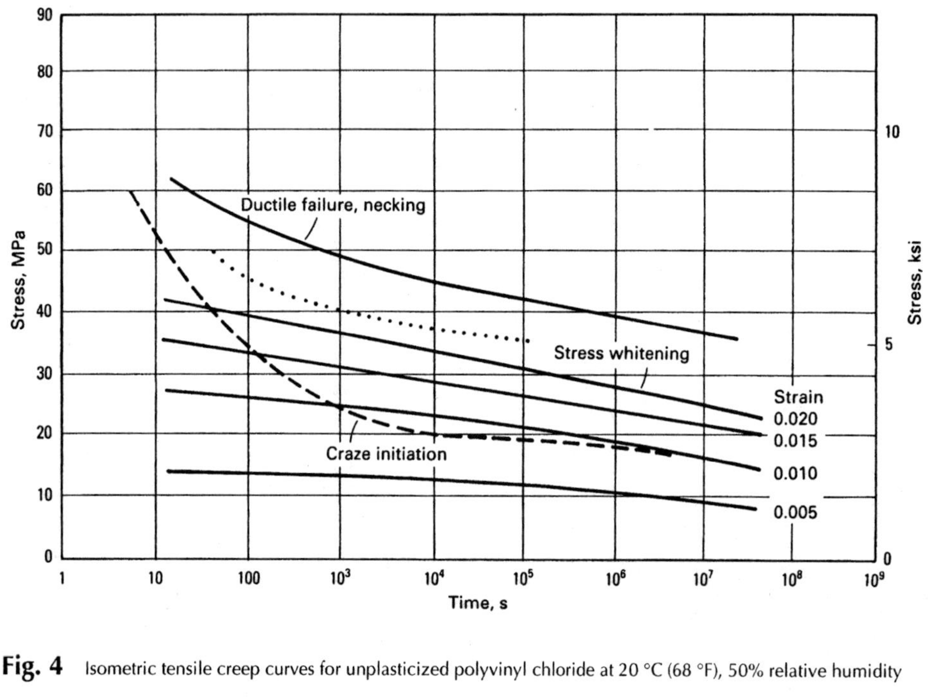

Fig. 4 Isometric tensile creep curves for unplasticized polyvinyl chloride at 20 °C (68 °F), 50% relative humidity

Example 1: Solvent-Induced Cracking Failure of PC Ophthalmic Lenses (Ref 10). A chemical laboratory accident occurred when a solvent splashed on metal-framed safety glasses with PC ophthalmic lenses. The metal-framed glasses had been substituted for plastic-framed safety glasses.

Investigation. Metal-framed PC ophthalmic lenses appeared to have shattered from acetone solvent-induced cracking. The failed lenses exhibited primary and secondary cracks (Fig. 9), which were associated with solvent swelling and crazing. All previous solvent splashes on plastic-framed ophthalmic safety glasses did not produce any fractures.

The lenses were specified to be unfilled PC, which exhibits glasslike clarity. The metal frames were a custom designer series, chosen exclusively for style. The frames gripped approximately two-thirds of the periphery of the lenses. The failed ophthalmic lenses were verified by Fourier transform infrared analyses to be the specified PC material. The source of the splashed liquid was found to contain primarily acetone.

Discussion. Polycarbonate plastics are often used in applications that require glasslike clarity with extreme toughness. One such application is safety glasses. Under most circumstances, when PC material is stressed in the absence of a solvent environment, no shattering will occur. The converse of this event is also true; that is, shattering does not occur in the absence of stress with solvent exposure. However, when PC material is stressed in the presence of a solvent environment, solvent-induced or environmentally induced cracking can occur, resulting in a catastrophic brittlelike failure. This phenomenon also occurs with other plastics and other solvents.

Conclusion. The most probable cause of failure is acetone solvent-induced cracking. Solvent-induced cracking of the metal-framed ophthalmic lenses (but not the plastic-framed ophthalmic lenses) occurred because the metal frames were exerting an uneven stress distribution on the polycarbonate lenses. The failure could have been prevented by using the correct type of safety glasses with plastic frames instead of metal-framed lenses. However, it was also emphasized that PC ophthalmic lenses are usually not designed or designated for chemical splash protection.

Recommendation. Polycarbonate ophthalmic lenses should not be exposed to service environments with solvents, such as acetone. In addition, marking pens, adhesives, or soaps, which contain undesirable solvents, should never be used on the ophthalmic lenses. Excessive stress can originate from working stress or from residual stress and may result in shattering in the presence of a solvent.

Crack Propagation (Adapted from Ref 4)

In most cases, failure of load-bearing structural components involves fracture—that is, complete or partial separation of a critical member of the component under service loading that renders the component nonfunctional. The role of failure analysis, therefore, is to reconstruct the sequence of processes leading to failure, aiming at determining the primary cause. This effort requires knowledge of the various mechanisms by which the material responds to a loading environment similar to that under which a failure occurred. Frequently, laboratory experiments must be conducted to establish such cause-and-effect relationships. Again, failure mechanisms observed in a particular field failure should bear specific similarity to that observed in the laboratory to render the comparison valid.

Fracture of load-bearing engineering components is generally a macroscopic phenomenon resulting from a series of microscopic and submicroscopic processes. Although molecular theories of polymer fracture are a subject of continuing research, practical fracture analysis is achievable from examining deformation events within the resolution of optical and scanning electron microscopy (SEM). Irreversible deformation mechanisms in polymers may fall into two basic categories: dilatational, such as crazes, voids, and microcracks; or nondilatational, such as shear bands. Commonly, a mixture of both mechanisms may be encountered.

In terms of their fracture behavior, polymers are generally classified as brittle or ductile, as discussed further in the next section in this article, "Fractography." Brittle polymers are those that are known to fracture at relatively low elongations in tension (2 to 4%). These include PS, PMMA, and rigid (unplasticized) PVC. Crazing is the dominant mechanism of failure in such polymers. Highly cross-linked polymers, such as epoxies and unsaturated polyesters, are also brittle, yet their fracture involves a microcracking mechanism. On the other hand, semicrystalline polymers, such as PE and nylons, and some amorphous polymers, such as PC and polyethylene terephthalate (PET), exhibit considerable postyield plastic deformation and are thus classified as ductile. In spite of this classification, however, relatively low loads applied over long periods of time are reported to cause failure to occur even in the most ductile polymers. Indeed, early investigations on PE, a ductile polymer, have shown that the failure mechanism switches from ductile to brittle at longer times and lower stress levels (Fig. 10, Ref 11).

The crack-propagation behavior of polymers resembles that in metals, because a propagating crack in a polymer is usually preceded by a zone of transformed material (i.e., a plastic zone). This may involve any of the known irreversible deformation mechanisms. In most cases, a zone of crazes precedes crack propagation, as in PP (Fig. 11) or in PS (Fig. 12). It is important to note that PP is a semicrystalline polymer that usually deforms by yielding and recrystallizing

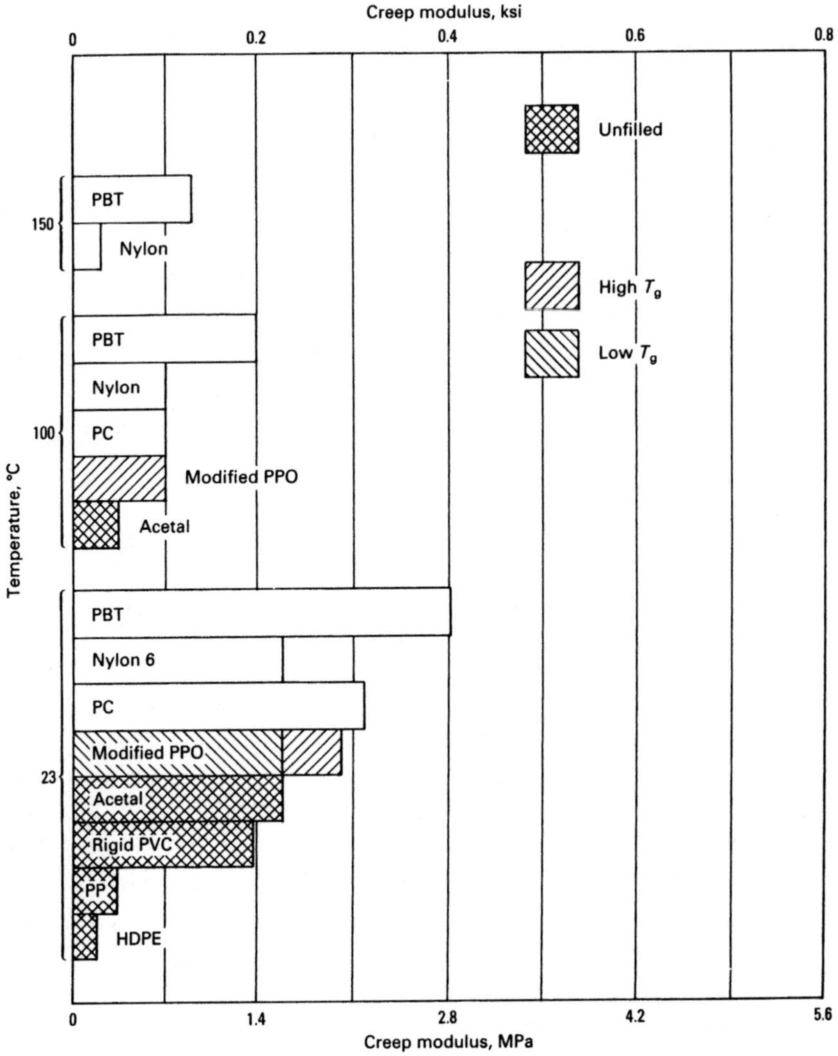

Fig. 5 A 1000 h creep modulus of several polymers as a function of temperature. PBT, polybutylene terephthalate; PC, polycarbonate; PPO, polyphenylene oxide; PVC, polyvinyl chloride; PP, polypropylene; HDPE, high-density polyethylene; T_g, glass transition temperature

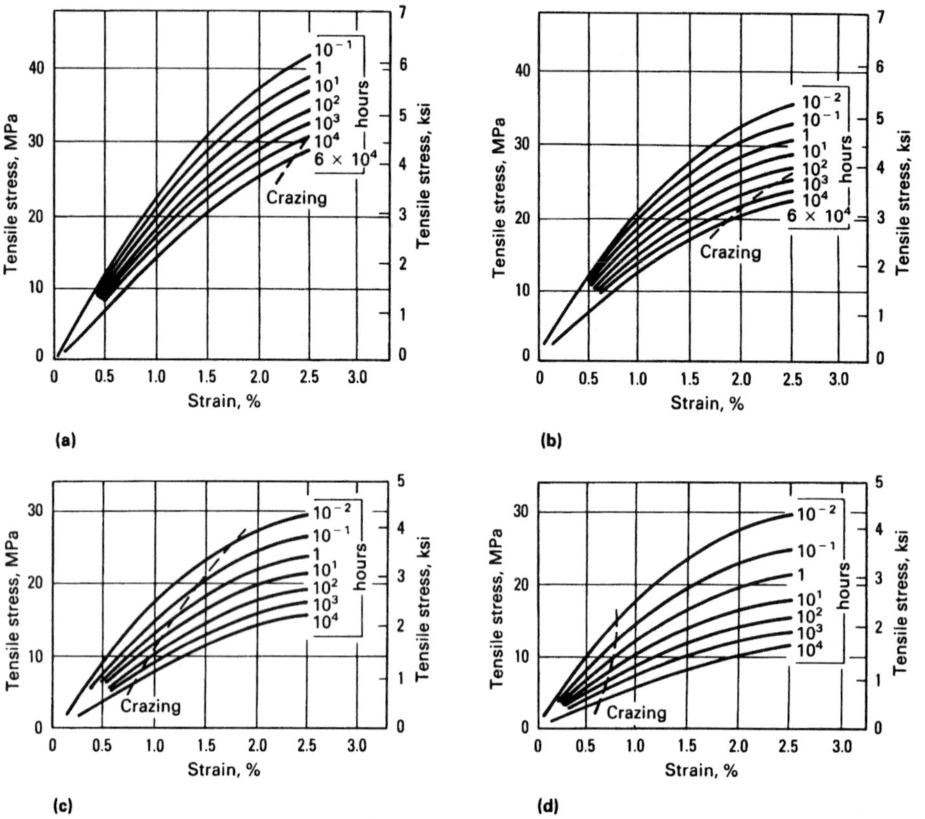

Fig. 6 Isochronous plot of polycarbonate stress-strain behavior as a function of temperature. Note that the crazing locus decreases in strain value with increasing temperature. (a) 23 °C (73.5 °F). Relative humidity, 50%. (b) 40 °C (104 °F). (c) 80 °C (176 °F). (d) 100 °C (212 °F). Courtesy of Mobay Chemical Company

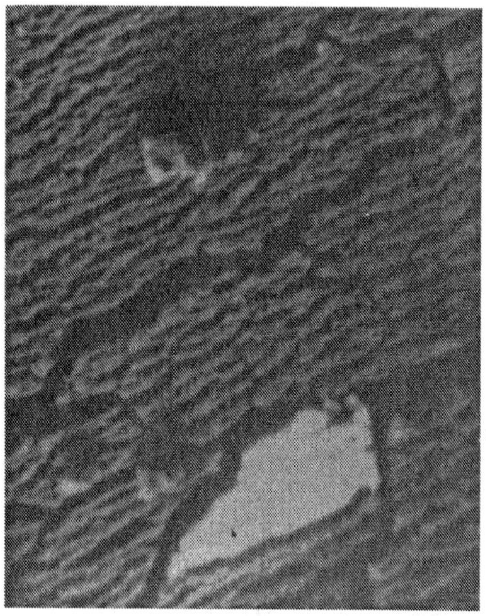

Fig. 8 Surface-microcracking network developed on polyoxymethylene due to ultraviolet exposure. 200×

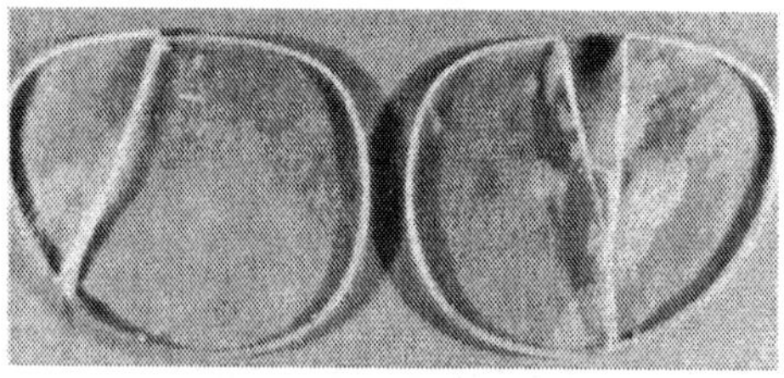

Fig. 9 Failed polycarbonate lenses exhibited primary and secondary cracking associated with solvent swelling and cracking

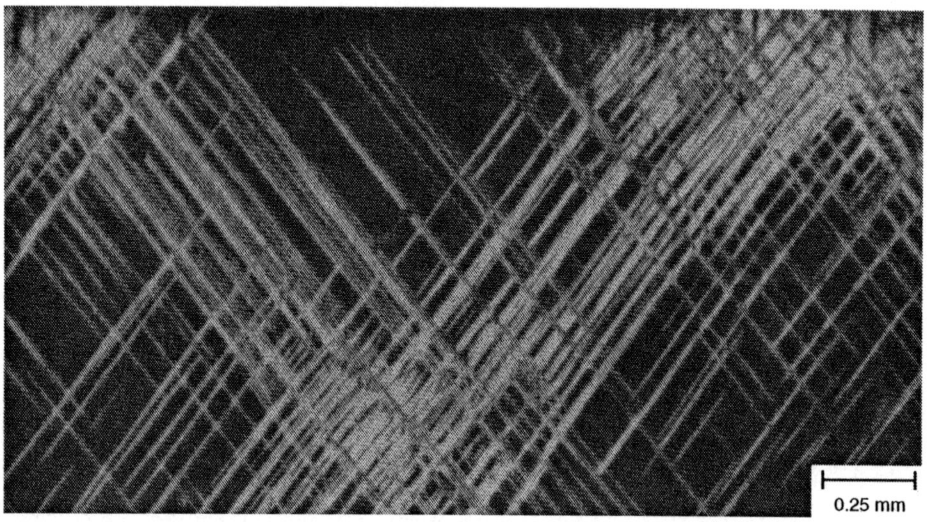

Fig. 7 Section from a polystyrene sample that was deformed past its compressive yield. The section is viewed between cross polars, showing shear bands. 50×

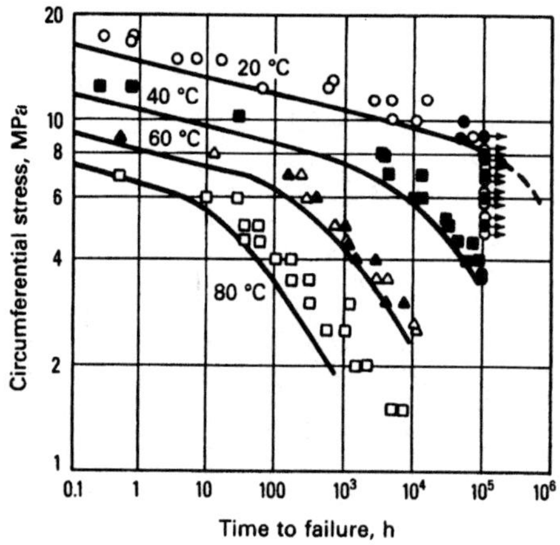

Fig. 10 Time-to-failure of high-density polyethylene pipes at different stresses and temperatures. Source: Ref 11

under monotonic loading. The transformation at the crack tip may appear as a yielded zone, as in thin PC sheet material (Fig. 13). A pair of shear bands are also reported to evolve at the tip of propagating cracks in PC (Ref 12). Recently, it was observed that when the thickness of the PC sheet is 6 mm (0.25 in.) or more, brittle microcracking ahead of the propagating crack becomes the dominant mechanism of fracture. Thus, one polymer displays more than one fracture mechanism. In all cases, the crack appears to propagate through a craze at its very tip.

The magnitude of transformation preceding the crack in a particular polymer is influenced by the loading history and consequently determines the resistance to crack propagation (Ref 13–15). Investigations show that damage evolution ahead of the crack tip accounts for the discrepancies in fracture toughness of polymers (Ref 16). These results suggest that examination of the material above and below the crack-propagation plane should be considered in failure analysis in addition to the commonly accepted fracture surface studies. As was briefly shown in preceding paragraphs, vital information about the resistance of the material to crack propagation and its loading history can easily be decoded from a thinned section normal to the crack-propagation plane.

Fractography (Adapted from Ref 17)

Fractography often reveals important clues about the cause of fracture and therefore plays an important role in the choice of subsequent testing or analyses to determine the cause of failure (Ref 18). Fractography involves the examination and interpretation of fracture surfaces, although the examination is not necessarily confined to the fracture surface alone. In most cases, the fracture surface examination is intended to reveal the location of the fracture origin, providing valuable information about the local service environment as well as the state of stress responsible for the crack initiation and growth that eventually led to fracture. The subsequent stress analysis of the failed part can be considerably simplified, because attention can be focused at the location of crack initiation (Ref 19).

The stereo zoom optical microscope is the instrument of choice for the preliminary fracture surface examination at moderate magnifications. A failed plastic part can be thoroughly inspected to assess the extent of cracking as well as the presence of other surface microcracks not revealed during the visual inspection. In addition, the origin of the fracture can frequently be identified under the optical microscope. Because virtually no specimen preparation is involved, the specimen is unaltered. This is sometimes an important consideration in situations where the specimen being examined must be preserved without contamination for subsequent analyses.

The SEM is frequently used to study fracture surfaces. Its superior depth of field is particularly useful in examining rough fracture surfaces at higher magnifications. However, because most polymers are electrically nonconductive, the specimen must be coated with a thin, conductive layer such as gold or carbon, which has a tendency to alter the surface appearance. If discoloration is evident on the failed part due to either environmental degradation or chemical attack, the specimen should be examined and documented first in the optical microscope. Because SEM is also more time-consuming than optical microscopy, specimens should be judiciously selected by preliminary examination under the optical microscope before a more detailed examination is conducted with the SEM. Additional information on polymer microscopy techniques and specimen preparation methods is available in Ref 20.

Modes of Fracture

When fracture occurs, a fracture surface is produced on a plane normal to the maximum principal tensile stress where the local stress exceeds the local strength of the material. Crack extension takes place along the path of least resistance to fracture. The characteristic fracture surface appearance depends on the complex interactions of the prevailing conditions of stress, materials properties, environment, and

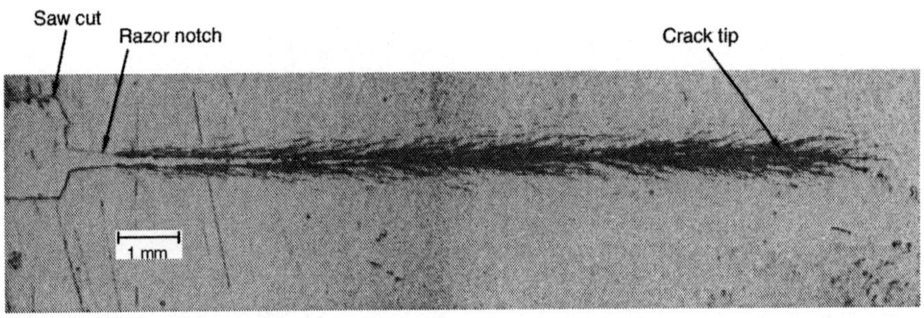

Fig. 11 A thinned section of fatigue-cracked polypropylene specimen. Crazes are visible surrounding and preceding the crack. 8×

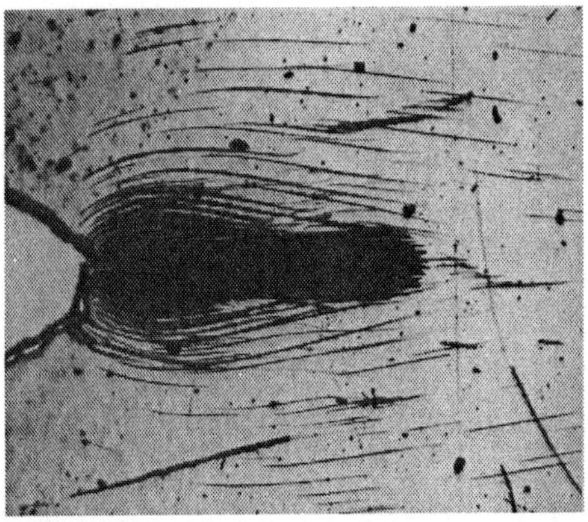

Fig. 12 Fatigue-crack initiation in polystyrene from a V-notch. Note crazes surrounding and preceding the crack. 37×

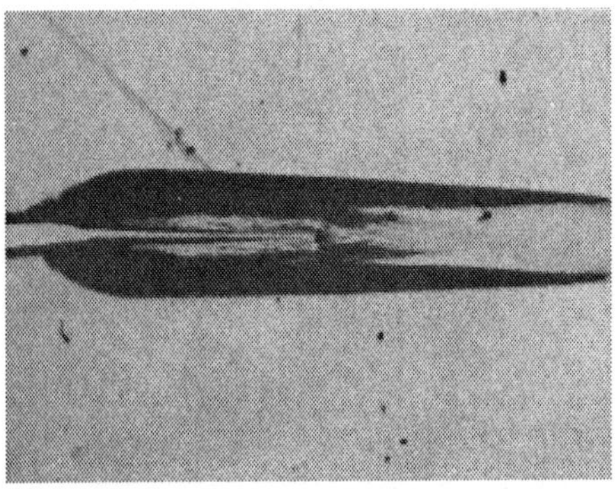

Fig. 13 Transmitted-light micrograph showing a yielded zone surrounding and preceding a fatigue crack in 0.25 mm (0.01 in.) thick polycarbonate sheet. 20×

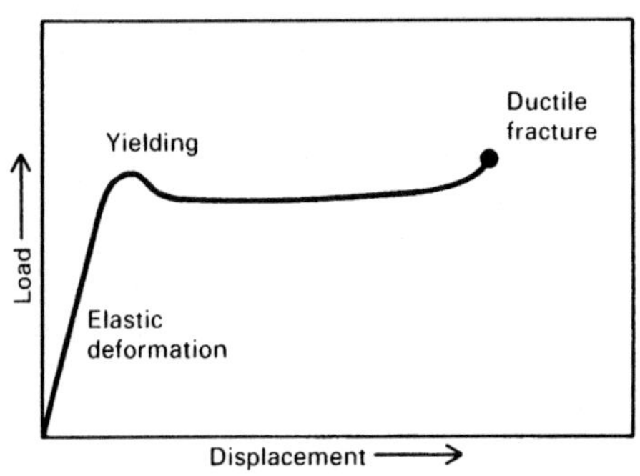

Fig. 14 Typical load-displacement curve for a ductile polymer tested in uniaxial tension

time. The distinction between the ductile and brittle fracture modes is generally made on the basis of macroscopic appearance.

Ductile fractures involve gross plastic deformation that is commonly described as yielding, tearing, plastic flow, necking, and shear—all of which involve shape changes or distortions of varying extent. Examples of commercial polymers that normally exhibit ductile fracture behavior are acrylonitrile-butadiene-styrene (ABS), polyethylene (PE), polypropylene (PP), high-impact polystyrene (HIPS), polyamide (PA), polybutylene, polycarbonate (PC), polyethylene terephthalate (PET), and polyvinyl chloride (PVC).

The most obvious example of a ductile fracture can be seen in the standard tensile specimen of a ductile polymer tested under ordinary conditions of temperature and strain rate. In a tensile test, a specimen containing no stress concentration is stretched at a constant rate until fracture. The stress in the specimen is essentially uniform uniaxial tension until the onset of yield and then again after the chains are oriented. Figure 14 shows the tensile load-displacement curve of a ductile polymer. Typically, the initial portion of the load-displacement record is nearly a straight line. This stress-strain response is termed linear elastic behavior because the stress is proportional to the strain, and, on unloading, the specimen will return to its original dimensions.

As the deformation increases, the stress is no longer linearly proportional to the strain. The slope of the curve begins to decrease, and a knee is eventually formed. For most engineering plastics, this typically occurs at a strain of a few percent. The rate of stress increase with deformation is substantially lower than the initial slope, which is a measure of the elastic modulus of the material. This sudden change in slope signifies the onset of net section plastic deformation or yielding. Even if the stress is removed, the specimen will not return to its original dimensions, in which case the specimen is permanently deformed. This suggests that ductile fractures occur with net section yielding after the applied stress has exceeded the yield strength of the material.

In practice, ductile fractures rarely become the subjects of failure analysis. The maximum stress normally expected in a well-designed component, or the maximum allowable design stress, should not exceed a small fraction of the yield strength of the material. Exceptions are those products that are intended to be broken in use, such as the tamper-evident rings found on the plastic closures of beverage bottles. When an unexpected ductile fracture occurs, it is usually found that the part was either grossly underdesigned or overloaded, so that the maximum stress it encountered not only exceeded the allowable design stress but also surpassed even the yield and tensile strengths of the material. Problems of underdesign or overloading are usually diagnosed early in the development of a product, such as during prototype testing. Ductile fractures can also result from other causes, such as excessive creep deformation or inadvertent exposure to elevated temperatures, which can substantially reduce yield strength.

Brittle fractures occur on a macroscopic level with little or no gross plastic deformation. Because shape changes or distortions are absent, the fracture faces can be precisely matched, although this practice is not recommended if the fracture surfaces are to be subsequently examined with a microscope. Inadvertent contact can produce artifacts that complicate the fracture surface examination. Examples of commercial plastics that normally fracture in a brittle manner are PS, styrene-acrylonitrile (SAN), PMMA, and thermosetting resins such as epoxy and polyester. A common example of a material that fractures in a truly brittle manner is inorganic glass.

The typical tensile stress-strain plot for a brittle plastic is essentially a straight line from the origin to the point of fracture, which coincidentally occurs at approximately the same strain as yielding in a ductile polymer. Except for a slight decrease in slope immediately before fracture, the deformation may be considered linearly elastic. Hooke's law for a homogeneous, isotropic material can be applied for most practical engineering analyses of stress and strain in a brittle plastic part.

When a ductile polymer fractures in a brittle manner, the stress analysis of the part may be similarly conducted, assuming linear elastic behavior. This is permissible because brittle fractures in a normally ductile polymer also occur at small strains before the onset of gross yielding. The stresses involved in producing the fracture are below the yield strength of the material and therefore lie within the linear elastic portion of the stress-strain plot.

Although no gross plastic yielding is evident in brittle fractures, plastic deformation is nevertheless involved in polymer fractures. Except for those rare cases in which polymer molecular motions are totally inhibited, such as at very low temperatures, brittle fractures of even brittle polymers are accompanied by plastic flow processes, which occur on a much more localized level than gross plastic yielding.

One form of plastic deformation that frequently leads to brittle fractures is crazing, sometimes called craze yielding (Ref 6). In many polymers, crack initiation is preceded by craze formation. In transparent polymers, the easily observed crazes appear as cracklike structures that are macroscopically indistinguishable from cracks. Geometrically, a craze is a planar defect similar to a crack, with its areal dimensions much larger than its thickness. A craze differs from a crack in that it contains an interpenetrating network of voids among highly drawn polymer fibrils bridging the craze faces.

Crazing begins with microvoid formation under the action of the hydrostatic tension component of the stress tensor. It is particularly sensitive to biaxial tensile stresses that frequently occur at sites of stress concentration with material constraints. After crazing has initiated, the voids increase in size and elongate along the direction of the maximum principal tensile stress. The polymer bulk material among the voids also undergoes gradual elongation to form thin fibrils. The strain in the craze fibrils depends on the amount of craze thickening but has been estimated to be approximately 50 to 100% in a well-developed craze section.

As the craze thickens, its growth in the lateral dimensions occurs by additional void nucleation at its leading edge. New craze matter is generated at the craze tip as a result of the triaxial tension there.

The deformation of a craze frequently leads to the initiation of a true crack. As the craze faces separate, the craze fibrils increase in length, while their diameters contract. The craze fibrils are not unlike microtensile specimens undergoing uniaxial extension. When the longitudinal strain in the fibrils exceeds the maximum extensibility of the molecular network, they rupture and form a crack. Figure 15 illustrates craze formation and its subsequent development into a crack.

Crack initiation is then followed by crack growth, in which more craze fibrils undergo extensive plastic deformation until they rupture. At the same time, craze growth continues, and new craze matter is generated at the craze tip. When the crack reaches a certain critical size, crack extension occurs in an uncontrollable man-

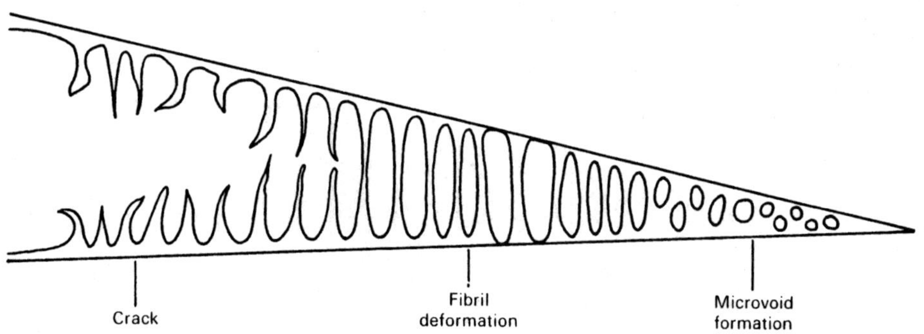

Fig. 15 Crack formation from a craze

ner, resulting in an unstable, brittle fracture. In terms of linear elastic fracture mechanics, the criterion of brittle fracture in the tensile crack opening mode is described by the stress-intensity factor, K_I, at the crack tip, reaching a critical value, K_{Ic}, for the plane-strain fracture toughness, which is considered to be a material property (Ref 21). Linear elastic fracture mechanics properties in accordance with ASTM D 5045 are appropriate for highly cross-linked thermosets or glassy thermoplastics incapable of significant plastic deformation (e.g., polystyrene.) Unlike shear yielding, which occurs at constant volume, craze yielding is a cavitation process that is accompanied by an increase in volume. Depending on the stage of craze development, the void volume in a craze has been calculated from refractive index measurements to be roughly 40 to 60%. Crazing, frequently described as the precursor of brittle fractures, is therefore favored by the presence of hydrostatic tensile stresses.

The influence of the state of stress on the mode of fracture is best illustrated by the deformation of the craze itself. The formation and growth of a craze, which frequently lead to brittle fractures, are promoted by a stress condition known as plane strain, in which triaxial tensile stresses exist. The plane-strain condition often results from the nonuniform stress distribution near stress raisers, such as cracks and other defects. Because of elastic constraints, lateral deformations near the crack tip or other stress concentrations are restricted, resulting in the development of lateral tensile stresses. The resultant hydrostatic tension tends to produce cavitation or void formation in the material, eventually leading to craze formation and brittle fractures.

On the other hand, the fibrils within the craze deform under a state of uniaxial tension in which lateral contraction is unconstrained, resulting in extensive shear band formation before rupture. A large amount of mechanical strain energy per unit volume is dissipated through craze yielding, which contributes to the toughness of the material. For this reason, even brittle polymers, such as PS and PMMA, are more impact resistant than ordinary window glass, in which brittle fractures occur without plastic deformation.

Despite the extensive plastic deformation that occurs within the craze, the total strain energy absorbed is small, because the plastic deformation takes place in a relatively small volume of material occupied by the craze. Therefore, only a small amount of strain energy is absorbed in a plane-strain brittle fracture. In contrast, a large amount of energy is absorbed in a ductile fracture, in which the plastic deformation takes place over a much larger volume of material.

Certain polymers that normally deform in a ductile manner in the standard tension test frequently sustain brittle fractures when a sharp notch or crack is introduced into the specimen. Some examples of notch-sensitive polymers are PC, PVC, PP, PET, PA, ABS, and HIPS. The transition from a ductile to a brittle fracture mode when a deep notch or crack is present is partly due to the change from a uniaxial to a triaxial tensile stress state.

Evidence of craze yielding can frequently be observed on the brittle fracture surface. As a result of craze fibril rupture, the fracture surface is lined with a thin layer of highly oriented polymer fibrils whose index of refraction differs from that of the underlying bulk polymer. As a result, colorful interference fringes similar to those seen on an oily, wet pavement can frequently be observed at the fracture origin on a brittle fracture surface.

Fracture Surface Features

The fracture origin is the point at which a crack is first nucleated. It usually coincides with the location of the maximum principal tensile stress or the minimum material resistance to fracture. Strictly speaking, because neither a uniform distribution of stress nor material homogeneity exists throughout an engineered component, certain points within a part are expected to be more probable sites of crack initiation due to either higher-than-average stresses or lower-than-normal crack resistance.

Local stress variation can result from a variety of factors, including loading configuration, part design, and discontinuities or inhomogeneities such as voids, cracks, inclusions, and other imperfections. The presence of holes, sharp corners, or sudden changes in wall thickness all contribute to stress concentration.

Other imperfections that can raise local stresses are introduced during the manufacture or the service life of the part. Some examples are shrinkage voids (Fig. 16) and contaminant inclusions resulting from poor molding practices. Improper techniques during postmolding operations, such as machining, decorating, bonding, or mechanical fastening, can produce partial fractures or cracks that escape detection and serve as sites of crack initiation.

The location of the fracture origin can also reveal points of material weakness. Material defects within a molded part include weak weld lines, inclusions of debris, poorly fused cross-linked gel particles, and other inhomogeneities such as agglomeration of pigment, filler, and reinforcement. Because these defects are caused by poor processing, their occurrence may be more sporadic than defects in design.

Material degradation caused by either natural aging or an aggressive environment is frequently revealed by examining the failed part in the vicinity of the crack origin. One of the leading causes of brittle fracture in polymers is environmental stress cracking that results from exposure to incompatible chemicals. Examples of potential stress crack agents are mold releases, cleaners, lubricating oils and greases, plasticizers, and solvents in paints and coatings. Prolonged exposure to ultraviolet radiation in sunlight can also embrittle the surface layers, resulting in microcracking. In these cases, fracture initiation is typically found at multiple locations on the affected areas and is frequently accompanied by surface microcracking.

Mirror Zone. Brittle fractures in many polymers are preceded by craze formation and its subsequent breakdown. The initial stage of crack growth results from the rupture of fibrils at the trailing edge of a craze. As the nucleated

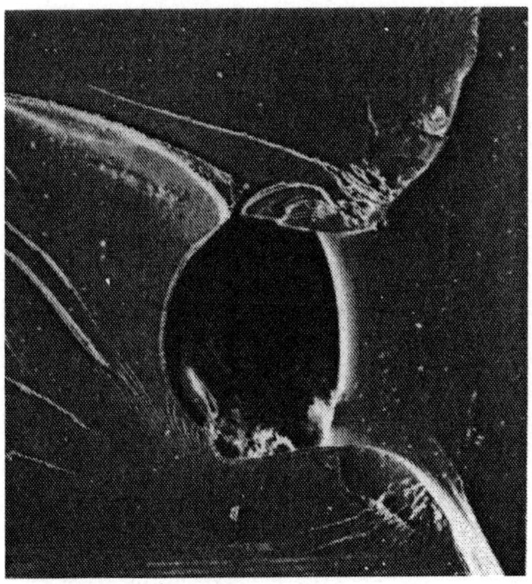

Fig. 16 Shrinkage void on field fracture surface of polycarbonate. 12×

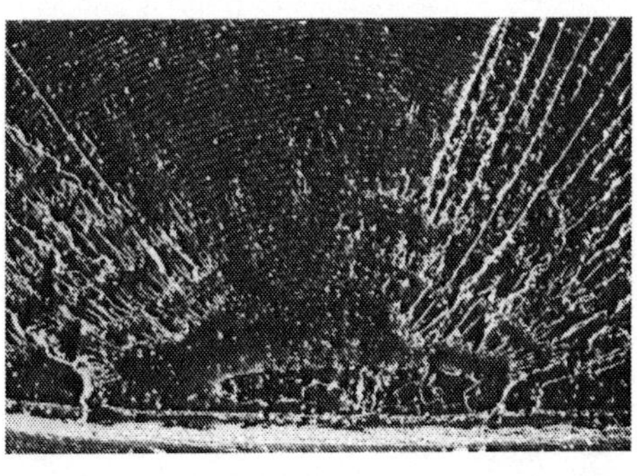

Fig. 17 Polycarbonate fracture surface showing mirror zone, mist and hackle regions, and Wallner lines. 14×

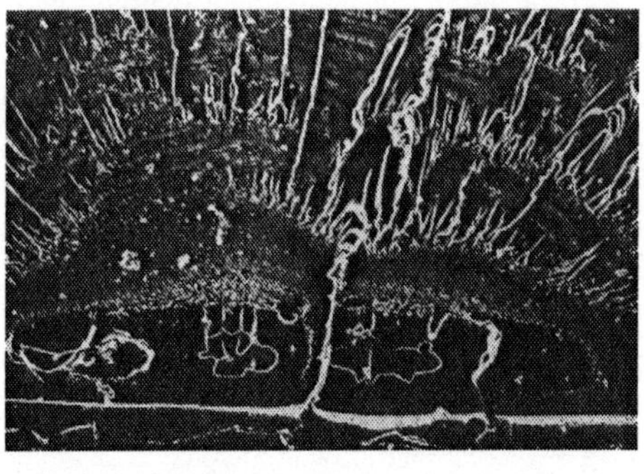

Fig. 18 Fracture initiation region of polycarbonate specimen after Izod impact showing mirror zone and mist region. 27×

crack increases in size, new craze matter is generated at the craze tip. Crazes are very thin, planar defects; therefore, they form a very flat and smooth fracture origin, commonly known as the mirror zone or mirror region, which actually contains craze remnants. Because of the presence of a thin layer of highly oriented polymer with a different refractive index from that of the bulk, interference color fringes are frequently observed in the mirror region when the specimen is viewed in visible light. Figure 17 shows the fracture surface area near the fracture origin of a PC specimen after Izod impact. The mirror zone appears in the lower central portion of Fig. 17 in the shape of an ellipse.

The crack initiation stage of fracture is sometimes referred to as nucleation, which is followed by slow, subcritical crack growth. As the newly formed crack increases in size, a critical size is eventually reached at the point when cracking becomes unstable overload fracture at a very rapid rate. The boundary of the mirror region marks the transition of crack velocity from a slow, stable extension to sudden acceleration, which leads to catastrophic fracture. In terms of fracture mechanics, the condition of crack instability can be described by:

$$K_{Ic} = Y\sigma_f\sqrt{a_c}$$

where K_{Ic} is the plane-strain fracture toughness dependent on temperature and environment, Y is the geometric factor dependent on loading and specimen and crack geometry, σ_f is the stress at fracture, and a_c is the critical crack size.

In other words, brittle fracture is said to occur when the stress-intensity factor, K_I, exceeds the fracture toughness, K_{Ic}, of the material. Because the stress-intensity factor depends on both the prevailing stress and crack geometry, the condition for brittle fracture can be satisfied by different combinations of stress and crack sizes. When the stress is high, the crack size at the moment of instability is expected to be smaller than when the stress is low. This suggests that the magnitude of the stress at fracture can be estimated by measuring the size of the slow-growth region.

According to this condition of instability, the mirror region size should be inversely proportional to the square of the fracture stress. In practice, however, this is of limited utility unless the other conditions of fracture, such as temperature and environment and their effects on fracture toughness, are also precisely known.

Mist Region. Surrounding the mirror zone is an area known as the mist region, commonly found in glass. It is also observed in some glassy polymers, such as PS, PMMA, and PC. This region is typically a flat, smooth area that is essentially featureless, except for a slight change in surface texture resembling a fine mist. Figure 18 shows the mist region immediately adjacent to the mirror zone at the fracture origin of a PC specimen after Izod impact. In polymers, mist regions are not necessarily confined to the vicinity of the fracture origin but can be observed elsewhere on the fracture surface.

Hackle Region. Unlike mirror and mist regions, which are relatively smooth surface features, hackle regions are particularly rough surfaces. They are easily recognized by the outward divergent lines pointing along the crack-propagation direction (Fig. 17, 18). Hackle regions are associated with a more violent stage of fracture in which a large amount of strain energy is absorbed through both plastic deformation and the generation of new fracture surface areas. The hackle region shown in Fig. 18 is slightly magnified in Fig. 19 to illustrate ductile shear yielding.

Hackle regions tend to appear in areas where the stress field is changing rapidly (either in direction or magnitude) or when the stress state changes from one of plane strain to plane stress. For example, they are frequently observed in the region of a specimen subjected to bending on the original compression side of the specimen but changes to tension as the crack approaches from the tension side. Figure 20 shows the hackle region in the last stage of fracture of a PC specimen after Izod impact. The stress state rapidly changes from plane strain to plane stress as the crack approaches the final ligament of the specimen.

As previously stated, cracking is a stress-relief mechanism. Elastic strain energy is released during crack extension, providing the crack driving force. As the driving force increases, the crack is driven to higher velocities. As the velocity of propagation approaches the limiting velocity in the material, the rapidly moving crack tends to branch into two or more cracks, thus increasing the rate of energy dissipation by creating additional fracture surface areas.

For a material incapable of plastic deformation, such as glass or a polymer at a very low temperature with totally inoperative flow processes, crack branching is the only mechanism for increasing the rate of energy dissipation. For this reason, an overpowering blow to a sheet of glass or a polymer at a very low temperature will produce a shattering, brittle fracture.

Another example involves the dicing fracture of tempered glass. Without any external stresses, a sheet of tempered glass is already under a state of high residual stress corresponding to a high level of stored elastic strain energy. If a crack develops in a sheet of tempered glass, a large amount of strain energy is suddenly released. To dissipate this energy, the material responds with profuse crack branching, which creates many small glass fragments that have many more new surfaces than if only a single crack is formed.

The rough appearance of hackle regions can be partially explained by the following description of a similar mechanism, but, in this case, the crack branching occurs on a much finer scale. As the crack driving force increases and the crack velocity is sufficiently high, a single crack front begins to split up into many smaller crack fronts. These continue to propagate and diverge outward alongside each other but at slightly different crack angles (Fig. 19, 21). Because this

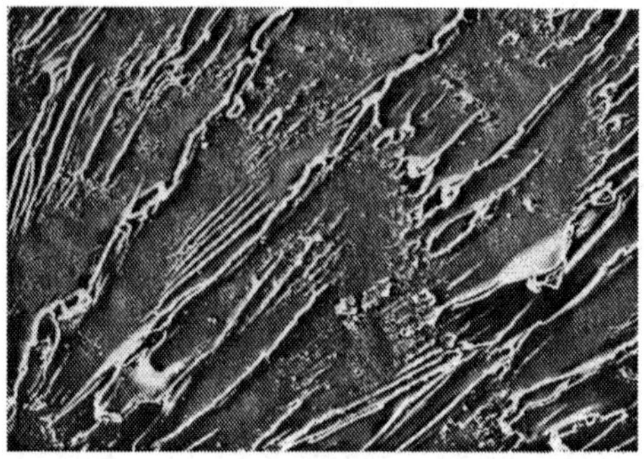

Fig. 19 Hackle region from Fig. 18 showing ductile shear yielding and crack-front branching. 65×

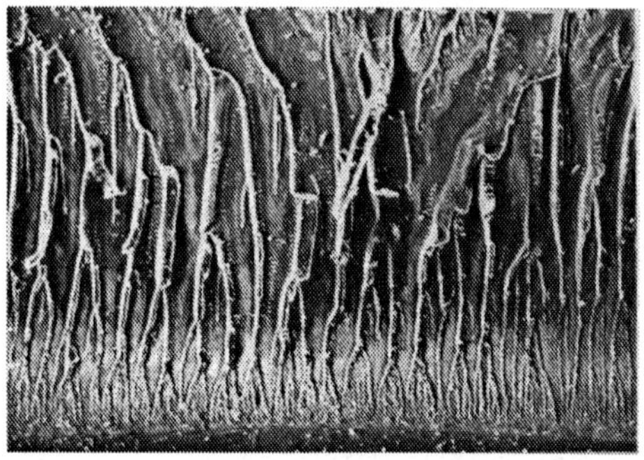

Fig. 20 Hackle region in final ligament of polycarbonate specimen after Izod impact. 14×

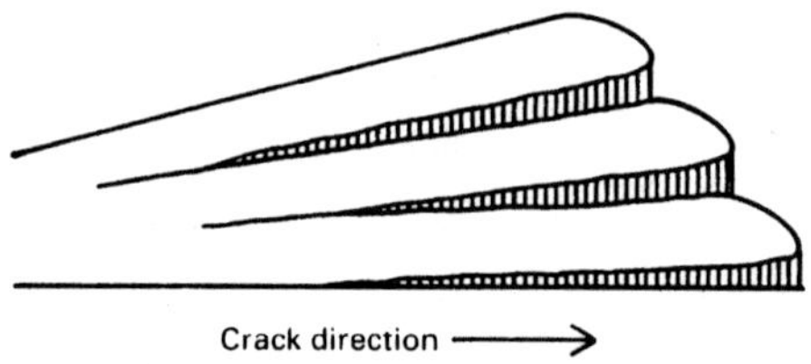

Fig. 21 Formation of hackle lines from crack-front branching

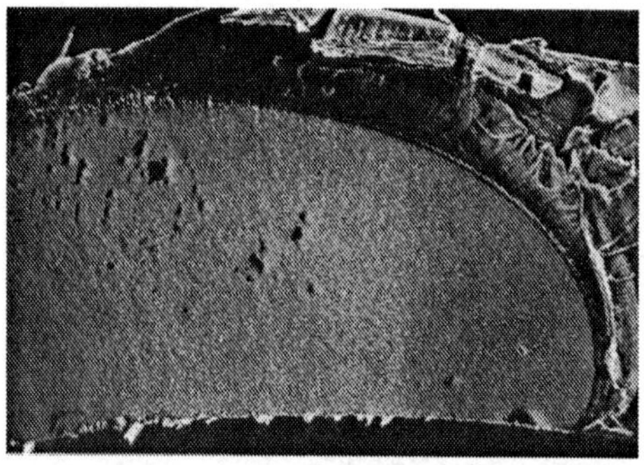

Fig. 22 Brittle fracture surface of a polyethylene gas pipe showing rib marking at crack arrest. 14.5×

results in a series of smaller crack fronts propagating on slightly different crack planes or elevations, their side boundaries may overlap or even undercut each other. Sharp slivers are frequently observed on glass fracture surfaces when two crack fronts whose side boundaries have previously undercut each other reemerge onto the same crack plane.

In polymeric materials, ductile plastic deformation is evident in the hackle regions of a brittle fracture surface (Fig. 18 to 20). The ductile tearing mode of deformation is typically observed in the side boundaries of adjacent cracks propagating on slightly different crack planes (Fig. 19). While the cracks are propagating on planes normal to the tensile stresses, the material at the crack boundaries is subjected to shear stresses. This results in shear yielding, a form of plastic deformation that occurs with no volume change. A large amount of strain energy is dissipated by this process. Energy absorption by plastic deformation has been estimated to be greater by several orders of magnitude than through the creation of new surfaces alone. The rough appearance of the hackle regions is due to both the occurrence of plastic flow on the fracture surface and the presence of non-coplanar crack surfaces.

Because hackle regions are prominent features on a fracture surface, the crack-propagation direction can be easily identified. The divergent nature of the hackle lines is advantageous in locating the crack origin. If the area of the fracture surface under examination is remote from its origin, some back tracing is necessary. Otherwise, hackle lines diverge from the fracture origin (Fig. 17, 18).

Wallner lines are sometimes observed as a faint ridged pattern on otherwise smooth fracture surfaces. They resemble fatigue striations with periodic spacing but are formed when stress waves reflected from the specimen boundaries interact with a propagating crack front. Very subtle changes in the fracture surface texture result when the stress waves produce a slight perturbation of the stress field ahead of the crack front. Wallner lines are shown near the central and upper portions of Fig. 17 and are interspersed among the hackle lines in Fig. 18 to 20.

Because Wallner lines are formed when reflected stress waves intersect a propagating crack, they are not true crack-front markings. In practice, however, stress wave velocities are so much higher than crack velocities that Wallner lines may be considered snapshots of the crack front during its propagation. Typically, they are curved markings similar to crack-front markings, with the fracture initiation site located on their concave side. As such, they may be useful in locating the fracture origin.

Rib Markings. A true crack-front marking is produced when a moving crack is stopped or arrested. These markings are commonly called crack-arrest lines or rib markings because of their resemblance to curved rib bones. Figure 22 shows the fracture surface area in a PE natural gas pipe where the crack has stopped.

Fatigue striations are also true crack-arrest markings, because they are formed during the repetition of crack extension and arrest. Figure 23 shows the fatigue fracture surface of a PC plumbing fixture subjected to the effect of a water hammer. In a laboratory specimen subjected to a well-controlled, uniform cyclic load, the fatigue striations are also well defined and nearly regularly spaced (Ref 7). This is generally not true of field fractures, in which the specimens are subjected to a more random loading history. The distance between crack arrests is generally more irregularly spaced, because it is entirely determined by the conditions affecting the progression of the crack.

Fig. 23 Fatigue striations on the fracture surface of a polycarbonate plumbing fixture after field failure. 32×

Fig. 24 Rib markings near the origin of polyethylene gas pipe fracture. 14×

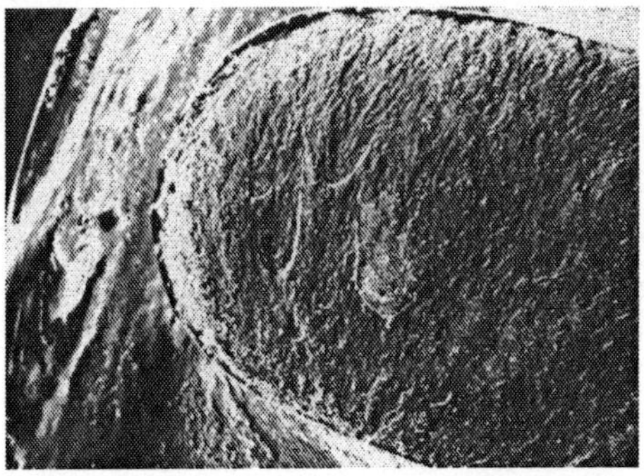

Fig. 25 Parabolic markings on acrylonitrile-butadiene-styrene fracture surface. 12.5×

Crack velocity changes also produce rib markings that are not as prominent as when the crack is totally stopped. Therefore, in this case, the term *crack-arrest marking* is not entirely accurate, even though it is more descriptive of the formation mechanism than the term *rib marking*. Figure 24 shows the fracture surface of the PE gas pipe in the vicinity of the origin near the inner pipe wall, where semicircular rib marks are seen expanding outward from the origin.

Because both fatigue striations and rib markings are true crack-front markings, the crack-propagation direction at any point on the crack

Fig. 26 SEM view of fatigue striations in medium-density polyethylene, laboratory tested at 0.5 Hz with maximum stress 30% of the yield strength. Crack growth is upward in this view. Original magnification 200×. Source: Ref 23

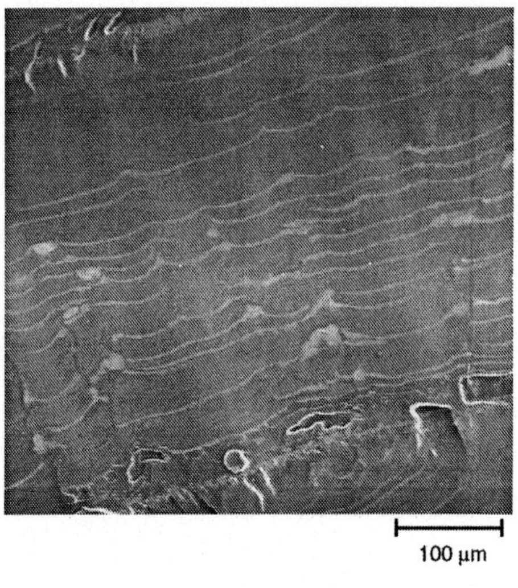

Fig. 27 Features observed on fatigue area of polymethyl methacrylate rotating beam specimen. Sample was sputter coated with platinum for SEM examination.

front can be determined by drawing the outward direction normal to the crack front. The fracture origin will be located by tracing back along the crack direction on the concave side of the curved markings.

Parabolic markings, which resemble the shape of a parabola, can also appear on the fracture surface of a plastic. Although often microscopic, they can be large enough to be observed with the unaided eye in certain plastics, such as PVC and ABS. Four parabolic markings are shown on the ABS fracture surface in Fig. 25. These markings are often useful surface features for determining the crack-propagation direction when more prominent features are lacking.

Parabolic markings are formed when a primary or main crack front intersects another crack that has initiated at a small distance just ahead of the main crack. Secondary crack origins arise at sites of local stress concentration because of material inhomogeneities and the rapidly rising stress field ahead of the primary crack. The intersection of the two crack fronts, frequently on slightly different crack elevations, produces a parabolic marking that diverges in the propagation direction of the main crack. The secondary crack origin can be found near the focus of the parabola. The primary crack origin, however, will be located on the convex side of the parabolic marking. The crack origin shown in Fig. 25 is near the lower central portion of the micrograph.

Fatigue (Ref 22). Exemplar SEM fractographs of polymers are provided in Ref 18, and fractography of fatigue in engineering plastics is included in Ref 7. Macroscopic features of fatigue of structural polymers can parallel those of metals in many circumstances: relatively flat fracture, beach marks, and ratchet marks. Some examples are given in Fig. 26 to 28.

The properties of polymers result in fatigue behavior different from that ordinarily encountered in metals. Except at elevated temperature or in corrosive environments, fatigue of structural metals depends on number (and magnitude) of loading cycles rather than time under load. However, because of time and rate sensitivity of many polymers at near-ambient temperatures, fatigue depends not only on the number of cycles at a given stress or stress-intensity level but also on frequency and time history of loading (Ref 24). Heat resulting from mechanical hysteresis during cyclic loading of plastics can cause thermal failure.

Additionally, depending on load level and time history, fatigue cracks in polymers may not propagate steadily (with an increment of growth for each fatigue cycle) but may grow in bursts or spurts (Ref 24). These spurts or discontinuous growth bands are associated with a large number of cycles (Ref 7). They can produce microscopic markings that appear very much like striations but which do not correspond to single load cycles. Consequently, without companion laboratory fatigue crack growth rate data and careful fractographic evaluation, estimation of service history from the spacing between fracture markings can be problematic (Ref 7). Some polymers, such as PC and PMMA, can exhibit either true striations or discontinuous growth bands, depending on load levels and loading history.

A brief discussion of fractography of fibrous organic-matrix structural composites is provided in the next article, "Fractography of Composites." Fractography of fatigue failures in composite materials can be more difficult than for other materials, such as metals. There may be little macroscopic difference between interlaminar fracture features formed by fatigue and those formed in overload. Composite materials may lack visual indications of the initiation site. Although microscopic fatigue striations may form, areas exhibiting striations are usually isolated and limited in extent.

Case Studies (Ref 4)

Several cases of field failure in various polymers are considered to illustrate the applicability of available analytical tools in conjunction with an understanding of failure mechanisms.

Example 2: Failure of an Irrigation Pipe. Figure 29 shows an optical micrograph of the fracture surface of an irrigation pipe made of medium-density PE that failed in service (Ref

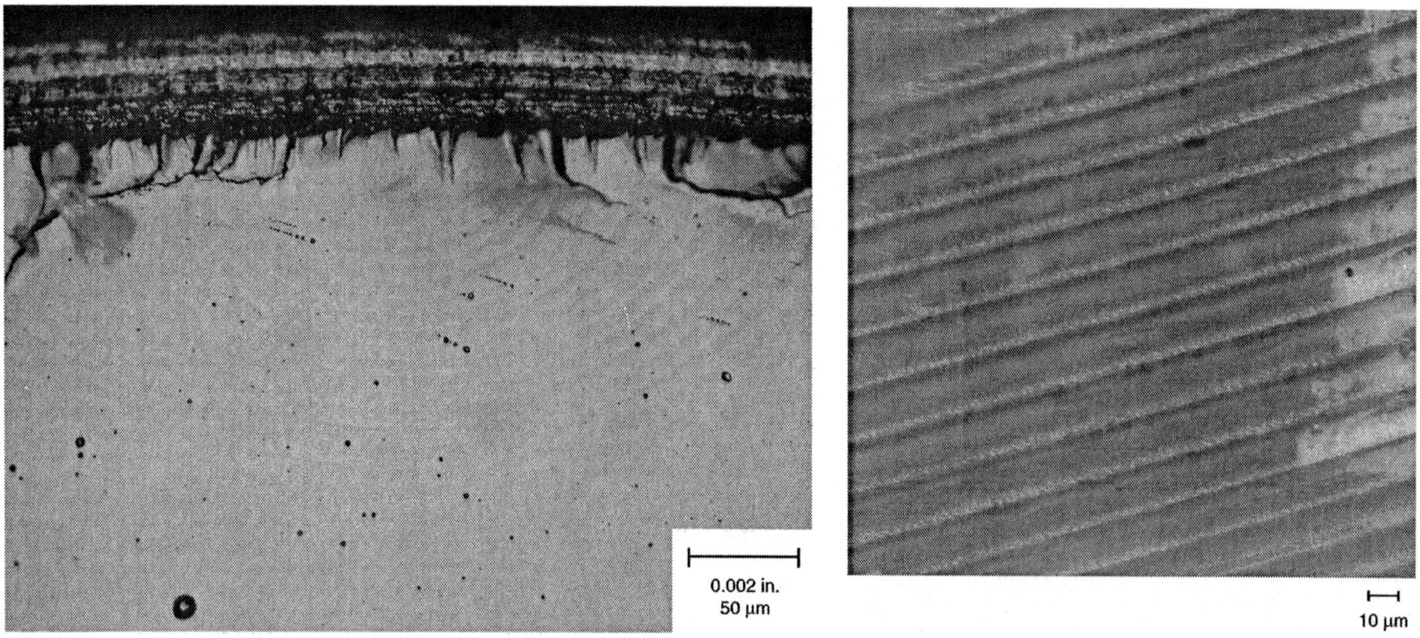

(a) (b)

Fig. 28 Features observed on fatigue area of polycarbonate rotating beam specimen. (a) Optical view at base of notch. (b) Higher-magnification electron fractograph. Sample was sputter coated with platinum for SEM examination.

Fig. 29 Reflected-light optical micrograph of the fracture surface of medium-density polyethylene pipe. Arrows indicate the direction of crack propagation.

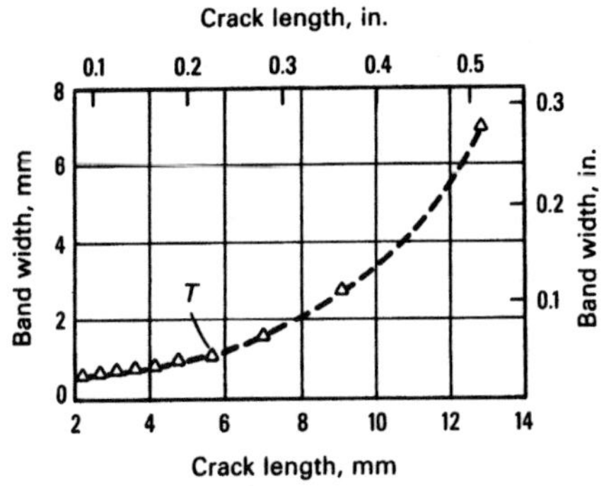

Fig. 30 Fracture band width as a function of crack length for the polyethylene pipe shown in Fig. 29. *T*, transition point

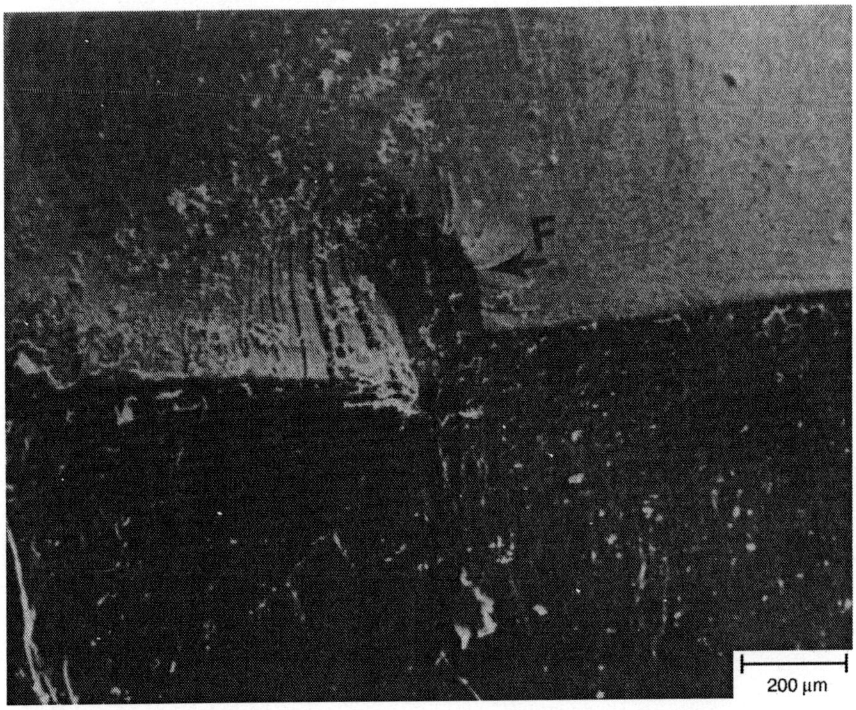

Fig. 31 Fracture in a polyvinyl chloride water filter. The fracture surface of the fatigue crack started from a fissure (arrow F). The lower dark zone is an artifact due to sectioning of the filter wall. 75×

25). This pipe was subjected to severe cyclic-bending strain of the order of 6% while under tensile stress of approximately 6.9 MPa (1000 psi) and a hoop stress of approximately 6.2 MPa (900 psi). These conditions of operation were far more stringent than those encountered in most applications of PE pipes. A subsurface imperfection in the pipe wall (dark, diamond-shaped spot, Fig. 29) acted as a crack starter. Contrary to the dominant belief that pipe failure initiates from surface defects, this example indicates that a critical-sized flaw within the pipe wall can also initiate failure. This crack starter (flaw) was located closer to the outside wall, where compressive residual stresses may be dominant.

Concentric circular striations originate from the crack starter and grow simultaneously in radial and circumferential directions. It is well established that such striations represent crack-arrest lines, where the distance between two striations (a band) is due to a crack excursion. Thus, evolution of the band width reflects the nature of crack propagation. The band width measured from larger micrographs is plotted as a function of crack length in Fig. 30. A smooth transition is observed at a point, *T*, as catastrophic failure (pipe separation) is approached. This transition is indicative of considerable increase in crack speed and coincides with a transition in band geometry from circular to elliptical. Plausibly, this occurred when the crack-tip stress field interacted with the inside wall of the pipe. It should also be noted that maximum residual tensile stress dominates close to the inside wall.

A similar transition has been noted to occur when the sufficient thermodynamic condition for crack instability is fulfilled. The major axis of the ellipse increases faster than the minor axis until no more striations are observed and ultimate failure results in large-scale yielding (approximately 50%) of the pipe wall (not shown in Fig. 30).

Whenever possible, similarity criteria should be established between the fracture behavior of a component in service and that observed in the laboratory. In this case, the band width appears to be a suitable candidate (Ref 26). This agreement indicates that discontinuous crack-growth band width, when available, can be employed as a similarity criterion to establish correspondence in loading history.

Example 3: Failure of a PVC Water-Filter Housing. Figure 31 shows an injection-molded PVC water-filter housing that fractured in service. An initial fissure (arrow F) is believed to have started first due to residual stresses developed during injection molding. Failure seems to have occurred due to fatigue crack propagation, as indicated by the presence of discontinuous crack-growth bands and their evolution. Although a tensile-stress component normal to the fracture surface was the dominant cause of failure, considerable triaxial stress seems evident in the early stages of fracture, as indicated by the successively smaller fissures to the left of the crack starter. As would be expected in PVC, catastrophic failure occurred by brittle failure as opposed to large-scale yielding, as in the PE pipe discussed in Example 2. This is evident from the relatively smooth appearance of the fracture surface beyond the last fatigue band in Fig. 31.

REFERENCES

1. I.M. Ward, *Mechanical Properties of Solid Polymers,* 2nd ed., John Wiley, 1983
2. R.P. Kambour and R.E. Robertson, Mechanical Properties of Plastics, *Polymer Science,* A.D. Jenkins, Ed., North-Holland Publishing Co., 1972, p 778
3. I.M. Ward, *Mechanical Properties of Solid Polymers,* John Wiley & Sons, 1982
4. A. Moet, Failure Analysis of Polymers, *Failure Analysis and Prevention,* Vol 11, *Metals Handbook,* 9th ed., American Society for Metals, 1986, p 758–765
5. A. Moet, Fatigue Failure, *Failure of Plastics,* W. Brostow and R.D. Coneliussen, Ed., Hanser Publishers, 1986
6. C.B. Bucknall, Toughened Plastics, *Appl. Sci.,* 1977
7. R.W. Hertzberg and J.A. Manson, *Fatigue of Engineering Plastics,* Academic Press, 1980
8. K. Matsushige, S.V. Radcliff, and E. Baer, *J. Mater. Sci.,* Vol 10, 1975, p 833
9. W. Schnabel, *Polymer Degradation,* Hanser International, 1981
10. E.C. Lochanski, Solvent-Induced Cracking Failure of Polycarbonate Ophthalmic Lenses, *Handbook of Case Histories in Failure Analysis,* Vol 2, ASM International, 1993, p 493
11. H.H. Kauch, *Polymer Fracture,* Springer-Verlag, 1978
12. M.T. Takemori and R.P. Kambour, *J. Mater. Sci.,* Vol 16, 1981, p 1108
13. F.W. Billmeyer, Jr., *Text Book of Polymer Science,* John Wiley & Sons, 1984
14. J.G. Williams, *Fracture Mechanics of Polymers,* John Wiley & Sons, 1984
15. A. Chudnovsky, A. Moet, R.J. Bankert, and M.T. Takemori, *J. Appl. Phys.,* Vol 54, 1983, p 5562
16. N. Haddaoui, A. Chudnovsky, and A. Moet, *Polym. Mater. Sci. Eng.,* Vol 49, 1983, p 117
17. P. So, Fractography, *Engineering Plastics,* Vol 2, *Engineered Materials Handbook,* ASM International, 1988, p 805
18. L. Engel, H. Klingele, G.W. Ehrenstein, and H. Schaper, *An Atlas of Polymer Damage,* Prentice-Hall, 1981
19. J.G. Williams, *Stress Analysis of Polymers,* John Wiley & Sons, 1973
20. L.C. Sawyer and D.T. Grubb, *Polymer Microscopy,* Chapman & Hall, 1987
21. J.G. Williams, *Fracture Mechanics of Polymers,* John Wiley & Sons, 1984
22. R. Lund and S. Sheybany, Fatigue Fracture Appearances, *Failure Analysis and Prevention,* Vol 11, *ASM Handbook,* ASM International, 2002, p 638–639
23. *Fractography,* Vol 12, *ASM Handbook,* ASM International, 1987
24. G.C. Pulos and W.G. Knauss, Nonsteady Crack and Craze Behavior in PMMA Under Cyclical Loading, *Int. J. Fract.,* Vol 93, 1998, p 145–207
25. K. Sehanobish, A. Moet, A. Chudnovsky, and P.P. Petro, *J. Mater. Sci. Lett.,* Vol 4, 1985, p 890
26. J.R. White and J.W. Teh, *Polymer,* Vol 20, 1979, p 764

SELECTED REFERENCES

- M. Ezrin, *Plastics Failure Guide: Causes and Prevention,* Hanser/Gardner Publications, Cincinnati, 1996
- J. Moalli, *Plastics Failure Analysis and Prevention,* Society of Plastics Engineers, 2001

Fractography of Composites*

FRACTURE SURFACES are examined during most investigations of failed structural components because these surfaces provide an actual physical record of the events at the time of failure. Fractographic analyses of the surfaces of metallic components reveal useful information about the cause and sequence of failure. Those surfaces reveal features that identify the crack origin, crack-propagation direction, failure mode, load, and environmental conditions at the time of failure. This information is extremely useful in the determination of failure cause. Hence, as composites developed into structural materials, a similar need arose to understand the fractographic evidence that these materials can provide.

The best method of developing an understanding of the fractographic evidence provided by those failures is to obtain pedigreed, fractographic data. These data are obtained by documenting the fractographic characteristics of specimens manufactured from different composite materials under different processes and exposed to different environmental and load conditions. The fracture surfaces of the pedigreed test specimens are examined, documented, and analyzed to determine which features are specific to a particular material, process, load, and/or environmental condition. The fractographs can then be used in the analysis of component failures.

This article depicts typical fractographic features for a number of different composite materials. Although not all-inclusive by any means, the fractographs depict a range of different, yet typical, fractographic features obtained from various composite materials that were manufactured and tested under different load and environmental conditions. It is hoped the fractographic data provided are useful for comparison with actual fractured surfaces to help determine the cause of component failures.

Material systems examined include epoxy resins with different fibers, such as carbon/epoxy (AS4/3501-6), fiberglass/epoxy (Hexcel E-glass/F155) and aramid/epoxy (Kevlar 49/3501-6), as well as fibers with different thermosetting resins, including carbon/bismaleimide (AS4/5250-3) and glass/polyimide (Celion 3K/PMR-15). Carbon-fiber and thermoplastic resin composite systems are also highlighted, mainly for comparison purposes, and include carbon thermoplastic (AS4/APC-2) and (AS4/KIII).

The specimens used for the fractographs depicted in this section were generally manufactured according to material supplier recommendations. Those specimens that were not manufactured according to manufacturer recommendations were processed with changes made solely to examine the effect of different material-processing conditions. Material-processing variations included changes in cure cycle, such as overcure or undercure conditions, surface contamination, and reduced resin content.

Environmental conditioning of the test specimens was conducted as noted in the fractographs provided. Environmental conditions examined included the effect of moisture in the laminate, moisture saturation followed by elevated-temperature exposure (i.e., hot/wet conditions), and elevated-temperature exposure without prior moisture conditioning.

Loading on the specimens was conducted using a variety of test specimens and load conditions. Mode I tension and tension fatigue failures were obtained using double-cantilever beam specimens; mode II shear and shear fatigue failures were obtained using end-notched flexural specimens. Translaminar tension and compression specimens used either the notched-bend bar specimens with four-point loading or the specimen configurations defined in ASTM D 3039, "Tensile Properties of Fiber-Resin Composites" and ASTM D 3410, "Compressive Properties of Unidirectional or Cross-Ply Fiber Resin Composites."

Following manufacture, environmental conditioning, and testing to failure, the fracture surfaces of the test specimens were examined in the scanning electron microscope (SEM). Typical fractographic features of each test specimen were then identified and documented. Further examination and analysis of the fractographs were then conducted in order to define the specific fractographic features that were indicative of a specific material, processing, environmental, or load condition at failure.

Interlaminar Fracture Features

An interlaminar fracture occurs when the load is applied perpendicular to the composite laminate and failure occurs in the plane of the reinforcement. Interlaminar fractures occur following mode I tension or fatigue loading, mode II shear or fatigue loading, flexural loading, and impact loading on the surface of the laminate.

Interlaminar Fracture of Composites with Brittle Thermoset Matrices. Most of the fractographic evidence in interlaminar fractures that would be indicative of the material, processing, load, and/or environmental conditions at failure are found in the matrix materials, rather than the fibers, of the composite. Analysis has shown that the fractographic features associated with these brittle thermoset matrices, including the epoxies, bismaleimides, and polyimides, are similar in nature. Because of this, most of the fractographic data presented in this section were obtained from epoxy-matrix materials; minimal fractographic data from the other brittle thermoset resin systems are presented.

Differences in Fracture Characteristics due to Different Loading Conditions. In general, brittle matrix composite materials tested under interlaminar, mode I tension loads fail in the plane of the reinforcement. Visually, these surfaces exhibit a glossy appearance, with some banding and resin covering most of the fibers on the fracture surfaces. On a microscopic level, the fractographic features evident in this failure mode consist mainly of river patterns on the surface of the matrix, as shown in Fig. 1 and 2, and matrix feathering, as shown in Fig. 3. River patterns are basically created by cleavage of the

*Adapted from the article by Patricia L. Stumpff, "Fractography," in *Composites,* Volume 21, *ASM Handbook,* ASM International, 2001, pages 977 to 987

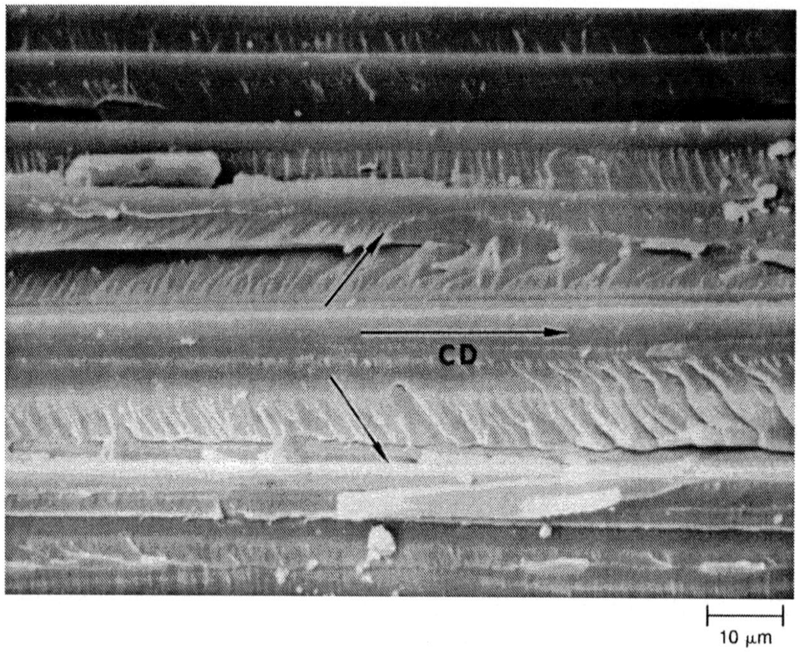

Fig. 1 River patterns on the surface of a mode I tensile failure in a carbon/epoxy (AS4/3501-6) composite laminate. Overall crack-growth direction is from left to right. 1000×

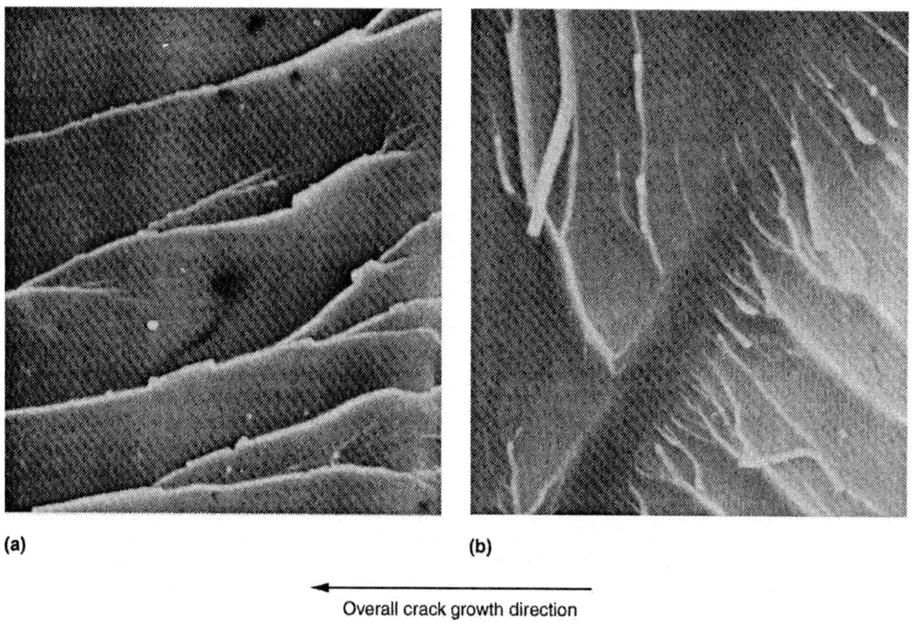

Fig. 2 Mode I tension interlaminar fractures that propagated at various angles to the direction of fiber reinforcement. (a) Fracture between adjacent 0° and 90° plies. (b) Fracture between 45° and –45° plies. 2000×. Source: Ref 1

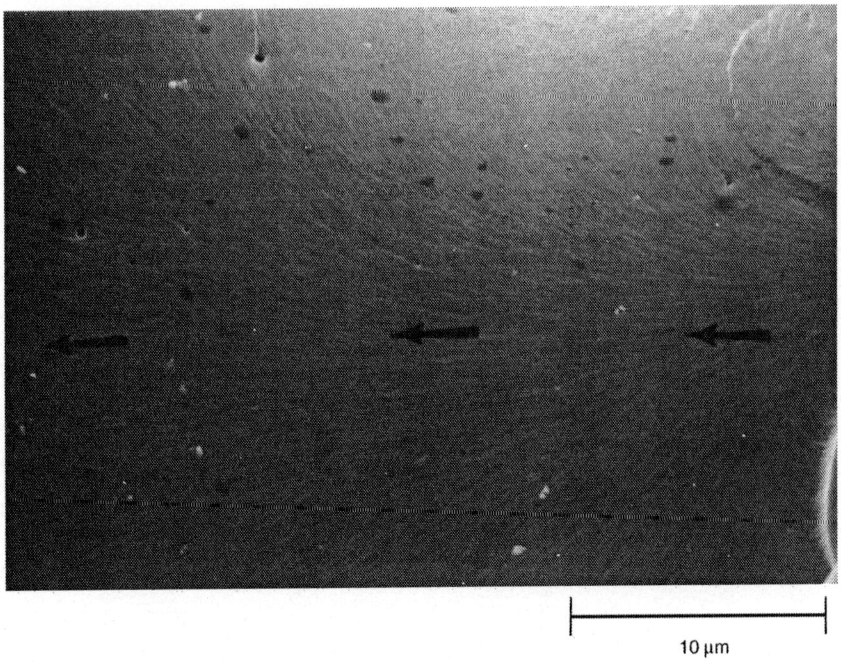

Fig. 3 Matrix feathering produced under interlaminar mode I tension. 3600×. Source: Ref 1

Fig. 4 Hackles in the resin of a carbon/epoxy (AS4/3501-6) laminate, indicative of mode II shear failure. 480×

matrix on different levels, resulting in what appear to be branches or small tributaries of a river. They can be found emanating from the resin at the surface of the fibers or fiber imprints, as shown in Fig. 2(b), or in the matrix between fibers. Matrix feathering, on the other hand, consists of small flow lines in the matrix that emanate from an imaginary centerline as the crack moves forward. Feathering is particularly evident in large, flat, resin-rich regions, where river patterns are not usually noted. Both river patterns and matrix feathering are not only indicative of mode I tension loading in brittle composite materials, but have also been noted to be indicative of crack-growth direction. Crack-growth direction can be ascertained by noting the direction in which the smaller rivers combine into the one large river, as shown in Fig. 1. In this figure, the crack-growth direction is from left to right. It has also been noted during fractographic examination that larger river patterns tend to give a better indication of the overall direction of crack growth in the specimen, because the larger river patterns are less influenced by the fibers themselves. In general, however, river patterns are indicative of mode I tensile loading and must be vectorially added across the majority of the fracture surface in order to obtain a definitive crack initiation site and crack-growth direction. Flow lines are also indicative of crack-growth direction, as shown in Fig. 3, where they are indicative of crack-growth direction from right to left.

Composites with brittle matrices tested under interlaminar mode II shear loading exhibit different fractographic features than those tested under interlaminar mode I tension loading. Visually, these surfaces exhibit milky white, dull fracture surfaces. Again, failure of the laminate generally occurs in the plane of the reinforcement, and SEM analysis reveals distinctive fractographic features. On these fracture surfaces, the appearance of the feature known as hackles becomes evident, as shown in Fig. 4 and 5. Hackles appear to form by the coalescence of numerous, small, 45°, tensile microcracks that form between the fibers under shear loading, as illustrated in Fig. 6 and 7. The size, shape, and form of the hackles are quite varied over the fracture surface, and the variation appears to be related to the actual percentage of mode I versus mode II loading, the amount of resin between the fibers, and the orientation of the fibers to the applied load. Under some mode II shear or mixed-mode load conditions, small river patterns are sometimes evident at the base of the hackle or on the surface of the hackle, as depicted in Fig. 5. These river marks can sometimes be used to help in determining the crack-growth direction.

Specimens tested under mode I tension and mode II shear fatigue loading generally result in fracture surfaces that contain fatigue striations. Fatigue striations, however, are not easily found in composite materials. This is partly because there may be little difference macroscopically between specimens that failed in fatigue and those that failed in overload by tension or shear. Unlike metallic materials, in which beach marks can often be found radiating outward from a visual fatigue initiation site, composite materials lack an apparent visual fatigue initiation site, which makes the diagnosis of fatigue failures at the macroscopic level somewhat more difficult. Additionally, fatigue failures can also be difficult to diagnose on the microscopic level. There are usually relatively few areas on the fracture surface that contain the fatigue striations. This lack of a significant number of striations on a fatigue fracture surface and the large separation between areas containing fatigue striations make locating them somewhat difficult and more time-consuming than in metals. The difficulty in finding these features is also enhanced by the fact that a certain amount of specimen tilt is often required in order to make them visible in the SEM. The amount of specimen tilt is of utmost importance in detecting the striations; higher tilt angles (>30°) often are required to find them. However, when fatigue striations are found, they can be found either in the matrix between two fibers, as shown in Fig. 7 and 8, or in the matrix itself, as shown in Fig. 9.

Impact damage is another form of loading that can result in specific interlaminar fracture characteristics in composite materials. In general, a delamination resulting from an impact load can often be ascertained by opening up the laminate in the plane of the delamination under mode I tension or mixed-mode loading. Visually, the delamination will generally exhibit a whitish, damaged surface, as compared to the darker, smoother, reflective surface of the manually fractured region. The delamination due to impact will also exhibit more evidence of shear (i.e., hackle formation in the damaged region) as compared to the surrounding area, which will generally have more river patterns indicative of the manually applied, mixed-mode, or mode I tensile loading. The impact-damaged region will also exhibit considerable matrix debris on the surface of the laminate, as compared to the surrounding area. For woven Kevlar/epoxy laminates, the fractographic evidence of impact damage can be found in both the matrix and in the fibers. Figures 10 and 11 depict the difference between an interlaminar fracture due to impact damage and an interlaminar fracture due to mode I tension loading for this particular material system. In addition to the visual differences noted previously, the microscopic features in the impact-damaged region (Fig. 10) include hackles, matrix debris, and significant fiber fibrillation and damage. The microscopic features in the mode I tensile region (Fig. 11) include the formation of river patterns, minimal matrix debris, and significantly less fiber fibrillation than in the impact-damaged zone.

Differences in Fracture Characteristics due to Different Material-Processing Conditions. Composite materials were then manufactured using material-processing conditions other than those recommended by the manufacturer. The purpose of manufacturing and testing these specimens was to determine if material-processing defects could be identified in the fracture

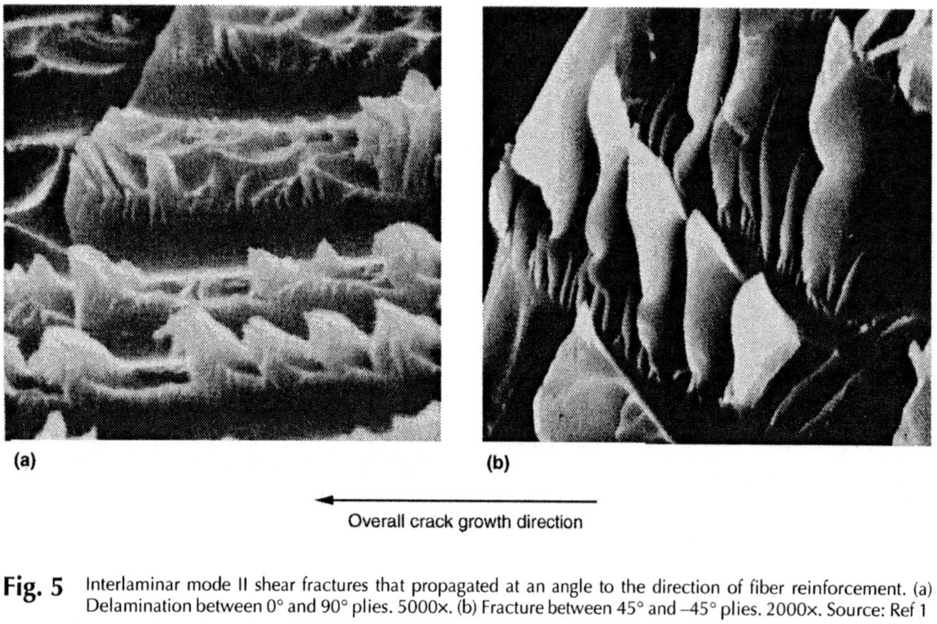

Fig. 5 Interlaminar mode II shear fractures that propagated at an angle to the direction of fiber reinforcement. (a) Delamination between 0° and 90° plies. 5000×. (b) Fracture between 45° and –45° plies. 2000×. Source: Ref 1

characteristics of the laminates. Brittle thermoset-matrix composite test specimens were either overcured or undercured during the processing of the laminates and then tested to failure under mode I tension loading. The fracture surfaces of these specimens exhibit variations in the fractographic features that appeared to coincide with the variations in processing conditions. In specimens that were undercured and then tested to failure under mode I tension, the river patterns generally exhibited a more feathery appearance than the specimens that had received a normal cure cycle. Specimens that were overcured and then tested to failure under mode I tension generally exhibited more brittle-looking and distinct river patterns in the matrix. Other material-processing variations, including the use of materials with inadequate resin content, generally resulted in interlaminar fracture surfaces with fewer matrix-rich regions and hence, fewer fracture features, such as river marks, matrix feathering, hackles, and fatigue striations. Inadequate resin content in laminates also generally reveals the fracture characteristic known as fiber splinters. Fiber splinters are fibers that separate readily from the fracture surface, because insufficient adherent-matrix cannot keep them attached to the rest of the specimen under mode I tension loading. These splinters are shown in Fig. 12 for a woven

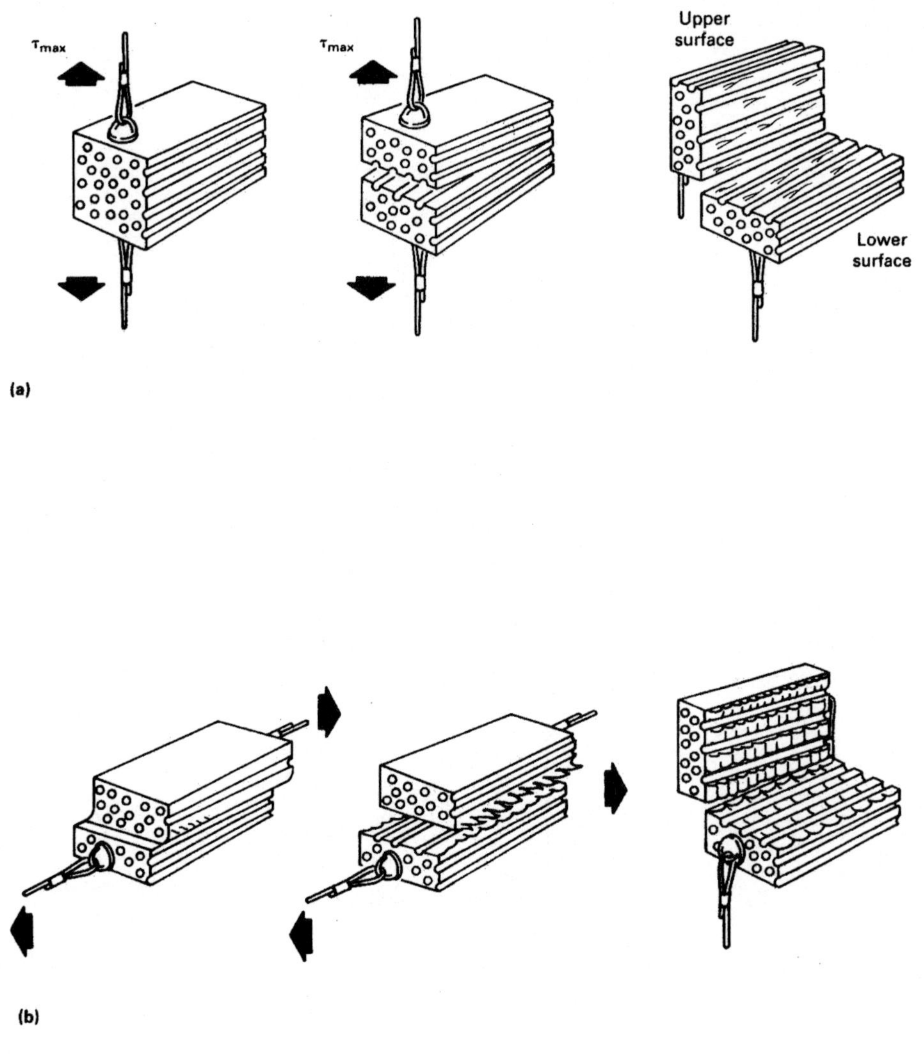

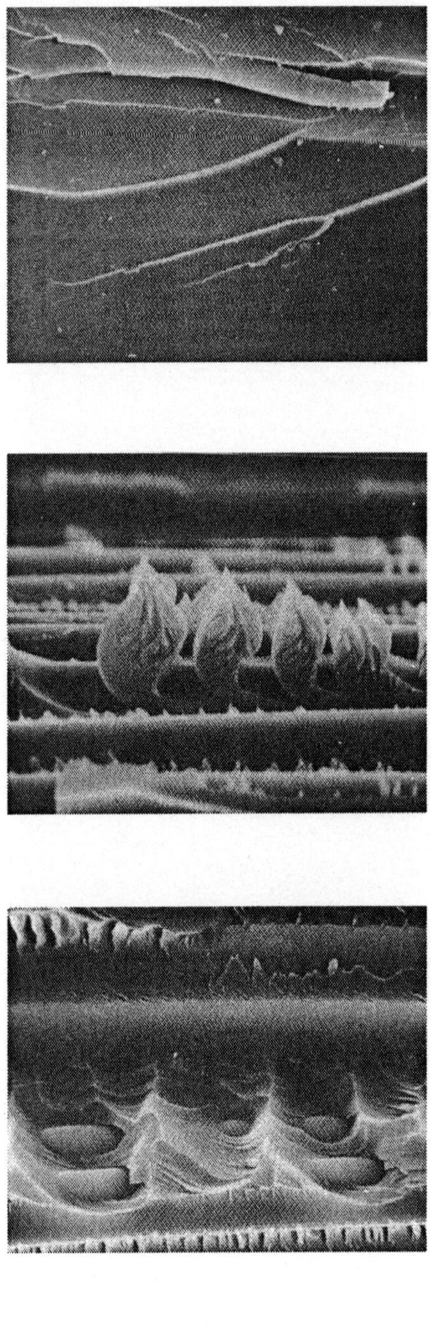

Fig. 6 Schematic of mode I (a) and mode II (b) failure

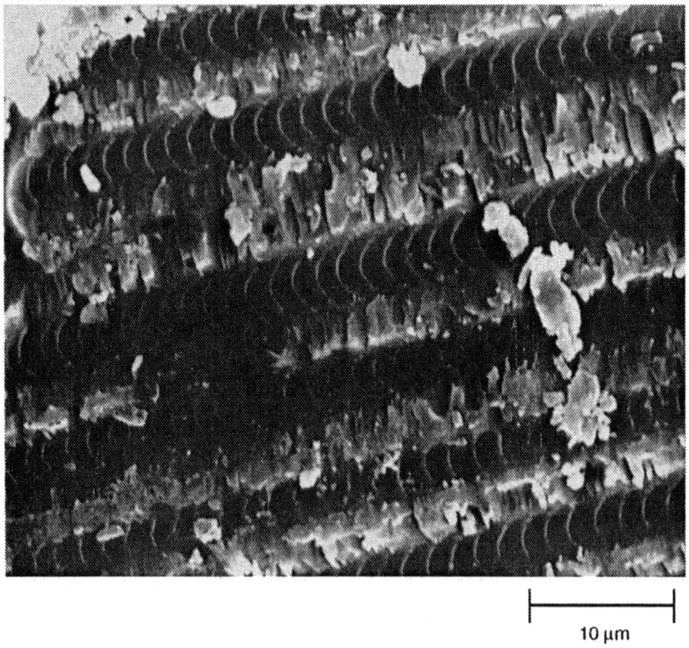

Fig. 7 Fatigue striations in a carbon-fiber composite. 2000×

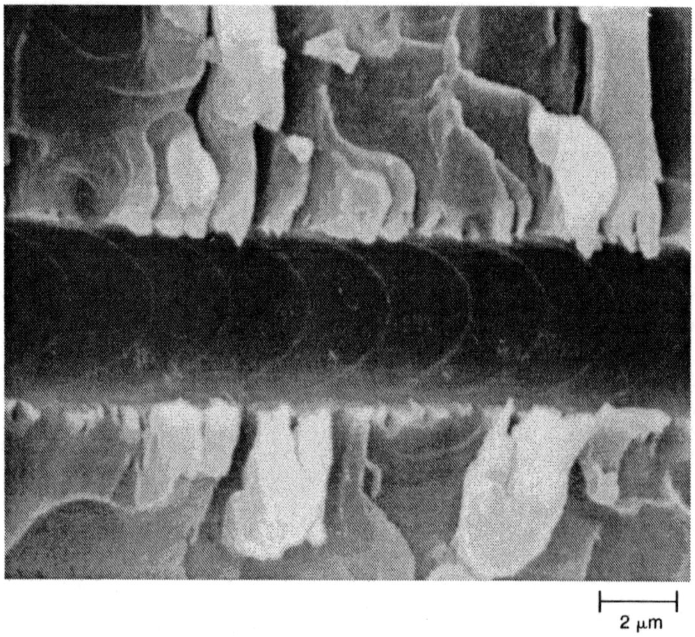

Fig. 8 Fatigue striations in the resin beneath a carbon fiber that was pulled out of a carbon/epoxy (AS4/3501-6) laminate following mode I fatigue loading. 5000×

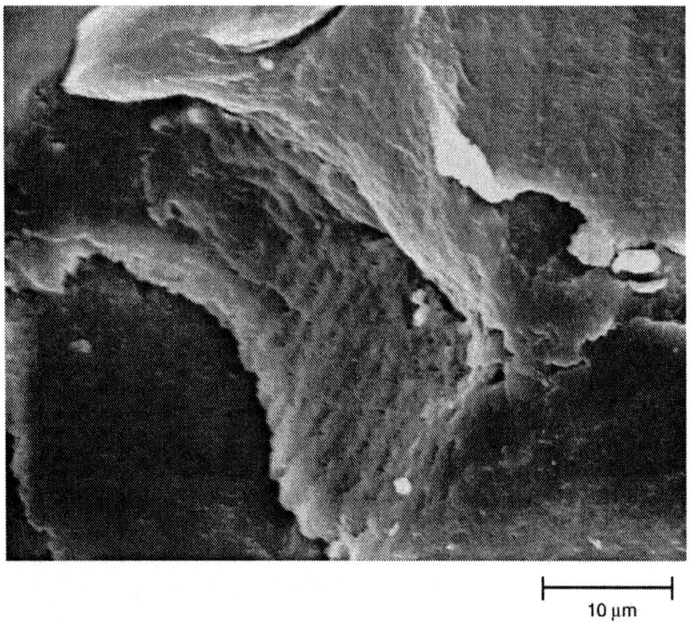

Fig. 9 Fatigue striations in the resin of a carbon-fiber composite laminate that failed in mode I fatigue loading. Striations cover the surfaces of several fibers. 1000×

Fig. 10 Impact damage in a Kevlar/epoxy composite laminate depicting hackle formation indicative of shear loading; resin debris indicative of impact loading and fiber fibrillation. 120×

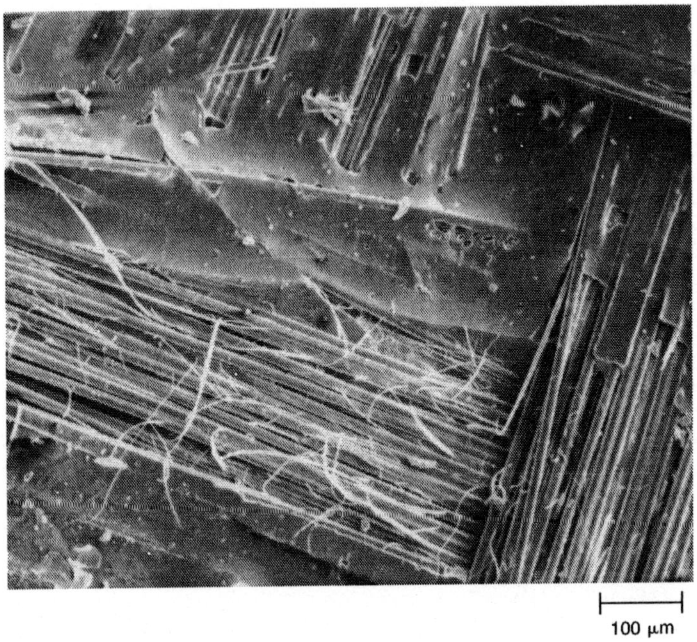

Fig. 11 River patterns and fiber fibrillation in a Kelvar/epoxy laminate in the region surrounding the impact damage, following peel failure of the laminate. 120×

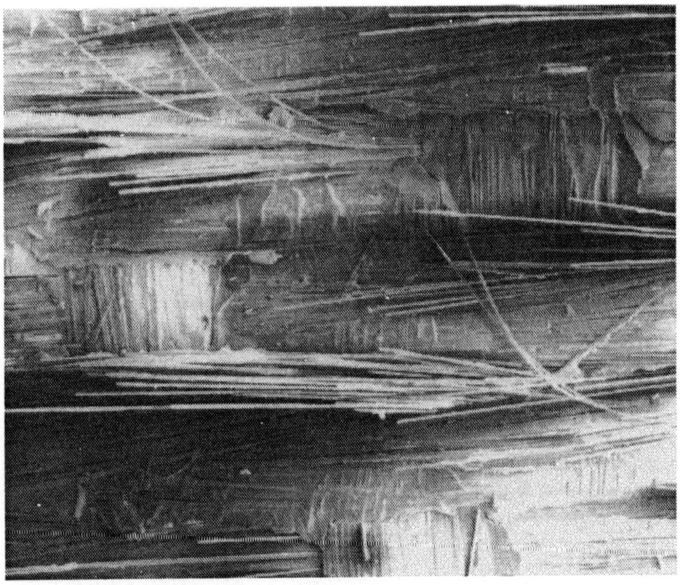

Fig. 12 Exposure of fiber splinters in a glass/polyimide laminate having inadequate resin content, following mode I tension loading of the specimen. 40×

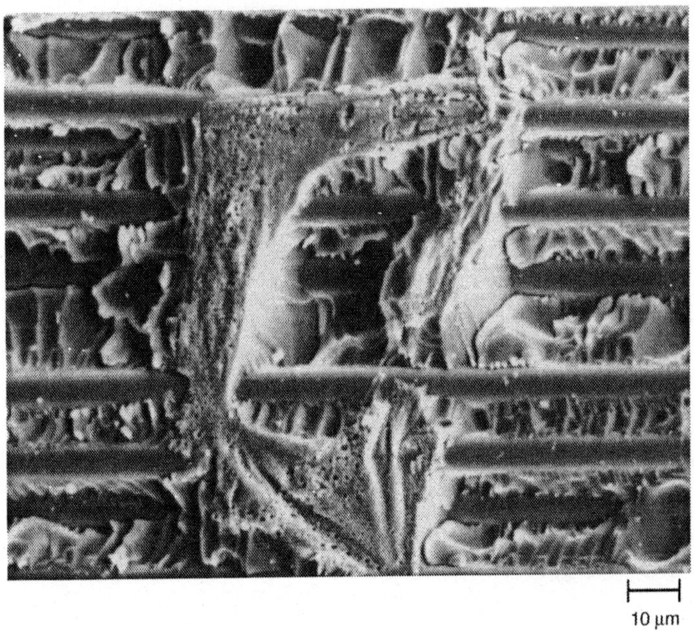

Fig. 13 Frekote contamination on the center portion of the fracture surface of a carbon/epoxy specimen, following mode II shear loading. 780×

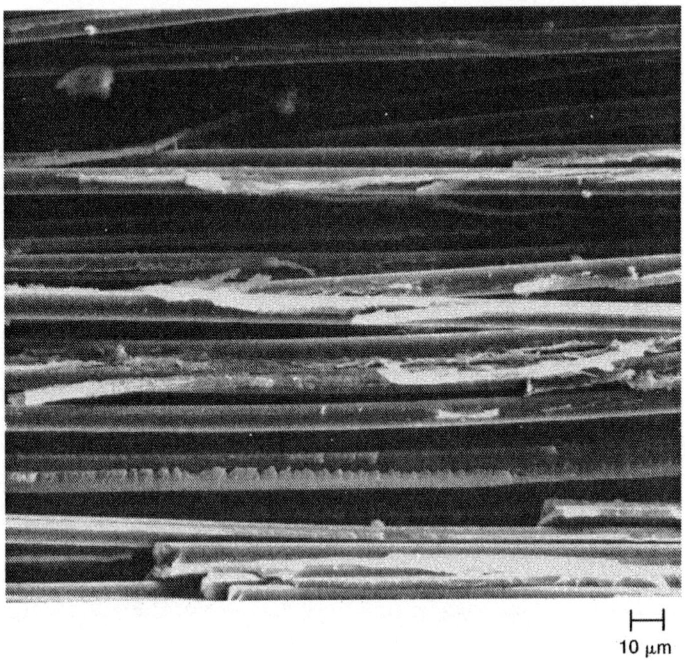

Fig. 14 Fiber/matrix interfacial failure in a carbon/epoxy (AS4/3501-6) test specimen after full moisture saturation and mode 1 tension loading at 130 °C (270 °F). 480×

glass/polyimide composite laminate. Other processing defects, such as contamination of an internal ply of the composite laminate with a release agent such as Frekote (Dexter Adhesive & Coating Systems), can also be found on the delaminated fracture surface during routine examination, as shown in Fig. 13.

Differences in Fracture Characteristics due to Different Environmental Conditions. Environmental exposure of the test specimens either before, during, or after loading also influences the fractographic features of brittle matrix composite materials. Specimens first exposed to moisture and then tested at room temperature or specimens exposed to moisture and then tested at elevated temperature revealed specific, identifiable fracture characteristics. Composites that were moisture-saturated and then tested at room temperature generally revealed more plasticity in their fracture characteristics than those specimens that were not moisture-conditioned. This effect, however, is very subtle and may or may not be evident, unless a similarly manufactured and tested dry specimen is available for comparison. For specimens that were moisture-conditioned and then exposed to elevated temperature, however, the fracture characteristics include not only an increase in matrix plasticity, but also an increase in the amount of fiber/matrix interfacial failure in the composite, as shown in Fig. 14. Although an increase in the amount of fiber/matrix interfacial failure may also be somewhat subjective in nature and difficult to discern, this effect is usually more significant, particularly at high moisture contents and temperatures near the wet, glass transition temperature of the resin. Environmental exposure of an organic composite laminate to high heat or fire, without prior moisture exposure, may also be determined from the fractographic evidence. Following exposure to elevated temperatures, the resin itself is often degraded or pyrolized. The carbon fibers themselves tend to become thinner and more distorted; they have a loss of fiber-end fracture features, and decomposition products appear on the surface, as shown in Fig. 15. The amount of degradation will depend on the glass transition and oxidation temperatures of the particular resin system used in the composite, as well as the time and temperature of the exposure. The result can be a partial or total loss of matrix fracture features, including river patterns, hackles, and striations, which makes analysis of composites exposed to high temperatures or fire significantly more difficult.

Interlaminar Fracture of Composites with Ductile Thermoplastic Matrices. In ductile thermoplastic resin systems, the interlaminar mode I and mode II fracture characteristics of composite materials are significantly different than for the brittle thermoset resin systems. In the case of carbon/thermoplastic (AS4/APC-2), the mode I tension fracture surfaces do not exhibit river patterns. These surfaces exhibit small matrix peaks, as shown in Fig. 16, or small, flat, radiating crystallite formations on the fiber surfaces, as shown in Fig. 17. Both of these formations are indicative of the semicrystalline nature of the polyetheretherketone (PEEK) resin system.

For thermoplastic composite laminates such as AS4/APC-2, tested under mode II shear, the fracture features are again unique and unlike the brittle thermoset matrices in which hackles are formed. However, in this case, a similar, repetitive formation of the resin occurs on the surface of the fibers, as shown in Fig. 18. These formations have been termed spikes. The tilt of the spikes and the flow of the material from the base to the tip can again be used as an indication of crack-growth direction.

Thermoplastic composite specimens, tested under mode I tension and mode II shear fatigue loading conditions, were also examined and fractographically documented. The fracture surfaces of these specimens also exhibit fatigue striations. These striations are similar to those

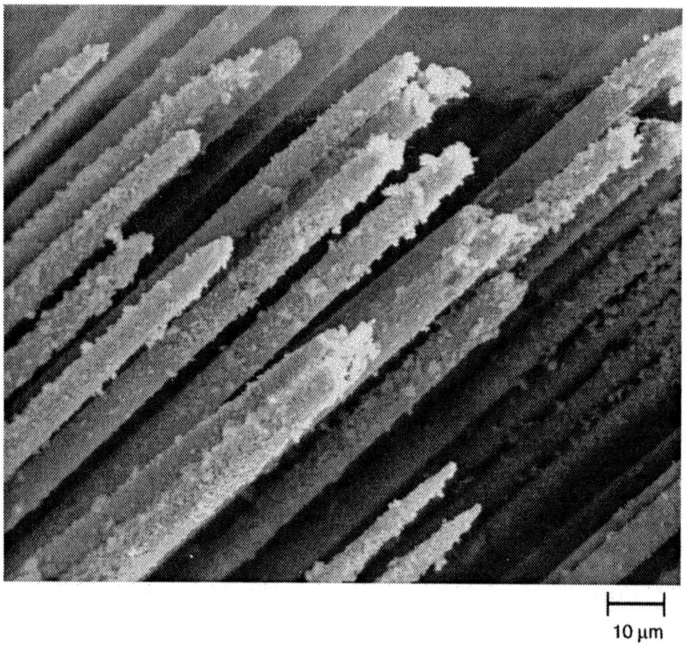

Fig. 15 Carbon fibers in a carbon/epoxy (AS4/3501-6) laminate, following exposure to fire for an unknown time period. 780×

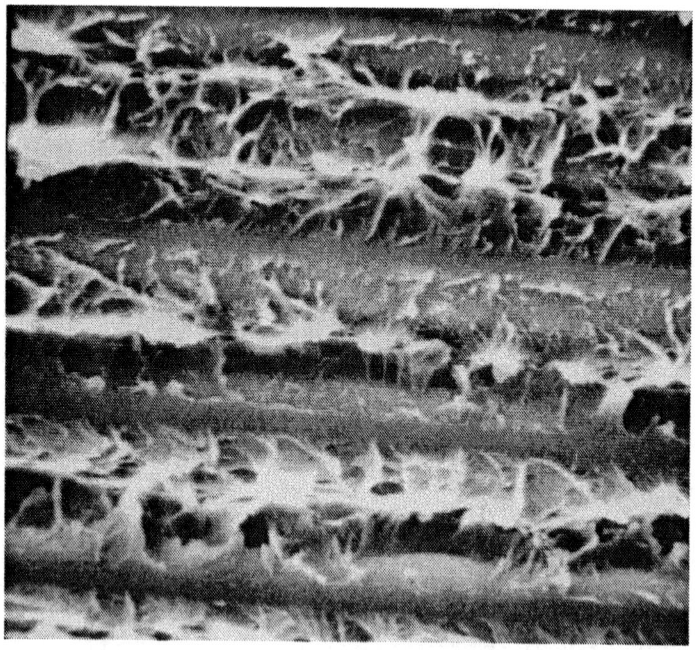

Fig. 16 Surface features of carbon/polyetheretherketone (AS4/APC-2), following mode I tensile fracture. 1500×

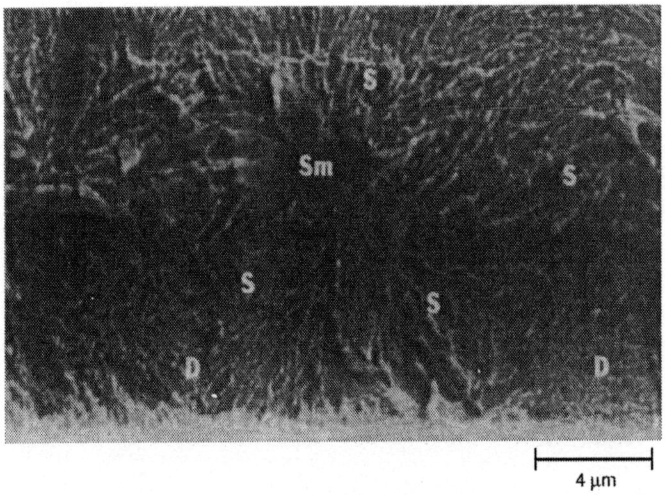

Fig. 17 Radial features thought to be crystallites in the matrix of the APC-2 material, indicative of the crystalline nature of the carbon/thermoplastic (AS4/APC-2) resin system, following failure due to mode I tension loading. 5000×

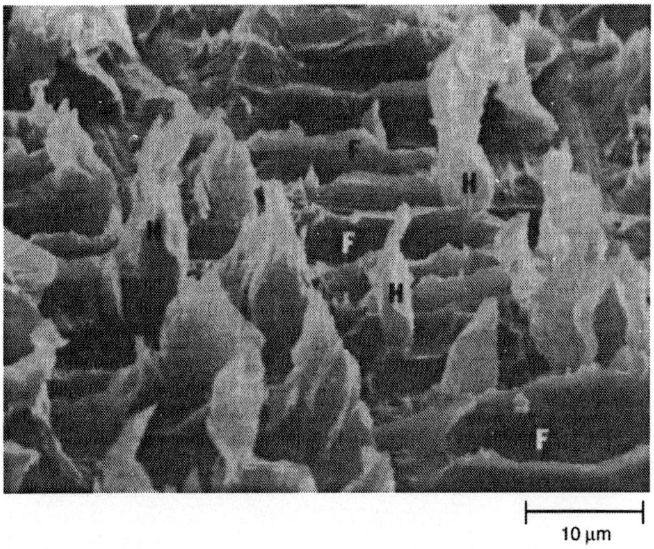

Fig. 18 The formation of the feature known as spikes in a mode II shear fracture surface of a carbon/PEEK composite laminate. 1700×

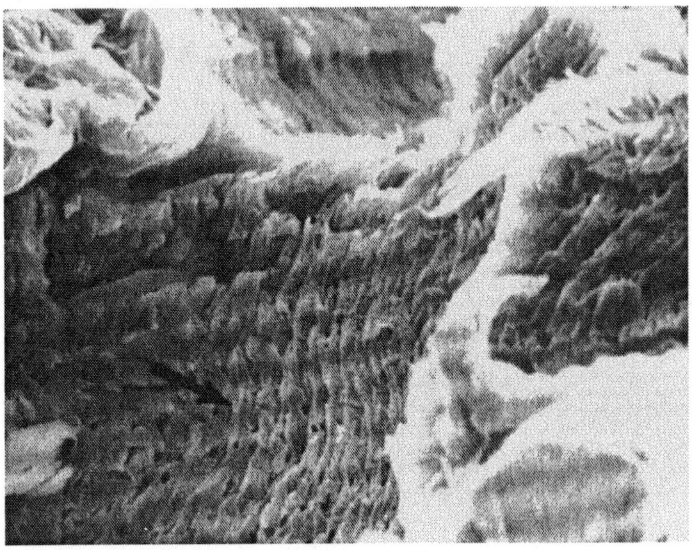

Fig. 19 Fatigue striations in the resin of an interlaminar failure, following mode I loading of a carbon/PEEK composite laminate. 900×

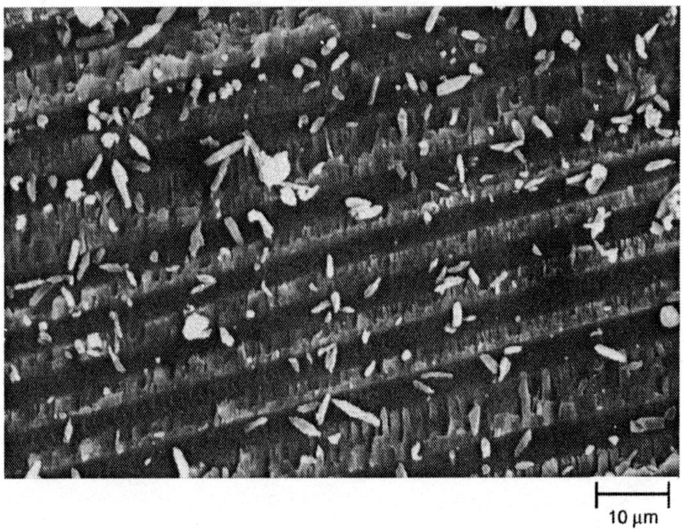

Fig. 20 Fracture feature known as matrix rollers on the surface of a carbon/KIII thermoplastic composite, following failure under mode II shear loading conditions. 1000×

formed in brittle thermoset-matrix composite systems but seemed to exhibit considerably more matrix plasticity, as shown in Fig. 19. The striations in these thermoplastic materials also tend to take on a somewhat irregular shape, often following the ductile matrix material; in brittle thermoset composites, the striations are generally sharp and regular and follow a relatively flat fracture path. Additionally, another feature indicative of fatigue has been noted on the fracture surface of other thermoplastic composite materials, including the carbon/KIII thermoplastic system; when tested under mode II shear fatigue loading conditions, the fracture surfaces of this material exhibit the feature known as matrix rollers. This feature consists of resin that tends to roll up on itself, as shown in Fig. 20. The appearance of these rollers, either between two fibers and/or on the top of the fracture surface, also indicates fatigue failure of the part.

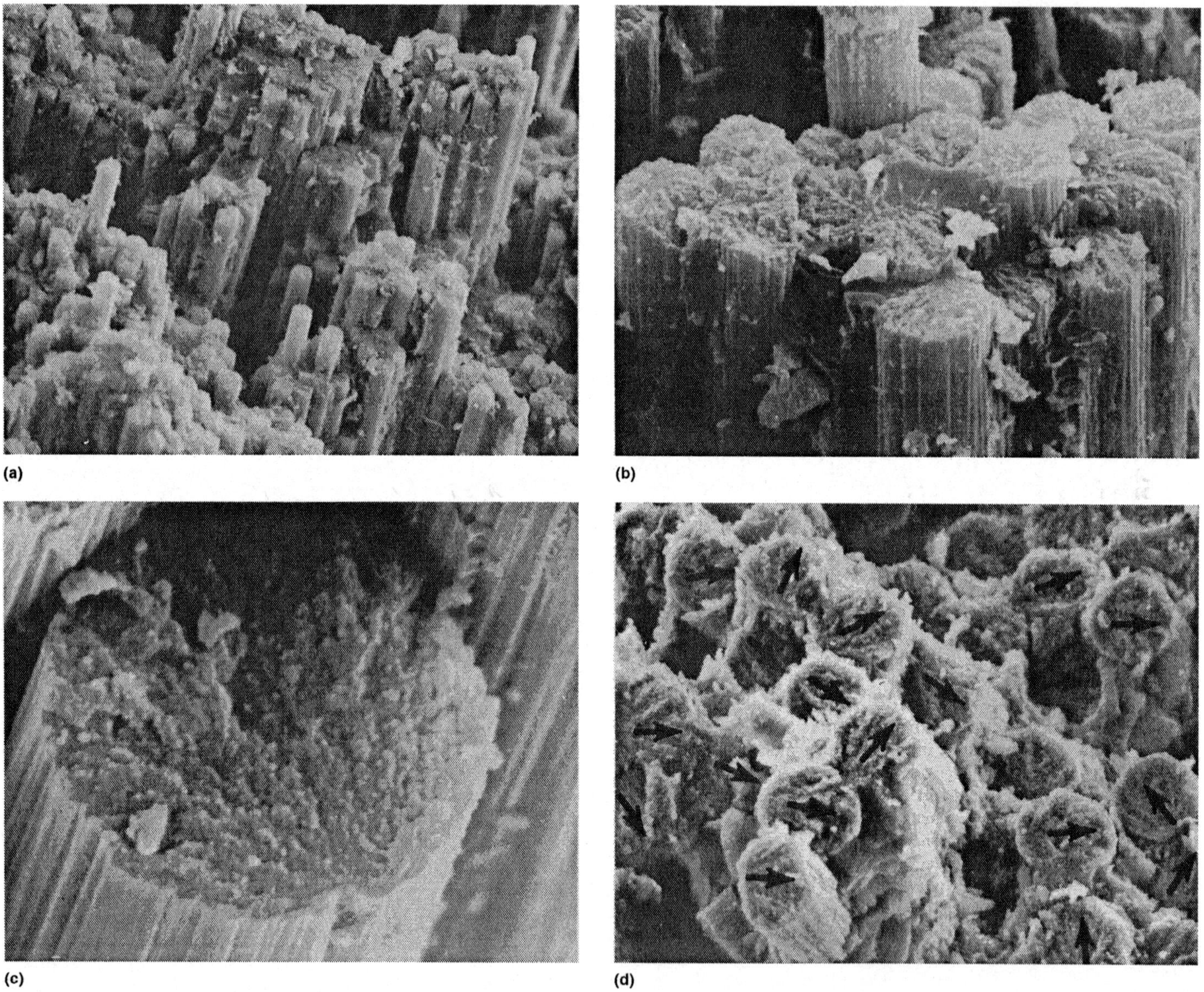

Fig. 21 Examples of translaminar tension fractures. (a) Translaminar tension fracture in a graphite/epoxy composite. Note fiber bundles and individual fiber pullout. 400×. Source: Ref 2. (b) Translaminar tension failure with localized area of flat fracture. 2000×. Source: Ref 2. (c) Radial fracture topography of an individual graphite-fiber failure under translaminar tension. 10,000×. Source: Ref 2. (d) Variations in fiber fracture mapped to determine overall crack-growth direction. 2000×. Source: Ref 3

Minimal changes in the manufacturing processes were explored and minimal environmental exposure was conducted for the thermoplastic composite systems. Moisture conditioning of the thermoplastic composite with room-temperature testing, however, did not appear to significantly alter fracture features of the carbon/PEEK material.

Translaminar Fracture Features

Translaminar fractures occur when loading of the composite specimen causes fracture perpendicular to the plane of fiber reinforcement. Unlike interlaminar failures, where most of the fractographic information is in the matrix material, translaminar fractures have the majority of the fractographic information in the fiber ends. Visually, translaminar fractures that fail under tensile loads, particularly those with some zero-degree or other off-axis plies, will exhibit considerable fiber pullout and have very irregular fracture surfaces. Scanning electron microscopic examination of the fiber-end fractures of carbon and glass fibers will often depict the feature known as fiber radials, as shown in Fig. 21 and 22. These fiber radial patterns can often be found in groups, particularly if failure occurred directionally across the laminate, as in the four-point, notched-bend specimens. In these directionally failed test specimens, the fiber radials can be used to determine the direction of the crack propagation. To do this, the direction of each fiber fracture must first be determined. The direction of fiber fracture is determined by creating a vector from the initiation point of the fiber, where the lines on the fiber ends radiate outward, to a point 180° across the fiber surface. Crack-growth direction in the laminate can then be determined from addition of these vectors on each fiber across the entire fracture surface.

For composites loaded under translaminar compression, the fracture surfaces are straighter and less jagged than those that failed under translaminar tension. The fracture surface of a composite failed in compression exhibits several different fracture layers, secondary cracking, and minimal fiber pullout. They often exhibit shear crippling due to microbuckling of the fibers, which occurs when the fibers kink under compressive loads (Fig. 23a, b). This shear crippling then results in the fracture feature known as chop marks, along with pieces of matrix debris on the surface, as shown for carbon fibers in Fig. 23(c) and (d) and for glass fibers in Fig. 24. Chop marks generally have three specific regions on the fiber ends: a tensile region, indicated by fiber radials; a compressive region, indicated by a flat, often angled fracture surface; and a neutral axis or line separating the two regions.

Translaminar flexural failure of composite laminates generally exhibits both tensile and compressive failure regions on the fracture surface. The amount of each depends on the loading conditions and the differences between the tensile and compressive strengths of the fibers. The differences between the two regions are generally quite visible, and the location of the neutral axis line can be easily identified from the differences in the fiber-end fractures.

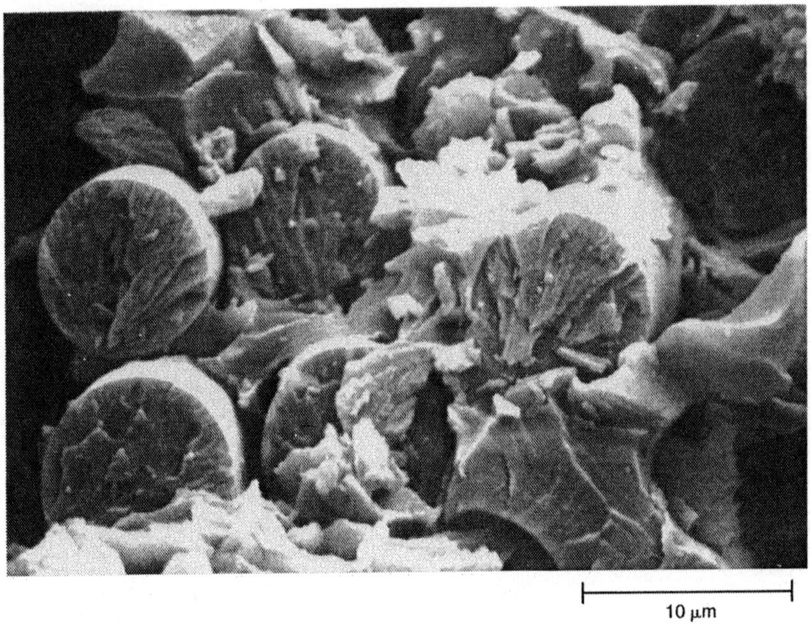

Fig. 22 Radial marks on the surfaces of glass fibers indicative of tensile failure in a glass/polyimide composite following failure of a notched four-point bend specimen. 3000×

Conclusion

In conclusion, the fracture surfaces of a number of composite test specimens have been examined and fractographically documented. It appears that, similar to metallic materials, composite materials have unique fractographic features that can be related to specific materials, processing methods, environmental exposures, and load conditions. These features have been catalogued by a number of researchers over the years and have been evaluated, based on the type of information they are able to provide about the fracture. The features can be used for determining information about component failures, particularly, information regarding crack initiation site, crack-growth direction, environmental conditioning, and failure mode. It should be noted, however, that even with the large amount of fractographic data that have become available in recent years, it still may not be possible to always determine the failure cause. This is because some of the fractographic information may have been obliterated, lost, or destroyed by some postfailure condition. Additionally, there are limitations as to the amount and type of information that can be obtained from a fractographic analysis, and additional techniques, such as mechanical testing and stress analysis, may be required to determine failure cause in some instances.

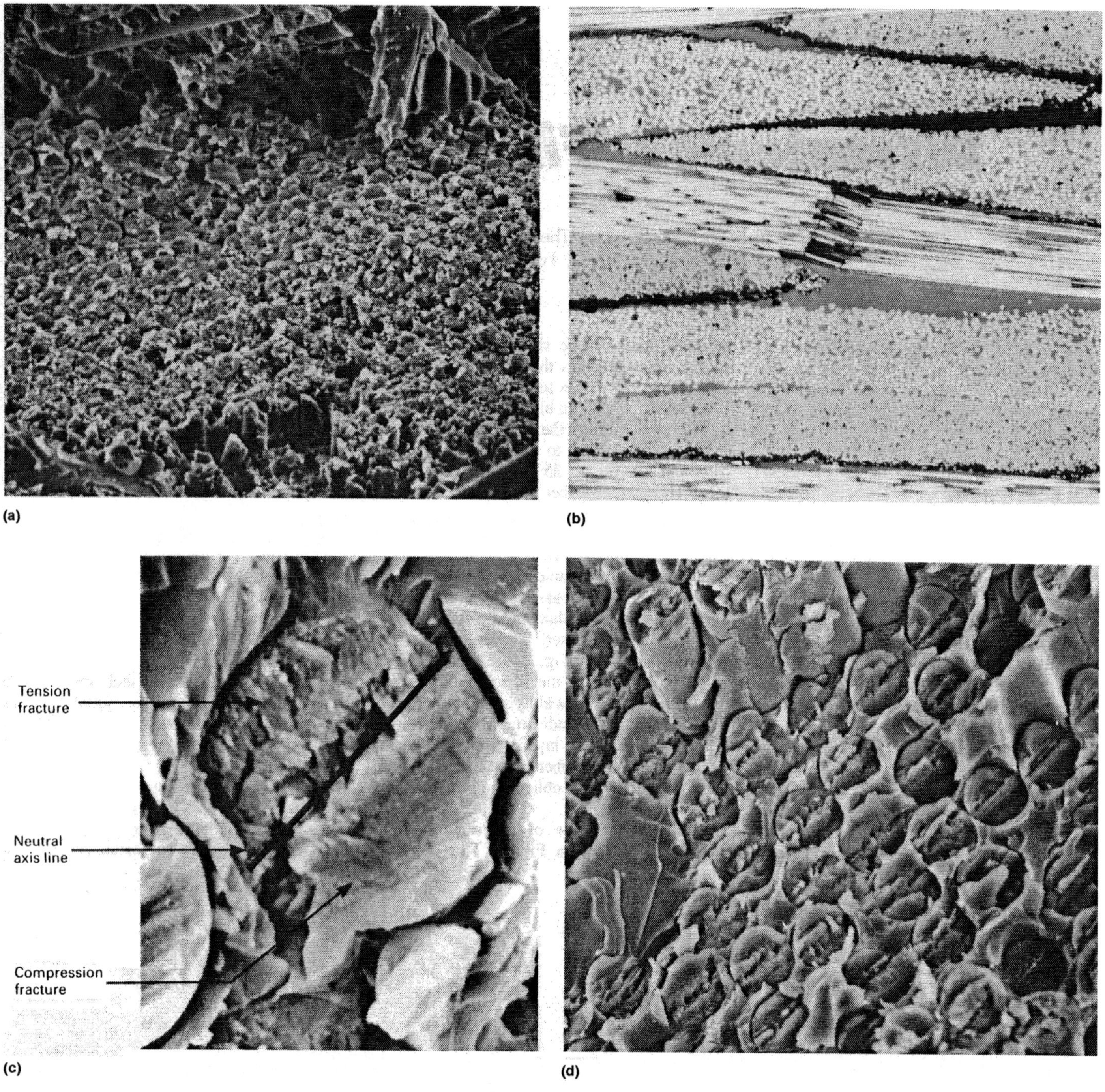

Fig. 23 Examples of translaminar compression fractures. (a) Translaminar compression fracture with extensive postfailure damage to fiber ends. 750×. (b) Translaminar compression-generated fiber kink in graphite/epoxy fabric. 100×. (c) Flexural fracture characteristics on fiber ends of a compression specimen. 10,000×. (d) Translaminar compression fracture illustrating parallel neutral axis lines representative of unified crack growth. 2000×

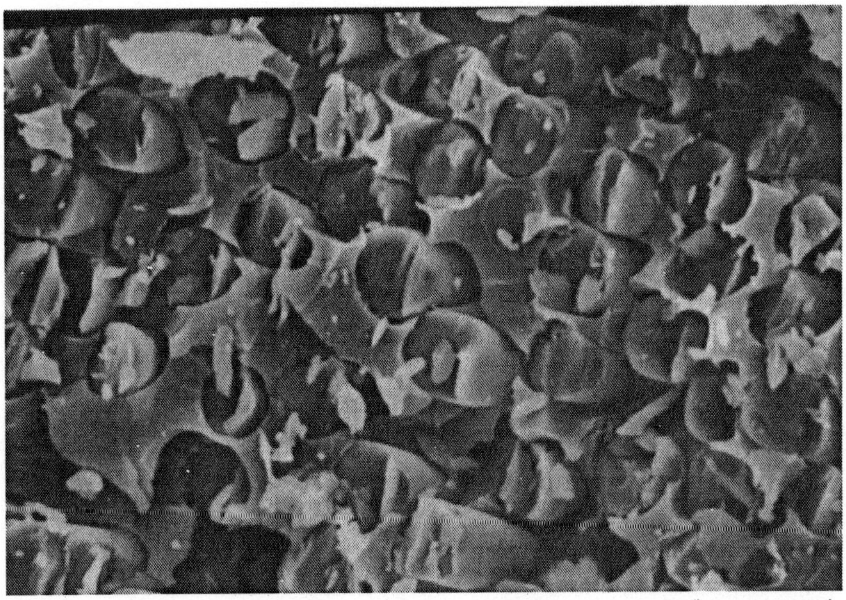

Fig. 24 Chop marks on the fracture surface of the glass fibers in a glass/polyimide composite tested as a notched four-point bend specimen that failed in compression. 1800×

ACKNOWLEDGMENTS

This article was done with the assistance of Boeing Military Airplane Company and Northrop Grumman, who provided some of the test specimens and fractographs for this work under Air Force Contracts F33615-84-C-5010 and F33615-87-C-5212.

REFERENCES

1. B.W. Smith et al., Fractographic Analysis of Interlaminar Fractures in Graphite-Epoxy Material Structures, *International Conference: Post Failure Analysis Techniques for Fiber Reinforced Composites,* Air Force Wright Aeronautical Laboratories, MLSE, July 1985
2. A.G. Miller et al., "Fracture Surface Characterization of Commercial Graphite/Epoxy Systems," *Nondestructive Evaluation and Flaw Criticality for Composite Materials,* STP 696, American Society for Testing and Materials, 1979, p 223–273
3. S.W. Tsai and H.T. Hahn, *Introduction to Composite Materials,* Technomic Publishing, 1980

Reference Information

Abbreviations and Symbols

a	crack length
a_c	critical crack length
a_i	initial crack size
ABS	acrylonitrile-butadiene-styrene
AC	acetals
AES	Auger electron spectroscopy
AF	aramid fiber
ANSI	American National Standards Institute
APP	ammonium polyphosphate
ASTM	American Society for Testing and Materials
ATH	alumina trihydrate
b	thickness
BD	butadiene
BMC	bulk molding compound
BOCA	Building Officials and Code Administrators
BPA	bisphenol A
BTDA	benzophenonetetracarboxylic acid dianhydride
c	velocity of light
CA	cellulose acetate
CAB	cellulose-acetate-butyrate
CAP	cellulose acetate-propionate
CD	compact disk
CN	cellulose nitrate
CN	coordination number
CSA	Canadian Standards Association
CTE	coefficient of thermal expansion
CTFE	chlorotrifluoroethylene
d	interplanar spacing
DAP	diallyl phthalate
DDA	dynamic dielectric analysis
DDS	diaminodiphenylsulfone
DDS	diaminodiphenyl sulfone
DFA	design for assembly
DFMA	design for manufacturing and assembly
DGEBA	diglycidyl ether of bisphenol A
DMA	dynamic mechanical analysis
DOP	dioctyl phthalate
DSC	differential scanning calorimetry
DTA	differential thermal analysis
DTUL	deflection temperature under load
DWI	drop weight index
E	elastic modulus (Young's modulus)
E_d	dissociation energy
EAA	poly(ethylene coacrylic acid)
EC	ethyl cellulose
E/CO	ethylene/carbon monoxide
EP	epoxy
EP	ethylene propylene
EPR	ethylene propylene rubber
ESC	environmental stress cracking
ESR	electron spin resonance
$E(t)$	time-dependent Young's modulus; stress relaxation modulus
FDA	Food and Drug Administration
FEP	fluorinated ethylene propylene
FID	flame ionization detector
FRP	fiber-reinforced polymer
FTIR	Fourier transform infrared
G^*	complex modulus
G'	elastic, or storage, modulus
G''	viscous, or loss, modulus
G'/G''	loss tangent, or dissipation factor
GPC	gel permeation chromatography
h	height
h	Planck's constant
HDPE	high-density polyethylene
HDT	heat-deflection temperature
HIPS	high-impact polystyrene
HMC	high-strength sheet molding compound
HPLC	high-performance liquid chromatography
HPPE	high-pressure polyethylene
HSLC	high-speed liquid chromatography
IEC	International Electrotechnical Commission
IR	infrared
IRHD	International Rubber Hardness Degrees
ISO	International Organization for Standardization
J	creep compliance
$\hat{J}$	normalized creep compliance
K	proportionality constant
ΔK	stress intensity range
K_c	critical stress-intensity factor
K_I	stress-intensity factor
K_{Ic}	plane-strain fracture toughness
l	length
LC	liquid chromatography
LCP	liquid crystal polymer
LDPE	low-density polyethylene
LLDPE	linear low-density polyethylene
LOI	limited oxygen index
LSC	liquid-solid chromatography
m	constant
M_c	critical weight average
$\overline{M}_n$	number-average molecular weight
M_T	modulus of toughness
$\overline{M}_w$	molecular weight average
$\overline{M}_w$	weight-average molecular weight
M_w/M_n	polydispersity index
$\overline{M}_z$	Z-average molecular weight

MDA methylene dianiline
MDAB 4,4′-bismaleimido-diphenylmethane
MDI methylene diphenylisocyanate
MF melamine-formaldehyde
MFR melt flow rate
MI melt index
MMA methyl methacrylate
MPDA metaphenylene diamine
M-PPE modified polyphenylene ether
M-PPO modified polyphenylene oxide
MW molecular weight
MWD molecular-weight distribution
N number of cycles
Nf number of cycles to failure
NASA National Aeronautics and Space Administration
NBS National Bureau of Standards
NFPA National Fire Protection Association
NIST National Institute of Standards and Technology
NMR nuclear magnetic resonance
NR nitrile resins
NRM nuclear magnetic resonance
P pressure
P_{ductile} maximum ductile load
P_{failure} failure load
PA polyamide
PAE polyaryl ether
PAEK polyaryl ether ketone (also abbreviated PEK)
PAI polyamide-imide
PAK polyaromatic ketones
PAN polyacrylonitrile
PB polybutylene
PBI polybenzimidazole
PBMA polybutyl methacrylate
PBT polybutylene terephthalate
PC polycarbonate
PCHDMT polycyclohexane dimethylene terephthalate
PCL polycaprolactone
PDMS polydimethyl siloxane
PE polyethylene
PEEK polyetheretherketone
PEEKK polyetheretherketoneketone
PEI polyether-imide
PEK polyetherketone
PEKK polyether ketone ketone
PEMA polyethyl methacrylate
PEN polyethernitrile
PES polyether sulfone (also abbreviated PESV, which is the preferred ASTM abbreviation)
PESU polyether sulfone
PESV polyether sulfone (or PES)
PET polyethylene terephthalate
PEVA polyethylene-vinyl acetate
PF phenol-formaldehyde
PHB poly(3-hydroxybutyrate)
PHBV poly(3-hydroxybutyrate-valerate)
PI polyimide
PIB polyisobutylene
P-iso-BMA polyisobutyl methacrylate
PMDA pyromellitic dianhydride
PMDI polymethylene diphenylene isocyanate
PMMA polymethyl methacrylate
PMP poly (4-methylpentene)
POB polyoxybutylene
POBT poly(polyoxybutylene terephthalate)
POM polyoxymethylene
PP polypropylene
PPDA paraphenylene diamine
PPE polyphenylene ether
pph parts per hundred
PPO polyphenylene oxide
PPO polypropylene oxide
PPS polyphenylene sulfide
PPSU polyphenylene sulfone
PS polystyrene
PSF polysulfone (or PSU)
PSU polysulfone (or PSF)
PTFCE polytrifluorochloroethylene
PTFE polytetrafluoroethylene
PU polyurethane
PUR polyurethane
PUR polyurethane resin
PVAC polyvinyl acetate
PVB polyvinyl butyral
PVC polyvinyl chloride
PVDC polyvinylidene chloride
PVDF polyvinylidene fluoride
PVF polyvinyl formal
PVFM polyvinyl formal
PVOH polyvinyl alcohol
Q activation energy
R gas constant
R_a roughness average
RIM reaction injection molding
RRIM reinforced reaction injection molding
RTI relative thermal index
RTM resin transfer molding
S fluctuating stress
SAE Society of Automotive Engineers
SAM scanning Auger microscope
SAN styrene-acrylonitrile
SB styrene butadiene
SBS styrene-butadiene-styrene
SEBS styrene-ethylene-butylene-styrene
SEC size-exclusion chromatography
SEM scanning electron microscopy
SFRP short-fiber-reinforced polymer
SIMS secondary ion mass spectroscopy
SMA styrene-maleic anhydride
SMC sheet molding compound
SMU sheet molding compound
SPE Society of Plastics Engineers
SRIM structural reaction injection molding
STEM scanning transmission electron microscope
t time
T temperature
T_g glass-transition temperature
T_m melting temperature
TDI toluene diisocyanate
TETA tetraethylenetriamine
TGA thermogravimetric analysis
TGMDA tetraglycidyl methylenedianiline
TLC thin-layer chromatography

TMA	thermomechanical analysis
TMS	tetramethyl silane
TOF-SIMS	time-of-flight secondary ion mass spectrometry
TPX	poly-(4-methyl-1-pentane)
TTT	time-temperature-transformation
UBC	Uniform Building Code
UHMWPE	ultrahigh-molecular-weight polyethylene
UL	Underwriter's Laboratory
UP	unsaturated polyester
UV	ultraviolet
VA	vinyl acetate
VC	vinyl chloride
VDC	vinylidene chloride
W	wear volume, wear rate
W^{-1}	wear resistance
WDS	wavelength-dispersive spectroscopy
XMC	extra-high-strength molding compound
XPS	x-ray photoelectron spectroscopy
XRD	x-ray diffraction analysis
γ	surface energy, shear strain
$\dot{\gamma}$	shear strain rate
δ	loss angle
ε	strain
η	melt viscosity
μ	coefficient of friction
σ	stress
$\Delta\sigma$	stress amplitude

Index

A

B

C

D

E

F

G

H

I

N

O

P

Q

R

S

T

U

V

W